The Entire Electromagnetic Spectrum

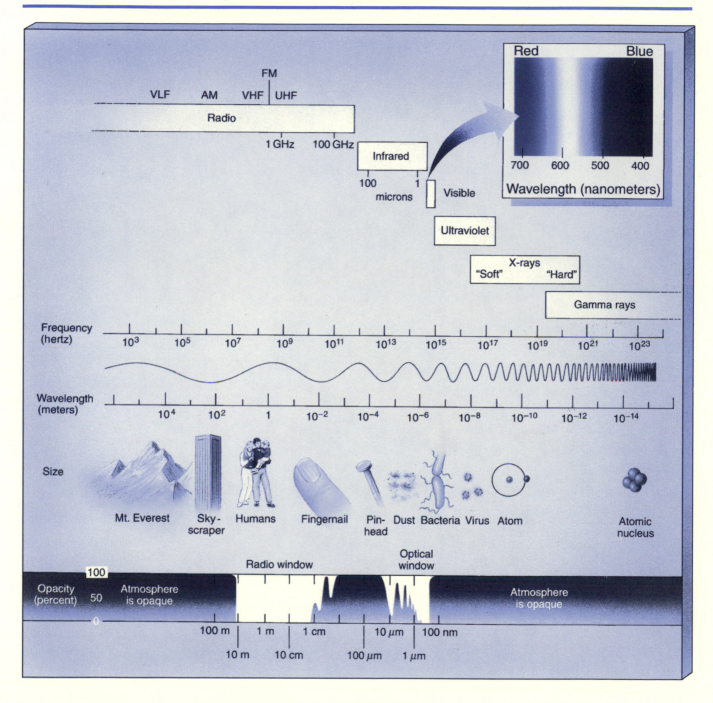

ASTRONOMY TODAY

ABOUT THE AUTHORS

ERIC CHAISSON

Eric holds a doctorate in Astrophysics from Harvard University, where he spent ten years on the faculty of Arts and Sciences. For five years, Eric was a Senior Scientist and Director of Educational Programs at the Space Telescope Science Institute and Adjunct Professor of Physics at Johns Hopkins University. He recently joined Tufts University, where he is now Professor of Physics, Professor of Education, and Director of the Wright Center for Innovative Science Education. He has written nine books on astronomy, which have received such literary awards as the Phi Beta Kappa Prize, two American Institute of Physics Awards, and Harvard's Smith Prize for Literary Merit. He has published more than 100 scientific papers in professional journals, and has also received Harvard's Bok Prize for original contributions to astrophysics.

STEVE McMILLAN

Steve holds a bachelor's and master's degree in Mathematics from Cambridge University and a doctorate in Astronomy from Harvard University. He held post-doctoral positions at the University of Illinois and Northwestern University, where he continued his research in theoretical astrophysics, star clusters, and numerical modeling. Steve is currently a Professor of Physics at Drexel University and a frequent visiting researcher at Princeton's Institute for Advanced Study, Cambridge's Institute of Astronomy, and the University of Tokyo. He has published over 40 scientific papers in professional journals.

ASTRONOMY TODAY

Second Edition

Eric Chaisson
Tufts University

Steve McMillan
Drexel University

Prentice Hall
Upper Saddle River, New Jersey 07458

Library of Congress Cataloging-in-Publication Data

Chaisson, Eric
     Astronomy today/Eric Chaisson, Steve McMillan— 2nd ed.
     1 v. (various pagings) : ill. (some col.), col. maps : 29 cm.
     Includes three pages of transparencies.
     Includes bibliographical references and index.
     ISBN 0-13-712382-5 (hardcover)
     1. Astronomy. I. McMillan, S. (Stephen), 1955- II. Title.
QB43.2.C44   1996
520—DC20   95-30896

Executive Editor: Alison Reeves
Editor in Chief: Paul Corey
Editorial Director: Tim Bozik
Development Editor: Dan Schiller
Production Editor: Susan Fisher/James Buckley
ESM Project Manager: Cynthia Dunn
New Media Production: Cindy Harford/David Moles/Grace Walkus
Special Projects Manager: Barbara A. Murray
CD-ROM Formatters: Jeff Henn/Ray Caramanna/Mike McLean/Richard Foster
Page Make-up and Composition: Shari Toron/Cynthia Dunn
    Kerry Reardon/Molly Pike Riccardi/Eric Hulsizer/Karen Noferi
Manager, Formatting and Art Production: John J. Jordan
Marketing Manager: Kelly McDonald/Leslie Cavaliere
Assistant Vice President ESM Production and Manufacturing: David W. Riccardi
Executive Managing Editor: Kathleen Schiaparelli
Art Director: Heather Scott
Creative Director: Paula Maylahn
Art Gurus: Rolando Corujo/Patrice Van Acker
Copy Editor: Sally Ann Bailey
Interior Designer: Lee Goldstein/Jeannette Jacobs
Cover Designer: Heather Scott
Cover Artist: Bill McClosky/Paul Gourhan
Manufacturing Buyer: Trudy Pisciotti
Editorial Assistant: Pam Holland-Moritz
Art Studio: Network Graphics
Photo Editor: Lorinda Morris-Nantz
Photo Researcher: Eloise Marion

 © 1997, 1996, 1993 Prentice Hall, Inc.
Simon & Schuster/A Viacom Company
Upper Saddle River, New Jersey 07458

Printed in the United States of America
10 9 8 7 6 5 4 3 2 1

ISBN 0-13-712382-5

Prentice Hall International (UK) Limited, *London*
Prentice Hall of Australia Pty. Limited, *Sydney*
Prentice Hall Canada, Inc., *Toronto*
Prentice Hall Hispanoamericana, S.A., *Mexico*
Prentice Hall of India Private Limited, *New Delhi*
Prentice Hall of Japan, Inc., *Tokyo*
Simon & Schuster Asia Pte. Ltd., *Singapore*
Editora Prentice Hall do Brasil, Ltda., *Rio de Janeiro*

BRIEF CONTENTS

1 CHARTING THE HEAVENS: *The Foundations of Astronomy* 1

2 THE COPERNICAN REVOLUTION: *The Birth of Modern Science* 28

3 RADIATION: *Information from the Cosmos* 52

4 SPECTROSCOPY: *The Inner Workings of Atoms* 72

5 TELESCOPES: *The Tools of Astronomy* 94

6 THE SOLAR SYSTEM: *An Introduction to Comparative Planetology* 126

7 THE EARTH: *Our Home in Space* 144

8 THE MOON AND MERCURY: *Scorched and Battered Worlds* 170

9 VENUS: *Earth's Sister Planet* 194

10 MARS: *A Near Miss for Life?* 214

11 JUPITER: *The Giant of the Solar System* 232

12 SATURN: *Spectacular Rings and Mysterious Moons* 252

13 URANUS, NEPTUNE AND PLUTO: *The Outer Worlds of the Solar System* 274

14 SOLAR SYSTEM DEBRIS: *Keys to Our Origin* 296

15 THE FORMATION OF THE SOLAR SYSTEM: *The Birth of Our World* 316

16 THE SUN: *Our Parent Star* 330

17 MEASURING THE STARS: *Giants, Dwarfs, and the Main Sequence* 356

18 THE INTERSTELLAR MEDIUM: *Gas and Dust between the Stars* 380

19 STAR FORMATION: *A Traumatic Birth* 398

20 STELLAR EVOLUTION: *From Middle Age to Death* 418

21 STELLAR EXPLOSIONS: *Novae, Supernovae, and the Formation of the Heavy Elements* 442

22 NEUTRON STARS AND BLACK HOLES: *Strange States of Matter* 462

23 THE MILKY WAY GALAXY: *A Grand Design* 486

24 NORMAL GALAXIES: *The Large Scale Structure of the Universe* 512

25 ACTIVE GALAXIES AND QUASARS: *Limits of the Observable Universe* 538

26 COSMOLOGY: *The Big Bang and the Fate of the Universe* 564

27 THE EARLY UNIVERSE: *Toward the Beginning of Time* 587

28 LIFE IN THE UNIVERSE: *Are We Alone?* 608

CONTENTS

PREFACE *xi*

1

CHARTING THE HEAVENS
The Foundations of Astronomy 2

1.1 Our Place in Space *4*
1.2 The Scale of Things *4*
1.3 The Obvious View *6*
1.4 The Motion of the Sun and the Stars *10*
1.5 Celestial Coordinates *13*
1.6 The Motion of the Moon *15*
1.7 Eclipses *17*
1.8 The Earth Moves *21*
1.9 The Measurement of Distance *20*
Chapter Review *26*

More Precisely... *Scientific Notation* 8
More Precisely... *Astronomical Measurement* 13
More Precisely... *Angular Measure* 14
More Precisely... *Earth Dimensions* 25

2

THE COPERNICAN REVOLUTION
The Birth of Modern Science 28

2.1 Ancient Astronomy *30*
2.2 The Geocentric Universe *32*
2.3 The Heliocentric Model of the Solar System *35*
2.4 The Birth of Modern Astronomy *36*
2.5 Kepler's Laws of Planetary Motion *40*
2.6 Dimensions of the Solar System *42*
2.7 Newton's Laws *44*
Chapter Review *50*

Interlude 2-1 *The Foundations of the Copernican Revolution* 35
Interlude 2-2 *The Scientific Method* 38
More Precisely... *Newton's Laws of Motion and Gravitation* 45

3

RADIATION
Information from the Cosmos 52

3.1 Information from the Skies *54*
3.2 Waves in What? *57*
3.3 The Electromagnetic Spectrum *59*
3.4 The Distribution of Radiation *62*
3.5 Another Inverse-Square Law *67*
3.6 The Doppler Effect *67*
Chapter Review *70*

More Precisely... *The Kelvin Temperature Scale* 63
More Precisely... *More About the Radiation Laws* 65

4

SPECTROSCOPY
The Inner Workings of Atoms 72

4.1 Spectral Lines *74*
4.2 The Particle Nature of Radiation *78*
4.3 Atomic Structure and Spectra *81*
4.4 Molecules *95*
4.5 Spectral-Line Analysis *86*
Chapter Review *92*

More Precisely... *The Photoelectric Effect* 80
More Precisely... *The Energy Levels of the Hydrogen Atom* 88

5

TELESCOPES
The Tools of Astronomy 94

5.1 Telescopes *96*
5.2 Telescope Size *101*
5.3 High-Resolution Astronomy *104*
5.4 Radio Astronomy *108*
5.5 Interferometry *113*
5.6 Other Astronomies *115*
5.7 Full-Spectrum Coverage *121*
Chapter Review *124*

Interlude 5-1 *The Hubble Space Telescope* 110

6

THE SOLAR SYSTEM
An Introduction to Comparative Planetology 126

6.1 Exploring Our Planetary System 128
6.2 The Overall Layout of the Solar System 129
6.3 Planetary Properties 130
6.4 Terrestrial and Jovian Planets 132
6.5 Interplanetary Debris 133
6.6 Spacecraft Exploration of the Solar System 134
6.7 Comparative Planetology 141
Chapter Review 142

Interlude 6-1 The Titius-Bode "Law" 135
Interlude 6-2 Interplanetary Navigation 139
More Precisely... Why Air Sticks Around 136

7

THE EARTH
Our Home in Space 144

7.1 The Earth in Bulk 146
7.2 The Tides 146
7.3 The Atmosphere 149
7.4 The Magnetosphere 153
7.5 The Interior 156
7.6 Surface Change 159
Chapter Review 168

More Precisely... Why is the Sky Blue? 152
More Precisely... Radioactive Dating 160

8

THE MOON AND MERCURY
Scorched and Battered Worlds 170

8.1 Orbital Properties 172
8.2 The Moon and Mercury in Bulk 174
8.3 Rotation Rates 177
8.4 Lunar Cratering 180
8.5 Lunar Surface Composition 183
8.6 The Surface of Mercury 185
8.7 Interiors 186
8.8 The Origin of the Moon 189
8.9 Evolutionary History of the Moon
and Mercury 190
Chapter Review 191

Interlude 8-1 Lunar Laser Ranging 180
Interlude 8-2 Lunar Exploration 188

9

VENUS
Earth's Sister Planet 194

9.1 Venus in Bulk 196
9.2 Long-Distance Observations of Venus 198
9.3 The Atmosphere of Venus 200
9.4 The Surface of Venus 203
9.5 Venus's Magnetic Field and Internal Structure 210
Chapter Review 211

10

MARS
A Near Miss for Life? 214

10.1 Mars in Bulk 216
10.2 Earth-Based Observations of Mars 217
10.3 The Surface of Mars 218
10.4 The Martian Atmosphere 225
10.5 Martian Internal Structure 226
10.6 The Moons of Mars 227
Chapter Review 230

Interlude 10-1 Canals on Mars? 223
Interlude 10-2 Life on Mars? 228

11

JUPITER
The Giant of the Solar System 232

11.1 Jupiter in Bulk 234
11.2 The Atmosphere of Jupiter 236
11.3 Internal Structure 240
11.4 Jupiter's Magnetosphere 241
11.5 The Moons of Jupiter 242
11.6 Jupiter's Ring 249
Chapter Review 250

Interlude 11-1 Almost a Star? 241

12

SATURN
Spectacular Rings and Mysterious Moons 252

12.1 Saturn in Bulk 254
12.2 Saturn's Atmosphere 254
12.3 Saturn's Interior and Magnetosphere 258
12.4 Saturn's Spectacular Ring System 259
12.5 The Moons of Saturn 265
Chapter Review 272

13

URANUS, NEPTUNE, PLUTO
The Outer Worlds of the Solar System 274

13.1 The Discovery of Uranus 276
13.2 The Discovery of Neptune 277
13.3 Uranus and Neptune in Bulk 278
13.4 The Atmosphere of Uranus and Neptune 279
13.5 Magnetospheres and Internal Structures 281
13.6 The Moon Systems of Uranus and Neptune 283
13.7 The Rings of the Outermost Jovian Planets 288
13.8 The Discovery of Pluto 290
13.9 Pluto in Bulk 292
13.10 The Origin of Pluto 293
Chapter Review 294

14

SOLAR SYSTEM DEBRIS
Keys to Our Origin 296

14.1 Asteroids 298
14.2 Comets 304
14.3 Meteoroids 310
Chapter Review 314

Interlude 14-1 What Killed the Dinosaurs? 300

15

THE FORMATION OF THE SOLAR SYSTEM
The Birth of Our World 316

15.1 Modeling the Origin of the Solar System 318
15.2 The Condensation Theory 319
15.3 The Differentiation of the Solar System 324
15.4 The Role of Catastrophes 326
15.5 The Angular Momentum Problem 327
Chapter Review 328

More Precisely... The Concept of Angular Momentum 321

16

THE SUN
Our Parent Star 330

16.1 The Sun in Bulk 332
16.2 The Solar Interior 334
16.3 The Solar Atmosphere 337
16.4 The Active Sun 341
16.5 The Heart of the Sun 348
16.6 Observations of Solar Neutrinos 351
Chapter Review 354

Interlude 16-1 Solar-Terrestrial Relations 349
More Precisely... Fundamental Forces 353

17

MEASURING THE STARS
Giants, Dwarfs, and the Main Sequence 356

17.1 The Distances to the Stars 358
17.2 Stellar Motion 359
17.3 Stellar Sizes 361
17.4 Luminosity and Brightness 362
17.5 Temperature and Color 363
17.6 The Classification of Stars 364
17.7 The Hertzsprung-Russell Diagram 367
17.8 Extending the Cosmic Distance Scale 371
17.9 Stellar Mass 372
17.10 Star Clusters 375
Chapter Review 378

More Precisely... The Magnitude Scale 366
Interlude 17-1 Stacks and Stacks of Photographs

18

THE INTERSTELLAR MEDIUM
Gas and Dust between the Stars 380

18.1 Interstellar Matter 382
18.2 Emission Nebulae 385
18.3 Dark Dust Clouds 389
18.4 21-Centimeter Radiation 392
18.5 Interstellar Molecules 394
Chapter Review 396

Interlude 18-1 A Satellite Named Iue 393

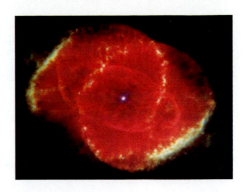

19

STAR FORMATION
A Traumatic Birth 398

19.1 Gravitational Competition 400
19.2 The Formation of Stars Like the Sun 401
19.3 Stars of Different Masses 406
19.4 Observations of Cloud Fragments and Protostars 407
19.5 Shock Waves and Star Formation 410
19.6 Emission Nebulae and Star Clusters 412
Chapter Review 416

Interlude 19-1 Evolution Observed 415

20

STAR EVOLUTION
From Middle Age to Death 418

20.1 Evolution Off the Main Sequence 420
20.2 Helium Fusion in Low-Mass Stars 424
20.3 The Death of a Low-Mass Star 428
20.4 High-Mass Stars 432
20.5 Observing Stellar Evolution in Star Clusters 433
20.6 The Evolution of Binary-Star Systems 436
Chapter Review 439

More Precisely... The CNO Cycle 422
Interlude 20-1 Mass Loss from Giant Stars 426
Interlude 20-2 Learning Astronomy from History 432

21

STELLAR EXPLOSIONS
Novae, Supernovae, and the Formation of the Heavy Elements 442

21.1 Life after Death for White Dwarfs 444
21.2 The End of a High-Mass Star 446
21.3 Supernova Explosions 448
21.4 The Formation of the Elements 454
21.5 The Cycle of Stellar Evolution 459
Chapter Review 460

Interlude 21-1 Nearby Supernovae 449
Interlude 21-2 Supernova 1987a 452

22

NEUTRON STARS AND BLACK HOLES
Strange States of Matter 462

22.1 Neutron Stars 464
22.2 Pulsars 465
22.3 Neutron-Star Binaries 467
22.4 Disappearing Matter 470
22.5 Properties of Black Holes 475
22.6 Space Travel Near Black Holes 477
22.7 Observational Evidence for Black Holes 480
Chapter Review 484

More Precisely... Einstein's Theories of Relativity 470
More Precisely... Tests of General Relativity 473
More Precisely... Black Hole Evaporation 479
Interlude 22-1 Gravity Waves 483

23

THE MILKY WAY GALAXY
A Grand Design 486

23.1 Our Parent Galaxy 488
23.2 Spiral Nebulae and Island Universes 489
23.3 The Structure of the Milky Way Galaxy 490
23.4 Disk and Halo Stars 493
23.5 The Galactic Disk 494
23.6 Galactic Dynamics 497
23.7 Spiral Structure 500
23.8 The Mass of the Galaxy 503
23.9 The Center of Our Galaxy 507
Chapter Review 510

Interlude 23-1 Early Computers 495
Interlude 23-2 Density Waves 502
Interlude 23-3 Cosmic Rays 508

24

NORMAL GALAXIES
The Large-Scale Structure of the Universe 512

24.1 Hubble's Galaxy Classification 514
24.2 The Distribution of Galaxies in Space 519
24.3 Galaxy Masses 525
24.4 Hubble's Law 527
24.5 The Formation and Evolution of Galaxies 532
       Chapter Review 536

Interlude 24-1 The Clouds of Magellan 522
Interlude 24-2 Colliding Galaxies 532

25

ACTIVE GALAXIES AND QUASARS
Limits of the Observable Universe 538

25.1 Beyond the Local Realm 540
25.2 Seyfert Galaxies 541
25.3 Radio Galaxies 542
25.4 The Central Engine of an Active Galaxy 546
25.5 Quasi-Stellar Objects 550
25.6 Active Galaxy Evolution 558
       Chapter Review 562

Interlude 25-1 Bl Lac Objects
Interlude 25-2 Could Quasars Be Local? 560
*More Precisely... Relativistic Red Shifts and Look-Back
                    Time 554*
More Precisely... Faster-Than-Light Velocities? 558

26

COSMOLOGY
The Big Bang and the Fate of the Universe 564

26.1 The Universe on the Largest Scale 566
26.2 The Expanding Universe 568
26.3 The Evolution of the Universe 573
26.4 Will the Universe Expand Forever? 577
26.5 The Geometry of Space 578
26.6 The Cosmic Microwave Background 579
       Chapter Review 584

Interlude 26-1 The Cosmological Constant 572
Interlude 26-2 The Steady-State Universe 583
More Precisely... Curved Space 580

27

THE EARLY UNIVERSE
Toward the Beginning of Time 587

27.1 Back to the Big Bang 588
27.2 Epochs in the Evolution of the Universe 590
27.3 The Formation of Nuclei and Atoms 593
27.4 The Inflationary Universe 597
27.5 The Formation of Structure in the Universe 601
27.6 Toward Creation 604
       Chapter Review 506

More Precisely... More on Fundamental Forces 592

28

LIFE IN THE UNIVERSE
Are We Alone? 608

28.1 Cosmic Evolution 610
28.2 Life in the Solar System 613
28.3 Intelligent Life in the Galaxy 615
28.4 The Search for Extraterrestrial Intelligence 619
       Chapter Review 622

Interlude 28-1 The Virus 613

APPENDICES A1

Some Useful Constants and Physical Measurements *A1*
The Periodic Table *A1*
Planetary Data *A2*
The Twenty Brightest Stars *A3*
The Thirty Nearest Stars *A4*

GLOSSARY G1

CREDITS FOR PHOTOGRAPHS C1

INDEX I1

STAR CHARTS S1

ANSWERS TO SELF-TEST ANS1

PREFACE

Astronomy continues to enjoy a golden age of exploration and discovery. Fueled by new technologies and novel theoretical insights, the study of the cosmos has never been more exciting. We are pleased to have the opportunity to present in this book a representative sample of the known facts, evolving ideas, and frontier discoveries in astronomy today.

This book is written for students who have taken no previous college science courses and who will likely not major in physics or astronomy. The text is suitable for both one-semester and two-semester courses. We present a broad view of astronomy, straightforwardly descriptive and without complex mathematics. The absence of sophisticated mathematics, however, in no way prevents discussion of important concepts. Rather, we rely on qualitative reasoning as well as analogies with objects and phenomena familiar to the student to explain the complexities of the subject without oversimplification. We have tried to impart the enthusiasm that we feel about astronomy, and to awaken students to the marvelous universe around us.

In teaching astronomy to non-scientists, as in writing this book, we are not seeking to convert students to careers in astronomy or even science. Instead, we strive to reach the wider audience of students who are majoring in many other worthwhile fields. We want to encourage these students to become scientifically literate members of modern society—to appreciate new developments in the world of science, to understand what scientists do for a living and why it is important, to make informed judgements regarding national initiatives in science and the public funding of scientific projects, and to vote intelligently in our democratic, increasingly technological world.

We are very gratified that the first edition of this text has been received so well by many in the astronomy education community. In using that earlier text, many of you—teachers and students alike—have sent us your helpful feedback and constructive criticisms. From these, we have learned to communicate better both the fundamentals and the excitement of astronomy. Many improvements inspired by your comments have been incorporated into this new edition.

ORGANIZATION AND APPROACH

The second edition of *Astronomy Today* is two chapters and 48 pages shorter than its predecessor. All chapters have been updated in content and several have seen significant internal reorganization. Specifically, the first three chapters of the previous edition have been restructured and condensed into two, resulting in a more concise and effective presentation of this important introductory material. Similarly, the solar system section has been reduced from 11 chapters to 10. The general introduction to the solar system and comparative planetology (Chapter 6) now, more logically, precedes the chapter on the Earth; and a single chapter is devoted to the Moon and Mercury, allowing for a discussion of these similar worlds that is at once more compact and more revealing.

Our overall organization follows the popular and effective "Earth-out" progression. We have found that most students, especially those with little scientific background, are much more comfortable studying the (relatively familiar) solar system before tackling stars and galaxies. Thus, Earth is the first object we discuss in detail. With the Earth and Moon as our initial planetary models, we move through the solar system, drawing on comparative planetology to provide an understanding of the many varied worlds we encounter. We conclude our coverage of the solar system with a discussion of its formation, a line of investigation that leads directly into a study of our Sun. With the Sun as our model star, we broaden the scope of our discussion to include stars in general—their properties, their evolutionary histories, and their varied fates. This journey naturally leads us to coverage of the Milky Way Galaxy, which in turn serves as an introduction to our treatment of other galaxies, both normal and active. Finally, we reach the subject of cosmology and the large-scale structure and dynamics of the universe as a whole. Throughout, we strive to emphasize the dynamic nature of the cosmos—virtually every major topic, from planets to quasars, includes a discussion of how those objects formed and how they evolve.

We continue to place much of the needed physics in the early chapters—an approach derived from years of experience teaching thousands of students. Additional physical principles are developed as needed later, both in the text narrative and in the boxed *More Precisely...* features (described below). We feel strongly that this is the most economical and efficient means of presentation. However, we acknowledge that not all instructors feel the same way. Accordingly, we have made the treatment of physics, as well as the more quantitative discussions, as modular as possible, so that these topics can be deferred to later stages of an astronomy course if desired. In addition, we have included as much modern astronomy as possible in the introductory chapters. These chapters are likely to engage students only if they are made to realize how simple physical principles provide the keys to our understanding of a vast and otherwise incomprehensible universe.

THE ILLUSTRATION PROGRAM

Visualization plays an important role in both the teaching and the practice of astronomy, and we continue to place strong emphasis on this aspect of our book. We have tried to combine aesthetic beauty with scientific accuracy in the artist's conceptions that adorn the text, and we have sought to present the best and latest imagery of a wide range of cosmic objects. More than 300 new photographs appear in the book, most of them not previously published in any text. Above all, each illustration has been carefully crafted to enhance student learning; each is pedagogically sound and tied tightly to nearby discussion of important scientific facts and ideas.

COMPOUND ART ▶

It is rare that a single image, be it a photograph or an artist's conception, can capture all aspects of a complex subject. Wherever possible, multiple-part figures are used in an attempt to convey the greatest amount of information in the most vivid way:

- Visible images are often presented along with their counterparts captured at other wavelengths.
- Interpretive line drawings are often superimposed on or juxtaposed with real astronomical photographs, helping students to really "see" what the photographs reveal.
- Breakouts—often multiple ones—are used to zoom in from wide-field shots to closeups, so that detailed images can be understood in their larger context.

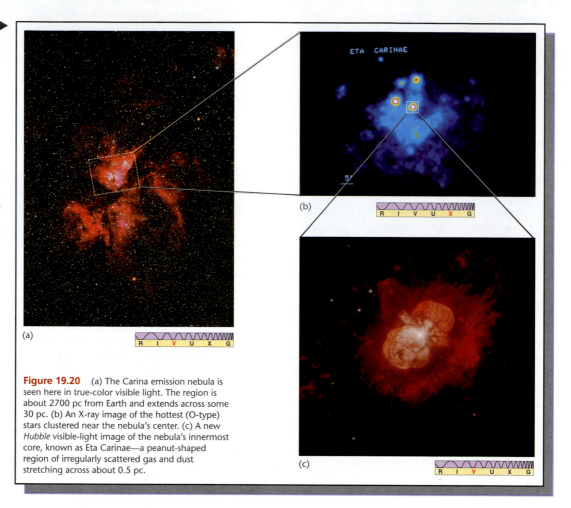

Figure 19.20 (a) The Carina emission nebula is seen here in true-color visible light. The region is about 2700 pc from Earth and extends across some 30 pc. (b) An X-ray image of the hottest (O-type) stars clustered near the nebula's center. (c) A new *Hubble* visible-light image of the nebula's innermost core, known as Eta Carinae—a peanut-shaped region of irregularly scattered gas and dust stretching across about 0.5 pc.

▲

EXPLANATORY CAPTIONS Students often review a chapter by "looking at the pictures." For this reason, the captions in this book are often a bit longer and more detailed than those in other texts.

FULL-SPECTRUM COVERAGE AND SPECTRUM ICONS

 Increasingly, astronomers are exploiting the full range of the electromagnetic spectrum to gather information about the cosmos. Throughout this book, images taken at radio, infrared, ultraviolet, X ray, or gamma ray wavelengths are used to supplement visible-light images. As it is sometimes difficult (even for a professional) to tell at a glance which images are visible-light photographs and which are false-color images created with other wavelengths, each photo in the text is provided with an icon that identifies the wavelength of electromagnetic radiation used to capture the image. This unique feature of the photographic program has now been enhanced by the addition of a running wave to the spectrum icon design, reinforcing the connection between wavelength and radiation properties.

H-R DIAGRAMS AND ACETATE OVERLAYS ▶

All of the book's H-R diagrams have been redone in uniform format, using real data. In addition, a unique set of transparent acetate overlays dramatically demonstrate to students how the H-R diagram helps us to organize our information about the stars and track their evolutionary histories.

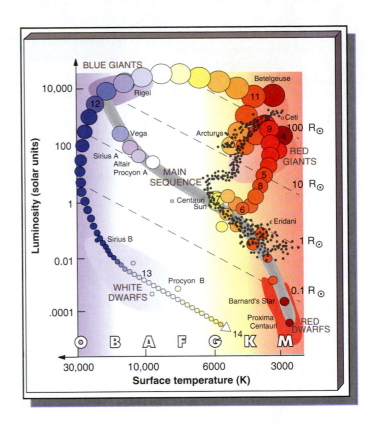

PEDAGOGICAL FEATURES

As with many other parts of our text, adopting instructors have helped guide us toward what is most helpful for effective student learning. With their assistance, we have revised both our in-chapter and end-of-chapter pedagogical apparatus to increase its utility to students.

LEARNING GOALS
Studying this chapter will enable you to:

1. Describe the basic properties of the main types of normal galaxies.

2. Discuss the distance-measurement techniques that enable astronomers to map the universe beyond our Milky Way.

3. Summarize what is known about the large-scale distribution of galaxies throughout the universe.

4. Describe some of the methods used to determine the masses of distant galaxies.

5. Explain why astronomers think that most of the matter in the universe is invisible.

6. State Hubble's law, and explain how it is used to derive distances to the most remote objects in the observable universe.

7. Discuss some theories of how galaxies form and evolve.

◀ **LEARNING GOALS** Studies indicate that beginning students often have trouble prioritizing textual material. For this reason, a few (typically 6 to 8) well-defined Learning Goals are provided at the start of each chapter. These help students to structure their reading of the chapter and then test their mastery of key facts and concepts. The Goals are numbered, and cross-referenced to key sections in the body of each chapter. This in-text highlighting of the most important aspects of the chapter also helps students to review. The Goals have also been reorganized and rephrased to make them more objectively testable, affording students better means of gauging their own progress.

KEY TERMS Like all subjects, astronomy has its own special vocabulary. To aid student learning, the most important astronomical terms are boldfaced at their first appearance in the text. Each boldfaced Key Term is also incorporated in the appropriate chapter summary, together with the page number where it was defined. In addition, a full alphabetical glossary, defining each Key Term and locating its first use in the text, appears at the end of the book.

The 12 constellations through which the Sun passes as it moves along the ecliptic—that is, the constellations we would see looking in the direction of the Sun, if they weren't overwhelmed by the Sun's light—had special significance for astrologers of old. These constellations are collectively known as the **zodiac**. The time required for the constellations to complete one cycle around the sky and to return to their starting points as seen from a given point on the Earth is one **sidereal year**. Earth completes exactly one orbit around the Sun in this time. One sidereal year is 365.256 solar days long,

BOXES The text of this edition is interspersed with a generous number of boxed features designed to add both breadth and depth to the book's coverage. These features are of two kinds:

"Interludes" explore a wide variety of interesting supplementary topics. ▶

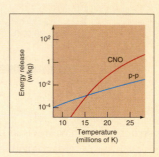
◀ **"More Precisely..."** boxes, a new element, provide quantitative treatments of subjects discussed qualitatively in the text, or explore the physics of astronomical processes in greater detail. Removing these more challenging topics from the main flow of the narrative and placing them within a separate modular element of the chapter design (so that they can be covered in class, assigned as supplementary material, or simply left as optional reading for those students who find them of interest) will allow instructors greater flexibility in setting the level of their coverage.

CROSS-LINKS In astronomy, as in many scientific disciplines, almost every topic seems to have some bearing on almost every other. In particular, the connection between the specifically astronomical material and the physical principles set forth early in the text is crucial. Practically everything in Chapters 6-28 of this text rests on the foundation laid in the first five chapters. It is important that students, when they encounter, say, the cosmological redshift in Chapter 24, recall what they learned about the Doppler shift in Chapter 3. Similarly, the discussions of the mass of binary star components (Chapter 17) and of galactic rotation (Chapter 23) both depend on the discussion of Kepler's and Newton's Laws in Chapter 2. It is therefore essential to remind students of these links, so that they can recall the principles on which later discussions rest, and if necessary review them. To reinforce these connections, "cross-links" have been inserted throughout the text—symbols that mark key intellectual bridges between material in different chapters. The links,

denoted by the symbol ∞, together with a page reference, signal to students that the topic under discussion is related in some significant way to ideas developed earlier, and direct them to material that they might need to review before proceeding. ▼

CHAPTER SUMMARIES The Chapter Summaries, a primary review tool for the student, have been expanded and improved for the second edition. All Key Terms introduced in each chapter are listed again, in context and in boldface, in these Summaries, along with page references to the text discussion.

▶

Chapter Review

SUMMARY

The **interstellar medium** (p. 382) occupies the space between stars. It is made up of cold (less than 100 K) gas, mostly atomic or molecular hydrogen and helium, and **dust grains** (p. 382). Interstellar dust is very effective at blocking our view of distant stars, even though the density of the interstellar medium is very low. The spatial distribution of interstellar matter is very patchy.

QUESTIONS, PROBLEMS, AND PROJECTS

Other elements of the end-of-chapter material have also seen substantial reorganization:

SELF-TEST: True or False?

The following 9 questions present properties of the solar system which a model of solar-system formation must explain. Which are correctly stated and which are not?

____ **1.** Each planet is relatively isolated in space.

____ **2.** The orbits of the planets are not very circular, nificantly elliptical.

____ **3.** The orbits of the planets all lie near the ecliptic

◀ Each chapter now incorporates 25-30 Self-Test Questions, about equally divided between "true/false" and "fill-in-the-blank" formats, designed to allow students to assess their understanding of the chapter material. Answers to all these questions appear at the end of the book. ▼

SELF-TEST: Fill in the Blank

1. The condensation theory, which currently is used to explain the formation of the solar system, is actually just a refined version of the older _____ theory.

2. In the condensation theory, astronomers realized the critical role played by _____ in starting the formation of small clumps of matter.

REVIEW AND DISCUSSION

1. List six properties of the solar system that any model of its formation must be able to explain.

2. Explain the difference between evolutionary theories and catastrophic theories of the solar system's origin.

3. Describe the basic features of the nebular theory of solar-system formation.

4. Give three examples of how the nebular theory explains some observed features of the present-day solar system.

5. Name two basic problems with the old nebular theory.

◀ Each chapter has 15-20 Review and Discussion Questions, which may be used for in-class review or for assignment. As with the Self-Test Questions, the material needed to answer Review Questions may be found within the chapter. The Discussion Questions explore particular topics more deeply, often asking for opinions, not just facts. As with all discussions, these questions usually have no single "correct" answer.

Several Problems in each chapter entail some numerical calculation; their answers are not contained verbatim within the chapter, but the information necessary to solve them has been presented in the text. ▶

PROBLEMS

1. The orbital angular momentum of a planet in a circular orbit is simply the product of its mass, its orbital velocity, and its distance from the Sun. Compare the orbital angular momenta of Jupiter, Saturn, and Earth.

2. A typical comet contains some 10^{13} kg of water ice. How many comets would have to strike the Earth in order to account for the roughly 2×10^{21} kg of water presently found

PROJECTS

1. Many amateur astronomers enjoy turning their telescopes the ninth-magnitude companion to Cygnus X-1, the sky most famous black hole candidate. Because none of us can see in X-rays, no sign of anything unusual can be seen. Still, it's fun to gaze toward this region of the heavens and contemplate Cygnus X-1's powerful energy emission strange proprerties. Even without a telescope, it is easy to locate the region of the

◀ Each chapter ends with a few Projects meant to get the student out of the classroom and looking at the sky, although some entail research in libraries or other extracurricular activities.

SUPPLEMENTARY MATERIAL

This edition of *Astronomy Today* is accompanied by an outstanding set of instructional aids.

COMETS A unique media subscription, available free to adopters. Each year, instructors will receive a coordinated kit with updates for their course materials, including:

• Computer-generated animations on VHS tape, created exclusively for *Astronomy Today* by renowned astronomical artist Dana Berry with James Palmer, and scripted by author Eric Chaisson to coordinate with figures in the text. The COMETS kit for Fall 1995 will contain seven animations created exclusively for this edition, including the bipolar jets and accretion disk of an active galaxy, the breakup of a comet or asteroid, and periodic nova outbursts in a binary system. Over the next two years, many additional animations will be supplied via the annual COMETS update package, eventually forming an extensive library.

• "Earth and Sky" Radio Broadcasts: highlights from this nationally syndicated radio program on audio casettes.

• New slides. 20 dramatic, current photographs are included in each issue, eliminating the inconvenience of having to track down the latest images every semester.

• COMETS Newsletter. This supplement includes an index to our entire library of supplemental materials in each issue, as well as suggestions regarding how each supplement might be used in your classroom or laboratory.

(Comets Fall 1995) ISBN: 0-13-376286-6
(Comets Fall 1996) ISBN: 0-13-376302-1

 PRENTICE HALL/*THE NEW YORK TIMES* THEMES OF THE TIMES Timely and relevant articles from recent editions of *The New York Times* compiled into a customized newspaper format. It enhances text coverage by helping students relate what they learn in class to exciting new theories and discoveries. Available free up to the quantity of adoption.

INSTRUCTOR'S EDITION The Instructor's Edition is a desk copy of the student text shrinkwrapped with the Instructor's Manual. The Instructor's Manual is also available as a stand-alone item (see below for a detailed description).

Instructor's Edition (student text and Instructor's Manual)
ISBN: 0-13-376328-5

INSTRUCTOR'S MANUAL, by Leo Connolly (California State University at San Bernardino). This manual provides an overview of each chapter; pedagogical tips, useful analogies, and suggestions for classroom demonstrations; answers to the end-of-chapter review and discussion questions and problems; and a list of selected readings.

Instructor's Manual (stand-alone item)
ISBN: 0-13-532151-4

ACETATES AND SLIDES An extensive set of color acetates and a comprehensive 35mm slide set are available free to qualified adopters.

(Slide set) ISBN: 0-13-376385-4
(Transparency pack) ISBN: 0-13-376393-5

TEST ITEM FILE An extensive file of test questions, newly compiled for the second edition by Leo Connolly, is offered free upon adoption. Available in both printed and electronic form (Macintosh or IBM-compatible formats).
ISBN: 0-13-376351-X

PRENTICE HALL CUSTOM TEST Prentice Hall Custom Test is based on the powerful testing technology developed by Engineering Software Associates, Inc. (ESA). Available for Windows, Macintosh, and DOS, Prentice Hall Custom Test allows educators to create and tailor the exam to their own needs. With the Online Testing option, exams can also be administered online and data can then be automatically transferred for evaluation. A comprehensive desk reference guide is included, along with on-line assistance.
(IBM) ISBN: 0-13-376377-3
(MAC) ISBN: 0-13-376369-2

DIRECTOR ACADEMIC
AUTHORWARE ACADEMIC

By Prentice Hall and Macromedia
These educational adaptations of the leading multimedia authoring tools, available exclusively through Prentice Hall, put the power of multimedia development into your hands at a fraction of the cost. Both products also contain templates designed specifically for academic use. Available for both Macintosh or Windows platforms. For further information and/or sales, contact Prentice Hall Multimedia Group at 1-800-887-9998.

STUDENT OBSERVATION GUIDE WITH LABORATORY EXERCISES, by Michael Seeds and Joseph Holzinger (Franklin and Marshall College). The 2nd edition of this useful supplement contains 42 classic labs and observational activities, along with cardboard cutout instruments that students can build and use for observations. Available for sale to students.
ISBN: 0-13-644196-3

BASIC ASTRONOMY LABS, by Jay Huebner and Terry Smith (University of North Florida, Jacksonville), and Michael Reynolds (Chabot Observatory and Science Center). A collection of 40 laboratory exercises, including a wide range of both traditional and innovative topics (such as the nature of human vision, radioactivity and time, and astronomy on the Internet). Detailed introductions provide a fully-developed context for each exercise. Available for sale to students.
ISBN: 0-13-376336-6

WORLD WIDE WEB PAGE We will be hosting a worldwide web homepage for this text that will include updated imagery from the world's best telescopes as they become available. These images, plus additional www resources, will be organized around the chapters and themes of the text. You may visit this area at *http://www.prenhall.com/~chaisson*

ACKNOWLEDGMENTS

Throughout the many drafts that have led to this book, we have relied on the critical analysis of many colleagues. Their suggestions ranged from the macroscopic issue of the book's overall organization to the minutiae of the technical accuracy of each and every sentence. We have also benefited from much good advice and feedback from many users of the first edition of the text and from our shorter book, *Astronomy: A Beginner's Guide to the Universe*. To these many helpful colleagues, we offer our sincerest thanks.

Robert H. Allen
University of Wisconsin, La Crosse

Timothy C. Beers
University of Evansville

Donald J. Bord
University of Michigan, Dearborn

William J. Boardman
Birmingham Southern College

Anne Cowley
Arizona State University

Bruce L. Cragin
Richland College

Norman Derby
Bennington College

Kimberly Engle
Drexel University

Donald Gudehus
Georgia State University

Marilynn Harper
Delaware County Community College

Joseph Heafner
Catawba Valley Community College

Steven D. Kawaler
Iowa State University

William Keel
University of Alabama

Mario Klairc
Midlands Technical College

Kristine Larsen
Central Connecticut State University

Robert J. Leacock
University of Florida

Larry A. Lebofsky
University of Arizona

M.A.K. Lodhi
Texas Tech University

Michael C. LoPresto
Henry Ford Community College

Phillip Lu
Western Conneticut State University

Milan Mijic
California State University, Los Angeles

Ronald Olowin
Saint Mary's College of California

Gregory W. Ojakangas
University of Minnesota, Duluth

Cynthia W. Peterson
University of Connecticut

James A. Roberts
University of North Texas

Malcolm P. Savedoff
University of Rochester

Harry L. Shipman
University of Delaware

C.G. "Pete" Shugart
Memphis State University

Maurice Stewart
Willamette University

Stephen R. Walton
California State University, Northridge

Peter Wehinger
Arizona State University

Louis Winkler
Pennsylvania State University

We would also like to acknowledge our gratitude to Leo Connolly for preparing the end-of-chapter questions and problems; to Ray Villard of the Space Telescope Science Institute for compiling the COMETS supplement; and to Don Neill of Columbia University for creating our world-wide web site.

The publishing team at Prentice Hall has assisted us at every step along the way. First and foremost, we owe heartfelt thanks to Tim Bozik, Editorial Director for Math and Science, for his whole-hearted support of this project, and to Ray Henderson, until recently Executive Editor for Physics and Astronomy, who has successfully navigated us through the tortuous maze of the publishing world, all the while managing the many variables that go into a multifaceted publication such as this. Ray has been a daily source of encouragement, sound advice, and good humor through three Chaisson/McMillan projects. We appreciate his dedication and wish him well in his new position in the Prentice Hall organization. His worthy successor, Alison Reeves, arriving at a crucial juncture, has played an essential role in helping us to see the work through to completion.

Dan Schiller, our Development Editor, has skillfully helped us revise numerous drafts of the manuscript and has been a constant source of knowledge and insight. Dan's critical eye, attention to detail, and bulldog-like tenacity have made him the linchpin holding the entire project together. We have also been fortunate to work once again with one of Prentice Hall's most experienced Production Editors, Susan Fisher, who has handled the innumerable details associated with this complex undertaking with customary efficiency and British imperturbability. The intricate and crucial task of page makeup was handled with aplomb by Shari Toron and Cindy Dunn, under the able leadership of John Jordan; they deserve the greatest credit for their extraordinary patience, diligence, and craftsmanship. A special word of thanks is due to Rolando Corujo and Patrice Van Acker for their invaluable help in refining the text's figures. Finally, we would like to express our gratitude to renowned space artist Dana Berry for allowing us to use many of his strikingly beautiful renditions of astronomical scenes, and to Lola Judith Chaisson for her painstaking work on the new H-R diagrams that she created especially for this edition.

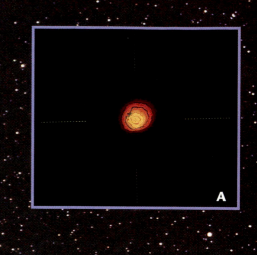

A

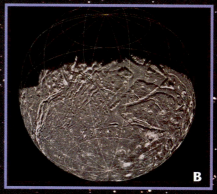

B

C

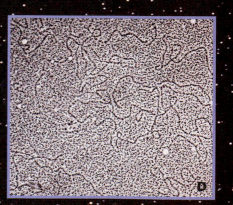

D

1

CHARTING THE HEAVENS

The Foundations of Astronomy

LEARNING GOALS

Studying this chapter will enable you to:

1 Explain the concept of the celestial sphere and the conventions of angular measurement that enable us to locate objects in the sky.

2 Describe how the Sun, the Moon, and the stars appear to change their positions from night to night and from month to month.

3 Account for these apparent motions in terms of the actual motions of the Earth and the Moon.

4 Show how the relative motions of the Earth, the Sun, and the Moon lead to eclipses.

5 Explain the simple geometric reasoning that allows astronomers to measure the distances of faraway objects.

(Opposite page, background) One of the most easily recognizable star fields in the winter nighttime sky, the familiar constellation Orion. This field of view spans roughly 100 light-years, or 10^{15} kilometers. (See also Figure 1.6.)

(Inset A) If we magnify the wide view of the Orion constellation, shown at left, by a million times, we enter into the realm of the largest stars, with sizes of about a billion kilometers. Such a star is seen in this false-color image of the red-giant star Betelguese (which is actually the bright star at the upper left of the Orion constellation).

(Inset B) Another magnification of a million brings us to the scale of typical moons—roughly 1,000 kilometers—represented here by Ariel, one of the many moons of Uranus.

(Inset C) With yet another million-times magnification, we reach scales of meters, represented here by an astronomer at the controls of her telescope.

(Inset D) At a final magnification of an additional million, we reach the scale of molecules (about 10^{-6} meter), represented by this coiled DNA molecule of a rat's liver.

Nature offers no greater splendor than the starry sky on a clear, dark night. Silent and jeweled with the constellations of ancient myth and legend, the night sky has inspired wonder throughout the ages—a wonder that leads our imaginations far from the confines of Earth and the pace of the present day, and out into the distant reaches of space and cosmic time itself. Astronomy, born in response to that wonder, is built on two of the most basic traits of human nature: the need to explore and the need to understand. Through the interplay of curiosity, discovery, and analysis—the keys to exploration and understanding—people have sought answers to questions about the universe since the earliest times. Astronomy is the oldest of all the sciences, yet never has it been more exciting than it is today.

1.1 *Our Place in Space*

In all of human history, there have been only two periods in which our understanding of the universe has been revolutionized within a single lifetime. The first spanned the years from the middle of the sixteenth century to the early part of the seventeenth, when the work of Copernicus, Kepler, and Galileo established beyond reasonable doubt the fact that our Earth is not the unmoving center of the entire universe, but in fact revolves about the Sun. The second is now underway. In the late twentieth century we have begun to break away from planet Earth, and in doing so we have achieved a whole new perspective on the universe in which we live.

Of all the scientific insights attained to date, one stands out boldly: Earth is neither central nor special. We inhabit no unique place in the universe. Astronomical research, especially within the past few decades, strongly suggests that we live on what seems to be an ordinary rocky *planet* called Earth, which is one of 9 known planets orbiting an average *star* called the Sun, which is one star near the edge of a huge collection of stars called the Milky Way *galaxy*, which is one galaxy among countless billions of others spread throughout the observable *universe*.

We are connected to these distant realms of space and time not only by our imaginations, but also through a common cosmic heritage: Most of the chemical elements in our bodies were created billions of years ago in the hot centers of long-vanished stars. Their fuel supply spent, these giant stars died in huge explosions, scattering afar the elements created deep within their cores. Eventually, this matter collected into clouds of gas that slowly collapsed to give birth to a new generation of stars. In this way, the Sun and its family of planets were formed nearly 5 billion years ago. Everything on Earth embodies atoms from other parts of the universe and from a past far more remote than the beginning of human evolution.

Although ours is the only planetary system we know of, others may orbit many of the billions upon billions of stars in the universe. Elsewhere, other beings, perhaps with an intelligence much greater than our own, might at this very moment be gazing in wonder at their own nighttime sky. Our own Sun might be nothing more than an insignificant point of light to them, if it is visible at all. If such beings exist, they too must share our cosmic origin.

1.2 *The Scale of Things*

Before going any further, let us clarify just what we mean by "the universe." We might define it poetically as the vast tracts of space and enormous stretches of time populated sparsely by stars and galaxies glowing in the dark. More scientifically, the **universe** is the totality of all space, time, matter, and energy. Consult Figures 1.1 through 1.4, and put some of these objects in perspective by studying Figure 1.5.

Take another look at the galaxy in Figure 1.3. This galaxy, whose catalog name is M83, is a swarm of about a hundred billion stars—more stars than people who have ever lived on Earth. The entire assemblage is spread across some 100,000 light years. A **light year** is the *distance* traveled by light, at a velocity of about 300,000 kilometers per second, in a year. It equals about 10 trillion kilometers (or around 6 trillion miles). Typical galactic systems are truly "astronomical" in size.

A thousand (1000), a million (1,000,000), a billion (1,000,000,000), and even a trillion (1,000,000,000,000)—these words occur regularly in everyday speech. But let's take a moment to understand the magnitude of these numbers and to appreciate the differences among them. One thousand is easy enough to understand; at the rate of one number per second, you could count to a thousand in about 16 minutes. However, if you wanted to count to a million, you would need more than 2 weeks of counting at the rate of one number per second, 16 hours per day (allowing 8 hours per day for sleep). To count from 1 to a billion at the same rate of one number per second and 16 hours per day would take nearly 50 years—the better part of an entire human lifetime.

In this text we will consider spatial domains spanning not just billions of kilometers, but billions of light years. We will discuss objects containing not just trillions of atoms, but trillions of stars. We will contemplate time intervals of not just billions of seconds or hours, but billions of years. You will need to become familiar with—and comfortable with—such enormous numbers. A good way to begin is to try and recognize just how much larger than a thousand is a million, and how much larger still is a billion. (The *More Precisely* feature on p. 8 explains the convenient method used by scientists for writing and manipulating very large and very small numbers.)

15,000 kilometers

1,500,000 kilometers

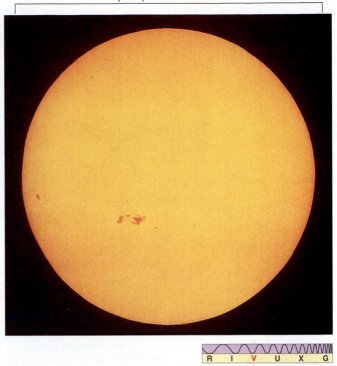

R I V U X G

R I V U X G

Figure 1.1 The Earth is a planet, a mostly solid object, though it has some liquid in its oceans and its core, and gas in its atmosphere. (In this view, you can clearly see the North and South American continents.)

Figure 1.2 The Sun is a star, a very hot ball of gas. Much bigger than the Earth, the Sun is held together by its own gravity.

About 100,000 light years

About 10,000,000 light years

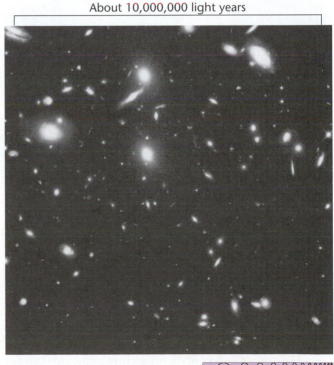

R I V U X G

R I V U X G

Figure 1.3 A typical galaxy is a collection of a hundred billion stars, each separated by vast regions of nearly empty space. This galaxy is the eighty-third entry in the catalog compiled by the eighteenth-century French astronomer Charles Messier—M83 for short. Our Sun is a rather undistinguished star near the edge of another such galaxy, called the Milky Way.

Figure 1.4 This photograph shows a portion of the Coma cluster of galaxies, some 300 million light years from Earth. Each galaxy contains hundreds of billions of stars, probably planets, and possibly living creatures.

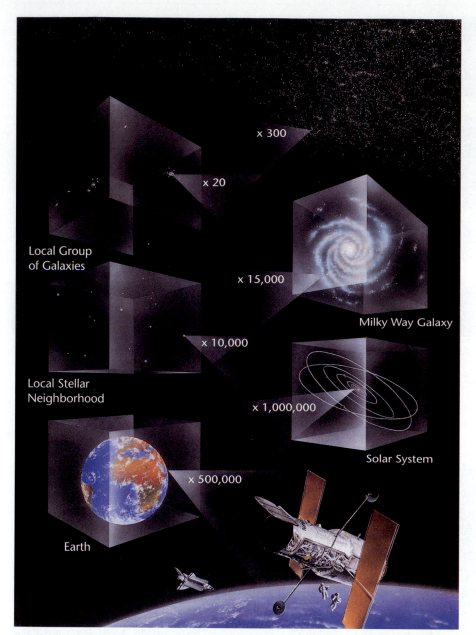

Local Group
of Galaxies

Local Stellar
Neighborhood

Earth

x 300

x 20

x 15,000

x 10,000

x 1,000,000

x 500,000

Milky Way Galaxy

Solar System

Figure 1.5 This artist's conception puts each of the previous figures in perspective. The bottom of this figure shows spacecraft (and astronauts) in Earth orbit, a view that widens progressively in each of the next five cubes drawn from bottom to top—the Earth, the planetary system, the local neighborhood of stars, the Milky Way Galaxy, and the closest cluster of galaxies. The numbers indicate approximately the increase in scale between each successive image.

1.3 *The Obvious View*

How have we come to know the universe around us? How do we know the proper perspective sketched in Figure 1.5? Our study of the universe, the subject of **astronomy**, begins by examining the sky.

CONSTELLATIONS IN THE SKY

Between sunset and sunrise on a clear night, we can see some 3000 points of light. Include the view from the opposite side of Earth, and nearly 6000 stars are visible to the unaided eye. A natural human tendency is to see patterns, and so people connected the brightest stars into configurations called **constellations**, which ancient astronomers named after mythological beings, heroes, and animals—whatever was important to them.

Figure 1.6 shows a constellation especially prominent in the nighttime sky from October through March: the "hunter" named Orion. Orion was a mythical Greek hero famed for his great beauty and stature, his hunting prowess, and his amorous pursuit of the Pleiades, the seven daughters of the giant Atlas. According to Greek mythology, in order to protect the Pleiades from Orion, the gods placed them among the stars, where Orion nightly stalks them across the sky but never catches them. Many constellations have similarly fabulous connections with ancient cultures.

Generally speaking, the stars that make up any particular constellation are not actually close to one another, even by astronomical standards. These stars merely are bright enough to observe with the naked eye and happen to lie in roughly the same direction in the sky as seen from

(a)

(b)

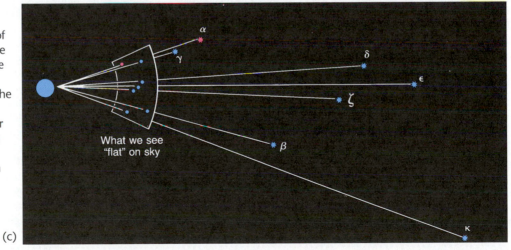

(c)

Figure 1.6 (a) A photograph of the group of bright stars that make up the constellation Orion. (b) The stars are connected to show the pattern visualized by the Greeks: the outline of a hunter. You can easily find this constellation in the winter sky by identifying the line of three bright stars in the hunter's "belt." (c) The true relationships between the stars, in three dimensions.

the Earth. That these stars have become associated with one another over the centuries is a tribute to the power of the human brain, which is extremely good at recognizing patterns and relationships between objects even when no true connection exists.

Perhaps not surprisingly, the patterns seen have a strong cultural bias—the astronomers of ancient China saw mythical figures different from those seen by the ancient Greeks, the Babylonians, and the people of other cultures, even though they were all looking at the same stars in the nighttime sky. Interestingly, different cultures often made the same basic *groupings* of stars, despite widely varying interpretations of what they saw.

For example, the group of seven stars usually known in North America as "the Big Dipper" is known as "the Wagon" or "the Plough" in Western Europe. The ancient Greeks regarded these same stars as the tail of "the Great Bear," the Egyptians saw them as the leg of an ox, the Siberians as a stag. Some Native Americans saw two mythical brothers, others an ermine, others still a funeral procession. The Chinese saw the pattern as a minor government official, dealing with the day-to-day concerns of the emperor.

The origins of most constellations, and of their names, date back to the dawn of recorded history. Some constellations served as navigational guides. For example, the star

More Precisely...   *Scientific Notation*

The objects studied by astronomers range in size from the smallest particles to the largest expanse of matter we know—the entire universe. Subatomic particles have sizes of about 0.000000000000001 meter, while galaxies (like that shown in Figure 1.3) typically measure some 1,000,000,000,000,000,000,000 meters across. The most distant known objects in the universe lie on the order of 100,000,000,000,000,000,000,000,000 meters from Earth.

Obviously, writing all those zeros is both cumbersome and inconvenient. More important, it is also very easy to make an error—write down one zero too many or too few and your calculations become hopelessly wrong! To avoid this, scientists always write large numbers using a shorthand notation in which the number of zeros following or preceding the decimal point is denoted by a superscript power, or *exponent,* of 10. The exponent is simply the number of places between the first significant (nonzero) digit in the number (reading from left to right) and the decimal point. Thus 1 is 10^0, 10 is 10^1, 100 is 10^2, 1000 is 10^3, and so on. For numbers less than 1, with zeros between the decimal point and the first significant digit, the exponent is negative: 0.1 is 10^{-1}, 0.01 is 10^{-2}, 0.001 is 10^{-3}, and so on. Using this notation we can shorten the number describing subatomic particles to 10^{-15} meter and write the number describing the size of a galaxy as 10^{21} meters.

More complicated numbers are expressed as a combination of a power of 10 and a multiplying factor. This factor is conventionally chosen to be a number between 1 and 10, starting with the first significant digit in the original number. For example, 150,000,000,000 meters (the distance from the Earth to the Sun, in round numbers) can be more concisely written as 1.5×10^{11} meters, 0.000000025 meters as 2.5×10^{-8} meter, and so on. The exponent is simply the number of places the decimal point must be moved *to the left* to obtain the multiplying factor.

Some other examples of scientific notation are

- the approximate distance to the Andromeda Galaxy = 2,000,000 light years = 2×10^6 light years
- the size of a hydrogen atom = 0.00000000005 meter = 5×10^{-11} meter

- the diameter of the Sun = 1,392,000 kilometers = 1.392×10^6 kilometers
- the U.S. national debt (as of July 1995) = \$4,924,917,000,000.00 = \$4.924917 trillion = 4.924917×10^{12} dollars.

In addition to providing a simpler way of expressing very large or very small numbers, this notation also makes it easier to do basic arithmetic. The rule for multiplication of numbers expressed in this way is simple: Just multiply the factors and add the exponents. Similarly for division: Divide the factors and subtract the exponents. Thus, 3.5×10^{-2} multiplied by 2.0×10^3 is simply $(3.5 \times 2.0) \times 10^{-2+3} = 7.0 \times 10^1$—that is, 70. Again, 5×10^6 divided by 2×10^4 is just $(5/2) \times 10^{6-4}$, or 2.5×10^2 (= 250). Applying these rules to unit conversions, 200,000 nanometers is $200,000 \times 10^{-9}$ meter (since 1 nanometer = 10^{-9} meter; see the *More Precisely* feature on p.13), or $2 \times 10^5 \times 10^{-9}$ meter, or $2 \times 10^{5-9} = 2 \times 10^{-4}$ meter. Verify these rules yourself with a few examples of your own. The advantages of this notation when considering astronomical objects will soon become obvious.

Scientists often use "rounded-off" versions of numbers, both for simplicity and for ease of calculation. For example, we will usually write the diameter of the Sun as 1.4×10^6 kilometers, instead of the more precise number given earlier. Similarly, the diameter of the Earth is 12,756 kilometers, or 1.2756×10^4 kilometers, but for "ballpark" estimates, we really don't need so many digits, and the more approximate number 1.3×10^4 kilometers will suffice. Very often, we perform rough calculations using only the first one or two significant digits in a number, and that may be all that is necessary to make a particular point. For example, to support the statement, "The Sun is much larger than the Earth," we need only say that the ratio of the two diameters is roughly 1.4×10^6 divided by 1.3×10^4. Since 1.4/1.3 is close to 1, the ratio is approximately $10^6/10^4 = 10^2$, or 100. The essential fact here is that the ratio is much larger than 1; calculating it to greater accuracy (to get 109.13) would give us no additional *useful* information. This technique of stripping away the arithmetic details to get to the essence of a calculation is very common in astronomy, and we will use it frequently throughout this text.

Polaris, part of the Little Dipper, signals north, and the near-constancy of its location in the sky, from hour to hour and night to night, has aided travelers for centuries. Other constellations served as primitive calendars to predict planting and harvesting seasons. For example, many cultures knew well that the appearance of certain stars on the horizon just before daybreak signaled the beginning of spring and the end of winter. Such knowledge provided the foundations for the science of *astronomy*.

In many societies, people came to believe that there were other benefits in being able to trace the regularly changing positions of heavenly bodies. The relative posi-

tions of stars and planets at a person's birth were carefully studied by *astrologers*, who used the data to make predictions about that person's destiny. Thus, in a sense, astronomy and astrology arose from the same basic impulse—the desire to "see" into the future—and indeed, for a long time they were indistinguishable from one another. Today most people recognize that astrology is nothing more than an amusing diversion (although millions still study their horoscopes in the newspaper every morning!). Nevertheless, the ancient astrological terminology—the names of the constellations and some terms used to describe the locations and motions of the planets, for exam-

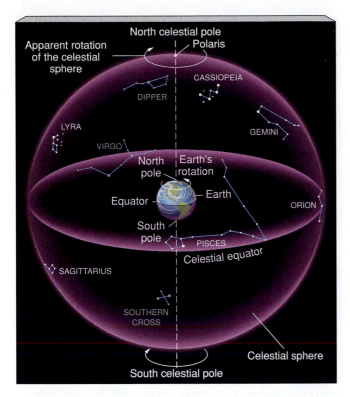

Figure 1.7   Planet Earth sits fixed at the hub of the celestial sphere, which contains all the stars. This is one of the simplest possible models of the universe, but it doesn't agree with all the facts that astronomers know about the universe.

THE CELESTIAL SPHERE

① Following the constellations nightly, ancient sky-watchers noted that the star patterns seemed unchanging. It was natural for them to conclude that the stars must be firmly attached to a **celestial sphere** surrounding Earth—a canopy of stars resembling an astronomical painting on a heavenly ceiling. Figure 1.7 shows how early astronomers pictured the stars as moving with this celestial sphere as it turned around a fixed, unmoving Earth.

For the most part, stars rise in the east, move across the sky, and set in the west each night. Some sweep out a large arc high above the horizon, but others appear to move very little. In fact, closer scrutiny (or time-lapse photography—see Figure 1.8) shows that all stars move in circles around a point in the sky very close to the star Polaris (better known as the Pole Star, or the North Star). To the ancients, this point represented the axis around which the entire celestial sphere turned.

From our modern standpoint, the apparent motion of the stars is the result of the spin, or **rotation**, not of the celestial sphere but of Earth. Polaris indicates the direction—due north—in which Earth's rotation axis points. Even though we now know that the celestial sphere is an incorrect description of the heavens, we still use the idea as a convenient fiction that helps us visualize the positions of stars in the sky. The point where Earth's axis intersects the celestial sphere in the Northern Hemisphere is known as the **north celestial pole**, and it is directly above Earth's North Pole. In the Southern Hemisphere, the extension of Earth's axis in the opposite direction defines the **south celestial pole**, directly above Earth's South Pole. Midway between the north and south celestial poles lies the **celestial equator**, representing the intersection of Earth's equatorial plane with the celestial sphere. These parts of the celestial sphere are marked on Figure 1.7.

ple—is still used throughout the astronomical world. The constellations still help astronomers to specify large areas of the sky, much as geologists use continents or politicians use voting precincts to identify certain localities on planet Earth. In all, there are 88 constellations, most of them visible from North America at some time during the year.

(a)

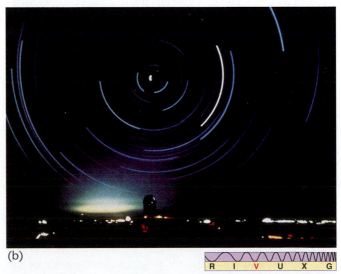

(b)

R   I   V   U   X   G

Figure 1.8   Time-lapse photographs of the northern sky. Each trail is the path of a single star across the nighttime sky. The duration of the exposure in (a) is about 1 hour, that in (b) is about 5 hours. How can you tell that this is so? The center of the concentric circles is near the north star, Polaris, whose short, bright arc is prominently visible in (b).

1.4 *The Motion of the Sun and the Stars*

DAY-TO-DAY CHANGES

2 We measure time by the Sun. Because the rhythm of day and night is central to our lives, it is not surprising that the period of time from one sunrise (or noon, or sunset) to the next, the 24-hour **solar day**, is our basic social time unit. The daily progress of the Sun and the other stars across the sky is known as *diurnal motion*. As we have just seen, it is a consequence of the rotation of Earth. But the star's positions in the sky do *not* repeat themselves exactly from one night to the next. In fact, each night, the whole celestial sphere appears to be shifted a little relative to the horizon, compared with the night before. The easiest way to confirm this difference is by noticing the stars that are visible just after sunset or just before dawn. You will find that they are in slightly different locations from the previous night. Because of this shift, a day measured by the stars—called a **sidereal day** after the Latin word *sidus*, meaning star—differs from a solar day. Evidently, there is more to the apparent motion of the heavens than just simple rotation. In fact, the motion of the Earth relative to the Sun—Earth's **revolution**—is also of great importance.

The reason for the difference between a solar day and a sidereal day is sketched in Figure 1.9. Each time Earth rotates once on its axis, it also moves a small distance along its orbit about the Sun. Earth therefore has to rotate through slightly more than 360° for the Sun to return to the same apparent location in the sky. Thus, the interval of time between noon one day and noon the next (a solar day) is slightly greater than one true rotation period (one sidereal day). Our planet takes 365 days to orbit the Sun, so the additional angle is 360°/365 = 0.986°. Because Earth takes about 3.9 minutes to rotate through this angle, the solar day is 3.9 minutes longer than the sidereal day (that is, one sidereal day is roughly 23^h56^m long.) From the point of view of the ancients, the "fact" that the Sun moved relative to the stars made it necessary to envisage not just one celestial sphere, but *two*: one for the stars and one for the Sun, each spinning about the Earth but at slightly different rates and with different orientations.

SEASONAL CHANGES

Seen from Earth, the Sun appears to move relative to the other stars. That is because Earth orbits the Sun, completing one orbit in 365.242 solar days—a period of time known as one *tropical year*. By the end of that time, the Sun has "traversed the sky" and returned to its starting position on the celestial sphere, and the cycle begins anew. The apparent motion of the Sun in the sky, expressed relative to the stars, follows a path known as the **ecliptic**. As illustrated in Figure 1.10(a), the ecliptic forms a great circle on the celestial sphere, inclined at an angle of about 23.5° to the celestial equator. This tilt is just a consequence of the inclination of Earth's rotation axis to the plane of its orbit, as shown in Figure 1.10(b).

The point on the ecliptic where the Sun is at its northernmost point above the celestial equator is known as the

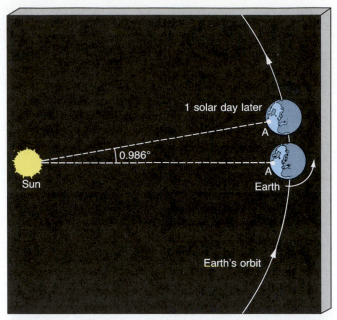

Figure 1.9 The difference between a solar and a sidereal day can be easily explained once we understand that Earth revolves around the Sun at the same time as it rotates on its axis. A solar day is the time from one noon to the next. In that time, Earth also moves a little in its solar orbit. Because Earth completes one circuit (360°) around the Sun in 1 year (365 days), it moves through nearly 1° in 1 day. Thus, between noon at point *A* on one day and noon at the same point the next day, Earth actually rotates through about 361°. Consequently, the solar day exceeds the sidereal day (360° rotation) by about 4 minutes. Note that the diagram is not drawn to scale, so that the true 1° angle is in reality much smaller than shown here.

summer solstice (from the Latin word *sol*, meaning sun). As shown in Figure 1.10, it represents the point in Earth's orbit where our planet's North Pole points closest to the Sun. This occurs on or near June 21—the exact date varies slightly from year to year because the actual length of a year is not a whole number of days. As Earth rotates, points north of the equator spend the greatest fraction of their time in sunlight on that date, so the summer solstice corresponds to the longest day of the year in the Northern Hemisphere and the shortest day in the Southern Hemisphere. Six months later, the Sun is at its southernmost point, and we have reached the **winter solstice** (December 21)—the shortest day in the Northern Hemisphere and the longest in the Southern Hemisphere. These two effects—the height of the Sun above the horizon and the length of the day— combine to account for the **seasons** we experience. In summer in the Northern Hemisphere, the Sun is high in the sky and the days are long, so temperatures are generally much higher than in winter, when the Sun is low and the days are short.

The two points where the ecliptic intersects the celestial equator are known as **equinoxes**. On those dates, day and night are of equal duration. (The word *equinox* derives from the Latin for "equal night.") In the fall (in the Northern Hemisphere), as the Sun crosses from the Northern into the Southern Hemisphere, we have the **autumnal equinox** (on September 21). The **vernal equinox** occurs in Northern spring, on or near

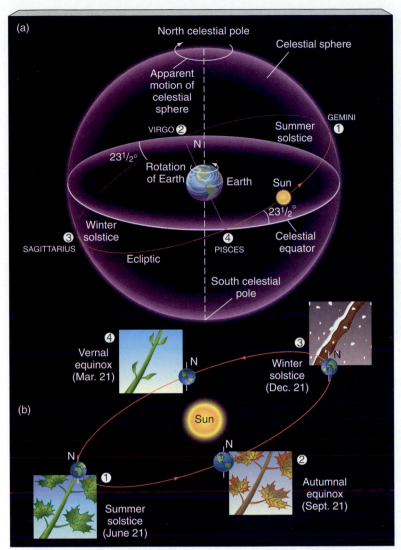

Figure 1.10 (a) The apparent path of the Sun on the celestial sphere and (b) its actual relation to Earth's rotation and revolution. The seasons result from the changing height of the Sun above the horizon. At the summer solstice (the points marked 1), the Sun is highest in the sky, as seen from the northern hemisphere, and the days are longest. In the "celestial sphere" picture (a), the Sun is at its northernmost point on its path around the ecliptic; in reality (b), the summer solstice corresponds to the point on Earth's orbit where our planet's north pole points most nearly toward the Sun. The reverse is true at the winter solstice (point 3). At the vernal and autumnal equinoxes, day and night are of equal length. These are the times when, as seen from Earth (a), the Sun crosses the celestial equator. They correspond to the points in Earth's orbit when our planet's axis is perpendicular to the line joining the Earth and Sun (b).

March 21, as the Sun crosses the celestial equator moving north. Because of its association with the end of winter and the start of a new growing season, the vernal equinox was particularly important to early astronomers and astrologers.

SUMMER AND WINTER CONSTELLATIONS

Figure 1.11 (a) illustrates the major stars visible from most locations in the United States on clear summer evenings. The brightest stars—Vega, Deneb, and Altair—form a conspicuous triangle high above the constellations Sagittarius and Capricornus, which are low on the southern horizon. In the winter sky, however, these stars have been replaced, as shown in Figure 1.11(b), by several well-known constellations that include Orion, Leo, and Gemini. In the constellation Canis Major lies Sirius (the Dog Star), the brightest star in the sky. Year after year, the same stars and constellations return, each in its proper season. Every winter evening, Orion is high overhead; every summer, it is gone. For more detailed maps of the sky at different seasons, consult the star charts at the end of the book.

The reason for these regular seasonal changes is, once again, the revolution of the Earth around the Sun. Earth's darkened hemisphere faces in a slightly different direction in space each evening. The change in direction is only about 1° per night (see Figure 1.9)—too small to be easily noticed with the naked eye from one evening to the next, but clearly noticeable over the course of weeks and months, as illustrated in Figure 1.12. After 6 months, the Earth has reached the opposite side of its orbit, and at night we face an entirely different group of stars and constellations. Ancient astronomers would have said that the Sun has moved to the opposite side of the celestial sphere, so that a different set of stars is visible at night.

The 12 constellations through which the Sun passes as it moves along the ecliptic—that is, the constellations we would see looking in the direction of the Sun, if they weren't overwhelmed by the Sun's light—had special significance for astrologers of old. These constellations are collectively known as the **zodiac**. The time required for the constellations to complete one cycle around the sky and to return to their starting points as seen from a given point on the Earth is one **sidereal year**. Earth completes exactly one orbit around the Sun in this time. One sidereal year is 365.256 solar days long, about 20 minutes longer than a tropical year. We will return to the reason for this slight discrepancy in a moment.

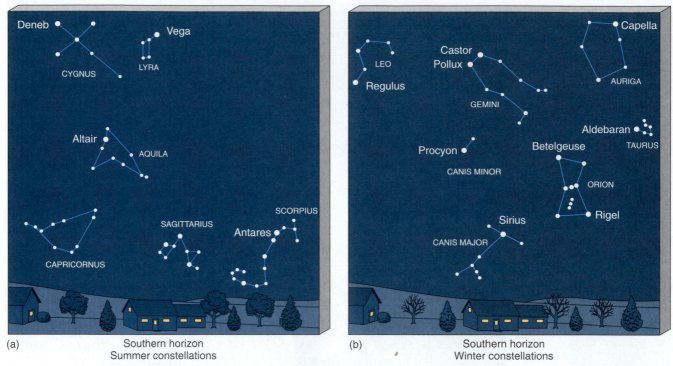

Figure 1.11 (a) A typical summer sky above the United States. Some prominent stars (labeled in larger print) and constellations (labeled in small capital letters) are shown. (b) A typical winter sky above the United States.

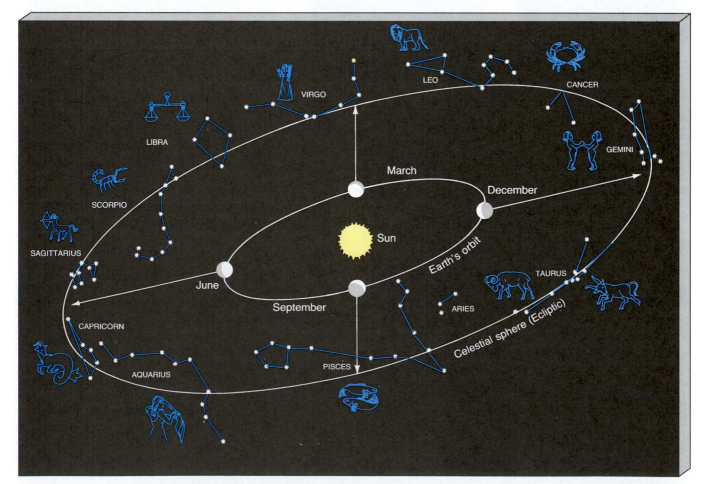

Figure 1.12 The view of the night sky changes as the Earth moves in its orbit about the Sun. As drawn here, the night side of Earth faces a different set of constellations at different times of the year.

More Precisely... *Astronomical Measurement*

Astronomers use many different kinds of units in their work, simply because no single system of units will do. Rather than the *Système Internationale* (SI), or meter-kilogram-second (MKS), metric system used in most high school and college science classes, many professional astronomers still prefer the older centimeter-gram-second (CGS) system. However, astronomers also commonly introduce new units when convenient. For example, when discussing stars, the mass and radius of the Sun are often used as reference points. The solar mass, written as $M_\odot$, is equal to 2.0×10^{33} g, or 2.0×10^{30} kg (since 1 kg = 1000 g). The solar radius, $R_\odot$, is equal to 700,000 km, or 7.0×10^8 m (1 km = 1000 m). The subscript $\odot$ always stands for Sun. Similarly, the subscript $\oplus$ always stands for Earth. In this book, we will use the units that astronomers commonly use in any given context, but we will also give the "standard" SI equivalents where appropriate.

Of particular importance are the units of length astronomers use. On small scales, the *angstrom* (1 Å = 10^{-10} m = 10^{-8} cm), the *nanometer* (1 nm = 10^{-9} m = 10^{-7} cm), and the *micron* (1 μm = 10^{-6} m = 10^{-4} cm) are used. Distances within the solar system are usually expressed in terms of the *astro-*nomical unit (A.U.), the mean distance between Earth and the Sun. One A.U. is approximately equal to 150,000,000 km, or 1.5×10^{11} m. On larger scales, the *light year* (1 ly = 9.5×10^{15} m = 9.5×10^{12} km) and the *parsec* (1 pc = 3.1×10^{16} m = 3.1×10^{13} km = 3.3 ly) are commonly used. Still larger distances use the regular prefixes of the metric system: *kilo* for one thousand and *mega* for one million. Thus, 1 *kiloparsec* (kpc) = 10^3 pc = 3.1×10^{19} m, 10 megaparsecs (Mpc) = 10^7 pc = 3.1×10^{23} m, and so on.

Astronomers use units that make sense within a context, and as contexts change, so do the units. For example, we might measure densities in grams per cubic centimeter (g/cm^3), in atoms per cubic meter (atoms/m^3), or even in solar masses per cubic megaparsec ($M_\odot$/Mpc^3), depending on the circumstances. The important thing to know is that once you understand the units, you can convert freely from one set to another. For example, the radius of the Sun could equally well be written as $R_\odot = 6.96 \times 10^8$ m, or 6.96×10^{10} cm, or 109 $R_\oplus$, or 4.65×10^{-3} A.U., or even 7.36×10^{-8} ly—whichever happens to be most useful. Some of the more common units used in astronomy, and the contexts in which they are most likely to be encountered, follow.

Length:

1 angstrom (Å)	= 10^{-10} m	atomic physics,
1 nanometer (nm)	= 10^{-9} m	spectroscopy
1 micron (μm)	= 10^{-6} m	interstellar dust and gas
1 centimeter (cm)	= 0.01 m	in widespread use
1 meter (m)	= 100 cm	throughout all
1 kilometer (km)	= 1000 m = 10^5 cm	astronomy
Earth radius ($R_\oplus$)	= 6378 km	planetary astronomy
Solar radius ($R_\odot$)	= 6.96×10^8 m	solar system,
1 astronomical unit (A.U.)	= 1.496×10^{11} m	stellar evolution
1 light year (ly)	= 9.46×10^{15} m = 63,200 A.U.	galactic astronomy,
1 parsec (pc)	= 3.09×10^{16} m = 3.26 ly	stars and star clusters
1 kiloparsec (kpc)	= 1000 pc	galaxies, galaxy clusters,
1 megaparsec (Mpc)	= 1000 kpc	cosmology

Mass:

1 gram (g)		in widespread use in
1 kilogram (kg)	= 1000 g	many different areas
Earth mass ($M_\oplus$)	= 5.98×10^{24} kg	planetary astronomy
Solar mass ($M_\odot$)	= 1.99×10^{30} kg	"standard" unit for all mass scales larger than Earth

Time:

1 second (s)		in widespread use throughout astronomy
1 hour (h)	= 3600 s	planetary and stellar
1 day (d)	= 86400 s	scales
1 year (yr)	= 3.16×10^7 s	virtually all processes occurring on scales larger than a star

1.5 *Celestial Coordinates*

❶ The simplest method of locating stars in the sky is to specify their constellation and then rank the stars in it in order of brightness. The brightest star is denoted by the Greek letter α (alpha), the second brightest by β (beta), and so on. Thus, the two brightest stars in the constellation Orion—Betelgeuse and Rigel—are also known as α Orionis and β Orionis, respectively. (Precise recent observations show that Rigel is actually brighter than Betelgeuse, but the names are now permanent.) Because there are many more stars in any given constellation than there are letters in the Greek alphabet, this method is of limited utility. However, for naked-eye astronomy, where only bright stars are involved, it is quite satisfactory.

For more precise measurements, astronomers find it helpful to lay down a system of **celestial coordinates** on

More Precisely... *Angular Measure*

The size and scale of things are often specified by measuring lengths and angles. The concept of *length measurement* is fairly intuitive. We've already noted the dimensions of some objects found in the universe, ranging from subatomic particles with sizes of 10^{-13} centimeter to whole galaxies with sizes of 10^{23} centimeters. The concept of *angular measurement* may be less intuitive, but it too can become second nature if you remember a few simple facts.

A full circle contains 360 *arc degrees* (or just 360°). Therefore, the half-circle that stretches from horizon to horizon, passing directly overhead and spanning the portion of the sky visible to one person at any one time, contains 180°.

Each 1° increment can be further subdivided into fractions of an arc degree, called *arc minutes*; there are 60 arc minutes in 1 arc degree. Both the Sun and the Moon project an angular size of 30 arc minutes on the sky. Your little finger, held at arm's length, does just about the same, covering a small (actually, about 40 arc minutes) slice of the entire 180 arc degrees from horizon to horizon.

Finally, an arc minute can be further sliced into 60 equal *arc seconds*. Put another way, an arc minute is 1/60 of an arc degree, and an arc second is $1/60 \times 1/60 = 1/3600$ of an arc degree. An arc second is an extremely small unit of angular measure. It is, in fact, the angle subtended (projected) by a centimeter-sized object at a distance of about 2 kilometers or approximately the angle subtended by a dime when seen from a mile away. The accompanying figure illustrates this subdivision of the circle into progressively smaller units.

Don't be confused by the units used to measure angles. Arc minutes and arc seconds have nothing to do with the measurement of time, and arc degrees have nothing to do with temperature. Arc degrees, arc minutes, and arc seconds are simply ways to measure the size and position of objects in the universe. An arc degree is usually written as 1°, 1 arc minute is denoted 1′, and 1 arc second is 1″.

The angular size of an object depends both on its actual size and on its distance from us. For example, the Moon, at its present distance from Earth, has an angular diameter of 0.5°, or 30′. If the Moon were twice as far away, it would appear half as big—15′ across—even though its actual size would be the same. Similarly, if the Moon were one-third as far away, its angular diameter would be three times as great—1.5°, and so on. Thus, *angular size by itself is not enough to determine the actual diameter of an object—the distance must also be known.*

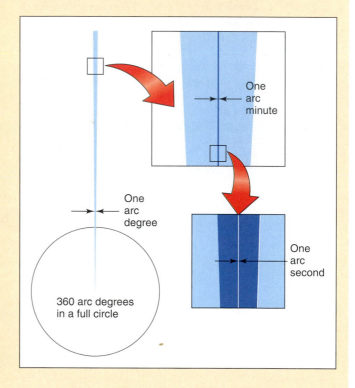

the sky. If we think of the stars as being attached to the celestial sphere centered on Earth, then the familiar system of latitude and longitude on Earth's surface extends quite naturally to cover the sky. The celestial analogs of latitude and longitude on the Earth's surface are known as *declination* and *right ascension*, respectively. Figure 1.13 illustrates the meanings of right ascension and declination on the celestial sphere and compares them with longitude and latitude on Earth.

Declination (dec) is measured in **degrees** (°) north or south of the celestial equator, just as latitude is measured in degrees north or south of Earth's equator. (See the *More Precisely* feature above for a discussion of angular measure.) Thus, the celestial equator is at a declination of 0°, the north celestial pole is at +90°, and the south celestial pole is at -90° (the minus sign here just means "south of the celestial equator").

Right as**cension** (RA) is measured in units called *hours*, *minutes*, and *seconds*, and it increases in the eastward direction. The angular units used to measure right ascension are constructed to parallel the units of time; the two sets of units are connected by the rotation of the Earth (or of the celestial sphere). In 24 hours, Earth rotates once on its axis, or through 360°. Thus, in a time period of 1 hour, Earth rotates through 360°/24 = 15°, or 1^{h}. In 1 minute of time, Earth rotates through 1^{m}; in 1 second, Earth rotates through 1^{s}. The choice of zero right ascension is quite arbitrary—it is conventionally taken to be the position of the Sun in the sky *at the instant of the vernal equinox*, when the Sun happens to lie between the constellations Pisces and Aquarius.

Although the units of right ascension were originally defined in this way to assist astronomical observation, their names are rather unfortunate, as these are angular

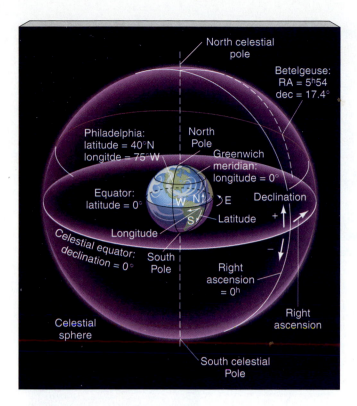

Figure 1.13 The right ascension and declination of a star on the celestial sphere are defined similarly to longitude and latitude on the surface of the Earth. Just as longitude and latitude allow us to locate a point on the surface of Earth, right ascension and declination specify locations on the sky. For example, to find Philadelphia on Earth, look 75° west of the Greenwich Meridian and 40° north of the Equator. Similarly, to locate the star Betelgeuse on the celestial sphere, look 5ʰ54ᵐ east of the vernal equinox (the line on the sky with a right ascension of zero) and 7°24′ north of the celestial equator.

measures, *not* units of time. Moreover, the angular units used for right ascension are *not* the same units as defined in the *More Precisely* feature on p. 14, which discusses angular measure. In fact, right ascension $1^m = 15°/60 = 0.25°$, or *15* arc minutes (15′). Similarly, right ascension $1^s = 15$ arc seconds (15″). If you remember that the units of right ascension are used *only* for that one purpose, and that all angular measurements except right ascension use *arc minutes* (′) and *arc seconds* (″), you should avoid undue confusion.

Just as latitude and longitude are tied to Earth, right ascension and declination are fixed on the celestial sphere. Although the stars appear to move across the sky because of the Earth's rotation, their celestial coordinates remain *constant* over the course of a night. Thus, we have a quantitative alternative to the use of constellations in specifying the positions of stars in the sky. For example, the stars Rigel and Betelgeuse mentioned earlier can be precisely located by looking in the directions $5^h13^m36^s$ (RA), -8°13′ (dec) and $5^h54^m0^s$ (RA), 07°24′ (dec), respectively. The coordinates of Betelgeuse on the celestial sphere are marked on Figure 1.13.

1.6 The Motion of the Moon

❷ ❸ The Moon is our nearest neighbor in space. Apart from the Sun, it is by far the brightest object in the sky. Unlike the Sun and the other stars, however, it emits no light of its own. Instead, it shines by reflecting sunlight. Another difference is that the Moon's appearance *changes* from night to night—in fact, on some nights it cannot be seen at all. Also, the Moon's daily rising and setting and its nightly motion through the sky differ from the motion of the celestial sphere. Like the Sun, the Moon appears to move relative to the stars—it crosses the sky at a rate of about 12° per day, which means it moves an angular distance equal to its own diameter—30′—in about an hour. Today, we explain these observations in terms of the Moon's revolution around the Earth. For ancient astronomers, however, the Moon's motion meant having to introduce yet another sphere in the sky, with a motion separate from either that of the stars or that carrying the Sun.

The Moon's appearance undergoes a regular cycle of changes, or **phases**, taking a little more than 29 days to complete. (The word *month* is actually derived from the word *Moon*.) Figure 1.14 illustrates the appearance of the Moon at different times in this monthly cycle. Starting from the so-called **new Moon**, which is all but invisible in the sky, the Moon appears to *wax* (or grow) a little each night and is visible as a growing *crescent* (panel 1 of Figure 1.14). One week after new Moon, half of the lunar disk can be seen (panel 2). This phase is known as a **quarter Moon**. During the next week, the Moon continues to wax, passing through the *gibbous* phase (panel 3) until, two weeks after new Moon, the **full Moon** (panel 4) is visible. During the next 2 weeks, the Moon *wanes* (or shrinks), passing in turn through the gibbous, quarter, and crescent phases (panels 5–7), eventually becoming new again. The waxing and waning phases are not merely time reversals of each other, however. The waxing Moon grows from the western edge of the disk, while the waning Moon shrinks toward the eastern edge.

The Moon doesn't actually change its size and shape on a monthly basis, of course; the full circular disk of the Moon is present at all times. Why then don't we always see a full Moon? As illustrated in Figure 1.14, half of the Moon's surface is illuminated by the Sun at any instant. However, not all of the Moon's sunlit face can be seen because of the Moon's position with respect to Earth and the Sun. When the Moon is full, we see the entire "daylit" face because the Sun and the Moon are in opposite directions from the Earth in the sky. In the case of a new Moon, the Moon and the Sun are in almost the same part of the sky, and the sunlit side of the Moon is oriented away from us. During the new Moon, the Sun must be almost behind the Moon, from our perspective.

As the Moon revolves around Earth, its position in the sky changes with respect to the stars. In one **sidereal month** (27.3 days), the Moon returns to the starting point

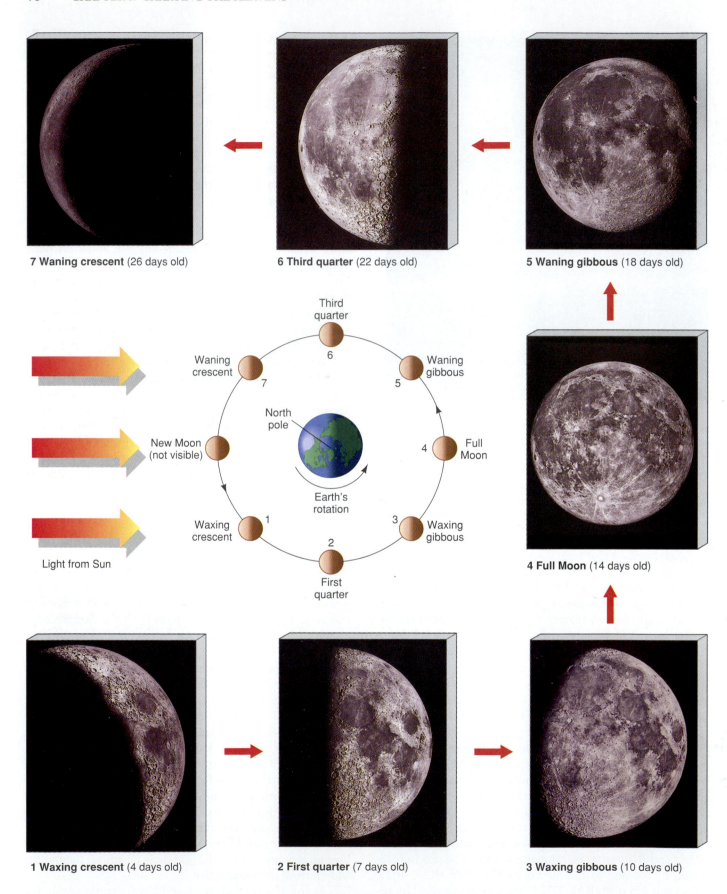

7 Waning crescent (26 days old)

6 Third quarter (22 days old)

5 Waning gibbous (18 days old)

Third
quarter

6

Waning
crescent

7

Waning
gibbous

5

North
pole

New Moon
(not visible)

4   Full
Moon

Earth's
rotation

Waxing
crescent

1

3   Waxing
gibbous

2

First
quarter

Light from Sun

4 Full Moon (14 days old)

1 Waxing crescent (4 days old)

2 First quarter (7 days old)

3 Waxing gibbous (10 days old)

Figure 1.14 Because the Moon orbits the Earth, the visible fraction of the sunlit face differs from night to night. The complete cycle of lunar phases takes 29 days to complete.

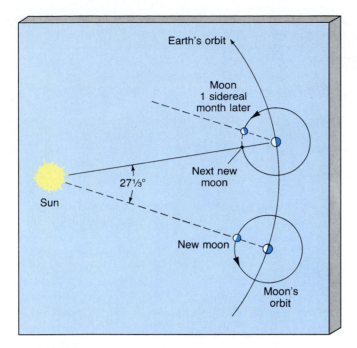

Figure 1.15 The difference between a synodic and a sidereal month stems from the motion of the Earth relative to the Sun. Because the Earth orbits the Sun in 365 days, in the 29.5 days from one new Moon to the next (one synodic month), Earth moves through an angle of approximately 29°. Thus the Moon must revolve more than 360° between new Moons. The sidereal month, which is the time taken for the Moon to revolve through exactly 360°, relative to the stars, is about 2 days shorter.

on its celestial sphere, having traced out a great circle in the sky. The time required for the Moon to complete a full cycle of phases, one **synodic month**, is a little longer—about 29.5 days. The synodic month is a little longer than the sidereal month for the same reason that a solar day is slightly longer than a sidereal day: Because of the motion of the Earth around the Sun, the Moon must complete slightly more than one full revolution to return to the same phase in its orbit (see Figure 1.15).

1.7 *Eclipses*

4 From time to time—but only at new or full Moon—the Sun and the Moon line up precisely as seen from Earth, and we observe the spectacular phenomenon known as an *eclipse*. When the Sun and the Moon are in exactly *opposite* directions, as seen from Earth, Earth's shadow sweeps across the Moon, temporarily blocking the Sun's light and darkening the Moon in a **lunar eclipse**, as illustrated in Figure 1.16. From Earth, we see the curved edge of Earth's shadow begin to cut across the face of the full Moon and slowly eat its way into the lunar disk. Usually, the alignment of the Sun, Earth, and Moon is imperfect, and the shadow never completely covers the Moon. Such an occurrence is known as a **partial eclipse**. Occasionally, however, the entire lunar surface is obscured in a **total eclipse**, as shown in Figure 1.16. Total lunar eclipses last only as long as is needed for the Moon to pass through Earth's shadow—no more than about 100 minutes. During that time, the Moon

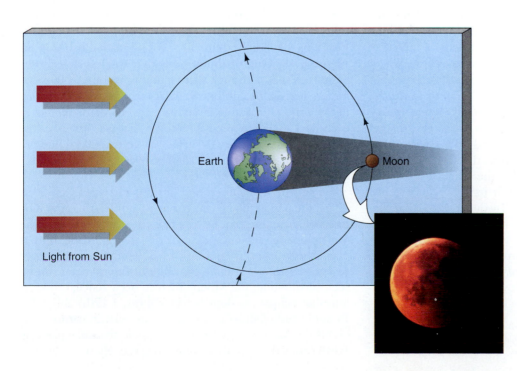

Figure 1.16 A lunar eclipse occurs when the Moon passes through Earth's shadow. At these times we see a darkened, copper-colored Moon, as shown in the inset photograph. The coloration is caused by sunlight deflected by the Earth's atmosphere onto the Moon's surface. (Note that this figure is not drawn to scale.)

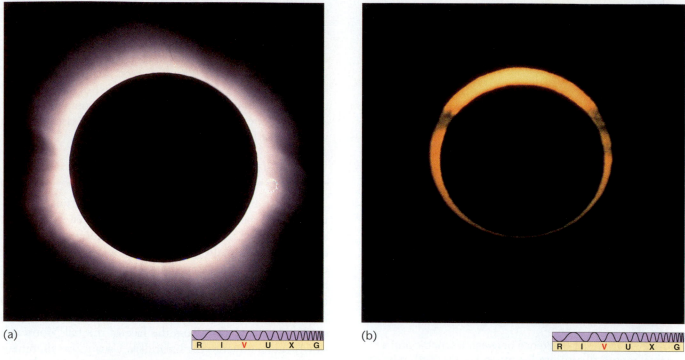

(a)

(b)

Figure 1.17 (a) During a total solar eclipse, the Sun's corona becomes visible as an irregularly shaped halo surrounding the blotted-out disk of the Sun. This was the July 1991 eclipse, as seen from the Baja Peninsula. (b) During an annular eclipse, the Moon fails to completely hide the Sun, so a thin ring of light remains. No corona is seen in this case because even the small amount of the Sun still visible completely overwhelms the corona's faint glow. This was the December 1973 eclipse, as seen from Algiers. (The gray fuzzy areas at top left and right are clouds in Earth's atmosphere.)

often acquires an eerie, deep red coloration—the result of a small amount of sunlight being refracted (bent) by Earth's atmosphere onto the lunar surface, preventing the shadow from being completely black. It is perhaps understandable that many primitive cultures interpreted lunar eclipses as harbingers of disaster.

When the Moon and the Sun are in exactly the *same* direction, an even more awe-inspiring event occurs. The Moon passes directly in front of the Sun, briefly turning day into night in a **solar eclipse**. In a total solar eclipse, when the alignment is perfect, planets and some stars become visible in the daytime as the Sun's light is reduced to nearly nothing. We can also see the Sun's ghostly outer atmosphere, or *corona* (Figure 1.17a).* In a partial solar eclipse, the Moon's path is slightly "off center," and only a portion of the Sun's face is covered. In either case, the sight of the Sun apparently being swallowed up by the black disk of the Moon is disconcerting even today. It must surely have inspired fear in early observers. Small wonder, then, that the ability to predict such events was a highly prized skill.

Unlike a lunar eclipse, which is simultaneously visible from all locations on the nighttime side of Earth, a total solar eclipse can be seen from only a small portion of the daytime side. The Moon's shadow on Earth's surface is

about 7000 kilometers wide—roughly twice the diameter of the Moon. Outside of that shadow, no eclipse is seen. However, only within the central region of the shadow, the **umbra**, is the eclipse total. Within the shadow but outside the umbra, in the **penumbra**, the eclipse is partial, with less and less of the Sun being obscured the farther one travels from the shadow's center. The connections between the umbra, the penumbra, and the relative locations of Earth, Sun, and Moon are illustrated in Figure 1.18. One of the reasons that total solar eclipses are rare is that, although the penumbra is some 7000 kilometers across, the umbra is always very small—even under the most favorable circumstances, its diameter never exceeds 270 kilometers. Moreover, because the shadow sweeps across the Earth's surface at over 1700 kilometers per hour, the duration of a total eclipse at any given point can never exceed 7.5 minutes.

The Moon's orbit around Earth is not exactly circular. Thus, the Moon may be far enough from Earth at the moment of an eclipse that its disk fails to cover the disk of the Sun completely, even though their centers coincide. In that case, there is no region of totality—the umbra never reaches Earth at all, and a thin ring of sunlight can still be seen surrounding the Moon. Such an occurrence, called an **annular eclipse**, is depicted in Figures 1.17(b) and 1.18. Roughly half of all solar eclipses are annular in nature.

If the Moon orbited Earth in exactly the same plane as Earth orbits the Sun, there would be precisely one solar and

---

Actually, although a total solar eclipse is undeniably a spectacular occurrence, the visibility of the corona is probably the most important astronomical aspect of such an event today. It enables us to study this otherwise hard-to-see part of our Sun.

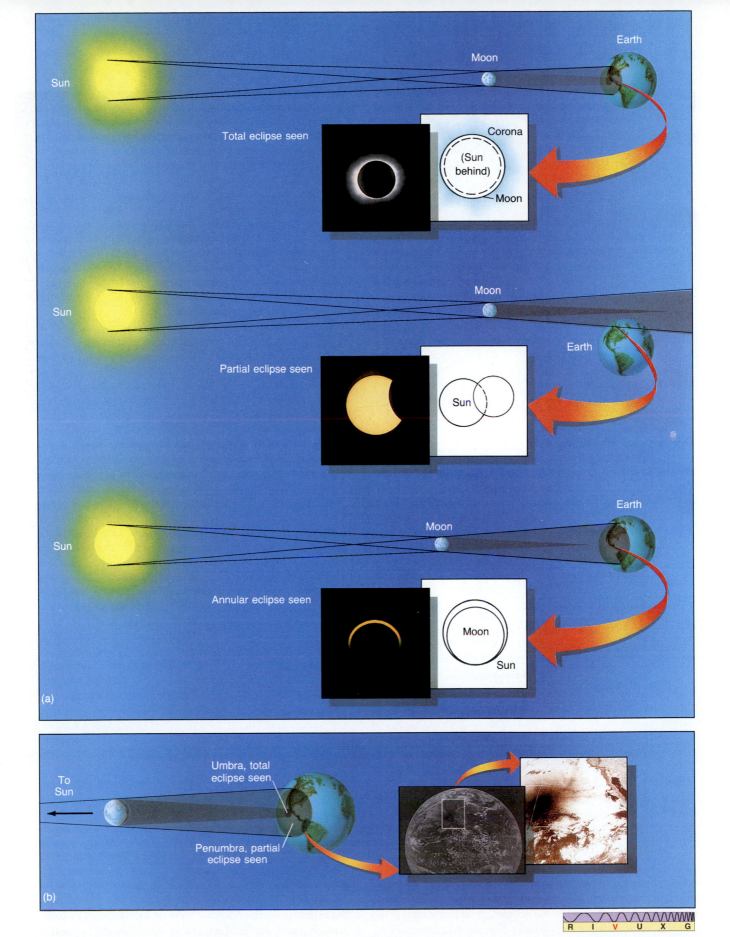

Figure 1.18 (a) The Moon's shadow on Earth during a solar eclipse consists of the umbra, where the eclipse is total, and the penumbra, where the Sun is only partially obscured. If the Moon is too far from Earth at the moment of the eclipse, there is no region of totality; instead, an annular eclipse is seen. (b) Actual photographs taken by an Earth-orbiting weather satellite of the Moon's shadow projected onto the Earth's surface (near the Baja Peninsula) during the total solar eclipse of July 11, 1991.

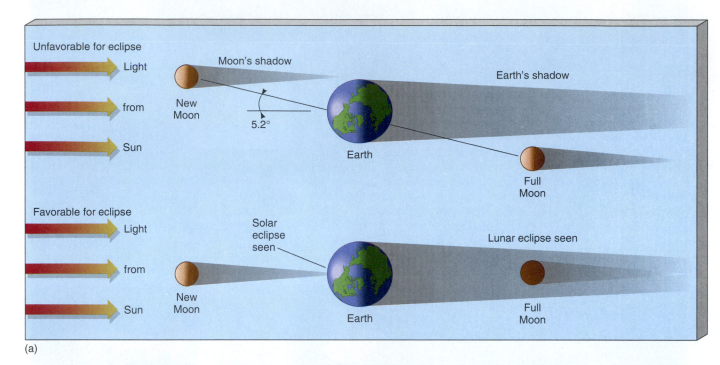

(a)

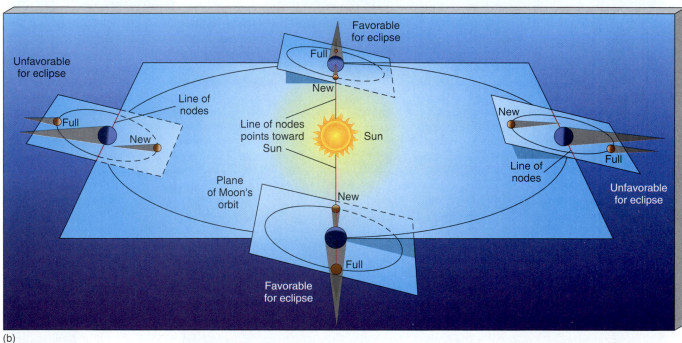

(b)

Figure 1.19 (a) An eclipse occurs when Earth, Moon, and Sun are precisely aligned. If the Moon's orbital plane lay in exactly the plane of the ecliptic, this alignment would occur once a month. However, the Moon's orbit is inclined at about 5° to the ecliptic, so not all configurations are actually favorable for producing an eclipse. (b) For an eclipse to occur, the line of intersection of the two planes must lie along Earth–Sun line. Thus, eclipses can occur only at specific times of the year.

one lunar eclipse every synodic month, with the two alternating at half-month intervals. However, the Moon's orbit is actually slightly inclined to the ecliptic, at an angle of 5.2°, so that the chance of an Earth–Moon–Sun alignment occurring just as the Moon crosses the ecliptic plane is greatly reduced. Figure 1.19 illustrates some possible configurations of the three bodies. If the Moon happens to lie above or below the plane of the ecliptic when new (or full), a solar (or lunar)

eclipse cannot occur. Such a configuration is termed *unfavorable* for producing an eclipse. In a *favorable* configuration, on the other hand, the Moon is new or full just as it crosses the ecliptic plane, and eclipses are seen. Unfavorable configurations are much more common than favorable ones.

Eclipses are relatively rare events. Moreover, they can occur only at certain times of the year. As indicated on Figure 1.19(b), the two points on the Moon's orbit where it

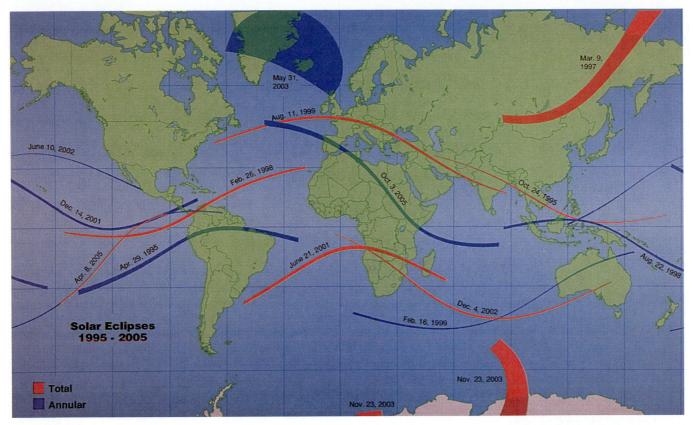

Figure 1.20   Regions of Earth that will see total or annular solar eclipses between the years 1995 and 2005. Each track represents the path of the Moon's umbra across Earth's surface during an eclipse.

crosses the ecliptic plane are known as the *nodes* of the orbit. The line joining them, which is also the line of intersection of Earth's and the Moon's orbital planes, is known as the *line of nodes*. Times when the line of nodes is not directed toward the Sun are unfavorable for eclipses. However, when the line of nodes briefly lies along the Earth–Sun line, eclipses are possible. These two periods, known as *eclipse seasons*, are the only times at which an eclipse can occur. Notice that there is no guarantee that an eclipse *will* occur. For a solar eclipse, we must have a new Moon during an eclipse season. Similarly, a lunar eclipse can occur only at full Moon during an eclipse season.

In fact, the gravitational tug of the Sun causes the Moon's orbital orientation, and hence the line of nodes, to change slowly with time. The result is that the eclipse seasons gradually progress backward through the calendar, occurring about 20 days earlier each year and taking 18.6 years to make one complete circuit. This phenomenon is known as the *regression of the line of nodes*. In 1991, the eclipse seasons were in January and July; on July 11, a total eclipse actually occurred, visible in Hawaii, Mexico, and parts of Central and South America. Three years later, in 1994, the eclipse seasons were in May and October; on May 10, an annular eclipse was visible across much of the continental United States. Because we know the orbits of Earth and the Moon to great accuracy, we can predict eclipses far into the future. Figure 1.20 shows the location and duration of all total and annular eclipses of the Sun from 1995 to 2005.

The solar eclipses that we do see highlight a remarkable cosmic coincidence. Although the Sun is many times farther away from Earth than is the Moon, it is also much larger. In fact, the ratio of distances is almost exactly the same as the ratio of sizes, so the Sun and the Moon both have roughly the *same* angular diameter—about half a degree seen from Earth. Thus, the Moon covers the face of the Sun almost exactly. If the Moon were larger, we would never see annular eclipses, and total eclipses would be much more common. If the Moon were a little smaller, we would see only annular eclipses.

1.8 *The Earth Moves*

❸ Earth has many motions—it spins on its axis, it travels around the Sun, and it moves with the Sun through the Galaxy. We have just seen how some of these motions can account for the changing nighttime sky and the changing seasons. In fact, the situation is even more complicated. Like a spinning top that rotates rapidly on its own axis while that axis slowly revolves about the vertical, Earth's axis changes its *direction* over the course of time (although the angle between the axis and a line perpendicular to the plane of the ecliptic remains close to 23.5°). This kind of change is called **precession**. Figure 1.21 illustrates Earth's precession, which is caused mostly by the gravitational pulls of the Moon and the Sun. During a complete cycle of precession, taking about 26,000 years, Earth's axis traces out a cone.

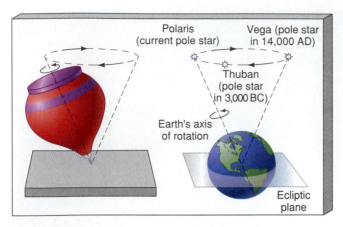

Figure 1.21 Earth's axis currently points nearly toward the star Polaris. Some 12,000 years from now—nearly half-way through one cycle of precession—Earth's axis will point toward a star called Vega, which will then be the "North Star". Five thousand years ago, the North Star was a star named Thuban in the constellation Draco.

Because of Earth's precession, the length of time from one vernal equinox to the next—one tropical year—is not quite the same as the time required for Earth to complete one orbit—one sidereal year. Recall that the vernal equinox occurs when Earth's rotation axis is perpendicular to the line joining Earth and Sun, and the Sun is crossing the celestial equator moving from south to north. In the absence of precession, this would occur exactly once per orbit, and the tropical and sidereal years would be identical. However, because of the slow precessional shift in the orientation of Earth's rotation axis, the instant when the axis is next perpendicular to the line to the Sun occurs slightly *sooner* than we would otherwise expect. Consequently, the vernal equinox drifts slowly around the zodiac over the course of the precession cycle. This is the cause of the 20-minute discrepancy between the two "years" mentioned earlier.

The tropical year is the year that our calendars measure. If our timekeeping were tied to the sidereal year, the seasons would slowly march around the calendar as the Earth precessed—13,000 years from now, summer in the Northern Hemisphere would be at its height in late February! By using the tropical year instead, we ensure that July and August will always be summer months. However, in 13,000 years' time, Orion will be a summer constellation.

There are many more complexities to Earth's motion. For example, the cone traced out by Earth's precession is not as clean as that drawn in Figure 1.21. Because of the combined gravitational influence of the Moon and the planet Jupiter, the tilt of Earth's axis with respect to the ecliptic varies back and forth between 22° and 24°. This variation is very slow, with one cycle completed only every 41,000 years. In addition, the regression of the line of nodes of the Moon's orbit in turn causes our planet's rotation axis to wobble ever so slightly, changing the angle between Earth's axis and the ecliptic by plus or minus 9 arc seconds every 18.6 years. This additional motion, superimposed on Earth's precession, is known as *nutation*. Finally, the Sun itself travels through

space. Currently, the Sun is moving at a speed of about 20 kilometers per second (relative to our neighboring stars) toward Vega, a bright star almost directly overhead in the early evening autumn sky. Along with the other planets that make up the solar system, the Earth is just along for the ride, tracing out a corkscrew path as it travels through space.

1.9 *The Measurement of Distance*

5 We have seen a little of how astronomers—ancient and modern— track and record the positions of the stars on the sky. But knowing the directions in which objects lie is only part of the information needed to locate them in space. Before we can make a systematic study of the heavens, we must find a way of measuring *distances*, too. One such method is called **triangulation**. It is based on the principles of Euclidean geometry and finds widespread application today in both terrestrial and astronomical settings. Today's engineers, especially surveyors, use these age-old geometrical ideas to measure indirectly the distance to faraway objects. In astronomical contexts, triangulation forms the foundation of the family of distance-measurement techniques that together make up the **cosmic distance scale**.

Imagine trying to measure the distance to a tree on the other side of a river. The most direct method is to lay a tape across the river, but that's not the simplest way. A smart surveyor would make the measurement by visualizing an *imaginary* triangle, sighting the tree on the far side of the river from two positions on the near side, as illustrated in Figure 1.22. The simplest possible triangle is a right triangle, in which one of the angles is exactly 90°, so it is usually conve-

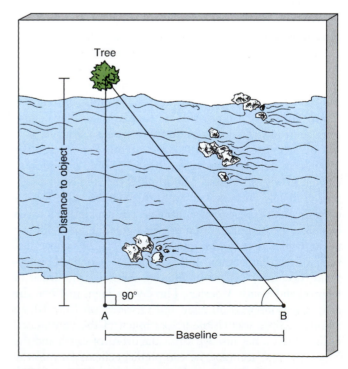

Figure 1.22 Surveyors often use simple geometry and trigonometry to estimate the distance to a faraway object.

nient to set up one observation position directly opposite the object, as at point *A*. The surveyor then moves to another observation position at point *B*, noting the distance covered between points *A* and *B*. This distance is the **baseline** of the imaginary triangle. Finally, the surveyor sights toward the tree whose distance is to be measured and notes the angle at point *B*. No further observations are required. The rest of the problem is a matter of calculation. Knowing the value of one side (*AB*) and two angles (the right angle itself, at point *A*, and the angle at point *B*) of the right triangle, the surveyor can geometrically construct the remaining sides and angles and so establish the distance to the tree.

To use triangulation to measure distances, a surveyor must be familiar with *trigonometry*, the mathematics of geometrical angles. However, even if we knew no trigonometry at all, we could still solve the problem by graphical means, as shown in Figure 1.23. Suppose that we pace off the baseline *AB*, measuring it to be 450 m, and measure the angle between the baseline and the line from *B* to the tree to be 52°, as illustrated in the figure. We can transfer the problem to paper by letting one box on our graph represent 25 m on the ground. Drawing the line *AB* on paper, completing the other two sides of the triangle, at angles of 90° (at *A*) and 52° (at *B*), we measure the distance on paper from *A* to the tree to be 23 boxes—that is, 575 m. We have solved the real problem by *modeling* it on paper. The point to remember here is this: Nothing more complex than basic geometry is needed to infer the distance, the size, and even the shape of an object too far away or too inaccessible for direct measurement.

Triangles with larger baselines are needed if we are to measure greater distances. Figure 1.24 shows a triangle having a fixed baseline between two observation positions at points *A* and *B*. Note how the triangle becomes narrower as an object's distance becomes progressively greater.

Narrow triangles cause problems because the angles at points *A* and *B* are hard to measure accurately. The measurements can be made easier by "fattening" the triangle—in other words, by lengthening the baseline.

Now consider an imaginary triangle extending from Earth to a nearby object in space, perhaps the Moon or a neighboring planet. The imaginary triangle is extremely long and narrow, even for the nearest cosmic objects. Figure 1.25(a) illustrates the case in which Earth's diameter, measured from point *A* to point *B*, is the baseline. In principle, two observers could sight the object from opposite sides of the Earth and thus measure the triangle's angles at points *A* and *B*. In practice, though, these angles cannot be accurately measured. It is actually easier to measure the third angle of the imaginary triangle, namely the very small one near the object. Here's how.

The observers on either side of Earth sight toward the object, taking note of its position *relative to some distant stars* seen on the plane of the sky. The observer at point *A* sees the object projected against a field of very distant stars. Call its apparent location *A'*, as indicated in Figure 1.25(a). Similarly, the object appears projected at point *B'* to the observer at point *B*. If each observer takes a photograph of the appropriate region of the sky, the object will appear at slightly different places in the two images. In other words, the object's photographic image is slightly displaced, or shifted, relative to the field of distant background stars, as shown in Figure 1.25(b). The background stars themselves appear undisplaced because of their much greater distance from the observer. This appar-

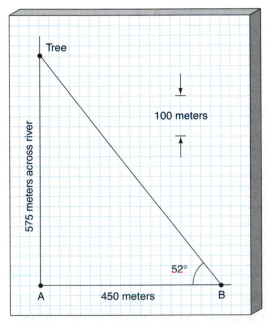

Figure 1.23 We don't even need trigonometry to estimate distances indirectly. Scaled estimates, like this one on a piece of paper, often suffice.

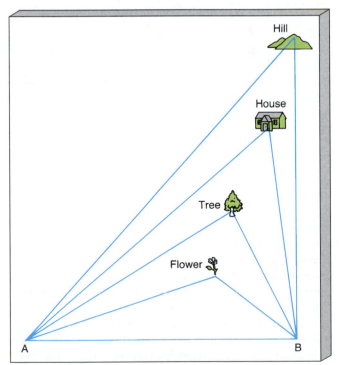

Figure 1.24 A triangle of fixed baseline (distance between points *A* and *B*) is narrower the farther away the object. As shown here, the imaginary triangle is much thinner when estimating the distance to a remote hill than it is when estimating the distance to a nearby flower.

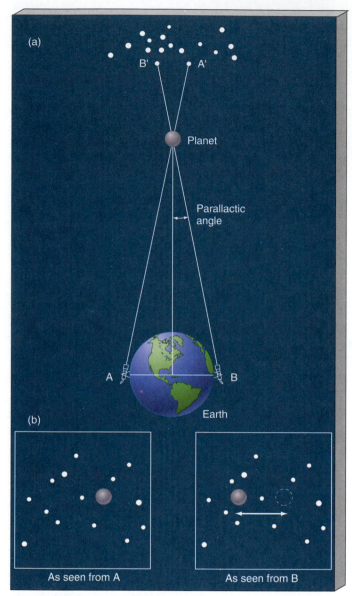

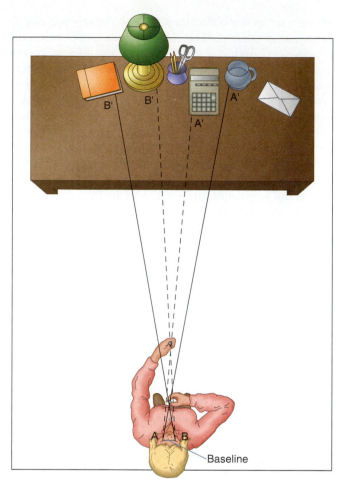

Figure 1.26 Parallax is inversely proportional to an object's distance. An object near your nose has a much larger parallax than an object held at arm's length.

Figure 1.25 (a) This imaginary triangle extends from Earth to a nearby object in space (such as a planet). The group of stars at the top represents a background field of very distant stars. (b) Hypothetical photographs of the same star field showing the nearby object's apparent displacement, or shift, relative to the distant, undisplaced stars.

ent displacement of a foreground object relative to the background as the observer's location changes is known as **parallax**. The size of the shift in Figure 1.25(b), measured as an angle on the celestial sphere, is equal to the very small angle shown in Figure 1.25(a). For historical reasons, *one-half* of this angle is called the **parallactic angle**.

The closer an object is to the observer, the larger is its parallax. To understand this concept, hold a pencil vertically in front of your nose, as sketched in Figure 1.26. Concentrate on some far-off object—say, a distant wall. Close one eye, then open it while closing the other. By blinking in this way, you should be able to see a large shift of the apparent position of the pencil projected onto the distant wall. In this example, one eye corresponds to point *A*, the other eye to point *B*, the

distance between your eyeballs to the baseline, the pencil to the nearby object, and the distant wall to a remote field of stars. If you now hold the pencil at arm's length, corresponding to a more distant object but one still not as far away as the distant stars, the apparent shift of the pencil will be less. By moving the pencil farther away, we are narrowing the triangle and decreasing the parallax (and, in the process, making its accurate measurement more difficult). If you were to paste the pencil to the wall, corresponding to the case where the object of interest is as far away as the background star field, blinking would produce no apparent shift of the pencil at all.

The amount of parallax is thus inversely proportional to an object's distance. Small parallax implies large distance. Conversely, large parallax implies small distance. Knowing the amount of parallax and the length of the baseline, we can easily derive that distance through triangulation.

As surveyors of the sky, we use the same basic information as the surveyor of the land. The calculation is basically the same. Only the means used to obtain the angles are different. The *More Precisely* feature on the next page illustrates application of simple geometrical logic to solve a distance-measurement problem—in this case, the determination of the radius of the Earth. We will see many instances of similar reasoning throughout this text.

More Precisely... Earth Dimensions

In about 200 B.C. a Greek philosopher named Eratosthenes (276–194 B.C.) used simple geometric reasoning to calculate the size of the Earth. The logic he employed still provides the basis for all measurements of distance outside of our own solar system.

Eratosthenes knew that, at noon on the first day of summer, observers in the city of Syene (now called Aswan), in Egypt, saw the Sun pass *directly overhead*. This was evident from the fact that vertical objects cast no shadows and sunlight reached to the very bottoms of deep wells, as shown in the figure here. However, at noon of the same day in Alexandria, a city 5000 *stadia* to the north, the Sun was *displaced* slightly from the vertical. (The *stadium* was a Greek unit of length, believed to have been about 0.16 km, although the exact value is uncertain—not all Greek stadia were the same size!) Using the simple technique of measuring the length of the shadow of a vertical stick and applying elementary trigonometry, Eratosthenes determined the angular displacement of the Sun from the vertical at Alexandria to be 7.2°.

What could have caused this discrepancy between the two measurements? It was not the result of measurement error—the same results were obtained every time the observations were repeated. Instead, as illustrated next, the explanation is simply that Earth's surface is not flat, but is actually *curved*. Our planet is a sphere. Eratosthenes was not the first person to realize that Earth is spherical—the philosopher Aristotle had done that over 100 years earlier (see *Interlude 2-2*)—but he was apparently the first to build on this knowledge, combining geometry with direct measurement to infer the size of our planet. Here's how he did it.

Rays of light reaching us from a very distant object, such as the Sun, travel almost parallel to one another. Consequently, as shown in the figure, the angle measured at Alexandria between the Sun's rays and the vertical (that is, the line joining Alexandria to the center of the Earth) is equal to the angle between Syene and Alexandria, as seen from Earth's center. (For the sake of clarity, this angle has been exaggerated in the drawing.) The size of this angle in turn is proportional to the fraction of Earth's circumference that lies between Syene and Alexandria. Because there are 360 degrees in a full circle, we see that 7.2° is 1/50 of a full circle, so Earth's entire circumference can be estimated by multiplying the distance between the two cities by a factor of 50. We can express this reasoning as follows:

$$\frac{7.2°}{360°} = \frac{5000 \text{ stadia}}{\text{Earth's circumference}}.$$

The circumference of the Earth is therefore 50×5000, or 250,000 stadia. If we take the stadium unit to be 0.16 km, we find that Eratosthenes estimated Earth's circumference to be about 40,000 km. In fact, he went one step further. Knowing that the circumference C of a circle is related to its radius r by the relation $C = 2\pi r$, he was also able to calculate Earth's radius to be $250,000/2\pi$ stadia, or 6366 km. The correct values for Earth's circumference and radius, now measured accurately by orbiting spacecraft, are 40,070 km and 6378 km, respectively.

Eratosthenes' reasoning was a remarkable accomplishment. More than 20 centuries ago, he estimated the circumference of Earth to within 1 percent accuracy, using only simple geometry. Even if our modern value for the size of one stadium turns out to be incorrect, the real achievement—that a person making measurements on only a small portion of Earth's surface was able to compute the size of the entire planet on the basis of observation and logic—is undiminished.

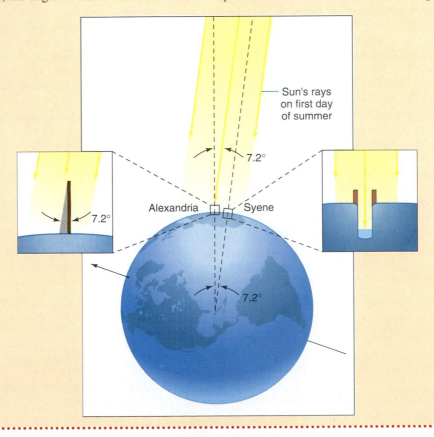

Chapter Review

SUMMARY

The **universe** (p. 4) is the totality of all space, time, matter and energy. **Astronomy** (p. 4) is the study of the universe. A widely used unit of distance in astronomy is the **light year** (p. 4), the distance traveled by a beam of light in one year.

Early observers grouped the thousands of stars visible to the naked eye into patterns called **constellations** (p. 6). These patterns have no physical significance, although they are a very useful means of labeling regions of the sky.

The nightly motion of the stars across the sky is the result of Earth's **rotation** (p. 9) on its axis. Early astronomers, however, imagined that the stars were attached to a vast **celestial sphere** (p. 9) centered on Earth, and that the motions of the heavens were caused by the rotation of the celestial sphere about a fixed Earth. The points where Earth's rotation axis intersects the celestial sphere are called the **north** and **south celestial poles** (p. 9). The line where Earth's equatorial plane cuts the celestial sphere is the **celestial equator** (p. 9).

The time from one sunrise to the next is called a **solar day** (p. 10). The time between successive risings of any given star is one **sidereal day** (p. 10). Because of Earth's **revolution** (p. 10) around the Sun, the solar day is a few minutes longer than the sidereal day.

The Sun's yearly path around the celestial sphere, or, equivalently, the plane of Earth's orbit around the Sun, is called the **ecliptic** (p. 10). Because Earth's axis is inclined to the ecliptic plane, we experience **seasons** (p. 10), depending on which hemisphere (northern or southern) happens to be "tipped" toward the Sun.

At the **summer solstice** (p. 10), the Sun is highest in the sky and the length of the day is greatest. At the **winter solstice** (p. 10), the Sun is lowest, and the day is shortest. At the **vernal** (p. 10) and **autumnal equinoxes** (p. 10), Earth's rotation axis is perpendicular to the line joining Earth to the Sun, and so day and night are of equal length.

Because Earth orbits the Sun, we see different constellations at different times of the year. The constellations lying along the ecliptic are collectively called the **zodiac** (p. 11). The time required for the same constellations to reappear at the same location in the sky, as viewed from a given point on Earth, is one **sidereal year** (p. 11).

It is possible to locate a star by specifying its constellation and its brightness relative to other stars in that constellation. A more precise method is to use **celestial coordinates** (p. 13) on the celestial sphere. **Declination** and **right ascension** (p. 14) are the celestial equivalents of latitude and longitude on Earth's surface. In general, angles on the sky are measured in **(arc) degrees** (p. 14).

The Moon emits no light of its own. It shines by reflected sunlight. As the Moon orbits Earth, we see **lunar phases** (p. 15) as the amount of the Moon's sunlit face visible to us varies. At **full Moon** (p. 15), we can see the entire illuminated side. At **quarter Moon** (p. 15), only half the sunlit side can be seen. At **new Moon** (p. 15), the sunlit face points away from us, and the Moon is all but invisible from Earth.

The time between successive full Moons is one **synodic month** (p 16). The time taken for the Moon to return to the same position in the sky, relative to the stars, is one **sidereal month** (p. 15). Because of Earth's motion around the Sun, the synodic month is about two days longer than the sidereal month.

A **lunar eclipse** (p. 17) occurs when the Moon enters Earth's shadow. The eclipse may be **total** (p. 17), if the entire Moon is (temporarily) darkened, or **partial** (p. 17), if only a portion of the Moon's surface is affected.

A **solar eclipse** (p. 18) occurs when the Moon passes between Earth and the Sun, so that a small part of Earth's surface is plunged into shadow. For observers in the **umbra** (p. 18), the entire Sun is obscured, and the solar eclipse is total. In the **penumbra** (p. 18), a **partial solar eclipse** (p. 18) is seen. If the Moon happens to be too far from Earth for its disk to completely hide the Sun, an **annular eclipse** (p. 18) occurs.

Because the Moon's orbit around Earth is slightly inclined with respect to the ecliptic, solar and lunar eclipses do not occur every month, but only a few times per year.

In addition to its rotation about its axis and its revolution around the Sun, Earth has may other motions. One of the most important of these is **precession** (p. 21), where the influence of the Moon causes Earth's axis to "wobble" slightly. As a result, the particular constellations that happen to be visible on any given night change slowly over the course of many years.

Surveyors on Earth use **triangulation** (p. 22) to determine the distances to distant objects. Astronomers use the same technique to measure the distances to planets and stars. The **cosmic distance scale** (p. 23) is the family of distance-measurement techniques by which astronomers chart the universe.

Parallax (p. 24) is the apparent motion of a foreground object relative to a distant background as the observer's position changes. One-half of the observed angular displacement is defined as the **parallactic angle** (p. 24). The larger the **baseline** (p. 23), the distance between the two observation points, the greater the parallax. Astronomers use parallax when measuring the distances to the planets by triangulation.

SELF TEST: True or False?

_____ **1.** The light-year is a measure of distance.

_____ **2.** The number 2×10^6 is equal to 2 billion.

_____ **3.** The stars in a constellation are physically close to each other.

_____ **4.** Some constellations, at one time, were used as simple calendars.

_____ **5.** Constellations are no longer of any use to astronomers.

_____ **6.** The solar day is longer than the sidereal day.

_____ **7.** The constellations lying immediately adjacent to the north celestial pole are collectively referred to as the zodiac.

_____ **8.** The seasons are caused by the precession of Earth's axis.

_____ **9.** The vernal equinox marks the beginning of Fall.

_____ **10.** The new phase of the moon cannot be seen because it always occurs during the daytime.

_____ **11.** A lunar eclipse can occur only during the full phase.

____ **12.** Solar eclipses are possible during any phase of the Moon.

____ **13.** An annular eclipse is a type of eclipse that occurs every year.

____ **14.** The parallax of an object is inversely proportional to its distance.

SELF TEST: Fill in the Blanks

1. A _____ is a collection of hundreds of billions of stars.
2. The angstrom, astronomical unit, parsec, and nanometer are all units of _____.
3. Rotation is the term used to describe the motion of a body around some _____.
4. To explain the daily and yearly motions of the heavens, ancient astronomers imagined that the Sun, Moon, stars, and planets were attached to a rotating _____.
5. The solar day is measured relative to the Sun; the sidereal day is measured relative to the _____.
6. The apparent path of the Sun across the sky is known as the _____.
7. On December 21, known as the _____, the Sun is at its _____

point on the celestial sphere.

8. Declination measures the position of an object north or south of the _____.
9. An arc second is _____ (give the fraction) of an arc minute.
10. When the Sun, Earth, and Moon are positioned to form a right angle at Earth, the Moon is seen in the _____ phase.
11. A _____ eclipse can be seen by about half the Earth at once.
12. As seen from Earth, the Sun and the Moon have roughly the same _____.
13. To measure distances to nearby stars by parallax, a baseline equal to the Earth's _____ is used.
14. The radius of the _____ was first measured by Eratosthenes in 200 B.C.

REVIEW AND DISCUSSION

1. Compare the size of the Earth with the Sun, the Milky Way, and the entire universe.
2. What does an astronomer mean by "the universe?"
3. What is a constellation?
4. Why does the Sun rise in the east and set in the west each day? Does the Moon also rise in the east and set in the west? Why? Do stars do the same? Why?
5. How many times in your life have you orbited the Sun?
6. Why are there seasons on Earth?
7. Why do we see different stars in summer and in winter?
8. If one complete hemisphere of the Moon is always lit by the sun, why do we see different phases of the Moon?

9. What causes a lunar eclipse? A solar eclipse?
10. Why aren't there lunar and solar eclipses every month?
11. What is precession, and what is its cause?
12. What is parallax? Give an everyday example.
13. Why is it necessary to have a long baseline when using triangulation to measure the distances to objects in space?
14. If you traveled to the outermost planet in our solar system, do you think the constellations would appear to change their shapes? What would happen if you traveled to the next-nearest star? If you traveled to the center of the Galaxy, could you still see the familiar constellations found in Earth's night sky?

PROBLEMS

1. In 1 second, light leaving Los Angeles will reach approximately as far as (a) San Francisco (about 500 km), (b) London (roughly 10,000 km), (c) the Moon (400,000 km), (d) Venus (0.3 AU from Earth at closest approach), or (e) the nearest star (about 1 pc from Earth). Which is correct?
2. Through how many degrees, arc minutes, or arc seconds does the Moon move in (a) one hour of time, (b) one minute, (c) one second? How long does it take for the Moon to move a distance equal to its own diameter?
3. Write these numbers in scientific notation: 100; 1000; 1,000,000,000,000,000; 0.01; 0.001; 123,000; 0.000456.

4. A surveyor wishes to measure the distance between two points on either side of a river, as illustrated in Figure 1.22. She measures the distance AB to be 250 m and the angle at B to be 30°. What is the distance between the two points?
5. At what distance is an object if its parallax, as measured from either end of a 1000 km baseline, is (a) 1°, (b) 1′, (c) 1″?
6. Given that the distance to the Moon is 384,000 km and its angular size is 0.5°, calculate the Moon's diameter.
7. What angle would Eratosthenes have measured (see the *More Precisely* feature on p. 25) had the Earth been flat?

PROJECTS

1. Go to a country location on a clear dark night. Imagine patterns among the stars, and name the patterns yourself. Note (or better yet, draw) these stars' location with respect to trees or buildings in the foreground. Do this every week or so for a couple of months. Be sure to look at the same time every night. What happens?
2. Find the star Polaris, also known as the North Star, in the evening sky. Identify any separate pattern of stars in the

same general vicinity of the sky. Wait several hours, at least until after midnight, and then locate Polaris again. Has Polaris moved? What has happened to the nearby pattern of stars? Why?

3. Hold your little finger out at arm's length. Can you cover the disk of the Moon? The Moon projects an angular size of 30′ (half a degree); your finger should more than cover it. How can you apply this fact in making sky measurements?

2

THE COPERNICAN REVOLUTION

The Birth of Modern Science

LEARNING GOALS

Studying this chapter will enable you to:

1 Relate how some ancient civilizations attempted to explain the heavens in terms of Earth-centered models of the universe.

2 Summarize the role of Renaissance science in the history of astronomy.

3 Explain how the observed motions of the planets led to our modern view of a Sun-centered solar system.

4 Sketch the major contributions of Galileo and Kepler to the development of our understanding of the solar system.

5 State Kepler's laws of planetary motion.

6 Explain how Kepler's laws enable us to construct a scale model of the solar system, and the technique used to determine the actual size of the planetary orbits.

7 State Newton's laws of motion and universal gravitation, and explain how they account for Kepler's laws.

8 Explain how the law of gravitation enables us to measure the masses of astronomical bodies.

(Opposite page, background) Most of our understanding of the universe has come within the last century. In an early twentieth-century classroom at Harvard College, as at other schools, astronomy came into its own as a viable and important subject of study.

(Inset A) Harlow Shapley (1885–1972), seen here at Harvard sitting at his huge circular desk that he could spin to work on different topics of astronomical interest. He discovered our place in the "suburbs" of the Milky Way, dispelling the notion that the Sun resides at the center of the universe.

(Inset B) Annie Cannon (1863–1941), one of the greatest astronomical cataloguers of all time, carefully analyzed photographic plates to classify nearly a million stars over the course of fifty years of work at Harvard Observatory.

(Inset C) Maria Mitchell (1818–1889) in her observatory on Nantucket Island. The first woman professional astronomer in the United States, she taught at Vassar College and made important contributions to several areas of astronomy, including the photographic study of the Sun's surface.

(Inset D) Edwin Hubble (1889–1953), here posing in front of one of the telescopes on Mount Wilson, in California, is often credited with having discovered the expansion of the universe.

Living in the Space Age, we have become accustomed to the modern view of our place in the universe. Images of our planet taken from space leave little doubt that Earth is round, and no one seriously questions the idea that we orbit the Sun. Yet there was a time, not so long ago, when our ancestors maintained that Earth was flat and lay at the center of all things. Our view of the universe—and of ourselves—has undergone a radical transformation since those early days. Earth has become a planet like many others, and humankind has been torn from its throne at the center of the cosmos and relegated to a rather unremarkable position on the periphery of the Milky Way Galaxy. But we have been amply compensated for our loss of prominence—we have gained a wealth of scientific knowledge in the process. The story of how all this came about is the story of the rise of the scientific method and the genesis of modern astronomy.

2.1 Ancient Astronomy

1 Many ancient cultures took a keen interest in the changing nighttime sky. The records and artifacts that have survived until the present make that abundantly clear. But unlike today, the major driving force behind the development of astronomy in those early societies was probably neither scientific nor religious in nature. Instead, it was decidedly practical and very down to earth. Seafarers needed to navigate their vessels, and farmers had to know when to plant their crops. In a very real sense, then, human survival depended on knowledge of the heavens. As a result, the ability to predict the arrival of the seasons, as well as other astronomical events, was undoubtedly a highly prized, and perhaps also jealously guarded, skill.

In Chapter 1 we saw that the human brain's ability to perceive patterns in the stars led to the "invention" of constellations as a convenient means of labeling regions of the celestial sphere. The realization that these patterns returned to the night sky at the same time each year met the need for a practical means of tracking the seasons. Many separate cultures, all over the world, built large and elaborate structures to serve, at least in part, as primitive calendars. In some cases, the keepers of the secrets of the sky eventually enshrined their knowledge in myth and ritual, so that these astronomical sites were often also used for religious rites.

Perhaps the best known such site is *Stonehenge*, located on Salisbury Plain, in England, and shown in Figure 2.1. This ancient stone circle, which today is one of the most popular tourist attractions in Britain, dates back to the Stone Age. Researchers believe it was an early astronomical observatory of sorts—not in the modern sense of the term (a place for making new observations and discoveries), but rather a kind of three-dimensional calendar or almanac, enabling its builders and their descendents to identify important dates by means of specific celestial events. Its construction apparently spanned a period of some 17 centuries, beginning around 2800 B.C. Additions and modifications continued up to about 1100 B.C., indicating its ongoing importance to the Stone Age and later Bronze Age people who built, maintained, and used Stonehenge. The largest stones shown in Figure 2.1 weigh up to 50 tons and were transported from quarries many miles away.

Many of the stones are aligned so that they point toward important astronomical events. For example, the line joining the center of the inner circle to the so-called *heel stone*, set off at some distance from the rest of the structure, points in the direction of the rising Sun on the summer solstice. Other alignments are related to the rising and setting of the Sun and the Moon at various other times of the year. Although the accurate alignments (within a degree or so) of the stones of Stonehenge were first noted in the eighteenth century, it is only relatively recently—in the second half of the twentieth century, in fact—that the scientific community has credited Stone Age technology with the ability to carry out such a precise feat of engineering. While some of Stonehenge's purposes remain uncertain and controversial, the site's function as an astronomical almanac seems well established. Although Stonehenge is the most impressive and the best preserved, other stone circles, found all over Europe, are believed to have performed similar functions.

Many ancient cultures are now known to have been capable of similarly precise accomplishments. The Big Horn Medicine Wheel in Wyoming (Figure 2.2a) is similar to Stonehenge in design—and, presumably, intent—although it is somewhat simpler in execution. The Medicine Wheel's alignments with the rising and setting Sun and with some bright stars indicate that its builders—the Plains Indians—had much more than a passing familiarity with the changing nighttime sky. Figure 2.2b shows the Caracol temple, built by the Mayans around 1000 A.D. in Mexico's Yucatán peninsula. This temple is much more sophisticated than Stonehenge, but it probably played a similar role as an astronomical observatory. Its many windows are accurately aligned with astronomical events, such as sunrise and sunset at the solstices and equinoxes and the risings and settings of the planet Venus. Astronomy was of more than mere academic interest to the Mayans, however. Caracol was also the site of countless human sacrifices, carried out when Venus appeared in the morning or evening sky.

The ancient Chinese too observed the heavens. Their astrology attached great importance to "omens" such as comets and "guest stars"—stars that appeared suddenly in the sky and then slowly faded away—and they kept careful and extensive records of such events. Twentieth-century astronomers still turn to the Chinese records to obtain obser-

Figure 2.1 Stonehenge was probably constructed as a primitive calendar and almanac. The fact that the largest stones were carried to the site from many miles away attests to the importance of this structure to its Stone Age builders. The inset shows sunrise at Stonehenge on the summer solstice. As seen from the center of the stone circle, the Sun rose directly over the "heel stone" on the longest day of the year.

vational data recorded during the Dark Ages (roughly from the fifth to the tenth century A.D.), when turmoil in Europe largely halted the progress of Western science. Perhaps the best-known guest star was one that appeared in 1054 A.D. and was visible in the daytime sky for many months. We now know that the event was actually a *supernova*: the explosion of a giant star, which scattered most of its mass into space. It left behind a remnant that is still detectable today, nine centuries later. The Chinese data are a prime source of historical information for supernova research.

(a)

(b)

Figure 2.2 (a) The Big Horn Medicine Wheel, in Wyoming, was built by the Plains Indians. Its spokes and other features are aligned with risings and settings of the Sun and other stars. (b) Caracol temple in Mexico. The many windows of this Mayan construct are aligned with astronomical events, indicating that at least part of Caracol's function was to keep track of the seasons and the heavens.

Figure 2.3   Arab astronomers at work, as depicted in a medieval manuscript.

A vital link between the astronomy of ancient Greece and that of medieval Europe was provided by Arab astronomers (Figure 2.3). For six centuries, from the depths of the Dark Ages to the beginning of the Renaissance, Islamic astronomy flourished and grew, preserving and augmenting the knowledge of the Greeks. The Arab influence on modern astronomy is subtle but quite pervasive. Many of the mathematical techniques involved in trigonometry were developed by Muslim astronomers in response to very practical problems, such as determining the precise dates of holy days or the direction of Mecca from any given location on Earth. Astronomical terms like "zenith" and "azimuth" and the names of many stars, such as Rigel, Betelgeuse, and Vega, all bear witness to this extended period of Muslim scholarship.

Astronomy, we see, is not the property of any one culture, civilization, or era. The same ideas, the same tools, and even the same misconceptions have been invented and reinvented by human societies all over the world, in response to the same basic driving forces. Astronomy came into being because people believed that there was a practical benefit in being able to predict the positions of the stars, but its roots go much deeper than that. The need to understand where we came from, and how we fit into the cosmos, is an integral part of human nature.

2.2 *The Geocentric Universe*

The Greeks of antiquity, and undoubtedly civilizations before them, built models of the universe. The study of the workings of the universe on the very largest scales is called **cosmology**. Today, cosmology entails looking at the universe on scales so large that even entire galaxies can be regarded as mere points of light scattered throughout space.

To the Greeks, however, the universe was basically the *solar system*—namely, the Sun, Earth, Moon, and the planets known at that time. The stars beyond were surely part of the universe, but they were considered to be fixed, unchanging beacons on a mammoth celestial dome. The Greeks did not consider the Sun, the Moon, and the planets to be part of the celestial sphere, however. Those objects had patterns of behavior that set them apart.

Over the course of a night, the stars slide smoothly across the sky. Over the course of a month, the Moon moves smoothly and steadily along its path on the sky relative to the stars, passing through its familiar cycle of phases. Over the course of a year, the Sun progresses along the ecliptic at an almost constant rate, varying little in brightness from day to day. In short, the behavior of both Sun and Moon seemed fairly simple and orderly. But the Greeks were also aware of five other bodies in the sky—the planets Mercury, Venus, Mars, Jupiter, and Saturn—whose behavior was not so easy to grasp. Their motions ultimately led to the downfall of an entire theory of the solar system and to a fundamental change in humankind's view of the universe.

Planets do not behave in as regular and predictable a fashion as the Sun, Moon, and stars. They vary in brightness, and they don't maintain a fixed position in the sky. Unlike the Sun and the Moon, with their regular paths, the planets seem to wander around the celestial sphere—indeed, the word planet derives from the Greek word *planetes*, meaning wanderer. Planets never stray far from the ecliptic and generally traverse the celestial sphere from west to east, as the Sun does. The planets, however, seem to speed up and slow down during their journeys. They even appear to loop back and forth relative to the stars, as shown in Figure 2.4. In other words, there are periods when a planet's eastward motion (relative to the stars) stops, and the planet appears to move westward in the sky for a month or two before reversing direction again and continuing on its eastward journey.

Motion in the eastward sense is usually referred to as *direct*, or *prograde*, motion; the backward (westward) loops are known as **retrograde motion**. Mars, Jupiter, and Saturn are always brightest—which means closest to Earth—during the retrograde portions of their orbits. Obviously this planetary behavior requires an explanation more complex than the relatively simple motions of the Moon and the Sun. The occasional retrograde loops of some planets, and the brightness variations of all of them, necessitated major modifications to the Greek's basic cosmological model describing the Sun, the Moon, and the stars.

The earliest models of the solar system followed the teachings of the Greek philosopher Aristotle, and were **geocentric** in nature. These geocentric models held that Earth lay at the *center* of the universe and that all other bodies moved around it. (Recall Figures 1.7 and 1.10a, which illustrate the basic geocentric view.) Aristotle's views were immensely influential—so much so, in fact, that his teachings carried great weight even centuries after his death. The geocentric model went largely unchallenged until the sixteenth century A.D.

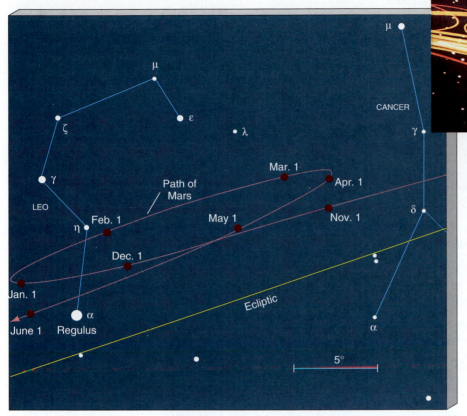

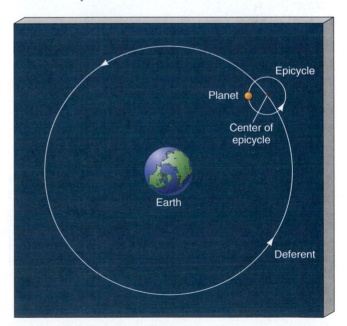

Figure 2.4 Most of the time, planets move from west to east relative to the background stars. Occasionally, however, they change direction and temporarily undergo retrograde motion before looping back. The image at left shows an actual retrograde loop in the motion of the planet Mars. The inset depicts the movements of several planets over the course of several years, as reproduced on the inside dome of a planetarium. The motion of the planets relative to the stars (represented as unmoving points) produces continuous streaks on the planetarium "sky."

Actually, history records that some ancient Greek astronomers reasoned differently about the motions of heavenly bodies. Foremost among them was Aristarchus of Samos (310–230 B.C.), who proposed that all the planets, including the Earth, revolve around the Sun and, furthermore, that Earth rotates on its axis once each day. This, he argued, would create an *apparent* motion of the sky—a simple idea that is familiar to anyone who has ridden on a merry-go-round and watched the landscape appear to move past them as they go. However, Aristarchus's description of the heavens, though essentially correct, did not gain widespread acceptance during his lifetime. Aristotle's influence was too strong, his followers too numerous, his writings too comprehensive.

The Aristotelian school presented some simple and (at the time) compelling arguments in favor of their views. First, of course, Earth doesn't *feel* as if it's moving. And if it were, wouldn't there be a strong wind as we moved at high speed around the Sun? Then again, considering that the vantage point from which we view the stars changes over the course of a year, why don't we see stellar parallax? Nowadays we might be inclined to dismiss the first two points as merely naive, but the third is a valid argument and the reasoning is essentially sound. We now know that there *is* stellar parallax as Earth orbits the Sun. However, because the stars are so distant, it amounts to less than 1″, even for the closest stars. Early astronomers simply would not have noticed it. We will encounter many other instances in astronomy where correct reasoning has led to the wrong conclusions because it was based on inadequate data.

The earliest geocentric models of the universe employed what Aristotle, and Plato before him, had taught was the perfect form: the circle. The simplest possible description—uniform motion around a circle having Earth at its center—provided a fairly good approximation of the orbits of the Sun and the Moon, but it could not possibly account for observed variations in planetary brightness or retrograde motion. A more complex model was needed to describe the planets.

The first step toward this new model modified the idea that the planets moved on circles centered on Earth.

Figure 2.5 In the geocentric model of the solar system, the observed motions of the planets made it impossible to assume that they moved on simple circular paths around Earth. Instead, each planet was thought to follow a small circular orbit (the epicycle) about an imaginary point that itself traveled in a large, circular orbit (the deferent) about Earth.

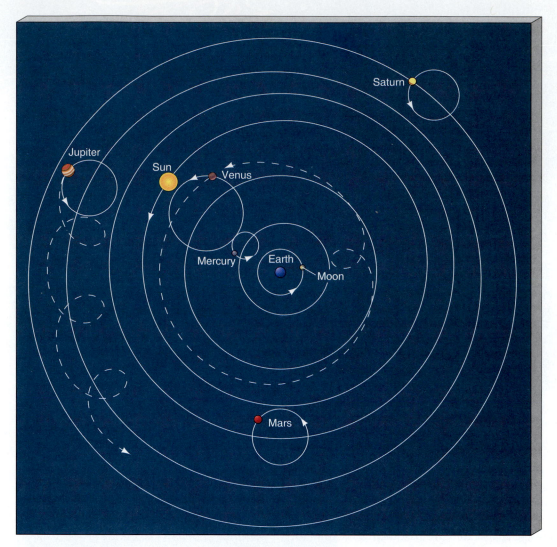

Figure 2.6 The basic features, drawn roughly to scale, of the geocentric model of the inner solar system that enjoyed widespread popularity prior to the Renaissance. To avoid confusion, we have drawn partial paths (dashed) of only two planets, Venus and Jupiter.

Instead, each planet was taken to move uniformly around a small circle, called an **epicycle**, whose *center* moved uniformly around Earth on a second and larger circle, known as the **deferent** (see Figure 2.5). The motion was now composed of two separate circular orbits, which created the possibility that, at some times, the planet's apparent motion in the sky could be retrograde. Also, the distance from the planet to Earth would vary, accounting for changes in brightness. By tinkering with the relative sizes of the epicycle and the deferent, with the planet's speed on the epicycle, and with the epicycle's speed along the deferent, this "epicyclic" motion could be brought into fairly good agreement with the observed paths of the planets in the sky. Moreover, this model had good predictive power, at least to the accuracy of observations at the time.

However, as time went by and the number and the quality of observations increased, it became clear that the simple epicyclic model was not perfect. Small corrections had to be introduced to bring it into line with new observations. The center of the deferents had to be shifted slightly from the center of Earth, and the motion of the epicycles had to be imagined uniform with respect not to Earth but to yet another point in space. Around A.D. 140, a Greek as-

tronomer named Ptolemy constructed perhaps the best geocentric model of all time. Illustrated in simplified form in Figure 2.6, it explained remarkably well the observed paths of the five planets then known, as well as the paths of the Sun and the Moon. However, to achieve its explanatory and predictive power, the full **Ptolemaic model** required a series of no fewer than 80 distinct circles. To account for the paths of the Sun, the Moon, and all the nine planets (and their moons) that we know today would require a vastly more complex set. Nevertheless, Ptolemy's text on the topic, *Syntaxis* (better known today by its Arabic name *Almagest*—"the greatest") provided the intellectual framework for all discussion of the universe for well over a thousand years.

Today, our scientific training leads us to seek simplicity, because simplicity in the physical sciences has so often proved to be an indicator of truth. We would regard the intricacy of a model as complicated as the Ptolemaic system as a clear sign of a fundamentally flawed theory. With the benefit of hindsight, we now recognize that the major error lay in the assumption of a geocentric universe. This was compounded by the insistence on uniform circular motion, whose basis was largely philosophical, rather than scientific, in nature.

2.3   The Heliocentric Model of the Solar System

The Ptolemaic picture of the universe survived, more or less intact, for almost 13 centuries. Given the scope of the cultural, scientific, and technological changes that have occurred in even the last 50 years, it may be difficult for us to grasp how *any* theory, and especially one so erroneous, could have persisted for such a long time. Whatever the reasons, the Ptolemaic model of the solar system became deeply embedded in European culture at all levels and lasted until the fifteenth century. Then came the *Renaissance*, a rebirth of artistic, philosophical, and scientific inquiry. Western thought moved away from the passive acceptance of ancient dogma and static beliefs toward critical thinking and observational testing. In astronomy, thinking (theory) and looking (observation) merged to produce a model of the solar system that was simpler than the geocentric one embraced by the ancients. A sixteenth-century Polish cleric, Nicholas Copernicus (see Figure 2.7), rediscovered Aristarchus's **heliocentric** model—one centered on the Sun—and showed how, in its harmony and organization, it better fit the observed facts than did the tangled geocentric cosmology.

Copernicus asserted that Earth spins on its axis every day and, like the other planets, orbits the Sun. Only the Moon, he said, orbits Earth. The observed daily and seasonal changes in the heavens can be understood in terms of these simple motions. The seven crucial statements that form the basis for what is now known as the **Copernican revolution** are summarized in *Interlude 2-1*. The Copernican view stands in stark contrast to the conventional beliefs of the preceding two millennia and presents a more ordered and natural explanation of the observed facts than any geocentric model could provide. We have already seen in Chapter 1 how this picture accounts for the motions of the Sun and the stars. Figure 2.8 shows how it explains both the varying brightness of the planets and their observed looping motions. If we suppose that Earth moves

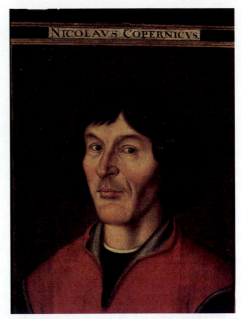

Figure 2.7   Nicholas Copernicus (1473–1543).

faster than the planet Mars, then every so often we will "overtake" Mars. Mars will appear to move backwards in the sky, in much the same way as a car we overtake on the highway seems to slip backwards relative to us. Note that the looping motions are now only apparent, not real, as they were in the Ptolemaic view.

Copernicus's major motivation for introducing the heliocentric model was simplicity. Even so, he was still influenced by Greek thinking and clung to the idea of circles to model the planets' motions. To bring his theory into agreement with observations, he was forced to retain the idea of epicyclic motion, though with the deferent centered on the Sun rather than on Earth, and with the epicycles being smaller than in the Ptolemaic picture. Thus, he retained unnecessary complexity and actually gained little in accuracy over the geocentric model. The heliocentric model

---

Interlude 2-1   The Foundations of the Copernican Revolution

The following seven points are essentially Copernicus's own words. The italicized material is additional explanation.

1. The celestial spheres do not have just one common center. *Specifically, Earth is not at the center of everything.*

2. The center of Earth is not the center of the universe, but is instead only the center of gravity and of the lunar orbit.

3. All the spheres revolve around the Sun. *By spheres, Copernicus meant the planets.*

4. The ratio of Earth's distance from the Sun to the height of the firmament is so much smaller than the ratio of Earth's radius to the distance to the Sun that the distance to the Sun is imperceptible when compared with the height of the firmament. *By firmament, Copernicus meant the distant stars. The point he was making is that the stars are very much farther away than the Sun.*

5. The motions appearing in the firmament are not its motions, but those of Earth. Earth performs a daily rotation around its fixed poles, while the firmament remains immobile as the highest heaven. *Because the stars are so far away, any apparent motion we see in them is the result of Earth's rotation.*

6. The motions of the Sun are not its motions, but the motion of the Earth. *Similarly, the Sun's apparent daily and yearly motion are actually due to the various motions of the Earth.*

7. What appears to us as retrograde and forward motion of the planets is not their own, but that of the Earth. *The heliocentric picture provides a natural explanation for retrograde planetary motion, again as a consequence of Earth's motion.*

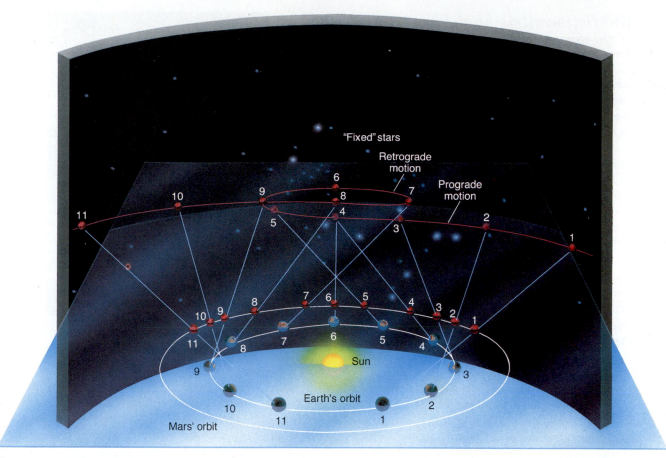

Figure 2.8   The Copernican model of the solar system explains the varying brightnesses of the planets, something the Ptolemaic system largely ignored. Here, for example, when Earth and Mars are relatively close to one another in their respective orbits (as at position 6), Mars seems brighter; when farther away (as at position 1), Mars seems dimmer. Also, because the line of sight from Earth to Mars changes as the two planets smoothly orbit the Sun, Mars would appear to loop back and forth, undergoing retrograde motion. The line of sight changes because Earth, on the inside track, moves faster in its orbit than Mars moves along its path.

did indeed rectify some small discrepancies and inconsistencies in the Ptolemaic system, but for Copernicus, the primary attraction of heliocentricity was its simplicity, its being "more pleasing to the mind." His theory was more something he *felt* than he could *prove*. To the present day, scientists still are guided by simplicity, symmetry, and beauty in modeling all aspects of the universe.

Despite the support of some observational data, neither his fellow scholars nor the general public easily accepted Copernicus's model. For the learned, heliocentricity went against the grain of much previous thinking and violated many of the religious teachings of the time, largely because it relegated Earth to a noncentral and undistinguished place within the solar system and the universe. In the heliocentric model, Earth became just one of several planets. And Copernicus's work had little impact on the general populace of his time, partly because it was published in Latin, which most people could not read. Only after Copernicus's death, when others—notably Galileo Galilei—popularized his ideas, did the Roman Catholic church take them seriously enough to bother banning them. Copernicus's writings on the heliocentric universe were placed on the *Index of Prohibited Books* in 1616, 73 years after they were first published. They remained there until the end of the eighteenth century.

2.4   *The Birth of Modern Astronomy*

❸   In the century following the death of Copernicus and the publication of his theory of the solar system, two scientists—Galileo Galilei and Johannes Kepler—made indelible imprints on the study of astronomy. Contemporaries, they were aware of each other's work, and corresponded from time to time about their theories. Each achieved fame for his discoveries and made great strides in popularizing the Copernican viewpoint. Yet, in their approach to astronomy—and in their personalities—they were as different as night and day.

GALILEO'S HISTORIC OBSERVATIONS

❹   Galileo Galilei (Figure 2.9) was an Italian mathematician and philosopher. By his willingness to perform experiments to test his ideas—a rather radical approach in those days (see Interlude 2-2)—and by embracing the brand-new technology of the telescope, he revolutionized the way science was done, so much so that he is now widely regarded as the father of experimental science. The telescope was invented in Holland in the early seventeenth century. Hearing of the invention—but without seeing one—Galileo built a telescope for himself in 1609 and aimed it at the sky. What he saw conflicted greatly with the philosophy of Aristotle, and provided much new data to support the ideas of Copernicus.*

Using his telescope, Galileo discovered that the Moon had mountains, valleys, and craters. Its surface was reminiscent more of Earth's than that of a perfect, unblemished celestial orb, which the Moon was conventionally held to be.

Looking at the Sun—something that should *never* be done directly and that eventually blinded Galileo—he saw dark blemishes. These blemishes, which we now call sunspots and which we will study in Chapter 16, directly contradicted the ancient Greek notion that the Sun was a perfect, jewellike body. Furthermore, he could see that the blemishes moved across the face of the Sun. From this he inferred that the Sun *rotates* on an axis roughly perpendicular to the ecliptic about once a month.

In studying the planet Jupiter, Galileo saw four small points of light, invisible to the naked eye, orbiting it. He recognized that Jupiter and its natural satellites (now called the Galilean moons) were a system similar to the Sun and its family of planets, but on a smaller scale. That another planet had moons—and thus that some body other than Earth could be the center of motion—conflicted directly with Aristotelianism. To Galileo, the moons of Jupiter provided the strongest support for the Copernican model. Clearly, the Earth was not the center of all things.

In fact, Galileo had already abandoned Aristotle in favor of Copernicus, although he had not published these beliefs at the time he began his telescopic observations.

Figure 2.9   Galileo Galilei (1564–1642).

Galileo also discovered that Venus showed a complete cycle of phases, like those of our Moon (see Figure 2.10). The insets show how the full and crescent views of Venus

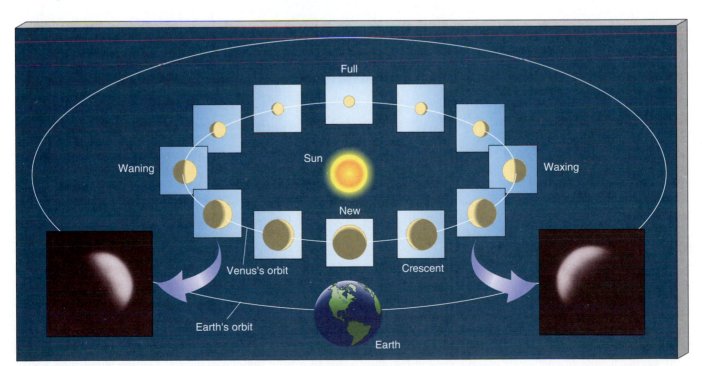

Figure 2.10   The phases of Venus, rendered at different points in the planet's orbit. If Venus orbits the Sun and is closer to the Sun than is Earth, as Copernicus maintained, then Venus should display phases, much as our Moon does. As shown here, when directly between Earth and the Sun, Venus's unlit side faces us, and the planet is invisible to us. As Venus moves in its orbit (at a faster speed than Earth moves in its orbit), progressively more of its illuminated face is visible from Earth. Note also the connection between orbital phase and the apparent size of the planet. Venus seems much larger in its crescent phase than when it is full because it is much closer to us during its crescent phase. (The insets at bottom left and right are actual photographs of Venus at two of its crescent phases.)

Interlude 2–2   *The Scientific Method*

Most ancient philosophers held firmly to the belief that, whatever the reasons for the motions of the heavens, Earth in general and humankind in particular were absolutely central to the workings of the universe. Modern science, by contrast, has arrived at a diametrically opposite view. Our present-day outlook is that Earth, the solar system, and (some would argue) humanity are ordinary in every way. This idea is often (and only half-jokingly) called the "principle of mediocrity," and it is deeply embedded in modern scientific thought. It is a natural extension of the Copernican principle discussed in Sections 2.3 and 2.4 (see also *Interlude 2–1*). Nowadays, any theory or observation that even appears to single out Earth, the solar system, or the Milky Way Galaxy as in some way special is immediately regarded with great suspicion in scientific circles.

The principle of mediocrity extends far beyond mere philosophical preference, however. Simply put, if we do not make this assumption, then we cannot make much headway in science, and we cannot do astronomy at all. Virtually every statement made in this text rests squarely on the premise that the laws of physics, as we know them here on Earth, apply everywhere else too, without modification and without exception.

This transformation in the perception of humanity's place in the universe went hand in hand with a gradual—but radical—shift in the way philosophers and scientists conducted their investigation of the cosmos. The earliest known models of the universe were based largely on imagination and pure reasoning, with little attempt to explain the workings of the heavens in terms of known earthly experience. However, history shows that some philosophers did come to realize the importance of careful observation and testing to the formulation of their theories. The success of their approach changed, slowly but surely, the way science was done, and opened the door to a fuller understanding of nature.

As knowledge from all sources was sought and embraced for its own sake, the influence of logic and reasoned argument grew, and the power of myth diminished. People began to inquire more critically about themselves and the universe. They realized that thinking about nature was no longer sufficient; looking at it is also necessary. Experiments and observations became a central part of the process of inquiry. To be effective, a theory—the framework of ideas and assumptions used to explain some set of observations and make predictions about the real world—must

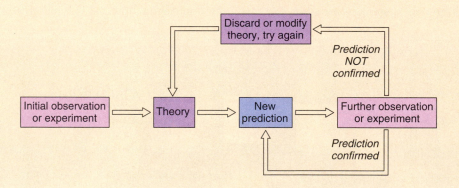

must be caused by the planet's motion around the Sun. The observations of the phases of Venus were further strong evidence that the Earth is not the center of things, and that at least one planet orbited the Sun.

Galileo published his findings, and his controversial conclusions supporting the Copernican theory, in 1610, in a book called *Sidereus Nuncius* (*The Starry Messenger*). In reporting these and other wondrous observations made with his new telescope, Galileo was challenging the scientific establishment and religious dogma of the time and aggressively urging people to change their basic view of the universe. He was (literally) playing with fire. He certainly was aware that only a few years earlier, in 1600, the astronomer Giordano Bruno had been burned at the stake in Rome for his heretical teaching that the Earth orbited the Sun. By all accounts, however, Galileo delighted in publicly ridiculing and irritating his Aristotelian colleagues. In 1616, his ideas were judged heretical, and Copernicus's works were

banned by the Roman Church. Galileo was instructed to abandon his cosmological pursuits.

But Galileo would not desist. In 1632, he raised the stakes by publishing *Dialogue Concerning the Two Chief World Systems*, which compared the Ptolemaic and Copernican models. The book presented a discussion among three people: one of them a dull-witted Aristotelian, whose views time and again were roundly defeated by the arguments of one of his companions, an articulate proponent of the heliocentric system. To make the book accessible to a wide popular audience, Galileo wrote it in Italian rather than Latin. These actions brought Galileo into direct conflict with the Church. Eventually, the Inquisition forced him, under threat of torture, to retract his claim that the Earth orbits the Sun, and he was placed under house arrest in 1633; he remained imprisoned for the rest of his life. Not until 1992 were Galileo's "crimes" publicly forgiven by the Church. But the damage to the orthodox view of the

be continually tested. If experiments and observations favor it, a theory can be further developed and refined, but if they do not, it must be rejected, no matter how appealing it originally seemed. The process is illustrated schematically in the accompanying figure. This new approach to investigation, combining thinking and doing—that is, theory and experiment—is known as the scientific method. It lies at the heart of modern science.

Notice, incidentally, that there is no "end-point" to the process depicted in the figure. A theory can be invalidated by a single wrong prediction, but no amount of observation or experimentation can ever prove it correct. Theories simply become more and more widely accepted as their predictions are repeatedly confirmed.

In astronomy, we are rarely afforded the luxury of performing experiments to test our theories, so observation becomes vitally important. One of the first documented uses of the scientific method in an astronomical context was performed by Aristotle (384–322 B.C.) nearly 25 centuries ago. He noticed that during a lunar eclipse, when Earth is positioned between the Sun and the Moon, it casts a curved shadow onto the surface of the Moon. The figure below shows a series of photographs taken during a recent lunar eclipse. The Earth's shadow, projected onto the Moon's surface, is indeed slightly curved. This is what Aristotle must have seen and recorded so long ago.

Because the observed shadow seemed always to be an arc of the same circle, Aristotle concluded that Earth, the cause of the shadow, must be round. On the basis of this hypothesis—

this possible explanation of the observed facts—he then went on to predict that any and all future lunar eclipses would show that Earth's shadow was curved, regardless of the orientation of our planet. That prediction has been tested every time a lunar eclipse has occurred. It has yet to be proved wrong. Aristotle was not the first person to argue that Earth is round, but he was apparently the first to offer a proof of it using the lunar-eclipse method.

The reasoning procedure Aristotle used forms the basis of all scientific inquiry today. He first made an observation. He then formulated a hypothesis to explain that observation. Finally, he tested the validity of his hypothesis by making predictions that could be confirmed or refuted by further observations. Observation, theory, and testing—these are the cornerstones of the scientific method, a technique whose power will be demonstrated again and again throughout our text.

Scientists throughout the world today use an approach that relies heavily on testing ideas. They gather data, form a working hypothesis that explains the data, and then proceed to test its predictions using experiment and observation. Experiment and observation are integral parts of the process of scientific inquiry. Theories unsupported by such evidence rarely gain any measure of acceptance in scientific circles. Used properly over a period of time, this rational, methodical approach enables us to arrive at conclusions that are mostly free of the personal bias and human values of any one scientist. The scientific method is designed to yield an objective view of the universe we inhabit.

universe was done, and the Copernican genie was out of the bottle once and for all.

THE ASCENDANCY OF THE COPERNICAN SYSTEM

Although Renaissance scholars were correct, none of them could *prove* that our planetary system is centered on the Sun, or even that Earth moves through space. Direct evidence for this was obtained only in the early eighteenth century, when astronomers discovered the *aberration of starlight*—a slight (20″) shift in the observed direction to a star, caused by Earth's motion perpendicular to the line of sight. Further proof came in the mid-nineteenth century, with the first unambigiuous measurement of stellar parallax. Further verification of the heliocentricity of the solar system came gradually, with innumerable observational tests that culminated with the expeditions of our unmanned space probes of the 1960s, 1970s, and 1980s. The Copernican world view is fully confirmed today.

The Copernican episode is a good example of how the scientific method, though affected at any given time by the subjective whims, human biases, and sheer luck of researchers, does ultimately lead to a definite degree of objectivity. Over time, many groups of scientists checking, confirming, and refining experimental tests will neutralize the subjective attitudes of individuals. Usually one generation of scientists can bring sufficient objectivity to bear on a problem, though some especially revolutionary concepts are so swamped by tradition, religion, and politics that more time is necessary. In the case of heliocentricity, objective confirmation was not obtained until about three centuries after Copernicus published his work and more than 2000 years after Aristarchus had proposed the concept. Nonetheless, that objectivity *did in fact* eventually prevail.

The development and eventual acceptance of the heliocentric model were milestones in our thinking.

Understanding our planetary system freed us from an Earth-centered view of the universe and eventually enabled us to realize that Earth orbits only one of myriad similar stars in the Milky Way Galaxy, which is itself one of myriad galaxies. This removal of the Earth from any position of great cosmic significance is generally known, even today, by the term *Copernican principle.*

2.5 *Kepler's Laws of Planetary Motion*

4 At about the same time that Galileo was becoming famous for his telescopic observations, Johannes Kepler (see Figure 2.11), a German mathematician and astronomer, announced his discovery of a set of simple empirical "laws" that accurately described the motions of the planets. While Galileo was the first "modern" observer, Kepler was a pure theorist; he based his work almost entirely on the observations of another (in part because of his own poor eyesight). Those observations, which predated the telescope by several decades, had been made by Kepler's employer, Tycho Brahe (1546–1601), arguably one of the greatest observational astronomers who ever lived.

BRAHE'S COMPLEX DATA

Tycho, as he is often called, was both an eccentric aristocrat and a talented observer. He was born in Denmark and educated at some of the best universities in Europe, where he studied astrology, alchemy, and medicine. By all accounts, he was impossibly rude and insulting to virtually everyone he met, an attitude that cost him dearly and ultimately resulted in his leaving Denmark in 1597. He moved to Prague, which happens to be fairly close to Graz, in Austria,

where Kepler lived. Kepler joined Tycho in Prague in 1600. There he was put to work trying to find a theory that could explain Brahe's planetary data. When Tycho died a year later, Kepler inherited not only Brahe's position as Imperial Mathematician of the Holy Roman Empire (then actually located in Eastern Europe), but also his most priceless possession: the accumulated observations of the planets, spanning several decades. Tycho's observations, though made with the naked eye, were nevertheless of very high quality. In most cases, his measured positions of stars and planets were accurate to within about 1'. Kepler set to work seeking a unifying principle to explain in detail the motions of the planets, without the need for epicycles. The effort was to occupy much of the remaining 29 years of his life.

Kepler and Tycho held different theories of the solar system. Brahe never fully accepted the Copernican view. Instead, he had his own "hybrid" cosmology, in which the Sun orbited the Earth, but the other planets orbited the Sun—a picture that, from a purely observational viewpoint, was just as consistent with observations as Copernicus's and philosophically more satisfying to Tycho.

In contrast, Kepler accepted the heliocentric picture of the solar system, but was careful not to say so in such a way as to antagonize his contemporaries. He was concerned about the relationship of his work to established Church doctrine, not so much from fear of retribution, but simply because he was a religious man. Kepler's goal was to find a simple, elegant description of the solar system that fit Tycho's complex mass of detailed observations. In the end, he found that it was necessary to abandon Copernicus's original simple idea of circular planetary orbits. As a result, an even greater simplicity emerged. After long years studying Brahe's planetary data, and after many false starts and blind alleys, Kepler developed the laws of planetary motion that now bear his name.

Kepler determined the shape of each planet's orbit by triangulation (see Section 1.9)—not from different points on Earth, but from different points on Earth's orbit. For the data, he used Brahe's detailed observations that had been made at many different times of the year. By using a portion of Earth's orbit as a baseline, Kepler was able to measure the relative sizes of the other planetary orbits. Noting where the planets were on successive nights, he also found the speeds at which the planets moved. We do not know how many geometrical shapes Kepler tried for the orbits before he hit upon the correct one. His difficult task was made even more complex because he had to determine Earth's own orbit, too. Nevertheless, he eventually succeeded in summarizing the motions of all the known planets, including Earth, in just three laws, the **laws of planetary motion**.

KEPLER'S SIMPLE LAWS

5 *Kepler's first law* has to do with the *shapes* of the planetary orbits:

> The orbital paths of the planets are elliptical (not circular), with the Sun at one focus.

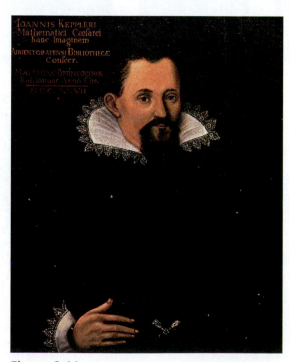

Figure 2.11 Johannes Kepler (1571–1630).

Figure 2.12 illustrates a means of constructing an **ellipse**, which is simply an elongated circle. Take a length of string and attach it to a piece of paper using two thumbtacks. Then, keeping the string taut at all times, use a pencil to trace out the curve shown in the diagram. Almost any ellipse can be drawn in this way by varying the length of the string or the distance between the tacks. The two points where the string is pinned are each called a **focus** (plural: *foci*) of the ellipse. The long axis of the ellipse, containing the two foci, is known as the *major axis*. We conventionally refer to half the length of this long axis—the **semi-major axis**—as a measure of the ellipse's "size." The **eccentricity** of the ellipse is the ratio of the distance between the foci to the length of the major axis. Notice that a circle is a special kind of ellipse in which the two foci coincide, so that the eccentricity is zero. The semi-major axis of a circle is simply its radius.

These two numbers—the semi-major axis and the eccentricity—are all that we need to describe the size and shape of the orbital path. From them, we can derive other useful quantities. For example, if a planet's orbit has semi-major axis a and eccentricity e, we can compute that its **perihelion** (closest approach to the Sun) is at a distance $a(1-e)$ and that its **aphelion** (greatest distance from the Sun) is $a(1+e)$. Thus, for example, a (hypothetical) planet with a semi-major axis of 400 million km and an eccentricity of 0.5 would range between $400 \times (1-0.5) = 200$ million km and $400 \times (1+0.5) = 600$ million km from the Sun over the course of one complete orbit.

None of the planets' elliptical orbits is nearly as elongated as the one shown in Figure 2.12. With two exceptions, the paths of Mercury and Pluto, the planetary orbits have such small eccentricities that our eyes would have trouble distinguishing them from true circles. Only because the orbits are so nearly circular were the Ptolemaic and Copernican models able to come as close as they did to describing reality.

Kepler's substitution of elliptical for circular orbits was no small advance. It amounted to abandoning an aesthetic bias—the belief in the perfection of the circle—that had governed astronomy since Greek antiquity. And it was another heavy blow to Aristotelian philosophy. Even Galileo, not known for his conservatism in these scholarly matters, clung to the idea of circular motion and never accepted that the planets move on elliptical paths.

Kepler's second law, illustrated in Figure 2.13, addresses the *speed* at which a planet traverses different parts of its orbit:

An imaginary line connecting the Sun to any planet sweeps out equal areas of the ellipse in equal intervals of time.

While orbiting the Sun, a planet traces the arcs labeled A, B, and C in Figure 2.13 in equal times. But notice that the distance traveled by the planet along arc C is greater than the distance traveled along arc A or arc B. Because the time is the same and the distance is different, the speed must vary. When a planet is close to the Sun, as in sector C, it moves much faster than when farther away, as in sector A. This law is not restricted to planets. It applies to any orbiting object. Spy satellites, for example, move very rapidly as they swoop close to Earth's surface—not because they are propelled with powerful on-board rockets, but because

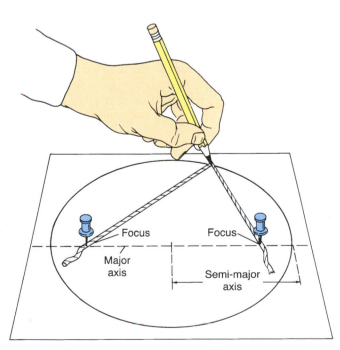

Figure 2.12 Any ellipse can be drawn with the aid of a string, a pencil, and two thumbtacks. The wider the separation of the foci, the more elongated, or eccentric, is the ellipse. In the special case where the two foci are at the same place, the drawn curve is a circle.

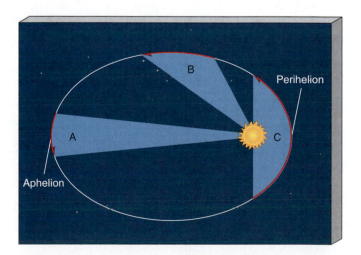

Figure 2.13 A diagram illustrating Kepler's second law: Equal areas are swept out in equal intervals of time. The three shaded areas (*A*, *B*, and *C*) are equal. Note that an object would travel the length of each of the three arrows in the same amount of time. Therefore, planets move faster when closer to the Sun.

any object in a highly elliptical orbit moves much faster as it approaches the focus where the object around which it orbits (in this case, the Earth) is located.

By taking into account the relative speeds and positions of the planets in their elliptical orbits about the Sun, Kepler's first two laws explained the variations in planetary brightness and some observed peculiar nonuniform motions that could not be accommodated within the assumption of circular motion, even with the inclusion of epicycles. Gone at last were the circles within circles that rolled across the sky. Kepler's modification of Copernicus's theory to allow the possibility of elliptical orbits both greatly simplified the model of the solar system and at the same time provided much greater predictive accuracy than had previously been possible.

Kepler published these two laws in 1609, stating that he had proved them only for the orbit of Mars. Ten years later, he extended the first and second laws to all the known planets and added a third law relating the size of a planet's orbit to its sidereal orbital **period**—the time needed for the planet to complete one circuit around the Sun. *Kepler's third law* states that

> The square of a planet's orbital period is proportional to the cube of its semi-major axis.

In other words, the planet's "year"—or, more technically, its (sidereal) orbital period P—increases more rapidly than does the size of its semi-major axis a, according to the rule $P^2 \propto a^3$ (the symbol $\propto$ means "is proportional to"). For example, Earth has, by definition, an orbital semi-major axis of 1 *astronomical unit* (1 *A.U.*—see the *More Precisely* feature on p. 13) and an orbital period of 1 Earth year. The planet Venus, orbiting at a distance of roughly 0.7 A.U., takes only 0.6 Earth years—about 225 days—to complete one circuit. By contrast, Saturn, almost 10 A.U. out, takes considerably more than 10 Earth years—in fact, nearly 30 years—to orbit the Sun just once.

Kepler's third law becomes particularly simple when we choose the astronomical unit as our unit of length and the (Earth) year as our unit of time. If we do this, the constant of proportionality in the above relation becomes equal to one, and we can replace proportionality ("$\propto$") by equality ("="). In other words, for any planet, we can write

$$P^2 \text{ (in Earth years)} = a^3 \text{ (in astronomical units)}.$$

Table 2-1 presents basic data describing the orbits of the nine planets now known. Renaissance astronomers knew these properties for the innermost six planets and used them to construct the currently accepted heliocentric model of the solar system. The second column presents each planet's orbital semi-major axis, measured in astronomical units; the third column gives the orbital period, in Earth years. The fourth column lists the planets' orbital eccentricities. For purposes of verifying Kepler's third law, the rightmost column lists the ratio P^2/a^3. As we have just seen, in the units used in the table, the third law implies that this number should equal 1 in all cases.

The main points to be grasped from Table 2-1 are these: (1) with the exception of Mercury and Pluto, the planets' orbits are very nearly circular (that is, their eccentricities are close to zero), and (2) the farther a planet is from the Sun, the greater is its orbital period, in precise agreement with Kepler's third law, to within the four-digit accuracy of the numbers in the table. For example, in the case of Pluto, verify for yourself that $39.53^3 = 248.6^2$ (at least, to three significant figures). Most important, note that Kepler's laws are exactly obeyed by *all* the known planets, *not just by the six on which he based his conclusions.*

2.6 Dimensions of the Solar System

❻ Kepler's laws allow us to construct a scale model of the solar system, with the correct shapes and *relative* sizes of all the planetary orbits, but they do not tell us the *actual* size of any orbit. We can express the distance to each planet only in terms of the distance from Earth to the Sun. Why is this? Because Kepler's triangulation measurements all used a portion of Earth's orbit as a baseline, distances could be expressed only relative to the size of that orbit, which was not itself determined. Thus our model of the solar system

TABLE 2.1 *Some Solar System Dimensions*

PLANET	ORBITAL SEMI-MAJOR AXIS, a (astronomical units)	ORBITAL PERIOD, P (Earth years)	ORBITAL ECCENTRICITY	P^2/a^3
Mercury	0.387	0.241	0.206	1.002
Venus	0.723	0.615	0.007	1.001
Earth	1.000	1.000	0.017	1.000
Mars	1.524	1.881	0.093	1.000
Jupiter	5.203	11.86	0.048	0.999
Saturn	9.539	29.46	0.056	1.000
Uranus	19.19	84.01	0.046	0.999
Neptune	30.06	164.8	0.010	1.000
Pluto	39.53	248.6	0.248	1.001

would be analogous to a road map of the United States showing the *relative* positions of cities and towns, but lacking the all-important scale marker indicating distances in kilometers or miles. For example, we would know that Kansas City is about three times more distant from New York than it is from Chicago, but we would not know the actual mileage between any two points on the map.

If we could somehow determine the value of the astronomical unit—in kilometers, say—we would be able to add the vital scale marker to our map of the solar system and compute the exact distances between the Sun and each of the planets. We might propose using triangulation to measure the distance from Earth to the Sun directly. However, we would find it impossible to measure the Sun's parallax using Earth's diameter as a baseline. The Sun is too bright, too big, and too fuzzy for us to distinguish any apparent displacement relative to the field of distant stars. To measure the Sun's distance from Earth, we must resort to some other method.

Before the middle of the twentieth century, the most accurate measurements of the astronomical unit were made using triangulation on the planets Mercury and Venus during their rare *transits* of the Sun—that is, during the brief periods when those planets passed directly between the Sun and the Earth (as shown for the case of Mercury in Figure 2.14). Because the time at which a transit occurs can be determined with great precision, astronomers can use this information to make very accurate measurements of a planet's position in the sky. They can then use simple geometry to compute the distance to the planet by combining observations made from different locations on Earth, just as discussed earlier in Chapter 1. ∞ (pp. 23-24) For example, the parallax of Venus at closest approach to Earth, as seen from two diametrically opposite points on Earth (separated by about 13,000 km), is about 1 arc minute—at the limit of naked-eye capabilities, but easily measurable telescopically. This parallax represents a distance of 45 million km.

Knowing the distance to Venus, we can immediately compute the magnitude of the astronomical unit. Figure 2.15 is an idealized diagram of the Sun–Earth–Venus orbital geometry. The planetary orbits are drawn as circles here, but in reality they are slight ellipses. This is a subtle difference, and we can correct for it using detailed knowledge of orbital motions. Assuming for the sake of simplicity that the orbits are perfect circles, we see from the figure that the distance from Earth to Venus at closest approach is approximately 0.3 A.U. Knowing that 0.3 A.U. is 45,000,000 km makes determining 1 A.U. straightforward—the answer is 45,000,000/0.3, or 150,000,000 km.

The modern method for deriving the absolute scale of the solar system uses *radar* rather than triangulation. The word **radar** is an acronym for **ra**dio **d**etection **a**nd **r**anging. In this technique, radio waves are transmitted toward an astronomical body, such as a planet. Their returning echo indicates the body's direction and range, or distance, in absolute terms (that is, in kilometers rather than in astronomical units). The calculations involved in using radar to mea-

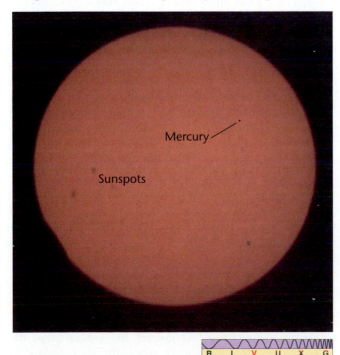

R I V U X G

Figure 2.14 A solar transit of Mercury. Such transits happen only about once per decade, because Mercury's orbit does not quite coincide with the plane of the ecliptic. Transits of Venus are even rarer, occurring only about twice per century.

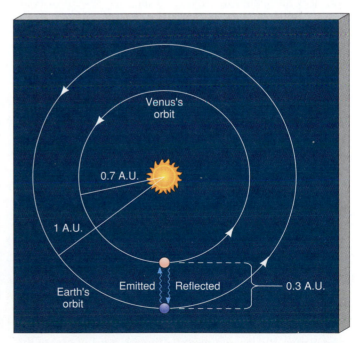

Figure 2.15 Simplified geometry of the orbits of Earth and Venus as they move around the Sun. The wavy lines represent the paths along which radar signals might be transmitted toward Venus and received back at Earth at the moment when Venus is at its minimum distance from Earth. Because the radius of Earth's orbit is 1 A.U. and that of Venus is about 0.7 A.U., we know that this distance is 0.3 A.U. Thus, radar measurements allow us to determine the astronomical unit in kilometers.

sure astronomical distances resemble those used to derive the distance between two cities if our car's speed and travel time are known. If we multiply the round-trip travel time of the radar signal (the time elapsed between transmission of the signal and reception of the echo) by the speed of light (300,000 km/s, which is also the speed of radio waves), we obtain twice the distance to the target planet (back and forth). The round-trip travel time can be measured with high precision—in fact, well enough to determine the dimensions of the orbit of Venus to an accuracy of 1 km. Through precise radar ranging, the astronomical unit is now known to be 149,597,870 km. In this text, we will round this number off to a value of 1.5×10^8 km.

Having determined the value of the astronomical unit, we can reexpress the sizes of the other planetary orbits in terms of more familiar units, such as miles or kilometers. The entire scale of the solar system can then be calibrated to high precision.

2.7 Newton's Laws

⑦ Kepler's three laws, which so simplified the solar system, were discovered *empirically*. In other words, they resulted solely from the analysis of observational data and were not derived from a theory or mathematical model. Indeed, Kepler did not have any appreciation for the physics underlying his laws. Nor did Copernicus understand the basic reasons *why* his heliocentric model of the solar system worked. Even Galileo, often called the father of modern physics, failed to understand why the planets orbit the Sun.

THE LAWS OF MOTION

What prevents the planets from flying off into space or from falling into the Sun? What causes the planets to revolve about the Sun, apparently endlessly? To be sure, the motions of the planets obey Kepler's three laws, but only by considering something more fundamental than those laws can we understand these motions. The heliocentric system was secured when, in the seventeenth century, the British mathematician Isaac Newton (Figure 2.16) developed a deeper understanding of the way *all* objects move and interact with one another as they do.

Isaac Newton was born in Lincolnshire, England, on Christmas Day in 1642, the year that Galileo died. Newton studied at Trinity College of Cambridge University, but when the bubonic plague reached Cambridge in 1665, he returned to the relative safety of his home for two years. During that time he made probably the most famous of his discoveries, the law of gravity (although it is but one of the many major scientific advances Newton is responsible for). However, either because he regarded the theory as incomplete or possibly because he was afraid that he would be attacked or plagiarized by his colleagues, he did not tell anyone of his monumental achievement for almost 20 years. It was not until 1684, when Newton was discussing with Edmund Halley (of Halley's comet fame) the leading astro-

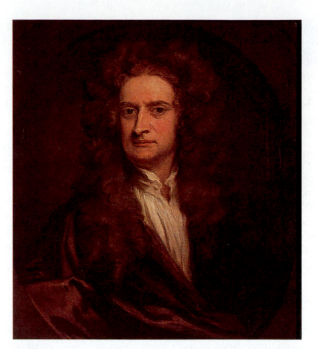

Figure 2.16 Isaac Newton (1642–1727).

nomical problem of the day—*Why* do the planets move according to Kepler's laws?—that he astounded his companion by remarking casually that he had solved the problem in its entirety nearly two decades before!

Prompted by Halley, Newton published his theories in perhaps the most influential physics book ever written: *Philosophiae Naturalis Principia Mathematica* (or *The Mathematical Principles of Natural Philosophy*—what we would today call "science"), usually known simply as Newton's *Principia*. The ideas expressed in that work form the basis for what today is known as **Newtonian mechanics**. Three basic laws of motion, the law of gravity, and the calculus (which Newton also invented) are sufficient to explain and quantify virtually all of the complex dynamic behavior we see on Earth and throughout the universe. Newton's laws are listed in the *More Precisely* feature on p. 45.

Figure 2.17 illustrates *Newton's first law of motion*. The first law simply states that a moving object will, in principle, move forever in a straight line unless some external **force** changes its direction of motion. For example, the object might glance off a brick wall or be hit with a baseball bat; in either case, a force changes the original motion of the object. The tendency of an object to keep moving at the same speed and in the same direction unless acted upon by a force is known as **inertia**. A familiar measure of an object's inertia is its **mass**—loosely speaking, the total amount of matter it contains. The greater an object's mass, the more inertia it has, and the greater is the force needed to change its state of motion.

Newton's first law contrasts sharply with the view of Aristotle, who incorrectly maintained that the natural state of an object was to be *at rest*—most probably an opinion based on Aristotle's observations of the effect of friction. In

More Precisely... Newton's Laws of Motion and Gravitation

THE THREE LAWS OF MOTION

1. Every body continues in a state of rest or in a state of uniform motion in a straight line unless it is compelled to change that state by a force acting on it.

   It requires no force to maintain motion in a straight line with constant velocity—that is, motion with constant speed and constant direction in space. The tendency of a body to remain in a state of uniform motion is usually called inertia. When velocity does vary (the speed increases or decreases, or the direction of motion changes), its rate of change is called acceleration. The relation of acceleration to any forces acting on a body is the subject of the second law of motion:

2. When a force F acts on a body of mass m, it produces in it an acceleration a equal to the force divided by the mass. Thus, $a = F/m$, or $F = ma$.

   In honor of Newton, the SI unit of force is named after him. By definition, 1 newton (N) is the force required to cause a mass of 1 kilogram to accelerate at a rate of 1 meter per second every second.

   Newton's third law relates the forces acting between separate bodies:

3. To every action, there is an equal and opposite reaction.

This law means, for example, that you attract Earth with exactly the same force as it attracts you (a force known as your weight). This attraction is governed by one final law (see p. 46):

THE LAW OF UNIVERSAL GRAVITATION ("NEWTON'S LAW OF GRAVITY")

Every particle of matter in the universe attracts every other particle with a force that is directly proportional to the product of the masses of the particles and inversely proportional to the square of the distance between them.

In other words, two bodies of masses m_1 and m_2, separated by a distance R, attract each other with a force F that is proportional to $(m_1 \times m_2)/R^2$. The constant of proportionality is known as the gravitational constant, or often simply as Newton's constant, and is always denoted by the letter G. We can then express the law of gravity as

$$F = \frac{Gm_1m_2}{R^2}.$$

The value of G has been measured in extremely delicate laboratory experiments. In SI units, its value is 6.67×10^{-11} newton meter^2/kilogram^2 (N m^2/kg^2).

our discussion, we will neglect the familiar concept of friction—the force that slows balls rolling along the ground, blocks sliding across tabletops, and baseballs moving through the air. In any case, it is not an issue for the planets because there is no appreciable friction in outer space. The fallacy in Aristotle's argument was first realized and exposed by Galileo, who conceived of the notion of inertia long before Newton formalized it into a law.

The rate of change of the velocity of an object—speeding up, slowing down, or simply changing direction—is called its **acceleration**. *Newton's second law* states that the acceleration of an object is directly proportional to the applied force and inversely proportional to its mass—that is, the greater the force acting on the object, or the smaller the mass of the object, the greater its acceleration. Thus, if two objects are pulled with the same force, the more massive one will accelerate less; if two identical objects are pulled with different forces, the one experiencing the greater force will accelerate more. Finally, *Newton's third law* simply tells us that forces cannot occur in isolation—if body A exerts a force on body B, then body B necessarily exerts a force on body A that is equal in magnitude, but oppositely directed.

Figure 2.17 An object at rest will remain at rest (a) until some force acts on it (b). It will then remain in that state of uniform motion until another force acts on it. The arrow in (c) shows a second force acting at a direction different from the first, which causes the object to change direction.

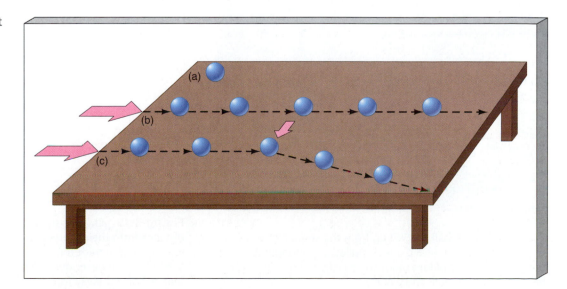

Only in extreme circumstances do Newton's laws break down, and this fact was not realized until the twentieth century, when Albert Einstein's theories of relativity once again revolutionized our view of the universe (see Chapter 22). Most of the time, however, Newtonian mechanics provides an excellent description of the motion of planets, stars, and galaxies through the cosmos.

GRAVITY

Forces can act *instantaneously* or *continuously*. To a good approximation, the force from a baseball bat that hits a home run can be thought of as being instantaneous, or momentary, in nature. A good example of a continuous force is the one that prevents the baseball from zooming off into space—**gravity**, the phenomenon that started Newton on the path to the discovery of his laws. Newton hypothesized that any object having mass always exerts an attractive *gravitational force* on all other massive objects. The more massive an object, the stronger its gravitational pull.

The continuous pull of Earth's gravity can be visualized if we consider a baseball thrown upward. Figure 2.18 illustrates how the baseball's path changes continuously. The baseball, having some mass of its own, also exerts a gravitational pull on Earth. By Newton's third law, this force is equal and opposite to the weight of the ball (the force with which Earth attracts it). But, by Newton's second law, the Earth has a much greater effect on the light baseball than the baseball has on the much more massive Earth. The ball and Earth each feel the same gravitational force, but the Earth's *acceleration* is much smaller.

Now consider the trajectory of a baseball batted from the surface of the Moon, which has much less mass than Earth has. Because the pull of gravity is about one-sixth as great on the Moon as on Earth, a baseball's path changes more slowly near the Moon. A typical home run in a ballpark on Earth would travel nearly half a mile on the Moon. The Moon, less massive than Earth, has less gravitational influence on the baseball. The magnitude of the gravitational force, then, depends on the *masses* of the attracting bodies. Theoretical insight, as well as detailed experiments, tells us that the force is in fact directly proportional to the product of the two masses.

Studying the motions of the planets around the Sun reveals a second aspect of the gravitational force. At all locations equidistant from the Sun's center, the gravitational force has the same strength, and it is always directed toward the Sun. Furthermore, in much the same way that temperature decreases with distance from a fire, gravity weakens with distance from any object that has mass.

Forces that decrease with distance from their source are encountered throughout all of science. Many of them, including gravity, decrease in proportion to the *square* of the distance. They are said to obey an **inverse-square law**. As shown in Figure 2.19, inverse-square forces decrease rapidly with distance from their source. For example, tripling the distance makes the force $3^2 = 9$ times weaker, while multiplying the distance by 5 results in a force that is $5^2 = 25$

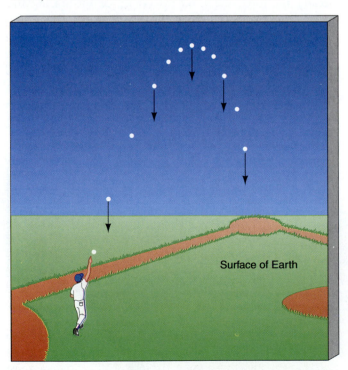

Surface of Earth

Figure 2.18 A ball thrown up from the surface of a massive object such as a planet is pulled continuously by the gravity of that planet (and, conversely, the gravity of the ball continuously pulls the planet).

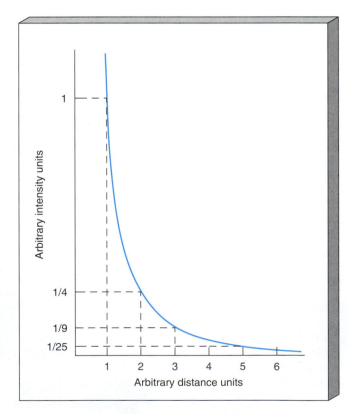

Arbitrary intensity units

Arbitrary distance units

Figure 2.19 Inverse-square forces rapidly weaken with distance from their source. The strength of the gravitational force decreases with the square of the distance from the Sun. The force never quite diminishes to zero, however, no matter how far away from the Sun we go.

times weaker. Despite this rapid decrease, the force never quite reaches zero. The gravitational pull of an object having some mass can never be completely extinguished.

We can combine the preceding statements about mass and distance to form a law of gravity that dictates the way in which *all* material objects attract one another. As a proportionality, Newton's law of gravity is

$$\text{gravitational force} \propto \frac{\text{mass of object \#1} \times \text{mass of object \#2}}{\text{distance}^2}.$$

This relationship is a compact way of stating that the gravitational pull between two objects is directly proportional to the product of their masses and inversely proportional to the square of the distance separating them. (See the *More Precisely* feature on p. 45 for a fuller statement of this law.)

To Newton, gravity was a force that acted at a distance, with no obvious way in which it was actually transmitted from place to place. Newton was not satisfied with this explanation, but he had none better. To appreciate the modern view of gravity, consider any piece of matter having some mass—it could be smaller than an atom or larger than a galaxy. Extending outward from this object in all directions is a **gravitational field** produced by the matter. We now regard such a field as a property of space itself—a property that determines the influence of one massive object on another. All other matter "feels" the field as a gravitational force.

PLANETARY MOTION

8 It is the mutual gravitational attraction of the Sun and the planets, as expressed by Newton's law of gravity, that produces the observed planetary orbits. Because the Sun is much more massive than any of the planets, it dominates the interaction. We might say that the Sun "controls" the planets, not the other way around. As the Sun pulls, it tends to draw the planets directly toward itself. As depicted in Figure 2.20, this inward gravitational force continuously pulls each planet toward the Sun, deflecting its forward motion into a curved orbital path.

The Sun–planet interaction sketched here is analogous to what occurs when you whirl a rock at the end of a string above your head. The Sun's gravitational field is your hand and the string, and the planet is the rock at the end of that string. The tension in the string provides the force necessary for the rock to move in a circular path. If you were suddenly to release the string—which would be like eliminating the Sun's gravity—the rock would fly away along a tangent to the circle, in accordance with Newton's first law. In the solar system, at this very moment, Earth is moving under the combined influence of these two effects: the competition between gravity and inertia. The net result is a stable orbit, despite our continuous rapid motion through space. In fact, Earth orbits the Sun at a speed of about 30 km/s, or some 70,000 miles per hour. (Verify this for yourself by calculating how fast Earth must move to complete a circle of radius 1 A.U.—and hence of circumference

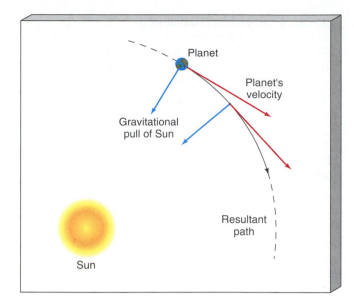

Figure 2.20   The Sun's inward pull of gravity on a planet competes with the planet's tendency to continue moving in a straight line. These two effects combine, causing the planet to move smoothly along an intermediate path, which continually "falls around" the Sun. This unending tug-of-war between the Sun's gravity and the planet's inertia results in a stable orbit.

2π A.U., or 940 million km—in 1 year, or 3.2×10^7 seconds. The answer is 9.4×10^8 km/3.1×10^7s, or 30.3 km/s.)

Before we leave this topic, let's make one final, very important point. We can use Newtonian mechanics to calculate the relationship between the distance (r) and the speed (v) of a planet moving in a circular orbit around the Sun (of mass m). By calculating the force required to keep the planet moving in a circle, and comparing it with the gravitational force due to the Sun, we can show that the circular speed is

$$v = \sqrt{\frac{Gm}{r}},$$

where the gravitational constant G is defined in the *More Precisely* feature on p. 45. Because we have measured G in the laboratory on Earth and because we know the length of a year and the size of the astronomical unit, we can use Newtonian mechanics to *weigh* the Sun. Inserting the known values of $v = 30$ km/s, $r = 1$ A.U. $= 1.5 \times 10^{11}$ m, and $G = 6.7 \times 10^{-11}$ Nm^2/kg^2, we can calculate the mass of the Sun to be 2×10^{30} kg—an enormous mass by terrestrial standards. Similarly, knowing the distance to the Moon and the length of the (sidereal) month, we can measure the mass of Earth to be 6×10^{24} kg.

In fact, this is how basically *all* masses are measured in astronomy. Because we can't just go out and weigh an astronomical object when we need to know its mass, we must look for its gravitational influence on something else. This principle applies to planets, stars, galaxies, and even clusters of galaxies—very different objects, but all subject to the same physical laws.

KEPLER'S LAWS RECONSIDERED

Newton's laws of motion and his law of universal gravitation provided a theoretical explanation for Kepler's empirical laws of planetary motion. Just as Kepler modified Copernicus's model by introducing ellipses rather than circles, so too did Newton make corrections to Kepler's first and third laws. It turns out that a planet does not orbit the exact center of the Sun. Instead, both the planet and the Sun orbit their common **center of mass**. Because the Sun and the planet feel equal and opposite gravitational forces (by Newton's third law), the Sun must also move (by Newton's first law), driven by the gravitational influence of the planet. The Sun is so much more massive than any planet that the center of mass is very close to the center of the Sun, which is why Kepler's laws are so accurate. Thus, Kepler's first law becomes

The orbit of a planet around the Sun is an ellipse, with the common center of mass of the planet and the Sun at one focus.

As shown in Figure 2.21, however, the center of mass for two objects of comparable mass does not lie within either object. For identical masses (Figure 2.21a), the orbits are identical ellipses, with a common focus located midway between the two objects. For unequal masses (as in Figure 2.21b), the elliptical orbits still share a focus and both have the same eccentricity, but the more massive object moves more slowly, and on a tighter orbit. (Note that Kepler's second law, as stated earlier, continues to apply without modification to each orbit separately, but the *rates* at which the two orbits sweep out area are different.) In the case of a planet orbiting the much more massive Sun (see Figure 2.21c), the path traced out by the Sun's center is smaller than the Sun itself.

The change to Kepler's third law is also small in the case of a planet orbiting the Sun but very important in other circumstances, such as when we consider the orbital motion of two stars that are bound to one another by gravity. Following through the mathematics of Newton's theory, we find that the true relationship between the semi-major axis a of the planet's orbit relative to the Sun and its orbital period P is

$$P^2 \propto \frac{a^3}{M_{\text{total}}},$$

where M_{total} is the *combined* mass of the two objects. Notice that Newton's restatement of Kepler's third law preserves the proportionality between P^2 and a^3, but now the proportionality includes M_{total}, so it is *not* quite the same for all the planets. The Sun's mass is so great, however, that the differences in M_{total} among the various combinations of the Sun and the other planets are almost unnoticeable, so Kepler's third law, as originally stated, is a very good approximation.

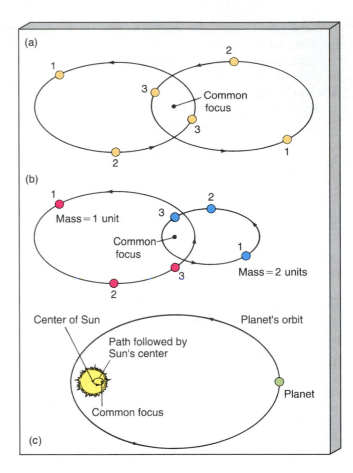

Figure 2.21   (a) The orbits of two bodies (stars, for example) with equal masses, under the influence of their mutual gravity, are identical ellipses with a common focus. That focus is not at the center of either star but instead is located at the center of mass of the pair, midway between them. The positions of the two bodies at three different times are indicated by the pairs of numbers. (Notice that a line joining the bodies always passes through the common focus.) (b) The orbits of two bodies, one of which is twice as massive as the other. Again, the elliptical orbits have a common focus, and the two ellipses have the same eccentricity. However, in accordance with Newton's laws of motion, the more massive body moves more slowly, and in a smaller orbit, staying closer to the center of mass (at the common focus). In this particular case, the larger ellipse is twice the size of the smaller one. (c) In this extreme case of a hypothetical planet orbiting the Sun, the common focus of the two orbits lies inside the Sun.

Just as before, we can once again choose units to simplify this proportionality. Expressing distance in astronomical units, time in years, and mass in units of the mass of the Sun, we can now say:

$$P^2 \text{ (in Earth years)} = \frac{a^3 \text{ (in astronomical units)}}{M_{\text{total}} \text{ (in solar masses)}}.$$

This modified form of Kepler's third law is true in *all* circumstances, inside or outside the solar system.

ESCAPING FOREVER

The law of gravity that describes the orbits of planets around the Sun applies equally well to natural moons and artificial satellites orbiting any planet. All of our Earth-orbiting, human-made satellites move along paths governed by a combination of the inward pull of Earth's gravity and the forward motion gained during the rocket launch. If the rocket initially imparts enough speed to the satellite, it can go into orbit. Satellites not given enough speed at launch (such as intercontinental ballistic missiles, ICBMs) fail to achieve orbit and fall back to Earth (see Figure 2.22). (Technically, ICBMs actually do orbit Earth's attracting center, but their orbits intersect Earth's surface.)

Some space vehicles, such as the robot probes that visit the other planets, attain enough speed to escape our planet's gravitational field and move away from Earth forever. This speed, known as the **escape velocity**, is about 41 percent greater (actually, $\sqrt{2} = 1.41421\ldots$ times greater) than the speed of a circular orbit at any given radius.* At less than escape velocity, the old adage "what goes up must come down" (or at least stay in orbit) still applies. At more than escape velocity, our spacecraft will leave Earth for good (neglecting the effect of air resistance on our way through Earth's atmosphere and assuming that we don't turn the craft around using an on-board rocket motor). Planets, stars, galaxies—all gravitating bodies—have escape veloci-

*In terms of our earlier formula, the escape velocity is just $v_{escape} = \sqrt{2Gm/r}$.

ties. No matter how massive the body, gravity decreases with distance. As a result, the escape velocity diminishes with increasing separation. The farther we go from Earth (or any gravitating body), the easier it becomes to escape.

The speed of a satellite in a circular orbit just above Earth's atmosphere is 7.9 km/s (roughly 18,000 mph). The satellite would have to travel at 11.2 km/s (about 25,000 mph) to escape from Earth altogether. In the case of an object exceeding the escape velocity, the motion is said to be **unbound**, and the orbit is no longer an ellipse. In fact, the path of the spacecraft relative to Earth is a related geometrical figure called a *hyperbola*. In the intermediate case, where the spacecraft has exactly the escape velocity and so has just enough energy to get away, the orbital trajectory is an intermediate geometrical shape—a *parabola*. If we simply change the word *ellipse* to *hyperbola* or *parabola*, the modified version of Kepler's first law still applies, as does Kepler's second law. (Kepler's third law does not extend to unbound orbits because it doesn't make sense to talk about a period in those cases.)

Newton's laws explain the paths of objects moving at any point in space near any gravitating body. These laws provide a firm physical and mathematical foundation for Copernicus's heliocentric model of the solar system and for Kepler's laws of planetary motions. But they do much more than that. Newtonian gravitation governs not only the planets, moons, and satellites in their elliptical orbits, but also the stars and galaxies in their motion throughout our universe—as well as apples falling to the ground.

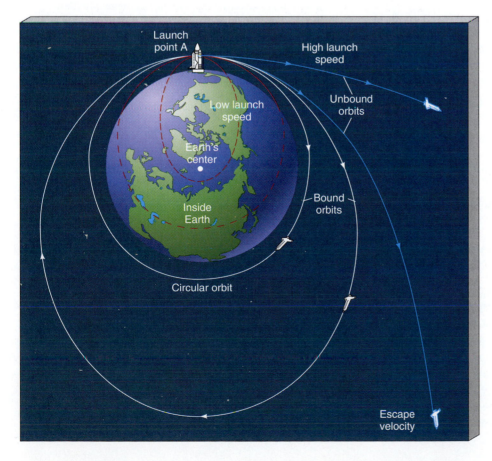

Figure 2.22 The effect of launch speed on the trajectory of a satellite. With too low a speed at point *A* the satellite will simply fall back to Earth. Given enough speed, however, the satellite will go into orbit—it "falls around the Earth." As the initial speed at point *A* is increased, the orbit will become more and more elongated. When the initial speed exceeds the escape velocity, the satellite will become unbound from Earth and will escape along a hyperbolic trajectory.

Launch point A
High launch speed
Low launch speed
Unbound orbits
Earth's center
Inside Earth
Bound orbits
Circular orbit
Escape velocity

Chapter Review

SUMMARY

Many ancient cultures constructed elaborate structures that served as calendars and astronomical observatories. The study of the universe on the very largest scales is called **cosmology** (p. 32).

Unlike the Sun and the Moon, planets sometimes appear to temporarily reverse their direction of motion (from night to night) relative to the stars, and then resume their normal "forward" course. This phenomenon is called **retrograde motion** (p. 32).

Geocentric (p. 32) models of the universe were based on the assumption that the Sun, the Moon, and the planets all orbit Earth. The most successful, and long-lived, of these was the **Ptolemaic model** (p. 34). To account for retrograde motion within the geocentric picture, it was necessary to suppose that planets moved on small circles called **epicycles** (p. 34), whose centers orbited Earth on larger circles called **deferents** (p. 34).

The **heliocentric** (p. 35) view of the solar system holds that Earth, like all the planets, orbits the Sun. This accounts for retrograde motion and the observed size and brightness variations of the planets in a much more natural way than the geocentric model. The widespread realization during the Renaissance that the solar system is Sun centered, and not Earth centered, is known as the **Copernican revolution** (p. 35), in honor of Nicholas Copernicus, who laid the foundations of the modern heliocentric model.

Galileo Galilei is often regarded as the father of experimental science. His telescopic observations of the Moon, the Sun, Venus, and Jupiter played a crucial role in supporting and strengthening the Copernican picture of the solar system.

Johannes Kepler improved on Copernicus's model with his three **laws of planetary motion** (p. 40): (1) Planetary orbits are **ellipses** (p. 41), with the Sun at one **focus** (p. 41). (2) A planet moves faster as its orbit takes it closer to the Sun. (3) The **semi-major axis** (p. 41) of the orbit is related in a simple way to

the planet's orbit **period** (p. 41). Most planets move on orbits whose **eccentricities** (p. 41) are quite small, so their paths differ only slightly from perfect circles. A planet's closest approach to the Sun is called its **perihelion** (p. 41); its farthest distance from the Sun is called its **aphelion** (p. 41).

The distance from Earth to the Sun is called the *astronomical unit*. Nowadays, the astronomical unit is determined by bouncing **radar** (p. 43) signals off the planet Venus and measuring the time taken for the signal to return.

Isaac Newton succeeded in explaining Kepler's laws in terms of a few general physical principles, now known as **Newtonian mechanics** (p. 44). The tendency of a body to keep moving at constant velocity is called **inertia** (p. 44). The greater the body's **mass** (p. 44), the greater its inertia. To change the velocity, a **force** (p. 44) must be applied. The rate of change of velocity, called **acceleration** (p. 45), is simply equal to the applied force divided by the body's mass.

To explain planetary orbits, Newton postulated that **gravity** (p. 46) attracts the planets to the Sun. Every object with any mass is surrounded by a **gravitational field** (p. 47), whose strength decreases with distance according to an **inverse-square law** (p. 46). This field determines the gravitational force exerted by the object on any other body in the universe.

Newton's laws imply that a planet does not orbit the precise center of the Sun, but instead that both the planet and the Sun orbit the common **center of mass** (p. 48) of the two bodies.

For an object to escape from the gravitational pull of another, its speed must exceed the **escape velocity** (p. 49) of the second body. In this case, the motion is said to be **unbound** (p. 49), and the orbital path is no longer an ellipse, although it is still described by Newton's laws.

SELF-TEST: True or False?

_____ **1.** Historical records show that it was Aristotle who first proposed that all planets revolve around the Sun.

_____ **2.** The teachings of Aristotle remained unchallenged until the eighteenth century A.D.

_____ **3.** Ptolemy was responsible for a geocentric model that was successful at predicting the positions of the planets, Moon, and the Sun.

_____ **4.** The heliocentric model of the universe held that the Earth was at the center, and everything else moved around it.

_____ **5.** Kepler's discoveries regarding the orbital motion of the planets were based on his own observations.

_____ **6.** The Sun's location in a planet's orbit is at the center.

_____ **7.** The semi-major axis of an orbit is half the major axis.

_____ **8.** A circle has an eccentricity of zero.

_____ **9.** The astronomical unit is a distance equal to the semi-major axis of Earth's orbit around the Sun.

_____ **10.** The speed of a planet orbiting the Sun is independent of the planet's position in its orbit.

_____ **11.** Kepler's laws work only for the six planets known in his time.

_____ **12.** Kepler never knew the true distances between the planets and the Sun, only their relative distances.

_____ **13.** Using his laws of motion and gravity, Newton was able to prove Kepler's Laws.

_____ **14.** You throw a baseball to someone. Before the ball is caught, it is temporarily in orbit around Earth's center.

SELF-TEST: Fill in the Blank

1. Stonehenge was used as a _____ by people in the Stone Age.

2. Accurate records of comets and "guest" stars were kept over many centuries by _____ astrologers.

3. The astronomical knowledge of ancient Greece was kept alive and augmented by _____ astronomers.

4. When the planets Mars, Jupiter, or Saturn appear to move "backwards" (westward) in the sky relative to the stars, this is known as _____ motion.

5. Observation, theory, and testing are the cornerstones of the _____.

6. The heliocentric model was reinvented by _____.

7. Central to the heliocentric model is the assertion that the observed motions of the planets and the Sun are the result of _____ motion around the Sun.

8. Galileo discovered _____ of Jupiter, the _____ of Venus, and the Sun's rotation from observations of _____.

9. Kepler discovered that the shape of an orbit is an _____ , not a _____, as had previously been believed.

10. Kepler's third law relates the _____ of the orbital period to the _____ of the semi-major axis.

11. The modern method of measuring the astronomical unit is by using _____ measurements of a planet or asteroid.

12. Newton's first law states that a moving object will continue to move in a straight line with constant speed unless acted upon by a _____.

13. Newton's law of gravity states that the gravitational force between two objects depends on the _____ of their masses and inversely on the _____ of their separation.

14. Newton discovered that, in Kepler's third law, the orbital period depends on the semi-major axis and on the the sum of the _____ of the two objects involved.

REVIEW AND DISCUSSION

1. What contributions to modern astronomy were made by Chinese and Islamic astronomers during the Dark Ages of medieval Europe?

2. Briefly describe the geocentric model of the universe.

3. The benefit of our current knowledge lets us see flaws in the Ptolemaic model of the universe. What is its basic flaw?

4. What was the great contribution of Copernicus to our knowledge of the solar system? What was still a flaw in the Copernican model?

5. When were Copernicus's ideas finally accepted?

6. What is the Copernican principle?

7. What discoveries of Galileo helped confirm the views of Copernicus?

8. Briefly describe Kepler's three laws of orbital motion.

9. If radio waves cannot be reflected from the Sun, how can radar be used to find the distance from Earth to the Sun?

10. What does it mean to say that Kepler's laws are empirical in nature?

11. List the two modifications made by Newton to Kepler's laws.

12. Why does a baseball fall toward the Earth and not Earth toward the baseball?

13. Why would a baseball go higher if it were thrown up from the surface of the Moon than if it were thrown with the same velocity from the surface of the Earth?

14. What is the meaning of the term escape velocity?

15. Explain why geometry has been so important in astronomy.

16. Is the climate for new ideas better today than it was during the time of Copernicus? Why or why not?

17. What would happen to Earth if the Sun's gravity were suddenly "turned off?"

PROBLEMS

1. How long does an Earth-Venus radar signal take to complete its round trip when Earth and Venus are at their closest to one another (0.3 A.U.)?

2. Using the data in Table 2-1 show that Pluto is closer to the Sun at perihelion (the point of closest approach to the Sun in its orbit), than Neptune is in its orbit. For simplicity, assume Neptune has a circular orbit, which is not far from the truth.

3. Jupiter's moon Callisto orbits it at a distance of 1.88 million km. Its orbital period about the planet is 16.7 days. What is the mass of Jupiter? (Assume that Callisto's mass is negligible compared with that of Jupiter.) Use the modified version of Kepler's Third Law (see Section 2.7).

4. Use Newton's law of gravity to calculate the force of gravity between you and the Earth. Convert your answer, which will be in newtons, to pounds using the conversion 4.45 N equals one pound. What would you normally call this force?

PROJECTS

1. Look in an almanac for the date of opposition of one or all of these bright planets: Mars, Jupiter, and Saturn. At opposition, these planets are at their closest points to the Earth, and are at their largest and brightest in the night sky. Observe these planets. How long before opposition does each planet's retrograde motion begin? How long afterwards does it end?

2. Draw an ellipse. (See Figure 2.12, p. 41.) You'll need two pins, a piece of string, and a pencil. Tie the string in a loop and place around the pins. Place the pencil inside the loop and run it around the inside of the string, holding the loop taut. The two pins will be at the foci of the ellipse. What is the eccentricity of the ellipse you have drawn?

3. Use a small telescope to replicate Galileo's observations of Jupiter's four largest moons. Note the moons' brightnesses and their locations with respect to Jupiter. If you watch over a period of several nights, draw what you see; you'll notice that these moons change their positions as they orbit the giant planet. Check the charts given monthly in *Astronomy* or *Sky & Telescope* magazines to identify each moon you see.

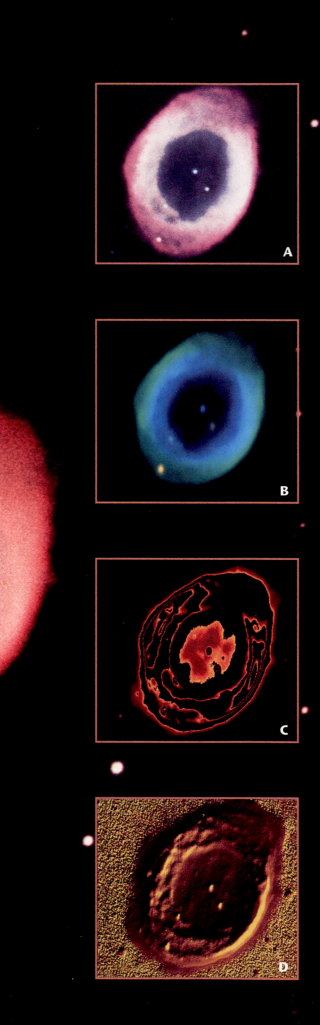

3

RADIATION

Information from the Cosmos

LEARNING GOALS

Studying this chapter will enable you to:

1 Discuss the nature of electromagnetic radiation and how that radiation transfers energy and information through interstellar space.

2 List the major regions of the electromagnetic spectrum, and explain how the properties of Earth's atmosphere affect our ability to make astronomical observations at different wavelengths.

3 Explain what is meant by the term "black-body radiation," and describe the basic properties of such radiation.

4 Tell how we can determine the temperature of an object by observing the radiation that it emits.

5 Describe the way in which the intensity of electromagnetic radiation varies with the distance from its source.

6 Show how the relative motion of a source of radiation and its observer can change the perceived wavelength of the radiation, and explain the importance of this phenomenon to astronomy.

(Opposite page, background) The Ring Nebula in the constellation Lyra is one of the most magnificent sights in the nighttime sky. Seen here glowing in the light of its own emitted radiation, the nebula is actually the expanding outer atmosphere of a nearly dead star.
(Inset A) This nearly true color view clearly shows the dying dwarf star at the center of the expanding gas cloud, which is really a three-dimensional shell and not a ring.
(Inset B) False-color images can sometimes enhance certain features. Here some of the fine structure in the shell of gas can be seen more clearly.
(Inset C) Contour images, like this one derived from red light emitted by hydrogen gas in the nebula's shell and star's lingering atmosphere, can be used to map regions of relative brightness.
(Inset D) Topographic images, created by taking two photographs of the same object, can highlight structural features, in this case enhancing subtle shadings in the nebula's brightness pattern.

*A*stronomical objects are more than just things of beauty in the night sky. Planets, stars, and galaxies are of vital significance if we are to understand fully our place in the big picture—the "grand design" of the universe. Each object is a source of information about the material aspects of our universe—its state of motion, its temperature, its chemical composition, even its past history. When we look at the stars, the light we see actually began its journey to Earth decades, centuries—even millennia—ago. The faint rays from the most distant galaxies have taken billions of years to reach us. The stars and galaxies in the night sky show us the far away and the long ago. In this chapter, we begin our study of how astronomers extract information from the light emitted by astronomical objects. These basic concepts of radiation are central to modern astronomy.

3.1 *Information from the Skies*

Figure 3.1 shows a neighboring spiral galaxy in the constellation Andromeda. On a dark, clear night, far from cities or other sources of light, this galaxy can be seen with the naked eye as a faint, fuzzy patch on the sky, comparable in diameter to the full Moon.

The fact that it is easily visible from the Earth belies Andromeda's enormous distance from us. The Andromeda Galaxy, as it is generally called, lies roughly 2 million light years away. We can begin to appreciate the enormous distances involved in astronomy when we realize that a light year is the *distance* traveled by light in a year and that light moves at the fastest speed known, nearly 300,000 kilometers *per second*. A single light year therefore equals about 10 trillion kilometers (or 6 trillion miles)—and Andromeda is *2 million times that far away*.

An object at such a distance is truly inaccessible, in any realistic human sense. Even if a space probe could miraculously travel at the speed of light, 2 million years would be needed for the probe to reach its destination and 2 million more for it to return with its findings. Considering that civilization has existed on Earth for fewer than 10,000 years (and its prospects for the next 10,000 are far from certain), even this unattainable technological feat would not provide us with a practical means of exploring other galaxies—or even the farthest reaches of our own, several tens of thousands of light years away.

LIGHT AND RADIATION

How do astronomers know *anything* about objects far from Earth? How do we obtain detailed information about any planet, star, or galaxy too distant for a personal visit or any kind of controlled experiment? The answer is that we use the laws of physics, as we know them here on Earth, to interpret the **electromagnetic radiation** emitted by these objects. *Radiation* is any way in which energy is transmitted through space from one point to another without the need for any physical connection between those two locations. The term *electromagnetic* just means that the energy is carried in the form of rapidly fluctuating *electric* and *magnetic fields* (to be discussed in more detail later in Section 3.2). Virtually all we know about the universe beyond the Earth's atmosphere has been gleaned from painstaking analysis of electromagnetic radiation received from afar.

Visible light is the particular type of electromagnetic radiation to which our human eyes happen to be sensitive. As light enters our eye, the cornea and lens focus it onto the retina, whereupon small chemical reactions triggered by the incoming energy send electrical impulses to the brain, producing the sensation of sight. But there is also *invisible* electromagnetic radiation, which goes completely unde-

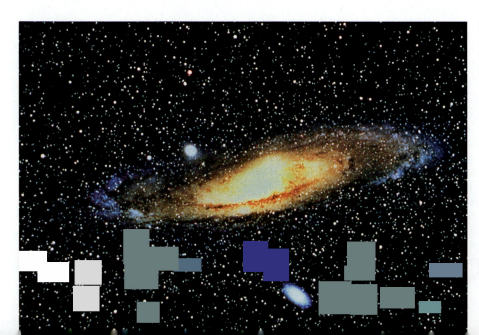

Figure 3.1 The pancake-shaped Andromeda Galaxy is about 2 million light years away and contains a few hundred billion stars.

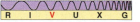

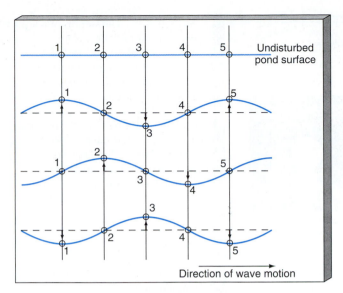

Figure 3.2 The passage of a wave across a pond causes the surface of the water to bob up and down, but there is no movement of water from one part of the pond to another. The points numbered 1 through 5 represent particles on the pond's surface.

tected by our eyes. **Radio, infrared,** and **ultraviolet** waves, as well as **X rays** and **gamma rays,** all fall into this category.

Recognize that, despite the different names, the words *light, rays, radiation,* and *waves* really all refer to the same thing. The names are just historical accidents, reflecting the fact that it took many years for scientists to realize that these apparently very different types of radiation are in reality one and the same physical phenomenon. We will use the general terms "light" and "electromagnetic radiation" more or less interchangeably throughout this text.

WAVE MOTION

Despite the early confusion still reflected in the modern terminology, scientists now know that all types of electromagnetic radiation travel through space in the form of **waves.** To understand the behavior of light, then, we must know a little about wave motion.

A wave is a way in which energy can be transferred from place to place without physical movement of material from one location to another. In wave motion, the energy is carried by a disturbance of some sort. This *disturbance,*

whatever its nature, occurs in a distinctive pattern that repeats itself cyclically both in time and in space. Ripples on the surface of a pond, sound waves in air, and electromagnetic waves in space, despite their many obvious differences, all share this basic defining property.

As a familiar example, imagine a twig floating in a pond. A pebble, thrown into the pond at some distance from the twig, disturbs the surface of the water, setting it into up-and-down motion. This disturbance will propagate outward from the point of impact in the form of waves. When the waves reach the twig, some of the pebble's energy will be imparted to it, causing the twig to bob up and down. In this way, both energy and *information*—the fact that the pebble entered the water—are transferred from the place where the pebble landed to the location of the twig. We could tell that a pebble (or, at least, some object) had entered the water just by observing the twig. With a little additional physics, we could even estimate the pebble's energy.

We must emphasize that the wave is *not* a physical object. No water traveled from the point of impact of the pebble to the twig—at any location on the surface, the water surface simply moved up and down as the wave passed. What, then, *did* move across the surface of the pond? As illustrated in Figure 3.2, the wave is the *pattern* of up-and-down motion, transmitted from one point to the next as the disturbance moves across the water. It is not a thing but a series of events—something that happens repeatedly at any one place and successively at different places as the wave progresses.

Figure 3.3 shows how wave properties are quantified and establishes some standard terminology. We characterize waves not only by the speed with which they move, but also by the length of their cycle. How many seconds does it take for a wave to repeat itself at some point in space? This is its **wave period.** How many meters does it take for the wave to repeat itself at a given moment in time? This is its **wavelength.** The wavelength (denoted by the Greek letter λ—lambda—in the figure) is defined as the length of an individual wave cycle. This can be measured as the distance between two adjacent wave *crests,* two adjacent wave *troughs,* or any other two similar points on adjacent wave cycles. Not surprisingly, wavelength has units of length, often expressed in centimeters or meters. The maximum departure of the wave from the

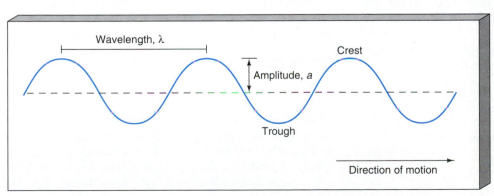

Figure 3.3 A typical wave, showing its direction of motion, wavelength (λ), and amplitude (*a*).

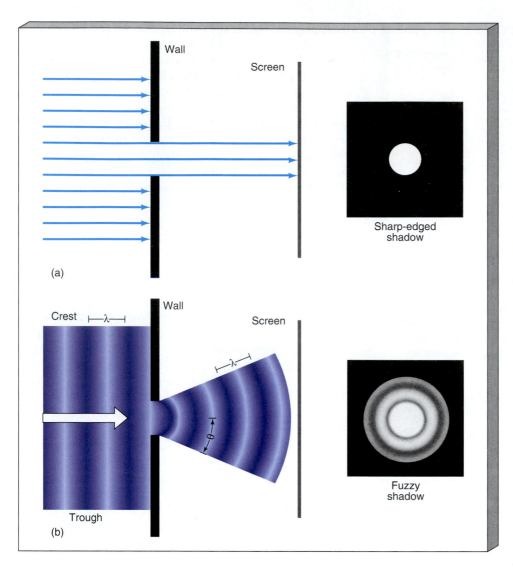

Wall
Screen
(a)
Sharp-edged shadow

Crest $\longmapsto\lambda\longmapsto$
Wall
Screen
θ
Trough
(b)
Fuzzy shadow

Figure 3.4 Diffraction of a light wave. (a) If radiation were composed of rays or particles moving in perfectly straight lines, no bending would occur as a beam of light passed through a circular hole in a barrier, and the outline of the hole, projected onto a screen, would have perfectly sharp edges. (b) In fact, light is diffracted through an angle θ (theta) that depends on the ratio of the wavelength of the wave to the size of the gap. The result is that the outline of the hole becomes "fuzzy," as shown in this actual photograph of the diffraction pattern produced by a small circular opening.

undisturbed state—still air or the flat pond surface—is called its **amplitude** (denoted by *a* in the figure).

If a wave moves at high speed, then the number of crests or cycles passing any given point per unit time—the wave's **frequency**—will be high. Conversely, if a wave moves slowly, with only a few crests passing per unit time, we say that it has a low frequency. The frequency, *f*, of a wave is just 1 divided by the wave's period, *T*. In symbols, $f = 1/T$. Frequency is expressed in units of inverse time, or cycles per second, termed hertz (Hz) in honor of the nineteenth-century German scientist Heinrich Hertz, who studied the properties of radio waves.

Wavelength and wave frequency are *inversely* related—$\lambda \propto 1/f$. In other words, doubling the frequency halves the wavelength, halving the frequency doubles the wavelength, and so on. The product of wavelength and frequency equals the *wave velocity, v*:

$$\text{wavelength} \times \text{frequency} = \text{velocity},$$

or

$$\lambda f = v.$$

A wave thus has a high frequency when its wavelength is small and a low frequency when its wavelength is large. This inverse relationship is easily understood. For a given wave velocity, if the wave crests are close together, more of them pass by a given point each second; when the crests are far apart, few of them pass by per unit time. Put another way, a wave crest moves a distance equal to one wavelength in one wave period. For example, a ripple on a pond might have a wavelength of 10 cm (0.1 m) and a velocity of 0.5 m/s. From the foregoing equation, we can calculate the wave's frequency to be (0.5 m/s)/0.1 m = 5 Hz—in other words, a stationary observer would see 5 wave crests pass by every second.

DIFFRACTION AND INTERFERENCE

Light exhibits two key properties that are characteristic of all forms of wave motion: *diffraction* and *interference*. **Diffraction** is the deflection, or "bending," of a wave as it passes a corner or moves through a narrow gap. As depicted in Figure 3.4, a sharp-edged hole in a barrier seems at first glance to produce a sharp shadow. Closer inspection, however, reveals that the shadow actually has a "fuzzy" edge. We

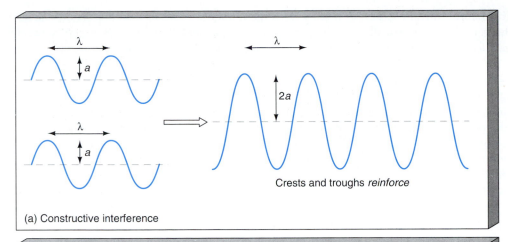

(a) Constructive interference

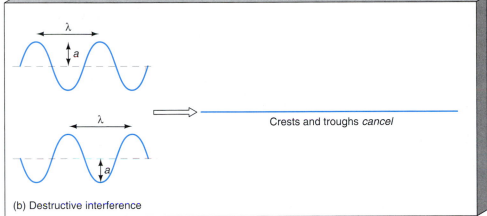

(b) Destructive interference

Figure 3.5 Interference of two identical waves: (a) constructive and (b) destructive. In constructive interference, the two waves reinforce each other to produce a larger-amplitude wave. However, in destructive interference, the two waves exactly cancel out.

are not normally aware of such effects in everyday life, because diffraction is generally very small for visible light. For any wave, the amount of diffraction is proportional to the ratio of the wavelength to the width of the gap. The longer the wavelength, and/or the smaller the gap, the greater the angle through which the wave is diffracted. Thus, visible light, with its extremely short wavelengths, shows perceptible diffraction only when passing through very narrow openings. (The effect is much more noticeable for sound waves, however—no one thinks twice about our ability to hear people even when they are around a corner and out of our sight.)

Interference is the ability of two or more waves to reinforce or cancel each other. Figure 3.5 shows two identical waves moving through the same region of space. In Figure 3.5(a), the waves are positioned so that their crests and troughs exactly coincide. The net effect is that the two wave motions reinforce each other, resulting in a wave of greater amplitude. This is known as *constructive interference*. In Figure 3.5(b), the two waves exactly cancel, so that no net motion remains. This is *destructive interference*. As with diffraction, interference between waves of visible light is not noticeable in everyday experience. However, it is easily measured in the laboratory.

In the chapters that follow we will find that both diffraction and interference play important roles in many areas of observational astronomy.

3.2 Waves in What?

① Waves of radiation differ fundamentally from water waves, sound waves, or any other waves that travel through a material medium. Radiation needs *no* such medium. When light radiation travels from a distant galaxy, or from any other cosmic object, it moves through the virtual vacuum of space. Sound waves, by contrast, cannot do this; if we were to remove all the air from a room, oral conversation would be impossible. Communication by flashlight or radio, however, would be entirely feasible.

The ability of light to travel through empty space was once a great mystery. The idea that light, or any other kind of radiation, could move as a wave through nothing at all seemed to violate common sense, yet it is now a cornerstone of modern physics.

INTERACTIONS BETWEEN CHARGED PARTICLES

To understand more about the nature of light, consider for a moment an *electrically charged* particle, such as an **electron** or a **proton**. Like mass, electrical charge is a fundamental property of matter. Electrons and protons are elementary particles—"building blocks" of atoms (as we will see in more detail in Chapter 4)—that carry the basic unit of charge. Electrons are said to carry a *negative* charge, while protons carry an equal and opposite *positive* charge. Just as a massive object exerts a gravitational force on any other massive body (as we saw in Chapter 2), an electrically charged particle exerts an *electrical* force on every other charged particle in the universe. ∞ (p. 46)

Unlike gravity, which is always attractive, electrical forces can be either attractive or repulsive. Particles with *like* charges (that is, both negative or both positive—for example, two electrons or two protons) repel one another. Particles with *unlike* charges (that is, having opposite signs—an electron and a proton, say) attract. Buildup of electrical charge (a net imbalance of positive over negative, or vice versa) is what causes "static cling" on your clothes when you take them out of a hot clothes dryer, or the shock you sometimes feel when you touch a metal door frame on a particularly dry day.

Extending outward in all directions from our charged particle is an **electric field**, which determines the electric force exerted by the particle on all other charged particles in the universe. The strength of the electric field, like the strength of the gravitational field, decreases with increasing distance from the charge according to an inverse-square law. By means of the electric field, the particle's presence is "felt" by other charged particles, near and far. Now suppose that our particle begins to move, perhaps because it becomes heated or collides with some other object. Its changing position causes its associated electric field to change, and this changing field in turn causes the force exerted on other charges to vary. If we measure the change in this force on these other charges, we learn about our original particle. Thus, *information about the particle's state of motion is transmitted through space via a changing electric field.*

ELECTROMAGNETIC WAVES

One other critical aspect of electric fields must be taken into account before we can fully understand the behavior of light radiation. The laws of physics tell us that a **magnetic field** must accompany every changing electric field. Magnetic fields govern the influence of *magnetized* objects on one another, much as electric fields govern interactions between charged particles. The fact that a compass needle always points to magnetic north is the result of the interaction between the magnetized needle and Earth's magnetic field. Magnetic fields also exert forces on *moving* electric charges (that is, electric currents)—electric meters and motors rely on this basic fact. Conversely, moving charges *create* magnetic fields (electromagnets are a fairly familiar example). In short, electric and magnetic fields are inextricably linked to one another: a change in either one necessarily creates the other.

Thus, as illustrated in Figure 3.6, the disturbance produced by our moving charge actually consists of rhythmically oscillating electric *and* magnetic fields, always oriented perpendicular to one another and moving together through space. These fields do not exist as independent entities; rather, they are different aspects of a single physical phenomenon: **electromagnetism**. Together, they constitute an *electromagnetic wave* that carries energy and information from one part of the universe to another.

How *quickly* does one charge feel the change in the electromagnetic field when another begins to move? This is an important question, because it is equivalent to asking how fast an electromagnetic wave travels. Does it propagate at some measurable speed, or is it instantaneous? Both theory and experiment tell us that all electromagnetic waves move a very specific speed—the **speed of light** (always denoted by the letter c). Its exact value is 299,792.458 km/s in a vacuum (and somewhat less in material substances such as air or water). However, it is usually rounded off for convenience, and we can say $c = 3.00 \times 10^5$ km/s to good accuracy. This is an extremely high speed. In the time needed to snap a finger—about a tenth of a second-light can travel three quarters of the way around the Earth. If the currently known laws of physics are correct, the speed of light is the fastest speed possible.

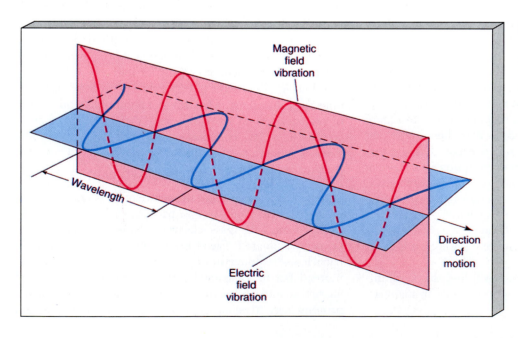

Figure 3.6 Electric and magnetic fields vibrate perpendicular to each other. Together they form an electromagnetic wave that moves through space at the speed of light.

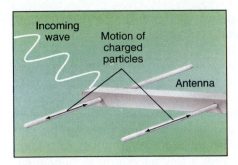

Figure 3.7   Charged particles in an ordinary household television antenna vibrate in response to electromagnetic radiation broadcast by a distant transmitter. The radiation is produced when electric charges are made to oscillate in the transmitter's emitting antenna. The vibrations in the receiving antenna "echo" the oscillations in the transmitter, allowing the original information to be retrieved.

Figure 3.7 shows a familiar example of information being transferred by electromagnetic radiation. A television transmitter causes electric charges to oscillate up and down a metal rod near the tower's top, thereby generating electromagnetic radiation. This radiation can be detected by rooftop antennas. In the metal rods of the receiving antenna, electric charges respond by vibrating in time with the transmitted wave frequency. The information carried by the pattern of vibrations is then converted into sound and pictures by your TV set.

Now consider a real cosmic object—a star, say. When some of its charged contents move around, their electric fields change, and we can detect that change. The resulting electromagnetic ripples travel outward in waves, requiring no material medium in which to travel. Small charged particles, either in our eyes or in our experimental equipment, eventually respond to the field changes by vibrating in tune with the received radiation. This response is how we detect the radiation—with our eyes or with our detectors.

The speed of light is very large, but it is still *finite*. That is, light does not travel instantaneously from place to place. This fact has some interesting consequences for our study of distant objects. It takes time—often lots of time—for light to travel through space. The light we see from the nearest large galaxy—the Andromeda Galaxy, shown in Figure 3.1—left that object about 2 million years ago—around the time our first human ancestors appeared on planet Earth. We can know nothing about this galaxy as it exists today. For all we know, it might no longer even exist! Only our descendents, 2 million years into the future, will know if it exists now. So, as we study objects in the cosmos, remember that the light now seen left those objects long ago. We can never observe the universe as it is—only as it was.

3.3   *The Electromagnetic Spectrum*

❷   White light is a mixture of many different colors, which we conventionally divide into six major hues—red, orange, yellow, green, blue, and violet. As shown in Figure

3.8, we can identify each of these basic colors by passing light through a prism.

THE COMPONENTS OF VISIBLE LIGHT

What determines the color of a beam of light? The answer is its *wavelength* (or, equivalently, its *frequency*). We see different colors because our eyes react differently to electromagnetic waves of different wavelengths. A prism splits a beam of light up into the familiar "rainbow" of colors because light rays of different wavelengths are bent, or **refracted**, slightly differently as they pass through the prism-red light the least, violet light the most. In principle, the original beam of white light could be produced once again by passing the entire red-to-violet range of colors—called a *spectrum* (plural, *spectra*)—through a second, oppositely oriented prism to recombine the colored beams. This experiment was first reported by Isaac Newton over 300 years ago.

Red light has a frequency of roughly 4.3×10^{14} Hz, corresponding to a wavelength of about 7.0×10^{-7} m. Violet light, at the other end of the visible range, has nearly double the frequency—7.5×10^{14} Hz—and (since the speed of light is the same in either case) just over half the wavelength—4.0×10^{-7} m. Wavelengths have been measured for the other colors, too, spanning the whole **visible spectrum** shown in Figure 3.8. Human eyes are insensitive to radiation of wavelength shorter than 4×10^{-7} m or longer than 7×10^{-7} m—radiation outside this range is invisible.

For convenience, astronomers often use a unit called the angstrom (Å) when describing the wavelength of light. This unit of length is named after the nineteenth-century Swedish physicist Anders Ångström (pronounced "Ongstrem"). There are 10^{10} Å in 1 m (see the *More Precisely* feature on p. 13). However, in SI units, the *nanometer* (1 nm = 10^{-9} m = 10 Å) is preferred. ∞ (p. 13) Thus, the visible spectrum covers the wavelength range from 400 to 700 nm, or from 4000 to 7000 Å. The radiation to which our eyes are most sensitive has a wavelength near the middle of this range, at about 550 nm (5500 Å), in the yellow-green region of the spectrum. It is no coincidence that this wavelength falls within the range of wavelengths at which the Sun emits most of its electromagnetic energy—our eyes have evolved to take greatest advantage of the available light.

THE FULL RANGE OF RADIATION

Figure 3.9 plots the entire range of all electromagnetic radiation, illustrating the relationships among the different "types" of electromagnetic radiation listed earlier. We see that the only characteristic that distinguishes one from another is its frequency—or, equivalently, its wavelength. To the low-frequency, long-wavelength side of visible light lie *radio* and *infrared* radiation. Radio frequencies include radar, microwave radiation, and the familiar AM, FM, and TV bands. We perceive infrared radiation as heat. At higher frequencies (shorter wavelengths) are the domains of *ultraviolet*, *X-ray*, and *gamma-ray* radiation. Ultraviolet radiation, lying just beyond the violet end of the visible spectrum, is responsible for suntans and sunburns. X rays are perhaps best known for their ability to penetrate human tissue and reveal the state of our insides without resorting to surgery. Gamma rays are the shortest-wavelength radiation. They are often associated with radioactivity and are invariably damaging to living cells they encounter.

All these spectral regions, including the visible spectrum, collectively make up the **electromagnetic spectrum**. Remember that, despite their greatly differing wavelengths and the very different roles they play in everyday life on Earth, all are basically the same phenomenon, and all move at the same speed—the speed of light, c.

Figure 3.9 is worth studying carefully, as it contains a great deal of information. Note that wave frequency (in hertz) increases from left to right, and wavelength (in meters) increases from right to left. These wave properties behave in opposite ways because, as noted earlier, they are inversely related: $\lambda \propto 1/f$. We will adhere to the convention that frequency increases toward the right when picturing wavelengths and frequencies in this book.

Notice that the wavelength and frequency scales in Figure 3.9 do not increase by equal increments of 10. Instead, successive values marked on the horizontal axis differ by *factors* of 10—each is 10 times greater than its neighbor. This type of scale, called a *logarithmic* scale, is often used in science in order to condense a very large range of some quantity into a manageable size. Had we used a linear scale for the wavelength range shown in Figure 3.9, the figure would have been many light years long!

Throughout the text we will often find it convenient to compress a wide range of some quantity onto a single easy-to-view plot by using such a scale.

Figure 3.9 shows that wavelengths extend from the size of mountains for radio radiation to the size of an atomic nucleus for gamma-ray radiation. The box at the upper right emphasizes how small the visible portion of the electromagnetic spectrum is. Most objects in the universe emit large amounts of invisible radiation. Indeed, many of them emit only a tiny fraction of their total energy in the visible range. A wealth of extra knowledge can be gained by studying the invisible regions of the electromagnetic spectrum.

ATMOSPHERIC BLOCKAGE

Our eyes are sensitive to only a minute portion of the many different kinds of radiation known. In addition, only a small fraction of the radiation produced by astronomical objects actually reaches our eyes, in part because of the *opacity* of Earth's atmosphere. **Opacity** is the extent to which radiation is blocked by the material through which it is passing—in this case, air. The more opaque an object is, the less radiation gets through it. Opacity is thus the opposite of transparency. At the bottom of Figure 3.9, the atmospheric opacity is plotted along the wavelength and frequency scales. The extent of shading is proportional to the opacity. Where the shading is greatest, no radiation can get in or out. Where there is no shading at all, the atmosphere is almost completely transparent, so that extraterrestrial radiation can reach the Earth's surface and

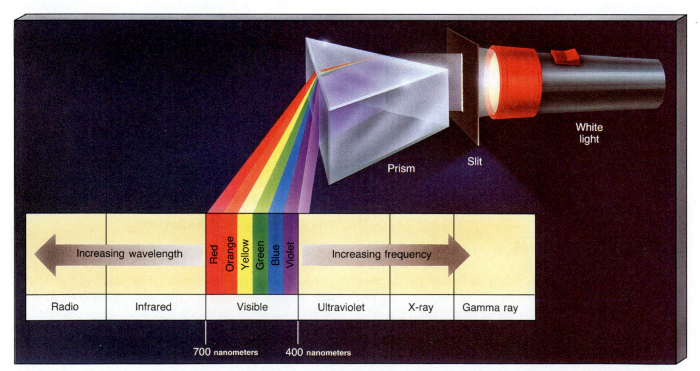

Figure 3.8   While passing through a prism, white light splits into its component colors, spanning red to violet in the visible part of the electromagnetic spectrum. The slit narrows the beam of radiation. The image on the screen is just a series of different-colored images of the slit.

terrestrial radiation from human transmissions can pass virtually unhindered into space.

What causes opacity to vary along the spectrum? Certain atmospheric gases are known to absorb radiation very efficiently at some wavelengths. For example, water vapor (H_2O) and oxygen (O_2) absorb radio waves having wavelengths less than about a centimeter, while water vapor and carbon dioxide (CO_2) are strong absorbers of infrared radiation. Ultraviolet, X-ray, and gamma-ray radiation are completely blocked by the *ozone layer* high in Earth's atmosphere (see Section 7.3). A passing but unpredictable source of atmospheric opacity in the visible part of the spectrum is the blockage of light by atmospheric clouds. In addition, the interaction between the Sun's ultraviolet radiation and the upper atmosphere produces a thin, electrically conducting layer at an altitude of about 100 km. The *ionosphere*, as this layer is known, reflects long-wavelength radio waves (wavelengths greater than about 10 m) as well as a mirror reflects visible light. In this way, extraterrestrial waves are kept out, and terrestrial waves—such as those produced by AM radio stations—are kept in. (That is why it is possible to transmit some radio frequencies beyond the horizon—the broadcast waves bounce off the ionosphere.)

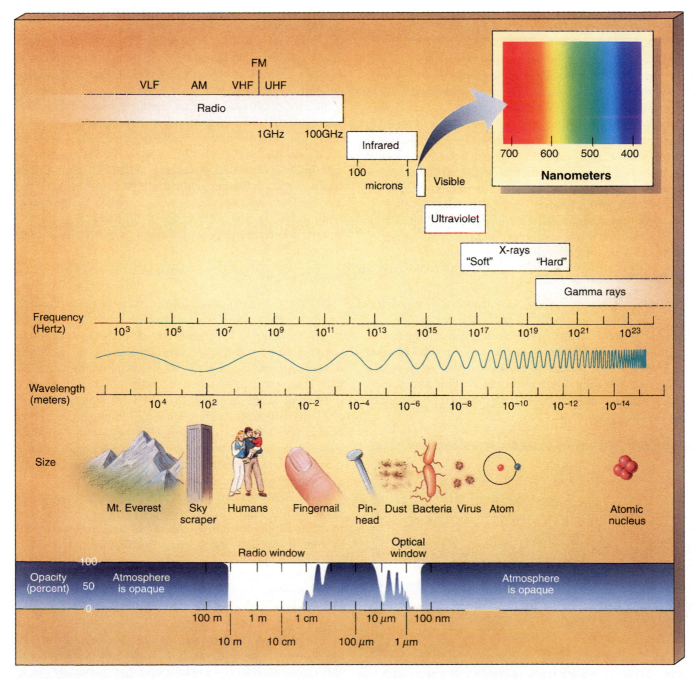

Figure 3.9 The entire electromagnetic spectrum.

The effect of all this blockage is that there are only a few *windows*, at well-defined locations in the electromagnetic spectrum, where Earth's atmosphere is transparent. In much of the radio and in the visible portions of the spectrum, the opacity is low, so we can study the universe at those wavelengths from ground level. In parts of the infrared range, the atmosphere is partially transparent, so we can make certain infrared observations from the ground. Moving to the tops of mountains, above as much of the atmosphere as possible, improves observations. In the rest of the spectrum, however, the atmosphere is opaque. Ultraviolet, X-ray, and gamma-ray observations can be made only from above the atmosphere, from orbiting satellites.

3.4 *The Distribution of Radiation*

③ *All* macroscopic objects—fires, ice cubes, people, stars—emit radiation at all times, regardless of their size, shape, or chemical composition. They radiate mainly because the microscopic charged particles they are made up of are in constantly varying random motion, and whenever charges change their state of motion, electromagnetic radiation is emitted. The **temperature** of an object is a direct measure of the amount of microscopic motion within it (see the *More Precisely* feature on p. 63). The hotter the object is, the faster its constituent particles move—and the more energy they radiate. **Intensity** is a term often used to specify the amount or strength of radiation at any point in space. Like frequency and wavelength, it is a basic property of radiation.

THE BLACK-BODY SPECTRUM

No natural object emits all of its radiation at just one frequency. Instead, the energy is generally spread out over some fairly broad portion of the electromagnetic spectrum. By studying the way in which the intensity of this radiation is distributed with respect to frequency (that is, to color), we can gain not only direct information about the object's temperature, but also much indirect information about the object's other properties.

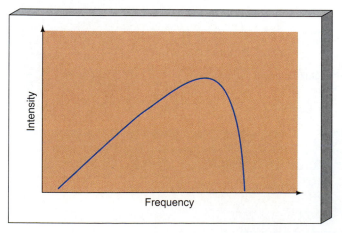

Figure 3.10 The black-body, or Planck, curve represents the distribution of the intensity of radiation emitted by any heated object.

The intensity of radiation emitted by a heated object is greatest at one particular frequency and falls off to lesser values above or below that frequency. Figure 3.10 shows the distribution of radiation emitted by such an object. It is called the "curve of a perfect thermal emitter," or a **black-body curve**, or sometimes the **Planck curve**, after Max Planck, whose mathematical analysis of such thermal emission in 1900 played a key role in modern physics. We will use the term *black-body curve* in this text. Notice that the curve is not shaped like a symmetrical bell that declines evenly on either side of the peak. Instead, the intensity falls off more slowly from the peak to lower frequencies than it does from the peak to high frequencies. This peculiar shape is characteristic of the emission of radiation from *all* heated objects, regardless of their temperatures.

The expression *black-body* refers to an idealized concept—an object that absorbs all radiation falling upon it. In a steady state, the object must reemit exactly the same amount of energy as it absorbs. The black-body curve describes the distribution of that reemitted radiation. No real body absorbs and radiates as a perfect black body, but in many cases the behavior of a heated object does approximate that of a black body quite closely. In this chapter, we will simply assume that the objects under discussion can be treated as black bodies, emitting and absorbing a broad spectrum of radiation that is well described by the black-body curve. In the next chapter, however, we will encounter some very important situations where this is definitely *not* the case.

WIEN'S LAW

④ The frequency at which the intensity peaks depends on the temperature of the emitting object. In fact, as illustrated in Figure 3.11, the entire black-body curve shifts toward higher frequencies and greater intensities as an object's temperature increases. Even so, the *shape* of the curve remains the same. This shifting of radiation's peak frequency with temperature is familiar to us all: Glowing objects, such as toaster filaments or stars, emit visible radiation. Cooler objects, such as warm rocks or household radiators, produce invisible radiation. The latter, warm to the touch but not glowing hot to the eye, emit their radiation most intensely as low-frequency infrared and radio waves.

As a further illustration of the shift of the black-body curve with temperature, imagine a piece of metal placed in a hot furnace. At first, the metal becomes warm, although it doesn't change its visible appearance. As it heats up, it begins to glow dull red, then orange, brilliant yellow, and finally white. How do we explain this? At first, when the metal was warm, it emitted invisible radio and infrared radiation. As the metal became hotter, the peak of the metal's emitted radiation shifted toward higher frequencies, producing more radiation in the visible domain. The black-body curve for an object of several thousand *kelvins*—the unit scientists generally use to measure temperature, as noted in the *More Precisely* feature on p.63—peaks in the visible part of the spectrum,

More Precisely...   *The Kelvin Temperature Scale*

The atoms and molecules that make up any piece of matter are in constant random motion. This motion represents a form of energy known as *thermal energy*—or, more commonly, *heat*. The quantity we call *temperature* is a direct measure of this internal motion: the higher an object's temperature, the faster, on average, the random motion of its constituent particles. More precisely, the temperature of a piece of matter specifies the average thermal energy of the particles it contains.

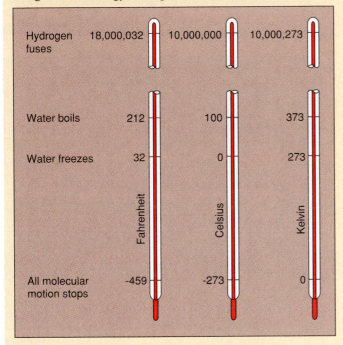

	Fahrenheit	Celsius	Kelvin
Hydrogen fuses	18,000,032	10,000,000	10,000,273
Water boils	212	100	373
Water freezes	32	0	273
All molecular motion stops	-459	-273	0

Our familiar Fahrenheit temperature scale, like the archaic English system in which length is measured in feet and weight in pounds, is of somewhat dubious value. In fact, the "degree Fahrenheit" is now a peculiarity of American society. Most of the world uses the Celsius scale of temperature measurement (also called the centigrade scale). In the Celsius system, water freezes at 0 degrees (0°C) and boils at 100 degrees (100°C), as illustrated in the accompanying figure.

There are, of course, temperatures below the freezing point of water. Although we know of no matter anywhere in the universe that is actually this cold, temperatures can in theory reach as low as -273.15°C. This is the temperature at which all atomic and molecular motion virtually ceases. It is convenient to construct a temperature scale based on this lowest possible temperature, or *absolute zero*. Scientists commonly use such a scale, called the *Kelvin scale* in honor of the nineteenth-century British physicist Lord Kelvin. Since it takes absolute zero as its starting point, the Kelvin scale differs from the Celsius scale by 273.15°. In this book, we round off the decimal places and simply use

$$\text{kelvins} = \text{degrees Celsius} + 273.$$

Thus,

- Translational motion ceases at 0 kelvins (0 K).
- Water freezes at 273 kelvins (273 K).
- Water boils at 373 kelvins (373 K).

Note that the unit is "kelvins," or "K," *not* "degrees kelvin" or "°K." (Occasionally, the term "degrees absolute" is used instead.)

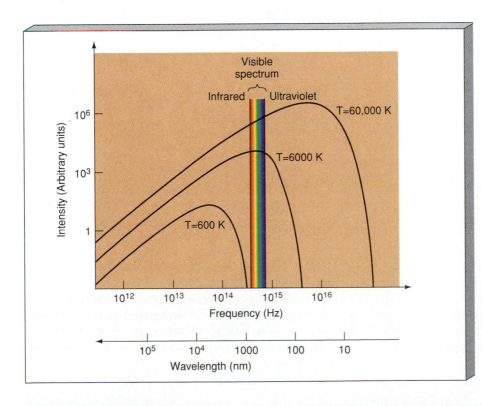

Figure 3.11 As an object is heated, the radiation it emits is still described by the black-body curve, but the curve shifts to peak at higher and higher frequency as the temperature rises. Shown here are the curves corresponding to temperatures of 600 K, 6000 K, and 60,000 K, which peak, respectively, in the infrared, the visible, and the ultraviolet regions of the electromagnetic spectrum.

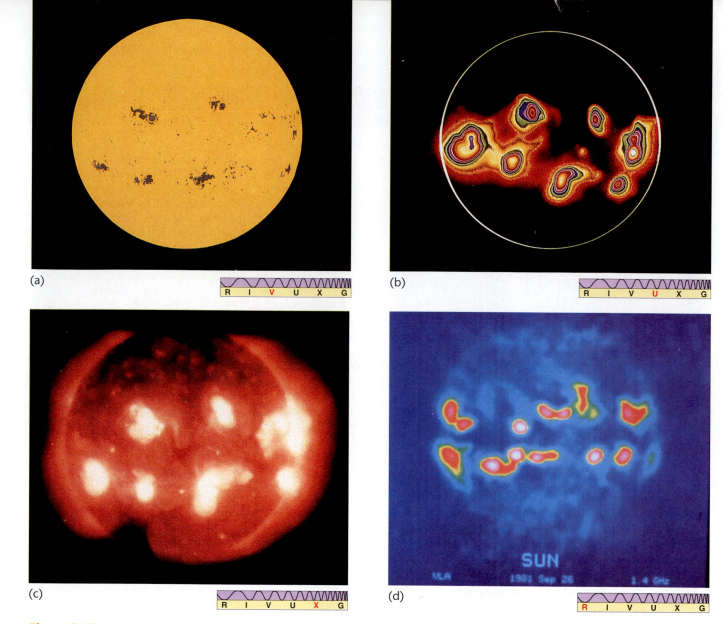

(a)

R I **V** U X G

(b)

R I V **U** X G

(c)

R I V U **X** G

(d)

SUN

VLA   1981 Sep 26   1.4 GHz

R I V U X G

Figure 3.12 Four images of the Sun, made using (a) visible light, (b) ultraviolet light, (c) X rays, and (d) radio waves. By studying the similarities and differences among these views of the same object, important clues to its structure and composition can be found.

regardless of its chemical composition. As the object becomes even hotter, its color gradually changes from red to orange to yellow to blue-white.

From detailed studies of the precise form of the blackbody curve, we obtain a very simple connection between the wavelength at which most radiation is emitted and the temperature T of the emitting object. These quantities are inversely related:

$$\text{maximum wavelength} \propto \frac{1}{T}.$$

This relationship, called **Wien's law**, is discussed in more detail in the *More Precisely* feature on p.65.

Simply put, Wien's law tells us that the hotter the object, the bluer its radiation (see Figure 3.11). For example, an object with a temperature of 6000 K emits most of its energy in the visible part of the spectrum, with a peak wavelength of 480 nm. At 600 K, the object's emission

would peak at a wavelength of 4800 nm, well into the infrared portion of the spectrum. At a temperature of 60,000 K, the peak would move all the way through the visible spectrum to a wavelength of 48 nm, in the ultraviolet range. Objects eventually become "white hot" because black-body curves that peak in the blue or violet part of the spectrum have a broad "tail" that extends through the longer-wavelength portion of the visible spectrum. This means that substantial amounts of green, yellow, orange, and red are also emitted. Together, all these colors combine to produce white.

STEFAN'S LAW

It is also a matter of everyday experience that as the temperature of an object increases, the *total* amount of energy it radiates (summed over all frequencies) increases rapidly. For example, the heat given off by an electric heater increases very sharply as it warms up and begins to emit visible light. Careful experimentation leads to the conclusion that the total amount of energy radiated per unit time is actually proportional to the fourth power of the object's temperature:

$$\text{energy radiated} \propto T^4 .$$

This relation is called **Stefan's law**. It is also discussed further in the *More Precisely* feature below. From the form of Stefan's Law we can see that the energy emitted by a body rises dramatically as its temperature increases. Doubling the temperature, for example, causes the total energy radiated to increase by a factor of 16.

ASTRONOMICAL APPLICATIONS

No known natural *terrestrial* objects can reach temperatures high enough to emit very-high-frequency radiation. Only human-made thermonuclear explosions are hot enough for their spectra to peak in the X-ray or gamma-ray range. (Most other human inventions that produce short-wavelength, high-frequency radiation, such as X-ray machines, are designed to emit only a specific range of wavelengths and do not operate at high temperatures.[*]) Many extraterrestrial objects, however, do emit copious quantities of ultraviolet, X-ray, and even gamma-ray radiation. Figure 3.12 shows a familiar object—our Sun—as it appears when viewed using radiation from different regions of the electromagnetic spectrum. While most sunlight is visible, a great deal of information about our parent star can be obtained by studying it in other parts of the electromagnetic spectrum.

Astronomers often use black-body curves as thermometers, to determine the temperatures of distant objects. For example, study of the solar spectrum makes it possible to measure the temperature of the Sun's surface. Observations of the radiation from our Sun at many different frequencies yield a curve shaped somewhat like the one shown in Figure 3.10. The Sun's curve peaks in the visible part of the electromagnetic spectrum; it also emits a lot of infrared and a little ultraviolet radiation. The overall curve indicates that the temperature of the *surface* of the Sun is approximately 6000 K.

Other cosmic objects have surfaces very much cooler or hotter than our Sun's. These objects emit most of their radiation in invisible parts of the spectrum (Figure 3.13). For example, much cooler regions, such as the surfaces of very young stars, measure about 600 K and emit mostly infrared radiation. The brightest stars, by contrast, have surface temperatures as high as 60,000 K and hence emit mostly ultraviolet radiation. By measuring the frequency distribution of emitted radiation, we can infer the surface temperature of *any* object—terrestrial or extraterrestrial. Provided that nothing interferes with the emitted radiation, the black-body curve yields a reasonably accurate temperature for whatever matter emits the radiation.

[*]*They are said to produce a* nonthermal *spectrum of radiation.*

More Precisely...   *More About the Radiation Laws*

As mentioned in the text, Wien's law relates the temperature of an object to the wavelength λ_{max} at which it emits the most radiation. Mathematically, if we measure T in kelvins and λ_{max} in centimeters, we find that

$$\lambda_{max} = \frac{0.29 \text{ cm}}{T} .$$

We could, of course, convert Wien's law into a statement about frequency, using the relation $f = c/\lambda$ to obtain $f_{max} = 1.0 \times 10^{11} \, T$ hertz, but the law is most commonly stated in terms of wavelength, and is probably easier to remember that way. Thus at 6000 K, the wavelength of maximum intensity is 0.29/6000 cm, or 480 nm (recall that 1 nanometer—1 nm—is equal to 10^{-9} m, or 10^{-7} cm). ∞ (p. 13)

We can also give Stefan's Law a more precise mathematical formulation. If we measure T in kelvins, then the total amount of energy emitted per square meter of its surface per second (a quantity known as the *energy flux, F*) is given by

$$F = \sigma T^4 .$$

The constant σ (the Greek letter sigma) is known as the *Stefan-Boltzmann constant*, or often just Stefan's constant, after Josef Stefan, the Austrian scientist who formulated the equation.

The SI unit of energy is the *joule* (J). Perhaps more familiar is the closely related unit called the watt (W), which measures power—the *rate* at which energy is emitted or expended by an object. One watt is the emission of one joule per second; for example, a 100 watt light bulb emits energy (mostly in the form of infrared and visible light) at a rate of 100 joules per second. In these units, the Stefan-Boltzmann constant has the value $\sigma = 5.67 \times 10^{-8}$ W/m$^2 \cdot$K^4.

Notice just how *rapidly* the energy flux increases with increasing temperature. A piece of metal in a furnace, when at a temperature of 3500 K, radiates energy at a rate of about 850 watts for every square centimeter of its surface area. Doubling its temperature to 7000 K (so that it becomes yellow to white hot, by Wien's law) increases the energy emitted by a factor of 16 (four "doublings"), to 13.6 *kilo*watts (13,600 watts) per square centimeter.

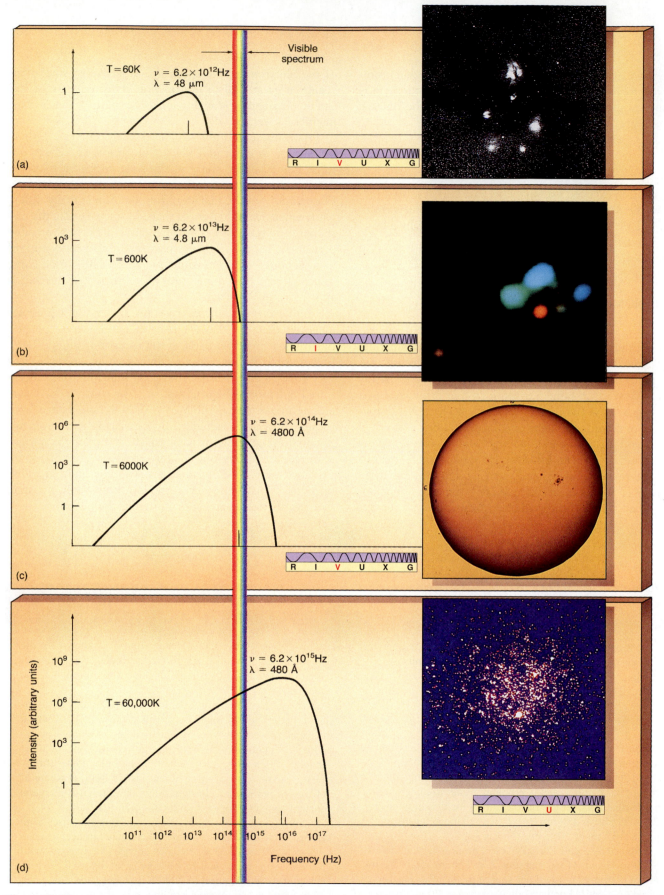

Figure 3.13 Comparison of black-body curves for four cosmic objects having different temperatures. (a) A cool, invisible galactic gas cloud called Rho Ophiuchi. At a temperature of 60 K, it emits mostly low-frequency radio radiation. (b) A dim, young star (shown here in red) near the center of the Orion Nebula. The star's atmosphere, at 600 K, radiates primarily in the infrared. (c) The Sun's surface, at 6000 K, is brightest in the visible region of the electromagnetic spectrum. (d) A cluster of very bright stars, called Omega Centauri, as observed by a telescope aboard the space shuttle. At a temperature of 60,000 K, these stars radiate strongly in the ultraviolet.

3.5 Another Inverse-Square Law

⑤ The farther away an object is, the fainter it appears. In order to explain and quantify this well-known observation, let's consider in a little more detail what happens when we look at an astronomical object such as a star. The star's energy is emitted in all directions, but only a tiny fraction of it actually arrives at the Earth. Our eyes and our detectors do not measure the total amount of energy leaving the star, of course. Instead, they measure the amount of energy striking a given area of some light-sensitive surface or device per unit time—a quantity called the *energy flux* produced by the star at our location. We will refer to this as the star's **apparent brightness**.

Consider Figure 3.14, which shows light leaving a star and traveling through space. Moving outward, the radiation passes through imaginary spheres of increasing radius surrounding the source. The amount of radiation leaving the star per unit time is constant, so the farther the light travels from the source, the less energy passes through each unit of area. You can think of the energy as being spread out over an ever-larger area, and therefore spread more thinly, or "diluted," as it expands into space. Because the area of a sphere grows as the square of the radius, the energy per unit area—the star's energy flux—is *inversely* proportional to the square of the distance from the star. Doubling the distance from a star makes it appear 2^2, or 4, times dimmer. Tripling the distance reduces the apparent brightness by a factor of 3^2, or 9, and so on.

The apparent brightness of a star is therefore *inversely* proportional to the square of its distance:

$$\text{apparent brightness (energy flux)} \propto \frac{1}{(\text{distance})^2}.$$

This relation applies to *any* source of radiation—the Sun, a distant star, an entire galaxy—so long it is viewed from a distance that is large compared to its diameter. Thus the planet Jupiter, at a distance of roughly 5 A.U. from the Sun, receives only $1/5^2 = 1/25$ the amount of sunlight per unit area that the Earth receives. Similarly, the star Alpha Centauri, which emits roughly the same amount of energy per unit time as the Sun, is nearly 270,000 times farther away (we will discuss how this enormous distance is measured in Chapter 17). As a result, Alpha Centauri's apparent brightness is roughly 70 *billion* (= 270,000²) times less than that of the Sun.

It is important to realize that the inverse-square law does not depend on *wavelength* in any way—the intensities of *all* wavelengths in the light emitted by an object are diminished by the same distance-dependent factor. Thus, the relative intensities of different wavelengths that we observe mirrors their relative intensities at the source. Increase your distance from a red star, and it will look fainter—but it will still look *red*. As a result (to take just one example), we can apply Wien's law to determine the star's temperature regardless of its distance. This simple fact—that when we observe a distant object, the distribution of apparent brightness at various wavelengths matches the actual distribution at the source—is central to virtually every analysis of starlight we will see in this book.

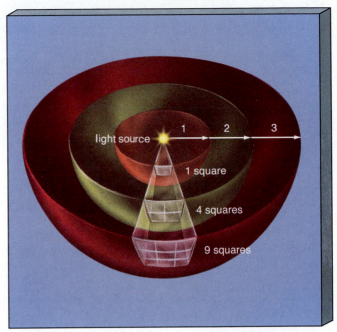

Figure 3.14 The amount of detectable radiation (apparent brightness) varies inversely as the square of the distance from an emitting object. As radiation moves away from a source, it is steadily diluted as it spreads over progressively larger surface areas (depicted here as sections of spherical shells).

3.6 The Doppler Effect

⑥ The speed of electromagnetic radiation never changes: it always equals the speed of light, *c*. In fact, as we will see later (in Chapter 22), we will *always* observe light to move at the same speed, *c*, regardless of our own state of motion, or that of the light source. The same is not true, however, for the *wavelength* of that light.

AN OBSERVATION

Imagine a rocket ship of the future launched from Earth with enough fuel to allow it to accelerate to speeds comparable to the speed of light. As the ship's speed increased, a remarkable thing would happen (see Figure 3.15). Passengers looking out from the front of the spacecraft would begin to notice that the light emitted from the star system toward which they were traveling seemed to be getting *bluer*. At first, they might think that the star's surface temperature was increasing, so that it was emitting more radiation in the blue part of the spectrum, but this would not be the reason for the change in color. As the travelers would soon realize, *all* stars in front of the ship would seem to be bluer than normal, and the greater the ship's velocity, the greater the color change would be. Furthermore, the stars behind the vessel would seem *redder* than normal. As the spacecraft slowed down and came to rest relative to the Earth, all stars would resume their "normal" appearance. The travelers would have to conclude that the stars had changed their colors not because of any real change in their physical properties, but because of the spacecraft's own *motion*.

Movement toward or away from a source of radiation does indeed change the way we perceive that radiation. Motion along any particular line of sight is known as **radial motion**. It should be distinguished from **transverse motion**, which is perpendicular to the line of sight. Radial motion—but *not* transverse motion—brings about changes in the observed properties of radiation.

Such changes are not restricted to electromagnetic radiation and fast-moving spacecraft. Waiting at a railroad crossing for an express train to pass, most of us have had the experience of hearing the pitch of the engine's horn change from high shrill to low blare as the train approaches and then recedes. High-pitched sound (treble) has shorter wavelengths than low-pitched sound (bass); this well-known sound change is similar in many respects to the light change observed by our space travelers. If the train had a light atop its engine, we could, in principle, witness the color change of this light, from bluer than normal to redder than normal, as the train passed by. (In practice, however, the light's color change would be impossible to discern because trains travel far too slowly for us to perceive the effect.)

This motion-induced change in the observed wavelength of a wave be it an electromagnetic (light) wave or an acoustical (sound) wave—is known as the **Doppler effect**, in honor of Christian Doppler, the nineteenth-century Austrian physicist who first explained it. Note, however, that in our example of the spaceship, the *observer* was in motion, whereas, in the train example, it was the *source* of

the wave that was moving. For electromagnetic radiation, the result is the same in either case—it is only the *relative* motion of source and observer that matters.

AN EXPLANATION

Imagine a wave moving from the place where it is generated toward an observer who is not moving with respect to the wave source, as shown in Figure 3.16(a). By counting the number of wave crests passing per unit time, the observer could determine the wavelength of the emitted wave.

Suppose now that the wave source begins to move. This is analogous to the train example given earlier, and in fact is also equivalent to the rocket ship example (since only *relative* motion is important). As illustrated in Figure 3.16(b), successive wave crests in the direction of motion of the source will be seen to be *closer together* than normal, simply because the source moves between the times of emission of one wave crest and the next. An observer in front of the source will therefore measure a *shorter* wavelength than normal. In the case of electromagnetic radiation, the observed wave is said to be **blueshifted**, because blue light has a shorter wavelength than red light. Similarly, an observer situated behind the source will measure a longer than normal wavelength—the wave is said to be **redshifted**. (Note that this terminology also holds even for invisible radiation, for which "red" and "blue" have no meaning. Any shift toward shorter wavelengths is called a *blue shift*, and any shift toward longer wavelengths is called a *red shift*. Thus, ultraviolet radiation might be blueshifted into the X-ray part of the spec-

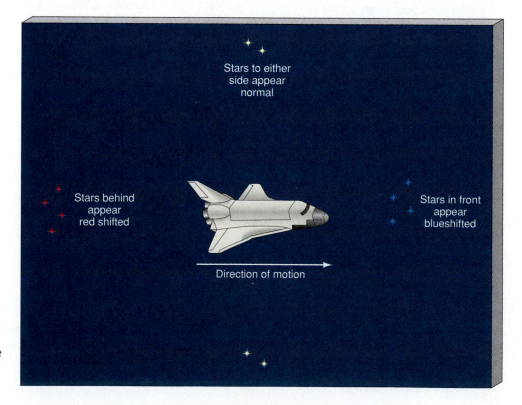

Figure 3.15 Observers in a fast-moving spacecraft will see the stars ahead of them seem bluer than normal, while those behind are reddened. The stars have not changed their properties—the color changes are the result of the motion of the spacecraft relative to the stars.

trum, infrared radiation could be redshifted into the microwave range, and so on.)

The greater the relative speed of the source and the observer, the greater the extent of the observed shift. In fact, the shift is *directly proportional to the net radial velocity*. If we (conventionally) denote by *v* the net velocity of *recession* between source and observer (so a positive value of *v* means that the two are moving farther apart from one another, a negative value that they are approaching), then the apparent wavelength (measured by the observer) is always related to the true wavelength (measured by an observer with no motion relative to the source) by

$$\frac{\lambda \text{ (apparent)}}{\lambda \text{ (true)}} = 1 + \frac{v}{c},$$

where *c* is the speed of the wave (the speed of light, in the case of radiation). Because the speed of light is so large—300,000 km/s—the effect is *extremely* small for everyday terrestrial velocities. Even with its source receding at Earth's orbital speed of 30 km/s, a beam of blue light would be redshifted by only 0.01 percent, from 400 nm to 400.04 nm—a very small change indeed, and one that the human eye could not distinguish (but one that is easily detectable with modern instruments). Notice, incidentally, that the Doppler effect depends only on the *net radial motion* of the observer relative to the source. It does not depend on distance in any way.

Of what value is the Doppler effect? The answer is that astronomers can use it to measure the *speed* of any cosmic object along the line of sight, simply by determining the extent to which its light is red- or blueshifted. The motions of nearby stars and distant galaxies—even the expansion of the universe itself—have *all* been measured in this way. (Motorists who have been stopped for speeding on the highway have experienced another, much more down-to-earth, application of the same basic law of physics! Police radar measures speed by means of the Doppler effect, as do the radar guns used to clock the velocity of a pitcher's fastball or a tennis player's serve.)

In practice, however, it is hard to measure Doppler shifts of an entire black-body curve, simply because it is widely spread over many wavelengths, making small shifts hard to determine with any accuracy. But if the radiation were more narrowly defined and took up just a narrow "sliver" of the spectrum, then precise measurements of Doppler effect *could* be made. We will see in the next chapter that, in some circumstances, this is precisely what does happen, making the Doppler effect one of the observational astronomer's most powerful tools.

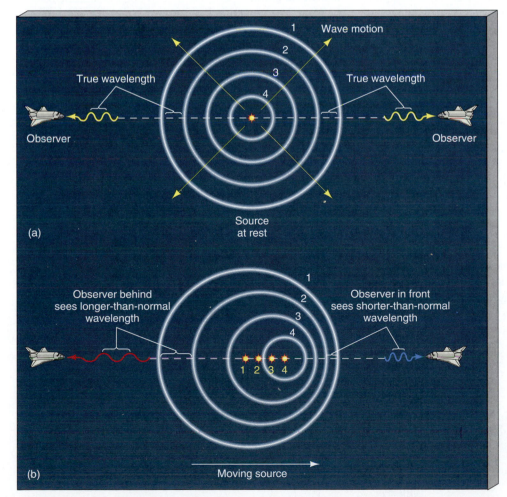

Figure 3.16 (a) Wave motion from a source toward an observer at rest with respect to the source. The four numbered circles represent successive wave crests emitted by the source; at the instant shown, the fifth wave crest is just about to be emitted. As seen by the observer, the source is not moving, so the wave crests are just concentric spheres (shown here as circles). (b) Waves from a moving source tend to "pile up" in the direction of motion and be "stretched out" on the other side. (The numbered points indicate the location of the source at the instant each wave crest was emitted.) As a result, an observer situated in front of the source measures a shorter than normal wavelength—a blueshift—while an observer behind the source sees a redshift. In this diagram, the source is shown in motion. However, the same general statements hold whenever there is any relative motion between source and observer.

Chapter Review

SUMMARY

Visible light (p. 54) is a particular type of **electromagnetic radiation** (p. 54) and travels through space in the form of a **wave** (p. 55). A wave is characterized by the **wave period** (p. 55), the length of time taken for one complete cycle; the **wavelength** (p. 55), the distance between successive wave crests; and the wave **amplitude** (p. 56), which measures the size of the disturbance associated with the wave. The wave **frequency** (p. 56) is simply 1 over the wave period—it counts the number of wave crests that pass a given point in one second. **Diffraction** (p. 56) is the tendency of a wave to spread out after passing through an opening or to bend around a corner. **Interference** (p. 57) is the ability of two waves to reinforce or partially cancel one another.

Electrons (p. 57) and **protons** (p. 57) are elementary particles that carry equal and opposite electrical charges. Any electrically charged object is surrounded by an **electric field** (p. 58) that determines the force it exerts on other charged objects. Like gravitational fields, electric fields decrease as the square of the distance from their source.

When a charged particle moves, information about that motion is transmitted throughout the universe by the particle's changing electric field. The information travels in the form of a wave at the **speed of light** (p. 58). Both electric and **magnetic fields** (p. 58) are involved, so the phenomenon is known as **electromagnetism** (p. 58).

A beam of white light is bent, or **refracted** (p. 60), as it passes through a prism. Different frequencies of light within the beam are refracted by different amounts, so the beam is split up into its component colors—the **visible spectrum** (p. 60). The color of visible light is simply a measure of its wavelength—red light has a longer wavelength than blue light.

The entire **electromagnetic spectrum** (p. 60) consists of (in order of increasing frequency) of *radio waves, infrared radiation, visible light, ultraviolet radiation, X rays,* and *gamma rays.*

The **opacity** (p. 60) of Earth's atmosphere—the extent to which it absorbs radiation—varies greatly with wavelength. Only radio waves, some infrared wavelengths, and visible light can penetrate the atmosphere and reach the ground from space.

The **temperature** (p. 62) of an object is a measure of the speed with which its constituent particles move. The **intensity** (p. 62) of radiation of different frequencies emitted by a hot object has a characteristic distribution, called a **black-body** (or **Planck**) **curve** (p. 62), that depends only on the temperature of the object. **Wien's law** (p. 64) tells us that the wavelength at which the object radiates most energy is directly proportional to its temperature. **Stefan's law** (p. 65) states that the total amount of energy radiated is proportional to the fourth power of the temperature.

The **apparent brightness** (p. 67) of a star, the energy flux measured by a telescope on Earth, falls off as the inverse square of the star's distance from the observer.

Our perception of the frequency of a beam of light can be altered by our net **radial motion** (p. 68) with respect to the source. **Transverse motion** (p. 68) has no effect on our measurements of the beam. This motion-induced change in the observed frequency of a wave is called the **Doppler effect** (p. 68). Any net motion away from the source will cause a **redshift** (p. 68)—a shift to lower frequencies—in the received beam. Motion toward the source causes a **blueshift** (p. 68). The extent of the shift is directly proportional to the observer's radial velocity relative to the source.

SELF-TEST: True or False?

_____ **1.** Light, radio, ultraviolet, and gamma rays are all forms of electromagnetic radiation.

_____ **2.** Sound is a familiar form of electromagnetic wave.

_____ **3.** The amount of diffraction increases with increasing wavelength.

_____ **4.** Interference occurs when one wave is brighter than another; the fainter wave can not be observed.

_____ **5.** Electromagnetic waves cannot travel through a perfect vacuum.

_____ **6.** Electromagnetic waves all travel at the same speed, the speed of light.

_____ **7.** Visible light makes up the greatest part of the entire electromagnetic spectrum.

_____ **8.** Ultraviolet light has the shortest wavelength of any electromagnetic wave.

_____ **9.** A black body emits all its radiation at one wavelength or frequency.

_____ **10.** A perfect black-body emits exactly as much radiation as it absorbs from outside.

_____ **11.** The inverse-square law of light affects short-wavelength radiation more than it does long-wavelength radiation.

_____ **12.** Objects moving away from an observer are redshifted because they actually turn red.

_____ **13.** The Doppler effect occurs for all types of wave motion.

_____ **14.** An object emitting radiation, moving transverse to the line of sight, produces no Doppler effect.

SELF-TEST: Fill in the Blank

1. The speed of light is _____ km/s.

2. The _____ of a wave is the distance between any two adjacent wave crests.

3. The _____ of a wave is measured in units of hertz (Hz).

4. When a charged particle moves, information about this motion is transmitted through space by means of its changing _____ and _____ fields.

5. The visible spectrum ranges from _____ to _____ in wavelength.

6. Light with a wavelength of 700 nm is perceived to be _____ in color.

7. Earth's atmosphere has low opacity for three forms of electromagnetic radiation. They are _____, _____, and _____.

8. The peak of an object's emitted radiation occurs at a frequency or wavelength determined by the object's _____.

9. The lowest possible temperature is _____ K.

10. Water freezes at _____ K.

11. Because the Sun emits its peak amount of radiation at about 480 nm, its temperature must be about _____ K.

12. Two identical objects have temperatures of 1000 K and 1200 K. It is observed that one of the objects emits twice as much radiation as the other. Which one is it? _____.

13. If an astronomical object is observed to emit X rays, it is reasonable to assume its temperature is very _____.

14. To make the Sun appear 9 times fainter than it now does, you would have to move _____ times farther way.

15. When an observer and/or an object emitting radiation move toward one other, the observer sees the radiation shifted to _____ wavelengths.

REVIEW AND DISCUSSION

1. Define the following wave properties: period, wavelength, amplitude, frequency.

2. What is the relationship between wavelength, wave frequency and wave velocity?

3. What is diffraction, and how does it relate to the behavior of light as a wave?

4. What's so special about c ?

5. Compare and contrast the gravitational force with the electric force.

6. Describe the way in which light radiation leaves a star, travels through the vacuum of space, and finally is seen by someone on Earth.

7. Name the colors that combine to make white light. What is it about the various colors that cause us to perceive them differently?

8. What do radio waves, infrared radiation, visible light, ultraviolet radiation, X rays, and gamma rays have in common? How do they differ?

9. In what regions of the electromagnetic spectrum is the atmosphere transparent enough to allow observations from the ground?

10. What is a black-body? What are the characteristics of the radiation emitted by a black-body?

11. What does Wien's Law reveal about stars in the sky?

12. What does Stefan's Law tell us about the radiation emitted by a black-body?

13. Why does the light coming from a star follow an inverse-square law?

14. What is the Doppler effect, and how does it alter the way in which we perceive radiation?

15. If Earth were completely blanketed with clouds and we couldn't see the sky, could we learn about the realm beyond the clouds? What other forms of radiation might be received?

16. In terms of its black-body curve, describe what happens when a red-hot glowing coal cools off.

PROBLEMS

1. What is the wavelength of a 100 MHz ("FM 100") radio signal?

2. According to Wien's law, how many times hotter is an object whose black-body emission spectrum peaks in the ultraviolet, at a wavelength of 200 nm, than an object whose spectrum peaks in the red, at 650 nm? According to Stefan's law, how much more energy does it radiate per unit area per second?

3. Normal human body temperature is about 37°C. What is this temperature in kelvins? What is the peak wavelength emitted by a person with this temperature? In what part of the spectrum does this occur?

4. The Sun has a temperature of almost 6000 K, and its black-body emission peaks at a wavelength of approximately 550 nm. At what wavelength does a star-forming cloud with a temperature of 1000 K radiate most strongly?

5. At what radial velocity, and in what direction, would a spacecraft have to be moving for a radio station transmitting at 100 MHz to be picked up by a radio tuned to 99.9 MHz?

PROJECTS

1. Locate the constellation Orion. Its two brightest stars are Betelgeuse and Rigel. Which of these is the hotter star? Which is cooler? How can you tell? Which of the other stars scattered across the night sky are hot, and which are cool?

2. Stand near (but not too near!) a train track or busy highway and wait for a train or traffic to pass by. Can you notice the Doppler effect in the sound of the engine noise or whistle blowing? How does the sound frequency depend on (a) speed, (b) the motion toward or away from you?

4

SPECTROSCOPY

The Inner Workings of Atoms

LEARNING GOALS

Studying this chapter will enable you to:

1 Describe the characteristics of continuous, emission, and absorption spectra, and the conditions under which each is produced.

2 Explain the relation between emission and absorption lines and what we can learn from these lines.

3 Discuss the observations that led scientists to conclude that light has particle as well as wave properties.

4 Specify the basic components of the atom and describe our modern conception of its structure.

5 Explain how electron transitions within atoms produce unique emission and absorption features in the spectra of those atoms.

6 Describe the general features of the spectra produced by molecules.

7 List and explain the kinds of information that can be obtained by analyzing the spectra of astronomical objects.

(Opposite page, background) When a prism is placed in the path of light captured by a telescope, the resulting photograph can look almost psychedelic. Each of the stars and other sources of radiation in the picture has its light split into its component colors, from red to violet. Here, in this strange image, we see a star-forming region (called Eta Carinae) containing many stars and much loose gas—all of whose light has been colorfully dispersed.

(Inset A) This is the same celestial object, now photographed without the prism. The result is Eta Carinae in its true color—mostly red, due to vast quantities of hydrogen gas spread across the 900-light-year-squared area of this photo.

(Inset B) Looking more carefully at spectra from individual stars, we often see two things: a colorful spectrum extending from red to violet, and a series of thin, dark lines across the spectrum. This is the spectrum of the brightest star in the sky, Sirius A.

(Inset C) This is the slightly different spectrum of Vega, the bright star in the constellation Lyra.

(Inset D) Examining spectra of stars even more closely, we find evermore dark lines. Here, in this spectrum of the star Arcturus, we can see many such lines in the yellow region alone. The tracing below records precise details about these lines that would be difficult or impossible to obtain just by looking at the spectrum. Spectra like theses, which tell astronomers virtually all that we know about stars, are the subject of this chapter.

In the previous chapter, we saw how light behaves as a continuous wave, and how this description of electromagnetic radiation allows us to begin to decipher the information reaching us from the cosmos in the form of visible and invisible light. However, early in the twentieth century, it became clear that the wave theory of electromagnetic phenomena was incomplete—some aspects of light simply could not be explained purely in wave terms. When radiation interacts with matter on atomic scales, it does so not as a continuous wave, but in a jerky, discontinuous way—in fact, as a particle. This so-called "wave-particle duality" of light is even today not fully understood; nevertheless, it is an undeniable fact of nature. With this discovery, scientists quickly realized that atoms too must behave in a discontinuous way, and the stage was set for a scientific revolution that has affected virtually every area of modern life. In astronomy, the observational and theoretical techniques that enable researchers to determine the nature of distant atoms by the way they emit and absorb radiation are now the indispensable foundation of modern astrophysics.

4.1 Spectral Lines

1 In Chapter 3, we saw something of how astronomers can analyze electromagnetic radiation received from space to obtain information about distant objects. A vital step in this process is the formation of a *spectrum*—splitting the incoming radiation into its component wavelengths. But in reality, *no* cosmic object emits a perfect black-body spectrum like those discussed in Section 3.4. ∞ (p. 62) All spectra deviate from this idealized form—some by only a little, others by a lot. Far from invalidating our earlier studies, however, these deviations contain a wealth of detailed information about physical conditions in the source of the radiation. Because spectra are so important, let's examine in a little more detail how astronomers obtain and interpret them.

Simple studies of radiation spectra can be performed with an instrument known as a **spectroscope**. In its most basic form, this might consist of an opaque barrier with a slit in it (to define a beam of light), a prism (to split the beam into its component colors), and an eyepiece or screen (to allow the user to view the resulting spectrum). Figure 4.1 shows such an arrangement. The research instruments (called *spectrographs*, or *spectrometers*) used by professional astronomers are rather more complex, consisting of a telescope (to capture the radiation), a dispersing device (to spread it out into a spectrum), and a detector (to record the result). Despite their greater sophistication, their basic operation is conceptually similar to the simple spectroscope shown in the figure.

In many large instruments, the prism is replaced by a device called a *diffraction grating*, consisting of a sheet of

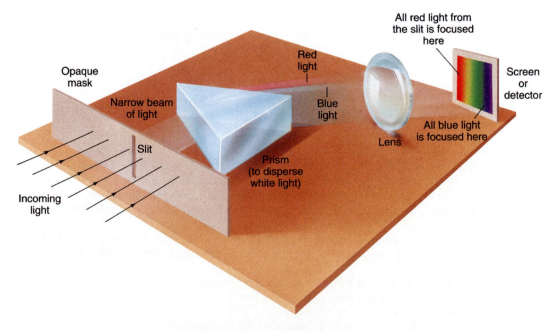

Figure 4.1 Diagram of a simple spectroscope. A small slit in the mask on the left allows a narrow beam of light to pass. The light passes through a prism and is split up into its component colors. The resulting spectrum can be viewed through an eyepiece or simply projected onto a screen.

transparent material with many closely spaced parallel lines ruled on it. The spaces between the lines act as many tiny openings, and light is diffracted as it passes through them. Because different wavelengths of electromagnetic radiation are diffracted by different amounts as they pass through a narrow gap (see Section 3.1), the effect of the grating is to split a beam of light into its component colors. ∞ (p. 56)

EMISSION LINES

The spectra we encountered in Chapter 3 are examples of **continuous spectra**. A white-hot metal bar emits radiation of all wavelengths (mostly in the visible range). The distribution of the intensity of the various wavelengths is described by the black-body curve corresponding to the bar's temperature. ∞ (p. 64) Viewed through a spectroscope, the spectrum of the light from the bar would show the familiar rainbow of colors, from red to violet, without interruption, as presented in Figure 4.2(a). The light from an ordinary incandescent light bulb is another example of a continuous spectrum.

Not all spectra are continuous, however. For instance, if we took a glass jar containing pure hydrogen gas and passed an electrical discharge through it (a little like a lightning bolt arcing through Earth's atmosphere), the gas would begin to glow—that is, it would emit radiation. If we were to examine that radiation with our spectroscope, we would find that its spectrum consisted of only a few bright lines on an otherwise dark background, quite unlike the continuous spectrum described for the incandescent light bulb. Figure 4.2(b) shows this schematically. An actual photograph of the spectrum of hydrogen appears in the top panel of Figure 4.3. The light produced by the hydrogen in this experiment does *not* consist of all possible colors but instead includes only a few narrow, well-defined **emission lines** that resemble narrow "slices" of the continuous spectrum.

After further experimentation, we would also find that although we could alter the *intensity* of the lines (for example, by changing the amount of hydrogen in the jar or the strength of the electrical discharge), we could not alter their *color* (in other words, their frequency or wavelength).

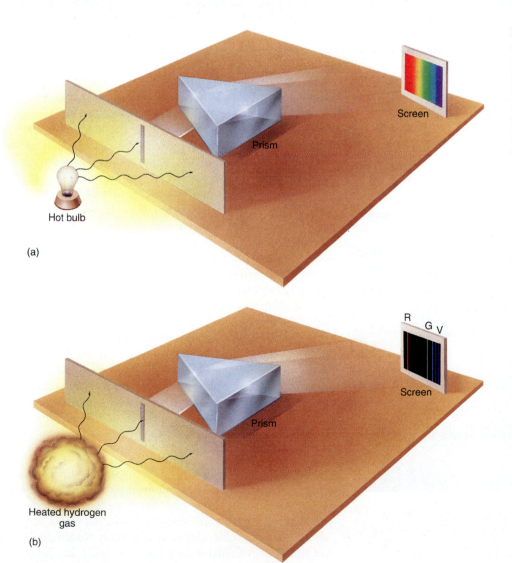

Figure 4.2   When passed through a slit and split up by a prism, light from a source of continuous radiation (a) gives rise to the familiar rainbow of colors. By contrast, the light from excited hydrogen gas (b) consists of a series of distinct spectral lines.

Hydrogen

Sodium

Helium

Neon

Mercury

Figure 4.3   The emission spectra of some well-known elements.

Figure 4.4   This visible spectrum of the Sun shows hundreds of dark absorption lines superimposed on a bright continuous spectrum. Here, the scale extends from long wavelengths (red) at the upper left to short wavelengths (blue) at the lower right.

The pattern of spectral emission lines is a property of the element hydrogen. Whenever we perform this experiment, the same characteristic colors for each gas result.

By the early nineteenth century, scientists had carried out similar experiments on many different gases. By vaporizing solids and liquids in a flame, they extended their inquiries to include materials that are not normally found in the gaseous state. Sometimes the pattern of lines was fairly simple, sometimes it was very complex. Always, though, it was *unique*. Even though the origin of the lines was a mystery, scientists quickly realized that the lines provided a one-of-a-kind "fingerprint" of the substance under investigation. Scientists could deduce the presence of a particular atom or molecule (a group of atoms held together by chemical bonds—see Section 4.4) solely through the study of the light it emitted.

Scientists have by now accumulated extensive catalogs of the specific wavelengths at which many different hot gases emit radiation. For gas of a given chemical composition, the particular pattern of the light it emits is known as its **emission spectrum**. Examples of the emission spectra of some common substances are shown in Figure 4.3.

ABSORPTION LINES

When sunlight is split by a spectroscope, at first glance it appears to produce a continuous spectrum. This is what Isaac Newton must have seen over three centuries ago when he used a prism to view sunlight. However, closer scrutiny shows that the solar spectrum is actually interrupted by a large number of narrow dark lines, as shown in Figure 4.4. We now know that many of these lines are formed when intervening gases remove light from the Sun's otherwise continuous spectrum. These gases are present in the outer layers of the Sun or in Earth's atmosphere, and they absorb certain wavelengths, which then do not appear in the spectrum. The lines are called **absorption lines**.

The English astronomer William Wollaston first noticed the solar absorption lines in 1802. They were studied in greater detail about 10 years later by the German physicist Joseph Fraunhofer, who measured and cataloged over 600 of them. They are now referred to collectively as **Fraunhofer lines**. Although the Sun is by far the easiest star to study, and so has the most extensive set of observed absorption lines, similar lines are now known to exist in the spectra of all stars.

At around the same time as the solar absorption lines were discovered, scientists found that absorption lines could also be produced in the laboratory by passing a beam of light from a continuous source through a cool gas, as shown in Figure 4.5. They quickly observed an intriguing connection between emission and absorption lines: The absorption lines associated with a given gas occur at precisely the *same* wavelengths as the emission lines produced when the gas is heated. Thus, if the emission lines form a unique fingerprint, so do the absorption lines.

As an example, consider the element sodium, whose emission spectrum appears in Figure 4.3. When sodium vapor is heated or energized in some other way, it emits almost all of its radiation at just two particular wavelengths in the visible range. The two characteristic yellow lines, with wavelengths of 589.0 nm and 589.6 nm, are clearly visible.

Now compare that result with what happens when a continuous spectrum is passed *through* some relatively *cool* sodium vapor. Two sharp, dark *absorption* lines appear in the spectrum at precisely the same wavelengths as the two lines that were emitted by hot sodium. The emission and absorption spectra of sodium are compared in Figure 4.6. The two spectra are shown to the same scale, clearly illustrating the relation between emission and absorption features.

KIRCHHOFF'S LAWS

❷ The analysis of the ways in which matter emits and absorbs radiation is called **spectroscopy**. One early spectro-

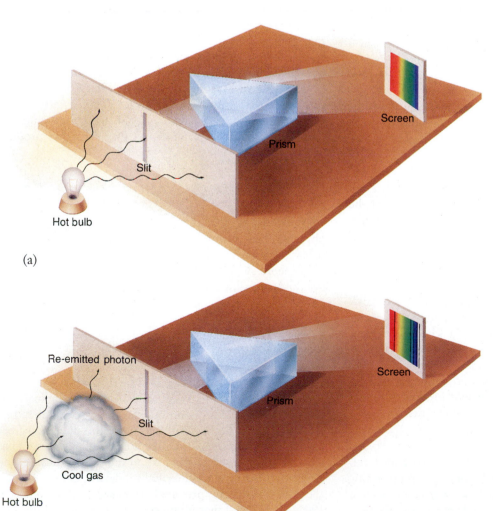

Figure 4.5 (a) A hot light bulb produces a continuous emission spectrum. (b) When some cool gas is placed between the bulb and the detector, the resulting emission spectrum is crossed by a series of dark absorption lines. These lines are formed when the intervening gas absorbs certain wavelengths (colors) from the original beam. The absorption lines appear at precisely the same wavelengths as the emission lines that would be produced if the gas were heated to high temperatures.

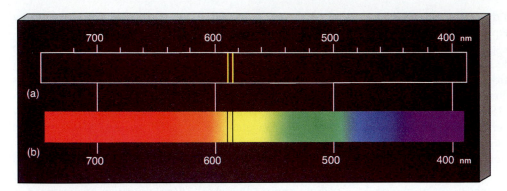

Figure 4.6 (a) The characteristic emission lines of sodium. The two bright lines in the center appear in the yellow part of the spectrum. (b) The absorption spectrum of sodium. The two dark lines appear at exactly the same wavelengths as the bright lines in the sodium emission spectrum.

scopist, the German physicist Gustav Kirchhoff, summarized the observed relationships between the three types of spectra—continuous, emission line, and absorption line—in 1859. He listed three spectroscopic rules, now known as **Kirchhoff's laws**, governing the formation of spectra:

1. A luminous solid or liquid, or a sufficiently dense gas, emits light of all wavelengths and so produces a *continuous spectrum* of radiation.

2. A low-density, hot gas emits light whose spectrum consists of a series of bright *emission lines*. These lines are characteristic of the chemical composition of the gas.

3. A cool, thin gas absorbs certain wavelengths from a continuous spectrum, leaving dark *absorption lines* in their place, superimposed on the continuous spectrum. Once again, these lines are characteristic of the composition of the intervening gas—they occur at precisely the same wavelengths as the emission lines produced by that gas at higher temperatures.

Figure 4.7 illustrates Kirchhoff's laws and the relationship between absorption and emission lines. When viewed directly, the light source, a hot solid (the filament of the bulb), has a continuous (black-body) spectrum. When viewed through a cloud of cool hydrogen gas, a series of dark absorption lines appear at wavelengths characteristic of hydrogen. The lines appear because the light at those wavelengths is absorbed by the hydrogen. As we will see later in this chapter, the absorbed energy is subsequently reradiated into space, but in all directions, not just the original direction of the beam. Consequently, when the cloud is viewed from the side against an otherwise dark background, a series of faint emission lines is seen. These lines contain the energy lost by the forward beam. If the gas were heated to incandescence, it would produce stronger emission lines at precisely the same wavelengths.

ASTRONOMICAL APPLICATIONS

By the late nineteenth century, spectroscopists had developed a formidable arsenal of techniques for interpreting the radiation received from space. Once astronomers knew that spectral lines were indicators of chemical composition, they set about identifying the observed lines in the solar spectrum. Almost all of the lines in light from extraterrestrial sources could be attributed to known elements (for example, many of the Fraunhofer lines in sunlight are associated with the element iron). However, some new lines also appeared in the solar spectrum. In 1868, astronomers realized that those lines must correspond to a previously unknown element. It was given the name helium, after the Greek word *helios*, meaning "Sun." Only in 1895, almost three decades after its detection in sunlight, was helium discovered on Earth. (A laboratory spectrum of helium is part of Figure 4.3.)

For all the information that nineteenth-century astronomers could glean from observations of stellar spectra, they still lacked a theory explaining how the spectra themselves arose. Despite their sophisticated spectroscopic equipment, they knew scarcely any more about the physics of stars than did Galileo or Newton. To understand how spectroscopy can be used to extract detailed information about astronomical objects from the light they emit, we must delve more deeply into the processes that produce line spectra.

4.2 *The Particle Nature of Radiation*

❸ By the start of the twentieth century, experimental physicists had accumulated evidence that light sometimes behaves in a manner that simply cannot be explained by the wave theory. As we have just seen, the production of absorption and emission lines involves only certain very specific frequencies or wavelengths (colors) of light. This would not be expected if light behaved like a continuous wave and matter always obeyed the laws of Newtonian mechanics. Other experiments conducted around the same time strengthened the conclusion that the notion of radiation as a wave was *incomplete*. It became clear that, when light interacts with matter on very small scales, it does so not in a continuous way but in a discontinuous, stepwise manner. The challenge was to find an explanation for this unexpected behavior. The eventual solution revolutionized our view of nature and now forms the foundation not just for physics and astronomy, but for virtually all of modern science.

Albert Einstein provided a major breakthrough in 1905. He realized that it was possible to explain a number of puzzling experimental results (especially the *photoelectric*

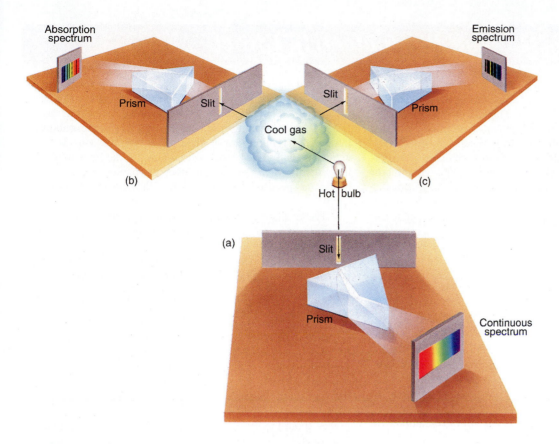

Figure 4.7 A source of continuous radiation, here represented by a light bulb, is used to illustrate Kirchhoff's laws of spectroscopy. (a) The unimpeded beam shows the familiar continuous spectrum of colors. (b) When viewed through a cloud of hydrogen gas, a series of dark hydrogen absorption lines appears in the continuous spectrum. These lines are formed when the gas absorbs some of the bulb's radiation and reemits it in random directions. Since most of the reemitted radiation does not go through the slit, the effect is to remove the absorbed radiation from the light that reaches the screen at left. (c) When the gas is viewed from the side, a fainter hydrogen emission spectrum is seen, consisting of reemitted radiation. The absorption lines in (b) and the emission lines in (c) thus have the same wavelengths. (If the gas were heated to incandescence, it would produce stronger emission lines at the same wavelengths.)

effect—see the *More Precisely* feature on p. 80) by assuming that light travels as individual packets of electromagnetic energy, now called **photons**. In this way, Einstein revised the entire theory of electromagnetic radiation, showing that although it sometimes behaves as a wave, at other times it acts as a stream of *particles*.

Further, Einstein was able to quantify the relationship between the two aspects of light's double nature. He found that the energy carried by a photon is proportional to the *frequency* of the radiation:

$$\text{photon energy} \propto \text{radiation frequency.}$$

Thus, for example, a "red" photon with a frequency of 4×10^{14} Hz (corresponding to a wavelength of approximately 750 nm, or 7500 Å) has 4/7 the energy of a "blue" photon with a frequency of 7×10^{14} Hz (and a wavelength 400 nm, or 4000 Å).

The realization that light can behave both as a wave and as a particle is another example of the scientific method at work. Despite the enormous success of the wave theory of radiation in the nineteenth century, the experimental evidence led scientists to the inevitable conclusion that the theory was incomplete—it had to be modified to allow for the fact that light exhibits particle characteristics. In addition to bringing about the birth of a whole new branch of physics—now known as *quantum mechanics*—this new theory radically changed the way physicists view light and all other forms of radiation.

Environmental conditions ultimately determine which description best fits the behavior of electromagnetic radiation—a wave or a stream of particles. As a general rule of thumb, in the macroscopic realm of everyday experience, radiation is more usefully described as a wave, while in the microscopic domain of atoms, it is best characterized as a series of particles. Many people are confused by the idea that light can behave in two such different ways. To be truthful, modern physicists don't yet fully understand *why* nature displays this wave-particle duality. Nevertheless, there is irrefutable experimental evidence for both of these aspects of radiation.

More Precisely... The Photoelectric Effect

Einstein developed his insight into the nature of radiation partly as a means of explaining a puzzling experimental result known as the *photoelectric effect*. This effect can be demonstrated by shining a beam of light on a metal surface (as shown in the accompanying Figure). When high-frequency ultraviolet light is used, bursts of electrons are dislodged from the surface by the beam. The effect is much like one billiard ball hitting another and knocking it off the table. Curiously, the speed with which the particles are ejected from the metal is found to depend only on the *color* of the light, not on its intensity. For lower-frequency light—blue, say—an electron detector will still record bursts of electrons, but now their speeds, and hence their *energies,* are less. For even lower frequencies—red or infrared light—*no* electrons are kicked out of the metal surface at all.

These results were very difficult to reconcile with a wave model of light. Einstein realized that the only way to explain the presence of an abrupt cutoff at the detector—that is, that the detector registered nothing when the frequency of the incoming radiation dropped below a certain level—and the increase in electron speed with light frequency was to envision radiation as traveling as "bullets," or particles, or *photons*. To account for the experimental findings, the energy of any photon must be proportional to the *frequency* of the radiation:

photon energy ∝ radiation frequency.

If we suppose that some minimum amount of energy is needed just to "unglue" the electrons from the metal, then we can see why no electrons are emitted below some critical frequency—red photons just don't carry enough energy. Above the critical frequency, photons have enough energy to dislodge the electrons; moreover, any energy they possess above the necessary minimum is imparted to the electrons as kinetic energy, the energy of motion. Thus, as the frequency of the radiation increases, so too does the photon energy, and hence the speed of the electrons that they liberate from the metal.

The constant of proportionality in the foregoing relation is now known as *Planck's* constant, in honor of the German physicist Max Planck, who first determined its numerical value. It is always denoted by the symbol *h*. The equation relating photon energy *E* to radiation frequency *f* is thus

$$E = hf.$$

As noted in Chapter 3, the SI unit of energy is the *joule*. ∞ (p. 65) The value of Planck's constant is a very small number—6.63×10^{-34} joule seconds. Consequently, the energy of a single photon is tiny—even a very-high-frequency gamma ray (the most energetic type of electromagnetic radiation) with a frequency of 10^{22} Hz has an energy of only 7×10^{-12} joule—about the same energy as is carried by a flying gnat. Nevertheless, this energy is more than enough to damage a living cell—the basic reason that gamma rays are so much more dangerous to life than, say, visible light, is that each gamma-ray photon carries billions of times more energy than a photon of visible radiation.

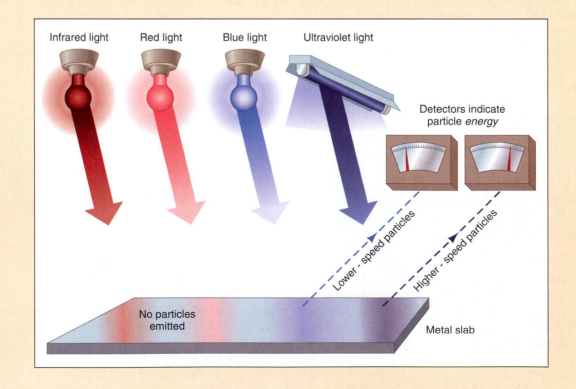

4.3 *Atomic Structure and Spectra*

4 To explain the formation of spectral lines, we must understand not just the nature of light, but also something of the structure of **atoms**—the microscopic building blocks from which all matter is constructed. Let us start with the simplest atom of all—hydrogen. A hydrogen atom consists of an *electron*, with a negative electrical charge, orbiting a *proton*, which carries a positive charge. The proton forms the central **nucleus** (plural: nuclei) of the atom. The hydrogen atom as a whole is electrically neutral. The equal and opposite charges of the proton and the orbiting electron produce an electrical attraction that binds them together within the atom.

How does this picture of the hydrogen atom relate to the characteristic emission and absorption lines associated with hydrogen gas? If an atom emits some energy in the form of radiation, that energy has to come from somewhere within the atom. Similarly, if energy is absorbed, it must cause some internal change. It is reasonable (and correct) to suppose that the energy emitted or absorbed by the atom is associated with changes in the motion of the orbiting electron.

THE BOHR ATOM

The first theory of the atom to provide an explanation of hydrogen's observed spectral lines was propounded by the Danish physicist Niels Bohr. This theory is now known simply as the **Bohr model** of the atom. Its essential features are as follows. First, there is a state of lowest energy—the **ground state**—which represents the "normal" condition of the electron as it orbits the nucleus. Second, there is a maximum energy that the electron can have and still be part of the atom. Beyond that energy, the electron is no longer bound to the nucleus, and the atom is said to be **ionized**. (An *ion* is an atom that has either lost or gained electrons.) Third, and most important (and also least intuitive), between those two energy levels, the electron can exist only in certain sharply defined energy states, sometimes referred to as *orbitals*. The orbital energies are said to be **quantized**. This description of the atom contrasts sharply with the predictions of Newtonian mechanics, which would permit orbits at *any* energy, not just at certain specific values. The rules of quantum mechanics are far removed from everyday experience.

In Bohr's model, each electron orbital was pictured as having a specific radius, much like a planetary orbit in the solar system, as shown in Figure 4.8. However, the modern view is not so simple. Although each orbital *does* have a precise energy, the electron is now envisioned as being smeared out in an "electron cloud" surrounding the nucleus, as illustrated in Figure 4.9. It is common to speak of the *mean* (average) distance from the cloud to the nucleus as the "radius" of the electron's orbit. When a hydrogen atom is in its ground state, the radius of the orbit is about 0.05 nm (0.5 Å). As the orbital energy increases, the radius

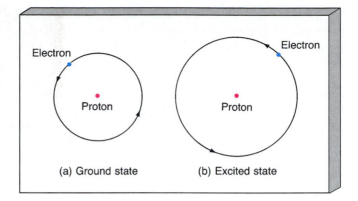

Figure 4.8   An early conception of the hydrogen atom pictured its electron orbiting the central proton in a well-defined orbit, rather like a planet orbiting the Sun. Two electron orbits of different energies are shown. The left-hand figure represents the ground state, the right-hand figure an excited state.

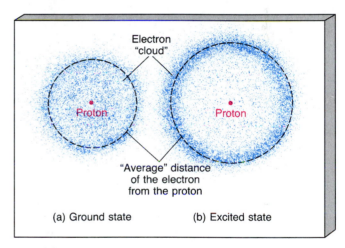

Figure 4.9   The modern view of the hydrogen atom sees the electron as a "cloud" surrounding the nucleus. The same two energy states are shown as in Figure 4.8.

increases, too. For the sake of clarity in the diagrams that follow, we will represent electron orbitals by solid lines, but bear in mind always that Figure 4.9 is a more accurate depiction of reality.

EXCITATION

Atoms do not always remain in their ground states. An atom is said to be in an **excited state** when an electron occupies an orbital at a greater than normal distance from its parent nucleus. An atom in such an excited state has a greater than normal amount of energy. The excited state with the lowest energy (that is, the one closest in energy to the ground state) is conventionally called the *first excited state*, that with the second-lowest energy the *second excited state*, and so on.

How can atoms become excited? Generally, in one of two ways: They can become *radiatively excited* by absorbing some light energy from a source of electromagnetic radiation, or they can become *collisionally excited* by colliding with another particle—an atom or a free electron, for example. However, the electron cannot stay "out of place" in this higher orbital forever—the ground state is the only level where it can remain indefinitely. After about 10^{-8} s, it returns to its normal ground state.

Here now is the crucial point that links atoms to radiation and allows us to interpret atomic spectra. Because electrons may exist only in orbitals having specific energies, atoms can absorb only specific amounts of energy as their electrons are boosted into excited states. Likewise, they can emit only specific amounts of energy as their electrons fall back to lower energy states. Thus, the amount of light energy absorbed or emitted in these processes *must correspond precisely to the energy difference between two orbitals*. The quantized nature of the atom's energy levels requires that light must be absorbed and emitted in the form of little "packets" of electromagnetic radiation, each carrying a very specific amount of energy—that is, in the form of photons.

In brief,

- Radiative excitation of an atom involves the absorption of a photon with a definite wavelength—or color—carrying an amount of energy precisely equal to the energy difference between the two orbitals involved.

- In the reverse process—radiation by an excited atom—the photon emitted similarly carries an amount of energy precisely equal to the energy difference between the two orbitals involved.

An atom therefore absorbs and emits radiation at the same characteristic wavelengths, which are determined by the atom's own internal structure. The basic process of absorption and emission of photons by a hydrogen atom are illustrated schematically in Figure 4.10(a). In the example shown here, where a hydrogen atom makes transitions between the ground state and the first excited state, the photon energy happens to correspond to an *ultraviolet* photon, of wavelength 121.6 nm (1216 Å).

Absorption can also boost an electron into higher excited states. From such a state, the electron may return to the ground state via several alternate paths. Thus, the absorption of a single high-energy photon may lead to the subsequent emission of two or more lower-energy photons as the atom returns to its ground state. For example, as illustrated in Figure 4.10(b), a hydrogen atom can be boosted from the ground state past the first excited state, all the way up into the second excited state. As before, the electron returns rapidly to the ground state, but this time it can do so in two possible ways:

- It can proceed directly back to the ground state, in the process emitting an ultraviolet photon with a wave-

length of 102.6 nm (1026 Å)—a photon identical to the one that excited the atom in the first place.

- Alternatively, the electron can *cascade* down one orbital at a time. If this occurs, the atom will emit *two* photons: one with an energy equal to the difference between the second and first excited states, and the other with an energy equal to the difference between the first excited state and the ground state. The second step of this cascade process produces a 121.6 nm ultraviolet photon, just as in Figure 4.10(a). However, the first transition of the cascade process—the one from the second to the first excited state—produces a photon with a wavelength of 656.3 nm (6563 Å), which is in the visible part of the spectrum. This photon is seen as red light. An individual atom—if one could be isolated—would emit a momentary red flash.

Absorption of additional energy could boost the electron to even higher orbitals within the atom. As the excited electron cascaded back down to the ground state, the atom could emit many photons, each with a different energy and hence a different wavelength. The resulting spectrum would therefore show many spectral lines. In a sample of heated hydrogen gas, at any instant, atoms are found in many different excited states. The complete emission spectrum therefore consists of wavelengths corresponding to all possible transitions between those states and states of lower energy. The visible hydrogen spectrum shown earlier (in Figure 4.3) is the result. In the case of hydrogen, it so happens that transitions from higher states back to the *first* excited state are the ones corresponding to spectral lines in the visible range. As we have just seen, transitions ending at the ground state produce ultraviolet photons. The energy levels and the spectrum of hydrogen are discussed in more detail in the *More Precisely* feature on p. 88.

KIRCHHOFF'S LAWS EXPLAINED

5 To appreciate further how spectral lines are produced, let's reconsider our earlier discussion of emission and absorption lines in terms of the model just presented. In Figure 4.7, a beam of continuous radiation shines through a cloud of hydrogen gas. The beam contains photons of all energies, but most of them cannot interact with the gas—the gas can absorb only those photons having precisely the right energy to cause a change in an electron's orbit from one state to another (an *electronic transition*). All other photons in the beam—with energies that cannot produce a transition—do not interact with the gas at all, but pass through it unhindered. Those photons that do have the right energy excite the gas and are removed from the beam.

The excited atoms return rapidly to their original states, each emitting one or more photons in the process. We might think, then, that although some photons from the beam are absorbed by the intervening cool gas, they would quickly be replaced by reemitted photons, so we would never observe the effects of absorption. This is not

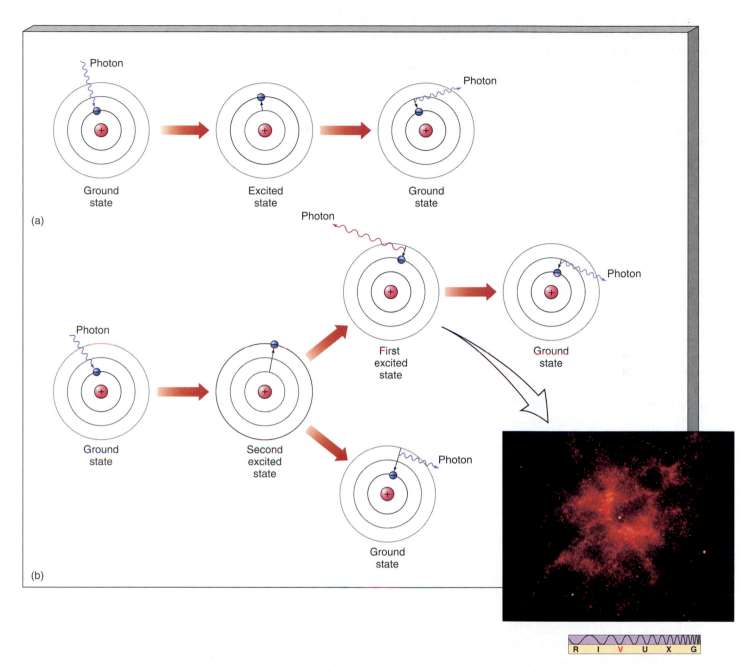

Figure 4.10   (a) Diagram of a photon being absorbed by a hydrogen atom (left), causing the momentary excitation of that atom (center) into its first excited state. Eventually, the atom returns to its ground state, accompanied by the emission of a photon of the same energy as the original photon (right). (b) Absorption of a photon might also boost the atom into a higher excited state, from which there may be several possible paths back to the ground state. (The sharp lines used for the orbitals here and in similar figures that follow are intended merely as a schematic representation of the electron energy levels, and are not meant to be taken literally. In actuality, electron orbitals are "clouds," as shown in Figure 4.9. As ultraviolet photons from a hot star pass through surrounding hydrogen gas, many are absorbed by the gas, boosting its atoms into excited states. Electrons in the second excited state can fall to the first excited state on their way back to the ground state (the upper path in part b). This transition produces radiation in the visible region of the spectrum—the 656.3 nm red glow that is characteristic of excited hydrogen gas. The object shown in the inset, designated NGC 2440, is an emission nebula: an interstellar cloud consisting largely of hydrogen gas excited by an extremely hot star (the white dot in the center).

the case, however, for two important reasons. First, although the photons not absorbed by the intervening gas follow a clear path to the detector, the reemitted photons can leave in *any* direction. Thus, many of the photons are absorbed and reemitted at other angles, and so are effec-

tively lost from the original beam. Second, as we have just seen, electrons can cascade back to the ground state, emitting several lower-energy photons instead of a single photon equal in energy to the one originally absorbed. The net result of these two processes is that the original energy is channeled into photons of many different colors, moving in many different directions.

A detector looking through the cloud at the source of the radiation (a star, say) records a continuous spectrum, except at those precise wavelengths where photons have been subtracted from the beam. The dark absorption lines thus produced are characteristic of the intervening gas. They are direct indicators of the energy differences between atomic orbitals. The regions of bright emission surrounding the dark lines is unchanged by the intervening cloud. It is produced by those photons in the beam that passed through the gas without being absorbed.

A detector looking at the cloud from the side records the energy that is reemitted from the cloud after absorption within the gas. Again, the spectrum is characteristic of the gas, not of the original beam. An example is shown in the inset of Figure 4.10, where the (mainly ultraviolet) radiation from a young, hot star excites the surrounding cool hydrogen gas out of which the star recently formed. The gas emits a characteristic red glow at a wavelength of precisely 656.3 nm—the wavelength corresponding to transitions from the second to the first excited states of hydrogen.

Absorption and emission spectra are created by the same atomic processes. They correspond to the same atomic transitions. They contain the same information about the composition of the cloud. In the lab, we can move our detector and can measure both. In astronomy, we are not able to change our vantage point—the type of spectrum we see depends on our chance location with respect to both the source and the cloud.

MORE COMPLEX ATOMS

All hydrogen atoms have basically the same structure—a single electron orbiting a single proton—but, of course, there are many other kinds of atoms, each kind having a unique internal structure. The number of protons in the nucleus of an atom determines the **element** that it represents. That is, just as all hydrogen atoms have a single proton, all oxygen atoms have 8 protons, all iron atoms have 26 protons, and so on.

The next simplest element after hydrogen is helium. The central nucleus of the most common form of helium is made up of two protons and two **neutrons** (another kind of elementary particle having a mass slightly larger than that of a proton but having no electrical charge at all). About this nucleus orbit two electrons. As with hydrogen and all other atoms, the "normal" condition for helium is to be electrically neutral, with the negative charge of the orbiting electrons exactly canceling the positive charge of the nucleus (Figure 4.11).

More complex atoms contain more protons (and neutrons) in the nucleus and have correspondingly more orbiting electrons. For example, an atom of carbon, shown in Figure 4.12, consists of six electrons orbiting a nucleus containing six protons and six neutrons. As we progress to

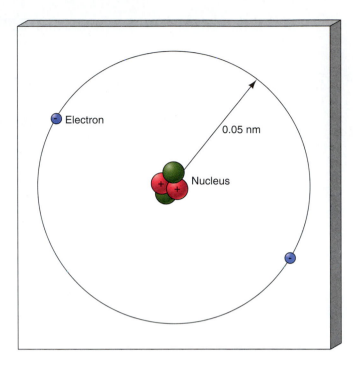

Figure 4.11 A helium atom in its normal, ground state. Two electrons occupy the lowest-energy orbital around a nucleus containing two protons and two neutrons.

heavier and heavier elements, the number of orbiting electrons increases, and the number of possible electronic transitions rises rapidly. The result is that very complicated spectra can be produced. The complexity of atomic spectra generally reflects the complexity of the atoms themselves. A good example is the element iron, which contributes several hundred of the Fraunhofer absorption lines seen in the solar spectrum. The many possible transitions of its 26 orbiting electrons yield an extremely rich line spectrum (see Figure 4.4). Even very heavy elements, such as gold (with 79 orbiting electrons) and lead (with 82) have been observed in some astrophysical settings.

A CHARACTERISTIC FINGERPRINT

The complex transitions among the various orbitals are *unique* to each element. Consequently, as electrons in a given atom return to lower orbitals from higher-energy states, or absorb radiation and jump to excited states, the photons that emitted or absorbed carry energies characteristic of that element—*and only that element*. This fact explains the power of spectroscopy. Even though many different kinds of atoms might be mixed together in a gas, spectroscopy enables us to study one kind of atom to the exclusion of all others simply by focusing on specific wavelengths of radiation. Thus, for example, a cool intervening gas cloud containing many elements will produce a very complicated absorption spectrum in the light received from a background continuous source. Nevertheless, by identifying the (superimposed) absorption spectra of many

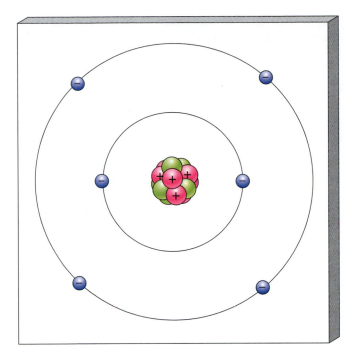

Figure 4.12 A carbon atom in its normal, ground state. Six electrons orbit a six-proton, six-neutron nucleus, two in an inner orbital, the other four at a greater distance from the center.

different atoms, we can determine the cloud's composition. Figure 4.13 shows an actual spectrum observed coming from a cosmic object, with some spectral lines labeled.

Spectral lines occur throughout the entire electromagnetic spectrum. Usually, electron transitions among the lowest orbitals of the lightest elements (such as hydrogen and helium) produce visible and ultraviolet spectral lines. Transitions among very highly excited states of hydrogen and other elements can produce spectral lines in the infrared and radio parts of the electromagnetic spectrum. Conditions on Earth make it all but impossible to detect these radio and infrared features in the laboratory, but they *can* be routinely observed coming from space. Electron transitions among lower energy levels in heavier, more complex elements produce X-ray spectral lines. These lines have been observ7ed in the laboratory; some of them are also observed in stars and other cosmic objects.

4.4 *Molecules*

6 A **molecule** is a tightly bound group of atoms held together by interactions between their orbiting electrons—interactions that we call chemical bonds. Much like atoms, molecules can exist only in certain well-defined energy states, and again like atoms, molecules produce emission or absorption spectral lines when they make a transition from one state to another. Because molecules are more complex than individual atoms, the rules of molecular physics are also much more complex. Nevertheless, as with atomic spectral lines, painstaking experimental work over many decades has determined the precise frequencies (or wavelengths) at which millions of molecules emit and absorb radiation.

In addition to the lines resulting from electron transitions, lines are produced by molecules because of two other kinds of changes not possible in atoms: molecules can ro-

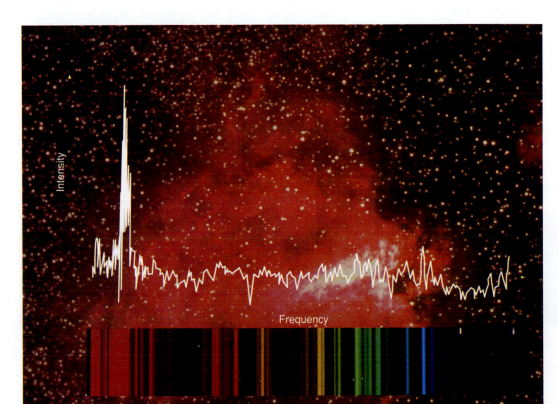

Figure 4.13 The visible spectrum of the hot gases in a nearby star-forming region known as the Omega nebula (M17). Shining by the light of several very hot stars, the nebula produces a complex spectrum of bright and dark lines (bottom), also shown here as an intensity trace from red to blue (center).

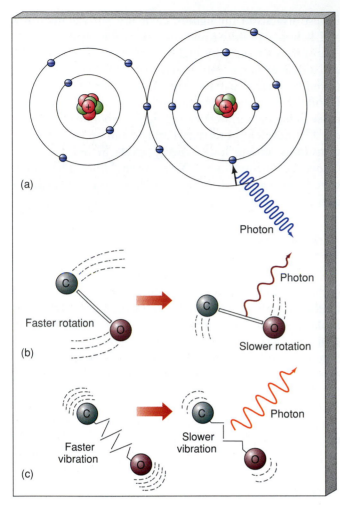

(a)

Photon

C

Faster rotation

Photon

C

O

Slower rotation

(b)

C

Faster vibration

Photon

C

Slower vibration

O

Photon

(c)

Figure 4.14   Molecules can change in three ways while emitting or absorbing electromagnetic radiation. Sketched here is the molecule carbon monoxide (CO) experiencing (a) a change in electron arrangement, in which an electron in the outermost orbital of the oxygen atom drops to a lower-energy state, (b) a change in rotational state, and (c) a change in vibrational state.

tate, and they can vibrate. Figure 4.14 illustrates these basic molecular motions. Molecules rotate and vibrate in specific ways. Only certain spins and vibrations are allowed by the rules of molecular physics. When a molecule *changes* its rotational state or its vibrational state, a photon is emitted or absorbed. Spectral lines characteristic of the specific kind of molecule result. These lines are molecular fingerprints,

just like their atomic counterparts, enabling researchers to identify and study one kind of molecule to the exclusion of all others.

As a rule of thumb, we can say that

- *Electron transitions* within molecules produce visible and ultraviolet spectral-line features.
- Changes in molecular *vibration* produce infrared spectral features.
- Changes in molecular *rotation* produce spectral lines in the radio part of the electromagnetic spectrum.

Molecular lines usually bear little resemblance to the spectral lines associated with their component atoms. For example, Figure 4.15(a) shows the emission spectrum of the simplest molecule known—molecular hydrogen. Notice how different it is from the spectrum of atomic hydrogen shown in part (b) of the figure.

4.5   *Spectral-Line Analysis*

Astronomers apply the laws of spectroscopy in analyzing radiation from beyond the Earth. A nearby star or a distant galaxy takes the place of the light bulb in our previous examples. A galactic cloud or a stellar (or even planetary) atmosphere plays the role of the intervening cool gas. And a spectrograph attached to a telescope replaces our simple prism and detector.

LINE WAVELENGTHS

Stars are very hot, especially deep down in their cores, where the temperature is measured in millions of kelvins. Because of the heat, the atoms are ionized and the spectrum of radiation is continuous. However, at the relatively cool surface of a star, some atoms retain a few, or even most, of their orbital electrons. By matching the spectral lines we see with the laboratory spectra of known atoms and molecules, the chemical *composition* of the star can be determined.

As we have already seen, literally thousands of dark absorption lines cover the Sun's visible spectrum; nearly 800 of them are produced by variously excited atoms and ions of just one element: iron. Atoms of a single element, such as iron, can yield many lines for two reasons. First, the 26 electrons of a normal iron atom can make an enormous number of different transitions among energy levels. Second, many iron atoms are ionized, with some of their

Figure 4.15   (a) The spectrum of molecular hydrogen. Notice how it differs from the spectrum of the simpler atomic hydrogen (b).

26 electrons stripped away. Because the removal of electrons alters an atom's electromagnetic structure, the energy levels of ionized iron are quite *different* from those of neutral iron. Each new level of ionization introduces a whole new set of spectral lines. Besides iron, many other elements, also in different stages of excitation and ionization, absorb photons at visible wavelengths. When we observe the entire Sun, all these atoms and ions absorb simultaneously to yield the rich spectrum we see.

The spectra of many atoms and ions are well known from laboratory measurements. Often, however, a familiar pattern of lines appears, but the lines are displaced from their expected locations. In other words, a set of spectral lines might be recognized as belonging to a particular element, except that they are all offset—blueshifted or redshifted—by the same amount from their normal wavelengths. Such shifts are produced by the Doppler effect, discussed in Section 3.6. They thus allow astronomers to find out how fast the source of the radiation is moving along the line of sight to the observer (its *radial velocity*). ∞ (p. 68)

LINE INTENSITY

Still more information can be obtained from detailed study of the lines themselves. Because the intensity of a line is proportional to the number of photons emitted or absorbed by the atoms, the intensity of a particular line depends in part on the number of atoms giving rise to the line. The more atoms present to emit or absorb the photons corresponding to a given line, the stronger (brighter or darker, depending on whether it is seen in emission or absorption) that line is.

But intensity also depends on the *temperature* of the atoms—that is, the temperature of the entire gas of which the atoms are members—because temperature determines what fraction of the atoms at any instant are in the right orbital to undergo any particular transition. Consider the absorption of radiation by hydrogen atoms in an interstellar gas cloud or in the outer atmosphere of a star. If all the hydrogen were in its ground state—as it would be if the temperature were relatively low—the only transitions that could occur would be the Lyman series (see the *More Precisely* feature on p. 88), resulting in absorption lines in the ultraviolet portion of the spectrum. Thus, astronomers would observe *no* visible hydrogen absorption lines (for example, the Balmer series) in the spectrum of this object, not because there is no hydrogen, but because there would be no hydrogen atoms in the first excited state (as is required to produce visible absorption features).

The spectrum of our own Sun is a case in point. Because the temperature of the Sun's atmosphere is a relatively cool 6000 K (as we saw in Chapter 3), few hydrogen atoms have electrons in any excited state. ∞ (p. 65) Hence, in the Sun, visible hydrogen lines are relatively weak—that is, of low intensity compared to the same lines in many other stars—even though hydrogen is by far the most abundant element there.

As the temperature rises, atoms move faster and faster. More and more energy becomes available in the form of collisions, and more and more electrons are boosted into an excited state. It takes a little time—about 10 nanoseconds (10^{-8} s), typically—for the electrons to fall back to the ground state. At any instant, then, some atoms are temporarily in an excited state and so are capable of absorbing at visible or longer wavelengths. As the number of atoms in the first excited state increases, lines in the Balmer series become more and more evident in the spectrum. Eventually, a temperature is reached where *most* of the atoms are in the first excited state, simply because of their frequent, energetic collisions with other atoms in the gas. At this point, the Balmer lines are at their strongest (while the Lyman lines are much weaker).

At even higher temperatures, most atoms are kicked beyond the first excited state into higher-energy orbitals, and new series of absorption lines are seen, while the strength of the Balmer series declines again. Eventually, the temperature becomes so high that most hydrogen is *ionized,* and no spectral lines are seen at all.

Over the years, astronomers have developed mathematical formulas that relate the number of emitted or absorbed photons to the temperature of the atoms, as well as to their number. Once an object's spectrum is measured, astronomers can interpret it by matching the observed intensities of the spectral lines with those predicted by the formulas. In this way, astronomers can refine their measurements of both the composition and the temperature of the gas producing the lines.

LINE BROADENING

Consider an emission line, such as the one shown in Figure 4.16(a). (The discussion that follows holds equally

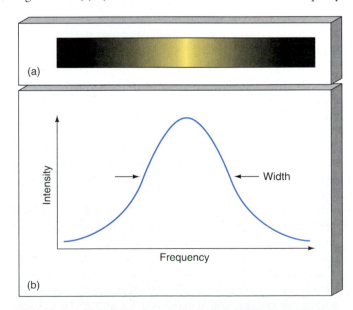

Figure 4.16   By tracing the changing brightness across a typical emission line (a) and expanding the scale, we obtain a graph of its intensity plotted against wavelength (b).

More Precisely...   The Energy Levels of the Hydrogen Atom

By observing the emission spectrum of hydrogen, Niels Bohr determined early in the twentieth century what the energy differences between the various energy levels must be. Using that information, he was then able to infer the actual energies of the excited states of hydrogen.

A unit of energy often used in atomic physics is the *electron volt* (eV). Its name derives from the amount of energy imparted to an electron by accelerating it through an electric potential of 1 volt. For our purposes, however, it is just a convenient quantity of energy, numerically equal to 1.6×10^{-19} joule—roughly half the energy carried by a single photon of red light. The minimum amount of energy needed to ionize hydrogen from its ground state is 13.6 eV. Bohr numbered the energy levels of hydrogen, with level 1 the ground state, level 2 the first excited state, and so on. He found that by assigning zero energy to the ground state, the energy of any state (the *n*th, say) could be written as follows:

$$E_n = 13.6\left(1 - \frac{1}{n^2}\right) \text{ eV.}$$

Thus, the ground state has energy $E_1 = 0$ (by our definition), the first excited state has an energy of $E_2 = 10.2$ eV, the second excited state has an energy of $E_3 = 12.1$ eV, and so on. Notice that there is an *infinite* number of excited states between the ground state and the energy at which the atom is ionized, crowding closer and closer together as *n* becomes large and E_n approaches 13.6 eV.

Knowing the energy of each electron orbital, we can calculate the energy associated with a transition between any two given states. For example, to boost an electron from the second state to the third, an atom must be supplied with $E_3 - E_2 = 1.9$ eV of energy, or 3.0×10^{-19} joule. Using the formula $E = hf$ given in the text, we find that this corresponds to a photon with frequency 4.6×10^{14} Hz, having a wavelength 656.3 nm (6563 Å) and lying in the red portion of the spectrum. Similarly, the jump from level 3 to level 4 requires $E_4 - E_3 = 0.66$ eV of energy, corresponding to an infrared photon with a wavelength 1890 nm (18,900 Å), and so on.

The accompanying diagram summarizes the structure of the hydrogen atom. The various energy levels are shown as a series of circles of increasing radius, representing increasing energy (but remember that the electron orbitals are actually rather fuzzy and do not really have well-defined radii). The electronic transitions between these levels (indicated by arrows) are conventionally grouped into families, named after their discoverers, which define the terminology used to identify specific spectral lines (Note that the spacings of the energy levels are not drawn to scale here, to provide room for all labels on the diagram. In reality, the circles should become more and more closely spaced as we move outward.)

Transitions down to (or up from) the ground state (level 1) from higher, excited levels form the *Lyman series*. The first is *Lyman alpha* (Lyα), corresponding to the transition between the first excited state (level 2) and the ground state. As we have seen, the energy difference is 10.2 eV, and the Lyα photon has a wavelength of 121.6 nm (1216 Å). The Lyβ (beta) transition, between level 3 (the second excited state) and the ground state, corresponds to an energy change of 12.1 eV and a photon of wavelength 102.6 nm (1026 Å). Lyγ (gamma) corresponds to a level 4 to level 1 jump, and so on. You can calculate the energies, frequencies, and wavelengths of all the photons in the Lyman series using the formulas given in the text. All Lyman series energies lie in the ultraviolet region of the spectrum.

The next series of lines, the *Balmer series,* involves transitions down to (or up from) level 2, the first excited state. All the Balmer series lines lie in or close to the visible portion of the electromagnetic spectrum. Because they form the most easily observable part of hydrogen's spectrum and were the first to be discovered, the Balmer series lines are often referred to simply as the "Hydrogen" series and signaled by the letter H. As with the Lyman series, the individual transitions are labeled with Greek letters. An Hα photon (level 3 to level 2) has a wavelength of 656.3 nm (6563 Å) and is red, Hβ (level 4 to level 2) has a wavelength of 486.1 nm (4861 Å) (green), Hγ (level 5 to level 2) has a wave-

---

well for absorption features.) The line seems uniformly bright, but more careful study shows that its brightness is greatest at the center and tapers off toward either side, as illustrated in Figure 4.16(b). We stressed earlier that photons are emitted and absorbed at very precise wavelengths. Why aren't spectral lines extremely narrow, occurring only at specific wavelengths? This *line broadening* is not the result of some inadequacy of our experimental apparatus. It is caused by the *environment* in which the emission or absorption occurs, which often changes our perception of a photon's energy, and it tells us a lot about the physical state of the gas involved.

Several physical mechanisms can broaden spectral lines. The most important of these involve the Doppler ef-

fect. ∞ (p. 68) To understand how the Doppler effect can broaden a spectral line, imagine a hot gas cloud. Individual atoms are in random, chaotic motion. The hotter the gas, the faster the random *thermal motions* of the atoms, as illustrated in Figure 4.17(a). If a photon is emitted by an atom *in motion*, the wavelength of the detected photon is changed by the Doppler effect. For example, if an atom is moving away from our eye or from our detector while in the process of emitting a photon, that photon is redshifted. The photon is not recorded at the precise wavelength predicted by atomic physics, but rather at a slightly longer wavelength. According to our previous discussion of the Doppler effect, the extent of this red shift is proportional to the velocity away from the detector. Similarly, if the

length of 434.1 nm (4341 Å) (blue), and so on. The most energetic Balmer series photons have energies that place them just beyond the blue end of the visible spectrum, in the near ultraviolet.

The classification continues with the *Paschen series* (transitions down to or up from the second excited state), the *Brackett series* (third excited state), and the *Pfund series* (fourth excited state). Beyond that point, infinitely many other families exist, moving farther and farther into the infrared and radio regions of the spectrum, but they are not commonly referred to by any special names. A few of the transitions making up the Lyman, Balmer, and Paschen series are marked on the figure. Astronomically, the most important series are the Lyman and Balmer (Hydrogen) sequences.

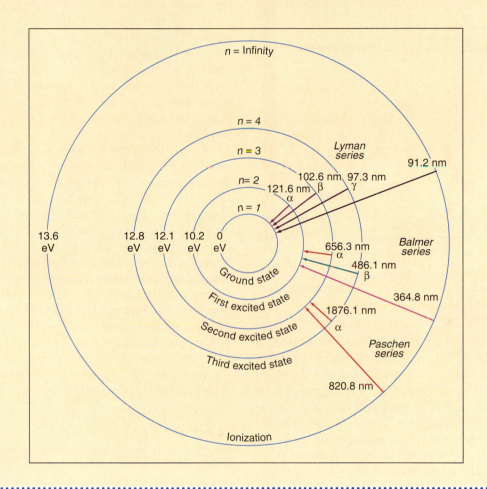

atom is moving toward us, its light is received at a shorter wavelength and so is blueshifted.

In a cloud of gas, atoms are in constant thermal motion. Some atoms move toward us, some away from us. Still others are moving transverse to our line of sight and are unaffected by the Doppler effect (at least, from our perspective). Throughout the whole cloud, atoms move in every possible direction. The result is that many atoms emit or absorb photons at slightly different wavelengths than would normally be the case if all the atoms were motionless. Most atoms in a typical cloud have very small thermal velocities. As a result, most atoms emit or absorb radiation that is Doppler-shifted only a little, and very few atoms have large shifts. So, the center of any spectral line is much more pronounced than either of its "wings." The result is a bell-shaped spectral feature like that in Figure 4.17(b). Thus, even if all atoms emitted and absorbed photons at only one specific wavelength, the effect of their thermal motion would be to smear the line out over a small range of wavelengths. The hotter the gas, the larger the spread of Doppler motions and the greater the width of the line. By measuring a line's width, astronomers can estimate the temperature of the gas producing it.

Actually, the situation is not so simple. Several other physical mechanisms can also produce line broadening. One such mechanism is gas *turbulence*, which exists when the gas in a cloud is not at rest or flowing smoothly, but in-

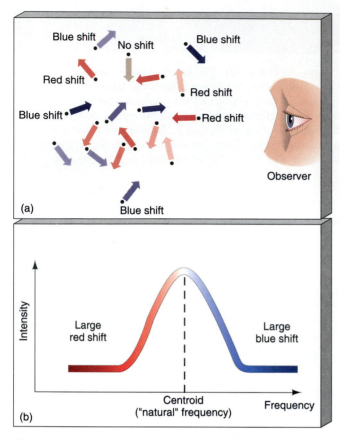

Figure 4.17 Atoms moving around randomly (a) produce broadened spectral lines (b) as their individual redshifted and blueshifted emission lines merge in our detector.

stead is seething and churning in eddies and vortices of many sizes. Motion of this type causes Doppler-shifting of spectral lines, but lines from different parts of the cloud are shifted more or less randomly. Very often, the cloud is so small or far away that our equipment cannot distinguish, or *resolve*, different parts from one another—the light from the entire cloud is blended together in our detector. When averaged over the whole cloud, the net effect appears rather similar to the thermal broadening just discussed. However, it has nothing to do with the temperature of the gas.

Rotation produces a similar effect. Consider a star or a gas cloud oriented so that we see it spinning. Photons emitted from the side spinning toward us are blueshifted by the Doppler effect. Photons emitted from the side spinning away from us are redshifted. As with turbulence, if our equipment is unable to resolve the object, a net broadening of its observed spectral lines results, as illustrated in Figure 4.18. Like the effect of turbulence, line broadening due to

rotation has nothing to do with the temperature of the gas producing the lines.

Other broadening mechanisms do not depend on the Doppler effect at all. For example, if electrons are moving between orbitals while their parent atom is colliding with another atom, the energy of the emitted or absorbed photons changes slightly, thus "blurring" the spectral lines. This mechanism occurs most often in dense gases, where collisions are most frequent. It is usually referred to as *collisional broadening*. The amount of broadening increases as the density of the emitting or absorbing gas rises.

Yet another cause of spectral-line broadening is *magnetism*. The electrons and nuclei within atoms behave as tiny, spinning magnets. As a result, the basic emission and absorption rules of atomic physics change slightly whenever atoms are immersed in a magnetic field, as is found in many stars to a greater or lesser degree. Generally, the greater the magnetic field, the more pronounced the spectral-line broadening.

Given sufficiently sensitive equipment, there is almost no end to the wealth of data contained in starlight. Table 4-1 lists some basic measurable properties of an incoming beam of radiation, and indicates what sort of information can be obtained from them. It is important to realize, however, that deciphering the extent to which each of the factors just described influences a spectrum can be a very difficult task. Typically, the spectra of many elements are superimposed on one another, and several competing physical effects are occurring simultaneously, each modifying the spectrum in its own way. The challenge facing astronomers is to unravel the extent to which each mechanism contributes to spectral-line profiles and so obtain meaningful information about the source of the lines. In the next chapter, we will discuss some the means by which astronomers obtain the data they need in their quest to understand the cosmos.

TABLE 4–1   *Spectral Information from Starlight*

OBSERVED SPECTRAL CHARACTERISTIC	INFORMATION PROVIDED
Peak frequency or wavelength (continuous spectra only)	Temperature (Wien's law)
Lines present	Composition
Line intensities	Composition, temperature
Line width	Temperature, turbulence, rotation speed, density, magnetic field
Doppler shift	Line-of-sight velocity

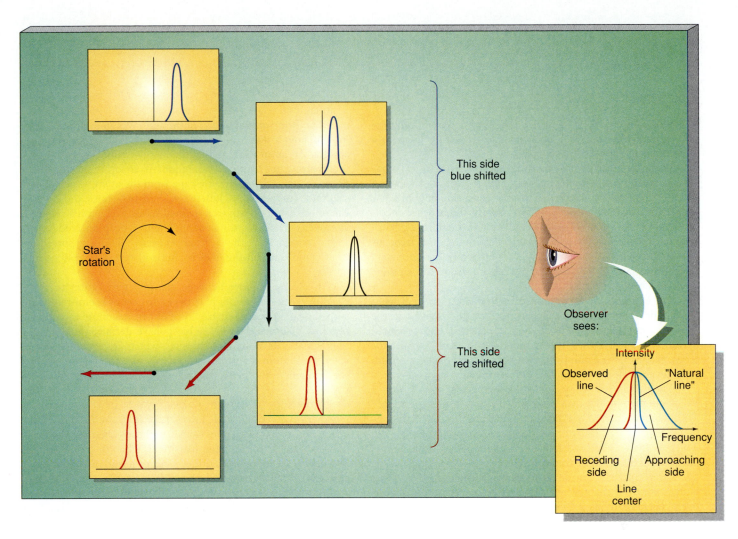

Figure 4.18   The rotation of a star can cause spectral line broadening. Since most stars are unresolved, light rays from all parts of the star merge together to produce wide lines.

Chapter Review

SUMMARY

A **spectroscope** (p. 74) is a device for splitting a beam of radiation into its component frequencies and delivering them to a screen or detector for detailed study.

Many hot objects emit a **continuous spectrum** (p. 75) of radiation, containing light of all wavelengths. A hot gas may instead produce an **emission spectrum** (p. 76), consisting only of a few well-defined **emission lines** (p. 75) of specific frequencies, or colors. Passing a continuous beam of radiation through cool gas will produce **absorption lines** (p. 77) at precisely the same frequencies as would be present in the gas's emission spectrum. **Kirchhoff's laws** (p. 78) describe the relationships between these different types of spectra.

The emission and absorption lines produced by each element are unique—they provide a "fingerprint" of that element in the light it emits or absorbs. The study of the spectral lines produced by different substances is called **spectroscopy** (p. 77). Spectroscopic studies of the **Fraunhofer lines** (p. 77) in the solar spectrum yield detailed information on the Sun's composition.

The *photoelectric effect*, which occurs when light strikes a metal surface and dislodges electrons from it, is strong evidence for the particle nature of radiation. The wave theory simply cannot explain it. Particles of radiation are called **photons** (p. 79). The energy of a photon depends on its color—it is directly proportional to the photon's frequency.

Atoms (p. 81) are made up of negatively charged electrons orbiting a positively charged heavy **nucleus** (p. 81) consisting of positively charged protons and electrically neutral **neutrons** (p. 84). In normal circumstances, the number of orbiting electrons equals the number of protons in the nucleus, and the atom as a whole is electrically neutral. The number of protons in the nucleus determines the type of **element** (p. 84) that the atom represents.

The **Bohr model** (p. 81) of the atom was an early attempt to explain how atoms can produce emission and absorption line spectra. An atom has a minimum-energy **ground state** (p. 81), representing its "normal" condition. If an orbiting electron is given enough energy, it can escape from the atom, which is then said to be **ionized** (p. 81). Between these two states, the electron can only exist in certain well-defined states, each with a very specific energy. The electron's energy is said to be **quantized** (p. 81). In the modern view, the electron is envisaged as being spread out in a "cloud" around the nucleus, but still with a sharply defined energy.

As electrons move between energy levels within an atom, the difference in the energy between the states is emitted or absorbed in the form of "packets" of electromagnetic radiation—photons. Because the energy levels have definite energies, the photons also have definite energies, which are characteristic of the type of atom involved. The energy of a photon determines the frequency, and hence the color, of the light emitted or absorbed.

Molecules (p. 85) are groups of two or more atoms bound together by electromagnetic forces. Like atoms, molecules exist in energy states that obey rules similar to those governing the internal structure of atoms. Again like atoms, when they make transitions between energy states, they emit or absorb a characteristic spectrum of radiation that identifies them uniquely.

Astronomers apply the laws of spectroscopy in analyzing radiation from beyond Earth. Several physical mechanisms can broaden spectral lines. The most important is the Doppler effect, which occurs because stars are hot and their atoms are in motion or because the object being studied is rotating or in turbulent motion.

SELF-TEST: True or False?

_____ **1.** Emission spectra are characterized by narrow, bright lines of different colors.

_____ **2.** Imagine an emission spectrum produced by a container of hydrogen gas. Changing the amount of hydrogen in the container will change the color of the lines in the spectrum.

_____ **3.** In the previous question, changing the gas in the container from hydrogen to helium will change the color of the lines occurring in the spectrum.

_____ **4.** An absorption spectrum appears as a continuous spectrum that is interrupted by a series of dark lines.

_____ **5.** The wavelengths of the emission lines produced by an element are different from the wavelengths of the absorption lines produced by the same element.

_____ **6.** Gustav Kirchhoff is credited with the discovery of black-body radiation.

_____ **7.** The density of the hot gas producing an emission spectrum must be very high.

_____ **8.** The energy of a photon is inversely proportional to the wavelength of the radiation.

_____ **9.** The ground state of an atom is when the electron is in its lowest energy level (orbital).

_____ **10.** An electron can have any energy within an atom, so long as it is above the ground state energy.

_____ **11.** An atom can remain in an excited state indefinitely.

_____ **12.** Emission and absorption lines correspond to the specific energy differences between orbitals in an atom.

_____ **13.** The number of electrons in an atom or ion determines the identity of the element it represents.

_____ **14.** More than one element or molecule can have the same emission or absorption spectrum.

SELF-TEST: Fill in the Blank

1. A _____ is a glass wedge that disperses light into a spectrum.

2. Black-body radiation is an example of a _____ spectrum.

3. Fraunhofer discovered absorption lines in the _____.

4. A continuous spectrum can be produced by a luminous solid, liquid, or _____ gas.

5. An absorption spectrum is produced when a _____ gas lies in front of a source of a continuous source.

6. Light behaves both as a wave and as a _____.

7. The experiment known as the _____ caused Einstein to realize that light does not always behave like a wave.

8. Protons carry a _____ charge; electrons carry a _____ charge.

9. Ionization is a process by which one or more _____ are stripped from the atom.

10. When an electron moves to a higher energy level in an atom it _____ a photon of a specific energy.

11. When an electron moves to a lower energy level in an atom it _____ a photon of a specific energy.

12. The "specific energy" of the photon, referred to in the last two questions, is exactly equal to the energy _____ between the two energy levels the electron moves.

13. Changes in molecular vibration states produce spectral features in the _____ part of the electromagnetic spectrum.

14. Changes in molecular rotational states produce spectral features in the _____ part of the electromagnetic spectrum.

REVIEW AND DISCUSSION

1. What is spectroscopy? Why is it so important to astronomers?

2. Describe the basic components of a simple spectroscope.

3. What is a continuous spectrum? An absorption spectrum?

4. Why are gamma rays generally harmful to life forms, but radio waves generally harmless?

5. In the particle description of light, what is color?

6. Give a brief description of a hydrogen atom.

7. What is the normal condition for atoms? What is an excited atom? What are orbitals?

8. Why do excited atoms absorb and reemit radiation at characteristic frequencies?

9. How are absorption and emission lines produced in a stellar spectrum? What information might absorption lines in the spectrum of a star reveal about a cloud of cool gas lying between us and the star?

10. According to Kirchhoff's Laws, what are the necessary conditions for a continuous spectrum to be produced?

11. Why is the Hα absorption line of hydrogen in the Sun relatively weak, even though the Sun has abundant hydrogen?

12. How do molecules produce spectral lines unrelated to electrons moving between energy levels?

13. How does the intensity of a spectral line yield information about the souce of the line?

14. How does the Doppler effect cause broadening of a spectral line?

15. List three properties of a star that can be determined from observations of its spectrum.

16. Suppose a luminous cloud of gas is discovered emitting an emission spectrum. What can be learned about this cloud from this observation?

PROBLEMS

1. What is the frequency of a 600 nm red photon?

2. How many times more energy has a 1 nm gamma ray than a 10 MHz radio photon?

3. Calculate the wavelength and frequency of the radiation emitted by the electronic transition from the 100th to the 99th excited state of hydrogen. In what part of the electromagnetic spectrum does this lie?

4. How many different photons (that is, photons of different frequencies) can be emitted as a hydrogen atom in the second excited state falls back, directly or indirectly, to the ground state? What about a hydrogen atom in the third excited state?

5. The Hα line of a certain star is received on Earth at a wavelength of 655 nm. What is the star's radial velocity with respect to Earth?

PROJECTS

1. Find a spectrum of the Sun that also has a wavelength scale alongside. Figure 16.8 is a good example; however, you may want to enlarge it on a copying machine. Select various absorption lines and determine their wavelengths by interpolation. Now, try to identify the element that has produced these lines. Use a reference of lines such as found in Moore's *A Multiplet Table of Astrophysical Interest*. Other references may be found in the astronomy, chemistry or physics sections of your library. Work with the darkest lines first before trying the fainter lines.

2. Use a hand-held spectroscope, available through Learning Technologies Inc. While in the shade, point the spectroscope at a white cloud or white piece of paper that is in direct sunlight. Look for the absorption lines in the Sun's spectrum. Note their wavelength from the scale inside the spectroscope. Compare your list to the Fraunhofer lines given in many physics, astronomy, or chemistry reference books.

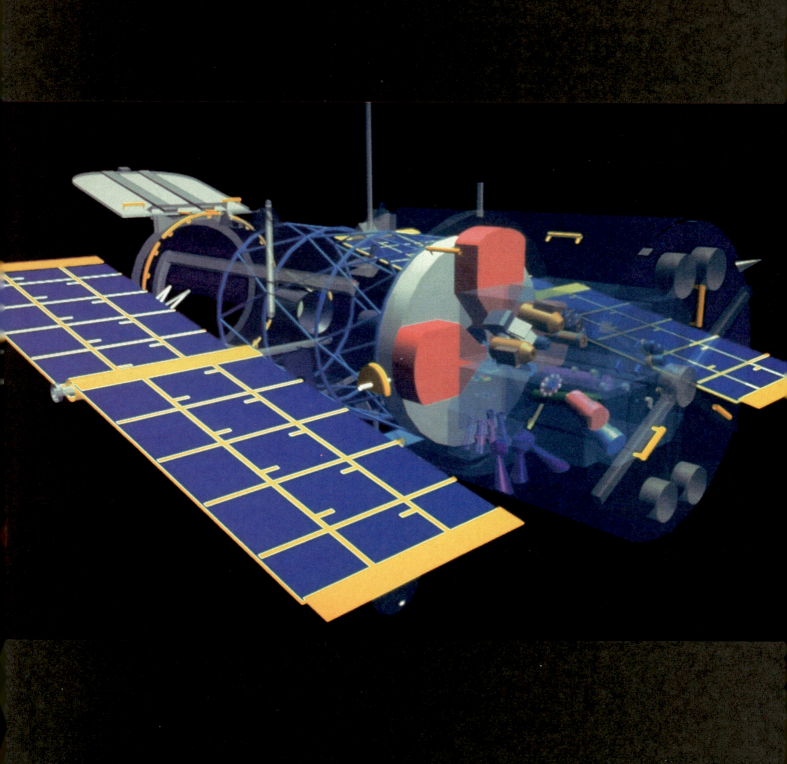

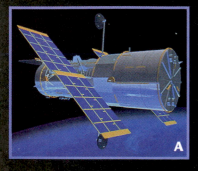

A

B

C

D

5

TELESCOPES

The Tools of Astronomy

LEARNING GOALS

Studying this chapter will enable you to:

1 Sketch and describe the basic designs of the major types of optical telescopes used by astronomers.

2 Explain why very large telescopes are needed for most astronomical study, and specify the particular advantages of reflecting telescopes for astronomical use.

3 Describe how the Earth's atmosphere affects astronomical observations, and discuss some of the current efforts to improve ground-based astronomy.

4 Discuss the specific advantages and disadvantages of radio astronomy.

5 Explain how interferometry can enhance the usefulness of radio observations.

6 List the other types of nonvisible radiation currently being exploited for astronomical observation and summarize the advantages, limitations, and chief uses of each.

7 Say why it is important to make astronomical observations in different regions of the electromagnetic spectrum.

(Opposite page, background) Called the "family portrait" of all the many devices aboard the *Hubble Space Telescope*, this semi-transparent illustration shows some of the main features of this spaceborne observatory. The large blue disk at center of the spacecraft is the primary mirror, and the red gadgets to its rear are the sensors that guide the pointing of the telescope. The open aperture door is at upper left. The huge solar panels are shown in yellow-checkered blue at left and partly obscured at right. Looking inside the aft bay of the vehicle, we can see key components of each of the science instruments—the spectrometers are shown in copper and blue (in the foreground), the cameras in pink, green and lavender (mostly in the background). **(Insets A, B, C, and D)** The four small insets are computer-rendered views of *Hubble* in orbit. Despite having the size of a city bus, *Hubble* is designed to move in space with the grace of a prima ballerina. All these illustrations are taken from video animations made by the astronomy artist and animator Dana Berry.

Observational knowledge of the cosmos normally advances in three phases: First, astronomers collect and measure radiation from space, using a device known as a telescope. Second, they store the resulting data for future use, usually on photographic film or in digital form. Finally, they analyze and interpret the data. The laws of physics are applied, and a model, or theory, that explains the data is developed and tested. Theoretical work plays a vital role in this process and often suggests what new data need to be collected. At its heart, however, astronomy is an observational science. More often than not, observations of cosmic phenomena precede any clear theoretical understanding of their nature. Our detecting instruments—our telescopes—have evolved to observe as broad a range of wavelengths as possible. Until the middle of the twentieth century, telescopes were limited to visible light. Since then, however, technological advances have broadened our view of the universe to all regions of the electromagnetic spectrum. Whatever the details of its design, a telescope is a device whose basic purpose is to collect electromagnetic radiation and deliver it to a detector for detailed study.

5.1 Telescopes

❶ In essence, a **telescope** is a "light bucket," whose primary function is to capture as many photons as possible from a given region of the sky and then to concentrate them into a focused beam for analysis. An *optical* telescope is one designed specifically to collect the wavelengths that are visible to the human eye. Optical telescopes have a long history, reaching back to the days of Galileo in the early seventeenth century. They are probably also the best-known type of telescope, so it is fitting that we begin our study of astronomical hardware with these devices.

Modern astronomical telescopes have evolved a long way from Galileo's simple apparatus. Their development over the years has seen a steady increase in *size* for one simple, but very important, reason: Large telescopes can gather and focus more radiation than can their smaller counterparts, allowing astronomers to study fainter objects and to obtain more detailed information about bright ones. This fact has played a central role in determining the design of contemporary instruments.

TYPES OF TELESCOPE

Optical telescopes fall into two basic categories—*reflectors* and *refractors*. Figure 5.1(a) shows how a **reflecting telescope** uses a carefully shaped, curved *mirror* to gather and concentrate a beam of light. The mirror is constructed so that all light rays arriving parallel to its axis, regardless of their distance from that axis, are reflected to pass through a single point, called the **prime focus**. The distance between the center of the mirror and the prime focus is the *focal length*.

A **refracting telescope** uses a *lens* to focus the incoming light. Refraction is simply the bending of a beam of light as it passes from one transparent medium (for example, air) into another (such as glass). Consider how a stick in water looks bent. The stick itself remains straight, of course, but the light by which we see it is bent—*refracted*—as it leaves the water and enters the air. As illustrated in Figure 5.1(b), we can think of a lens as a

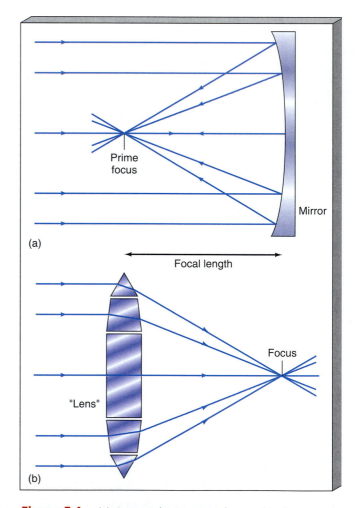

Figure 5.1   (a) A curved mirror can be used to focus rays of light parallel to the axis to a point. Light rays traveling along the axis of the mirror are simply reflected back along the axis. Off-axis rays are reflected through greater and greater angles the farther they are from the axis, so that they all pass through the same focal point. (b) A lens can be thought of as a series of prisms. A light ray traveling along the axis of the lens is undeflected as it passes through the lens. Rays arriving at progressively greater distances from the axis are deflected by increasing amounts, so all rays parallel to the axis are focused to a point.

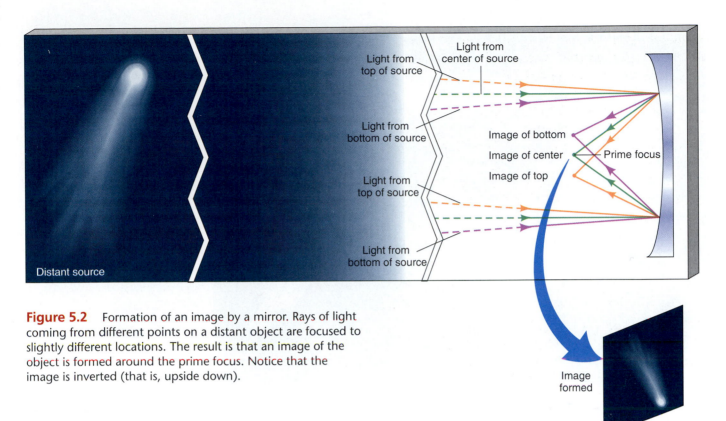

Figure 5.2 Formation of an image by a mirror. Rays of light coming from different points on a distant object are focused to slightly different locations. The result is that an image of the object is formed around the prime focus. Notice that the image is inverted (that is, upside down).

series of prisms, chosen in such a way that all rays of light striking the lens parallel to the axis are refracted to pass through the focus.

Astronomical telescopes are often used to make **images** of their field of view. Figure 5.2 illustrates how this is accomplished, in this case by the mirror in a reflecting telescope. (Lenses also form images; however, since all of the world's large optical telescopes are reflectors, we will concentrate on mirrors here.) Light from a distant object (such as a star) reaches us as parallel, or very nearly parallel, rays. Any ray of light entering the instrument parallel to the telescope's axis strikes the mirror and is reflected through the prime focus. Light coming from a slightly different direction—inclined slightly to the axis—is focused to a slightly different point. In this way, an image is formed near the prime focus. Each point on the image corresponds to a different point in the field of view.

The prime-focus images produced by large telescopes are actually quite small—the image of the entire field of view may be as little as 1 centimeter across. Often, the image is magnified with a lens known as an *eyepiece* before being observed by eye or, more likely, recorded as a photograph or digital image. Figure 5.3(a) shows the basic design of a simple reflecting telescope, illustrating how a small secondary mirror and eyepiece are used to view the image. Figure 5.3(b) shows how a refracting telescope accomplishes the same function. In principle, there is no limit to the magnification that can be achieved just by using more and more powerful eyepieces. However, as we will see in a moment, there are

important practical restrictions on how much detail can be extracted in this way.

These two telescope designs—reflecting and refracting—achieve the same basic result: A beam of light, initially close to the axis of the instrument, is focused to form an image. On the face of it, then, it might appear that there is little reason to prefer either mirrors or lenses in telescope construction. However, there are some important factors to consider when deciding which type to buy or build:

- The fact that light must pass through the lens is a major disadvantage of refracting telescopes. Large lenses cannot be constructed in such a way that light passes through them uniformly. Just as a prism disperses white light into its component colors, the lens in a refracting telescope focuses red and blue light differently. This deficiency is known as **chromatic aberration**. Figure 5.4 shows how chromatic aberration occurs and indicates how it affects the image of a star. Careful design and choice of materials can largely correct chromatic aberration, but it is very difficult to eliminate entirely. And as the lens diameter increases, so do the aberration and the number of problems that need to be solved to correct it. Mirrors do not suffer from this defect.

- As light passes through a lens, some light is absorbed by the glass. This absorption is a particular problem for nonoptical radiation because glass is opaque to (that is, it blocks completely) much of the infrared and ultraviolet regions of the spectrum. The problem of opacity obviously does not affect mirrors.

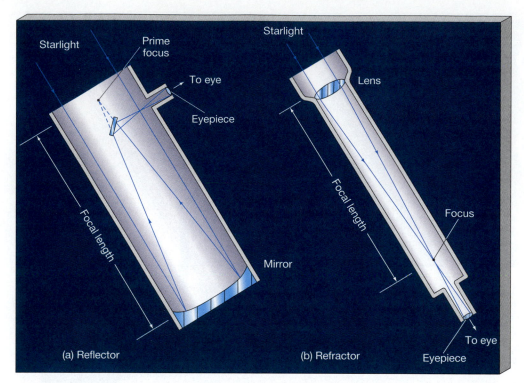

Figure 5.3 Comparison of (a) reflecting and (b) refracting telescope systems. Both types are used to gather and focus cosmic radiation—to be observed by human eyes or recorded on photographs or in computers. In either case, the image formed at the focus is viewed with a small magnifying lens called an eyepiece.

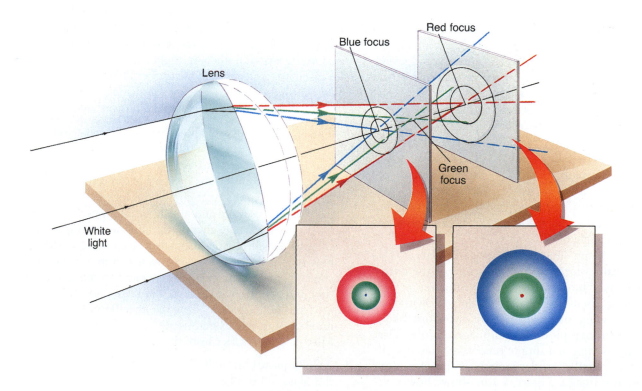

Figure 5.4 Chromatic aberration. A prism bends blue light more than it bends red light, so the blue component of light passing through a lens is focused slightly closer to the lens than is the red component. As a result, the image of an object acquires a colored "halo," no matter where we place our detector.

- A large lens can be quite heavy. Because it can be supported only around its edge (so as not to block the incoming radiation), the lens tends to deform under its own weight. A mirror, conversely, can be supported over its entire back surface.

- A lens has two surfaces that must be accurately machined and polished—which can be very difficult—but a mirror has only one.

For these reasons, *all* large modern telescopes use mirrors as their primary light gatherers. Figure 5.5 shows the world's largest refractor, installed in 1897 at the Yerkes Observatory in Wisconsin and still in use today. It has a lens diameter of 1 m (about 40 inches). By contrast, some new reflecting telescopes have mirror diameters in the 10 m range, and larger instruments are on the way.

TELESCOPE DESIGN

Figure 5.6 presents a diagram of some basic reflecting telescope designs. Radiation from a star enters the instrument, passes down through the main telescope tube, strikes the curved surface of the *primary mirror*, and is reflected back toward the prime focus, near the top of the tube. Sometimes astronomers simply place their instruments right at the prime focus, as in Figure 5.6(a). However, it can be very inconvenient, or even impossible, to suspend bulky pieces of equipment there. More often, the light is intercepted on its path to the focus by a smaller *secondary mirror* and then redirected to a more convenient location. Three such arrangements are shown in Figure 5.6(b) through (d).

Figure 5.5 Photograph of the Yerkes Observatory's 1 m diameter refracting telescope.

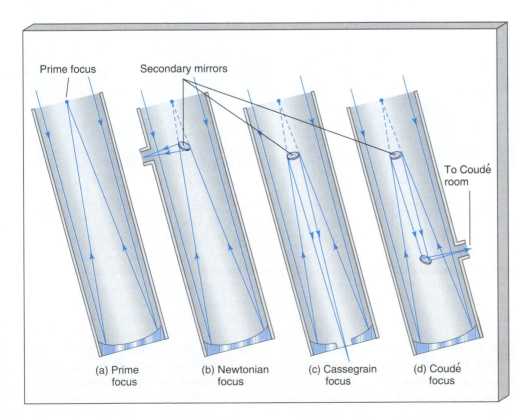

Prime focus

Secondary mirrors

To Coudé room

(a) Prime focus

(b) Newtonian focus

(c) Cassegrain focus

(d) Coudé focus

Figure 5.6 The essential features of the apparatus used to collect, focus, and record information from cosmic objects. Shown here are four different reflecting telescope designs: (a) prime focus, (b) Newtonian focus, (c) Cassegrain focus, and (d) coudé focus. Each uses a primary mirror at the bottom of the telescope to capture radiation, which is then directed along different paths for analysis. Notice that the secondary mirrors shown in (c) and (d) are actually slightly diverging, moving the focus outside the telescope.

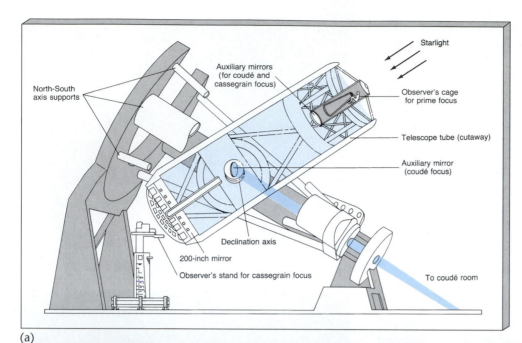

Starlight

Auxiliary mirrors
(for coudé and
cassegrain focus)

North-South
axis supports

Observer's cage
for prime focus

Telescope tube (cutaway)

Auxiliary mirror
(coudé focus)

Declination axis

200-inch mirror

Observer's stand for cassegrain focus

To coudé room

(a)

Figure 5.7 (a) An artist's illustration of the 5 m diameter Hale optical telescope on Palomar Mountain in California. (b) A photograph of the telescope. (c) Astronomer Edwin Hubble in the observer's cage at the Hale prime focus.

(b)

(c)

In a **Newtonian telescope** (named after Sir Isaac Newton, who invented this particular design), the light is intercepted before it reaches the prime focus and is deflected by 90°, usually to an eyepiece at the side of the instrument. This is a particularly popular design for smaller reflecting telescopes, such as those used by amateur astronomers.

Alternatively, astronomers may choose to work on a rear platform where they can use equipment, such as a spectroscope, that is too heavy to hoist to the prime focus. In this case, light reflected by the primary mirror toward the prime focus is intercepted by a smaller secondary mirror, which reflects it back down through a small hole at the center of the primary mirror. This arrangement is known as a **Cassegrain telescope** (after Guillaume Cassegrain, a French lensmaker), and the point behind the primary mirror where the light from the star finally converges is called the Cassegrain focus.

Another, more complex, observational configuration requires starlight to be reflected by several mirrors. As in the Cassegrain design, light is first reflected by the primary mirror toward the prime focus and reflected back down the tube by a secondary mirror. A third, much smaller mirror then deflects the light into an environmentally controlled laboratory. Known as the *coudé* room (from the French word for "bent"), this laboratory is separate from the telescope itself, enabling astronomers to use very heavy and finely tuned equipment that could not possibly be lifted to either the prime focus or to an elevated platform at the rear of the telescope. The light path to the coudé focus lies along the axis of the telescope's mount—that is, the axis around which the telescope rotates as it tracks objects across the sky—so that it does not change as the telescope moves.

To illustrate some of the points we have discussed, let us briefly consider an instrument that has been at or near the forefront of astronomical research for much of the last half-century. Figure 5.7 depicts the Hale 5 m (200 inch) diameter optical telescope on California's Palomar Mountain, dedicated in 1948. As the size of the figure in the observer's cage at the prime focus indicates, this is indeed a very large telescope. In fact, for almost three decades the Hale telescope was the largest in the world. Observations can be made at the prime, the Cassegrain, or the coudé focus, depending on the needs of the user. The coudé room itself is out of the picture, to the lower right.

IMAGES AND DETECTORS

Large reflectors are good at forming images of narrow fields of view, where all the light that strikes the mirror surface moves almost parallel to the axis of the instrument. However, if the light enters at an appreciable angle, it cannot be accurately focused, degrading the overall quality of the image. The effect (called *coma*) worsens as we move farther from the center of the field of view. Eventually, the image quality is reduced to the point where it is no longer usable. The distance from the center to where the image becomes unacceptable defines the useful field of view of the telescope—typically, only a few arc minutes for large instruments.

A design that overcomes this problem is the *Schmidt telescope*, named after its inventor, Bernhard Schmidt, who built the first such instrument in the 1930s. The telescope uses a correcting lens, which sharpens the final image of the entire field of view. Consequently, a Schmidt telescope is well suited to producing wide-angle photographs, covering several degrees of the sky. Because the design of the Schmidt telescope results in a curved image that is not suitable for viewing with an eyepiece, the image is recorded on a specially shaped piece of photographic film. For this reason, the instrument is often called a *Schmidt camera* (Figure 5.8). The Palomar Observatory Schmidt camera, one of the largest in the world (with a 1.8 m mirror and a 1.2 m lens), performed a survey of the entire northern sky in the 1950s. The Palomar Observatory Sky Survey, as it is known, has for decades been an invaluable research tool for professional observers. A second, digital survey is now nearing completion.

When a photographic plate is placed at the focus to record an image of the field of view, the telescope is acting in effect as a high-powered camera. However, this is by no means the only light-sensitive device that can be placed at the focus to analyze the radiation received from space. When very accurate and rapid measurements of light intensity are required, a device known as a **photometer** is

Figure 5.8 The Schmidt camera of the European Southern Observatory in Chile.

used. A photometer measures the total amount of light received in all or part of the image. When only part of the image is under study, the region of interest is selected simply by masking out the rest of the field of view. Using a photometer often means "throwing away" spatial detail, but in return more information is obtained about the intensity and time variability of a particular source, such as a pulsating star or a supernova explosion.

Often, astronomers want to study the *spectrum* of the incoming light. Large **spectrometers** frequently work in tandem with optical telescopes. Light radiation collected by the primary mirror may be redirected to the underground coudé room, defined by a narrow slit, passed through a prism, and projected onto a screen—a process not so different from the operation of the simple spectroscope described in Chapter 4. ∞ (p. 74) The spectrum can be studied in real time (that is, as it happens) or stored on a photographic plate (or, more commonly nowadays, on a computer disk) for later analysis.

5.2 *Telescope Size*

② As we have already noted, astronomers generally prefer large telescopes over small ones. This preference has to do both with the amount of light the telescopes collect and with the amount of detail that can be seen.

LIGHT-GATHERING POWER

For optical work, the main reason for using a larger telescope is simply that it has a greater **collecting area**—the total area of a telescope capable of capturing radiation. The larger the telescope's reflecting mirror or refracting lens, the more light it collects, and the easier it is to measure and study an object's radiative properties. Astronomers spend a large fraction of their time observing very distant—and hence, by the inverse-square law, very *faint*—cosmic sources. ∞ (p. 67) If we wish to make detailed observations of ob-

Figure 5.9 Effect of increasing telescope size on an image of the Andromeda Galaxy. Both photographs had the same exposure time; the bottom image was taken with a telescope twice the size of that used to make the top image. Fainter detail can be seen as the diameter of the telescope mirror increases because larger telescopes are able to collect more photons per unit time.

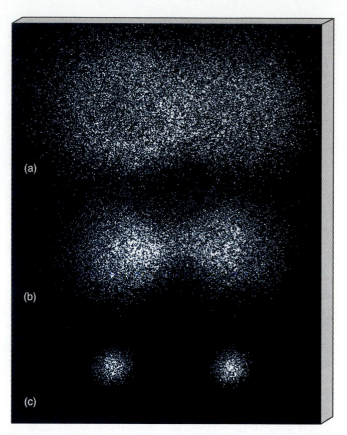

Figure 5.10 Two comparably bright light sources become progressively clearer when viewed at finer and finer angular resolution. When the angular resolution is much poorer than the separation of the objects, as at the top, the objects appear as a single fuzzy "blob." As the resolution improves, the two sources become discernible as separate objects.

jects far from our own cosmic neighborhood, very large telescopes are essential tools. Figure 5.9 illustrates the effect of increasing telescope size by comparing images of the Andromeda Galaxy taken with different instruments.

The observed brightness of an astronomical object is directly proportional to the area of our telescope's mirror, and therefore to the *square* of the mirror diameter. Thus, a 5 m telescope will produce an image 25 times as bright as a 1 m instrument because a 5 m mirror has $5^2 = 25$ times the collecting area of a 1 m mirror. We can also think of this relationship in terms of the length of *time* required for a telescope

to collect enough energy to create a recognizable image on a photographic plate. Our 5 m telescope will produce an image 25 times faster than the 1 m device because it gathers energy at a rate 25 times greater. Expressed in another way, a 1-hour time exposure with a 1-m telescope is roughly equivalent to a 2.4-minute time exposure with a 5 m instrument.

RESOLVING POWER

A second advantage of large telescopes is their resolution. **Angular resolution** refers to the ability to distinguish between two adjacent objects in the sky. The finer the angular resolution, the better we can make such a distinction, and the better we can see the details of any given object. Figure 5.10 illustrates how the appearance of two nearby objects might change as the angular resolution varies.

What limits a telescope's resolution? One important factor is *diffraction*, the tendency of light, and all other waves for that matter, to bend around corners. ∞ (p. 56) As light enters the telescope, the rays are bent slightly, and this bending makes it impossible to focus the light to a sharp point, even with a perfectly constructed mirror. The angle through which the beam bends is pro-

(a)

(b)

(c)

(d)

Figure 5.11 Detail becomes clearer in the Andromeda Galaxy as the angular resolution is improved some 600 times, from (a) 10′, to (b) 1′, (c) 5″, and (d) 1″.

R I V U X G

portional to the wavelength of the radiation divided by the width of the opening (in this case, the diameter of the mirror). In convenient units,

$$\text{angular resolution (arc sec)} = 0.25 \, \frac{\text{wavelength (μm)}}{\text{diameter (m)}},$$

where 1 μm (micron) = 10^{-6} m. ∞ (p. 13) This angle determines the angular resolution of the telescope.

Thus, for a given telescope size, the amount of diffraction increases in proportion to the wavelength used, so that observations in the infrared or radio range are often limited by its effects. For example, in an otherwise perfect observing environment, the best possible angular resolution of blue light (with a wavelength of 400 nm) that could be obtained using a 1-m telescope observing would be about 0.1″. This quantity is known as the *diffraction-limited* resolution of the telescope. But if we were to use our 1-m telescope to make observations in the near infrared range, at a wavelength of 10 microns, the best resolution we could obtain would be only 2.5″. A 1 m radio telescope operating at a wavelength of 1 cm would have an angular resolution of slightly under 1°.

Similarly, for light of any given wavelength, large telescopes produce less diffraction than small ones. A 5 m telescope observing in blue light would have a diffraction-limited resolution five times finer than the 1-m telescope just discussed—about 0.02″. A 0.1 m (10 cm) telescope would have a diffraction limit of 1″ and so on. For comparison, the angular resolution of the human eye is about 0.5′. Figure 5.11 shows how the Andromeda Galaxy would appear in greater detail with progressively higher resolution, when viewed in visible light through a hypothetical series of telescopes. In fact, as we will see in a moment, no large ground-based telescope actually comes close to its diffraction limit, because of the blurring effects of Earth's atmosphere.

BUILDING VERY LARGE TELESCOPES

We have seen that in the push toward larger and larger telescopes, refractors were abandoned long ago, in large part because of the difficulty in constructing them. But, while large refracting telescopes are certainly hard to build, large reflectors are not exactly easy. Conventional telescope mirrors are made from large blocks of quartz, glass, or some other type of polishable material capable of withstanding

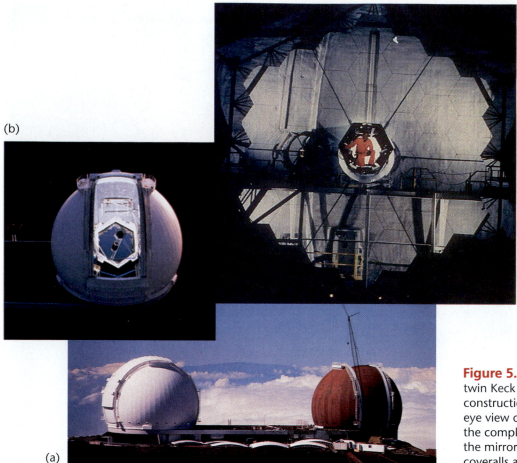

(b)

(c)

(a)

Figure 5.12   (a) Photograph of the twin Keck telescopes, one still under construction as of mid-1995. (b) A bird's-eye view of the segmented mirror inside the completed dome. (c) A closer view of the mirror. Note the technician in orange coveralls at the center.

large temperature changes with little expansion or contraction. Workers begin the construction process by pouring the molten material into a large cast, then cooling it slowly over the course of several years to keep it from cracking or developing internal stresses while it changes from liquid to solid. It takes years more to grind and polish the surface to the required curvature. Finally, the surface is coated with a thin film of aluminum to provide a reflecting surface.

All these stages are slow, painstaking, and, unfortunately, not always successful. Engineers encounter severe difficulties in building very large telescopes by these means. Indeed, since the construction of the 5 m Palomar instrument in 1948, only one larger single-mirror telescope (a 6 m instrument in Russia) has been completed; this telescope, regrettably, suffers from several optical defects and does not focus light well, resulting in poor-quality images.

Until the 1980s, the conventional wisdom was that telescopes with mirrors larger than 5 or 6 m in diameter were simply too expensive and impractical to build. However, new manufacturing techniques, coupled with radically new mirror designs, now make the construction of telescopes in the 8- to 12-m range almost a routine matter. Experts can now make large mirrors much lighter for their size than had previously been believed feasible, and can combine many smaller mirrors into the equivalent of a much larger single-mirror telescope. Several large-diameter instruments are now under construction, and many more are planned.

The California Institute of Technology and the University of California are cooperating to build two 10 m telescopes atop Mauna Kea, in Hawaii (see Figure 5.26 later in this chapter). These telescopes, known as the Keck telescopes, and shown in Figure 5.12, employ a segmented design, each combining 36 1.8 m six-sided mirrors into the equivalent area of a single 10 m reflector. The first Keck telescope saw "first light" in 1990. It became fully operational in 1992.

5.3   *High-Resolution Astronomy*

3   In recent years, new technologies have made it possible to improve both light collection and image formation. Telescopes are being moved out beyond much of the atmosphere, and computers are playing an increasingly important role in astronomy.

ATMOSPHERIC BLURRING

Even large telescopes have their limitations. For example, consider again the 5 m optical telescope on Palomar Mountain. According to our earlier discussion of diffraction, this telescope should achieve an angular resolution of around 0.02″. In practice, it cannot exceed 1″. In fact, apart from instruments using special techniques developed to examine some bright stars, *no* ground-based optical telescope

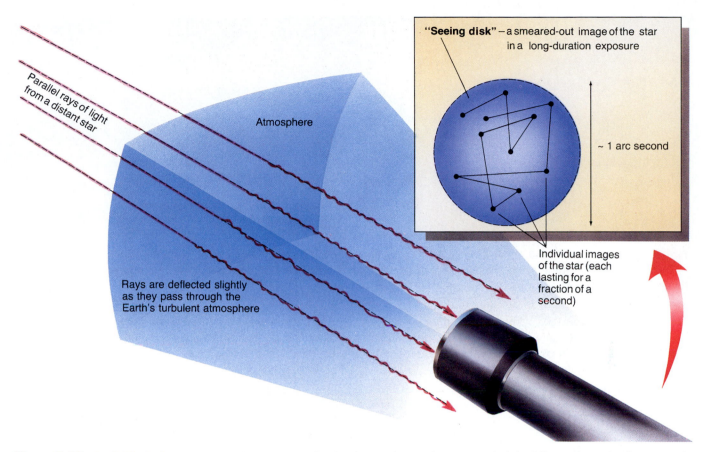

Figure 5.13 Individual photons from a distant star strike the detector in a telescope at slightly different locations because of turbulence in the Earth's atmosphere. Over time, the individual photons cover a roughly circular region on the detector, and even the pointlike image of a star is recorded as a small disk, called the seeing disk.

built before 1990 can presently resolve astronomical objects to much better than an arc second. Why?

As we observe a star, atmospheric turbulence produces continual small changes in the optical properties of the air between the star and our telescope. The light from the star is refracted slightly, and the stellar image dances around on the detector (or on our retina). This continual deflection is the cause of the well-known "twinkling" of stars. It occurs for the same reason that objects appear to shimmer when viewed across a hot roadway on a summer day.

On a good night at the best observing sites, the maximum amount of deflection produced by the atmosphere is slightly less than 1″. Consider taking a photograph of a star. After a few minutes' exposure time (long enough for the intervening atmosphere to have undergone many small, random changes), the image of the star has been smeared out over a roughly circular region an arc second or so in diameter. Astronomers use the term **seeing** to describe the effects of atmospheric turbulence. The circle over which a star's light (or the light from any other astronomical source) is spread is called the **seeing disk**. Figure 5.13 illustrates the formation of the seeing disk for a small telescope.*

In fact, for a large instrument—more than about 1 m in diameter—the situation is more complicated, because rays striking different parts of the mirror have actually passed through different turbulent atmospheric regions. The end result is still a seeing disk, however.

To achieve the best possible seeing, telescopes are sited on mountaintops (to get above as much of the atmosphere as possible) in regions of the world where the atmosphere is known to be fairly stable and relatively free of dust, moisture, and light pollution from cities. In the continental United States, these sites tend to be in the desert Southwest. The U.S. National Observatory for optical astronomy in the Northern Hemisphere, completed in 1973, is located high on Kitt Peak near Tucson, Arizona. The site was chosen because of its many dry, clear nights. Seeing of 1″ from such a location is regarded as good, and seeing of a few arc seconds is tolerable for many purposes. Even better conditions are found on Mauna Kea, Hawaii, and at Cerro Tololo and La Silla in the Andes Mountains of Chile (Figure 5.14)—which is why many large telescopes have recently been located at those two exceptionally clear locations.

An optical telescope placed in orbit about the Earth or on the Moon could obviously overcome the limitations imposed by the atmosphere on ground-based instruments. Without atmospheric blurring, extremely fine resolution—close to the diffraction limit—can be achieved, subject only to the engineering restrictions of building or placing large structures in space. The *Hubble Space Telescope (HST)* (named for one of America's most notable astronomers, Edwin Hubble) was launched into the Earth's orbit by NASA's space shuttle *Discovery* in 1990. This telescope has a 2.4 m mirror, with a

Figure 5.14   Located in the Andes Mountains of Chile, the European Southern Observatory at La Silla is run by a consortium of European nations. Numerous domes house optical telescopes of different sizes, each with varied support equipment, making this one of the most versatile observatories south of the equator.

diffraction limit of only 0.05″. Thus, this orbiting observatory can give us a view of the universe as much as 20 times sharper than is normally available from even the much larger ground-based instruments. (See *Interlude 5-1* on p. 110.)

IMAGE PROCESSING

Most large telescopes today are controlled by computers or by operators who rely heavily on their assistance, and the images themselves are recorded in a form that can be easily read and manipulated by computer programs. It is becoming fairly rare for photographic equipment to be used as the primary means of data acquisition at large observatories. Rather, electronic detectors known as **charge-coupled devices**, or **CCDs**, are now in widespread use. Their output goes directly to a computer.

A CCD (see Figure 5.15) consists of a wafer of silicon divided into a two-dimensional array of many tiny picture elements, known as **pixels**. When light strikes a pixel, an electric charge builds up on the device. The amount of charge is directly proportional to the number of photons striking each pixel—in other words, to the intensity of the light at that point. The charge buildup is monitored electronically, so that a two-dimensional image can be obtained.

The entire device is typically only a few square centimeters in area and may contain several million pixels, generally arranged on a square grid. As the technology improves, both the areas of CCDs and the number of pixels they contain are steadily increasing. Incidentally, the technology is not limited to astronomy—many home video cameras contain CCD chips quite similar in basic design to those in use at the great astronomical observatories of the world.

Charge-coupled devices have many advantages over photographic plates, which were the staple of astronomers for over a century. CCDs are much more efficient than photographs. They record as many as 75 percent of the photons striking them, while photographic methods record less than 5 percent. This fact alone means that a CCD image can show objects ten to twenty times fainter as can a photograph taken using the same telescope and the same exposure time. Alternatively, CCDs can record the same level of detail in less than a tenth of the time required by photographs, or record that detail with a much smaller telescope. CCDs also produce a faithful representation of an image in a digital format that can be stored on magnetic tape, or disk, or even sent directly across a computer network to an observer's home institution for detailed study.

(a)

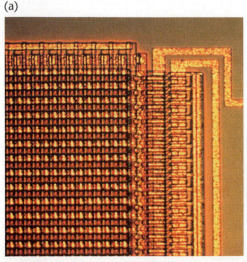

(b)

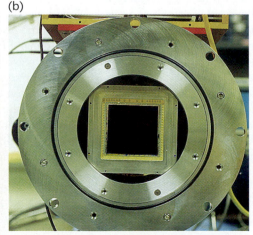

Figure 5.15   A charge-coupled device consists of hundreds of thousands, or even millions, of tiny light-sensitive cells, or pixels, usually arranged in a square array. Light striking a pixel causes an electrical charge to build up on it. By electronically reading out the charge on each pixel, a computer can reconstruct the pattern of light—the image—falling on the chip. Photograph (a) is a detail of a CCD array; (b) shows a CCD chip mounted for use at the focus of a telescope.

(a)   (b)   (c)   (d)

Figure 5.16   (a) A ground-based view of the star cluster R136, a group of stars in the Large Magellenic Cloud (a nearby galaxy). (b) The "raw" image of this same region as seen by the *Hubble Space Telescope* in 1990, before the repair mission. (c) The same image after computer processing that partly compensated for imperfections in the mirror. (d) The same region as seen by the repaired *HST* in 1994.

With the aid of high-speed computers, the background noise found in the "raw" image from a telescope can be greatly reduced, allowing astronomers to see features that would otherwise remain hidden. Noise is anything that corrupts the integrity of a message, such as static on an AM radio or "snow" on a television screen. The noise corrupting telescopic images has many causes. In part, it results from faint, unresolved sources in the telescope's field of view and from light scattered into the line of sight by the Earth's atmosphere. It can also be caused by electronic "hiss" within the detector itself. Whatever the origin of noise, its characteristics can be determined (for example, by observing a part of the sky where there are no known sources of radiation), and then the noise can be partially removed.

Using computer processing, astronomers can also compensate for known instrumental defects and even correct some effects of bad seeing. In addition, the computer can often carry out many of the relatively simple, but tedious and time-consuming, chores that must be performed before an image (or spectrum) reaches its final "clean" form. Figures 5.16(b) and (c) illustrate how computerized image-processing techniques were used to correct for known instrumental problems in the *Hubble Space Telescope*, allowing much of the planned resolution of the telescope to be recovered even before its repair in 1993. For comparison, Figure 5.16(a) shows the best image obtainable from the ground.

NEW TELESCOPE DESIGN

An exciting development, promising to bring about striking improvements in the resolution of ground-based optical telescopes, takes these ideas of computer control and image processing one stage farther. If an image could be analyzed while the light was still being collected (a process that can take many minutes, or even hours, in some cases), it might be possible to adjust the telescope from moment to moment to correct for the effects of mirror distortion, temperature changes, and bad seeing. Perhaps the telescope could come close to the theoretical (diffraction-limited) resolution. Some of these techniques, collectively known as **active optics**, are already in use in the New Technology Telescope (NTT), located at the European Southern Observatory in Chile (the most prominent instrument in Figure 5.14). This 3.5 m instrument, employing the latest in real-time telescope controls, achieves res-

olution of about 0.5″ by making minute modifications to the tilt of the mirror as its temperature and orientation change, thus maintaining the best possible focus at all times. From its very first observing run, NTT became the highest-resolution optical telescope on Earth (see Figure 5.17a). The Keck 10-m instruments, one of whose mirrors is shown in Figure 5.17b, also employ these methods, and may ultimately achieve resolution as fine as 0.25″.

Figure 5.17   (a) These false-color infrared photographs of part of the star cluster R136—the same object shown in Figure 5.16 above—contrast the resolution obtained without the active optics system (left image) with that achievable when the active optics system is in use (right image). (b) A hexagonal mirror segment destined for one of the Keck telescopes undergoes shaping and polishing. The unusually thin glass will be backed by push-pull pistons that can adjust the precise configuration of the segment during observations so as to attain improved resolution.

(a)

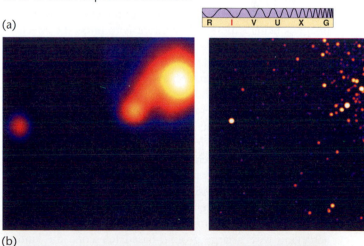

(b)

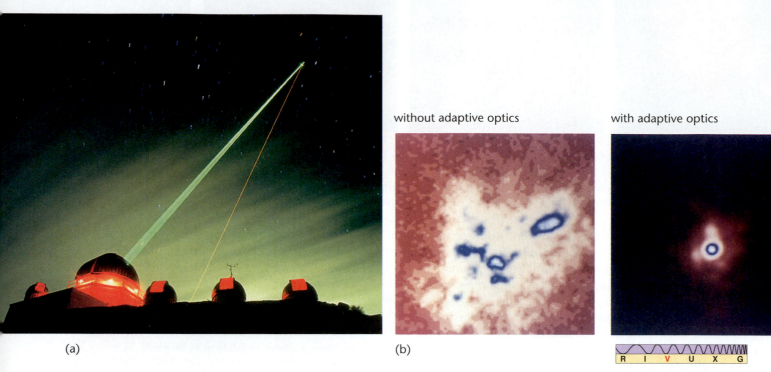

without adaptive optics   with adaptive optics

Figure 5.18   (a) Until mid-1991, the Starfire Optical Range at Kirtland Air Force Base in New Mexico was one of the U.S. Air Force's most closely guarded secrets. Here, beams of laser light probe the atmosphere above the system's 1.5 m telescope, allowing minute computer-controlled changes to be made to the mirror surface thousands of times each second. (b) The improvement in seeing produced by such systems can be dramatic, as can be seen in these images acquired at another military observatory atop Mt. Haleakala in Maui, Hawaii, employing similar technology. The uncompensated image (left) of the bright star Procyon is a blur spread over several arc seconds. With adaptive compensation applied (right), the resolution is improved to a mere 0.2 arc second.

An even more ambitious undertaking is known as **adaptive optics**. This technique actually deforms the shape of the mirror's surface, under computer control, while the image is being exposed. The intent is to undo the effects of atmospheric turbulence. In the experimental system shown in Figure 5.18(a), lasers probe the atmosphere above the telescope, returning information about the air's swirling motion to a computer that modifies the mirror thousands of times per second to compensate for poor seeing. Adaptive optics presents formidable theoretical and technological problems, but the rewards are so great that they are presently the subject of intense research. Recently declassified SDI ("Star Wars") technology has provided an enormous boost to this effort. Already, impressive improvements in image quality have been obtained (see Figure 5.18b). In the next decade, it may well be possible to have the "best of both worlds" by achieving with a large ground-based telescope the kind of resolution presently attainable only from space.

5.4 Radio Astronomy

④ Optical telescopes are limited to those wavelengths that the human eye can detect. However, information also comes to us at other wavelengths. In addition to the visible radiation that normally penetrates the Earth's atmosphere on a clear day, radio radiation also reaches the ground. In fact, the radio window in the electromagnetic spectrum is much wider than the optical window, as indicated in Figure 3-9. ∞ (p. 60) As the atmosphere is no hindrance to long-wavelength radiation, radio astronomers have built many ground-based **radio telescopes** capable of detecting cosmic radio waves. These devices have all been constructed since the 1950s—radio astronomy is a much younger subject than optical astronomy.

The field originated with the work of Karl Jansky at Bell Labs in 1931, but only after the technological push of World War II did it grow into a distinct branch of astronomy. Jansky was engaged in a study of shortwave radio interference when he discovered a faint static "hiss" that had no apparent terrestrial source. He noticed that the strength of the hiss varied in time and that its peak occurred about 4 minutes earlier each day. He soon realized that the peaks were coming exactly one *sidereal day* apart, and correctly inferred that the hiss was not of terrestrial origin but came from a definite direction in space. That direction is now known to correspond to the center of our Galaxy. It took over a decade, and the realization by astronomers that interstellar gas could actually be observed at radio wavelengths, for the full importance of his work to be appreciated, but today Jansky is widely regarded as the father of radio astronomy.

Figure 5.19   The 43-m-diameter radio telescope at the National Radio Astronomy Observatory in Green Bank, West Virginia.

ESSENTIALS OF RADIO TELESCOPES

Figure 5.19 shows a fairly typical radio telescope, the large 43 m (140 foot) diameter telescope located at the National Radio Astronomy Observatory in West Virginia.

Although much larger than any reflecting optical telescope, most radio telescopes are built in basically the same way. They have a large horseshoe-shaped mount that supports the large, curved metal "dish," or mirror. The collecting area captures cosmic radio waves and reflects them to the focus, where a receiver detects the signals and channels them to the computer. The signals may be partially analyzed before being stored digitally for later processing and full analysis. Unlike optical instruments, which can simultaneously detect all visible wavelengths, radio detectors normally register only a narrow band of wavelengths at any one time. To observe radiation at another radio frequency, we must retune the equipment, much as we tune a television set to a different channel.

Large radio telescopes are very sensitive instruments and can detect even very faint radio sources. Indeed, the *total* amount of radio energy detected by *all* the radio telescopes on Earth since the first receiver was built would barely be enough to keep a 100-W light bulb burning for just 10 billionths of a second. However, their angular resolution is generally poor compared to their optical counterparts. The major disadvantage of *all* radio telescopes is their relatively low resolving power, despite the enormous size of many radio dishes. It is not our atmosphere that is to blame—the radio wavelengths normally studied pass through air without any significant distortion. The problem is that the typical wavelengths of radio waves are about a million times longer than those of visible light, and these longer wavelengths impose a corresponding crudeness in angular resolution because of the effects of diffraction. Recall from Section 5.2 that the longer the wavelength, the greater the diffraction. ∞ (p. 102)

The best angular resolution obtainable with a single radio telescope is about 10″ (for the largest instruments operating at millimeter wavelengths)—at least 10 times coarser than the capabilities of the largest optical mirrors. The resolution varies widely, depending on the wavelength being observed. The 43 m radio telescope shown in Figure 5.19 can achieve resolution of about 1′ when receiving radio waves having wavelengths of around 1 cm. However, it was designed to operate most efficiently (that is, it is most sensitive to radio signals) at wavelengths closer to 5 cm, where the resolution is only about 6′, or 0.1°.

Radio telescopes are large not only because that is the only way they can achieve good resolution, but also because the total amount of energy arriving at the Earth in the form of radio radiation is extremely small—less than a trillionth of a watt spread over the entire surface of our planet. Radio telescopes can be built so much larger than their optical counterparts because their reflecting surface need not be as smooth as is needed for shorter-wavelength light waves. Provided that surface irregularities (dents, bumps, and the like) are much smaller than the wavelength of the waves to be detected, the surface will reflect them without distortion. Because the wavelength of visible radiation is short (approximately 10^{-6} m), very smooth mirrors are needed to reflect the waves properly, and it is difficult to construct very large mirrors to such exacting tolerances.

For example, when visible light shines on a curved surface having irregularities of even a fraction of a millimeter, the light scatters and does not focus well; the image of the light source is severely distorted. You could test this statement by looking at your own blurred reflection in a piece of unpolished metal. But radio waves of a centimeter or longer wavelength are not scattered at all by slightly rough surfaces. Instead, they are reflected to an accurate focus. Very long radio waves of, say, 1 m wavelength can reflect perfectly well from surfaces having irregularities even as large as your fist. The situation is somewhat analogous to trying to bounce a ball off an irregular surface. If the surface irregularities are much smaller than the radius of the ball (the wavelength of the photon, in this analogy), the bounce will be true, and the ball will travel in the intended direction (that is, the photon will be reflected toward a focal point). However, if the irregularities are comparable in size to the radius of the ball, the bounce will be unpredictable and erratic (that is, the focus will be blurred).

Interlude 5–1   *The Hubble Space Telescope*

The *Hubble Space Telescope (HST)* is the largest, most complex, and most sensitive observatory ever deployed in space. At nearly $2.5 billion, it is also the most expensive scientific instrument ever constructed. Built jointly by NASA and the European Space Agency, *HST* is designed to allow astronomers to probe the universe with at least 10 times finer resolution and with some 30 times greater sensitivity to light than existing Earth-based devices. It is operated remotely from the ground. There are no astronauts aboard the telescope, which orbits the Earth about once every 95 minutes, at an altitude of about 600 kilometers (380 miles).

The telescope's overall dimensions approximate those of a city bus or railroad tank car—13 m (43 feet) long, 12 m (39 feet) across with solar arrays extended, and 11,000 kg (12.5 tons when weighed on the ground). The heart of *HST* is a 2.4 m (94.5 inch) diameter mirror designed to capture optical, ultraviolet, and infrared radiation before it reaches the Earth's murky atmosphere. The accompanying figure shows the telescope being lifted out of the cargo bay of the space shuttle *Discovery* in the spring of 1990.

The optical system and scientific instruments aboard *HST* are compact and pioneering. The telescope reflects light from its large mirror back to a smaller, 0.3 m (12 inch) secondary mirror, which in turn sends the light through a hole in the doughnut-shaped main mirror and into the aft bay of the spacecraft. There, any of five major scientific instruments wait to analyze the incoming radiation. Most of these instruments are about the size of a telephone booth. They include two cameras to image (or electronically photograph) various regions of the sky, two spectrographs to split the radiation into its component wavelengths, and a group of fine guidance sensors to measure the positions of stars in the sky.

Although not the largest ever built, *HST*'s mirror is assuredly the most finely polished mirror of its size. If *HST*'s mirror were scaled up to equal the width of the continental United States, the highest hill or lowest valley would be less than 2 inches from the average surface. By contrast, skyscraper-sized imperfections would result if ordinary eyeglass lenses were scaled up to reach from coast to coast. Unfortunately, soon after launch, astronomers discovered that the mirror had been polished to the wrong shape. The mirror is too flat by 2 μm, or about 1/50 the width of a human hair. Even though it *is* the smoothest mirror ever made, its imperfect shape makes it impossible to focus all the captured light as well as expected. This optical flaw (known as *spherical aberration*) meant that *HST* was not as sensitive as designed, although it could still see many objects in the universe with unprecedented resolution.

Figure 5.20 shows the world's largest radio telescope, located in Arecibo, Puerto Rico. Approximately 300 m (1000 feet) in diameter, the surface of the Arecibo telescope spans nearly 20 acres. Constructed in 1963 in a natural depression in the hillside, the dish was originally surfaced with chicken wire, which was lightweight and cheap. Although fairly rough, the chicken wire was adequate for proper reflection because the openings between adjacent strands of wire were much smaller than the long-wavelength radio waves to be detected. The entire Arecibo dish was resurfaced in 1974 with thin metal plates, so it can now be used to study shorter-wavelength radio radiation. Even so, useful observations are still restricted to radiation of wavelength greater than about 10 cm, making the telescope's angular resolution no better than that of some smaller radio telescopes, despite its enormous size. The huge size of the dish, in fact, creates one distinct disadvantage: The Arecibo telescope cannot be pointed very well to follow cosmic objects across the sky. The dish is literally strung among several limestone hills, restricting its observations to those objects that happen to pass within about 20° of overhead.

In late 1993 (see the next figure), astronauts aboard the space shuttle *Endeavour* visited *HST* and succeeded in repairing some of its ailing equipment. For example, they replaced *Hubble*'s gyroscopes to help the telescope point more accurately and installed sturdier versions of the solar arrays that power the telescope's electronics. They also partly corrected the flawed vision of *HST*'s instruments by inserting an intricate set of small mirrors (each about the size of the coins in your pocket) to compensate for the faulty main mirror—in much the same way that we use eyeglasses or contact lenses to help nearsighted humans see better.

Given that *Hubble* was so expensive to build, was meant to be the flagship of a whole new generation of *NASA* spacecraft, and was greeted with such public fanfare, it is perhaps understandable that the news media sensationalized so many aspects of the repair mission. The bottom line, however, is that *HST* was never really "broken," nor is it now completely "fixed." Despite the fact that it still does not operate as originally designed, *Hubble* has always been a superb telescope since its deployment in space in 1990, and it remains so today. In fact, especially due to the heroic efforts of the astronaut repair crew, *Hubble* is probably the best telescope built by humans to date.

Following the repair, *Hubble* has regained much of its lost sensitivity, enabling it to see very faint objects, and its resolution is now somewhat better than the best resolution attainable (with image processing) before the repair. A good example of its scientific capabilities can be seen by comparing the two images of the spiral galaxy M100. On the left is perhaps the best ground-based photograph of this beautiful galaxy, showing rich detail and color in its spiral arms. On the right is an *HST* image, showing improvement in both resolution and sensitivity. (The odd, chevron-shaped field of view is caused by the corrective optics inserted into the telescope by the shuttle astronauts; an additional trade-off is *Hubble*'s smaller field of view compared with those of ground-based telescopes.) Many examples of *HST*'s remarkable new data appear throughout this book.

Arecibo is an example of a roughly surfaced telescope capable of detecting long-wavelength radio radiation. At the other extreme, Figure 5.21 shows the 36 m diameter Haystack dish in northeastern Massachusetts. It is constructed of polished aluminum and maintains a parabolic curve to an accuracy of about a millimeter all the way across its solid surface. It can reflect and accurately focus radio radiation with a wavelength as short as a few millimeters. The telescope is contained within the protective shell, or radome, which protects the surface from the harsh wind and weather of New England. It acts much like the protective dome of an optical telescope, except that there is no slit through which the telescope "sees." Incoming cosmic radio signals pass virtually unimpeded through the radome's fiberglass construction.

THE VALUE OF RADIO ASTRONOMY

Despite the inherent disadvantage of relatively poor angular resolution, radio astronomy enjoys many advantages. Radio telescopes can observe 24 hours a day—darkness is not needed for receiving radio signals. The reason for this is simply that the Sun is a relatively weak

Figure 5.20   An aerial photograph of the 300 m diameter dish at the National Astronomy and Ionospheric Center near Arecibo, Puerto Rico. The receivers that detect the focused radiation are suspended nearly 300 m (about 80 stories) above the center of the dish. One insert shows a close-up of the radio receivers hanging high above the dish. The other insert shows technicians adjusting the surface of the dish.

source of radio energy, so its emission does not swamp radio signals from elsewhere. Astronomers can often make radio observations through cloudy skies, and radio telescopes can detect the longest-wavelength radio waves even during rain or snow storms. Poor weather causes few problems because the wavelength of most radio waves is much larger than the typical size of atmospheric raindrops or snowflakes. Optical astronomy cannot be done under these conditions because the wavelength of visible light is smaller than a raindrop, a snowflake, or even a minute water droplet in a cloud.

However, perhaps the greatest value of radio astronomy (and, in fact, of all other invisible astronomies) is that it opens up a whole new window on the universe. Objects that are bright in the optical region of the spectrum are not necessarily strong radio emitters, and very often, strong radio sources are completely undetectable in visible wavelengths. Thus, radio observations do not just afford us the opportunity of studying the same objects at different wavelengths. They have also allowed us to see new classes of objects that had been completely unknown.

For example, Figure 5.22(a) shows an optical photograph of the Orion Nebula (a huge cloud of interstellar gas) taken with the 4 m telescope on Kitt Peak. Figure 5.22(b) shows a radio contour map of the same region superimposed on the optical image. By aiming a radio telescope at

the nebula, radio astronomers determine the strength of its radio emission. Scanning back and forth across the nebula and taking many measurements, the astronomers construct a radio map of the entire region. The map is drawn as a series of contour lines connecting locations of equal radio brightness. These radio contours are similar to pressure contours drawn by meteorologists on weather maps and height contours drawn by cartographers on topographic maps. The inner contours usually represent relatively strong radio signals, the outside contours weak signals.

The radio map shown in Figure 5.22(b) has many similarities to the visible image of the nebula—the radio emission is clearly strongest near the center of the optical image and declines toward the nebular edge. But there are also some subtle differences between the radio and optical images. The two maps differ mainly toward the upper left of the main cloud, where visible light seems to be absent, despite the existence of radio waves. How can radio waves be detected from locations not showing any light emission? This particular nebular region is known to be especially dusty in its top left quadrant. The dust obscures the short-wavelength visible radiation but not the long-wavelength radio radiation.

Thus, our radio map allows us to see the true extent of this cosmic source. Optical images are often distorted by intervening dust somewhere along our line of sight. In fact,

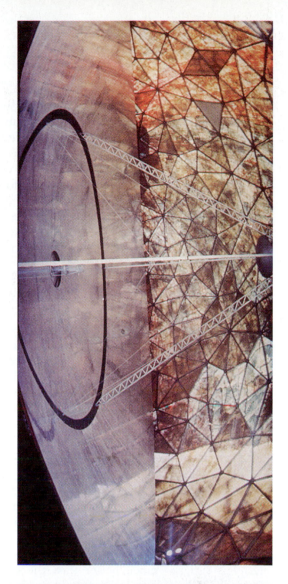

Figure 5.21 Photograph of the Haystack dish, taken from inside the radome. For scale, note the engineer standing at the bottom. Also note the dull shine on the telescope surface, indicating its smooth construction. Haystack is a poor optical mirror but a superb radio telescope. Accordingly, it can be used to reflect and accurately focus radiation having short radio wavelengths, even as small as a fraction of a centimeter.

many important objects and regions of the universe cannot be seen at all by optical astronomy; the very center of our Milky Way Galaxy is a prime example of a totally invisible region. Our knowledge of such regions results almost entirely from analyses of their longer-wavelength radio and infrared emissions.

5.5 *Interferometry*

⑤ The main disadvantage of radio astronomy compared with optical work is its lack of good angular resolution. However, radio astronomers have invented ways to overcome this problem. By using a technique known as **interferometry**, they can improve the angular resolution of some radio maps enormously. In fact, using interferometry, it is actually possible to produce radio images of much higher angular resolution than can be achieved with even the best optical telescopes.

In interferometry, two or more radio telescopes are used in tandem to observe the *same* object at the *same* wavelength and at the *same* time. The combined instruments to-

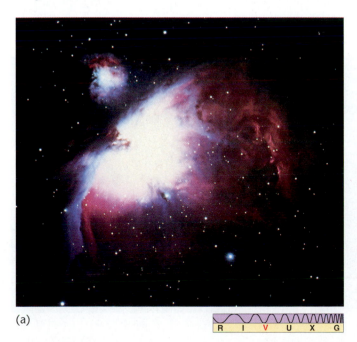

(a)

R I V U X G

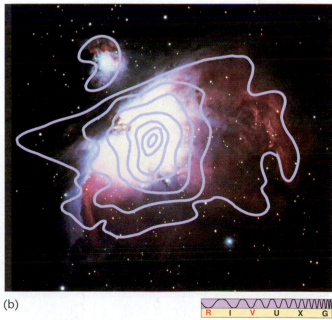

(b)

R I V U X G

Figure 5.22 (a) Optical photograph of the Orion Nebula, a star-forming region some 1500 light years distant. The bright regions in this photograph are stars and clouds of glowing gas. The dark regions are not empty but are simply obscured by interstellar matter. The nebula can be seen in Figure 1.6 of the hunter's sword in the constellation Orion. (b) Superposition of a radio contour map onto the optical photograph. Each curve represents a different intensity of radio emission. The resolution of the optical image is about 1 arc second; that of the radio map 1 arc minute.

gether make up an **interferometer**. Figure 5.23 shows an interferometer—several separate radio telescopes working together as a team. By means of electronic cables or radio links, the signals each antenna receives are sent to a central computer that analyzes how the waves interfere with each other when added together. If the detected waves are in step when added, they combine positively to form a strong radio signal. If the signals are not in step, they destructively interfere and cancel each other. As the antennas track their target, a pattern of peaks and troughs emerges, which, after extensive computer processing, translates into an image of the observed object.

An interferometer is essentially a substitute for a single huge antenna. With respect to resolving power, the effective telescope diameter of an interferometer equals the distance between its outermost dishes. In other words, two small dishes can act as opposite ends of an imaginary but huge single radio telescope, dramatically improving the angular resolution. For example, resolution of a few arc seconds can be achieved at typical radio wavelengths (such as 10 cm), either by using a single radio telescope 5 km in diameter (which is quite impossible to build) or by using two or more much smaller dishes *separated* by 5 km and connected electronically.

Large interferometers made up of many dishes, like the instrument shown in Figure 5.23, now routinely attain radio resolution comparable to that of optical images. Figure 5.24 compares an interferometric radio map of a nearby galaxy with a photograph of that same galaxy made using a large optical telescope. The radio clarity is superb—much better than the radio map of Figure 5.22(b).

The larger the distance separating the telescopes—the longer the *baseline* of the interferometer—the better the resolution attainable. Astronomers have created radio interferometers spanning very great distances, first across North America and later between continents. A typical very-long-baseline interferometry experiment (usually known by the acronym VLBI) might use radio telescopes in North America, Europe, Australia, and Russia to achieve angular resolution on the order of 0.001″, about 1000 times better than images produced by most current optical telescopes. It seems that even the Earth's diameter is no limit. Radio astronomers have successfully used an antenna in orbit, together with several antennas on the ground, to construct an even longer baseline and achieve still better resolution. Proposals even exist to place interferometers entirely in Earth orbit, and even on the Moon.

Nowadays, interferometry is no longer restricted to the radio domain. Radio interferometry became feasible when electronic equipment and computers achieved speeds great enough to combine and analyze radio signals from separate radio detectors without loss of data. As the technology has improved, it has become possible to apply the same methods to higher-frequency radiation. Millimeter-wave interferometry has already become an established and important observational technique, and it is very likely that infrared interferometry will become commonplace in the coming few years. Interferometry is not yet widely used in optical work because of the technical difficulties involved, but optical interferometry is the subject of intensive research. The new Keck telescopes on Mauna Kea will be used for infrared—and perhaps someday for optical—interferometric work.

(a)

(b)

Figure 5.23   This large interferometer is made up of 27 separate dishes spread along a Y-shaped pattern about 30 km across on the Plain of San Augustin in New Mexico. The most sensitive radio device in the world, it is called the Very Large Array or VLA, for short. (b) A close-up view from ground level of some of the VLA antennas. Notice that the dishes are mounted on railroad tracks so that they can be repositioned easily.

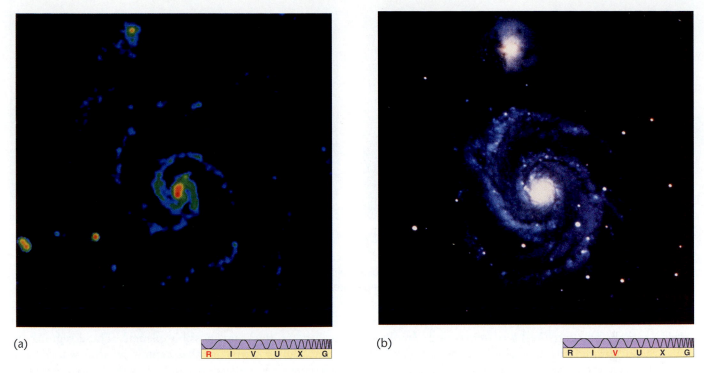

(a) R I V U X G   (b) R I V U X G

Figure 5.24 VLA radio "photograph" (or radiograph) of the spiral galaxy M51, observed at radio frequencies with an angular resolution of a few arc seconds (a); shows nearly as much detail as an actual (light) photograph of that same galaxy (b) made with the 4 m Kitt Peak optical telescope.

5.6 *Other Astronomies*

6 The electromagnetic spectrum consists of far more than just visible light and radio waves. Optical and radio astronomy are the oldest and best-established branches of astronomy, but since the 1970s there has been a virtual explosion of observational techniques covering the many other types of electromagnetic radiation. Today all portions of the spectrum are studied, from radio waves to gamma rays, to maximize the amount of information available about astronomical objects.

As we have already noted in the context of radio astronomy, the *types* of astronomical objects that can be observed may differ quite markedly from one wavelength range to another. Thus, full-spectrum coverage is essential, not only to see things more clearly, but even to see some things at all.

Because of the transmission characteristics of the Earth's atmosphere, astronomers must study most wavelengths (other than optical and radio) from space. The rise of these "other astronomies" has therefore been closely tied to the development of the space program.

INFRARED ASTRONOMY

Infrared studies are an important component of modern observational astronomy. Generally, **infrared telescopes** resemble optical telescopes (indeed, many optical telescopes are also used for infrared work), but their detectors are sensitive to longer-wavelength radiation. Although most infrared radiation is absorbed by the atmosphere (primarily by water vapor), in a few windows in the high-frequency part of the infrared spectrum (see Figure 3.9) the opacity is low enough to allow ground-based observations. Indeed, some of the most useful infrared observing is done from the ground, even though the radiation is somewhat diminished in intensity by our atmosphere.

As with radio observations, the longer wavelength of infrared radiation often enables us to perceive objects partially hidden from optical view. As an example of the penetrating properties of infrared radiation, Figure 5.25 shows a dusty and hazy region in California, hardly viewable optically, but easily seen using infrared radiation.

Figure 5.26 is a photograph of the world's highest ground-based observatory, perched more than 4 km (about 14,000 feet) above sea level on top of an extinct volcano at Mauna Kea, Hawaii. Despite its remoteness, this site draws a full schedule of astronomers throughout the year. The thin air at this high altitude guarantees less atmospheric absorption of incoming radiation, and hence a clearer view, than is possible from sea level. Mauna Kea is one of the finest locations on the Earth for ground-based optical and infrared astronomy, but the air is so thin that astronomers must occasionally wear oxygen masks while performing their observations.

Astronomers can make still better infrared observations if they can place their instruments above most or all

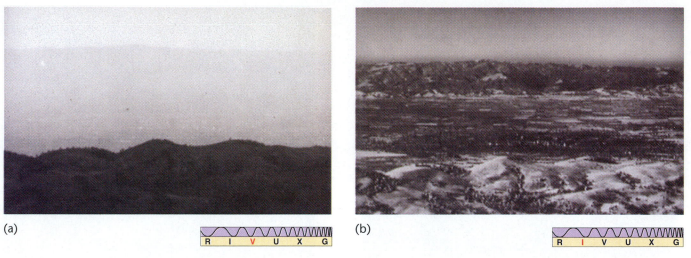

(a)

R I V U X G

(b)

R I V U X G

Figure 5.25   An optical photograph (a) taken near San Jose, California, and an infrared photo (b) of the same area taken at the same time. Longer wavelength infrared radiation can penetrate smog much better than short-wavelength visible light.

of the Earth's atmosphere. Improvements in balloon-, aircraft-, rocket-, and satellite-based telescope technologies have made infrared research a very powerful tool with which to study the universe (see Figure 5.27). As might be expected, the infrared telescopes that can be carried above the atmosphere are considerably smaller than massive ground-based instruments.

The most advanced facility to function in this part of the spectrum is the *Infrared Astronomy Satellite (IRAS)*, shown in Figure 5.27(b). Launched into Earth orbit in 1983 but now inoperative, this British–Dutch–U.S. satellite housed a 0.6 m mirror with an angular resolution as fine as 30″. (As usual, the resolution depended on the precise wavelength observed.) Its sensitivity was greatest for radiation in the 10 to 100 μm range. During its 10-month lifetime (and long afterwards—the data archives are still heavily used even today), *IRAS* contributed greatly to our

knowledge of clouds of galactic matter that seem destined to become stars, and possibly planets. These regions are composed of warm gas that cannot be seen with optical telescopes or adequately studied with radio telescopes. Throughout the text we will encounter many findings made by this satellite about comets, stars, galaxies, and the scattered dust and rocky debris found between the stars. All these objects "glow" in the infrared. For example, because much of the material between the stars has a temperature between a few tens and a few hundred kelvins, Wien's law (Section 3.4) tells us that the infrared domain is the natural portion of the spectrum in which to study it. ∞ (p. 64)

Figure 5.28(a) shows an *IRAS* image of the Orion Nebula. At about 1′ angular resolution, the fine details of Orion visible in the earlier optical image (Figure 5.22) cannot be perceived. Nonetheless, astronomers can ex-

Figure 5.26   Photograph of the world's highest ground-based observatory at Mauna Kea, Hawaii. Among the domes visible in the picture are those that house the Canada–France–Hawaii 3.6 m telescope, the 2.2 m telescope of the University of Hawaii, Britain's 3.8 m infrared facility, and the twin Keck telescopes (see Figure 5.12).

(a) (b)

Figure 5.27 (a) A gondola containing a 1-m infrared telescope (lower left) is readied for its balloon-borne ascent to an altitude of about 30 km (100,000 feet), where it will capture infrared radiation that cannot penetrate the atmosphere. (b) An artist's conception of the *Infrared Astronomy Satellite,* placed in orbit in 1983. This 0.6 m telescope surveyed the infrared sky at wavelengths ranging from 10 to 100 μm. During its 10 months of operation, it greatly increased astronomers' understanding of many different aspects of the universe, from the formation of stars and planets to the evolution of galaxies.

tract useful information about this object and others like it from such observations. For example, clouds of warm dust and gas, believed to play a critical role in the process of star formation, and extensive groups of bright young stars, completely obscured at visible wavelengths, are seen.

Unfortunately, by Wien's law, telescopes themselves also radiate strongly in the infrared unless they are cooled to nearly absolute zero. The end of *IRAS*'s mission came not because of any equipment malfunction or unexpected mishap but simply because its supply of liquid helium coolant ran out. *IRAS*'s own thermal emission then overwhelmed the radiation it was built to detect. The European Space Agency (ESA) and NASA both plan to launch infrared instruments into Earth orbit in the late 1990s. The first of these, ESA's *Infrared Space Observatory (ISO)*, is now in orbit, refining and extending the groundbreaking work begun by *IRAS*.

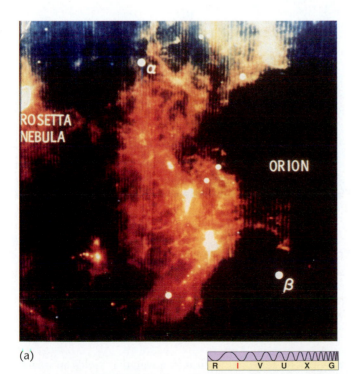

(a)

(b)

Figure 5.28 (a) This infrared image of the Orion Nebula and its surrounding environment was made by the *Infrared Astronomy Satellite.* The whiter regions denote greater strength of infrared radiation; the false colors denote different temperatures, descending from white to red to black. (b) The same region photographed in visible light.

ULTRAVIOLET ASTRONOMY

On the short-wavelength side of the visible spectrum lies the ultraviolet domain. This region of the spectrum, extending in wavelength from 400 nm (4000 Å, blue light) down to a few nanometers ("soft" X rays), has only recently begun to be explored. Because Earth's atmosphere is partially opaque to radiation below 400 nm and is totally opaque below about 300 nm (in part because of the ozone layer), astronomers cannot conduct any useful ultraviolet observations from the ground, not even from the highest mountaintop. Rockets, balloons, or satellites are therefore essential to any **ultraviolet telescope**—a device designed to capture and analyze this high-frequency radiation.

One of the most successful ultraviolet space missions is the *International Ultraviolet Explorer*, called *IUE* for short. This satellite was placed in Earth orbit in 1978 and is still functioning as designed (see *Interlude 18-1*). Like all ultraviolet telescopes, its basic appearance and construction are quite similar to optical and infrared devices. Several hundred astronomers from all over the world have used *IUE* to explore a variety of phenomena in planets, stars, and galaxies. In subsequent chapters, we will learn what this relatively new window on the universe has shown us about the activity and even the violence that seems to pervade the cosmos. The *Hubble Space Telescope*, described in Interlude 5-1, is also a superb ultraviolet instrument.

An alternative means of placing astronomical payloads into (temporary) Earth orbit is provided by NASA's space shuttle. In December 1990 and March 1995, a shuttle carried aloft the *Astro* package of three ultraviolet telescopes (see Figure 5.29). Astronomical shuttle missions offer a potentially very flexible way for astronomers to get instruments into space, without the long lead times and great expense of permanent satellite missions like the *Hubble* telescope.

HIGH-ENERGY ASTRONOMY

High-energy astronomy studies the universe as it presents itself to us in X rays and gamma rays—the types of radiation whose photons have the highest frequencies, and hence the greatest energies. How do we detect radiation of such short wavelengths? First, it must be captured high above the Earth's atmosphere because none of it reaches the ground. Second, its detection requires the use of equipment basically different in design from that used to capture the relatively low-energy radiation discussed up to this point.

The basic difference in the design of **high-energy telescopes** comes about because X and gamma rays cannot be reflected easily by any kind of surface. Rather, these rays tend to pass straight through, or be absorbed by, any material they strike. When X rays barely graze a surface, however,

(a)

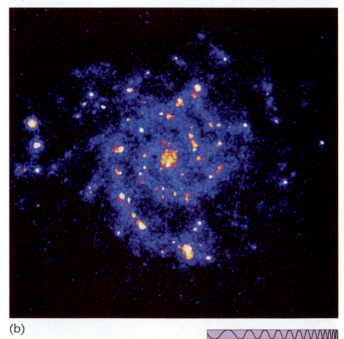

(b)

| R | I | V | U | X | G |

Figure 5.29   (a) The *Astro* payload, carried by the space shuttle in 1990 and 1995, performed ultraviolet and X-ray observations from orbit for the 10-day duration of each mission. (b) This false-color image of the spiral galaxy M74 was made by an ultraviolet telescope aboard *Astro*.

they can be reflected from it in a way that yields an image, although the mirror design is fairly complex (see Figure 5.30). For gamma rays (with wavelengths less than about 0.01 nm), no such method of producing an image has yet

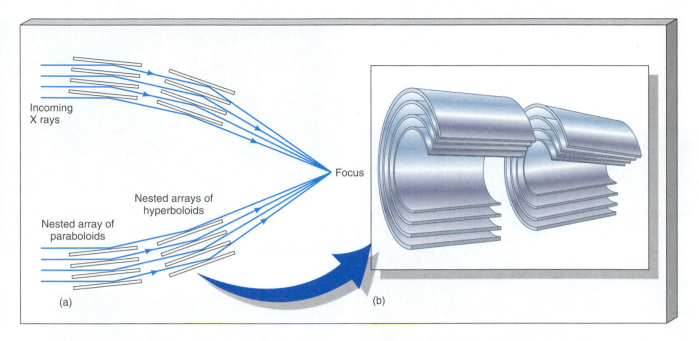

Figure 5.30 The arrangement of mirrors in an X-ray telescope allows X rays to be reflected at grazing angles and focused into an image.

been devised. Present-day gamma-ray telescopes simply point in a specified direction and count photons received.

In addition, X-ray and gamma-ray detection methods using photographic plates or CCD devices do not work well. Instead, individual X-ray and gamma-ray photons are counted by electronic detectors on board an orbiting device, and the results are then transmitted to the ground for further processing and analysis. Furthermore, the number of photons in the universe seems to be inversely related to frequency. Billions of visible (starlight) photons arrive at the Earth each second, but hours or even days are often needed for a single gamma-ray photon to be recorded. Not only are these photons hard to focus and measure, they are also few and far between.

Toward the end of the 1970s, a new generation of X-ray and gamma-ray telescopes was launched into Earth orbit. Called the *High-Energy Astronomy Observatories*, or *HEAO* for short, these spacecraft made major advances in our understanding of high-energy phenomena throughout the universe. Having greater accuracy and sensitivity than all earlier high-energy satellites, these spacecraft did for X-ray astronomy what the first large optical and radio telescopes did for longer-wavelength radiation. Figure 5.31 is a photograph of the *HEAO-2* spacecraft, the first X-ray telescope capable of forming an image of its field of view. In 1979, the year when it first came on-line, the satellite was renamed the *Einstein Observatory*, in honor of the birth centenary of the great scientist.

Although the collecting diameter of *Einstein* was only 0.6 m, its angular resolution was a mere 3″. Accordingly, this spacecraft could produce images of quality comparable

Figure 5.31 HEAO-2, also known as the *Einstein Observatory*, the first imaging X-ray telescope.

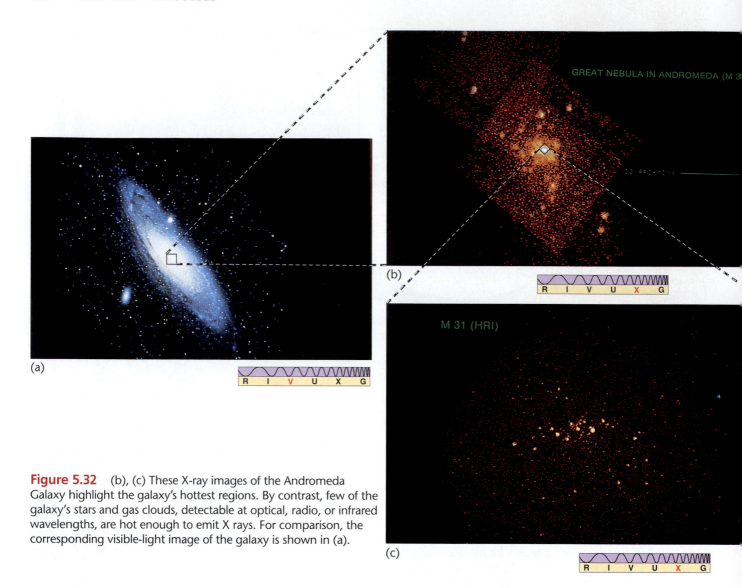

GREAT NEBULA IN ANDROMEDA (M 3

(b)

R I V U X G

M 31 (HRI)

(a)

R I V U X G

(c)

R I V U X G

Figure 5.32 (b), (c) These X-ray images of the Andromeda Galaxy highlight the galaxy's hottest regions. By contrast, few of the galaxy's stars and gas clouds, detectable at optical, radio, or infrared wavelengths, are hot enough to emit X rays. For comparison, the corresponding visible-light image of the galaxy is shown in (a).

to that of optical photographs. Figures 5.32(b) and (c) are *Einstein* X-ray images, showing some of the many hot regions in and around the center of the Andromeda Galaxy. In this image, the hottest regions stand out most clearly because the black-body curve of their emission peaks well into the high-energy domain. Thus, they shine brightly in X rays compared to the much cooler surrounding material, which emits primarily in the infrared and visible regions of the spectrum and hardly at all at higher energies.

The most recent major X-ray satellite is the German *ROSAT* (short for *Röntgen Satellite*, after Wilhelm Röntgen, the discoverer of X rays). Launched in 1990 by a European *Ariane* rocket, it began its mission with a detailed survey of the X-ray sky and is now making detailed observations of specific astronomical objects (see Figure 5.33). With more sensitivity, a wider field of view, and better resolution than *Einstein*, *ROSAT* is providing high-energy astronomers with new levels of observational detail. Even more powerful will be NASA's planned *Advanced X-ray Astrophysics Facility (AXAF)*, another long-duration orbiting observa-

tory (in the spirit of *IUE* and *HST*) that may become operational by the end of the 1990s.

The youngest entrant into the observational arena is *gamma-ray astronomy*. As mentioned earlier, true imaging gamma-ray telescopes do not exist, so only fairly coarse (1° resolution) observations can be made. Nevertheless, even at that resolution, there is much to be learned. Cosmic gamma rays were originally detected in the 1960s by the U.S. *Vela* series of satellites, whose primary mission was to monitor illegal nuclear detonations on Earth. Since then, several X-ray telescopes have also been equipped with gamma-ray detectors. By far the most advanced instrument is the *Gamma Ray Observatory (GRO)*, launched by the space shuttle in 1991. This satellite can scan the sky and study individual objects in much greater detail than previously attempted. Figure 5.34 shows *GRO* on station in low Earth orbit, along with a (false color) gamma-ray image of a highly energetic outburst in the nucleus of a distant galaxy. Figure 5.35(e) below shows another GRO image, this time of our own galaxy.

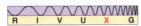

Figure 5.33   An X-ray image of the Orion region, taken by the *ROSAT* X-ray satellite. (Compare with Figures 1.6, 5.22, and 5.28.) Note the three stars of Orion's belt and the glowing nebula below them at bottom left in the photograph.

5.7 *Full-Spectrum Coverage*

⑦ In this chapter we have studied some of the basic techniques and equipment used by astronomers to study the universe. Besides the familiar optical telescopes that collect light from cosmic objects, other tools are needed to capture the invisible radiation emitted by a variety of celestial sources. Often, these "invisible astronomies" are crucial in our study of objects that are totally obscured from view in the visible range or simply do not emit any visible light. Radio astronomy is the oldest of the nonvisible subjects, high-energy astronomy the newest. In the end, they all supplement one another, helping us accumulate a growing store of astronomical knowledge. Table 5-1 lists the basic regions of the electromagnetic spectrum and describes some of the objects that are typically studied in each frequency range. Bear in mind, though, that the list is far from exhaustive—many important astronomical objects are now routinely observed at many different electromagnetic wavelengths, and are not listed here.

As we proceed through the text, we will discuss more fully the wealth of information that high-precision astronomical instruments can provide us. It is reasonable to suppose that the future holds many further improvements in both the quality and the availability of astronomical data

(a)

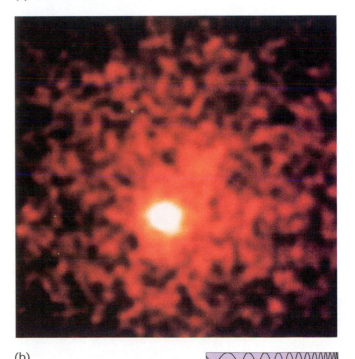

(b)

Figure 5.34   (a) This photograph of the 17-ton *Gamma-Ray Observatory* (also called the *Compton Observatory,* after an American gamma-ray pioneer) was taken by an astronaut during the satellite's deployment from the space shuttle *Atlantis* over the Pacific Coast of the United States. (b) A typical false-color gamma-ray image—this one showing a violent event in the distant galaxy 3C279, also known as a "gamma-ray blazar."

TABLE 5-1 *Astronomy at Many Wavelengths*

WAVELENGTH/FREQUENCY RANGE	GENERAL CONSIDERATIONS	COMMON APPLICATIONS	CHAPTER REFERENCE
Radio	Radio radiation can penetrate dusty regions of interstellar space.	Radar studies of planets	2, 9
		Planetary magnetic fields	11
	Earth's atmosphere is largely transparent to radio wavelengths.	Interstellar gas clouds	18
		Center of the Milky Way Galaxy	23
	Radio emissions can be detected in the daytime as well as a night.	Galactic structure	23, 24
		Active galaxies	25
	High resolution at long wavelengths requires very large telescopes.	Cosmic background radiation	27
Infrared	IR radiation can penetrate dusty regions of interstellar space.	Star formation	19
		Cool stars	
	Earth's atmosphere is only partially transparent to IR radiation: some observations must be made from space.	Center of the Milky Way Galaxy	23
		Active galaxies	25
		Large-scale structure in the universe	24, 27
Visible	Earth's atmosphere is transparent to visible light.	Planets	7–14
		Stars and stellar evolution	17, 20, 21
		Galactic structure	23, 24
		Large-scale structure in the universe	24, 27
Ultraviolet	Earth's atmosphere is opaque to UV radiation: must be observed from space.	Interstellar medium	19
		Hot stars	
X ray	Earth's atmosphere is opaque to X rays: must be observed from space.	Stellar atmospheres	16
		Neutron stars and black holes	22
	Special mirror configurations are needed to form images.	Hot gas in galaxy clusters	24
		Active galactic nuclei	25
Gamma ray	Earth's atmosphere is opaque to gamma rays: must be observed from space.	Neutron stars	22
		Active galactic nuclei	25
	Cannot form images.		

and that many new discoveries will be made. The current and proposed pace of technological progress presents us with the following very exciting prospect: In the twenty-first century, if all goes according to plan, it will be possible, for the first time ever, to make *simultaneous* high-quality measurements of any astronomical object at *all* wavelengths, from radio to gamma ray. The consequences of this development for our understanding of the workings of the universe may be little short of revolutionary.

As a preview of the sort of comparison that full-spectrum coverage allows, Figure 5.35 shows a series of images of our own Milky Way Galaxy. They were made by several different instruments, at wavelengths ranging from radio to gamma ray, over a period of about five years. By comparing the features visible in each, we immediately see how multiwavelength observations can complement each other, greatly extending our perception of the universe around us.

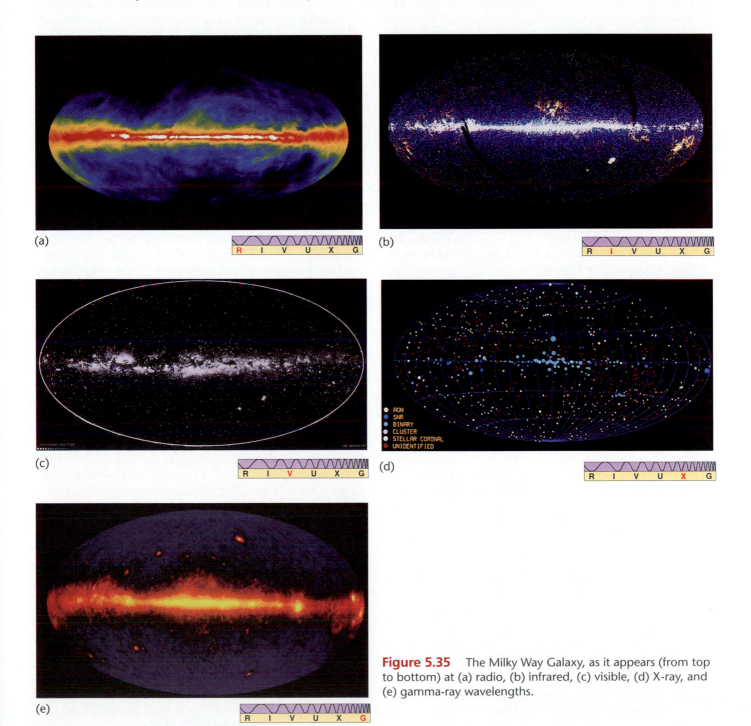

Figure 5.35 The Milky Way Galaxy, as it appears (from top to bottom) at (a) radio, (b) infrared, (c) visible, (d) X-ray, and (e) gamma-ray wavelengths.

Chapter Review

SUMMARY

A **telescope** (p. 96) is a device designed to collect as much light as possible from some distant source and deliver it to a detector for detailed study. **Reflecting telescopes** (p. 96) use a mirror to concentrate and focus the light. **Refracting telescopes** (p. 96) use a lens. The **prime focus** (p. 96) of the telescope is the point where the incoming beam is focused and where analysis instruments may be placed. The **Newtonian** (p. 100) and **Cassegrain telescope** (p. 100) designs employ secondary mirrors to avoid placing heavy equipment at the prime focus.

The lenses used in refracting telescopes suffer from a number of problems, among them **chromatic aberration** (p. 97), the tendency of lenses to focus different colors to different prime foci. These problems become more difficult to correct the larger the lens is, with the result that all astronomical telescopes larger than about 1 m in diameter use mirrors in their design.

The light collected by a telescope may be processed in a number of ways. It can be made to form an **image** (p. 97), a **photometer** (p. 101) may be used to make detailed measurements of the energy received, or a **spectrometer** (p. 101) may study its spectrum.

The light-gathering power of a telescope depends on its **collecting area** (p. 101), which is proportional to the square of the mirror diameter. To study the faintest sources of radiation, astronomers must use large telescopes.

An important aspect of a telescope is its **angular resolution** (p. 102), the ability to distinguish between light sources lying close together on the sky. One limitation on the resolution of a telescope is diffraction, which makes it impossible to focus a beam perfectly. The amount of diffraction is proportional to the wavelength of the radiation under study, and inversely proportional to the size of the mirror. Thus, at any given wavelength, larger telescopes suffer least from the effects of diffraction.

The resolution of most ground-based optical telescopes is actually limited by **seeing** (p. 105)—the blurring effect of Earth's turbulent atmosphere, which smears the pointlike images of stars out into **seeing disks** (p. 105) a few arc seconds in diameter. Radio and space-based telescopes do not suffer from atmospheric effects, so their resolution is determined by the effects of diffraction.

Most modern telescopes now use **charge-coupled devices**, or **CCDs** (p. 106), instead of photographic plates to collect their data. The field of view is divided into an array of millions of **pixels** (p. 106) that accumulate an electric charge when light strikes them. CCDs are many times more sensitive than photographic plates, and the resultant data are easily saved directly on disk or tape for later image processing.

Using **active optics** (p. 107), in which a telescope's environment and focus are carefully monitored and controlled, and **adaptive optics** (p. 108), in which the blurring effects of atmospheric turbulence are corrected for in real time, it may soon be possible to achieve diffraction-limited resolution in ground-based optical instruments.

Radio telescopes (p. 108) are conceptually similar in construction to optical reflectors. However, radio telescopes are generally much larger than optical instruments, for two reasons. First, the amount of radio radiation reaching Earth from space is tiny compared to optical wavelengths, so a large collecting area is essential. Second, the long wavelengths of radio waves mean that diffraction severely limits the resolution unless large instruments are used.

In order to increase the effective area of a telescope, and hence improve its resolution, several separate instruments may be combined into a device called an **interferometer** (p. 114). Using **interferometry** (p. 113), radio telescopes can produce images much sharper than those from the best optical equipment. Infrared interferometers are under construction, and optical interferometric systems are under active development.

Infrared telescopes (p. 115) and **ultraviolet telescopes** (p. 118) are similar in their basic design to optical systems. Infrared studies in some parts of the infrared range can be carried out using large ground-based systems. Ultraviolet astronomy must be carried out from space.

High-energy telescopes (p. 118) study the X- and gamma-ray regions of the electromagnetic spectrum. X-ray telescopes can form images of their field of view, although the mirror design is more complex than for lower-energy instruments. Gamma-ray telescopes simply point in a certain direction and count photons received. Because the atmosphere is opaque at these short wavelengths, both types of telescope must be placed in space.

Radio and other nonoptical telescopes are essential to studies of the universe because they allow astronomers to probe regions of space that are completely opaque to visible light and to study the many objects that emit little or no optical radiation at all.

SELF-TEST: True or False?

_____ **1.** The primary purpose of any telescope is to collect as much radiation as possible and magnify the image.

_____ **2.** A Newtonian telescope has no secondary mirror.

_____ **3.** A Cassegrain telescope has a hole in the middle of the primary mirror to allow light reflected from its secondary mirror to reach a focus behind the primary mirror.

_____ **4.** The term "seeing" is used to describe how faint an object can be detected by a telescope.

_____ **5.** The primary advantage to using the _Hubble Space Telescope_ is the increased amount of "night" time available to it.

_____ **6.** One of the primary advantages of CCDs over photograph plates is their high efficiency in detecting light.

_____ **7.** The _Hubble Space Telescope_ can observe objects in the optical, infrared, and ultraviolet parts of the spectrum.

_____ **8.** The Keck Telescope has the largest single mirror ever produced.

_____ **9.** Radio telescopes are large, in part to improve their angular resolution, which is poor because of the long wavelengths at which they observe.

_____ **10.** Radio telescopes are large, in part because the sources of radio radiation they observe are very faint.

____ **11.** Radio telescopes have to have surfaces as smooth as those in optical telescope mirrors.

____ **12.** Infrared astronomy must be done from space.

____ **13.** Because the ozone layer absorbs ultraviolet light, as-

tronomers must make observations in the ultraviolet from the highest mountain tops.

____ **14.** X-ray and gamma-ray telescopes employ the same basic design as optical instruments.

SELF-TEST: Fill in the Blank

1. A telescope that uses a lens to focus light is called a _____ telescope.

2. A telescope that uses a mirror to focus light is called a _____ telescope.

3. All large modern telescopes are of the _____ type.

4. The light-gathering power of a telescope is determined by the _____ of its mirror or lens.

5. The angular resolution of a telescope is limited by the _____ of the telescope and the _____ of the radiation being observed.

6. The angular resolution of ground-based optical telescopes is more seriously limited by Earth's _____ than by diffraction.

7. Optical telescopes on Earth can see angular detail down to about _____ arc second.

8. CCDs produce images in _____ form that can be easily transmitted, stored, and later processed by computers.

9. Active optics and adaptive optics are both being used to improve the _____ of ground-based optical telescopes.

10. All radio telescopes are of the _____ design.

11. An _____ is two or more telescopes used in tandem to observe the same object, in order to improve angular resolution.

12. An object with a temperature of 300 K would be best observd with an _____ telescope.

REVIEW AND DISCUSSION

1. Cite two reasons why astronomers are continually building larger and larger telescopes.

2. What are three advantages of reflecting telescopes over refracting telescopes?

3. How does the Earth's atmosphere affect what is seen by an optical telescope?

4. What advantages does the *Hubble Space Telescope* have over ground-based telescopes? List some disadvantages.

5. What are the advantages of a CCD over a photographic plate?

6. What is image processing?

7. Describe some ways in which optical astronomers can compensate for the blurring effects of Earth's atmosphere.

8. Why do radio telescopes have to be very large?

9. What kind of astronomical objects can we best study with radio techniques?

10. What is interferometry, and what problem in radio astronomy does it address?

11. Compare the highest resolution attainable with optical telescopes with the highest resolution attainable with radio telescopes (including interferometers).

12. What special conditions are required to conduct observations in the infrared?

13. What is the main advantage of studying objects at different wavelengths of radiation?

14. Our eyes can see light with an angular resolution of 1'. Suppose our eyes detected only infrared radiation, with 1° angular resolution. Would we be able to make our way around on Earth's surface? To read? To sculpt? To create technology?

PROBLEMS

1. A 2-m telescope can collect a given amount of light in 1 hour. Under the same observing conditions, how much time would be required for a 6-m telescope to perform the same task?

2. A certain space-based telescope can achieve (diffraction-limited) angular resolution of 0.05" for red light (of wavelength 700 nm). What would its resolution be (a) in the infrared, at 3.5 μm; and (b) in the ultraviolet, at 140 nm?

3. The photographic equipment on a telescope is replaced by a CCD. If the photographic plate records 5 percent of the light

reaching it, while the CCD records 75 percent, how much time would the new system take to collect as much information as the old detector recorded in a 1-hour exposure?

4. The Andromeda galaxy lies about 700,000 pc away. To what distances do the angular resolutions of *HST* (0.05") and a radio interferometer (0.001") correspond to at that distance?

5. What would be the equivalent single-mirror area of a telescope constructed from six separate 4-m mirrors?

PROJECTS

1. Here's how to take some easy pictures of the night sky. You will need a location with a clear, dark sky, a 35mm camera with a standard 50 mm lens, tripod, and cable release, a watch with a seconds display visible in the dark, and a role of high-speed color slide film.

Set your camera to the "bulb" setting for the exposure and attach the cable release so you can take a long exposure. Set the focus on infinity. Point the camera to a favored constellation, seen through your viewfinder, and take a 20 to 30

second exposure. Don't hold on to the cable release during the exposure; minimize all vibration. Keep a log of your shots. When finished, have the film developed in the standard way.

2. For some variations, vary your exposure times, use different films, take hours-long exposures for star trails, use different lenses such as wide-angle or telephoto, place the camera piggyback on a telescope that is tracking and take exposures that are a few minutes long. Experiment and have fun!

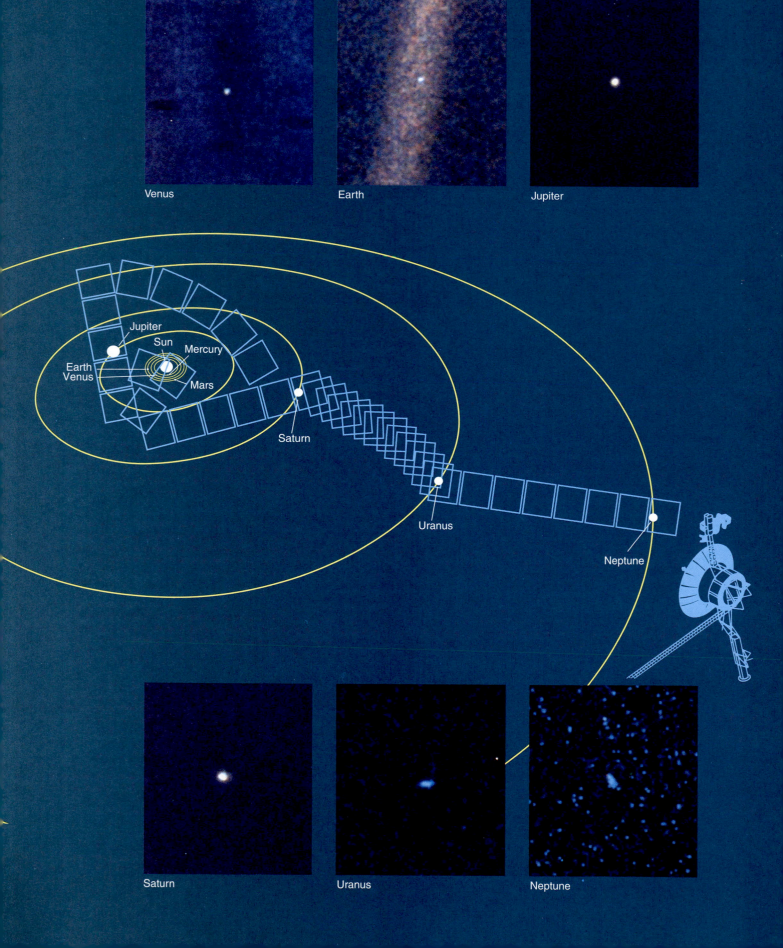

Venus

Earth

Jupiter

Jupiter
Sun
Mercury
Earth
Venus
Mars

Saturn

Uranus

Neptune

Saturn

Uranus

Neptune

6

THE SOLAR SYSTEM

An Introduction to Comparative Planetology

LEARNING GOALS

Studying this chapter will enable you to:

1 Describe the overall scale and structure of the solar system.

2 Summarize the basic differences between the terrestrial and the jovian planets.

3 Identify and describe the major nonplanetary components of the solar system.

4 Describe some of the spacecraft missions that have contributed significantly to our knowledge of the solar system.

5 Discuss the importance of comparative planetology to solar-system studies.

(Opposite page) *Voyager*, one of the most remarkable robotic space probes to travel through the solar system, sent back a wealth of scientific data about several of our neighboring planets. As it sped toward interstellar space after its close encounter with Neptune, it also took a "family portrait" of our planetary system. This diagram depicts the positions at which numerous photographs were taken, superposed on a sketch of the solar system. The inserts show some of the actual images radioed back to Earth.

*L*ooking up at the nighttime sky, we see an almost bewildering array of stars. Thousands are visible to the naked eye, and even a small telescope increases the number we can observe into the millions. Because astronomers can see and study so many stars, they have come to recognize patterns in their properties and have classified the stars accordingly. Ultimately, this classification has led to an understanding of the birth, maturity, and death of stars everywhere. In studying the planets, we are not so fortunate. We are aware of only a single planetary system—our own solar system—and each of its nine planets differs significantly from all the others. We don't have tens of thousands of Earthlike planets to compare, so we cannot perform the statistical analyses that have taught us so much about the stars. Every piece of information we can glean about the structure and history of the planets and smaller bodies orbiting our Sun plays a potentially vital role in helping us understand not just our own solar system, but countless undiscovered planetary systems throughout the Galaxy.

6.1 *Exploring Our Planetary System*

In less than a single generation, we have learned more about our **solar system**—the Sun and everything that orbits it—than in all the centuries that went before. By studying the **planets**—the major bodies that orbit the Sun and reflect its light—and their **moons**—which orbit the planets—astronomers have gained a richer outlook on our own home in space. Instruments aboard unmanned robots have taken close-up photographs of the planets and their moons and in some cases have made on-site measurements. The discoveries of the past few decades have revolutionized our understanding not only of our cosmic neighborhood but also its history, for our solar system is filled with clues to its own origin and evolution.

As we describe the solar system in the next few chapters, we will use the powerful and still emerging perspective of **comparative planetology**—comparing and contrasting the properties of the diverse worlds we encounter—to un-

derstand better the conditions under which planets form. Having made some important stops along the way, we will conclude our tour in Chapter 10 with a look at the modern theory of how our planetary system came into being and how it developed over time.

Before we start our journey, let us pause for a moment to consider the solar system as a whole, to try to place the planets in perspective and to see what patterns we can discern in their orbits and properties. We will begin by making a short historical survey of our cosmic neighborhood.

The Greeks and other astronomers of old were aware of five planets in the nighttime sky—Mercury, Venus, Mars, Jupiter, and Saturn. We saw in Chapter 2 how observations of the apparently erratic motions of those wanderers across the celestial sphere ultimately led to the Copernican revolution and the birth of our modern view of the cosmos. ∞ (p. 34) In addition to the Sun and the Moon, the ancients also knew of two other types of heavenly objects that were clearly neither stars nor planets.

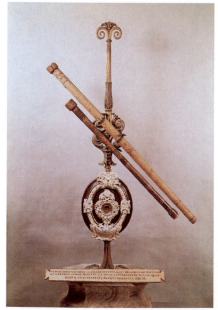

Figure 6.1 The telescope with which Galileo made his first observations was simple, but its influence on astronomy was immeasurable.

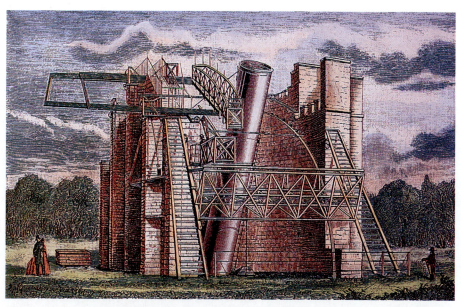

Figure 6.2 By the midnineteenth century, telescopes had improved enormously in both size and quality. Shown here is the telescope built and used by Irish nobleman and amateur astronomer the Earl of Rosse.

Figure 6.3 An *Apollo* astronaut doing some lunar geology—prospecting near a huge boulder near the Mare Serenitatus.

Figure 6.4 The launch of the space probe *Galileo* on a mission to explore in detail the moons and the atmosphere of Jupiter.

Comets appear as long, wispy strands of light in the night sky that remain visible for periods of up to several weeks, then slowly fade from view. *Meteors*, or "shooting stars" are sudden bright streaks of light that flash across the sky, usually vanishing less than a second after they first appear. While these transient phenomena were familiar to ancient astronomers, their role in the "big picture" of the solar system was not understood until much later.

With the invention of the telescope, more detailed observations of the known planets could be made. Galileo Galilei was the first to capitalize on this new technology (his simple telescope is shown in Figure 6.1). His discovery of the phases of Venus and of four moons around Jupiter early in the seventeenth century played a large part in changing forever humankind's vision of the universe. From that time on, Earth was considered a planet like all the others.

With continuing technological advances, knowledge of the solar system improved rapidly. Astronomers began discovering objects invisible to the unaided human eye. By the end of the nineteenth century, astronomers had found Saturn's rings (1659), Uranus (1758), Neptune (1846), many planetary moons, and the first of **asteroids**, "minor planets" that orbit the Sun mostly in a broad band, the **asteroid belt**, lying between Mars and Jupiter. Ceres, the largest asteroid and the first to be sighted, was discovered in 1801. A large telescope of midnineteenth-century vintage is shown in Figure 6.2.

The twentieth century has brought continued improvements in optical telescopes. One more planet (Pluto) has been discovered, along with three more ring systems, dozens of moons, and thousands of asteroids. The century has also seen the rise of both nonoptical astronomy—especially radio and infrared—and spacecraft exploration, both making vitally important contributions to the field of planetary science. Astronauts have carried out experiments on the Moon (see Figure 6.3), and numerous unmanned probes have left Earth and traveled to all but one of the other planets. Figure 6.4 shows the 1989 launch of *Galileo*, which was carried in the cargo bay of the space shuttle *Atlantis*.

As currently explored, we know that our solar system contains 1 star (the Sun), 9 planets, 63 moons (at last count), 6 asteroids larger than 300 km in diameter, more than 4000 smaller (but well-studied) asteroids, myriad comets a few kilometers in diameter, and countless meteoroids less than a meter across. The list may grow as we continue to explore our neighborhood. The near-void between all these objects is termed **interplanetary space**.

6.2 *The Overall Layout of the Solar System*

❶ By terrestrial standards, the solar system is immense. The distance from the Sun to Pluto is about 40 A.U., almost a million times the radius of Earth and roughly 15,000 times the distance from Earth to the Moon. Despite its vast extent, though, the entire solar system lies very close to its parent Sun, astronomically speaking. Even the diameter of Pluto's orbit is less than 1/1000 of a light year, and the next nearest star is several light years distant from the Sun.

The planet closest to the Sun is Mercury. Moving outward, we encounter in turn Venus, Earth, Mars, Jupiter, Saturn, Uranus, Neptune, and Pluto. In Chapter 2 we saw the basic properties of the planets' orbits. Their paths are all ellipses, with the Sun near one focus. ∞ (p. 41) Most orbits have low eccentricities, the exceptions being the innermost and the outermost worlds, Mercury and Pluto. We can reasonably think of most planets' orbits as circles centered on the Sun. Figure 6.5 is an artist's rendition of the planetary system as future generations of space voyagers might perceive it from a distant vantage point.

All the planets orbit the Sun in nearly the same plane as Earth (the ecliptic plane). Again, Mercury and Pluto deviate somewhat from this rule—their orbital planes lie at 7° and 17° to the ecliptic, respectively. Still, we can think of the solar system as being quite flat. Its "thickness" perpendicular to the plane of the ecliptic is less than 1/50 the diameter of Pluto's orbit. If we were to view the planets' orbits from a vantage point in the ecliptic plane about 50 A.U. from the Sun, only Pluto's orbit would be noticeably tilted. All planets orbit the Sun in the same direction—counterclockwise, seen

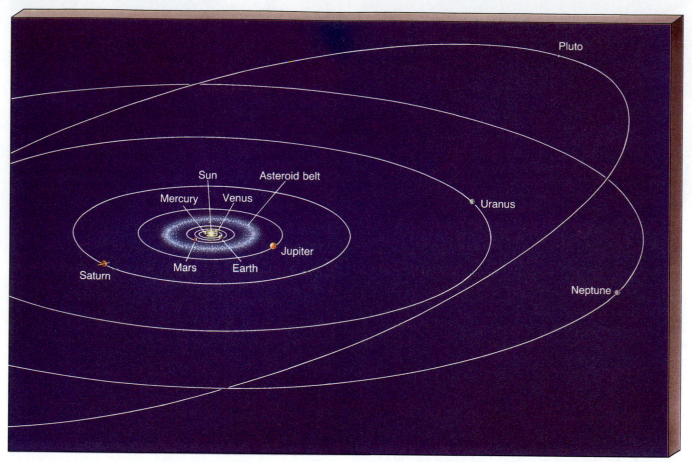

Figure 6.5 Might future space voyagers travel far enough from Earth to gain this perspective on our solar system? Except for Mercury and Pluto, the orbits of the planets lie nearly in the same plane. As we move out from the Sun, the distance between the orbits of the planets increases. The entire solar system spans nearly 80 A.U.

from above Earth's North Pole. Figure 6.6 is a photograph of planets Mercury, Venus, Mars, and Jupiter taken during the July 1991 solar eclipse. These four planets happen to be visible in this one photograph in large part because their orbits lie nearly in the same plane.

The planetary orbits are not evenly spaced: The orbits get farther and farther apart as we move farther out from the Sun. Nevertheless, there is a certain regularity in their spacing. In the eighteenth century a fairly simple rule, now known as the Titius-Bode law (see *Interlude 6-1* on p. 135), seemed to "predict" the radii of the planetary orbits remarkably well. Even the asteroid belt between Mars and Jupiter had a place in the scheme, which excited great interest among astronomers and numerologists alike. There is apparently no simple explanation for this empirical "law." Today, it is regarded more as a curiosity than as a fundamental property of the solar system.

6.3 Planetary Properties

The solar system presents us with a sense of orderly motion. The planets move nearly in a plane, on almost concentric elliptical paths, in the same direction around the Sun, at steadily increasing orbital intervals. However, the individual properties of the planets themselves are less reg-

Figure 6.6 Taken from Hawaii during the July 1991 eclipse of the Sun, this single photograph shows four planets—Mercury, Venus, Mars, and Jupiter. Because they all orbit in nearly the same plane, it is possible for them all to appear (by chance) in the same region of the sky, as seen from Earth.

TABLE 6-1 *Properties of Some Solar System Objects*

OBJECT	ORBITAL SEMI-MAJOR AXIS (A.U.)	ORBIT PERIOD (Earth years)	MASS (Earth masses)	RADIUS (Earth radii)	NUMBER OF KNOWN MOONS	ESCAPE VELOCITY (km/s)	ROTATION PERIOD (days)	AVERAGE DENSITY (kg/m³)	(g/cm³)
Mercury	0.39	0.24	0.055	0.38	0	4.3	59	5400	5.4
Venus	0.72	0.62	0.81	0.95	0	10.4	−243	5200	5.2
Earth	1.0	1.0	1.0	1.0	1	11.2	1.0	5500	5.5
Moon	1.0	—	0.012	0.27	—	2.4	27.3	3300	3.3
Mars	1.5	1.9	0.11	0.53	2	5.0	1.0	3900	3.9
Ceres (asteroid)	2.8	4.7	0.0002	0.07	0	0.6	0.38	2700	2.7
Jupiter	5.2	11.9	318	11.2	16	60	0.41	1300	1.3
Saturn	9.5	29.5	95	9.5	20	36	0.43	700	0.7
Uranus	19.2	84	15	4.0	15	21	−0.69	1200	1.2
Neptune	30.1	165	17	3.9	8	24	0.72	1700	1.7
Pluto	39.5	249	0.003	0.2	1	1.3	−6.4	2300	2.0
Sun	—	—	332,000	109	—	617	25.8	1400	1.4

ular. Some selected properties of various solar system objects are presented in Table 6-1.

Table 6-1 lists, for each object, its distance from the Sun, the length of its year, its mass and radius, the number of moons known to orbit it, its escape velocity, its rotation period, and finally its *density*—a very important property of any object, to which we will return in more detail in a moment. In most cases, all these quantities can be determined by straightforward methods already described in this book. As we saw in Chapter 2, the distance of each planet from the Sun is known from Kepler's laws once the scale of the solar system is set by radar-ranging on Venus. ∞ (p. 43) The masses of planets with moons may be calculated by application of Newton's laws of motion and gravity, again as described in Chapter 2, just by observing the moons' orbits around the planets. The sizes of those orbits, like the sizes of the planets themselves, are found by measuring their *angular* sizes and then applying elementary geometry, as we saw in Chapter 1. ∞ (p. 22)

The masses of Mercury and Venus, as well as our own Moon and the asteroid Ceres, are a little harder to determine accurately because they have no natural satellites of their own. Nevertheless, it is possible to measure their masses by careful observations of their influence on other planets or nearby bodies. Mercury and Venus produce small but measurable effects on each other's orbits, as well as that of the Earth. The Moon causes small "wobbles" in Earth's motion as the two bodies orbit their common center of mass. Only in the case of Ceres is the mass poorly known, primarily because that asteroid's gravity is so slight. All the techniques necessary for determining mass were available to astronomers well over a century ago. Today, the masses of most of the objects in Table 6-1 have been very accurately measured through their gravitational interaction with artificial satellites and space probes launched from Earth.

Once a planet's mass and radius are known, its escape velocity—the minimum speed required for an object to escape forever from its gravitational grasp—can be calculated from the formula given in Chapter 2. ∞ (p. 47) As explained in the *More Precisely* feature on p. 136, escape velocity is the primary factor in determining the composition, and even the existence, of a planet's atmosphere. A planet's rotation period is determined simply by watching surface features appear and disappear again as the planet rotates. For some planets this is difficult to do, as their surfaces are hard to see or may even be nonexistent! Venus's surface is completely obscured by clouds, while Jupiter, Saturn, Uranus, and Neptune have no solid surfaces at all—their atmospheres simply thicken, and eventually become liquid, as we descend deeper and deeper below the visible clouds. We will describe the methods used to measure the rotation periods of these planets in later chapters.

The final column in Table 6-1 lists a property called **density**. This is a measure of the "compactness" of the matter within an object. It is computed by dividing the object's mass (in kilograms, say) by its volume (in cubic meters, for example). Dividing Earth's mass (determined by observations of the Moon's orbit) by its volume (which we know because we know Earth's radius—see the *More Precisely* feature on p. 25), we obtain an average density of approximately 5500 kg/m³. *On average*, then, there are about 5500 kilograms of Earth matter in every cubic meter of Earth volume. For comparison, the density of ordinary water is 1000 kg/m³, rocks on the Earth's surface have densities in the range 2000–3000 kg/m³, and iron has a density of some 8000 kg/m³. Earth's atmosphere has a density of only a few kilograms per cubic meter. Because most astronomers are more familiar with the CGS units of density (grams per cubic centimeter—g/cm³—where 1 kg/m³ = 1000 g/cm³), Table 6-1 lists density in both SI and CGS units.

A clear distinction can be drawn between the inner and the outer members of our planetary system based on densities and other physical properties. In short, the inner planets—Mercury, Venus, Earth, and Mars—are small, dense, and *rocky* in composition. The outer worlds—Jupiter, Saturn, Uranus, and Neptune (but not Pluto)—are large, of low density, and *gaseous* in nature.

Because their physical and chemical properties are somewhat similar to Earth's, the four innermost planets—Mercury, Venus, Earth, and Mars—are often called the **terrestrial planets**. (The word *terrestrial* derives from the Latin word *terra*, meaning "land," or "earth.") The larger, outer planets—Jupiter, Saturn, Uranus, and Neptune—are often labeled the **jovian planets** because of their physical and chemical resemblance to Jupiter. (The word *jovian* comes from *Jove*, another name for the Roman god Jupiter.) The jovian worlds are all much larger than Earth and quite different from it in both composition and structure.

Finally, beyond the outermost gas giant lies one more small world, frozen and mysterious. Pluto doesn't fit well into either planetary category. It might once have been a moon of Neptune, or perhaps even a large cometlike body and not originally a planet at all.

6.4 *Terrestrial and Jovian Planets*

The four terrestrial planets are close to the Sun in astronomical terms, all being within 1.5 A.U. of their parent star. All are small and of low mass, and all have generally rocky composition and solid surfaces. Beyond that, however, the similarities end. When we take into account the differing compression of their interiors by the weight of the overlying layers (greatest for Earth, least for Mercury), we find that the average *uncompressed densities* of the terrestrial worlds—that is, the densities in the absence of any compression—decrease steadily as we move farther from the Sun. This indicates that their overall compositions differ significantly. In addition, the planets' present-day surface conditions are quite distinct. All but Mercury have atmospheres, but the atmospheres are about as dissimilar as we could imagine, ranging from a near-perfect vacuum on Mercury to a hot, dense inferno on Venus. Earth alone has oxygen in its atmosphere (as well as liquid water on its surface). Earth and Mars spin at roughly the same rate—one rotation every 24 (Earth) hours—but Mercury and Venus both take months to rotate just once. Earth and Mars have Moons, but Mercury and Venus do not. Finding the common threads in the evolution of four such worlds is no sim-

Figure 6.7   Diagram, drawn to scale, of the relative sizes of the planets and our Sun. Notice how much larger the jovian planets are than the Earth and the other terrestrials and how much larger still is the Sun.

ple task! Comparative planetology will be our indispensable guide as we proceed through the coming chapters.

For all the differences among the terrestrial worlds, they still seem very similar when compared with the jovian planets. Perhaps the simplest way to express the major differences between the terrestrial and jovian worlds is to say that the jovian planets are everything the terrestrial planets are not: The terrestrial worlds lie close together, near the Sun; the jovian worlds are widely spaced through the outer solar system. The terrestrial worlds are small, dense, and rocky; the jovian worlds are large and gaseous, being made up predominantly of hydrogen and helium (the lightest elements), which are quite rare on the inner planets for reasons to be described in the *More Precisely* feature on p. 136. The terrestrial worlds have solid surfaces; the jovian worlds have none (their dense atmospheres thicken with depth, eventually merging with their liquid interiors). The terrestrial worlds have weak magnetic fields, if any; the jovian worlds all have strong magnetic fields. The terrestrial worlds have only three moons among them; the jovian worlds each have many moons, no two of them alike and none of them like our own. In addition, all the jovian planets have rings, a feature unknown on the inner planets. Despite their greater size, the jovian worlds all rotate much faster than any terrestrial planet.

Figure 6.7 presents a diagram of the sizes of the planets relative to the Sun. The Sun is clearly the largest object in our solar system. It has over 1000 times the mass of the

TABLE 6-2   *Properties of Terrestrial and Jovian Planets*

TERRESTRIAL PLANETS	JOVIAN PLANETS
close to the Sun	far from the Sun
closely spaced orbits	widely spaced orbits
small masses	large masses
small radii	large radii
predominantly rocky	predominantly gaseous
high density	low density
slower rotation	faster rotation
weak magnetic fields	strong magnetic fields
few moons	many moons

next largest object, the planet Jupiter. The Sun in fact contains about 99.9 percent of all solar system material. The terrestrial planets—including our home planet—are insignificant in comparison. Table 6-2 compares and contrasts the key properties of the terrestrial and jovian worlds.

6.5 Interplanetary Debris

3 In the vast space among the nine known major planets move countless chunks of rock and ice, all orbiting the Sun, many of them on highly eccentric paths. This final

---

Interlude 6–1   The Titius-Bode "Law"

A close look at the list of semi-major axes of the known planets (see Table 6-1) reveals considerable regularity in their orbits. The *spacing* of the orbits increases more or less geometrically as we move out from the Sun: At any point in the list, the distance to the next planet out is about twice that to the next planet in.

Is there some underlying structure to the solar system? In 1766, in search of an answer to that question, the German astronomer Johann Titius came up with a simple formula that "predicted" quite well the orbits of the then-known planets, Mercury through Saturn. Johann Bode, a better-known astronomer of the day, later popularized this relationship among the planets' orbits, so that it now is usually known as the Titius-Bode law, or even just Bode's law. We must emphasize that it is not a law at all, in the scientific sense, but rather just a rule for determining the approximate orbital semi-major axes of the planets. Nevertheless, the *law* part of the name has stuck.

The rule for determining the planets' orbits is as follows. Start with 0.4, the distance (in A.U.) from the Sun to Mercury. Then add to it 0.0 (Mercury: 0.4 A.U.), 0.3 (Venus: 0.7 A.U.), 0.6 (Earth: 1.0 A.U.), 1.2 (Mars: 1.6 A.U.), 2.4 (?: 2.8 A.U.), 4.8 (Jupiter: 5.2 A.U.), and 9.6 (Saturn: 10.0 A.U.) to arrive at the (approximate) orbital distances of the known planets, plus one extra "planet" between Mars and Jupiter. The relation between successive orbits is easily seen—after Venus, the added

term simply doubles at each step. Even the fictitious extra "planet" is acceptable—it corresponds to the middle of the asteroid belt! Thus it was that this rule appeared to contain some deep insight into our planetary system.

The test of any theory is its predictive power. According to the Titius-Bode rule, the next three planets beyond Saturn should lie at 19.6, 38.8, and 77.2 A.U. When Uranus was discovered in 1781, it fell close to the prediction, at 19.2 A.U. Speculation was rife that the next planet would also lie where the "law" decreed it should. Sadly, the rule fails for Neptune, which lies only 30 A.U. from the Sun. However, if we ignore Neptune, Pluto is in just about the right place!

Few people today take the rule sufficiently seriously to bother to look for a scenario in which Neptune started out in the "right" orbit and somehow got moved farther in toward the Sun, perhaps leaving Pluto behind. Still, the regularity of the "law" was appealing to astronomers of the eighteenth and nineteenth centuries, as it suggested some fundamental harmony in the structure of the solar system, and, to some extent, that appeal remains today.

The Titius-Bode rule is not the result of any known planetary interaction, nor is it even very accurate. Yet many astronomers suspect that it may be telling us *something* about the formation of the solar system—they just don't know quite what.

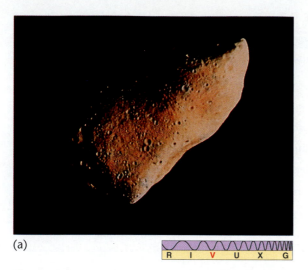

(a)

(b)

R I V U X G   R I V U X G

Figure 6.8 (a) Asteroids, like meteoroids, are generally composed of rocky material. This asteroid, Gaspra, is about 20 km long; it was photographed by the *Galileo* spacecraft on its way to Jupiter. (b) Comet West, seen as it approached the Sun in 1976. Most comets are composed largely of ice, and so tend to be relatively fragile. Shortly after this photograph was taken, the comet split into several fragments.

component of the solar system is the collection of *interplanetary matter*—debris that ranges in size from the relatively large asteroids, through the smaller comets and even smaller meteoroids, down to the smallest grains of interplanetary dust that litter our cosmic environment. Larger bodies collide and break apart into smaller bodies. In turn, smaller bodies collide and are ground into dust, which eventually settles into the Sun or is swept away by the *solar wind*, a stream of energetic charged particles that continually flows outward from the Sun. The dust is generally quite difficult to detect in visible light, but studies in the infrared range reveal that interplanetary space contains surprisingly large amounts of it. Our solar system is an extremely good vacuum by terrestrial standards, but positively dirty by the standards of interstellar or intergalactic space.

Asteroids (Figure 6.8a) and meteoroids are generally rocky, a little like the outer layers of the terrestrial planets. Their total mass is much less than that of Earth's Moon. They are important because many of them are made of material that has scarcely changed since the early days of the solar system. In addition, they often conveniently deliver themselves right to our doorstep (in the form of meteorites), allowing us to study them in detail without having to fetch them from space.

The comets are quite distinct from the other small bodies in the solar system. They are icy rather than rocky—in fact, they are quite similar in composition to some of the icy moons of the outer planets—but they too represent truly ancient material. Since their formation long ago, along with the rest of the solar system, most comets have probably not changed or interacted with anything at all. Comets striking Earth's atmosphere do not reach the surface intact, so we do not have actual samples of cometary material. However, they do vaporize and emit radiation as their highly elongated orbits take them near the Sun (Figure 6.8b). We can,

therefore, determine their makeup by spectroscopic study of the radiation they emit as they are destroyed. ∞ (p. 84) In this way we can obtain information on what the solar system was like soon after its birth.

6.6 Spacecraft Exploration of the Solar System

4 Since the 1960s, there have been dozens of unmanned space missions to the other planets in the solar system. All the planets but Pluto have been visited and probed at close range by U.S. or Soviet craft. The impact of these missions on our understanding of our planetary system has been nothing short of revolutionary. In the next few chapters, we will see many examples of the marvelous images radioed back to Earth. Here, we focus on just a few of these remarkable technological achievements.

THE *MARINER 10* FLYBYS OF MERCURY

In 1974, the U.S. spacecraft *Mariner 10* came within 10,000 km of the surface of Mercury, sending back high-resolution images of the planet. These photographs, which showed surface features as small as 150 m across, revolutionized our knowledge of the planet. For the first time, we saw Mercury as a heavily cratered world, in many ways reminding us of our own Moon.

Mariner 10 (see Figure 6.9) was launched from Earth in November 1973 and was placed in an eccentric 176-day orbit about the Sun, aided by a gravitational assist (see *Interlude 6–2* on p. 139) from the planet Venus. In that orbit, *Mariner 10*'s nearest point to the sun (perihelion) is close to Mercury's path, and its farthest point away (aphelion) lies between the orbits of Venus and Earth. The 176-day period is exactly two Mercury years, so the spacecraft revisits Mercury roughly every 6 months. However, only

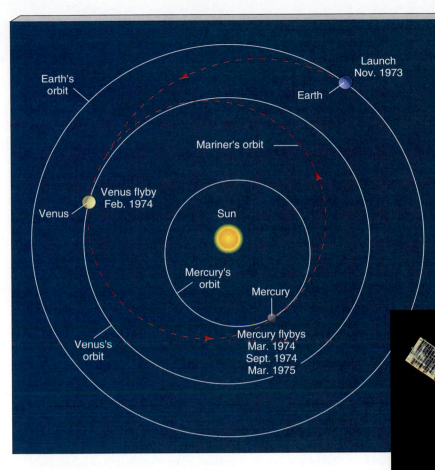

Figure 6.9   The path of the *Mariner 10* probe to Mercury included a gravitational boost from Venus. The spacecraft (inset) returned data from March 1974 until March 1975 providing astronomers with a wealth of information on the planet Mercury.

on the first three encounters—in March 1974, September 1974, and March 1975—did the spacecraft return data. After that, the craft's supply of maneuvering fuel was exhausted. In total, over 4000 photographs, covering about 45 percent of the planet's surface, were radioed back to Earth during the mission's active lifetime. The remaining 55 percent of Mercury is still unexplored.

EXPLORATION OF VENUS

In all, some 20 spacecraft have visited Venus since the 1970s, far more than have spied on any other planet. The Soviet space program took the lead role in exploring Venus's atmosphere and surface, while American spacecraft have performed extensive radar mapping of the planet from orbit. The American *Mariner 2* and *Mariner 5* missions passed within 35,000 km of the planet in 1962 and 1967, while *Mariner 10* grazed Venus at a distance of 6000 km en route to Mercury.

During roughly the same period, the Soviet *Venera* (derived from the Russian word for Venus) program got under way, and the Soviet *Venera 4* through *Venera 12* probes parachuted into the planet's atmosphere between 1967 and 1978. The early *Venera* probes were destroyed by enormous atmospheric pressures before reaching the surface. Then, in 1970, *Venera 7* (sketched in Figure 6.10) became the first spacecraft to soft-land on the planet. During the 23 minutes it survived on the surface, it radioed back information on atmospheric pressure and temperature. Since that time, a

Figure 6.10   One of the Soviet *Venera* landers that reached the surface of Venus. The design was essentially similar for all the surface missions. Note the heavily armored construction, necessary to withstand the harsh conditions on the planet's surface.

number of *Venera* landers have transmitted photographs of the surface back to Earth and have analyzed the atmosphere and the soil. None of them survived for more than an hour in the planet's hot, dense atmosphere. The data they sent

More Precisely... **Why Air Sticks Around**

Some planets and moons in the solar system—for example, Venus and Titan (the largest moon of Saturn)—have atmospheres thicker than Earth's. Other objects—such as Mars—have thinner atmospheres. Still others—for example, our own Moon and the planet Mercury—have virtually no atmosphere at all. The jovian planets have atmospheres rich in hydrogen, while gaseous hydrogen is rare on the terrestrial worlds.

Why do planets have atmospheres, and what determines their composition? Why does a layer of air, made up mostly of nitrogen and oxygen, lie just above the surface of Earth? After all, experience shows that most gas naturally expands to fill all the volume available. Perfume in a room, fumes from a poorly running engine, and steam from a tea kettle all rapidly disperse until we can hardly sense them. Why, then, doesn't our atmosphere similarly disperse by just floating away into outer space?

The answer is that *gravity* holds it down. The gravitational field of the Earth exerts a pull on all the atoms and molecules in our atmosphere, preventing them from escaping. However, gravity can't be the only influence on the air. Other agents must compete with Earth's gravity to keep the atmosphere buoyant. Otherwise, gravity would eventually pull all the air down to the ground.

It is *heat* that competes with gravity. All gas molecules are in constant random motion. The *temperature* of the gas is a direct measure of this motion—the hotter the gas, the faster the molecules are moving. (See the *More Precisely* feature on p. 63.) The Sun continually supplies heat to our planet's atmosphere, and the resulting rapid movement of heated molecules produces pressure. This pressure tends to oppose the force of gravity, exerting a net upward force and preventing our atmosphere from collapsing under its own weight. If we are to understand planetary atmospheres, we must explore this competition between gravity and heat in a little more detail.

An important measure of the strength of a body's gravity is its *escape velocity*—the speed needed for any object to escape forever from its surface. As we saw in Chapter 2, this speed increases with increased mass or decreased radius of the parent body (often a moon or a planet). ∞ (p. 47) Mathematically, it can be expressed in the following way:

$$\text{escape velocity} \propto \sqrt{\frac{\text{mass of parent object}}{\text{radius of parent object}}}.$$

Thus, if the *mass* of the parent were to quadruple, the escape velocity would double. If the parent's *radius* were to quadruple, then the escape velocity would be halved.

Earth's escape velocity is about 11 km/s. This is the minimum velocity needed for any object—a rocket, a baseball, or a molecule—to escape from our planet's surface. Civilian rockets designed to propel robot space probes toward the outer planets must attain even greater velocities because they must overcome the gravitational pull of the Sun in addition to escaping from our planet. Military intercontinental ballistic missiles are designed to achieve less than escape velocity (actually, about 6 km/s), so that they follow a curved path back to their designated target on Earth. A speeding bullet attains a maximum velocity of only about 1 km/s—far less than escape velocity—which explains why bullets don't leave the Earth when fired from guns. Table 6-1 lists the escape velocities for some prominent members of the solar system.

To determine whether or not a planet will retain an atmosphere, we must compare the planet's escape velocity with the *molecular velocity,* which is the average speed of the gas particles making up the atmosphere. This speed actually depends not only on the *temperature* of the gas, as described earlier, but also on the *mass* of the individual molecules—the hotter the gas, or the smaller the molecular mass, the larger will be the average velocity of the molecules:

$$\text{average molecular velocity} \propto \sqrt{\frac{\text{temperature of gas}}{\text{molecular mass}}}.$$

Thus, increasing the temperature of a sample of gas by a factor of 4—from, for example, 100 K to 400 K—doubles the average speed of its constituent molecules. And, at a given temperature, molecules of hydrogen in air move, on average, 4 times faster than molecules of oxygen, which are 16 times heavier.

For nitrogen and oxygen in Earth's atmosphere, where the temperature near the surface is nearly 300 K, the typical molecular velocity is about 0.5 km/s, far smaller than the 11 km/s needed for a molecule to escape into space. As a result, Earth is able to retain its nitrogen–oxygen atmosphere. On the

back make up the entirety of our direct knowledge of Venus's surface. In 1983, the *Venera 15* and *Venera 16* orbiters sent back detailed radar maps (at about 2 km resolution) of large portions of Venus's northern hemisphere.

The United States' *Pioneer Venus* mission in 1978 placed an orbiter at an altitude of some 150 km above Venus's surface and dispatched a "multiprobe," consisting of five separate instrument packages, into the planet's atmosphere. During their hour-long descent to the surface, the probe returned information on the variation of density, temperature, and chemical composition with altitude in the atmosphere. The orbiter's radar produced images of most of the planet's surface.

The most recent U.S. mission was the *Magellan* probe (shown in Figure 6.11), which entered orbit around Venus in August 1990. The spacecraft began sending back spectacular data (see Chapter 9) in September 1990. It completed its first 243-day mapping cycle (the time required for Venus to rotate once beneath the probe's orbit) in May 1991. *Magellan's* spatial resolution was at least 10 times better than the best images previously obtained. It could distinguish objects as small as 120 m across and measure vertical distances to within less than 50 m. The probe covered the entire surface of Venus with unprecedented clarity, rendering all previous data virtually obso-

whole, the gravity of our planet simply has more influence than the heat of our atmosphere.

The situation is a little more complicated than this simple comparison of velocities would suggest. Atmospheric molecules can gain or lose energy (and, therefore, velocity) by bumping into one another or by colliding with objects near the ground. Thus, we can characterize a gas by its average molecular velocity, but, as indicated in the figure below, the molecules do not *all* move at the same speed. Actually, only a tiny fraction of the molecules in any gas has velocity much greater than average—one molecule in 2 million has a velocity more than three times the average, and only one in 10^{16} exceeds the average by more than a factor of 5. Nevertheless, this means that at any instant, *some* molecules are moving fast enough to escape, even when the average molecular speed is much less than the escape velocity. The result is that all planetary atmospheres slowly leak away into space.

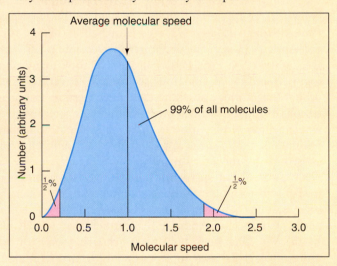

Don't be alarmed—the leakage is usually very gradual! As a rule of thumb, if the escape velocity from a planet exceeds the average velocity of a given type of molecule by a factor of 6 or more, then molecules of that type will not have escaped from the planet's atmosphere in significant quantities since the solar system formed. Conversely, if the escape velocity is less than six times the average velocity of molecules of a given type, then most of them will have escaped by now, and we should not expect to find them in the atmosphere.

This line of reasoning explains the lack of any significant atmosphere on the Moon and the planet Mercury. Mercury's escape velocity is 4.3 km/s. Its peak surface temperature is around 700 K, corresponding to an average molecular velocity for nitrogen or oxygen of about 0.8 km/s. Thus, the escape velocity is less than six times the average molecular velocity (4.8 km/s), and there has been ample time for those gases to escape. Similarly, if our own Moon originally had an Earthlike atmosphere, it would have been heated by the Sun to much the same temperature as Earth's air today, so the average molecular velocity would have been about 0.5 km/s. However, the Moon's escape velocity is only 2.4 km/s. Thus, any original lunar atmosphere has long ago dispersed into interplanetary space.

We can also use these arguments to understand some aspects of atmospheric *composition*. For example, hydrogen molecules move, on average, at about 2 km/s in Earth's atmosphere at sea level, so they have had plenty of time to escape since our planet formed (as 6×2 km/s = 12 km/s, which is greater than Earth's 11 km/s escape velocity). Consequently, we find very little hydrogen in Earth's atmosphere today. On Jupiter, where the temperature is lower (about 100 K, much less than Earth's 300 K), the velocity of hydrogen molecules is correspondingly lower—about 1.2 km/s. At the same time, Jupiter's escape velocity is 60 km/s, over five times higher than on Earth. For those reasons, Jupiter has retained its hydrogen—in fact, hydrogen is the dominant ingredient of Jupiter's atmosphere.

The distinction between those properties of a planet (such as the existence and chemical makeup of an atmosphere) that can be explained by *evolutionary* processes and those properties (for example, mass and overall planetary composition) that were laid down when the planet formed is very important to understanding the formation of the solar system. By categorizing observed planetary properties into "evolutionary" and "original" in nature, we will compile a list of solar-system characteristics that must have been determined at the time the solar system formed. In doing so, we will lay the foundation for the theory of how stars and planets are born.

lete. Many theories of the processes shaping the planet's surface have been radically altered or abandoned completely because of *Magellan*.

SPACECRAFT EXPLORATION OF MARS

Both NASA and the Soviet (now Russian) space agency have Mars exploration programs that began back in the 1960s. However, the Soviet effort was plagued by a string of technical problems, along with a liberal measure of plain bad luck. As a result, almost all of the detailed planetary data we have on Mars has come from unmanned U.S. probes launched in the 1960s and 1970s.

The first spacecraft to reach the red planet was *Mariner 4*, which flew by Mars in July 1965. The images sent back by the craft showed large numbers of craters caused by meteoroids impacting the planet's surface, but nothing of the Earthlike terrain some scientists had expected to find. Flybys in 1969 by *Mariner 6* and *Mariner 7* confirmed these findings, leading to the conclusion that Mars was a geologically dead planet having a heavily cratered, old surface. Studies of Mars received an enormous boost with the arrival in November 1971 of the *Mariner 9* orbiter. The craft mapped the entire Martian surface at a resolution of about 1 km, and it rapidly became clear that here was a world far

Figure 6.11 The U.S. *Magellan* spacecraft is launched from the space shuttle *Atlantis* in May 1989.

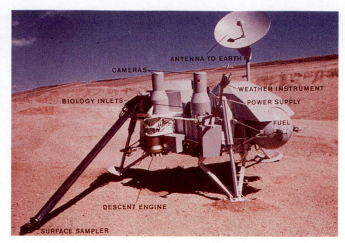

Figure 6.12 A *Viking* lander, here being tested in the Mojave Desert prior to launch.

MISSIONS TO THE OUTER PLANETS

Two pairs of U.S. spacecraft have revolutionized our knowledge of Jupiter and the jovian planets. The first pair, *Pioneer 10* and *Pioneer 11*, were launched in March 1972 and April 1973, arriving at Jupiter in December 1973 and December 1974. The *Pioneer* spacecraft took many photographs and made numerous scientific discoveries. Their orbital trajectories also allowed them to observe the polar regions of Jupiter in much greater detail than later missions would achieve. In addition to their many scientific accomplishments, the *Pioneer* craft also played an important role as "scouts" for the later *Voyager* missions. The *Pioneer* series demonstrated that spacecraft could travel the long route from Earth to Jupiter without colliding with debris in the solar system. They also discovered—and survived—the perils of Jupiter's extensive radiation belts (somewhat like Earth's Van Allen belts, but on a much larger scale). In addition, *Pioneer 11* used Jupiter's gravity to propel it along the same trajectory to Saturn that the *Voyager* controllers planned for *Voyager 2*'s visit to Saturn's rings.

Figure 6.13 The *Voyager* spacecraft. *Voyager 1* and *Voyager 2* (shown here) were identical.

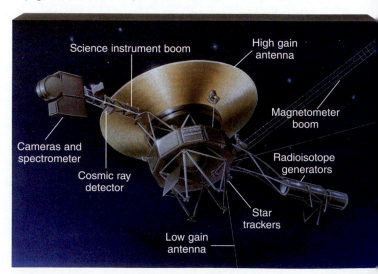

more complex than the dead planet imagined only a year or two previously. *Mariner 9*'s maps revealed vast plains, volcanoes, drainage channels, and canyons. All these features were completely unexpected, given the data provided by the earlier missions. These new findings paved the way for the next step—actual landings on the planet's surface.

The last U.S. spacecraft to visit Mars were the two *Viking* missions, which reached the planet in mid-1976. *Viking 1* and *Viking 2* each consisted of two parts. An orbiter mapped the surface at a resolution of about 100 m (about the same as the resolution achieved by *Magellan* on Venus), and a lander (see Figure 6.12) descended to the surface and performed a wide array of geological and biological experiments. *Viking 1* touched down on Mars on July 20, 1976. *Viking 2* arrived in September of the same year.

By any standards, the mission was a complete success, as the *Viking* orbiters and landers returned a wealth of data on the Martian surface and atmosphere. In August 1993, the first U.S. probe since *Viking*—the *Mars Observer*, which was designed to radio back detailed images of the planet's surface and provide data on the Martian atmosphere—mysteriously fell silent, for reasons that are still unclear, just before entering orbit . As a result, the U.S. *Mars* program is now on hold. Although most space scientists would agree that more exploration of Mars is needed, they disagree as to the best means of achieving that goal. NASA has ambitious plans for manned missions to Mars. However, the enormous expense of such an undertaking, coupled with the belief of many astronomers that unmanned missions are economically and scientifically preferable to manned missions, make the future of these projects uncertain at best.

Interlude 6–2   *Interplanetary Navigation*

Celestial mechanics is the study of the motions of gravitationally interacting objects, such as planets and stars. These days, people studying the subject bring powerful computers to bear on Newton's laws of motion (see Chapter 2) to understand the intricate movements of astronomical bodies. Computerized celestial mechanics enables astronomers to calculate the orbits of the planets with high precision, taking their small gravitational influences on one another into account. Even before the computer age, the discovery of one of the outermost planets, Neptune, came about almost entirely through studies of the distortions of Uranus's orbit that were caused by Neptune's gravity.

Celestial mechanics has become an essential tool as scientists and engineers navigate manned and unmanned spacecraft throughout the solar system. Robot probes can now be sent on stunningly accurate trajectories, expressed in the trade with such slang phrases as "sinking a corner shot on a billion-kilometer pool table." In fact, near-flawless rocket launches, aided by occasional midcourse changes in flight paths, now enable interplanetary navigators to steer remotely controlled spacecraft through an imaginary "window" of space just a few kilometers wide and a billion kilometers away.

Sophisticated knowledge of celestial mechanics also aids navigation of a single space probe toward several planets. For example, in 1974 the *Mariner 10* spacecraft was guided into the part of Venus's gravitational field that would swing the ship around to precisely the right path for an additional trek toward Mercury. In other words, Venus itself propelled the probe in a new direction, a course alteration that required no fuel.

The accompanying figure illustrates how such a *gravitational slingshot* works. From the point of view of the planet, the spacecraft arrives along a hyperbolic (unbound) trajectory, passes close by, and then escapes along the same trajectory, in a new direction and with the same speed relative to the planet. However, if the planet itself is moving, some of the planet's momentum is transferred to the spacecraft as it passes by. If the orbit of the spacecraft is chosen correctly, the craft can gain energy, and speed up, as a result of the encounter. Of course, there is no "free lunch"—the extra energy acquired by the spacecraft comes from the planet's motion, causing its orbit to change ever so slightly. However, since planets are so much more massive than spacecraft, the effect is tiny—*Mariner 10* had no measurable effect on Venus's orbit.

Such a slingshot maneuver has been used several times. For example, the *Voyager 2* spacecraft (launched in late 1977) closely passed by Jupiter (in 1979), Saturn (in 1981), Uranus

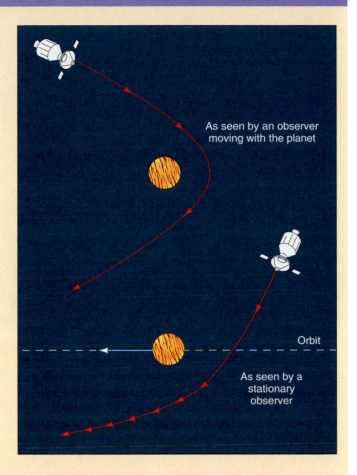

As seen by an observer moving with the planet

Orbit

As seen by a stationary observer

(in 1986), and Neptune (in 1989). It is now moving through the outermost reaches of the solar system. The gravitational fields of the giant planets whipped the craft around at each visitation, enabling flight controllers to get considerable extra "mileage" out of the probe.

The *Galileo* mission to Jupiter which was launched in 1989 and arrived at its target in 1995, received *three* gravitational assists en route—one from Venus and two from Earth. Once in the Jupiter system, one of its probes used the gravity of Jupiter and its moons to propel it through a complex series of maneuvers designed to bring it close to all the major moons as well as to the planet itself. Every encounter with a moon had a slingshot effect—sometimes accelerating and sometimes slowing the probe, but each time moving it onto a different orbit—and every one of these effects was carefully calculated long before *Galileo* ever left the Earth.

The two *Voyager* spacecraft (see Figure 6.13) left Earth in 1977, reaching Jupiter in March (*Voyager 1*) and July (*Voyager 2*) of 1979 to study the planet and its major satellites in detail. Each craft carried sophisticated equipment to study the planet's magnetic field, as well as radio, visible-light, and infrared sensors to analyze its reflected and emitted radiation.

Both *Voyager 1* and *Voyager 2* used Jupiter's gravity to send them on to Saturn. *Voyager 1* visited Titan, Saturn's largest moon, and so did not come close enough to the planet to receive a gravity-assisted boost to Uranus. However, *Voyager 2* went on to visit both Uranus and Neptune in a spectacularly successful "Grand Tour" of the outer planets. The data returned by the two craft are still

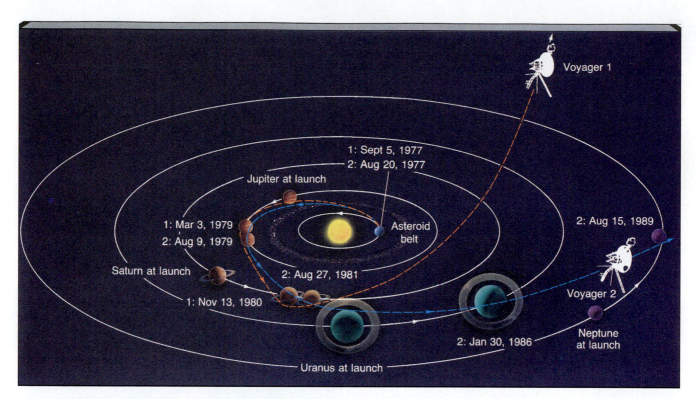

Figure 6.14   The paths taken by the two *Voyager* spacecraft to reach the outer planets. *Voyager 1* is now high above the plane of the solar system, having been deflected up and out of the ecliptic plane following its encounter with Saturn. *Voyager 2* continued on for a "Grand Tour" of the four jovian planets. It is now outside the orbit of Pluto.

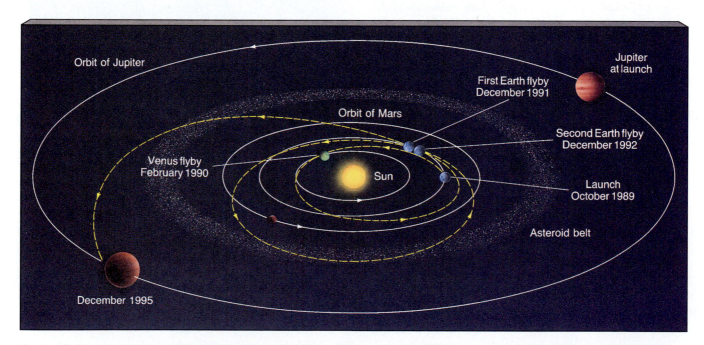

Figure 6.15   *Galileo*'s path to Jupiter included one flyby of Venus, two of Earth, and two trips through the asteroid belt, before reaching its destination in 1995.

being analyzed today. Like *Pioneer 11*, the two *Voyager* craft are now headed out of the solar system, still sending data as they race toward interstellar space. Figure 6.14 shows the past and present trajectories of the *Voyager* spacecraft.

The most recent mission to Jupiter is the U.S. *Galileo* probe, launched by NASA in 1989 (see Figure 6.4). It arrived at its target in 1995 after a rather roundabout route (shown in Figure 6.15) involving a gravity assist from Venus and two from Earth itself. The mission consists of an orbiter and an atmospheric probe. The probe will descend into the atmosphere of Jupiter, slowed by a heat shield and a parachute, making measurements and chemical analyses as it goes. The orbiter will execute a complex series of gravity-assisted maneuvers through Jupiter's moon system, returning to some moons already studied by *Voyager* and visiting others for the first time. Scientists around the world eagerly anticipate the results of the mission. Though *Galileo*'s findings may answer many old riddles, they are sure to pose many new ones too.

Shortly after launch, *Galileo* mission controllers discovered that the craft's main antenna, needed to radio its findings back to scientists on Earth, had failed to fully deploy. Despite repeated efforts, the instrument remains jammed in an almost closed position. As a result, all data must be transmitted via a small secondary antenna, greatly reducing the amount of information that can be returned. Nevertheless, with better data processing techniques, engineers believe that most of the scientific objectives of the mission can still be met. Already, *Galileo* has provided astronomers with spectacular views of Earth and Moon, as well as the only close-up photographs of asteroids ever obtained.

6.7 Comparative Planetology

5 As we proceed through the solar system, we will seek to understand how each planet compares with our own Earth–Moon system, and we will learn what each planet contributes to our knowledge of the solar system as a whole. On the basis of our studies of our own backyard—the Earth and Moon—we will identify some important questions to ask. For example, does the planet have a magnetic field, an atmosphere, or geological activity? Does it have a rocky surface? Does it have a liquid core? But, as we will see, each planet will also present us with new questions and insights of its own. Whatever the answers, the comparison enriches our knowledge of the ways planets work.

As we catalog the similarities and differences among the planets, terrestrial and jovian, what can we learn? To understand the solar system, we must try to answer basic questions such as, "Why did planet X evolve in one way, while planet Y turned out completely different?" "Why are the planets' orbits so orderly, when their individual properties are not?" "What can we learn from the makeup of the debris that orbits among the planets?" Our main problem in finding answers is that we don't have many objects to work with. Thus, we must study the properties of all known planets, moons, and larger asteroids in great detail to determine their common features and their differences and to find the reasons for both. Our goal is to develop a comprehensive theory of the *origin* and *evolution* of the solar system that explains all, or at least most, of its observed properties. As we unravel the origin of our own solar system, perhaps we will learn about the origin of planetary systems beyond our own.

Chapter Review

SUMMARY

The **solar system** (p. 128) consists of the Sun and everything that orbits it, including the nine major **planets** (p. 128), the **moons** (p. 128) that orbit them, and the many small bodies found in **interplanetary space** (p. 129).

The **asteroids** (p. 129), or "minor planets," are small bodies, none of them larger than Earth's Moon, most of which orbit in a broad band called the **asteroid belt** (p. 129) between the orbits of Mars and Jupiter.

Comets (p. 129) are chunks of ice found mostly in the outer solar system. Their importance to planetary astronomy lies in the fact that they are thought to be "leftover" material from the formation of the solar system and therefore contain clues to the very earliest stages of its development.

The major planets orbit the Sun in the same sense—counter-clockwise as viewed from above Earth's north pole—on roughly circular orbits that lie close to the ecliptic plane. The orbits of the innermost planet, Mercury, and the outermost, Pluto, are the most eccentric and have the greatest orbital inclination. The spacing between planetary orbits increases as we move outward from the Sun.

Density (p. 131) is a convenient measure of the compactness of any object. The average density of a planet is obtained by dividing the planet's total mass by its volume. The innermost four planets in the solar system have average densities compara-ble to the Earth and are generally rocky in composition. The outermost planets have much lower densities and, with the exception of Pluto, are made up mostly of gaseous or liquid hydrogen and helium.

All planetary atmospheres slowly leak away into space. The higher the temperature, or the weaker the planet's gravity, the faster this occurs. Based on densities and composition, planetary scientists divide the eight large planets (excluding Pluto) in the solar system into the rocky **terrestrial planets** (p. 132)—Mercury, Venus, Earth, and Mars—which lie closest to the Sun, and the gaseous **jovian planets** (p. 132)—Jupiter, Saturn, Uranus, and Neptune—which lie at greater distances. Compared to the terrestrial worlds, the jovian planets are larger and more massive, rotate more rapidly, and have stronger magnetic fields. In addition, the jovian planets all have ring systems and many moons orbiting them.

All the major planets, with the exception of Pluto, have been visited by unmanned space probes. Spacecraft have landed on Venus and Mars. In many cases, the spacecrafts' trajectories have included "gravitational assists" from one or more planets to reach their destinations.

The science of **comparative planetology** (p. 128) compares and contrasts the properties of the diverse bodies found in the solar system, to understand better the conditions under which planets form and develop.

SELF-TEST: True or False?

____ **1.** Moons are small bodies that orbit the Sun.

____ **2.** Most planets orbit the Sun in nearly the same plane as the Earth.

____ **3.** The Titius-Bode Law, giving the spacing of planetary orbits, is not a true law at all and is not very accurate.

____ **4.** The largest planets also have the largest densities.

____ **5.** A planet with a density of 5,000 kg/m^3 most likely has a gaseous composition.

____ **6.** The larger the radius of a body (of constant mass), the higher its escape velocity.

____ **7.** The solar wind occurs on Earth during days of exceptional solar heating.

____ **8.** The *Mars Observer* was successful in mapping the surface of Mars to a resolution of 100 m.

____ **9.** All planets have moons.

____ **10.** The landing of the Soviet *Venera* 7 on the planet Venus was notable because it was the first time any spacecraft had ever landed on a planet.

SELF-TEST: Fill in the Blank

1. The major bodies orbiting the Sun are known as _____.

2. The _____ are bodies that orbit the Sun between the orbits of Mars and Jupiter

3. The two planets with the highest eccentricities and orbital tilts are _____ and _____.

4. The temperature of a gas is a measure of the _____ of its molecules.

5. In order for Earth's atmosphere to retain hydrogen, the Earth's escape velocity would have to be much _____.

6. Asteroids and meteoroids have a _____ composition, in contrast to comets, which have an _____ composition.

7. Asteroids are similar in composition to the _____ planets.

8. Comets have compositions similar to the _____ moons of the _____ planets.

9. The *Mariner 10* spacecraft was sent by _____ (country) to photograph the planet _____ .

10. The U.S *Magellan* probe mapped the entire surface of Venus using _____.

11. The *Viking 1* and *Viking 2* missions sent orbiters and landers to the planet _____.

12. Jupiter was first visited by the U.S. spacecraft _____.

13. The U.S. spacecraft _____ is the only probe to have visited each of the Jovian planets.

14. Spacecraft visiting one planet often use a _____ assist to visit another planet.

REVIEW AND DISCUSSION

1. Name and describe all the different types of objects found in the solar system. Give one distinguishing characteristic for each. Include a mention of interplanetary space!

2. What is the order of the planets, from the closest to the farthest from the Sun?

3. What is comparative planetology? Why is it useful? What is its ultimate goal?

4. Why has our knowledge of the solar system increased greatly in recent years?

5. Compare and contrast Kepler's Laws with the Titius-Bode "Law" (Interlude 6-1). Why are Kepler's Laws considered to be true natural laws, while the Titius-Bode "Law" is not?

6. According to the Titius-Bode "Law," where should the ninth, tenth, and eleventh planets in the solar system lie?

7. Which are the terrestrial planets? Why are they given this name? Repeat the question for the jovian planets. What property actually defines these two different types of planets?

8. Name three differences between the terrestrial planets and the jovian planets.

9. Compare the properties of Pluto given in Table 6-1 to the properties of the terrestrial and jovian planets in Table 6-2. What do you conclude regarding the classification of Pluto as either a terrestrial or jovian planet?

10. Discuss two mechanisms that operate to keep a planet's atmosphere in place. What would happen to the atmosphere if a planet's surface temperature increased or if its mass decreased?

11. Why are asteroids and meteoroids important to planetary scientists?

12. Comets generally vaporize upon striking Earth's atmosphere. How then do we know their composition?

13. How and why do scientists use gravity assists to propel spacecraft through the solar system?

14. Which planets have been visited by spacecraft from Earth? On which ones have spacecraft actually landed?

PROBLEMS

1. Choose one of the objects in Table 6-1 and calculate its density. Compare your result with that given in the table. Assume that all objects are spherical.

2. How long would it take for a radio signal to complete the round trip between Earth and Jupiter? Assume that Jupiter is at its closest point to Earth. Why can't Mission Control on Earth maneuver spacecraft in real time?

3. Choose one of the objects in Table 6-1 and calculate its escape velocity using the relationship given in the *More Precisely* feature on p. 136. (Hint: Think in terms of Earth units and use the fact that Earth's escape velocity is 11.2 km/s.)

4. Suppose the average mass of each of 6000 asteroids (which includes a few yet-to-be-discovered asteroids!) in the solar system is about 10^{17} kg. Compare the total mass of all asteroids to the mass of the Earth.

5. A spacecraft has an orbit that just grazes Earth's orbit at perihelion and that of Mars at aphelion. What is its orbital eccentricity and semi-major axis? What is its orbital period? (This is the so-called *minimum energy orbit* for a craft leaving Earth and reaching Mars. Assume circular orbits for simplicity.)

6. Assume that a planet will have lost its atmosphere by the present time if the molecular velocity exceeds one-sixth of the escape velocity. What would Jupiter's temperature have to be for its hydrogen to have escaped by now?

PROJECTS

1. You can begin to visualize the ecliptic—the plane of the planets' orbits—just by noticing the path of the Sun throughout the day and of the full Moon in the course of a single night. It helps if you watch from one spot, such as your backyard or a rooftop. It's also good to have a general notion of direction (west is where the Sun sets). You will see that the movements of the Sun and Moon are confined to a narrow pathway across our sky. The planets also travel along this path. The motion of the Sun, Moon, and planets is a two-dimensional reflection of the three-dimensional plane of our solar system.

2. Once you get a feeling for the whereabouts of the ecliptic, try locating the North Star. Knowing the direction to celestial north makes it easier to imagine the motion of the planets in the plane of the solar system. Don't worry about being too precise. Just get a sense of the ecliptic as a kind of merry-go-round of planets—that we on Earth also ride!

A

B

C

D

7

THE EARTH

Our Home in Space

LEARNING GOALS

Studying this chapter will enable you to:

1 Summarize the overall physical properties of the Earth.

2 Describe how the Moon and Sun influence Earth's oceans.

3 Explain how Earth's atmosphere helps to heat us as well as protect us.

4 Discuss the nature and origin of Earth's magnetosphere.

5 Outline our current model of the Earth's interior structure and describe some of the experimental techniques used to establish this model.

6 Summarize the evidence for the phenomenon of "continental drift," and discuss the physical processes that drive it.

(Opposite page, background) This beautiful image of the whole Earth was taken by a weather satellite hovering over the equator. One can clearly see Africa and the Middle East (with desert areas prominent in both), as well as Europe (at top).

(Inset A) A band of cirrus clouds cuts across Egypt, the Nile River, and the Red Sea (upper left) in this (southward) view taken by a crewed spaceship in Earth orbit.

(Inset B) The Red Sea, separating Egypt (right) from Saudi Arabia (left) and the Sinai Peninsula is a new ocean being formed as the African continent begins to drift away from Asia Minor.

(Inset C) Here, the southern tip of the Sinai Peninsula is centered, with the Red Sea at the top, the Gulf of Aqaba at lower left, and the Gulf of Suez at right.

(Inset D) An even closer view of desolate sand dunes in Arabia. The long dunes parallel the prevailing wind direction and can be up to 150 meters high and 600 kilometers long without a break.

*E*arth is the best-studied terrestrial planet. From the matter of our world sprang life, intelligence, culture, and all the technology we now use to explore the cosmos. We ourselves are made of "earthstuff" as much as are rocks, trees, and air. Now, as humanity begins to explore the solar system, we can draw on our knowledge of the Earth to aid our understanding of the other planets. By cataloging Earth's properties and attempting to explain them, we set the stage for our comparative study of the solar system. Every piece of information we can glean about the structure and history of our own world plays a potentially vital role in helping us understand the planetary system in which we live. If we are to appreciate the universe, we must first come to know our own planet. Our study of astronomy begins at home.

7.1 The Earth in Bulk

① The radius of our planet has been known since the time of the ancient Greeks to be about 6500 km (see the *More Precisely* feature on p. 25). More precise measurements are now routinely made by spacecraft, leading to the result $R_\oplus = 6378$ km.* Table 1 in the Appendix lists in detail the physical properties of the Earth and other planets of the solar system; we will use rounded-off numbers throughout the body of the text.

We saw in Chapter 2 how we can measure the mass of Earth, or any other astronomical body, by observing its gravitational influence on some other nearby object and applying Newton's law of gravity. ∞ (p. 47) By studying the dynamical behavior of objects near our planet—be they baseballs, rockets, or the Moon—we can compute Earth's mass, $M_\oplus$. In effect, we can weigh the Earth. We find, approximately, $M_\oplus = 6.0 \times 10^{24}$ kg—6000 billion billion metric tons.

Dividing Earth's mass by its volume yields a density of 5500 kg/m³, or 5.5 g/cm³. ∞ (p. 131) This simple measurement of Earth's average density allows us to make a very important deduction about the interior of our planet. The water that makes up much of Earth's surface has a density of 1000 kg/m³ (1 g/cm³), and the rock beneath us on the continents, as well as on the seafloor, has a density in the range of 2000 to 4000 kg/m³. We can immediately conclude that because the surface layers have densities much less than the average, much denser material must lie deeper in. Hence we should expect that much of Earth's interior is made up of very dense matter, far more compact than the densest continental rocks on the surface.

Based on measurements made in many different ways—using aircraft in the atmosphere, satellites in orbit, gauges on the land, submarines in the ocean, and drilling gear below the rocky crust—scientists have built up the following overall picture of our planet. Earth can be divided into six main regions, shown in Figure 7.1. A zone of charged particles trapped by Earth's magnetic field forms the **magnetosphere**. This lies high above the **atmosphere** of air. At the surface, we have the **hydrosphere**, which contains the liquid oceans and accounts for some 70 percent of our planet's total surface area, and a relatively thin **crust**, comprising the solid continents and the seafloor. In the interior, a large **mantle** surrounds a smaller **core**. Virtually all of our planet's mass is contained within the surface and interior. The gaseous atmosphere and the magnetosphere contribute hardly anything at all—less than 0.1 percent—to the mass.

7.2 The Tides

Earth is unique among the planets in that it has large quantities of liquid water on its surface. Approximately three quarters of Earth's surface is covered by water, to an average depth of about 3.6 km. Only 2 percent of the water is contained within lakes, rivers, clouds, and glaciers. The remaining 98 percent is in the oceans.

GRAVITATIONAL DEFORMATION

② A familiar hydrospheric phenomenon is the daily fluctuation in ocean level known as the **tides**. At most coastal locations on Earth, there are two low tides and two high tides each day. The "height" of the tides—the magnitude of the variation in sea level—can range from a few centimeters to many meters, depending on the location on Earth and time of year. The height of a typical tide on the open ocean is about a meter, but if this tide is funneled into a narrow opening such as the mouth of a river, it can become much higher. At the Bay of Fundy, on the Maine–Canada border, the high tide can reach nearly 20 m (approximately 60 feet, or the height of a six-story building!), above the low-tide level. An enormous amount of energy is contained in the daily motion of the oceans. This energy is constantly eroding and reshaping our planet's coastlines. In some locations, it has been harnessed as a source of electrical power for human activities.

What causes the tides? A clue comes from the fact that tides have daily, monthly, and yearly cycles. The tides are a direct result of the gravitational influence of the Moon and the Sun on the Earth. We have already seen how gravity keeps the Earth and Moon in orbit about one another, and both in orbit around the Sun. For simplicity, let us first consider just the interaction between Earth and the Moon.

Recall that the strength of gravity depends on the distance separating any two objects. Thus, different parts of

---

Recall that the symbol ⊕ is commonly used to represent the Earth (see the More Precisely feature on p.13). Apart from the naming of the constellations, one of the few remnants of astrology in modern astronomy is the use of such symbols for the Sun, the Moon, and the planets. Thus, any symbol with a ⊕ attached refers specifically to the Earth. Similarly, when the Sun sign ⊙ is used, the symbol relates to the Sun, and so on.

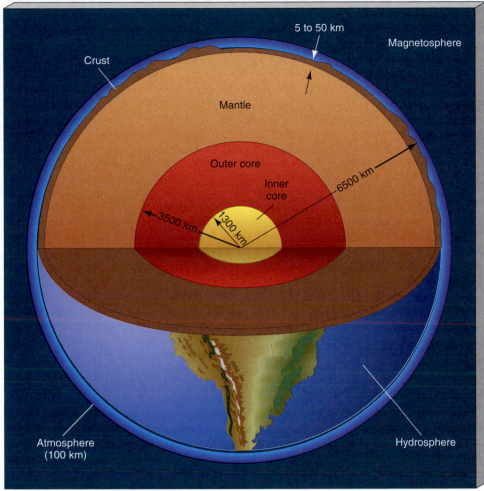

Figure 7.1 The main regions of planet Earth. At Earth's center lies our planet's inner core, about 2600 km in diameter. Surrounding the inner core is an outer core, some 7000 km across. Most of the rest of Earth's 13,000 km diameter is taken up by the mantle, which is topped by a thin crust only a few tens of kilometers thick. The liquid portions of Earth's surface make up the hydrosphere. Above the hydrosphere and solid crust lies the atmosphere, most of it within 50 km of the surface. Earth's outermost region is the magnetosphere, extending thousands of kilometers out into space.

the Earth feel slightly different pulls due to the Moon's gravity, depending on their distance from the Moon. For example, the Moon's gravitational attraction is greater on the side of Earth that faces the Moon than on the opposite side, some 13,000 km farther away. This difference in the gravitational force is small—only about 3 percent—but it produces a noticeable effect, a **tidal bulge**. The Earth be-

comes slightly elongated, with the long axis of the distortion pointing toward the Moon, as illustrated in Figure 7.2.

The liquid portions of Earth undergo the greatest deformation, because they can most easily move around on the surface. (A bulge *is* actually raised in the solid material of Earth, but it is about a hundred times smaller than the oceanic bulge.) Thus, the ocean becomes a little deeper in

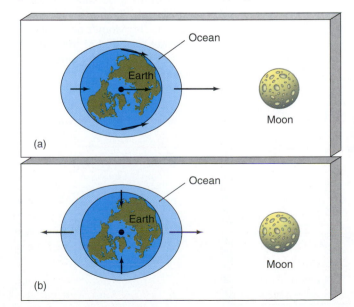

Figure 7.2 This exaggerated illustration shows how the Moon induces tides on both the near and the far sides of the Earth. The lengths of the straight arrows indicate the relative strengths of the Moon's gravitational pull on various parts of Earth. (a) The lunar gravitational forces acting on several different locations on and in Earth. The force is greatest on the side nearest the Moon and smallest on the opposite side. (b) The difference between the lunar forces experienced at those same locations and the force acting on Earth's center. The arrows represent the force with which the Moon tends to pull matter away from, or squeeze it toward, the center of our planet. Material on the side of the Earth nearest the Moon tends to be pulled away from the center, and material on the far side is "left behind," so that a bulge is formed. High and low tides result, twice per day, as the Earth rotates beneath the bulges in the oceans.

some places (along the line joining Earth to the Moon) and shallower in others (perpendicular to this line). The daily tides result as Earth rotates beneath this deformation. As noted in Figure 7.2, the side of the Earth *opposite* the Moon also experiences a tidal bulge. The different gravitational pulls—greatest on that part of Earth closest to the Moon, weaker at Earth's center, and weakest of all on Earth's opposite side—cause average tides on either side of our planet to be approximately equal in height. On the side nearer the Moon, the ocean water is pulled slightly toward the Moon. On the opposite side, the ocean water is literally left behind as the Earth is pulled closer to the Moon. Thus, at any point on Earth high tide occurs twice, not once, every day.

In fact, both the Moon and the Sun exert tidal forces on our planet. Even though the Sun is roughly 375 times farther away from Earth than is the Moon, its mass is so much greater (by a factor of about 27 million) that its tidal influence is still significant—about half that of the Moon. Thus, instead of one tidal bulge, there are actually two— one pointing toward the Moon, the other toward the Sun—and the interaction between them accounts for the changes in the height of the tides over the course of a month or a year. When Earth, the Moon, and the Sun are roughly lined up, the gravitational effects reinforce one another, and so the highest tides are generally found at times of new and full moons. These tides are known as *spring*

tides. When the Earth–Moon line is perpendicular to the Earth–Sun line (at the first and third quarters), the daily tides are smallest. These are termed *neap tides*. The relative orientations of the Earth, the Moon, and the Sun at times of spring and neap tides are illustrated in Figure 7.3.

The variation of the Moon's gravity across Earth is an example of a *differential force*, or **tidal force**. The *average* gravitational interaction between two bodies determines their orbit around one another. However, the *tidal* force, superimposed on that average, tends to deform the bodies themselves. The tidal influence of one body on another diminishes very rapidly with increasing distance (in fact, as the inverse *cube* of the separation). For example, if the distance from Earth to the Moon were to double, the tides resulting from the Moon's gravity would decrease by a factor of 8. We will see many examples in this book of situations where tidal forces are critically important in understanding astronomical phenomena. Notice that we still use the word *tidal* in these other contexts, even though we are not discussing oceanic tides and possibly not even planets at all.

EFFECT OF TIDES ON THE EARTH'S ROTATION

Earth's rotation is gradually slowing down. As the spin slows, the length of the day increases. We now have direct evidence that the day is lengthening by about 0.002 seconds every century. That's a small time increase on the

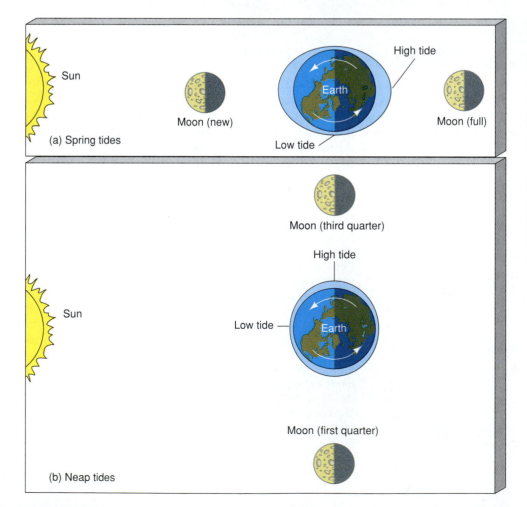

Figure 7.3 The combined effects of the Sun and the Moon produce variations in the high and low tides. (a) When the Moon is either full or new, Earth, Moon, and Sun are approximately aligned, and the tidal bulges raised in Earth's oceans by the Moon and the Sun reinforce one another. (b) At first or third quarter Moon, the tidal effects of the Moon and the Sun partially cancel, and the tides are smallest. The Moon's tidal effect is greater than that of the Sun, and so the net bulge points toward the Moon.

scale of a human lifetime, but over millions of years, this steady slowing of the Earth's rotation adds up.

A number of natural biological clocks lead us to the conclusion that Earth's spin rate is decreasing. For example, each day a growth mark is deposited on a certain type of coral in the reefs off the Bahamas. These growth marks are similar to the annual rings found in tree trunks, except that in the case of coral, the marks are made daily, in response to the day–night cycle of solar illumination. However, they also show yearly variations, as the coral's growth responds to Earth's seasonal changes, allowing us to perceive annual cycles. Coral growing today shows 365 marks per year, but ancient coral shows many more growth deposits per year. In fact, fossilized reefs that are several hundred million years old contain coral with nearly 400 deposits per year of growth. Nearly 500 million years ago, the day was only 22 hours long and the year contained 400 days.

Why is Earth's spin slowing? The main reason is the tidal effect of the Moon. In reality, the tidal bulge raised by the Moon does *not* point directly at it, as shown in Figure 7.2. Because of the effects of friction, both between the crust and the oceans and within Earth itself, the bulge actually points slightly *ahead* of our satellite, as the rotating Earth tries to drag the tidal bulge around with it. The net effect of the Moon's pull on this slightly offset bulge is that our planet's rotation rate decreases. At the same time the Moon spirals slowly away from us, increasing its average distance from Earth by about 4 cm per century. This process will continue until Earth rotates on its axis at exactly the same rate as the Moon orbits Earth. At that time the Moon will al-

ways be above the same point on Earth and will no longer lag behind the bulge it raises. The Earth's rotation period will be 47 of our present days, and the distance to the Moon will be 550,000 km (about 43 percent greater than at present). However, this will take a very long time to occur—many billions of years!

7.3   *The Atmosphere*

From a human perspective, probably the most important aspect of Earth's atmosphere is that we can breathe it. Air is a mixture of gases, the most common of which are nitrogen (78 percent by volume), oxygen (21 percent), argon (0.9 percent), and carbon dioxide (0.03 percent). Water vapor is a variable constituent of the atmosphere, making up anywhere from 0.1 percent to 3 percent, depending on location and climatic conditions. The presence of a large amount of oxygen makes our atmosphere unique in the solar system, and the presence of even trace amounts of water and carbon dioxide play a vital role in the workings of our planet.

Our atmosphere does much more than just provide the oxygen we breathe, however. It protects us from most of the harsh radiation emitted by the Sun and other cosmic objects—particularly the high-frequency and often harmful ultraviolet and X-ray radiation. It also guards us from most rocky debris, or "meteoroids," falling in from space—friction with the air causes all but the largest to burn up long before they reach the ground. Furthermore, it helps to keep us warm. In short, the atmosphere also acts as a pro-

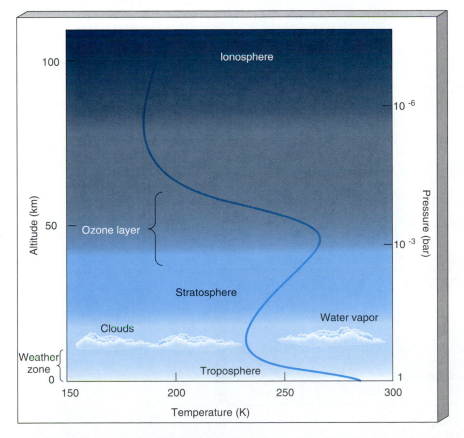

Figure 7.4   Diagram of Earth's atmosphere, showing the changes of temperature and pressure from the surface to the bottom of the ionosphere. (One bar is the pressure at sea level.)

tective blanket, making the surface of our planet a relatively comfortable place on which to live.

ATMOSPHERIC STRUCTURE

3 Figure 7.4 shows a cross section of Earth's atmosphere. Compared with the overall dimensions of our planet, our atmosphere is not thick. Half of it lies within 5 km of the surface, and all but 1 percent of it is found below 30 km. The temperature and the density both decrease with altitude in the lowest-lying atmospheric zones, where virtually all weather occurs. Climbing even a modest mountain, 4 or 5 km high, clearly demonstrates this cooling and thinning of the air. The portion of the atmosphere below about 15 km is called the **troposphere**. Above it, extending up to an altitude of nearly 100 km, lies the **stratosphere**.

The troposphere is the region of Earth's (or any other planet's) atmosphere where **convection** occurs, driven by the heat of Earth's warm surface. Convection is the constant upwelling of warm air and the concurrent downward flow of cooler air to take its place, a process that physically transfers heat from a lower (hotter) to a higher (cooler) level. In Figure 7.5(a), part of Earth's surface is heated by the Sun. The air immediately above the warmed surface is heated, expands a little, and becomes less dense. As a result, the hot air becomes buoyant and starts to rise. At higher altitudes, the opposite effect occurs: The air gradually cools, grows denser, and sinks back to the ground. Cool air at the surface rushes in to replace the hot buoyant air. In this way, a circulation pattern is established. These *convection cells* of rising and falling air not only contribute to atmospheric heating, but are also responsible for surface winds. This constant churning motion is responsible for all the weather we experience.

Atmospheric convection can also create clear-air turbulence—the bumpiness we sometimes experience on aircraft flights. Ascending and descending parcels of air, especially below fluffy clouds (themselves the result of convective processes), can cause a choppy ride. For this reason, passenger aircraft tend to fly above most of the turbulence, or in the lower stratosphere, where the atmosphere is stable and the air is calm.

Within the stratosphere lies the **ozone layer** where, at an altitude of around 50 km, the air temperature increases as incoming solar ultraviolet radiation is absorbed by atmospheric oxygen, ozone, and nitrogen. The ozone layer is one of the insulating spheres that serve to protect life on Earth from the harsh realities of outer space. Not long ago,

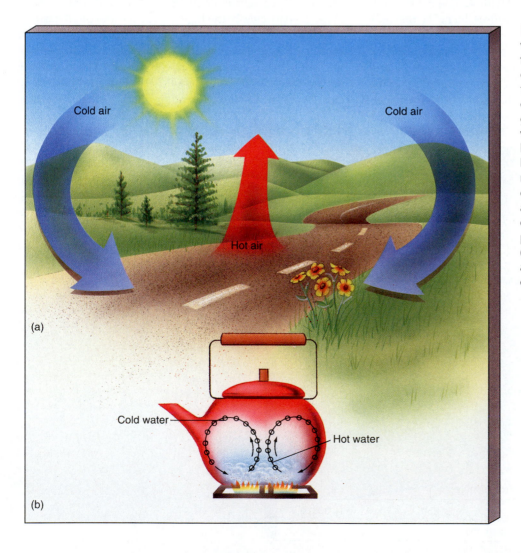

(a)

(b)

Figure 7.5 Convection occurs whenever cool fluid overlies warm fluid. The resulting circulation currents are familiar to us (a) as the winds in Earth's atmosphere caused by the solar-heated ground and (b) as the upwelling water motions in a pot of water heated on a stove. Hot air or liquid rises, cools, and falls repeatedly. Eventually, steady circulation patterns with rising and falling currents are established and maintained, provided that the source of heat (the Sun in the case of the atmosphere, the stove in the case of the water) remains intact.

scientists judged space to be hostile to advanced life forms because of what is missing out there: breathable air and a warm environment. Now, most scientists regard outer space harsh because of what is *present* out there: fierce radiation and energetic particles, both of which are injurious to human health. The ozone layer is one of our planet's umbrellas. Without it, advanced life (at least on Earth's surface) would be at best unlikely and at worst impossible.

Above about 100 km, the atmosphere is significantly ionized by the high-energy portion of the Sun's radiation spectrum, which breaks down molecules into atoms and atoms into ions, and the degree of ionization increases with altitude. This electrically conducting portion of the upper atmosphere is known as the **ionosphere**. Its conductivity makes it highly reflective to certain radio wavelengths. The reason that AM radio stations can be heard beyond the horizon is that their signal actually bounces off the ionosphere before reaching the receiver.

Despite these absorbing and reflecting layers, much of the Sun's radiation manages to penetrate Earth's atmosphere, eventually reaching the ground. Most of this energy is in the form of visible and infrared radiation—ordinary sunlight. (Solar radio waves, in the other portion of the electromagnetic spectrum to which Earth's atmosphere is transparent, also reach Earth's surface, but the hot Sun emits comparatively little of this radiation.) All the solar radiation not absorbed or reflected from clouds in the upper atmosphere shines directly onto Earth's surface. The result is that our planet's surface and most objects on it heat up considerably during the hours of daylight. But the Earth can't absorb this solar energy indefinitely. If it did, surface objects would soon become hot enough to melt, and all life forms would be cooked.

SURFACE HEATING

3 As it heats up, the surface reradiates much of its absorbed energy. The reradiated radiation follows the usual Planck curve, shown in Chapter 3. ∞ (p. 62) As the surface temperature rises, the amount of energy radiated increases rapidly, according to Stefan's law. Eventually, Earth radiates as much energy back into space as it receives from the Sun, and a stable balance is struck. In the absence of any complicating effects, this balance would be achieved at an average surface temperature of about 250 K (−23°C). At that temperature, Wien's law tells us that most of the reemitted energy is in the form of infrared radiation.

But there are complications. Infrared radiation is partially blocked by Earth's atmosphere. The primary reason for this is the presence of molecules of water vapor and carbon dioxide, which absorb very efficiently in the infrared portion of the spectrum. Even though these two gases are only trace constituents of our atmosphere, they manage to absorb a large fraction of all the infrared radiation emitted from the surface. Consequently, only some of that radiation escapes back into space. The remainder is trapped within our atmosphere, and the temperature rises.

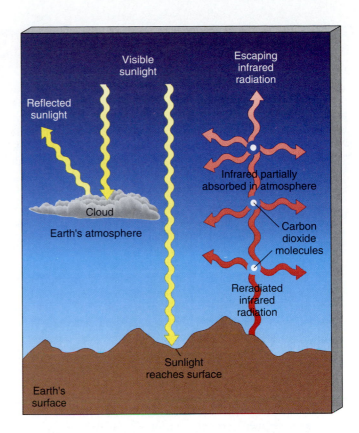

Figure 7.6 The greenhouse effect. Sunlight that is not reflected by clouds reaches the Earth's surface, warming it up. Infrared radiation reradiated from the surface is partially absorbed by water vapor and carbon dioxide in the atmosphere, causing the overall surface temperature to rise.

This partial trapping of solar radiation is known as the **greenhouse effect**. The name comes from the fact that a very similar process operates in a greenhouse. Sunlight passes relatively unhindered through glass panes, but much of the infrared (heat) radiation reemitted by the plants is blocked by the glass and cannot get out. Consequently, the interior of the greenhouse heats up, and flowers, fruits, and vegetables can grow even on cold wintry days.* The radiative processes that determine the temperature of Earth's atmosphere are shown in Figure 7.6. Earth's greenhouse effect makes our planet about 40K hotter than would otherwise be the case.

The magnitude of the effect is very sensitive to the concentration of the so-called greenhouse gases (that is, those that absorb infrared radiation efficiently) in the atmosphere. Of greatest importance among these is carbon dioxide, although water vapor too plays a significant role. The amount of carbon dioxide in Earth's atmosphere is increasing, largely as a result of the burning of fossil fuels (principally oil and coal) in the industrialized world. Carbon dioxide

---

*Actually, we now know that this is only a minor part of the reason that greenhouses work as they do—the interior stays warm mainly because the glass prevents convection from carrying heat away—but the name has stuck.

More Precisely...   *Why Is the Sky Blue?*

Is the sky blue because it reflects the color of the ocean, or is the ocean blue because it reflects the color of the surrounding sky? The answer is the latter, and the reason has to do with the way that light is *scattered* by air molecules and minute dust particles. By scattering we mean the process by which radiation is absorbed and then reradiated by the material through which it passes.

As sunlight passes through our atmosphere, it is scattered by gas molecules in the air. This process turns out to be very sensitive to the wavelength of the light involved. The British physicist Lord Rayleigh first investigated this phenomenon about a century ago, and today it bears his name—it is known as Rayleigh scattering.

Rayleigh found that blue light is much more easily scattered than red light, in essence because the wavelength of blue light (400 nm) is closer to the size of air molecules than is the wavelength of red light (700 nm). He went on to prove that the amount of scattering is actually inversely proportional to the *fourth* power of the wavelength, so that blue light is scattered about nine times (that is, $[700/400]^4$) more efficiently than red light. Dust particles also preferentially scatter blue light, but the amount of scattering by dust is just inversely proportional to the wavelength, so the contrast between red and blue scattering is only a factor of (700/400), or 1.75.

Consequently, with the Sun at a reasonably high elevation, the blue component of incoming sunlight will scatter much more than any other color. Thus, some blue light is re-moved from the line of sight between us and the Sun, and may scatter many times in the atmosphere before eventually entering our eyes, as shown in the figure. Red or yellow light is scattered relatively little, and arrives at our eyes predominantly along the line of sight to the Sun. The net effect of all this is that the Sun is "reddened" slightly, because of the removal of blue light, while the sky away from the Sun appears blue. In outer space, where there is no atmosphere, there is no Rayleigh scattering of sunlight, and the sky is black (although, as we will see in Chapter 18, light from distant stars is reddened in precisely the same way as it passes through clouds of interstellar gas and dust.)

At dawn or dusk, with the Sun near the horizon, sunlight must pass through much more atmosphere before reaching our eyes—so much so, in fact, that the blue component of the Sun's light is almost entirely scattered out of the line of sight, and even the red component is diminished in intensity. Accordingly, the Sun itself appears orange—a combination of its normal yellow color and a reddishness caused by the subtraction of virtually all of the blue end of the spectrum—and dimmer than at noon. At the end of a particularly dusty day, when weather conditions or human activities during the daytime hours have raised excess particles into the air, short-wavelength Rayleigh scattering can be so heavy that the Sun appears brilliantly red. Reddening is often especially evident when we look at the westerly "sinking" summer Sun over the ocean, where seawater molecules have evaporated into the air, or during the weeks and months after an active volcano has released huge quantities of gas and dust particles into the air—as was the case in North America when the Philippine volcano Mt. Pinatubo erupted in 1991.

levels have increased by over 20 percent in the last century, and they are continuing to rise at a present rate of 4 percent per decade. In Chapter 9, we will see how a runaway increase in carbon dioxide levels in the atmosphere of the planet Venus has radically altered conditions on its surface, causing its temperature to rise to over 700 K. Although no one is predicting that Earth's temperature will ever reach that of Venus, many scientists now believe that this increase, if left unchecked, may result in global temperature increases of several kelvins over the next half-century—enough to cause dramatic, and possibly catastrophic, changes in our planet's climate.

ORIGIN OF THE EARTH'S ATMOSPHERE

Why is our atmosphere made up of its present constituents? Why is it not composed entirely of nitrogen, say, or of carbon dioxide, like the atmospheres of Venus and Mars? The origin and development of the Earth's atmosphere was a fairly complex and lengthy process. The main evolutionary stages are outlined in the paragraphs that follow.

When Earth first formed, any *primary atmosphere* it might have had would have consisted of the gases most common in the early solar system. These were light gases, such as hydrogen, helium, methane, ammonia, and water vapor—a far cry from the atmosphere we enjoy today. Almost all of this light material, and especially any hydrogen or helium, escaped into space during the first half-billion or so years after Earth was formed. (For more information on how planets retain or lose their atmospheres, consult the *More Precisely* feature on p. 136.)

Subsequently, Earth developed a *secondary atmosphere*, which was *outgassed* from the planet's interior as a result of volcanic activity. Volcanic gases are rich in water vapor, methane, carbon dioxide, sulfur dioxide, and compounds containing nitrogen (such as nitrogen gas, ammonia, and nitric oxide). Solar ultraviolet radiation decomposed the lighter, hydrogen-rich gases, allowing the hydrogen to escape, and liberated much of the nitrogen from its bonds with other elements. As Earth's surface temperature fell and the water vapor condensed, oceans formed. Much of the carbon dioxide and sulfur dioxide became dissolved in the oceans or combined with surface rocks. Oxygen is such a reactive gas that any free oxygen that appeared at early times was removed as quickly as it formed. An atmosphere consisting largely of nitrogen slowly appeared.

The final major development in the story of our planet's atmosphere is known so far to have occurred only on Earth. *Life* appeared in the oceans more than 3.5 billion years ago, and organisms eventually began to produce atmospheric oxygen. The ozone layer formed, shielding the surface from the Sun's harmful radiation. Eventually, life spread to the land and flourished. The fact that oxygen is a major constituent of the present-day atmosphere is a direct consequence of the evolution of life on the Earth.

7.4 *The Magnetosphere*

4 Another part of Earth also helps protect us from the harsh realities of outer space. Discovered by artificial satellites launched in the late 1950s, the magnetosphere lies far above the atmosphere. Simply put, the magnetosphere is the region around Earth that is influenced by our planet's magnetic field. As shown in Figure 7.7, Earth's field is similar to that of a bar magnet. Having a north and a south pole, the magnetic field surrounds our planet three dimensionally. The magnetic field lines, which indicate the strength and direction of the field at any point in space, run from geographic south to geographic north, as indicated. The north and south *magnetic poles*, where the axis of the imaginary bar magnet intersects Earth's surface, are roughly aligned with the spin axis of Earth. The magnetic north pole is currently in northern Canada, at a latitude of about 80°N, almost due north of the center of North America. The magnetic south pole lies at a latitude of about 60°S, just off the coast of Antarctica south of Adelaide, Australia.

Instrumented spacecraft have found that the magnetosphere contains two doughnut-shaped zones of high-energy

Figure 7.7 Earth's magnetic field resembles that of an enormous bar magnet situated inside our planet. The arrows on the field lines indicate the direction in which a compass needle would point.

particles, one about 3000 and the other 20,000 km above Earth's surface. Called the **Van Allen belts**, these zones are named for the American physicist whose instruments on board one of the first artificial satellites detected them. We call them "belts" because they are most pronounced near Earth's equator and because they completely surround the planet. Figure 7.8 shows how these invisible regions envelop our planet, except near the North and South Poles.

We could never survive unprotected in the Van Allen belts. Unlike the lower atmosphere, on which humans and other life forms rely for warmth and protection, much of the magnetosphere is subject to intense bombardment by large numbers of high-velocity, and potentially very harmful, charged particles. Colliding violently with an unprotected human body, these particles would deposit large amounts of energy wherever they made contact, causing severe damage to living organisms. Without sufficient shielding on the *Apollo* spacecraft, for example, the astronauts might not have survived the passage through the magnetosphere on their journey to the Moon.

Of course, Earth's magnetism is not really produced by a huge bar magnet lying within our planet. What then is the origin of the magnetosphere and the Van Allen belts within it? In fact, Earth's magnetic field is not a permanent part of the planet, but is thought to be generated within Earth's core. Like the dynamos that run industrial machines, the spin of electrically conducting metal deep inside Earth induces powerful electrical currents that produce our planet's magnetism. The theory that explains planetary (and other) magnetic fields in terms of rotating, conducting material flowing in the planet's interior is known as **dynamo theory**.

Why do the particles that make up the Van Allen belts stay in Earth's magnetosphere? Traveling through space,

neutral particles and electromagnetic radiation are unaffected by Earth's magnetism, but electrically *charged* particles are strongly influenced by it. A magnetic field exerts a force on a moving charged particle, causing it to spiral around the magnetic field lines, as illustrated in Figure 7.9. Charged particles headed toward Earth, especially protons and electrons from the Sun—the so-called *solar wind*—can become trapped by Earth's magnetism. In this way, the Earth's magnetic field, sketched in Figure 7.7, exerts electromagnetic control over the particles, herding them into the shape of the Van Allen belts shown in Figure 7.8.

The charged particles often escape from the magnetosphere near the North and South Poles, where the field lines intersect the atmosphere. Their collisions with the air rip apart some atmospheric molecules, creating a spectacular light show like the one shown in Figure 7.10. Called an **aurora**, this colorful display results when atmospheric molecules, excited by collision with the charged particles, fall back to their ground states and emit visible light. Many different colors are produced because each type of atom or molecule can take one of several possible paths as it returns to its ground state (see Figure 4.10). ∞ (p. 83) Aurorae are most brilliant at high latitudes, especially within the Arctic and Antarctic circles. In the north, the spectacle is called the *aurora borealis*, or northern lights. In the south, it is called the *aurora australis*, or southern lights.

Occasionally, particularly after a storm on the Sun, the Van Allen belts can become distorted by the solar wind and overloaded with many more particles than normal, allowing some particles to escape prematurely and at lower latitudes. For example, in North America, the aurora borealis is normally seen with any regularity only in northern Canada and Alaska. However, at times of great-

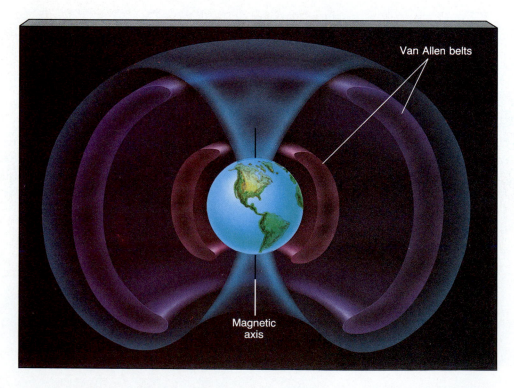

Figure 7.8 High above the Earth's atmosphere, the magnetosphere (lightly shaded blue area) contains at least two doughnut-shaped regions (heavily shaded violet areas) of magnetically trapped charged particles. These are the Van Allen belts.

Van Allen belts

Magnetic axis

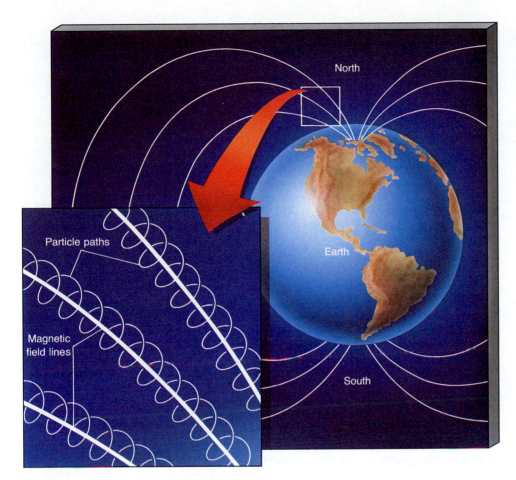

Figure 7.9 A charged particle in a magnetic field spirals around the field lines. Thus, charged particles tend to become "trapped" by strong magnetic fields.

North

Particle paths

Magnetic field lines

Earth

South

(a)

(b)

R   I   V   U   X   G

Figure 7.10   (a) A colorful aurora rapidly flashes across the sky like huge wind-blown curtains glowing in the dark. The aurora is created by the emission of light radiation after magnetospheric particles collide with atmospheric molecules. The colors are produced as excited atoms and molecules return to their ground states. (b) The aurora high above the Earth, as photographed from a space shuttle (visible at left).

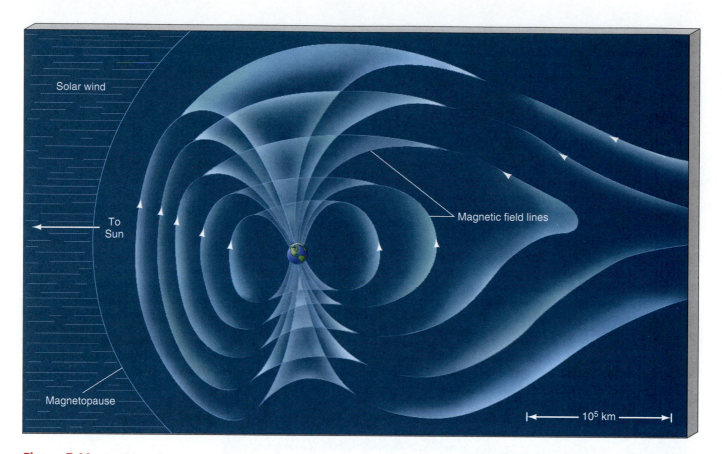

Figure 7.11   Earth's real magnetosphere is actually greatly distorted by the solar wind, with a long tail extending from the nighttime side of the Earth well into space.

est solar activity, the display has been seen as far south as the southern United States.

Actually, Earth's magnetosphere is not nearly as symmetrical as depicted in Figure 7.7. Satellites have mapped its true shape. As shown in Figure 7.11, the entire region of trapped particles is quite distorted, forming a teardrop-shaped cavity. On the sunlit (daytime) side of Earth, the magnetosphere is compressed by the flow of high-energy particles in the solar wind. The boundary between the magnetosphere and this flow is known as the *magnetopause*. It is found at about 10 Earth radii from our planet. On the side opposite the Sun, the belts are extended, with a long tail often reaching beyond the orbit of the Moon.

Earth's magnetic field plays an important role in controlling many of the potentially destructive charged particles that venture near our planet. Without the magnetosphere, the Earth's atmosphere—and perhaps the surface, too—would be bombarded by harmful particles, possibly damaging many forms of life on our planet. Some researchers have even suggested that had the magnetosphere not existed in the first place, life might never have arisen at all on planet Earth.

7.5   The Interior

5   Now let's shift our attention from the sky above our heads to the ground beneath our feet. Although we reside on Earth, we cannot easily probe our planet's interior.

Drilling gear can penetrate rock only so far before breaking. No substance used for drilling—even diamond, the hardest known material—can withstand the pressure below a depth of about 10 km. That's rather shallow compared with the Earth's radius of (approximately) 6500 km. Fortunately, geologists have developed other techniques that indirectly probe the deep recesses of our planet.

SEISMIC WAVES

A sudden dislocation of rocky material near Earth's surface—an **earthquake**—causes the entire planet to vibrate a little. The whole planet rings like a bell. These vibrations are not random. They are systematic waves, called **seismic waves**, that move outward from the site of the quake. Like all waves, they carry information. This information can be detected and recorded using sensitive equipment—a seismograph—designed to monitor Earth tremors.

Decades of earthquake research have demonstrated the existence of many kinds of seismic waves. Two types are of particular importance to the study of Earth's internal structure. First to arrive at a monitoring site after a distant earthquake are the primary waves, or *P-waves*. These are *pressure* waves, a little like ordinary sound waves in air, that alternately expand and compress the material medium through which they move. Seismic P-waves usually travel at velocities ranging from 5 to 6 km/s and can travel through both liquids and solids. Some time later (the actual delay depending on the

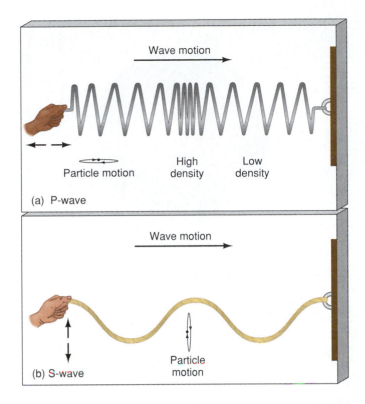

Wave motion

Particle motion   High density   Low density

(a) P-wave

Wave motion

Particle motion

(b) S-wave

Figure 7.12   (a) A pressure (P-) wave traveling through Earth's interior causes material to vibrate in a direction parallel to the direction of motion of the wave. Material is alternately compressed and expanded. (b) A shear (S-) wave produces motion perpendicular to the direction in which the wave travels, pushing material from side to side. Also shown is the motion of one typical particle. In case (a), the particle oscillates forward and backward about its initial position. In (b), the particle moves from side to side.

distance from the earthquake site), secondary waves, or *S-waves*, arrive. These are *shear* waves. Unlike P-waves, which vibrate the material through which they pass back and forth along the direction of travel of the wave, S-waves cause side-to-side motion, more like waves in a guitar string. The two types of waves are illustrated in Figure 7.12. S-waves normally travel through Earth's interior at 3 to 4 km/s. Seismic S-waves cannot travel through liquid, which absorbs them.

The velocity of both P- and S-waves depends on the density of the matter through which they are traveling. Consequently, if we can measure the time taken for the waves to move from the site of an earthquake to one or more monitoring stations on Earth's surface, we can determine the density of matter in the interior. Figure 7.13 illus-

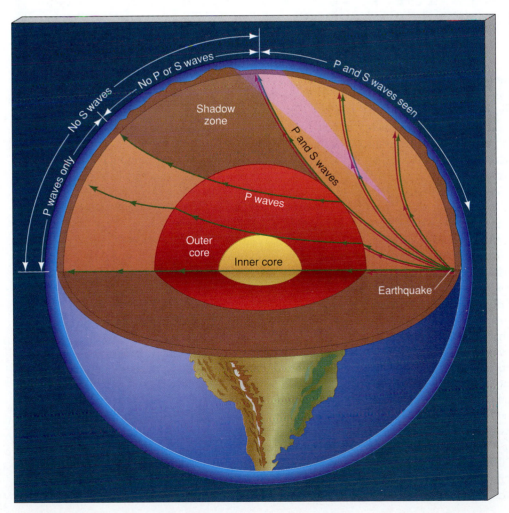

Figure 7.13   Earthquakes generate pressure (P, or primary) and shear (S, or secondary) waves that can be detected at seismographic stations around the world. S-waves are not detected by stations "shadowed" by the liquid core of the Earth. P-waves do reach the side of Earth opposite the earthquake, but their interaction with the Earth's core produces another shadow zone, where no P-waves are seen.

trates some paths followed by P- and S-waves from the site of an earthquake. Seismographs located around the world measure the times of arrival as well as the strengths of the seismic waves. Both observations contain much useful information—both about the earthquake itself and about the Earth's interior through which the waves pass. Notice that the waves do not travel in straight lines through the planet. Because the wave velocity varies with depth, the waves bend as they move through the interior.

A particularly important result emerged after numerous quakes were monitored several decades ago: Seismic stations on the side of the world opposite a quake never detect S-waves—some shear waves seem to be blocked by material within Earth's interior. Further, while P-waves never fail to arrive at stations diametrically opposite the quake, there are parts of Earth's surface that they cannot reach (see Figure 7.13). Most geologists believe that S-waves are absorbed by a liquid core at the center of the Earth and that P-waves are "refracted" at the core boundary, much as light is refracted by a lens. The result is the S- and P-wave "shadow zones" we observe. The fact that every earthquake exhibits these shadow zones is the best evidence that the core of our planet is hot enough to be in the liquid state.

The radius of the core, as determined from seismic data, is about 3500 km. In fact, very faint P-waves *are* observed in the P-wave shadow zone indicated in Figure 7.13. These are believed to be reflected off the surface of a solid *inner core*, of radius 1300 km, lying at the center of the liquid outer core.

MODELING EARTH'S INTERIOR

Because earthquakes occur often and at widespread places across the globe, geologists have accumulated a large amount of data about shadow zones and seismic-wave properties. They have used these data, along with direct knowledge of surface rocks, to build mathematical models of Earth's interior. Figure 7.14 presents a recent model that most scientists accept. As the graphs show, the density and temperature both increase with depth. Specifically, from Earth's surface to its very center, the density increases from about 3000 kg/m³ to a little more than 12,000 kg/m³. These densities suggest to geologists that the inner parts of Earth must be rich in nickel and iron. Under the heavy pressure of the overlying layers, these metals (whose densities under surface conditions are around 8000 kg/m³) can be compressed to the high densities predicted by the model. The sharp density increase at the mantle–core boundary results from the difference in composition between the two regions. The mantle is composed of dense but *rocky* material, compounds of silicon and oxygen. The core consists primarily of even denser *metallic* elements. There is no similar jump in density or temperature at the inner core boundary—the material there simply changes from the liquid to the solid state.

The model indicates that the core must be a mixture of nickel, iron, and some other lighter element, possibly sulfur. Without direct observations, it is difficult to be absolutely certain of the light component's identity. All geologists agree that much of the core must be in the liquid state. The exis-

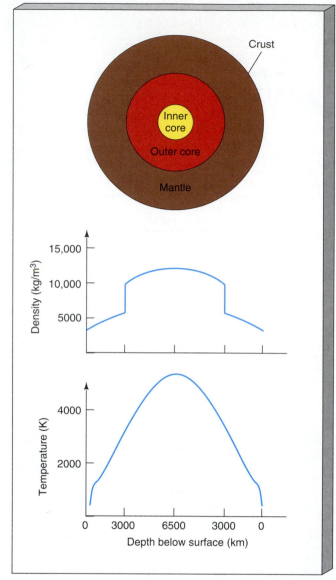

Figure 7.14   Computer models of the Earth's interior imply that the density and temperature vary considerably through the mantle and the core. Note the sharp density discontinuity between Earth's core and mantle.

tence of the shadow zone demands that, and our current explanation of the Earth's magnetic field relies on it. The outer parts of the core must be liquid, because of their high temperature. However, the pressure near the center, about 4 million times atmospheric pressure at the Earth's surface, is high enough to force the material there into the solid state.

The outer core is surrounded by a thick mantle and topped with a thin crust. The mantle is about 3000 km thick and accounts for the bulk (80 percent) of our planet's volume. Models of Earth's interior suggest that much of the mantle has a density midway between the densities of the core and crust: about 5000 kg/m³. The crust has an average thickness of only 15 km—a little less (around 8 km) under the oceans and somewhat more (20 to 50 km) under the continents. The average density of the material making up the Earth's crust is around 3000 kg/m³.

Despite the fact that no experiment has yet succeeded in piercing Earth's crust to recover a sample of the mantle, we are not entirely ignorant of the mantle's properties. In a *volcano*, hot lava upwells from below the crust, bringing a little of the mantle to us and providing some inkling of Earth's interior. The chemical makeup and physical state of the newly emerged lava are generally consistent with predictions based on the model sketched in Figure 7.14.

The chemical composition of the upper mantle is probably similar to the iron-magnesium-silicate mixtures known as *basalt*. You may have seen some dark gray basaltic rocks scattered across Earth's surface, especially near volcanoes. Basalt is formed as mantle material upwells from Earth's interior as lava, then cools, and solidifies. With a density between 3000 and 3300 kg/m^3, basalt contrasts with the lighter *granite* (density 2700–3000 kg/m^3) that constitutes much of the rest of Earth's crust. Granite is richer than basalt in the light elements silicon and aluminum, which explains why the surface continents do not sink into the interior. Their low-density composition lets the crust "float" atop the denser matter of the mantle and core below..

DIFFERENTIATION OF THE EARTH

Earth, then, is not a homogeneous ball of rock. Instead, it has a layered structure, with a low-density crust at the surface, intermediate-density material in the mantle, and a high-density core. Such variation in density and composition is known as **differentiation**. Why isn't our planet just one big, rocky ball of uniform density? The answer appears to be that much of Earth was *molten* at some time in the past. As a result, the higher-density matter sank to the core, and the lower-density material was displaced toward the surface. A remnant of this ancient heating exists today: Earth's central temperature is nearly equal to the surface temperature of the Sun. What processes were responsible for heating the entire planet to this extent? To answer this question, we must try to visualize the past.

We will see in Chapter 15 that when the Earth formed, it did so by capturing material from its surroundings, growing in mass as it swept up "preplanetary" chunks of matter in its vicinity. As the young planet grew, its gravitational field strengthened, and the speed with which newly captured matter struck its surface increased. This process generated a lot of heat—so much, in fact, that Earth may already have been partially or wholly molten by the time it reached its present size. As Earth began to differentiate and heavy material sank to the center, even more gravitational energy was released, and the interior temperature must have increased still further. Later, Earth continued to be bombarded with debris left over from the formation process. At its peak some 4 billion years ago, this secondary bombardment was probably intense enough to keep the surface molten, but only down to a depth of a few tens of kilometers. Erosion by wind and water has long since removed all trace of this early period from the surface of Earth, but our Moon still bears visible scars of the onslaught.

Another important process for heating Earth soon after its formation was **radioactivity**—the release of energy by certain rare, heavy elements, such as uranium, thorium, and plutonium (see the *More Precisely* feature on p. 160). These elements emit energy as their complex, heavy nuclei decay into simpler, lighter nuclei. While the energy produced by the decay of a single radioactive atom is tiny, Earth contained a lot of radioactive atoms, and a lot of time was available. Rock is such a poor conductor of heat that the energy would have taken a very long time to reach the surface and leak away into space, so the heat built up in the interior, adding to the energy left there by Earth's formation. Provided that enough radioactive elements were originally spread throughout the primitive Earth, rather like raisins in a cake, the entire planet—from crust to core—could have melted and remained molten for about a billion years. That's a long time by human standards, but not so long in the cosmic scheme of things. Measurements of the ages of some surface rocks indicate that Earth finally began to solidify roughly a billion years after it originally formed. Radioactive heating did not stop after the first billion years. It continued even after the Earth's surface cooled and solidified. But radioactive decay works in only one direction, always producing lighter elements from heavier ones. Once decayed, the heavy and rare radioactive elements cannot be replenished.

The early source of heat diminished with time, allowing the planet to cool over the past 3.5 billion years. In so doing, it cooled from the outside in, much like a hot potato. The reason for this is that it cooled by radiating its energy into space, and regions closest to the surface could most easily unload their excess heat. In this way, the surface developed a crust, and the differentiated interior attained the layered structure now implied by seismic studies. Today, radioactive heating continues throughout Earth. However, there is probably not enough of it to melt any part of our planet. The high temperatures in the core are merely the trapped remnant of a much hotter Earth that existed eons ago. The surface has completely cooled and solidified, making it a habitable place from which intelligent living beings can begin to unravel the history of our planet.

7.6 Surface Change

ACTIVE SITES

Earth is geologically alive today. Its interior seethes and its surface changes. Figure 7.15 shows two indicators of surface geological activity: volcanoes, where molten rock and hot ash upwell through fissures or cracks in the surface, and earthquakes, which occur when the crust suddenly dislodges under great pressure. Catastrophic volcanoes and earthquakes are relatively rare events these days, but geological studies of rocks, lava, and other surface features imply that surface activity must have been more frequent, and probably more violent, long ago.

More Precisely...   *Radioactive Dating*

In Chapter 4 we saw that atoms are made up of electrons and nuclei, and that nuclei are composed of protons and neutrons. ∞ (p. 81) The number of protons in a nucleus determines which element it represents. However, the number of neutrons can vary. In fact, most elements can exist in several *isotopic* forms, all containing the same number of protons but different numbers of neutrons in their nuclei. The particular nuclei we have encountered so far—the most common forms of hydrogen, helium, carbon, iron—are all *stable*. For example, left alone, a carbon-12 nucleus, consisting of six protons and six neutrons, will remain unchanged forever. It will not break up into smaller pieces, nor will it turn into anything else.

Not all nuclei are stable, however. Many nuclei—for example, carbon-14 (containing 6 protons and 8 neutrons), thorium-232 (90 protons, 142 neutrons), uranium-235 (92 protons, 143 neutrons), uranium-238 (92 protons, 146 neutrons), and plutonium-241 (94 protons, 147 neutrons)—are inherently *unstable*. Left alone, they will eventually break up into lighter "daughter" nuclei, in the process emitting some elementary particles and releasing some energy. The change happens spontaneously, without any external influence. This instability is known as *radioactivity*. The energy released by the disintegration of the radioactive elements we just listed is the basis for nuclear fission reactors (and nuclear bombs).

Unstable heavy nuclei achieve greater stability by disintegrating into lighter nuclei, but they do not do so immediately. Each type of "parent" nucleus takes a characteristic amount of time to decay. The *half-life* is the name given to the time required for half of a sample of parent nuclei to disintegrate. Notice that this is really a statement of probability. We cannot say *which* nuclei will decay in any given half-life interval, only that half of them are expected to do so. Thus, if we start with a billion radioactive nuclei embedded in a sample of rock, a half-billion nuclei would remain after one half-life, a quarter-billion after two half-lives, and so on.

Every kind of radioactive element has its own characteristic half-life, and most of them are now well known from studies conducted since the 1950s. For example, the half-life of uranium-235 is 713 million years, while that of uranium-238 is 4.5 billion years. Some radioactive elements decay much more rapidly, others much more slowly, but these two types of uranium are particularly important to geologists because their half-lives are comparable to the age of the solar system. The accompanying figure illustrates the half-lives and decay reactions for four unstable heavy nuclei.

The decay of unstable radioactive nuclei into more stable "daughter" nuclei provides us with a useful tool to measure the ages of any rocks we can get our hands on. The first step is to measure the amount of stable nuclei of a given kind (for example, lead-206, which results from the decay of uranium-238). This amount is then compared with the amount of remaining unstable parent nuclei (in this case, uranium-238) from which the daughter nuclei descended. Knowing the rate (or half-life) at which the disintegration occurs, the age of the rock then follows directly. For example, if half of the parent nuclei of some element have decayed, so that the number of daughter nuclei equals the number of parents, the age of the rock must be equal to the half-life of the radioactive nucleus studied. Similarly, if only a quarter of the parent nuclei remains (three times as many daughters as parents), the rock's age is twice the half-life of that element, and so on. In practice, ages can be determined by these means to within an accuracy of a few percentage points. The most ancient rocks on Earth are dated at nearly 4 billion years old (3.9 billion years old, to be more precise). These rare specimens have been found in Greenland and Labrador.

The radioactive-dating technique rests on the assumption that the rock has remained *solid* while the radioactive decays have been going on. If the rock melts, there is no reason to expect the daughter nuclei to remain in the same locations their parents had occupied, and the whole method fails. Thus, ra-

(a)

(b)

Figure 7.15 (a) An active volcano on Mount Kilauea in Hawaii. Kilauea seems to be a virtually ongoing eruption. Other eruptions, such as that of Mount St. Helens in Washington State on May 18, 1980, are catastrophic events that can release more energy than the detonation of a thousand nuclear bombs. (b) The aftermath of the earthquake that claimed more than 5000 lives and caused billions of dollars' worth of damage in Kobe, Japan in January 1995.

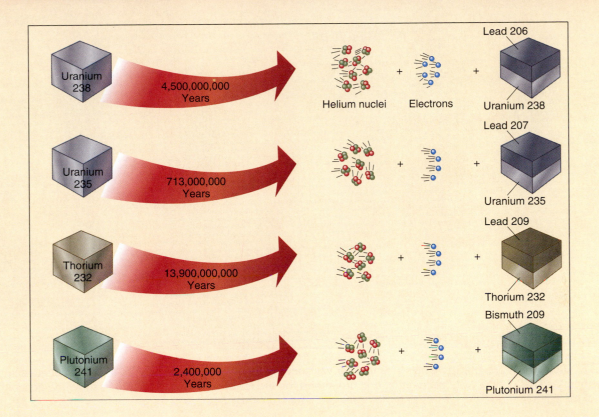

dioactive dating indicates the time that has elapsed since the *last* time the rock in question solidified. Hence this 4 billion year value represents only a portion of the true age of our planet. It does not measure the duration of its molten existence.

A variety of indirect arguments suggests that 4.5 billion years ago is a very special date. For example, all meteorites are radioactively dated to be 4.5 billion years old. Furthermore, the oldest Moon rocks have ages close to 4.5 billion years. Clearly, something important happened in the vicinity of the Earth, and perhaps throughout the entire solar system, 4.5 billion years ago. That something, astronomers reason, was the formation of the Sun and planets.

Many traces of past geological events are scattered across our globe. Erosion by wind and water has wiped away much of the evidence for ancient activity, but modern exploration has documented the sites of most of the recent activity, such as earthquakes and volcanic eruptions. Figure 7.16 is a map of the currently active areas of our planet. The red dots represent sites of volcanism or earthquakes. Nearly all these sites have experienced surface activity within this century, some of them suffering much damage and the loss of many lives. These sites are especially abundant along the western coast of the United States, throughout the Aleutian Islands off the coast of Alaska, down along the Andes Mountains, across the Japanese Islands, up through India, and throughout much of Turkey, Greece, and the Aegean Sea.

CONTINENTAL DRIFT

6 The intriguing aspect of Figure 7.16 is that the active sites are not spread evenly across our planet. Instead, they trace well-defined lines of activity, where crustal rocks dislodge (as in earthquakes) or mantle material upwells (as in volcanoes). In the mid-1960s, it became clear that these lines are really the outlines of gigantic "plates," or slabs of Earth's surface.

Most startling of all, the plates are slowly moving—literally drifting around the surface of our planet. These plate motions have created the surface mountains, the oceanic trenches, and the other large-scale features across the face of planet Earth. In fact, plate motions have shaped the continents themselves. The process is popularly known as "continental drift." The technical term for the study of plate movement and its causes is **plate tectonics**.

The plates are not simply slowing to a stop after some ancient initial movements. Rather, they are still drifting today, although at an extremely slow rate. Typical velocities of the plates amount to only a few centimeters per year—about the same rate as your fingernails grow. This change is

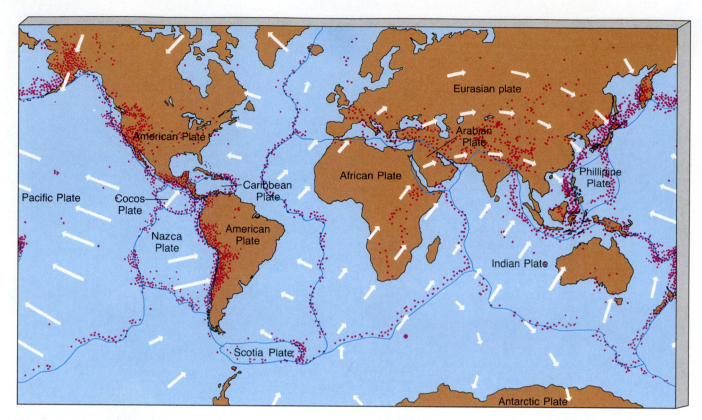

Figure 7.16   Red dots represent active sites where major volcanoes or earthquakes have occurred in the twentieth century. Taken together, the sites outline vast "plates" that drift around on the surface of our planet. The arrows show the general directions of the plate motions.

well within the measuring capabilities of modern equipment. Curiously, one of the best ways of monitoring plate motion on a global scale is by making accurate observations of very distant astronomical objects. Quasars (see Chapter 25), lying many hundreds of millions of light years from Earth, will never show any measurable apparent motion on the sky stemming from their own motion in space. Thus any apparent change in their position (after correction for Earth's motion, of course) can be interpreted as arising from the motion of the telescope—or of the plate on which it is located. On smaller scales, laser-ranging and other techniques now routinely track the relative motion of plates in many populated areas, such as California.

During the course of Earth history, each plate has had plenty of time to move large distances, even at their sluggish pace. For example, a drift rate of only 2 cm per year can cause two continents (for example, Europe and North America) to separate by some 4000 km over the course of 200 million years. That may be a long time by human standards, but it represents only about 5 percent of the age of the Earth.

A common misconception is that the plates are the continents themselves. Some plates are indeed made mostly of continental landmasses, but other plates are made of a continent plus a large part of an ocean. For example, the Indian plate includes all of India, much of the Indian Ocean, and all of Australia and its surrounding south seas (see Figure 7.16). Still other plates are mostly ocean. The seafloor itself is the slowly drifting plate, and the oceanic water merely fills in the

depressions between continents. The southeastern portion of the Pacific Ocean, called the Nazca plate, contains no landmass at all. For the most part, the continents are just passengers riding on much larger plates. Taken together, the plates make up Earth's **lithosphere**, which contains both the crust and a small part of the upper mantle. The lithosphere is the portion of Earth that undergoes tectonic activity. The semisolid part of the mantle over which the lithosphere slides is known as the **aesthenosphere**. The relationships between these regions of Earth are shown in Figure 7.17.

The major plates of the world are marked on Figure 7.16. Note that the major boundary separating the North American plate from the Eurasian plate is marked by a thin strip in the middle of the Atlantic Ocean. Discovered after World War II by oceanographic ships studying the geography of the seafloor, this giant fault is called the Mid-Atlantic Ridge. It extends, like a seam on a giant baseball, all the way from Scandinavia in the North Atlantic to the latitude of Cape Horn at the southern tip of South America. This ridge thus also separates the South American plate from the African plate. The entire ridge is a region of seismic and volcanic activity, but the only major part of it that rises above sea level is the island of Iceland.

As the plates drift around, we might expect collisions to be routine. Indeed, plates do collide. But unlike two automobiles that collide and then stop, the surface plates are driven by enormous forces. They do not stop easily. Instead, they just keep crunching into one another. Figure

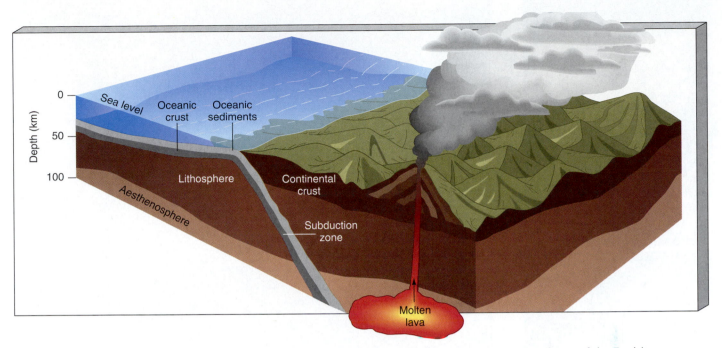

Figure 7.17   The outer layers of Earth's interior. The rocky lithosphere comprises both the crust and part of the Earth's upper mantle. It is typically between 50 and 100 km thick. Below it lies the aesthenosphere, a relatively soft part of the mantle over which the lithosphere slips.

7.18(a) shows a collision currently occurring between two continental landmasses: the subcontinent of India, on the prow of the northward-moving Indian plate, is crashing into the landmass of Asia, located on the Eurasian plate (see Figure 7.16). The resulting folds of rocky crust, create mountains—in this case the Himalayas snow-covered mountain range at upper right. A peak like Mt. Everest (Figure 7.18b) represents a portion of the Earth's crust that has been lifted over 8800 m by the slow but inexorable force of one plate plowing into another.

(b)

Figure 7.18   Mountain building results largely from plate collisions. (a) The subcontinent of India, imaged here in infrared light from orbit, lies at the northernmost tip of the Indian plate. As this plate drifts northwards, the Indian landmass collides with Asia, on the Eurasian plate. The impact causes the Earth's crust to buckle and fold, thrusting up the Himalayan mountain range (covered with snow at the upper right). (b) The results of this process can be seen in this view of Mt. Everest.

(a)

Figure 7.19 The San Andreas and associated faults in California result from the North American and Pacific plates sliding roughly past one another. The Pacific plate, which includes a large slice of the California coast, is drifting to the northwest with respect to the North American plate. (a) A small part of the fault line separating the two plates. (b) A satellite photo of the San Francisco Bay area, showing the location of two fault lines. (c) A larger-scale view of the fault system. The numbers represent estimates of the probability of a major earthquake occurring during a 30-year period at various locations along the faults.

Not all colliding plates produce mountain ranges. At other collision locations, called *subduction zones*, one plate slides under the other, ultimately to be destroyed as it sinks into the mantle. Subduction zones are responsible for most of the deep trenches in the world's oceans.

Nor do all plates experience head-on collisions. As noted by the arrows of Figure 7.16, many plates slide or shear past one another. A good example is the most famous active region in North America—the San Andreas fault in California (Figure 7.19). The site of much earthquake activity, this fault marks the boundary where the Pacific and North American plates are rubbing past one another. The motion of these two plates, like that of moving parts in a poorly oiled machine, is neither steady nor smooth. The sudden jerks that occur when they do move against each other are often strong enough to cause major earthquakes.

At still other locations, such as down the center of the Atlantic Ocean, the plates are moving apart. As they re-

cede, new mantle material wells up between them, forming midocean ridges. Today, hot mantle material is rising through a crack all along the Mid-Atlantic Ridge, and radioactive dating techniques indicate that material has been upwelling along the ridge more or less steadily for the past 200 million years. The Atlantic seafloor is slowly growing, as the North and South American plates move away from their Eurasian and African counterparts. Similar events are occurring elsewhere on our planet.

WHAT DRIVES THE PLATES?

What process is responsible for the enormous forces that drag plates apart in some locations and ram them together in others? The answer is probably convection—the same physical process we encountered earlier in our study of the atmosphere.

Figure 7.20 is a cross-sectional diagram of the top few hundred kilometers of our planet's interior. It depicts roughly the region in and around a midocean ridge. There the ocean floor is covered with a layer of sediment—dirt, sand, and dead sea organisms that have fallen through the seawater for millions of years. Below the sediment lies about 10 km of granite, the low-density rock that makes up the crust. Deeper still lies the upper mantle, whose temperature increases with depth. Below the base of the lithosphere, at a depth of perhaps 50 km, the temperature is sufficiently high that the mantle is soft enough to flow, very slowly, although it is not molten. This region is the aesthenosphere.

This is a perfect setting for convection—warm matter underlying cool matter. The warm mantle rock rises, just as hot air rises in our atmosphere. Sometimes, the rock squeezes up through cracks in the granite crust. Every so often, such a fissure may open in the midst of a continental landmass, producing a volcano such as Mount St. Helens or possibly a geyser like those at Yellowstone National Park. However, most such cracks are under water. The Mid-Atlantic Ridge is a prime example.

Not all the rising warm rock in the upper mantle can squeeze through cracks and fissures. Some warm rock cools and falls back down to lower levels. In this way, large circulation patterns become established within the upper mantle, as depicted in Figure 7.20. Riding atop these convection patterns are the plates. The circulation is extraordinarily sluggish. Semisolid rock takes millions of years to complete one convection cycle. Although the details are far from certain and remain controversial, many researchers believe that it is the large-scale circulation patterns near plate boundaries that ultimately drive the motion of the plates. Some scientists also conjecture that the aesthenosphere acts like a lubricant, enabling the thin plates to slide across the surface of our planet.

EVIDENCE FOR PAST CONTINENTAL DRIFT

Several pieces of data support the theory of plate tectonics. The first is geographical. Figure 7.21 shows how all the continents nearly fit together like pieces of a puzzle, suggesting the existence of a single huge landmass at some time in the past. Note especially how the Brazilian coast meshes nicely with the Ivory Coast of Africa. In fact, most of the continental landmasses in the Southern Hemisphere fit together remarkably well. Following the arrows on Figure 7.16 backwards, we can see that the fits are roughly consistent with the present motions of the plates involved. The fit is not so good in the Northern Hemisphere, but we shouldn't be terribly surprised, given all the geological "debris" in the North Atlantic—Iceland, Greenland, and the British Isles. However, if we were to consider the entire continental shelves (the continental borders, which are under water) instead of just the portions that happen to stick up above sea level, the fit would improve markedly.

The idea of continental drift was first suggested in 1912 by a German meteorologist named Alfred Wegener, and part of the evidence he cited was this geographical fit of the continents. No one took him seriously, however, in part because there was then no known mechanism that could drive the plates' motions. Nearly all scientists thought it preposterous that large segments of rocky crust could be drifting across the surface of our planet. These skeptical views persisted until the mid-1960s, when the accumulation of data in support of continental drift became overwhelming.

We now believe that, sometime in the past, a single gargantuan landmass dominated our planet. Geologists call this ancestral supercontinent Pangaea, meaning "all lands," as illustrated in Figure 7.21(a). The rest of the planet was presumably covered with water. The current locations of the continents, along with measurements of their current drift rates, suggest that Pangaea was the major land feature on the Earth approximately 200 million years ago. Dinosaurs, which were then the dominant form of life, could have sauntered from Russia to Texas via Boston without getting their feet wet. The other frames

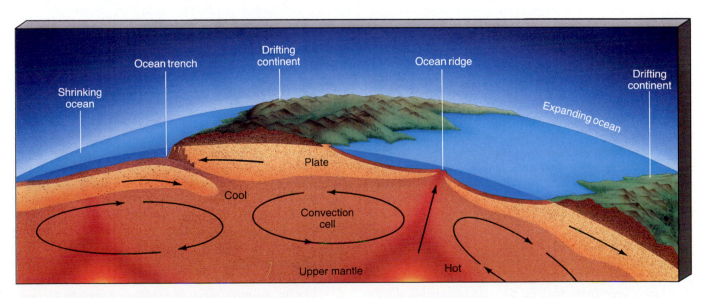

Figure 7.20   Plate drift is probably caused by convection—in this case, giant circulation patterns in the upper mantle that drag the plates across the surface.

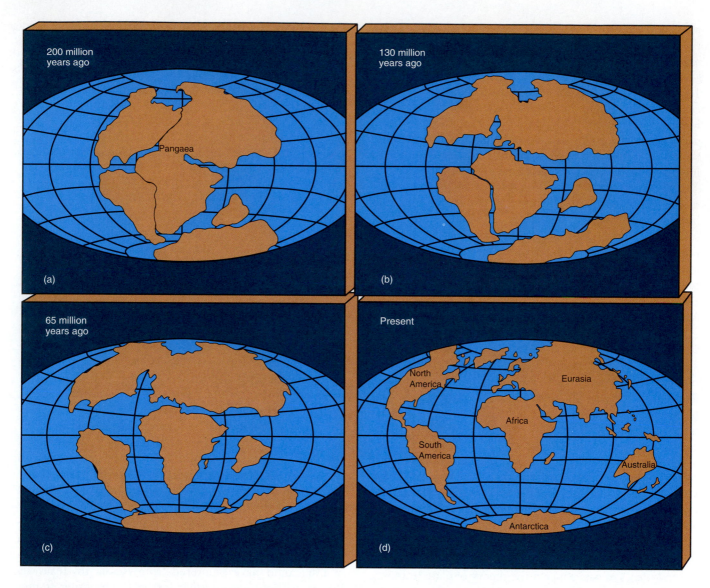

Figure 7.21 Given the currently estimated drift rates and directions of the plates, we can trace their movements back into the past. About 200 million years ago, they would have been at the approximate positions shown in (a). The continents' current positions are shown in (d).

in Figure 7.21 show how Pangaea split apart, eventually leading to the familiar continents of today.

A second piece of evidence favoring the theory of plate tectonics came when fossils of a reptile extinct for nearly 200 million years were uncovered at only two locations on Earth, one on the Brazilian coast and the other on the west coast of Africa. These two places are precisely where the continents apparently once meshed as part of the ancestral supercontinent of Pangaea. If Africa and South America had always been separated, these creatures could hardly have survived the swim between the coasts. Even if they had, the chances are slim that they would have departed and landed at exactly those parts of the continents that geographically mesh. A much more reasonable conclusion is that Africa and South America were once joined in the region where this reptile happened to live.

A third piece of evidence comes from active sites submerged beneath the ocean, forming a giant system of undersea cracks. One example is the Mid-Atlantic Ridge, mentioned earlier. During the 1970s, robot submarines retrieved samples of the ocean floor at a variety of locations on either side of this mountain range. As depicted in Figure 7.22, the ocean floor closest to the underwater ridge is relatively young, while material farther away, on either side, is noticeably older. This is exactly what would be expected if hot molten matter had been upwelling and solidifying as the Eurasian and North American plates were drifting apart. The plates on either side of the Atlantic Ocean must have been drifting apart for the past 200 million years, the oldest age found for any part of the seafloor.

Finally, a fourth piece of evidence comes from *paleomagnetism*—the study of ancient, or fossilized, magnetism.

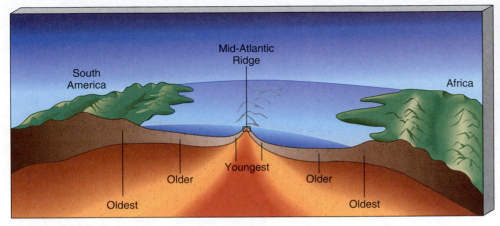

Figure 7.22 Samples of ocean floor retrieved by oceanographic vessels are youngest close to the Mid-Atlantic Ridge and progressively older farther away. The insert shows some underwater vents near the ridge, as photographed by an undersea robot.

As hot mantle material (carrying traces of iron) upwells from cracks in the oceanic ridges and solidifies, it becomes slightly magnetized, retaining an imprint of the Earth's magnetic field *at the time of cooling*. Thus, the ocean-floor matter has preserved within it a record of Earth's magnetism during past times, rather like a tape recording.

Figure 7.23 is a schematic diagram of a small portion of the ocean floor around the Mid-Atlantic Ridge. As shown, the current magnetism of Earth is oriented in the familiar north–south fashion. When samples of ocean floor are examined close to the ridge, the iron deposits are oriented just as expected—north–south. This is the "young" basalt that upwelled and cooled fairly recently. Samples retrieved far from the ridge, corresponding to older material that upwelled at earlier times, however, are often magnetized with the opposite orientation. As we move farther from the ridge, the imprinted magnetic field flips back and forth, more or less regularly, and does so symmetrically on either side of the ridge.

Scientists believe that these different magnetic orientations were caused by reversals in Earth's magnetic field as the plates drifted away from the central ridge. Working backwards, we can use the fossil magnetic field to infer the past positions of the plates, as well as the orientation of the Earth's magnetism. These paleomagnetic data provide strong support for the idea of seafloor spreading and, when taken in conjunction with the data on seafloor age, also allow us to time our planet's magnetic reversals. On average, the Earth's magnetic field reverses itself roughly every half-million years.

There is nothing particularly special about a time 200 million years in the past. We do not suppose that Pangaea remained intact for some 4 billion years after the crust first formed, only to break up so suddenly and so (relatively) recently. Much more likely, Pangaea itself came into existence after an earlier period during which other plates, carrying widely separated continental masses, were driven together by tectonic forces, merging their landmasses to form a single supercontinent. It is quite possible that there has been a long series of "Pangaeas" stretching back in time over much of the Earth's history. There will probably be many more.

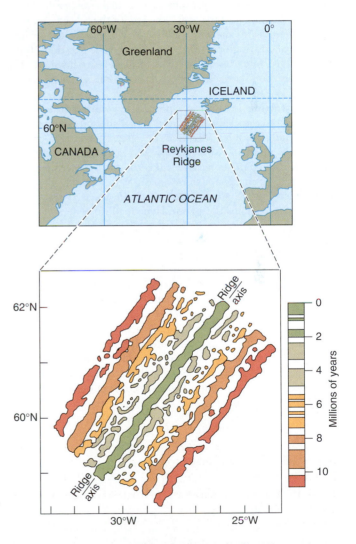

Figure 7.23 Samples of basalt retrieved from the ocean floor often show Earth's magnetism to have been oriented oppositely from the current north–south magnetic field. This simplified diagram shows the ages of some of the regions in the vicinity of the Mid-Atlantic Ridge (compare Figure 7.16), together with the direction of the fossil magnetic field. The colored areas have the current orientation; they are separated by regions of reversed magnetic polarity.

Chapter Review

SUMMARY

The six main regions of Earth are (from inside to outside) a central metallic **core** (p. 146), which is surrounded by a thick rocky **mantle** (p. 146) and topped with a thin **crust** (p. 146). The liquid oceans on our planet's surface make up the **hydrosphere** (p. 146). Above the surface is the **atmosphere** (p. 146), which is composed primarily of nitrogen and oxygen. Higher still lies the **magnetosphere** (p. 146), where charged particles from the Sun are trapped by Earth's magnetic field.

The daily **tides** (p. 146) in Earth's oceans are caused by the gravitational effect of the Moon and the Sun, which raise **tidal bulges** (p. 147) in the hydrosphere. The tidal effect of the Moon is almost twice that of the Sun. The size of the tides depends on the orientations of the Sun and the Moon relative to Earth. A differential gravitational force is always called a **tidal force** (p.148), even when no oceans or even planets are involved. The tidal interaction between Earth and the Moon is causing Earth's spin to slow.

Convection (p. 150) is a process by which heat is moved from one place to another by the actual motion (upwelling or downwelling) of a fluid, such as air or water. Convection occurs in the **troposphere** (p. 150), the lowest region of Earth's atmosphere. It is the cause of surface winds and the weather we experience. Earth's atmosphere thins rapidly with altitude. Above the troposphere, in the **stratosphere** (p. 150), the air is calm. The stratosphere contains the **ozone layer** (p. 150), where incoming solar ultraviolet radiation is absorbed. At even higher altitudes is the **ionosphere** (p. 151), where the atmosphere is kept ionized by high-energy radiation and particles from the Sun.

The **greenhouse effect** (p. 151) is the absorption and trapping by atmospheric gases (primarily carbon dioxide and water vapor) of infrared radiation emitted by Earth's surface. Incoming visible light from the Sun is not significantly absorbed by these gases. By making it more difficult for Earth to radiate its energy back into space, the greenhouse effect makes our planet's surface some 40K warmer than would otherwise be the case.

The air we breathe is not Earth's original atmosphere. It originated in material that was outgassed from our planet's interior by volcanoes, then altered by solar radiation, and finally by the emergence of life.

Earth's magnetic field extends far beyond the surface of our planet. Charged particles from the solar wind are trapped by Earth's magnetic field lines to form the **Van Allen belts** (p. 154) that surround our planet. When particles from the Van Allen belts hit Earth's atmosphere, they heat and ionize the atoms there, causing them to glow in an **aurora** (p. 154). According to **dynamo theory** (p. 154), planetary magnetic fields are produced by the motion of rapidly rotating, electrically conducting fluid (such as molten iron) in the planet's core.

Studies of the interior of our planet are carried out by observing how **seismic waves** (p. 156), produced by **earthquakes** (p. 156) just below Earth's surface, travel through the mantle. We can also study the upper mantle by analyzing the material brought to the surface when a volcano erupts. Seismic studies and mathematical modeling indicate that Earth's iron core consists of a solid inner core surrounded by a liquid outer core. Earth's center is extremely hot—about the same temperature as the surface of the Sun.

The density at the center of the Earth is much greater than the density of surface rocks. The process by which heavy material sinks to the center of a planet while lighter material rises to the surface is called **differentiation** (p. 159). The differentiation of Earth implies that our planet must have been at least partially molten in the past. One way in which this could have occurred is by the heat released during Earth's formation and subsequent bombardment by material from interplanetary space. Another possibility is the energy released by the decay of **radioactive** (p. 159) elements present in the material from which Earth formed.

Earth's surface is made up of about a dozen enormous slabs, or plates. The slow movement of these plates across the surface is called continental drift, or **plate tectonics** (p. 161). Earthquakes, volcanism, and mountain building are associated with plate boundaries, where plates may collide, move apart, or rub against one another. The motion of the plates is thought to be driven by convection in Earth's mantle. The rocky upper layer of Earth that makes up the plates is the **lithosphere** (p. 162). The semisolid region in the upper mantle over which the plates slide is called the **aesthenosphere** (p. 162). Evidence for past plate motion can be found in the geographical fit of continents, in the fossil record, and in the ages and magnetism of surface rocks.

SELF-TEST: True or False?

_____ **1.** The average density of the Earth is less than the density of water.

_____ **2.** There is one high tide and one low tide per day at any given coastal location on Earth.

_____ **3.** Because of the tides, Earth's rate of rotation is slowing down.

_____ **4.** Other than at the surface of the Earth, the ozone layer is the warmest part of the atmosphere.

_____ **5.** Earth's atmosphere is composed primarily of oxygen.

_____ **6.** Most of Earth's atmosphere is within 30 km of the surface.

_____ **7.** P-waves can travel through both liquid and solid material; S-waves travel only through solid material.

_____ **8.** Water vapor and nitrogen are the primary greenhouse gases in Earth's atmosphere.

_____ **9.** Earth's magnetic field is the result of our planet's large, permanently magnetized iron core.

_____ **10.** Motion of the crustal plates is driven by convection in Earth's upper mantle.

_____ **11.** Samples of Earth's core are available from volcanoes.

SELF-TEST: Fill in the Blank

1. The radius of the Earth is roughly _____ km.
2. Of Earth's crust, mantle, outer core, and inner core, which layer is the thinnest?
3. Earth is unique among the planets in that it has _____ on its surface.
4. The tidal force is due to the _____ in the gravitational force from one side of the Earth to the other.
5. 78 percent of Earth's atmosphere is _____; 21 percent is _____.
6. The troposphere is where the process of _____ occurs.
7. Sunlight is absorbed by Earth's surface and is reemitted in the form of _____ radiation.
8. A continued rise in the level of carbon dioxide in Earth's atmosphere would _____ our planet's temperature.
9. Earth's secondary atmosphere was outgassed by _____.
10. Oxygen in Earth's atmosphere is the result of the appearance of _____.
11. When trapped electrons and protons in the magnetosphere eventually collide with the upper atmosphere, they produce an _____.
12. Observations of S- and P-waves have confirmed that Earth's inner core is _____ and the outer core is _____.
13. Crustal rocks are made up primarily of low-density _____ ; the upper mantle is composed of slightly higher density _____.
14. For differentiation to have occurred, Earth's interior must, at some time in the past, have been largely _____.
15. Continental drift, volcanism, earthquakes, faults, and mountain building can all be explained by the process known as _____.

REVIEW AND DISCUSSION

1. By comparison with Earth's average density, what do the densities of the water and rocks in Earth's crust tell us about Earth's interior?
2. What is Rayleigh scattering? What is its most noticeable effect for us on Earth?
3. Give a brief description of Earth's magnetosphere, and tell how it was discovered.
4. Compare and contrast P-waves and S-waves, and say how they are useful.
5. How would our knowledge of Earth's interior change if our planet were geologically dead, like the Moon?
6. What conditions are needed to create a dynamo in Earth's interior? What effect does this dynamo have?
7. Give two reasons why geologists believe that part of Earth's core is in a liquid state.
8. What clue does Earth's differentiation provide to our planet's history?
9. What is convection? What effect does it have on (a) Earth's atmosphere, and (b) Earth's interior?
10. How did radioactive decay heat the Earth early in its history? When did this heating end?
11. What process has created the surface mountains, oceanic trenches, and other large-scale features on Earth's surface?
12. Discuss how distant quasars, lying hundreds of millions of light-years from Earth, are used to monitor the motion of Earth's tectonic plates.
13. How do we know that Earth's magnetic field has undergone reversals in the past? How might Earth's magnetic field reversals have affected the evolution of life on our planet?
14. If the Moon had oceans like the Earth's, what would the tidal effect be like there? How many high and low tides are there during a "day?" How would the variations in height compare to those on the Earth?
15. Is the greenhouse effect operating in Earth's atmosphere helpful or harmful? Give examples. What are the consequences of an enhanced greenhouse effect?

PROBLEMS

1. Approximating Earth's atmosphere as a layer of its gas 10 km thick, with uniform density 1.2 kg/m^3, calculate total mass. Compare this with the mass of the Earth.
2. What is the tidal effect on Earth due to Jupiter, relative to the tidal effect of the Moon? Assume an Earth-Jupiter distance of 4.2 A.U.
3. At 2 cm/yr, how long would it take a typical plate to traverse the present width of the Atlantic Ocean, about 6000 km?
4. Following an earthquake, how long would it take a P-wave moving in a straight line with speed 5 km/s to reach the opposite side of the Earth?

PROJECTS

1. Go to a sporting goods store and get a tide table; many stores near the ocean provide them free. Choose a month and plot the height of one high and one low tide versus the day of the month. Now mark the dates when the primary phases of the Moon occur. How well does the phase of the Moon predict the tides?
2. Measure the radius of the Earth. You will need a friend or colleague (or another astronomy student with a project assignment!) who lives a few hundred kilometers due north or south of you. On the day of the first quarter moon, right at sunset, you should both estimate, to within a tenth of a degree, the angular distance of the moon above your southern horizon. Compare the angles you obtain; they should be different. Call this difference θ. Determine the exact distance between your two locations using a map; call this d. The radius of the Earth can then be computed from the equation r = 57.3° (d/θ). Many details of how to do this experiment have been left for you to figure out. While you are at it, show where the formula comes from!

8

THE MOON AND MERCURY

Scorched and Battered Worlds

LEARNING GOALS

Studying this chapter will enable you to:

1 Specify the general characteristics of the Moon and Mercury, and compare them with those of the Earth.

2 Explain how the Moon's rotation is influenced by its orbit around the Earth, and Mercury's by its orbit around the Sun.

3 Describe the surface features of the Moon and Mercury, and recount how they were formed by dynamic events early in their history.

4 Explain how observations of cratering can be used to estimate the age of a body's surface.

5 Compare the Moon's interior structure with that of Mercury.

6 Summarize the various theories for the formation of the Moon, and indicate which is presently considered most likely.

7 Discuss how astronomers have pieced together the story of the Moon's evolution, and compare its evolutionary history with that of Mercury.

(Opposite page, background) The *Apollo 15* mission to the Moon explored a geological fault, called Hadley Rille, where molten lava once flowed. In the large photograph printed here, the rille—a system of valleys—runs along the base of the Apennine Mountains (lower right) at the edge of Mare Imbrium (to the left). For scale, the lower of the two large craters, Autolycus, spans 40 km. The shadow-sided, most prominent peak at lower right, Mount Hadley, rises almost 5 km high.

Inset A) An astronaut embarks from his lunar rover, preparing to explore Hadley Rille in the distance. The width of the rille is about 1.5 km and its depth averages 300m.

(Inset B) Photograph of an astronaut and a lunar rover in front of Mount Hadley.

(Inset C) An astronaut collects samples at the edge of Hadley Rille.

(Inset D) This photographic mosaic of a small part of an interior wall of Hadley Rille shows evidence for subsurface horizontal layering. Each distinct layer (denoted A, B, C) is a few meters deep and presumably represents successive lava flows that helped form the extensive lava plain called Mare Imbrium.

*T*he Moon is the only natural satellite of planet Earth. Mercury, the smallest terrestrial world, is the planet closest to the Sun. What do these bodies have in common that leads us to consider them together? In fact, quite a lot. They are comparable in both size and appearance—indeed, at first glance, you might even mistake one for the other. Both have heavily cratered, ancient surfaces, littered by boulders and pulverized dust. They have no air, no sound, no water. Clouds, rainfall, and blue sky are all absent. With no atmosphere to moderate the variations in solar heating, each experiences wild temperature swings from day to night. Why, then, bother to explore and study such hostile, barren, desolate worlds? Precisely because they are so very different from our own. It is the job of the planetary scientist to explain why the Earth and the Moon differ so greatly, despite their proximity to one another, and why Mercury, a terrestrial planet, has so much in common with Earth's Moon. In this chapter we explore the similarities and differences between these two worlds as we begin our comparative study of the planets and the moons that make up our solar system.

8.1 Orbital Properties

Let's begin our study of the Moon and Mercury by examining their orbits. Knowledge of these will, in turn, aid us in determining many of the other properties of these worlds.

THE MOON

We can determine the orbit of the Moon rather easily. Parallax methods, described in Chapter 1, can provide us with quite accurate measurements if we use Earth's diameter as the baseline. ∞ (pp. 23–24) Radar yields a more accurate lunar distance; the Moon is much closer than any of the planets, and the radar echo bounced off the Moon's surface is strong. A radio telescope receives the echo after about a 2.56-second wait. Dividing this time by 2 (to account for the round trip taken by the signal) and multiplying it by the speed of light (300,000 km/s) gives us a mean distance of 384,000 km. (The actual distance at any specific time depends on the Moon's location in its slightly elliptical orbit around the Earth.) Current technology allows astronomers to measure the radar's round-trip time to submicrosecond accuracy. Repeated radar measurements allow us to determine the Moon's orbit to within a few meters (see also *Interlude 8-1* on p. 180). This precision is needed for programming unmanned spacecraft to land successfully on the lunar surface.

As a result of these measurement techniques, the Moon's orbit (see Figure 1.19) is very well known. Its sidereal period (that is, relative to the stars) is 27.3 days; its synodic period (as seen from Earth—full Moon to full Moon, say) is 29.5 days. The orbit is *prograde*, meaning that the Moon revolves around Earth in the same sense as Earth revolves around the Sun; has an eccentricity of 0.055; and lies close to, but not exactly in, the plane of the ecliptic. The inclination of the orbital plane to the plane of the ecliptic is about 5.2°.

MERCURY

Mercury was well known to ancient astronomers, but was misinterpreted because it never strays far from the Sun. As viewed from Earth, the angular distance between the Sun and Mercury is never greater than 28°. Consequently,

Mercury is visible to the naked eye only when the Sun's light is blotted out—just before dawn or just after sunset (or, much less frequently, during a total solar eclipse). It is, therefore, impossible to follow Mercury through a full cycle of phases. In fact, the ancients originally believed that this companion to the Sun was two different objects, and the connection between the planet's morning and evening appearances took some time to establish. The early Greek astronomers, for example, had two separate names for Mercury—Hermes in the evening and Apollo in the morn-

Figure 8.1 Four planets, together with the Moon, are visible in this photograph taken shortly after sunset. To the right of the Moon (top left) is the brightest planet, Venus. A little farther to the right, at top center, is Mars, with the star Regulus just below and to its left. At the lower right, at the edge of the Sun's glare, is Jupiter, with Mercury just below it. (The moon appears round rather than crescent-shaped because the "dark" portion of its disk is indirectly illuminated by sunlight reflected from Earth. This "earthshine," relatively faint to the naked eye, is exaggerated in the overexposed photographic image.)

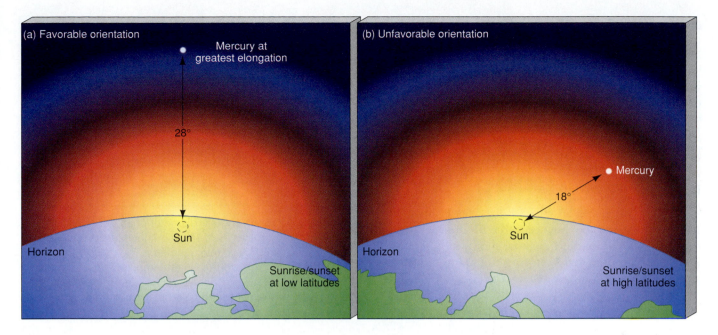

Figure 8.2   Favorable and unfavorable orientations of Mercury's orbit result from different Earth orientations and observer locations. At the most unfavorable orientations, Mercury is close to both the Sun and the horizon.

ing. (Mercury is the Roman name for the Greek god Hermes.) However, later Greek astronomers were certainly aware that the "two planets" were really different alignments of a single body. Figure 8.1, a photograph taken just after sunset, shows Mercury above the western horizon, along with three other planets and the Moon.

Because Earth rotates at a rate of 15° per hour, Mercury is visible for at most 2 hours on any given night, even under the most favorable circumstances. For most observers at most times of the year, Mercury is considerably less than 28° above the horizon, so it is generally visible for a much shorter period (see Figure 8.2). Nowadays, large telescopes can filter out the Sun's glare and observe Mercury even during the daytime, when the planet is higher in the sky and the atmosphere's effects are reduced.

(The amount of air that the light from the planet has to traverse before reaching our telescope decreases as the height above the horizon increases.) In fact, some of the best views of Mercury have been obtained in this way. The naked-eye or amateur astronomer, however, is generally limited to nighttime observations. In all cases, it becomes progressively more difficult to view Mercury the closer (in the sky) its orbit takes it to the Sun. The best images of the planet therefore show a "half Mercury," close to its maximum angular separation from the Sun, or *elongation*, as illustrated in Figure 8.3.

Mercury orbits the Sun in the same sense as Earth—counterclockwise, as seen from above the ecliptic looking down on Earth's North Pole. Its orbital plane is inclined approximately 7° to the plane of the ecliptic. The orbital semi-

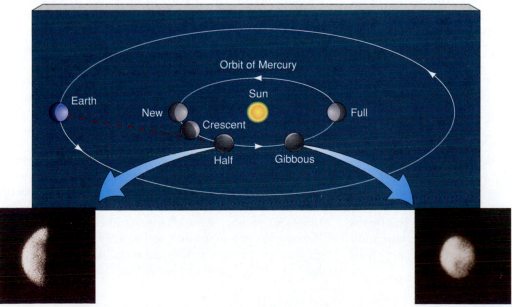

Figure 8.3   Some views of Mercury at different points along its orbit. The best images of the planet are taken when it is at its maximum elongation (greatest apparent distance from the Sun) and show a "half Mercury." (Compare Figure 2.10.)

R I V U X G

major axis is 0.39 A.U. (58 million km), and its orbital eccentricity is 0.21. Because of its relatively large eccentricity, its distance from the Sun varies substantially around its orbit, ranging from just 0.31 A.U. (46 million km) to 0.47 A.U. (70 million km). In accordance with Kepler's third law, Mercury's orbital period is 88 days. (Note that whenever we use the word day, we will mean an Earth day unless otherwise stated.)

8.2 *The Moon and Mercury in Bulk*

PHYSICAL PROPERTIES

① The Moon has an angular diameter of about 0.5°. Knowing that and the distance to the Moon, we can easily calculate its true size. The method, shown in Figure 8.4, is essentially the same that Eratosthenes used to measure the size of the Earth (see the *More Precisely* feature on p. 25). The ratio of the Moon's diameter to the circumference of its orbit ($2\pi \times 384{,}000$ km) is equal to the ratio of its angular diameter to 360°. The Moon's radius is therefore ($0.5 \times 2\pi \times 384{,}000$ km $\times 0.5°/360°$), or about 1700 km, roughly 1/4 the radius of Earth. More precise calculations yield a lunar radius of 1738 km. We can determine Mercury's radius by similar reasoning. At its closest approach to the Earth, at a distance of about 0.52 A.U., Mercury's angular diameter is measured to be $\approx13''$ (arc seconds), so its radius is about 2450 km. More accurate measurements by unmanned space probes yield a result of 2439 km, or 0.38 Earth radii.

As with all astronomical objects, we determine the masses of the Moon and Mercury by studying their gravitational pull on other objects in their vicinity, such as Earth and humanmade spacecraft. For example, if we know the period of a spacecraft's orbit around the Moon and its distance from the Moon, we can find the Moon's mass from Kepler's third law (as modified by Newton), which relates any satellite's orbital period and semi-major axis to the mass of the body it orbits. ∞ (p. 48) In fact, as mentioned in Chapter 6, even before the Space Age, the masses of both the Moon and Mercury were already quite well known from studies of the motion of the Earth. ∞ (p. 131) The mass of the Moon is 7.4×10^{22} kg, approximately 1/80 the mass of the Earth. The mass of Mercury is 3.3×10^{23} kg—about 0.055 Earth masses.

The Moon's average density of 3300 kg/m^3 contrasts with the average Earth value of about 5500 kg/m^3, suggesting that the Moon contains fewer heavy elements (such as iron) than Earth. Despite its many other similarities to the Moon, Mercury's mean density is 5400 kg/m^3, only slightly less than that of Earth. Assuming that the surface rock on Mercury is of similar density to the surface rock on Earth and the Moon, we are led to the conclusion that the interior of Mercury must contain a lot of high-density material, most probably iron. In fact, since Mercury is considerably less massive than Earth, its interior is squeezed less by the weight of overlying material, so Mercury's iron core must actually contain a much larger fraction of the planet's mass than is the case for our own planet.

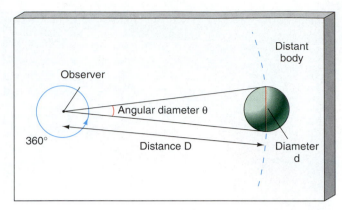

Figure 8.4 If the angular diameter of an observed object can be measured and its distance is known, its true diameter can be calculated by simple geometry. The ratio of the object's diameter (d) to the circumference of an imaginary circle centered on the observer and passing through the object ($2\pi D$) is equal to the ratio of its observed angular diameter θ to 360°.

Because the Moon and Mercury are so much less massive than Earth, their gravitational fields are also weaker. The force of gravity on the lunar surface is only about 1/6 that on Earth; Mercury's gravity is a little stronger—about 0.4 Earth gravities. Thus, an astronaut weighing 180 lbs. on Earth would weigh a mere 30 lbs. on the Moon and 72 lbs. on Mercury. Those bulky spacesuits used by the *Apollo* astronauts on the Moon were not as heavy as they appeared!

ATMOSPHERES

Astronomers have never observed any appreciable atmosphere on either the Moon or Mercury, either spectroscopically from the Earth or during close approaches by spacecraft. As discussed in the *More Precisely* feature on pp. 136–137, this is a direct consequence of these bodies' weak gravitational fields. Massive objects have a better chance of retaining their atmospheres, because the more massive the object, the larger are the velocities needed for atoms and molecules to escape. The Moon's escape velocity is only 2.4 km/s, compared with 11 km/s for Earth; Mercury's escape velocity is 4.3 km/s. Any primary atmospheres these worlds had initially, or secondary atmospheres that appeared later, have gone forever.

Although *Mariner 10* found a trace of what was at first thought to be an atmosphere on Mercury, this gas is now known to be temporarily trapped hydrogen and helium from the solar wind. Mercury captures this gas and holds it for just a few weeks. Earth-based observations have also found an extremely tenuous envelope of sodium and potassium around the planet. Scientists believe that these atoms are torn out of the surface rocks by impacts with high-energy particles in the solar wind; they do not constitute a "true" atmosphere in any sense. Thus, neither the Moon nor Mercury has any protection against the harsh environment of interplanetary space. This fact is crucial in understanding their surface evolution and present-day appearance.

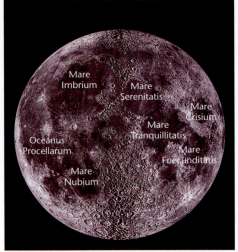

(a)

Figure 8.5 (a) A photographic mosaic of the full Moon, north pole at the top. Because the Moon emits no visible radiation of its own, we can see it only by the reflected light of the Sun. (b) The Moon near third quarter. Notice that surface features are much more visible near the terminator (the line separating light from dark), where sunlight strikes at a sharp angle and shadows highlight the topography. The light-colored crater just below the lower left corner of the breakout box is Copernicus. (c) Magnified view of a region near the terminator, as seen from Earth through a large telescope. The central dark area is Mare Imbrium, ringed at bottom right by the Apennine mountains. (d) An enlargement of a portion of (c). The crater at the bottom left is Eratosthenes; Archimedes is at the top center.

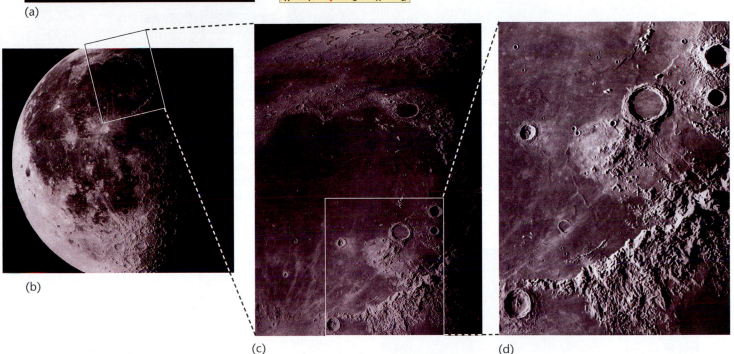

(b)

(c)

(d)

Lacking the moderating influence of an atmosphere, both the Moon and Mercury experience wide variations in surface temperature. Noontime temperatures on the Moon can reach 400 K, well above the boiling point of water. Because of its proximity to the Sun, Mercury's daytime temperature is even higher—radio observations of the planet's thermal emission indicate that it can reach 700 K. ∞ (p. 64) But at night or in the shade, temperatures on both worlds fall to about 100 K, well below water's freezing point. Mercury's 600 K temperature range is the largest of any planet or moon in the solar system.

SURFACE FEATURES

The first observers to point their telescopes at the Moon, most notable among them Galileo Galilei, noted large dark areas, resembling (they thought) the Earth's oceans. They also saw light-colored areas resembling the continents. Both types of region are clear in Figure 8.5, a mosaic of the full Moon, made possible by sunlight reflected toward Earth from the lunar surface. The light and dark surface features are also evident to the naked eye, creating the face of the familiar "Man-in-the-Moon."

Today we know that the dark areas are not oceans but extensive flat areas that resulted from lava flows during a much earlier period of the Moon's evolution. Nevertheless, they are still called **maria**, a Latin word meaning "seas" (singular, *mare*). There are 14 maria, all roughly circular. The largest of them (Mare Imbrium) is about 1100 km in diameter. The lighter areas, originally dubbed *terrae*, from the Latin word for "land," are now known to be elevated several kilometers above the maria. Accordingly, they are usually called the lunar **highlands**.

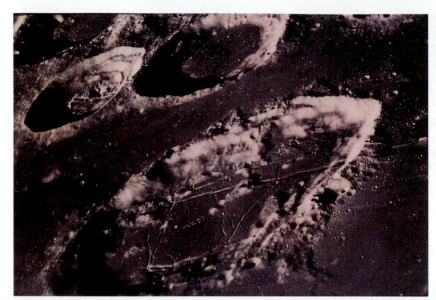

Figure 8.6   The Moon, as seen from the *Apollo 8* orbiter during the first human circumnavigation of the Moon in 1968. Craters ranging in size from 50 km to 500 m (also the width of the long fault lines) can be seen.

The smallest lunar features we can distinguish with the naked eye are about 200 km across. Telescopic observations further resolve the surface into numerous bowl-shaped depressions, or **craters**. Most craters apparently formed eons ago primarily as the result of meteoritic impact. Craters are particularly clear in Figures 8.5(b) and (c) near the *terminator* (the line that separates day from night on the surface), where the Sun is low in the sky and casts long shadows that enable us to distinguish quite small surface details. Because of the effects of our atmosphere, the smallest lunar objects that telescopes on the Earth's surface can resolve are about 1 km across (see Figure 8.5c).

Much more detailed photographs have been taken by orbiting spacecraft and, of course, by visiting astronauts. Figure 8.6 shows a view of the Moon taken from an orbiting spacecraft. The smallest features that can be distinguished in this image are about 500 m across. Lunar craters are now known to come in all sizes—the largest are hundreds of kilometers in diameter, the smallest are microscopic. Craters are found everywhere on the Moon's surface, although they are much more prevalent in the highlands.

All of the Moon's significant surface features have names. The 14 maria bear fanciful Latin names—Mare Imbrium ("Sea of Showers"), Mare Nubium ("Sea of Clouds"), Mare Nectaris ("Sea of Nectar"), and so on. Most mountain ranges in the highlands bear the names of terrestrial mountain ranges—the Alps, the Carpathians, the Apennines, the Pyrenees, and so on. Most of the craters are named after great scientists or philosophers, such as Plato, Aristotle, Eratosthenes, and Copernicus.

To the surprise of most astronomers, when the far side of the Moon was mapped, first by Soviet and later by

Figure 8.7   The western hemisphere of the Moon, as seen by the *Galileo* probe en route to Jupiter. The large, dark region is part of the face visible from Earth. This image shows only one or two small maria on the far side.

Figure 8.8   Photograph of Mercury taken from Earth with one of the largest ground-based optical telescopes. Only a few surface features are discernible.

Figure 8.9   (a) Mercury is imaged here as a mosaic of photographs taken by the *Mariner 10* spacecraft in the mid-1970s during its approach to the planet. At the time, the spacecraft was some 200,000 km away. (b) *Mariner 10*'s view of Mercury as it sped away from the planet after each encounter. Again, the spacecraft was about 200,000 km away when the photographs making up this mosaic were taken.

American spacecraft, no major maria were found there. The lunar far side (see Figure 8.7) is composed almost entirely of highlands. This fact has great bearing on our theory of how the Moon's surface terrain came into being, for it implies that the processes involved could *not* have been entirely internal in nature. The location of the Earth must somehow have played a role.

Mercury is difficult to observe from Earth because of its closeness to the Sun. Even with a fairly large telescope, we see it only as a slightly pinkish disk. Figure 8.8 is one of the few photographs of Mercury taken from Earth that shows any evidence of surface markings. Astronomers could only speculate about the faint, dark markings in the days before the robot spacecraft *Mariner 10* provided clearer images; we now know that these markings are much like those one sees when gazing casually at Earth's Moon. The largest telescopes can resolve features on the surface of Mercury no better than we can perceive features on our Moon with our unaided eyes.

In 1974, the U.S. spacecraft *Mariner 10* (see Chapter 6) approached within 10,000 km of the surface of Mercury, sending back high-resolution images of the planet. ∞ (p. 134) These photographs, which showed surface features as small as 150 m across, revolutionized our knowledge of the planet. Figures 8.9 (a) and (b) show views of Mercury radioed back to Earth from a distance of 200,000 km by *Mariner 10*. Together, these two mosaics cover the known surface of Mercury. No similar photographs exist of the hemisphere that happened to be in shadow during the encounters. Figure 8.10 shows a higher-resolution photograph of the planet from a distance of 20,000 km. The similarities to the Moon are very striking. We see no sign of clouds, rivers, dust storms, or other aspects of weather. Much of the cratered surface bears a strong resemblance to the Moon's highlands. Mercury, however, shows few extensive lava flow regions akin to the lunar maria.

Figure 8.10   Another photograph of Mercury by *Mariner 10*, this time from about 20,000 km above the planet's surface. The "double-ringed crater" at the upper left, named C. Bach, is about 100 km across; it exemplifies how many of the large craters on Mercury tend to form double, rather than single, rings. The reason is not yet understood.

8.3 *Rotation Rates*

THE ROTATION OF THE MOON

❷   The Moon's rotation period about its axis is precisely equal to its period of revolution about the Earth (27.3 days), so the Moon keeps the same side facing the Earth at all times (see Figure 8.11). To an astronaut standing on the Moon's surface, Earth would appear almost stationary in the sky (though its daily rotation would be obvious). Sunrise and sunset on the Moon happen roughly two weeks apart. Is this just a remarkable coincidence, or is there some reason for this state of affairs?

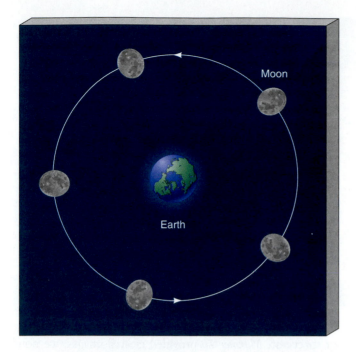

Figure 8.11   The Moon is slightly elongated in shape, with its long axis perpetually pointing toward Earth. (The elongation is highly exaggerated in this diagram.)

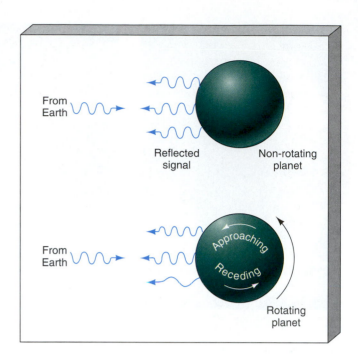

Figure 8.12   A radar beam reflected from a rotating planet yields information about both the planet's overall motion and its rotation rate.

That the Moon's rotation and revolution periods are equal is no accident. Just as the Moon raises tides on Earth, Earth also produces a tidal bulge in the Moon. In fact, because Earth is so much more massive, the tidal force on the Moon is about 20 times greater than that on Earth and the Moon's tidal bulge is correspondingly larger. In Chapter 7, we saw how lunar tidal forces are causing Earth's spin to slow and how, as a result, Earth will eventually rotate on its axis at the same rate as the Moon revolves around Earth. ∞ (p. 149) This condition—when the spin of one body is precisely equal to (or *synchronized* with) the period of its orbit around another—is known as a **synchronous orbit.**

The Earth's rotation will not become synchronous with the Earth–Moon orbital period for hundreds of billions of years. In the case of the Moon, however, the process has already gone to completion. The Moon's much larger tidal deformation caused it to evolve into a synchronous orbit long ago—the Moon is said to have become *tidally locked* to Earth (see Figure 8.11). As a consequence, the Moon has a "near" side, which is always visible from Earth, and a "far" side, which never is. Thus, until very recently, no one on Earth had any idea what half of our satellite looked like. It was only when spacecraft actually flew around the Moon that we finally saw the far side. Most of the moons in the solar system are similarly locked by the tidal fields of their parent planets.

In fact, the size of the lunar bulge is too great to be produced by Earth's present-day tidal influence. The explanation seems to be that, long ago, the distance from Earth to the Moon may have been as little as two-thirds of its current value, or about 250,000 km. ∞ (p. 149) Earth's tidal force on the Moon would then have been

more than three times greater than it is today, and could have accounted for the Moon's elongated shape. The resulting distortion would have "set" when the Moon solidified, thus surviving to the present day, while at the same time accelerating the synchronization of the Moon's orbit.

MEASUREMENT OF MERCURY'S SPIN

In principle, the ability to discern surface features on Mercury should allow us to measure its rotation rate simply by watching the motion of a particular region around the planet. In the mid-nineteenth century, an Italian astronomer named Giovanni Schiaparelli did just that. He concluded that Mercury always keeps one side facing the Sun, much as our Moon perpetually presents only one face to Earth. The explanation suggested for this synchronous rotation was the same as for the Moon—the tidal bulge raised in Mercury by the Sun had modified the planet's rotation rate until the bulge always pointed directly at the Sun. While the surface features could not be seen clearly, the combination of Schiaparelli's observations and a plausible physical explanation was enough to convince most astronomers, and the belief that Mercury rotates synchronously with its revolution about the Sun persisted for almost half a century.

In 1965, astronomers making radar observations of Mercury from the Arecibo radio telescope in Puerto Rico (see Figure 5.20) discovered that this long-held view was in error. The technique they used is illustrated in Figure 8.12, which shows a radar signal reflecting from the surface of a hypothetical planet. Let us imagine, for the purpose of this discussion, that the pulse of outgoing radiation is of a single frequency.

The returning pulse bounced off the planet is very much weaker than the outgoing signal. Beyond this change, the reflected signal can be modified in two important ways. First, the signal as a whole might be redshifted or blueshifted as a consequence of the Doppler effect, depending on the overall radial velocity of the planet with respect to Earth. ∞ (p. 68) Let's assume for simplicity that this velocity is zero, so, on average, the frequency of the reflected signal is the same as the outgoing beam. Second, if the planet is rotating, the radiation reflected from the side of the planet moving toward us returns at a slightly higher frequency than the radiation reflected from the receding side, simply. (Think of the two hemispheres as being separate sources of radiation and moving at slightly different velocities, one toward us and one away.) The effect is very similar to the rotational line broadening discussed in Chapter 4 (see Figure 4.18), except that, in this case, the radiation we are measuring was not emitted by the planet, only reflected from its surface. ∞ (p. 90) What we see in the reflected signal is a spread of frequencies on either side of the original frequency. By measuring the extent of that spread, we can determine the planet's rotational speed.

The Arecibo researchers found that the rotation period of Mercury is not 88 days, as astronomers had previously believed, but actually 59 days. Astronomers quickly realized that the 59-day rotation period is very close to two-thirds of the 88-day orbital period. More detailed measurements confirmed that the rotation period is in fact *exactly* two-thirds of a Mercury year. Because there are exactly three rotations for every two revolutions, we say that there is a 3:2 **spin-orbit resonance** in Mercury's motion. In this context, the term *resonance* just means that two characteristic times—here Mercury's day and year—are related to one another in a simple way. An even simpler example of a spin-orbit resonance is the Moon's orbit around the Earth. In that case, the rotation is synchronous with the revolution, so the resonance is said to be 1:1.

Figure 8.13 illustrates the implications of Mercury's curious rotation for a hypothetical inhabitant of the planet. Mercury's solar day—the time from noon to noon, say—is actually two Mercury years long! The Sun stays "up" in the black Mercury sky for almost three Earth months at a time, after which follows nearly three months of darkness. At any given point in its orbit, Mercury presents the same face to the Sun not *every* time it revolves, but *every other* time.

EXPLANATION OF MERCURY'S ROTATION

Mercury's 3:2 spin-orbit resonance did not occur by chance. What mechanism establishes and maintains it? In the case of the Moon orbiting the Earth, we explain the 1:1 resonance as the result of tidal forces. In essence, the lunar rotation period, which probably started off much shorter than its present value, has lengthened so that the tidal bulge created by Earth is fixed relative to the body of the Moon. We now know that tidal forces (this time due to the Sun) are also responsible for Mercury's 3:2 resonance, but in a much more subtle way.

Mercury could not settle into a 1:1 resonance because its orbit around the Sun is quite eccentric. By Kepler's second law, Mercury's orbital speed is greatest at perihelion (closest approach to the Sun) and least at aphelion (greatest distance from the Sun). ∞ (p. 41) A moment's thought shows that there is no way that the planet (rotating at a constant rate) can remain in a synchronous orbit—if its rotation were synchronous near perihelion, it would be too rapid at aphelion, while synchronism at aphelion would be too slow at perihelion.

Tidal forces always act to try to synchronize the rotation rate with the instantaneous orbital speed, but such synchronization cannot be maintained over Mercury's entire orbit. What happens? The answer is found when we realize that tidal effects diminish very rapidly with increasing distance, so the tidal forces acting on Mercury at perihelion are much greater than those at aphelion. As a result, perihelion "won" the struggle to determine the rotation rate. In the 3:2 reso-

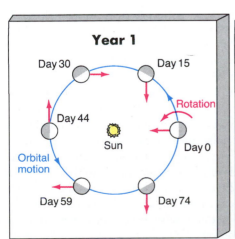
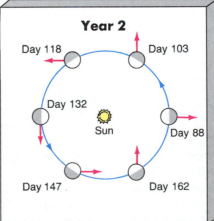
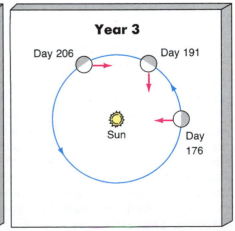

Figure 8.13 Mercury's orbital and rotational motions combine to produce a day that is two years long. The arrow represents an observer standing on the surface of the planet. At day 0, it is noon for our observer, and the Sun is directly overhead. By the time Mercury has completed one full orbit around the Sun and moved from day 0 to day 88, it has rotated on its axis exactly 1.5 times, so that it is now midnight at the observer's location. After another complete orbit, it is noon once again.

Interlude 8-1   *Lunar Laser Ranging*

Several manned *Apollo* (U.S.) and unmanned *Luna* (USSR) missions left equipment on the Moon. One of the most interesting devices the Americans left is an array of mirrors designed to intercept light pulses launched from Earth and to reflect them back toward Earth. Each small mirror resembles those used on highway posts and signs to reflect automobile headlight beams. The whole array of reflectors is not much larger than this book; it is at center right in the photo.

Astronomers can use this equipment to measure the distance to the Moon by a method that is similar to the radar technique described in the text but differs in two important ways. First, the lunar-laser-ranging method does not use radio signals. Instead, it uses a *laser,* which transmits highly concentrated *light* pulses. (The word *laser* is an acronym for *light amplification by stimulated emission of radiation.*) Second, because the reflectors are positioned at well-known locations on the Moon, researchers know precisely which point on the surface is being measured. Radar echoes bounce from much larger areas on the Moon's surface and are thus a little less accurate.

Aiming the laser at the reflector on the Moon is comparable to hitting a dime with a rifle at a distance of about a kilometer. And like the radar technique, the return echo of light is a lot weaker than the transmitted light. Only about one returning photon is detected for every billion billion (10^{18}) photons in the burst of light sent toward the target.

Astronomers can currently determine the distance to the Moon—and so gather data to model its orbit—to an accuracy of about 6 cm. Further refinements in equipment during the next few years should enable researchers to decrease the error to within a centimeter or two.

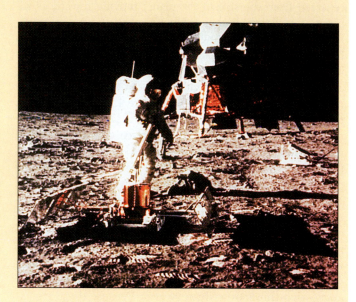

nance, Mercury's orbital and rotational motion are almost exactly synchronous *at perihelion,* so that particular rotation rate was naturally "picked out" by the Sun's tidal influence on the planet. Notice that, even though Mercury rotates through only 180° between one perihelion and the next (see Figure 8.13), the appearance of the tidal bulge is the *same* each time around.

The motion of Mercury is one of the simplest nonsynchronous resonances known in the solar system. Astronomers now believe that these intricate dynamical interactions are responsible for much of the fine detail observed in the motion of the solar system. Examples of resonances can be found in the orbits of many of the planets, their moons, their rings, and in the asteroid belt.

The Sun's tidal influence also causes Mercury's rotation axis to be exactly perpendicular to its orbit plane. As a result, and because of Mercury's eccentric orbit and the spin-orbit resonance, some points on the surface get much hotter than others. In particular, the two (diametrically opposite) points on the surface where the Sun is directly overhead at perihelion get hottest of all. They are called the *hot longitudes.* The peak temperature of 700 K quoted earlier occurs at noon at these two locations. At the *warm longitudes,* where the Sun is directly overhead at aphelion, the peak temperature is about 150 K cooler—a mere 550 K! By contrast, the Sun is always on the horizon as seen from the

planet's poles. Recent Earth-based radar studies suggest that Mercury's polar temperatures may be as low as 125 K and that, despite the planet's scorched equator, the poles may be covered with extensive sheets of water ice.

8.4 *Lunar Cratering*

❸ On Earth, the combined actions of wind and water erode our planet's surface and reshape its appearance almost on a daily basis. Coupled with the never-ending motion of the Earth's surface plates, this means that most of the ancient history of our planet's surface is lost to us. The Moon, on the other hand, has no air, no water, no plate tectonics, and no ongoing volcanic or seismic activity. Consequently, features dating back almost to the formation of the Moon itself are still visible today. The lunar surface is not entirely changeless, however. There is ample evidence for erosion—for example, the soft edges of the craters visible in the foreground of Figure 8.14. In the absence of erosion, these features would still be as jagged and angular today as they were when they formed. Something must have worn them down to their present condition.

METEORITIC IMPACTS

The primary source of erosion on the Moon is interplanetary debris, in the form of small *meteoroids* that collide with the lu-

Figure 8.14 Despite the complete lack of wind and water on the airless Moon, the surface has still eroded a little under the constant "rain" of impacting meteoroids, especially micrometeoroids. The twin tracks were made by the *Apollo* lunar rover.

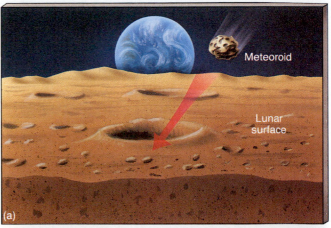

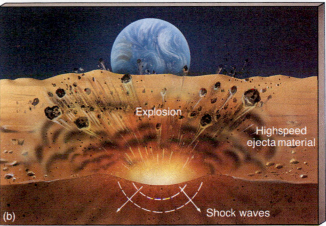

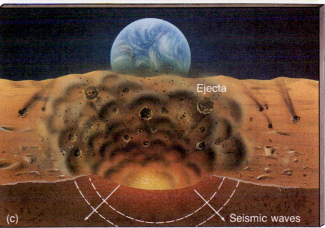

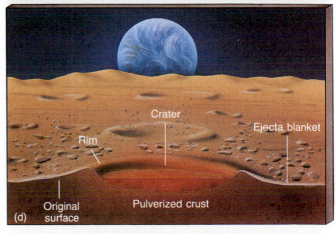

nar surface. This material, much of it rocky or metallic in composition, is strewn throughout the solar system. It wanders around in interplanetary space, perhaps for billions of years, until it collides with some planet or moon. On Earth, most meteoroids burn up in the atmosphere, producing the streaks of light known as *meteors*, or "shooting stars." But the Moon, without an atmosphere, has no protection against this onslaught. Large and small meteoroids just zoom in and collide with the surface, sometimes producing huge craters. Over billions of years, these collisions have scarred, cratered, and sculpted the landscape. Craters are still being formed today—even as you read this—all across the lunar surface.

Meteoroids generally strike the Moon at velocities of several kilometers per second. At these speeds, even a small piece of matter carries an enormous amount of energy—for example, a 1-kg object hitting the Moon's surface at 10 km/s would release as much energy as the detonation of 10 kg of TNT. As illustrated in Figure 8.15, impact of a meteoroid with the surface causes sudden and tremendous pressures to build up, heating the normally brittle rock and deforming the ground like heated plastic. The ensuing explosion pushes previously flat layers of rock up and out, forming a crater.

The diameter of the eventual crater is typically 10 times that of the incoming meteoroid; the crater depth is about twice the meteoroid's diameter. Thus, our 1 kg meteoroid, measuring perhaps 10 cm across, would produce a crater about 1 m in diameter and 20 cm deep. Shock waves from the impact pulverize the lunar surface to a depth many times that of the crater itself. The material thrown out by the explosion surrounds the crater in a layer called an *ejecta blanket*. The ejected debris ranges in size from

Figure 8.15 Several stages in the formation of a crater by meteoritic impact. (a) A meteoroid strikes the surface, releasing a large amount of energy. (b, c) The resulting explosion ejects material from the impact site and sends shock waves through the underlying surface. (d) Eventually, a characteristic crater surrounded by a blanket of ejected material results.

(a)

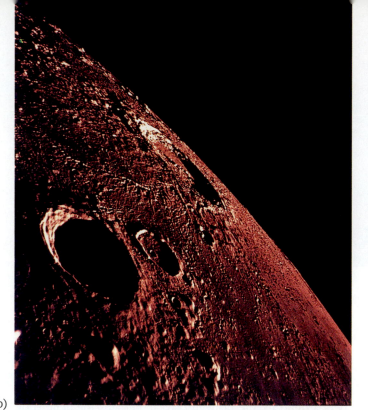

(b)

R I V U X G

Figure 8.16 (a) A large lunar crater, called the Orientale basin. The impact that produced this crater thrust up much surrounding matter, which can be seen as concentric rings of cliffs called the Cordillera Mountains. The outermost ring is nearly 1000 km in diameter. (b) Two smaller craters called Reinhold and Eddington sit amid the secondary cratering resulting from the impact that created the 90 km wide Copernicus crater (near the horizon) about a billion years ago. The ejecta blanket from crater Reinhold, 40 km across, and in the foreground, can be seen clearly. View is looking northeast from the Lunar Module during the *Apollo 12* mission.

fine dust to large boulders. Figure 8.16(a) shows the result of one large meteoritic impact on the Moon. As shown in Figure 8.16(b), the larger pieces of ejecta may themselves form secondary craters.

CRATERING RATES

④ In addition to bombardment by meteoroids with masses of a gram or more, a constant "rain" of *micrometeoroids* (debris with masses ranging from a few micrograms up to about 1 gram) also eat away at the structure of the lunar surface, contributing to the overall erosion process. Some examples can be seen in Figure 8.17, a photomicrograph of glassy "beads" brought back to Earth by the *Apollo* astronauts. The beads themselves were formed during the explosion following a meteoroid impact, when surface rock was melted, ejected, and rapidly cooled. However, several of them also display fresh, miniature craters caused by micrometeoroids that struck them after they cooled and solidified. The *rate* of cratering decreases rapidly with crater size—fresh large craters are very few and far between, but small craters are very common. The reason for this is simple: There just aren't very many large chunks among the interplanetary debris, so their collisions with the Moon are rare. At the present average rates, one new 10 km (diameter) lunar crater is formed

roughly every 10 million years, a new meter-sized crater is created about once a month, and centimeter-sized craters are formed every few minutes.

Despite this constant barrage from space, the Moon's present-day erosion rate is still very low—about 1/10,000 that on Earth. Wind and water on Earth are far more effective erosive agents than meteoritic bombardment on the Moon. For example, the Barringer Meteor Crater (shown in Figure 8.18) in the Arizona desert, one of the largest meteoroid craters on Earth, is only 25,000 years old, but has already undergone noticeable erosion. It will probably disappear completely in a mere million years, quite a short time geologically. If a crater that size had formed on the Moon even 4 billion years ago, it would still be plainly visible today.

When the *Apollo* astronauts visited several lunar sites and brought back rock samples, it became possible to measure the ages of the highlands and the maria using radioactive dating techniques, and astronomers can now use the known ages of Moon rocks to estimate the rate of cratering in the past. The highlands are typically more than 4 billion years old, while the maria have ages ranging from 3.2 to 3.9 billion years. Thus the much more heavily cratered highlands are indeed older than the less cratered maria, but the difference in cratering is not sim-

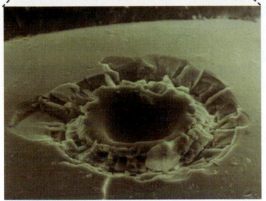

Figure 8.17   Craters of all sizes litter the lunar landscape. Some shown here, embedded in glassy beads retrieved by *Apollo* astronauts, measure only 0.01 mm across. (The scale at the top is in millimeters.)

Figure 8.18   The Barringer Meteor Crater, near Winslow, Arizona, is 1.2 km in diameter and 0.2 km deep. Geologists think a large meteoroid made it about 25,000 years ago. The meteoroid probably was about 50 m across and weighed around 300,000 tons.

ply a matter of exposure time. Astronomers now believe that the Moon, and presumably the entire inner solar system, experienced a sudden sharp drop in meteoritic bombardment about 3.9 billion years ago. The highlands solidified and received most of their craters before that time, while the maria solidified afterward. The rate of cratering has been roughly constant ever since.

8.5 *Lunar Surface Composition*

SURFACE ROCKS

❸ The *Apollo* program demonstrated clear differences in composition between the lunar highlands and the maria. The highlands are made largely of rocks rich in aluminum, making them lighter in color and lower in density (2900 kg/m^3). The maria's basaltic matter contains more iron, giving it a darker color and greater density (3300 kg/m^3). Loosely speaking, the highlands represent the Moon's crust, while the maria are made of mantle material. Many of the rock samples brought back by the *Apollo* astronauts show evidence for repeated shattering and melting—direct evidence of the violent shock waves and high temperatures produced in meteoritic impacts.

Geologists believe that the type of rock in the maria arose on the Moon much as basalt did on Earth, through the upwelling of molten material through the lunar crust. The great basins that formed the maria are thought to have been created during the final stages of the heavy meteoritic bombardment just described, between about 4.1 and 3.9 billion years ago. Subsequent volcanic activity filled the craters with lava, creating the formations we see today. In a sense, then, the maria *are* oceans—ancient seas of molten lava, now solidified.

Not all of these great craters became flooded with lava, however. One of the youngest craters is the Orientale basin (Figure 8.16a), which formed about 3.9 billion years ago. It did not undergo much subsequent volcanism, and so we can recognize its structure as an impact crater rather than as another mare. On the far side of the Moon, similar "unflooded" basins can be seen.

LUNAR DUST

Meteoroid collisions with the Moon are the primary cause of the layer of pulverized ejecta—also called lunar dust or *regolith* (meaning "fine rocky layer")—that covers the lunar landscape to an average depth of about 20 m. This microscopic dust has a typical particle size of about 0.01 mm. In consistency, it is rather like talcum powder or ready-mix

Figure 8.19 Photograph of an *Apollo* astronaut's bootprint in the lunar dust. The astronaut's weight has compacted the regolith to a depth of a few centimeters.

dry mortar. Figure 8.19 shows an astronaut's bootprint in the regolith. Owing to the very low rate of lunar erosion, even those shallow bootprints will remain intact for millions of years. The regolith is thinnest on the maria (10 m) and thickest on the highlands (over 100 m in places).

In contrast to Earth's soil, the lunar regolith contains no organic matter like that produced by biological organisms. No life whatsoever exists on the Moon. Nor were any fossils found in *Apollo* samples. Lunar rocks are barren of life and apparently always have been. NASA was so confident of this that the astronauts were not even quarantined on their return from the last few *Apollo* landings. Furthermore, all the lunar samples returned by the American and Soviet Moon programs were bone dry. Lunar rock doesn't even contain minerals having water molecules locked within their crystal structure. Terrestrial rocks, conversely, are almost always 1 or 2 percent water.

VOLCANISM

Only a few decades ago, debate raged in scientific circles as to the origin of lunar craters, with most scientists holding the opinion that the craters were the result of volcanic activity. We now know that almost all lunar craters are actually meteoritic in origin. However, a few of them apparently are not. For example, Figure 8.20 shows an intriguing alignment of several craters in a *crater-chain* pattern so straight that it is very unlikely to have been produced by meteoroids randomly colliding with the surface. Instead, the crater chain probably marks the location of a subsurface fault—a place where cracking or shearing of the surface once allowed molten matter to well up from below. As the lava cooled, it formed a solid "dome" above each fissure. Subsequently, the underlying lava receded and the centers of the domes collapsed, forming the

Figure 8.20 This "chain" of well-ordered craters was photographed by an *Apollo 14* astronaut. The largest crater, called Davy, is located on the western edge of Mare Nubium. The entire field of view measures about 100 km across.

Figure 8.21 A volcanic rille, photographed from the *Apollo 15* spacecraft orbiting the Moon, can be seen clearly here (bottom and center) winding its way through one of the maria. Called Hadley Rille, this system of valleys runs along the base of the Apennine Mountains (lower right) at the edge of the Mare Imbrium (to the left). Autolycus, the large crater closest to the center, spans 40 km. The shadow-sided, most prominent peak at lower right, Mount Hadley, rises almost 5 km high. (See also the chapter–opening image on p. 170.)

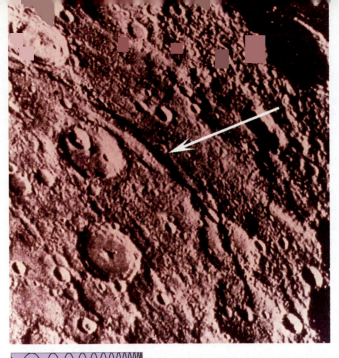

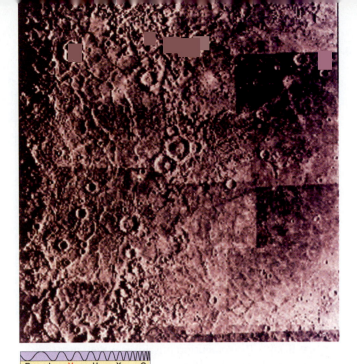

Figure 8.22 Discovery Scarp on Mercury's surface. This appears to be a compressional feature that formed when the planet's crust cooled and contracted early in its history, causing a crease in the surface. This scarp, running diagonally across the center of the frame, is several hundred kilometers long and up to 3 km high in places. (Another scarp can be seen in the top left of the figure, close to the horizon.)

Figure 8.23 Mercury's most prominent geological feature—the Caloris basin—measures about 1400 km across and is ringed by concentric mountain ranges that reach more than 3 km high in places. This huge circular basin, only half of which can be seen (at the left) in this *Mariner 10* photo, is similar in size to the Moon's Mare Imbrium and spans more than half of Mercury's radius.

craters we see today. Similar features have been observed on Venus by the orbiting *Magellan* probe (see Chapter 9).

Many other examples of lunar volcanism are known, both in telescopic observations from the Earth and in the close-up photographs taken during the *Apollo* missions. Figure 8.21 shows a volcanic **rille**, a ditch where molten lava once flowed. There is good evidence for surface volcanism early in the Moon's history, and the volcanism in turn explains the presence of the lava that formed the maria. However, whatever volcanic activity once existed on the Moon ended long ago. The low-density lunar highlands are dated to be *at least* 4 billion years old (and some are as old as 4.4 billion years). The high-density maria are in all cases found to be only a little younger. No rocks on the Moon are known to be younger than 3 billion years. (Recall from Chapter 7 that the radioactivity clock doesn't start "ticking" until the rock solidifies.) ∞ (p. 160) Apparently, the maria solidified more than 3 billion years ago, and the Moon has been dormant ever since.

8.6 The Surface of Mercury

❸ Like the Moon's craters, almost all those on Mercury are the result of meteoritic bombardment. But the craters are less densely packed than their lunar counterparts, and there are extensive **intercrater plains.** The crater walls are generally not as high as those on the Moon, and the ejected material appears to have landed closer to the impact site, exactly as we would expect on the basis of the greater surface gravity on Mercury. One likely explanation for Mercury's relative lack of

craters is that the older craters have been filled in by volcanic activity, in much the same way as the Moon's maria filled in craters as they formed. However, the plains do not look much like maria—they are much lighter in color and not as flat. Still, most geologists believe that volcanism did occur in Mercury's past, obscuring the old craters. The details of how Mercury's landscape came to look the way it does remain unexplained. The apparent absence of rilles or other obvious features associated with very-large-scale lava flows, along with the light color of the lava-flooded regions, suggest that Mercury's volcanic past was different from the Moon's.

Mercury has at least one type of surface feature not found on the Moon. Figure 8.22 shows a **scarp**, or cliff, on the surface that does not appear to be the result of volcanic or other familiar geological activity. The scarp cuts across several craters, which indicates that whatever produced it occurred *after* most of the meteoritic bombardment was over. Mercury shows no evidence for crustal motions. The scarps, of which several are known from the *Mariner* images, probably formed when the planet's interior cooled and shrank long ago, much as wrinkles form on the skin of an old shrunken apple. If we can apply to Mercury the cratering age estimates we use for the Moon, the scarps probably formed about 4 billion years ago.

Figure 8.23 shows what was probably the last great event in the geological history of Mercury—an immense bull's-eye crater called the Caloris basin, formed eons ago by the impact of a large asteroid. (The basin is so called because it lies in Mercury's "hot longitudes"—see Section 8.3—close to the planet's equator; *calor* is the Latin word for heat.) Because of

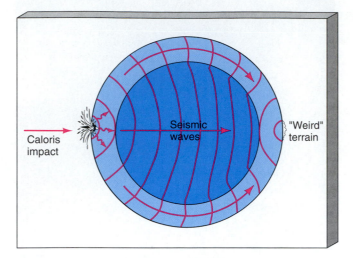

Figure 8.24 The refocusing of seismic waves after the Caloris basin impact may have created the weird terrain on the opposite side of the planet.

the orientation of the planet during *Mariner 10*'s flybys, only half of the basin was visible. The center of the crater is off the left-hand side of the photograph. Compare this basin with the Orientale basin on the Moon (Figure 8.16a). The impact-crater structures are quite similar, but even here there is a mystery: The patterns visible on the Caloris floor are unlike any seen on the Moon. Their origin, like the composition of the floor itself, is unknown.

So large was the impact that created the Caloris basin that it apparently sent strong seismic waves reverberating throughout the entire planet. On the opposite side of Mercury from Caloris there is a region of oddly rippled and wavy surface features, usually referred to as *weird* (or *jumbled*) terrain. Scientists believe that this terrain was produced by seismic waves from the Caloris impact traveling around the planet and converging on the diametrically opposite point, causing large-scale disruption of the surface there, as illustrated in Figure 8.24.

8.7 *Interiors*

INTERIOR OF THE MOON

5 The Moon's average density, about 3300 kg/m³, is similar to that of lunar surface rock, virtually eliminating any chance that the Moon has a large, massive, and very dense nickel–iron core like that of the Earth. In fact, the low density implies that the entire Moon is actually deficient in iron and other heavy metals relative to our planet. There is presently no evidence for any lunar magnetic field. Researchers believe that planetary magnetism requires a rapidly rotating liquid metal core, like Earth's. Thus, the absence of a lunar magnetic field could result from the Moon's slow rotation, from the absence of a liquid core, or from both.

Our knowledge of the details of the Moon's interior is very limited. Models based on the available data indicate that the Moon's interior is of rather uniform density. As depicted schematically in Figure 8.25, the models suggest

a core of perhaps 200 km radius surrounded by a roughly 500 km thick inner mantle of semisolid rock having properties similar to Earth's aesthenosphere. The core is probably somewhat more iron-rich than the rest of the Moon (although it is still iron-poor compared with Earth's core). Near the center, the current temperature may be as low as 1500 K, too cool to melt rock. However, seismic data collected by sensitive equipment left on the surface by *Apollo* astronauts (see *Interlude 8-2* on p. 188) suggest that the inner parts of the core may be at least partially molten, implying a somewhat higher temperature.

Above these regions lies an outer mantle of solid rock, some 900–950 km thick, topped by a 60–150 km crust (considerably thicker than that of Earth). Together, these layers constitute the Moon's lithosphere. The crust material, which forms the lunar highlands, is lighter than the mantle, which is similar in chemical composition to the lunar maria.

The crust on the lunar far side is *thicker* than that on the side facing Earth. If we assume that lava takes the line of least resistance in getting to the surface, then we can readily understand why the far side of the Moon has no large maria—volcanic activity did not occur on the far side simply because the crust was too thick to allow it to occur there. But *why* is the far-side crust thicker? The answer is probably related to the gravitational pull of Earth. Just as heavier material tends to sink to the center of Earth, the denser lunar mantle tended to sink below the lighter crust in Earth's gravitational field. The effect of this was that the crust and the mantle became slightly off center with respect to one another. The mantle was pulled a little closer to Earth, while the crust moved slightly away. Thus, the crust became thinner on the near side and thicker on the far side.

INTERIOR OF MERCURY

Mercury's magnetic field, discovered by *Mariner 10*, is about 1/100 that of Earth. Actually, the discovery that Mercury had any magnetic field at all came as a surprise to planetary scientists. Having detected no magnetic field in the Moon (and, in fact, none in Venus or Mars, either), they had expected Mercury to have no measurable magnetism. Mercury certainly does not rotate rapidly, and it may lack a liquid metal core. Yet, a magnetic field undeniably surrounds it. Although weak, the field is strong enough to deflect the solar wind and create a small magnetosphere around the planet.

To be honest, scientists have no clear understanding of the origin of Mercury's magnetic field. If it is produced by ongoing dynamo action, as in the Earth, Mercury's core must be at least partially molten. Yet the absence of any recent surface geological activity suggests that the outer layers are solid to a considerable depth, as on the Moon. It is difficult to reconcile these two considerations in a single theoretical model of Mercury's interior. If the field is being generated dynamically, Mercury's slow rotation might at least account for the field's weakness. Alternatively, Mercury's current weak magnetism might be a mere remnant of an extinct dynamo. Given Mercury's rather small size, its metallic core might well have long since solidified. The models are

Figure 8.25 Cross-sectional diagram of the Moon. Unlike that of Earth, the Moon's rocky lithosphere is very thick—about 1000 km. Below the lithosphere is the inner mantle, or lunar aesthenosphere, similar in properties to that of Earth. At the center lies the core, which may be partly molten.

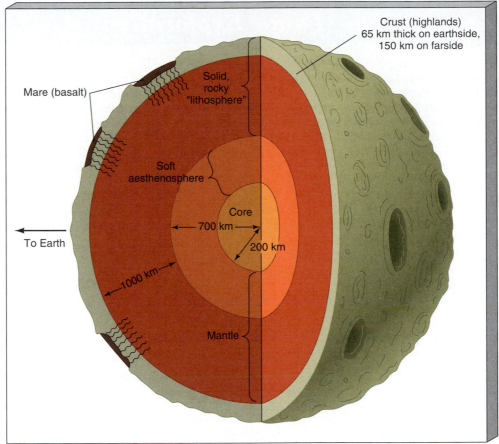

inconclusive on this issue, and no spacecraft is scheduled to revisit Mercury in the foreseeable future.

Mercury's magnetic field and large average density together imply that the planet is differentiated. Even without the luxury of seismographs on the surface, we can infer that most of its interior must be dominated by a large, heavy, iron-rich core with a radius of perhaps 1800 km. Whether that core is solid or liquid remains to be deter-

mined. Probably a less dense, lunarlike mantle lies above this core, to a depth of about 500 to 600 km. Thus, about 40 percent of the volume of Mercury, or 60 percent of its mass, is contained in its iron core. The ratio of core volume to total planet volume is greater for Mercury than for any other object in the solar system. Figure 8.26 illustrates the relative sizes and internal structures of Earth, the Moon, and Mercury.

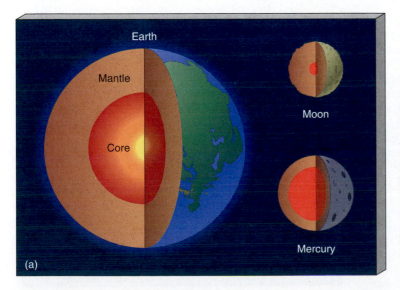

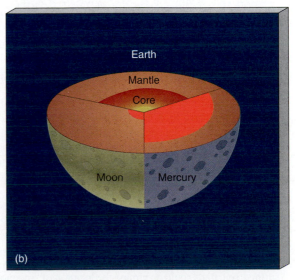

Figure 8.26 (a) The internal structures of the Earth, the Moon, and Mercury, drawn to the same scale. Note how large a fraction of Mercury's interior is core. (b) Enlarging the Moon and Mercury to the same size as Earth makes it easier to compare the relative sizes of their cores.

Interlude 8–2   *Lunar Exploration*

The Space Age began in earnest on October 4, 1957, with the launch of the Soviet satellite *Sputnik 1.* Thirteen months later, on January 4, 1959, the Soviet *Luna 1,* the first humanmade craft to escape Earth's gravity, passed the Moon. *Luna 2* crash-landed on the surface in September of that year, and *Luna 3* returned the first pictures of the far side a month later. The long-running *Luna* series established a clear Soviet lead in the early "space race" and returned volumes of detailed information about the lunar surface. Several of the *Luna* missions landed and returned lunar surface material to Earth.

The U.S. lunar exploration program got off to a rocky start. The first six attempts in the *Ranger* series, between 1961 and 1964, failed to accomplish their objective of just hitting the Moon. The last three were successful, however. *Ranger 7* collided with the lunar surface (as intended) on June 28, 1964. Five U.S. *Lunar Orbiter* spacecraft, launched in 1966 and 1967, were successfully placed in orbit around the Moon, and they relayed back to Earth high-resolution images of much of the lunar surface. For regions not visited by later landers, those images are often the best available even today. Between 1966 and 1968, seven *Surveyor* missions soft-landed on the Moon and performed detailed analyses of the surface.

Many of these unmanned U.S. missions were performed in support of the manned *Apollo* program. On May 25, 1961, at a time when the U.S. space program was in great disarray, President John F. Kennedy declared that the United States would "send a man to the Moon and return him safely to Earth" before the end of the decade, and the *Apollo* program was born. On July 20, 1969, less than 12 years after *Sputnik* and only 8 years after the statement of the program's goal, *Apollo 11* commander Neil Armstrong became the first human to set foot on the Moon, in Mare Tranquilitatis (Sea of Tranquility). Three and a half years later, on December 14, 1972, scientist-astronaut Harrison Schmitt, of *Apollo 17,* was the last.

The astronauts who traveled in pairs to the lunar surface in each lunar lander (see the accompanying photograph) performed numerous geological and other scientific studies on the surface. The later landers brought with them a "lunar rover"—a small golf cart–sized vehicle that greatly expanded the area the astronauts could cover. Probably the most important single aspect of the *Apollo* program was the collection of samples of surface rock from various locations on the Moon. In all, some 382 kg of material was returned to Earth. Chemical analysis and radioactive dating of these samples revolutionized our understanding of the Moon's surface history—no amount of Earth-based observations could have achieved the same results.

Each *Apollo* lander also left behind a nuclear-powered package of scientific instruments called *ALSEP* (*Apollo* Lunar Surface Experiments Package) to monitor the solar wind, measure heat flow in the Moon's interior, and, perhaps most important, record lunar seismic activity. With several ALSEPs on the surface, scientists could determine the location of "moonquakes" by triangulation and could map out the Moon's inner structure, obtaining information critical to our understanding of the Moon's evolution. NASA turned off the ALSEPs in 1978 to save money.

By any standards, the *Apollo* program was a spectacular success. It represents one of the most towering achievements of the human race. The project goals were met on schedule and within budget, and our knowledge of the Moon, Earth, and the solar system increased enormously. But the "Age of Apollo" was short-lived. Public interest quickly waned. Over half a billion people breathlessly watched television as Neil Armstrong set foot on the Moon; yet barely three years later, when the program was abruptly canceled for largely political (rather than technological, scientific, or economic) reasons, the landings had become so routine that they no longer excited the interest of the American public. Unmanned space science moved away from the Moon and toward the other planets, and the manned space program floundered. Perhaps one of the most amazing—and saddest—aspects of *Apollo* is that today, over two decades later, *no* nation on Earth (including the United States) has the desire, the capability, or the money to repeat the feat.

Currently there are no ongoing, long-term lunar exploration programs. The most recent spacecraft to study the Moon is the small U.S. military satellite *Clementine,* placed in lunar orbit in December 1993. Plans do exist to establish permanent human colonies on the Moon, either for commercial ventures, such as mining, or for scientific research. Proposals have also been made to site large optical, radio, and other telescopes on the lunar surface. Such instruments, which could be constructed larger than Earth-based devices, would enjoy perfect seeing and no light pollution. However, none of these projects is scheduled to become reality in the foreseeable future. After a brief encounter with humankind, the Moon is once again a lifeless, unchanging world.

8.8 *The Origin of the Moon*

6 The origin of the Moon is uncertain, although several theories have been advanced to account for it. As we will see, both the similarities between the Moon and the Earth *and* their differences conspire to confound many promising attempts to explain the Moon's existence.

One theory (the *sister*, or *coformation*, theory) suggests that the Moon formed as a separate object near Earth and in much the same way as our own planet—the "blob" of material that eventually coalesced into Earth gave rise to the Moon at about the same time. The two objects thus formed as a double-planet system, each revolving about the common center of mass. Although once favored by many astronomers, this idea suffers from a major flaw: The Moon differs in both density and composition from the Earth, which makes it hard to understand how both could have originated from the same material.

A second theory (the *capture* theory) maintains that the Moon formed far from Earth and was later captured by it. In this way, the density and composition of the two objects need not be similar, for the Moon presumably materialized in a quite different region of the early solar system. The objection to this theory is that the Moon's capture would be an extraordinarily difficult event; it might even be an impossible one. Why? Because the mass of our Moon is so large relative to that of Earth. It's not that our Moon is the largest natural satellite in the solar system, but it is unusually large compared with its parent planet. Mathematical modeling suggests that it is unreasonable to expect Earth's gravity to have attracted our Moon in exactly the right way to capture it during a close encounter sometime in the past. Furthermore, there are too many similarities in composition between the mantles of the Earth and the Moon to admit the possibility that the two bodies formed entirely independently of one another.

A third theory (the *daughter*, or *fission*, theory) speculates that the Moon originated out of Earth itself. The Pacific Ocean basin has often been mentioned as the place from which protolunar matter may have been torn—the result of the young Earth's rapid spin or even of tidal effects (from the Sun) on the young, mostly molten Earth. Indeed, the matter constituting both the Moon and the Pacific basin is basalt, which is largely devoid of iron. However, there remains the fundamental mystery of how Earth could have possibly have been spinning fast enough to eject an object as large as our Moon. Also, computer simulations indicate that the ejection of the Moon into a stable orbit simply would not have occurred.

Today, many astronomers favor a hybrid of the capture and fission themes. This idea—often called the *impact* theory—postulates a collision by a large, Mars-sized object with a youthful and molten Earth. Such collisions may have been quite frequent in the early solar system (see Chapter 15). The collision presumed by the impact theory would have been more a glancing blow than a direct impact. The matter dislodged from our planet then assembled to form the Moon.

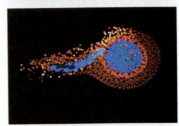
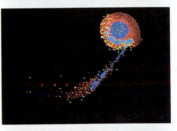

Figure 8.27 This sequence shows a simulated collision between the Earth and an object the size of Mars. The sequence proceeds top to bottom, and zooms out dramatically. The arrow in the final frame shows the newly formed Moon.

Computer simulations of such a catastrophic event show that most of the bits and pieces of splattered Earth could have coalesced into a stable orbit. Figure 8.27 shows some of the stages of one such simulation. If the Earth had already formed an iron core by the time the collision occurred, the Moon would indeed have ended up with a composition similar to the Earth's mantle. During the collision, any iron core in the impacting object itself would have been left behind in

Earth, eventually to become part of the Earth's core. Thus both the Moon's overall similarity to that of the Earth's mantle and its lack of a dense central core are naturally explained. Over the past decade, planetary scientists have come to realize that collisions such as this probably played very important roles in the formation of all the terrestrial planets.

8.9 *Evolutionary History of the Moon and Mercury*

EVOLUTION OF THE MOON

7 Given all the data, can we construct a reasonably consistent history of the Moon? The answer seems to be "Yes". Many specifics are still debated, but a consensus exists. Refer to Figure 8.28 while studying the following details.

The Moon apparently formed about 4.5 billion years ago. The approximate date of the oldest rocks discovered in the lunar highlands is 4.4 billion years, so we know that at least part of the crust must already have solidified by that time. At formation, the Moon was already depleted in heavy metals compared with Earth.

During the earliest phases of the Moon's existence— roughly the first half billion years or so—meteoritic bombardment must have been frequent enough to heat and remelt most of the *surface* layers of the Moon, perhaps to a depth of 400 km. The early solar system was surely populated with lots of interplanetary matter, much of it in the form of boulder-sized fragments, capable of generating large amounts of energy on collision with planets and their moons. But the intense heat derived from such collisions could not have penetrated very far into the lunar interior. Rock simply does not conduct heat well.

This situation resembles the surface melting we suspect occurred on Earth from meteoritic impacts during the first billion years or so. But the Moon is much less massive than Earth, and it does not contain enough radioactive elements to heat it much further. Radioactivity probably heated the Moon a little, but not sufficiently to transform it from a warm, plastic object to a completely liquid one. The

Moon must have differentiated during this period. If the Moon has a small iron core, it also formed at this time.

About 3.9 billion years ago, around the time that Earth's crust solidified, the heaviest phase of the meteoritic bombardment ceased. The Moon was left with a solid crust, ultimately to become the highlands, dented with numerous large basins, soon to flood with lava and become the maria. Between 3.9 and 3.2 billion years ago, lunar volcanism filled the maria with the basaltic material we see today. The age of the youngest maria—3.2 billion years—apparently indicates the time when the volcanic activity subsided. The maria are the sites of the last extensive lava flows on the Moon, over 3 billion years ago. Their smoothness, compared with the older, more rugged highlands, disguises their great age.

Small objects cool more rapidly than large ones—their interior is closer to the surface, on average. Being so small, the Moon rapidly lost its internal heat to space. As a consequence, it cooled much faster than did Earth. As the Moon cooled, the volcanic activity ended as the thickness of the solid surface layer increased. With the exception of a few meters of surface erosion from eons of meteoritic bombardment, the lunar landscape has remained more or less structurally frozen for the past 3 billion years. The Moon is dead now, and it has been dead for a long time.

EVOLUTION OF MERCURY

7 Like the Moon, Mercury seems to have been a geologically dead world for much of the past 4 billion years. On both the Moon and Mercury, the lack of ongoing geological activity results from a thick solid mantle that prevents volcanism or tectonic motion. Because of the *Apollo* program, the Moon's early history is much better understood than Mercury's, which remains somewhat speculative. What we do know about Mercury's history is learned mostly through comparison with the Moon.

When Mercury formed some 4.5 billion years ago, it was already depleted in lighter, rocky material. We will see later that this was largely a consequence of its location in

Figure 8.28 Paintings of the Moon (a) about 4 billion years ago, after much of the meteoritic bombardment had subsided and the surface had somewhat solidified; (b) about 3 billion years ago, after molten lava had made its way up through surface fissures to fill the low-lying impact basins and create the smooth maria; and (c) today, with much of the originally smooth maria now heavily pitted with craters formed at various times within the past 3 billion years.

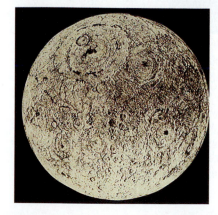

the hot inner regions of the early solar system, although possibly a collision stripped away some of its light mantle. During the next half-billion years, it melted and differentiated, like the other terrestrial worlds. It suffered the same intense meteoritic bombardment as the Moon. Being more massive than the Moon, Mercury cooled more slowly, so its crust was thinner and volcanic activity more common at early times. More craters were erased, leading to the intercrater plains found by *Mariner 10*.

As the planet's large iron core formed and then cooled, the planet began to shrink, compressing the crust. This compression produced the scarps seen on Mercury's surface, and may have prematurely terminated volcanic activity by squeezing shut the cracks and fissures on the surface. Thus Mercury did not experience the extensive volcanic outflows that formed the lunar maria. Despite its larger mass and greater internal temperature, Mercury has probably been geologically inactive for even longer than the Moon.

Chapter Review

SUMMARY

Both the Moon and Mercury are airless, virtually unchanging worlds that experience extremes in temperature. Mercury has no permanent atmosphere, although it does have a thin envelope of gas temporarily trapped from the solar wind. The main surface features on the Moon are the dark **maria** (p. 175) and the lighter-colored **highlands** (p. 175). The surfaces of both the Moon and Mercury are covered with **craters** (p. 175) of all sizes, caused by impacting meteoroids. Meteoritic impacts are the main source of erosion on the surfaces of both worlds. The lunar highlands are older than the maria, and are much more heavily cratered. The rate at which craters are formed decreases rapidly with increasing crater size.

The high day-side temperatures and cold night-side temperatures on the Moon and Mercury result from the absence of significant heat conduction or atmospheric blanketing on the planet. Sunlight strikes Mercury's polar regions at such an oblique angle that temperatures there may be very low, and the planet may have frozen polar caps of water ice.

The tidal interaction between Earth and the Moon is causing Earth's spin to slow, and is responsible for the Moon's **synchronous orbit** (p. 178), in which the same side of the Moon always faces our planet. The large lunar equatorial bulge probably indicates that the Moon once rotated more rapidly and orbited closer to Earth. Mercury's rotation rate is strongly influenced by the tidal effect of the Sun. Because of Mercury's eccentric orbit, the planet rotates not synchronously but exactly three times for every two orbits around the Sun. The condition in which a body's rotation rate is simply related to its orbit period around some other body is known as a **spin-orbit resonance** (p. 179).

The Moon's surface consists of both rocky and dusty material. Highland rocks, which are less dense than rocks from the maria, are believed to represent the Moon's crust. Maria rocks are believed to have originated in the mantle of the Moon. Lunar dust, called regolith, is made mostly of pulverized lunar rock, mixed with a small amount of material from impacting meteorites. Evidence for past volcanic activity on the Moon is found in the form of solidified lava channels called **rilles** (p. 185).

Mercury's surface features bear a striking similarity to those of the Moon. The planet is heavily cratered, much like the lunar highlands. Among the differences between Mercury and the Moon are Mercury's lack of lunarlike maria, its extensive **intercrater plains** (p. 185), and the great cracks, or **scarps** (p. 185), in its crust. The plains were caused by extensive lava flows early in Mercury's history. The scarps were apparently formed when the planet's core cooled and shrank, causing the surface to crack. Mercury's evolutionary path was similar to that of the Moon for half a billion years after they both formed. Mercury's volcanic period probably ended even before that of the Moon.

The absence of a lunar atmosphere and any present-day lunar volcanic activity are both consequences of the Moon's small size. Lunar gravity is too weak to retain any gases, while lunar volcanism was stifled by the Moon's cooling mantle shortly after extensive lava flows formed the maria more than 3 billion years ago. The crust on the near side of the Moon is substantially thicker than the crust on the far side. As a result, there are almost no maria on the lunar far side.

Mercury has a large impact crater called the Caloris basin, whose diameter is comparable to the radius of the planet. The impact that formed it apparently sent violent shock waves around the entire planet, buckling the crust on the opposite side.

The Moon's average density is not much greater than that of its surface rocks, probably because the Moon cooled more rapidly than the larger Earth and solidified sooner, so there was insufficient time for significant differentiation to occur. The lunar crust is too thick and the mantle too cool for plate tectonics to occur. Mercury's average density is considerably greater—similar to that of Earth—implying that Mercury contains a large high-density core, probably composed primarily of iron.

The Moon has no measurable magnetic field, a consequence of its slow rotation and lack of a molten metallic core. Mercury's weak magnetic field seems to have been "frozen in" long ago when the planet's iron core solidified.

The most likely explanation for the formation of the Moon is that the newly formed Earth was struck by a large (Mars-sized) object. Part of the impacting body remained behind as part of our planet. The rest ended up in orbit as the Moon.

SELF-TEST: True or False?

_____1. Radar-ranging can determine the distance to the Moon to an accuracy of a few meters.

_____2. Mercury has a very small orbital eccentricity.

_____3. Neither Mercury nor the Moon has any detectable atmosphere.

_____4. Both the Moon and Mercury have nighttime low temperatures of 250 K, well below the freezing point of water.

_____5. Telescopes on Earth could see the astronauts and the lunar landers on the Moon during the Apollo missions.

_____6. Unlike the lunar maria on the moon, Mercury has few lava-flow regions.

_____7. Mercury's solar day is actually longer than its solar year.

_____8. Large lunar craters are formed frequently today on the surface of the Moon.

_____9. Some volcanic activity continues today on the surface of the Moon.

_____10. Mercury's craters are more densely packed than craters on the Moon.

_____11. Scarps are found only on Mercury, not on the Moon.

_____12. Unlike the Moon, Mercury is differentiated.

SELF-TEST: Fill in the Blank

1. The most accurate method for determining the distance to the Moon is by _____.

2. Mercury can be seen only just before _____ or just after _____.

3. The radius of the Moon is about _____ Earth's radius; the radius of Mercury is about _____ that of Earth. (Give your answers as simple fractions, not decimals.)

4. Because the Moon's average density is so much lower than Earth's average density, the Moon must contain less _____.

5. Mercury's iron core contains a _____ fraction of the planet's mass than does Earth's core.

6. Mercury's daytime temperature is higher than the Moon's because it is _____.

7. The _____ on the Moon are dark, flat, roughly circular regions hundreds of kilometers in diameter.

8. Craters on the Moon and Mercury are primarily due to _____.

9. Mercury's rate of rotation was first measured using _____.

10. Although Mercury's daytime temperatures are always very hot, it may still be possible for it to have sheets of water ice at its _____.

11. The crater produced by the impact of a meteoroid on the Moon has a typical diameter about _____ times the diameter of the meteoroid.

12. The lunar maria's dark, dense rock originally was part of the _____ of the Moon.

13. Mercury's _____ is about 1/100 that of the Earth, and was originally thought not to exist at all.

REVIEW AND DISCUSSION

1. How far away is the Moon? How do we know this?

2. Why is Mercury seldom seen with the naked eye?

3. How big is the Moon's equatorial bulge, and how does its size compare with what one would expect on theoretical grounds?

4. Employ the concept of escape velocity to explain why the Moon and Mercury have no significant atmospheres.

5. What does it mean to say that Mercury has a 3:2 spin-orbit resonance? Why didn't Mercury settle into a 1:1 spin-orbit resonance with the Sun like the Moon did with the Earth?

6. Why is the surface of Mercury often compared to that of the Moon? List two similarities and two differences between the surfaces of Mercury and the Moon.

7. What is a scarp? How are scarps thought to have formed? Why do scientists believe that the scarps formed after most of the meteoritic bombardment ended?

8. What is weird terrain on Mercury? How might it have formed?

9. What is the primary source of erosion on the Moon? Why is the average rate of lunar erosion so much less than on Earth?

10. Name two pieces of evidence indicating that the lunar highlands are older than the maria.

11. How does lunar soil differ from earthly soil?

12. In contrast to the Earth, the Moon and Mercury undergo extremes in temperature. Why?

13. What do Mercury's magnetic field and high average density imply about the planet's interior?

14. How is Mercury's evolutionary history like that of the Moon? How is it different?

15. Describe the theory of the Moon's origin favored by many astronomers.

16. Because the Moon always keeps one face toward Earth, an observer on the moon's near side would see Earth appear almost stationary in the lunar sky. Still, Earth would change its appearance as the Moon orbited Earth. How would Earth's appearance change? Why?

17. The best place to aim a telescope or binoculars on the Moon is along the terminator line, the line between the Moon's light and dark hemispheres. Why? If you were standing on the lunar terminator, where would the Sun be in your sky? What time of day is it when you're standing on Earth's terminator line?

18. Where on the Moon would be the best place from which to make astronomical observations? What would be this locations' advantage over locations on the Earth?

19. Explain why Mercury is never seen overhead at midnight in Earth's sky.

20. How is the varying thickness of the lunar crust related to the presence or absence of maria on the Moon?

21. Mercury is smaller than three moons in the solar system. Why then, is it called a planet instead of a moon? Think about the names and terms for other objects in the solar system. Asteroids orbiting the Sun are even smaller than Mercury. They are sometimes called "minor planets." Do you think this is a good way to describe them? Why or why not?

PROBLEMS

1. How long does it take a radar signal to travel from Earth to Mercury and back?

2. What is the angular diameter of the Sun, as seen from Mercury? Compare this with the angular diameter of the Sun as seen from Earth.

3. Using the rate given in the text for the formation of 10-km craters on the Moon, how long would it take to cover the Moon with new craters of this size? How much higher must the cratering rate have been in the past to produce the Moon we see today?

4. The Moon's mass is 1/80 the mass of the Earth, and the lunar radius is 1/4 Earth's radius. Calculate the total weight on the Moon of a 100-kg astronaut with a 50-kg spacesuit and backpack, relative to his weight on Earth.

5. The *Hubble Space Telescope* has a resolution of about 0.05 arc seconds. What is the smallest object it can see on the surface of the Moon? Give your answer in meters.

PROJECTS

1. Observe the Moon during an entire cycle of phases. When does the Moon rise, set, and appear highest in the sky at each major phase? What is the interval of time between each phase?

2. If you have binoculars, turn them on the Moon when it appears at twilight and when it appears overhead. Draw pictures of what you see. What differences do you notice in your two drawings? What color is the Moon seen near the horizon? What color is the Moon seen overhead? Why is there a difference?

3. Watch the Moon over a period of hours on a night when you can see one or more bright stars near it. Estimate how many moon diameters it moves per hour. Knowing the Moon is about 0.5° in diameter, how many degrees per hour does it move? What is your estimate of its orbital period?

4. Try to spot Mercury in morning or *evening twilight*. (Hint: As seen from the Northern Hemisphere, the best evening apparitions of the planet take place in the spring and the best morning apparitions take place in the fall.)

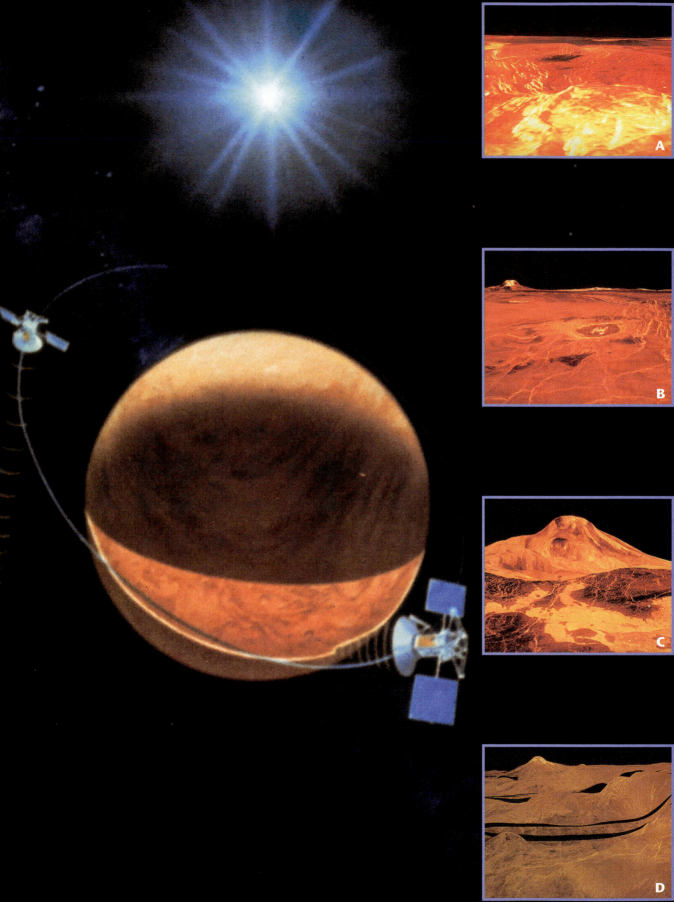

9

VENUS

Earth's Sister Planet

LEARNING GOALS

Studying this chapter will enable you to:

1. Summarize Venus's general orbital and physical properties.

2. Explain why Venus is hard to observe from Earth, and how we have obtained more detailed knowledge of the planet.

3. Describe the characteristics of Venus's atmosphere, and contrast it with that of Earth.

4. Explain why the greenhouse effect has produced conditions on Venus very different from those on Earth.

5. Compare the surface features and geology of Venus with those of Earth and the Moon.

6. Describe Venus's magnetic field and internal structure.

(Opposite page, background) The *Magellan* robot spacecraft is seen here in artist's conception orbiting Venus in the early 1990s. When close to the veiled planet, its radar system scans the surface; when farther away, it sends data home to Earth.

(Inset A) Each of the insets shows a three-dimensional, computer-enhanced view of surface features on Venus, artificially colored and vertically exaggerated to enhance small-scale structure. This is the volcanic Gula Mons area, with extensive lava flows across otherwise fractured plains.

(Inset B) A distant view of Maat Mons (at left rear), an 8-kilometer high volcano, now apparently dormant.

(Inset C) A closer view of Maat Mons, showing clearly more lava flows across the fractured terrain in the foreground.

(Inset D) The Danu Montes comprise a mountain range rising 1 to 3 kilometers above the plains—probably formed ages ago by surface compression, much like the Andes and Appalachian mountain ranges on Earth.

In the previous chapter, we gained insight into the nature and history of Mercury by comparing it with Earth's Moon. Similarly, we can learn about Venus by comparing it with Earth itself. In its bulk properties at least, Venus is almost a carbon copy of our own world. The two planets are similar in size, density, and chemical composition. They orbit at comparable distances from the Sun. At formation, they must have been almost indistinguishable from one another. Yet they are now about as different as two terrestrial planets could be. While Earth is a vibrant world, teeming with life, Venus is an uninhabitable inferno, with a dense, hot atmosphere of carbon dioxide, lacking any trace of oxygen or water. Somewhere along their respective evolutionary paths, Venus and Earth diverged, and diverged radically. How did this occur? What were the factors leading to Venus's present condition? Why are Venus's surface, atmosphere, and interior so different from Earth's? In answering these questions, we will discover that a planet's environment, as well as its composition, can play a critical role in determining its future.

9.1 *Venus in Bulk*

ORBITAL PROPERTIES

① Venus is the second planet from the Sun. Its orbit lies within Earth's, so Venus, like Mercury, is always found fairly close to the Sun in the sky—Venus is never seen more than 47° from the Sun. Given Earth's rotation rate of 15° per hour, this means that Venus is visible above the horizon for at most three hours before the Sun rises or after it sets. Because we can see Venus from Earth only just before sunrise or just after sunset, the planet is often called the "morning star" or the "evening star," depending on where it happens to be in its orbit. Figure 9.1 shows Venus in the western sky just after sunset. The planet appears much brighter than any star.

Venus's orbital semi-major axis is 0.72 A.U. (108 million km). Its orbital eccentricity is only 0.007, so its path around the Sun is essentially circular. In fact, of all the planets, the orbit of Venus is closest to being a perfect circle. The planet's sidereal orbital period is 225 days, or about seven and a half Earth months. Like all the planets, Venus orbits the Sun in the same sense as Earth—counterclockwise, as seen from above the north celestial pole. Its orbit is inclined just 3° to the plane of the ecliptic.

Venus is the third brightest object in the entire sky (after the Sun and the Moon)—it appears more than 10 times brighter than the brightest star, Sirius. You can see Venus even in the daytime, if you know just where to look. On a moonless night away from city lights, Venus casts a faint shadow. The planet's brightness stems from the fact that Venus is highly reflective. More than 70 percent of the sunlight reaching Venus is reflected back into space (compare this with roughly 10 percent in the case of Mercury and the Moon). Most of the sunlight is reflected from clouds high in the planet's atmosphere.

We might expect Venus to appear brightest when it is "full"—that is, when we can see the entire sunlit side. However, because Venus orbits between Earth and the Sun, Venus is full when it is at its greatest distance from us, 1.7 A.U. away on the other side of the Sun (an alignment known as *superior conjunction*). Actually, when Venus is exactly on the other side of the Sun, we can't see it at all—it is lost in the Sun's glare—but we can see an almost full Venus within a few degrees of this configuration. When Venus is closest to us, the planet is at the new phase, lying between Earth and the Sun (*inferior conjunction*). At this time we again can't see it, because now the sunlit side faces away from us. Only a thin ring of sunlight, caused by refraction in Venus's atmosphere, surrounds the planet. As Venus moves away from this location, more and more of it becomes visible, but its distance from us also increases. Venus's maximum apparent brightness actually occurs about 36 days before or after closest approach to Earth. At that time, the planet is about 39° away from the Sun, and we see it as a rather fat crescent. Figure 9.2 illustrates the relation between Venus's orbit and its brightness.

RADIUS, MASS, AND DENSITY

We can determine Venus's radius from simple geometry, just as we did for Mercury. At closest approach, when Venus is only 0.28 A.U. from us, its angular diameter is 64″. From this observation, we can determine its radius to be about 6000 km. More accurate measurements from spacecraft lead to a value of 6052 km, or 0.95 Earth radii.

Like Mercury, Venus has no moon. Before the Space Age, astronomers calculated its mass by indirect means—through studies of its small gravitational effect on the orbits of the other planets, especially Earth. Now that spacecraft have orbited the planet, we know its mass very accurately: 4.9×10^{24} kg, or 0.82 times the mass of Earth.

From its mass and radius we find that Venus's average density is 5300 kg/m^3 (5.3 g/cm^3). As far as these bulk properties are concerned, then, Venus appears to be very similar to Earth. Is Venus's overall composition similar to Earth's? If so, we could reasonably conclude that the planet's internal structure and evolution are basically Earthlike. Later in this chapter we will review what little evidence there is on this subject.

Figure 9.1 The Moon and Venus in the western sky just after sunset. Venus clearly outshines even the brightest stars in the sky.

ROTATION RATE

The same clouds whose reflectivity make Venus so easy to see in the night sky also make it impossible for us to discern any surface features, at least in visible light. Even when viewed through a large optical telescope, the clouds themselves show few features. Until the advent of suitable radar techniques in the 1960s, astronomers did not know the rotation period of Venus. Attempts to determine its rotation by watching motions of the cloud markings as seen from Earth were frustrated by the rapidly changing nature of the clouds themselves. Some astronomers argued for a 25-day period, while others favored a 24-hour cycle. Controversy raged until, to the surprise of all, radar observers announced that the Doppler broadening of their returned echoes implied a sluggish 243-day rotation period. Furthermore,

Venus's spin was found to be *retrograde*—that is, in a sense opposite that of Earth and most other solar system objects, and opposite that of Venus's orbital motion.

Planetary astronomers conventionally define "north" and "south" for individual members of the solar system by the condition that planets *always* rotate from west to east. With this definition, Venus's retrograde spin means that the planet's north pole actually lies *below* the ecliptic plane, unlike any of the other terrestrial worlds. Venus's axial tilt—the angle between its equatorial and orbital planes—is 177.4° (compared to 23.5° in the case of the Earth). Note, however, that astronomical images of solar-system objects conventionally place objects lying above the ecliptic plane at the top of the frame. Thus, with the above definition of north and south, all the images of Venus shown in this chapter have the *south* pole at the top.

Figure 9.3 illustrates Venus's retrograde rotation and compares it with the rotation of its neighbors, Mercury, Earth, and Mars. Because of Venus's slow, retrograde rotation, the planet's solar day (noon to noon) is quite different from its sidereal rotation period of 243 Earth days (the time for one "true" rotation relative to the stars). ∞(p. 10) In fact, one Venus day is a little more than half a Venus year

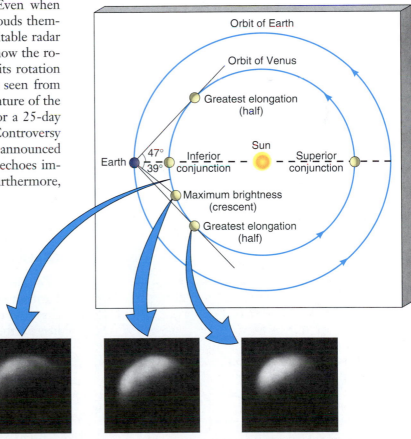

Figure 9.2 Venus appears full when it is at its greatest distance from Earth, on the opposite side of the Sun from us (superior conjunction). As its distance decreases, less and less of its sunlit side becomes visible. When it is closest to Earth, it lies between us and the Sun (inferior conjunction), so we cannot see the sunlit side of the planet at all. Venus appears brightest when it is about 39° from the Sun. (Compare Figure 2.10.)

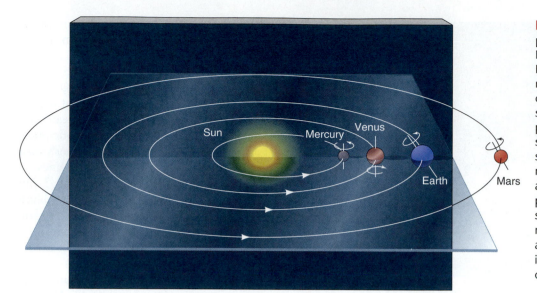

Figure 9.3 The inner four planets of the solar system—Mercury, Venus, Earth, and Mars—display widely differing rotational properties. While all orbit the Sun in the same sense and in nearly the same plane, Mercury's rotation is slow and prograde (in the same sense as the orbital motion), the rotations of Earth and Mars are fast and prograde, while Venus's is slow and retrograde. Venus rotates clockwise as seen from above the plane of the ecliptic in the direction of the north celestial pole.

(225 Earth days). Figure 9.4 depicts this interplay between Venus's orbital and rotational motion.

Why is Venus rotating "backwards," and why so slowly? At present, the best explanation planetary scientists can offer is that, early in Venus's evolution, the planet was struck by a large body, much like the one that may have hit the Earth and formed the Moon, and that impact was sufficient to reduce the planet's spin almost to zero. Whatever its cause, Venus's rotation poses practical problems for Earthbound observers. It turns out that the planet rotates almost exactly five times between one closest approach to Earth and the next. As a result, *Venus always presents nearly the same face to Earth at closest approach.* This means that radar observations of the planet's surface cover one side—the one facing us at

closest approach—much more thoroughly than the other side, which we can see only when the planet is close to its maximum distance from Earth.

Astronomers hate to appeal to coincidence to explain their observations, but the case of Venus's rotation appears to be just that. No known interaction between Earth and Venus can explain the near-perfect 5:1 resonance between Venus's rotation and orbital motion—Earth's tidal effect on Venus is tiny, and is much less than the Sun's tidal effect in any case. Furthermore, the key word in the preceding sentence is *near*. A resonance, if it existed, would require that the number of rotations per orbit be *exactly* five. The slight discrepancy—which amounts to less than 3 hours in 584 days—appears to be real, and if that is so, no resonance exists. For now we are simply compelled to accept this coincidence without adequate explanation.

9.2 *Long-Distance Observations of Venus*

GENERAL FEATURES

2 Because Venus, of all planets, most nearly matches Earth in size, mass, and density, and because its orbit is closest to us, it is often called Earth's sister planet. But unlike Earth, Venus has a dense atmosphere that is nearly opaque to visible radiation, making its surface completely invisible from the outside at optical wavelengths. Even cloud patterns are very difficult to detect. Figure 9.5 shows one of the best photographs of Venus taken with a large telescope on Earth. The planet resembles a white-yellow disk and shows rare hints of cloud circulation.

The atmospheric patterns are much more evident when examined with equipment capable of detecting ultraviolet radiation. Some of Venus's atmospheric constituents absorb this high-frequency radiation, thereby in-

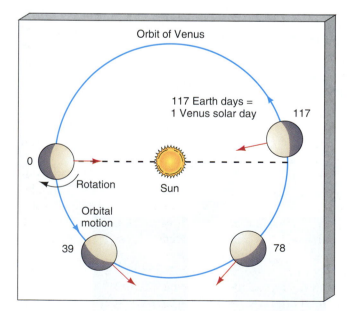

Figure 9.4 Venus's orbit and retrograde rotation combine to produce a Venus solar day equal to 117 Earth days, or slightly more than half a Venus year. The numbers in the figure mark time in Earth days.

R I V U X G

Figure 9.5   This photograph, taken from Earth, shows Venus with its creamy yellow mask of clouds.

R I V U X G

Figure 9.6   Venus as it was photographed by the *Pioneer* spacecraft's cameras some 200,000 km away from the planet. This image was made by capturing solar ultraviolet radiation reflected from the planet's clouds, which are probably composed mostly of droplets of sulfuric acid (the highly corrosive acid used in car batteries).

creasing the cloud contrast. Figure 9.6 is an ultraviolet image taken in 1979 by America's *Pioneer Venus* spacecraft at a distance of 200,000 km from the planet. The large, fast-moving cloud patterns resemble Earth's high-altitude jet stream more than the great cyclonic whirls characteristic of Earth's low-altitude clouds. In fact, the upper deck of clouds on Venus rotates around the planet in just four days, which is much faster than the underlying surface. The upper-level winds reach speeds of 400 km/hr relative to the planet.

Spectroscopic examination of sunlight reflected from the clouds shows the presence of large amounts of carbon dioxide, with little evidence for any other atmospheric gases. ∞ (p. 85) The first studies to reveal carbon dioxide as a major constituent of Venus's atmosphere were performed in the 1930s. Until the 1950s, astronomers generally believed that observational difficulties alone prevented them from seeing other atmospheric components. The hope lingered that Venus's clouds were actually predominantly water vapor, like those on Earth, and that below the cloud cover Venus might be a habitable planet similar to our own. Indeed, in the 1930s scientists had measured the temperature of the atmosphere spectroscopically at about 240 K, not much different from our own upper atmosphere. ∞ (p. 64) Calculations of the surface temperature—taking into account the cloud cover and Venus's proximity to the Sun, and assuming an atmosphere much like our own—suggested that Venus should have a surface temperature only 10 or 20 degrees higher than Earth's.

RADIO MEASUREMENTS BELOW THE CLOUDS

These hopes for an Earthlike Venus were dashed in the 1950s, when radio observations of the planet were used to measure its thermal energy emission. Unlike visible light, radio waves easily penetrate the cloud layer, and they gave the first indication of conditions on or near the surface. The radiation emitted by the planet has a black-body spectrum characteristic of a temperature near 750 K. ∞ (p. 64) Almost overnight, the popular conception of Venus changed from that of a lush tropical jungle to an arid, uninhabitable desert.

Radar observations of the surface of Venus are routinely carried out from Earth using the Arecibo radio telescope. ∞ (p. 110) This instrument can achieve a resolution of a few kilometers, but it can adequately cover only a small fraction (roughly 25 percent) of the planet. This telescope's observation of Venus is limited because of the planet's peculiar near-resonance and because radar reflections from regions near the "edge" of the planet are hard to obtain. However, the results from Arecibo have been combined with information from probes orbiting Venus to build up a detailed picture of the planet's surface. Only very recently, with the arrival of the *Magellan* probe, were more accurate data obtained.

9.3   *The Atmosphere of Venus*

ATMOSPHERIC STRUCTURE

❸   The data returned from the *Venera* and *Pioneer* spacecraft have allowed us to paint a fairly detailed picture of Venus's atmosphere. Figure 9.7 shows the run of temperature and pressure with height. Compare this figure with Figure 7.4, which gives similar information for Earth. The atmosphere of Venus is about 90 times more massive than Earth's, and it extends to a much greater height above the surface. On Earth, 90 percent of the atmosphere lies within about 10 km of sea level. On Venus the corresponding (90 percent) level is found at an altitude of 50 km instead. The surface temperature and pressure of Venus's atmosphere are much greater than Earth's. However, the temperature drops more rapidly with altitude, and the upper atmosphere of Venus is actually colder than our own.

The troposphere on Venus extends up to an altitude of nearly 100 km. The reflective clouds that block our view of the surface lie between 50 and 70 km above the surface. The *Pioneer* multiprobe data indicate that the clouds may actually be separated into three distinct layers within that altitude range. Below the clouds, extending down to an altitude of some 30 km, is a layer of haze. Below 30 km, the air is clear.

Above the clouds, a high-speed "jet stream" blows from west to east at about 300–400 km/h, fastest at the equator and slowest at the poles. This high-altitude flow is responsible for the rapidly moving cloud patterns seen in ultraviolet light. Figure 9.8 shows a sequence of three ultraviolet images of Venus in which the variations in the cloud patterns can be seen. Note the characteristic V-shaped appearance of the clouds—a consequence of the fact that, despite their slightly lower speeds, the winds near the poles have a shorter distance to travel in circling the planet, and so are always forging ahead of winds at the

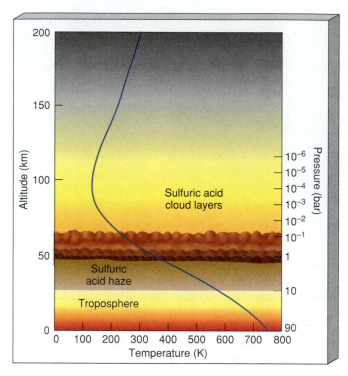

Figure 9.7   The structure of the atmosphere of Venus, as determined by U.S. and Soviet probes.

equator. Near the surface, by contrast, the dense atmosphere moves more sluggishly—indeed, the fluid flow bears more resemblance to Earth's oceans than to its air. Surface wind speeds are typically less than 2 m/s (roughly 4 mph).

ATMOSPHERIC COMPOSITION

As we have seen, observations from Earth revealed the presence of carbon dioxide in Venus's atmosphere but were inconclusive about other possible constituents. We now know that carbon dioxide is in fact the dominant component of the atmosphere, accounting for 96.5 percent of it

Figure 9.8   Three ultraviolet views of Venus, taken by the *Pioneer Venus* orbiter, showing the changing cloud patterns in the planet's upper atmosphere. The wind flow is in the direction opposite the "V" in the clouds. Notice the motion of the dark region marked by the arrow. Venus's retrograde rotation means that north is at the bottom of these images, and west is to the right. The time difference between the left and right photographs is about 20 hours.

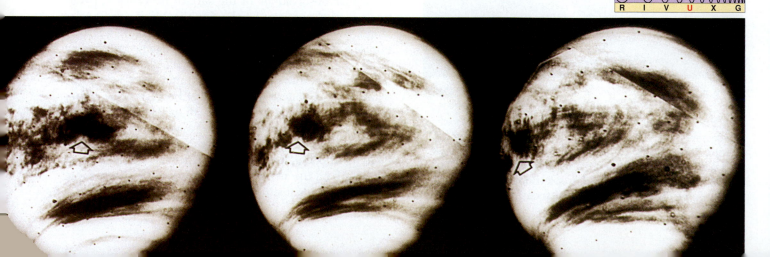

by volume. Almost all of the remaining 3.5 percent is nitrogen. Trace amounts of other gases, such as water vapor, carbon monoxide, sulfur dioxide, and argon, are also present. This composition is clearly radically different from Earth's atmosphere. The absence of oxygen is perhaps not surprising, given the absence of life (recall our discussion of Earth's atmosphere in Chapter 7). ∞ (p. 153) However, there is no sign of the water vapor that we might expect to find if a volume of water equivalent to Earth's oceans had evaporated. If Venus started off with Earthlike composition, something has happened to its water—it is now a very dry planet. The lack of water and the dominance of carbon dioxide in the atmosphere of Venus are closely related, as we will see.

For a long time, the chemical makeup of the highly reflective cloud layer surrounding Venus was unknown. At first scientists assumed the clouds were water vapor or ice, as on Earth, but the reflectivity of the clouds at different wavelengths didn't match that of water ice. Infrared observations carried out in the 1970s showed that the clouds (or at least the top layer of clouds) are actually composed of sulfuric acid, created by reactions between water and sulfur dioxide. Sulfur dioxide is an excellent absorber of ultraviolet radiation and could be responsible for some of the cloud patterns seen in ultraviolet light. Spacecraft observations have since confirmed the presence of all these compounds in the atmosphere. They also indicate that there may be particles of sulfur suspended in and near the cloud layers, which may account for Venus's characteristic yellowish hue.

THE GREENHOUSE EFFECT ON VENUS

4 Given the distance of Venus from the Sun, the planet was not expected to be such a pressure cooker. Calculations based on Venus's orbit and its surface reflectivity indicated a temperature not much different from Earth's, and early measurements of the cloud temperatures seemed to concur. Certainly, scientists reasoned, Venus could be no hotter than the sunward side of Mercury, and it should probably be much cooler. This reasoning was obviously seriously in error.

Why is Venus's atmosphere so hot? And if, as we believe, Venus started off like Earth, why is it now so different? The answer to the first question is fairly easy: Given the present composition of its atmosphere, Venus is hot because of the greenhouse effect. Recall from our discussion in Chapter 7 that the "greenhouse gases" in Earth's atmosphere, particularly water vapor and carbon dioxide, serve to trap heat from the Sun. ∞ (p. 151) By inhibiting the escape of infrared radiation reradiated from Earth's surface, these gases serve to increase the planet's equilibrium temperature, in much the same way as an extra blanket keeps you warm on a cold night. Continuing the analogy a little further, the more blankets you place on the bed, the warmer you will become. Similarly, the more greenhouse gases there are in the atmosphere, the hotter

the surface will be. (This is precisely the mechanism thought to be responsible for global warming on Earth as atmospheric carbon dioxide levels rise.)

The same effect naturally occurs on Venus, whose dense atmosphere is made up almost entirely of a primary greenhouse gas, carbon dioxide. As illustrated schematically in Figure 9.9, the thick carbon dioxide blanket absorbs 99 percent of all the infrared radiation released from the surface of Venus, and it is the immediate cause of the planet's sweltering 750 K surface temperature. Furthermore, the temperature is nearly as high at the poles as at the equator, and there is not much difference between the temperatures on the "night side" (facing directly away from the Sun) and the "day side." The circulation of the atmosphere spreads energy very efficiently around the planet, making it impossible to escape the blazing heat, even at night.

THE RUNAWAY GREENHOUSE EFFECT

But why is Venus's atmosphere so different from Earth's? Why is there so much carbon dioxide in the atmosphere of Venus, and why is the atmosphere so dense? To address these questions, we must consider the processes that created the atmospheres of the terrestrial planets and then determined their evolution. In fact, we can turn the question around and ask instead, "Why is there so little carbon dioxide in Earth's atmosphere compared with that of Venus?"

We believe that Earth's atmosphere has evolved greatly since it first appeared. Any primary atmosphere escaped soon after Earth formed and was replaced by a secondary atmosphere outgassed from the interior by volcanic activity 4 billion years ago. Since then, the atmosphere has been reprocessed, in part by living organisms, into its present form. On Venus, the initial stages probably took place in more or less the same way, so that at some time in the past, Venus may well have had an atmosphere similar to the primitive secondary atmosphere on Earth, containing water, carbon dioxide, sulfur dioxide, and nitrogen-rich compounds. What happened on Venus to cause such a major divergence from subsequent events on our own planet?

On Earth, nitrogen was released into the air by the action of sunlight on the chemical compounds containing it. Meanwhile, the water condensed into oceans, and much of the carbon dioxide and sulfur dioxide eventually became dissolved in them. Most of the remaining carbon dioxide combined with surface rocks. Thus, much of the secondary outgassed atmosphere quickly became part of the surface of the planet. If all the dissolved or chemically combined carbon dioxide were released back into Earth's present-day atmosphere, its new composition would be 98 percent carbon dioxide and 2 percent nitrogen, and it would have a pressure about 70 times its current value. In other words, apart from the presence of oxygen (which appeared on Earth only after the development of life) and water (the absence of which on Venus will be explained shortly), Earth's atmosphere would be a lot like that of Venus! The real dif-

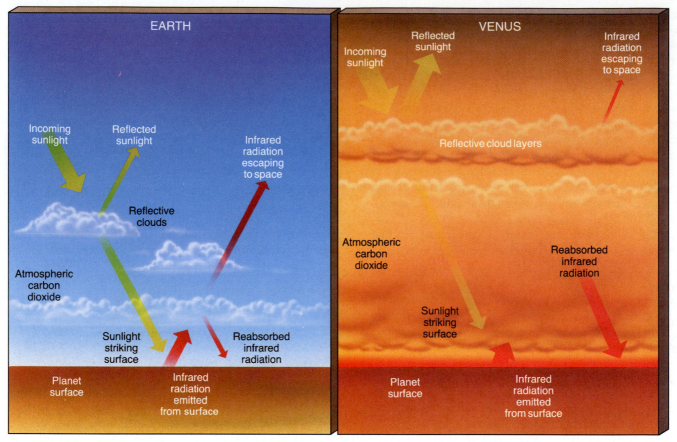

Figure 9.9 Because Venus's atmosphere is much thicker and denser than Earth's, a much smaller fraction of the infrared radiation leaving the planet's surface actually escapes into space. The result is a much stronger greenhouse effect than on Earth and a correspondingly hotter planet. The outgoing infrared radiation is not absorbed at a single point in the atmosphere; instead, absorption occurs at all atmospheric levels. (The arrows are meant to indicate only that absorption occurs, not that it occurs at one specific level.)

ference between Earth and Venus is that the greenhouse gases of Venus never left the atmosphere.

When Venus's secondary atmosphere appeared, the temperature was higher than on Earth, simply because Venus is closer to the Sun. However, the Sun was probably somewhat dimmer then (see Chapter 22)—perhaps only half its present brightness—so there is some uncertainty as to exactly how much hotter than Earth Venus actually was. If the temperature was already so high that no oceans condensed, the outgassed water vapor and carbon dioxide would have remained in the atmosphere, and the full greenhouse effect would have gone into operation immediately. If oceans did form and most of the greenhouse gases left the atmosphere, the temperature must still have been sufficiently high that a process known as the **runaway greenhouse effect** came into play.

To understand the runaway greenhouse effect, imagine that we took Earth from its present orbit and placed it in Venus's orbit, about 30 percent closer to the Sun. At its new distance from the Sun, the inverse square law tells us that the amount of sunlight striking Earth's surface would be about twice its present level, so the planet

would warm up. ∞ (p. 67) More water would evaporate from the oceans, leading to an increase in atmospheric water vapor. At the same time more carbon dioxide would escape from the oceans and surface rocks to the atmosphere. This would increase the greenhouse heating, so the planet would warm still further, leading to a further increase in atmospheric greenhouse gases, and so on. Once started, then, the process would "run away" and lead to the complete evaporation of all Earth's oceans, in effect restoring all the original greenhouse gases to the atmosphere. Although the details are quite complex, basically the same thing would have happened on Venus long ago.

Venus's atmosphere never lost its greenhouse gases to the surface. The greenhouse effect on Venus was even more extreme in the past, when the atmosphere contained water vapor. By intensifying the blanketing effect of the carbon dioxide, the water vapor helped the surface of Venus reach temperatures perhaps twice as hot as at present. At those high temperatures, the water vapor was able to rise high into the planet's upper atmosphere—so high, in fact, that it was broken up by solar ultraviolet radiation into its components, hydrogen and oxygen. The light hydrogen rapidly escaped, and the reactive oxygen quickly combined with other atmospheric constituents. In this way, essentially *all* of the water in Venus's atmosphere and surface layers was lost forever.

Although it is highly unlikely that global warming will ever send Earth down the path taken by Venus, this episode highlights the relative fragility of the planetary environment. No one knows how close to the Sun Earth could have formed before a runaway greenhouse effect would have occurred. But in comparing our planet with Venus, we have come to understand that there is an orbital limit, presumably between 0.7 and 1.0 A.U., inside of which Earth would have suffered a similar catastrophic runaway. We must consider this "greenhouse limit" when assessing the likelihood that planets harboring life formed elsewhere in the Galaxy.

9.4 *The Surface of Venus*

⑤ Although the clouds are extremely thick and the surface totally shrouded, we are by no means ignorant of Venus's topography. As mentioned earlier, radar astronomers have bombarded the planet with radio signals, both from Earth and from the *Pioneer, Venera,* and, most recently, *Magellan* spacecraft. Analysis of the radar echoes yields a map of the planet's surface. As Figure 9.10 illustrates, the early maps of Venus suffered from poor resolution; however, the more recent probes, especially *Magellan*, have provided much sharper views.

LARGE-SCALE TOPOGRAPHY

Figure 9.11(a) shows basically the same *Pioneer* data of Venus as Figure 9.10(a), except that it has been flattened out into a more conventional map. The altitude of the surface relative to the average radius of the planet is indicated by the use of color, with red representing the highest elevations, blue the lowest. Figure 9.11(b) shows a map of Earth to the same scale and at the same spatial resolution. Figure 9.11(c) is the same as Figure 9.11(a), except that some of the main features on the planet have been labeled. The planet's surface appears to be mostly smooth, perhaps resembling rolling plains with modest highlands and lowlands. Only two or three continental-sized features adorn the landscape, and these contain mountains comparable in height to those on Earth. The highest peaks rise some 14 km above the level of the deepest surface depressions. For comparison, the highest point on Earth (the summit of Mount Everest) lies about 20 km above the deepest section of the ocean floor (Challenger Deep, at the bottom of the Marianas trench on the eastern edge of the Philippines plate).

The *Pioneer* maps allow us to discern the largest-scale features of the surface of Venus. The planet has two elevated continental-sized regions on its surface. In the southern high latitudes, at the *top* of Figure 9.11c (recall our discussion of the planet's retrograde rotation on p. 197), we find an extensive uplifted plateau, called Ishtar Terra (or "Land of Ishtar," after the Babylonian counterpart of the Roman Venus and the Greek Aphrodite). The projection of the map makes Ishtar Terra appear much larger than it re-

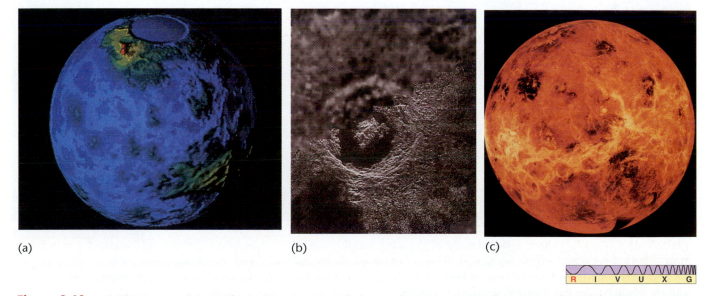

(a)   (b)   (c)

| R | I | V | U | X | G |

Figure 9.10   (a) This image of the surface of Venus was made by a radar transmitter and receiver onboard the *Pioneer* spacecraft, still in orbit about the planet. The two continent-sized landmasses are named Ishtar Terra (upper left) and Aphrodite (lower right). The spatial resolution is about 25 km. (b) At even higher resolution, more surface features are detected. Here the resolution of the Soviet *Venera 15* orbiter (previously the best available, about 1–2 km) is compared with that of *Magellan* (which can resolve features almost 20 times smaller). The view shows an impact crater named Golubkina, about 34 km in diameter. The lower right side is the *Magellan* image, the upper left came from *Venera*. (c) A planetwide mosaic of *Magellan* images. The largest "continent" on Venus, Aphrodite Terra, is at the center of the image.

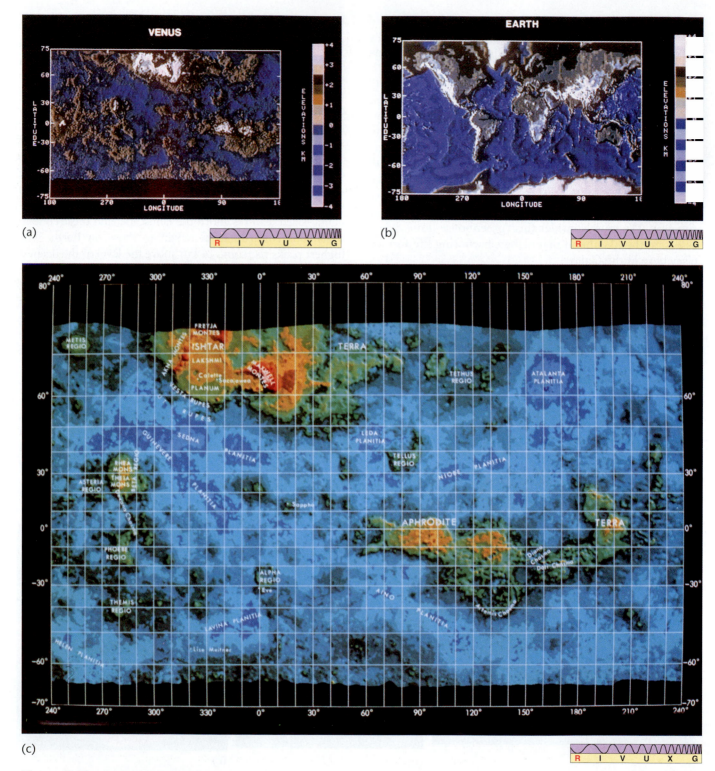

Figure 9.11 (a) Radar map of the surface of Venus, based on the *Pioneer Venus* data. Color represents elevation according to the scale at the right. (b) A similar map of Earth, at the same spatial resolution. (c) Another version of (a), with the major surface features labeled. Note that North is down in this image.

ally is—it is actually about the same size as Australia. This landmass is dominated by a great plateau known as Lakshmi Planum (see Figure 9.12), which is some 1500 km across at its widest point. This plain is ringed by mountain ranges that include the highest peak on the planet, Maxwell Mons,

which reaches an altitude higher than Mount Everest above sea level on Earth. The region also houses a great crater, called Cleopatra, about 100 km across.

Figure 9.12(a) shows a large-scale *Venera* image of Lakshmi Planum, at a resolution of about 2 km. The "wrin-

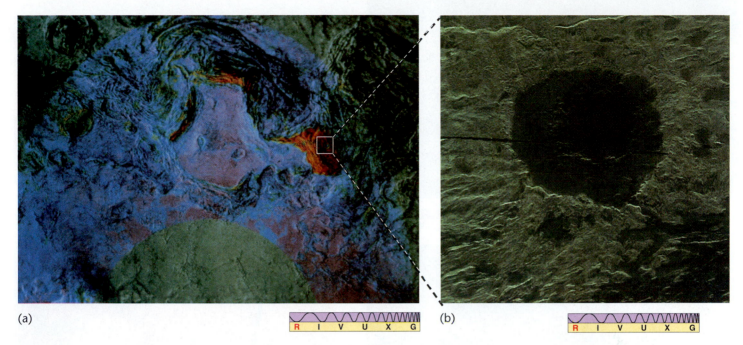

Figure 9.12 (a) A *Venera* orbiter image of a plateau in Ishtar Terra known as Lakshmi Planum. Maxwell Mons (red) lies on the western margin of the plain, near the right-hand edge of the image. A meteor crater named Cleopatra is visible on the western slope of the Maxwell mountain range. Note the two larger craters in the center of the plain itself. (b) A *Magellan* image of Cleopatra showing a double-ringed structure that identifies it to geologists as an impact crater.

kles" are actually chains of mountains, hundreds of kilometers long and tens of kilometers apart. The red area immediately to the right of the plain is Maxwell Mons. The crater Cleopatra lies on the western (right-hand) slope of the Maxwell range. Figure 9.12(b) shows a *Magellan* image of Cleopatra itself. As with all the *Magellan* images, the light areas in Figure 9.12(b) represent regions where the surface is rough and efficiently scatters *Magellan*'s sideways-looking radar beam back to the detector. Smooth areas tend to reflect the beam off into space instead and so appear dark. Close-up views of the crater's structure have led planetary scientists to conclude that it is meteoritic in origin, although there was some volcanism associated with its formation. Notice the dark lava flow emerging from within the inner ring and cutting across the outer rim at the upper right.

The other continental-sized formation is called Aphrodite Terra. It is located on the planet's equator and is comparable in size to Africa. Before *Magellan*'s arrival, some researchers had speculated that Aphrodite Terra might have been the site of something equivalent to seafloor spreading on Earth—a region where two lithospheric plates moved apart and molten rock rose to the surface in the gap between them, forming an extended ridge. With the low-resolution data then available, the issue could not be settled at the time.

The *Magellan* images now seem to rule out any plate tectonic activity on Venus, and the Aphrodite region shows no signs of spreading. The crust appears buckled and fractured, suggesting large compressive forces, and there seem to have been repeated periods when extensive lava flows

occurred. Figure 9.13(a) shows a portion of Aphrodite Terra called Ovda Regio. The ridges running in two distinct directions across the image attest to both the magnitude and the variability of the forces compressing and distorting the crust. The dark (smooth) regions are probably solidified lava flows. Some narrow lava channels, akin to rilles on the Moon, also appear. Figure 9.13(b) shows a series of angular cracks in the crust, which are thought to have formed when lava welled up from a deep fissure, flooded the surrounding area, and then retreated back below the planet's surface. As the molten lava withdrew, the thin new crust of solidified material that had formed on top of it was unable to support its own weight, and the surface collapsed, forming the cracks we now see. Even taking into account the temperature and composition differences between Venus's crust and Earth's, this terrain is not at all what we would expect at a spreading site similar to the Mid-Atlantic Ridge.

Two other highland regions, lying close to the equator of Venus and generally north of Ishtar, bear the names Alpha Regio and Beta Regio. These two mountainous areas, which are both somewhat smaller than Ishtar, were identified in the early radar images of Venus made from Earth and designated by the first two letters of the Greek alphabet (alpha and beta; *regio* just means "region"). It is now conventional to name features on Venus after famous women—Aphrodite, Ishtar, Cleopatra, and so on. However, the early nonfemale names (Maxwell Mons, named after the Scottish physicist James Clerk Maxwell, for example) have stuck, and they are unlikely to change.

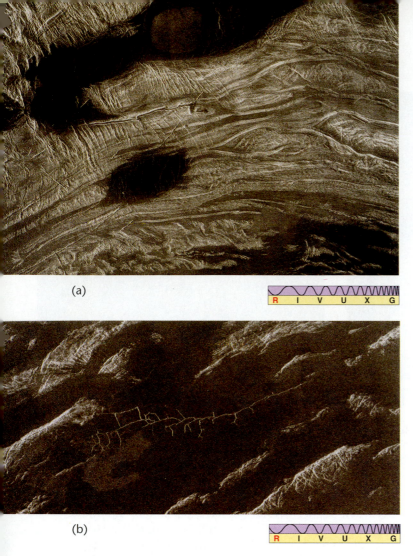

(a)

R I V U X G

(b)

R I V U X G

Figure 9.13 (a) A *Magellan* image of Ovda Regio in Aphrodite Terra. The intersecting ridges indicate repeated compression and buckling of the surface. The dark areas represent regions that have been flooded by lava upwelling from cracks like those shown in (b), detected by *Magellan* in another part of Aphrodite Terra. This network of fissures is about 50 km long.

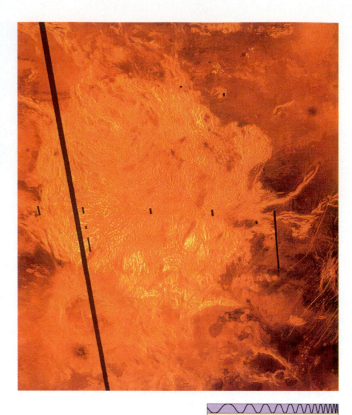

R I V U X G

Figure 9.14 *Magellan* image of Alpha Regio. The region shown here is about 1300 km across. The image is made up of over 600 separate strips, each about 20 km across, corresponding to consecutive orbits of the spacecraft. The black regions represent periods when data were lost due to technical problems with the spacecraft. The bright dot in the lower left, near the longest black line, is the crater of the volcano, Eve.

A large-scale *Magellan* image of Alpha Regio is shown in Figure 9.14. A wide variety of terrains are evident. Alpha Regio itself is the light-colored (that is, rough), somewhat diamond-shaped region in the center of the frame. It consists of mountainous ridges and troughs. Numerous fractures (the diagonal "stripes" at right) can be seen in the surrounding lava-flooded (dark) lowlands. The large, oval dark region near the southwest edge of Alpha is a volcanic feature known as Eve, which is probably responsible for much of the flooding in the southern portion of the image. Lava flow channels can be seen emerging from its rim. Alpha and Beta Regio are among the many areas of Venus that are now known to have extensive volcanic features. Scientists suspect they are also the sites of currently active volcanoes, although no direct evidence has yet been found.

The elevated "continents" occupy only 8 percent of Venus's total surface area. (For comparison, continents on Earth make up about 25 percent of the surface.) The remainder of Venus's surface is classified as lowlands (27 percent) or rolling plains (65 percent), although there is probably little geological difference between the two terrains. Although there is no evidence for any large-scale plate tectonic activity on Venus, there is plenty of evidence for volcanism in the recent past. It is likely that the stresses in the crust that led to the large mountain ranges were caused by convective motion within Venus's mantle, the same basic process that drives Earth's plates. ∞ (p. 165) Lakshmi Planum, for example, is probably the result of a "plume" of upwelling mantle material that raised and buckled the planet's surface.

VOLCANISM AND CRATERING

There are numerous craters on Venus's surface. They cannot be seen at the 25 km resolution of the *Pioneer* maps, but more detailed observations by *Venera 15* and *16* and from Arecibo on Earth have revealed their presence. The *Magellan* data allow them to be studied in great detail. Although some craters appear to have arisen from meteoritic impact, most are volcanic in origin.

Figure 9.15, an enlargement of part of the southeast portion of Figure 9.14, shows a series of seven pancake-

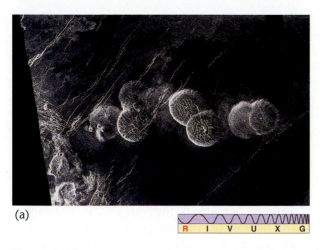

(a)

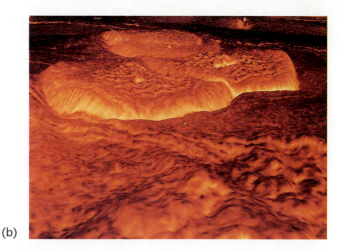

(b)

Figure 9.15   (a) A series of dome-shaped structures on the southeast edge of Alpha Regio. They are the result of viscous molten rock bulging out of the ground and then retreating, leaving behind a thin solid crust that subsequently cracked and subsided. *Magellan* has found features like this in several locations on Venus. (b) A three-dimensional representation of four of the domes. The computer view is looking toward the right from near the center of the image in part (a). Color is based on data returned by Soviet landers.

shaped lava domes, each about 25 km across. They probably formed when lava oozed out of the surface, formed the dome, and then withdrew, leaving the crust to crack and subside. Lava domes such as these are found in sev-

eral locations on Venus, but the most common volcanoes on the planet are of the type known as **shield volcanoes**. Two large shield volcanoes, called Sif Mons and Gula Mons, are shown (in false color) in Figure 9.16. Shield

(a)

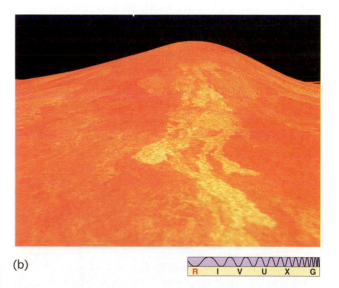

(b)

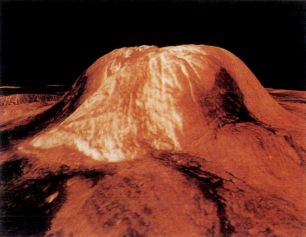

(c)

Figure 9.16   (a) Two larger volcanoes, known as Sif Mons (left) and Gula Mons, appear in this *Magellan* image of a region east of Beta Regio. Color indicates height, ranging from purple (the level of the surrounding plain) to orange (corresponding to an altitude of about 4 km). The two volcanic calderas at the summits are about 100 km across. (b) A computer-generated view of Sif Mons, as seen from ground level. (c) Gula Mons, as seen from ground level. In (b) and (c), the colors are based on data returned from Soviet landers, and the vertical scales have been greatly exaggerated; Venus is actually a remarkably flat place.

volcanoes on Earth, like the Hawaiian Islands, are associated with lava welling up through a "hot spot" in the crust. They are built up over long periods of time by successive eruptions and lava flows. A characteristic of shield volcanoes is the formation of a *caldera*, or crater, at the summit when the underlying lava withdraws and the surface collapses.

The largest volcanic structures on Venus are huge, roughly circular regions known as **coronae**. A large corona, called Aine, can be seen in Figure 9.17, another large-scale mosaic of *Magellan* images. Coronae are unique to Venus. They appear to have been caused by upwelling mantle material, similar to the uplift that resulted in Lakshmi Planum, but on a somewhat smaller scale. They generally have volcanoes both in and around them, and closer inspection of the rims usually shows evidence for extensive lava flows into the plains below.

Not all of the craters found on Venus are volcanic in origin. Some, like the crater Cleopatra on the slopes of Maxwell Mons, were formed by meteoritic impact instead. The largest impact craters on Venus are generally circular, but those less than about 15 km in diameter can be quite asymmetric in appearance. Figure 9.18(a) shows a *Magellan* image of a relatively small meteoritic impact crater, about 10 km across, in Venus's southern hemisphere. Geologists believe the light-colored region to be the ejecta blanket— material ejected from the crater following the impact. The odd shape may be the result of a large meteoroid breaking up just before impact into pieces that hit the surface near one another. This seems to be a fairly common fate for medium-sized bodies (1 km or so in diameter) plowing through Venus's dense atmosphere. Numerous impact craters, again identifiable by their ejecta blankets, can also be discerned in Figure 9.17.

Venus's atmosphere is sufficiently thick that small meteoroids do not reach the ground, so there are no impact craters smaller than about 3 km across. Atmospheric effects probably also account for the observed deficiency in impact craters less than 25 km in diameter. Overall, the rate of formation of large-diameter craters on Venus's surface seems to be only about 1/10 that in the lunar maria. Applying the same crater-age estimates to Venus as we do to Earth and the Moon suggests that much of the surface of Venus is quite young—around a billion years old. Some planetary scientists have suggested that some areas, such as the region shown in Figure 9.16 are even younger—perhaps as little as 200 or 300 million years. Although erosion by the planet's atmosphere may play some part in obliterating surface features, the main agent is volcanism, which appears to "resurface" the planet every few hundred million years.

There is now overwhelming evidence for past surface activity on Venus. Has this activity now stopped, or is it still going on? Two pieces of indirect evidence suggest that volcanism continues. First, the level of sulfur dioxide above Venus's clouds shows large and fairly frequent fluctuations. It is quite possible that these variations result from volcanic eruptions on the surface. If so, volcanism may be the primary cause of Venus's thick cloud cover. Second, both the *Pioneer* and the *Venera* orbiters observed bursts of radio energy from the Beta and Aphrodite regions. These bursts are similar to those produced by lightning discharges that often occur in the plumes of erupting volcanoes on Earth, again suggesting ongoing activity. These pieces of evidence may be persuasive, but they are still only circumstantial. No "smoking gun" (an erupting volcano) has yet been seen, so the case for active volcanism is not yet complete.

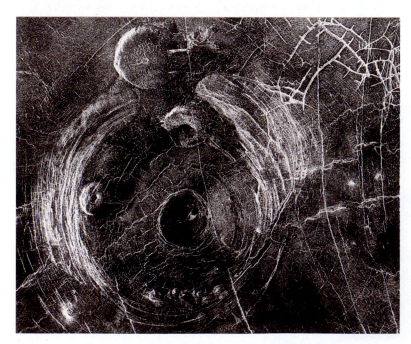

Figure 9.17   This corona, called Aine, lies in the plains south of Aphrodite Terra; it is about 300 km across. It is probably the result of mantle material causing the surface to bulge outward. Note the fractures in the crust, and the many large impact craters with their surrounding white (rough) ejecta blankets that stud the region.

(a)

(b)

(c)

Figure 9.18   (a) A *Magellan* image of an impact crater in Venus's southern hemisphere. The peculiar kidney shape seems to be the result of a meteoroid that fragmented just prior to impact. The dark regions in the crater may be pools of solidified lava. (b) Some more regular looking craters. (c) Three-dimensional computer-generated representations of the craters shown in part (b).

DATA FROM THE SOVIET LANDERS

The 1975 soft-landings of the Soviet *Venera 9* and *Venera 10* spacecraft established directly that Venus's surface is dry and dusty. Figure 9.19 shows one of the first photographs of the surface of Venus radioed back to Earth. Each craft lasted only about an hour before overheating, their electronic circuitry literally melting in this planetary oven. Typical rocks in the photo measure about 50 cm by 20 cm across—a little like flagstones on Earth. Having sharp edges and a slablike character, they show little evidence of erosion. Apparently they are quite young rocks, again supporting the idea of ongoing surface activity of some kind.

Later *Venera* missions took more detailed photographs, as shown in Figure 9.20. The presence of small rocks and finer material indicates the effects of erosive processes. These later missions also performed simple chemical analyses of the surface of Venus. The samples studied by *Venera 13* and *Venera 14* were predominantly basaltic in nature, again implying a volcanic past. However, not all of the rocks were found to be volcanic. The *Venera 17* and *Venera 18* landers also found surface material resembling terrestrial granite, probably (as on Earth) part of the planet's ancient crust.

Figure 9.19 The first direct view (in false color) of the surface of Venus, radioed back to Earth from the Soviet *Venera 9* spacecraft, which made a soft-landing in 1975 in Beta Regio. The amount of sunlight penetrating Venus's cloud cover apparently resembles that on a heavily overcast day on Earth.

Figure 9.20 Another view of Venus, this time in true color from *Venera 14*. Flat rocks like those visible in Figure 9.19 are present, but there are also many smaller rocks and even fine soil on the surface. These craft also landed in Beta Regio, not far from the earlier landing sites. The peculiar filtering effects of whatever light does penetrate the clouds make Venus's air and ground peach-colored.

9.5 *Venus's Magnetic Field and Internal Structure*

6 In 1962, *Mariner 2* flew by Venus, carrying, among other instruments, magnetometers to measure the strength of the planet's magnetic field. None was detected, and subsequent Soviet and U.S. missions, carrying more sensitive detectors, have confirmed this finding. Venus, with an average density similar to Earth's, probably has a similar overall composition and a partially molten iron-rich core. The lack of any detectable magnetic field on Venus, then, is almost surely the result of the planet's extremely slow rotation and consequent lack of dynamo action. ∞ (p. 154)

Having no magnetosphere, Venus has no protection from the solar wind. Its upper atmosphere is continually bombarded by high-energy particles from the Sun, keeping the topmost layers permanently ionized. However, the great thickness of the atmosphere prevents any of these particles from reaching the surface.

None of the *Venera* landers carried seismic equipment, so no direct measurements of the planet's interior have been made, and theoretical models of the interior have very little hard data to constrain them. However, to many geologists the surface of Venus resembles the young Earth, at an age of perhaps a billion years. At that time, volcanic ac-

tivity had already begun, but the crust was still relatively thin, and the convective processes in the mantle that drive plate tectonic motion were not yet established.

Why has Venus remained in that immature state and not developed plate tectonics like Earth? That question remains to be answered. Some planetary geologists have speculated that the high surface temperature on Venus has inhibited the planet's evolution by slowing the planet's cooling. Possibly the high surface temperature has made the crust too soft for Earth-style plates to develop. Or perhaps the high temperature and soft crust led to more volcanism, tapping the energy that might otherwise go into convective motion. It may also be that the presence of water plays an important role in lubricating mantle convection and plate motion, so that arid Venus could not evolve along the same path as Earth. Without more detailed data, it is difficult to say. It is hoped that the detailed gravity maps of Venus to be constructed from in-depth studies of *Magellan's* orbit will shed more light on the planet's internal structure.

As further data are gathered, astronomers will eagerly compare the interior of Venus with that of Earth. The two planets have nearly equal masses and radii but very different environmental conditions. In determining how and why these two near-identical twins diverged in later life, we will surely achieve a much more comprehensive understanding of planetary physics.

Chapter Review

SUMMARY

The interior orbit of Venus means that it never strays far from the Sun in the sky. Because of its highly reflective cloud cover, Venus is brighter than any star in the sky, as seen from Earth. It is so bright that it can be seen even in the daytime.

Venus's rotation is slow and retrograde, most likely because of a collision between Venus and some other solar-system body during the late stages of the planet's formation.

The extremely thick atmosphere of Venus is nearly opaque to visible radiation, making the planet's surface invisible from the outside. Spectroscopic examination of sunlight reflected from the planet's cloud tops shows the presence of large amounts of carbon dioxide. Venus's atmosphere is nearly 100 times denser than Earth's. The temperature of the upper atmosphere is much like that of Earth's upper atmosphere, but the surface temperature is a sizzling 750K.

Venus is comparable in both mass and radius to Earth, suggesting that the two planets started off with fairly similar surface conditions. However, the atmospheres of Earth and Venus are now very different. The total mass of Venus's atmosphere is about 90 times greater than Earth's. The greenhouse effect stemming from the large amount of carbon dioxide in Venus's atmosphere is the basic cause of the planet's current high temperatures.

Almost all of the water vapor and carbon dioxide initially present in Earth's early atmosphere quickly became part of the oceans or surface rocks. Because Venus orbits closer to the Sun than Earth, surface temperatures were initially higher, and the planet's greenhouse gases never left the atmosphere. On Venus, the **runaway greenhouse effect** (p. 202) resulted in all of the planet's greenhouse gases—carbon dioxide and water vapor—ending up in the atmosphere, leading to the extreme conditions we observe today.

Venus's surface cannot be seen in visible light from Earth, but it has been thoroughly mapped by radar from Earth-based ra-

dio telescopes and orbiting satellites. The most recent and most thorough survey has been carried out by the U.S. *Magellan* satellite. Its radar images of Venus are at least 10 times better than the best images previously obtained.

Venus's surface is mostly smooth, resembling rolling plains with modest highlands and lowlands. Two elevated continental-sized regions are called Ishtar Terra and Aphrodite Terra.

Many lava domes and **shield volcanoes** (p. 207) have been found by *Magellan* on Venus's surface, but none of the volcanoes has yet been proven to be currently active. The planet's surface shows no sign of plate tectonics. Features called **coronae** (p. 208) are thought to have been caused by an upwelling of mantle material that, for unknown reasons, never developed into full convective motion. The surface of the planet appears to be relatively young, resurfaced by volcanism every few hundred million years. Some craters on Venus are due to meteoritic impact, but the majority are volcanic in origin.

The evidence for currently active volcanoes on Venus includes surface features resembling those produced in earthly volcanism, fluctuating levels of sulfur dioxide in Venus's atmosphere, and bursts of radio energy similar to those produced by lightning discharges that often occur in the plumes of erupting volcanoes on Earth. However, no actual eruptions have been seen.

Soviet spacecraft that landed on Venus photographed surface rocks with sharp edges and a slablike character. Some rocks on Venus appear predominantly basaltic in nature, implying a volcanic past. Other rocks resemble terrestrial granite and are probably part of the planet's ancient crust.

Venus has no detectable magnetic field, almost certainly because the planet's rotation is too slow for any appreciable dynamo effect to occur.

To some planetary geologists, Venus's interior structure suggests that of the young Earth, before convection became established in the mantle.

SELF-TEST: True or False?

_____ 1. Venus appears brighter in the sky than any star.

_____ 2. Venus has the least eccentric orbit of any of the planets in the solar system.

_____ 3. Venus is brightest when it is in its full phase.

_____ 4. Although Venus is slightly smaller than Earth, its atmosphere is significantly larger.

_____ 5. The average surface temperature of Venus is about 260 K.

_____ 6. Numerous large surface features on Venus can be seen from Earth-based observations made in the ultraviolet part of the spectrum.

_____ 7. The top cloud layers of Venus are composed of sulfuric acid.

_____ 8. Venus has roughly the same temperature at its equator as at its poles.

_____ 9. The surface of Venus is relatively rough, compared to the surface of the Earth, with higher highs and lower lows.

_____ 10. Images from *Magellan* show signs of tectonic activity on Venus.

_____ 11. Lava flows are common on the surface of Venus.

_____ 12. There is strong circumstantial evidence that active volcanism continues on Venus.

_____ 13. Venus has a magnetic field similar to that of the Earth.

_____ 14. Soviet spacecraft are still sending back data from the surface of Venus.

SELF-TEST: Fill in the Blank

1. Venus's mass has now been well determined through the use of _____ orbiting it.

2. Because Venus has a mass and average _____ only slightly lower than Earth's, we might expect that its internal structure and evolution should be Earthlike.

3. Venus's rotation is unusual because it is _____.

4. The most abundant gas in the atmosphere of Venus is _____.

5. Water vapor is found in _____ amounts in the atmosphere of Venus.

6. The process which makes Venus so hot is known as the _____.

7. The runaway greenhouse effect on Venus was a result of the planet being _____ to the Sun than is the Earth.

8. The surface of Venus has been mapped using _____.

9. Ishtar Terra and Aphrodite Terra are two _____ on the surface of Venus.

10. Most craters on the surface of Venus are the result of _____.

11. Huge, roughly circular regions on the surface of Venus are known as _____.

12. A meteoroid less than about 1 km in diameter will most likely _____ as it passes through Venus's atmosphere.

13. The deficiency in large impact craters on the surface of Venus indicates that its surface is quite _____.

14. The surface of Venus appears to have been resurfaced by _____ every few hundred million years.

15. The main difficulties in using landers to study Venus's surface are the planet's extremely high _____ and _____.

REVIEW AND DISCUSSION

1. Why does Venus appear so bright to the eye? Upon what factors does the brightness of Venus depend?

2. Explain why Venus is always found in the same region of the sky as the Sun.

3. If you were standing on Venus, how would Earth look?

4. How was the radius of Venus determined using observations from Earth?

5. How did radio observations of Venus made in the 1950s change our conception of the planet?

6. What two features of Venus's atmosphere make Venus extremely hostile to earthly life?

7. What did ultraviolet images returned by *Pioneer Venus* show about the planet's high-level clouds?

8. Name three ways in which the atmosphere of Venus is different from that of Earth.

9. What are the main constituents of Venus's atmosphere? What are clouds in the upper atmosphere made of?

10. What characteristics of Venus's atmosphere cause Venus to be so hot?

11. Explain why there is so much carbon dioxide in the atmosphere of Venus, compared to Earth. What happened to all the water that Venus must have had when formed?

12. What is the runaway greenhouse effect, and how might it have altered the climate of Venus?

13. If Venus had formed at Earth's distance from the Sun, what do you imagine its climate would be like today? Why do you think so?

14. How do the "continents" of Venus differ from earthly continents?

15. How are the impact craters of Venus different from those found on other bodies?

16. What evidence exists that there has been volcanism of various types changing the surface of Venus?

17. What is the evidence for active volcanoes on Venus?

18. How might an orbiting spacecraft with sophisticated imaging abilities, such as *Magellan*, be used to discover active volcanoes on Venus?

19. Given that Venus probably has a partially molten iron-rich core, why doesn't it have a magnetic field?

20. Earth and Venus are nearly alike in size and density. What primary fact caused one planet to evolve as an oasis for life, while the other became a dry and inhospitable inferno?

21. Might there be life on Venus? Explain your answer.

PROBLEMS

1. Approximating Venus's atmosphere as a layer of gas 50 km thick, with uniform density 21 kg/m^3, calculate its total mass. Compare it with the mass of Earth's atmosphere and with the mass of Venus.

2. Starting from the angular size of Venus when it is at its closest to Earth, given in the text, calculate Venus's angular size when it is farthest from Earth.

3. *Pioneer Venus* observed high-level clouds moving around Venus in 4 days. What would be their velocity in km/hr? In mi/hr? Is this true for all parts of Venus?

4. According to Stefan's Law (see Chapter 3), how much more radiation—per square meter, say—is emitted by Venus's surface at 750 K than is emitted by the Earth's surface at 300 K?

5. Given that HST orbits 400 km above Earth's surface with a period of 95 minutes, what is the orbit period of the *Magellan* spacecraft if it orbits 500 km above the surface of Venus?

PROJECTS

1. Is Venus in the morning or evening sky right now? Look for it every few days, over the course of several weeks. Draw a picture of the planet with respect to foreground trees or buildings. If you always observe at the same time every day, you might begin to notice that the planet is getting higher or lower in the sky.

2. Consult an almanac to determine the next time Venus will pass between the Earth and Sun. How many days before and after this event can you glimpse the planet with the eye alone?

3. Consult the almanac again to find out the next time Venus will pass on the far side of the Sun from Earth. How many days before and after this event can you see the planet with the naked eye?

4. When Venus ornaments the predawn sky, try keeping track of the planet with your eye alone until it appears in a blue sky, after sunrise. As always, be careful not to look at the Sun!

5. Using a powerful pair of binoculars or a small telescope, examine Venus as it goes through its phases. Note the phase and the relative size of it. (You can compare its size to the field of view in a telescope; always use the same eyepiece for this.) Look at it every few days or once a week. Make a table of the shape of the phase, the size, and the relative brightness to the naked eye. After you have observed it through a significant change in phase, can you see the correlations between these three properties first recognized by Galileo?

10

MARS

A Near Miss for Life?

LEARNING GOALS

Studying this chapter will enable you to:

1 Summarize the general orbital and physical properties of Mars.

2 Describe the observational evidence for seasonal changes on Mars.

3 Discuss the evidence that Mars once had a much denser atmosphere and running water on its surface.

4 Compare the atmosphere of Mars with those of Earth and Venus, and explain why the evolutionary histories of these three worlds diverged so sharply.

5 Compare the surface features and geology of Mars with those of the Moon and Earth, and account for these characteristics in terms of Martian history.

6 Describe the characteristics of Mars's moons, and explain their probable origin.

(Opposite page, background) The red planet, Mars, shown in true color here, displays an array of fascinating surface features. Most prominent in this *Viking*-mission image is the Mariner Valley, a vast "canyon" named after an earlier spacecraft, *Mariner*, that discovered it. The valley extends for approximately 4000 km, roughly the width of the United States.

(Inset) A close-up of the Martian surface, rock-strewn and flat, seen through the eyes of the *Viking 2* robot that soft-landed on the northern Utopian plains. The discarded canister is about 20 cm long. The scars in the "dirt" were made by the robot's shovel.

As we continue beyond Venus on our outward journey from the Sun, crossing the orbit of Earth–Moon system, we come to the fourth terrestrial planet—Mars. This red world, named by the ancient Romans for their bloody god of war, is for some people the most intriguing of all celestial objects. Over the years it has inspired much speculation that life—perhaps intelligent, and possibly hostile—may exist there. With the dawn of the Space Age, those notions had to be abandoned. Visits by robot spacecraft have revealed no signs of life of any sort, even at the microbial level, on Mars. Even so, the planet's properties are close enough to those of Earth that Mars is still widely regarded as the next most hospitable environment for the appearance of life in the solar system, after the Earth itself. At about the same time as Earth's "twin," Venus, was evolving into a searing inferno, the Mars of long ago may have had running water and blue skies. If life ever arose there, however, it must be long extinct. The Mars of today appears to be a dry, dead world.

10.1 *Mars in Bulk*

ORBITAL PROPERTIES

① Mars is the fourth planet from the Sun and the outermost of the four terrestrial worlds in the solar system. It lies outside Earth's orbit, as illustrated in Figure 10.1(a), which shows the orbits of both planets drawn to scale. Because of its exterior orbit, Mars ranges in our sky from a position close to the Sun (for example, when Earth and Mars are at the points marked A in the figure) to one far from the Sun (with the planets at points B). Contrast this orbit with the nighttime appearances of Mercury and Venus, whose interior orbits ensure that we never see them far from the Sun. From our earthly viewpoint, Mars appears to traverse a great circle in the sky, keeping close to the ecliptic and occasionally executing retrograde loops, as we saw in Chapter 2. ∞ (p. 33) Because Mars never comes within Earth's orbit, it can never pass between the Earth and the Sun—there can be no Martian transit of the Sun, although a careful Martian astronomer might occasionally glimpse Earth crossing the solar disk.

Mars revolves around the Sun with an orbital semimajor axis of 1.52 A.U. (228 million km). Its orbital eccentricity is 0.093, much larger than that of most other planets—only the innermost and the outermost planets, Mercury and Pluto, have more elongated orbits. Because of the ellipticity of its orbit, Mars's perihelion distance from the Sun—1.38 A.U. (206 million km)—is substantially smaller than its aphelion distance—1.66 A.U. (249 million km), resulting in a large variation in the amount of sunlight striking the planet over the course of its year. The inverse square law tells us that the intensity of sunlight on the Martian surface is almost 45 percent greater when the planet is at perihelion than when it is at aphelion. ∞ (p. 67) As we will see, this has a substantial effect on the Martian climate.

Mars's sidereal orbital period is 687 Earth days. The planet is at its largest and brightest when it is at *opposition*— that is, when Earth lies between Mars and the Sun (location B on Figure 10.1). If this happens to occur near a Martian perihelion, the two planets can come within 0.38

A.U. (56 million km) of one another. Mars's angular size under those circumstances is about 25″. Ground-based observations of the planet at those times can distinguish surface features as small as 100 km across—about the same resolution as the unaided human eye can achieve when viewing the Moon. Such a coincidence of opposition with Martian perihelion is relatively rare. It last happened in September 1988 and will next occur in August 2003, although Mars will not be quite so close to perihelion then as it was in 1988. The dates and configurations of some oppositions of Mars are illustrated in Figure 10.1(b), which shows the locations of the Earth and Mars at six oppositions (at 780-day intervals) between September 1988 and April 1999.

Although Mars is quite bright and easily seen at opposition, the planet is still considerably fainter than Venus. This faintness results from a combination of three factors. First, Mars is more than twice as far from the Sun as is Venus, so each square meter on the Martian surface receives less than one-quarter the amount of sunlight that strikes each square meter on Venus. Second, Mars's surface area is only about 30 percent that of Venus, so there are fewer square meters to intercept the sunlight. Finally, Mars is much less reflective than Venus—only about 15 percent of the sunlight striking the planet is reflected back into space, compared with over 70 percent in the case of Venus. Still, at its brightest, Mars is brighter than any star. Its characteristic red color, visible even to the naked eye, makes it easily identifiable in the night sky.

RADIUS, MASS, AND DENSITY

As with Mercury and Venus, we can determine Mars's radius by means of simple geometry. From the data given earlier for the planet's size and distance, we can calculate a radius of about 3400 km. More accurate measurements give a result of 3397 km, or 0.53 Earth radii.

Unlike Mercury and Venus, Mars has two small moons in orbit around it, which are visible (through telescopes) from Earth. Named Phobos (Fear) and Deimos (Panic) for the horses that drew the Roman war god's chariot, these moons are little more than large rocks trapped by the planet's gravity. We will return to their individual properties

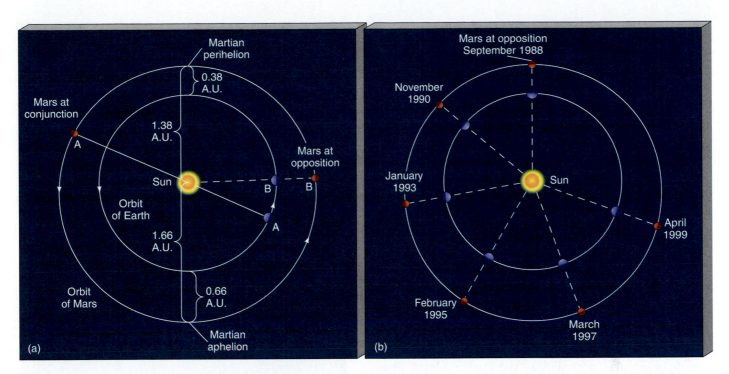

Figure 10.1 (a) The orbit of Mars compared with that of Earth. Notice that Mars's orbit is noticeably elliptical, unlike Earth's, whose eccentricity is barely perceptible. When the planets are on opposite sides of the Sun, as at the points marked A, Mars is said to be at conjunction. The planets are at their closest at opposition, when Earth and Mars are aligned and on the same side of the Sun, as at the points marked B. (Note that to get from point A to point B, Earth must travel for nearly 13 months—all the way around its orbit and then some.) (b) Several oppositions of Mars, from the particularly favorable configuration of September 1988 to the unfavorable opposition of March 1997.

in a moment. The larger of the two, Phobos, orbits at a distance of only 9380 km from the center of the planet once every 459 minutes. Applying the modified version of Kepler's third law (which states that the square of a moon's orbital period is proportional to the cube of its orbital semimajor axis divided by the mass of the planet it orbits), we find that the mass of Mars is 6.4×10^{23} kg, or 0.11 times that of the Earth. ∞ (p. 48) Naturally, the orbit of Deimos yields the same result.

From the mass and radius, we find that the average density of Mars is 3900 kg/m³(3.9 g/cm³), only slightly greater than that of the Moon. If we assume that the Martian surface rocks are similar to those on the other terrestrial planets, this average density suggests the existence of a substantial core of higher than average density within the planet. Scientists now believe that this core is composed largely of iron sulfide (a compound about five times denser than surface rock) and has a diameter of about 2500 km.

ROTATION RATE

Surface markings easily seen on Mars allow astronomers to track the planet's rotation. Mars rotates once on its axis every 24.6 hours. One Martian day is thus very similar in length to one Earth day. The planet's equator is inclined to the orbit plane at an angle of 24.0°, again very similar to Earth's inclination of 23.5°. Thus, as Mars orbits the Sun,

we find both daily and seasonal cycles, just as on the Earth. In the case of Mars, however, the seasons are complicated somewhat by variations in solar heating due to the planet's eccentric orbit.

10.2 Earth-Based Observations of Mars

➋ At opposition, when Mars is closest to us and most easily observed, it is also full, so the Sun's light strikes the surface almost vertically, casting few shadows and preventing us from seeing any topographic detail, such as craters or mountains. Even through a large telescope Mars appears only as a reddish disk, with some light and dark patches and prominent polar caps. These surface features undergo slow seasonal changes over the course of a Martian year—a consequence of Mars's axial tilt and somewhat eccentric orbit, which combine to produce large changes in the amount of sunlight striking the planet's surface. We saw in Chapter 1 how the inclination of Earth's axis produces similar seasonal changes. ∞ (p. 10) Figure 10.2 shows some of the best images of Mars ever made from the Earth, along with a photograph taken by one of the U.S. *Viking* spacecraft en route to the planet.

Viewed from Earth, the most obvious Martian surface features are its bright polar caps (see Figure 10.2a). The caps are mostly frozen carbon dioxide (that is, dry ice), not

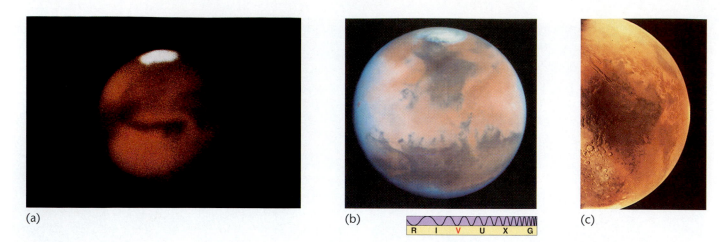

(a)   (b)   (c)

R I U V X G

Figure 10.2 (a) A deep-red (800 nm) image of Mars, taken in 1991 at Pic du Midi, an exceptionally clear site in the French Alps. One of the planet's polar caps appears at the top. (b) A visible-light *Hubble Space Telescope* image of Mars, taken while the planet was near opposition in 1995. (c) A view of Mars taken from a *Viking* spacecraft during its approach in 1976. Some of the planet's surface features can be clearly seen at a level of detail completely invisible from the Earth.

water ice as at the Earth's North and South poles. They grow or diminish according to the seasons, almost disappearing at the time of Martian summer.

The dark surface features on Mars also change from season to season, although their variability probably has little to do with the melting of the polar ice caps. To the more fanciful observers around the start of the twentieth century, these changes suggested the seasonal growth of vegetation on the planet. It was but a small step from seeing polar ice caps and speculating about teeming vegetation to imagining a planet harboring intelligent life, perhaps not unlike us.

But those speculations and imaginings were not to be confirmed. The polar caps do contain water, but it remains permanently frozen, and the dark markings seen in Figures 10.2, once claimed to be part of a network of "canals" dug by Martians for irrigation purposes (see *Interlude 10-1* on p. 223), are actually highly cratered and eroded areas around which surface dust occasionally blows. Repeated covering and uncovering of these landmarks gives the impression (from a distance) of surface variability, but it's only the thin dust cover that changes. A powdery material, the surface dust is borne aloft by strong winds that often reach hurricane proportions (that is, hundreds of kilometers per hour). In fact, when the American *Mariner 9* spacecraft went into orbit around Mars in 1971, a planetwide dust storm obscured the entire landscape. Had the craft been on a flyby mission (for a quick look) instead of an orbiting mission (for a longer view), its visit would have been a failure. Fortunately, the storm subsided, enabling the craft to radio home detailed information about the surface.

10.3 *The Surface of Mars*

The maps of the surface of Mars returned by the *Mariner* and *Viking* orbiters show a wide range of geological features. Mars has huge volcanoes, deep canyons, vast dune fields, and many other geological wonders. The orbiters

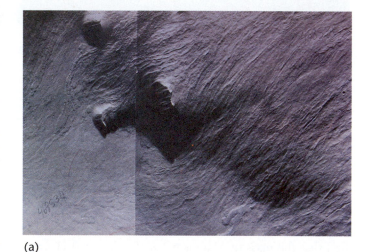

(a)

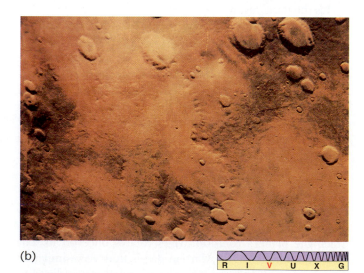

(b)

R I U V X G

Figure 10.3 (a) The northern hemisphere of Mars consists of rolling, volcanic plains (false-color image). (b) The southern Martian highlands are heavily cratered (true-color).

R I V U X G

Figure 10.4 The Tharsis region of Mars, some 5000 km across, bulges out from the planet's equatorial region, rising to a height of about 10 km. The three large volcanoes on the left mark the approximate peak of the bulge. One of the plains flanking the Tharsis bulge, Chryse Planitia, is toward the right. Dominating the center of the field of view is a vast "canyon" known as Valles Marineris. (See also the chapter-opening image on p. 214.)

performed large-scale surveys of much of the planet's surface; the lander data complemented these planetwide studies with detailed information on two specific sites.

LARGE-SCALE TOPOGRAPHY

A striking feature of the terrain of Mars is the marked difference between the northern and southern hemispheres. The northern hemisphere is made up largely of rolling volcanic plains, not unlike the lunar maria. These extensive northern lava plains—much larger than those found on Earth or the Moon—were formed by eruptions involving enormous volumes of lava. The plains are strewn with blocks of volcanic rock, as well as with boulders blasted out of impact areas by infalling meteoroids (the Martian atmosphere is too thin to offer much resistance to incoming debris). The southern hemisphere consists of heavily cratered highlands lying several kilometers above the level of the lowland north. Most of the dark regions visible from the Earth are mountainous regions in the south. Figure 10.3 contrasts typical terrains in the two hemispheres.

The northern plains are cratered much less than the southern highlands. On the basis of arguments presented in

Chapter 8, this smoother surface suggests that the northern surface is younger. ∞ (p. 182) Its age is perhaps 3 billion years, compared with 4 billion in the south. In places, the boundary between the southern highlands and the northern plains is quite sharp. The surface level can drop by as much as 4 km in a distance of 100 km or so. Most scientists assume that the southern terrain is the original crust of the planet. How most of the northern hemisphere could have been lowered in elevation and flooded with lava remains a mystery.

The major geological feature on the planet is the Tharsis bulge (shown in Figure 10.4, a large-scale view of the planet). It is a region, roughly the size of North America, on the Martian equator that rises some 10 km higher than the rest of the Martian surface. To the east of Tharsis lies Chryse Planitia (the "Plains of Gold"), and to its west lies a region known as Isidis Planitia (the "Plains of Isis," an Egyptian goddess). These features are wide depressions, hundreds of kilometers across and up to 3 km deep.

Tharsis appears to be even less heavily cratered than the north, making it the youngest region on the planet. It is estimated to be only 2 to 3 billion years old. If we wished to extend the idea of "continents" from Earth and Venus to Mars, we would have to conclude that Tharsis is the only continent on the Martian surface. However, as on Venus, there is no sign of plate tectonics—the continent of Tharsis is not drifting as are its earthly counterparts. ∞ (p. 161)

VOLCANISM

Mars contains the largest known volcanoes in the solar system. Four very large volcanoes are found on the Tharsis bulge, three of them visible in Figure 10.4. The largest volcano of all is Olympus Mons (shown in Figure 10.5), on the

Figure 10.5 Olympus Mons, the largest volcano known on Mars or anywhere else in the solar system. Nearly three times taller than Mount Everest on Earth, this Martian mountain measures about 700 km across the base and extends 25 km at the peak. It seems currently inactive and may have been extinct for at least several hundred million years. By comparison, the largest volcano on Earth, Hawaii's Mauna Loa, measures a mere 120 km across and peaks about 9 km above the Pacific Ocean floor.

R I V U X G

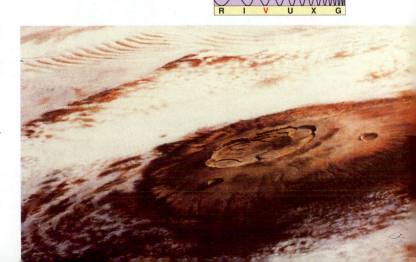

northwestern slope of Tharsis, lying just off the left-hand edge of Figure 10.4. It measures some 700 km in diameter at its base—only slightly smaller than the state of Texas—and rises to a height of 25 km above the surrounding plains. The caldera, or crater, at its summit, measures 80 km across. The other three volcanoes (on the left of Figure 10.4) are a little smaller—a mere 18 km high—and lie near the top of the bulge. Like Maxwell Mons on Venus, these volcanoes are not associated with plate motion but instead are shield volcanoes (see Chapter 9), sitting atop a hot spot in the underlying Martian mantle. All four volcanoes show distinctive lava channels and other flow features very similar to those found on shield volcanoes on the Earth. The *Viking* images of the Martian surface reveal many hundreds of volcanoes. Most of the largest are associated with the Tharsis bulge, but many smaller volcanoes are also found in the northern plains.

The great height of Martian volcanoes is a direct consequence of the planet's low surface gravity. As lava flows and spreads to form a shield volcano, its eventual height depends on the new mountain's ability to support its own weight. The lower the gravity, the less the weight and the higher the mountain. It is no accident that Maxwell Mons on Venus and the Hawaiian shield volcanoes on the Earth rise to roughly the same height (about 10 km) above their respective bases—Earth and Venus have similar surface gravity. Mars has surface gravity only 40 percent that of Earth, and volcanoes rise roughly 2.5 times as high.

Are these volcanoes still active? Scientists have found no direct evidence for recent or ongoing eruptions. However, if these volcanoes have been around since the Tharsis uplift (as the formation of the Tharsis bulge is often known) and have been active as recently as 100 million years ago (an age estimate based on the extent of impact cratering on their slopes), some of them may still be at least intermittently active. Millions of years, though, may pass between eruptions.

IMPACT CRATERING

The *Mariner* series of spacecraft found that the surfaces of Mars and its two moons are pitted with impact craters formed by meteoroids falling in from space. As on our Moon, the smaller craters are often filled with surface matter—mostly dust—confirming that Mars is a dry, desert world. However, Martian craters are filled in considerably faster than their lunar counterparts. On the Moon, ancient craters less than 20 m deep (corresponding to about 100 m in diameter) have been obliterated, primarily by meteoritic erosion, as discussed in Chapter 8. ∞ (p. 183) On Mars, there is also a deficit of small craters, but extending to craters 5 km in diameter. The Martian atmosphere is an efficient erosive agent, transporting dust from place to place and erasing surface features much faster than meteoritic impacts alone. As on the Moon, the extent of large impact cratering (that is, craters too big to have been filled in by erosion since they formed) can be used as an age indicator for the Martian surface. Ages derived in this way range from 4 billion years for the southern highlands to a few hundred million years in the youngest volcanic areas.

The detailed appearance of Martian impact craters provides an important piece of information about conditions just below the planet's surface. The ejecta blankets

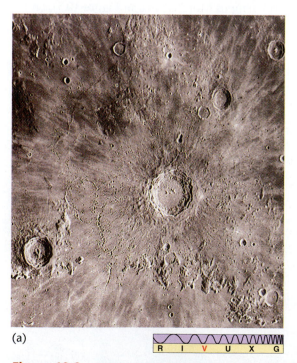

(a)

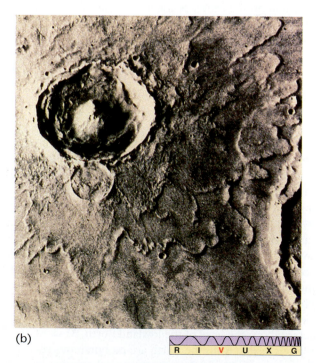

(b)

Figure 10.6   (a) The large lunar crater Copernicus is quite typical of those found on Earth's Moon. Its ejecta blanket appears to be composed of dry, powdery material. (b) The ejecta from Mars's crater Yuty (18 km in diameter) evidently was liquid in nature. This type of crater is sometimes called a "splosh" crater.

surrounding many Martian craters look quite different from their lunar counterparts. Figure 10.6 compares the Copernicus crater on the Moon with the (fairly typical) crater Yuty on Mars. The material surrounding the lunar crater is just what one would expect from an explosion ejecting a large volume of dust, soil, and boulders. The ejecta blanket on Mars gives the distinct impression of a liquid that has splashed or flowed out of the crater. Geologists believe that this **fluidized ejecta** crater indicates that a layer of **permafrost**, or water ice, lies under the surface. The explosive impact heated and liquefied the ice, resulting in the fluid appearance of the ejecta.

THE MARTIAN "GRAND CANYON"

Yet another feature associated with the Tharsis bulge is a great "canyon" known as Valles Marineris (the Mariner Valley). Shown in its entirety in Figure 10.4 and in more detail in Figure 10.7, it is not really a canyon in the terrestrial sense, because running water played no part in its formation. Planetary astronomers believe that it was formed by the same crustal forces that forced the entire Tharsis region to bulge outward, causing the surface to split and crack. These cracks, called *tectonic fractures*, are found all around the Tharsis bulge. The Mariner Valley is the largest of them. Cratering studies suggest that the cracks are at least 2 billion years old. Similar (but smaller) cracks, with

similar causes, have been found in the Aphrodite Terra region of Venus, as we saw in Chapter 9.

Valles Marineris runs for almost 4000 km along the Martian equator, about one-fifth of the way around the planet. At its widest, it is some 120 km across, and it is as deep as 7 km in places. Like many Martian surface features, it simply dwarfs Earthly competition. The Grand Canyon in Arizona would easily fit into one of its side "tributary" cracks. Valles Marineris is so large that it can even be seen from Earth—in fact, it was one of the few canals observed by nineteenth-century astronomers (see *Interlude 10-1* on p. 223) that actually corresponded to a real feature on the planet's surface (it was known as the Coprates canal). We must reemphasize, however, that this Martian feature was *not* constructed by intelligent beings, nor was it carved by a river, nor is it a result of Martian plate tectonics. For some reason, the crustal forces that formed it never developed into full-fledged plate motion as on the Earth.

EVIDENCE FOR RUNNING WATER

Although the great surface cracks in the Tharsis region are not really canyons and were not formed by running water, photographic evidence reveals that liquid water once existed in great quantity on the surface of Mars. Two types of flow features are seen: the **runoff channels** and the **outflow channels**.

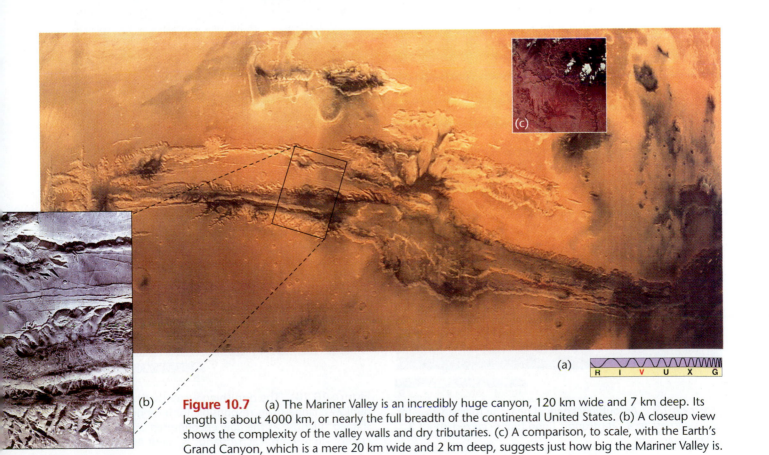

(a)

R I V U X G

(b)

Figure 10.7 (a) The Mariner Valley is an incredibly huge canyon, 120 km wide and 7 km deep. Its length is about 4000 km, or nearly the full breadth of the continental United States. (b) A closeup view shows the complexity of the valley walls and dry tributaries. (c) A comparison, to scale, with the Earth's Grand Canyon, which is a mere 20 km wide and 2 km deep, suggests just how big the Mariner Valley is.

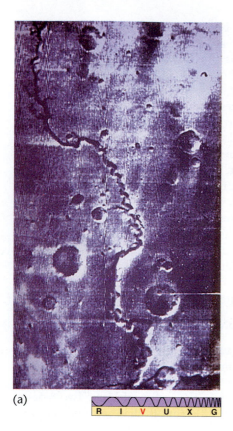

(a)

R I V U X G

(b)

R I V U X G

Figure 10.8 (a) This runoff channel on Mars measures about 400 km long and 5 km wide. Here, we compare it with a photograph of the Red River (b) running from Shreveport, Louisiana, to the Mississippi River. The two differ mainly in that there is currently no liquid water in this, or any other, Martian valley.

The runoff channels (one of which is shown in Figure 10.8) are found in the southern highlands. They are extensive systems—sometimes hundreds of kilometers in total length—of interconnecting, twisting channels that seem to merge into larger, wider channels. They bear a strong resemblance to river systems on Earth, and it is believed by geologists that this is just what they are—the dried-up beds of long-gone rivers that once carried rainfall on Mars from the mountains down into the valleys. These runoff channels speak of a time 4 billion years ago (the age of the

R I V U X G

Figure 10.10 The onrushing water that carved out the outflow channels was responsible for forming these oddly shaped "islands" as the flow encountered obstacles—impact craters—in its path. Each "island" is about 40 km long.

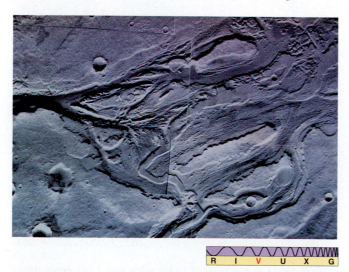

R I V U X G

Figure 10.9 An outflow channel near the Martian equator bears witness to a catastrophic flood that occurred about 3 billion years ago.

Interlude 10–1 *Canals on Mars?*

The year 1877 was an important one in the human study of the planet Mars. The red planet came unusually close to Earth, affording astronomers an especially good view. Of particular note was the discovery, by U.S. Naval Observatory astronomer Asaph Hall, of the two moons circling Mars. But most exciting was the report of the Italian astronomer Giovanni Schiaparelli (the man responsible for erroneously reporting the synchronous orbit of Mercury) on his observation of a network of linear markings that he termed *canali*. In Italian *canali* means simply "grooves" or "channels," but the word was translated into English as "canals," suggesting that the grooves had been constructed by intelligent beings. Observations of these features became sensationalized in the world's press (especially in the United States), and some astronomers began drawing elaborate maps of Mars, showing oases and lakes where canals met in desert areas.

Percival Lowell (see photo below), a successful Boston businessman (and brother of the poet Amy Lowell and Harvard president Abbott Lawrence Lowell), became fascinated by these reports. He abandoned his business and purchased a clear-sky site at Flagstaff, Arizona, where he built a major observatory. He devoted his life to achieving a better understanding of the Martian "canals." In doing so, he championed the idea that Mars was drying out and that an intelligent society had constructed the canals to transport water from the wet poles to the arid equatorial deserts.

Alas, the Martian valleys and channels photographed by robot spacecraft during the 1970s are far too small to be the "canali" that Schiaparelli, Lowell, and others thought they saw on Mars. The entire episode represents a classic case in the history of science—a case where well-intentioned observers, perhaps obsessed with the notion of life on other worlds, let their personal opinions and prejudices seriously affect their interpretations of reasonable data. The two figures of Mars below show how surface features (which were probably genuinely observed by astronomers at the turn of the century) might be imagined to be connected. The figure at center is a photograph of how Mars actually looked in a telescope at the end of the nineteenth century. The sketch at right is an interpretation (done at the height of the canal hoopla) of the pictured view. The human eye, under physiological stress, tends to connect dimly observed yet distinctly separated features. Humans saw patterns and canals where none in fact existed.

The chronicle of the Martian canals illustrates how the scientific method requires scientists to acquire new data to sort out sense from nonsense, fact from fiction. Rather than simply believing the claims about the Martian canals, other scientists demanded further observations to test Lowell's hypothesis. Eventually, improved observations, climaxing in the *Mariner* and *Viking* exploratory missions to the red planet nearly a century after all the fuss began, totally disproved the existence of canals. Although it often takes time, the scientific method does in fact lead to progress toward understanding reality.

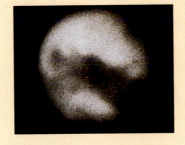
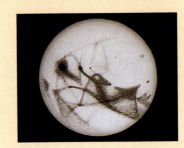

Martian highlands), when the atmosphere was thicker, the surface warmer, and liquid water widespread.

The outflow channels (see Figure 10.9) are probably relics of catastrophic flooding on Mars long ago. They appear only in equatorial regions and generally do not form the extensive interconnected networks that characterize the runoff channels. Instead, they are probably the paths taken by huge volumes of water draining from the southern highlands into the northern plains. The onrushing water arising from these flash floods probably also formed the odd teardrop-shaped "islands" (resembling the miniature versions seen in the wet sand of our beaches at low tide) that have been found on the plains close to the ends of the outflow channels (Figure 10.10). Judging from the width and depth of the channels, the flow rates must have been truly enormous—perhaps as much as a hundred times greater

than the 10^5 tons per second carried by the Amazon river, the largest river system on Earth. Flooding shaped the outflow channels about 3 billion years ago, about the same time as the northern volcanic plains formed.

Planetary astronomers find no evidence for liquid water anywhere on Mars today, and the amount of water vapor in the Martian atmosphere is tiny. Yet the extent of the outflow channels indicates that a huge total volume of water existed on Mars in the past. Where did all that water come from? And where did it all go? The answer may be that virtually all of the water on Mars is now locked in the permafrost layer under the surface, with perhaps a little more contained in the polar caps. Four billion years ago, as climatic conditions changed, the running water that formed the runoff channels began to freeze, forming the permafrost and drying out the river beds. Mars remained frozen for about a billion years, until volcanic (or some other) activity heated large regions of the surface, melting the subsurface ice and causing the flash floods that created the outflow channels. Subsequently, volcanic activity subsided, the water refroze, and Mars once again became a dry world.

POLAR CAPS

We have already noted that Mars's polar caps are composed predominantly of carbon dioxide frost—dry ice—and show seasonal variations. Each cap in fact consists of two distinct parts—the **seasonal cap**, which grows and shrinks each Martian year, and the **residual cap**, which remains permanently frozen. At maximum size, in southern midwinter, the southern seasonal cap is some 4000 km across, about one-fifth the circumference of the planet. Half a Martian year later, the northern cap is at its largest, reaching a diameter of roughly 3000 km. The two seasonal polar caps do not have the same maximum size because of the eccentricity of Mars's orbit around the Sun. During southern winter, Mars is considerably farther from the Sun than half a year later, in northern winter. Thus, the southern winter season is longer and colder than in the north, and the polar cap grows correspondingly larger.

The seasonal caps are composed entirely of carbon dioxide. Their temperatures are never greater than about 150 K (–120°C), the point at which dry ice can form. During the Martian summer, when sunlight striking a cap is most intense, carbon dioxide evaporates into the atmosphere, and the cap shrinks. In the winter, atmospheric carbon dioxide refreezes, and the cap reforms. As the caps grow and shrink, they cause substantial variations (up to 30 percent) in the Martian atmospheric pressure. A large fraction of the atmosphere condenses out and evaporates again on a yearly basis. From studies of these atmospheric fluctuations, scientists can estimate the amount of carbon dioxide in the seasonal polar caps. They believe that the maximum thickness of the seasonal caps is about 1 m.

The residual caps, shown in Figure 10.11, are smaller and brighter than the seasonal caps and show an even more marked north–south asymmetry. The southern residual cap is about 350 km across and, like the seasonal caps, is made mostly of carbon dioxide, although it may contain some

water ice. Its temperature remains below 150 K at all times. The northern residual cap is much larger—about 1000 km across—and warmer, with a temperature that can exceed 200 K in northern summertime. Planetary scientists believe that the northern cap is made mostly of water ice, an opinion strengthened by the observed increase in the concentration of water vapor above the north pole in northern summer as some small fraction of its water ice evaporates in the Sun's heat. It is quite possible that the northern residual polar cap is a major storehouse for water on Mars.

Why is there such a temperature difference (at least 50 K) between the two residual polar caps? The reason seems to be related to the giant dust storms that envelop the planet during southern summer. These storms, which last for a quarter of a Martian year (about six Earth months), tend to blow the dust from the warmer south into the northern hemisphere. As a result, the northern ice cap becomes dusty. Since dust absorbs much more sunlight than ice, which is highly reflective, the cap tends to warm up.

THE VIEW FROM *VIKING*

Viking 1 landed in Chryse Planitia, the broad depression to the east of Tharsis. The view that greeted its cameras (Figure 10.12) was a windswept, gently rolling, rather desolate plain, littered with rocks of all sizes, not unlike a high desert on Earth. *Viking 2* landed somewhat farther north, in a region of Mars called Utopia. Mission planners chose Utopia in part because they anticipated greater seasonal climatic variations there. The plain on which *Viking 2* landed was flat and featureless. From space, the landing site appeared smooth and dusty. In fact, the surface turned out to be very rocky, even rockier than the Chryse site, and without the dust layer the mission directors had expected. The surface rocks visible in Figure 10.12 are probably part of the ejecta blanket of a nearby impact crater. The views that the two landers recorded may turn out to be quite typical of the low-latitude northern plains.

The *Viking* landers performed numerous chemical analyses of the regolith of Mars. One important finding of these studies was the high iron content of the planet's surface. Chemical reactions between the iron-rich surface soil

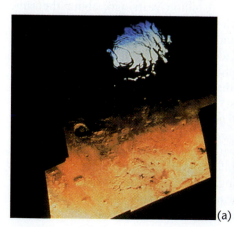
(a)

(b)

Figure 10.11 The southern (a) and northern (b) polar caps of Mars are shown to scale in these mosaics of *Mariner 9* images. These are the residual caps, seen here during their respective summers half a Martian year apart. The southern cap is only 350 km across and made up mostly of frozen carbon dioxide. The northern cap is about 1000 km across and composed mostly of water ice.

Figure 10.12   Panoramic view from the perspective of the *Viking 1* spacecraft now parked on the surface of Mars. The fine-grained soil and the rock-strewn terrain stretching toward the horizon are reddish. Containing substantial amounts of iron ore, the surface of Mars is literally rusting away. The sky is a pale yellow-pink color, the result of airborne dust.

and free oxygen in the atmosphere is responsible for the iron oxide ("rust") that gives Mars its characteristic color. Although the surface layers are rich in iron relative to Earth's surface, the overall abundance is similar to the Earth's average iron content. On Earth, much of the iron has differentiated to the center. Chemical differentiation does not appear to have been nearly so complete on Mars.

10.4   *The Martian Atmosphere*

COMPOSITION

❸   Even before the arrival of the *Mariner* and *Viking* spacecraft, astronomers knew from Earth-based spectroscopy that the Martian atmosphere was quite thin and composed primarily of carbon dioxide. In 1964 *Mariner 4* confirmed these results, finding that the atmospheric pressure is only about 1/150 the pressure of Earth's atmosphere at sea level and that carbon dioxide makes up at least 95 percent of the total atmosphere. With the arrival of *Viking*, more detailed measurements of the Martian atmosphere were made. Its composition is now known to be 95.3 percent carbon dioxide, 2.7 percent nitrogen, 1.6 percent argon, 0.13 percent oxygen, 0.07 percent carbon monoxide, and about 0.03 percent water vapor. The level of water vapor is quite variable.

As the *Viking* landers descended to the surface, they made measurements of the temperature and pressure at various heights. The results appear in Figure 10.13. The Martian atmosphere contains a troposphere (the lowest-lying atmospheric zone, where convection and "weather" occur), which varies both from place to place and from season to season. The variability of the troposphere arises from the variability of the Martian surface temperature. At noon in the summertime, surface temperatures may reach 300 K. Atmospheric convection is strong, and the top of the troposphere can reach an altitude of 30 km. At night, the atmosphere retains little heat, and the temperature can drop by as much as 100 K. Convection then ceases entirely, and the troposphere vanishes. Temperatures in the stratosphere are low enough for carbon dioxide to solidify, giving rise to a high-level layer of carbon dioxide clouds and haze.

On average, surface temperatures on Mars are about 50 K cooler than on Earth. Weather, such as dust storms and most clouds, is confined to the troposphere, as it is on Earth. Only a few thin carbon dioxide clouds are occasionally found in the lower stratosphere. The low early-morning temperatures often produce water–ice "fog" in the

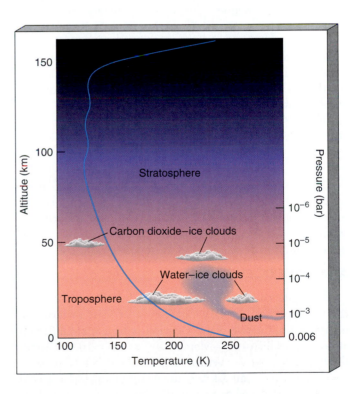

Figure 10.13   Structure of the Martian atmosphere, as determined by *Viking*. The troposphere rises up to an altitude of about 30 km in the daytime. It occasionally contains clouds of water ice or, more frequently, dust during the planetwide dust storms that occur each year. Above the troposphere lies the stratosphere. Note the absence of a higher-temperature zone in the stratosphere, indicating the absence of an ozone layer.

R I V U X G

Figure 10.14   Fog in the Martian canyons, photographed by the *Viking* orbiter. As the Sun's light reaches and heats the canyon floor, it drives water vapor from the surface. When this vapor comes in contact with the colder air above the surface, it condenses again, and a temporary water–ice "fog" results.

Martian canyons, as can be seen in Figure 10.14. For most of the year, there is little day-to-day variation in the weather: The Sun rises, the surface warms up, and light winds blow until sunset, when the temperature drops again. Only in the southern summer does the daily routine change. Strong surface winds (without rain or snow) sweep up the dry dust, carry it aloft, and eventually deposit it elsewhere on the planet. At its greatest fury, a Martian storm floods the atmosphere with dust, making the worst storm we could imagine on Earth's Sahara Desert seem inconsequential by comparison. The dust can remain airborne for months at a time. The blown dust forms systems of sand dunes similar in appearance to those found on the Earth.

ATMOSPHERIC EVOLUTION

4 While there is some superficial similarity between the atmospheres of Mars and Venus, at least in terms of composition, the two planets obviously have quite different atmospheric histories—Mars's "air" is over 10,000 times thinner than that on Venus. As with the other planets we have studied, we can ask *why* the Martian atmosphere is as it is. Presumably, Mars acquired a secondary outgassed atmosphere quite early in its history, just as the other terrestrial worlds did. Around 4 billion years ago, as indicated by the runoff channels in the highlands, Mars may have had a fairly dense atmosphere, complete with blue skies and rain. Despite Mars's distance from the Sun, the greenhouse effect would have kept conditions fairly comfortable, and a surface temperature of over 0°C seems quite possible.

Sometime in the next billion years, most of the Martian atmosphere disappeared. Possibly some of it was lost because of impacts with other large (Moon-sized) bodies in the early solar system. More likely, the Martian atmosphere became unstable, in a kind of reverse runaway greenhouse effect. On Venus, as we have seen, the familiar greenhouse effect ran away to high temperatures and pressures. Planetary scientists theorize that the early Martian atmosphere also ran away, but in the opposite direction. The presence of liquid water would have caused much of the atmospheric carbon dioxide to have dissolved in Martian rivers and lakes (and oceans, if any), ultimately to combine with Martian surface rocks. Recall that, on the Earth, most carbon dioxide is presently found in surface rocks. Calculations show that much of the Martian atmospheric carbon dioxide could have been depleted in this way in a relatively short period of time, perhaps as little as a few hundred million years, although some of it may have been replenished by volcanic activity, possibly extending the "comfortable" lifetime of the planet to a half-billion years or so.

As the level of carbon dioxide declined and the greenhouse-heating effect diminished, the planet cooled. The water froze out of the atmosphere, lowering still further the level of atmospheric greenhouse gases, and so accelerating the cooling. (Recall from Chapter 9 that water vapor also contributes to the greenhouse effect. (p. 99) Since that time, the overall density of the atmosphere has steadily declined through a steady loss of carbon dioxide, nitrogen, and water vapor. Solar ultraviolet radiation in the upper Martian atmosphere splits the molecules of these gases into their component atoms and provides some with enough energy to escape. The present level of water vapor in the Martian atmosphere is the maximum possible, given the atmosphere's present density and temperature. Estimates of the total amount of water stored as permafrost or in the polar caps are quite uncertain, but it is likely that, if all the water on Mars were to become liquid, it would cover the surface to a depth of several meters.

10.5 *Martian Internal Structure*

5 The *Viking* landers carried seismometers to probe the internal structure of the planet, but one failed to work and the other was unable to distinguish seismic activity from the buffeting of the Martian wind. Thus, as yet no seismic studies of Mars's interior have been carried out. On the basis of studies of the stresses that occurred during the Tharsis uplift, astronomers estimate the thickness of the Martian crust to be about 100 km.

Mariner 4 detected no Martian magnetic field. Thus, the strength of Mars's field must be less than 1/1000 that of Earth—the level of sensitivity of *Mariner*'s instruments. Because Mars is rotating rapidly, we must conclude that either its core is nonmetallic, or not liquid, or both.

The small size of Mars means that any radioactive (or other internal) heating of its interior would have been less effective at heating and melting the planet than heating on Earth. The heat was able to reach the surface and escape more easily than in a larger planet such as Earth or Venus. The evidence we noted earlier for ancient surface activity, especially volcanism, suggests that at least parts of the planet's interior must have melted and possibly differentiated at some time in the past. But the lack of current activity, the absence of any significant magnetic field, the relatively low density (4000 kg/m³), and an abnormally high abundance of iron at the surface all suggest that Mars never melted as extensively as did the Earth.

The history of Mars appears to be that of a planet where large-scale tectonic activity almost started but was stifled by the planet's rapidly cooling outer layers. The large upwelling of material that formed the Tharsis bulge might have developed on a larger, warmer planet into full-fledged plate tectonic motion, but the Martian mantle became too rigid and the crust too thick for that to occur. Instead, the upwelling continued to fire volcanic activity, almost up to the present day, but geologically much of the planet apparently died 2 billion years ago.

10.6   *The Moons of Mars*

6 Unlike Earth's moon, Mars's moons are tiny compared with their parent planet and orbit very close to it (relative to the planet's radius). Discovered by American astronomer Asaph Hall in 1877, the two Martian moons—Phobos and Deimos—are only a few kilometers across. Their composition is quite dissimilar to that of the planet. Astronomers generally believe that Phobos and Deimos did not form along with Mars but instead are asteroids that were slowed and captured by the outer fringes of the early Martian atmosphere (which, as we have just seen, was probably much denser than the atmosphere today). It is even possible that they are remnants of a single object that broke up during the capture process. They are quite difficult to study from the Earth because their proximity to Mars makes it hard to distinguish them from their much brighter parent. The *Mariner* and *Viking* orbiters, however, have studied them in great detail.

Both moons, shown in Figure 10.15, are quite irregularly shaped and heavily cratered. The larger of the two is Phobos (Figure 10.15a), which is about 28 km long and 20 km wide and dominated by an enormous 10 km wide crater named Stickney (after Angelina Stickney, Asaph Hall's wife, who encouraged him to persevere in his observations). The smaller Deimos (Figure 10.15b) is only 16 km long by 10 km wide. Its largest crater is 2.3 km in diameter. The fact that both moons have quite dark surfaces, reflecting no more than 6 percent of the light falling on them, contributes to the difficulty in observing them from the Earth.

Phobos and Deimos move in circular, equatorial orbits, and they rotate synchronously (that is, they keep the same face permanently turned toward the planet). All of

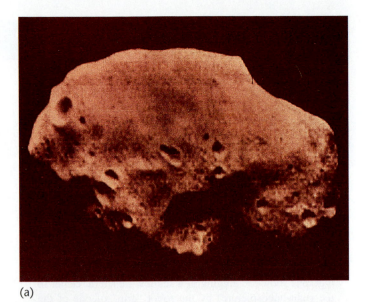

(a)

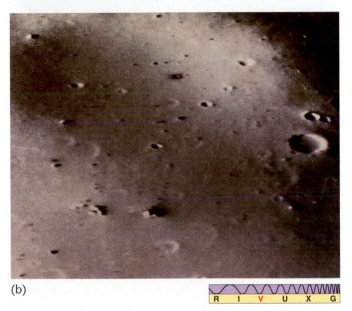

(b)

R   I   V   U   X   G

Figure 10.15   Mars is accompanied in its trek around the Sun by two tiny moons. Phobos (Greek for "fear") and Deimos ("panic") measure about 28 and 16 km long, respectively. Both are shaped like potatoes, with lots of cratering. (a) This is a *Mariner 9* photograph of the irregularly shaped Phobos, not much larger than Manhattan Island. (b) Like Phobos, the smaller moon Deimos has a composition unlike that of Mars. Both moons are probably captured asteroids. This photograph was taken by the *Viking* orbiter. The field of view is only 2 km across, with most of the boulders shown about the size of a house.

these characteristics are direct consequences of the tidal influence of Mars. Both moons orbit Mars in the prograde sense—that is, in the same sense (counterclockwise, as seen from above the north celestial pole) as the planet orbits the Sun and rotates on its axis. Phobos lies only 9378 km (less than three planetary radii) from the center of Mars and, as

Interlude 10–2 *Life on Mars?*

Even before *Viking* reached Mars, astronomers had abandoned hope of finding intelligent, or even animal, life on the planet. Scientists knew there were no large-scale canal systems, no surface water, almost no oxygen in the atmosphere, and no seasonal vegetation changes. The present lack of liquid water on Mars especially dims the chances for life there now. Might life have existed once? Running water and possibly a dense atmosphere in the past may have created conditions suitable for the emergence of life long ago. In the hope that some form of microbial life might have survived to the present day, the *Viking* landers carried out experiments designed to detect biological activity.

The accompanying pair of before and after photographs show the digging of a shallow trench (at center in the right-hand picture) by the robot "arm" of one of the *Vikings*. In this way, soil samples were scooped up and taken inside the lander, where instruments tested them for chemical composition and any sign of life.

All three *Viking* biological experiments assumed some basic similarity between hypothetical Martian microbes and those found on Earth. For example, it was assumed that any living organisms present would probably use nutrients to grow and perform metabolic activities, in the process releasing gaseous waste or by products, as most earthly organisms do carbon dioxide. Thus, a *gas exchange experiment* offered a nutrient broth to any residents of a sample of Martian soil and looked for gases that might signal metabolic activity. The *labeled release experiment* added compounds containing radioactive carbon to the soil and then waited for results indicating that Martian organisms had either eaten or breathed in this carbon. The *pyrolitic release experiment* added radioactively tagged carbon dioxide to a sample of Martian soil and atmosphere, waited a while, then removed the gas, and tested the soil (by heating it) for signs that something had absorbed the tagged gas. In all cases, contamination by terrestrial bacteria was a major concern. Indeed, any release of Earth organisms would have invalidated these and all future such experiments on Martian soil. Both *Viking* landers were carefully sterilized prior to launch, and international agreement presently protects the Martian environment from contamination by future terrestrial probes. (How we will sterilize a manned mission is still a little unclear.)

Initially, all of the experiments appeared to be giving positive signals! However, subsequent careful studies showed that all of the results could be explained by inorganic (that is, nonliving) chemical reactions. Thus, we have no clear evidence for even microbial life on the surface. The *Viking* robots detected peculiar reactions that mimic in some ways the basic chemistry of living organisms, but they did not detect life itself.

Two criticisms have been aimed at these remarkable experiments, suggesting that they did not comprise a definitive search for life on Mars. First, NASA sent robots to Mars to search for the only kind of life we know—carbon-based life, operating in a water-based medium, with higher forms utilizing oxygen to metabolize their food. The assumption underlying this strategy could well be wrong. Water and carbon are central to the biochemistry of life on Earth (see Chapter 28), but are we being chauvinistic in searching only for our own kind of life? The element silicon, for example, can form

(Before)   (After)

we saw earlier, has an orbital period of 7 hours and 59 minutes. This orbit period is much less than a Martian day, so an observer standing on the Martian surface would actually see Phobos move backwards across the Martian sky—that is, in a direction opposite to the apparent daily motion of the Sun. Because the moon moves faster than the observer, it overtakes the planet's rotation, rising in the west and set-

ting in the east, crossing the sky from horizon to horizon in about 5.5 hours. Deimos lies a little farther out, at 23,459 km, or slightly less than seven planetary radii, and orbits in 30 hours and 18 minutes. Because it completes its orbit in more than a Martian day, it moves "normally," as seen from the ground (that is, from east to west), taking almost 3 days to traverse the sky.

complex, long-chain molecules similar to those of carbon (which make up the key building blocks of life here on Earth), and silicon is very abundant on Mars—in fact, more abundant than carbon. Furthermore, water might not be the only possible medium in which biological molecules can interact. Ammonia has often been suggested as a possible alternative. In fact, with its lower melting point, ammonia might enable strange forms of life to prevail on a cold planet like Mars, where water is normally frozen.

Most chemists and biologists believe that silicon and ammonia could not provide the basis for the numbers and diversity of species we see on Earth. Yet it is well to remember that all the textbooks labeled "General Chemistry" or "General Biology" on our shelves are not really general at all; they are books about Earth chemistry and biology. Could those sciences be different on another world? In part because of this possibility, however remote, a camera was added to the *Viking* landers at the last moment, to search visually for some obvious yet unexpected form of life (see Figure 6.12). Conceivably, trees that thrive on ammonia (rather than water) might have been swaying on the horizon, and a camera-less *Viking* would have missed them. Or if a 17-foot silicon-based giraffe were to have sauntered past a blind *Viking*, its scientific instruments would not have detected it. In the end, however, the *Viking* cameras saw nothing suggestive of life.

The other criticism of the *Viking* experiments is that they searched for life now living. What about fossilized life? Despite the lack of clear evidence for life on today's Mars, this most intriguing planet might well harbor evidence for life

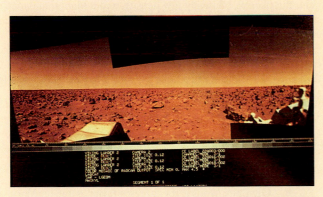

that flourished during some previous epoch. Today, Mars seems locked in an ice age—the kind of numbing cold that would prohibit sustained life as we know it. But if bacterial life did arise on an Earth-like early Mars, then we might be able to find its fossil remains preserved at or near its surface.

Viking could not have found fossilized microbial life; none of its experiments nor its camera would have detected it. In any case, neither of the *Viking* landers touched down in the most interesting sites on Mars, such as near the dried-up tributaries of the many canyons, or the mouths of ancient river beds. Spacecraft controllers were intent on navigating the robots to the safest—that is, the smoothest—terrain. They accordingly landed the *Viking* robots on the Martian equivalent of Earth's Sahara Desert—not exactly the most hospitable place for any kind of life. (The panoramic view below suggests how desolate were the landing sites chosen.)

Later attempts to send machines to Mars have failed. Mistakes by Soviet ground controllers caused the loss of two craft of the *Phobos* mission during their final approach pattern to Mars in the late 1980s, and the billion-dollar U.S. *Mars Observer* spacecraft exploded just as it was entering Martian orbit in 1993. Despite these setbacks, future missions are planned to explore some of the more interesting places on the Martian surface and to check for any trace of fossils buried in its rusting dirt and red rocks. If all goes well, by the end of the 1990s humans will have resumed the exploration of Mars with a whole new class of semi-intelligent robots.

To give just one preview of an ambitious U.S.–European space mission, the *Mars Pathfinder* vehicle is specifically designed to lay the groundwork for future efforts to find fossil remains of living things. Now under construction for a 1998 launch, *Pathfinder* will enter the Martian atmosphere and parachute an instrument package toward the surface. As it approaches the ground, the parachute will fall away, huge airbags will deploy, and the robot craft will bounce—perhaps as high as a ten-story building, and probably several times. Once settled, side panels will open and a solar-powered rover will emerge. Controlled from Earth, *Pathfinder* will roam the Martian dunes, valleys, and especially the mouths of its old flood channels, seeking evidence that life—any kind of life—once existed (or perhaps still exists) on this neighboring planet that has always captured human imagination.

On the basis of measurements of their gravitational effect on the *Viking* orbiters, astronomers have estimated the masses of the two moons. Their densities are around 2000 kg/m^3, far less than that of any world we have yet encountered in our outward journey through the solar system. This is an important reason why astronomers do not believe these moons formed along

with Mars. If they are indeed captured asteroids, Phobos and Deimos represent material left over from the earliest stages of the solar system. Astronomers study them not to gain insight into Martian evolution, but rather because the moons contain information about the very early solar system, before the major planets had formed.

Chapter Review

SUMMARY

Mars lies outside Earth's orbit, so it traverses the entire ecliptic plane, as seen from Earth. Its orbit is more elliptical than Earth's, so its distance from the Sun varies more. Mars rotates at almost the same rate as Earth, and its rotation axis is inclined to the ecliptic at almost the same angle as Earth's axis. Because of its axial tilt, Mars has daily and seasonal cycles much like those on our own planet, but they are more complex than those on Earth because of Mars's eccentric orbit.

From Earth, the most obvious Martian surface features are the polar caps, which grow and diminish as the seasons change on Mars. The appearance of the planet also changes because of seasonal dust storms that obscure its surface.

Like the atmosphere of Venus, Mars's atmosphere is composed primarily of carbon dioxide. However, unlike Venus, the density of the cool Martian atmosphere is less than 1 percent that of Earth's. Mars may once have had a dense atmosphere, but it was lost, partly to space and partly to surface rocks and subsurface **permafrost** (p. 221). Even today, the thin atmosphere is slowly leaking away. Surface temperatures on Mars average about 50 K cooler than those on Earth. Otherwise, Martian weather is reminiscent of that on Earth, with dust storms, clouds, and fog.

The two polar caps on Mars consist of a **seasonal cap** (p. 224), composed of carbon dioxide, which grows and shrinks, and a **residual cap** (p. 224), of water ice, which remains permanently frozen. Much of the original Martian atmosphere has escaped into space; most of the rest is now stored in the permafrost and polar caps.

There is a marked difference between the two Martian hemispheres, with rolling plains in the north and heavily cratered highlands in the south. The northern hemisphere consists of rolling volcanic plains and lies several kilometers below the level of the heavily cratered southern hemisphere. The lack of craters in the north suggests that this region is younger. The cause of the north–south asymmetry is not known.

In 1971, *Mariner 9* mapped the entire Martian surface, revealing plains, volcanoes, channels, and canyons. *Viking 1* and *Viking 2* reached Mars in 1976 and returned a wealth of data on the planet's surface and atmosphere. Experiments on board the *Viking* landers detected no evidence for Martian life.

Mars's craters differ from those on the Moon by the presence of **fluidized ejecta** (p. 221), providing direct evidence for the permafrost layer beneath the surface.

Mars's major surface feature is the Tharsis bulge, located on the planet's equator. It may have been caused by a "plume" of upwelling material in the youthful Martian mantle. Associated with the bulge is Olympus Mons, the largest known volcano in the solar system, and a huge crack, called the Mariner Valley, in the planet's surface. The height of the Martian volcanoes is a direct consequence of Mars's low surface gravity. No evidence for recent or ongoing eruptions has been found.

There is clear evidence that water once existed in great quantity on Mars. Mars may have had a brief "Earthlike" phase early on in its evolution. The **runoff channels** (p. 221) are the remains of ancient Martian rivers. The **outflow channels** (p. 221) are the paths taken by flash floods that cascaded from the southern highlands into the northern plains. Today, a large amount of that water may be locked up in the polar caps and in the layer of permafrost lying under the Martian surface.

Convection in the Martian interior seems to have been stifled 2 billion years ago by the planet's rapidly cooling and solidifying mantle.

Mars has no magnetic field. Since the planet rotates rapidly, this implies that its core is nonmetallic, nonliquid, or both. The lack of current volcanism, the absence of any significant magnetic field, the planet's relatively low density, and a high abundance of surface iron all suggest that Mars never melted and differentiated as extensively as Earth.

Mars's moons, Phobos and Deimos, are probably asteroids captured by Mars early in its history. Their densities are far less than any planet in the inner solar system. They may be representative of conditions in the early solar system.

SELF-TEST: True or False?

_____ **1.** It is possible, over time, to see Mars at any angular distance from the Sun, at any time of night.

_____ **2.** Seen from Earth, Mars goes through phases, just like Venus and Mercury.

_____ **3.** Mars, at times, is engulfed by global dust storms.

_____ **4.** Because Mars has such a thin atmosphere, its winds can never blow any faster than a few kilometers per hour.

_____ **5.** Seasonal changes in the appearance of Mars are caused by vegetation on the surface.

_____ **6.** Mars has the largest volcanoes in the solar system.

_____ **7.** The northern hemisphere of Mars is much older than the southern hemisphere.

_____ **8.** Olympus Mons is the largest impact crater on Mars.

_____ **9.** There are many indications of past plate tectonics on Mars.

_____ **10.** Valles Marineris is similar in size to Earth's Grand Canyon.

_____ **11.** The orange-red color of the surface of Mars is primarily due to rust (iron oxide) in its soil.

_____ **12.** Daytime temperatures on Mars can reach 300 K.

_____ **13.** The *Viking* landers are still sending data back to Earth.

_____ **14.** One of the two moons of Mars is larger than Earth's Moon.

SELF-TEST: Fill in the Blank

1. When Mars is on the other side of the Sun from the Earth, it is said to be at _____.
2. The radius of Mars is about ____ that of the Earth. (Give an answer as a simple fraction.)
3. The length of the Martian day is about _____ hours.
4. The seasonal polar ice caps of Mars are composed of _____ .
5. The southern hemisphere of Mars consists of heavily _____ highlands.
6. The Tharsis region of Mars is a large equatorial _____ .
7. Several large _____ lie at approximately the center of Tharsis.
8. The great height of Martian volcanoes is a direct result of the planet's low _____ .
9. The fluidized ejecta surrounding Martian impact craters is evidence of a layer of _____ just under the surface.
10. Runoff channels carried _____ from the southern mountains into the valleys.
11. Outflow channels are the result of catastrophic _____ .
12. Water flowed on the surface of Mars a few _____ years ago, but not now.
13. Mars's northern residual polar ice cap is consists mostly of _____ .
14. Most of the carbon dioxide on Mars is now found in the planet's _____ .

REVIEW AND DISCUSSION

1. When is the best time to see Mars from Earth?
2. What is the evidence that water once flowed on Mars?
3. Is there water on Mars today?
4. For a century, there was speculation that intelligent life had constructed irrigation canals on Mars. What did the "canals" turn out to be?
5. Imagine that you will be visiting the southern hemisphere of Mars during its summer. Describe the atmospheric conditions you might face.
6. Describe the two Martian polar caps, their seasonal and permanent composition, and the differences between them.
7. Why is Mars red?
8. Why couldn't you breathe on Mars?
9. Why were Martian volcanoes able to grow so large?
10. How were the masses of Mars's moons measured, and what did these measurements tell us about their origin?
11. What is the evidence that Mars never melted as extensively as did Earth?
12. How would Earth look from Mars?
13. If humans were sent to Mars to live, what environmental factors would have to be considered? In particular, what resources might Mars provide, and which would have to be supplied from Earth?
14. Since Mars has an atmosphere and its composition is mostly carbon dioxide, why isn't there a significant greenhouse effect to warm its surface?
15. Compare and contrast the evolution of the atmospheres of Mars, Venus, and Earth.

PROBLEMS

1. Verify that the surface gravity on Mars is 40 percent that of Earth.
2. What is the maximum elongation of the Earth, as seen from Mars (assuming circular orbits for both planets)?
3. How much less sunlight, on average, does each square meter of Martian surface receive than each square meter on Earth?
4. A certain star, observed from Earth, has a parallax of 0.1″. What would be its parallax as seen from Mars? How might observations from Mars-based telescopes be superior to those made from Earth?

PROJECTS

1. Track the motion of the red planet in front of the stars for several months following its return to the predawn sky. (Consult an almanac to determine where Mars will be in the sky this year.) You will see that Mars moves rapidly in front of the stars, crossing many constellation boundaries.
2. Several months before opposition, Mars begins retrograde motion. Chart the planet's motion in front of the stars to determine when it stops moving eastward and begins moving toward the west.
3. Notice the increase in Mars's brightness as it approaches opposition. Why is it getting brighter? What other planets appear in the sky now? How do their brightnesses compare with that of Mars?
4. Look at Mars with as large a telescope as is available to you; binoculars will not be of use. Prepare ahead of time and find out the Martian season that will be occurring at the time of your observation, which hemisphere will be tilted in Earth's direction, and what longitude will be pointing towards Earth at the time of observation. This information can be obtained from many different computer almanacs. Sketch what you see. Look very carefully and take your time. Afterward, try to identify the various features you have seen with known objects on Mars.

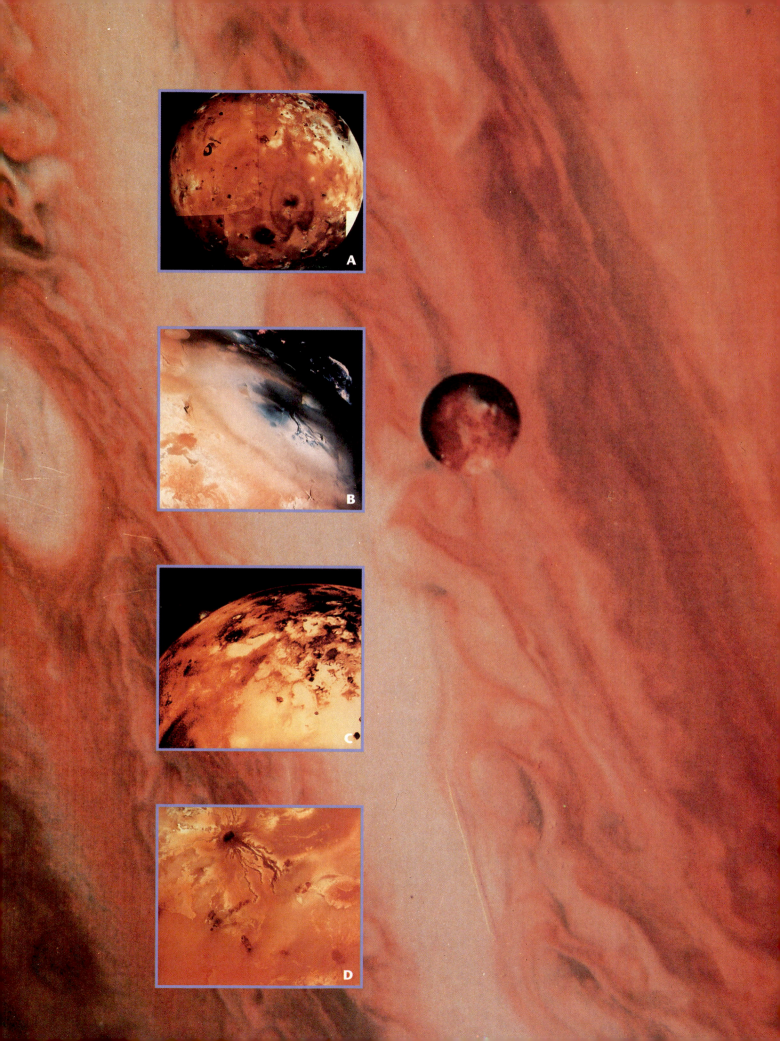

11

JUPITER

The Giant of the Solar System

LEARNING GOALS

Studying this chapter will enable you to:

1 Specify the ways in which Jupiter differs from the terrestrial planets in its physical and orbital properties.

2 Discuss the processes responsible for the appearance of Jupiter's atmosphere.

3 Describe Jupiter's internal structure and composition, and explain how these are inferred from external measurements.

4 Summarize the characteristics of Jupiter's magnetosphere.

5 Discuss the orbital properties of the Galilean moons of Jupiter, and describe the appearance and physical properties of each.

6 Explain how tidal forces can produce enormous internal stresses in a jovian moon, and discuss the results of those stresses.

(Opposite page, background) In this close-up image of Jupiter, taken by the *Voyager 2* spacecraft, the moon Io is clearly visible. One of the most peculiar moons in the solar system, and certainly the most active, Io's multi-hued surface is caused by various chemical compounds deposited by currently active volcanos.
(Inset A) Color-enhanced photo of Io taken by *Voyager 1* with a remarkable resolution of 7 km. The red color results largely from sulfur compounds. The white areas are probably covered with sulfur dioxide frost. And the smaller dark areas are "lava lakes," possibly composed of liquid sulfur.
(Inset B) This region on Io, known as Pele, shows the largest of its volcanos. The field of view is only several hundred km across.
(Inset C) At top left, a plume from an erupting volcano, known as Prometheus, is clearly seen at the limb of Io against the blackness of space.
(Inset D) The absence of impact craters on Io demonstrates the youth of its surface deposits. A variety of lava flows can be seen at several places on Io, like this one stretching for several hundred km from the large volcano Ra Patera.

*B*eyond the orbit of Mars, the solar system is very different from our own backyard. The outer solar system presents us with a totally unfamiliar environment—huge gas balls, peculiar moons, ringlike structures, and a wide variety of physical and chemical phenomena, many of which are still only poorly understood. Although the jovian planets—Jupiter, Saturn, Uranus, and Neptune—differ from one another in many ways, we will find that they have much in common, too. As with the terrestrial planets, we will learn from their differences as well as from their similarities. Our study of these alien places begins with the jovian planet closest to Earth—Jupiter, the largest planet in the solar system and a model for the other jovian worlds.

11.1 *Jupiter in Bulk*

ORBITAL PROPERTIES

❶ Named after the most powerful god of the Roman pantheon, Jupiter is by far the largest planet in the solar system. Ancient astronomers could not have known the planet's true size, but their choice of names was very apt.

Jupiter is the third-brightest object in the night sky (after the Moon and Venus), making it very easy to locate and study. Its sidereal orbital period is 11.9 Earth years; the orbital semi-major axis is 5.20 A.U. (778 million km), with an eccentricity of 0.05. As in the case of Mars, Jupiter is brightest when it is near opposition. Its closest approach to Earth is about 3.95 A.U., when opposition happens to occur near perihelion. Jupiter's angular size is greatest then—the planet can be up to 50″ across, and a lot of detail can be discerned through even a small telescope.

Figure 11.1 shows two of the best views of Jupiter ever obtained from Earth—one from the ground and one from space. Figure 11.1(a) shows a ground-based photograph of the planet, and Figure 11.1(b) shows a *Hubble Space Telescope* image taken during the opposition of December 1990. Notice the alternating light and dark bands that cross the planet parallel to its equator and the prominent dark spot at the lower right of Figure 11.1(a).

MASS AND RADIUS

Unlike any of the terrestrial planets, Jupiter has many moons, with a wide range of sizes and properties. The four largest are visible from Earth with a small telescope (or even with the naked eye). They are known as the **Galilean moons**, after Galileo Galilei, who first observed them early in the seventeenth century. Astronomers have been able to study the motion of the Galilean moons for quite some time. Consequently, Jupiter's mass has long been known. It is 1.9×10^{27} kg, or 318 Earth masses. Jupiter has more than twice the mass of all the other planets combined.

Jupiter is such a large planet that many celestial mechanicians—those researchers concerned with the motions of interacting cosmic objects—regard our solar system as containing only two important objects—the Sun and Jupiter. To be sure, in this age of sophisticated and precise spacecraft navigation, the gravitational influence of all the planets must be considered, but in the broadest sense, our solar system is a two-object system with a lot of debris. As massive as Jupiter is, though, it is still some 1000 times less massive than the Sun. This makes studies of Jupiter all the more important, for here we have an object intermediate in size between the Sun and the terrestrial planets.

Knowing Jupiter's distance and angular size, we can easily determine its radius. It is 71,400 km, or 11.2 Earth radii. More dramatically stated, more than 1400 Earths would be needed to equal the volume of Jupiter. From the size and mass, we derive an average density of 1300 kg/m³ (1.3 g/cm³) for the planet. Here (as if we needed it) is yet another indicator that Jupiter is radically different from the terrestrial worlds. It is clear that, whatever Jupiter's composition, it is not made up of the same material as the inner planets (recall from Chapter 7 that the Earth's average density is 5500 kg/m³). Studies of the planet's internal structure indicate that Jupiter must be composed primarily of hydrogen and helium. The enormous pressures in the planet's interior greatly compress these light gases, producing the average density we observe—very high for hydrogen, but still considerably lower than that of the terrestrial planets.

ROTATION RATE

As with other planets, we can attempt to determine Jupiter's rotation rate simply by timing a surface feature as it moves around the planet. However, in the case of Jupiter (and, indeed, all the gaseous outer planets), there is a catch—Jupiter has no solid surface. All we see are cloud features in the planet's atmosphere. Unlike Earth's atmosphere, different parts of Jupiter's atmosphere, with no solid surface to "tie them down," move independently. Visual observations and Doppler-shifted spectral lines prove that the equatorial zones rotate a little faster ($9^{h}50^{m}$ period) than the higher latitudes ($9^{h}56^{m}$ period). Thus, Jupiter exhibits **differential rotation**—the rotation rate is not constant from one location to another. Differential rotation is not possible in solid objects like the terrestrial planets, but it is normal for fluid bodies such as Jupiter.

Observations of Jupiter's magnetosphere provide a more meaningful measurement of the rotation period. The planet's magnetic field is strong and emits radiation at radio wavelengths. Careful studies show a periodicity of $9^{h}56^{m}$ in this radio emission. Scientists assume that this measure-

Figure 11.1   (a) Photograph of Jupiter made with a large ground-based telescope, showing several of its moons. (b) A *Hubble Space Telescope* image of Jupiter, in true color. Features as small as a few hundred kilometers across are resolved.

ment matches the rotation of the planet's interior, where the magnetic field arises. We see that the planet's interior rotates at the same rate as the clouds at its poles. The equatorial zones rotate more rapidly.

A rotation period of 9^h56^m is fast for such a large object. In fact, Jupiter has the fastest rotation rate of *any* planet in the solar system, and this rapid spin has al-

tered Jupiter's shape. As illustrated in Figure 11.2, a spinning object tends to develop a bulge around its midsection. The more loosely the object's matter is bound together, or the faster it spins, the larger the bulge becomes. In objects like Jupiter, which are made up of gas or loosely packed matter, high spin rates can produce a quite pronounced bulge. Jupiter's equatorial

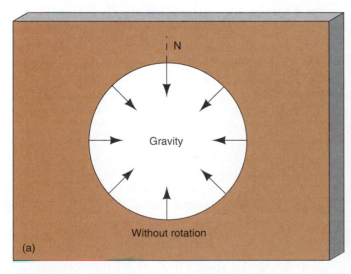

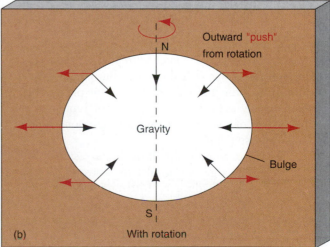

Figure 11.2   All spinning objects tend to develop an equatorial bulge because rotation causes matter to push outward against the inward-pulling gravity. The size of the bulge depends on the mechanical strength of the matter and the rate of rotation. The inward-pointing arrows denote gravity, the outward arrows the "push" due to rotation.

radius (71,400 km) exceeds its polar radius (66,800 km) by about 6.5 percent.*

But there is more to the story of Jupiter's shape. Jupiter's observed equatorial bulge also tells us something very important about the planet's deep interior. Careful calculations indicate that Jupiter would be *more* flattened than it actually is if its core were composed of hydrogen and helium alone. To account for the planet's observed shape, we must assume that Jupiter has a small, dense, probably rocky core, between 10 and 20 times the mass of the Earth. This is one of the few pieces of data we have on Jupiter's internal structure.

11.2   *The Atmosphere of Jupiter*

② Jupiter is visually dominated by two features, one a series of ever-changing atmospheric bands arranged parallel to the equator and the other an oval atmospheric blob called the **Great Red Spot**, or often just the "Red Spot." The cloud bands, seen clearly in Figure 11.1, display many colors—pale yellows, light blues, deep browns, drab tans, and vivid reds among others. Scientists believe that chemical compounds in Jupiter's atmosphere create these different colors, but the detailed chemistry is still not completely understood. The Red Spot (shown in more detail in Figure 11.3) is one of many features associated with Jupiter's weather. It seems to be an Earth-sized hurricane that has persisted for hundreds of years.

ATMOSPHERIC COMPOSITION

Spectroscopic studies of sunlight reflected from Jupiter gave scientists their first look at the planet's atmospheric composition. Radio, infrared, and ultraviolet observations later provided more details. The most abundant gas is molecular hydrogen (86.1 percent by number), followed by helium (13.8 percent). Together they make up over 99 percent of Jupiter's atmosphere. Scientists generally accept that these two gases also make up the bulk of the planet's interior. This belief is based not on direct evidence of the interior (until the recent collision of a comet with Jupiter, described in Chapter 14, there was virtually none), but largely on theoretical studies of the internal structure of the planet, such as we have already seen in the discussion of Jupiter's shape. Small amounts of atmospheric methane, ammonia, and water vapor are also found.

Unlike the gravitational pull of the terrestrial planets, the gravity of the larger jovian planets is strong enough to have retained even hydrogen. Little, if any, of Jupiter's original atmosphere has escaped. Because of the great abundance of hydrogen, all of the common elements other

Earth also bulges slightly at the equator because of rotation. However, our planet is much more rigid than Jupiter, and the effect is much smaller—the equatorial diameter is only about 40 km larger than the distance from pole to pole, a tiny difference compared with the Earth's full diameter of nearly 13,000 km. Relative to its overall dimensions, Earth is smoother and more spherical than a billiard ball.

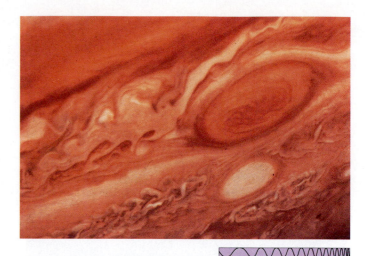

Figure 11.3 *Voyager 1* took this photograph of Jupiter's Red Spot (upper right) from a distance of about 100,000 kilometers. The resolution is about 100 km. Note the complex turbulence patterns to the left of both the Red Spot and the smaller white oval vortex below it. (For scale, planet Earth is about the size of the white oval.)

than helium (in particular, carbon, nitrogen, and oxygen) are chemically combined with it.

ATMOSPHERIC STRUCTURE AND COLOR

None of the atmospheric gases just listed can, by itself, account for Jupiter's observed coloration. For example, frozen ammonia and water vapor would simply produce white clouds, not the many colors actually seen. We now believe that Jupiter's clouds are arranged in several layers and are the product of complex and continuous chemical processes occurring in the planet's turbulent atmosphere. The various visible clouds lie at different levels—specifically, the white ammonia clouds generally overlie the more brightly colored layers, whose composition we will discuss in a moment. Above the clouds themselves there is a thin, faint layer of haze, created by chemical reactions similar to those that cause smog on the Earth. When we observe Jupiter's many colors, we are actually looking down to many different depths in the planet's atmosphere.

Figure 11.4 shows a diagram of Jupiter's atmosphere. The planet lacks a solid surface to use as a reference level for measuring altitude, so by convention scientists take the top of the troposphere to lie at 0 km. As on other planets, the weather on Jupiter is the result of convection in the troposphere, so the colored clouds, which are associated with planetary weather systems, all lie at negative altitudes in the diagram. The haze layer lies at the upper edge of Jupiter's troposphere, at an altitude of zero. The temperature at this level is about 110 K. Above the troposphere, as on Earth, the temperature rises as the atmosphere absorbs solar ultraviolet light. Below the haze layer, at a depth of about 40 km (shown as -40 km in Figure 11.4), lie white

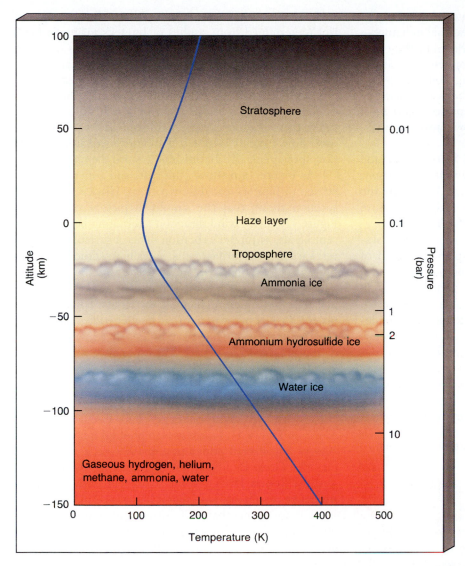

Figure 11.4 The vertical structure of Jupiter's atmosphere. Jupiter's clouds are arranged in three main layers, each with quite different colors and chemistry. The colors we see in photographs of the planet depend on the cloud cover. The white regions are the tops of the upper ammonia clouds. The yellows, reds, and browns are associated with the second cloud layer, which is composed of ammonium hydrosulfide ice. The lowest cloud layer is water ice and bluish in color. However, the overlying layers are sufficiently thick that this level is not seen in visible light.

wispy clouds made up of ammonia ice. At these cloud tops, the temperature is approximately 125–150 K. It increases quite rapidly with increasing depth.

A few tens of kilometers below the ammonia clouds, the temperature is a little warmer—over 200 K—and the clouds are probably made up mostly of droplets or crystals of ammonium hydrosulfide, produced by reactions between ammonia and hydrogen sulfide in the planet's atmosphere. However, instead of being white (the color of ammonium hydrosulfide on Earth), these clouds are tawny. This is the level at which atmospheric chemistry begins to play its part in determining Jupiter's appearance. Many planetary scientists believe that molecules containing the element sulfur, and perhaps even sulfur itself, play an important role in influencing the cloud colors—particularly the reds, browns, and yellows, all colors associated with sulfur or sulfur compounds. It is also possible that compounds containing the element phosphorus contribute to the coloration. At deeper levels in the atmosphere, the ammonium hydrosulfide clouds give way to clouds of water ice or water vapor. This

lowest cloud layer, which is not seen in visible-light images of Jupiter, lies some 80 km below the top of the troposphere.

Deciphering the detailed causes of Jupiter's distinctive colors is a difficult task. The cloud chemistry is complex, and it is very sensitive to small changes in atmospheric conditions, such as pressure and temperature, as well as to chemical composition. The atmosphere is in incessant, churning motion, causing these conditions to change from place to place and from hour to hour. In addition, the energy that powers the reactions comes in many different forms: the planet's own interior heat, solar ultraviolet radiation, aurorae in the planet's magnetosphere, and lightning discharges within the clouds themselves. All these factors combine to keep the complete explanation of Jupiter's appearance beyond our present grasp.

ATMOSPHERIC BANDS

Astronomers generally describe Jupiter's banded appearance—and, to a lesser extent, the appearance of the other jovian worlds as well—as a series of bright **zones** and dark **belts** crossing the planet. The zones and belts vary in both latitude and intensity during the year, but the general pattern remains. These variations are not seasonal in nature—Jupiter has no seasons—but instead appear to be the result of dynamic motion in the planet's atmosphere. The light-colored zones lie above upward-moving convective currents in Jupiter's atmosphere. The dark belts are caused by the other part of the convection cycle, representing regions where material is generally sinking downward, as illustrated schematically in Figure 11.5.

Because of the upwelling material below them, the zones are regions of high pressure; the belts, conversely, are low-pressure regions. Thus, the belts and zones are the planet's equivalents of the familiar high- and low-pressure systems that cause our weather on Earth. A major difference is that Jupiter's rapid rotation has caused these systems to wrap all the way around the planet, instead of forming localized circulating storms, as on our own world. Because of the pressure difference, the zones lie slightly higher in the atmosphere than the belts. The temperature difference between the two (recall that the temperature decreases with altitude), and the associated changes in chemical reactions, is the basic reason for their different colors.

Underlying the bands is an apparently very stable pattern of eastward and westward wind flow, often referred to as Jupiter's **zonal flow**. This zonal flow is evident in Figure 11.6, which shows the wind speed at different planetary lat-

itudes measured relative to the rotation of the planet's interior (determined from studies of Jupiter's magnetic field). As we have already seen, the equatorial regions of the atmosphere rotate faster than the planet, with an average flow speed of some 85 m/s, or about 300 km/h, in the easterly direction. The speed of this equatorial flow is quite similar to that of the jet stream on Earth. At higher latitudes, there are alternating regions of westward and eastward flow, roughly symmetric about the equator, with the flow speed generally diminishing toward the poles.

As Figure 11.6 shows, the belts and zones are closely related to Jupiter's zonal flow pattern. However, closer inspection shows that the simplified picture presented in Figure 11.5, with wind direction alternating between adjacent bands, as Jupiter's rotation deflects surface winds into eastward or westward streams, is really too crude to explain the actual flow. Scientists now believe that the interaction between convective motion in Jupiter's atmosphere and the planet's rapid rotation channels the largest eddies into the observed zonal pattern, but that smaller eddies cause irregularities in the flow. Near the poles, where the zonal flow disappears, the band structure vanishes also.

THE WEATHER ON JUPITER

In addition to the zonal flow pattern, Jupiter has many "small-scale" weather patterns. The Great Red Spot, shown in Figure 11.3—a close-up photograph taken as the *Voyager 1* spacecraft glided past in 1979—is a prime example. The Great Red Spot was first reported by the British scientist Robert Hooke in the midseventeenth century, so we can be reasonably sure that it has existed in one form or another for more than 300 years, and it might well be much older. *Voyager* observations show the spot to be a region of swirling, circulating winds, rather like a whirlpool or a terrestrial hurricane—a persistent and vast atmospheric storm. The size of the Spot varies, although it averages about twice the diameter of Earth. Its present dimensions are roughly 25,000 km by 15,000 km. It rotates around Jupiter at a rate similar to the planet's interior, perhaps suggesting that its roots lie far below the atmosphere.

The origin of the Spot's red color is unknown, as is its source of energy, although it is generally supposed that the Spot is somehow sustained by Jupiter's large-scale atmospheric motion. Repeated observations show that the gas flow around the Spot is counterclockwise, with a period of about six days. Turbulent eddies form and drift away from its edge. The Spot's center, however, remains quite tranquil in appearance, like the eye of a hurricane on Earth. The zonal motion north of the Spot is westward, while that to the south is eastward (see Figure 11.7), supporting the idea that the Spot is confined and powered by the zonal flow. However, the details of how this occurs are still a matter of conjecture. Computer simulations of the complex fluid dynamics of Jupiter's atmosphere are only now beginning to hint at answers.

Actually, storms, which as a rule are much smaller than the Red Spot, may be quite common on Jupiter.

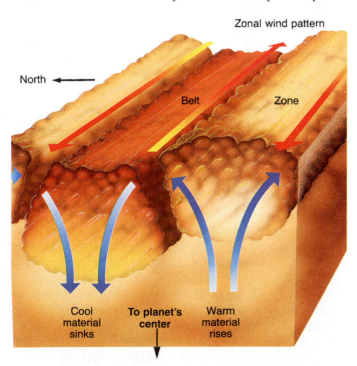

Figure 11.5   The colored bands in Jupiter's atmosphere are associated with vertical convective motion. Upwelling warm gas results in the lighter-colored zones; the darker bands lie atop lower-pressure regions where cooler gas is sinking back down into the atmosphere. As on Earth, surface winds tend to blow from high- to low-pressure regions. Jupiter's rotation channels these winds into an east–west flow pattern, as indicated.

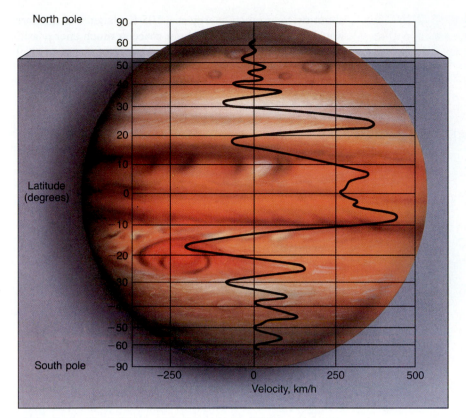

North pole

Latitude
(degrees)

South pole

Velocity, km/h

Figure 11.6 The wind speed in Jupiter's atmosphere, measured relative to the planet's internal rotation rate. The alternations in wind direction are associated with the atmospheric band structure.

Spacecraft photographs of the dark side of the planet reveal both auroral activity and bright flashes resembling lightning. The *Voyager* mission discovered many smaller light- and dark-colored spots that are also apparently circulating storm systems. Note the several **white ovals** in Figures 11.3 and 11.7, south of the Red Spot. Like the Red Spot, they rotate counterclockwise. Their high cloud tops give them their color. These particular white ovals are known to be at least 40 years old. Figure 11.8 shows a **brown oval**, a "hole" in the clouds that allows us to look down into the lower atmosphere. For unknown reasons, brown ovals appear only in latitudes around 20°N. Although not as long-lived as the Red Spot, these systems can persist for many years or even decades.

We cannot explain their formation, but we can offer at least a partial explanation for the longevity of these storm

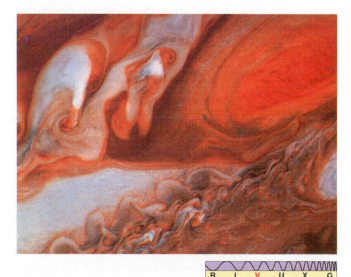

R   I   V   U   X   G

Figure 11.7 These *Voyager 2* close-up views of the Red Spot, taken four hours apart, show clearly the turbulent flow around its edges. The general direction of motion of the gas north of (above) the Spot is westward (to the left), while gas south of the Spot flows east. The Spot itself rotates counterclockwise, suggesting that it is being "rolled" between the two oppositely directed flows. The colors have been exaggerated somewhat to enhance the contrast.

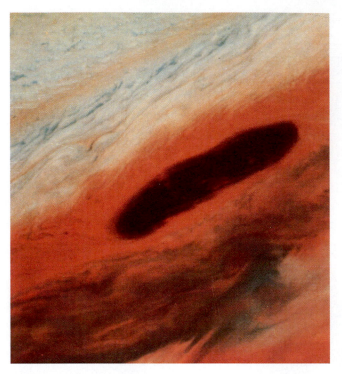

Figure 11.8   A brown oval in Jupiter's northern hemisphere. Its color comes from the fact that it is actually a break in the upper cloud layer, allowing us to see deeper in. The oval's length is approximately equal to the diameter of the Earth.

systems on Jupiter. On Earth a large storm, such as a hurricane, forms over the ocean and may survive for many days, but it dies quickly once it encounters land. The Earth's continental landmasses disrupt the flow patterns that sustain the storm. Jupiter has no continents, so once a storm is established and has reached a size at which other storm systems cannot destroy it, apparently little affects it. The larger the system, the longer its lifetime.

11.3   *Internal Structure*

3 On the basis of Jupiter's distance from the Sun, astronomers had expected to find that the temperature of the cloud tops was around 105 K. At that temperature, they reasoned, Jupiter would radiate back into space exactly the same amount of energy as it received from the Sun. When radio and infrared observations were first made of the planet, however, astronomers found that its Planck spectrum corresponded to a temperature of 125 K instead. Subsequent measurements, including those of *Voyager*, as we have just seen, have verified that finding. Although a difference of 20 K may seem small, recall from Chapter 5 that the energy emitted by a planet grows as the *fourth* power of the surface temperature (in Jupiter's case, the temperature of the cloud tops). A planet at 125 K radiates $(125/105)^4$, or

about twice as much energy as a 105 K planet. Put another way, Jupiter actually emits about twice as much energy as it receives from the Sun. Thus, unlike any of the terrestrial planets, Jupiter must have its own internal heat source.

What is responsible for Jupiter's extra energy? It is not the decay of radioactive elements within the planet—that must be occurring, but not at nearly the rate necessary to produce the temperature we record. Nor is it the process that generates energy in the Sun, nuclear fusion—the temperature in Jupiter's interior, high as it is, is far too low for that. (See *Interlude 11-1*.) Instead, astronomers theorize that the source of Jupiter's excess energy is the slow escape of gravitational energy released during the planet's formation. As the planet took shape, some of its energy was converted into heat in the interior. That heat is still slowly leaking out through the planet's heavy atmospheric blanket, resulting in the excess emission we observe. Despite the huge amounts of energy involved—Jupiter's energy emission is about 4×10^{17} watts more than it receives from the Sun—the energy loss is quite slight compared with the planet's total energy. A simple calculation indicates that the average temperature of the interior of Jupiter falls by only about a millionth of a kelvin per year.

Jupiter's clouds, with their complex chemistry, are probably less than 200 km thick. Below them, the temperature and pressure steadily increase, as the atmosphere becomes the "interior" of the planet. Much of our knowledge of Jupiter's interior comes from theoretical modeling. Planetary scientists use all available bulk data on the planet—mass, radius, composition, rotation, temperature, and so on—to construct a model of the interior that agrees with observations. Our statements about Jupiter's interior are, then, really statements about the model that best fits the facts. However, because the interior consists largely of hydrogen and helium—two simple gases whose physics we think we understand well—we can be fairly confident that Jupiter's internal structure is now understood.

Figure 11.9 shows that both the temperature and the density of Jupiter's atmosphere increase with depth below the cloud cover. However, no "surface" of any kind exists anywhere inside. Instead, Jupiter's atmosphere just becomes denser and denser, because of the pressure of the overlying layers. At a depth of a few thousand kilometers, the gas makes a gradual transition into the liquid state. By a depth of about 20,000 km, the pressure is about 3 million times greater than atmospheric pressure on Earth. Under those conditions, the hot liquid hydrogen is compressed so much that it undergoes another transition, this time to a "metallic" state, with properties in many ways similar to a liquid metal. Of particular importance for Jupiter's magnetic field, this metallic hydrogen is an excellent conductor of electricity.

As we have already mentioned, Jupiter's observed flattening requires that there be a small, dense core at its center, containing perhaps 15 times the mass of the Earth. The core's exact composition is unknown, but planetary scientists think that it contains heavier materials than the rest of the planet. Present best estimates indicate that it consists of "rocky" ma-

Interlude 11-1 *Almost a Star?*

Jupiter has a starlike composition—predominantly hydrogen and helium, with a trace of heavier elements. Did Jupiter ever come close to becoming a star itself? Might the solar system have formed as a double-star system? Probably not. Unlike a star, Jupiter is cold. Its central temperature is far too low to ignite the nuclear fires that power our Sun. Jupiter's mass would have to increase 80-fold before its central temperature rose to the point where nuclear reactions could begin, converting Jupiter into a small, dim star. Even so, it is interesting to note that although Jupiter's present-day energy output is very small (by solar standards, at least), it must have been much greater in the distant past, while the planet was still contracting rapidly toward its present size. For a brief period of time—a few hundred million years—Jupiter may actually have been as bright as a faint star, although its brightness never came

within a factor of 100 of the Sun's. Seen from the Earth at that time, Jupiter would have been about 100 times brighter than the Moon!

What might have happened had our solar system formed as a double-star system? Conceivably, had Jupiter been massive enough, its radiation might have produced severe temperature fluctuations on all the planets, perhaps to the point of making life on Earth impossible. Even if Jupiter's brightness were too low to cause us any problems, its gravitational field (which would be 1/12 that of the Sun if its mass were 80 times greater) might have made the establishment of stable, roughly circular planetary orbits a fairly improbable event. The size of the "Jupiter," or second largest body, in a new-born planetary system may be a very important factor in determining the likelihood of the appearance of life.

terials, similar to those found in the terrestrial worlds. In fact, it now appears that all of the jovian planets contain large rocky cores and that the formation of such a large "terrestrial" planetary core is a necessary stage in the process of building up a gas giant. Because of the high pressure at the

center of Jupiter—approximately 50 million times that on Earth's surface, or 10 times that at its center—the core must be compressed to quite high densities (perhaps twice the core density of the Earth). It is probably not much more than 20,000 km in diameter, and the central temperature may be as high as 40,000 K.

11.4 *Jupiter's Magnetosphere*

④ For decades, ground-based radio telescopes monitored radiation leaking from Jupiter's magnetosphere, but only when the *Pioneer* and *Voyager* spacecraft reconnoitered the planet in the mid-1970s did astronomers realize the full extent of the planet's magnetic field. Jupiter, as it turns out, is surrounded by a vast sea of energetic charged particles, mostly electrons and protons, somewhat similar to Earth's Van Allen belts but much larger. The radio radiation detected on Earth is emitted when these particles are accelerated to very high speeds—close to the speed of light—by Jupiter's powerful magnetic field. This radiation is several thousand times more intense than that produced by Earth's magnetic field, and represents a potentially serious hazard for manned and unmanned space vehicles alike. Sensitive electronic equipment (not to mention even more sensitive human bodies) would require special protective shielding to operate for long in this hostile environment.

Direct spacecraft measurements show Jupiter's magnetosphere to be almost 30 million km across, roughly a million times more voluminous than Earth's magnetosphere, and far larger than the entire Sun. As in the case of Earth, the size of Jupiter's magnetosphere is determined by the interaction between the planet's magnetic field and the solar wind. Outside, the solar wind particles flow freely away from the Sun, past the planet. Inside, their motions are governed by the planetary magnetic field. Jupiter's magnetosphere has a long tail extending away from the Sun at least as

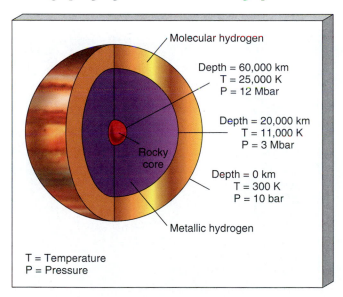

Figure 11.9 Jupiter's internal structure, as deduced from *Voyager* measurements and theoretical modeling. The outer radius represents the top of the cloud layers, some 70,000 km from the planet's center. The density and temperature increase with depth, and the atmosphere gradually liquefies over the outermost few thousand kilometers. Below a depth of 20,000 km, the hydrogen behaves like a liquid metal. At the center of the planet lies a large rocky core, somewhat terrestrial in composition but much larger than any of the inner planets. Although very uncertain, the temperature and pressure at the center are probably about 40,000 K and 50 Mbar, respectively. (One bar is the pressure of Earth's atmosphere at sea level. One Mbar is one million bars.)

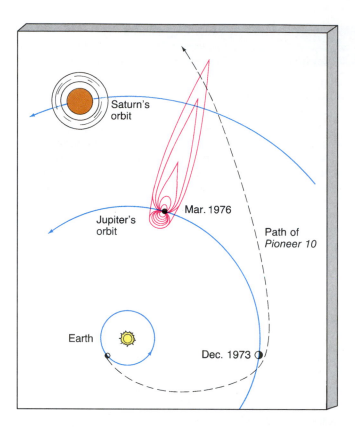

Figure 11.10 The *Pioneer 10* spacecraft did not detect any solar particles while moving behind Jupiter. Accordingly, as sketched here, Jupiter's magnetosphere apparently extends beyond the orbit of Saturn.

far as Saturn's orbit (over 4 A.U. farther out from the Sun), as sketched in Figure 11.10. On the sunward side, the magnetopause—the boundary of Jupiter's magnetic influence on the solar wind—lies about 3 million km from the planet. The outer magnetosphere appears to be quite unstable, sometimes deflating in response to "gusts" in the solar wind, then growing back. In the inner magnetosphere, Jupiter's rapid rotation has forced most of the charged particles into a flat *current sheet*, lying on the planet's magnetic equator. The portion of the magnetosphere close to Jupiter is sketched in Figure 11.11. Notice that the planet's magnetic axis is not exactly aligned with its rotation axis, but is inclined to it at an angle of approximately 10°.

Ground-based and spaceborne observations of the radiation emitted from Jupiter's magnetosphere imply that the *intrinsic* strength of the planet's magnetic field is nearly 20,000 times greater than Earth's. The existence of such a strong field further supports our theoretical model of the interior; the conducting liquid interior that is thought to

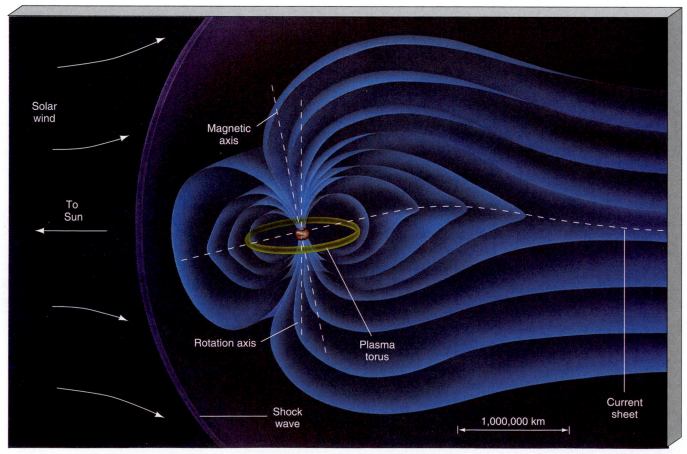

Figure 11.11 Jupiter's inner magnetosphere is characterized by a flat current sheet, consisting of charged particles squeezed into the magnetic equatorial plane by the planet's rapid rotation. The plasma torus is a ring of charged particles associated with the moon Io; it is discussed on p.246.

make up most of the planet should combine with Jupiter's rapid rotation to produce a large dynamo effect and a strong magnetic field, just as observed.

11.5 The Moons of Jupiter

5 Jupiter has at least 16 moons. In many ways, the entire Jupiter system resembles a miniature solar system. Its four largest moons—the **Galilean satellites**—are each comparable in size to Earth's Moon. Moving outward from Jupiter, the four are named Io, Europa, Ganymede, and Callisto, after the mythical attendants of the Roman god Jupiter. They move in nearly circular orbits about their parent planet. When the *Voyager 1* spacecraft passed close to the Galilean moons in 1979, it sent some remarkably detailed photographs back to Earth, allowing planetary scientists to discern fine surface detail on each moon and greatly expanding our knowledge of these small, distant worlds. We will consider the Galilean satellites in more detail in a moment.

Within the orbit of Io lie four small satellites, all but one discovered by *Voyager* cameras. The largest of the four, Amalthea, is less than 300 km across and is irregularly shaped. E. E. Barnard discovered it in 1892. It orbits at a distance of 181,000 km from Jupiter's center—only 110,000 km above the cloud tops. Its rotation, like that of most of Jupiter's satellites, is synchronous with its orbit because of Jupiter's strong tidal field. Amalthea rotates once per orbit period, every 11.7 hours.

Beyond the Galilean moons lie eight more small satellites, all discovered in the twentieth century, but before the *Voyager* missions. They fall into two groups of four moons each. The moons in the inner group move in eccentric, inclined orbits, about 11 million km from the planet. The outer four moons lie about 22 million km from Jupiter. Their orbits too are fairly eccentric, but *retrograde*, moving in a sense opposite to all the other moons (and to Jupiter's rotation). It is very likely that each group represents a single body that was captured by Jupiter's strong gravitational field long after the planet and its larger moons originally formed. Both bodies subsequently broke up, either during or after the capture process, resulting in the two families of similar orbits we see today. The masses, and hence the densities, of these small worlds are unknown. However, their appearance and sizes suggest compositions more like asteroids than their larger Galilean companions. Table 11-1 presents the general properties of Jupiter's moons.

THE GALILEAN MOONS AS A MODEL OF THE INNER SOLAR SYSTEM

If we think of Jupiter's moon system as a scaled-down solar system, the Galilean moons correspond to the terrestrial planets. Their orbits are direct (that is, in the same sense as Jupiter's rotation), roughly circular, and lie close to Jupiter's equatorial plane. They range in size from slightly smaller than Earth's Moon (Europa) to slightly larger than Mercury (Ganymede). The parallel with the inner solar system contin-

TABLE 11-1 *The Moons of Jupiter*

NAME	DISTANCE FROM JUPITER (km)	(planet radii)	ORBIT PERIOD (days)	SIZE (longest diameter, km)	MASS (Earth Moon masses)	DENSITY (kg/m³)	(g/cm³)
Metis	128,000	1.79	0.29	40			
Adastea	129,000	1.81	0.30	20			
Amalthea	181,000	2.54	0.50	200			
Thebe	222,000	3.11	0.67	90			
Io	422,000	5.91	1.77	3630	1.22	3600	3.6
Europa	671,000	9.40	3.55	3140	0.65	3000	3.0
Ganymede	1,070,000	15.0	7.16	5260	2.02	1900	1.9
Callisto	1,880,000	26.4	16.7	4800	1.47	1900	1.9
Leda	11,100,000	155	239	15			
Himalia	11,500,000	161	251	180			
Lysithea	11,700,000	164	259	40			
Elara	11,700,000	164	260	80			
Ananke	21,200,000	297	631*	30			
Carme	22,600,000	317	692*	40			
Pasiphae	23,500,000	329	735*	70			
Sinope	23,700,000	332	758*	40			

*Indicates a retrograde orbit.

Figure 11.12 The *Voyager 1* spacecraft photographed each of the four Galilean moons of Jupiter. Shown here to scale, as they would appear from a distance of about 1 million km, they are, clockwise from upper left, Io, Europa, Callisto, and Ganymede.

ues with the realization that their densities decrease with increasing distance from Jupiter. It is quite likely that the inner two Galilean moons, Io and Europa, have a rocky composition, possibly similar to the crusts of the terrestrial planets.

Figure 11.13 *Voyager 1* took this photo of Jupiter with ruddy Io on the left and pearl-like Europa toward the right. Note the scale of objects here: Io and Europa are each comparable in size to our Moon, and the Red Spot (seen here to the left bottom) is roughly twice as big as Earth. (See also the chapter-opening image on p.232.)

The two outer Galilean moons, Ganymede and Callisto, are clearly deficient in rocky materials. Lighter materials, such as water ice, may account for as much as half of their total mass. Figure 11.12 compares the appearances and sizes of the four Galilean satellites. Figure 11.13 shows two of the moons photographed against the background of their parent planet.

Many astronomers think that the formation of Jupiter and the Galilean satellites may in fact have mimicked on a small scale the formation of the Sun and the inner planets. For that reason, studies of the Galilean moon system may provide us with valuable insight into the processes that created our own world. We will return to this parallel in Chapter 15. But let us point out that not all of the properties of the Galilean moons find analogs in the inner solar system. For example, because of Jupiter's tidal effect, all four Galilean satellites are in states of synchronous rotation, so that they all keep one face permanently pointing toward their parent planet. By contrast, of the terrestrial planets, only Mercury is strongly influenced by the Sun's tidal force, and even its orbit is not synchronous. And, of course, the Jupiter system has no analogs of the jovian planets. Finally, inspection of Table 11-1 shows a remarkable coincidence in the orbit periods of the three inner Galilean moons: Their periods are almost exactly in the ratio 1:2:4—a kind of "Bode's law" for Jupiter. This is most probably the result of a complex, but poorly understood, three-body resonance in the Galilean moon system, something not found among the terrestrial worlds.

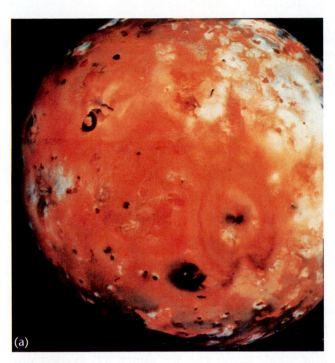

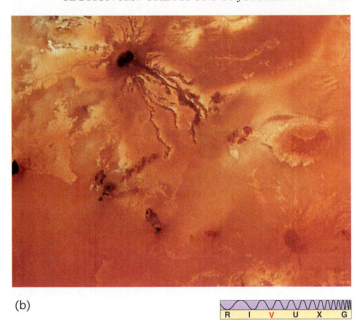

(b)

(a)

Figure 11.14   Jupiter's innermost moon, Io, is quite different in character from the other three Galilean satellites. Its surface is kept smooth and brightly colored by the moon's constant volcanism. The resolution of the photograph in (a) is about 7 km. In the more detailed image (b), features as small as 2 km across can be seen.

IO

6 Io, the densest of the Galilean moons, is the most geologically active object in the entire solar system. Its mass and radius are fairly similar to those of Earth's own Moon, but there the resemblance ends. Shown in Figure 11.14, Io's surface is a collage of reds, yellows, and blackish browns—resembling a giant pizza in the minds of some startled *Voyager* scientists. As the space-craft glided past Io, an outstanding discovery was made: Io has active volcanoes! *Voyager 1* photographed eight erupting volcanoes, and six were still erupting when *Voyager 2* passed by 4 months later. In Figure 11.15, one volcano is seen ejecting matter to an altitude of over 200 km. The gases are spewed forth at speeds up to 2 km/s, quite unlike the (relatively) sluggish ooze that emanates from the Earth's insides.

Figure 11.15   One of Io's volcanoes was caught in the act of erupting while the *Voyager* spacecraft flew past this fascinating moon. Surface features here are resolved to within a few kilometers. In (a), the volcano's umbrellalike profile shows clearly against the darkness of space. The plume measures about 100 km high and 300 km across. In (b), several jets of volcanic ejecta (dark against Io's brighter surface) can be discerned as *Voyager* prepares to "overfly" another volcano.

(a)

(b)

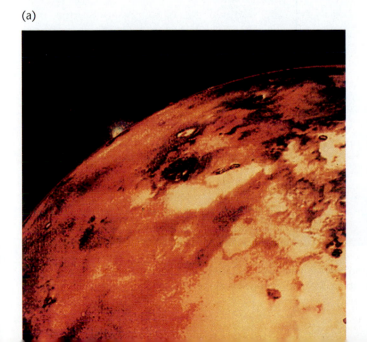

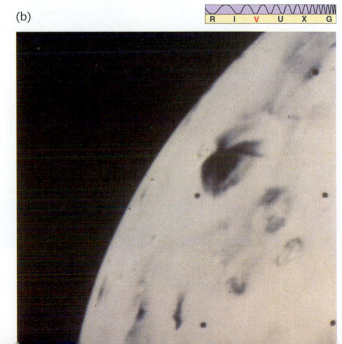

The orange color immediately surrounding the volcano most likely results from sulfur compounds in the ejected material. In stark contrast to the other Galilean moons, Io's surface is neither cratered nor streaked. (The circular features visible in Figures 11.14 and 11.15 are volcanoes.) Its surface is exceptionally smooth, apparently the result of molten matter constantly filling in any "dents and cracks." Accordingly, we can conclude that this remarkable moon has the youngest surface of any known object in the solar system. Of further significance, Io also has a thin, temporary atmosphere made up primarily of sulfur dioxide, presumably the result of gases ejected by volcanic activity.

Io's volcanism has a major effect on Jupiter's magnetosphere. All of the Galilean moons orbit within the magnetosphere and play some part in modifying its properties, but Io's influence is particularly marked. Although many of the charged particles in Jupiter's magnetosphere come from the solar wind, there is strong evidence that Io's volcanism is the primary source of heavy ions in the inner regions. Jupiter's magnetic field continually sweeps past Io, gathering up the particles its volcanoes spew into space and accelerating them to high speed. The result is the *Io plasma torus* (Figure 11.16; see also Figure 11.11), a doughnut-shaped region of energetic heavy ions that follows Io's orbital track, completely encircling Jupiter. (A plasma is a gas that has been heated to such high temperatures that all of its atoms are ionized.) It is quite easily detectable from Earth, but before *Voyager* its origin was unclear. Spectroscopic analysis shows that sulfur is indeed one of the torus's major constituents, strongly implicating Io's volcanoes as its source. As a hazard to spacecraft—manned or unmanned—the plasma torus is formidable. The radiation levels there are lethal.

What causes such astounding volcanic activity on Io? Surely that moon is too small to have geological activity like the Earth. Io should be long dead, like our own Moon. At one time, some scientists suggested that Jupiter's magnetosphere might be the culprit—perhaps the (then-unknown) processes creating the plasma torus were somehow also stressing the moon. We now know that this is not the case. The real source of Io's energy is gravity—Jupiter's gravity. Io orbits very close to Jupiter—only 422,000 km, or 5.9 Jupiter radii, from the center of the planet. As a result, Jupiter's huge gravitational field produces strong tidal forces on the moon. If Io were the only satellite in the Jupiter system, it would long ago have come into a state of synchronous rotation with the planet, just like our own Moon, for the reasons discussed in Chapter 8. In that case, Io would move in a perfectly circular orbit, with one face permanently turned toward Jupiter. The tidal bulge would be stationary with respect to the moon, and there would be no internal stresses and hence no volcanism.

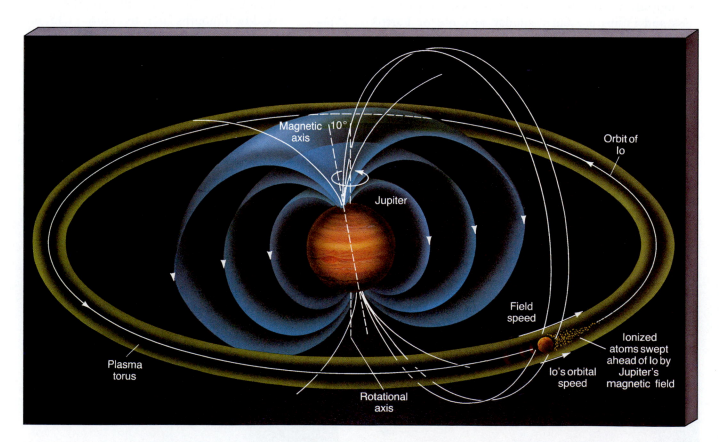

Figure 11.16 The Io plasma torus is the result of material being ejected from Io's volcanoes and swept up by Jupiter's rapidly rotating magnetic field. Spectroscopic analysis indicates that the torus is composed primarily of sodium and sulfur atoms.

But Io is not alone. As it orbits, it is constantly tugged by the gravity of its nearest large neighbor, Europa. These tugs are small and not enough to cause any great tidal effect in and of themselves, but they are sufficient to make Io's orbit slightly noncircular, preventing the moon from settling into a precisely synchronous state. The reason for this effect is exactly the same as in the case of Mercury, as discussed in Chapter 8. ∞ (p. 179) In a noncircular orbit, the moon's speed varies from place to place as it revolves around its planet, but its rate of rotation on its axis remains constant. Thus it cannot keep one face always turned toward Jupiter. Instead, as seen from Jupiter, Io rocks or "wobbles" slightly from side to side as it moves. The large tidal bulge, however, always points directly toward Jupiter, so it moves back and forth across Io's surface as the moon wobbles. These conflicting forces result in enormous tidal stresses that continually flex and squeeze Io's interior.

Just as repeated back-and-forth bending of a piece of wire can produce heat through friction, Io is constantly energized by the ever-changing distortion of its interior. This generation of large amounts of heat within Io ultimately causes huge jets of gas and molten rock to squirt out of the surface. It is likely that much of Io's interior is soft or molten, with only a relatively thin solid crust overlying it. In fact, Io's volcanoes are probably more like geysers on the Earth, but the term volcano has stuck. Researchers estimate that the total amount of heat generated in Io as a result of tidal flexing is about 100 million megawatts. This phenomenon makes Io one of the most fascinating objects in our solar system.

EUROPA

Europa (Figure 11.17) is a very different world from Io. Lying outside Io's orbit, 670,000 km (9.4 Jupiter radii) from Jupiter, it has relatively few craters on its surface, suggesting geologic youth. Recent activity must have erased the scars of ancient meteoritic impacts. Europa's surface does display a vast network of lines crisscrossing bright, clear fields of water ice. Some of these linear "bands," or fractures, appear to extend halfway around the satellite and resemble in some ways the pressure ridges that develop in ice floes on the Earth's polar oceans.

Some researchers have theorized that Europa is covered completely by an ocean of liquid water whose top is frozen at the low temperatures that prevail so far from the Sun. The cracks are attributed to the tidal influence of Jupiter and the gravitational pulls of the other Galilean satellites, although these forces are weaker than those powering Io's volcanic activity. Other planetary scientists suggest that Europa's fractured surface is instead related to some form of tectonic activity, one involving ice rather than rock. If the markings truly are fault lines of ice, then this moon is probably still quite active. If Europa does have a liquid ocean below the ice, it opens up many interesting avenues of speculation into the possible development of life there.

(a)

(b)   (c)

Figure 11.17 The second Galilean moon is Europa. Its icy surface is only lightly cratered, indicating that some ongoing process must be obliterating impact craters soon after they are formed. The origin of the cracks criss-crossing the surface is uncertain. The resolution of the *Voyager* mosaic in (a) is about 5 km. The two images below it (b and c) display even finer detail.

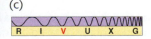

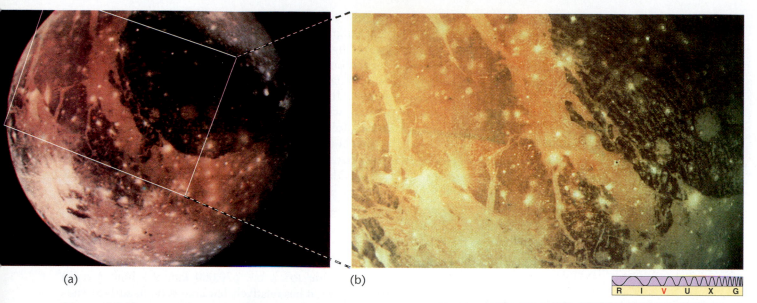

(a)　　　　　　　　　　　　　　　(b)

R I V U X G

Figure 11.18 Jupiter's largest moon, Ganymede, is also the largest satellite in the solar system. The dark regions on the surface are the oldest and probably represent the original icy crust of the moon. The largest dark region visible in the *Voyager 2* image in (a) is called Galileo Regio. It spans some 320 km. The lighter, younger regions are the result of flooding and freezing that occurred within a billion years or so of Ganymede's formation. The light-colored spots are recent impact craters. The resolution of the detailed image in (b) is about 3 km.

GANYMEDE AND CALLISTO

The two outermost Galilean moons are Ganymede (at 1.1 million km, or 15 planetary radii, from the center of Jupiter) and Callisto (at 1.9 million km, or 26 Jupiter radii). Their densities are each only about 2000 kg/m³, suggesting that they harbor substantial amounts of ice throughout and are not just covered by thin icy or snowy surfaces. Ganymede, shown in Figure 11.18, is the largest moon in the solar system, exceeding not only Earth's Moon but also the planets Mercury and Pluto in size. It has many impact craters on its surface and patterns of dark and light markings that are reminiscent of the highlands and maria on Earth's own Moon. In fact, Ganymede's history has many parallels with that of the Moon (with water ice replacing lunar rock). The large, dark region clearly visible in Figure 11.18 is called Galileo Regio.

As with the inner planets, we can estimate ages on Ganymede by counting craters. We learn that the darker regions, like Galileo Regio, are the oldest parts of Ganymede's surface. These regions are the original icy surface of the moon, just as the ancient highlands on our own Moon are

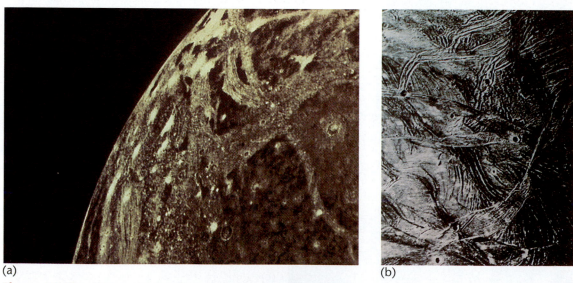

(a)　　　　　　　　　　　　　　　(b)

R I V U X G

Figure 11.19 "Grooved terrain" on Ganymede may have been caused by a process similar to plate tectonics on Earth. The resolution of the detailed image (b) is about 3 km.

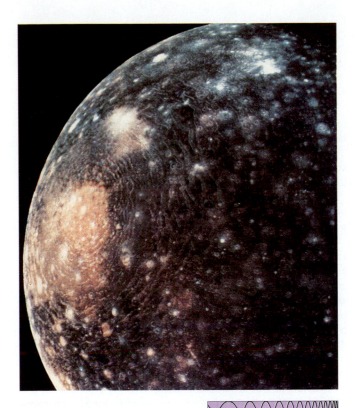

R   I   V   U   X   G

Figure 11.20   Callisto, the outermost Galilean moon of
Jupiter, is similar to Ganymede in composition but is more
heavily cratered. The large series of concentric ridges visible
on the left of the image is known as Valhalla. Extending nearly
1500 km from the basin center, they formed when "ripples"
from a large meteoritic impact froze before they could
disperse completely. The resolution here is around 10 km.

its original crust. The surface darkens with age as micromete-
orite dust slowly covers it. The light-colored parts of
Ganymede are much less heavily cratered, so they must be
younger. They are Ganymede's "maria" and probably formed
in a manner similar to the maria on the Moon. Intense meteoritic bombardment caused liquid water—Ganymede's counterpart to our own Moon's molten lava—to upwell from the
interior, flooding the impacting regions before solidifying.

Not all of Ganymede's surface features follow the lunar
analogy. Ganymede has a system of grooves and ridges
(shown in Figure 11.19) that may have resulted from crustal
tectonic motion, much as the Earth's surface undergoes
mountain building and faulting at plate boundaries.
Ganymede's large size indicates that its original radioactivity
probably helped to heat and differentiate its partly rocky interior, after which the moon cooled and the crust cracked.
Ganymede seems to have had some early plate tectonic activity, but the process stopped about 3 billion years ago when
the cooling crust became too thick.

Callisto, shown in Figure 11.20, is in many ways similar in appearance to Ganymede, although it has more
craters and fewer fault lines. Its most obvious feature is a
huge series of concentric ridges surrounding each of two

large basins. The larger of the two, on Callisto's Jupiter-
facing side, is named Valhalla and measures some 3000 km
across. It is clearly visible in Figure 11.20. The ridges resemble the ripples made as a stone hits water, but on
Callisto, they probably resulted from a cataclysmic impact
with an asteroid or comet. The upthrust ice was partially
melted, but it resolidified quickly, before the ripples had a
chance to subside. Today, both the ridges and the rest of
the crust are frigid ice and show no obvious signs of geological activity (such as the grooved terrain on Ganymede).
Apparently, Callisto froze before plate tectonic or other activity could start. The density of impact craters on the
Valhalla basin indicates that it formed long ago, perhaps 4
billion years in the past.

11.6 *Jupiter's Ring*

Yet another remarkable finding of the 1979 *Voyager* missions
was the discovery of a faint ring of matter encircling Jupiter in
the plane of the planet's equator (see Figure 11.21). This ring
lies roughly 50,000 km above the top cloud layer of the planet,
inside the orbit of the innermost moon. A thin sheet of material may extend all the way down to Jupiter's cloud tops, but
most of the ring is confined within a region only a few thousand kilometers across. The outer edge of the ring is quite
sharply defined. In the direction perpendicular to the equatorial plane, the ring is only a few tens of kilometers thick. The
small, dark particles that make up the ring may have been
chipped off by meteorite impacts on two small moons—Metis
and Adastrea, discovered by *Voyager*—that lie very close to the
ring itself. Despite its different appearance and structure,
Jupiter's ring can perhaps be best understood by studying the
most famous ringed planet—Saturn—so we will postpone further discussion of ring properties until the next chapter.

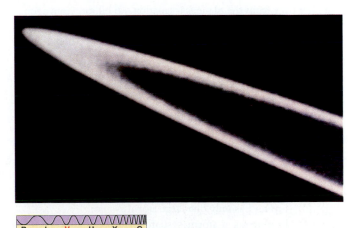

R   I   V   U   X   G

Figure 11.21   Jupiter's faint ring, as photographed (nearly
edge-on) by *Voyager 2*. The ring, made up of dark fragments of
rock and dust, possibly chipped off the innermost moons by
meteorites, was unknown before the two *Voyager* spacecraft
arrived at the planet. It lies in Jupiter's equatorial plane, only
50,000 km above the cloud tops.

Chapter Review

SUMMARY

Jupiter is the largest planet in the solar system. Its mass is more than twice the mass of all the other planets combined, although it is still about 1000 times less massive than the Sun. It is composed primarily of hydrogen and helium gas.

Jupiter rotates very rapidly, producing a pronounced equatorial flattening. The amount of flattening allows astronomers to infer the presence of a large rocky cores in its interior. Jupiter displays **differential rotation** (p. 234)—having no solid surface, the rotation rate varies from place to place in the atmosphere. Measurements of radio emission from Jupiter's magnetospheres provide a measure of the planet's interior rotation rate.

Jupiter's atmosphere consists of three main cloud layers. The colors we see are the result of chemical reactions, fueled by the planet's interior heat, solar ultraviolet radiation, auroral phenomena, and lightning, at varying depths below the cloud tops, seen through "holes" in the overlying clouds.

The cloud layers on all the jovian worlds are arranged into bands of bright **zones** (p. 237) and darker **belts** (p. 237) crossing the planets parallel to the equator. The bands are the result of convection in the planets' interiors and the planets' rapid rotation. The lighter zones are the tops of upwelling, warm currents, and the darker bands are cooler regions where gas is sinking. Underlying them is a stable pattern of eastward or westward wind flow called the **zonal flow** (p. 238). The wind direction alternates as we move north or south away from the equator.

The main weather pattern on Jupiter is the **Great Red Spot** (p. 236), an Earth-sized hurricane that has been raging for at least three centuries. Other, smaller, weather systems—the **white** (p. 239) and **brown ovals** (p. 239)—are also observed. They can persist for decades.

Jupiter radiates about twice as much energy into space as it receives from the Sun. The source of this energy is most likely heat left over from the planet's formation 4.6 billion years ago.

Jupiter's atmosphere becomes hotter and denser with depth, eventually becoming liquid. Interior pressures are so high that the hydrogen is "metallic" in nature near the center. The planet has a large "terrestrial" core 10–20 times the mass of the Earth.

The magnetosphere of Jupiter is about a million times more voluminous than Earth's magnetosphere, and the planet has a long magnetic "tail" extending away from the Sun to at least the distance of Saturn's orbit. Energetic particles spiral around magnetic field lines, accelerated by Jupiter's rotating magnetic field, producing intense radio radiation.

Jupiter and its system of moons resemble a small solar system. Sixteen moons have been discovered so far. The outermost eight moons resemble asteroids and have retrograde orbits, suggesting that they may have been captured by Jupiter's gravity long after the planets and largest moons formed. Jupiter's four major moons are called the **Galilean satellites** (p. 242), after their discoverer, Galileo Galilei. Their densities decrease with increasing distance from the planet.

The innermost Galilean moon, Io, has active volcanoes powered by the constant flexing of the moon by Jupiter's tidal forces. As Io orbits Jupiter, it "wobbles" because of the gravitational pull of Europa. The ever-changing distortion of its interior energizes this moon, and geyserlike volcanoes keep its surface smooth with constant eruptions. The material ejected by these volcanoes forms the Io plasma torus in Jupiter's inner magnetosphere.

Europa has a cracked, icy surface that may possibly conceal an ocean of liquid water. Its fields of ice are nearly devoid of craters, but have extensive fractures most likely due to the tidal influence of Jupiter and the gravitational effects of the other Galilean satellites. Ganymede and Callisto have ancient, heavily cratered surfaces. Ganymede, the largest moon in the solar system, shows some evidence for past geological activity, but it is now unmoving rock and ice. Callisto apparently froze before tectonic activity could start there.

Jupiter has a faint, dark ring extending down to the planet's cloud tops. It was discovered in 1979 by *Voyager 1*.

SELF-TEST: True or False?

_____ **1.** Jupiter has over 300 times the mass of Earth, and twice the mass of all the other planets combined.

_____ **2.** The solid surface of Jupiter lies just below the cloud layers visible from Earth.

_____ **3.** There is no evidence to suggest that Jupiter has a rocky core.

_____ **4.** Jupiter has only one large storm system.

_____ **5.** In general, a storm system in Jupiter's atmosphere is much longer-lived than storms in Earth's atmosphere.

_____ **6.** The element helium plays an important role in producing the colors in Jupiter's atmosphere.

_____ **7.** The magnetosphere of Jupiter is similar in intensity to Earth's magnetosphere.

_____ **8.** Most of Jupiter's moons rotate synchronously with their orbits.

_____ **9.** The densities of the Galilean moons increase with increasing distance from Jupiter.

_____ **10.** Io has a noticeable lack of impact craters on its surface.

_____ **11.** The surface of Europa is completely covered by water ice.

_____ **12.** Ganymede shows evidence for ancient plate tectonics.

_____ **13.** Jupiter's ring is made up of icy particles a few meters across.

_____ **14.** Most of the small moons of Jupiter have diameters of several hundred kilometers.

_____ **15.** The thickness of Jupiter's cloud layer is less than 1 percent of the planet's radius.

SELF-TEST: Fill in the Blank

1. The _____ of Jupiter indicates that its overall composition differs greatly from that of the terrestrial planets.
2. The main constituents of Jupiter are _____ and _____.
3. Jupiter's rapid _____ produces a significant equatorial bulge.
4. Jupiter's Great Red Spot has similarities to _____ on Earth.
5. The diameter of the Great Red Spot is about _____ that of the Earth .
6. Jupiter's clouds consist of a series of bright _____ and dark _____.
7. Jupiter emits about _____ more radiation than it receives from the Sun.
8. Although often referred to as a gaseous planet, Jupiter is mostly _____ in its interior.
9. Jupiter's magnetic field is generated by its rapid rotation and the element _____, which becomes metallic in the interior of the planet.
10. The Galilean moons make up _____ of the _____ moons of Jupiter. (Give the numbers.)
11. The Galilean moon _____ is larger than the planet Mercury.
12. In contrast to the inner Galilean moons, the outer two Galilean moons have compositions that include significant amounts of _____.
13. Io is the only moon in the solar system with active _____.

REVIEW AND DISCUSSION

1. In what sense does our solar system consist of only two important objects?
2. What is differential rotation, and how is it observed on Jupiter?
3. Describe some of the ways in which the *Voyager* mission changed our perception of Jupiter.
4. What is the Great Red Spot? What is known about the source of its energy?
5. What is the cause of the colors in Jupiter's atmosphere?
6. Briefly describe the weather on Jupiter.
7. Why has Jupiter retained most of its original atmosphere?
8. Explain the theory that accounts for Jupiter's internal heat source.
9. What is Jupiter thought to be like beneath its clouds? Why do we think this?
10. What is responsible for Jupiter's enormous magnetic field?
11. Why might we say that Jupiter was nearly a star?
12. In what sense are Jupiter and its moons like a miniature solar system?
13. How does the density of the Galilean moons vary with increasing distance from Jupiter? Is there a trend to this variation? If so, why?
14. What is the source of Io's volcanic activity?
15. How does the amount of cratering vary among the Galilean moons? Does it depend on their location? If so, why?
16. Why is there speculation that the Galilean moon Europa might be an abode for life?
17. Water is relatively uncommon among the terrestrial planets. Is it common among the moons of Jupiter?

PROBLEMS

1. How does the force of gravity at Jupiter's cloudtops compare with that at Earth's surface?
2. Calculate the ratio of Jupiter's mass to the total mass of the Galilean moons. Compare this with the ratio of Earth's mass to that of the Moon.
3. Io orbits Jupiter at a distance of six planetary radii in 42 hours. At what distance would a satellite orbit Jupiter in the time taken for Jupiter to rotate exactly once (10 hours, say), so that the satellite would appear "stationary" above the planet? (Use Kepler's Third Law; see Chapter 2.)
4. Compare the apparent sizes of the Galilean moons, as seen from Jupiter's cloudtops, with the angular diameter of the Sun at Jupiter's distance. Would you expect ever to see a total solar eclipse from Jupiter's cloudtops?

PROJECTS

1. Are there any stars in the night sky that look as bright as Jupiter? What other difference do you notice between Jupiter and the stars?
2. Use binoculars to peer at Jupiter. Be sure to hold them steadily (try propping your arms up on the hood of a car, or sitting down and bracing them against your knees). Can you see any of Jupiter's four largest moons? If you come back the following evening, the moons' relative positions will have changed. Have some changed more than others?
3. Through a telescope, you should be able to see the red-and-tan cloud bands of Jupiter, and you can clearly see some moons. Do the moons orbit equatorially? Before observing, look up the positions of the Galilean moons in a current magazine such as *Astronomy* or *Sky & Telescope*. Identify each of the moons. Watch Io over a period of at least an hour or more. Can you see its motion? Do the same for Europa.

12

SATURN

Spectacular Rings and Mysterious Moons

LEARNING GOALS

Studying this chapter will enable you to:

1 Summarize the orbital and physical properties of Saturn, and compare them with those of Jupiter.

2 Describe the composition and structure of Saturn's atmosphere and interior.

3 Explain why Saturn's internal heat source and magnetosphere differ from those of Jupiter.

4 Describe the structure and composition of Saturn's rings.

5 Define the Roche limit, and explain its relevance to the origin of Saturn's rings.

6 Summarize the general characteristics of Titan, and discuss the chemical processes in its atmosphere.

7 Discuss some of the orbital and geological properties of Saturn's smaller moons.

(Opposite page, background) Through the cameras onboard the *Hubble Space Telescope*, the planet Saturn appears much as our naked eyes would see it if it were only twice as far away as the Moon. Resolution here is 670 km, good enough to see clearly the band structure on the ball of the planet, as well as several gaps in the rings. True-color images like this one are made by combining separate images taken in red, green, and blue light, as shown in some of the smaller inserts.

(Inset A) Saturn as seen through a red filter, centered at a wavelength of 718 nm.

(Inset B) Saturn as seen through a green filter, centered at a wavelength of 547 nm.

(Inset C) Saturn as seen through a blue filter, centered at a wavelength of 439 nm.

(Inset D) Sometimes odd combinations of photos can provide unique insight. This false-color image of Saturn sums individual images taken at blue and infrared wavelengths. The resulting "psychedelic cloudscape" enables studies of the vertical growth of the planets' white cirrus–cloudlike features.

To many people, Saturn is the most beautiful and enchanting of all astronomical objects. Its rings are a breathtaking sight when viewed through even a small telescope, and they are probably the planet's best-known feature. Aside from its famous rings, however, Saturn presents us with another good example of a giant gaseous planet. Saturn is in many ways similar to its larger neighbor, Jupiter, in terms of composition, size, and structure. Yet when we study the two planets in detail, we find that there are important differences as well. A comparison between Saturn and Jupiter provides us with valuable insight into the structure and evolution of all the jovian worlds.

12.1 Saturn in Bulk

PHYSICAL PROPERTIES

❶ Saturn was the outermost planet known to ancient astronomers. Named after the father of Jupiter in Greek and Roman mythology, Saturn orbits the Sun at almost twice the distance of Jupiter, with an orbital semi-major axis of 9.54 A.U. (1430 million km) and an orbital eccentricity of 0.06. The planet's sidereal orbital period of 29.5 Earth years was the longest natural unit of time known to the ancient world. At opposition, when Saturn is at its brightest, it can lie within 8 A.U. of the Earth. However, its great distance from the Sun still makes it considerably fainter than either Jupiter or Mars. Saturn ranks behind Jupiter, the inner planets, and several of the brightest stars in the sky in terms of apparent brightness.

Less than one-third the mass of Jupiter, Saturn is still an enormous planet, at least by terrestrial standards. As with Jupiter, Saturn's many moons allowed an accurate determination of the planet's mass long before the arrival of the *Pioneer* and *Voyager* missions. Saturn's mass is 5.6×10^{26} kg, or 95 times the mass of Earth. From Saturn's distance and angular size, the planet's radius—and hence the average density—quickly follow. Saturn's equatorial radius is 60,000 km, or 9.4 Earth radii. The average density then is 700 kg/m^3 (0.7 g/cm^3)—less than the density of water (which is 1000 kg/m^3). Here we have a planet that would float in the ocean—if Earth had one big enough! Saturn's low average density indicates that, like Jupiter, it is composed primarily of hydrogen and helium. Saturn's lower mass, however, results in lower interior pressure, so that these gases are less compressed than in Jupiter's case.

ROTATION RATE

Saturn, like Jupiter, rotates very rapidly and differentially. The rotation period of the interior (as measured from magnetospheric outbursts, which trace the rotation of the planet's core) and at high planetary latitudes (determined by tracking weather features observed in Saturn's atmosphere) is 10^{h}40^m. The rotation period at the equator is 10^{h}14^m, about 26 minutes shorter. Because of Saturn's lower density, this rapid rotation makes Saturn even more flattened than Jupiter. Saturn's polar radius is only 54,000 km, about 6000 km less than the equatorial radius. Careful calculations show that this degree of flattening is less than would be expected for a planet composed of hydrogen and

helium alone. Astronomers believe that Saturn also has a rocky core, perhaps twice the mass of Jupiter's.

RINGS

Saturn's best-known feature is its spectacular *ring system*. Because the rings lie in the equatorial plane, their appearance (as seen from Earth) changes in a seasonal manner, as shown in Figure 12.1. Saturn's rotation axis is significantly tilted with respect to the planet's orbit plane—the axial tilt is 27°, similar to that of both Earth and Mars. Consequently, as Saturn orbits the Sun, the angles at which the rings are illuminated, and at which we can view them, vary. When the planet's north or south pole is tipped toward the Sun, during Saturn's summer or winter, the highly reflective rings are at their brightest. During Saturn's spring and fall, the rings are close to being edge-on, both to the Sun and to us, so that they seem to disappear altogether. One important deduction that we can make from this simple observation is that the rings are very thin. In fact, we now know that their thickness is less than a few hundred meters, even though they are over 200,000 km in diameter.

12.2 Saturn's Atmosphere

❷ Saturn is much less colorful than Jupiter. Figure 12.2 shows yellowish and tan cloud belts that parallel the equator, but these regions display less atmospheric structure than do the belts on Jupiter. No obvious large "spots" or "ovals" adorn Saturn's cloud decks. Bands and storms do exist, but the color changes that distinguish them on Jupiter are largely absent on Saturn.

COMPOSITION AND COLORATION

Astronomers first observed methane in the spectrum of sunlight reflected from Saturn in the 1930s, about the same time that it was discovered on Jupiter. However, it was not until the early 1960s, when more sensitive observations became possible, that gaseous ammonia was finally detected. In Saturn's cold upper atmosphere, most ammonia is in the solid or liquid form, with relatively little of it present as a gas to absorb sunlight and create spectral lines. Astronomers finally made the first accurate determinations of the hydrogen and helium content in the late 1960s. These Earth-based measurements were later confirmed with the arrival of the *Pioneer* and *Voyager* spacecraft in the 1970s.

Figure 12.1 Over a period of several years, Saturn's rings change their appearance to terrestrial observers as the tilted ring plane orbits the Sun. The first two photos are separated by about 1 year; the one at right was taken a decade later. At some times during Saturn's orbital period of 29.5 Earth years, the rings seem to disappear altogether as Earth passes through their plane and we view them edge-on.

Saturn's atmosphere consists of molecular hydrogen (92.4 percent), helium (7.4 percent), methane (0.2 percent), and ammonia (0.02 percent). As on Jupiter, hydrogen and helium dominate—these most abundant elements never escaped from Saturn's atmosphere because of the planet's large mass and low temperature (see the *More Precisely* feature on p. 136). However, the fraction of helium on Saturn is far less than is observed on Jupiter (where, as we saw, helium ac-

counts for nearly 14 percent of the atmosphere) or in the Sun. It is extremely unlikely that the processes that created the outer planets preferentially stripped Saturn of nearly half its helium or that the missing helium somehow escaped from the planet while the hydrogen remained behind. Instead, astronomers believe that at some time in Saturn's past, the heavier helium began to sink toward the center of the planet, reducing its abundance in the outer layers and leaving them

Figure 12.2 Saturn as seen by the *Hubble Space Telescope* in December 1994. At the time, a rare storm was visible near the planet's equator. The bland colors are approximately true—that is, as the human eye sees things. The insert shows the northern polar hood at an earlier time.

relatively hydrogen-rich. We will return to the reasons for this differentiation and its consequences in a moment.

Figure 12.3 illustrates Saturn's atmospheric structure (compare with the corresponding diagram for Jupiter, Figure 11.4). In many respects, Saturn's atmosphere is quite similar to Jupiter's, except that the temperature is a little lower because of its greater distance from the Sun, and because its clouds are somewhat thicker. Since Saturn, like Jupiter, lacks a solid surface, we take the top of the troposphere as our reference level and set it to 0 km. The top of the visible clouds lies about 50 km below this level. As on Jupiter, the clouds are arranged in three distinct layers, composed (in order of increasing depth) of ammonia, ammonium hydrosulfide, and water ice. Above the clouds lies a layer of haze formed by the action of sunlight on Saturn's upper atmosphere.

The total thickness of the three cloud layers in Saturn's atmosphere is roughly 200 km, compared with about 80 km on Jupiter, and each layer is itself somewhat thicker than its counterpart on Jupiter. The reason for this difference is Saturn's weaker gravity. At the haze level, Jupiter's gravitational field is nearly two and a half times stronger than Saturn's, so Jupiter's atmosphere is pulled much more powerfully toward the center of the planet. Thus Jupiter's atmos-phere is compressed more than Saturn's, and the clouds are squeezed more closely together. The colors of Saturn's cloud layers, as well as the planet's overall butterscotch hue, are due to the same basic cloud chemistry as on Jupiter. However, because Saturn's clouds are thicker, there are few holes and gaps in the top layer, so we rarely glimpse the more colorful levels below. Instead, we see only different levels in the topmost layer, which leads to Saturn's rather uniform appearance.

WEATHER

Saturn has atmospheric wind patterns that are in many ways reminiscent of those on Jupiter. There is an overall east–west zonal flow, which is apparently quite stable. Computer-enhanced images of the planet that bring out more cloud contrast (see Figure 12.4) clearly show the existence of bands, oval storm systems, and turbulent flow patterns looking very much like those seen on Jupiter. Scientists believe that Saturn's bands and storms have essentially the same cause as does Jupiter's weather. Ultimately, the large-scale flows and small-scale storm systems are powered by convective motion in Saturn's interior and the planet's rapid rotation.

The zonal flow on Saturn is considerably faster than on Jupiter and shows fewer east–west alternations, as can be

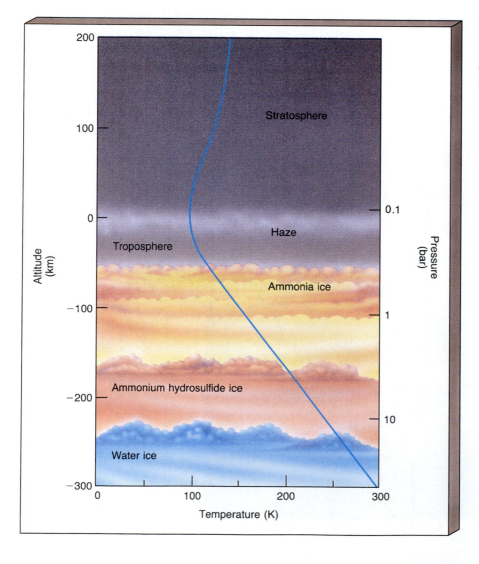

Figure 12.3 The vertical structure of Saturn's atmosphere. As with Jupiter, there are several cloud layers, but Saturn's weaker gravity results in thicker clouds and a more uniform appearance.

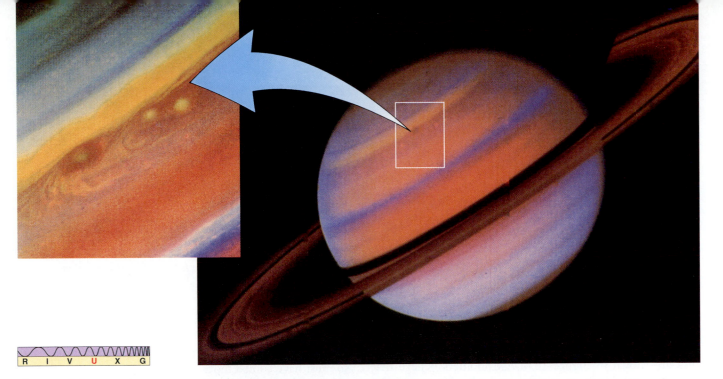

R I V U X G

Figure 12.4 We see more structure in Saturn's cloud cover when computer processing and artificial color are used to enhance the image contrast, as in these *Voyager* images of the entire gas ball and of a smaller, magnified piece of it.

seen from Figure 12.5 (compare Figure 11.6). The equatorial eastward jet stream, which reaches a speed of about 400 km/h on Jupiter, moves at a brisk 1500 km/h on Saturn, and extends to much higher latitudes. Not until latitudes 40° north and south of the equator are the first westward flows found. Latitude 40°N also marks the strongest bands on

Saturn and the most obvious ovals and turbulent eddies. Astronomers still do not fully understand the reasons for the differences between Jupiter's and Saturn's flow patterns.

In September 1990, amateur astronomers detected a large white spot in Saturn's southern hemisphere, just below the equator. In November of that year, when the

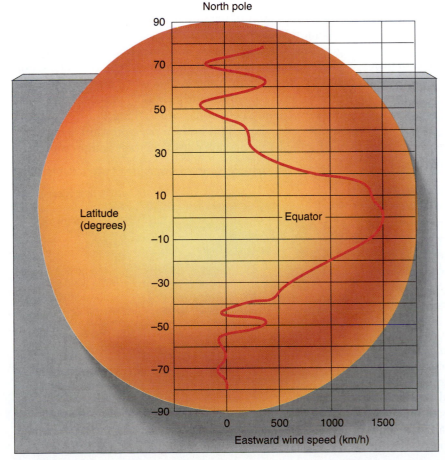

Figure 12.5 Winds on Saturn reach speeds even greater than those on Jupiter. As on Jupiter, the visible bands appear to be associated with variations in wind speed.

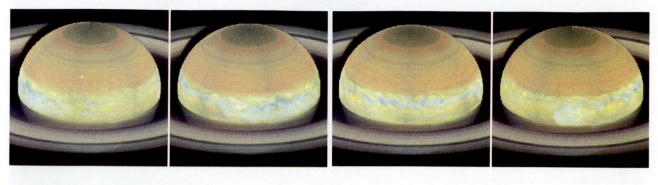

Figure 12.6   Circulating and evolving cloud systems on Saturn, imaged by the *Hubble Space Telescope* in 1990.

R I V U X G

Hubble Space Telescope imaged the phenomenon in more detail, the spot had developed into a band of clouds completely encircling the planet's equator. Some of these images are shown in Figure 12.6. Astronomers believe that the white coloration arose from crystals of ammonia ice formed when an upwelling plume of warm gas penetrated the cool upper cloud layers. Because the crystals were freshly formed, they had not yet been affected by the chemical reactions that color the planet's other clouds.

Such spots are rare on Saturn. The last one visible from Earth appeared in 1933, but it was much smaller than the 1990 system and much shorter-lived, lasting for only a few weeks. The turbulent flow patterns seen around the 1990 white spot have many similarities to the flow around Jupiter's Great Red Spot. Scientists speculate that these white spots may represent long-lived weather systems on Saturn and hope that routine observations of such temporary atmospheric phenomena on the outer worlds will enable them to gain greater insight into the dynamics of planetary atmospheres.

12.3 *Saturn's Interior and Magnetosphere*

INTERIOR STRUCTURE AND INTERNAL HEATING

❸ Figure 12.7 depicts Saturn's internal structure (compare Figure 11.9 for Jupiter). This picture has been pieced together by planetary scientists using the same tools—*Voyager* observations and theoretical modeling—that they used to infer Jupiter's inner workings. Saturn has the same basic internal parts as Jupiter, but their relative proportions are somewhat different: Saturn's metallic hydrogen layer is thinner, and its core is larger. Because of its lower mass, Saturn has a less extreme core temperature, density, and pressure than Jupiter. The central pressure is around a tenth of Jupiter's—not too different from the pressure at the center of the Earth.

Infrared measurements indicate that Saturn's surface (that is, cloud-top) temperature is 97 K, substantially higher than the temperature at which Saturn would reradiate all the energy it receives from the Sun. In fact, Saturn radiates away almost three times more energy than it absorbs. Thus Saturn, like Jupiter, has an internal energy source. But the

explanation behind Jupiter's excess energy—that the planet has a large reservoir of heat left over from its formation—doesn't work for Saturn. Saturn is smaller than Jupiter and so must have cooled more rapidly—rapidly enough that its original supply of energy has long ago been used up. What then is happening inside Saturn to produce this extra heat?

The explanation for this strange state of affairs also explains the mystery of Saturn's apparent helium deficit. At the temperatures and high pressures found in Jupiter's interior, liquid helium *dissolves* in liquid hydrogen. In Saturn, where the internal temperature is lower, the helium doesn't dissolve so easily and tends to form droplets instead. The phenomenon is familiar to cooks, who know that it is generally much easier to dissolve ingredients in hot liquids than in cold ones. Saturn probably started out with a fairly uniform solution of helium dissolved in hydrogen, but the helium tended to condense out of the surrounding hydrogen, much as water vapor condenses out of Earth's atmosphere to form a mist. The amount of helium condensation was greatest in the planet's cool outer layers, where the mist turned to rain about 2 billion years ago. A light shower of liquid helium has been falling through Saturn's interior ever since. This **helium precipitation** is responsible for depleting the outer layers of their helium content.

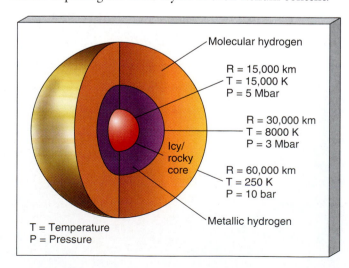

Figure 12.7   Saturn's internal structure, as deduced from *Voyager* observations and computer modeling.

Labels in figure:
Molecular hydrogen
R = 15,000 km
T = 15,000 K
P = 5 Mbar
R = 30,000 km
T = 8000 K
P = 3 Mbar
Icy/ rocky core
R = 60,000 km
T = 250 K
P = 10 bar
Metallic hydrogen
T = Temperature
P = Pressure

So we can account for the unusually low abundance of helium in Saturn's atmosphere—much of it has rained down to lower levels. But what about the excess heating? The answer is simple: As the helium sinks toward the center, the planet's gravitational field compresses it and heats it up. The gravitational energy thus released is the source of Saturn's internal heat. In the distant future, the helium rain will stop, and Saturn will cool until its outermost layers radiate only as much energy as they receive from the Sun. When that happens, the temperature at Saturn's cloud tops will be 74 K. As Jupiter cools, it too may someday experience helium precipitation in its interior, causing its surface temperature to rise once again.

MAGNETOSPHERIC ACTIVITY

Saturn's electrically conducting interior and rapid rotation produce a strong magnetic field and an extensive magnetosphere. Probably because of the considerably smaller mass of Saturn's metallic hydrogen zone, the planet's basic magnetic field strength is only about 1/20 that of Jupiter, or about 1000 times greater than that of the Earth. The magnetic field at Saturn's cloud tops (roughly 10 Earth radii from the planet's center) is approximately the same as at Earth's surface. *Voyager* measurements indicate that, unlike Jupiter and Earth, whose magnetic axes are slightly tilted, Saturn's magnetic field is not inclined with respect to its rotation axis.

Saturn's magnetosphere extends about 1 million km toward the Sun and is large enough to contain the planet's ring system and the innermost 16 small moons. Saturn's largest moon, Titan, orbits about 1.2 million km from the planet, so that it is sometimes found just inside the outer magnetosphere and sometimes just outside, depending on the intensity of the solar wind (which tends to push the sunward side of the magnetosphere closer to the planet). Because no major moons lie deep within Saturn's magnetosphere, the details of its structure are different from those of Jupiter's magnetosphere. For example, there is no equivalent of the Io plasma torus. Like Jupiter, Saturn emits radio waves, but as luck would have it, they are reflected from the Earth's ionosphere (they lie in the AM band) and were not detected until the *Voyager* craft approached the planet.

12.4 *Saturn's Spectacular Ring System*

THE VIEW FROM EARTH

④ The most obvious aspect of Saturn's appearance is, of course, its **planetary ring system**. Astronomers now know that all the jovian planets have rings, but Saturn's are by far the brightest, the most extensive, and the most beautiful. Galileo saw them first in 1610, but he did not recognize what he saw as a planet with a ring. At the resolution of his small telescope, the rings looked like bumps on the planet, or perhaps components of a triple system of some sort. In 1659, the Dutch astronomer Christian Huygens realized what the "bump" was—a thin, flat ring, completely encircling the planet.

In 1675, the French–Italian astronomer Giovanni Domenico Cassini discovered the first ring feature, a dark band about two-thirds of the way out from the inner edge. From Earth, the band looks like a gap in the ring (an observation that is not too far from the truth, although we now know that there is actually some ring material within it). This "gap" is named the **Cassini Division**, in honor of its discoverer. Careful observations from Earth show that the inner "ring" is in reality also composed of two rings. From the outside in, the three rings are known somewhat prosaically as the **A, B, and C rings**. The Cassini Division lies between A and B. A smaller gap, known as the **Encke Division**, is found in the outer part of the A ring. Its width is 270 km. These ring features are marked on Figure 12.8.

Encke division
A ring
Cassini division
B ring
C ring

Figure 12.8 Much fine structure, especially in the rings, appears in this image of Saturn taken while the *Voyager 2* spacecraft approached the planet. (One of Saturn's moons appears at bottom, and a second casts a black shadow on the cloud tops.) The banded structure of Saturn's atmosphere is also more evident in this photograph. Note the absence of the vivid colors that characterize Jupiter's atmospheric cloud layers.

R   I   V   U   X   G

TABLE 12-1   *The Rings of Saturn*

RING	INNER RADIUS (km)	(planet radii)	OUTER RADIUS (km)	(planet radii)	WIDTH (km)
D	60,000	1.00	74,000	1.23	14,000
C	74,000	1.23	92,000	1.53	18,000
B	92,000	1.53	117,600	1.96	25,600
Cassini Division	117,600	1.96	122,200	2.04	4600
A	122,200	2.04	136,800	2.28	14,600
Encke Division*	135,500	2.26	135,900	2.27	400
F	140,500	2.34	140,600	2.34	100
E	210,000	3.50	300,000	5.00	90,000

The Encke Division lies within the A ring.

No finer ring details are visible from our earthly vantage point. Of the three main rings, the B ring is brightest, followed by the somewhat fainter A ring, and then by the almost translucent C ring. A more complete list of ring properties appears in Table 12-1. (The E and F rings are discussed later in this section.)

WHAT *ARE* SATURN'S RINGS?

A fairly obvious question—and one that perplexed the best scientists and mathematicians on Earth for centuries—is "What are Saturn's ring made of?" By the middle of the nineteenth century, various dynamical and thermodynamic arguments had conclusively proved that the rings could not be solid, liquid, or gas! What is left? In 1857, Scottish physicist James Clerk Maxwell, after proving that a solid ring would become unstable and break up, suggested that the rings are composed of a great number of small particles, all independently orbiting Saturn, like so many tiny moons. That inspired speculation was verified in 1895, when Lick Observatory astronomers measured the Doppler shift of sunlight reflected from the rings and showed that the velocities thus determined were exactly what would be expected from separate particles moving in circular orbits in accordance with Newton's law of gravity.

What sort of particles make up the rings? The fact that they reflect most (over 80 percent) of the sunlight striking them had long suggested to astronomers that they were made of ice, and infrared observations in the 1970s confirmed that water ice is indeed a prime ring constituent. Radar observations and later *Voyager* studies of scattered sunlight showed that the diameters of the particles range from fractions of a millimeter to tens of meters, with most

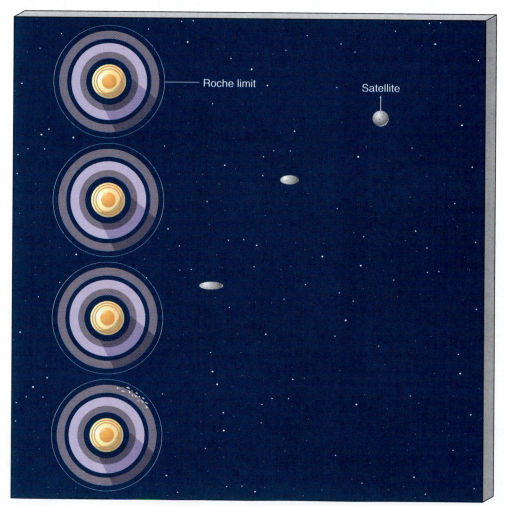

Figure 12.9   The increasing tidal field of a planet first distorts, and then destroys, a moon that strays too close. (The distortion is exaggerated in the second and third panels.)

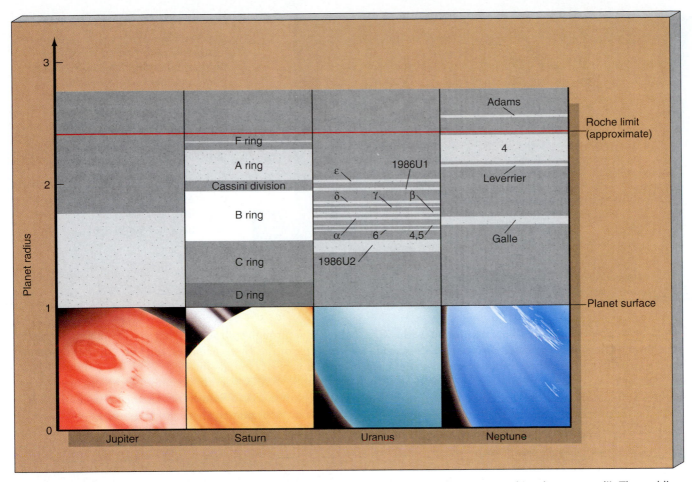

Figure 12.10   The rings of Jupiter, Saturn, Uranus, and Neptune. All distances are expressed in planetary radii. The red line represents the Roche limit. In all cases, the rings lie within the Roche limit of the parent planet.

particles being about the size (and composition) of a large snowball on Earth. We now know that the rings are truly thin—perhaps only a few tens of meters thick in places. Stars can occasionally be seen through them, like automobile headlights penetrating a snowstorm. Why are the rings so thin? The answer seems to be that collisions between ring particles tend to keep them all moving in circular orbits in a single plane. Any particle that tries to stray away from this orderly motion finds itself in an orbit that soon runs into other ring particles. Over long periods of time, the ensuing jostling serves to keep all particles moving in circular, planar orbits. The asymmetric gravitational field of Saturn (the result of its flattened shape) sees to it that the rings lie in the plane of the planet's equator.

THE ROCHE LIMIT

⑤   But why a ring of particles at all? What process produced the rings in the first place? To answer these questions, consider the fate of a small moon orbiting close to a massive planet such as Saturn. The moon is held together by internal forces—its own gravity, for example. As we bring our hypothetical moon closer to the planet, the tidal force on it increases. Recall from Chapter 7 that the effect of such a tidal force is to stretch the moon along the direction to the planet—that is, to create a tidal bulge. Recall

also that the tidal force increases rapidly with decreasing distance from the planet. ∞ (p. 147) As the moon is brought closer to the planet, it reaches a point where the tidal force tending to stretch it out becomes *greater* than the internal forces holding it together. At that point, the moon is torn apart by the planet's gravity, as shown in Figure 12.9. The pieces of the satellite then pursue their own individual orbits around that planet, eventually spreading all the way around it in the form of a ring.

For any given planet and any given moon, this critical distance, inside of which the moon is destroyed, is known as the *tidal stability limit*, or the **Roche limit**, after the nineteenth-century French mathematician Edouard Roche, who first calculated it. If our hypothetical moon is held together by its own gravity and its average density is the same as that of the parent planet (both reasonably good approximations for Saturn's larger moons), then the Roche limit is just 2.4 times the radius of the planet. Thus, for Saturn, no moon can survive within a distance of 144,000 km of the planet's center, about 7000 km beyond the outer edge of the A ring. The rings of Saturn occupy the region inside Saturn's Roche limit.

These considerations apply equally well to the other jovian worlds. Figure 12.10 shows the location of the ring system of each jovian planet relative to the planet's Roche limit. Given the approximations in our assumptions, we

Figure 12.11 The *Voyager 2* cameras took this close-up of the ring structure just before plunging through the tenuous outer rings of Saturn. The ringlets in the B ring, spread over several thousand kilometers, are resolved here to about 10 km. As *Voyager* approached Saturn, more and more of these tiny ringlets became noticeable in the main rings. The (false) color variations probably indicate different sizes and compositions of the particles making up the thousands of rings. Earth is superposed, to proper scale, for a size comparison.

can conclude that all of the rings are found within the Roche limit of their parent planet. Notice that the calculation of this limit applies only to moons massive enough for their own gravity to be the dominant force binding them together. Sufficiently small moons can survive even within the Roche limit because they are held together mostly by interatomic (electromagnetic) forces, not by gravity.

THE VIEW FROM *VOYAGER*

Thus it was, as *Voyager* approached Saturn, that scientists on Earth were fairly confident that they understood the nature of the rings. However, there were many surprises in store. The *Voyager* mission changed forever our view of this spectacular region in our cosmic backyard, revealing the rings to be vastly more complex than astronomers had imagined.

As the *Voyager* probes approached Saturn, it became obvious that the main rings are actually composed of tens of thousands of narrow **ringlets** (shown in Figure 12.11). Although *Voyager* cameras did find several new gaps in the rings, the ringlets are generally not separated from each other by empty space. Instead, the rings contain concentric regions of alternating high and low concentrations of ring particles—the ringlets are just the high-density peaks. Although the process is not fully understood, it seems that the mutual gravitational attraction of the ring particles (as well as the effects of Saturn's inner moons) enables waves of matter to form and move in the plane of the rings, rather like ripples on the surface of a pond. The wave crests typically wrap around the rings, forming tightly wound spiral patterns called *spiral density waves* that resemble grooves in a huge celestial phonograph record.

Although the ringlets are probably the result of spiral waves in the rings, the true gaps are not. The narrower gaps—about 20 of them—are most likely swept clean by the action of small moonlets embedded in them. These moonlets are larger (perhaps 10 or 20 km in diameter) than the largest true ring particles, and they simply "sweep up" ring

material through collisions as they go. Despite many careful searches of the *Voyager* images, only one of these moonlets has so far been found—in 1991, after five years of exhaustive study, NASA scientists confirmed the discovery of the eighteenth moon of Saturn (now named Pan) in the Encke Division. Astronomers have found indirect evidence for embedded moonlets, in the form of "wakes" that they leave behind them in the rings, but no other direct sightings have occurred. Despite their elusiveness, however, moonlets are still regarded as the best explanation for the small gaps.

Voyager found a series of faint rings, now known collectively as the **D ring**, inside the inner edge of the C ring, stretching down almost to Saturn's cloud tops. The D ring contains relatively few particles and is so dark that it is completely invisible from Earth. Another faint ring, also a *Voyager* discovery, lies well outside the main ring structure. Known as the **E ring**, it appears to be associated with volcanism on the moon Enceladus. The *Voyager* cameras revealed one other completely unexpected feature. A series of dark radial "spokes" formed on the B ring, moved around the planet for about one ring orbit period, and then disappeared (Figure 12.12). Careful scrutiny of these peculiar drifters showed that they were composed of very fine (micron-sized) dust hovering a few tens of meters *above* the plane of the rings. Scientists believe that this dust was held in place by electrostatic forces generated in the ring plane, perhaps resulting from particle collisions there. The electrical fields slowly dispersed, and the spokes faded as the ring revolved. We expect that the creation and dissolution of such spokes is a regular occurrence in the Saturn ring system.

ORBITAL RESONANCES AND SHEPHERD SATELLITES

Voyager images show that the largest gap in the rings, the Cassini Division, is not completely empty of matter. In fact, as shown in Figure 12.13, the Division contains a series of faint ringlets and gaps (and, presumably, embedded moonlets too). The overall concentration of ring particles

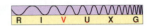

Figure 12.12   Saturn's B ring showed a series of dark temporary spokes as *Voyager 2* flew by at a distance of about 4 million km. The spokes were caused by small particles suspended just above the ring plane.

Figure 12.13   Close inspection by *Voyager* revealed that the Cassini Division (shown here as the darker color) is not completely empty. It contains a series of faint ringlets and gaps, assumed to be caused by unseen embedded satellites. The density of material in the division is very low, accounting for its dark appearance from the Earth.

in the division as a whole is, however, much lower than in the A and B rings. Although its small internal gaps probably result from embedded satellites, the division itself does not. It owes its existence to another solar-system resonance, this time involving particles orbiting in the division and Saturn's innermost major moon, Mimas.

A ring particle moving in an orbit within the Cassini Division has an orbital period exactly half that of Mimas. Particles in the division thus complete exactly two orbits around Saturn in the time taken for Mimas to orbit once— a configuration known as a 2:1 resonance. The effect of this resonance is that particles in the division feel a gravitational tug from Mimas at exactly the same location in their orbit every other time around. Successive tugs reinforce one another, and the initially circular trajectories of the ring particles soon get stretched out into ellipses. In their new orbits these particles collide with other particles and eventually find their way into new circular orbits at other radii. The net effect is that the number of ring particles in the Cassini Division is greatly reduced.

Particles in "nonresonant" orbits (that is, at radii where the orbital period is not simply related to the period of Mimas) also experience Mimas's gravitational pull. But the times when the force is greatest are spread uniformly around the orbit, and the tugs cancel out. It's a little like pushing a child on a swing—pushing at the same point in the swing's motion each time produces much better results than do random shoves. Thus, Mimas (or any other moon) has a large effect on the ring at those radii where a resonance exists and little or no effect elsewhere. We now know that resonances between ring particles and moons play an important role in shaping the fine structure of Saturn's

rings. For example, the sharp outer edge of the A ring corresponds precisely to a 3:2 resonance with Mimas (three ring orbits in two Mimas orbital periods). Most theories of planetary rings predict that the ring system should spread out with time, basically because of collisions among ring particles. The A ring's outer edge is "patrolled" by a small satellite named Atlas, held in place by the resonance with Mimas, which prevents ring particles from diffusing away.

Outside the A ring lies the strangest ring of all. The faint, narrow **F ring** (shown in Figure 12.14) was discov-

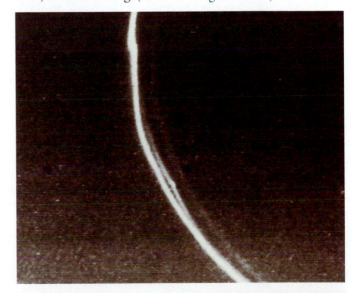

Figure 12.14   Saturn's narrow F ring appears to contain kinks and braids, making it unlike any of Saturn's other rings.

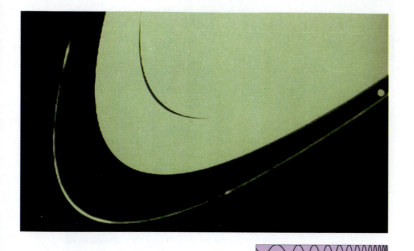

Figure 12.15 The F ring's thinness, and possibly its other peculiarities too, can be explained by the effects of two shepherd satellites that orbit a few hundred kilometers inside and outside the ring. This photo shows one of the shepherding satellites, roughly 100 km in length, at right.

ered by *Pioneer* in 1979, but its full complexity became evident only when *Voyager* took a closer look. Unlike the inner major rings, the F ring is narrow, less than a hundred kilometers wide. It lies just inside Saturn's Roche limit, separated from the A ring by about 3500 km. Its narrowness by itself is unusual, as is its slightly eccentric shape, but its oddest feature is that it looks for all the world as though it is made up of several separate strands braided together! This remarkable discovery sent dynamicists scrambling in search of an explanation. It now seems as though the ring's intricate structure, as well as its thinness, arise from the in-

fluence of two small moons, known as **shepherd satellites**, that orbit on either side of it (Figure 12.15).

These two small, dark satellites, each about 50 km in diameter, are called Prometheus and Pandora. They orbit about 1000 km on either side of the F ring, and their gravitational influence on the F-ring particles keeps the ring tightly confined in its narrow orbit. As shown in Figure 12.16, any particle straying too far out of the F ring is gently guided back into the fold by one or the other of the moons. (The moon Atlas confines the A ring in a somewhat similar way.) However, the details of how Prometheus and Pandora produce the braids in the F ring and why the two moons are there at all, in such similar orbits, remain unclear. There is some evidence that other eccentric rings found in the gaps in the A, B, and C rings may also result from the effects of shepherding moonlets.

THE ORIGIN OF THE RINGS

Two possible origins have been suggested for Saturn's rings. Astronomers estimate that the total mass of ring material is no more than 10^{15} tons—enough to make a satellite about 250 km in diameter. If such a satellite strayed inside Saturn's Roche limit or was destroyed (perhaps by a collision) near that radius, a ring could have resulted. An alternative view is that the rings represent material left over from Saturn's formation stage 4.5 billion years ago. In this scenario, Saturn's tidal field prevented any moon from forming inside the Roche limit, and so the material has remained a ring ever since. Which is correct?

All the dynamic activity observed in Saturn's rings suggests to many researchers that the rings must be quite young—perhaps no more than 50 million years old, or 100 times younger than the solar system. There is just too much going on for them to have remained stable for bil-

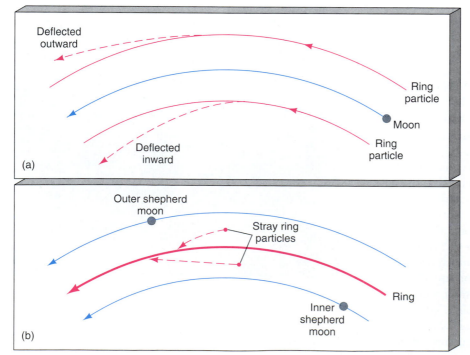

Figure 12.16 (a) Strange as it may seem, the net effect of the interactions between a moon and ring particles is that the moon tends to push those particles *away* from it. (b) The F ring shepherd satellites operate by forcing errant F ring particles back into the main ring. Each moon operates as in part (a) so that the ring is confined between the two moons. As a consequence of Newton's Third Law of Motion, the satellites themselves slowly drift away from the ring.

lions of years, so they probably aren't left over from the planet's formative stages. If this is so, then either the rings must be continuously replenished, perhaps by fragments of Saturn's moons chipped off by meteorites, or they are the result of a relatively recent, possibly catastrophic, event in the planet's system—perhaps a small moon that was hit by a large comet, or even by another moon. Astronomers normally prefer not to invoke catastrophic events to explain observed phenomena, but the more we learn of the universe, the more we realize that catastrophe probably plays an important role. For now, the details of the formation of Saturn's ring system simply aren't known.

12.5 *The Moons of Saturn*

GENERAL FEATURES

Saturn has the most extensive, and, in many ways, the most complex, system of natural satellites of all the planets. Saturn's 18 named moons are listed in Table 12–2. In addition, at least 2, and possibly as many as 4, still unnamed small moons were discovered by the *Hubble Space Telescope* in May 1995, when Earth passed through the planet's ring plane and the normally bright glare from the rings was greatly reduced. These new moons all orbit in the vicinity of Saturn's F ring. The reflected light from Saturn's moons suggests that most are covered with snow and ice. Many of them are probably made almost entirely of water ice. Even so, they are a curious and varied lot.

The satellites fall into three fairly natural groups. First, there are the "small" moons—irregularly shaped chunks of ice, all less than 300 km across—which exhibit a bewildering variety of complex and fascinating motion. Second, there are six "medium-sized" moons—spherical bodies with diameters ranging from about 400 to 1500 km—that offer clues to the past and present state of the environment of Saturn, at the same time presenting many puzzles in their own appearance and history. Finally, there is Saturn's single "large" moon—Titan—which, at 5150 km in diameter, is the second-largest satellite in the solar system (Jupiter's Ganymede is a little bigger). It has an atmosphere denser than Earth's and (some scientists think) surface conditions possibly conducive to life. Notice, incidentally, that Jupiter has no "medium" moons, as just defined. The Galilean satellites are large, like Titan, and all of Jupiter's other satellites are small—no more than 200 km in diameter.

TITAN

6 Perhaps the most intriguing of all Saturn's moons is Titan, discovered by Christian Huygens in 1655. Even through a large Earth-based telescope, Titan is visible only as a barely resolved reddish disk. Long before the *Voyager* missions, astronomers already knew (from spectroscopic

TABLE 12–2   *The Moons of Saturn*

NAME	DISTANCE FROM SATURN		ORBIT PERIOD	SIZE	MASS	DENSITY	
	(km)	(planet radii)	(days)	(longest diameter, km)	(Earth Moon masses)	(kg/m³)	(g/cm³)
Pan	134,000	2.23	0.58	20	4×10^{-8}		
Atlas	138,000	2.30	0.60	40			
Prometheus	139,000	2.32	0.61	80			
Pandora	142,000	2.37	0.63	100			
Janus	151,000	2.52	0.69	190			
Epimetheus	151,000	2.52	0.69	120			
Mimas	186,000	3.10	0.94	394	0.00054	1200	1.2
Enceladus	238,000	3.97	1.37	502	0.0011	1200	1.2
Tethys	295,000	4.92	1.89	1050	0.010	1300	1.3
Telesto	295,000	4.92	1.89	25			
Calypso	295,000	4.92	1.89	25			
Dione	377,000	6.28	2.74	1120	0.015	1400	1.4
Helene	377,000	6.28	2.74	30			
Rhea	527,000	8.78	4.52	1530	0.034	1300	1.3
Titan	1,220,000	20.3	16.0	5150	1.83	1900	1.9
Hyperion	1,480,000	24.7	21.3	270			
Iapetus	3,560,000	59.3	79.3	1440	0.026	1200	1.2
Phoebe	13,000,000	217	550*	220			

*Indicates a retrograde orbit.

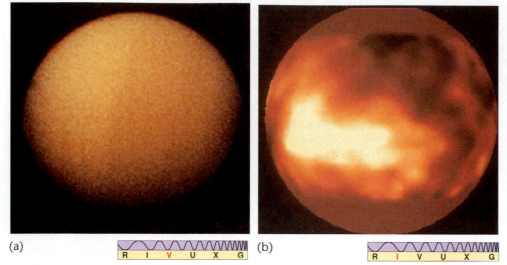

(a)

(b)

Figure 12.17 (a) Titan, larger than the planet Mercury and roughly half the size of Earth, was photographed in visible light from only 4000 km away as the *Voyager 1* spacecraft passed by in 1980. All we can see here is Titan's upper cloud deck. For unknown reasons, the northern hemisphere appears slightly brighter than the southern. (b) In infrared, however, as captured by the *Hubble Space Telescope*, we can see large-scale surface features. The most prominent bright area shown here is nearly 400 km across, about the size of Australia.

observations) that the moon's reddish coloration is caused by something quite special—an atmosphere. So anxious were mission planners to obtain a closer look that they programmed *Voyager 1* to pass very close to Titan, even though that meant the spacecraft could not then use Saturn's gravity to continue on to Uranus and Neptune. (Instead, *Voyager 1* left the Saturn system on a path taking the craft out of the solar system well above the ecliptic plane.) A *Voyager 1* image of Titan is shown in Figure 12.17.

Scientists believe that Titan's internal composition and structure must be similar to Ganymede and Callisto because these three moons have quite similar masses and radii and hence average densities (Titan's density is 1900 kg/m³). Thus, Titan probably contains a rocky core surrounded by a thick mantle of water ice. Despite *Voyager 1*'s close pass, the moon's surface remains a mystery. A thick, uniform haze layer, similar to the photochemical smog found over many cities on the Earth, envelops the moon and completely obscured the spacecraft's view.

Voyager 1 was able to provide mission specialists with detailed atmospheric data, however. Titan's atmosphere is thicker and denser even than Earth's, and it is certainly more substantial than that of any other moon. Prior to *Voyager 1*'s arrival in 1980, only methane and a few other simple hydrocarbons had been conclusively detected on Titan. (Hydrocarbons are molecules consisting solely of hydrogen and carbon atoms, of which methane, CH_4, is the simplest). But radio and infrared observations from the spacecraft showed that the atmosphere is actually made up mostly of nitrogen (roughly 90 percent) and argon (at most 10 percent), with a few percent of methane. In addition, the complex chemistry in Titan's atmosphere maintains steady (but trace) levels of hydrogen gas, the hydrocarbons ethane and propane, and carbon monoxide.

In fact, Titan's atmosphere seems to act like a gigantic chemical factory. Powered by the energy of sunlight, it is undergoing a complex series of chemical reactions that ultimately result in the observed smog and trace chemical composition. The upper atmosphere is thick with aerosol

haze, and the unseen surface may be covered with organic sediment that has settled down from the clouds. Speculation runs the gamut from oceans of liquid hydrocarbons, especially ethane, to icy valleys laden with petrochemical sludge. Future spacecraft exploration of Titan may present scientists with an opportunity to study the kind of chemistry thought to have occurred billions of years ago on Earth—the prebiotic chemical reactions that eventually led to life on our own planet.

Based largely on *Voyager* measurements, Figure 12.18 shows the probable structure of Titan's atmosphere. Despite Titan's low mass (a little less than twice that of Earth's Moon) and hence its low surface gravity (one-seventh of Earth's), the atmospheric pressure at ground level is 60 percent *greater* than on Earth. Titan's atmosphere contains about 10 times more gas than Earth's. Because of Titan's weaker gravitational pull, the atmosphere extends some 10 times farther into space than does our own. The top of the main haze layer lies some 200 km above the surface, although there are additional layers, seen primarily through their absorption of ultraviolet radiation, at about 300 km and 400 km (Figure 12.19). Below the haze, the atmosphere is reasonably clear, although rather gloomy, because so little sunlight gets through. The surface temperature is a frigid 94 K, roughly what we would expect simply on the basis of Titan's distance from the Sun. At the temperatures typical of the lower atmosphere, methane and ethane may behave rather like water on Earth, raising the possibility of methane rain, snow, and fog and even ethane oceans! At high altitudes, the temperature rises, the result of photochemical absorption of solar radiation.

Why does Titan have such a thick atmosphere, when similar moons of Jupiter such as Ganymede and Callisto have none? The answer seems to be a direct result of Titan's greater distance from the Sun. The moons of Saturn formed at considerably lower temperature than those of Jupiter. Those conditions would have enhanced the ability of the water ice that makes up the bulk of Titan's interior to absorb methane and ammonia, both of which

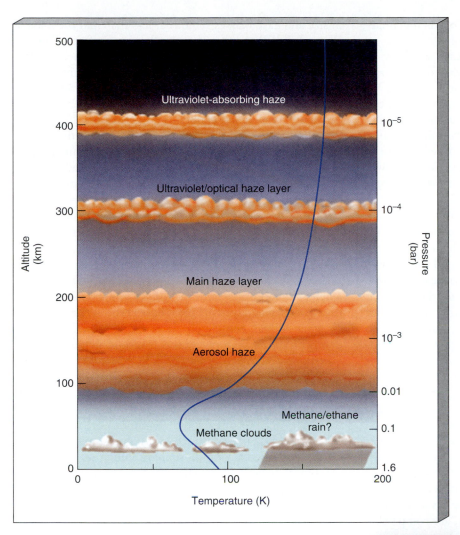

Figure 12.18 The structure of Titan's atmosphere, as deduced from *Voyager 1* observations.

were present in abundance at those early times. As a result, Titan was initially laden with much more methane and ammonia gas than either Ganymede or Callisto. As Titan's internal radioactivity warmed the moon, the ice released the trapped gases, forming a thick methane–ammonia atmosphere. Sunlight split the ammonia into hydrogen, which escaped into space, and nitrogen, which remained in the atmosphere. The methane, which was less easily broken apart, survived intact. Together with argon outgassed from Titan's interior, these gases form the basis of the atmosphere we see today.

SATURN'S MEDIUM-SIZED MOONS

7 Saturn's complement of midsized moons consists (in order of increasing distance from the planet) of Mimas (at 3.1 planetary radii), Enceladus (at 4.0), Tethys (at 4.9), Dione (at 6.3), Rhea (at 8.8), and Iapetus (at 59.3). All six were known from Earth-based observations long before the space age. The inner five move on circular trajectories and are tidally locked by Saturn's gravity into synchronous rotation (so that one side always faces the planet). They therefore all have permanently "leading" and "trailing" faces as they move in their orbits, a fact that is important in understanding their often asymmetrical surface markings.

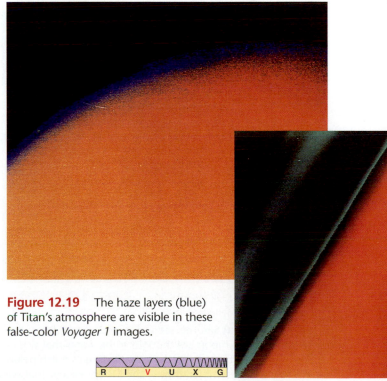

Figure 12.19 The haze layers (blue) of Titan's atmosphere are visible in these false-color *Voyager 1* images.

R I V U X G

The sixth member of the group, Iapetus, orbits much farther out on an inclined, eccentric path.

Unlike the densities of the Galilean satellites of Jupiter, the densities of these six moons do not show any correlation with distance from Saturn. Their densities are all about 1200 kg/m³, implying that nearness to the central planetary heat source was a less important influence during their formation than it was in the Jupiter system. Scientists believe that the midsized moons are composed largely of rock and water ice, as is Titan. Their densities are lower than Titan's, primarily because their lower masses produce less compression of their interiors.

The largest of the six, Rhea, is shown in Figure 12.20. Its mass is only 1/30 that of Earth's Moon, and its icy surface is very reflective and heavily cratered. At the low temperatures found on its surface, water ice is very hard and behaves rather like rock on the inner planets. For that reason, Rhea's surface craters look very much like craters on the Moon or Mercury. The crater density is similar to that in the lunar highlands, indicating that the surface is old, and there is no evidence of extensive geological activity. Rhea's only real riddle is the presence of so-called *wispy terrain*—prominent light-colored streaks—on its trailing side. The leading face, by contrast, shows no such markings, only craters. Astronomers believe that the "wisps" were caused by some event in the distant past during which water was released from the interior and condensed on the surface. Any similar markings on the leading side have presumably been long since obliterated by cratering, which should be much more frequent on the satellite's forward-facing surface.

Inside Rhea's orbit lie the orbits of Tethys and Dione. These two moons, shown in Figure 12.21, are comparable in size and have masses somewhat less than half the mass of Rhea. Like Rhea, they have reflective surfaces that are heavily cratered, but each shows signs of surface activity, too. Dione's trailing face has prominent bright streaks, which are probably similar to Rhea's wispy terrain. Dione also has "maria" of sorts, where flooding appears to have obliterated the older craters. The cracks on Tethys may have been caused by cooling and shrinking of the surface layers or, more probably, by meteoritic bombardment.

The innermost, and smallest, medium-sized moon is Mimas, shown in Figure 12.22. Despite its low mass—only 1 percent the mass of Rhea—its closeness to the rings causes resonant interactions with the ring particles, resulting most notably in the Cassini Division, as we have already seen. Possibly because of its proximity to the rings, Mimas is heavily cratered. The moon's chief surface feature, known as Herschel, is an enormous crater on the leading face. Its diameter is almost one-third that of the moon itself. The impact that formed Herschel must have come very close to destroying Mimas completely. It is quite possible that the debris produced by such impacts is responsible for creating or maintaining the spectacular rings we see.

Enceladus (Figure 12.23) orbits just outside Mimas. Its size, mass, composition, and orbit are so similar to those of Mimas that one might guess that the two moons would also be very similar to one another in appearance and history. This is not so. Enceladus is so bright and shiny—it reflects virtually 100 percent of the sunlight falling on it—that astronomers believe its surface must be completely coated with fine crystals of pure ice, which may be the icy "ash" of water "volcanoes" on Enceladus. The moon bears visible evidence for large-scale volcanic activity. Much of its sur-

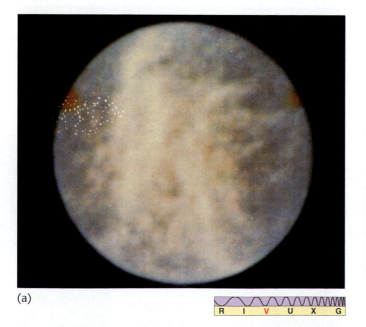

(a)

R I V U X G

(b)

R I V U X G

Figure 12.20   (a) Saturn's second-largest moon, Rhea, has a heavily cratered, old surface. The light-colored "wisps" on the trailing face of the moon are thought to be water that was released from the moon's interior during some long-ago period of activity and then froze on the surface. (b) Rhea's north polar region, seen here at a superb resolution of only 1 km. The heavily cratered surface resembled that of the Moon, except that we see here craters in bright ice, rather than dark lunar rock.

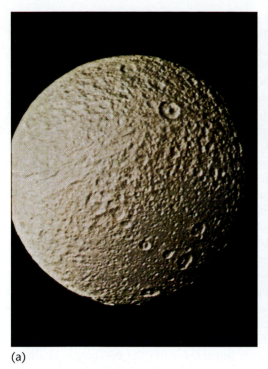

(a)

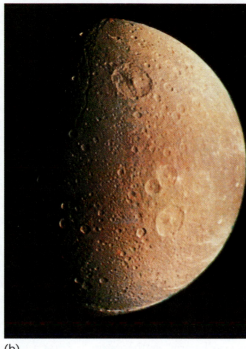

(b)

Figure 12.21 Tethys (a) and Dione (b) are similar to each other and are each about 1000 km across, smaller than Rhea. They show evidence for ancient geological activity of some sort, in addition to extensive meteoritic cratering. Resolution here is just a few kilometers.

face is devoid of impact craters, which seem to have been erased by what look like lava flows, except that the "lava" is water, temporarily liquefied during recent internal upheavals and now frozen again. Although no geysers or vol-

canoes have actually been observed on Enceladus, there seems to be strong circumstantial evidence for volcanism on the satellite. In addition, the nearby thin cloud of small, reflective particles that makes up Saturn's E ring is known

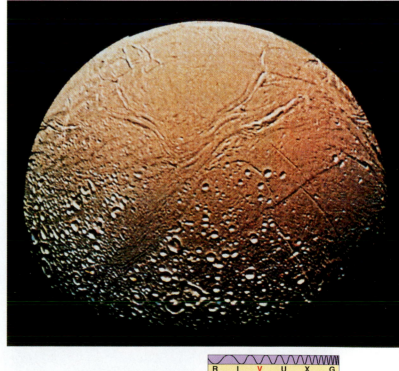

Figure 12.22 Mimas's main surface feature is the large crater Herschel, plainly visible in this *Voyager* image. The impact that caused Herschel must have come very close to shattering Mimas. The moon is about 400 km in diameter.

Figure 12.23 Despite its similarity in size and location to Mimas, Enceladus, with its highly reflective icy surface and apparent water volcanism, looks very different. Apparently, the moon is still active but the cause of the volcanism is unexplained.

to be densest near Enceladus. Calculations indicate that the E ring is unstable because of the disruptive effects of the solar wind, supporting the view that volcanism on Enceladus continually supplies new particles to maintain the ring.

Why is there so much activity on a moon so small? No one knows. Attempts have been made to explain Enceladus's water volcanism in terms of tidal stresses. (Recall the role that Jupiter's tidal stresses play in creating volcanism on Io.) ∞ (p. 247) However, Saturn's tidal force on Enceladus is only one quarter the force exerted by Jupiter on Io, and there are no nearby large satellites to force Enceladus away from a circular trajectory. Thus, the ingredients that power Io's volcanoes may not be present on Enceladus. For now, the mystery of Enceladus's internal activity remains unsolved.

The outermost midsized moon is Iapetus (Figure 12.24). It orbits Saturn on a somewhat eccentric, inclined orbit with a semi-major axis of 3.6 million km. Its mass is about three quarters that of Rhea. Iapetus is a two-faced moon. The dark, leading face reflects only about 3 percent of the sunlight reaching it, while the icy trailing side reflects 50 percent. Spectral studies of the dark regions seem to indicate that the material originates on Iapetus, so the moon is not simply sweeping up dark material as it orbits. Similar dark deposits seen elsewhere in the solar system are thought to be organic (carbon-containing) in nature; they can be produced by the action of solar radiation on hydrocarbon (for example, methane) ice. But how the dark markings can adorn only one side of Iapetus in that case is still unknown.

THE SMALL SATELLITES

7 Finally we come to Saturn's dozen or so small moons. Their masses are mostly unknown, but scientists believe their composition to be similar to the small moons of Jupiter. The outermost small moons, Hyperion and Phoebe, were discovered in the nineteenth century, in 1848 and 1898, respectively. The others were first detected in the second half of the twentieth century. Only the moons in or near the rings themselves were actually discovered by *Voyager*.

Just 10,000 km beyond the F ring lie the so-called *co-orbital satellites* Janus and Epimetheus. As the name implies, these two satellites "share" an orbit, but in a very strange way. At any given instant, both moons are in circular orbits about Saturn, but one of them has a slightly smaller orbital radius than the other. Each satellite obeys Kepler's laws, so the inner satellite orbits slightly faster than the outer one and slowly catches up to it. The inner moon takes about four Earth years to "lap" the outer one. As the inner satellite gains ground on the outer one, a strange thing happens. As illustrated in Figure 12.25, when the two get close enough to begin to feel each others' weak gravity, they switch orbits—the new inner moon (which used to be the outer one) begins to pull away from its companion, and the whole process begins again! No one knows why the co-orbital satellites are engaged in this curious dance. Possibly they are portions of a single moon that broke up, perhaps after a meteoritic impact, leaving the two pieces in almost the same orbit.

In fact, several of the other small moons also share orbits, this time with larger moons. Telesto and Calypso have orbits that are synchronized with the orbit of Tethys, always remaining fixed relative to the larger moon, lying precisely 60° ahead and 60° behind it as it travels around Saturn (see Figure 12.26). The moon Helene is similarly tied to Dione. These 60° points are known as **Lagrange points**, after the French mathematician Joseph Louis Lagrange, who first studied them. Later we will see further examples of this special 1:1 orbital resonance in the motion of some asteroids about the Sun, trapped in the Lagrange points of Jupiter's orbit.

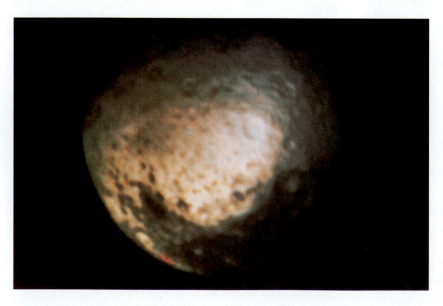

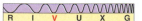

Figure 12.24   Iapetus is one of the most peculiar moons known. The contrast is clearly evident here between its light (icy) trailing surface at top and center and the black leading hemisphere at bottom. The large, circular, black region is called Cassini Regio. Its makeup and origin are unknown.

Figure 12.25   The peculiar motion of Saturn's co-orbital satellites, Janus and Epimetheus, which play a never-ending game of tag as they move on their orbits around the planet. The labeled points represent the locations of the two moons at a few successive times. From A to C, satellite 2 gains on satellite 1. However, before it can overtake it, the two moons swap orbits, and satellite 1 starts to pull ahead of satellite 2 again, through points D and E. The whole process is repeated ad infinitum.

The strangest motion of all is that of the moon Hyperion, which orbits between Titan and Iapetus, at a distance of 1.5 million km from the planet. Unlike most of Saturn's moons, its rotation is not synchronous with its orbital motion. Because of the gravitational effect of Titan, Hyperion's orbit is not exactly circular, so synchronous rotation cannot occur. In response to the competing gravitational influences of Titan and Saturn, this irregularly shaped satellite constantly changes both its rotation speed and its rotation axis, in a condition known as **chaotic rotation**. As Hyperion orbits Saturn, it tumbles apparently at random, never stopping and never repeating itself, in a completely unpredictable way. Since the 1970s, the study of chaos on Earth has revealed new classes of unexpected behavior in even very simple systems. Hyperion is one of the few other places in the universe where this behavior has been unambiguously observed.

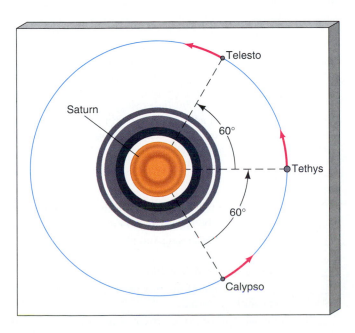

Figure 12.26   The orbits of the moons Telesto and Calypso are tied to the motion of the moon Tethys. The combined gravitational pulls of Saturn and Tethys keep the small moons exactly 60° ahead and behind the larger moon at all times, so that all three moons share an orbit and never change their relative positions.

Chapter Review

SUMMARY

Saturn was the outermost planet known to ancient astronomers. Its rings and moons were not discovered until after the invention of the telescope. Saturn is smaller than Jupiter, but still much larger than any of the terrestrial worlds. Like Jupiter, Saturn rotates rapidly, producing a pronounced flattening, and displays differential rotation. Strong radio emission from the planet's magnetosphere allows the rotation rate of the interior to be determined.

Like all the jovian planets, Saturn has many moons and a planetary *ring system* around it. The rings lie in the planet's equatorial plane, so their appearance from Earth changes as Saturn orbits the Sun.

Weather systems are seen on Saturn, as on Jupiter, although they are less distinct. Short-lived storms are occasionally seen. Saturn has weaker gravity and a more extended atmosphere than Jupiter. The planet's overall butterscotch hue is due to cloud chemistry similar to that occurring in Jupiter's atmosphere. Saturn, like Jupiter, has bands, ovals, and turbulent flow patterns powered by convective motion in the interior.

Like Jupiter, Saturn emits far more radiation into space than it receives from the Sun. Unlike Jupiter, Saturn's excess energy emission is the result of **helium precipitation** (p. 259) in the planet's interior, where helium liquefies and forms droplets, which then fall toward the center of the planet. This process is also responsible for Saturn's observed helium deficit.

Saturn's interior is theoretically similar to that of Jupiter, but it has a thinner layer of metallic hydrogen and a larger core. Its lower mass gives Saturn a less extreme core temperature, density, and pressure than Jupiter. Saturn's conducting interior and rapid rotation produce a strong magnetic field and an extensive magnetosphere large enough to contain the planet's ring system and the innermost 16 moons.

From Earth, the main visible features of Saturn's rings are the **A**, **B**, and **C rings** (p. 259) and the **Cassini** (p. 259) and **Encke Divisions** (p. 259). The Cassini Division is a dark region between the A and B rings. The Encke Division lies near the outer edge of the A ring. The rings are made up of trillions of icy particles ranging in size from dust grains to boulders, all orbiting Saturn like so many tiny moons. Their total mass is comparable to that of a small moon. Both divisions are dark because they are almost empty of ring particles.

The **Roche limit** (p. 261) of a planet is the distance within which the planet's tidal field would overwhelm the internal gravity of an orbiting moon, tearing the moon apart and forming a ring. All known planetary ring systems lie inside their parent planets' Roche limits.

When the *Pioneer* and *Voyager* probes reached Saturn, they found that the rings are actually made up of tens of thousands of narrow **ringlets** (p. 262). Interactions between the ring particles and the planet's inner moons are responsible for much of the fine structure observed in the main rings.

Saturn's narrow **F ring** (p. 263), discovered by the *Pioneer* probe, lies just outside the A ring. It has a kinked, braided structure, apparently caused by two small **shepherd satellites** (p. 264) that orbit close to it and prevent it from breaking up. The *Voyager* probes also discovered the faint **D ring** (p. 262), lying between the C ring and Saturn's cloud layer, and the **E ring** (p. 262), associated with the moon Enceladus.

Many scientists believe that planetary rings have a lifetime of only a few tens of millions of years. The fact that we see rings around all four jovian planets means that they must constantly be reformed or replenished, perhaps by material chipped off moons by meteoritic impact or by the tidal destruction of entire moons.

Saturn's large moon Titan is the second-largest moon in the solar system. Its thick atmosphere obscures the moon's surface, and may be the site of complex cloud and surface chemistry. The existence of Titan's atmosphere is a direct consequence of the cold conditions that prevailed at the time of the moon's formation.

The medium-sized moons of Saturn are made up predominantly of rock and water ice. They show a wide variety of surface terrains, and are also heavily cratered. They are all tidally locked by the planet's gravity into synchronous orbits. The innermost midsized moon Mimas exerts influence over the structure of the rings. The Cassini Division, now known to contain faint ringlets and gaps, is the result of resonance between its particles and Mimas. The moon Iapetus has a marked contrast between its leading and trailing faces, while Enceladus has a highly reflective appearance, possibly the result of water "volcanoes" on its surface.

Saturn's small moons exhibit a wide variety of complex motion. Several moons "share" orbits, in some cases lying at the **Lagrange points** (p. 270) 60° ahead and behind the orbit of a larger moon. The moon Hyperion undergoes **chaotic rotation** (p. 271)— constantly tumbling, in an unpredictable way, as it orbits the planet.

SELF-TEST: True or False?

_____ **1.** Saturn's orbit is almost twice the size of Jupiter's orbit.

_____ **2.** The rotation of Saturn is unlike that of Jupiter; it is slow and shows little differential rotation.

_____ **3.** Saturn's large axial tilt produces strong seasonal variations in the planet's appearance.

_____ **4.** Saturn probably does not have a rocky core.

_____ **5.** The atmosphere of Saturn contains only half the helium of Jupiter's atmosphere but overall, Saturn probably has a normal helium abundance.

_____ **6.** The magnetic field of Saturn is much less than Jupiter's and overall, is about as strong as Earth's magnetic field.

_____ **7.** Saturn is unique among the planets in having a ring system.

_____ **8.** A typical ring particle is 100 meters in diameter.

_____ **9.** The composition of the ring particles of Saturn is predominantly water ice.

_____ **10.** Although Saturn's ring system is tens of thousands of kilometers wide, it is only a few tens of meters thick.

_____ **11.** Saturn has small and mid-sized moons, but no large moons.

_____ **12.** Water ice predominates in Saturn's moons.

_____ **13.** Titan's atmosphere is 10 times denser than Earth's.

_____ **14.** Titan's surface is obscured by thick clouds of water ice.

SELF-TEST: Fill in the Blank

1. Saturn is best known for its spectacular _____ system.

2. Features in the atmosphere of Saturn, as on Jupiter, are mostly hidden by an upper layer of frozen _____ clouds.

3. Saturn's cloud layers are thicker than those of Jupiter because of Saturn's weaker _____.

4. Saturn's excess energy emission is caused by _____.

5. As viewed from Earth, Saturn's ring system is conventionally divided into _____ broad rings.

6. The Cassini Division lies between the _____ and _____ rings.

7. The _____ ring of Saturn is the brightest.

8. The rings exist because they lie within Saturn's _____.

9. Two small moons, known as _____ satellites, are responsible for the unusually complex form of the F ring.

10. The composition of Titan's atmosphere is 90 percent _____.

11. There may be a large amount of _____ sediment covering the surface of Titan.

12. Except for Titan, most of the moons of Saturn are heavily cratered, even though their surfaces are mostly water ice. The low _____ of the ice makes it hard as rock.

13. The icy, highly reflective surface of Enceladus and Saturn's _____ ring are probably related, possibly through volcanism on the moon.

14. When small moons share an orbit with a large moon, they are found at the _____ points of the large moon's orbit.

REVIEW AND DISCUSSION

1. Seen from Earth, Saturn's rings sometimes appear broad and brilliant, but at other times seem to disappear. Why?

2. Why does Saturn have a less varied appearance than Jupiter?

3. Compare and contrast the atmospheres of Saturn and Jupiter, describing how the differences affect the appearance of each planet.

4. Compare the thicknesses of Saturn's various layers (clouds, molecular hydrogen, metallic hydrogen, and core), to the equivalent layers in Jupiter.

5. What is the Roche limit, and what is its relevance to planetary rings?

6. What evidence supports the idea that a relatively recent catastrophic event was responsible for Saturn's rings?

7. What effect does Mimas have on the rings?

8. When *Voyager* passed Saturn in 1980, why didn't it see the surface of Saturn's largest moon, Titan?

9. Compare and contrast Titan with Jupiter's Galilean moons.

10. Why does Titan have a dense atmosphere when other large moons in the solar system don't?

11. What is the evidence for geological activity on Enceladus?

12. What mystery is associated with Iapetus?

13. Describe the behavior of Saturn's co-orbital satellites.

14. Imagine what the sky would look like from Saturn's moon Hyperion. Would the Sun rise and set in the same way it does on Earth? How do you imagine Saturn might look? On what sort of schedule would the planet rise and set?

PROBLEMS

1. How long does it take for Saturn's equatorial flow, moving at 1500 km/h, to encircle the planet?

2. The text states that the total mass of ring material is about 10^{15} tons. Suppose the average ring particle is 6 cm in radius (a large snowball) and has a density of 1000 kg/m^3. How many ring particles would there be?

3. What is the orbital speed of ring particles in the inner part of the B ring? Give your answer in km/s. Compare with the speed of a satellite in low Earth orbit. Why are these speeds so different?

4. Show that Titan's surface gravity is about one-seventh of Earth's, as stated in the text.

PROJECTS

1. Saturn moves more slowly among the stars than any other visible planet. How many degrees per year does it move? Look in an almanac to see where the planet is now. What constellation is it in now? Where will it be in one year?

2. Binoculars may not reveal the rings of Saturn, but most small telescopes will. Use a telescope to look at Saturn. Does Saturn appear oblate? Examine the rings. How are they tilted? Can you see a dark line in the rings? This is the Cassini Division.

It once was thought to be a gap in the rings, but the *Voyager* spacecraft discovered that it is filled with tiny ringlets. Can you see the shadow of the rings on Saturn?

3. While looking at Saturn through a telescope, can you see any of its moons? The moons line up with the rings; Titan is often the farthest out, and always the brightest of them. How many moons can you see? Use an almanac to identify each one you find.

13

URANUS, NEPTUNE, AND PLUTO

The Outer Worlds of the Solar System

LEARNING GOALS

Studying this chapter will enable you to:

1 Describe how both calculation and chance played a major role in the discoveries of the outer planets.

2 Summarize the similarities and differences between Uranus and Neptune, and compare these planets with the other two jovian worlds.

3 Explain what the moons of the outer planets tell us about their past.

4 Contrast the rings of Uranus and Neptune with those of Jupiter and Saturn.

5 Summarize the orbital and physical properties of Pluto, and explain how the Pluto–Charon system differs fundamentally from all the other planets.

(Opposite page, background) This odd looking photo, taken by the *Hubble* telescope orbiting Earth, shows the planet Uranus with its many rings and five of its inner moons. The picture seems odd because each of the moons appears as a string of three dots. It is actually a composite of three images, taken 6 minutes apart, thereby showing the moons' rather rapid revolution just beyond the planet's rings.

(Inset A) The largest of Uranus's inner moons is Miranda, seen here from the perspective of the oncoming *Voyager* spacecraft. Also visible at lower right in the *Hubble* photo, Miranda has less than one percent of the mass of Earth's Moon.

(Insets B, C, and D) Progressively closer views of Miranda, radioed back as *Voyager* skirted to within 30,000 kilometers of this peculiar moon, show several of its prominent features. So many major terrain types are evident in these photos that some scientists have likened Miranda to a cosmic geology museum. Currently, astronomers have no clear consensus about how Miranda's unusual surface features originated, but it seems clear that some very violent events must have been involved, probably about 4 billion years ago. Consult Figure 13.14 for further details about Miranda's cliffs, grooves, and greatly puzzling terrain.

*T*he three outermost planets were unknown to the ancients. All were discovered by telescopic observations—Uranus in 1781, Neptune in 1846, and Pluto in 1930. Uranus and Neptune have very similar masses and radii, so it is natural to consider them together. They are part of the jovian family of planets. Pluto, by contrast, is not a jovian world. It is very much smaller than even the terrestrial planets and generally seems much more moonlike than planetlike in character. Indeed, it may well be a one-time moon that has escaped from one of the outer planets, most likely Neptune. Because of Pluto's similarity to the jovian moons and its possible (although unproven) connection with Neptune, we study it here along with its larger jovian neighbors.

13.1   *The Discovery of Uranus*

①   The planet Uranus was discovered by British astronomer William Herschel in 1781. Herschel was engaged in charting the faint stars in the sky when he came across an odd-looking object that he described as "a curious either nebulous star or perhaps a comet." Repeated observations showed that it was neither. The object appeared as a disk in Herschel's 6-inch telescope and moved relative to the stars, but it traveled too slowly to be a comet. Herschel soon realized that he had found the seventh planet in the solar system. Since this was the first new planet discovered in well over 2000 years, the event caused quite a stir at the time. The story goes that Herschel's first instinct was to name the new planet "Georgium Sidus" (Latin for "George's star") after his king, George III of England. The solar system was saved from a planet named George by the wise advice of another astronomer, Johann Bode. He suggested instead that the tradition of using names from Greco–Roman mythology be continued and that the planet be named after Uranus, the father of Saturn.

Careful observations since its discovery have allowed astronomers to determine the orbital properties of Uranus. Its orbit has a semi-major axis of 19.2 A.U., an eccentricity of 0.05, and a sidereal orbital period of 84.0 years. Since its discovery in 1781, Uranus has completed only two and a half revolutions about the Sun. The planet is in fact just barely visible to the naked eye, if you know exactly where to look. At opposition, it has a maximum angular diameter of 4.1″ and shines just above the unaided eye's threshold of visibility. It looks like a faint, undistinguished star. No wonder it went unnoticed by the ancients. Even today, few astronomers have seen it without a telescope.

Through a large Earth-based optical telescope (see Figure 13.1), Uranus appears hardly more than a tiny pale

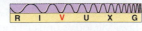

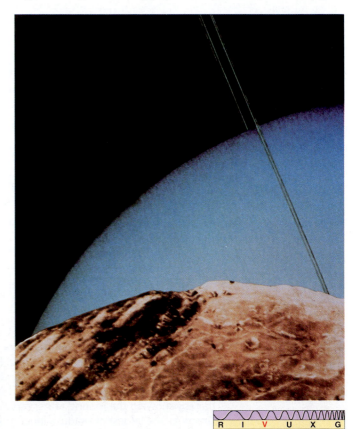

Figure 13.1   Details are barely visible on photographs of Uranus made with large Earth-based telescopes. (Arrows point to three of its moons.)

Figure 13.2   A montage of Uranus and one of its satellites, composed of photos sent back to Earth by the *Voyager 2* spacecraft as it whizzed past this giant planet at 10 times the speed of a rifle bullet. The image of Uranus, taken from a distance of about 100,000 km, shows the planet's blue yet featureless upper atmosphere. The image in the foreground is Miranda, one of Uranus's moons. (Uranus's rings, not visible in the photograph, have been added by an artist.)

greenish disk. With the flyby of *Voyager 2* in 1986, our detailed knowledge of Uranus increased dramatically. Figure 13.2 is a combination of close-up visible-light images of the planet and one of its moons. The apparently featureless atmosphere of Uranus contrasts sharply with the bands and spots visible on all the other jovian worlds.

13.2 *The Discovery of Neptune*

1 Once Uranus was discovered, astronomers set about charting its orbit. The figures we have just listed do indeed describe Uranus's orbital motion, but eighteenth-century astronomers quickly discovered a small discrepancy between the planet's predicted position and where they actually observed it. Try as they might, astronomers could not find an elliptical orbit that fit the planet's trajectory to within the accuracy of their measurements. Half a century after Uranus's discovery, the discrepancy had grown to a quarter of an arc minute, far too big to be explained away as observational error.

The logical conclusion was that an unknown body must be exerting a gravitational force on Uranus—much weaker than that of the Sun, but still measurable. But what body could this be? Astronomers realized that there had to be *another* planet in the solar system perturbing Uranus's motion.

In the 1840s, two mathematicians independently solved the difficult problem of determining this new planet's mass and orbit. A British astronomer, John Adams, reached the solution in September 1845; in June of the following year, the French mathematician Urbain Leverrier came up with essentially the same answer. British astronomers seeking the new planet found nothing during the summer of 1846. In September, a German astronomer named Johann Galle began his own search from the Berlin Observatory, using a newly completed set of more accurate sky charts. He found the new planet within one or two degrees of the predicted position—on his first attempt. After some wrangling over names and credits, the new planet was named Neptune, and Adams and Leverrier (but not Galle!) are now jointly credited with its discovery.

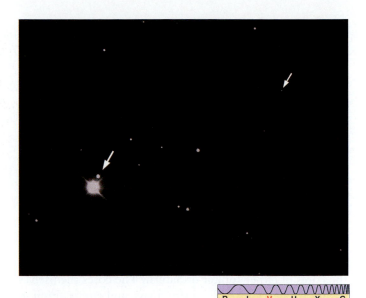

Figure 13.3 Neptune and two of its moons, Triton (large arrow) and Nereid (small arrow), imaged with a large Earth-based telescope.

Neptune orbits the Sun with a semi-major axis of 30.1 A.U. (4.5 billion km) and an eccentricity of just 0.01. Since its sidereal orbital period is 164.8 years, it has not yet completed one revolution since its discovery. Unlike Uranus, Neptune cannot be seen with the naked eye, although it can be seen with a small telescope—in fact, Galileo may actually have seen Neptune, although he had no idea what it really was at the time. Through a large telescope, Neptune appears as a bluish disk, with a maximum angular diameter of 2.4″ at opposition.

Figure 13.3 shows a long Earth-based exposure of Neptune and its largest moon, Triton. Neptune is so distant that surface features are virtually impossible to discern. Even under the best observing conditions, only a few markings can be seen. These are suggestive of multicolored cloud bands—light bluish hues seem to dominate. With *Voyager 2*'s arrival, much more detail emerged, as shown in Figure 13.4. Superficially, at least, Neptune resembles a blue-tinted Jupiter, with atmospheric bands and spots clearly evident.

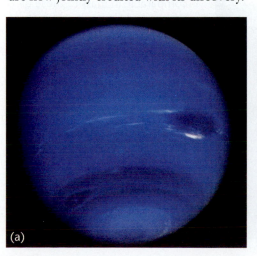

(a)   (b)

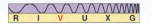

Figure 13.4 (a) Neptune as seen by *Voyager 2,* from a distance of roughly 1 million km. A closer view (b), resolved to about 10 km, shows cloud streaks ranging in width from 50 to 200 km.

13.3 *Uranus and Neptune in Bulk*

MASSES AND RADII

❷ Figure 13.5 shows Uranus and Neptune to scale, along with the Earth for comparison. The two giant planets are quite similar in their bulk properties, too. The diameter of Uranus is 51,100 km (4.0 times that of the Earth); that of Neptune, 49,500 km (3.9 Earth diameters). Their masses (first determined from terrestrial observations of their larger moons and later refined by *Voyager 2*) are 8.7×10^{25} kg (14.5 Earth masses) for Uranus and 1.0×10^{26} kg (17.1 Earth masses) for Neptune. Thus, Uranus's average density is 1200 kg/m³ (1.2 g/cm³), and Neptune's is 1700 kg/m³. These densities imply that large rocky cores constitute a greater fraction of the planets' masses than do the cores of either Jupiter or Saturn. The cores themselves are probably comparable in size, mass, and composition to those of the two larger giants.

ROTATION RATES

Like the other jovian planets, Uranus has a short rotation period. Earth-based observations of the Doppler shifts in spectral lines first indicated that Uranus's "day" was between 10 and 20 hours long. The precise value of the planet's rotation period—accurately determined when *Voyager 2* timed radio signals associated with its magnetosphere—is now known to be 17.2 hours. Again as with Jupiter and Saturn, the planet's atmosphere rotates differentially, but Uranus's atmosphere actually rotates *faster* at the poles (where the period is 14.2 hours) than near the equator (where the period is 16.5 hours).

Each planet in our solar system seems to have some outstanding peculiarity, and Uranus is no exception. Unlike all the other planets, which have their spin axes roughly perpendicular to the ecliptic plane, Uranus's rotation axis lies almost *within* that plane—98° from the perpendicular, to be precise. Relative to the other planets, we might say that Uranus lies tipped over on its side. As a result, the "north" (spin) pole* of Uranus, at some time in its orbit, points almost directly toward the Sun. Half a "year" later, its "south" pole faces the Sun, as illustrated in Figure 13.6. When *Voyager 2* encountered the planet in 1986, the north pole happened to be pointing nearly at the Sun, so it was midsummer in the northern hemisphere.

The strange orientation of Uranus's rotation axis produces some extreme seasonal effects. Starting at the height of northern summer, when the north pole points closest to the Sun, an observer near that pole would see the Sun move in gradually increasing circles in the sky, completing one circuit (counterclockwise) every 17 hours and dipping slightly lower in the sky each day. Eventually the Sun would begin to set and rise again in a daily cycle, and the nights would grow progressively longer with each passing day. Twenty-one years after the summer solstice, the autumnal equinox would occur, with day and night each 8.5 hours long.

As in Chapter 9, we adopt the convention that a planet's rotation is always counterclockwise as seen from above the north pole (that is, planets always rotate from west to east). ∞ (p. 197)

Figure 13.5 Jupiter, Saturn, Uranus, Neptune, and Earth, drawn to scale. Uranus and Neptune are quite similar in their bulk properties. Each probably contains a core about 10 times more massive than the Earth. Jupiter and Saturn are each substantially larger, but their rocky cores are probably comparable in mass to those of Uranus and Neptune.

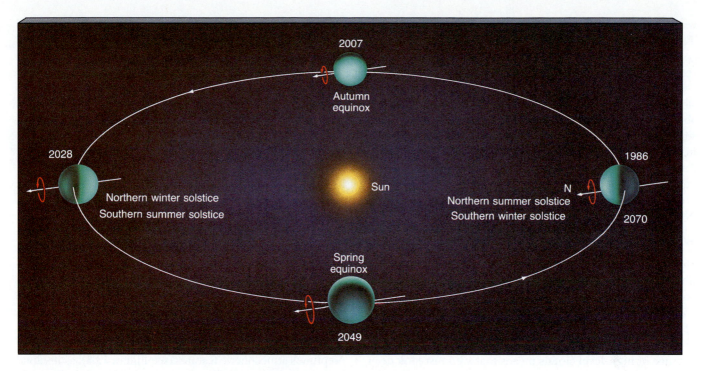

Figure 13.6   Uranus's 98° axial tilt places its equator almost perpendicular to the ecliptic. As a result, the planet experiences the most extreme seasons known in the solar system. The equatorial regions experience two warm seasons (around the two equinoxes) and two cold seasons (at the solstices) each year, while the poles are alternately plunged into darkness for 42 years at a time.

The days would continue to shorten until one day the Sun would fail to rise at all. The period of total darkness that followed would be equal in length to the earlier period of constant daylight, plunging the northern hemisphere into the depths of winter. Eventually, the Sun would rise again; the days would lengthen through the vernal equinox and beyond, and in time the observer would again experience a summer of uninterrupted (though dim) sunshine.

From the point of view of an observer on the equator, by contrast, summer and winter would be almost equally cold seasons, with the Sun never rising far above the horizon. Spring and fall would be the warmest times of year, with the Sun passing almost overhead each day.

No one knows why Uranus is tilted in this way. Some scientists have speculated that a catastrophic event, such as a grazing collision between the planet and another planet-sized body, might have altered the planet's spin axis. There is no direct evidence for such an occurrence, however, and no theory to tell us how we should seek to confirm it.

Neptune's clouds show more variety and contrast than do those of Uranus, and Earth-based astronomers studying them determined a rotation rate for Neptune even before *Voyager 2*'s flyby in 1989. The average rotation period of Neptune's atmosphere is 17.3 hours (virtually identical to that of Uranus). Measurements of Neptune's radio emission by *Voyager 2* showed that the magnetic field of the planet, and presumably also its interior, rotates once every 16.1 hours. Thus, Neptune is unique among the jovian worlds in that its atmosphere rotates *more slowly* than its in-terior. Neptune's rotation axis is inclined 29.6° to a line perpendicular to its orbital plane, quite similar to the 27° tilt of Saturn.

13.4   *The Atmospheres of Uranus and Neptune*

COMPOSITION

Spectroscopic studies of sunlight reflected from Uranus's and Neptune's dense clouds indicate that the two planets' outer atmospheres (the parts we actually measure spectro-scopically) are similar to those of Jupiter and Saturn. The most abundant element is molecular hydrogen (84 percent), followed by helium (about 14 percent) and methane, which is more abundant on Neptune (about 3 percent) than on Uranus (2 percent). Ammonia, which plays such an impor-tant role in the Jupiter and Saturn systems, is not present in any significant quantity in the outermost jovian worlds.

Comparing this information with that given previously for Jupiter and Saturn, we can see that the abundances of gaseous ammonia and methane vary systematically among the jovian planets. Jupiter has much more gaseous ammonia than methane, but as we proceed outward, we find that the more distant planets have steadily decreasing amounts of ammonia and relatively greater amounts of methane. The reason for this variation is temperature. Ammonia gas freezes into ammonia ice crystals at about 70 K. This is cooler than the cloud-top temperatures of Jupiter and Saturn but warmer than those of Uranus (58 K) and Neptune (59 K). Thus, the outermost jovian planets have little or no *gaseous* ammonia in their atmospheres, so their spectra (which record atmospheric gases only) show just traces of ammonia.

The increasing amounts of methane are largely responsible for the outer jovian planets' blue coloration. Methane absorbs long-wavelength red light quite efficiently, so that sunlight reflected from the planet's atmospheres is deficient in red and yellow photons and appears blue-green or blue. As the concentration of methane increases, the reflected light should appear bluer. This is just the trend observed: Uranus, with less methane, looks blue-green, while Neptune, with more methane, looks distinctly blue.

WEATHER

Voyager 2 detected only a few features in Uranus's atmosphere, and even these became visible only after extensive computer enhancement (see Figure 13.7). Uranus apparently lacks any significant internal heat source, and because of the planet's low surface temperature, its clouds are found only at lower-lying, warmer levels in the atmosphere. The absence of high-level clouds means that we must look deep into the planet's atmosphere to see any structure, and the bands and spots that characterize flow patterns on the other jovian worlds are largely "washed out" on Uranus by intervening stratospheric haze.

With computer-processed images, astronomers have learned that Uranus has atmospheric clouds and flow patterns that move around the planet in the same sense as the planet's rotation, with wind speeds in the range 200–500 km/h. Tracking these clouds allowed the measurement of the differential rotation mentioned earlier. Despite the odd angle at which sunlight is currently striking the surface (recall that it is just after midsummer in the northern hemi-

sphere), the planet's rapid rotation still channels the wind flow into bands reminiscent of those found on Jupiter and Saturn. Even though the predominant wind flow is in the east–west direction, Uranus's atmosphere seems to be quite efficient at transporting energy from the heated north to the unheated southern hemisphere. Although the south is currently in total darkness, the temperature there is only a few kelvins less than in the north.

Neptune's cloud and band structure is much more easily seen. Although it lies at a greater distance from the Sun, Neptune's upper atmosphere is actually slightly warmer than Uranus's. Like Jupiter and Saturn, but unlike Uranus, Neptune has an internal energy source—in fact, it radiates 2.7 times more heat than it receives from the Sun. The cause of this heating is still uncertain. Some scientists have suggested that Neptune's excess methane has helped "insulate" the planet, tending to maintain its initially high internal temperature. If that is so, then Neptune's internal heat has the same basic explanation as Jupiter's—it is energy left over from the planet's formation. The combination of extra heat and less haze may be responsible for the greater visibility of Neptune's atmospheric features (see Figure 13.8), as its cloud layers lie at higher levels in the atmosphere than do Uranus's.

Neptune sports several storm systems similar in appearance to those seen on Jupiter (and assumed to be produced and sustained by the same basic processes). The largest such storm, known simply as the **Great Dark Spot**, is shown in Figure 13.8(a). Discovered by *Voyager 2* in 1989, the Spot was about the size of the Earth, was located near the planet's equator, and exhibited many of the same general characteristics as the Great Red Spot on Jupiter. The flow around it was counterclockwise, as with the Red Spot, and there appeared to be turbulence where the winds associated with the Great Dark Spot interacted with the zonal

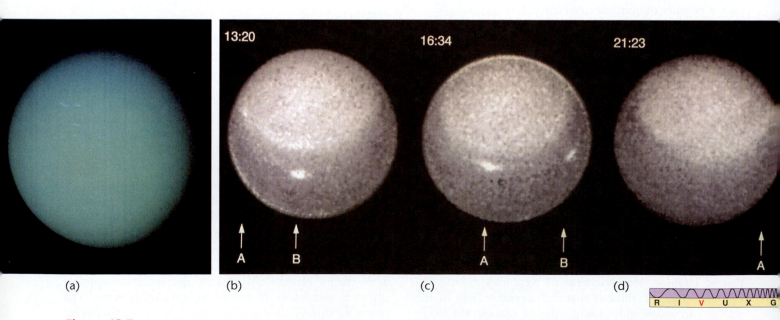

(a) (b) 13:20 A B (c) 16:34 A B (d) 21:23 A

R I V U X G

Figure 13.7   (a) This *Voyager* view of Uranus approximates the planet's true color but shows little else. Parts (b), (c), and (d) are *Hubble Space Telescope* photographs made at roughly four-hour intervals, showing the motion of a pair of bright clouds (labeled A and B) in the planet's southern hemisphere. (The numbers at the top give the time of each photo.)

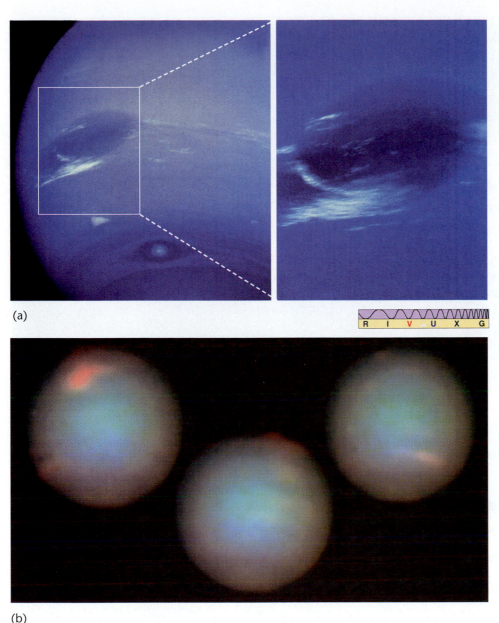

(a)

(b)

Figure 13.8 (a) Close-up views, taken by *Voyager 2* in 1989, of the Great Dark Spot of Neptune, a large storm system in the planet's atmosphere, possibly similar in structure to Jupiter's Great Red Spot. Resolution in the photo on the right is about 50 km; the entire dark spot is roughly the size of planet Earth. (b) These three *Hubble Space Telescope* views of Neptune were taken about 10 days apart in 1994, when the planet was some 4.5 billion km from the Earth. The aqua color is caused by absorption of red light by methane; the cloud features (mostly methane ice crystals) are tinted pink here because they were imaged in the infrared, but are really white in visible light. Neptune apparently has a remarkably dynamic atmosphere that changes over just a few days. Notice that the Great Dark Spot has now disappeared.

flow to its north and south. The flow around this and other dark spots may drive updrafts to high altitudes, where methane crystallizes out of the atmosphere to form the high-lying cirrus clouds. Astronomers did not have long to study the Dark Spot's properties, however. As shown in Figure 13.8(b), when the *Hubble Space Telescope* viewed Neptune in 1994, the Spot had disappeared.

13.5 *Magnetospheres and Internal Structure*

Voyager 2 found that both Uranus and Neptune have fairly strong internal magnetic fields—about 100 times stronger than Earth's field and 1/10 as strong as Saturn's. However, because the radii of Uranus and Neptune are so much larger than the radius of Earth, the magnetic fields at the cloud tops, spread out over a far larger volume than the field on Earth, are actually comparable in strength to

Earth's field. Uranus and Neptune each have a substantial magnetosphere, populated largely by electrons and protons either captured from the solar wind or created from ionized hydrogen gas escaping from the planets themselves.

When *Voyager 2* arrived at Uranus, it discovered that the planet's magnetic field is tilted at about 60° to the axis of rotation. On Earth, such a tilt would put the North (magnetic) Pole somewhere in the Caribbean. Furthermore, the magnetic field lines are *not* centered on the planet. It is as though Uranus's field were due to a bar magnet that is tilted with respect to the planet's rotation axis and displaced from the center by about one-third the radius of the planet. Figure 13.9 compares the magnetic field structures of the four jovian planets. The locations and orientations of the bar magnets represent the observed planetary fields, and the sizes of the bars indicate magnetic field strength.

Because dynamo theories generally predict that the magnetic axis should be roughly aligned with the rotation

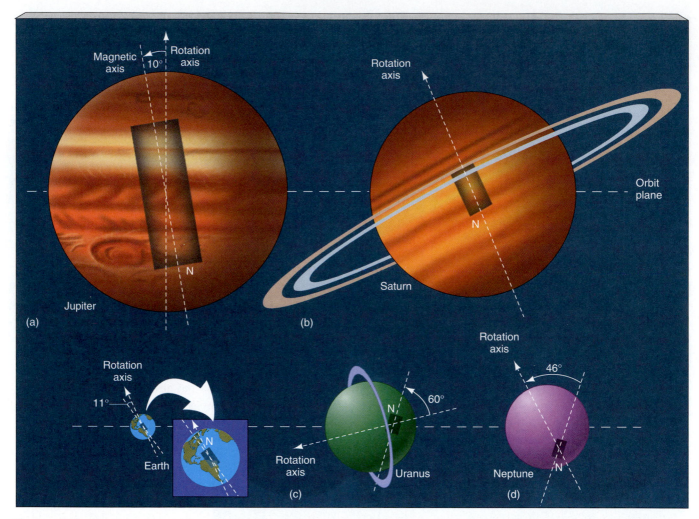

Figure 13.9   A comparison of the magnetic field strengths, orientations, and offsets in the four jovian planets: (a) Jupiter, (b) Saturn, (c) Uranus, (d) Neptune. The planets are drawn to scale, and in each case the magnetic field is represented as though it came from a simple bar magnet. The size and location of each magnet represent the strength and orientation of the planetary field. Notice that the fields of Uranus and Neptune are significantly offset from the center of the planet and are significantly inclined to the planet's rotation axis. The Earth's magnetic field is shown for comparison.

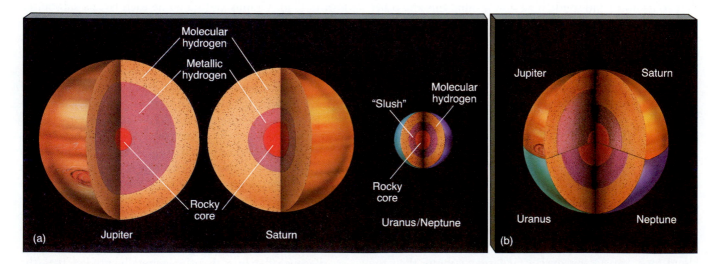

Figure 13.10   A comparison of the interior structures of the four jovian planets. (a) The planets drawn to scale. (b) The relative proportions of the various internal zones.

axis—as on Earth, Jupiter, Saturn, and the Sun—the misalignment on Uranus suggested to some researchers that perhaps the planet's field had been caught in the act of reversing. Another possibility was that the oddly tilted field was in some way related to the planet's axial tilt—perhaps one catastrophic collision skewed both axes at the same time. Those ideas evaporated in 1989 when *Voyager 2* found that Neptune's field is also inclined to the planet's rotation axis, at an angle of 46° (see Figure 13.9d), and also substantially offset from the center. It now appears that the internal structures of Uranus and Neptune are different from those of Jupiter and Saturn, and this difference somehow changes the nature of the field-generation process.

Theoretical models indicate that Uranus and Neptune have rocky cores similar to those found in Jupiter and Saturn—about the size of Earth and perhaps 10 times more massive. However, the pressure outside the cores of Uranus and Neptune (unlike the pressure within Jupiter and Saturn) is too low to force hydrogen into the metallic state, so hydrogen stays in its molecular form all the way in to the planets' cores. Astronomers theorize that, deep below the cloud layers, Uranus and Neptune may have high-density, "slushy" interiors containing thick layers of water clouds. It is also possible that much of the planets' ammonia could be dissolved in the water, accounting for the absence of ammonia at higher cloud levels. Such an ammonia solution would provide a thick, electrically conducting ionic layer that could conceivably explain the planets' magnetic fields. At the present time, however, we don't know enough about the interiors to assess the correctness of this picture. Our current state of knowledge is summarized in Figure 13.10, which compares the internal structures of the four jovian worlds.

13.6 *The Moon Systems of Uranus and Neptune*

3 Like Jupiter and Saturn, both Uranus and Neptune have extensive moon systems, each consisting of a few large moons, long known from the Earth, and many smaller moonlets, discovered by *Voyager 2*.

URANUS'S MOONS

William Herschel discovered and named the two largest of Uranus's five major moons, Titania and Oberon, in 1789. British astronomer William Lassell found the next largest, Ariel and Umbriel, in 1851. Gerard Kuiper found the smallest, Miranda, in 1948. In order of increasing distance from the planet, they are Miranda (at 5.1 planetary radii), Ariel (at 7.5), Umbriel (at 10.4), Titania (at 17.1), and Oberon (at 22.8). The 10 smaller moons discovered by *Voyager 2* all lie inside the orbit of Miranda. Many of them are intimately related to Uranus's ring system. All the moons revolve in Uranus's skewed equatorial plane, almost perpendicular to the ecliptic, in circular, tidally locked orbits. Their properties are listed in Table 13-1. Because the satellites share Uranus's odd orientation, they experience the same extreme seasons as their parent planet.

The five largest moons are similar in many respects to the six midsized moons of Saturn. Their densities lie in the range 1300–1600 kg/m^3, suggesting composition of ice and rock, like Saturn's moons, and their diameters range from 1600 km for Titania and Oberon, to 1200 km for Umbriel and Ariel, to 490 km for Miranda. Uranus has no moons

TABLE 13-1   *The Moons of Uranus*

NAME	DISTANCE FROM URANUS		ORBIT PERIOD	SIZE	MASS	DENSITY	
	(km)	(planet radii)	(days)	(longest diameter, km)	(Earth Moon masses)	(kg/m^3)	(g/cm^3)
Cordelia	49,700	1.95	0.34	40			
Ophelia	53,800	2.11	0.38	50			
Bianca	59,200	2.32	0.44	50			
Cressida	61,800	2.42	0.46	60			
Desdemona	62,700	2.45	0.48	60			
Juliet	64,600	2.53	0.50	80			
Portia	66,100	2.59	0.51	80			
Rosalind	69,900	2.74	0.56	60			
Belinda	75,300	2.95	0.63	60			
Puck	86,000	3.37	0.76	170			
Miranda	130,000	5.09	1.41	485	0.0011	1300	1.3
Ariel	191,000	7.48	2.52	1160	0.018	1600	1.6
Umbriel	266,000	10.4	4.14	1190	0.018	1400	1.4
Titania	436,000	17.1	8.71	1610	0.048	1600	1.6
Oberon	583,000	22.8	13.5	1550	0.040	1500	1.5

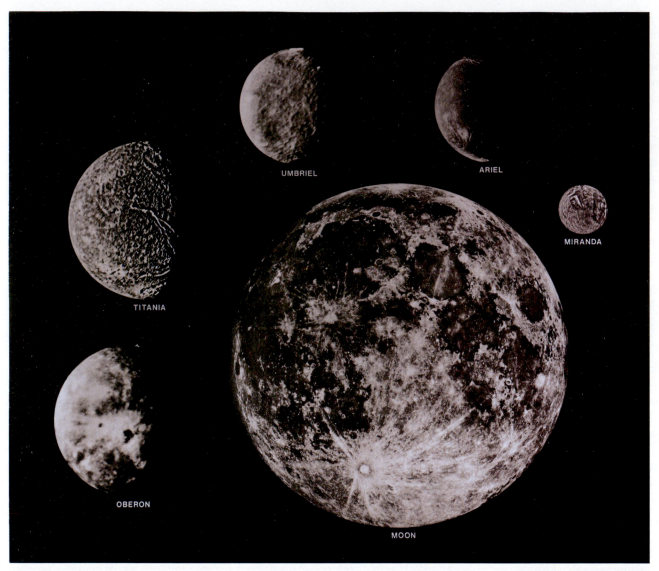

Figure 13.11 The five largest moons of Uranus, to scale. In order of increasing distance from the planet, they are Miranda, Ariel, Umbriel, Titania, and Oberon. Earth's moon is shown for comparison.

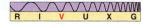

(a)

(b)

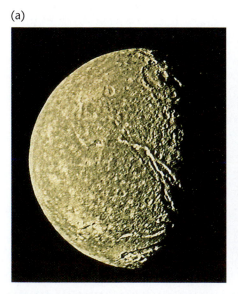

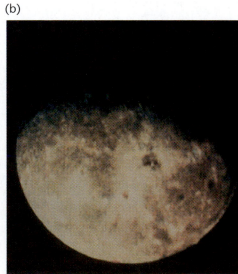

Figure 13.12 Close-up comparison of Uranus's two largest moons, Titania (a) and Oberon (b). Their appearance, structure, and history may be quite similar to those of Saturn's moon Rhea. Smallest details visible on both moons are about 15 km across.

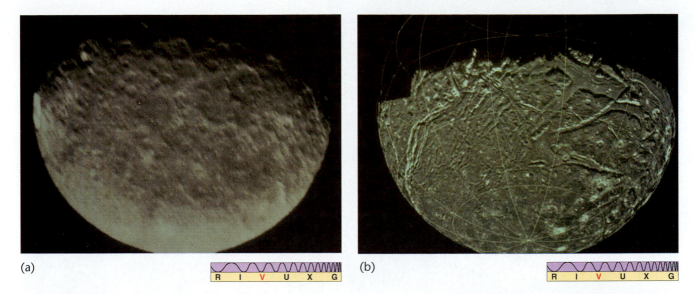

(a)   (b)

Figure 13.13 (a) The moon Umbriel is one of the darkest bodies in the solar system. Its most noteworthy feature is a bright white spot on its sunward side. (b) Ariel is similar in size but has a brighter surface. Unlike Umbriel, its surface shows signs of past geological activity. Resolution is approximately 10 km.

comparable to the Galilean satellites of Jupiter, nor to Saturn's single large moon, Titan. Figure 13.11 shows Uranus's large moons to scale, along with the Earth's Moon for comparison.

The two outermost moons, Titania and Oberon (shown in Figure 13.12), are heavily cratered and show little indication of any geological activity. Their overall appearance (and quite possibly their history) is comparable to that of Saturn's moon Rhea, except that they lack Rhea's wispy streaks. Also, like all Uranus's moons, they are considerably less reflective than Saturn's satellites, suggesting that their icy surfaces are quite dirty. One possible reason for this might simply be that the planetary environment in the vicinity of Uranus and Neptune contains more small "sooty" particles than does the solar system closer to the Sun. An alternative explanation, now considered more likely by many planetary scientists, cites the effects of radiation and high-energy particles striking the surfaces of these moons. These impacts tend to break up the molecules on the moons' surfaces, eventually leading to chemical reactions that slowly build up a layer of dark, organic material. This **radiation darkening** is thought to contribute to the generally darker coloration of many of the moons and rings in the outer solar system. In either case, the longer a moon has been inactive and untouched by meteoritic impact, the darker its surface should be.

The darkest of the moons of Uranus is Umbriel (Figure 13.13a). It displays little evidence for any past surface activity; its only mark of distinction is a bright spot about 30 km across, of unknown origin, on its sunward side. By contrast, Ariel (Figure 13.13b), similar in size to Umbriel but closer to Uranus, does appear to have experienced some activity in the past. It shows signs of resurfacing in places and exhibits surface cracks a little like those seen on another of Saturn's moons, Tethys. However, unlike Tethys, whose cracks are probably due to meteoritic impact, Ariel's activity probably occurred as internal forces and external tidal stresses (due to the gravitational pull of Uranus) distorted the moon and cracked its surface.

Strangest of all Uranus's icy moons is Miranda, shown in Figure 13.14. Before the *Voyager 2* encounter, astronomers thought that Miranda would most resemble Mimas, the moon of Saturn whose size and location it most closely approximates. However, instead of being a relatively uninteresting cratered, geologically inactive world, Miranda displays a wide range of surface terrains, including ridges, valleys, large oval faults, and many other tortuous geological features. In order to explain why Miranda seems to combine so many different types of surface features, some researchers have hypothesized that this baffling object has been catastrophically disrupted several times (from within or without), with the pieces falling back together in a chaotic, jumbled way. Certainly, the frequency of large craters on the outer moons suggests that destructive impacts may once have been quite common in the Uranus system. It will be a long time, though, before we can obtain more detailed information to test this theory.

NEPTUNE'S MOONS

From Earth, we can see only two moons orbiting Neptune. William Lassell discovered the inner moon, Triton, in 1846. The outer moon, Nereid, was located by Gerard Kuiper in 1949. *Voyager 2* discovered six additional moons, all less than a few hundred kilometers across and all lying within Nereid's orbit. Neptune's known moons are listed in Table 13-2.

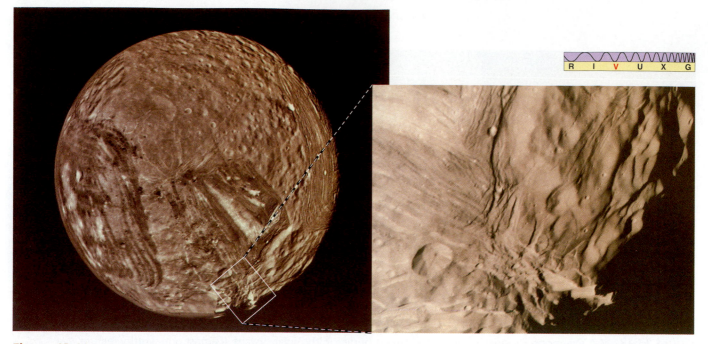

Figure 13.14 Miranda, an asteroid-sized moon of Uranus photographed by *Voyager 2,* has a strange, fractured surface suggestive of a violent past, but the cause of the grooves and cracks is presently unknown. The resolution here is about 1 km.

In its moons we find Neptune's contribution to our list of solar system peculiarities. Unlike the other jovian worlds, Neptune has no regular moon system. The larger moon, Triton, is 2800 km in diameter and occupies a circular retrograde orbit 354,000 km (14.2 planetary radii) from the planet, inclined at about 20° to Neptune's equatorial plane. It is the only large moon in our solar system to have a retrograde orbit. The other moon visible from Earth, Nereid, is only 200 km across. It orbits Neptune in the prograde sense, but on an elongated trajectory that brings it as close as 1.4 million km to the planet and as far away as 9.7 million km. Nereid is probably similar in both size and composition to Neptune's small inner moons.

Voyager 2 approached to within 24,000 km of Triton's surface, providing us with essentially all that we now know about that distant, icy world. Astronomers redetermined the moon's radius (which was corrected downward by about 20 percent) and measured its mass for the first time.

Along with Saturn's Titan and the four Galilean moons of Jupiter, Triton is one of the six large moons in the outer solar system. Triton is the smallest of them, with about half the mass of the next smallest, Jupiter's Europa.

Lying 4.5 billion km from the Sun, and with a fairly reflective surface, Triton has a surface temperature of just 37 K. It has a tenuous nitrogen atmosphere, perhaps a hundred thousand times thinner than Earth's, and a surface that most likely consists primarily of water ice. A *Voyager 2* mosaic of Triton's south polar region is shown in Figure 13.15. The moon's low temperatures produce a layer of nitrogen frost that forms and evaporates over the polar caps, a little like the carbon dioxide frost responsible for the seasonal caps on Mars. The frost is visible as the pinkish region on the left of Figure 13.15. Overall, there is a marked lack of cratering on Triton, presumably indicating that surface activity has obliterated the evidence of most impacts. There are many other signs of an active past. Triton's face is scarred by large fis-

TABLE 13-2 *The Moons of Neptune*

NAME	DISTANCE		ORBIT PERIOD	SIZE	MASS	DENSITY	
	(km)	(planet radii)	(days)	(longest diameter, km)	(Earth Moon masses)	(kg/m³)	(g/cm³)
Naiad	48,200	1.94	0.30	60			
Thalassic	50,100	2.02	0.31	80			
Designate	52,500	2.12	0.33	180			
Galatea	62,000	2.50	0.43	150			
Larissa	73,600	2.97	0.55	190			
Proteus	118,000	4.76	1.12	415			
Triton	354,000	14.3	5.88*	2760	0.291	2100	2.1
Nereid	5,520,000	223	360.2	200	0.0000034	2000	2.0

*Indicates a retrograde orbit

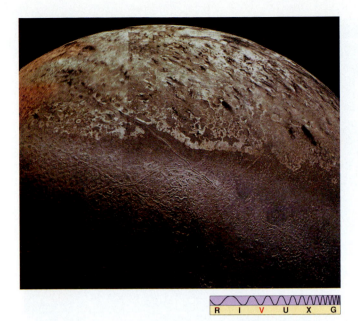

R I V U X G

Figure 13.15   The south polar region of Triton, showing a variety of terrains, ranging from deep ridges and gashes to what appear to be frozen water lakes, all indicative of past surface activity. The pinkish region at the left is nitrogen frost, forming the moon's polar cap. Resolution is about 4 km.

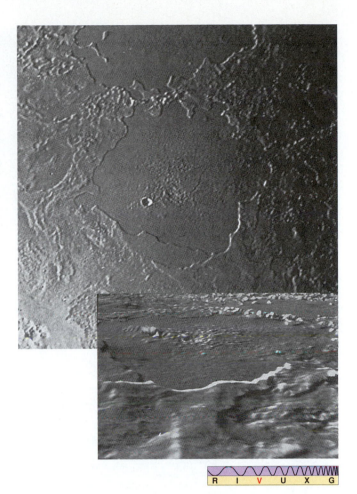

R I V U X G

Figure 13.16   Scientists believe that this lakelike feature on Triton may have been caused by the eruption of an ice volcano. The water "lava" has since solidified, leaving a smooth surface. The absence of craters indicates that this eruption was a relatively recent event in Triton's past. The nearly circular feature at the center of this image spans some 200 km in diameter; its details are resolved to a remarkable 1 km. The insert is a computer-generated view illustrating the topographic relief of the same area.

sures similar to those seen on Ganymede, and the moon's odd cantaloupe-like terrain may indicate repeated faulting and deformation over the moon's lifetime. In addition, Triton has numerous frozen "lakes" of water ice (Figure 13.16), which are believed to be volcanic in origin.

Triton's surface activity is not just a thing of the past. As *Voyager 2* passed the moon, its cameras detected two great jets of nitrogen gas erupting from below the surface, rising several kilometers into the sky. It is thought that these "geysers" result when liquid nitrogen below Triton's surface is heated and vaporized by some internal energy source, or perhaps even by the Sun's feeble light. Vaporization produces high pressures, which force the gas through cracks and fissures in the crust, creating the displays *Voyager 2* saw. Scientists conjecture that nitrogen geysers may be very common on Triton and are perhaps responsible for much of the moon's thin atmosphere.

The event or events that placed Triton on a retrograde orbit and Nereid on such an eccentric path are unknown, but they are the subject of considerable speculation. Triton's peculiar orbit and surface features suggest to some astronomers that the moon did not form as part of the Neptune system but instead was captured, perhaps not too long ago. Other astronomers, basing their views on Triton's chemical composition, maintain that it formed as a "normal" moon but was later kicked into its abnormal orbit by some catastrophic event, such as an interaction with another similar-sized body. It has even been suggested that the planet Pluto may have played a role in this process, although no really convincing demonstration of such an encounter has ever been presented. The surface deformations

on Triton certainly suggest fairly violent and relatively recent events in the moon's past. However, they were most likely caused by the tidal stresses produced in Triton as Neptune's gravity circularized its orbit and synchronized its spin, and they give little indication of the processes leading to the orbit in the first place.

Whatever its past, Triton's future is fairly clear. Because of its retrograde orbit, the tidal bulge Triton raises on Neptune tends to make the moon spiral *toward* the planet rather than away from it (as our Moon moves away from Earth). Thus, Triton is doomed to be torn apart by Neptune's tidal gravitational field, probably in no more than 100 million years or so, the time required for the moon's inward spiral to bring it inside Neptune's Roche limit (see Chapter 12). ∞ (p. 261) By that time, it is conceivable that Saturn's ring system may have disappeared, so that Neptune will then be the planet in the solar system with spectacular rings!

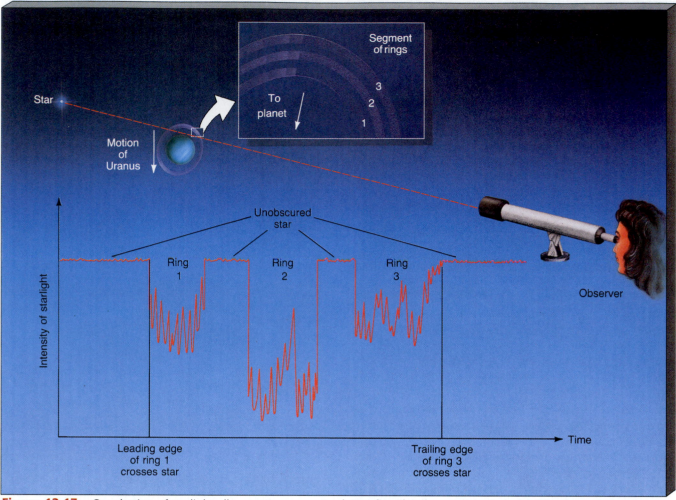

Figure 13.17   Occultation of starlight allows astronomers to detect fine detail on a distant planet. The rings of Uranus were discovered using this technique.

13.7 *The Rings of the Outermost Jovian Planets*

THE RINGS OF URANUS

④ All of the jovian planets have rings. The ring system surrounding Uranus was discovered in 1977, when astronomers observed it passing in front of a bright star, momentarily dimming its light. Such a **stellar occultation** (see Figure 13.17) happens a few times per decade and allows astronomers to measure planetary structures that are too small and faint to be detected directly. The 1977 observation was actually aimed at studying the planet's atmosphere by watching how it absorbed starlight. However, 40 minutes before and after Uranus itself occulted the star, the flickering starlight revealed the presence of a set of rings. The discovery was particularly exciting because, at the time, only Saturn was known to have rings. Jupiter's rings went unseen until *Voyager 1* arrived there in 1979, and those of Neptune were unambiguously detected only in 1989, by *Voyager 2*.

The ground-based observations revealed the presence of a total of nine thin rings. The main rings, in order of increasing radius, are named Alpha, Beta, Gamma, Delta,

Figure 13.18   The main rings of Uranus, as imaged by *Voyager 2*. All of the rings known before *Voyager 2*'s arrival can be seen in this photo. From the inside out, they are 6, 5, 4, Alpha, Beta, Eta, Gamma, Delta, and Epsilon. Resolution is about 10 km, which is just about the width of most of these rings. The two rings discovered by *Voyager 2* are too faint to be seen here.

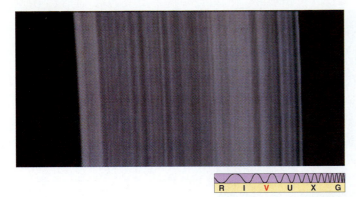

Figure 13.19 A close-up of Uranus's Epsilon ring, showing some of its internal structure. The width of the rings averages 30 km; special image processing has magnified the resolution to about 100 meters.

TABLE 13-3   *The Rings of Uranus*

RING	INNER		OUTER RADIUS*		WIDTH
	(km)	(planet radii)	(km)	(planet radii)	(km)
1986U2R	37,000	1.45	39,500	1.55	2500
6	41,900	1.64			2
5	42,200	1.65			2
4	42,600	1.67			2
Alpha	44,700	1.75			10
Beta	45,700	1.79			10
Eta	47,200	1.85			1
Gamma	47,600	1.86			3
Delta	48,300	1.89			6
1986U1R	50,000	1.96			2
Epsilon	51,200	2.00			20–90

Most of Uranus's rings are so thin that there is little difference between their inner and outer radii.

and Epsilon, and they range from 44,000 to 51,000 km from the planet's center. All of these lie within the Roche limit of Uranus, which is about 62,000 km. A fainter ring, known as the Eta ring, lies between the Beta and Gamma rings, and three other faint rings, known as 4, 5, and 6, lie between the Alpha ring and the planet itself. In 1986, *Voyager 2* discovered two more even fainter rings, one between Delta and Epsilon and one between ring 6 and Uranus. The main rings are shown in Figure 13.18. More details on the rings are provided in Table 13-3.

Uranus's rings are quite different from those of Saturn. While Saturn's rings are bright and wide with relatively narrow gaps between them, the rings of Uranus are dark, narrow, and widely spaced. With the exception of the Epsilon ring, which is about 100 km wide, the rings of Uranus are all less than 10 km wide, and the spacing between them ranges from a few hundred to about a thousand kilometers. However, like Saturn's rings, all of Uranus's rings are less than a few tens of meters thick (that is, measured in the direction perpendicular to the ring plane).

The density of particles within the rings themselves is comparable to that found in Saturn's A and B rings. The particles that make up Saturn's rings range in size from dust grains to boulders, but in the case of Uranus, the particles show a much smaller spread—few if any are smaller than a centimeter or so in diameter. The ring particles are also considerably less reflective than Saturn's ring particles, possibly because they are covered with the same dark material as Uranus's moons. The Epsilon ring (shown in detail in Figure 13.19), the widest of the 11, exhibits properties a little like those of Saturn's F ring. It is slightly eccentric (its eccentricity is 0.008) and of variable width, although no braids are found. It also appears to be composed of ringlets.

Like the F ring of Saturn, Uranus's narrow rings require shepherding satellites to keep them from diffusing away. In fact, the theory of shepherd satellites was first worked out to explain the rings of Uranus, which had been detected by stellar occultation even before *Voyager 2*'s Saturn encounter. Thus, the existence of the F ring did not come as quite such a surprise as it might otherwise have

done! Presumably many of the small inner satellites of Uranus play some role in governing the appearance of the rings. *Voyager 2* detected the shepherds of the Epsilon ring, Cordelia and Ophelia (see Figure 13.20). Many other, undetected shepherd satellites must also exist.

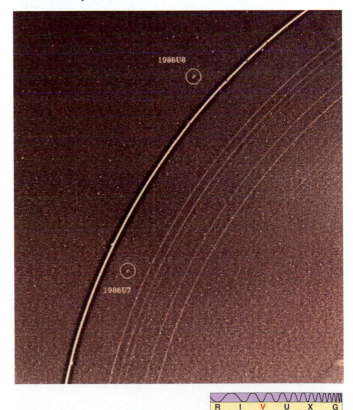

Figure 13.20 These two small moons, discovered by *Voyager 2* in 1986 and named Cordelia and Ophelia, tend to "shepherd" Uranus's Epsilon ring, thus keeping it from diffusing away into space.

Figure 13.21 Neptune's faint rings. In this long-exposure image, the planet (center) is heavily overexposed and has been artificially blotted out to make the rings easier to see. One of the two fainter rings lies between the inner bright ring and the planet. The other lies between the two bright rings.

THE RINGS OF NEPTUNE

As shown in Figure 13.21 and presented in more detail in Table 13-4, Neptune is surrounded by four dark rings. Three are quite narrow, like the rings of Uranus, and one is quite broad and diffuse, more like Jupiter's ring. The dark coloration probably results from radiation darkening, as discussed earlier in the context of the moons of Uranus. All the rings lie within Neptune's Roche limit. The outermost (Adams) ring is noticeably clumped in places. From Earth we see not a complete ring, but only partial arcs—the unseen parts of the ring are simply too thin (unclumped) to be detected. The connection between the rings and the planet's small inner satellites has not yet been firmly established, although many astronomers believe that the clumping is caused by shepherd satellites.

While all the jovian worlds have ring systems, the rings themselves differ widely from planet to planet. This leads us to ask: Is there some "standard" way in which rings form around a planet? And is there a standard manner in which ring systems evolve? Or do the processes of ring formation and evolution depend entirely on the particular planet in question? If, as now appears to be the case, ring systems are relatively short-lived, their formation must be a fairly common event. Otherwise, we would not expect to find rings around all four jovian planets at once. There are also many indications that the individual planetary environment plays an important role in determining a ring system's appearance and longevity. Although many aspects of ring formation and evolution are now understood, it must be admitted that no comprehensive theory yet exists.

13.8 The Discovery of Pluto

❶ By the end of the nineteenth century, observations of the orbits of Uranus and Neptune suggested that Neptune's influence was not sufficient to account for all of the irregularities in Uranus's motion. Further, it seemed that Neptune itself might be affected by some other unknown body. Following their success in the discovery of Neptune, astronomers hoped to pinpoint the location of this new planet using similar techniques. One of the most ardent searchers was Percival Lowell, a capable, persistent observer and one of the best-known astronomers of his day. (Recall that he was the leading proponent of the theory that the "canals" on Mars were constructed by an intelligent race of Martians—see *Interlude 10-1*.) ∞ (p. 223)

TABLE 13-4   *The Rings of Neptune*

RING	INNER		OUTER RADIUS*		WIDTH
	(km)	(planet radii)	(km)	(planet radii)	(km)
Galle (1989N3R)*	41,900	1.69			15
Leverrier (1989N2R)*	53,200	2.15			30
1989N4R	53,200	2.15	59,000	2.38	5800
Adams (1989N1R)*	62,900	2.54			50

Three of Neptune's rings are so thin that there is little difference between their inner and outer radii.

Basing his investigation primarily on the motion of Uranus (Neptune's orbit was still relatively poorly determined at the time), Lowell set about calculating where the supposed ninth planet should be. He searched for it, without success, during the decade preceding his death in 1916. Not until 14 years later did American astronomer Clyde Tombaugh, working with improved equipment and photographic techniques at the Lowell Observatory, finally succeed in finding Lowell's ninth planet, only 6° away from Lowell's predicted position. The new planet was named Pluto for the Roman god of the dead who presided over eternal darkness (and also because its first two letters and its astrological symbol ♇ are Lowell's initials). Its discovery was announced on March 13, 1930, Percival Lowell's birthday.

On the face of it, the discovery of Pluto looks like another spectacular success for celestial mechanics. Unfortunately, it now appears that the supposed irregularities in the motions of Uranus and Neptune did not exist and that the mass of Pluto, not measured accurately until the 1980s, is far too small to have caused them anyway. The discovery of Pluto owed much more to simple luck than to elegant mathematics!

Pluto's orbital semi-major axis is 39.5 A.U. (5.9 billion km). But unlike the paths of the other outer planets, Pluto's orbit is quite elongated, with an eccentricity of 0.25. It is also inclined at 17.2° to the plane of the ecliptic. Already we have some indications that Pluto is unlike its jovian neighbors. Because of its substantial orbital eccentricity, Pluto's distance from the Sun varies considerably. At perihelion, it lies 29.7 A.U. (4.4 billion km) from the Sun, inside the orbit of Neptune. At aphelion, the distance is 49.3 A.U. (7.4 billion km), well outside Neptune's orbit. Pluto last passed perihelion in 1989, and it will remain inside Neptune's orbit until 1999. Its sidereal period is 248.6

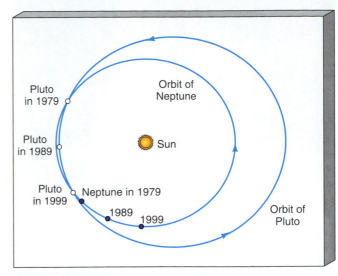

Figure 13.22 The orbits of Neptune and Pluto cross, although Pluto's orbital inclination and a 3:2 resonance prevent the planets from actually coming close to one another. Between 1979 and 1999, Pluto is inside Neptune's orbit, making Neptune the most distant planet from the Sun.

years, so the next perihelion passage will not occur until the middle of the twenty-third century. Pluto's orbital period is apparently exactly 1.5 times that of Neptune—the two planets are locked into a 3:2 resonance (two orbits of Pluto for every three of Neptune) as they orbit the Sun. The orbits of these two outer planets are sketched in Figure 13.22.

At nearly 40 A.U. from the Sun, Pluto is often hard to distinguish from the background stars. As the two photographs of Figure 13.23 indicate, the planet is actually considerably fainter than many stars in the sky. Like Neptune, it is never visible to the naked eye.

Figure 13.23 These two photographs, taken one night apart, show motion of the planet Pluto (arrow) against a field of much more distant stars. Most of Pluto's apparent motion in these two frames is actually due to the orbital motion of the Earth rather than that of Pluto.

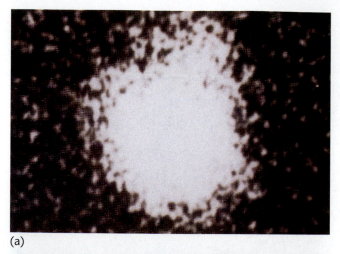

(a)

(b)

R I V U X G

Figure 13.24 (a) The discovery photograph of Pluto's moon, Charon. The moon is the small bump on the top right portion of the image. (b) The Pluto–Charon system, to the same scale, as seen by the *Hubble Space Telescope*. The angular separation of the planet and its moon is about 0.9 arc second.

13.9 *Pluto in Bulk*

⑤ Pluto is so far away that little is known of its physical nature. Until the late 1970s, studies of its reflected light variations suggested a rotation period of nearly a week, but measurements of its mass and diameter were very uncertain. All this changed in July 1978, when astronomers at the U.S. Naval Observatory discovered that Pluto has a satellite. It is now named Charon, after the mythical boatman who ferried the dead across the river Styx into Hades, Pluto's domain. The discovery photograph of Charon is shown in Figure 13.24(a). Charon is the small bump near the top of the image. Knowing the moon's orbital period of 6.4 days, astronomers could determine the mass of Pluto to much greater accuracy. It is 0.0025 Earth masses (1.5 × 10^{22} kg), far smaller than any earlier estimate—more like the mass of a moon than of a planet. In 1990, the *Hubble Space Telescope* imaged the Pluto–Charon system (Figure

13.24b). The improved resolution of that instrument clearly separates the two bodies and allowed even more accurate measurements of their properties.

Before Charon was discovered, Pluto's radius was also poorly known. Pluto's angular size is much less than 1″, so its true diameter is blurred by the effects of Earth's turbulent atmosphere. But Charon's orbital orientation has given astronomers new insight into the system. By pure chance, Charon's orbit over the 6-year period from 1985 to 1991 (less than 10 years after the moon was discovered) has produced for Earth viewers a series of eclipses. Pluto and Charon repeatedly passed in front of one other, as seen from our vantage point. Figure 13.25 sketches this orbital configuration. With more good fortune, these eclipses took place while Pluto was closest to the Sun, making for the best possible Earth-based observations.

Basing their calculations on the variations in light as Pluto and Charon periodically hid each other, astronomers have computed their masses and radii and have determined their orbit plane. Additional studies of sunlight reflected from Pluto's surface indicate that the two objects are tidally locked as they orbit each other. Pluto's diameter is 2250 km, about one-fifth the size of the Earth. Charon is about 1300 km across and orbits at a distance of 19,700 km from Pluto. If planet and moon have the same composition (probably a reasonable assumption), Charon's mass must be about one-sixth that of Pluto, giving the Pluto–Charon system by far the largest satellite-to-planet mass ratio in the solar system. As shown in Figure 13.26, Charon's orbit is inclined at an angle of 118° to the plane of Pluto's orbit around the Sun. Since the spins of both planet and moon are perpendicular to the plane of Charon's orbit around Pluto, the geographic "north" poles of both bodies lie below the plane of Pluto's orbit. Thus, Pluto is the third planet in the solar system found to have retrograde rotation.

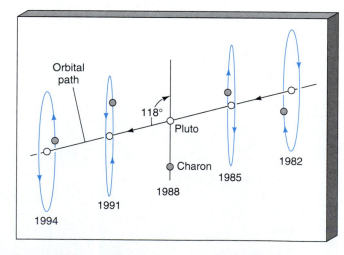

Figure 13.25 The orbital orientation of Charon produced a series of eclipses between 1985 and 1991. Observations of eclipses of Charon by Pluto and of Pluto by Charon have provided detailed information about both bodies' sizes and orbits.

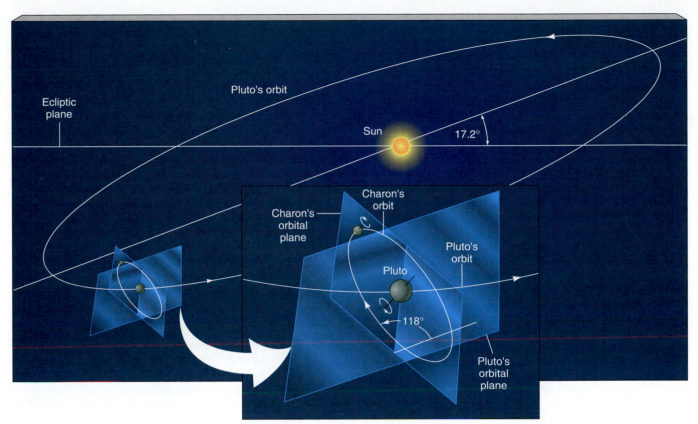

Figure 13.26   Charon's path around Pluto is circular, synchronous, and inclined at 118° to the orbit plane of the Pluto–Charon system about the Sun. The Pluto–Charon orbit plane is itself inclined at 17° to the plane of the ecliptic.

The known mass and radius of Pluto allow us to determine its average density, which is 2300 kg/m³—too low for a terrestrial planet, but far too high for a mixture of hydrogen and helium of that mass. Instead, the mass, radius, and density of Pluto are just what we would expect for one of the icy moons of a jovian planet. In fact, Pluto is quite similar in mass and radius to Neptune's large moon, Triton. The planet is almost certainly made up mostly of water ice. In addition, spectroscopy reveals the presence of *frozen* methane as a major surface constituent. Pluto is the only planet in the solar system on which methane exists in the solid state, implying that the surface temperature on Pluto is no more than 50 K. Pluto may also have a thin methane atmosphere, associated with the methane ice on its surface. Recent computer studies indicate that Charon may have bright polar caps, but their composition and nature are as yet unknown.

13.10   *The Origin of Pluto*

Because Pluto is neither terrestrial nor jovian in its makeup, and because of its similarity to the ice moons of the outer planets, some researchers suspect that Pluto is not a "true" planet at all. Pluto may be an escaped planetary moon or a large icy chunk of debris left over from the formation of the solar system. This idea is bolstered by Pluto's eccentric, inclined orbit, which is quite unlike the orbits of the other known planets. Since 1978, the explanation of Pluto's origin has been greatly complicated by the presence of Charon. It was much easier to suppose that Pluto was an escaped moon before we learned that it had a moon of its own. There is still no clear or easy answer to the puzzle of Pluto's origin.

Pluto may be just what it seems—a planet that formed in its current orbit, possibly even with its own moon right from the outset. Because we know so little about the environment in the outer solar system, we cannot rule out the possibility that planets beyond Neptune should simply look like Pluto. There is evidence for large chunks of ice circulating in interplanetary space beyond the orbits of Jupiter or Saturn (see Chapter 14), and some researchers have even suggested that there might have been thousands of Pluto-sized objects initially present in the outer solar system. The capture of a few of these objects by the giant planets would explain the strange moons of the outer worlds, especially Triton. And if there were enough moon-sized chunks originally orbiting beyond Neptune, it is quite plausible that Pluto could have captured Charon following a collision (or near-miss) between the two. At present, our scant knowledge of the compositions of the two bodies does not allow us to confirm or disprove either the coformation or the capture theory of the Pluto–Charon system. Unfortunately, this uncertainty may persist for some time—there is no present or proposed space mission that might suddenly and radically improve our understanding of these distant worlds.

Chapter Review

SUMMARY

The outer planets Uranus, Neptune, and Pluto were unknown to ancient stargazers. Uranus was discovered in the eighteenth century, by chance. Neptune was discovered after mathematical calculations of Uranus's slightly non-Keplerian orbit revealed the presence of an eighth planet.

At opposition, Uranus is barely visible to the unaided eye. It appears as a pale green disk through a telescope. Neptune cannot be seen with the naked eye, but a telescope shows it as a tiny bluish disk. Today we know the giant planets Uranus and Neptune mainly through data taken by *Voyager 2*. Small, remote Pluto has not been visited by a spacecraft, and our knowledge of it stems from painstaking Earth-based observations.

The masses of the outer planets are determined from measurements of their orbiting moons. The radii of Uranus and Neptune were relatively poorly known until the *Voyager 2* flybys in the 1980s. Uranus and Neptune have similar bulk properties; their densities imply large, rocky cores making up a greater fraction of the planets' masses than in either Jupiter or Saturn.

Surface features are barely discernible on Uranus, but computer-enhanced images from *Voyager 2* revealed atmospheric clouds and flow patterns moving beneath Uranus's haze. Neptune, although farther away from us, has atmospheric features that are clearer because of warmer temperatures and less haze. The **Great Dark Spot** (p. 280) on Neptune had many similarities to Jupiter's Red Spot. It disappeared in 1994.

For unknown reasons, Uranus's spin axis lies nearly in the ecliptic plane, leading to extreme seasonal variations in solar heating on the planet as it orbits the Sun.

Unlike the other jovian planets, Uranus has no excess heat emission. The source of Neptune's excess energy, like that of Jupiter's, is most likely heat left over from the planet's formation.

Both Uranus and Neptune have substantial magnetospheres. *Voyager 2* discovered that the magnetic fields of both planets are tilted at large angles to the planets' rotation axes. The reason for this is not known.

All of Uranus's moons revolve in the planet's equatorial plane, almost perpendicular to the ecliptic, in circular, synchronous orbits. Like the moons of Saturn, the medium-sized moons of Uranus are made up predominantly of rock and water ice. Many of them are heavily cratered and in some cases must have come close to being destroyed by the meteoritic impacts whose craters we now see. The strange moon Miranda has geological features that suggest repeated violent impacts in the past.

Neptune's moon Triton has a fractured surface of water ice and a thin atmosphere of nitrogen, probably produced by nitrogen "geysers" on its surface. Triton is the only large moon in the solar system to have a retrograde orbit around its parent planet. This orbit is unstable and will eventually cause Triton to be torn apart by Neptune's gravity.

Uranus has a series of dark, narrow rings, first detected from Earth by **stellar occultation** (p. 288)—their obscuration of the light received from background stars. Shepherd satellites are responsible for the rings' thinness. Neptune has three narrow rings like Uranus's and one broad ring, like Jupiter's. They were discovered by *Voyager 2*. The dark coloration of both the rings and the moons of the outer giant planets may be due to **radiation darkening** (p. 285), where exposure to solar high-energy radiation slowly causes a dark hydrocarbon layer to build up on a body's icy surface.

Pluto was discovered in the twentieth century after a laborious search for a planet that was supposedly affecting Uranus's orbital motion. We now know that Pluto is far too small to have any detectable influence on Uranus's path.

Pluto has a moon, Charon, whose mass is about one-sixth that of Pluto itself. Studies of Charon's orbit around Pluto have allowed the masses and radii of both bodies to be accurately determined. Pluto is too small for a terrestrial planet. Its properties are far more moonlike than planetlike. It may be an escaped moon, or an icy asteroid. The origin of the Pluto–Charon system is unknown.

SELF-TEST: True or False?

_____ **1.** Uranus was discovered by Galileo.

_____ **2.** After the discovery of Uranus, astronomers started looking for other planets, and quickly discovered Neptune.

_____ **3.** The semi-major axis of Neptune's orbit is about 30 A.U.

_____ **4.** Neptune has a smaller radius than Uranus and therefore a smaller mass and density.

_____ **5.** The rotation rates of Uranus and Neptune are almost identical.

_____ **6.** During the northern summer of Uranus, an observer near the north pole would observe the Sun high and almost stationary in the sky.

_____ **7.** Both Uranus and Neptune have a layer of metallic hydrogen surrounding their central cores.

_____ **8.** Relative to the size of the planet, the rocky cores of Uranus and Neptune are larger than the cores of Jupiter and Saturn.

_____ **9.** The system of moons around Uranus have orbits that share the tilt of the planet.

_____ **10.** Uranus has no large moons.

_____ **11.** The surfaces of the largest Uranian moons are darker than similar moons of Saturn.

_____ **12.** Triton's surface has a marked lack of cratering, indicating significant amounts of surface activity.

_____ **13.** Pluto's moon, Charon, is relatively small compared to the size of Pluto.

_____ **14.** Pluto is larger than Earth's Moon.

SELF-TEST: Fill in the Blank

1. Uranus is about _____ as distant from the Sun as is Saturn.
2. Uranus's rotation axis is almost _____ to the ecliptic.
3. Whereas the abundance of hydrogen and helium for the jovian planets remains more or less constant among the four planets, the relative abundance of _____ increases with increasing distance from the Sun.
4. The planet _____ is blue-green in color, and virtually featureless.
5. The planet _____ is dark blue in color, with white cirrus clouds and visible storm systems.
6. The _____ of both Uranus and Neptune are highly tilted relative to their rotation axis and significantly offset from the planets' centers.
7. The Uranian moon that shows the greatest amount of geological activity and disruption over time is _____.
8. *Voyager 2* discovered six of the _____ known moons of Neptune.
9. Triton's orbit is unusual because it is _____.
10. Nereid's orbit is unusual because it has a high orbital _____ .
11. Like only one other moon in the solar system, Triton has an _____.
12. Nitrogen geysers are found on _____.
13. The orbit of Pluto is locked into a 3:2 resonance with _____.
14. Overall, Pluto is most similar to which object in the solar system? _____.

REVIEW AND DISCUSSION

1. How was Uranus discovered?
2. Why did astronomers suspect an eighth planet beyond Uranus?
3. What is unusual about the rotation of Uranus?
4. What is responsible for the colors of Uranus and Neptune?
5. How are the interiors of Uranus and Neptune thought to differ from those of Jupiter and Saturn?
6. What is odd about the magnetic fields of Uranus and Neptune?
7. What is unique about Miranda? Give a possible explanation for this uniqueness.
8. How does Neptune's moon system differ from those of the other jovian worlds? What do these differences suggest about the origin of the moon system?
9. What is the predicted fate of Triton?
10. The rings of Uranus are dark, narrow, and widely spaced. Which of these properties makes them different from the rings of Saturn?
11. How do the rings of Neptune differ from those of Uranus and Saturn?
12. How was Pluto discovered?
13. How was the mass and radius of Pluto determined?
14. In what respect is Pluto more like a moon than a jovian or terrestrial planet?
15. Why was the discovery of Uranus in 1781 so surprising? Might there be similar surprises in store for today's astronomers?
16. From what is known today, which planet or moon in the solar system is the most mysterious? Why?

PROBLEMS

1. Compare the amount of sunlight (per square meter) reaching Uranus (at 19.2 A.U.) and Neptune (at 30.1 A.U.) with the amount of sunlight reaching Earth.
2. What is the angular diameter of the Sun, as seen from Neptune? Compare it with the angular diameter of Triton. Would you expect solar eclipses to be common on Neptune?
3. How close is Charon to Pluto's Roche limit?
4. What is the round-trip travel time of light from Earth to Pluto (at a distance of 40 A.U.)? How far would a spacecraft orbiting Pluto at a speed of 0.5 km/s travel in that time?
5. Add up the masses of all the moons of Uranus, Neptune, and Pluto. (Just estimate the masses of the smallest moons—they will contribute little to the result.) How does this compare to the mass of Earth's Moon, and to the mass of Pluto?

PROJECTS

1. The major astronomy magazines *Sky and Telescope* and *Astronomy* print charts showing the whereabouts of the planets in their January issues. Consult one of these charts, and locate Neptune and Uranus in the sky. Uranus may be visible to the naked eye, but binoculars make the search much easier. (Hint: Uranus shines more steadily than the background stars.) With the eye alone, can you detect a color to Uranus? Through binoculars?
2. The search for Neptune requires a much more determined effort! A telescope is best, but high-powered binoculars mounted on a steady support will do. If you can see both planets through a telescope—and they will remain close together on the sky for the rest of this century—contrast their colors. Which planet appears bluer? Through a telescope, does Uranus show a disk? Can you see that Neptune shows a disk, or does it look more like a point of light?

14

SOLAR SYSTEM DEBRIS

Keys to Our Origin

LEARNING GOALS

Studying this chapter will enable you to:

1 Describe the orbital properties of the major groups of asteroids.

2 Explain the effect of orbital resonances on the structure of the asteroid belt.

3 Summarize the composition and physical properties of a typical asteroid.

4 Discuss the characteristics of cometary orbits and what they tell us about the probable origin of comets.

5 Detail the composition and structure of a typical comet, and explain the formation and appearance of its tail.

6 Distinguish among the terms meteor, meteoroid, and meteorite.

7 Summarize the orbital and physical properties of meteoroids, and explain what these suggest about their probable origin.

(Opposite page, background) Wind and rain have eroded much of the evidence for meteoritic impacts on our home planet. Such craters are especially hard to find at ground level. In recent years, satellite observations have made more clear several huge impact craters on Earth—like this photo of the nearly 100–kilometer diameter Manicougan crater lake in Quebec, Canada. (See also Figure 14.16.)
(Inset A) Partly obscured by clouds, this million-year-old, 10-kilometer-wide crater lake is located in Bosumtwi, Ghana.
(Inset B) This extremely circular crater, with a diameter of only 2 kilometers, is situated in the southern Namib Desert, near Roter Kamm, Namibia.
(Inset C) Located at Kara-Kul, Russia, this 40-kilometer-wide crater is located some 6 kilometers above sea level near the Afghan border.
(Inset D) These highly eroded twin craters, forming the 30-kilometer-wide Clearwater Lakes in Quebec, were caused by two separate but presumably related meteoritic impacts some 300 million years ago.
(Inset E) A much closer view of a much more recent crater, near Lake Lejia in Chile. Note how little erosion is evident. The age of this crater, like that of the Barringer Crater in Arizona (Figure 8.18), can probably be measured in thousands rather than millions of years.
All photographs on this page with the exception of E were taken by astronauts aboard U.S. space shuttles.

According to classical definitions, there are only nine planets in the solar system. But several thousand other celestial bodies are also known to revolve around the Sun in well-determined orbits. These minor objects—called asteroids and comets—are small and of negligible mass compared with the planets and their major moons. Yet each is a separate world, with its own story to tell about the early solar system. On the basis of statistical deductions, astronomers estimate that there are more than a billion such objects still to be discovered. These objects may seem to be only rocky and icy "debris," but, more than the planets themselves, they hold a record of the formative stages of our planetary system. Many are nearly "pure," unevolved bodies that may one day teach us much about our local origins.

14.1 Asteroids

Asteroids are relatively small, rocky objects that revolve around the Sun. Their name literally means "starlike bodies," but asteroids are definitely not stars. They are too small even to be classified as planets. Astronomers often refer to them as "minor planets" or sometimes "planetoids."

Asteroids differ from planets both by their orbits and by their size. They generally move on highly eccentric trajectories between Mars and Jupiter, quite unlike the almost circular paths of the major planets. Few are larger than 300 km in diameter, and most are far smaller—as small as a tenth of a kilometer across. The largest known asteroid, Ceres, is just 1/10,000 the mass of Earth and measures only 940 km across. Together, the 4000 or so cataloged asteroids amount to less than 1/10 the mass of the Moon, so they do not contribute significantly to the total mass of the solar system.

ORBITAL PROPERTIES

❶ European astronomers discovered the first asteroids early in the nineteenth century as they searched for an additional planet in the region between Mars and Jupiter, where the Titius–Bode "law" (see *Interlude 6-1* on p. 135) suggested one might be found. Italian astronomer Giuseppe Piazzi was the first to discover an asteroid. He detected Ceres in 1801 and measured its orbital semi-major axis to be 2.8 A.U., exactly as the "law" predicted. Within a few years, three more asteroids—Pallas (at 2.8 A.U.), Juno (at 2.7 A.U.), and Vesta (at 3.4 A.U.)—were discovered. By the start of the twentieth century, astronomers had cataloged several hundred asteroids with well-determined orbits. In this last decade of the twentieth century, the list numbers over 4000. The vast majority are found in a region of the solar system known as the **asteroid belt**, located between 2.1 and 3.3 A.U. from the Sun— roughly midway between the orbits of Mars (at 1.5 A.U.) and Jupiter (at 5.2 A.U.). All but one of the known asteroids revolve about the Sun in prograde orbits, in the same sense as the planets. The overall layout of the asteroid belt is sketched in Figure 14.1.

Such a compact concentration of asteroids in a well-defined belt suggests that they are either the fragments of a planet broken up long ago or primal rocks that never managed to accumulate into a genuine planet. On the basis of the best evidence currently available, researchers favor the latter view. There is far too little mass in the belt to constitute a planet, and the marked chemical differences between individual asteroids strongly suggest that they could not all have originated in a single planet. Instead, astronomers believe that the strong gravitational field of Jupiter continuously disturbs the motions of these chunks of primitive matter, nudging and pulling at them, preventing them from aggregating into a planet.

EARTH-CROSSING ASTEROIDS

The orbits of most asteroids have eccentricities lying in the range 0.05—0.3, ensuring that they always remain between the orbits of Mars and Jupiter. Very few asteroids have eccentricities greater than 0.4. Those that do, of course, are of great interest to us, as their orbits may intersect that of the Earth, leading to the possibility of a collision with our planet. The few stray asteroids with very elliptical orbits have probably been influenced by the gravitational fields of nearby Mars and especially Jupiter. These planets can disturb the normal asteroid orbits, deflecting them into the inner solar system. Asteroids whose paths cross the orbit of the Earth are termed *Apollo asteroids* (after the first-known Earth-crossing asteroid, Apollo). Those crossing only the orbit of Mars are known as *Amor asteroids*.

Although we are currently aware of only a few dozen Apollo asteroids, they are among the best known because of their occasional close encounters with Earth. For example, the Apollo asteroid Icarus (Figure 14.2) has perihelion within 0.2 A.U. of the Sun. On its way past Earth in 1968, it missed our planet by "only" 6 million km—a close call by cosmic standards. More recently, in 1989 an unnamed asteroid (designated 1989FC) came even closer, passing only 800,000 km from Earth, only twice the distance to the Moon. In 1991, asteroid 1991BA missed us by a mere 170,000 km.

The potential for collision with Earth is real. Calculations imply that most Earth-crossing asteroids will eventually collide with Earth and, on average during any given million-year period, our planet is struck by about three asteroids. Because Earth is largely covered with water, on average two of those impacts should occur in the ocean and only one on land. Several dozen large land basins and eroded craters on our planet are suspected to be sites of ancient asteroid collisions (see, for example, Figure 14.16 later

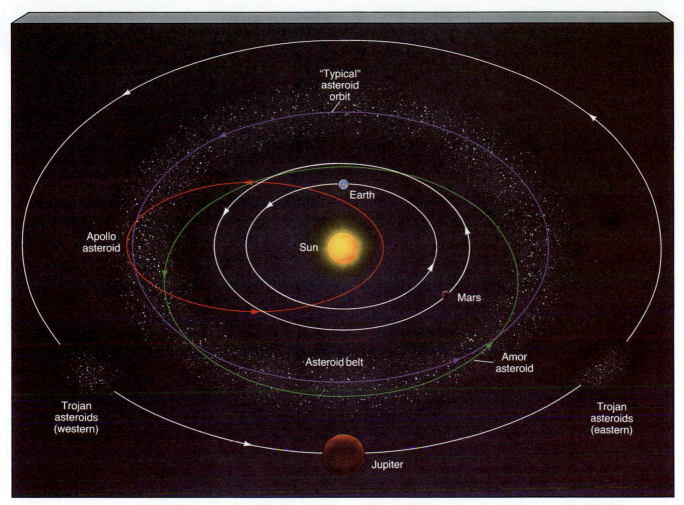

Figure 14.1 The asteroid belt, along with the orbits of Earth, Mars, and Jupiter. The main belt, the Trojan asteroids, and some Apollo (Earth-crossing) and Amor (Mars-crossing) orbits are shown.

in this chapter). The many large-impact craters on the Moon, Venus, and Mars are direct evidence of similar events on those worlds.

Most known Apollo asteroids are relatively small—about 1 km in diameter (although one of 10 km in diameter has been identified). Even so, a visit of even a kilometer-sized asteroid to Earth could be catastrophic by human standards. Such an object packs enough kinetic energy to devastate an area some 100 km in diameter. The explosive power would be equivalent to about a million 1-megaton nuclear bombs, a hundred times more than all the nuclear weapons currently in existence on Earth. A fatal blast wave would doubtless affect a much larger area still. Should an asteroid hit our planet hard enough, it might even cause the extinction of entire species—indeed, many scientists think that the extinction of the dinosaurs was the result of just such an impact (see *Interlude 14-1* on p. 300). Some people take the prospect of an asteroid impact seriously enough to advocate an "asteroid watch"—an effort to catalog and monitor all Earth-crossing asteroids in order to maximize our warning time of any impending collision.

Figure 14.2 The asteroid Icarus has an orbit that passes within 0.2 A.U. of the Sun, well within the orbit of Earth. Icarus occasionally comes close to our own planet, making it one of the best-studied asteroids in the solar system. The asteroid's motion relative to the stars makes it appear as a streak (marked) in this long-exposure photograph, taken in the 1970s.

Interlude 14–1   *What Killed the Dinosaurs?*

The name *dinosaur* derives from the Greek words, *deinos* (terrible) and *sauros* (lizard). Dinosaurs were no ordinary reptiles. In their prime, roughly 100 million years ago, the dinosaurs were the all-powerful rulers of the Earth. Their fossilized remains have been uncovered on all the world's continents. Despite their dominance, according to the fossil record, these creatures vanished from Earth quite suddenly about 65 million years ago. What happened to them?

Until fairly recently, the prevailing view among paleontologists—scientists who study prehistoric life—was that dinosaurs were rather small-brained, cold-blooded creatures. In chilly climates, or even at night, the metabolisms of these huge reptiles would have become sluggish, making it difficult for them to move around and secure food. The suggestion was that they were poorly equipped to adapt to sudden changes in Earth's climate, so that they eventually died out.

However, a competing, and still controversial, view of dinosaurs has emerged. Recent fossil evidence suggests that many of these monsters may in fact have been warm-blooded and relatively fast-moving creatures—not at all the dull-witted, slow-moving giants of earlier conception. In any case, no species able to dominate the Earth for more than 100 million years could have been too poorly equipped for survival. By comparison, humans have thus far dominated for little more than 2 million years.

If the dinosaurs didn't die out simply because of stupidity and inflexibility, then what happened to cause their sudden and complete disappearance? Many explanations have been offered for the extinction of the dinosaurs. Devastating plagues, magnetic field reversals, increased tectonic activity, severe climate changes, and supernova explosions have all been proposed. In the 1980s, it was suggested that a huge extraterrestrial object collided with Earth 65 million years ago, and this is now (arguably) the leading explanation for the demise of the dinosaurs, although it is by no means universally accepted. According to this idea, a 10- to 15-km-wide asteroid or comet struck the Earth, releasing as much energy as 10 million or more of the largest hydrogen bombs humans have ever constructed and kicking huge quantities of dust (including the pulverized remnants of the asteroid itself) high into the atmosphere. The dust may have shrouded our planet for many years, virtually extinguishing the Sun's rays during this time. On the darkened surface, plants could not survive. The entire food chain was disrupted, and the dinosaurs, at the top of that chain, eventually died out.

Although we have no direct astronomical evidence to confirm or refute this idea, we can estimate the chances of a large asteroid or comet striking the Earth today, on the basis of observations of the number of objects presently on Earth-crossing orbits. The next figure in this interlude shows the likelihood of an impact as a function of the size of the impacting body. The horizontal scale indicates the energy released by the collision, measured in *megatons* of TNT. (The megaton—4.2×10^{16} joules, the explosive yield of a large nuclear warhead—is the only common terrestrial measure of energy adequate to describe the violence of these occurrences.) We see that 100 million megaton events, like the planetwide catastrophe that supposedly wiped out the dinosaurs, are very rare, occurring only once every 10 million years or so. However, smaller impacts, equivalent to "only" a few tens of kilotons of TNT (about one MX missile warhead from the U.S. nuclear arsenal), could happen every few years—we may be long overdue for one. The most recent large impact was the Tunguska explosion in Siberia, in 1908, which packed a roughly 1-megaton punch (see Figure 14.17).

The main geological evidence supporting this theory is a layer of clay enriched with the element iridium. This layer is found in 65-million-year-old rocky sediments all around our planet. Iridium on the Earth's surface is rare because most of it sank into our planet's interior long ago. The abundance of iridium in this one layer of clay is about 10 times greater than in other terrestrial rocks, but it matches closely the abundance of iridium found in meteorites (and, we assume, in asteroids and comets too).

ORBITAL RESONANCES

2   Although most asteroids orbit in the main belt, between 2 and 3 A.U. from the Sun, an additional class of asteroids, known as the **Trojan asteroids**, orbit at the distance of Jupiter. These asteroids are locked into a 1:1 orbital resonance with Jupiter by that planet's strong gravity, just as some of the small moons of Saturn share orbits with the medium-sized moons Tethys and Dione, as described in Chapter 12. ∞ (p. 271)

Calculations first performed by the French mathematician Joseph Louis Lagrange in 1772 show that there are exactly five places in the solar system where a small body can orbit the Sun in synchrony with Jupiter (or, for that matter, any other planet) subject to the combined gravitational influence of both large bodies. These places are known as the *Lagrange points* of Jupiter's orbit. As illustrated in Figure 14.3, three of these points (referred to as L_1, L_2, and L_3) lie on the line joining Jupiter and the Sun (or its extension in either direction). The other two—L_4 and L_5—are located on Jupiter's orbit, exactly 60° ahead of and behind the planet. All five Lagrange points revolve around the Sun at the same rate as Jupiter.

In principle, an asteroid placed at any of the Lagrange points will circle the Sun in lockstep with Jupiter, always maintaining the same position relative to the planet. However, the three Lagrange points in line with Jupiter and the Sun are known to be *unstable*—a body displaced, however slightly, from any of those points will tend to drift slowly away from it, not back toward it. Since matter in the solar system is constantly subjected to small perturbations—by the planets, the asteroids, and

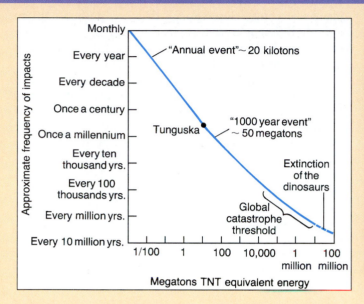

This theory has its problems, however. The amount of iridium in this clay layer varies greatly from place to place across the globe, and there is no complete explanation of why that should be so. And if this body was so massive as to kick up enough dust to darken the entire planet, then where is its impact crater? One leading candidate has recently been identified in the Yucatan Peninsula in Mexico, where possible evidence for a heavily eroded, but not completely obliterated, crater has been found. However, its identity as the site of the dinosaur-killing collision has not yet been conclusively proven. Perhaps, some scientists argue, the iridium layer was laid down by volcanoes and had nothing to do with an extraterrestrial impact at all.

Another potential difficulty concerns the speed at which the dinosaurs disappeared. The fossil and geological records are rather imprecise in the ages they yield, and this translates into uncertainties in just how long the extinction process took. It seems to be no more than a million or so years, but that is still a very long time. If the dinosaurs vanished as a direct result of a collision and subsequent explosion, we might expect their disappearance to have been complete in a matter of decades, or maybe centuries. It is hard to see how the process could have spanned tens or hundreds of millennia. Greatly improved geological age-measurement techniques will be needed to settle this issue once and for all.

Whatever killed the dinosaurs, dramatic environmental change of some sort was almost surely responsible. It is important that we continue the search for the cause of their extinction, for there's no telling if and when that sudden change might strike again. As the dominant species on the Earth, we are the ones who now stand to lose the most.

even the solar wind—matter does not accumulate in these regions. No asteroids orbit near the L_1, L_2, and L_3 points of Jupiter's orbit.

Not so for the other two Lagrange points, L_4 and L_5. They are both *stable*—matter placed near them tends to remain in the vicinity. Consequently, asteroids tend to accumulate near these points. The 50 or so Trojan asteroids are roughly equally divided between Jupiter's leading (L_4) and trailing (L_5) Lagrange points. Astronomers know of no asteroids orbiting in the Lagrange points of any other planet. However, space colony enthusiasts frequently tout the stable L_5 point of the Earth–Moon system as an excellent site for a permanent human presence in space.

The main asteroid belt also has structure—not as obvious as the Trojan orbits or the prominent gaps and ringlets in Saturn's ring system, but nevertheless of great dynamical significance. A graph of the number of asteroids having various orbital semi-major axes (Figure 14.4a) shows that there are several prominent underrepresented regions in the distribution. These "holes" are known as the **Kirkwood gaps**, after their discoverer, the nineteenth-century American astronomer Daniel Kirkwood.

The Trojan asteroids share an orbit with Jupiter—they orbit in 1:1 resonance with that planet. The Kirkwood gaps result from other, more complex orbital resonances with Jupiter. For example, an asteroid with a semi-major axis of 3.3 A.U. would (by Kepler's third law) orbit the Sun in exactly half the time taken by Jupiter. (p. 42) The gap at 3.3 A.U., then, corresponds to the 2:1 resonance. An asteroid at that particular resonance feels a regular, periodic tug from Jupiter at the same point in every other orbit (Figure 14.4b). The cumulative effect of

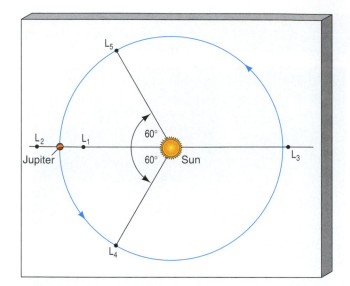

Figure 14.3 The Lagrange points of the Jupiter–Sun system, where a third body could orbit in synchrony with Jupiter on a circular trajectory. Only the L_4 and L_5 points are stable. They are the locations of the Trojan asteroids.

those tugs is to deflect the asteroid into an elongated orbit—one that crosses the orbit of Mars or Earth. Eventually, the asteroid collides with one of those two planets or comes close enough that it is pushed onto an entirely different trajectory. In this way, Jupiter's gravity creates the Kirkwood gaps, while some of the cleared-out asteroids become Apollo or Amor asteroids.

Notice that, while there are many similarities between this mechanism and the resonances that produce the gaps in Saturn's rings (see Chapter 12), there are differences too. ∞ (p. 263) Unlike Saturn's rings, where eccentric orbits are rapidly circularized by collisions among ring particles, there are no *physical* gaps in the asteroid belt. The in-and-out motion of the belt asteroids as they travel in their eccentric orbits around the Sun means that no part of the belt is actually empty. Only when we look at semi-major axes (or, equivalently, at orbital *energies*) do the gaps become apparent.

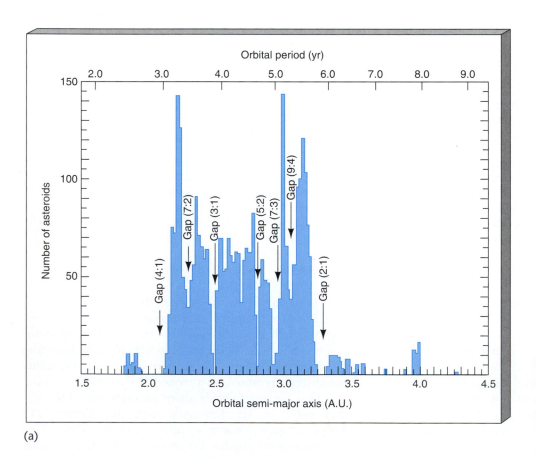

(a)

(b)

Figure 14.4 (a) The distribution of asteroid semi-major axes shows some prominent gaps caused by resonances with Jupiter's orbital motion. Note, for example, the prominent gap at 3.3 A.U., which corresponds to the 2:1 resonance—the orbital period is 5.9 years, exactly half that of Jupiter. (b) An asteroid in a 2:1 resonance with Jupiter receives a strong gravitational tug from the planet each time they are closest together (as in panels 1 and 3). Because the asteroid's period is precisely half that of Jupiter, the tugs come at exactly the same point in every other orbit, and their effects reinforce each other.

PHYSICAL PROPERTIES

3 With few exceptions, asteroids are too small to be resolved by Earth-based telescopes. We must rely on indirect methods to find their sizes, shapes, and compositions. Consequently, only a few of their physical and chemical properties are accurately known.

To the extent that astronomers can determine their compositions, asteroids have been found to differ not only from the nine known planets and their many moons, but also among themselves.

Asteroids are classified in terms of their spectroscopic properties. The darkest, or least reflective, asteroids contain a large fraction of carbon in their makeup. They are known as *C-type* (or *carbonaceous*) asteroids. The more reflective *S-type* asteroids contain silicate, or rocky, material. Generally speaking, S-type asteroids predominate in the inner portions of the asteroid belt, and the fraction of C-type bodies steadily increases as we move outward. Overall, about 15 percent of all asteroids are S-type, 75 percent are C-type, and 10 percent are other types (such as those containing large fractions of iron). Many planetary scientists believe that the carbonaceous asteroids consist of very primitive material, representative of the earliest stages of the solar system, and have not experienced significant heating or chemical evolution since they first formed 4.6 billion years ago.

In most cases, astronomers estimate the sizes of asteroids from the amount of sunlight they reflect and the amount of heat they radiate. These observations are difficult, but we now have size measurements for more than 1000 asteroids. In rare cases, astronomers witness an asteroid occulting a star, which allows them to determine its size and shape with great accuracy. The largest asteroids are roughly spherical, but the smaller bodies can be highly irregular.

Astronomers have measured masses only for the largest asteroids. The computed densities are consistent with the rocky or carbonaceous compositions just described.

The three largest asteroids, Ceres, Pallas, and Vesta, have diameters of 940, 580, and 540 km, respectively. Only two dozen or so asteroids are more than 200 km across, and most are much smaller. Almost assuredly, there exist hundreds of thousands more asteroids awaiting discovery. However, observers estimate that they are mostly small. Probably 99 percent of all asteroids larger than 100 km are known and cataloged, and at least 50 percent of those asteroids larger than 10 km are accounted for. While most asteroids are probably less than a few kilometers across, most of the *mass* in the asteroid belt resides in objects greater than a few tens of kilometers in diameter. The total number of potentially visible asteroids (that is, visible through telescopes if we knew just where and when to look) may exceed 100,000.

The first close-up views of asteroids were provided by the Jupiter probe *Galileo* (see Chapter 6), which on its rather roundabout path to the giant planet passed twice through the asteroid belt, making close encounters with asteroid Gaspra in October 1991 and asteroid Ida in August 1993 (see Figure 14.5). ∞ (p. 141) Technical problems limited the amount of data that could be sent back from the spacecraft during the asteroid flybys. Nevertheless, the images produced by *Galileo* show far more detail than any photographs made from Earth.

Gaspra and Ida are irregularly shaped bodies with maximum diameters of about 20 and 50 km, respectively. They are pitted with craters ranging in size from a few hundred meters to 2 km across and are covered with a layer of dust of variable thickness. Ida is much more heavily cratered than Gaspra, in part because it resides in a denser part of the asteroid belt. Also, scientists believe that Ida has

(a)

(b)

R I V U X G

Figure 14.5   (a) The asteroid Gaspra as seen from a distance of 1600 km by the probe *Galileo* on its way to Jupiter. (b) The asteroid Ida, photographed by *Galileo* from a distance of 3400 km. (Ida's moon, Dactyl, is visible at the right of the photo.) The resolution in these photographs is on the order of 100 m. True-color images showed the surfaces of both bodies to be a fairly uniform shade of gray. Sensors on board the spacecraft indicated that the amount of infrared radiation absorbed by these surfaces varies from place to place, probably as a result of variations in the thickness of the dust layer blanketing them.

suffered more from the ravages of time. Ida is about a billion years old, far more aged than Gaspara, which is estimated to be a mere 100 million years old based on the extent of cratering. Both asteroids are thought to be fragments of much larger objects that broke up into many smaller pieces following violent collisions long ago.

To the surprise of most mission scientists, closer inspection of the Ida image (Figure 14.5b) revealed the presence of a tiny *moon*, just 1.5 km across, orbiting the asteroid. A few such *binary asteroids* had been observed from Earth (for example, the second-largest asteroid, Pallas, is a binary). However, the rare binary systems known before the Ida flyby were all much larger than Ida—a moon the size of Ida's cannot be detected from the ground. Scientists believe that, given the relative congestion of the asteroid belt, collisions between asteroids may be quite common, providing a source of both interplanetary dust and smaller asteroids, and possibly deflecting one or both of the bodies involved onto eccentric, Earth-crossing orbits. The less violent collisions may be responsible for the binary systems we see. By studying *Galileo's* images of Ida and its moon (now christened Dactyl), astronomers hope to determine the asteroid's mass, allowing them to gain more insight into the composition and structure of these tiny worlds.

14.2 Comets

Known for their long wispy tails, **comets** derive their name from the Greek *kome*, meaning "hair." They are usually discovered as faint, fuzzy patches of light on the sky while still several astronomical units away from the Sun, in whose direction they are racing. Far from the Sun, comets are visible only by reflected sunlight, but as they near the Sun, they can emit radiation of their own, or at least reprocess the Sun's light into other forms. Traveling in a highly elliptical orbit with the Sun at one focus, a comet brightens and develops a tail as its icy matter becomes heated and sublimes* away, as depicted in Figure 14.6. As a comet departs from the Sun's vicinity, its brightness and its tail diminish until it becomes once again a faint point of light receding into the distance.

COMET ORBITS

4 Comets that survive a close encounter with the Sun—some break up entirely—continue their outward journey to the edge of the solar system. Their highly elliptical orbits take many comets far beyond Pluto, perhaps even as far as 50,000 A.U., where, in accord with Kepler's second law, they move more slowly and so spend most of their time. ∞ (p. 41) Most take hundreds of thousands, some even millions, of years for a round trip. However, a few "short-period" comets (conventionally defined as those with periods less than 200 years) return for another

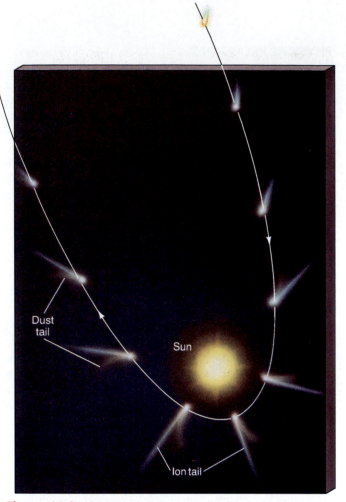

Figure 14.6   Diagram of part of the orbit of a typical comet. As the comet approaches the Sun, it develops an ion tail, which is always directed away from the Sun. Closer in, a curved dust tail, also directed generally away from the Sun, may also appear. Notice that although the ion tail always points directly away from the Sun on both the inbound and the outgoing portions of the orbit, the dust tail has a marked asymmetry, always tending to "lag behind" the ion tail.

encounter within a human life span. The short-period comets do not venture far beyond the distance of Pluto at aphelion. Notice that, during all the time they spend far from the Sun's warmth, comets are nothing more than small icy chunks. Only when they enter the vicinity of the Sun do they acquire their spectacular tails.

Unlike the orbits of the other solar-system objects we have studied so far, the orbits of comets are *not* necessarily confined to within a few degrees of the ecliptic plane. Short-period comets do tend to have prograde orbits lying close to the ecliptic. Their long-period counterparts, however, exhibit all inclinations and all orientations, both prograde and retrograde, roughly uniformly distributed in all directions from the Sun.

Astronomers believe that the short-period comets originate in a region of the solar system beyond the orbit of Neptune called the **Kuiper Belt** (after Gerard Kuiper, a pioneer in infrared and planetary astronomy). A little like the asteroids in the inner solar system, most Kuiper Belt comets move in roughly circular orbits between about 30 and 100

---

* *Sublimation is the process by which a solid changes directly into a gas without passing through the liquid phase. Frozen carbon dioxide—dry ice—is an example of a solid that undergoes sublimation rather than melting and subsequently evaporating. In space, sublimation is the rule, rather than the exception, for the behavior of ice when exposed to heat.*

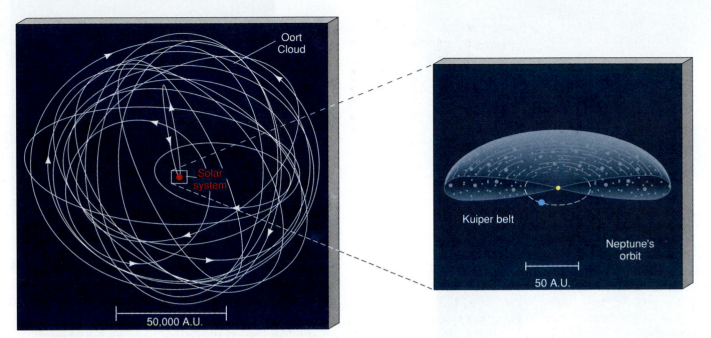

Figure 14.7   Diagram of the Oort Cloud, showing some cometary orbits. Most Oort Cloud comets never come close to the Sun. Of all the orbits shown, only the most elongated ellipse represents a comet that will actually enter the solar system (which is smaller than the dot at the center of the figure on this scale) and possibly become visible from Earth. The insert shows the Kuiper Belt, believed to be the source of the short-period comets.

A.U. from the Sun, never venturing inside the orbits of the jovian planets. Occasionally, however, a close encounter between two comets, or (more likely) the cumulative gravitational influence of one of the outer planets, "kicks" a Kuiper Belt comet into an eccentric orbit that brings it into the inner solar system, and into our view. The observed orbits of these comets reflect the flattened structure of the Kuiper Belt.

What of the long-period comets? How do we account for their apparently random orbital orientations? Only a tiny portion of a typical long-period cometary orbit lies within the inner solar system, so it follows that for every comet we see, there must be many more similar objects at great distances from the Sun. On these general grounds, many astronomers reason that there must be a huge "cloud" of comets far beyond the orbit of Pluto, completely surrounding the Sun. It is named the **Oort Cloud**, after the Dutch astronomer Jan Oort, who first wrote (in the 1950s) of the possibility of such a vast and distant reservoir of inactive, frozen comets. The Kuiper Belt and the orbits of some typical Oort Cloud comets are sketched in Figure 14.7.

Based on the observed orbital properties of long-period comets, researchers believe that the Oort Cloud may be up to 100,000 A.U. in diameter. Like those of the Kuiper Belt, however, most of the comets of the Oort Cloud never come anywhere near the Sun. Indeed, Oort-Cloud comets rarely approach even the orbit of Pluto, let alone that of the Earth. Only when the gravitational field of a passing star happens to deflect a comet into an extremely eccentric orbit that passes through the inner solar system do we actually get to see one of these objects. Because the Oort Cloud surrounds the Sun in all directions, instead of being confined to the ecliptic plane like the Kuiper Belt, the long-period comets

we see can come from any direction in the sky. But despite their great distances and long orbital periods, the Oort Cloud comets are still gravitationally bound to the Sun. Their orbits are governed by precisely the same laws of motion that control the planets.

COMET STRUCTURE

5 Even through a large telescope, the **nucleus**, or main solid body, of a comet is no more than a minute point of light. A typical cometary nucleus is extremely small—only a few kilometers in diameter. During most of the comet's orbit, far from the Sun, only this frozen nucleus exists. When a comet comes within a few astronomical units of the Sun, however, its icy surface becomes too warm to remain stable. Part of it becomes gaseous and expands into space, forming the exciting display we see from Earth. In response to the Sun's heat, a diffuse **coma** (or "halo") of dust and evaporated gas begins to form, surrounding the nucleus and growing in size as the comet nears the Sun. At maximum size, the coma can measure 100,000 km in diameter—almost as large as Saturn or Jupiter.

Engulfing the coma, an invisible **hydrogen envelope**, usually distorted by the solar wind, stretches across millions of kilometers of space. The comet's **tail**, most pronounced when the comet is closest to the Sun and the rate of sublimation of material from the nucleus is greatest, is much larger still, sometimes spanning as much as an astronomical unit. From Earth, only the coma and tail of a comet are visible to the naked eye. Despite the size of the tail, most of the light comes from the coma; most of the comet's mass resides in the nucleus.

(a)

(b)

(c)

Figure 14.8 (a) A comet with a primarily ion tail. Called Comet Giacobini–Zinner and seen here in 1959, its coma measured 70,000 km across; its tail was well over 500,000 km long. (b) Photograph of a comet having (mostly) a dust tail, showing both its gentle curvature and inherent fuzziness. This is Comet West, in 1976, whose tail stretched 13° across the sky. (c) Photograph of Comet Kohoutek, whose tail in 1975 was a mixture of the two categories. The "tail" is not a sudden streak in time across the sky, as in the case of meteors or fireworks. Instead, it travels along with the head of the comet as long as the comet is close enough to the Sun.

R   I   V   U   X   G

Two types of comet tails may be distinguished. The **ion tails** are approximately straight, often made of glowing, linear streamers like those seen in Figure 14.8a. Their spectra show emission lines of numerous *ionized* molecules—molecules that have lost some of their normal complement of electrons—including carbon monoxide, nitrogen, and water among many others. ∞ (p. 85) The **dust tails** are usually broad, diffuse, and gently curved (Figure 14.8b). They are rich in microscopic dust particles that reflect sunlight, making the tail visible even though it emits no light of its own. Some comets' tails are mixtures of the two types. Comet Kohoutek (Figure 14.8c), which appeared in 1975, is a typical example.

The tails are in all cases directed *away* from the Sun by the solar wind (the invisible stream of matter and radiation escaping the Sun). Consequently, as depicted in Figure 14.6, the tail always lies outside the comet's orbit and actually *leads* the comet during the portion of the orbit that is outbound from the Sun. Every tiny particle in space in our solar system—including those in comet tails—follows an orbit determined by gravity and the solar wind. If gravity alone were acting, the particle would follow the same curved path as its parent comet, in accordance with Newton's laws of motion. If the solar wind were the only influence, the tail would be swept up by it and would trail radially outward from the Sun. The ion tails are much more strongly influenced by the solar wind than by the Sun's gravity, so those tails always point directly away from the Sun. The heavier dust particles have more of a tendency to follow the comet's orbit, giving rise to the slightly curved dust tails.

A VISIT TO HALLEY'S COMET

Probably the most famous comet of all is Halley's Comet. In 1705, the British astronomer Edmund Halley realized that the 1682 appearance of this comet was not a one-time event. Basing his work on previous sightings of the comet, Halley calculated its path and found that the comet orbited the Sun with a period of 76 years. He predicted its reappearance in 1758. His successful determination of the comet's trajectory and his prediction of its return was an early triumph of Newton's laws of motion and gravity. Although Halley did not live to see his calculations proved correct, the comet was named in his honor.

Once astronomers knew the comet's period, they traced its appearances back in time. Historical records from many ancient cultures show that Halley's Comet has been observed at every passage since 240 B.C. A spectacular show, the tail of Halley's Comet can reach almost a full astronomical unit in length, stretching many tens of degrees across the sky. Figure 14.9(a) shows Halley's Comet as seen from the Earth in 1910. Its most recent visit, in 1986 (see Figure 14.9b), was not ideal for terrestrial viewing, but the comet was closely scrutinized by spacecraft. The comet's orbit is shown in Figure 14.10; its next scheduled visit to the inner solar system is in 2061.

When Halley's Comet rounded the Sun in 1986, a small armada of spacecraft launched by the USSR, Japan, and a group of western European countries went to meet it. One of the Soviet craft, *Vega 2*, traveled through the comet's coma, coming to within some 8000 km of the

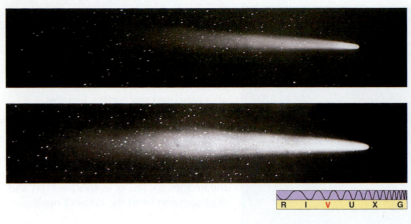

Figure 14.9 (a) A series of photographs of Halley's Comet as it appeared in 1910; top, on May 10, with a 30° tail, bottom, on May 12, with a 40° tail. (b) Halley's Comet in 1986, about one month before perihelion.

nucleus. Using positional knowledge of the comet gained from the Soviet craft encounter, the European *Giotto* spacecraft (named after the Italian artist who painted an image of Halley's Comet not long after its appearance in the year A.D.1301) was navigated to within 600 km of the nucleus. This was a daring trajectory, since at 70 km/s—the speed of the craft relative to the comet—a colliding dust particle becomes a devastating bullet. Debris did in fact damage *Giotto*'s camera, but not before it sent home a wealth of data. Figure 14.11 shows two images of the comet's nucleus radioed back to the Earth by these spacecraft.

The results of the Halley encounters were somewhat surprising. Halley's nucleus is an irregular, potato-shaped object, larger than astronomers had estimated. Spacecraft measurements showed it to be 15 km long by as much as 10 km wide. Also, the nucleus appeared almost jet black—as dark as finely ground charcoal. This solid nucleus was enveloped by a cloud of dust, which scattered light throughout the coma of the comet. Partly because of this scattering and partly because of dimming by the dust, no spacecraft was able to discern much surface detail on the nucleus.

Figure 14.12 is a composite of several findings concerning Halley's nucleus. The visiting spacecraft found direct ev-

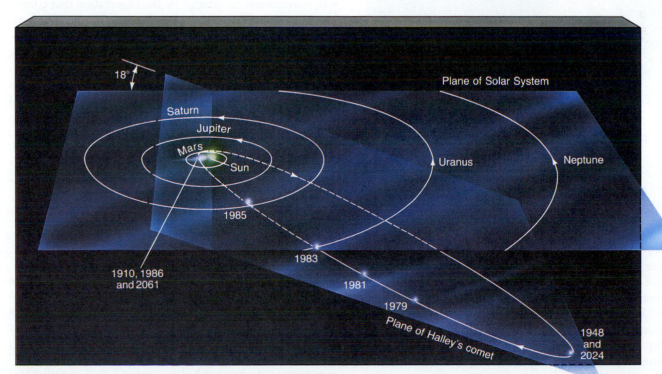

Figure 14.10 Halley's Comet has a smaller orbital path and a shorter period than most comets. Sometime in the past the comet must have encountered a jovian planet (probably Jupiter itself), which threw it into a tighter orbit that extends not to the Oort Cloud but merely a little beyond Neptune. Halley applied Newton's law of gravity to predict this comet's return.

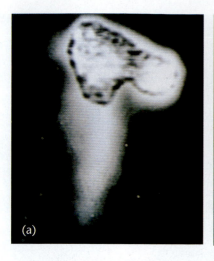

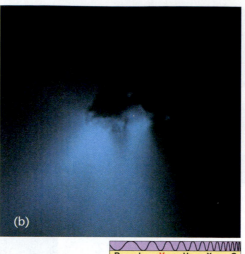

Figure 14.11 (a) *Vega 2's* camera imaged Halley's nucleus, coming within a mere 8000 km of the spinning bundle of dirty ice. (b) The *Giotto* spacecraft better resolved the comet, showing its nucleus to be very dark, although heavy dust in the area obscured any surface features. Resolution here is about 50 m—half the size of a football field. At the time of both pictures, March 1986, the comet was within days of perihelion, and the Sun was toward the bottom. The brightest parts of the images are jets of evaporated gas and dust spewing from the comet's nucleus.

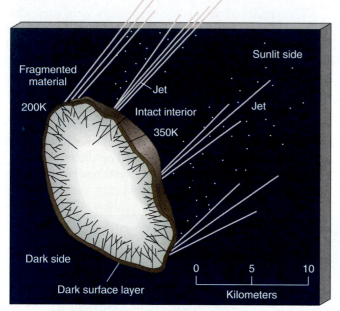

Figure 14.12 A diagram of Halley's nucleus, showing its size, shape, jets, and other physical and chemical properties.

idence for several jets of matter streaming from the nucleus. Instead of evaporating uniformly from the whole surface to form the comet's coma and tail, gas and dust apparently vent from small areas on the sunlit side of Halley's nucleus. The force of these jets may be largely responsible for the comet's 53-hour rotation period. Like maneuvering rockets on a spacecraft, such jets can cause a comet to change its rotation rate and even to veer away from a perfectly elliptical orbit. Astronomers had supposed the existence of these nongravitational forces for comets in general on the basis of the slight deviations from Kepler's laws observed in some cometary trajectories. However, only on the Halley encounter did astronomers actually see these jets at work.

PHYSICAL PROPERTIES OF COMETS

The mass of a comet can sometimes be estimated by watching how it interacts with other solar system objects, or by de-

termining the size of the nucleus and assuming a density characteristic of icy composition. These methods yield typical cometary masses ranging from 10^{12} to 10^{16} kg, comparable to the masses of small asteroids. A comet's mass decreases with time because some of it is lost each time it rounds the Sun. For comets that travel within an astronomical unit of the Sun, this evaporation rate can reach as high as 10^{30} molecules per second—about 10 tons of cometary material lost for *every second* the comet spends near the Sun. Astronomers have estimated that this loss of material will destroy Halley's Comet in about 5000 orbits, or 40,000 years.

In seeking the physical makeup of a cometary body itself, astronomers are guided by the observation that comets have dust that reflects light, as well as gas that emits spectral lines of hydrogen, nitrogen, carbon, and oxygen. Even as the atoms, molecules, and dust particles boil off, creating the coma and tail, the nucleus itself remains a cold mixture of gas and dust, hardly more than a ball of loosely packed ice with a density of about 100 kg/m^3 (0.1 g/cm^3) and a temperature of only a few tens of kelvins. Experts now consider cometary nuclei to be largely made of dust particles trapped within a mixture of methane, ammonia, and ordinary water ice. (These constituents should be fairly familiar to you as the main components of most of the small moons in the outer solar system, discussed in Chapters 12 and 13.) Comets are often described as "dirty snowballs," a term first coined by Fred Whipple of Harvard University.

A COMETARY IMPACT

In July 1994, skywatchers were treated to an exceedingly rare event that greatly increased our knowledge of comet composition and structure—the collision of a comet (known as Shoemaker–Levy 9, after its discoverers) with the planet Jupiter! When it was discovered in March 1993, Shoemaker–Levy 9 appeared to have a curious, "squashed" appearance. Higher-resolution images (see Figure 14.13a) revealed that the comet's flattened nucleus was really made up of several large pieces, each following the same orbit, but spread out along the comet's path like a string of pearls.

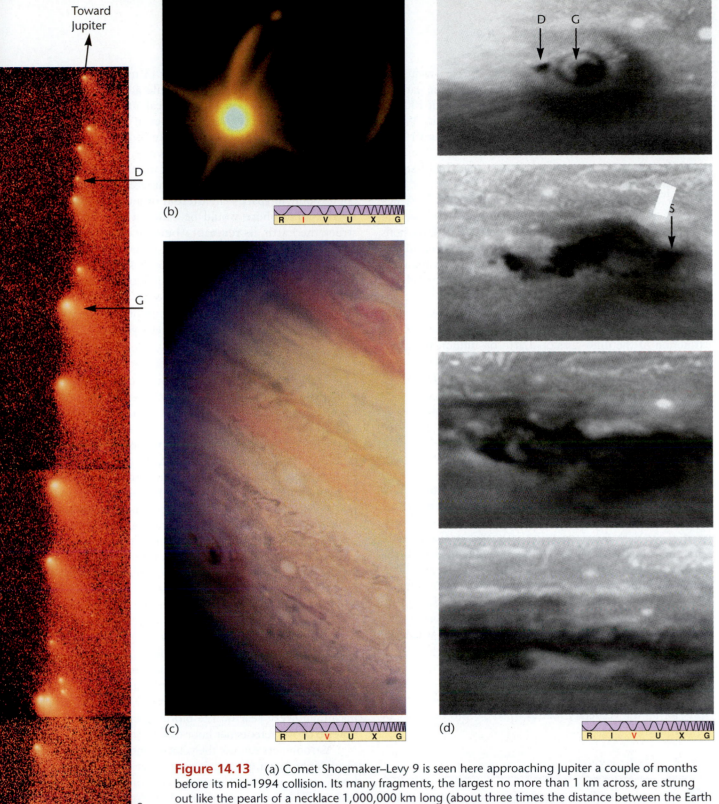

Figure 14.13 (a) Comet Shoemaker–Levy 9 is seen here approaching Jupiter a couple of months before its mid-1994 collision. Its many fragments, the largest no more than 1 km across, are strung out like the pearls of a necklace 1,000,000 km long (about three times the distance between the Earth and Moon). These fragments were created when the parent comet shattered during a close encounter with Jupiter years ago. (b) One of the largest parts of SL-9, fragment G, produced this fireball on the southwest limb of Jupiter; it is seen here some 10 minutes after impact, radiating strongly in the infrared (that is, giving off lots of heat). Also visible is the small, warm cloud on the southeast limb left over from the impact of fragment A, the first to hit the planet on the previous day. (c) The collisions caused several "black eyes" roughly the size of the Earth in Jupiter's southern hemisphere. One of the most prominent impact sites, again caused by fragment G, is shown clearly in this true-color, visible-light photo. Taken nearly 2 hours after impact, it also shows a large dark arc some 6000 km from the impact site—the result of plume material falling back onto Jupiter. (d) One small area on Jupiter got hit with three of SL-9's fragments. This series of photographs shows the month-long evolution of several of the impact sites. Comet fragments D and G hit on July 17 and 18, followed by fragment S on July 21. Thereafter, the impact debris was transported around the planet by the jovian winds.

What could have caused such an unusual object? Tracing the orbit backward in time, researchers calculated that, early in July 1992, the comet had approached within about 100,000 km of Jupiter, well inside the planet's Roche limit. ∞ (p. 261) They surmise that the objects shown in Figure 14.13a are the fragments produced when a previously "normal" comet was captured by Jupiter and torn apart by its strong gravitational field.

On its next approach, Shoemaker–Levy 9 struck Jupiter's upper atmosphere, plowing into it at a speed of over 60 km/s and causing a series of enormous explosions (see Figure 14.13b). Each impact created, for a period of a few minutes, a brilliant fireball hundreds of kilometers across, with a temperature of many thousands of kelvins. The energy released in each of these explosions was comparable to a billion terrestrial nuclear detonations, rivaling in violence the prehistoric impact suspected of causing the extinction of the dinosaurs on Earth 65 million years ago (see *Interlude 14-1* on p. 300). Every major telescope on Earth, as well as the *Hubble Space Telescope, Galileo* (which was only 1.5 A.U. from the planet at the time) and even *Voyager 2* were watching. The effects on the planet's atmosphere, and the vibrations produced throughout Jupiter's interior, were observable for days after the impact.

As best we can determine, none of the cometary fragments breached the jovian clouds. Only Galileo had a direct view of the impacts on the back side of Jupiter, and in every case the explosions seemed to occur high in the atmosphere, above the uppermost cloud deck. Consequently, most of the dark debris seen in the images probably arose from the comet itself. Spectral lines from metals such as silicon, magnesium, and iron were detected in the aftermath of the collisions, and these might explain the dark "smoke" near some of the impact sites (see Figure 14.13c and d). Water vapor was also detected spectroscopically, again apparently from the melted and vaporized comet—which really did resemble a loosely packed snowball.

The fallen debris from the impacts evolved slowly around Jupiter's bands, eventually spreading, after five months, completely around the planet. It will probably take years for all the cometary matter to settle into the body of Jupiter itself.

14.3 *Meteoroids*

6 On a clear night, it is possible to see a few **meteors**, or "shooting stars," every hour. As a small piece of interplanetary debris enters Earth's atmosphere, it heats and excites air molecules through friction. These molecules emit light as they return to their ground states, producing a sudden bright streak (Figure 14.14). The bright streak is a meteor. Before encountering the atmosphere, the chunk of debris is known as a **meteoroid**. Any part of a meteoroid that survives its fiery passage through our atmosphere and finds its way to the ground is a **meteorite**. Notice that the sudden flash of light produced by a meteor is in no way similar to the broad, steady swath of

light associated with a comet's tail. A meteor is a fleeting event in our own atmosphere, while a comet tail is truly enormous, possibly an astronomical unit or more long, and visible in the sky for weeks or even months.

Meteoroids and asteroids are both interplanetary fragments. What separates them is size. The dividing line between them is a little fuzzy, but meteoroids are usually taken to be less than 100 m in diameter. The distinction applies only in space—an asteroid or a comet striking the Earth's atmosphere would be called a meteor, and if it struck the Earth, its remnant would be called a meteorite. Some meteoroids are the rocky remains of broken-up comets. Others seem to be small bodies that have strayed from the asteroid belt, or pieces left over from asteroid collisions. Like comets and asteroids, many meteoroids are composed of ancient material and provide us with evidence of conditions in the early solar system.

COMETARY FRAGMENTS

Each time a comet passes near the Sun, some cometary fragments dislodge from the main body. The fragments initially travel in a tightly knit group of dust or pebble-sized objects, called a **meteoroid swarm**, moving in nearly the same orbit as the parent comet. Over the course of time, the swarm gradually disperses along the orbit, so that eventually the **micrometeoroids**, as these small meteoroids are known, become more or less smoothly spread all the way around the parent comet's orbit. If Earth's orbit happens to intersect the orbit of such a young cluster of meteoroids, a spectacular **meteor shower** can result. Earth's motion takes it across a given comet's orbit at most twice a year (depending on the precise orbit of each body). Intersection occurs at the same time each year (see Figure 14.15), so the appearance of certain meteor showers is a regular and (fairly) predictable event.

Meteor showers are composed of large numbers of meteors burning their way through the atmosphere. The showers are usually named for their *radiant*, the constellation from whose direction they appear to come. For example, the Perseid shower is seen to emanate from the constellation Perseus. It can last for several days, but reaches maximum every year on the morning of August 12, when upward of 50 meteors per hour can be observed.

Astronomers can use the velocity and direction of a meteor's flight to compute its interplanetary trajectory. This is how certain meteoroid swarms have come to be identified with well-known comet orbits. For example, the Perseid shower shares the same orbit as Comet 1862III, the third comet discovered in the year 1862 (also known as Comet Swift–Tuttle). Meteoroids travel around the Sun in elliptical orbits just as comets, asteroids, and planets do. All these objects obey Kepler's laws, as they all move under the gravitational influence of the Sun. Table 14-1 lists some prominent meteor showers, the dates they are visible from Earth, and the comet from which they are thought to originate. Notice in the table that the last "parent comet," Phaeton, is actually an asteroid. This object shows no sign of cometary activity, but its orbit matches the meteoroid paths very well.

(a)

(b)

Figure 14.14 A bright streak called a meteor is produced when a fragment of interplanetary debris plunges into the atmosphere, heating the air to incandescence. (a) A small meteor photographed against a backdrop of stars. (b) An auroral display provides the background for a brighter meteor trail.

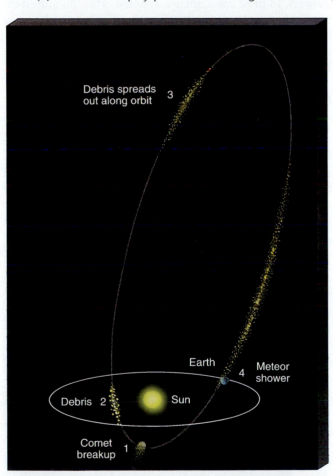

Figure 14.15 A meteoroid swarm associated with a given comet intersects the Earth's orbit at specific locations, giving rise to meteor showers at specific times of the year. We imagine that a portion of the comet breaks up near perihelion, at the point marked 1. The fragments continue along the original comet orbit, gradually spreading out as they go (points 2 and 3). The rate at which the debris disperses around the orbit is actually much slower than depicted here—it takes many orbits for the material to spread out as shown. Eventually, the fragments will extend all around the orbit, more or less uniformly. If the orbit happens to intersect Earth's orbit, a meteor shower is seen each time Earth passes through the intersection (point 4).

STRAY ASTEROIDS

Larger meteoroids—more than a few centimeters in diameter—are generally *not* associated with swarms of cometary debris. Regarded as stray asteroids or as the result of asteroid collisions, these objects have produced most of the cratering on the surfaces of the Moon, Mercury, Venus, Mars, and some of the moons of the jovian planets. When these large meteoroids enter Earth's atmosphere with a typical velocity of nearly 20 km/s, they produce energetic shock waves, or "sonic booms," as well as a bright sky streaks and dusty trails of discarded debris. Such large meteors are sometimes known as *fireballs*. The greater the velocity of the incoming object, the hotter its surface becomes and the faster it burns up. A few large meteors enter the atmosphere with such great velocity (about 75 km/s) that they either fragment or disperse entirely at high altitudes.

More massive meteors (at least a ton in mass and a meter across) do make it to the surface, converting their kinetic energy (motion) into mechanical energy (damage), thermal energy (heat), and acoustical energy (sound). This combination is inevitably explosive and produces a crater such as the kilometer-wide Barringer Crater of Figure 8.18. From the size of this crater, we can estimate that the meteoroid responsible must have had a mass of about 50,000 tons. Only 25 tons of iron meteorite fragments have been found at the crash site. The remaining mass must have been scattered by the explosion at impact, broken down by subsequent erosion, or buried in the ground.

Currently, Earth is scarred with nearly 100 craters larger than 0.1 km in diameter. Most of these are so heavily eroded by weather and distorted by crustal activity that they can be identified only in satellite photography, as shown in Figure 14.16. Fortunately, such major collisions between Earth and large meteoroids are thought to be rare events now. Researchers believe that, on average, they occur only once every few hundred thousand years (see *Interlude 14-1* on p. 300).

TABLE 14.1   *Some Prominent Meteor Showers*

MORNING OF MAXIMUM ACTIVITY	SHOWER NAME	ROUGH HOURLY COUNT	PARENT COMET
Jan. 3	Quadrantid	40	—
Apr. 21	Lyrid	10	1861I (Thatcher)
May 4	Eta Aquarid	20	Halley
June 30	Beta Taurid	25	Encke
July 30	Delta Aquarid	20	—
Aug. 11	Perseid	50	1862III (Swift–Tuttle)
Oct. 9	Draconid	up to 500	Giacobini–Zinner
Oct. 20	Orionid	30	Halley
Nov. 7	Taurid	10	Encke
Nov. 16	Leonid	12[*]	1866I (Tuttle)
Dec. 13	Geminid	50	3200 Phaeton?

[*]*Every 33 years, as the Earth passes through the densest region of this meteoroid swarm, we see intense showers that can reach 1000 meteors per minute for brief periods of time.*

Figure 14.16   This photograph, taken from orbit by the U.S. *Skylab* space station, clearly shows the ancient impact basin that forms Quebec's Manicouagan Reservoir. A large meteorite landed there about 200 million years ago. The central floor of the crater rebounded after the impact, forming an elevated central peak. The lake, 70 km in diameter, now fills the resulting ring-shaped depression. (See also the chapter opening photograph.)

Not all meteoroid encounters with the Earth result in an impact. One of the most recent meteoritic events occurred in central Siberia on June 30, 1908 (Figure 14.17). The presence of only a shallow depression as well as a complete lack of fragments implies that this Siberian intruder exploded several kilometers above the ground, leaving a blasted depression at ground level but no well-formed crater. Recent calculations suggest that the object in question was a rocky meteoroid about 30 m across. The explosion, estimated to have been equal in energy to a 10-megaton nuclear detonation, was heard hundreds of kilometers away and produced measurable increases in atmospheric dust levels all across the Northern Hemisphere.

The orbits of large meteorites that survive their plunge through Earth's atmosphere can be reconstructed in a manner similar to that used to determine the orbits of meteor showers. In most cases, their computed orbits do indeed intersect the asteroid belt, providing the strongest evidence we have that they were once part of the belt before being redirected, probably by a collision with another asteroid, into the Earth-crossing orbit that led to the impact with our planet.

METEORITE PROPERTIES

7 One feature that distinguishes the small micrometeoroids that burn up in Earth's atmosphere from the large meteoroids that manage to reach the ground is their composition. The average density of meteoritic fireballs too small to reach the ground is about 500–1000 kg/m³. Such a low density is typical of comets, which are made of loosely packed ice and dust. By contrast, the meteorites that reach Earth's surface are often much denser—up to 5000 kg/m³—suggesting a composition more like that of the asteroids.

Figure 14.17   The Tunguska event of 1908 leveled trees over a vast area. Although the impact of the blast was tremendous and its sound audible for hundreds of kilometers, the Siberian site was so remote that little was known about the event until scientific expeditions arrived to study it many years later.

Figure 14.18 (a) A large meteorite on display at the Hayden Planetarium in New York serves as a jungle gym for curious children in this photograph from the 1930s. (b) The Wabar meteorite, discovered in the Arabian desert. Although small fragments of the original meteor had been collected more than a century before, the 2000-kg main body was not found until 1965.

Meteorites like the ones shown in Figure 14.18 have received close scrutiny from planetary scientists—prior to the Space Age, they were the only type of extraterrestrial matter we could touch and examine in terrestrial laboratories.

Most meteorites are rocky in composition (Figure 14.19a), although a few percent are composed mainly of iron and nickel (Figure 14.19b). When meteorites are heated and broken down into their component elements, we learn that their basic composition is much like the stony inner planets and the Moon, except that some of their lighter elements—such as hydrogen and oxygen—appear to have boiled away into space when they were molten long ago. Some meteorites show clear evidence of strong heating at some time in their past, indicating that they originated on a larger body that either experienced some geological activity or was partially melted during the collision that liberated the meteorite. Others show no such evidence and probably date back to the formation of the solar system.

Most primitive of all are the *carbonaceous* meteorites, so called because of their relatively high carbon content. They are black or dark gray, and they may well be related to the carbon-rich asteroids that populate the outer asteroid belt. (Similarly, silicate-rich stony meteorites are associated with the inner asteroids.) Many contain significant amounts of ice and other volatile substances, and they are usually rich in organic molecules.

Finally, almost all meteorites are *old*. Direct radioactive dating shows most of them to be between 4.4 and 4.6 billion years old—roughly the age of the oldest lunar rocks. Meteorites, along with some lunar rocks, comets, and perhaps the planet Pluto, provide essential clues to the original state of matter in the solar neighborhood.

Figure 14.19 (a) A stony meteorite often has a dark fusion crust, created when its surface is melted by the tremendous heat generated during passage through the atmosphere. (b) Iron meteorites, much rarer than stony ones, usually contain some nickel as well. Most such meteorites show characteristic crystalline patterns when their surfaces are cut, polished, and etched with weak acid.

Chapter Review

SUMMARY

More than 4000 **asteroids** (p. 298) have been cataloged. Most orbit in a broad band called the **asteroid belt** (p. 298) between the orbits of Mars and Jupiter. They are probably primal rocks that never clumped together to form a planet. A few *Earth-crossing asteroids* have orbits that intersect Earth's orbit and will probably collide with our planet one day. The 50 or so **Trojan asteroids** (p. 300) share Jupiter's orbit, remaining 60° ahead or behind that planet as it moves around the Sun. The **Kirkwood gaps** (p. 301) in the main asteroid belt have been cleared by Jupiter's gravity.

The masses of the asteroids are small, totaling less than 1/10 the mass of Earth's Moon. The largest asteroids are a few hundred kilometers across. Most are much smaller. Asteroids are classified according to the properties of their reflected light. Brighter S-type (silicate) asteroids dominate the inner asteroid belt, while darker C-type (carbonaceous) asteroids are more plentiful in the outer regions. They are believed to have changed little since the solar system formed. The smaller asteroids are generally irregular in shape, and may have experienced violent collisions in the past.

The best views of asteroids have come from the *Galileo* probe as it passed through the asteroid belt on its way to Jupiter. *Galileo* found that the asteroid Ida has a tiny moon orbiting it. Several such binary asteroids have been observed from Earth.

Comets (p. 304) are fragments of icy material that normally orbit far from the Sun. Unlike most other bodies in the solar system, their orbits are often highly elongated and not confined to the ecliptic plane. Most comets are thought to reside in the **Oort Cloud** (p. 305), a vast "reservoir" of cometary material, tens of thousands of astronomical units across, completely surrounding the Sun. A very small fraction of comets happen to have highly elliptical orbits that bring them into the inner solar system. Comets with orbital periods less than about 200 years are thought to originate not in the Oort cloud, but in the **Kuiper Belt** (p. 304), a broad band lying roughly in the ecliptic plane, beyond the orbit of Neptune.

As a comet approaches the Sun, its surface ice begins to vaporize. We see the comet by the sunlight reflected from the dust and vapor released. The **nucleus** (p. 305), or core, of a comet may be only a few kilometers in diameter. It is surrounded by a **coma** (p. 305) of dust and gas. Surrounding this is an extensive invisible **hydrogen envelope** (p. 305). Stretching behind the comet is a long **tail** (p. 305), formed by the interaction between the cometary material and the solar wind. The **ion tail** (p. 306) consists of ionized gas particles and always points directly away from the Sun. The **dust tail** (p. 306) is less affected by the solar wind and has a somewhat curved shape.

Spacecraft visited Halley's Comet in 1986 and studied its nucleus. All other comet studies have been indirect, usually based on spectroscopic measurements. Comets are icy, dusty bodies, sometimes called "dirty snowballs," that are believed to be "leftover" material unchanged since the formation of the solar system. Their masses are comparable to the masses of small asteroids.

Comets and stray asteroids are responsible for most of the cratering on the various worlds in the solar system. Earth is still subject to these sorts of collisions. The most recent Earth impact occurred in 1908, when an asteroid apparently exploded several miles above Siberia. Comet Shoemaker–Levy 9 struck Jupiter in 1994, causing violent explosions in that planet's atmosphere.

Meteors (p. 310), or "shooting stars," are bright streaks of light that flash across the sky as a **meteoroid** (p. 310), a piece of interplanetary debris, enters Earth's atmosphere. If any of the meteoroid reaches the ground, it is called a **meteorite** (p. 310). The major difference between meteoroids and asteroids is their size. The dividing line between them is usually taken to be around 100 m.

Each time a comet rounds the Sun, some cometary material becomes dislodged, forming a **meteoroid swarm** (p. 310)—a group of small **micrometeoroids** (p. 310) following the comet's original orbit. If Earth happens to pass through the comet's orbit, a **meteor shower** (p. 310) occurs.

Larger meteoroids are probably pieces of material chipped off asteroids following collisions in the asteroid belt. Meteorite composition is thought to mirror the composition of the asteroids, and the few orbits that have been determined are consistent with an origin in the asteroid belt. Some meteorites show evidence of heating, but the oldest ones do not. Most meteorites are between 4.4 and 4.6 billion years old.

SELF-TEST: True or False?

____ **1.** Asteroids generally move on almost circular orbits.

____ **2.** The Apollo asteroids have perihelion distances less than 1 A.U.

____ **3.** The C-type asteroids are so named because of their heavily cratered surfaces.

____ **4.** S-type asteroids are more common in the inner part of the asteroid belt.

____ **5.** The least reflective asteroids are C-type.

____ **6.** Some comets travel up to 50,000 A.U. from the Sun.

____ **7.** Cometary orbits always lie close to the ecliptic plane.

____ **8.** The Oort cloud is the large cloud of gas surrounding a comet while it is near the Sun.

____ **9.** Tails of comets always lie along the path of the orbit.

____ **10.** The light emitted from a meteor is due to a meteoroid literally burning up in the Earth's atmosphere.

____ **11.** Some meteorites found on Earth originally came from the Moon or Mars.

____ **12.** Comets are the sources of meteor showers.

SELF-TEST: Fill in the Blank

1. Asteroids have a _____ composition.

2. The asteroid belt, where most asteroids are found, lies between the orbits of _____ and _____.

3. The largest asteroids are _____ kilometers in diameter; the smallest are only _____ meters across.

4. The Trojan asteroids share an orbit with _____.

5. The Kirkwood gaps are caused by an orbital _____ with Jupiter.

6. Comets have an _____ composition.

7. The process of sublimation is one in which a solid turns into a _____.

8. Comets have orbits that are highly _____.

9. The nucleus of a comet is typically _____ kilometers across; its tail may be up to _____ long.

10. Passage of a comet near the Sun may leave a _____ moving in the comet's orbit.

11. The Kuiper belt is the source of the short-period _____.

12. Meteoroids are mostly fragments of _____ and _____.

13. When a meteoroid enters Earth's atmosphere, you see a _____.

14. Meteoroids that impact a planet or moon are called _____.

15. The oldest meteorites have ages of _____.

REVIEW AND DISCUSSION

1. Describe the differences among the Trojan, Apollo, and Amor asteroids.

2. How have the best photographs of asteroids been obtained?

3. What are the Kirkwood gaps? How did they form?

4. Compare and contrast the C-type and S-type asteroids.

5. What are comets like when they are far from the Sun? What happens when they enter the inner solar system?

6. Where are most comets found?

7. Describe the various parts of a comet while it is near the Sun.

8. What are the typical ingredients of a comet nucleus?

9. What are some possible fates of comets?

10. Explain the difference between a meteor, a meteoroid, and a meteorite.

11. What causes a meteor shower?

12. What are the most primitive meteorites?

13. What do meteorites reveal about the age of the solar system?

14. Why can comets approach the Sun from any direction, while asteroids generally orbit close to the ecliptic plane?

15. Why do meteorites contain information about the early solar system, yet Earth does not?

16. What might be the consequences of a 10 km meteorite striking the Earth today?

PROBLEMS

1. The largest asteroid, Ceres, has a radius 0.073 times the radius of the Earth and a mass of 0.0002 Earth masses. How much would a 100 kg astronaut weigh on Ceres?

2. (a) Using Kepler's Laws of Planetary Motion (see Chapter 2), calculate the orbital period of a comet with a perihelion distance of 0.5 A.U. and aphelion in the Oort cloud, at a distance of 50,000 A.U. from the Sun. (b) A short-period comet has a perihelion distance of 1 A.U. and an orbital period of 125 yr. What is its maximum distance from the Sun?

3. A particular comet has a total mass of 10^{13} kg, 95 percent of which is ice and dust. The remaining 5 percent is in the form of rocky fragments of average mass 100 g. How many meteoroids would you expect to find in the swarm formed by the breakup of this comet?

4. You are standing on the surface of a spherical, 10 km diameter asteroid whose density is 3000 kg/m^3. Could you throw a small rock into circular orbit around it? Give the velocity required in km/s and mph.

PROJECTS

1. The only way to tell an asteroid from a star is to watch it over several nights; you can detect its movement in front of the star background. The astronomy magazines *Sky and Telescope* and *Astronomy* often publish charts for especially prominent asteroids. Look for the asteroids Ceres, Pallas, or Vesta. They are the brightest asteroids. Use the chart to locate the appropriate star field. Aim binoculars at that location in the sky; you may be able to pick out the asteroid from its location in the chart. If you can't, make a rough drawing of the entire field. Come back a night or two later, and look again. The "star" that has moved is the asteroid.

2. Although a spectacular naked-eye comet comes along only about once a decade, fainter comets can be seen with binoculars and telescopes in the course of every year. *Sky and Telescope* often runs a "Comet Digest" column announcing the whereabouts of comets. A comprehensive list of periodic comets expected to return in a given year can be found in Guy Ottewell's *Astronomical Calendar*. This calendar contains a wealth of other sky information as well, including monthly star charts. At this writing, it costs $15 a year and can be purchased from:

Astronomical Workshop
Furman University
Greenville, South Carolina 29613
(803) 294-2208

3. There are a number of major meteor showers every year, but, if you plan to watch one, be sure to notice the phase of the Moon. Bright moonlight or city lights can obliterate a meteor shower. A common misconception about meteor watching is that most meteors are seen in the direction of the shower's radiant point. It's true that if you trace the paths of the meteors backward in the sky, they all can be seen to come from the radiant. But most meteors don't become visible until they are 20 or 30 degrees from the radiant. Meteors can appear in all parts of the sky! Just relax and let your eyes rove among the stars. You will generally see many more meteors in the hours before dawn than in the hours after sunset. Why do you suppose meteors have different brightnesses? Can you detect their variety of colors? Watch for meteors that appear to "explode" as they fall. Watch for vapor trails that linger after the meteor itself has disappeared.

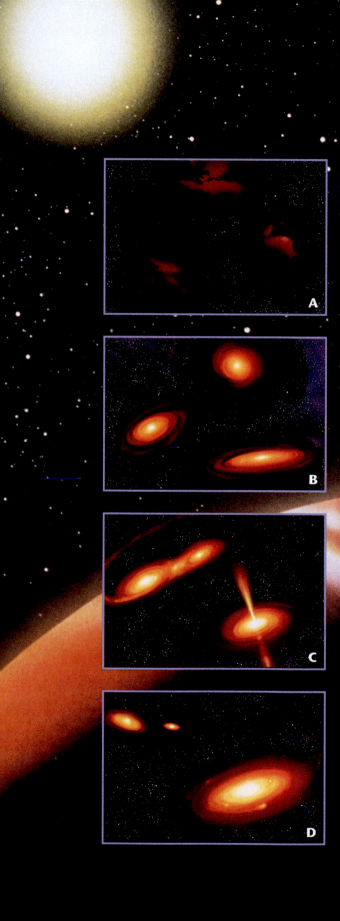

15

THE FORMATION OF THE SOLAR SYSTEM

The Birth of Our World

LEARNING GOALS

Studying this chapter will enable you to:

1 Summarize the major features that a theory of solar-system origins has to explain.

2 Outline the process by which planets form as natural by-products of star formation.

3 Explain the role played by dust in the currently accepted model of solar system formation.

4 Account for the differences between the terrestrial and the jovian planets.

5 Discuss the possible role of collisions in determining specific characteristics of the solar system.

6 Explain the angular momentum problem, and discuss some proposed solutions.

(Opposite page, background) An imaginary triple-star system, viewed from the vicinity of a Jupiter-like planet. The planet's parent star resembles our Sun, hence its yellowish appearance. The other two stars are much hotter, and appear bluish. The four inserts show, from top to bottom, a possible evolutionary sequence that gave rise to the triple-star system.
(Inset A) Neighboring, dark galactic clouds contracting, their central regions heating enough to emit a reddish glow.
(Inset B) Formation of huge, rotating disks of glowing matter around young stars ("protostars") foster further heating at their cores.
(Inset C) Two of the protostars become gravitationally bound to each other, interacting somewhat violently; the third (foreground) protostar drifts farther away, allowing it to evolve more calmly.
(Inset D) The third protostar continues to evolve slowly, with planet-sized clumps ("protoplanets") beginning to emerge within the disk.

Completing these chapters on the planets, you might be struck by the vast range of physical and chemical properties found in the solar system. The planets present a long list of interesting features and bizarre peculiarities, and the list grows even longer when we consider the characteristics of their moons. Every object has its idiosyncrasies, some of them due to particular circumstances, many others the result of planetary evolution. Each time a new discovery is made, we learn a little more about the properties and history of our planetary system. Our astronomical neighborhood may seem more like a great junkyard than a smoothly running planetary system. Can we really make any sense of the entire collection of solar system matter? Is there some underlying principle that unifies the knowledge we have gained? The answer, as we will see, is "yes."

15.1 Modeling the Origin of the Solar System

The origin of the planets and their moons is a complex and as yet incompletely solved puzzle, although the basic outlines of the process are quite well understood. Most of our knowledge of the solar system's formative stages has emerged from studies of interstellar gas clouds, fallen meteorites, and Earth's Moon, as well as from the various planets observed with ground-based telescopes and planetary space probes. Ironically, studies of Earth itself do not help much because information about our planet's early stages eroded away long ago. Meteorites provide perhaps the most useful information, for nearly all have preserved within them traces of solid and gaseous matter from the earliest times.

Astronomers still have no firm evidence of the existence of planets like our own anywhere beyond our solar system. For that reason, we must concentrate on the origin of the planetary system in which we live. Bear in mind, however, that no part of the scenario we will describe in the paragraphs that follow is in any way unique to our own system. The same basic processes could have occurred—and, many astronomers believe, probably *did* occur—during the formative stages of most of the stars in our Galaxy. It is part of the job of the planetary scientist to distinguish between those properties of the solar system that are inherent—that is, that were imposed at formation—and those that must have evolved since the solar system formed. In this chapter, we draw together all the planetary data we have amassed and show how the regularities—and the irregularities—of the solar system can be explained by a single comprehensive theory.

MODEL REQUIREMENTS

❶ Any theory of the origin and architecture of our planetary system must adhere to the known facts. We know of nine outstanding properties of our solar system as a whole. They may be summarized as follows.

1. *Each planet is relatively isolated in space.* The planets exist as independent bodies at progressively larger distances from the central Sun; they are not bunched together. In very rough terms, each planet tends to be twice as far from the Sun as its next inward neighbor.

2. *The orbits of the planets are nearly circular.* In fact, with the exceptions of Mercury and Pluto, which we will argue are special cases, each planetary orbit closely describes a perfect circle. The slight orbital eccentricity of Mercury, the innermost planet, is perhaps related to the influence of the Sun's intense gravity. As for the outermost Pluto, some researchers regard this planet as a comet-like body, or perhaps even an escaped moon of Neptune, thereby accounting for its large orbital eccentricity.

3. *The orbits of the planets all lie in nearly the same plane.* The planes swept out by the planets' orbits are accurately aligned to within a few degrees. Put another way, the solar system has the shape of a very thin disk. Again, Mercury and Pluto are slight exceptions, probably for the same reasons just noted.

4. *The direction of the planets' revolution in their orbits about the Sun (counterclockwise as viewed from Earth's north) is the same as that of the Sun's rotation on its axis.* Virtually all the large-scale motions in the solar system—such as the planets' orbits and the Sun's spin—are in the same plane and in the same sense. The plane is that of the Sun's equator, and the sense is that of the Sun's rotation.

5. *The direction in which most planets rotate on their axes also mimics that of the Sun's spin.* This property is less general, as three planets (Venus, Uranus, and Pluto) do not share it. Venus's slow spin is opposite (retrograde) to that of the Sun and the other planets. Uranus apparently lies on its side, with its poles almost in the plane of its orbit. The Pluto–Charon system also rotates almost on its side. (Pluto, as we have seen, is an exception to almost all our rules.)

6. *Most of the known moons revolve about their parent planets in the same direction that the planets rotate on their axes.* Some moons, like those of Jupiter, resemble miniature solar systems, revolving about their parent planet, roughly in the plane of the planet's equator. This motion suggests uniformity in our planetary system and must have been determined by the environment in which the planets and moons formed.

7. *Our planetary system is highly differentiated.* The inner terrestrial planets are characterized by high densities, moderate atmospheres, slow rotation rates, and few or no moons. By contrast, the outer jovian planets (Pluto excepted, as usual) have low densities, thick atmospheres, rapid rotation rates, and many moons.

8. *The asteroids are very old and exhibit a range of properties not characteristic of either the inner or the outer planets or their moons.* The asteroid belt shares, in rough terms, the bulk orbital properties of the planets. However, it appears to be made of primitive, unevolved material, and the meteorites that strike the Earth are the oldest rocks known.

9. *The comets are primitive, icy fragments that do not orbit in the ecliptic plane and probably reside primarily at large distances from the Sun.* The probable existence of the Oort Cloud, surrounding the Sun at tens of thousands of astronomical units, is thought to be directly related to the processes that formed the solar system long ago.

All these observed facts, when taken together, strongly suggest a high degree of order within our solar system. The whole system is not a random assortment of objects spinning or orbiting this way or that. Consequently, it hardly seems possible that our solar system could have formed by the slow accumulation of already made interstellar "planets" casually captured by our Sun over the course of billions of years. The overall architecture of our solar system is too neat, and the ages of its members too uniform, to be the result of random chaotic events. The overall organization points toward a single formation, an ancient but one-time event, 4.6 billion years ago. A convincing theory that explains all of the nine features just listed has been a goal of astronomers for many centuries.

ADDITIONAL CONSIDERATIONS

It is equally important to recognize what our theory of the solar system does *not* have to explain. There is plenty of scope for planets to evolve after their formation, so circumstances that have developed since the initial state of the solar system was established need not be included in our list. Examples are Mercury's 3:2 spinorbit coupling, Venus's runaway greenhouse effect, the Moon's synchronous rotation, the emergence of life on Earth and its absence on Mars, the Kirkwood gaps in the asteroid belt, and the rings and atmospheric appearance of the outer planets. There are many more. Indeed, all of the properties of the planets for which we have already provided an *evolutionary* explanation need not be included as items that our theory must account for at the outset.

In addition to its many regularities, our solar system also has many notable *irregularities*, some of which we have already mentioned. Far from threatening our theory, however, these irregularities are important facts for us to consider in shaping our explanations. For example, it is necessary that the explanation for the solar system not insist that *all* planets rotate in the same sense or have *only* prograde moons, because that is not what we observe. Instead, the theory of the solar system should provide strong reasons for the observed planetary characteristics, yet be flexible enough to allow for and explain the deviations, too. And, of course, the existence of the asteroids and comets that tell us so much about our past must be an integral part of the picture. That's quite a tall order, yet many researchers now believe that we are close to that level of understanding.

15.2 *The Condensation Theory*

NEBULAR CONTRACTION

2 One of the earliest heliocentric models of solar system formation is termed the **nebular theory**. It is often attributed to the eighteenth-century German philosopher Immanuel Kant, but he merely elaborated upon a proposal made a century earlier by the French philosopher René Descartes. In this model, a large cloud of interstellar gas began to collapse under the influence of its own gravity. As it contracted, it became denser and hotter, eventually forming a star—the Sun—at its center. While all this was going on, the outer, cooler, parts of the cloud formed a giant swirling region of matter, creating the planets and their moons essentially as by-products of the star-formation process. We have called this swirling mass "the primitive solar system," but it is more usually referred to as the **solar nebula**.

The nebular theory is an example of an **evolutionary theory**, which describes the development of the solar system as a series of gradual and natural steps, understandable in terms of well-established physical principles. Evolutionary theories may be contrasted with **catastrophic theories**—theories that invoke accidental or unlikely celestial events in order to interpret observations.* Scientists usually do not like to invoke catastrophes to explain the universe. However, as we will see, there are instances where pure chance has played a role in determining the present state of the solar system.

In 1796 the French mathematician-astronomer Pierre Simon de Laplace tried to develop the nebular model in a quantitative way. He was able to show mathematically that the conservation of angular momentum

---

A good example of such a theory is the collision hypothesis, which imagines that the planets were torn from the Sun by a close encounter with a passing star. This hypothesis enjoyed some measure of popularity during the nineteenth century, in large part due to the inability of other theories to account for the observed properties of the solar system, but no scientist takes it seriously today. Aside from its extreme improbability, it is completely unable to explain the orbits, the rotations, or the composition of the planets and their moons.

(consult the *More Precisely* feature on p. 321.) demands that an interstellar cloud like the hypothetical solar nebula must spin faster as it contracts. A decrease in the size of a rotating mass must be balanced by an increase in its rotational speed.

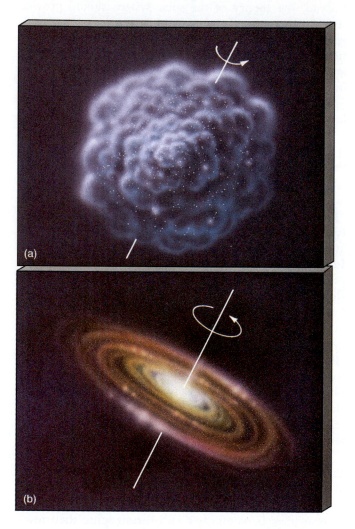

Figure 15.1   Conservation of angular momentum demands that a contracting, rotating cloud (a) must spin faster as its size decreases. Eventually, (b) the primitive solar system came to resemble a giant pancake. The large blob at the center would ultimately become the Sun.

The increase in rotation speed, in turn, must have caused the nebula's *shape* to change as it collapsed. In Chapter 11, we saw how a spinning body tends to develop a bulge around its middle. The rapidly spinning nebula behaved in exactly this way. As shown in Figure 15.1, the fragment eventually flattened into a pancake-shaped primitive solar system. If we now suppose that planets formed out of this spinning material, we can already begin to understand the origin of some of the architecture observed in our planetary system today, such as the circularity of the planets' orbits and the fact that they move in nearly the same plane.

Laplace imagined that as the spinning nebula contracted, it left behind a series of concentric rings, each of which would eventually become a planet orbiting a central **protosun**—a hot ball of gas well on its way to becoming the Sun. Each ring then clumped into a **protoplanet**—a forerunner of a genuine planet. Figure 15.2 is an artist's illustration of this scenario. In this diagram, several outer planets have already begun to develop, while the interior of the primitive solar system is still contracting to shape the inner planets and the central Sun.

THE ROLE OF DUST

❸   Two centuries ago, Laplace got the description of the collapse and flattening of the nebula essentially correct. However, in modern times, when computers were used to study more subtle aspects of the problem, some fatal flaws were found in the simple nebular picture. Detailed computations show that a ring of the sort assumed in the theory would probably not form, and even if it did, would not necessarily condense to form a planet in any case. In fact, computer calculations predict just the opposite: The rings would tend to disperse. The protoplanetary matter is too warm, and, moreover, no one ring would have enough mass to bind its own matter into a ball.

The model currently favored by most astronomers is really just a more sophisticated version of the nebular theory. Known as the **condensation theory**, it combines the good features of the old nebular theory with new information about interstellar chemistry to avoid most of the old theory's problems. The key new ingredient in the modern picture is the presence of **interstellar dust** in the solar nebula. Astronomers now recognize that the space

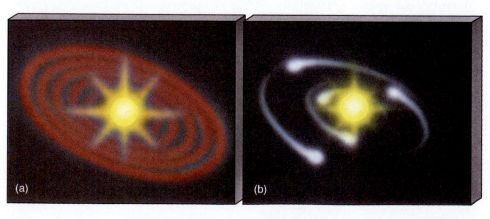

Figure 15.2   Laplace's nebular theory envisioned the formation of (a) rings of gaseous matter at various distances from the central protosun. Eventually, (b) the rings clump into planets.

between the stars is strewn with microscopic dust grains, an accumulation of the ejected matter of many long-dead stars (see Chapter 22). These dust particles probably formed in the cool atmospheres of old stars, then grew by accumulating more atoms and molecules from the interstellar gas within the Milky Way Galaxy. The end result is that our entire Galaxy is littered with miniature chunks of icy and rocky matter having typical sizes of about 10^{-5} m. Figure 15.3 shows one of many such dusty regions found in the vicinity of the Sun.

Dust grains play an important role in the evolution of any gas. Dust helps to cool warm matter by efficiently radiating its heat away in the form of infrared radiation, allowing it to collapse more easily. Furthermore, the dust grains greatly speed up the process of collecting enough atoms to form a planet. They act as **condensation nuclei**—microscopic platforms to which other atoms can attach, forming larger and larger balls of matter. This is similar to the way that raindrops form in Earth's atmosphere; dust and soot in the air act as condensation nuclei around which water molecules cluster. The presence of dust virtually guarantees that gaseous matter will clump, both because the dust "seeds" it with sites for the condensation process to start and because it cools the matter below the point at which outward-pushing pressure (which is just proportional to the gas temperature) can effectively compete with inward-pulling gravity.

More Precisely...   The Concept of Angular Momentum

It seems that most celestial objects rotate. Planets, moons, stars, galaxies—virtually all have some *angular momentum,* which we can define as the tendency of a body to keep spinning or moving in a circle. Angular momentum is as important a property of an object as its mass or its energy.

Consider, first, a simpler motion—*linear momentum,* which is the tendency of an object to keep moving in a straight line in the absence of external forces. Consider a truck and a bicycle rolling equally fast down a street. Each has some linear momentum, but you would obviously find it easier to stop the less massive bicycle. Although the two vehicles have the same speed, the truck has more momentum. We see that the linear momentum of an object depends on the mass of that object. It also depends on the velocity. If two bicycles were rolling down the street at different speeds, the slower one could be stopped more easily. Linear momentum is defined as the product of mass and velocity:

$$\text{linear momentum} = \text{mass} \times \text{velocity}$$

Angular momentum is an analogous property of objects that are rotating or revolving. However, in addition to mass and velocity, the angular momentum also depends on the way in which the object's mass is distributed. The farther the mass is from the object's axis of rotation, the greater the angular momentum. For example, if you whirl a ball with constant velocity at the end of a string, its angular momentum will depend directly on the length of the string—the longer the string, the greater the angular momentum. For the same reason, a doughnut-shaped object will have more angular momentum than a flat disc of the same mass spinning at the same speed, because more of the doughnut's mass is concentrated away from the center of rotation.

We can therefore say, loosely, that

$$\text{angular momentum} = \text{mass} \times \text{velocity} \times \text{"size"}$$

where, as we have seen, the "size" is a quantity that depends not only on the object's dimensions but also on the distribution of its mass. For the simple case of the ball on a string, it is just the length of the string (the radius of the ball's "orbit"). The "velocity" for linear momentum is velocity in a straight line. For angular momentum, "velocity" refers to the circular velocity of spinning or orbital motion.

According to Newton's laws of motion, both types of momentum—linear and angular—must be conserved at all times. In other words, both linear and angular momentum must remain constant before, during, and after a physical change in any object (again, as long as no external forces act). For example, if a spherical object having some spin begins to contract, the relationship above demands that it spin faster. After all, the object's mass does not change during the contraction, yet the size of the object clearly decreases. The circular velocity of the spinning object must therefore increase in order to keep the total angular momentum unchanged.

Figure skaters use the principle of angular-momentum conservation. They spin faster by drawing in their arms (as shown below) and slow down by extending them. Here, the mass of the human body remains the same, but its lateral size changes, causing the body's circular velocity to change to conserve angular momentum.

R I V U X G

Figure 15.3 Interstellar gas and dark dust lanes mark this region of star formation. The dark cloud known as Barnard 86 (left) flanks a cluster of young blue stars called NGC 6520 (right). Barnard 86 may be part of a larger interstellar cloud that gave rise to these stars.

Modern models trace the formative stages of our solar system along the following broad lines. Imagine a dusty interstellar cloud fragment measuring about a light year across. Intermingled with the preponderance of hydrogen and helium atoms, the cloud harbors some heavy-element gas and dust. Some external influence, such as the passage of another interstellar cloud or perhaps the explosion of a nearby star, starts the fragment contracting, down to a size of about 100 A.U. As the cloud collapses, it rotates faster and begins to flatten (just as described in the old nebular theory). By the time it has shrunk to 100 A.U., the solar nebula has already formed an extended, rotating disk.

Astronomers are fairly confident that our own solar nebula formed such a disk because similar disks have been observed (or inferred) around other stars. Figure 15.4 shows a visible-light image of the region around a star called Beta Pictoris, lying about 50 light years from the Sun. When the light from the star itself is suppressed and

the resulting image enhanced by a computer, a faint disk of warm matter (viewed almost edge-on here) can be seen. This particular disk is roughly 1000 A.U. across—about 10 times the diameter of Pluto's orbit. Astronomers believe that Beta Pictoris is a very young star, perhaps only 100 million years old, and that we are witnessing it pass through an evolutionary stage similar to that our own Sun experienced some 4.6 billion years ago.

ACCRETION AND FRAGMENTATION

Unlike the nebular theory sketched earlier, the condensation theory predicts no rings. As depicted in Figure 15.5, the process of planet formation took place in three stages. Early on in the solar nebula, dust grains formed condensation nuclei around which matter began to accumulate. This vital step greatly hastened the critical process of forming the first small clumps of matter. Once these clumps formed, they grew quite rapidly by sticking to other clumps or dust grains they encountered. (Imagine a snowball thrown through a fierce snowstorm, growing bigger as it encounters more snowflakes.) As the clumps grew larger, their surface areas increased, and the rate at which they swept up new material accelerated. They gradually grew into objects of pebble size, baseball size, basketball size, and larger.

Eventually, this process of **accretion**—this gradual growth by collision and sticking—created objects a few hundred kilometers across. By that time, their gravity was strong enough to sweep up material that would otherwise not have collided with them, and their rate of growth became faster still. At the end of this first stage, the solar system was made up of hydrogen and helium gas and millions of **planetesimals**—objects the size of small moons, whose gravitational fields were just strong enough to affect their neighbors.

In the second phase of the accretion process, gravitational forces between the planetesimals caused them to collide and merge, again forming larger and larger objects.

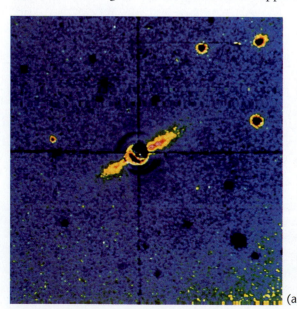

(a)

(b)

Figure 15.4 (a) A computer-enhanced photograph (taken from Las Campanas Observatory in Chile) of a disk of warm matter surrounding the star Beta Pictoris. Most of the light from the star itself is blocked by an instrument called a coronagraph, designed to detect faint halos around bright objects. The full extent of the disk, seen almost edge-on here, is about 1000 A.U. (b) An artist's conception of the disk of clumped matter.

R I V U X G

Figure 15.5 Planetesimals and planets forming in a collapsing, dusty cloud. Dust grains act as condensation nuclei, forming clumps of matter that collide, stick, and grow. Large eddies also come and go as the gas in the disk swirls around. According to the condensation theory, the larger eddies and clumps of matter will eventually become planets. The large blob in the center will become the Sun.

Because larger objects have stronger gravity, the rich became richer in the early solar system, and eventually almost all the planetesimal material was swept up into a few large protoplanets—the accumulations of matter that would eventually evolve into the planets we know today. Figure 15.6 shows a computer simulation of accretion in the inner solar system. Notice how, as the number of bodies decreases, the orbits of the remainder become more widely spaced and more nearly circular.

The four largest protoplanets became large enough that they were able to enter a third phase of planetary development, sweeping up large amounts of gas from the solar nebula to form what would ultimately become the jovian planets. The smaller, inner protoplanets never reached that point, so their masses remained relatively low.

There is an alternative explanation for the significant difference in size between the jovian and the terrestrial planets. Turbulent eddies in the cool outer solar nebula may themselves have become unstable and begun to contract, in some ways mimicking on a smaller scale the collapse of the initial interstellar cloud. In this way, protoplanets could have formed directly, skipping the initial accretion stage. These first protoplanets would have had gravitational fields strong enough to scoop up more of the remaining gas and dust in the solar nebula, allowing them to grow into the gas giants we see today. Their large size reflects the "head start" they obtained in the accretion process.

As the protoplanets grew, another process became important. The strong gravitational fields produced many high-speed collisions between planetesimals and protoplanets. These collisions led to **fragmentation**, as small objects broke into still smaller chunks, which were then swept up by the protoplanets. Not only did the rich get richer, but the poor were mostly driven to destruction! Some of these fragments produced the intense meteoritic bombardment we know occurred during the early evolution of the planets and moons, as we have seen in the last few chapters. ∞ (pp. 159, 190) Only a relatively small number of 10–100 km fragments escaped capture by a planet or a moon and became the asteroids and comets. Assuming that the accretion process was reasonably efficient throughout the disk, we can thus understand how our present solar system has come to exist as a collection of rather small planets orbiting throughout an otherwise empty region of space.

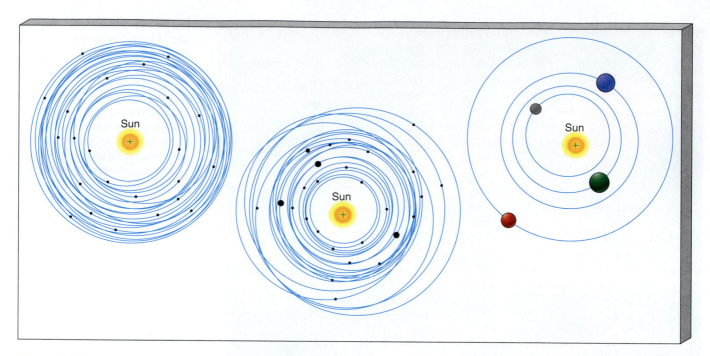

Figure 15.6 Accretion in the inner solar system: Initially, many moon-sized planetesimals orbited the Sun. Over the course of a hundred million years or so, they gradually collided and coalesced, forming a few large planets in roughly circular orbits.

Many of the moons of the planets (but not our own—see Chapter 8) presumably also formed through accretion but on a smaller scale, in the gravitational field of their parent planets. ∞ (p. 189) Once the nebular gas began to accrete onto the large jovian protoplanets, conditions probably resembled a miniature solar nebula, with condensation and accretion continuing to occur. The large moons of the outer planets almost certainly formed in this way. Some of the smaller moons may have been "chipped off" their parent planets during collisions with asteroids; others may be captured asteroids themselves.

Mathematical modeling, like the computer simulation shown in Figure 15.6, indicates that, after about 100 million years, the primitive solar system had evolved into nine protoplanets, dozens of protomoons, and the big protosolar mass at the center. Computer simulations generally reproduce the increasing spacing between the planets ("Bode's law"), although the reasons for the regularity seen in the actual planetary spacing remain unclear. ∞ (p. 132) Roughly a billion years more were required to sweep the system reasonably clear of interplanetary trash. This was the billion-year period that saw the heaviest meteoritic bombardment, tapering off as the number of planetesimals decreased.

15.3 *The Differentiation of the Solar System*

④ We can understand the basic differences in content and structure between the terrestrial and the jovian planets using the condensation theory of the solar system's origin. Indeed, it is in this context that the adjective *condensation* derives its true meaning. To see why the planets' composi-

tion depends on location in the solar system, it is necessary to consider the temperature structure of the solar nebula.

THE ROLE OF HEAT

As the primitive solar system contracted under the influence of gravity, it heated up as it flattened into a disk. The density and temperature were greatest near the central protosun and much lower in the outlying regions. Detailed calculations indicate that the gas temperature near the core of the contracting system was several thousand kelvins. At a distance of 10 A.U., out where Saturn now resides, the temperature was only about 100 K.

In the warmer regions of the cloud, dust grains broke apart into molecules, and they in turn split into excited atoms. Because the extent to which the dust was destroyed depended on the temperature, it also depended on location in the solar nebula. Most of the original dust in the inner solar system disappeared at this stage, but the grains in the outermost parts probably remained largely intact.

The destruction of the dust in the hot inner portion of the solar nebula introduced an important new ingredient into the theoretical mix, one that we omitted from our earlier account of the accretion process. With the passage of time, the gas radiated away its heat and the temperature decreased at all locations, except in the very core, where the Sun was forming. Everywhere beyond the protosun, new dust grains began to condense (or crystallize) from their hotter gas phase to their cooler solid phase, much as raindrops, snowflakes, and hailstones condense from moist, cooling air here on Earth. It may seem strange that although there was plenty of interstellar dust early on, it was mostly destroyed, only to form again later. However, a critical change had occurred. Initially, the nebular gas was uniformly peppered with dust grains. When the dust reformed later, the distribution of grains was very different.

Figure 15.7 plots the temperature gradient across the primitive solar system just prior to the onset of the accretion stage. At any given location, the only materials to condense out were those able to survive the temperature there. As marked on the figure, in the innermost regions, around Mercury's present orbit, only metallic grains could form. It was simply too hot for anything else to exist. A little farther out, at about 1 A.U., it was possible for rocky, silicate grains to form, too. Beyond about 3 or 4 A.U., water ice could exist, and so on, with the condensation of more and more material possible at greater and greater distances from the Sun. The composition of the material that could condense out at any given radius would ultimately determine the types of planets that formed there.

THE JOVIAN PLANETS

In the middle and outer regions of the primitive planetary system, beyond about 5 A.U. from the center, the temperature was low enough for the condensation of several abundant gases into solid form. After hydrogen and helium, the most common materials in the solar nebula (as they are today in the universe as a whole) were the elements carbon, nitrogen, and oxygen. The most common chemical compounds were those containing those elements—specifically, water vapor, ammonia, and methane. As we have seen, these compounds are still the primary constituents of jovian atmospheres. At temperatures of a few hundred kelvins or less, these gases condensed out of the nebula. Consequently, the ancestral fragments destined to become the cores of the jovian planets were formed under cold conditions out of low-density, icy material. The planetesimals that formed at these distances were predominantly composed of ice. Because more material could condense out of the solar nebula at these radii than in the inner regions near the protosun, accretion began sooner, with more resources to draw on. The outer planets grew rapidly to the point where they could accrete nebular gas, not just grains, and eventually formed the hydrogen-rich jovian worlds we see today.

With the formation of the four giant jovian planets, the remaining planetesimals were subject to those planets' strong gravitational fields. Over a period of hundreds of millions of years and after repeated "gravity assists" from the giant planets, many of the interplanetary fragments in the outer solar system were flung into orbits taking them far from the Sun. Astronomers believe that those fragments now make up the Oort Cloud, whose members occasionally visit the inner solar system as comets. During this period, many icy planetesimals were also deflected into the inner solar system, where they played an important role in the evolution of the inner planets. A key prediction of this model is that some of the original planetesimals should

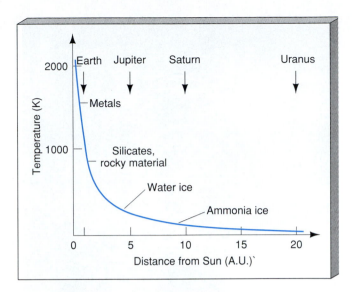

Figure 15.7 Theoretically computed variation of temperature across the primitive solar nebula. In the hot central regions, only metals could condense out of the gaseous state to form grains. At greater distances from the central protosun, the temperature was lower, so rocky and icy grains could also form. The labels indicate the minimum radii at which grains of various types could condense out of the nebula.

have remained behind, in the broad band called the *Kuiper belt*, lying beyond the orbit of Neptune. ∞ (p. 304) In 1993, several such asteroid-sized objects were discovered, lying between 30 and 35 A.U. from the Sun, lending strong support to the condensation theory.

THE TERRESTRIAL PLANETS

In the inner regions of the primitive solar system, condensation from gas to solid began when the average temperature was about 1000 K. The environment there was too hot for ices to survive. Many of the abundant heavier elements, such as silicon, iron, magnesium, and aluminum, combined with oxygen to produce a variety of rocky materials. Planetesimals in the inner solar system were therefore rocky in nature, as were the protoplanets and planets they ultimately formed.

These heavier materials condensed into grains in the outer solar system, too, of course. However, they would have been vastly outnumbered by the far more abundant light elements there. The outer solar system is not deficient in heavy elements. The inner solar system is underrepresented in light material. Here we have another reason why the jovian planets grew so much bigger than the terrestrial worlds. The inner regions of the nebula had to wait for the temperature to drop so that a few rocky grains could appear and begin the accretion process, but the outer regions may not have had to wait at all. The accretion process in the outer solar system began almost with the formation of the disk itself.

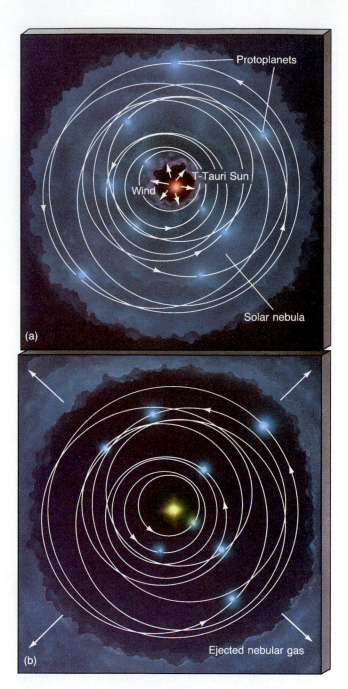

Figure 15.8 (a) Strong stellar winds from newly born stars are responsible for sweeping away any dust and gas left over from the star formation process, (b) leaving only planets and planetesimals behind.

Very abundant light elements such as hydrogen and helium, as well as any other gases that failed to condense into solids, would have escaped from the terrestrial protoplanets, or, more likely, they were simply never accreted from the solar nebula. The inner planets' surface temperature was too high, and their gravity too low, to capture and retain those gases. Where then did the Earth's volatile gases, particularly water, come from? The answer seems to be that icy fragments—comets—from the outer solar system, deflected into eccentric orbits by the jovian planets'

gravity, participated in the meteoritic bombardment of the newly born inner planets, supplying them with water *after* their formation.

The myriad rocks of the asteroid belt between Mars and Jupiter failed to accumulate into a planet. Probably nearby Jupiter's huge gravitational field caused them to collide too destructively to coalesce. Strong Jupiter tides on the planetesimals in the belt would have also hindered the development of a protoplanet. The result is a band of planetesimals, still colliding and occasionally fragmenting, but never coalescing into a larger body—surviving witnesses to the birth of the planets.

CLEANING UP THE DEBRIS

Most of the planetesimals left over after the major planets formed eventually collided with a planet or were ejected into the Oort Cloud. Little solid material remained. But what of the gas that made up most of the original cloud? Why don't we see it today throughout the planetary system? In the outer solar system, some (but not all) of that gas was swept up into planets. But that did not occur in the inner regions, where the terrestrial protoplanets never became massive enough to accrete such light material. Instead, the newly formed Sun took a hand. All young stars apparently experience a highly active evolutionary stage known as the *T-Tauri* phase (Figure 15.8; see also Chapter 19), during which their radiation and stellar winds are very intense. Any gas remaining between the planets was blown away into interstellar space by the solar wind and the Sun's radiation pressure when the Sun entered this phase, just before nuclear burning started at its center. Afterwards, all that remained were protoplanets and planetesimal fragments, ready to continue their long evolution into the solar system we know today.

15.4 *The Role of Catastrophes*

⑤ The condensation theory accounts for the nine "characteristic" points listed at the start of this chapter. Specifically, the planets' orbits are circular (2), in the same plane (3), and in the same direction as the Sun's rotation on its axis (4) as a direct consequence of the nebula's shape and rotation. The rotation of the planets (5) and the orbits of the moon systems (6) are due to the tendency of the smaller-scale eddies to inherit the nebula's overall sense of rotation. The growth of planetesimals throughout the nebula, with each protoplanet ultimately sweeping up the material near it, accounts for point (1), which is the fact that the planets are widely spaced (even if the theory does not explain the regularity of the spacing). The heating of the nebula and the Sun's ignition resulted in the observed differentiation (7), while the debris from the accretion–fragmentation stage naturally accounts for the asteroids (8) and comets (9).

We stressed earlier that an important aspect of any solar system theory is its ability to allow for the possibility of imperfections—deviations from the otherwise well-ordered scheme of things. In the condensation theory, that capacity is provided by the randomness inherent in the encounters that ultimately combined the planetesimals into protoplanets. As the numbers of large bodies decreased and their masses increased, individual collisions acquired greater and greater importance. The effects of these collisions can still be seen today in many parts of the solar system—the large craters on many of the moons we have studied thus far are a case in point. Having started with nine regular points to explain, we end with seven irregular solar-system features that still fall within the theory's scope. It is impossible to test any of these assertions directly, but it is reasonable to suppose that some (or even all) of the following "odd" aspects of the solar system can be explained in terms of collisions late in the formative stages of the protoplanetary system. Not all astronomers believe all of these explanations; however, most would accept at least some.

1. Mercury's exceptionally large nickel–iron core may be the result of a collision between two partially differentiated protoplanets. The cores may have merged, and much of the mantle material may have been lost.

2. Two large bodies could have merged to form Venus, giving it its abnormally low rotation rate.

3. The Earth–Moon system may have formed from a collision between the proto-Earth and a Mars-sized object.

4. A late collision with a large planetesimal may have caused Mars's curious north–south asymmetry and ejected much of the planet's atmosphere.

5. The tilted rotation axis of Uranus might have been caused by a grazing collision with a sufficiently large planetesimal, or by a merger of two smaller planets.

6. Uranus's moon Miranda may have been almost destroyed by a planetesimal collision, accounting for its bizarre surface terrain.

7. Interactions between the proto-Neptune and one or more planetesimals might account for Triton's retrograde motion, Nereid's eccentric orbit, and perhaps even the Pluto–Charon system.

15.5 The Angular Momentum Problem

6 A possible weak link in the condensation theory is known as the **angular momentum problem**. Although our Sun contains about 1000 times more mass than all the planets combined, it possesses a mere 0.3 percent of the total angular momentum of the solar system. Jupiter, for example, has a lot more angular momentum than does our Sun. Because of its large mass and great distance from the Sun, Jupiter possesses about 60 percent of the solar system's angular momentum. All told, the four jovian planets account for well over 99 percent of the total angular momentum of the solar system. By comparison, the lighter (and closer) terrestrial planets have negligible angular momentum.

The problem here is that all mathematical models predict that the Sun should have been spinning very rapidly during the earliest epochs of the solar system, and should command most of the solar system's angular momentum, basically because it contains most of the mass. However, as we have just seen, the reverse is true. Indeed, if all the planets, with their large amounts of orbital angular momentum, were placed inside the Sun, it would spin on its axis about 100 times as fast as it does at present. Somehow, the Sun must have lost most of its original angular momentum. Although the precise way it did so is unknown, we can surmise that the Sun probably transferred much of its spin angular momentum to the orbital angular momentum of the planets.

Many researchers speculate that the solar wind, moving away from the Sun into interplanetary space, carried away much of the Sun's initial angular momentum. The early Sun probably produced more of a dense solar gale than the relatively gentle "breezes" now measured by our spacecraft. High-velocity particles leaving the Sun followed the solar magnetic field lines. As the rotating magnetic field of the Sun tried to drag those particles around with it, they acted as a brake on the Sun's spin. Although each particle boiled off the Sun carries only a tiny amount of the Sun's angular momentum with it, over the course of nearly 5 billion years the vast numbers of escaping particles could have robbed the Sun of most of its initial spin momentum. Even today, our Sun's spin continues to slow.

Other researchers prefer to solve the Sun's momentum problem by assuming that the primitive solar system was much more massive than the present-day system. They argue that the accretion process was not entirely successful during the system's formative stages. Matter not captured by the Sun or the planets may well have transported much angular momentum while escaping back into interstellar space. This proposal is difficult to test, because the escaped matter would be well beyond the range of our current robot space probes. Perhaps the remote Oort Cloud of innumerable comets is the "escaped" matter.

Despite some minor controversy as to how this angular momentum quandary can best be resolved, nearly all astronomers agree that some version of the condensation theory is correct. The details have yet to be fully worked out, but the broad outlines of the processes involved are quite firmly established. Our planet is a by-product of the formation of the Sun. We might reasonably now ask what preceded the Sun, and what circumstances led to the collapse of the solar nebula in the first place. To answer these questions, we must widen the scope of our studies. We will find it necessary to understand the workings not only of the stars and the gas between them, but also of the Galaxy in which they reside. In the next chapter, we will begin this expanded inquiry with a closer look at our own parent star—the Sun.

Chapter Review

SUMMARY

Our solar system is an orderly place, making it unlikely that the planets were simply captured by the Sun. The overall organization points toward formation as the product of an ancient, one-time event, 4.6 billion years ago. An ideal theory of the solar system should provide strong reasons for the observed characteristics of the planets, yet be flexible enough to allow for deviations.

In the **nebular theory** (p. 319) of the formation of the solar system, a large cloud of dust and gas—the **solar nebula** (p. 319)—began to collapse under its own gravity. As it did so, it began to spin faster, to conserve angular momentum, eventually forming a disk. **Protoplanets** (p. 320) formed in the disk and became planets, while the central **protosun** (p. 320) eventually evolved into the Sun.

The nebular theory is an example of an **evolutionary theory** (p. 319), in which the properties of the solar system evolved smoothly into their present state. In a **catastrophic theory** (p. 319), changes occur abruptly, as the result of accident or chance.

The **condensation theory** (p. 321) builds on the nebular theory by the incorporation of the effects of particles of **interstellar dust** (p. 321), which help to cool the nebula and act as **condensation nuclei** (p. 321), allowing the planet-building process to begin.

Small clumps of matter grew by **accretion** (p. 322), gradually sticking together and growing into moon-sized **planetesimals** (p. 322), whose gravitational fields were strong enough to accelerate the accretion process. Competing with accretion in the solar nebula was **fragmentation** (p. 323), where small bodies were broken up following collisions with larger ones. Eventually, only a few planet-sized objects remained. The planets in the outer solar system became so large that they could capture the hydrogen and helium gas in the solar nebula, forming the jovian worlds.

The condensation theory can explain the basic differences between the jovian and terrestrial planets because the temperature of the solar nebula would be expected to decrease with increasing distance from the Sun. At any given location, the temperature would determine which materials could condense out of the nebula and so control the composition of any planets forming there. The terrestrial planets are rocky because they formed in the hot inner regions of the solar nebula, near the Sun, where only rocky and metallic materials condensed out. Farther out, the nebula was cooler and ices of water and ammonia could also form, leading to the observed differences in composition between the inner and outer solar system.

When the Sun became a star, its strong winds blew away any remaining gas in the solar nebula. Many leftover planetesimals were ejected into the Oort Cloud by the gravitational fields of the outer planets. They now occasionally revisit our part of the solar system as comets. In the inner solar system, light elements such as hydrogen and helium would have escaped into space. Much, if not all, of Earth's water was carried to our world by comets deflected from the outer solar system.

The asteroid belt is a collection of planetesimals that never managed to form a planet, probably because of Jupiter's gravitational influence. Many "odd" aspects of the solar system may conceivably be explained in terms of collisions late in the formation stages of the protoplanetary system.

The **angular momentum problem** (p. 327) is the fact that, while the Sun contains virtually all of the solar system's mass, it accounts for almost none of the angular momentum. It is believed that the solar wind or ejected planetesimals has carried off the Sun's initially high angular momentum, allowing its spin to slow to the rate observed today.

SELF-TEST: True or False?

The following 9 questions present properties of the solar system which a model of solar-system formation must explain. Which are correctly stated and which are not?

_____ **1.** Each planet is relatively isolated in space.

_____ **2.** The orbits of the planets are not very circular but significantly elliptical.

_____ **3.** The orbits of the planets all lie near the ecliptic plane.

_____ **4.** The direction of planetary revolution is in the same direction as the Sun's rotation.

_____ **5.** Planetary rotation is always in the same direction as the Sun's rotation.

_____ **6.** Moons do not usually revolve in the same direction as their parent planet rotates.

_____ **7.** The planetary system is highly differentiated.

_____ **8.** Asteroids were recently formed from the collision and breakup of an object orbiting within the asteroid belt.

_____ **9.** Most comets have short periods and orbit close to the ecliptic plane.

True or False?

_____ **10.** The terrestrial planets formed out of the original dust that made up the solar nebula.

_____ **11.** Water could not have condensed out any closer than 3 or 4 A.U. from the Sun.

_____ **12.** The accretion process occurred faster in the inner part of the solar system than it did in the outer regions.

_____ **13.** The condensation theory does not offer an explanation of the highly tilted rotation axis of Uranus.

_____ **14.** The condensation theory does not offer an explanation of the formation of the Moon.

_____ **15.** Random collisions, inherently a part of the condensation theory, can explain many of the odd properties found among some solar-system objects.

SELF-TEST: Fill in the Blank

1. The condensation theory, which currently is used to explain the formation of the solar system, is actually just a refined version of the older _____ theory.

2. In the condensation theory, astronomers realized the critical role played by _____ in starting the formation of small clumps of matter.

3. Initially, the accretion of matter into larger bodies occurred through _____ between particles in the solar nebula.

4. By the time planetesimals had formed, the accretion process was accelerated by the effect of _____.

5. In the final stage of accretion, the largest protoplanets were able to attract large quantities of _____ from the solar nebula.

6. The temperature of the inner part (out to 1 or 2 A.U.) of the solar nebula was _____.

7. Unlike the terrestrial planets, the planetesimals that formed the jovian planets were made up of _____ material.

8. High-speed collisions between planetesimals often led to _____ rather than accretion.

9. The large number of left-over planetesimals formed beyond 5 A.U. were destined to become _____.

10. The water now found on Earth was probably brought here by _____.

11. The reason the planetesimals of the asteroid belt did not form a larger object may was probably the gravitational influence of _____.

12. The _____ phase of the early Sun cleared out excess gas not used in planet formation.

13. Angular momentum depends on the mass, velocity, and _____ of an object.

14. The Sun's angular momentum is now much _____ than it was when the Sun was a protostar.

15. The total orbital angular momentum of all the planets is _____ than the Sun's angular momentum.

REVIEW AND DISCUSSION

1. List six properties of the solar system that any model of its formation must be able to explain.

2. Explain the difference between evolutionary theories and catastrophic theories of the solar system's origin.

3. Describe the basic features of the nebular theory of solar-system formation.

4. Give three examples of how the nebular theory explains some observed features of the present-day solar system.

5. Name two basic problems with the old nebular theory.

6. Explain the difference between angular momentum and linear momentum.

7. What is the key ingredient in the modern condensation theory of the solar system's origin that was missing or unknown in the nebular theory?

8. What are the two types of accretion that played a role in forming the planets?

9. Why are the jovian planets so much larger than the terrestrial planets?

10. What solar-system objects, still observable today, resulted from the process of fragmentation?

11. What influence did Earth's location in the solar nebula have on its final composition?

12. How did the temperature structure of the solar nebula determine planetary composition?

13. Why could Earth not have formed out of material containing water? How might Earth's water have gotten here?

14. What happened in the early solar system when the Sun became a T Tauri star?

15. What is the modern explanation for the formation of the Oort Cloud?

16. How do modern astronomers attempt to explain the angular momentum problem in light of modern theories of solar system formation?

17. Give at least two reasons why we can't simply look to a distant solar system that is in the process of formation to understand how our own solar system came to be.

18. Discuss the possibility that other stars have planets in light of two different theories of solar system formation: the collision theory and the condensation theory.

19. Describe the possible history of a single comet now visible from Earth, starting with its birth in the solar nebula somewhere near the planet Jupiter.

PROBLEMS

1. The orbital angular momentum of a planet in a circular orbit is simply the product of its mass, its orbital velocity, and its distance from the Sun. Compare the orbital angular momenta of Jupiter, Saturn, and Earth.

2. A typical comet contains some 10^{13} kg of water ice. How many comets would have to strike the Earth in order to account for the roughly 2×10^{21} kg of water presently found on our planet? If this amount of water accumulated over a period of 0.5 billion years, how often would Earth be hit by a comet during that time?

3. Consider a planet growing by accretion of material from the solar nebula. As it grows, its density remains roughly constant. Does the force of gravity at its surface to increase, decrease, or stay the same? Specifically, what would happen to the surface gravity as the radius of the planet doubled? Give reasons for your answer.

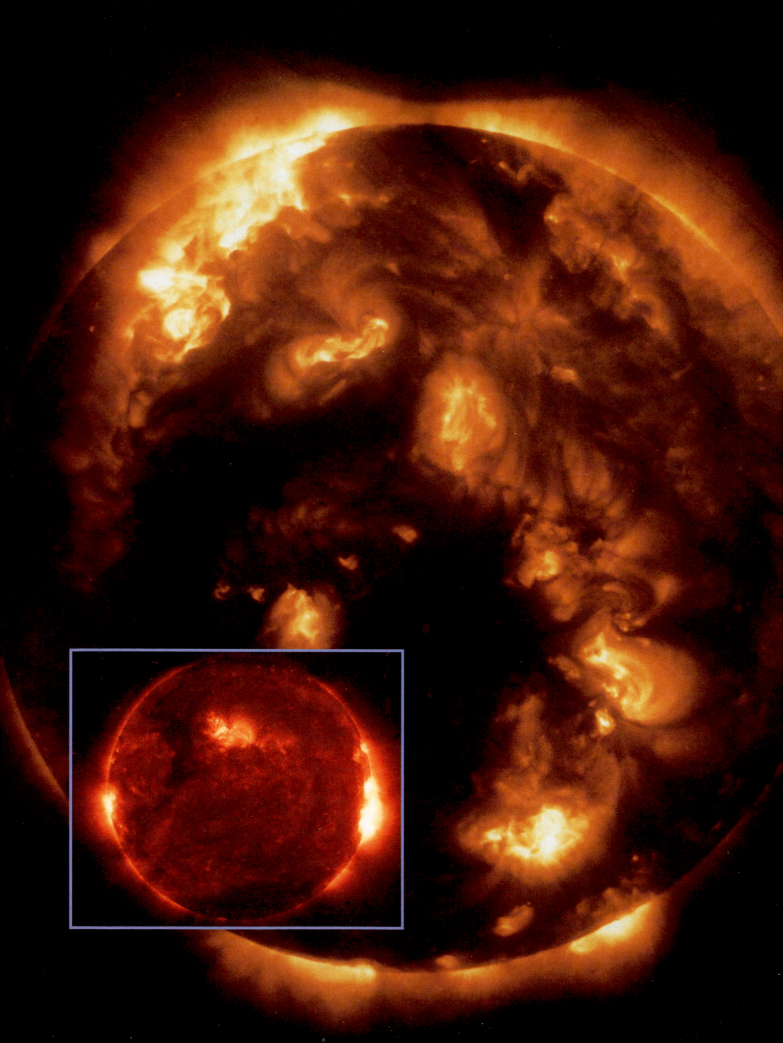

16

THE SUN

Our Parent Star

LEARNING GOALS

Studying this chapter will enable you to:

1 Summarize the overall properties of the Sun.

2 Explain how energy travels from the solar core, through the interior, and out into space.

3 Describe the Sun's outer layers, and say what those layers tell us about the Sun's surface composition and temperature.

4 Discuss the nature of the Sun's magnetic field and its relationship to the various types of solar activity.

5 Outline the process by which energy is produced in the Sun's interior.

6 Explain how observations of the Sun's core challenge our present understanding of solar physics.

(Opposite page, background) This spectacular image of the Sun was made by capturing X rays emitted by our star's most active regions. It was taken by a camera on a rocket lofted shortly before the total solar eclipse of July 1991. (Note the shadow of the Moon approaching from the west, at top.) The brightest regions in all these images have temperatures of about 3 million kelvins.
(Inset) An exquisitely resolved X-ray image of the Sun, made (again via rocket) in 1993.

Living in the solar system, we have the chance to study, at close range, perhaps the most common type of cosmic object—a star. Our Sun is a star, and a fairly average star at that, but with one unique feature: It is very close to us—some 300,000 times closer than our next nearest neighbor, Alpha Centauri. While Alpha Centauri is 4.3 light years distant, the Sun is only 8 light minutes away from us. Consequently, astronomers know far more about the properties of the Sun than about any of the other distant points of light in the universe. A good fraction of all our astronomical knowledge is based on modern studies of the Sun. Just as we studied our parent planet, Earth, to set the stage for our exploration of the solar system, we now study our parent star, the Sun, as the next step in our exploration of the universe.

16.1 *The Sun in Bulk*

❶ The Sun is the sole source of light and heat for the maintenance of life on Earth. It is a **star**, a glowing ball of gas held together by its own gravity and powered by nuclear fusion at its center. In its physical and chemical properties, the Sun is very similar to most other stars, regardless of when and where they formed. Our Sun appears to be a rather "typical" star, lying right in the middle of the observed ranges of stellar mass, radius, brightness, and composition. Far from detracting from the interest in the Sun, this very mediocrity is one of the main reasons that astronomers study the Sun—they can apply knowledge of solar phenomena to so many other stars in our Galaxy.

STRUCTURE

The Sun has a surface of sorts—not a solid surface (the Sun contains no solid material), but the part of the brilliant gas ball we see with our eyes or through a heavily filtered telescope—the so-called *solar disk*. This "surface" is known as the **photosphere**. Just above the photosphere is the Sun's lower atmosphere, called the **chromosphere**. Stretching far beyond that is a tenuous (thin) outer atmosphere, the solar **corona**. At still greater distances, the corona turns into the *solar wind*, which flows away from the Sun and permeates the entire solar system. Figure 16.1 shows the approximate dimensions of each of these regions.

Figure 16.1 also shows three more regions, all lying below the photosphere. Just beneath the surface, extending down some 200,000 km, lies the **convection zone**, a region where the material of the Sun is in constant convective motion. The region at the heart of the Sun, about 200,000 km in radius, is known as the **solar core**. This region is the site of powerful nuclear reactions that generate the Sun's enormous energy output. In between lies a region called the **radiation zone**, where energy is transferred by radiation rather than by convection. The term *solar interior* is often used to include both the radiation and convection zones. Table 16-1 summarizes the characteristics of the main regions of the Sun.

PHYSICAL PROPERTIES

Knowing that the Earth–Sun distance—the astronomical unit—is about 1.5×10^8 km, we can find the physical size of the solar photosphere. The solar disk is about 0.5° across, so we compute the diameter of the Sun to be about 1.4×10^6 km, a little over 100 times the diameter of the Earth. The Sun's volume is therefore more than 100^3, or 1 million times that of Earth. The Sun is truly vast by terrestrial standards.

Just as with the planets, we can determine the Sun's mass by studying the motion of objects (planets, asteroids, spacecraft) influenced by its gravity. ∞ (p. 47) It is about 2.0×10^{30} kg, about 300,000 times greater than the mass of Earth. However, the Sun is not especially massive considering its huge volume. In fact, the average solar density—roughly 1400 kg/m³—is quite similar to that of the jovian planets and only about one-quarter the average density of the Earth.

Again as with the planets, we can determine the Sun's rotation rate by watching features on the solar surface (in this case, disturbances such as sunspots—see Chapter 2 and the paragraphs that follow) traverse the solar disk. ∞ (p. 37) These observations indicate that the Sun does not rotate as a solid body. Instead, it spins *differentially*—faster

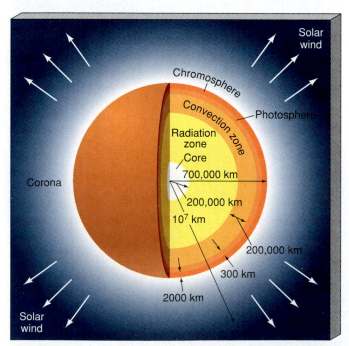

Figure 16.1 Six main regions of the Sun, not drawn to scale, with physical dimensions labeled.

TABLE 16-1 *The Standard Solar Model*

REGION	INNER RADIUS (km)	TEMPERATURE (K)	DENSITY (kg/m³)	DEFINING PROPERTIES
Core	0	15,000,000	150,000	Energy generated by nuclear fusion
Radiation zone	200,000	7,000,000	15,000	Energy transported by electromagnetic radiation
Convection zone	500,000	2,000,000	150	Energy carried by convection
Photosphere	696,000*	5800	2×10^{-4}	Electromagnetic radiation can escape—the part of the Sun we see
Chromosphere	696,500*	4500	5×10^{-6}	Cool lower atmosphere
Transition zone	698,000*	8000	2×10^{-10}	Rapid temperature increase
Corona	706,000	1,000,000	10^{-12}	Hot, low-density upper atmosphere
Solar wind	10,000,000	2,000,000	10^{-23}	Solar material escapes into space and flows outward through the solar system

These radii are based on the accurately determined radius of the photosphere. The other radii quoted are approximate, round numbers.

at the equator and slower at the poles, like Jupiter and Saturn. The solar photosphere rotates once every 27 days at the equator, but only once in 31 days at the poles.

To measure the Sun's temperature, we can apply the ideas discussed in Chapter 3. ∞ (p. 64) The distribution of solar radiation with respect to wavelength has the approximate shape of a black-body curve for an object at about 6000 K. This is the average temperature—sometimes called the *effective temperature*—of the Sun's photosphere. (A more careful calculation yields a temperature of 5800 K, but we will use the rounded-off value here.) Notice, by the way, that this measurement tells us nothing whatever about the temperature in the solar interior. Later in this chapter we will consider conditions deeper within the Sun.

LUMINOSITY

The properties of size, mass, density, rotation rate, and temperature are familiar from our study of the planets. But the Sun has an additional property, perhaps the most important of all from the point of view of life on Earth—it *radiates* a great deal of energy into space, uniformly (we assume) in all directions.

What is the total energy output of the Sun? If we were to hold a light-sensitive device—perhaps a solar cell of known dimensions—perpendicular to the Sun's rays, we could answer this question by measuring the amount of solar energy it received per square meter per second. The amount of solar energy reaching Earth per unit area per unit time (above the atmosphere) is known as the **solar constant**. Its value is approximately 1400 watts per square meter (W/m²). A sunbather's body, with a total surface area of about 0.5 m², receives solar energy at a rate of nearly 700 watts, roughly

equivalent to the output of a typical electric room heater, or about 10 household light bulbs.

Let us now ask about the *total* amount of energy radiated in all directions from the Sun, not just the small fraction intercepted by the Earth. Figure 16.2 shows how this

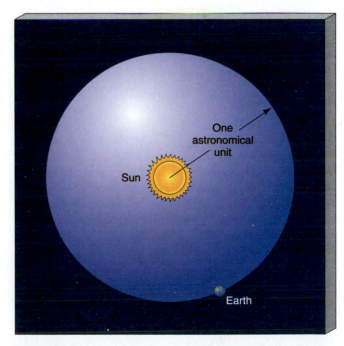

Figure 16.2 We can draw an imaginary sphere around the Sun so that the sphere's edge passes through Earth's center. The radius of this imaginary sphere equals one astronomical unit. By multiplying the sphere's surface area by the solar constant, we can measure the Sun's luminosity, the amount of energy it emits each second.

can be measured. Imagine a three-dimensional sphere centered on the Sun. The surface of this sphere intersects Earth, so the sphere's radius is 1 A.U. and its surface area is approximately 2.8×10^{24} m². Multiplying the amount of solar energy falling on each square meter (that is, the solar constant) by the total surface area of our imaginary sphere, we can determine the total *rate* at which energy leaves the Sun's surface. It turns out to be about 4×10^{26} W. (A more precise measurement gives 3.9×10^{26} W, but we will use the rounded-off figure here.) This quantity is known as the **luminosity** of the Sun—the total energy radiated by the Sun each second.

The Sun is a very powerful source of energy. *Every second,* it produces an amount of energy equivalent to the detonation of about 100 billion 1-megaton nuclear bombs. Put another way, the solar luminosity is equivalent to 4 trillion trillion l00-watt light bulbs shining simultaneously—about 10^{19} dollars' worth of energy radiated per second (at 1995 rates).

16.2 *The Solar Interior*

MODELING THE STRUCTURE OF THE SUN

Lacking any direct measurements of the solar interior, astronomers must use more indirect means to probe the inner workings of our parent star. To accomplish this, they construct mathematical models of the Sun. The physical processes that are believed to be important in determining the Sun's internal structure are incorporated into a computer program, and estimates are made about any unknown quantities. The program then calculates the internal properties of this model star and predicts how it would look from the outside. These predictions are compared with observations, the estimates are modified, and the cycle is repeated until the solar model agrees with reality. By requiring our theoretical model to agree with observations when comparisons can be made, and trusting our knowledge of the laws of physics when comparisons are impossible, we arrive at a self-consistent picture of the Sun. The model that has gained widespread acceptance is often referred to as the **Standard Solar Model**.

To test and refine the predictions of the Standard Solar Model, astronomers are eager to obtain information about the solar interior. However, we have little or no direct information about conditions below the solar photosphere, so we must rely on more indirect techniques. In the 1960s, it was discovered that the surface of the Sun vibrates like a complex set of bells. These vibrations, illustrated in Figure 16.3, are the result of internal pressure ("sound") waves that reflect off the photosphere and repeatedly cross the solar interior. Because these waves can penetrate deep inside the Sun, analysis of their surface patterns allows scientists to study conditions far below the Sun's surface. This process is similar to the way in which seismologists study the interior of the Earth by observing the P- and S-waves produced by earthquakes. ∞ (p. 157) For this reason, study of solar surface patterns is usually called **helioseismology**, even though solar pressure waves have nothing whatever to do with solar seismic activity—there is no such thing.

The observed surface vibrations can be extremely complex. Many separate patterns are superimposed on one another, so deciphering the observations is a difficult task. The most recent and most extensive study of solar oscillations is the ongoing "GONG" (short for Global Oscillations Network Group) project. By making continuous observations of the Sun from many clear sites around the Earth, solar astronomers can obtain uninterrupted high-quality solar data spanning many days and even weeks—almost as though Earth were not rotating and the Sun never set. Analysis of these data provides important additional information about the temperature, density, rotation, and the convective state of the solar interior, allowing detailed comparisons between theory and reality to be made.

Figure 16.4 shows the solar density and temperature plotted as functions of distance from the core of the Sun. Notice how the density drops rather sharply at first, then decreases more slowly near the solar photosphere, some 700,000 km from the center. The variation in density is large, ranging from a core value of about 150,000 kg/m³, 20 times the density of iron, to an intermediate value (at 350,000 km) of about 1000 kg/m³, the density of water, to an extremely small photospheric value of 2×10^{-4} kg/m³, about 10,000 times less dense than air at the surface of Earth. The *average* density, as we have seen, of the entire

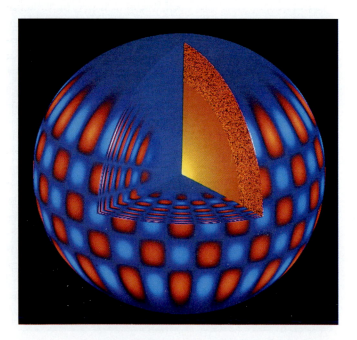

Figure 16.3 The Sun has been found to vibrate in a very complex way. By observing the motion of the solar surface, scientists can determine the wavelength and the frequencies of the individual waves and deduce information about the solar interior not obtainable by other means. The alternating patches represent gas moving down (red) and up (blue).

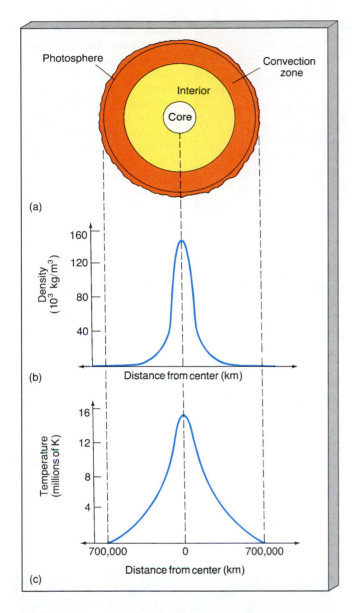

(a)

(b)

(c)

Figure 16.4 Theoretically modeled profiles of density (b) and temperature (c) for the interior of the Sun, presented for perspective in (a). All three parts describe a cross-sectional cut through the center of the Sun.

Sun is 1400 kg/m^3. Because the density is so high in the core, roughly 90 percent of the Sun's mass is contained within the inner half of its radius. The solar density continues to decrease out beyond the photosphere, reaching values as low as 10^{-23} kg/m^3 in the far corona—about as thin as the best vacuum that physicists can create in laboratories on the Earth.

As shown in Figure 16.4(c), the solar temperature also decreases with increasing radius, but not as rapidly as the density. Computer models indicate a temperature of about 15,000,000 K at the core, consistent with the minimum 10,000,000 K needed to initiate the nuclear reactions known to power all stars, decreasing to the observed value of about 6000 K at the photosphere.

CONVECTION

2 Models of the solar interior predict the existence of an extensive convection zone lying just below the photosphere. As in Earth's interior and atmosphere, and elsewhere in the solar system, convection can occur whenever cooler material overlies warmer material. The result is a rather well-defined circulation pattern that tends to even out the temperature. Because the process of solar convection is critically important for transporting energy to the Sun's surface, let's study it in a little more detail.

The very hot solar interior ensures violent and frequent collisions among gas particles. Particles move in all directions with high velocities, bumping into one another unceasingly. In and near the core, the extremely high temperatures guarantee that the gas is completely ionized. Recall from Chapter 4 that under less extreme conditions, atoms absorb photons that can boost their electrons to more excited states. ∞ (p. 82) With no electrons left on atoms to capture the photons, however, the deep solar interior is quite transparent to radiation. Only occasionally does a photon encounter and scatter off a free electron or proton. As a result, the energy produced by nuclear reactions in the core travels outward toward the surface in the form of radiation with relative ease.

As we move outward from the core, the temperature falls, atoms collide less frequently and less violently, and more and more electrons manage to remain bound to their parent nuclei. With more and more atoms retaining electrons that can absorb the outgoing radiation, the gas in the interior changes from being relatively transparent to being almost totally opaque. By the outer edge of the radiation zone, 200,000 km below the photosphere, *all* of the photons produced in the Sun's core have been absorbed. Not one of them reaches the surface. But what happens to the energy they carry?

The photons' energy must travel beyond the Sun's interior. If it did not, the Sun would have exploded long ago. That we see sunlight—visible energy—proves that some energy escapes. That energy reaches the surface by *convection*. Hot solar gas, beginning at the top of the radiation zone, physically moves upward, while cooler gas above it sinks, creating a characteristic pattern of convection cells. This action is the solar analog of the warm, rising bubbles in a saucepan of boiling soup. All through the upper interior, energy is transported to the surface by physical motion of the solar gas. Figure 16.5 is a diagram of solar convection cells, where columns of hot gas rise, cool, and descend. Throughout this discussion, it is important to keep in mind that *convection and radiation are fundamentally different mechanisms of energy transport.*

In reality, the zone of convection is much more complex than we have just described. There is a hierarchy of convection cells, organized in tiers at different depths. The deepest tier, about 200,000 km below the surface, is thought to contain large cells some tens of thousands of kilometers in diameter. Heat is then successively carried

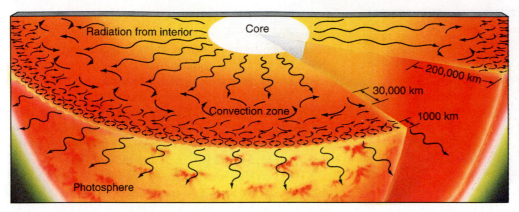

Figure 16.5 Physical transport of energy in the Sun's convection zone. We can visualize the upper interior as a boiling, seething sea of gas. Each convective loop is about 1000 km across. The convective cells are arranged in tiers with cells of progressively smaller size as the surface is neared. (This is a highly simplified diagram; there are many different cell sizes, and they are not so neatly arranged.)

upward through a series of progressively smaller-sized cells, stacked one upon another until, at a depth of about 1000 km, the individual cells are about 1000 km across. The top of this uppermost tier of convection is the visible surface, where astronomers can directly observe the cell sizes. Information about convection below that level is inferred mostly from computer models of the solar interior.

At some distance from the core, the solar gas becomes too thin to sustain further upwelling by convection. Theory suggests that this distance roughly coincides with the photospheric surface we see. Convection does *not* proceed into the solar atmosphere. There is simply not enough gas there. The density is so low that the gas becomes transparent once again, and radiation once again becomes the mechanism of energy transport. Photons reaching the photosphere escape more or less freely into space, and the photosphere emits thermal radiation, like any other hot object.

GRANULATION

Figure 16.6 is a high-resolution photograph of the solar surface taken with instruments aboard NASA's *Skylab* space station as it orbited above much of Earth's atmosphere in the mid-1970s. The visible surface is highly mottled, or **granulated**, with regions of bright and dark gas known as *granules*. Each bright granule measures about 1000 km across—comparable in size to a continent on Earth—and has a lifetime of between 5 and 10 minutes. Together, several million granules constitute the top layer of the convection zone, immediately below the photosphere.

Each granule forms the topmost part of a solar convection cell. Spectroscopic observation of the photosphere within and around the bright regions shows direct evidence for the upward motion of gas as it "boils" up from within. This evidence proves that convection really does occur at or below the photosphere. Spectral lines detected from the bright granules appear slightly bluer than normal, indicating Doppler-shifted matter coming toward us with a velocity of about 1 km/s. ∞ (p. 69) Conversely, spectroscopes focused on the darker portions of the granulated photosphere show the same spectral lines to be redshifted, indicating matter moving away.

The brightness variations of the granules result strictly from differences in temperature. The upwelling gas is hotter and therefore emits more radiation than the cooler downwelling gas. The adjacent bright and dark gases appear to contrast considerably, but in reality their temperature difference is less than about 500 K.

Careful measurements also reveal a much larger-scale flow on the solar surface. **Supergranulation** is a flow pattern quite similar to granulation except that supergranulation cells measure some 30,000 km across. As with granulation, material upwells at the center of the cells, flows across the surface, then sinks down again at the edges. Scientists believe that supergranules are the imprint on the photosphere of a deeper tier of large convective cells, like those depicted in Figure 16.5.

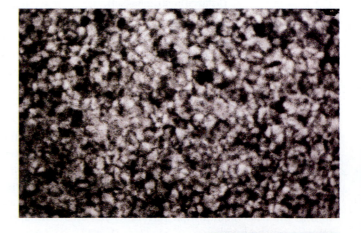

R I V U X G

Figure 16.6 *Skylab* photograph of the granulated solar photosphere. Typical solar granules are comparable in size to the Earth's continents. The bright portions of the image are regions where hot material is upwelling from below. The dark regions correspond to cooler gas that is sinking back down into the interior.

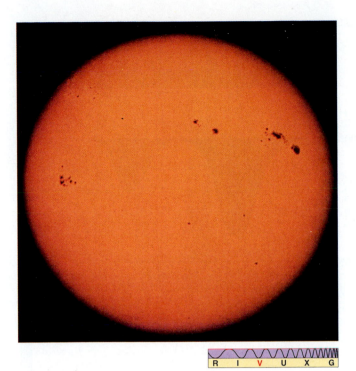

Figure 16.7 This photograph of the Sun shows a sharp solar limb, although our star, like all stars, is made of a gradually thinning gas. The edge appears sharp because the solar photosphere is so thin.

THE "EDGE" OF THE SUN

Figure 16.7 shows the entire solar disk photographed through a heavily filtered telescope. Despite the steady decrease in density and temperature from the interior to the atmosphere, the Sun displays a reasonably sharp edge, or *limb*. The limb is sharp because the overwhelming majority of visible photons arise in the extremely shallow photosphere. Slightly below the photosphere, the gas is still convective, and the radiation does not reach us directly. Slightly above, the gas is too thin to emit appreciable amounts of radiation. Recent estimates suggest that the depth of the photosphere is no more than 500 km, very small compared with the size of the Sun. The photosphere's thickness is less than 1/10 of 1 percent of the solar radius, which is why we perceive the Sun as having a well-defined edge.

16.3 *The Solar Atmosphere*

COMPOSITION

❸ Astronomers can glean an enormous amount of information about the Sun from an analysis of the *absorption* lines that arise in the photosphere and lower atmosphere. ∞ (p. 76) Figure 16.8 (see also Figure 5.4) is a detailed spectrum of the Sun, obtained for a small portion of the range of visual wavelengths, from 360 to 690 nm. Notice the intricate dark Fraunhofer absorption lines superposed on the background continuous spectrum.

As discussed in Chapter 4, spectral lines arise when electrons in atoms or ions make transitions between states of well-defined energies, emitting or absorbing photons of specific energies (that is, wavelengths or colors) in the process. ∞ (p. 83) To explain the observed solar spectrum, however, we must modify slightly our earlier description of the formation of absorption lines. We explained these lines in terms of cool foreground gas intercepting light from a hot background source. In actuality, both the bright background and the dark absorption lines in Figure 16.8 form in roughly the *same* location in the Sun—the solar photosphere and lower chromosphere. To understand how line formation occurs, let's reconsider the solar energy-emission process in a little more detail.

Below the photosphere, the solar gas is sufficiently dense, and interactions among photons, electrons, and ions sufficiently common, that radiation cannot escape directly into space. Photons are absorbed and reemitted many times as the Sun's energy travels outward from the core—a typical "parcel" of energy may take millions of years to reach the surface. In the solar atmosphere, however, the probability that a photon will escape without further interaction with matter depends on its energy. If that energy happens to correspond to some electronic transition in one of the atoms or ions present in the gas, then the photon may be absorbed again before it can travel very far—the more elements present of the type suitable for absorption, the lower the escape probability. Conversely, if the photon's energy does not coincide with any such transition,

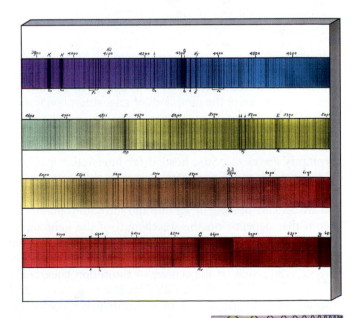

Figure 16.8 A detailed spectrum of our Sun in a portion of the visible domain shows thousands of Fraunhofer spectral lines, which indicate the presence of some 67 different elements in various stages of excitation and ionization in the lower solar atmosphere.

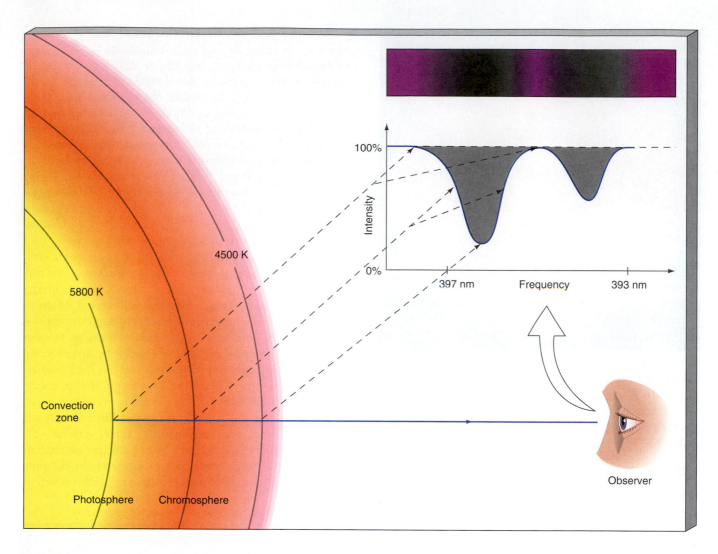

Figure 16.9 Formation of solar absorption lines. Photons with energies well away from any atomic transition can escape from relatively deep in the photosphere, but those with energies close to a transition are more likely to be reabsorbed before escaping, so the ones we see on Earth tend to come from higher, cooler levels in the solar atmosphere. The inset shows a close-up tracing of two of the thousands of solar absorption lines, those produced by calcium at about 395 nm.

then the photon cannot interact further with the gas, and it promptly leaves the Sun, headed for interstellar space, or perhaps the detector of an astronomer on the Earth.

As illustrated in Figure 16.9, when we look at the Sun, we are actually peering down into the solar atmosphere to a depth that depends on the wavelength of the light under consideration. Photons with wavelengths well away from any absorption feature tend to come from deep in the photosphere, while those at the centers of absorption lines, being so much more likely to interact with matter as they travel through the solar gas, mainly escape from higher, cooler levels. The lines are darker than their surroundings because the *temperature* of the atmosphere where they form is lower than the 5800 K temperature of the photosphere, where most of the continuous emission originates. (Recall that by Stefan's law, the brightness of a radiating object depends on its temperature—the cooler the gas, the less energy it radiates.) ∞ (p. 65) Thus, the existence of the Fraunhofer lines is direct

evidence that the temperature in the Sun's atmosphere decreases with height above the photosphere.

Tens of thousands of spectral lines have been observed and cataloged in the solar spectrum, although there are not nearly this many elements in the Sun. The reason for this is that most elements are present in many different states of excitation and ionization. ∞ (p. 87) They therefore absorb photons having many different energies, even in the relatively narrowly defined visible range, giving rise to many different lines. The more complex the element, the more lines it can produce. For example, hundreds of lines are attributed to just the element iron. In all, some 67 elements have been identified in the Sun. More elements probably exist there, but they are present in such small quantities that our instruments are simply not sensitive enough to detect them. Table 16-2 lists the 10 most common elements in the Sun. Notice that hydrogen is by far the most abundant element, followed by helium. This dis-

TABLE 16-2   *The Composition of the Sun*

ELEMENT	ABUNDANCE (percentage of total number of atoms)	ABUNDANCE (percentage of total mass)
Hydrogen	91.2	71.0
Helium	8.7	27.1
Oxygen	0.078	0.97
Carbon	0.043	0.40
Nitrogen	0.0088	0.096
Silicon	0.0045	0.099
Magnesium	0.0038	0.076
Neon	0.0035	0.058
Iron	0.0030	0.14
Sulfur	0.0015	0.040

R I V U X G

Figure 16.10 This photograph of a total solar eclipse shows the solar chromosphere, a few thousand kilometers above the Sun's surface.

Spicules

R I V U X G

Figure 16.11 Solar spicules, short-lived narrow jets of gas that typically last mere minutes, can be seen sprouting up from the solar chromosphere in this Hα image of the Sun. The spicules are the thin, dark, spikelike regions. They appear dark against the face of the Sun because they are cooler than the solar photosphere.

tribution is just what we saw on the jovian planets, and it is what we will find for the universe as a whole.

THE CHROMOSPHERE

Above the photosphere lies the cooler chromosphere, the inner part of the solar atmosphere. This region emits very little light of its own and cannot be observed visually under normal conditions. The photosphere is just too bright, dominating the chromosphere's radiation. The relative dimness of the chromosphere results from its low density—large numbers of photons simply cannot be emitted by a tenuous gas containing very few atoms per unit volume. Still, although it is not normally seen, astronomers have long been aware of the chromosphere's existence. Figure 16.10 shows the Sun during an eclipse in which the photosphere—but not the chromosphere—is obscured by the Moon. The chromosphere's characteristic reddish hue is plainly visible. This coloration is due to the Hα ("hydrogen alpha") emission line of hydrogen, which dominates the chromospheric spectrum. (Recall from Chapter 4 that the wavelength of this line is 656.3 nm, or 6563 Å, right in the middle of the red portion of the spectrum.) ∞ (p. 82)

The chromosphere is far from tranquil. Every few minutes, small solar storms erupt, expelling jets of hot matter known as *spicules* into the Sun's upper atmosphere (see Figure 16.11). These long, thin spikes of matter leave the Sun's surface at typical velocities of about 100 km/s, reaching several thousand kilometers above the photosphere. Spicules are not spread evenly across the solar surface. Instead, they cover only about 1 percent of the total area, tending to accumulate around the edges of supergranules. The Sun's magnetic field is also known to be somewhat stronger than average in those regions. Scientists speculate that the downwelling material there tends to strengthen the solar magnetic field and that spicules are the result of magnetic disturbances in the Sun's churning outer layers.

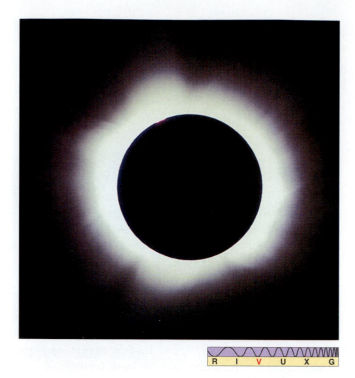

R I V U X G

Figure 16.12   When both the photosphere and the chromosphere are obscured by the Moon during a solar eclipse, the faint corona becomes visible. This photograph shows clearly the emission of radiation from the solar corona.

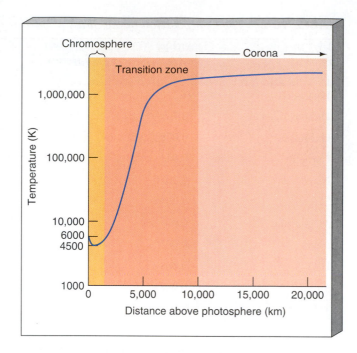

Figure 16.13   The change of gas temperature in the lower solar atmosphere is dramatic. The minimum temperature marks the outer edge of the chromosphere. Beyond that, the temperature rises sharply in the transition zone, finally leveling off at over 1,000,000 K in the corona.

THE CORONA

During the brief moments of an eclipse, if the Moon's angular size is large enough that both the photosphere and the chromosphere are blocked, the ghostly solar corona can be seen, as in Figure 16.12. With the photospheric light removed, the pattern of spectral lines changes dramatically. The intensities of the usual lines alter, suggesting changes in elemental abundances or gas temperature or both. The spectrum shifts from absorption to emission, and an entirely new set of spectral lines suddenly appears. These new coronal (and in some cases chromospheric) lines were first observed during eclipses in the 1920s. For years afterward, some researchers (for want of any better explanation) attributed them to a nonterrestrial element, which they called "coronium."

We now recognize that these new spectral lines do not indicate any new kind of atom. Coronium does not exist. Rather, the new lines arise because atoms in the corona have lost several more electrons than atoms in the photosphere—that is, the coronal atoms are much more highly ionized. For example, astronomers have identified coronal lines corresponding to iron ions with as many as 13 of their normal 26 electrons missing. In the photosphere, most iron atoms have lost only 1 or 2 of their electrons. The cause of this extensive electron stripping is the high coronal temperature. The degree of ionization inferred from spectra observed during solar eclipses tell us that the gas temperature of the upper chromosphere exceeds that of the photo-

sphere. Furthermore, the temperature of the solar corona, where even more ionization is seen, is still higher.

Based on many observations of conditions at different distances from the limb of the Sun, from the photosphere outward into the corona, Figure 16.13 plots the variation of gas temperature with altitude. The temperature decreases to a minimum of about 4500 K some 500 km above the photosphere, after which it rises steadily. About 1500 km above the photosphere, the gas temperature begins to rise rapidly, reaching more than 1,000,000 K at an altitude of 10,000 km. Thereafter, it remains roughly constant. Using this temperature profile, we can draw a clear distinction between the chromosphere and the corona: The chromosphere extends from the top of the photosphere for approximately 1500 km. The region in which the temperature rises rapidly—from about 1500 km to 10,000 km—is called the **transition zone**. At 10,000 km, the corona begins.

The cause of this rapid temperature rise is not fully understood. The temperature profile runs contrary to intuition—moving away from a heat source, we would normally expect the heat to diminish, but this is not the case for the lower atmosphere of the Sun. The corona must have another energy source. Astronomers now believe that magnetic disturbances in the solar photosphere—a little like spicules, but on a much larger scale—are ultimately responsible for heating the corona. We will return to these disturbances in more detail in the next section.

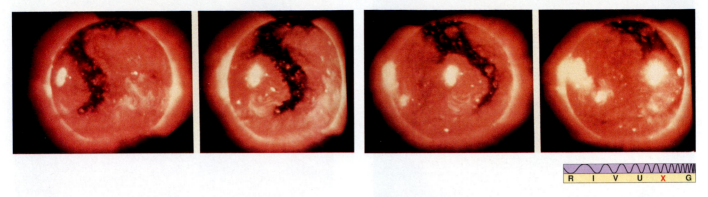

R I V U X G

Figure 16.14 Images of X-ray emission from the Sun observed by the *Skylab* space station. These frames were taken at one-day intervals. Note the dark, boot-shaped coronal hole traveling from left to right, where the X-ray observations outline in dramatic detail the abnormally thin regions through which the high-speed solar wind streams forth.

THE SOLAR WIND

Electromagnetic radiation and fast-moving particles—mostly protons and electrons—escape from the Sun all the time. The radiation moves away from the photosphere at the speed of light, taking 8 minutes to reach Earth. The particles travel more slowly, although at the still considerable speed of about 500 km/s, reaching the Earth in a few days. This constant stream of escaping solar particles is the **solar wind**.

The solar wind results from the high temperature of the solar corona. About 10,000,000 km above the photosphere, the coronal gas is hot enough to escape the Sun's gravity, and it begins to flow outward into space. At the same time, the solar atmosphere is continually replenished from below. If that were not the case, the corona would disappear in about a day. The Sun is, in effect, "evaporating"—constantly shedding mass through the solar wind. But the wind is an extremely thin medium. Although it carries away about a million tons of solar matter each second, less than 0.1 percent of the Sun has been lost since the solar system formed billions of years ago. Our star is indeed evaporating, but it is losing only a negligible fraction of its huge bulk.

THE SUN IN X RAYS

What sort of radiation is emitted by a gas of 1,000,000 K? Unlike the 6000 K photosphere, which emits most strongly in the visible part of the electromagnetic spectrum, the hotter coronal gas radiates at much higher frequencies—primarily in X rays. For this reason, X-ray telescopes have become important tools in the study of the solar corona. Figure 16.14 shows an X-ray image of the Sun. The full corona extends well beyond the regions shown, but the density of coronal particles emitting the radiation diminishes rapidly with distance from the Sun. The intensity of X-ray radiation farther out is too dim to be seen here.

In the mid-1970s, instruments aboard NASA's *Skylab* space station revealed that the solar wind escapes mostly through solar "windows" called **coronal holes**. The dark area moving from left to right in Figure 16.14 represents a coronal hole. Not really holes, such structures are simply deficient in matter—vast regions of the Sun's atmosphere where the density is about 10 times lower than the already tenuous, normal corona. Coronal holes are underabundant in matter because the gas there is able to stream freely into space at particularly high speeds, driven by disturbances in the Sun's atmosphere and magnetic field. In coronal holes, the solar magnetic field lines extend from the surface far out into interplanetary space. Charged particles tend to follow the field lines, so they can escape. In other regions of the corona, the solar magnetic field lines stay close to the Sun, keeping charged particles near the surface and inhibiting the outward flow of the solar wind (just as Earth's magnetic field tends to prevent the incoming solar wind from striking Earth), and the density remains (relatively) high. The largest coronal holes can be hundreds of thousands of kilometers across. Structures of this size are seen only a few times each decade. Smaller holes—perhaps only a few tens of thousand kilometers in size—are much more common, appearing every few hours.

16.4 *The Active Sun*

④ Most of the Sun's luminosity results from continuous emission from the photosphere. This radiation arises from what we call the **quiet Sun**—the underlying predictable star that blazes forth day after day. This steady behavior contrasts with the sporadic, unpredictable radiation of the **active Sun**, a much more irregular component of our star's energy output, characterized by explosive, unpredictable behavior. This aspect of solar radiation contributes little to the Sun's total luminosity, and has little effect on the evolution of the Sun as a star, but it does affect us directly here on Earth (see *Interlude 16–1* on p. 349). The size and duration of coronal holes are strongly influenced by the level of solar activity, as is the strength of the solar wind.

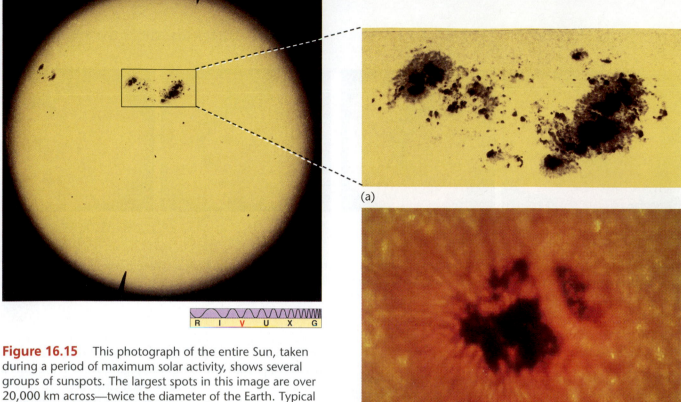

R I V U X G

Figure 16.15 This photograph of the entire Sun, taken during a period of maximum solar activity, shows several groups of sunspots. The largest spots in this image are over 20,000 km across—twice the diameter of the Earth. Typical sunspots are only about half this size.

(a)

(b)

R I V U X G

Figure 16.16 (a) An enlarged photograph of the largest pair of sunspots in Figure 16.15. Each spot consists of a cool, dark inner region called the umbra, surrounded by a warmer, brighter region called the penumbra. The spots appear dark because they are slightly cooler than the surrounding photosphere. (b) A high-resolution, true-color image of a single sunspot shows details of its structure as well as much surface granularity surrounding it. The spot is about the size of the Earth.

SUNSPOTS

Figure 16.15 is an optical photograph of the entire Sun, showing numerous dark blemishes on the surface. First studied in detail by Galileo, these "spots" provided one of the first clues that the Sun was not a perfect unvarying creation, but a place of constant change. These dark areas are called **sunspots**. They typically measure about 10,000 km across, about the size of the Earth. As shown in the figure, they often occur in groups. At any given time, the Sun may have hundreds of sunspots, or it may have none at all.

Studies of sunspots show an *umbra*, or dark center, surrounded by a grayish *penumbra*. The close-up view of a pair of sunspots in Figure 16.16 shows each of these dark areas and the brighter undisturbed photosphere nearby. This gradation in darkness is really a gradual change in photospheric temperature—sunspots are sim-

ply *cooler* regions of the photospheric gas. The temperature of the umbra is about 4500 K, compared with the penumbra's 5500 K. The spots, then, are certainly composed of hot gases. They seem dark only because they appear against an even brighter background (the hotter, 6000 K photosphere). If we could magically remove a

Figure 16.17 The evolution of some sunspots and lower chromospheric activity over a period of 12 days. The sequence runs from left to right. An Hα filter was used to make these photographs, taken from the *Skylab* space station. An arrow follows one set of sunspots over the course of a week as they are carried around the Sun by its rotation.

R I V U X G

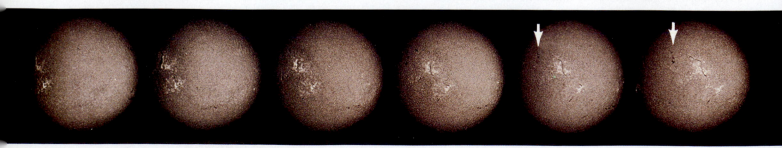

sunspot from the Sun (or just block out the rest of the Sun's emission), the spot would glow brightly, just like any emitter at roughly 5000 K.

Sunspots are not steady. Most change their size and shape, and all come and go. Figure 16.17 shows a time sequence in which several spots vary—sometimes growing, sometimes dissipating—over a period of several days. Individual spots may last anywhere from 1 to 100 days. A large group of spots typically lasts 50 days.

SOLAR MAGNETISM

What causes a sunspot? Why is it cooler than the surrounding photosphere? The answers to these questions involve the Sun's magnetism. As we saw in Chapter 4, analysis of spectral lines can yield information about the magnetic field where they originate. ∞ (p. 90) Such analysis reveals that the magnetic field in a typical sunspot is about 1000 times greater than the field in neighboring, undisturbed photospheric regions (which is itself several times stronger than the Earth's field). Scientists believe that sunspots are cooler than their surroundings because these abnormally strong fields tend to block (or redirect) the normal convective flow of hot gas toward the surface of the Sun.

Another indicator of the magnetic nature of sunspots is their grouping. They almost always come in pairs, and the magnetic fields observed in the two members of any pair are always opposite to one another—the members of the pair are said to have opposite magnetic *polarities*. As illustrated in Figure 16.18(a), magnetic field lines emerge from the interior through one member of a sunspot pair, loop through the solar atmosphere, then reenter the solar surface through the other spot. What's more, *all* the sunspot pairs in the same solar hemisphere (north or south) at any instant have the same magnetic configuration—if the magnetic field lines are directed into the Sun in one leading spot (measured in the direction of the Sun's rotation), they are inwardly directed in all leading spots in that hemisphere, as shown in Figure 16.18(b). In the other hemisphere at the same time, all sunspot pairs have the *opposite* polarity. Despite the irregularity of the sunspots themselves, these correlations suggest a high degree of order in the solar magnetic field.

The Sun is gaseous and rotates *differentially*, and these facts radically affect the character of solar magnetism. As il-

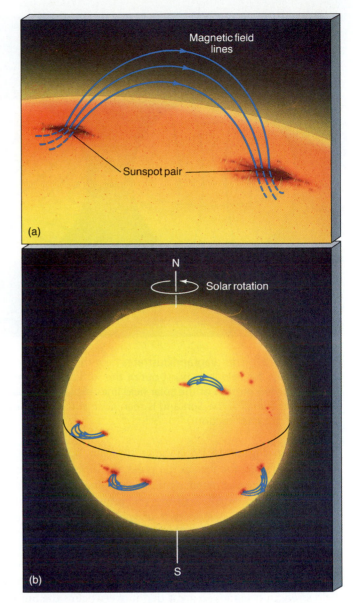

Figure 16.18 (a) Sunspot pairs are linked by magnetic field lines. The Sun's magnetic field emerges from the surface through one member of the pair and reenters the Sun through the other. (b) The leading members of all sunspot pairs in the solar northern hemisphere have the same polarity—if the magnetic field lines are directed into the Sun in one leading spot, they are inwardly directed in all leading spots in that hemisphere. The same is true in the southern hemisphere, except that the polarities are always opposite to those in the north.

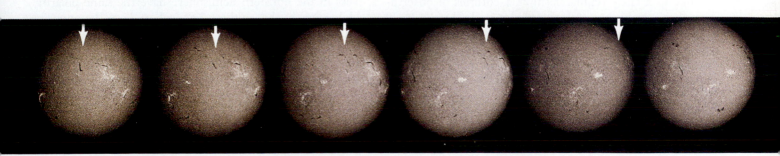

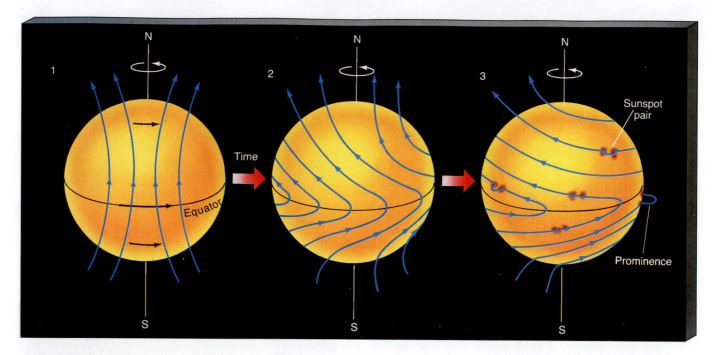

Figure 16.19 This diagram illustrates how the Sun's differential rotation wraps and distorts the solar magnetic field. Occasionally, the field lines burst out of the surface and loop through the lower atmosphere, thereby creating a sunspot pair. The underlying pattern of the solar field lines explains the observed pattern of sunspot polarities. (If the loop happens to occur on the limb of the Sun and is seen against the blackness of space, we see a phenomenon called a prominence, described in a later section.)

lustrated in Figure 16.19, because the Sun rotates more rapidly at the equator than at the poles, the differential rotation distorts the solar magnetic field, wrapping it around the solar equator, eventually causing the original north–south magnetic field to reorient itself in an east–west direction. Convection then causes the magnetized gas to upwell toward the surface, twisting and tangling the magnetic field pattern. In some places, the field becomes kinked like a knot in a garden hose, causing it to increase in strength. Occasionally, the field strength becomes so great that it overwhelms the Sun's gravitational field and a "tube" of field lines bursts out of the surface and loops through the lower atmosphere, forming a sunspot pair. The general east–west organization of the underlying solar field accounts for the observed polarities of the pairs in each hemisphere.

THE SOLAR CYCLE

Not only do sunspots come and go with time, but their numbers and distribution across the face of the Sun also change in a fairly regular fashion. Centuries of observations have established a clear **sunspot cycle**. Figure 16.20 shows the number of sunspots observed each year during the twentieth century. The average number of spots reaches a maximum every 11 or so years and then falls off almost to zero before the cycle begins afresh. The repetition is not exact, however. While the *average* time from one maximum to the next is 11 years, there is quite a spread in individual cycles, which can vary in length from 7 to 15 years.

The latitudes at which sunspots appear vary as the sunspot cycle progresses. Individual sunspots do not move up or down in latitude, but new spots appear closer to the equator as older ones at higher latitudes fade away over the course of the 11-year cycle. Figure 16.21 is a plot of observed sunspot latitude as a function of time. At the start of each cycle, at the time of **solar minimum**, only a few spots are seen. They are generally confined to two narrow zones about 25° to 30° north and south of the solar equator. Approximately 4 years into the cycle, around the time of **solar maximum**, the number of spots has increased markedly. They are found within about 15° to 20° of the equator. Finally, by the end of the cycle, at solar minimum again, the total number has fallen again, and most sunspots lie within about 10° of the solar equator. The beginning of each new cycle appears to overlap the end of the last.

Complicating this picture further, the 11-year sunspot cycle is actually only half of a longer 22-year **solar cycle**. During any given sunspot cycle, the leading spots of all the pairs in the northern hemisphere have the same polarity, while spots in the southern hemisphere have the opposite polarity (see Figure 16.18b). However, these polarities reverse their signs on successive 11-year cycles. The full cycle takes 22 years to repeat, when we take the Sun's magnetism into account. In fact, it has been found that between one 11-year segment and the next, the *entire* solar magnetic field reverses itself.

Astronomers now believe that the Sun's magnetic field is both generated and amplified by the constant stretching,

Figure 16.21 Sunspots cluster at high latitudes when solar activity is at a minimum. They appear at lower latitudes as the number of sunspots peaks. Finally, they are prominent near the Sun's equator as solar minimum is again approached. The most recent solar maximum occurred in 1990.

Figure 16.20 This graph presents the annual number of sunspots throughout the twentieth century, showing the 5-year average of the annual data to make long-term trends more evident. The (roughly) 11-year solar cycle is clearly visible. At the time of solar minimum, hardly any sunspots are seen. About 4 years later, at solar maximum, as many as 100–200 spots are observed per year.

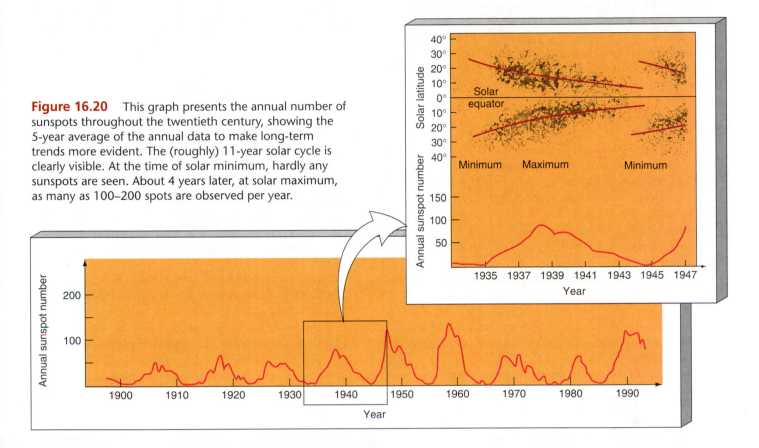

twisting, and folding of magnetic field lines that results from the combined effects of differential rotation and convection. The theory is essentially the same dynamo theory that accounts for the Earth's changing magnetic field (see Chapter 7), except that the solar dynamo operates much faster, and on a much larger scale. ∞ (p. 154) One prediction of this theory is that the Sun's magnetic field should rise to a maximum, then fall to zero and reverse itself in a more or less periodic way, just as observed. Solar surface activity, such as the sunspot cycle, simply follows the variations in the magnetic field. The changing numbers of sunspots and their migration to lower latitudes are both consequences of the strengthening and eventual decay of the field lines as they become more and more tightly wrapped around the solar equator.

Figure 16.22 plots the full extent of all the sunspot data recorded since the invention of the telescope. It is a simple extension of the data presented in Figures 16.20 and 16.21. As can be seen, the 11-year "periodicity" of the solar sunspot cycle is far from regular. Not only does the period vary from 7 to 15 years, but the sunspot cycle has

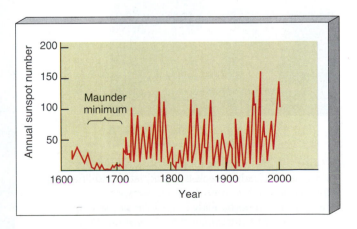

Figure 16.22 This graph plots the number of sunspots occurring each year. Note the approximate 11-year "periodicity" and the absence of spots during the late seventeenth century.

disappeared entirely in the relatively recent past. In honor of the British astronomer who drew attention to these historical records, the lengthy period of solar inactivity that extended from 1645 to 1715 is called the *Maunder minimum*. The corona was apparently also less prominent during total solar eclipses, and Earth auroras were sparse throughout the late seventeenth century. Lacking a complete understanding of the solar cycle, we cannot easily explain how it could shut down entirely. Most astronomers suspect changes in the Sun's convection zone and/or its rotation pattern, but the specific cause of the Sun's century-long variations remains a mystery.

ACTIVE REGIONS

Sunspots are relatively quiescent aspects of solar activity. However, the photosphere surrounding them occasionally erupts violently, spewing forth large quantities of energetic particles into the surrounding corona. The sites of these explosive events are known simply as **active regions**. Most pairs or groups of sunspots have active regions associated with them. Like all aspects of solar activity, these phenomena tend to follow the solar cycle and are most frequent and violent around the time of solar maximum.

Figure 16.23 shows two solar **prominences**. Prominences are loops or sheets of glowing gas ejected from an active region on the solar surface, moving through the inner parts of the corona under the influence of the Sun's magnetic field. Magnetic instabilities in the strong fields found in and near sunspot groups may cause the prominences, although the details are still not completely un-

derstood. Many observations, as in Figure 16.23(a), clearly show streams of hot ionized gas soaring high into the solar atmosphere, following the arching magnetic field lines between members of a sunspot pair (see also Figure 16.18).

Quiescent prominences persist for days or even weeks, hovering high above the photosphere, suspended by the Sun's magnetic field. *Active prominences* come and go much more erratically, changing their appearance in a matter of hours or surging up from the solar photosphere, then immediately falling back on themselves. A typical solar prominence measures some 100,000 km in extent, nearly 10 times the diameter of planet Earth. Prominences as large as that shown in Figure 16.23(b) (which traversed almost half a million kilometers of the solar surface) are less common and usually appear only at times of greatest solar activity. The largest prominences can release up to 10^{25} joules of energy, counting both particles and radiation—not much compared with the total solar luminosity of 4×10^{26} W (joule/s), but still enormous by terrestrial standards. (All the power plants on Earth would take a billion years to produce this much energy.)

Flares are another type of solar activity observed low in the Sun's atmosphere near active regions. Also the result of magnetic instabilities, flares, like that shown in Figure 16.24, are even more violent (and even less well understood) than prominences. They often flash across a region of the Sun in minutes, releasing enormous amounts of energy as they go. Observations made by the *Solar Maximum* satellite demonstrated that X-ray and ultraviolet emissions are especially intense in the extremely compact hearts of

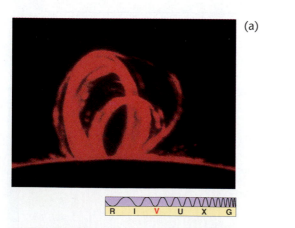

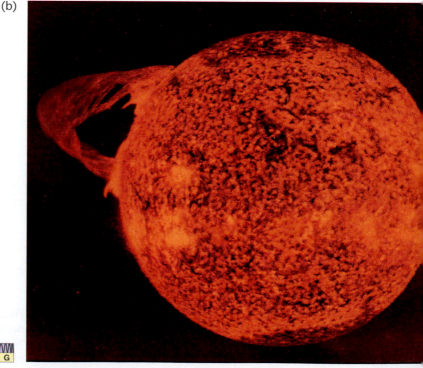

(a) (b)

Figure 16.23 (a) The looplike structure of this prominence clearly reveals the magnetic field lines connecting the two members of a sunspot pair. (b) This image of a particularly large solar prominence was observed by ultraviolet detectors aboard the *Skylab* space station in 1979. (See also Figures 16.18 and 16.19.)

R I V U X G

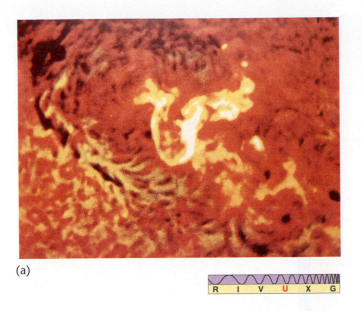

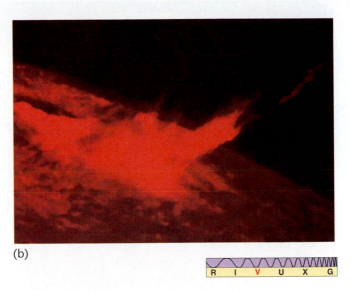

(a)

(b)

Figure 16.24 (a) Much more violent than a prominence, a solar flare is an explosion on the Sun's surface that sweeps across an active region in a matter of minutes, accelerating solar material to high speeds and blasting it into space. Visible here as the white, snakelike feature at the center, this flare extends across some 30,000 km. (b) A similar flare is seen from the side in this dramatic photograph, taken through a red Hα filter.

flares, where temperatures can reach 100,000,000 K. So energetic are these cataclysmic explosions that some researchers have likened flares to bombs exploding in the lower regions of the Sun's atmosphere. A major flare can release as much energy as the largest prominences, but in a matter of minutes or hours rather than days or weeks. Unlike the gas that makes up the characteristic loop of a prominence, the particles produced by a flare are so energetic that the Sun's magnetic field is unable to hold them and shepherd them back to the surface. Instead, the particles are simply blasted into space by the violence of the explosion.

THE CHANGING SOLAR CORONA

The solar corona also varies in step with the sunspot cycle. The photograph of the corona in Figure 16.12 shows the quiet Sun, at sunspot minimum. The corona is fairly regular in appearance and appears to surround the Sun more or less uniformly. Compare this image with Figure 16.25, which was taken in 1991, close to the most recent peak of the sunspot cycle. The active corona is much more irregular in appearance and extends farther from the solar surface. The "streamers" of coronal material pointing away from the Sun are characteristic of this phase. Astronomers now believe that the corona is heated primarily by solar surface activity, particularly prominences and flares, which can inject large amounts of energy into the upper solar atmosphere, greatly distorting its shape. Extensive disturbances often move through the corona above an active site in the photos-

phere, distributing the energy throughout the coronal gas. Given this connection, it is hardly surprising that both the appearance of the corona and the strength of the solar wind are closely correlated with the solar cycle.

Figure 16.25 Photograph of the solar corona during the July, 1991 eclipse, at the peak of the sunspot cycle. At these times, the corona is much less regular and much more extended than at sunspot minimum (compare Figure 16.12). Astronomers believe that coronal heating is caused by surface activity on the Sun. The changing shape and size of the corona are the direct result of variations in prominence and flare activity over the course of the solar cycle.

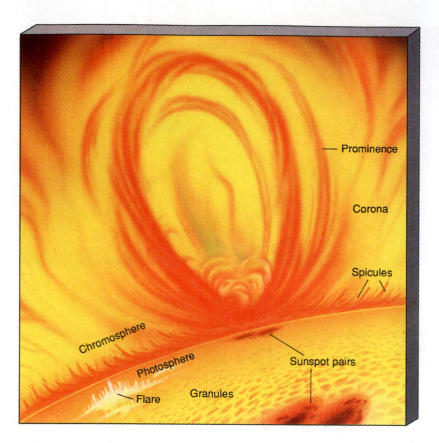

Figure 16.26 This summary piece of art illustrates many of the salient features of our Sun—from prominences in its upper atmosphere, to spicules and flares in its lower atmosphere, to granules and spots on the surface.

Figure 16.26 summarizes many of the features of the active Sun.

16.5 *The Heart of the Sun*

⑤ Compared with familiar terrestrial events—storms, tidal waves, tornados, even violent volcanic eruptions and earthquakes—the spots and flares of the active Sun are enormously energetic. The much greater *steady* emission from the inactive quiet Sun simply staggers the imagination. The Sun is somehow able to produce huge amounts of energy, and, according to Earth's fossil record, it has been doing so for the last several billion years. What powers the Sun? What forces are at work in the Sun's core to produce such energy? By what process does the Sun shine, day after day, year after year, eon after eon? Answers to these questions are central to all of astronomy. Without them, we can understand neither the physical existence of stars and galaxies in the universe nor the biological existence of life on Earth.

SOLAR ENERGY PRODUCTION

The Sun's luminosity is 4×10^{26} W, and its mass is 2×10^{30} kg. We can quantify how efficiently the Sun generates energy by dividing the solar luminosity by the solar mass:

$$\frac{\text{solar luminosity}}{\text{solar mass}} = 2 \times 10^{-4} \text{ W/kg}.$$

This simply means that, on average, every kilogram of solar material yields about 0.2 milliwatts of energy—0.0002 joules of energy every second. This is not a great deal of energy—a piece of burning wood generates about a million times more energy per unit mass per unit time than does our Sun. But there is an important difference: The wood will not burn for billions of years.

To appreciate the magnitude of the energy generated by our Sun, we must consider not the ratio of the solar luminosity to the solar mass, but instead the *total* amount of energy generated by each gram of solar matter *over the entire lifetime of the Sun as a star*. This is easy to do. We simply multiply the rate at which the Sun generates energy by the age of the Sun, about 5 billion years. We obtain a value of 3×10^{13} joule/kg. This is the average amount of energy radiated by every gram of solar material since the Sun formed. It represents a *minimum* value for the total energy radiated by the Sun, for more energy will be needed for every additional day the Sun shines. Should the Sun endure for another 5 billion years (as is predicted by theory), we would have to double this value.

This energy-to-mass ratio value is very large. Thirty trillion joules of energy must arise from *every* kilogram of solar matter (on average) to power the Sun throughout its lifetime. But the generation of energy is not explosive, releasing large amounts of energy in a short period of time. Instead, it is slow and steady, providing a *uniform* and long-lived rate of energy production. Only one known energy-generation mechanism can conceivably power the Sun in this way. That process is nuclear **fusion**—the combining of light nuclei into heavier ones.

Interlude 16-1   *Solar–Terrestrial Relations*

Our Sun has often been worshipped as a god with power over human destinies. Obviously, the steady stream of solar energy arriving at our planet every day is essential to our lives. But, over the past century there have also been repeated claims of a correlation between the Sun's *activity* and the Earth's weather. Only recently, however, has the subject become scientifically respectable—that is, more natural than supernatural.

In fact, there do seem to be some correlations between the 22-year solar cycle (two sunspot cycles, with oppositely directed magnetic fields) and periods of climatic dryness here on Earth. For example, near the start of the past eight cycles, there have been droughts in North America—at least within the middle and western plains from South Dakota to New Mexico. The most recent of these droughts, which typically last three to six years, came in the late 1950s. The one expected in the 1980s, however, did not occur as clearly as anticipated.

Other possible Sun–Earth connections include a link between solar activity and increased atmospheric circulation on our planet. As circulation increases, terrestrial storm systems deepen, extend over wider ranges of latitude, and carry more moisture. The relationship is complex, and the subject controversial, because no one has yet shown any physical mechanism (other than the Sun's heat, which does not vary much during the solar cycle) that would allow solar activity to stir our terrestrial atmosphere. Without a better understanding of the physical mechanism involved, none of these effects can be incorporated into our weather forecasting models.

Solar activity might also influence long-term climate on Earth. For example, the Maunder Minimum seems to correspond fairly well to the coldest years of the so-called "Little Ice Age" that chilled northern Europe during the late 1600s (p. 346). How the active Sun, and its abundance of sunspots, might affect the Earth's climate is a frontier problem in terrestrial climatology.

One correlation that *is* definitely established, and also better understood, is that between solar flares and geomagnetic activity at Earth. The extra radiation and particles thrown off by flares impinge on Earth's environment, overloading the Van Allen Belts, causing brilliant auroras in our atmosphere, and degrading our communication networks. These disturbances have been known for many years, but only recently have we associated them with the solar wind. We are only beginning to understand how the radiation and particles emitted by solar flares also interfere with terrestrial radars, power networks, and other technological equipment. Some power outages on the Earth are actually caused not by increased customer demand or malfunctioning equipment, but by flares on the Sun!

We cannot now predict when and where solar flares will occur. However, it would certainly be to our advantage to be able to do so, as this aspect of the active Sun affects our lives. This is a very fertile area of astronomical research, and one for which there are clear terrestrial applications.

NUCLEAR FUSION

We can represent a typical fusion reaction symbolically as

$$\text{nucleus 1 + nucleus 2} \rightarrow \text{nucleus 3 + energy.}$$

For powering the Sun, the most important piece of this equation is the energy produced. Let's see what gives rise to this energy.

The key point is that during a fusion reaction, the total mass *decreases*: The mass of nucleus 3 is *less* than the combined masses of nuclei 1 and 2. To understand the consequences of this, we can use a very important law of modern physics—the law of **conservation of mass and energy**. Albert Einstein showed at the beginning of the twentieth century that matter and energy are interchangeable. One can be converted into the other, in accordance with Einstein's famous equation, $E = mc^2$. To determine the amount of energy corresponding to a given mass, simply multiply it by the square of the speed of light (c in the equation). For example, the energy equivalent of 1 kg of matter is $1 \times (3 \times 10^8)^2$, or 9×10^{16} joules. The speed of light is so large that even small amounts of mass translate into enormous amounts of energy.

The law of conservation of mass and energy states that the *sum* of mass and energy must always remain constant in any physical process. There are no known exceptions. According to this law, an object can literally disappear, provided that some energy appears in its place. If magicians really made rabbits disappear, the result would be a flash of energy equaling the product of the rabbit's mass and the square of the speed of light—enough to destroy the magician, everyone in the audience, and probably all of the surrounding city as well! In the case of a fusion reaction, the lost mass is converted into energy, primarily in the form of electromagnetic radiation. The light energy we see coming from the Sun means that the Sun's mass must be slowly decreasing.

THE PROTON–PROTON CHAIN

All atomic nuclei are positively charged, so they repel one another. Furthermore, the closer two nuclei come to one another, the greater is the repulsive force between them. How then do nuclei—two protons, say—ever manage to fuse into anything heavier? The answer is that, if they collide at high enough speeds, one proton can momentarily plow deep into the other, eventually coming within the exceedingly short range of the *strong nuclear force*. (See the *More Precisely* feature on p. 353.) At distances less than about 10^{-15} m, the attraction of the nuclear force overwhelms the electromagnetic repulsion, and fusion occurs. Speeds in excess of a few hundred kilometers per second, corresponding to a gas temperature of 10^7 K or more, are needed to slam protons together fast enough to initiate fusion. Such conditions are found in the core of the Sun and at the centers of all stars.

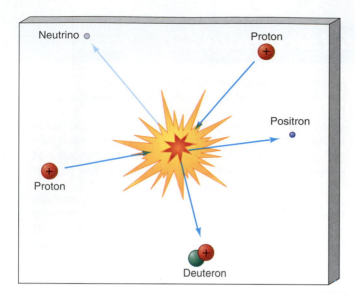

Figure 16.27   Two protons collide violently, initiating the chain of nuclear fusion that powers the Sun.

At these temperatures, two protons can interact to produce another proton, a neutron, and two additional elementary particles. Figure 16.27 is a diagram of this event. We can also represent it by the following equation:

$$\text{proton 1} + \text{proton 2} \rightarrow \text{proton 3} + \text{neutron} + \text{positron} + \text{neutrino}.$$

The **positron** particle in this reaction is a positively charged electron. Its properties are identical to those of a normal negatively charged electron, except for its positive charge. Scientists call the electron and the positron a "matter–antimatter pair"—the positron is said to be the *antiparticle* of the electron. These newly created positrons find themselves in the midst of a sea of electrons, with which they interact immediately and violently. The particles and antiparticles annihilate one another, producing pure energy in the form of gamma rays.

The final product of the reaction is a particle known as a **neutrino**, a word derived from the Italian for "little neutral one." Neutrinos carry no electrical charge, and are of very low mass—at most 1/10,000 the mass of an electron, which itself has only 1/2000 the mass of a proton. (Physicists are still unsure if the neutrino mass is actually zero.) They move at (or nearly at) the speed of light and interact with hardly anything. They can penetrate, without stopping, several light years of lead. Their interactions with matter are governed by the *weak force*, described in more detail in the *More Precisely* feature on p. 352. Despite their elusiveness, neutrinos can be detected with carefully constructed instruments (see p. 353).

The neutron and proton produced in the collision merge to form a **deuteron**, the nucleus of a special form of hydrogen—deuterium, also referred to as "heavy hydrogen." Deuterium differs from ordinary hydrogen by virtue of an extra neutron in the nuclei of its atoms. Nuclei containing the same number of protons but different numbers of neutrons represent different forms of the same element—they are known as **isotopes** of that element. Usually, there are about as many neutrons in a nucleus as protons, but the exact number of neutrons can vary, and most elements can exist in a number of isotopic forms.

To avoid confusion when talking about isotopes of the same element, nuclear physicists attach a number to the symbol representing the element. This number indicates the total number of particles (protons plus neutrons) in its nucleus. Ordinary hydrogen is denoted by ^1H (or sometimes simply by p, for proton), deuterium is ^2H, normal helium (2 protons plus 2 neutrons) is ^4He (also referred to as helium-4), and so on. We will adopt this convention for the remainder of this book. We can now write the net effect of the proton–proton reaction, including the formation of the deuteron, as

$$^1\text{H} + {}^1\text{H} \rightarrow {}^2\text{H} + \text{positron} + \text{neutrino}. \qquad \text{(I)}$$

This equation is labeled (I) because the production of a deuteron by the fusion of two protons is the first step in the fusion process powering most stars. It is the start of the **proton–proton chain**, whereby gargantuan quantities of protons are fused within the core of the Sun each second.

The next step in solar fusion is the formation of an isotope of helium. A proton interacts with the deuteron particle produced in step (I), as symbolized by the equation

$$^2\text{H} + {}^1\text{H} \rightarrow {}^3\text{He} + \text{energy}. \qquad \text{(II)}$$

Step (II) begins as soon as deuterons appear. The main product is an isotope of helium—helium-3—lacking one of the neutrons contained in the normal helium-4 nucleus (which has two protons and two neutrons). Energy is also emitted, again in the form of gamma-ray photons.

The third and final step in the proton–proton chain, also verified by direct laboratory experiments, involves the production of nuclei of helium-4. Helium-4 (^4He) comes about most often through the fusion of two of the helium-3 nuclei created in step (II):

$$^3\text{He} + {}^3\text{He} \rightarrow {}^4\text{He} + {}^1\text{H} + {}^1\text{H} + \text{energy}. \qquad \text{(III)}$$

The result is a helium-4 nucleus plus two more protons.

The net effect of steps (I) through (III) is this: Four hydrogen nuclei (protons) combine to create one helium-4 nucleus, plus some gamma-ray radiation, and two neutrinos. The whole process is illustrated in Figure 16.28. Symbolically, we have

$$4\,(^1\text{H}) \rightarrow {}^4\text{He} + \text{energy} + 2\,\text{neutrinos}.$$

The gamma-ray photons originally produced in the heart of the Sun are slowly reduced in energy as they pass through the solar interior. Photons are repeatedly absorbed by electrons and ions, then reemitted at a wavelength that reflects the temperature of the surrounding gas, in accordance with Wien's Law. ∞ (p. 64) Thus, as the radiation slowly makes its way to the surface through ever-cooler layers, its wavelength decreases. Eventually, the electromagnetic energy leaves the surface in the form of visible light. The neutrinos escape unhindered into space. The

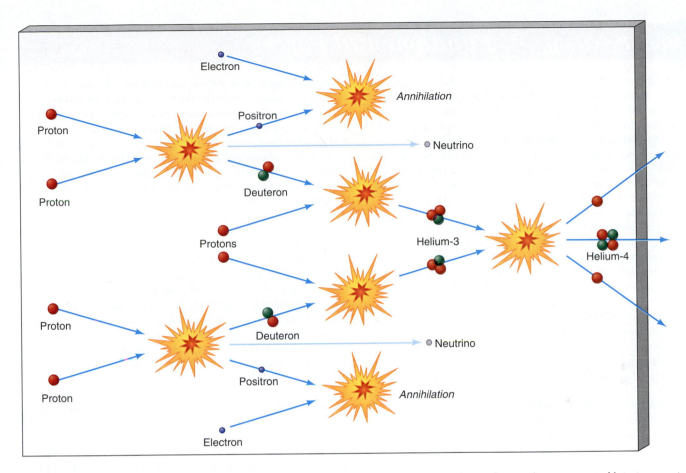

Figure 16.28   Diagram of the entire proton–proton chain. A total of six protons (and two electrons) are converted into two protons, one helium-4 nucleus, and two neutrinos. The two leftover protons are available as fuel for new proton–proton reactions, so the net effect is that four protons are fused to form one helium-4 nucleus. Energy, in the form of gamma rays, is produced in each reaction.

helium stays put in the core. Alternative reactions leading to the same final result exist (see the *More Precisely* feature on p. 420), but they are relatively rare in stars like the Sun.

ENERGY GENERATED BY THE PROTON–PROTON CHAIN

Let us now calculate the energy produced in the fusion process and compare it with the 3×10^{13} joule/kg needed to account for the Sun's luminosity. Careful laboratory experiments have determined the masses of all the particles involved in the conversion of four protons into a helium-4 nucleus: The total mass of the protons is 6.6943×10^{-27} kg, the mass of the helium-4 nucleus is 6.6466×10^{-27} kg, and the neutrinos are virtually massless. The difference between the total mass of the protons and the helium nucleus, 0.048×10^{-27} kg—not much mass, but easily measurable in laboratory experiments—is lost in the reaction. It is transformed into energy. Multiplying the vanished mass by the square of the speed of light yields 4.3×10^{-12} joule—this is the energy produced by the fusion of 6.7×10^{-27} kg (the rounded-off mass of the 4 protons) of hydrogen into helium. It follows that fusion of 1 kg of hydrogen would generate 6.4×10^{13} joule—more than enough to power the Sun. To fuel the Sun's present energy output, hydrogen must be fused into helium in the core at a rate of 600 million tons per second.

16.6   *Observations of Solar Neutrinos*

6   Theorists are quite sure that the proton–proton chain operates in the Sun. Based on detailed "numerical experiments"—that is, simulations on large computers—their models of the Sun's interior temperature, density, composition, and nuclear burning rates predict bulk properties that agree very well with observations. The observations, however, are almost exclusively confined to the solar exterior—the photosphere, chromosphere, and corona. Astronomers have little *direct* evidence of the nuclear reactions occurring at the center of the Sun. The reason for this is simple: Electromagnetic radiation cannot escape from the Sun's interior. The powerful gamma-ray photons created in the solar core are absorbed and reemitted many times, bouncing their way through the interior for hundreds of thousand of years before they finally leave the solar surface as visible or infrared photons.

By contrast, the neutrinos that arise as by-products of the proton–proton cycle *do* travel cleanly out of the Sun, interacting with virtually nothing. They leave at or near the speed of light, escaping into space a few seconds after being created at the core. Unlike the gamma rays, which heat the solar core as they interact with the matter there, the neutrinos do not heat the solar gas at all. On the contrary, they serve to cool it by swiftly carrying energy away. Thus neu-

More Precisely...   *Fundamental Forces*

Our studies of nuclear reactions have uncovered new ways in which matter can interact with matter at a subatomic level. Let's pause to consider in a slightly more systematic fashion the relationships among the various forces of nature.

As best we can tell, the behavior of all matter in the universe—from elementary particles to clusters of galaxies—is ruled by just four (or fewer) basic forces, which are *fundamental* to everything in the universe. In a sense, the search to understand the nature of the universe is the quest to understand the nature of these forces.

The *gravitational force* is probably the best known. Gravity binds galaxies, stars, and planets together and holds humans on the surface of the Earth. As we saw in Chapter 2, its magnitude decreases with distance, according to an inverse-square law. ∞ (p. 46) Its strength is also proportional to the masses of each of the two objects involved. Thus, the gravitational field of an atom is extremely weak, but that of a galaxy, consisting of huge numbers of atoms, is very powerful. Gravity is by far the weakest of the forces of nature, but its effect accumulates as we move to larger and larger volumes of space, and nothing can cancel its attractive pull. As a result, gravity is the dominant force in the universe on all scales larger than that of the Earth.

The *electromagnetic force* is another of nature's basic agents. Any particle having a net electric charge, such as an electron or a proton in an atom, exerts an electromagnetic force on any other charged particle. The everyday things we see around us are held together by this force. Like gravity, its strength also decreases with distance according to an inverse-square law. However, for subatomic particles, electromagnetism is much stronger than gravity. For example, the electromagnetic force between two protons exceeds their gravitational attraction by a factor of about 10^{36}. Unlike gravity, electromagnetic forces can repel (between like charges) as well as attract (between opposite charges). Positive and negative charges tend to neutralize each other, greatly diminishing their net electromagnetic influence. Above the microscopic level, most objects are in fact very close to being electrically neutral. Thus, except in unusual circumstances, the electromagnetic force is relatively unimportant on macroscopic scales.

A third fundamental force of nature is simply termed the *weak force*. It is much weaker than electromagnetism, and its influence is somewhat more subtle. The weak force governs the emission of radiation from some radioactive atoms; the emission of a neutrino during the first stage of the proton–proton reaction is also the result of a weak interaction. It is now known that the weak force is not really a separate force at all, but just a form of the electromagnetic force. Thus, physicists often speak of the "electroweak force." However, when acting in its "weak" mode, the electroweak force does not obey the inverse-square law. Its effective range is less than the size of an atomic nucleus, about 10^{-15} m.

Strongest of all the forces is the *strong* (or *nuclear*) *force*. It binds atomic nuclei together and governs the generation of energy in the Sun and all other stars. Like the weak force, and unlike the forces of gravity and electromagnetism, the strong force operates only at very close range. It is unimportant outside a distance of a hundredth of a millionth of a millionth (10^{-14}) of a meter. However, within this range (for example, in atomic nuclei), it binds particles with enormous strength. In fact, it is the range of the strong force that determines the typical sizes of atomic nuclei. Only when two protons are brought within about 10^{-15} m of one another can the attractive strong force overcome their electromagnetic repulsion.

Not all particles are subject to all types of force. All particles interact through gravity because all have mass. However, only *charged* particles interact electromagnetically. Protons and neutrons are affected by the strong force, but electrons are not. Under the right circumstances, the weak force can affect any type of subatomic particle, regardless of its charge.

---

trinos offer, at least in principle, the possibility of probing directly the conditions at the heart of the Sun.

Of course, the fact that they can pass through the entire Sun without interacting also makes neutrinos fairly difficult to detect on Earth! Nevertheless, they do interact a little more strongly with some elements—chlorine and gallium, for example—than with others, and this knowledge can be used in the construction of Earth-based neutrino-detection devices. Occasionally, a neutrino from the Sun will encounter a chlorine-37 nucleus, converting it into a nucleus of argon-37, or will interact with a nucleus of gallium-31, turning it into germanium-31 instead.

In the late 1960s, a team of researchers from Brookhaven National Laboratory built a large tank near the bottom of the Homestake gold mine in South Dakota and filled it with 400,000 liters (about 100,000 gallons) of a chlorine-containing chemical—the common cleaning fluid used by dry cleaners. Figure 16.29 shows a photograph of the apparatus. At 1.5 km below ground level, the experimenters could be reasonably sure of avoiding interference from other sources, as most subatomic particles are unable to penetrate Earth to such a depth. They left their tank in the mine for months at a time, periodically checking to see if any of the chlorine had been converted into argon, which would signal the absorption of a neutrino.

Given the size of the detector and the expected physical conditions at the Sun's core, theory predicts that about one solar neutrino of the roughly 10^{16} that streamed through the tank each day should have been detected. While the experiment succeeded in detecting *some* neutrinos, the numbers were not as great as predicted. Over the course of the entire experiment, neutrinos were detected about two or three times per week, on average, not once per day. Apparently, the only way we have of peering directly into the core of the Sun presents us with a problem. It is generally known as the **solar neutrino problem**.

Although the detection of solar neutrinos is an incredibly exacting task, it is unlikely that the Homestake experimental apparatus was at fault. The neutrino deficit persisted over two decades of almost continuous monitoring (until the experiment was terminated in 1993) and, though scores of technicians examined every facet of the Homestake instrument for instrumental flaws, they found none. The experimental gear was well designed and well built. The conclusion that the Homestake detector worked properly was bolstered by a more recent experiment (with a quite different detector design) conducted at Kamioka, Japan, which reported a similar neutrino deficit.

One drawback of the Homestake and Kamioka experiments was that they were sensitive to only a tiny fraction of the neutrinos actually produced by fusion in the Sun's core. The particular reaction creating the neutrinos they could detect is not reaction (I), but a much less probable sequence of events that occurs only about 0.25 percent of the time. Uncertainties in the exact rate at which these reactions occur once led astronomers to suspect that the theoretical calculation of "Homestake-detectable" neutrinos might simply be wrong. Because the beryllium–boron reaction chain accounts for only a tiny fraction of the Sun's total energy output, it has a negligible effect on the solar luminosity and surface temperature. As a result, modifications to its reaction rate might have made the Homestake and Kamioka results consistent with the Standard Model without significantly affecting any other aspect of the Sun's appearance.

That possibility has been all but eliminated by two recent neutrino detectors—the *Soviet–American Gallium Experiment* (or SAGE, for short) and the U.S.–European GALLEX collaboration—each of which uses the element gallium to capture solar neutrinos. Unlike Homestake and Kamioka, both SAGE and GALLEX can detect the neutrinos produced by reaction (I)—the initial step in the proton–proton chain—so they provide a much more direct probe of energy generation in the solar core. Like Homestake and Kamioka, each of them indicates a significant shortfall of neutrinos below the predicted number.

The four neutrino-detection experiments we have just described disagree somewhat in their measurements of the precise extent of the deficit, but *each* of them sees less than the expected number of solar neutrinos. It is hard to avoid the conclusion that there is a real discrepancy between the Sun's theoretical neutrino output and the neutrinos we actually observe on Earth. How can we explain this contradiction? If, as we think, the detectors are working correctly, there are really only two possibilities. Either neutrinos are not produced as frequently as we think, or not all of them make it to the Earth. Let us now consider these alternatives in turn.

If the temperature in the solar core were lower, the number of neutrinos predicted by theory would be lower. If the center of the Sun were about 10 percent cooler than in the Standard Solar Model—about 13,500,000 K—helium-4 would still be produced, but it would be accompanied by fewer neutrinos detectable by the Homestake experiment. But lowering the temperature would also lower the Sun's luminosity, and most theorists agree that the numerical

models could not be in error by as much as 1,500.000 K while remaining consistent with all other solar observations. In addition, observations by the GONG group (discussed earlier) seem to rule out a central temperature below 15,000,000 K. Most astronomers regard it as quite unlikely that the resolution of the solar neutrino problem will be found in the nuclear physics of the Sun's interior.

Instead, the properties of the neutrinos themselves may provide the answer. If neutrinos do have a minute amount of mass, it may be possible for them to change their properties, even to transform into other particles, during their 8-minute flight from the solar core to Earth, through a process generally known as **neutrino oscillations**. In this picture, neutrinos are produced in the Sun at the rate required by the Standard Solar Model, but some of them turn into something else (they are said to "oscillate" into other particles) on their way to the Earth and so go undetected. Proposed experiments near neutrino-producing nuclear reactors on Earth may be able to test this idea within the next few years.

Where are the missing neutrinos? Is the proton–proton chain operating as we think? Do we *really* know what processes are at work deep in the hearts of stars? For now, the mystery of the solar neutrinos remains unsolved, although most physicists favor the neutrino-oscillation explanation. With continued observations of the neutrinos from the main reaction in the proton–proton chain, and improved calculations of the nuclear reactions occurring in the solar core, astronomers are hopeful that the solution will be in hand by the end of the century. Virtually all researchers concur—or at least hope—that the correct interpretation of the solar neutrino problem will not tear apart the theoretical fabric of the proton–proton chain. Most believe that the description of solar fusion we have presented is this chapter is basically right; our understanding of neutrino physics just needs to be fine-tuned. But should drastic measures be needed to solve the solar puzzle, we may yet have to return to the drawing board to answer one of the most fundamental scientific questions of all: How does a star shine?

Figure 16.29 This swimming-pool-sized detector is a "neutrino telescope" of sorts, buried underground in a South Dakota gold mine.

Chapter Review

SUMMARY

A **star** (p. 332) is a glowing ball of gas held together by its own gravity and powered by nuclear fusion at its center. The main interior regions of the Sun are the **core** (p. 332), where nuclear reactions generate energy, the **radiation zone** (p. 332), where the energy travels outward in the form of electromagnetic radiation, and the **convection zone** (p. 332), where the Sun's matter is in constant convective motion.

The sharp *solar disk* that we see from Earth marks the solar **photosphere** (p. 332)—the thin region at the Sun's surface from which essentially all of the visible light is emitted. Above the photosphere lies the **chromosphere** (p. 332), which is separated from the solar **corona** (p. 332) by a thin **transition zone** (p. 341) in which the temperature increases from a few thousand to around a million kelvins.

The Sun's **luminosity** (p. 334) is the total amount of energy radiated from the solar surface per second. It is determined by measuring the **solar constant** (p. 333)—the amount of solar radiation reaching each square meter at Earth's distance from the Sun—and multiplying that amount by the area of an imaginary sphere of radius 1 A.U.

Much of our knowledge of the solar interior comes from mathematical models. The model that best fits the observed properties of the Sun is the **Standard Solar Model** (p. 334). Studies of **helioseismology** (p. 334)—oscillations of the solar surface caused by sound waves in the interior—provide further insight into the Sun's structure.

The effect of the solar convection zone can be seen on the surface in the form of **granulation** (p. 336) of the photosphere. As hotter (and therefore brighter) gas rises and cooler (dimmer) gas sinks, a characteristic "mottled" appearance results. Lower levels in the convection zone also leave their mark in the form of larger transient patterns called **supergranulation** (p. 336).

Most of the absorption lines seen in the solar spectrum are produced in the upper photosphere and the chromosphere. Studies of these allow scientists to determine the Sun's composition and the temperature structure of the solar atmosphere.

At about 10–15 solar radii, the gas in the corona is hot enough to escape the Sun's gravity, and the corona begins to flow outward as the **solar wind** (p. 341). Most of the solar wind flows from low-density regions of the corona called **coronal holes** (p. 341).

The steady component of the Sun's energy production is known as the **quiet Sun** (p. 341). Superimposed on that is the much more erratic emission of the **active Sun** (p. 341). Solar activity is generally associated with disturbances in the Sun's magnetic field.

Sunspots (p. 342) are Earth-sized regions on the solar surface that are a little cooler than the surrounding photosphere. They are regions of intense magnetism. They appear to move across the solar disk as the Sun rotates and usually survive for 1–2 months.

Both the numbers and locations of sunspots vary in an 11-year **sunspot cycle** (p. 344). At **solar minimum** (p. 344), only a few spots are typically seen, and they lie far from the solar equator. At **solar maximum** (p. 344), the number of spots is much greater, and they generally lie much closer to the equator. The sunspot cycle is quite irregular. Its length varies from 7 to 15 years. There have been times in the past when no sunspots were seen for long periods. The overall direction of the solar magnetic field reverses from one sunspot cycle to the next. The 22-year cycle that results when the direction of the field is taken into account is called the **solar cycle** (p. 344).

Solar activity tends to be concentrated in **active regions** (p. 346) associated with sunspot groups. **Prominences** (p. 346) are loop- or sheetlike structures produced when hot gas ejected by activity on the solar surface interacts with the Sun's magnetic field. The more intense **flares** (p. 346) are violent surface explosions that blast particles and radiation into interplanetary space.

The Sun generates energy by "burning" hydrogen into helium in its core by the process of nuclear **fusion** (p. 348). When four protons are converted into a helium nucleus in the **proton–proton chain** (p. 350), some mass is lost. The law of **conservation of mass and energy** (p. 349) requires that this mass appear as energy, eventually resulting in the light we see.

Some particles produced during the solar fusion process are the **positron** (p. 350), or antielectron, which quickly annihilates with electrons in the Sun's core to generate gamma rays, the **deuteron** (p. 350), an **isotope** (p. 350) of hydrogen consisting of a proton and a neutron, and the **neutrino** (p. 350), a near-massless particle that escapes from the Sun without any further interactions once it is created in the core.

Despite their elusiveness, it is possible to detect a small fraction of the neutrinos streaming from the Sun. The observations lead to the **solar neutrino problem** (p. 352)—substantially fewer neutrinos are observed than are predicted by theory. The resolution to this problem is unclear. A leading explanation is that **neutrino oscillations** (p. 353) convert some neutrinos into other (undetected) particles en route from the Sun to Earth.

SELF-TEST: True or False?

_____ **1.** The Sun is a rather normal star.

_____ **2.** The average density of the Sun is significantly greater than the density of the Earth.

_____ **3.** The Sun's diameter is about 10 times that of Earth.

_____ **4.** The Sun's differential rotation indicates that it is not solid.

_____ **5.** In the solar radiation zone, the gas is partly ionized.

_____ **6.** Convection involves cool gas rising to the solar surface, and hot gas sinking into the interior.

_____ **7.** Absorption lines in the solar spectrum are produced mainly in the corona.

_____ **8.** There are as many absorption lines in the solar spectrum as there are elements present in the Sun.

_____ **9.** The faintness of the chromosphere is a direct result of its low temperature.

_____ **10.** The temperature of the solar corona decreases with increasing radius.

_____ **11.** Sunspots are regions of intense magnetic fields.

_____ **12.** Prominences are large flames erupting from the burning surface of the Sun.

_____ **13.** Neutrinos are neutrons traveling close to the speed of light.

SELF-TEST: Fill in the Blank

1. The part of the Sun we actually see is called the _____.
2. Traveling outward from the surface, the two main regions of the solar atmosphere are the ____ and the _____.
3. Below the solar surface, in order of increasing depth, lie the _____ zone, the _____ zone, and the _____.
4. The _____ seen on the surface of the Sun is evidence of convective cells.
5. The Sun appears to have a well-defined edge because the thickness of the _____ is only 0.1 percent of the solar radius.
6. _____ is the most abundant element in the Sun.
7. _____ is the second most abundant element in the Sun.
8. The two most abundant elements in the Sun make up about _____ percent of its composition.

9. The gas in the corona is highly _____.
10. Sunspots appear dark because they are _____ than the surrounding gas of the photosphere.
11. The sunspot cycle is _____ years long; the solar cycle is _____ as long.
12. The entire solar luminosity is produced in the _____ (give the region) of the Sun.
13. The net result of the proton-proton chain is that ____ protons are fused into a nucleus of _____ , 2 _____ are emitted, and energy is released in the form of _____.
14. The solar neutrino "problem" is the fact that astronomers observe too _____ neutrinos coming from the Sun.

REVIEW AND DISCUSSION

1. Name and briefly describe the main regions of the Sun.
2. How massive is the Sun, compared with the Earth?
3. How hot is the solar surface? The solar core?
4. How do scientists construct models of the Sun?
5. Describe how energy generated at the center of the Sun reaches Earth.
6. Why does the Sun appear to have a sharp edge?
7. Give the history of "coronium," and tell how it increased our understanding of the Sun.
8. What is the solar wind?

9. What is the cause of sunspots, flares, and prominences?
10. What fuels the Sun's enormous energy output?
11. What are the ingredients and the end result of the proton-proton chain in the Sun? Why is energy released in the process?
12. Why are scientists trying so hard to detect solar neutrinos?
13. What would we observe on Earth if the Sun's internal energy source suddenly shut off? Would the Sun darken instantaneously? If not, how long do you think it might take—minutes, days, years, millions of years—for the Sun's light to begin to fade? Repeat the question for solar neutrinos.

PROBLEMS

1. Use Wien's Law (peak wavelength = 0.29 cm / T, where T is the temperature in kelvins; see Chapter 3) to determine the wavelength corresponding to the peak of the black-body curve (a) in the core of the Sun, where the temperature is 10^7 K, (b) in the solar convection zone (10^5 K), and (c) just below the solar photosphere (10^4 K). What form (visible, infrared, X-ray, etc.) does the radiation take in each case?
2. If convected solar material moves at 1 km/s, how long does it take to flow across the 1000-km expanse of a typical granule?

Compare this with the roughly 10-minute lifetimes observed for most solar granules.
3. Use Stefan's Law (flux $\propto T^4$, where T is the temperature in kelvins; see Chapter 3) to calculate how much less energy (as a fraction) is emitted per unit area of a 4500-K sunspot than from the surrounding 6000-K photosphere.
4. How long does it take for the Sun to convert 1 Earth mass of hydrogen into helium?

PROJECTS

The projects given here all require a special solar filter. Such filters are easily purchased from various sources.
NEVER LOOK DIRECTLY AT THE SUN WITHOUT A FILTER!

1. An appropriately filtered telescope will easily show you sunspots, although the numbers will vary considerably during the sunspot cycle. Count the number of sunspots you see on the Sun's surface. Notice that sunspots often come in pairs or groups. Come back and look again a few days later and you'll see that the Sun's rotation has caused spots to move and the spots themselves have changed. If a sufficiently large sunspot (or, more likely, sunspot group) is seen, continue to watch it as the Sun rotates. It will be out of view for about two weeks. Can you determine the rotation of the Sun from these observations?

2. Solar granulation is not too hard to see. The atmosphere of

the Earth is most stable in the morning hours. Observe the Sun on a cool morning, one or two hours after it has risen. Use high magnification and look initially at the middle of the Sun's disk. Can you see changes in the granulation pattern? They are there, but are not always obvious or easy to see.

3. View some solar prominences and flares. Hydrogen-alpha filters are commercially available for small telescopes. Although rather expensive, many science departments will have one. You can often see prominences and flares even during times of sunspot minimum. You are actually viewing the chromosphere rather than the photosphere, so the Sun looks quite different, and can look very impressive.

A

B

C

17

MEASURING THE STARS

Giants, Dwarfs, and the Main Sequence

LEARNING GOALS

Studying this chapter will enable you to:

1 Explain how stellar distances are determined.

2 Discuss the motions of the stars through space and how these motions are measured from Earth.

3 Explain how physical laws are used to estimate stellar sizes.

4 Distinguish between luminosity and brightness, and explain how stellar luminosity is determined.

5 Explain how stars are often classified according to their colors, surface temperatures, and spectral characteristics, and the usefulness of such classification.

6 State how an H-R diagram is constructed, and summarize the properties of the different types of stars that such a diagram helps us to identify.

7 Explain how the masses of stars are measured and how mass is related to other stellar properties.

8 Distinguish between open and globular star clusters, and explain why the study of clusters is important to astronomers.

(Opposite page, background) Even in our own local region of the cosmos, the number of stars is virtually beyond our ability to count. Relatively few of them have actually been studied in detail. Yet we have learned an enormous amount about the stars in general and the range of properties they exhibit: their masses, their temperatures, their luminosities, even their ages and (as we shall see in subsequent chapters) their destinies.
(Inset A) A stellar nursery: the star-forming region NGC 2024, just below Orion's belt. This infrared image shows many recently-formed stars that are obscured in visible-light photos by clouds of dust.
(Inset B) Young neighbors: The Pleiades star cluster. These stars are relatively nearby (about 400 ly) and still quite youthful: a mere 20 million years old, they had not yet been born when dinosaurs roamed the Earth.
(Inset C) A close-up look at an average star in mid-life would reveal an object very like this artist's depiction of our own Sun.

*U*p to this point, we have studied Earth, the Moon, the solar system, and the Sun. To continue our inventory of the contents of the universe, we must move away from our local environment, into the depths of space. In this chapter we take a great leap in distance and consider stars in general. Our primary goal is to comprehend the nature of the stars that make up the constellations as well as the myriad more distant stars we cannot perceive with our unaided eyes. Rather than studying their individual peculiarities, we will concentrate on determining the physical and chemical properties they share. There is order in the legions of stars scattered across the sky. Like comparative planetology in the solar system, comparing and cataloging the stars play vital roles in furthering our understanding of the galaxy and the universe we inhabit.

17.1 *The Distances to the Stars*

STELLAR PARALLAX

❶ In Chapter 1 we studied how we can use *parallax* to measure distances to terrestrial and solar system objects. Recall that parallax is an object's apparent shift relative to some more distant background as the observer's point of view changes. ∞ (p. 24) In order to measure parallax, we must observe the object from either end of some baseline and measure the angle through which the line of sight to the object shifts. In astronomical contexts, we determine the parallax by comparing photographs made from the two ends of the baseline. As the distance to the object increases or the baseline shrinks, the parallax becomes smaller and harder to measure. Accordingly, a large baseline is essential for measuring the distance to a very remote object.

The stars are so far away from us that even Earth's diameter is too short to use as a baseline in determining their distance. Their apparent shift, as seen from different points on Earth, is too small to measure. However, by comparing observations made of a star at different times of the year, as shown in Figure 17.1, we effectively extend the baseline to the diameter of Earth's orbit around the Sun, 2 A.U. Only with this enormously longer baseline do some stellar parallaxes become measurable. Other, more distant, stars do not reveal any apparent shifts, even with this 2-A.U. baseline.

The parallactic angle *p* is always very small. Even for the closest stars, it is less than 1 arc second, so the imaginary triangle formed by Earth, the Sun, and the star is actually much longer and narrower than is suggested by Figure 17.1. Astronomers generally find it convenient to measure parallax in arc seconds rather than in degrees. If we ask at what distance a star must lie in order for its parallax to measure exactly 1″ (arc second), we get an answer of 206,265 A.U., or 3.1×10^{16} m. Astronomers call this distance 1 **parsec** (1 pc), from "*par*allax in arc *sec*onds." Because parallax decreases as distance increases, we can relate the parallactic angle to a star's distance by the following simple formula:

$$\text{distance (in parsecs)} = \frac{1}{\text{parallax (in arc seconds)}}.$$

Thus, a star with a measured parallax of 1″ lies at a distance of 1 pc from the Sun. The parsec is defined so as to make the conversion between distance and parallactic angle easy. An object with a parallax of 0.1″ lies at a distance of 10 pc; an object with a parallax of 10″ lies at 0.1 pc, and so on. One parsec is approximately equal to 3.3 light years.

We saw in Chapter 2 how Aristotle and other geometers of ancient Greece used the apparent absence of stellar parallax to argue that Earth was stationary. ∞ (p. 33) They were eventually proved wrong, but only in the nineteenth century, when a German astronomer named Friedrich Bessel (1784–1846) succeeded in measuring the parallax of a nearby star. The observation of stellar parallax proves that Earth moves about the Sun. However, the lack of easily observed stellar parallax proves that even "nearby" stars lie at great distances from us.

OUR NEAREST NEIGHBORS

The closest star to Earth (excluding the Sun) is called Proxima Centauri. This star is a member of a triple-star system (three separate stars orbiting one another, bound together by gravity) known as the Alpha Centauri complex. Proxima Centauri displays the largest known stellar parallax, 0.76″, which means that it is about 1.3 pc away—about 270,000 A.U., or 4.3 light years. That's the *nearest* star to Earth—at almost 300,000 times the distance from Earth to the Sun! This is a fairly typical interstellar distance in the Milky Way Galaxy.

Vast distances can sometimes be grasped by means of analogies. Imagine Earth as a grain of sand that orbits a golfball-sized Sun at a distance of about 1 m. The nearest star, also a golfball-sized object, is then more than 100 *kilo*meters away. Except for the other planets in our solar system, themselves ranging in size from grains of sand to small marbles and all lying within 50 m of the "Sun," nothing else of consequence exists in the 100 km separating the two stars. Such is the void of interstellar space.

The next nearest neighbor to the Sun beyond the Alpha Centauri system is called Barnard's Star. Its parallax is 0.55″, so it lies at a distance of 1.8 pc, or 6.0 light years. The farther into space astronomers look, the greater the volume of space they observe and more stars they find.

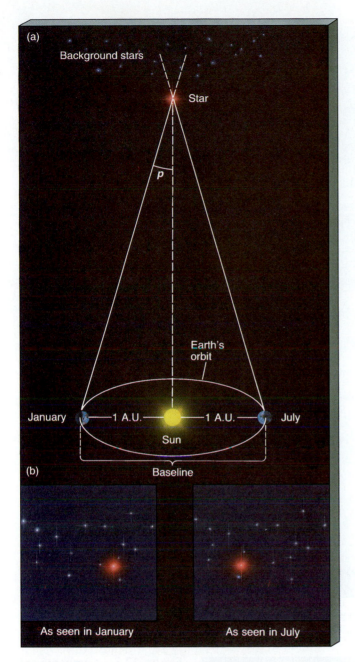

Figure 17.1 (a) The geometry of stellar parallax. For observations made 6 months apart, the baseline is twice the Earth–Sun distance, or 2 A.U. (b) The parallactic angle p is usually measured photographically.

About 30 stars are known to lie within 4 pc of Earth. Figure 17.2 is a map of our nearest galactic neighbors.

As we mentioned in Chapter 5, ground-based images of stars are generally smeared out into a disk of radius 1″ or so by turbulence in Earth's atmosphere. ∞ (p. 105) However, astronomers have special equipment that can routinely measure stellar parallaxes of 0.03″ or less, corresponding to stars within about 30 pc (100 ly) of Earth. Several thousand stars lie within this range. The majority of them are dimmer than the Sun and invisible to the naked eye. Most of the bright stars in the night sky are really much brighter than our Sun, so they are visible from Earth despite their great distance. The vast majority of stars in our Galaxy are far more distant than 30 pc.

17.2 Stellar Motion

❷ In addition to the apparent motion caused by parallax, stars have real motion, too. In other words, stars travel through space. The annual movement of a star across the sky, as seen from Earth (and corrected for parallax), is called **proper motion**. Like parallax, it is measured in terms of angular displacement, and the angles involved are typically very small. Proper motion is usually expressed in arc seconds per year. Stars' velocities can be quite large. Because of their great distances; however, it usually takes many years for us to be able to discern their movement.

Figure 17.3 compares two photographs of the sky around Barnard's Star. They were made on the same day of the year, but 22 years apart. As the photographs show, Barnard's Star moved during this interval. If the two photographs were superimposed, the two images of the star would not coincide. Because Earth was at the same point in its orbit when these photographs were taken, the displacement cannot be due to parallax caused by Earth's motion around the Sun. We conclude that the observed displacement indicates real space motion of Barnard's Star relative to the Sun.

Careful measurements show that Barnard's Star moved 227″ the 22-year interval. The proper motion—*the annual displacement*—of Barnard's Star is 227″/22 years, or 10.3″/yr. This is the largest known proper motion of any star. Only a few hundred stars have proper motions greater than 1″/yr.

Proper motion is one part of the total space motion of a star—namely, the *transverse* component, perpendicular to the line of sight. ∞ (p. 68) A star's transverse velocity is easily calculated once its proper motion and its distance are known. Figure 17.4 is a sketch of the Alpha Centauri star system in relation to our solar system. Its proper motion has been measured, relative to more distant background stars, at about 3.5″/yr. At Alpha Centauri's distance of 1.3 pc, an angle of 3.5″ corresponds to a physical displacement of 0.00002 pc, about 700 million km. Alpha Centauri takes a year to travel this distance, so its transverse velocity is (700 million km)/(3.2×10^7 s/yr), or 22 km/s. We can determine the other component of motion—the *radial velocity*, along the line of sight—using the Doppler effect, as discussed in Chapter 3. ∞ (p. 68) Spectral lines from Alpha Centauri are slightly blueshifted, allowing astronomers to measure the star system's radial velocity (relative to the Sun) as 20 km/s toward us.

What is the true space motion of Alpha Centauri? Will this alien system collide with our own some time in the future? The answer is no—Alpha Centauri's

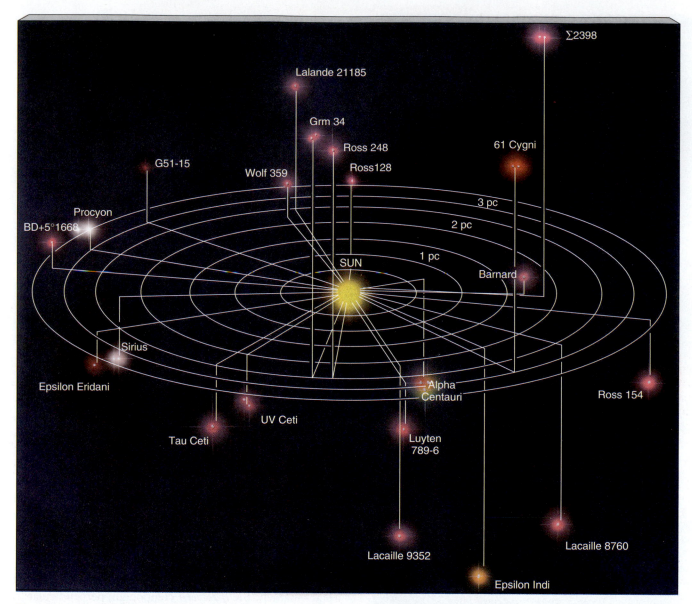

Figure 17.2 A plot of the 30 closest stars to the Sun, projected so as to reveal their three-dimensional relationships. Notice that many are members of multiple-star systems. All lie within 4 pc (about 13 light years) of Earth.

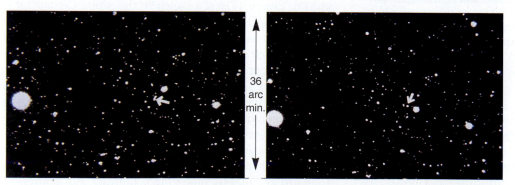

Figure 17.3 Comparison of two photographic plates taken 22 years apart shows evidence of real space motion for Barnard's star (denoted by an arrow).

transverse velocity will steer it well clear of the Sun. We can combine the transverse and radial velocities according to the Pythagorean theorem. The total velocity is $\sqrt{22^2 + 20^2}$, or about 30 km/s, in the direction shown by the horizontal red arrow in Figure 17.4. As that figure indicates, Alpha Centauri will get no closer to us than about 1 pc, and that won't happen until 280 centuries from now.

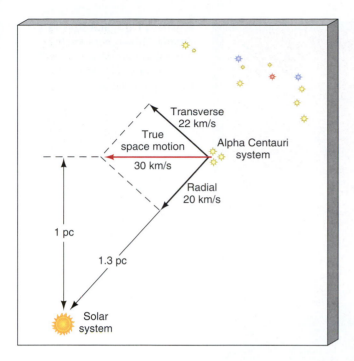

Figure 17.4 The motion of the Alpha Centauri star system drawn relative to our solar system. The transverse component of the velocity has been determined by observing the system's proper motion. The radial component is measured using the Doppler shift of lines in Alpha Centauri's spectrum. The true space velocity, indicated by the red arrow, results from the combination of the two.

17.3 Stellar Sizes

DIRECT AND INDIRECT MEASUREMENTS

3 Most stars are unresolvable points of light in the sky, even when viewed through the largest telescopes. Still, a few are big enough, bright enough, and close enough to Earth to allow their sizes to be *directly* measured. Until fairly recently, astronomers accomplished this task by using a special optical technique known as *speckle interferometry*. Using this technique, astronomers combine many short-exposure images of a star, each image too brief for Earth's turbulent atmosphere to smear it out into a seeing disk, to make a very-high-resolution map of the star's surface. In some cases, the results are detailed enough to allow surface features to be distinguished (see Figure 17.5). As adaptive optics techniques continue to improve (see Chapter 5), it is becoming possible to combine the individual images in real time, again allowing very-high-resolution stellar images to be made. ∞ (p. 108)

Using these techniques, optical astronomers have directly measured the sizes of a few dozen stars. By measuring a star's angular size and knowing its distance from the Earth, astronomers can determine its physical radius by simple geometry. In general, however, the sizes of most stars must be inferred by more indirect means.

Recall from Chapter 4 that the radiation emitted by a hot body is governed by *Stefan's law*, which states that the

Figure 17.5 The swollen star Betelgeuse (shown here in false color) is close enough for us to directly resolve its size, along with some surface features thought to be storms similar to those that occur on the Sun. Betelgeuse is such a huge star (300 times the size of the Sun) that its photosphere spans roughly the size of Mars's orbit. Most of the surface features discernible here are larger than the entire Sun.

energy emitted per unit area per unit time by a hot body —the body's *energy flux*—increases proportionally to the fourth power of the temperature. ∞ (p. 65) This law applies to a hot piece of metal, a glowing light bulb, or a star. The energy flux is measured per square meter. To extend this measure of energy emission to incorporate the entire surface of a star—in other words, to determine the *luminosity* of the star—we must multiply by the star's surface area. ∞ (p. 67) Because the energy flux is proportional to the fourth power of the stellar surface temperature and the area is proportional to the square of the stellar radius, it follows that

$$\text{luminosity} \propto \text{radius}^2 \times \text{temperature}^4,$$

where $\propto$ is the symbol representing proportionality. This **radius–luminosity–temperature relationship** is important because it demonstrates that knowledge of a star's luminosity and temperature can yield an estimate of its radius—an *indirect* determination of stellar size.

GIANTS AND DWARFS

Let's consider some examples to clarify the foregoing ideas. The star known as Mira (Omicron Ceti) has a surface temperature of about 3000 K and a luminosity of 1.6×10^{29} W. Thus, its surface temperature is half, and its luminosity about 400 times, the corresponding quantities for our Sun. The radius–luminosity–temperature relationship then implies that the star's radius is $\sqrt{400}/0.5^2 = 80$ times that of our Sun. If our Sun were this large, its photosphere would extend as far as the orbit of Mercury. A star as large as Mira is known as a *giant*. More precisely, **giants** are stars with radii between 10 and 100 times that of the Sun. Even larger stars, ranging up to 1000 solar radii in size, are known as **supergiants**. Since the color of any 3000 K object is red, Mira is a **red giant**.

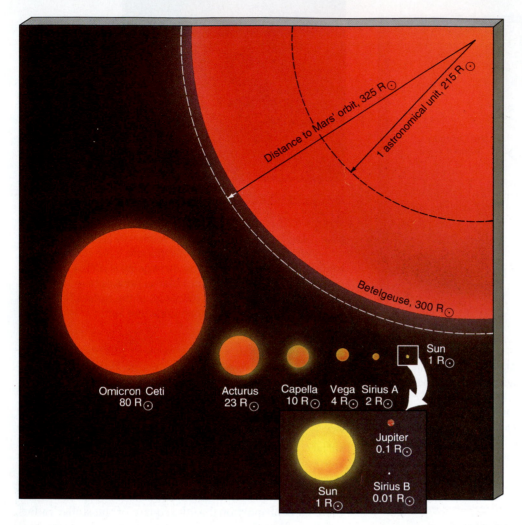

Figure 17.6 Star sizes vary greatly. Shown here are the estimated sizes of several well-known stars, including a few of those discussed in this chapter.

Now consider Sirius B—a faint companion to Sirius A, the brightest star in the night sky. Sirius B's surface temperature is roughly 24,000 K, four times that of the Sun. Its total luminosity is 10^{25} W, about 0.04 times the solar value. Substituting these quantities into our equation, we obtain a radius of $\sqrt{0.04}/4^2 = 0.01$ times the solar radius. Sirius B is much hotter but smaller and dimmer than our Sun. In fact, it is roughly the size of the Earth. Such a star is known as a **dwarf**. In astronomical parlance, the term *dwarf* refers to any star of radius comparable to or smaller than the Sun (including the Sun itself). Because any 24,000 K object glows white, Sirius B is an example of a **white dwarf**.

Stellar sizes determined by these methods range from 0.01 $R_\odot$ to 100 $R_\odot$. (Recall that $\odot$ is the symbol for the Sun, so $R_\odot$ is the radius of the Sun, 7.0×10^8 m.) ∞ (p. 13) We will encounter some exceptions to this statement as we proceed through this text; however, it is valid for the vast majority of stars. Figure 17.6 shows the estimated sizes of some well-known stars.

In order to determine a star's size, we need to know its luminosity and temperature. How do we determine these quantities? We now consider, in turn, each of these basic stellar properties. By studying them, we will learn a lot more about the nature of stars.

17.4 *Luminosity and Brightness*

4️⃣ Luminosity is the total amount of energy radiated into space each second from a star's surface. It is the *rate* of emission of energy—not just the energy of visible light, but the energy of any type of electromagnetic radiation—from radio waves to gamma rays. For the majority of stars, the spectrum peaks in or near the visible range, so much of the energy is in fact emitted in the form of visible light, although many stars emit a large fraction of their energy in the form of infrared or ultraviolet radiation. Luminosity is an intrinsic property of a star; it does not depend in any way on the location or motion of the observer. It is sometimes referred to as the star's **absolute brightness**.*

As explained in Chapter 3, when we look at a star, we do not "see" its luminosity. Instead, our eyes and astronomers' detectors actually measure *apparent brightness*—the amount of energy striking some light-sensitive surface or device per unit time. ∞ (p. 67) Apparent brightness is a measure not of a star's luminosity, but of the *energy flux* produced by the star, as seen from Earth.

Many astronomers prefer to define absolute brightness in a more "observational" manner—see the More Precisely feature on p. 366—but the definition is conceptually equivalent to the simpler one adopted here.

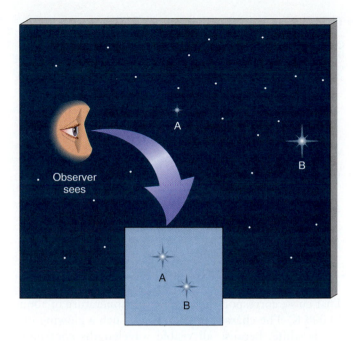

Figure 17.7 Two stars A and B of different luminosity can appear equally bright to an observer on Earth if the brighter star B is more distant than the fainter star A.

The apparent brightness of a light source decreases with distance according to an inverse-square law (see Section 3.5. ∞ (p. 67) But the source's luminosity also enters into the calculation—doubling the luminosity would double the energy crossing any spherical shell surrounding the source, and hence would double the apparent brightness. We can therefore extend our earlier statement of the inverse-square law for radiation and say that the apparent brightness of a star is *directly* proportional to the star's luminosity and *inversely* proportional to the square of its distance:

$$\text{apparent brightness (energy flux)} \propto \frac{\text{luminosity}}{(\text{distance})^2}.$$

Thus, two identical stars can have the same apparent brightness if (and only if) they lie at the same distance from Earth. However, as illustrated in Figure 17.7, two different stars can appear equally bright if the more luminous one lies farther away. A bright star (that is, a star with large apparent brightness) is a powerful emitter of radiation, is near Earth, or both. A dim star is a weak emitter, is far from Earth, or both.

Finding a star's luminosity is a twofold task. First, the astronomer must determine the star's apparent brightness by measuring the flux of energy detected through a telescope. Second, the distance must be measured—by parallax for nearby stars, and by other means (to be discussed later) for more distant stars. The luminosity can then be found using the inverse-square law.

17.5 *Temperature and Color*

Astronomers can obtain the surface temperature of a star from measurements of its brightness (radiation intensity) at different frequencies. Figure 17.8 shows three black-body curves describing the emission of radiation from three different stars. As we saw in Chapter 16, the surface ("effective") temperature of the Sun is found by measuring its radiation at many different frequencies, then matching the observations to the appropriate black-body curve. ∞ (p. 333) The theoretical curve that best fits the Sun's emission describes a 5800 K emitter.

We can use the same technique for any star, regardless of its distance from Earth. Actually, we need not measure every wavelength. Because the basic *shape* of the black-body curve is so well understood, we need only a few data points to determine the correct curve and thus determine the surface temperature (which is fortunate, as detailed spectra of faint stars are often difficult and time consuming to obtain). In practice, an astronomer observes a star's intensity at only a few selected wavelengths, using telescope filters that block out all radiation except that within specific wavelength ranges. For example, a B (blue) filter rejects all radiation except for a certain range of violet to blue light. Defined by international agreement to extend from 380 to 480 nm (3800–4800 Å), this range corresponds to wavelengths to which photographic film happens to be most sensitive. Similarly, a V (visual) filter passes only radiation within the

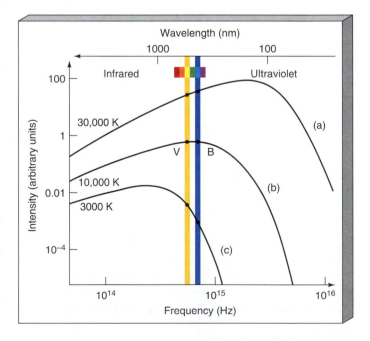

Figure 17.8 Black-body curves for three for different temperatures, along with the locations of the B (blue) and V (visual) filters. Star (a) is very hot—30,000 K—so its B intensity is considerably greater than its V intensity. Star (b) has B and V readings about the same and appears white. Its temperature is about 10,000 K. Star (c) is red. Its V intensity greatly exceeds the B value, and its temperature is 3000 K.

Figure 17.9 The constellation Orion, as it would appear through a small telescope or binoculars. The different colors of the member stars are easily distinguished. The bright red star at the upper left is Betelgeuse (α); the bright blue-white star at the lower right is Rigel (β). (Compare with Figure 1.6) The scale of the photograph is about 3° across.

TABLE 17-1 *Stellar Colors and Temperatures*

COLOR INDEX	SURFACE TEMPERATURE	COLOR	FAMILIAR EXAMPLES
(B intensity/V intensity)	(K)		
1.7	30,000	electric blue	
1.3	20,000	blue	Rigel
1.0	10,000	white	Vega, Sirius
0.8	7,000	yellow-white	Canopus
0.6	6,000	yellow	Sun, Alpha Centauri
0.4	4,000	orange	Arcturus, Aldebaran
0.2	3,000	red	Betelgeuse, Barnard's Star

490 to 590 nm range (green to yellow), corresponding to that part of the spectrum to which human eyes are particularly sensitive. Many other filters are also in routine use—a U (ultraviolet) filter covers the near ultraviolet, and infrared filters span longer-wavelength parts of the spectrum.

Temperature can be determined with as few as two filters. Figure 17.8 shows how the B and V filters admit different amounts of light for objects of different temperatures. These measurements, or those made in any two suitably separated wavelength ranges, are enough to specify the black-body curve and thus yield the surface temperature. This technique works for all objects that emit radiation through heat, be they glowing pokers, flashlight bulbs, or distant stars.

Let's consider a few examples. Suppose that we observe a particular star and find that the light passing through the B filter is almost twice as intense as that passing through the V filter, as shown in curve (a) of Figure 17.8. We can construct the entire black-body curve on the basis of only those two measurements—no other black-body curve could be drawn through *both* measured points. By determining the temperature associated with that particular curve, we can measure the temperature of the star to be 30,000 K. Such a hot star would have a distinct bluish tint when observed without any restricting filters because the radiation it emits is most intense in the blue-light range.

When the intensity measured through the B filter equals that through the V filter, the two measurements define the black-body curve labeled (b) in Figure 17.8.

That curve corresponds to a surface temperature of about 10,000 K. The characteristic color of such a glowing object is white, because all visible wavelengths contribute nearly equally. Finally, curve (c) of Figure 17.8 shows a star whose intensity measured through the B filter is only 1/5 that measured through the V filter. These two intensity values describe a black-body curve for a relatively cool, and distinctly red, 3000-K emitter.

The **color index** (often, just the "color") of a glowing object is the ratio of its B to V intensities. It is equivalent to the object's surface temperature to the extent that the object's radiation is well described by a black-body spectrum (which it often is), and it is easily measurable by telescopic means. Looking up at the night sky, you can tell at a glance which stars are hot and which are cool. In Figure 17.9, which shows the constellation Orion as it appears through a small telescope, the colors of the cool red star Betelgeuse and the hot blue star Rigel are evident. However, to obtain the actual temperatures (3000 K for Betelgeuse and 15,000 K for Rigel), at least two intensity measurements at different wavelengths are required. (Note that these colors are intrinsic properties of the stars and have *nothing* to do with Doppler redshifts or blueshifts.) Table 17-1 lists a more detailed description of the color index, the surface temperature derived from it, the dominant color perceived in the absence of filters, and some stellar examples. This type of nonspectral-line analysis, in which a star's intensity is measured through each of a set of standard filters, is known as **photometry** (literally meaning "light measurement").

17.6 *The Classification of Stars*

5 Astronomers can use color and temperature to classify stars reasonably well, but they often use a more detailed classification scheme. This scheme incorporates additional knowledge of stellar physics obtained through spectroscopy, the study of spectral-line radiation.

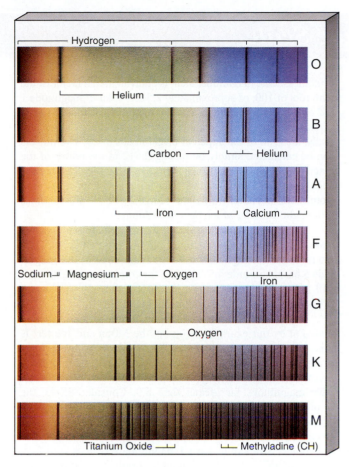

Figure 17.10 Comparison of spectra observed for seven different stars having a range of surface temperatures. The hottest stars, at the top, show lines of helium and multiply-ionized heavy elements. In the coolest stars, at the bottom, helium lines are not seen, but lines of neutral atoms and molecules are plentiful. At intermediate temperatures, hydrogen lines are strongest. The actual compositions of all seven stars are about the same.

DETAILED SPECTRA

Figure 17.10 compares the spectra of several different stars, arranged in order of decreasing surface temperature (as determined from measurements of their colors). All the spectra extend from 400 to 700 nm (4000–7000 Å), and each shows a series of dark absorption lines superimposed on a background of continuous color, like those of the Sun shown in Chapters 4 and 16. ∞ (p. 76, 337) However, the precise patterns of lines show many differences. Some stars display strong lines in the long-wavelength part of the spectrum (to the left in Figure17.10). Other stars have their strongest lines at short wavelengths (to the right). Still others show strong absorption lines spread across the whole visible spectrum. What do these differences tell us?

Although spectral lines of many elements are present with widely varying strengths, the differences among the spectra in Figure 17.10 are not due to differences in com-

position. Detailed spectral analysis indicates that the seven stars shown have very similar elemental abundances—all are more or less solar in makeup. Instead, the differences are due almost entirely to the stars' *temperatures*. The spectrum at the top of Figure 17.10 is exactly what we would expect from a star with solar composition and a surface temperature of about 30,000 K, the second from a 20,000 K star, and so on, down to the 3000 K star at the bottom of the figure. Let's now discuss these spectra in a little more detail.

The spectra of stars with surface temperatures exceeding 25,000 K usually show intense absorption lines of singly ionized helium and multiply ionized heavier elements, such as oxygen, nitrogen, and silicon. These lines are not seen in the spectra of cooler stars. Only very hot (blue) stars have surface temperatures high enough to excite and ionize these tightly bound atoms. In contrast, hydrogen produces only weak absorption lines in these stars. The reason is not a lack of hydrogen—it is still by far the most abundant element in stars. At these high temperatures, however, hydrogen is mostly ionized, so there are few intact hydrogen atoms to produce a characteristic spectrum. What's more, the few hydrogen atoms that have managed to retain their single electron are mostly in such highly excited states that their spectral lines are invisible (see the *More Precisely* feature on p. 88).

In cooler stars, hydrogen lines are more intense. In (whitish) stars with surface temperatures of about 10,000 K, hydrogen is responsible for the strongest absorption line. This temperature is just right for electrons to move frequently between hydrogen's second and third orbitals, producing the characteristic Hα (hydrogen alpha) line at 656.3 nm. ∞ (p. 339) Tightly bound atoms—elements such as helium, oxygen, and nitrogen, which need lots of energy for excitation or ionization—are rarely observed, whereas ions of more loosely bound atoms—such as calcium and titanium—are fairly common. The spectrum of a yellow star such as the Sun, with a surface temperature of about 6000 K, shows few strong lines from ionized elements—the Sun is too cool for that. The Hα line is no longer the most intense because, as with the hottest stars, the gas in the Sun does not have much electron traffic between the second and the third atomic orbitals. But unlike the hottest stars, in which most hydrogen atoms are highly excited or ionized, the Sun's hydrogen lines are weak because most of the electrons reside in the ground-level orbital.

Cool red stars, with surface temperatures of only a few thousand kelvins, show extremely weak hydrogen lines. The most intense lines in their spectra are due to neutral heavy atoms—and weakly excited ones at that. Astronomers observe no lines from ionized elements. In fact, the average energy of the photons leaving the surface of the coolest stars is less than that needed to destroy some molecules, and many absorption lines observed in red stars are produced by molecules rather than by elements.

More Precisely... *The Magnitude Scale*

Not all stars in the night sky have the same apparent brightness—some are bright, others faint. This fact is clear even to the casual observer, without any specialized astronomical measuring equipment. Astronomers working at any wavelength speak of luminosity, flux, and brightness, but optical astronomers also describe brightness in terms of an ancient notion called the *magnitude* of an object.

In the second century B.C., the Greek astronomer Hipparchus ranked the naked-eye stars into six groups. The brightest stars were placed in the first group: He categorized them as *first magnitude*. The next brightest stars were labeled *second magnitude*, and so on, down to the faintest stars visible to the naked eye, which were classified as *sixth magnitude*. The range 1 (brightest) through 6 (faintest) spanned all the stars known to the ancients. Notice that a *large* magnitude means a *faint* star!

When modern astronomers began using telescopes with sophisticated detectors to measure the light received from the stars, they quickly discovered two important facts about the magnitude scale. First, the 1–6 magnitude range defined by Hipparchus spans about a factor of 100 in energy received per unit area per unit time—energy flux, or apparent brightness—from stars: A first-magnitude star is approximately 100 times brighter than a star of the sixth magnitude. Second, the physiological characteristics of the human eye are such that each magnitude change of 1 corresponds to a factor of about 2.5 in brightness. A first-magnitude star is roughly 2.5 times brighter than a second-magnitude star, which is roughly 2.5 times brighter than a third-magnitude star, and so on. By combining factors of 2.5, we confirm that a first-magnitude star is indeed $(2.5)^5 \approx 100$ times brighter than a sixth-magnitude star.

Astronomers have found it convenient to retain the ancient magnitude scale. They have formalized it by *defining* 5 magnitudes to be equivalent to a factor of exactly 100 in apparent brightness. One magnitude, then, is equivalent to a factor of the fifth root of 100, or approximately 2.512. A star of magnitude 1 is 2.512 times brighter than a star of magnitude 2 and 6.31 (that is, 2.512^2) times brighter than a star of magnitude 3. More generally, the ratio of the apparent brightnesses of any two objects can be calculated simply by raising 2.512 to the power of the magnitude difference between them. For example, the Moon is about 10 magnitudes brighter than the planet Jupiter, as seen from Earth—in other words, we receive $2.512^{10} = 10,000$ times more light from the Moon than from Jupiter.

Astronomers have broadened the magnitude scale in two ways. First, it no longer includes only whole numbers. A star intermediate in brightness between magnitude 4 and magnitude 5 has magnitude 4.5. The rule for comparing brightnesses still holds. A star of magnitude 4.0 is about $\sqrt{2.512} \approx 1.6$ times brighter than a star measuring magnitude 4.5.

The second extension of the magnitude system (already hinted at by our example involving the Moon and Jupiter) is that magnitudes outside the range 1–6 are allowed. Very bright objects can have magnitudes between 1 and 0, or even less than 0 (negative magnitudes), and very faint objects can have magnitudes far greater than 6.0. Consider Sirius, the brightest star in our sky. Its energy flux, as measured by modern telescopes, actually places it brighter than magnitude 1. Because increases in flux are associated with decreases in magnitude, Sirius must have a magnitude less than 1. In fact, its magnitude is -1.5, 2.5 magnitudes brighter (10 times more flux) than the magnitude assigned to it by Hipparchus. Other very bright objects also have negative magnitudes—the Sun at -26.8, the full Moon at -12.5, Venus at -4.4 (at its brightest), and Jupiter at -2.7.

Modern telescopes can collect and focus light from stars much dimmer than were visible to the ancient Greeks. For example, the large Hale telescope on Mount Palomar can detect

---

The differences among stellar spectra, then, stem mainly from differences in temperature, not in composition. Most stars are made from the same elements, in roughly the same proportion. Observations of the spectrum of a star provide us with another means of measuring the star's surface temperature. Spectroscopy generally requires more telescope time than does photometry, because the available photons must be spread out over the entire visible spectrum instead of being divided into just a few wide bands, as discussed earlier. Nevertheless, so much more information is available from a complete spectrum than from a simple color index that astronomers prefer to work with spectra whenever possible. We now discuss the method actually used by astronomers to classify the stellar spectra they observe.

SPECTRAL CLASSIFICATION

Stellar spectra like those shown in Figure 17.10 had been obtained for numerous stars even before the start of the twentieth century. Observatories around the world amassed spectra for several hundred thousand stars, in both hemispheres of the sky. From 1880 to 1920, researchers correctly identified some of the observed spectral lines on the basis of comparisons between stellar lines and those obtained in the laboratory. Those workers, though, had no firm understanding of how the lines were produced. Modern atomic theory had not yet been developed, so the correct interpretation of the line strengths, as described in this chapter and in Chapter 4, was impossible at the time. Lacking full understanding of the atom, early workers classified stars primarily according to their hydrogen-line intensities. They adopted an A, B, C, D,… scheme in which A stars, with the strongest hydrogen lines, were thought to have more hydrogen than did B stars, and so on. The classification extended as far as the letter P.

In the 1920s, scientists began to understand the intricacies of atomic structure and the causes of spectral lines. Astronomers quickly realized that stars could be more

objects as faint as magnitude 27—equivalent to seeing a candle at a distance of about 50,000 km—from which we receive only about 1 photon per square centimeter per *hour*. The orbiting *Hubble Space Telescope* and the huge Keck telescope atop Mauna Kea can study even fainter objects—nearly as dim as magnitude 30, about as faint as a firefly seen from a distance equal to the diameter of the Earth. The accompanying figure illustrates the magnitudes of some astronomical objects.

So far, the magnitude we have been discussing is equivalent to radiation flux, or apparent brightness. It is more commonly known as *apparent magnitude*. Apparent magnitude denotes a star's brightness when viewed at its actual distance from Earth. When you compare intrinsic, or absolute, properties of stars, imagine looking at all stars from a "standard" distance of 10 pc. There is no particular reason to use 10 pc—it is simply convenient. A star's *absolute magnitude* is the apparent magnitude of a star placed at a distance of 10 pc. Because the distance is fixed in this definition, the absolute magnitude of a star depends *only* on its luminosity. Absolute magnitude is therefore a measure of the intrinsic brightness, or luminosity, of the star itself.

A star with a distance of less than 10 pc would be diminished in apparent brightness, and so *increased* in apparent magnitude, if it were moved to a point 10 pc away from us. Stars less than 10 pc from Earth therefore have apparent magnitudes that are less than their absolute magnitudes. For stars more distant than 10 pc, the reverse is true. An extreme example is our Sun. Because of its proximity to Earth, it appears very bright and thus has a large negative apparent magnitude. However, the Sun's absolute magnitude is 4.8. If the Sun were moved to a distance of 10 pc, it would be only slightly brighter than the faintest stars visible in the night sky.

Despite its intuitive underpinnings, the magnitude system is a quaint and, to many students, highly confusing heirloom of

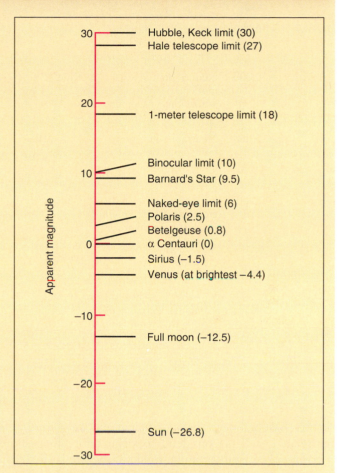

traditional astronomy. Nevertheless, in its modern form it is deeply embedded in optical astronomy and is unlikely to be replaced in the foreseeable future.

---

meaningfully classified according to their surface temperature. Instead of adopting an entirely new scheme, however, they chose to shuffle the existing alphabetical categories—those based on the strength of hydrogen lines—into a new sequence based on temperature. In the modern scheme, the hottest stars are designated O, because they have very weak absorption lines of hydrogen and they were classified toward the end of the original scheme. In order of decreasing temperature, the original letters now run O, B, A, F, G, K, M. (The other letter classes have been dropped.) These stellar designations are called **spectral classes**. Use the mnemonic "**O**h, **B**e **A** **F**ine **G**uy (**G**irl), **K**iss **M**e" to remember them in the correct order.

Astronomers further subdivide each lettered spectral classification into 10 subdivisions, denoted by the numbers 0–9. By convention, the lower the number, the hotter the star. Thus, for example, our Sun is classified as a G2 star (a little cooler than G1 and a little hotter than G3), Vega is a type A0, Barnard's Star is M5, Betelgeuse is M2, and so on.

Table 17-2 lists the main properties of each stellar spectral class for the stars presented in Table 17-1.

We should not underestimate the importance of the early work in classifying stellar spectra. Even though the original classification was based on erroneous assumptions, the painstaking accumulation of large quantities of accurate data paved the way for rapid improvements in understanding once a theory came along that explained the observations.

17.7 *The Hertzsprung–Russell Diagram*

6 We have now studied the two most important, most basic properties of any star: its luminosity (or absolute brightness or absolute magnitude) and its surface temperature (or color or spectral class). Astronomers use these two quantities to classify stars in much the same way that height and weight serve to classify the bulk properties of human

TABLE 17-2 *Stellar Spectral Classes*

SPECTRAL CLASS	SURFACE TEMPERATURE (K)	PROMINENT ABSORPTION LINES	FAMILIAR EXAMPLES
O	30,000	Ionized helium strong; multiply ionized heavy elements; hydrogen faint	
B	20,000	Neutral helium moderate; singly ionized heavy elements; hydrogen moderate	Rigel (B8)
A	10,000	Neutral helium very faint; singly ionized heavy elements; hydrogen strong	Vega (A0), Sirius (A1)
F	7,000	Singly ionized heavy elements; neutral metals; hydrogen moderate	Canopus (F0)
G	6,000	Singly ionized heavy elements; neutral metals; hydrogen relatively faint	Sun (G2), Alpha Centauri (G2)
K	4,000	Singly ionized heavy elements; neutral metals strong; hydrogen faint	Arcturus (K2), Aldebaran (K5)
M	3,000	Neutral atoms strong; molecules moderate; hydrogen very faint	Betelgeuse (M2), Barnard's Star (M5)

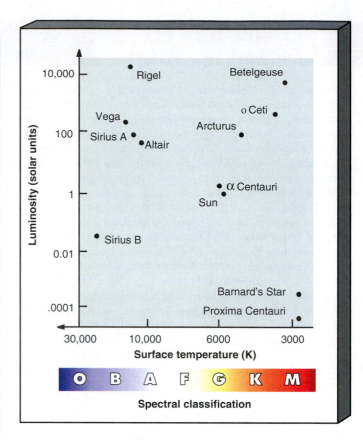

Figure 17.11 A plot of luminosity against surface temperature (or spectral classification), known as an H–R diagram, is a useful way to compare stars. Plotted here are the data for some stars mentioned earlier in the text. The Sun, of course, has a luminosity of 1 solar unit. Its temperature, read off the bottom scale, is 5800 K—a G type star. Similarly, the B-type star Rigel, at top left, has a temperature of about 15,000 K and a luminosity more than 10,000 times that of the Sun. The M-type star Proxima Centauri, at bottom right, has a temperature of 3000 K and a luminosity less than 1/10,000 that of the Sun.

beings. We know that people's height and weight are well correlated—tall persons tend to weigh more than short persons. We might naturally wonder if the two basic stellar properties are also related in some way. In the second decade of the twentieth century, Danish astronomer Ejnar Hertzsprung and the U.S. astronomer Henry Norris Russell independently discovered just such a relationship.

Figure 17.11 shows the way that Hertzsprung and Russell originally plotted the temperatures and luminosities of stars. In honor of these two scientists, such a plot is now known as a Hertzsprung–Russell diagram, or **H–R diagram** for short. The vertical scale, expressed in units of solar luminosity ($L_\odot = 3.9 \times 10^{26}$ W), extends over a large range, from 10^{-4} to 10^4; the Sun appears right in the middle of the luminosity range, at a luminosity of 1. Surface temperature is plotted on the horizontal axis, although in the unconventional sense of temperature increasing to the *left* (so that the spectral sequence O, B, A, … reads left to right). To change the horizontal scale so that temperature increases conventionally to the right would play havoc with historical precedent.

As we have just seen, astronomers often use a star's *color* to measure its temperature. Indeed, the spectral classes plotted along the horizontal axis of spectral class in Figure 17.11 are equivalent to the B/V color index. Also, because astronomers commonly express a star's luminosity as an absolute magnitude (see the *More Precisely* feature on p. 366),

stellar *magnitude* instead of stellar luminosity could be plotted vertically. For these reasons, many astronomers refer to diagrams such as Figure 17.11 as **color–magnitude diagrams**. In this book, however, we will cast our discussion strictly in terms of temperature and luminosity measurements.

CONSTRUCTING AN H–R DIAGRAM

The first step in making an H–R diagram for any given collection of stars is to determine the surface temperature (or spectral class) of each. We can do this in one of two ways: (1) We can measure the star's intensity in the B and V bands and fit a black-body curve to those measurements, or (2) we can observe the star's spectrum and so determine its spectral type. As we have seen, both methods provide a means of determining stellar temperature. Neither requires any knowledge of the star's distance.

Interlude 17–1   *Stacks and Stacks of Photographs*

Several of the world's observatories have stacks not only of books, filed by author and title, but also of celestial photographs, filed by sky location and date. The Harvard College Observatory houses the largest collection of stellar photographs—nearly a million glass plates that are cataloged and stored in a building suspended on springs to guard against earthquake damage.

Prior to the invention of photography in the midnineteenth century, astronomical observations were made by visual impressions and hand-drawn sketches. Viewed by most astronomers at its inception as an idle diversion from the real study of the skies, photography soon became a major tool for recording and quantifying observations. In effect, photography transformed observational astronomy from an art into a science. Foresight on the part of E. C. Pickering, director of the Harvard Observatory in the 1880s, who assigned funds and staff to this new pursuit, resulted in the large Harvard collection. That collection includes photographs regularly taken with several telescopes in both hemispheres over the past century. Pickering is shown in the accompanying photograph making his way to the summit of El Misti in Arequipa, Peru, the site of one of the most advanced observatories south of the equator around the turn of the century.

A small fraction of all stars in the Galaxy are known to vary in brightness over a period of days or weeks. Early photographic surveys concentrated mostly on the luminosity fluctuations of these *variable stars*. The cataloged plates are the best way to monitor long-term changes in stellar brightness. Toward the end of the nineteenth century, researchers realized that the spectroscopy of stars contains even more information than luminosity alone. Detailed spectral observations were made for tens of thousands of stars by 1900 and for millions of stars since then.

The Harvard Photographic Collection is a rich source of astronomical lore, only part of which has yet been tapped. Astronomers from around the world apply to analyze the plates, much as they routinely request observing time on telescopes or orbiting satellites. The cataloged plates allow astronomers to study the brightness of almost any visible cosmic object and the spectra of the brighter ones. Even in these days of charge-coupled devices and orbiting observatories, the long time span of the Harvard archival data makes the plate collection invaluable.

The second step is to determine each star's luminosity. This step is easy if the star's distance is known, because we can then easily convert the measured apparent brightness (energy flux) into luminosity using the inverse-square law. If the distance is unknown, however, the star's luminosity cannot be measured, so it cannot be used in the construction of the diagram. Thus, the first H–R diagrams we can construct can show only nearby stars, whose distances are measured by parallax.

THE MAIN SEQUENCE

As Hertzsprung and Russell plotted more and more stellar temperatures and luminosities, they found that stars are *not* uniformly scattered across the H–R diagram. Instead, most are confined to a fairly well-defined band stretching diagonally across the diagram, from the top left (high-temperature, high-luminosity) to the bottom right (low-temperature, low-luminosity). In other words, cool stars tend to be faint, and hot stars tend to be bright. This band of stars spanning the H–R diagram is known as the **main sequence**.

Figure 17.12 shows a more systematic study of stellar properties, covering the 80 or so stars that lie within 5 pc of the Sun. As more points are included in the diagram, the main sequence "fills up," and the pattern becomes more evident. The vast majority of stars in the immediate vicinity of the Sun lie on the main sequence.

The surface temperatures of main sequence stars range from about 3000 K (spectral type M) to over 30,000 K (spectral type O). This relatively small temperature range—only a factor of 10—is determined mainly by the rates at which nuclear reactions occur in stellar cores. In contrast, the observed range in luminosities is very large, covering some eight orders of magnitude (that is, a factor of 100 million) and ranging from 10^{-4} to 10^4 times the luminosity of the Sun.

Astronomers can use the radius–luminosity–temperature relationship ($L \propto R^2 T^4$) given earlier in this chapter to estimate the radii (R) of main-sequence stars from their temperatures (T) and luminosities (L). They find that in order to account for the observed range in luminosities, stellar radii must also vary along the main sequence. The faint, red M-type stars in the bottom right of the H–R diagram are only about 1/10 the size of the Sun, whereas the bright, blue O-type stars in the upper left are about 10 times larger. The oblique dashed lines in Figure 17.12 represent stars with the same radii; $L \propto T^4$ along these lines. By including such lines, we can indicate stellar temperatures, luminosities, and radii on a single H–R diagram.

We see a very clear trend as we traverse the main sequence from top to bottom. At one end, the stars are large, hot, and bright. Because of their size and color, they are referred to as **blue giants**. The very largest are called **blue supergiants**. At the other end, stars are small, cool, and faint. They are known as **red dwarfs**. Our Sun lies right in the middle.

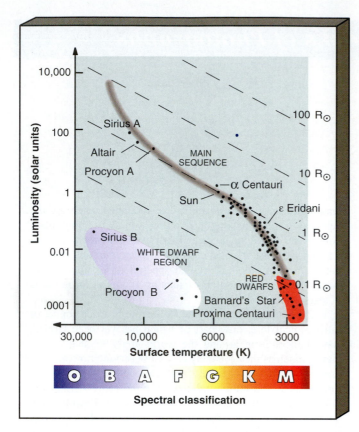

Figure 17.12 Most stars have properties within the shaded region known as the main sequence. The points plotted here are for stars lying within about 5 pc of the Sun. The diagonal lines correspond to constant stellar radius, so that stellar size can be represented on the same diagram as luminosity and temperature. (Recall that ⊙ stands for the Sun.)

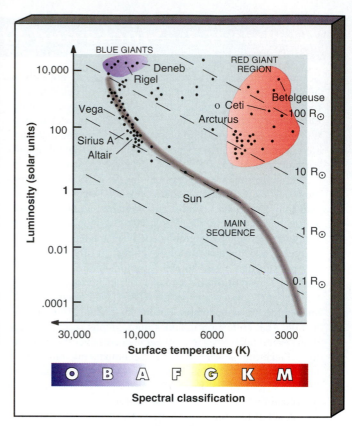

Figure 17.13 An H–R diagram for the 100 brightest stars in the sky. Such a plot is biased in favor of the most luminous stars—which appear toward the upper left—because we can see them more easily than we can the faintest stars. (Compare this with Figure 17.12, which shows only the closest stars.)

Figure 17.13 shows an H–R diagram for a different group of stars—the 100 stars of known distance having the greatest apparent brightness, as seen from Earth. Notice the much larger number of very luminous stars at the upper end of the main sequence than at the lower end in Figure 17.13. The reason for this excess of blue giants is simple—we can see very luminous stars a long way off. The stars shown in Figure 17.13 are scattered through a much greater volume of space than those in Figure 17.12, but the sample is heavily biased toward the brightest objects. In fact, of the 20 brightest stars in the sky, only 6 lie within 10 pc of us; the rest are visible, despite their great distances, because of their high luminosities.

If very luminous blue giants are overrepresented in Figure 17.13, low-luminosity red dwarfs are surely underrepresented. In fact, no dwarfs appear on that diagram. This absence is not surprising, because such low-luminosity stars are very difficult to observe from Earth. They are just too faint, and they radiate much of their energy in the invisible, infrared part of the electromagnetic spectrum. However, in the 1970s astronomers began to realize that they had greatly underestimated the number of red dwarfs in the Galaxy. As hinted at by the H–R diagram in

Figure 17.12, which shows an unbiased sample of stars in the solar neighborhood, red dwarfs are actually the most common type of star in the sky. They probably account for upward of 80 percent of all stars in the universe.

WHITE DWARFS AND RED GIANTS

Most stars lie on the main sequence. However, some of the points plotted in Figures 17.11–17.13 clearly do not. One such point in Figure 17.11 represents Sirius B, a white dwarf whose surface temperature (24,000 K) is about four times that of the Sun and whose luminosity is about 0.04 $L_\odot$. A few more such faint A-type stars can be seen in Figure 17.12 in the bottom left-hand corner of the H–R diagram. This region, known as the **white-dwarf region**, is marked on Figure 17.12. It represents a class of stars on the H–R diagram quite distinct from the stars of the main sequence.

Also shown in Figure 17.11 is Mira (Omicron Ceti), whose surface temperature (3000 K), is about half that of the Sun and whose luminosity is some 400 times greater than the Sun's. Another point represents Betelgeuse (Alpha Orionis), the ninth brightest star in the sky, a little cooler than Mira but more than 30 times brighter. The upper right-hand corner of the H–R diagram, where these stars plot (marked on Figure 17.13), is called the **red-giant region**. No red giants are found within 5 pc of the Sun (Figure 17.12), but many of the brightest stars seen in the sky are in fact red giants (Figure 17.13). Red giants are relatively rare, but they are so bright that they are visible to very great distances. They form a third distinct class of stars in the H–R diagram, very different in their properties from both main-sequence stars and white dwarfs.

Although dwarfs and giants give some feeling for the extreme properties of stars, most stars have properties much more like our Sun and lie on the main sequence in the H–R diagram. About 90 percent of all stars in our solar neighborhood, and probably a similar percentage elsewhere in the universe, are main-sequence stars. About 9 percent of stars are white dwarfs, and 1 percent are red giants.

17.8 *Extending the Cosmic Distance Scale*

SPECTROSCOPIC PARALLAX

In Chapter 2 we introduced the first "rung" on a ladder of distance-measurement techniques that will ultimately carry us to the edge of the observable universe. That rung is radar-ranging on the inner planets. ∞ (p. 43) It establishes the scale of the solar system to great accuracy and, in doing so, defines the astronomical unit. Earlier in this chapter, we discussed a second rung in the cosmic distance ladder—stellar parallax—which is based on the first. Now, having used the first two rungs to determine the distances and other physical properties of many nearby stars, we can use that knowledge to construct a third rung in the ladder. That new rung, illustrated schematically in Figure 17.14, is known as **spectroscopic parallax**.* It expands our cosmic field of view still deeper into space.

We have already discussed the connections between absolute brightness (luminosity), apparent brightness (energy flux), and distance. Knowledge of a star's apparent brightness and distance allows us determine its luminosity using the inverse-square law. But we can also turn the problem around. If we somehow knew a star's luminosity and then measured its apparent brightness, the inverse-square law would give us its distance from the Earth.

Consider another analogy. Most of us have a rough idea of the approximate brightness and size of a red traffic signal. Suppose we are driving down an unfamiliar street and see a red traffic light in the distance. Our knowledge of the intrinsic luminosity of the light often enables us immediately to make a mental estimate of its distance. A normal traffic light that is relatively dim must be quite distant (assuming it's not just dirty); a bright one must be relatively close. *A measurement of the apparent brightness of a light source, com-*

bined with some knowledge of its intrinsic properties, can yield an estimate of the source's distance. For a star, the trick is to find an independent measure of the luminosity without knowing the distance. The H–R diagram can provide just that.

The main sequence represents a fairly close correlation between temperature and luminosity for most stars, with the exception of a few giants and dwarfs. As such, the main sequence can be used to specify the *average* properties of most stars. Let's imagine for a moment that the main sequence really is a *line*, rather than a somewhat fuzzy band, in the H–R diagram and that *all* stars lie on the main sequence. From a star's spectrum, we can determine its surface temperature or spectral type. If the star lies on the main sequence, then there is only one possible luminosity corresponding to that temperature. We can read the star's luminosity directly off a graph, such as Figure 17.12, and then determine its distance by measuring the energy flux at Earth and using the inverse-square law. The existence of the main sequence allows us to make a connection between an easily measured quantity (temperature) and the star's luminosity, which would otherwise be unknown.

Spectroscopic parallax can be used for distances out to several thousand parsecs. Beyond that, spectra and colors of individual stars are difficult to obtain. The "standard" main sequence is obtained from H–R diagrams of stars whose distances can be measured by (geometric) parallax, so the method of spectroscopic parallax is calibrated using nearby stars. In using this method, we are applying the "principle of mediocrity" that we discussed in Chapter 1. ∞ (p. 38) Specifically, we are assuming (without proof) that distant stars are basically similar to nearby stars—in particular, that they *fall on the same main sequence as nearby stars*. Only by making this assumption can we expand the boundaries of our distance-measurement techniques.

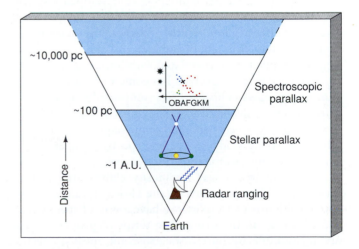

Figure 17.14 Knowledge of a star's luminosity and apparent brightness can yield an estimate of its distance. Astronomers use this third "rung" in our distance ladder, called spectroscopic parallax, to measure distances as far out as individual stars can be clearly discerned—several thousand parsecs.

This unfortunate name is rather misleading, as the method has nothing in common with trigonometric parallax other than its use as a means of determining stellar distances.

Of course, the main sequence is not really a line in the H–R diagram: It has some thickness. For example, the luminosity of a main-sequence G2-type star (such as the Sun) can range from about 0.5 to 1.5 $L_\odot$. The main reason for this range is the variation in stellar composition and age from place to place in the Galaxy. As a result, there is an uncertainty in the luminosity obtained by this method and so some uncertainty in the distance. Distances obtained by spectroscopic parallax are probably accurate to no better than 25 percent. Although this may not seem very accurate—a cross-country traveler in the United States would hardly be impressed to be told that the best estimate of the distance between Los Angeles and New York is somewhere between 3000 and 5000 km—it illustrates the point that in astronomy even something as simple as the distance to another star can be very difficult to measure. Still, an estimate with an uncertainty of ±25 percent is far better than no estimate at all.

LUMINOSITY CLASS

If, by chance, the star in question happens to be a giant or a dwarf, the determination of distance by spectroscopic parallax will be incorrect. But roughly 90 percent of all stars are on the main sequence, so we could argue that the assumption that a star is a main-sequence star will be valid 9 out of 10 times. In fact, astronomers can do much better than that. Recall from Chapter 4 that the *width* of a spectral line can provide information on the *density* of the gas where the line formed. ∞ (p. 91) The atmosphere of a red giant is much less dense than that of a main-sequence star, and this in turn is much less dense than the atmosphere of a white dwarf. By studying the width of a star's spectral lines, astronomers can usually tell with a high degree of confidence whether or not a star is on the main sequence.

Over the years, astronomers have developed a system for classifying stars according to the width of their spectral lines. Because the line width is particularly sensitive to density in the stellar photosphere, and the atmospheric density in turn is well correlated with luminosity, the class in which a star is categorized has come to be known as its **luminosity class**. This classification provides a means for astronomers to distinguish supergiants from giants, giants from main-sequence stars, and main-sequence stars from white dwarfs by studying a single spectral property—the line broadening—of the radiation received.

The standard stellar luminosity classes are given in Table 17.3. Their locations on the H–R diagram are indicated in Figure 17.15. Now we have a way of specifying a star's location in the diagram in terms of properties that are measurable by purely spectroscopic means; spectral type and luminosity class locate a star just as surely as do temperature and luminosity. The full specification of a star's spectral properties includes its luminosity class. For example, the Sun, on the main sequence, is of class G2V, Vega is A0V, the red dwarf Barnard's Star is M5V, the red supergiant Betelgeuse is M2Ia, and so on.

Consider, for example, a K7-type star (see Table 17.4).

TABLE 17–3   *Stellar Luminosity Classes*

CLASS	DESCRIPTION
Ia	Bright supergiants
Ib	Supergiants
II	Bright giants
III	Giants
IV	Subgiants
V	Main-sequence stars/dwarfs

If the star lies on the main sequence (that is, it is a K7V star), its luminosity is about 0.1 $L_\odot$. If its spectral lines are observed to be narrower than lines normally found for main-sequence stars, the star may be recognized as a giant, with a luminosity of 10 $L_\odot$. If the lines are very narrow, the star might instead be classified as a supergiant, brighter by a further factor of 100, at 1000 $L_\odot$. In this way, the observed width of the star's spectral lines translates directly into a measure of the physical state of the star. Knowledge of luminosity classes allows us to use spectroscopic parallax with some confidence that we are not accidentally counting a red giant or a white dwarf as a main-sequence star and making a huge error in our distance estimate as a result.

17.9 *Stellar Mass*

7   What ultimately determines a star's position on the main sequence? The answer is its *mass* and its *composition*. Mass and composition are fundamental properties of any star. They are set once and for all at the time of a star's birth. Together, they uniquely determine both the star's internal structure and its external appearance and even (as we will see in Chapter 20) its future evolution. The ability to measure these two key stellar properties is of the utmost importance if we are to understand how stars work. We have already seen how spectroscopy is used to determine a star's composition. Now we turn to the problem of finding a star's mass.

As with all other objects, we must measure a star's mass by observing its gravitational influence on some body that is near it—another star, perhaps, or a planet. If the distance between the two bodies is known, Newton's laws may be used to calculate the masses involved. In this way, we can derive the mass of Earth by watching the Moon or artificial

TABLE 17–4   *Variation in Stellar Properties Within a Spectral Class*

SURFACE TEMPERATURE (K)	LUMINOSITY ($L_\odot$)	RADIUS ($R_\odot$)	OBJECT
4000	0.1	0.7	K main-sequence star
4000	10	7	K giant star
4000	1000	70	K supergiant star

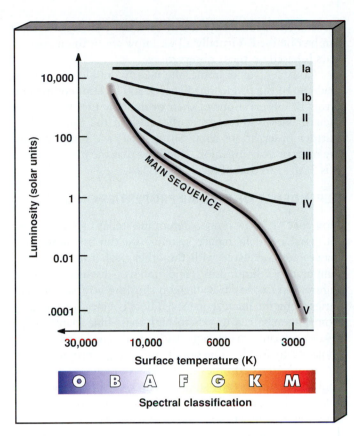

Figure 17.15 Stellar luminosity classes in the H–R diagram. Note that a star's location could be specified by its spectral type and luminosity class instead of by its temperature and luminosity.

satellites orbit it, and we can derive the mass of the Sun by studying the orbital motions of the planets. Unfortunately, astronomers have not yet been able to detect planets around any main-sequence star other than the Sun, much less measure their orbits. Even so, there is a way to estimate the masses of many stars.

BINARY STARS

Most stars are members of multiple-star systems—groups of two or more stars in orbit around one another. The majority of stars are found in **binary-star systems**, which consist of two stars in orbit about their common center of mass, held together by their mutual gravitational attraction. Other stars are members of triple, quadruple, or even more complex systems. Most complex of all are the star clusters discussed below. The Sun is not part of a multiple-star system; if it has anything at all uncommon about it, it may be its lack of stellar companions.

Astronomers classify binary-star systems (or *binaries*) according to their appearance from Earth and the ease with which they can be observed. **Visual binaries** have widely separated members that are bright enough to be observed and monitored separately, as shown in Figure 17.16(a). Others, known as **spectroscopic binaries**, are too distant to be resolved into separate stars, but they can be indirectly perceived by monitoring the back-and-forth Doppler shifts of their

spectral lines as the stars orbit one another. Recall that motion toward an observer blueshifts the lines, and motion away from the observer redshifts them. ∞ (p. 68) In a *double-line* spectroscopic binary, two distinct sets of spectral lines—one for each component star—shift back and forth as the stars move. Because we see particular lines alternately approaching and receding, we know that the objects emitting the lines are in orbit. In the more common *single-line* systems, such as that shown in Figure 17.16(b), one star is too faint for its spectrum to be distinguished, so only one set of lines is observed to shift back and forth. This shifting means that the detected star must be in orbit around another star, even though the companion cannot be directly observed.

In the much less common **eclipsing binaries**, the orbital plane of the pair of stars is almost edge-on to our line of sight. In this situation, depicted in Figure 17.16(c), we observe a periodic decrease of starlight as one component passes in front of the other. By studying the variation of the light from the binary system—the binary's **light curve**—astronomers can derive detailed information not only about the stars' orbits and masses but also about their radii.

These categories of binary-star systems are not mutually exclusive. For example, a single-line spectroscopic binary may also happen to be an eclipsing system. In that case astronomers can use the eclipses to gain extra information about the fainter member of the pair.

Occasionally, two unrelated stars just happen to lie close together in the sky, even though they are actually widely separated. These *optical doubles* are just chance superpositions and carry no useful information about stellar properties.

A great deal of data can be obtained from repeated observations of a binary system. By observing the actual orbit of the stars, or the back-and-forth motion of the spectral lines, or the dips in the light curve—whatever information is available—we can measure the binary's orbital period. Observed periods span a broad range—from hours to centuries. Doppler-shift measurements give us information on the orbital velocities of the member stars. In addition, if the distance to a visual binary is known, the size (semi-major axis) of its orbit can be determined directly, by simple geometry.

Knowledge of the binary period and orbit size is all we need to determine the combined mass of the component stars, using the modified form of Kepler's third law discussed in Chapter 2. ∞ (p. 48) Additional observations are needed to determine the individual masses of the components. For example, in any system of orbiting objects, each object orbits the common center of mass. Measuring the distance from each star to the center of mass of a visual binary yields the ratio of the stellar masses. Knowing both the sum of the masses and the ratio of the masses, we can then find the mass of each star. The individual component masses of a single-line spectroscopic binary system generally cannot be determined (unless the binary happens also to be an eclipsing system, in which case the extra information allows the mass ratio to be inferred). If a binary is too distant or if only one component is visible, we cannot determine the component masses, only their sum.

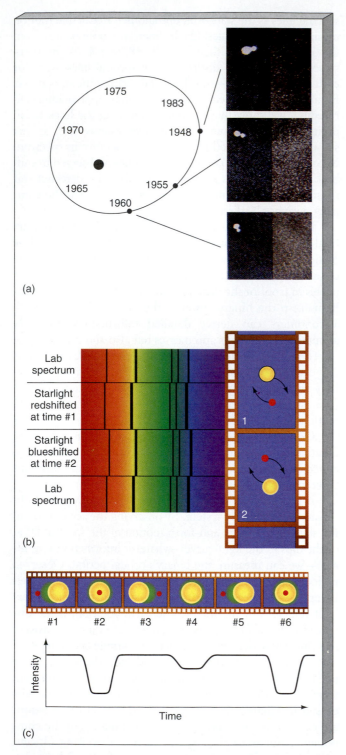

Figure 17.16   (a) The periods and separations of binary stars can be observed directly if each star is clearly seen. This binary star is known as *Kruger 60*. (b) Binary properties can also be found indirectly by measuring the periodic Doppler shift of one star relative to the other as they move in their orbits. The diagram shows a so-called single-line system, in which only one spectrum (from the brighter component) is visible. The observer is situated to the left of the diagram. Reference lab spectra are shown at top and bottom. (c) If the two stars happen to eclipse one another, additional information on their radii and masses can be obtained by observing the periodic decrease in starlight as one passes in front of the other.

Nevertheless, for many nearby systems, individual masses can be obtained. Virtually all we know about the masses of stars is based on those observations.

Consider, for example, the nearby double-star system made up of the bright star Sirius A and its faint companion Sirius B. Observations of their orbit show that the sum of their masses is three times the mass of the Sun—3 $M_\odot$. Further observations show Sirius A to have roughly twice the mass of its companion. It follows that the mass of Sirius A is 2 $M_\odot$ and that of Sirius B is 1 $M_\odot$.

DEPENDENCE OF STELLAR PROPERTIES ON MASS

Now that we know how to determine stellar masses, at least for stars found in binary systems, we can ask how these masses are correlated with the other properties of stars—temperature, luminosity, and radius—discussed earlier. Figure 17.17 is an H–R diagram showing how stellar mass varies along the main sequence. There is a clear progression from low-mass red dwarfs to high-mass blue giants. With few exceptions, main-sequence stars range in mass from about 0.1 to 20 $M_\odot$. The hot O- and B-type stars are generally about 10 to 20 times more massive than our Sun. The coolest K- and M-type stars contain only a few tenths of a solar mass. Because all other stellar properties are set once a star's mass is known, we can say that *the mass of a star at the time of formation determines its location on the main sequence.*

Figure 17.18 illustrates in a little more detail how a main-sequence star's radius and luminosity depend on its mass. The two curves are based on observations of binary-star systems. The curves show the **mass–radius** and **mass–luminosity** relations for main-sequence stars. Along the main sequence, both radius and luminosity increase with mass. As a rough rule of thumb, radius rises in direct proportion to mass, whereas luminosity increases much faster—more like the *cube* of the mass. For example, a 2 $M_\odot$ main-sequence star has a radius twice that of the Sun and a luminosity of 8 $L_\odot$ (2^3 solar luminosities); a 0.2 $M_\odot$ main-sequence star has a radius of about 0.2 $R_\odot$ and a luminosity of 0.008 $L_\odot$ (0.2^3 solar luminosities); and so on.

STELLAR LIFETIMES

The rapid rate of nuclear burning deep inside a star releases vast amounts of energy per unit time. How long can the fire continue to burn? We can estimate a star's lifetime simply by dividing the amount of fuel available (the mass of the star) by the rate at which the fuel is being consumed (the star's luminosity):

$$\text{stellar lifetime} \propto \frac{\text{stellar mass}}{\text{stellar luminosity}}.$$

Because the mass–luminosity relation tells us that a star's luminosity is roughly proportional to the cube of its mass, we can rewrite this expression to obtain, approximately,

$$\text{stellar lifetime} \propto \frac{1}{(\text{stellar mass})^2}.$$

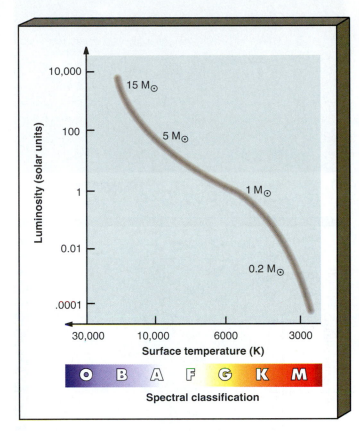

Figure 17.17   Mass, more than any other stellar property, determines a star's position on the main sequence. Stars that form with low mass will be cool and faint; they lie at the bottom of the main sequence. Very massive stars are hot and bright; they lie at the top of the main sequence.

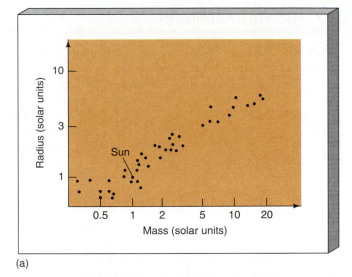

(a)

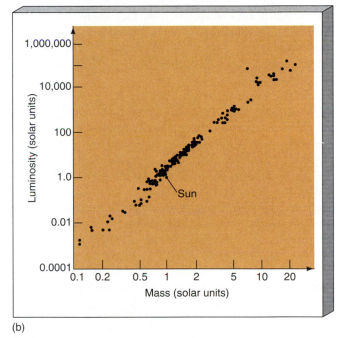

(b)

Figure 17.18   (a) Dependence of stellar radius on mass for main-sequence stars. The radius increases roughly in proportion to the mass over much of the range. (b) Dependence of luminosity on mass. The luminosity increases much faster than the mass.

For example, O and B stars have masses 10 to 20 times that of the Sun and luminosities thousands of times higher than the solar luminosity. Accordingly, these massive stars can survive only for short times. Their nuclear reactions proceed so rapidly that their fuel is quickly depleted, despite their large mass. We can be sure that all the O and B stars now observable in the sky are relatively young objects. Most of them are less than 20 million years old. Massive stars older than that have already exhausted their fuel and no longer emit large amounts of energy. They have, in effect, died.

At the opposite end of the main sequence, the cooler K- and M-type stars have less mass than our Sun. With their low core densities and temperatures, their proton–proton reactions churn away rather sluggishly, much more slowly than those in the Sun's core. The small energy release per unit time leads to low luminosities and surface temperatures for these stars, so they have very long lifetimes. Many of the K- and M-type stars now seen in the sky will shine for at least another trillion years.

Table 17-5 compares some key properties of several well-known main-sequence stars, arranged in order of decreasing mass. Notice how little the central temperature differs from one star to another and how large is the spread in stellar luminosities and lifetimes.

TABLE 17-5   *Key Properties of Some Well-Known Stars*

STAR	SPECTRAL TYPE	MASS $(M_\odot)$	CENTRAL TEMPERATURE $(10^6\,K)$	LUMINOSITY $(L_\odot)$	ESTIMATED LIFETIME $(10^6\,\text{years})$
Rigel	B8Ia	10	30	44,000	20
Sirius	A1V	2.3	20	23	1,000
Alpha Centauri	G2V	1.1	17	1.4	7,000
Sun	G2V	1.0	15	1.0	10,000
Proxima Centauri	M5V	0.1	0.6	0.00006	>1,000,000

17.10 Star Clusters

8 Although an H–R diagram can be drawn for any group of stars, astronomers are usually interested in comparing the properties of stars that have something in common. Very often, that common feature is the fact that the stars all lie in the same region of space (Figure 17.12). The H–R diagram then becomes a means of describing the average properties of stars in a particular part of the Galaxy. The diagram can be compared with similar plots made for stars found elsewhere in the Galaxy, allowing astronomers to test their theories about the way stars (and the Galaxy itself) formed and have evolved.

When trying to obtain an H–R diagram for a distant region of the Galaxy, however, astronomers face a problem. In order to plot the diagram, we must know luminosities; and to know luminosities, we must know distances. Thus it would seem impossible to construct H–R diagrams for stars more distant than 100 pc or so, the maximum distance measurable by stellar parallax. We can't use the method of spectroscopic parallax because that method *assumes* the properties of the main sequence.

In some circumstances, however, it is possible to plot an H–R diagram for very distant stars even though their distances are not known. If we observe a group of stars that all lie at the same distance from us, then comparing *apparent* brightnesses is equivalent to comparing *absolute* brightnesses. Why? Because as the radiation travels toward Earth, the brightness of every star in the group is diminished by the same amount, according to the inverse-square law. By measuring and plotting apparent brightnesses, we can create a "relative" H–R diagram for the group that looks (apart from the numbers on the vertical axis) exactly the same as a real H–R diagram based on luminosities. Such an easily recognizable group of distant stars is called a **star cluster**.

Star clusters can include anywhere from a few dozen to a million stars in a region a few parsecs across. Astronomers believe that all the stars in a given cluster formed at the same time, out of the same cloud of interstellar gas, and under the same conditions. Thus, when we look at a star cluster, we are looking at a group of stars that all have the same age, are similar in composition, and lie in the same region of space, at essentially the same distance from Earth. Unlike the stars plotted in Figures 17.11–17.13, which differ in mass, age, and (to a lesser extent) chemical composition, the only factor distinguishing one cluster star from another is its mass.

Clusters are therefore almost ideal "laboratories" for stellar studies—not in the sense that astronomers can perform experiments on the stars in them, but because the properties of the stars are very tightly constrained. Hence theoretical models of star formation and evolution can be compared with reality without the major complications introduced by broad spreads in age, composition, and place of origin. Clusters are of central importance to astronomers who wish to understand how stars evolve in time.

(a)

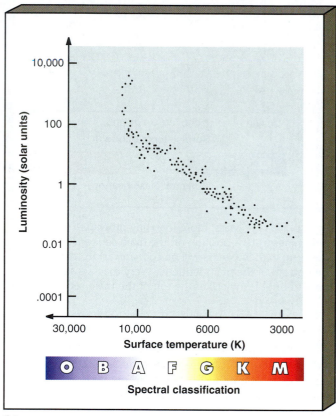

(b)

Figure 17.19 (a) The Pleiades cluster (also known as the Seven Sisters, or M45) lies about 120 pc from the Sun. The naked eye can see only its brightest stars. (b) The stars of this well-known open cluster yield an H–R diagram.

OPEN CLUSTERS

Figure 17.19(a) shows a rather loose cluster—the Pleiades, or Seven Sisters—a well-known naked-eye object in the constellation Taurus. Individual stellar colors provide an estimate of the surface temperature of each star in that cluster. The luminosities follow directly from measurement of the apparent brightness and the cluster's distance (which in this case is known to be about 120 pc). Figure 17.19(b) shows the cluster H–R diagram obtained from these data. This type of cluster, found mainly in the strip across the sky known as the Milky Way, is called an **open cluster** (or, sometimes, a *galactic cluster*). Open clusters typically contain from a few tens to a few hundred stars and are a few parsecs across.

The H–R diagram in Figure 17.19(b) shows stars throughout the main sequence—stars of all colors are represented. The blue stars must be relatively young, for, as we have seen, they burn their fuel rapidly. If all the stars in the cluster formed at the same time, then the red stars must be young too. Thus, even though we have no direct evidence of the cluster's birth, we can estimate its age as less than 20 million years, the lifetime of an O star. Other factors also hint at the cluster's youth. It contains a large amount of interstellar gas and dust not yet processed into stars or lost from the cluster. And it is abundant in heavy elements that (as we will see) can have been created only within the cores of many generations of ancient stars long since perished.

GLOBULAR CLUSTERS

A second type of stellar cluster, of which a representative is shown in Figure 17.20(a), is called a **globular cluster**. Globular clusters are much more tightly knit than the loose groups of stars that make up open clusters. All globular clusters are roughly spherical (which accounts for their name) and contain hundreds of thousands, and sometimes millions, of stars spread out over about 50 pc. As with open clusters, the entire assemblage is held together by gravity.

Figure 17.20(b) shows an H–R diagram for this cluster, which is called Omega Centauri. Notice its many differences from Figure 17.19(b)—globular clusters are a very different stellar environment from open clusters like the Pleiades. The distance to this cluster has been determined by a variation on the method of spectroscopic parallax, but applied to the entire cluster rather than to individual stars. By calculating the distance at which the apparent brightnesses of the cluster's stars taken as a whole best match theoretical models, the cluster is found to lie about 5,000 pc from Earth.

The most outstanding feature of globular clusters is their lack of O- and B-type stars. This deficit is clear from Figure 17.20(b). Low-mass red stars and intermediate-mass yellow stars abound, but high-mass white or blue stars are very rare—in fact, there are no main-sequence stars with masses greater than about 0.8 $M_\odot$. (The blue supergiants in this plot are actually stars at a much later stage in their evolution that happen to be "passing through" the location of the upper main sequence. Their internal structure is quite different from main-sequence giants.) Apparently, globular clusters formed long ago; the more massive O- through F-type stars have long since exhausted their nuclear fuel and disappeared from the main sequence. Other factors confirm that globular clusters are old. For example, their spectra show few heavy elements, implying that these stars formed in the distant past when heavy elements were much less abundant than they are today.

On the basis of these and other observations, astronomers estimate that all globular clusters are at least 10 billion years old. They contain the oldest known stars in the Milky Way Galaxy. As such, globular clusters are considered to be remnants of the earliest stages of our Galaxy's existence.

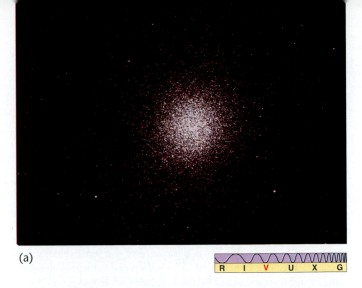

(a)

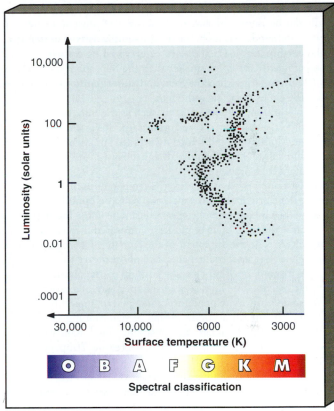

(b)

Figure 17.20   (a) The globular cluster Omega Centauri is approximately 5000 pc from Earth and spans some 40 pc in diameter. (b) An H–R diagram for many (but not all) of its stars.

We will never be able to watch a single star move through all its evolutionary phases. The lifetimes of humans—even of human civilizations—are far too short compared with the lifetimes of even the shortest-lived O and B stars. Instead, we must observe stars as they presently exist—through snapshots taken at specific moments in their life cycles. The H–R diagram is just such a snapshot. By studying stars of different ages or, even better, by studying stars in clusters, in which the ages are known to be the same, we can patch together an understanding of a star's "life story" without having to follow a few individuals from birth to death. Such evolutionary studies will be the subject of the next few chapters

Chapter Review

SUMMARY

The distances to the nearest stars can be measured by trigono-metric parallax. A star with a parallax of one arc second is 1 **parsec** (p. 358)—about 3.3 light years—away.

Stars have real motion through space as well as apparent motion as Earth orbits the Sun. A star's **proper motion** (p. 359), its true motion across the sky, is a measure of the star's velocity perpendicular to our line of sight. The star's radial velocity—along the line of sight—is measured by the Doppler shift of its spectral lines.

Only a few stars are large enough and close enough that their radii can be measured directly. The sizes of most stars are estimated indirectly through the **radius–luminosity–temperature relationship** (p. 361). Stars are categorized as **dwarfs** (p. 362) comparable in size to or smaller than the Sun, **giants** (p. 361) up to 100 times larger than the Sun, and **supergiants** (p. 361) more than 100 times larger than the Sun.

In addition to "normal" stars such as the Sun, two other important classes of star are **red giants** (p. 361), which are large, cool, and luminous, and **white dwarfs** (p. 362), which are small, hot, and faint. The **absolute brightness** (p. 362) of a star is equivalent to its luminosity.

Astronomers often measure the temperatures of stars by measuring their brightnesses through two or more optical filters, then fitting a black-body curve to the results. The **color index** (p. 364) of a star is the ratio of its brightnesses measured through two standard filters. The measurement of the amount of starlight received through each of a set of filters is called **photometry** (p. 364).

Astronomers classify stars according to the absorption lines in their spectra. The lines seen in the spectrum of a given star depend mainly on its temperature, and spectroscopic observations of stars provide an accurate means of determining both stellar temperatures and stellar composition. The standard stellar **spectral classes** (p. 367), in order of decreasing temperature, are OBAFGKM.

A plot of stellar luminosities versus stellar spectral classes (or temperatures) is called an **H–R diagram** (p. 368), or a **color–magnitude diagram** (p. 368). About 90 percent of all stars plotted on an H–R diagram lie on the **main sequence** (p. 369), stretches from hot, bright **blue supergiants** (p. 369) and **blue giants** (p. 369), through intermediate stars such as the Sun, to cool, faint **red dwarfs** (p. 369). Most main-sequence stars are red

dwarfs; blue giants are quite rare. About 9 percent of stars are in the **white dwarf region** (p. 370), and the remaining 1 percent are in the **red giant region** (p. 371).

By careful spectroscopic observations, astronomers can determine a star's **luminosity class** (p. 372), allowing them to distinguish main-sequence stars from red giants or white dwarfs of the same spectral type (or color). Once a star is known to be on the main sequence, measurement of its spectral type allows its luminosity to be estimated and its distance to be measured. This method of distance determination, which is valid for stars up to several thousand parsecs from Earth, is called **spectroscopic parallax** (p. 371).

Most stars are not isolated in space but instead orbit other stars in **binary-star systems** (p. 373). In a **visual binary** (p. 373), both stars can be seen and their orbit charted. In a **spectroscopic binary** (p. 373), the stars cannot be resolved, but their orbital motion can be detected spectroscopically. In an **eclipsing binary** (p. 373), the orbit is oriented in such a way that one star periodically passes in front of the other as seen from Earth and dims the light we receive. The binary's **light curve** (p. 373) is a plot of its apparent brightness as a function of time.

Studies of binary stars often allow stellar masses to be measured. The mass of a star determines its size, temperature, and brightness. Fairly well-defined **mass–radius** (p. 374) and **mass–luminosity** (p. 374) relations exist for main-sequence stars. Hot blue giants are much more massive than the Sun; cool red dwarfs are much less massive.

The lifetime of a star can be estimated by dividing its mass by its luminosity. High-mass stars burn their fuel rapidly and have much shorter lifetimes than the Sun. Low-mass stars consume their fuel slowly, and may remain on the main sequence for trillions of years.

Many stars are found in compact groups known as **star clusters** (p. 376). **Open clusters** (p. 376), with a few hundred to a few thousand stars, are found mostly in the plane of the Milky Way. They typically contain many bright blue stars, indicating that they formed relatively recently. **Globular clusters** (p. 377) are found mainly away from the Milky Way plane and may contain millions of stars. They include no main-sequence stars much more massive than the Sun, indicating that they formed long ago. Globular clusters are believed to date back to the formation of our Galaxy.

SELF-TEST: True or False?

_____ **1.** One parsec is a little over 200,000 A.U.

_____ **2.** There are no stars within one parsec of the Sun.

_____ **3.** Parallax can be used to measure stellar distances out to about 1000 pc.

_____ **4.** Most stars have radii from 0.1 $R_\odot$ to 10 $R_\odot$.

_____ **5.** Star A appears brighter than star B, as seen from Earth. Therefore, star A must be closer to Earth than star B.

_____ **6.** Star A and star B have the same absolute brightness, but star B is twice as distant as star A. Therefore, star A appears 4 times brighter than star B.

_____ **7.** A magnitude 5 star looks brighter than a magnitude 2 star.

_____ **8.** Differences among stellar spectra are mainly due to dif-

ferences in composition.

_____ **9.** Cool stars have very strong lines of hydrogen in their spectra.

_____ **10.** A G9 star is cooler than a G5 star.

_____ **11.** Red dwarfs lie in the lower left part of the H-R diagram.

_____ **12.** The brightest stars visible in the night sky are all found in the upper part of the H-R diagram.

_____ **13.** In a spectroscopic binary, the orbital motion of the component stars appears as variations in their radial velocities.

_____ **14.** It is impossible to have a one-billion-year-old O or B main-sequence star.

SELF-TEST: Fill in the Blank

1. Parallax measurements of the distances to the nearest stars use a baseline of _____.
2. The radial velocity of a star is determined by observing its _____ and using the _____ effect.
3. To determine the true space velocity of a star, its _____ , radial velocity, and _____ must all be known.
4. The radius of a star can be indirectly determined if the _____ and _____ of the star are known.
5. The smallest stars normally plotted on the H-R diagram are _____.
6. Observations of stars through B and V filters are used to determine stellar _____.
7. The hottest stars show little evidence of hydrogen in their spectra because hydrogen is mostly _____ at these temperatures.
8. The Sun has a spectral type of _____.
9. The H-R diagram is a plot of _____ on the horizontal scale versus _____ on the vertical scale.
10. The band of stars extending from the top left of the H-R diagram to its bottom right is known as the _____.
11. _____ star systems are important for providing measurements of stellar masses.
12. Going from spectral type O to M along the main sequence, stellar masses _____.

REVIEW AND DISCUSSION

1. How is parallax used to measure the distances to stars?
2. What is a parsec? Compare it to the astronomical unit.
3. Explain two ways in which a star's real space motion translates into motion observable from Earth.
4. Describe some characteristics of red giant and white dwarf stars.
5. What is the difference between absolute and apparent brightness?
6. How do astronomers measure the temperatures of stars?
7. Briefly describe how stars are classified according to their spectral characteristics.
8. What information is needed to plot a star on the Hertzsprung-Russell diagram?
9. What is the main sequence? What basic property of a star determines where it lies on the main sequence?
10. How are distances determined using spectroscopic parallax?
11. Which stars are most common in the Galaxy? Why don't we see many of them in H-R diagrams?
12. How can stellar masses be determined by observing binary star systems?
13. If a high-mass star starts off with much more fuel than a low-mass star, why doesn't the high-mass star live longer?
14. Compare and contrast the properties of open star clusters and globular star clusters.
15. In general, is it possible to determine the age of an individual star simply by noting its position on an H-R diagram?
16. Describe how the H-R diagram of a star cluster changes with time.

PROBLEMS

1. How far away is the star Spica, whose parallax is 0.013″?
2. A certain star has a temperature twice that of the Sun and a luminosity 64 times greater than the solar value. What is its radius, in units of the solar radius?
3. Two stars—A and B, of luminosities 0.5 and 4.5 times the luminosity of the Sun, respectively—are observed to have the same apparent brightness. Which one is more distant, and how much farther away is it than the other?
4. Astronomical objects visible to the naked eye range in apparent brightness from faint sixth-magnitude stars to the Sun, with magnitude -27. What is the range in flux corresponding to this magnitude range?
5. Given that the Sun's lifetime is about 10 billion years, estimate the life expectancy of (a) a 0.2-solar mass, 0.01-solar luminosity red dwarf, (b) a 3-solar mass, 30-solar luminosity star, (c) a 10-solar mass, 1000-solar luminosity blue giant.

PROJECTS

1. Every winter, you can find an astronomy lesson in the evening sky. The Winter Circle is an asterism—or pattern of stars—made up six bright stars in five different constellations: Sirius, Rigel, Betelgeuse, Aldebaran, Capella, and Procyon. These stars span nearly the entire range of colors (and therefore temperatures) possible for normal stars. Rigel is a B star. Sirius is an A. Procyon is an F star. Capella is a G star. Aldebaran a K star. Betelgeuse is an M star. The color differences of these stars are easy to see. Why do you suppose there is no O star in the Winter Circle?
2. Summer is a good time to search with binoculars for open star clusters. Open clusters are generally found in the plane of the Galaxy. If you can see the hazy band of the Milky Way arcing across your night sky—in other words, if you are far from city lights and looking at an appropriate time of night and year—you can simply sweep with your binoculars along the Milky Way. Numerous "clumps" of stars will pop into view. Many will turn out to be open star clusters.
3. Globular star clusters are harder to find. They are intrinsically larger, but they are also much farther away and therefore appear smaller in the sky. The most famous globular cluster visible from the Northern Hemisphere is M13 in the constellation Hercules, visible on spring and summer evenings. This cluster contains half a million or so of the Galaxy's most ancient stars. It may be glimpsed in binoculars as a little ball of light, located about one-third of the way from the star Eta to the star Zeta in the Keystone asterism of the constellation Hercules. Telescopes reveal this cluster as a magnificent, symmetrical grouping of stars.

18

THE INTERSTELLAR MEDIUM

Gas and Dust Between the Stars

LEARNING GOALS

Studying this chapter will enable you to:

1 Summarize the composition and physical properties of the interstellar medium.

2 Describe the characteristics of emission nebulae, and explain their significance in the life cycle of stars.

3 Discuss the nature of dark interstellar clouds.

4 Specify the radio techniques used to probe the nature of interstellar matter.

5 Discuss the nature and significance of interstellar molecules.

(Opposite page, background) The Tarantula Nebula, known also as 30 Doradus, is the largest and richest stellar nursery in the known universe. Stretching for several hundred parsecs, its interstellar clouds are illuminated by the strong ultraviolet radiation from the young blue-white stars shining amidst the darkness of the nighttime sky.

(Inset A) The Cone Nebula is part of a large and messy complex of nebulosity in the constellation Monoceros. The cone itself seems to be a clump of dense clouds of dust, thick enough to keep out, for now, the penetrating radiation surrounding it.

(Inset B) The big, overexposed star at left is Antares—a red supergiant star shedding gas and particles from its swollen surface. The different colors here result from the scattering of Antares' light among dusty interstellar grains of many sizes.

(Inset C) A better example of a reflection nebula is this bluish molecular cloud. It has the innocuous name CG4, and is about 300 pc away. The only part of the nebula that is aglow itself is the reddish matter near the center; all the rest of the blue color is caused by scattered light, much like that causing our blue sky on Earth.

(Inset D) This emission nebula, IC2944, is a parsec-sized cloud of mostly hydrogen gas illuminated by a few very hot stars. Embedded within the nebula are several tiny dark clouds, known as Bok globules, that might now be contracting to become more stars.

In addition to stars and planets, our Galaxy also harbors matter throughout the invisible regions of interstellar space, in the dark voids between the stars. The density of this interstellar matter is extremely low—approximately a trillion trillion times less dense than matter in either stars or planets, far more tenuous than the best vacuum attainable on Earth. Only because the volume of interstellar space is so vast does its mass amount to anything at all. So why bother to study this near-perfect vacuum? We do so for three important reasons. First, there is as much mass in the "voids" between the stars as there is in the stars themselves. Second, interstellar space is the region out of which new stars are born. Third, it is the region into which some old stars explode at death. It is one of the most significant crossroads through which matter passes in our universe.

18.1 *Interstellar Matter*

① Figure 18.1 shows a large region of space, a much greater expanse of universal real estate than anything we have studied thus far. The bright regions are congregations of innumerable stars, some of whose properties we have just studied in Chapter 17. However, the dark areas are not simply "holes" in the stellar distribution. They are regions of space that obscure or extinguish light from the stars beyond. These regions house *interstellar matter*, consisting of great clouds of gas and dust. Their very darkness means that they cannot be easily studied by the optical methods used for stellar matter. There is, quite simply, very little to see!

Figure 18.1 shows that the dark interstellar matter is rather patchy. It is spread very irregularly through space. In some directions, the obscuring matter is largely absent, and astronomers can study objects literally billions of parsecs from Earth. In other directions, the obscuration is moderate, prohibiting visual observation of objects beyond more than a few thousand parsecs, but still allowing us to study nearby stars. Still other regions are so heavily obscured that starlight from even relatively nearby stars is completely absorbed before reaching the Earth.

GAS AND DUST

The matter between the stars is called the **interstellar medium**. It is made up of two components—gas and dust—intermixed throughout all of space.

Interstellar gas is made up mainly of individual atoms, of average size 10^{-10} m (1 Å) or so. The gas also contains some small molecules, no larger than about 10^{-9} m across. Regions with such small particles are transparent to nearly all types of radiation, including ultraviolet, visible, infrared, and radio waves. Apart from numerous narrow atomic and molecular absorption lines, gas alone does not block radiation to any great extent.

Interstellar dust is more complex. It consists of clumps of atoms and molecules—not unlike chalk dust and the microscopic particles that make up smoke, soot, or fog. Light from distant stars cannot penetrate the densest accumulations of interstellar dust any more than a car's headlights can illuminate roadside objects in a thick Earth-bound fog.

Comparisons of how starlight is diminished in interstellar space with the scattering of light in terrestrial fog indicate that the typical size of an interstellar dust particle—or **dust grain**—is about 10^{-7} m. The grains are thus comparable in size to the wavelength of visible light, and about 1000 times larger than interstellar gas particles.

The ability of a particle to scatter a beam of light depends on (1) the size of the particle and (2) the wavelength of the radiation involved. (See the *More Precisely* feature on p. 152.) As a rule of thumb, only particles with diameters comparable to or larger than the wavelength can significantly influence the beam. Because the wavelength of radio waves greatly exceeds the size of the dust grains, dusty interstellar regions are completely transparent to long-wavelength radio radiation. In other words, radio waves pass through dusty regions unimpeded. These regions are also partially transparent to infrared radiation. Conversely, interstellar dust is very effective at blocking short-wavelength optical, ultraviolet, and X-ray radiation. This general dimming of starlight by interstellar matter is called **extinction**.

Because the interstellar medium is more opaque to short-wavelength radiation of than to radiation of longer wavelengths, light from distant stars is preferentially robbed of its higher-frequency ("blue") components. Hence stars also tend to appear redder than they really are, an effect known as **reddening**. It is similar to the process that produces spectacular red sunsets on Earth.

As illustrated in Figure 18.2, extinction and reddening change a star's apparent brightness and color, but they have no effect on its spectral type. Absorption lines in the star's spectrum are largely unaffected by interstellar dust. Astronomers can use this fact to study the interstellar medium. By determining a main-sequence star's spectral type, astronomers first learn its true luminosity and color. ∞ (p. 372) They can then measure the degree to which the starlight has been affected by extinction and reddening en route to Earth. This, in turn, allows them to estimate both the numbers and the sizes of interstellar dust particles along the line of sight to the star. By repeating these measurements for stars in many different directions and at many different distances from the Earth, a picture of the distribution and properties of the interstellar medium in the surrounding solar neighborhood can be built up.

TEMPERATURE AND DENSITY

The temperature of the interstellar gas and dust ranges from a few kelvins to a few hundred kelvins, depending on its proximity to a star or some other source of radiation. Generally, we can take 100 K as an average temperature of a typical dark region of interstellar space. Compare this with 273 K, at which water freezes, and 0 K, at which atomic and molecular motions all but cease. Interstellar space is very cold.

The density of interstellar matter is extremely low. It averages roughly 10^6 atoms per cubic meter—just 1 atom per cubic centimeter—but densities as great as 10^9 atoms/m^3 (1000 atoms/cm^3) and as small as 10^4 atoms/m^3 have been found. Matter of this low density is far more tenuous than the best vacuum—about 10^{10} molecules/m^3— that we can make in laboratories here on Earth. Interstellar matter is about a trillion trillion times less dense than water. Interstellar dust is even rarer than interstellar gas. On average, there are only about 10^{-6} dust particles per cubic meter—that is, 1000 per cubic *kilo*meter. Interstellar space is populated with gas so thin that harvesting all the matter in an interstellar region the size of Earth would yield barely enough matter to make a pair of dice.

How can such fantastically sparse matter diminish light radiation so effectively? The key is size—interstellar space is vast. The typical distance between stars (1 pc or so in the vicinity of the Sun) is much, much greater than the

Figure 18.1   A wide-angle photograph of a great swath of space, showing regions of brightness (vast fields of stars) as well as regions of darkness (obscuring interstellar matter). The field is roughly 30° across.

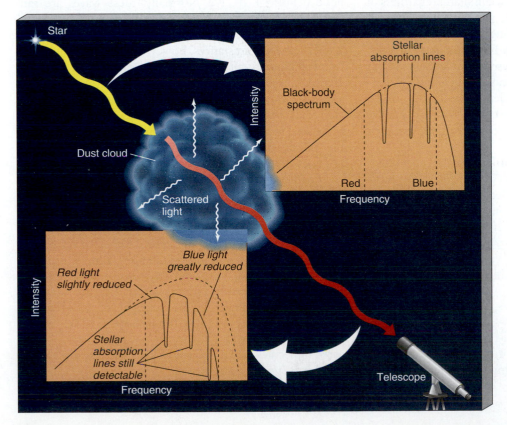

Figure 18.2   Starlight passing through a dusty region of space is both dimmed and reddened, but spectral lines are still recognizable in the light that reaches Earth.

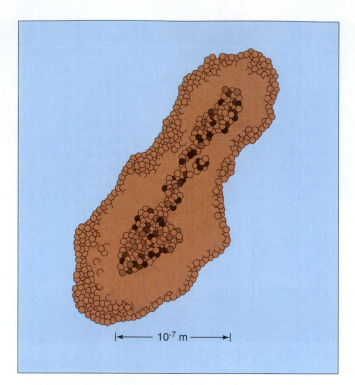

Figure 18.3 A diagram of a typical interstellar dust particle. The average size of such particles is only 1/10,000 of a millimeter, yet space contains enough of them to obscure our view in certain directions.

typical size of the stars themselves (around 10^{-7} pc). Stellar and planetary sizes pale in comparison to the vastness of interstellar space. Thus, matter can accumulate, regardless of how thinly spread. For example, an imaginary cylinder 1 m^2 in cross section and extending from Earth to Alpha Centauri would contain more than 10 billion dust particles. Over huge distances, dust particles accumulate slowly but surely, to the point at which they can effectively block visible light and other short-wavelength radiation. Even though the density of matter there is so low, interstellar space in the vicinity of the Sun contains about as much mass as exists in the form of stars.

Despite their rarity, dust particles make interstellar space a *relatively* dirty place. Earth's atmosphere, by comparison, is about a million times cleaner. Our air is tainted by only one dust particle for about every billion billion (10^{18}) atoms of atmospheric gas. If we could compress a typical parcel of interstellar space to equal the density of air on Earth, this parcel would contain enough dust to make a fog so thick that we would be unable to see our hand held at arm's length in front of us.

COMPOSITION

The composition of interstellar gas is reasonably well understood. Spectroscopic studies of interstellar absorption lines provide astronomers with comprehensive information on its elemental abundances. ∞ (p. 86) Most of it—about 90 percent of all particles—is atomic and molecular

hydrogen; some 9 percent is helium, and the remaining 1 percent consists of heavier elements. The abundances of several of the heavy elements, such as carbon, oxygen, silicon, magnesium, and iron, are much lower in interstellar gas than in our solar system or in stars. The most likely explanation for this finding is that substantial quantities of these elements have been used to form the interstellar dust, taking them out of the gas and locking them up in a form that is much harder to observe.

In contrast to interstellar gas, the composition of interstellar dust is currently not very well known. We have some infrared evidence for silicates, graphite, and iron—the same elements that are underabundant in the gas—lending support to the theory that interstellar dust forms out of interstellar gas. The dust probably also contains some "dirty ice," a frozen mixture of ordinary water ice contaminated with trace amounts of ammonia, methane, and other chemical compounds. This composition is quite similar to that of cometary nuclei in our own solar system (see Section 14.2).

DUST SHAPE

Curiously, astronomers know the *shape* of interstellar dust particles better than their composition. Although the minute atoms in the interstellar gas are basically spherical, the dust particles are not. Individual dust grains are apparently elongated or rodlike, as shown in Figure 18.3. We can infer this because the light emitted by stars is dimmed and partially **polarized**, or aligned, by the dust.

Recall from Chapter 3 that light consists of electromagnetic waves composed of vibrating electric and magnetic fields. ∞ (p. 58) Normally, these waves are randomly oriented, and we say the radiation is *unpolarized*. Stars emit unpolarized radiation from their photospheres. Under some circumstances, however, the electric fields can become aligned—all vibrating in the same plane as the radiation moves through space. We then say the radiation is *polarized*. Polarization of starlight does not occur by chance. If the light detected by our telescope is polarized, it is because some interstellar matter lies between the emitting object and Earth. The polarization of starlight, then, provides another way to study the interstellar medium.

On Earth we can produce polarized light by passing unpolarized light through a Polaroid filter, which has specially aligned elongated molecules that allow the passage of only those waves with their electric fields oriented in some specific direction (see Figure 18.4a). Other waves are absorbed and do not pass through the filter. The alignment of the molecules determines which waves will be transmitted. In interstellar space, dust grains can act like the molecules in the Polaroid filter. If the starlight is polarized, astronomers can conclude that the interstellar dust particles must have an *elongated* shape (by analogy with the elongated molecules of the Polaroid filter) and that these molecules are aligned, as shown in Figure 18.4(b). Only then can the dust preferentially absorb certain waves, leaving the remainder (the ones we observe) polarized.

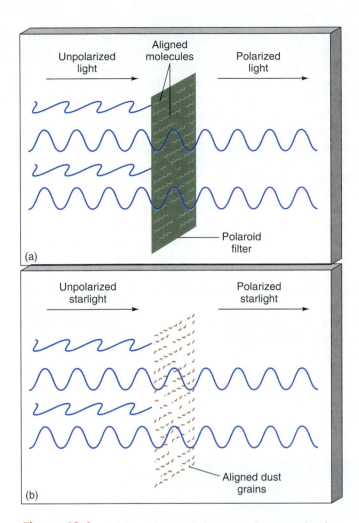

Figure 18.4 (a) Unpolarized light waves have randomly oriented electric fields. When the light passes through a Polaroid filter, only waves with their electric fields oriented in a specific direction are transmitted, and the resulting light is polarized. (b) Aligned dust particles in interstellar space polarize radiation in a similar manner. Observations of the degree of polarization allow astronomers to infer the size, shape, and orientation of the particles.

The alignment of the interstellar dust is the subject of intense research among astronomers. The current view, accepted by most, holds that the dust particles are affected by a weak interstellar magnetic field, perhaps a million times weaker than Earth's field. Each dust particle responds to the field in much the same way that small iron filings are aligned by an ordinary bar magnet. Measurements of the blockage and polarization of starlight thus yield information about the size and shape of interstellar dust particles, as well as about magnetic fields in interstellar space.

Now that we have a general idea of the basic contents and properties of interstellar space, let us examine some typical regions in more detail. We'll note especially how astronomers have unraveled the nature of the matter contained within them.

18.2 *Emission Nebulae*

Figure 18.5 is a mosaic of photographs showing a region of space where stars, gas, and dust seem to congregate. The bright areas are made up of myriad unresolved stars. The dark areas are vast pockets of dust, blocking from our view what would otherwise be a rather smooth distribution of bright starlight. From our vantage point within the solar system, this assemblage of stellar and interstellar matter follows a bright band extending across the sky. On a clear night, this band of patchy light is visible to the naked eye as the Milky Way. In Chapter 23 we will come to recognize this band as the flattened disk, or *plane*, of our own Galaxy.

Figure 18.6 shows a 12°-wide swath of the galactic plane in the general direction of the constellation Sagittarius, as photographed from Earth. The view is rather mottled, with a patchy distribution of stars and interstellar debris. In addition, several large fuzzy patches of light are clearly visible. These fuzzy objects, labeled M8, M16, M17, and M20, correspond to the 8th, 16th, 17th, and 20th objects in a catalog compiled by Charles Messier, an eighteenth-century French astronomer.* The stars, the

Messier was actually more concerned with making a list of celestial objects that might possibly be confused with comets, his main astronomical interest. However, the catalog of 109 "Messier objects" is now regarded as a much more important contribution to astronomy than any comets Messier discovered.

Figure 18.5 A mosaic of the plane of the Milky Way Galaxy. Photographed almost from horizon to horizon, and thus extending over nearly 180°, this band contains high concentrations of stars as well as interstellar gas and dust. The field of view is several times wider than that of Figure 18.1, whose outline is superimposed on this image.

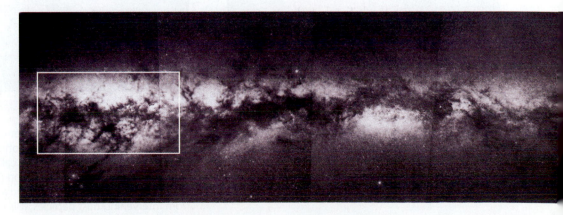

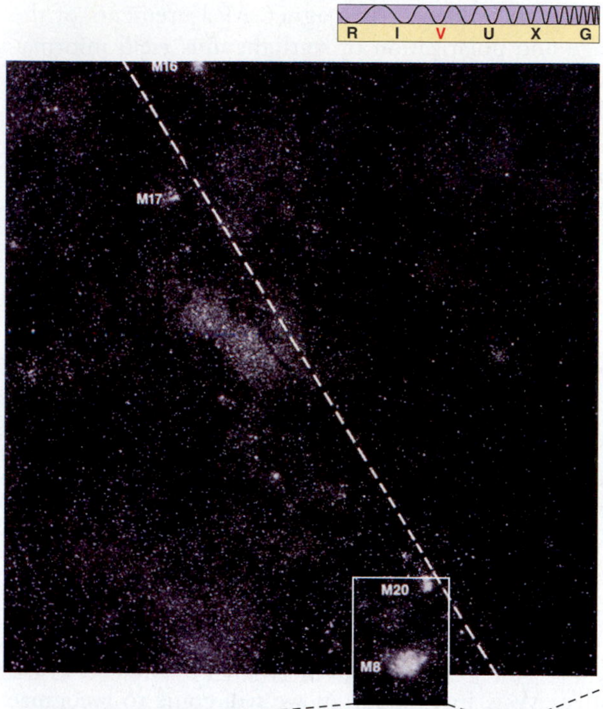

Figure 18.6 A photograph of a small portion (about 12° across) of the galactic plane shown in Figure 18.1, displaying higher-resolution evidence for stars, gas, and dust as well as several distinct fuzzy patches of light, known as emission nebulae. The plane of the Milky Way Galaxy is marked with a dashed line.

fuzzy objects, and the dark obscuring matter are all concentrated around and along the galactic plane. Indeed, this plane is the site of greatest concentration of almost all astronomical objects within our Galaxy.

Historically, astronomers used the term **nebula** to refer to any "fuzzy" patch (bright or dark) on the sky—any region of space that was clearly distinguishable through a telescope but not sharply defined, unlike a star or a planet. We now know that many (though not all) nebulae are clouds of interstellar dust and gas. If they happen to obscure stars lying behind them, we see a dark patch. If something within the cloud—a group of hot young stars, for example—causes it to glow, the nebula appears bright instead. The four fuzzy objects labeled in Figure 18.6 are **emission nebulae**—glowing clouds of hot interstellar gas. The method of spectroscopic parallax applied to stars visible within the nebulae indicates that their distances from Earth range from 900 pc (M20) to 1800 pc (M16). ∞ (p. 371) Thus, all four are near the limit of visibility for any object embedded in the dusty galactic plane. M16, at the top left, is approximately 1000 pc from M20, near the bottom.

We can gain a better appreciation of these nebulae by examining progressively smaller fields of view. Figure 18.7 is an enlargement of the region near the bottom of Figure 18.6. M20 is at the top, and M8 is at the bottom, only a few degrees away. Figure 18.8 is yet another enlargement of the

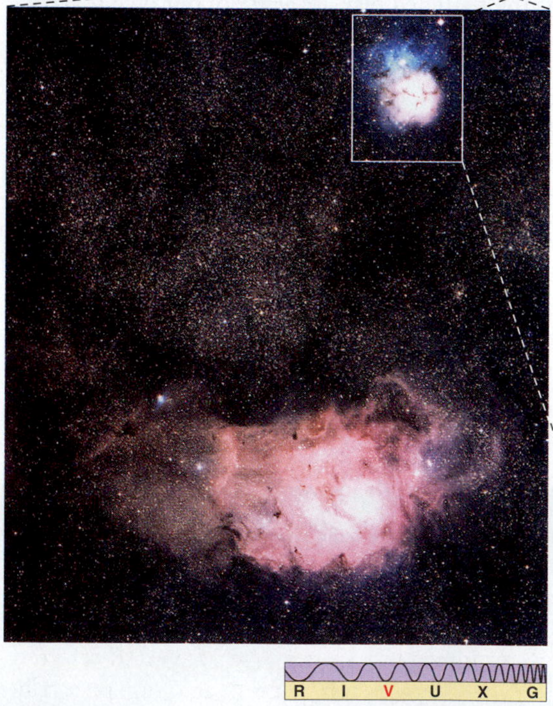

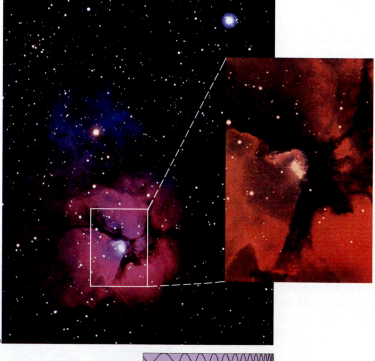

Figure 18.7 An enlargement of the bottom of Figure 18.6, showing M20 (top) and M8 (bottom) more clearly.

Figure 18.8 Further enlargements of the top of Figure 18.7, showing only M20 and its interstellar environment. The nebula itself (in red) is about 4 pc in diameter. It is often called the Trifid Nebula because of the dust lanes that trisect its midsection (insert). The blue region is unrelated to the red emission nebula and is caused by starlight reflected from intervening dust particles. It is called a *reflection nebula*.

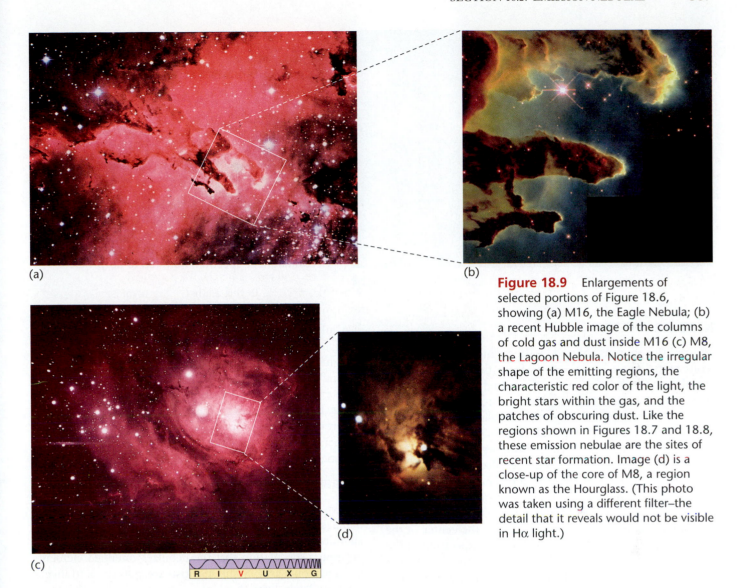

(a)

(b)

(c)

(d)

Figure 18.9 Enlargements of selected portions of Figure 18.6, showing (a) M16, the Eagle Nebula; (b) a recent Hubble image of the columns of cold gas and dust inside M16 (c) M8, the Lagoon Nebula. Notice the irregular shape of the emitting regions, the characteristic red color of the light, the bright stars within the gas, and the patches of obscuring dust. Like the regions shown in Figures 18.7 and 18.8, these emission nebulae are the sites of recent star formation. Image (d) is a close-up of the core of M8, a region known as the Hourglass. (This photo was taken using a different filter–the detail that it reveals would not be visible in Hα light.)

R I V U X G

top of Figure 18.7. This is a close-up of M20 and of its immediate environment—a real jewel of the night. The total area displayed measures some 10 pc across.

Emission nebulae such as M8 and M20 are among the most spectacular objects in the entire universe. Yet they appear only as small, undistinguished patches of light when viewed in the larger context of the entire galactic plane, as in Figure 18.5. Perspective is crucial in astronomy.

These nebulae are regions of glowing, ionized gas. At or near the center of each is at least one newly formed hot O- or B-type star producing copious amounts of ultraviolet light. As they travel outward from the star, ultraviolet photons collide with the surrounding gas, ionizing much of it. As electrons recombine with nuclei, they emit visible radiation, causing the gas to glow. Figure 18.9 completes our set of enlargements of the four nebulae visible in Figure 18.6. Notice in all cases the predominant red coloration of the emitted radiation and the hot bright stars embedded within the glowing nebular gas.

The reddish hue of these four nebulae—and in fact of any emission nebula—results from hydrogen atoms emit-

ting light in the red part of the visible spectrum. Specifically, it is caused by the emission of radiation at 656.3 nm (6563 Å)—the Hα line that we encountered in Chapter 4. ∞ (p. 82) Other elements in the nebula also emit radiation as their electrons recombine, but because hydrogen is so plentiful, its emission overwhelms that from all other atoms. Tinged here and there with other colors, red usually dominates in emission nebulae.

Scientists often refer to the *ionization state* of an atom by attaching a roman numeral to its chemical symbol—I for the neutral atom, II for a singly ionized atom (missing one electron), III for a doubly ionized atom (missing two electrons), and so on. Because emission nebulae are composed mainly of ionized hydrogen, they are often referred to as *HII regions*. Regions of space containing primarily neutral (atomic) hydrogen are known as *HI regions*.

Woven through the glowing nebular gas, and plainly visible in Figures 18.7–18.9, are lanes of dark obscuring dust. Recent studies have demonstrated that these **dust lanes** are part of the nebulae and are not just unrelated dust clouds that happen to lie along our line of sight.

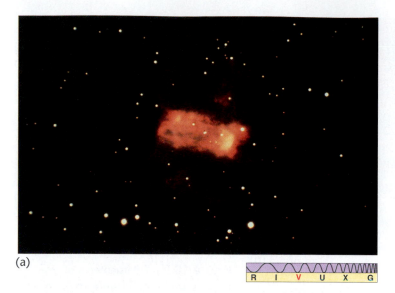

(a)

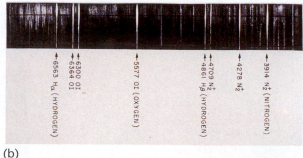

(b)

Figure 18.10 (a) The Butterfly Nebula, a glowing patch of gas a few parsecs across. (b) Its emission spectrum, showing light intensity as a function of frequency over the entire visible portion of the electromagnetic spectrum from red to deep violet.

R I V U X G

NEBULAR SPECTRA

Most of the photons emitted by the recombination of electrons with atomic nuclei escape from the nebula. Unlike the ultraviolet photons originally emitted by the embedded stars, they do not carry enough energy to ionize the nebular gas, and they pass through the nebula relatively unhindered. Some eventually reach the Earth, where we can detect them. Only through these lower-energy photons do we learn anything about emission nebulae.

Just as studies of the spectra of ordinary stars contain a wealth of information about stellar atmospheres, nebular spectra tell us a great deal about ionized interstellar gas. Because at least one hot star resides near the center of the nebula, we might think that the combined spectrum of the star and the nebula would be hopelessly confused. In fact, it is not. We can easily distinguish nebular spectra from stellar spectra because the physical conditions in stars and nebulae differ so greatly. In particular, emission nebulae are made of hot thin gas that, as we saw in Chapter 4, yields detectable *emission* lines. ∞ (p. 75) When our spectroscope is trained on a star, we see a familiar stellar spectrum, consisting of a black-body-like continuous spectrum and absorption lines, together with superimposed emission lines from the nebular gas. When no star appears in the field of view, only the emission lines are seen.

Figure 18.10(b) is a typical nebular emission spectrum spanning part of the visible and near-ultraviolet wavelength interval. Numerous emission lines can be seen, and information on the nebula shown in Figure 18.10a can be extracted from all of them. The results of analyses of many nebular spectra show abundances close to those derived from observations of the Sun and other stars: Hydrogen is 90 percent abundant by number, followed by helium at about 9 percent; the heavier elements together make up the remaining 1 percent.

Unlike stars, nebulae are large enough for their actual sizes to be measurable by simple geometry. Coupling this size information with estimates of the amount of matter along our line of sight (as revealed by the nebula's emission of light), we can find the nebula's density. Generally, emission nebulae have only a few hundred particles, mostly protons and electrons, in each cubic centimeter—a density some 10^{22} times lower than that of a typical planet. Spectral-line widths imply that the gas atoms and ions have temperatures around 8000 K. Table 18-1 lists some vital statistics for each of the nebulae shown in Figure 18.6.

"FORBIDDEN" LINES

When astronomers first studied the spectra of emission nebulae, they found many lines that did not correspond to anything observed in terrestrial laboratories. For a time, this prompted speculation that the nebulae contained elements unknown on Earth. Some scientists went so far as to invent the term "nebulium" for a new element, much as the name helium came about when that element was first discovered in the Sun (recall also "coronium" from Chapter 16). ∞ (p. 340) With a fuller understanding of the workings of the atom, astronomers realized that these lines did in fact result from electron transitions within the atoms of familiar elements. However, these transitions occurred under unfamiliar conditions not reproducible in laboratories.

For example, in addition to the dominant red coloration just discussed, many nebulae also emit light with a

TABLE 18–1 *Nebular Properties*

OBJECT	APPROX. DISTANCE (pc)	AVERAGE DIAMETER (pc)	DENSITY (10^6 particles/m³)	MASS ($M_\odot$)	TEMPERATURE (K)
M8	1200	14	80	2600	7500
M16	1800	8	90	600	8000
M17	1500	7	120	500	8700
M20	900	4	100	150	8200

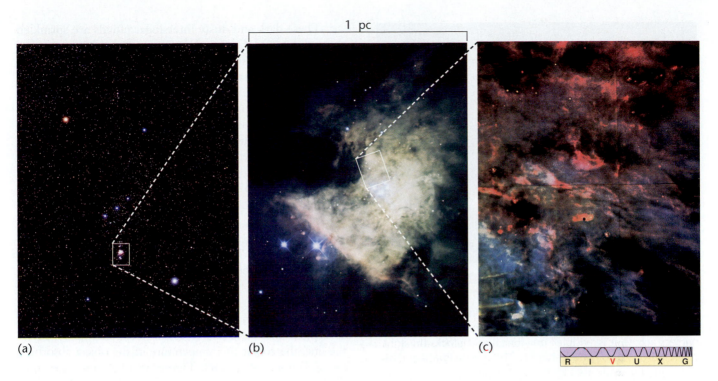

1 pc

(a) (b) (c)

R   I   V   U   X   G

Figure 18.11 (a) The Orion Nebula (M42) lies some 450 pc from Earth. It is visible to the naked eye as the fuzzy middle "star" of Orion's sword. (b) Like all emission nebulae, it consists of hot, glowing gas powered by a group of bright stars in the center. In addition to the red Hα emission, parts of the nebula show a slight greenish tint, caused by a so-called forbidden transition in ionized oxygen. (c) A high-resolution, approximately true-color image shows rich detail in a region about a light year across. Structural details are visible down to a level of 0.1 arc second, or 6 light *hours*—a scale comparable to our solar system.

characteristic green color (see Figure 18.11). The greenish tint of portions of this nebula greatly puzzled astronomers in the early twentieth century and defied explanation in terms of the spectral lines known at the time.

Astronomers now understand that the color is caused by a particular electron transition in doubly ionized oxygen. The structure of oxygen is such that an ion in the higher-energy state for this transition tends to remain there for a very long time—many hours, in fact—before dropping back to the lower state and emitting a photon. Only if the ion is left undisturbed during this time, and not kicked into another energy state, will the transition actually occur and the photon be emitted.

In a terrestrial experiment, no atom or ion is left undisturbed for long. Even in a "low-density" laboratory gas, there are many trillions of particles per cubic meter, and each particle experiences millions of collisions every second. The result is that an ion in the particular energy state that produces the peculiar green line in the nebular spectrum never has time to emit its photon—collisions kick it into some other state long before that occurs. For this reason, the line is usually called **forbidden**, even though it violates no laws of physics. It simply occurs on Earth with such low probability that it is never seen.

In an emission nebula, the density is so low that collisions between particles are very rare. There is plenty of time for the excited ion to emit its photon, and the forbidden line is produced. Numerous forbidden lines are known in nebu-

lar spectra. They remind us once again that the environment in the interstellar medium is very different from conditions on Earth and warn us of the problems of extending our terrestrial experience to the study of interstellar space.

18.3 *Dark Dust Clouds*

❸ Emission nebulae are only one small component of interstellar space. Most of space—in fact, more than 99 percent of it—is devoid of nebular regions, and contains no stars. It is simply dark. Look again at Figure 18.5, or just ponder the evening sky. The dark regions are by far the most representative regions of interstellar space. The gas density there is only about a million atoms per cubic meter—100 times lower than the density in an emission nebula. The temperature of this diffuse gas is typically around 100 K.

Within these dark voids between the nebulae and the stars lurks another type of astronomical object, the **dark dust cloud**. Such clouds are cooler than their surroundings and thousands or even millions of times denser. These clouds bear little resemblance to terrestrial clouds. Most of them are bigger than our solar system, and some are many parsecs across. (Yet even so, they make up no more than a few percent of the entire volume of interstellar space.) Despite their name, these clouds are actually made up primarily of *gas*, just like the rest of the interstellar medium. As just discussed, however, their absorption of starlight is due almost entirely to the dust they contain.

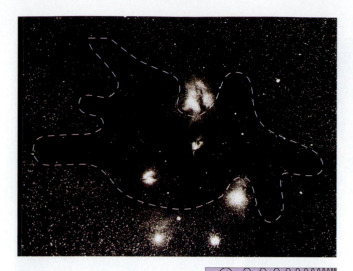

Figure 18.12 Photograph of a typical interstellar dark dust cloud. This cloud, known as Rho Ophiuchi, is several parsecs across. It is "visible" only because it blocks the light coming from stars lying behind it. The approximate outline of the cloud is indicated by the dashed line.

OBSCURATION OF VISIBLE LIGHT

Figure 18.12 is an optical photograph of a typical interstellar dust cloud. Pockets of intense blackness mark regions where the dust and gas are especially concentrated and the light from background stars is completely obscured. This cloud takes its name from a nearby star, Rho Ophiuchi, and resides at the relatively nearby distance of about 300 pc. Measuring several parsecs across, this cloud is only a tiny part of the grand mosaic shown in Figure 18.5. Note especially the long "streamers" of (relatively) dense dust and gas. This cloud clearly is not spherical. Indeed, most interstellar clouds are very irregularly shaped. The bright patches within the dark region are emission nebulae in the foreground. Some of them are part of the cloud itself, where newly formed stars near the surface have created a "hot spot" in the cold, dark gas. Others have no connection to the cloud and just happen to lie along our line of sight.

These dark and dusty interstellar clouds are sprinkled throughout our Galaxy. We can study them at optical wavelengths only if they happen to block the light emitted by more distant stars or nebulae. The dark outline of Rho Ophiuchi and the dust lanes visible in Figures 18.8 and 18.9 are good examples of this obscuration. The dust is apparent only because it blocks the light coming from behind it. Figure 18.13 shows another striking example of a dark cloud—the Horsehead Nebula in Orion. This curiously shaped finger of gas and dust projects out from the much larger dark cloud in the bottom half of the image and stands out clearly against the red glow of a background emission nebula.

ABSORPTION SPECTRA

Astronomers first became aware of the true extent of dark interstellar clouds in the 1930s as they studied the optical spectra of distant stars. In addition to the wide absorption lines normally formed in stars' lower atmospheres, much narrower absorption lines were also detected. Recall that the narrower the line, the cooler the temperature of the object absorbing the radiation. ∞ (p. 88) Figure 18.14(a) illustrates how light from a star may pass through several interstellar clouds on its way to Earth. These clouds need not be close to the star, and indeed they usually are not. Each absorbs some of the stellar radiation in a manner that depends on its own temperature, density, and elemental abundance. Figure 18.14(b) depicts part of a typical spectrum produced in this way.

The narrow absorption lines contain information about dark interstellar clouds, just as stellar absorption lines reveal the properties of stars and nebular emission lines tell us about conditions in hot nebulae. By studying these lines, astronomers can probe the cold depths of interstellar space. In most cases, the elemental abundances detected in interstellar clouds mirror those found in other astronomical objects—which is perhaps not surprising, because interstellar clouds are the regions that spawn nebulae and stars. It appears that most of the matter in the Galaxy has become fairly well mixed by repeated processing in and out of stars, nebulae, and clouds. Most objects in the Galaxy have fairly similar composition.

Figure 18.13 The Horsehead Nebula in Orion is a striking example of a dark dust cloud, silhouetted against the bright background of an emission nebula. The "neck" of the horse is about 0.25 pc across. The nebular region is roughly 1500 pc from the Earth.

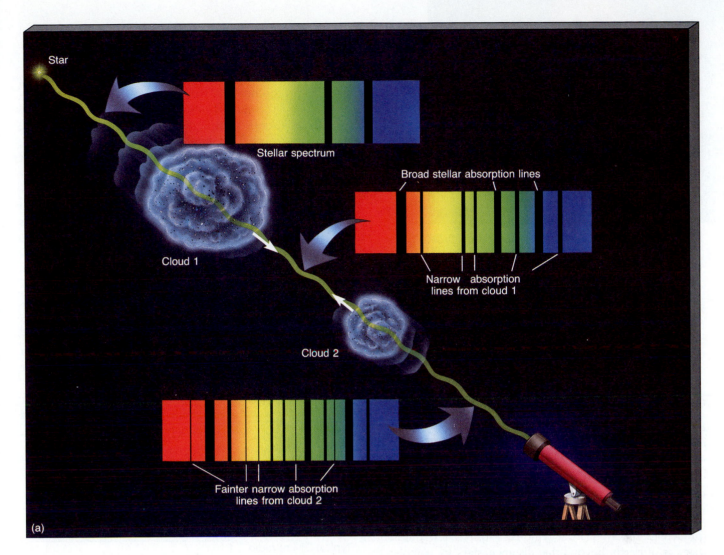

(a)

Figure 18.14 (a) A simplified diagram of some interstellar clouds between a hot star and Earth. Optical observations might show an absorption spectrum like that traced in (b). The wide, intense lines are formed in the star's hot atmosphere; narrower, weaker lines arise from the cold interstellar clouds. The smaller the cloud, the weaker the lines. The redshifts or blueshifts of the narrow absorption lines provide information on cloud velocities. The widths of all the spectral lines depicted here are greatly exaggerated for the sake of clarity.

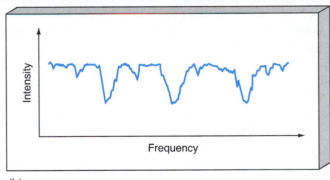

(b)

Dark dust clouds are much cooler than the thousands of kelvins that characterize emission nebulae and stellar atmospheres. Their temperatures, as determined from their absorption line widths, are usually less than about 100 K, and values as low as 10 or 20 K are common. None of this interstellar gas can be ionized. It consists of neutral matter—atoms and molecules. This is why the clouds are invisible—they are just too cold to emit any visible light. They do, however, emit strongly at longer wavelengths. Compare the optical photograph of Rho Ophiuchi in Figure 18.12 with the infrared image of the same dark dust cloud in Figure 18.15, captured by sensitive detectors aboard the *Infrared Astronomy Satellite*. ∞ (p. 116)

Along any given line of sight, cloud densities can range from 10^7 atoms/m³ to more than 10^{12} atoms/m³. These latter clouds are generally called *dense* interstellar clouds by researchers, but even these densest interstellar regions are about as tenuous as the best laboratory vacuum. Still, it is because their density is larger than the average value of 10^6 atoms/m³ that we can distinguish clouds from the surrounding expanse of interstellar space.

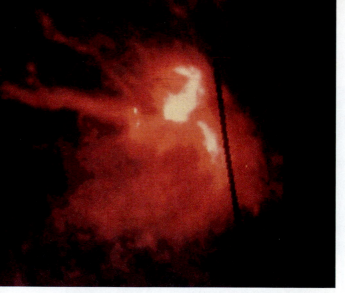

Figure 18.15 A filled contour map of infrared radiation detected from the dark interstellar cloud Rho Ophiuchi. The infrared radiation, and therefore the dust that emits this radiation, displays a structure similar to the cloud's visual image (Figure 18.12). The very bright source of infrared radiation near the top of the cloud comes from a hot emission nebula, which can also be seen in the optical image. (The scale of the image is 3pc on a side; the black diagonal streak at right is an instrumental artifact.)

18.4 *21-Centimeter Radiation*

④ A basic problem with the optical technique just described is that we can examine interstellar clouds *only* along the line of sight to a distant star. To form an absorption line, there has to be a background source of radiation to absorb. The need to see stars through clouds also restricts this approach to relatively local regions, within a few thousand parsecs of Earth. Beyond that distance, stars are completely obscured, and no optical observations are possible. Only the denser, dustier clouds emit enough infrared radiation for astronomers to study them in that part of the spectrum.

To probe interstellar space more thoroughly, we need a more general, more versatile observational method—one that does not rely on conveniently located stars and nebulae. In short, we need a way to detect cold, neutral interstellar matter anywhere in space through its *own* radiation.

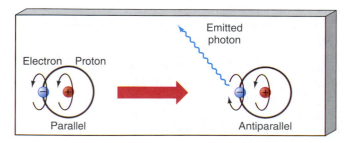

Figure 18.16 Diagram of a ground-level hydrogen atom changing from a higher-energy state (electron and proton spins are parallel) to a lower-energy state (spins are antiparallel). The emitted photon carries away an energy equal to the energy difference between the two spin states.

This may sound impossible, but such an observational technique does in fact exist. The method relies on low-energy *radio* emissions produced by the interstellar gas itself.

Recall that a hydrogen atom has one electron orbiting a single-proton nucleus. Besides the electron's orbital motion around the central proton, electrons also have some rotational motion—that is, *spin*—about their own axis. The proton also spins. This model parallels a planetary system, in which, in addition to the orbital motion of a planet about a central star, both the planet (electron) and the star (proton) rotate about their axis.

The laws of physics dictate that there are exactly two possible spin configurations for a hydrogen atom in its ground state. The electron and proton can rotate in the same direction, with their spin axes parallel, or they can rotate with their axes antiparallel (that is, parallel, but oppositely oriented). Figure 18.16 shows these two configurations. The antiparallel configuration has slightly less energy than the parallel state.

All matter in the universe tends to achieve its lowest possible energy state, and interstellar gas is no exception. Even slightly excited hydrogen atoms eventually end up in their least energetic state. After a certain amount of time, hydrogen atoms in the higher-energy state relax back to the lower level. That is, the electron suddenly and spontaneously flips its spin, which then becomes opposite to that of the proton.

Figure 18.17
Typical 21-cm radio spectral lines observed toward several different regions of interstellar space. The peaks do not all occur at a wavelength of exactly 21 cm, corresponding to a frequency of 1.4 GHz (gigahertz), because the gas in the Galaxy is moving with respect to Earth.

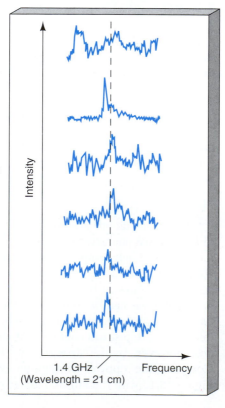

Interlude 18-1 *A Satellite Named IUE*

One of the most successful satellites ever launched—placed in orbit in 1978 and still going strong—is the *International Ultraviolet Explorer,* or *IUE.* ∞ (p. 118) *IUE* contains several on-board spectrometers capable of resolving narrow ultraviolet spectral lines. This orbiting observatory is controlled from the ground, where scientists point the telescope using the satellite's gyroscopes. Automatic on-board star trackers then take over the job of "fine tuning" the craft's orientation, using a navigational map based on the known positions of the stars to lock in on specific regions of the sky. Large solar panels provide the electrical power needed to operate the instruments that send the data back to Earth. The accompanying figure shows the *IUE* control center at NASA's Goddard Space Flight Center, outside Washington, D.C.

The line-of-sight technique for analyzing interstellar clouds, described in the text for optical observations, can also be applied to ultraviolet radiation. We search the ultraviolet spectrum of a hot star for particularly narrow absorption lines caused by intervening interstellar clouds. By this means, many interstellar gases can be studied, including hydrogen (both atomic and molecular), carbon, nitrogen, oxygen, iron, and several others. As we have seen, some of these gases are somewhat less abundant in galactic clouds than in our solar system or in stars, probably because substantial quantities of these elements have been used to form interstellar dust particles. This idea is supported by additional infrared observations.

IUE confirmed earlier findings (made by an ultraviolet satellite named *Copernicus*) that certain clouds were speeding through space. Usually, interstellar clouds move at velocities of 20 km/s or less relative to the Sun. However, in some re-gions of space, the ultraviolet observations have clocked clouds and thin "sheets" of material moving at nearly 100 km/s. The origin of this high-speed matter is not well understood, but it may have been pushed around by explosions of old stars or by stellar winds like that escaping the Sun.

IUE observations of weak spectral lines from highly excited atoms have shown that some regions of interstellar space are much thinner (5000 atoms/m³) and hotter (500,000 K) than expected. Although we still lack a complete inventory, it appears that some of the space between the dust clouds and the emission nebulae may contain extremely dilute yet seething plasma, probably the result of the concussion and expanding debris from ancient stars that exploded long ago. These superheated interstellar "bubbles," or *intercloud medium,* may extend far into interstellar space beyond our local neighborhood and conceivably into the even vaster spaces among the galaxies.

As with any such change, the transition from a high-energy state to a low-energy state releases a photon with energy equal to the energy difference between the two levels.

The energy difference between the two states is very small, so the energy of the emitted photon is very low. Consequently, the wavelength of the radiation is rather long—in fact, about 21 cm, roughly the width of this book. That wavelength lies in the radio portion of the electromagnetic spectrum. Researchers refer to the spectral line that re-sults from this hydrogen-spin-flip process as the **21-centimeter line**. Figure 18.17 shows typical spectral profiles of 21-cm radio signals observed toward several different regions of space. These tracings are the characteristic signatures of cold, atomic hydrogen in our Galaxy. Needing no visible starlight to help calibrate their signals, radio astronomers can observe *any* interstellar region that contains enough hydrogen gas to produce a detectable signal. Even the low-density regions between the dark clouds can be studied.

As can be seen in Figure 18.17, actual 21-cm lines are quite jagged and irregular, somewhat like nebular emission lines in appearance. These irregularities arise because there are usually numerous clumps of interstellar gas along any given line of sight. Each has its own density, temperature, ra-dial velocity, and internal motion, so the intensity, width, and Doppler shift of the resultant 21-cm line vary from place to place. All these different lines are superimposed in the signal we eventually receive at Earth, and sophisticated computer analysis is generally required to disentangle them. The figures quoted earlier for the temperatures (100 K) and densities (10^6 atoms/m³) of the regions between the dark dust clouds are based on 21-cm measurements; observations of the dark clouds themselves yield densities and temperatures in good agreement with those obtained by optical spectroscopy.

All interstellar atomic hydrogen emits 21-cm radiation. But if all atoms eventually fall into their lowest-energy configuration, why isn't all the hydrogen in the Galaxy in the lower-energy state by now? Why do we see 21-cm radiation today? The answer is that the energy difference between the two states is comparable to the energy of a typical atom at a temperature of 100 K or so. As a result, atomic collisions in the interstellar medium are energetic enough to boost the electron up into the higher-energy configuration and so maintain comparable numbers of hydrogen atoms in either state. Any sample of interstellar hydrogen at any instant will contain many atoms in the upper level, and 21-cm radiation will always be emitted.

Of great importance, the wavelength of this characteristic radiation is much larger than the typical size of interstellar dust particles. Accordingly, this radio radiation reaches Earth completely unscattered by interstellar debris. The opportunity to observe interstellar space well beyond a few thousand parsecs, and in directions lacking background stars, makes 21-cm observations among the most important and useful in all of astronomy. We will see this technique used both in studies of our own Galaxy and in observations of truly distant astronomical objects.

18.5 *Interstellar Molecules*

④ ⑤ In certain interstellar regions of cold (typically 20 K) neutral gas, densities can reach as high as 10^{12} particles/m^3. Until the late 1970s, astronomers regarded these regions simply as abnormally dense interstellar clouds, but it is now recognized that they belong to an entirely new class of interstellar matter. The gas particles in these regions are not in atomic form at all; they are molecules. Because of the predominance of molecules in these dense interstellar regions, they are known as **molecular clouds**. Only within recent years have astronomers begun to appreciate the vastness of these clouds. They literally dwarf even the largest emission nebulae, which were previously thought to be the most massive residents of interstellar space.

MOLECULAR SPECTRAL LINES

As noted in Chapter 4, molecules can become collisionally or radiatively excited, much like atoms. ∞ (p. 86) Furthermore, again like atoms, molecules relax back to their ground states whenever the opportunity arises. In the process, they emit radiation. The energy states of molecules are much more complex than those of atoms, however. Molecules, like atoms, can undergo internal electron transitions, but unlike atoms, they can also rotate and vibrate. They do so in specific ways, obeying the laws of quantum physics.

Figure 18.18 illustrates a simple molecule that is rotating rapidly. After a length of time that depends on its internal makeup, the molecule relaxes back to a slower rotational rate (a state of lower energy). This change causes a photon to be emitted, carrying an energy equal to the energy difference

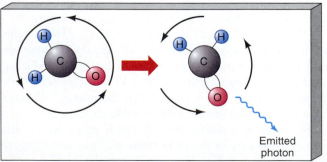

Figure 18.18   As a molecule changes from a rapid rotation (left) to a slower rotation (right), a photon is emitted that can be detected with a radio telescope. Depicted here is the formaldehyde molecule, H_2CO. The length of the curved arrows is proportional to the spin rate of the molecule.

between the two rotational states. The energy differences between rotational states are generally very small, so the emitted radiation is usually in the radio range.

We are fortunate that molecules emit radio radiation, because they are invariably found in the densest and dustiest parts of interstellar space. These are regions where the absorption of shorter-wavelength radiation is enough to prohibit the use of ultraviolet, optical, and most infrared techniques that might ordinarily detect changes in the energy states of the molecules. Only low-frequency radio radiation can escape, eventually to be detected on Earth.

Why are molecules found only in the densest and darkest of the interstellar clouds? One possible reason is that the dust serves to protect the fragile molecules from the normally harsh interstellar environment—the same absorption that prevents high-frequency radiation from getting out to our detectors also prevents it from getting in to destroy the molecules. Another possibility is that the dust acts as a catalyst that helps form the molecules. The grains provide both a place where atoms can stick and react and a means of dissipating any heat associated with the reaction, which might otherwise destroy the newly formed molecules. Probably the dust plays both roles. The close association between dust grains and molecules in dense interstellar clouds argues strongly in favor of this picture, but the details are still being debated.

MOLECULAR TRACERS

By far the most common constituent of molecular clouds is molecular hydrogen (H_2). Unfortunately, despite its prevalence, this molecule does not emit or absorb radio radiation. It emits only short-wavelength ultraviolet radiation, so it cannot easily be used as a probe of cloud structure. Nor are 21-cm observations helpful—they are sensitive only to atomic hydrogen, not to the molecular form of the gas. Theorists had expected H_2 to abound in these dense, cold pockets of interstellar space, but proof of its existence was hard to obtain. Only when spacecraft measured the ultraviolet spectra of a few stars located near the edges of some dense clouds was the presence of molecular hydrogen confirmed.

With hydrogen effectively ruled out as a probe of molecular clouds, astronomers had to find other ways to study the dark interiors of these dusty regions. Fortunately, there are plenty to choose from. Spectral emissions caused by rotational changes of many different heavy molecules have been detected by radio telescopes. Molecules such as carbon monoxide (CO), hydrogen cyanide (HCN), ammonia (NH_3), water (H_2O), methyl alcohol (CH_3OH), formaldehyde (H_2CO), and about 60 others, some of them quite complex, are now known to exist in interstellar space.* Their abundances are very small—they are generally 1 million to

---

* *Some remarkably complex organic molecules have been found in the densest of the dark interstellar clouds, including formaldehyde (H_2CO), ethyl alcohol (CH_3CH_2OH), methylamine (CH_3NH_2), and formic acid (H_2CO_2). Their presence has fueled speculation about the origins of life, both on Earth and in the interstellar medium—especially since the recent discovery by radio astronomers of evidence that glycine (NH_2CH_2COOH), one of the key amino acids that form the large protein molecules in living cells, may also be present in interstellar space.*

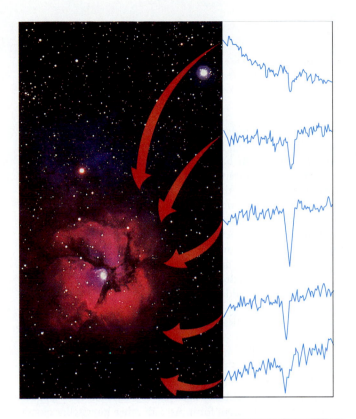

Figure 18.19 Spectra indicate that formaldehyde molecules exist around M20, as indicated by the arrows. The lines are most intense both in the dark dust lanes trisecting the nebula and in the dark regions beyond the nebula.

1 billion times less abundant than H_2. But their parent molecular clouds are large and dense enough that photons emitted by many molecules accumulate to yield detectable signals.

These molecules are unimportant in terms of the overall properties of molecular clouds, but they play a vital role as *tracers* of a cloud's structure and physical properties. We believe that such molecules are produced by chemical reactions within molecular clouds—so when we observe high densities of formaldehyde, for example, we know that the regions under study also contain high densities of molecular hydrogen, dust, and other important constituents. Spectral studies of formaldehyde can thus provide physical information on the entire cloud. Moreover, the rotational properties of different molecules often make them suitable as probes of regions with different physical properties. Formaldehyde may provide the most useful information on one region, carbon monoxide on another, and water on yet another, depending on the densities and temperatures of the regions involved. These data equip astronomers with a sophisticated spectroscopic "toolbox" for studying the interstellar medium.

For example, Figure 18.19 shows some of the sites where formaldehyde molecules have been detected near M20. At practically every dark area sampled between M16 and M8, this molecule is present in surprisingly large abundance (although it is still far less common than H_2). Analyses of spec-

tral lines at many locations along the 12°-wide swath shown in Figure 18.6 indicate that the temperature and density are much the same in all the molecular clouds studied (50 K and 10^{11} molecules/m^3, on average). Figure 18.20 shows a contour map of the distribution of formaldehyde molecules in the immediate vicinity of the M20 nebula. It was made by observing radio spectral lines of formaldehyde at various locations and then drawing contours connecting regions of similar abundance. Notice that the amount of formaldehyde (and, we assume, the amount of hydrogen) peaks in a dark region, well away from the visible nebula.

Radio maps of interstellar gas and infrared maps of interstellar dust reveal that molecular clouds do not exist as distinct and separate objects in space. Rather, they make up huge **molecular cloud complexes**, some spanning as much as 50 pc across and containing enough gas to make millions of stars like our Sun. There are about 1000 such giant complexes known in our Galaxy.

The very existence of molecules has forced astronomers to rethink and reobserve interstellar space. In doing so, they have begun to realize that this active and interesting domain is far from the void suspected by theorists not so long ago. Regions of space recently thought to contain nothing more than galactic "garbage"—the cool, tenuous darkness among the stars—now play a critical role in our understanding of stars and the interstellar medium from which they are born.

Figure 18.20 Contour map of the amount of formaldehyde near the M20 nebula, demonstrating how formaldehyde is especially abundant in the darkest interstellar regions. Other kinds of molecules have been found to be similarly distributed. The contour values increase from the outside to the inside, so the maximum density of formaldehyde lies just to the bottom right of the visible nebula. The green and red contours outline the intensity of the formaldehyde absorption lines at different rotational frequencies. The nebula itself is about 4 pc across.

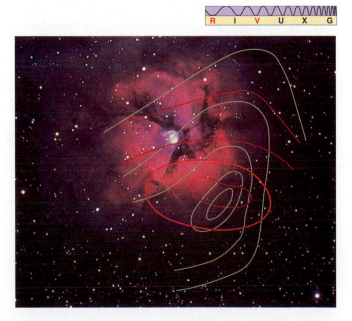

Chapter Review

SUMMARY

The **interstellar medium** (p. 382) occupies the space between stars. It is made up of cold (less than 100 K) gas, mostly atomic or molecular hydrogen and helium, and **dust grains** (p. 382). Interstellar dust is very effective at blocking our view of distant stars, even though the density of the interstellar medium is very low. The spatial distribution of interstellar matter is very patchy. The general diminution of starlight by dust is called **extinction** (p. 382). In addition, the dust preferentially absorbs short-wavelength radiation, leading to a distinct **reddening** (p. 382) of light passing interstellar clouds.

Interstellar dust is thought to be composed of silicates, graphite, iron, and "dirty ice." Interstellar dust particles are apparently elongated or rodlike. The **polarization** (p. 384) of starlight provides a means of studying them.

A **nebula** (p. 386) is a general term for any fuzzy bright or dark patch on the sky. **Emission nebulae** (p. 386) are extended clouds of hot, glowing interstellar gas. Associated with star formation, they are caused by hot O- and B-type stars heating and ionizing their surroundings. Studies of the emission lines produced by excited nebular atoms allow astronomers to measure the nebula's properties. Nebulae are often crossed by dark **dust lanes** (p. 387)—part of the larger cloud from which they formed.

Some excited atomic states take so long to emit a photon that the spectral lines associated with these transitions are never seen in terrestrial laboratories, where collisions always knock the atom into another energy state before it can emit any radiation. When these lines are seen in nebular spectra, they are called **forbidden lines** (p. 388).

Dark dust clouds (p. 389) are cold, irregularly shaped regions in the interstellar medium that diminish or completely obscure the light from background stars. Astronomers can learn about these clouds by studying the absorption lines they produce in starlight that passes through them.

Another way to observe cold, dark regions of interstellar space is through **21-centimeter radiation** (p. 393). Such radiation is produced whenever the electron in an atom of hydrogen reverses its spin, changing its energy very slightly in the process. This radio radiation is important because it is emitted by all cool atomic hydrogen gas, even if the gas is undetectable by other means. In addition, 21-centimeter radiation is not appreciably absorbed by the interstellar medium, so radio astronomers making observations at this wavelength can "see" to great distances.

The interstellar medium also contains many cold dark **molecular clouds** (p. 394), which are observed mainly through the radio radiation emitted by the molecules they contain. Dust within these clouds probably both protects the molecules and acts as a catalyst to help them form. As with other interstellar clouds, hydrogen is by far the most common constituent, but molecular hydrogen is very hard to observe. Astronomers usually study these clouds through observations of other "tracer" molecules that are less common but much easier to detect.

Astronomers believe that molecular clouds are likely sites of future star formation. Often, several molecular clouds are found close to one another, forming an enormous **molecular cloud complex** (p. 395) millions of times more massive than the Sun.

SELF-TEST: True or False?

_____ **1.** Interstellar matter is evenly distributed throughout the Milky Way Galaxy.

_____ **2.** In the vicinity of the Sun, there is about as much mass in the form of interstellar matter as in the form of stars.

_____ **3.** There is a lack of heavy elements in interstellar gas because they go into making interstellar dust.

_____ **4.** The fact that starlight becomes polarized as it passes through the interstellar medium tells us that interstellar dust particles are spherical in shape.

_____ **5.** A typical region of dark interstellar space has a temperature of about 500 K.

_____ **6.** An emission nebula is a cloud of dust reflecting the light of a nearby star cluster.

_____ **7.** Emission nebulae display spectra almost identical to those of the stars embedded in them.

_____ **8.** "Forbidden" emission lines can occur in emission nebulae because the density of interstellar gas there is extremely low.

_____ **9.** Because of the obscuration of visible light by interstellar dust, we can observe stars only within a few thousand parsecs of Earth.

_____ **10.** A typical dark dust cloud is many hundreds of parsecs across.

_____ **11.** Because of their low temperatures, dark dust clouds radiate mainly in the radio part of the electromagnetic spectrum.

_____ **12.** 21-centimeter radiation provides astronomers with information on the density, temperature, and internal motions of interstellar gas.

_____ **13.** 21-centimeter radiation can pass unimpeded through the entire Milky Way Galaxy.

_____ **14.** Water, formaldehyde, carbon monoxide, and numerous organic molecules are all commonly found in molecular clouds.

SELF-TEST: Fill in the Blank

1. The interstellar medium is made up of _____ and _____.
2. To scatter a beam of radiation, a particle must be _____ in size to the wavelength of the radiation.
3. Extinction is the _____ of starlight by interstellar _____.
4. The density of interstellar matter can be characterized as being very _____.
5. Interstellar gas is composed of 90 percent _____ and 9 percent _____.
6. The temperature of a typical emission nebula is about _____ K.
7. An HII region is another name for an _____.
8. Dark dust clouds can have temperatures as low as _____ K.
9. 21-centimeter radiation is emitted by _____ hydrogen.
10. 21-centimeter radiation results from a change in the _____ of the electron in a _____ atom.
11. Molecular clouds typically have temperatures of about _____ K.
12. Emissions from molecular clouds are in the _____ part of the electromagnetic spectrum.
13. The most common constituent of molecular clouds is molecular _____.
14. A molecular cloud complex may contain as much as _____ solar masses of gas.

REVIEW AND DISCUSSION

1. Give a brief description of the interstellar medium.
2. What is the composition of interstellar gas? What about interstellar dust?
3. How dense is the matter between the stars, on average?
4. If space is a near-perfect vacuum, how can there be enough dust in it to block light?
5. How is interstellar matter distributed through space?
6. What are some methods that astronomers use to study interstellar dust?
7. What is an emission nebula?
8. Why do emission nebulae appear red in color photographs?
9. Give a brief description of a dark dust cloud.
10. Dark dust clouds are said to obscure and redden light, yet starlight reflected off them appears blue. Why is this?
11. What is 21-centimeter radiation? With what element is it associated?
12. Why can't 21-centimeter radiation be used to probe the interiors of molecular clouds?
13. How does a molecular cloud differ from other interstellar matter?
14. If our Sun were surrounded by a cloud of gas, would this cloud be an emission nebula? Why or why not?
15. Compare the reddening of stars by interstellar dust with the reddening of the setting Sun.
16. Explain what it means for a star's light to be polarized. How does the polarization of starlight provide a means of studying the interstellar medium?

PROBLEMS

1. Calculate the total mass of interstellar matter (of density 10^7 hydrogen atoms/m³, each atom with a mass of 1.7×10^{-27} kg) contained in a volume equal to the volume of the Earth.
2. In order to carry enough energy to ionize a hydrogen atom, a photon must have a wavelength less than 9.12×10^{-8} m (91.2 nm). Using Wien's Law (Chapter 3), calculate the temperature a star must have for the peak wavelength of its black-body curve to equal this value.
3. Calculate the frequency of 21-cm radiation.
4. A beam of light shining through a dense molecular cloud is diminished in intensity by a factor of 2.5 for every 3 parsecs it travels. By what total factor is it reduced if the total thickness of the cloud is 60 pc?
5. Calculate the radius of a spherical molecular cloud whose total mass equals the mass of the Sun. Assume a cloud density of 10^{12} hydrogen atoms per cubic meter.

PROJECTS

1. The constellation Orion the Hunter is prominent in the evening sky of winter. Its most noticeable feature is a short, straight row of three medium-bright stars: the famous Belt of Orion. A line of stars extends from the east-most star of the Belt, towards the south. This line represents Orion's Sword. Towards the bottom of the sword is the sky's most famous emission nebula, M42, the Orion Nebula.

   Observe the Orion Nebula with your eye, with binoculars, and with a telescope. What is its color? How can you account for this? With the telescope, try to find the Trapezium, a grouping of four stars in the center of M42. These are hot, young stars; their energy causes the Orion Nebula to glow.

2. Observe the Milky Way on a dark, very clear night. Is it a continuous band of light across the sky or is it mottled? The parts of the Milky Way that appear missing are actually dark dust clouds that are relatively near the Sun. Identify the constellations in which you see these clouds. Make a sketch and compare to a star atlas. Find other small clouds in the atlas and try to find them with your eye or with binoculars.

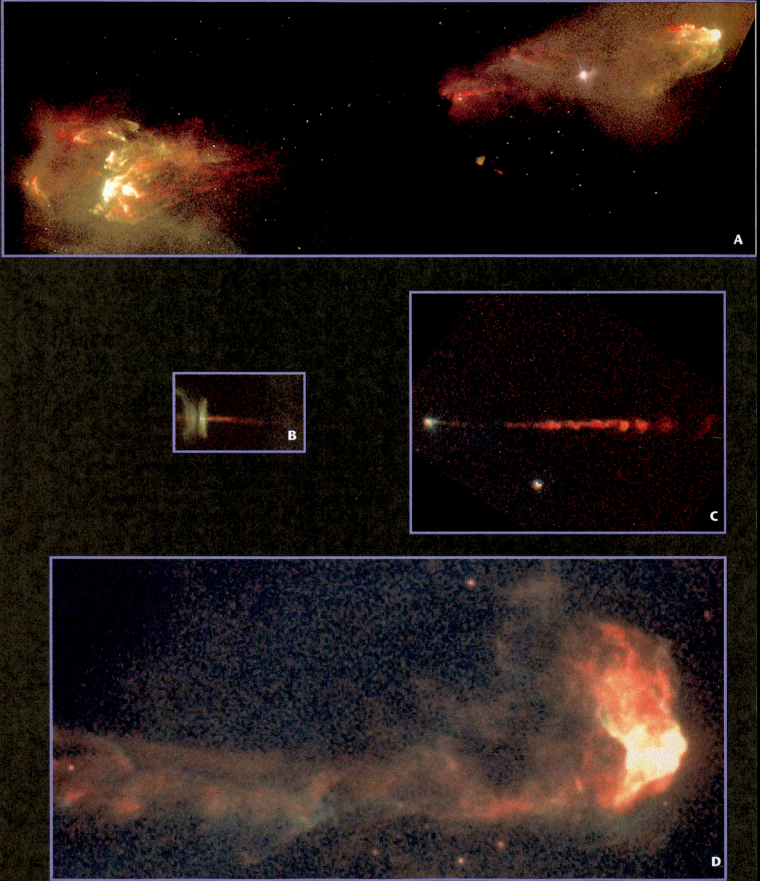

19

STAR FORMATION

A Traumatic Birth

LEARNING GOALS

Studying this chapter will enable you to:

1 Discuss the factors that compete against gravity in the process of star formation.

2 Summarize the sequence of events leading to the formation of a star like our Sun.

3 Explain how the formation of a star is affected by its initial mass.

4 Describe some of the observational evidence supporting the modern theory of star formation.

5 Explain the nature of interstellar shock waves, and discuss their possible role in the formation of stars.

(Photo A) Disks and jets pervade the universe on many scales. Here, they seem to be a natural result of a rotating cloud of gas contracting to form a star. Matter falling onto the embryonic star creates a pair of high-speed jets of gas perpendicular to the star's flattened disk, carrying away heat and angular momentum that might otherwise prevent the birth of the star. This image shows a small region near the Orion Nebula known as HH1/HH2, whose twin jets have blasted outward for several trillion km (nearly half a light-year) before colliding with interstellar matter. (HH stands for Herbig-Haro, after the discoverers of the first such objects.) The next three photos show stellar jets ejected from three different very young stars. Reproduced here to scale, these images collectively depict the propagation of a jet through space.

(Photo B) This image of HH30, spanning approximately 250 billion km, or about 0.01 pc, shows a thin jet (in red) emanating from a circumstellar disk (at left in grey) encircling a nascent star.

(Photo C) One of HH34's jets is longer, reaching some 600 billion km, yet remains narrow, with a beaded structure.

(Photo D) HH47 is more than a trillion km in length, or nearly 0.1 pc. This photo shows one of its jets plowing through interstellar space, creating bow shocks in the process.

We now move from the interstellar medium—the gas and dust between the stars—to the stars themselves. The next four chapters discuss the formation and evolution of stars. We have already seen that stars must evolve as they consume their fuel supply, and we have extensive observational evidence of stars at many different evolutionary stages. With the help of these observations, scientists have developed a good understanding of stellar evolution—the complex changes experienced by stars as they form, mature, grow old, and die. We begin by studying the process of star formation, through which interstellar clouds of gas and dust are transformed into the myriad stars we see in the night sky.

19.1 Gravitational Competition

GRAVITY AND HEAT

1 How do stars form? What factors determine the masses, luminosities, and distribution of stars in our Galaxy? In short, what basic processes are responsible for the appearance of our nighttime sky? Simply stated, star formation begins when part of the interstellar medium—one of the cold, dark clouds discussed in Chapter 18—starts to collapse under its own weight. ∞ (p. 389) The cloud fragment heats up as it shrinks, and eventually its center becomes hot enough for nuclear burning to begin. At that point, the contraction stops, and a star is born. But what determines which interstellar clouds collapse? For that matter, since all clouds exert a gravitational pull, why didn't they all collapse long ago? To begin to answer these questions, let us consider a small portion of a large cloud of interstellar gas. Concentrate first on just a few atoms, as shown in Figure 19.1.

Each atom has some random motion because of the cloud's heat, even if the cloud's temperature is very low. Each atom is also influenced by the gravitational attraction of all its neighbors. The gravitational force is not large, however, because the mass of each atom is so small. Even when a few atoms accidentally cluster for an instant, as shown in Figure 19.1(b), their combined gravity is insufficient to bind them into a lasting, distinct clump of matter. This accidental cluster would disperse as quickly as it formed. The effect of heat—the random motion of the atoms—is much stronger than the effect of gravity.

Now let's concentrate on a larger group of atoms. Imagine, for example, 50, 100, 1000, even a million atoms, each gravitationally pulling on all the others. The force of gravity is now stronger than before. Would this many atoms exert a combined gravitational attraction strong enough to prevent the clump from dispersing again? The

answer—at least under the conditions found in interstellar space—is still no. The gravitational attraction of this mass of atoms is still far too weak to overcome the effect of heat.

We have already seen numerous instances of the competition between heat and gravity (see, for example, the *More Precisely* feature on p. 136). Recall that the temperature of a gas is simply a measure of the average speed of the atoms or molecules in it. ∞ (p. 63) The higher the temperature, the greater the average speed, and thus the higher the pressure of the gas. This is the main reason that the Sun and other stars don't collapse. The outward pressure of their heated gases exactly balances gravity's inward pull.

SOME COMPLICATING FACTORS

Heat is not the only factor that tends to oppose gravitational contraction. Rotation—that is, spin—can also compete with gravity's inward pull. As we saw in Chapter 15, a contracting cloud having even a small spin tends to develop a bulge around its midsection. ∞ (p. 320) As the cloud contracts, it must spin faster (to conserve its angular momentum), and the bulge grows—material on the edge tends to fly off into space. Figure 19.2 illustrates this important feature of rotation. (Consider as an analogy mud flung from a rapidly rotating bicycle wheel.) Eventually, the cloud forms a flattened, rotating disk.

For material to remain part of the cloud and not be spun off into space, a force must be applied—in this case, the force of gravity. The more rapid the rotation, the greater the tendency for the gas to escape, and the greater the gravitational force needed to retain it. It is in this sense that we can regard rotation as opposing the inward pull of gravity. Should the rotation of a contracting gas cloud overpower gravity, the cloud would simply disperse. Thus, rapidly rotating interstellar clouds need more mass for contraction into stars than clouds having no rotation at all.

Magnetism can also hinder a cloud's contraction. Just as

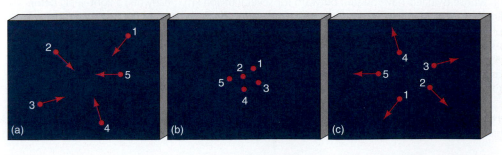

Figure 19.1 Motions of a few atoms within an interstellar cloud are influenced by gravity so slightly that their paths are hardly changed (a) before, (b) during, and (c) after an accidental, random encounter.

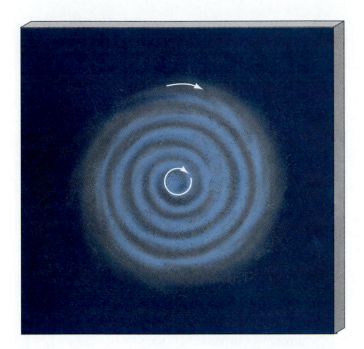

Figure 19.2   A rapidly rotating gas cloud tends to resist contraction. The spin tends to fling matter from the cloud, like mud spinning off the rim of a bicycle wheel. Spin thus competes with the inward pull of gravity. An extended disk of matter forms around the edge of the cloud.

Earth, most of the other planets, and the Sun all have some magnetism, magnetic fields permeate most interstellar clouds. As a cloud contracts, it heats up, and atomic encounters become violent enough to ionize the gas. As we noted in Chapter 7 when discussing Earth's Van Allen belts and in Chapter 16 when discussing activity on the Sun, magnetic fields can exert electromagnetic control over charged particles. ∞ (pp. 154, 343) In effect, the particles tend to become "tied" to the magnetic field—free to move *along* the field lines but inhibited from moving *perpendicular* to them. As a result, interstellar clouds may contract in distorted ways. Because the charged particles and the magnetic field are linked, the field itself follows the contraction of a cloud, as indicated in Figure 19.3. The charged particles literally pull the magnetic field toward the cloud's center in the direction perpendicular to the field lines. As the field lines are compressed, the magnetic field strength increases. In this way, the strength of magnetism in a cloud can become much larger than that normally permeating general interstellar space. The primitive solar nebula may have contained a strong magnetic field created in just this way.

Theory suggests that even small quantities of rotation or magnetism can compete quite effectively with gravity and can greatly alter the evolution of a typical gas cloud. Unfortunately, the interplay of these factors is not very well understood—both can lead to very complex behavior as a cloud contracts, and the combination of the two is extremely difficult to study theoretically. In this chapter we will gain an appreciation for the broad outlines of the star-formation process by neglecting these two complicating

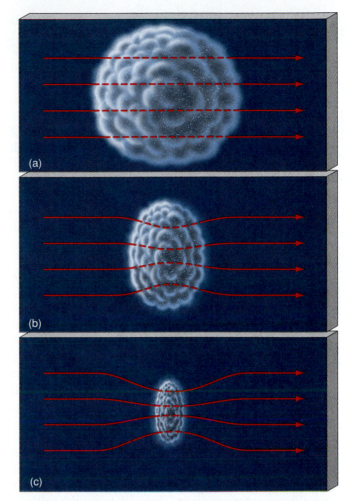

Figure 19.3   Magnetism can hinder the contraction of a gas cloud, especially in directions perpendicular to the magnetic field (solid lines). Frames (a), (b), and (c) trace the evolution of a slowly contracting interstellar cloud having some magnetism.

factors. Bear in mind, however, that both are probably important in determining the details.

We now return to our original question: How many atoms need to be accumulated for their collective pull of gravity to prevent them from dispersing back into interstellar space? The answer, even for a typical cool (100 K) cloud having no rotation or magnetism, is a truly huge number. Nearly 10^{57} atoms are required—much larger than the 10^{25} grains of sand on all the beaches of the world, even larger than the 10^{51} elementary particles that constitute all the atomic nuclei in the entire Earth. There is simply nothing on Earth comparable to a star.

19.2   *The Formation of Stars Like the Sun*

❷ We can best study the specific steps of star formation by considering the Hertzsprung–Russell (H–R) diagram studied earlier in Chapter 17. ∞ (p. 368) Recall that an H–R diagram is a plot of two key stellar properties—

surface temperature (increasing to the left) and luminosity (increasing upward). The luminosity scale in Figure 19.4 is expressed in terms of the solar luminosity ($L_\odot \approx 4 \times 10^{26}$ W). Our G2-type Sun is plotted at a temperature of 6000 K and a luminosity of 1 unit. We can also indicate the *size* of a star represented by any point on the diagram because of the radius–luminosity–temperature relationship. ∞ (p. 361) As before, the dashed diagonal lines in the H–R diagrams mark stellar radius, allowing us to follow the changes in a star's size as it evolves.

As we saw earlier, most stars plotted on the H–R diagram fall along the main sequence. For roughly 90 percent of their lifetimes, stars burn rather quietly and their physical properties do not change much. Data points representing such stable, full-fledged stars remain almost stationary on the H–R diagram.

Near the beginning and end of its existence, however, a star's properties change drastically and rapidly. The H–R diagram is a useful aid in describing these phases of its life. At each phase of a star's evolution, its surface temperature and luminosity can be represented by a point on the H–R diagram. The motion of that point about the diagram as the star evolves is known as the star's **evolutionary track**. It is a graphical representation of a star's life.

Table 19-1 lists seven evolutionary stages that an interstellar cloud goes through prior to becoming a main-sequence star like our Sun. These stages are characterized by varying central temperatures, surface temperatures, central densities, and radii of the prestellar object. They trace its progress from a quiescent interstellar cloud to a genuine star. The numbers given in Table 19-1 and the following discussion are valid *only* for stars of approximately the same mass as the Sun. In the next section we will relax this restriction and consider the formation of other stars.

STAGE 1—AN INTERSTELLAR CLOUD

The first stage in the star-formation process is an ordinary dense interstellar cloud, like those studied in Chapter 18. These clouds are truly vast, sometime spanning tens of parsecs (10^{14}–10^{15} km) across. Typical temperatures are about 10 K throughout, and densities are usually not much more

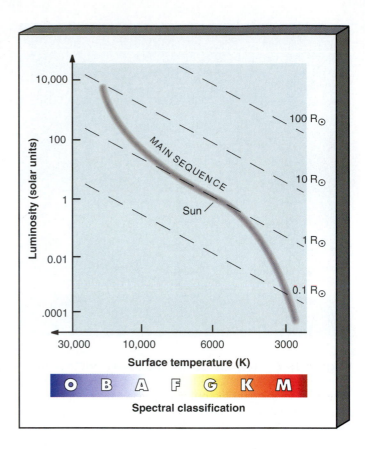

Figure 19.4   The H–R diagram is a useful way to summarize the observed properties of both stars and prestellar objects. The diagonal lines correspond to stars of the same radius.

than 10^9 particles/m³. Stage-1 clouds contain thousands of times the mass of the Sun in the form of cold atomic and molecular gas.

If such a cloud is to be the birthplace of stars, it must become unstable and eventually break up into smaller pieces. The initial collapse occurs when a pocket of gas becomes gravitationally unstable. Perhaps it is squeezed by some external event, such as the pressure wave produced when a nearby O- or B-type star forms and ionizes its surroundings. Or possibly its supporting magnetic field leaks

TABLE 19-1   *Prestellar Evolution of a Solar-Type Star*

STAGE	APPROXIMATE TIME TO NEXT STAGE	CENTRAL TEMPERATURE	SURFACE TEMPERATURE	CENTRAL DENSITY	DIAMETER	OBJECT
	(yr)	(K)	(K)	(particles/m³)	(km)	
1	2×10^6	10	10	10^9	10^{14}	Interstellar cloud
2	3×10^4	100	10	10^{12}	10^{12}	Cloud fragment
3	10^5	10,000	100	10^{18}	10^{10}	Cloud fragment/protostar
4	10^6	1,000,000	3000	10^{24}	10^8	Protostar
5	10^7	5,000,000	4000	10^{28}	10^7	Protostar
6	3×10^7	10,000,000	4500	10^{31}	2×10^6	Star
7	10^{10}	15,000,000	6000	10^{32}	1.5×10^6	Main-sequence star

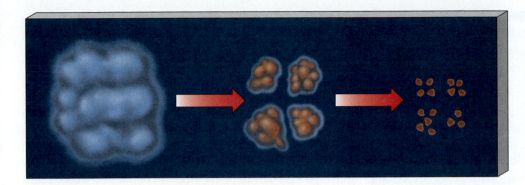

Figure 19.5 As an interstellar cloud collapses, gravitational instabilities cause it to fragment into smaller pieces. The pieces themselves continue to collapse and fragment, eventually to form many tens or hundreds of separate stars.

away as charged particles slowly drift across the confining field lines. Whatever the specific cause, theory suggests that once the collapse begins, fragmentation into smaller and smaller clumps of matter naturally follows, as gravitational instabilities continue to operate in the gas. As illustrated in Figure 19.5, a typical cloud can break up into tens, hundreds, even thousands, of fragments, each imitating the shrinking behavior of the parent cloud and contracting ever faster. The whole process, from a single quiescent cloud to many collapsing fragments, takes a few million years.

In this way, depending on the precise conditions under which fragmentation takes place, an interstellar cloud can produce either a few dozen stars, each much larger than our Sun, or a whole cluster of hundreds of stars, each comparable to or smaller than our Sun. There is *little* evidence for stars born in isolation, one star from one cloud. Most stars—perhaps even all stars—appear to originate as members of multiple systems or star clusters. The Sun, which is now found alone and isolated in space, probably escaped from the multiple-star system where it formed after an encounter with another star or some much larger galactic object (such as a molecular cloud).

STAGE 2—A COLLAPSING CLOUD FRAGMENT

The second stage in our evolutionary scenario represents the physical conditions in just one of the many fragments that develop in a typical interstellar cloud. A fragment destined to form a star like the Sun contains between 1 and 2 solar masses of material at this stage. Estimated to span a few hundredths of a parsec across, this fuzzy, gaseous blob is still about 100 times the size of our solar system. Its central density is now some 10^{12} particles/m^3.

Even though it has shrunk substantially in size, the fragment's average temperature is not much different from that of the original cloud. The reason is that the gas constantly radiates large amounts of energy into space. The material of the fragment is so thin that photons produced within it easily escape without being reabsorbed by the cloud, so virtually all the energy released in the collapse is radiated away and does not cause any significant increase in temperature. Only at the center, where the radiation must traverse the greatest amount of material in order to escape, is there any appreciable temperature increase. The gas there might be as warm as 100 K by this stage. For the most part, however, the fragment stays cold as it shrinks.

The process of continued fragmentation is eventually stopped by the increasing density within the shrinking cloud. As stage-2 fragments continue to contract, they eventually become so dense that radiation cannot get out easily. The trapped radiation causes the temperature to rise, the pressure to increase, and the fragmentation to cease.

STAGE 3—FRAGMENTATION CEASES

Several tens of thousands of years after it first began contracting, a typical stage-2 fragment has shrunk by the start of stage 3 to a gaseous sphere with a diameter roughly the size of our solar system (still 10,000 times the size of our Sun). The inner regions of the fragment have just become opaque to their own radiation and so have started to heat up, as noted in Table 19-1. The central temperature has reached about 10,000 K—hotter than the hottest steel furnace on Earth. However, the temperature at the fragment's periphery has not increased much. It is still able to radiate its energy into space and so remains cool. The density increases much faster in the core of the fragment than at its periphery, so the outer portions of the cloud are both cooler and thinner than the interior. The central density by this time is approximately 10^{18} particles/m^3 (still only 10^{-9} kg/m^3 or so).

For the first time, our fragment is beginning to resemble a star. The dense, opaque region at the center is called a **protostar**—an embryonic object perched at the dawn of star birth. Its mass grows as more and more material rains down on it from outside, although its radius continues to shrink because its pressure is still unable to overcome the relentless pull of gravity. After stage 3, we can distinguish a "surface" on the protostar—its *photosphere*. Inside the photosphere, the protostellar material is opaque to the radiation it emits.* From here on, the surface temperatures listed in Table 19-1 refer to the photosphere and not to the "periphery" of the collapsing fragment, whose temperature remains low.

STAGE 4—A PROTOSTAR

As the protostar evolves, it shrinks, its density grows, and its temperature rises, both in the core and at the photosphere. Some 100,000 years after the fragment began to form, it reaches stage 4, where its center seethes at about 1,000,000 K. The electrons and protons ripped from atoms

*Note that this is the same definition of "surface" that we used for the Sun in Chapter 16. ∞ (p. 337)

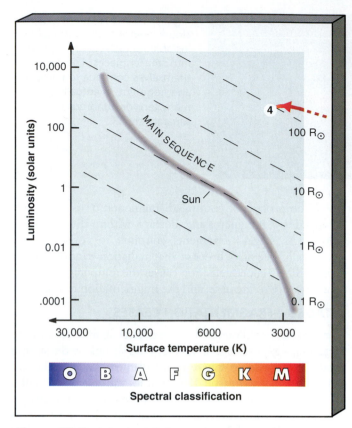

Figure 19.6 Diagram of the approximate evolutionary track followed by an interstellar cloud fragment prior to reaching the end of the Kelvin–Helmholtz contraction phase as a stage-4 protostar. (The circled numbers on this and subsequent plots refer to the prestellar evolutionary stages listed in Table 19-1 and described in the text.)

whiz around at hundreds of kilometers per second, yet the temperature is still short of the 10^7 K needed to ignite the proton–proton nuclear reactions that fuse hydrogen into helium. Still much larger than the Sun, our gassy heap is now about the size of Mercury's orbit. Heated by the material falling on it from above, its surface temperature has risen to a few thousand kelvins.

By the time stage 4 is reached, our protostar's physical properties can be plotted on the H–R diagram, as shown in Figure 19.6. Knowing the protostar's radius and surface temperature, we can calculate its luminosity. Surprisingly, it turns out to be several thousand times the luminosity of the Sun. Even though the protostar has a surface temperature only about half that of the Sun, it is hundreds of times larger, making its total luminosity very large indeed—in fact, much greater than the luminosity of most main-sequence stars. Because nuclear reactions have not yet begun, this luminosity is due entirely to the release of gravitational energy as the protostar continues to shrink in size and nebular material rains down on its surface.

Figure 19.6 depicts the approximate path followed by our interstellar cloud fragment since it became a protostar at stage 3 (which itself lies off the right-hand edge of the figure). This early evolutionary track is known as the *Kelvin–Helmholtz contraction phase*, after two European physicists (Lord Kelvin and Hermann von Helmholtz) who first studied the subject. Figure 19.7 is an artist's sketch of an interstellar gas cloud proceeding along the evolutionary path outlined so far.

Our protostar is still not in equilibrium. Even though its temperature is now so high that outward-directed pressure has become a powerful countervailing influence against gravity's continued inward pull, the balance is not yet perfect. The protostar's internal heat gradually diffuses out from the hot center to the cooler surface, where it is radiated away into space. As a result, the overall contraction slows, but it does not stop completely. From our perspective on Earth, this is quite fortunate: If the heated gas were somehow able to counteract gravity completely before the star reached the temperature and density needed to start nuclear burning in its core, the protostar would simply radiate away its heat and never become a true star. The night sky would be abundant in faint protostars, but completely lacking in the genuine article. Of course, there would be no Sun either, so it is unlikely that we, or any other intelligent life form, would exist to appreciate these astronomical subtleties.

After stage 4, the protostar on the H–R diagram moves down (toward lower luminosity) and slightly to the left (toward higher temperature), as shown in Figure 19.8. Its surface temperature remains almost constant, and it becomes less luminous as it shrinks. This portion of our protostar's evolutionary path running from point 4 to point 6 in Figure 19.8 is called the *Hayashi track*, after C. Hayashi, a twentieth-century Japanese researcher. Protostars on the Hayashi track often exhibit violent surface activity. As a consequence, they can have extremely strong protostellar winds, much denser than that of our own Sun. The **T Tauri stars** discussed in *Interlude 19–1* on p. 415 may well be direct observational evidence of this phase of stellar evolution.

STAGE 5—PROTOSTELLAR EVOLUTION

By stage 5 on the Hayashi track, the protostar has shrunk to about 10 times the size of the Sun, its surface temperature is about 4000 K, and its luminosity has fallen to about 10 $L_\odot$. At this point, the central temperature has reached about 5,000,000 K. The gas is completely ionized by now, but the protons still do not have enough thermal energy to overwhelm their mutual electromagnetic repulsion and enter the realm of the nuclear binding force. The core is still too cool for nuclear burning to begin.

Events in a protostar's development proceed more slowly as the protostar approaches the main sequence. The initial contraction and fragmentation of the interstellar cloud occurred quite rapidly, but as the protostar nears the status of a full-fledged star, its evolution slows. The cause of this slowdown is heat—even gravity must struggle to compress a hot object. The contraction is governed largely by the rate at which the protostar's internal energy can be radiated away into space. The greater this radiation of internal energy—that is, the more energy that moves through the star to escape from its surface, the faster the contraction occurs. As the luminosity decreases, so too does the contraction rate.

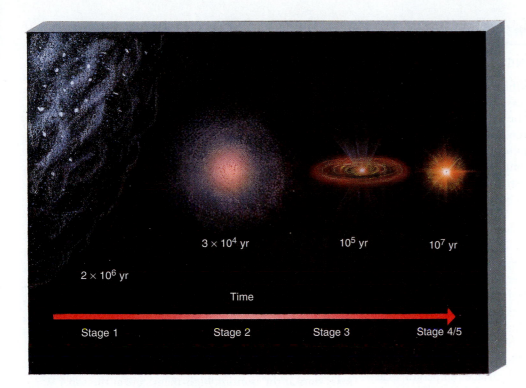

Figure 19.7 Artist's conception of the changes in an interstellar cloud during the early evolutionary stages outlined in Table 19-1. Shown are a stage-1 interstellar cloud; a stage-2 fragment; a smaller, hotter stage-3 fragment; and a stage-4/stage-5 protostar. (Not drawn to scale.) The duration of each stage, in years, is also indicated.

STAGE 6—A NEWBORN STAR

Some 10 million years after its first appearance, the protostar finally becomes a true star. By the bottom of the Hayashi track, at stage 6, when our roughly 1-solar-mass object has shrunk to a radius of about 1,000,000 km, the contraction has raised the central temperature to 10,000,000 K, enough to ignite nuclear burning. Protons begin fusing into helium nuclei in the core, and a star is born. As shown in Figure 19.8, the star's surface temperature at this point is about 4500 K, still a little cooler than the Sun. Even though the newly formed star is slightly larger in radius than our Sun, its lower temperature means that its luminosity is somewhat less than (actually, about two-thirds) the solar value.

STAGE 7—THE MAIN SEQUENCE AT LAST

Over the next 30 million years or so, the stage-6 star contracts a little more. In making this slight adjustment, the central density rises to about 10^{32} particles/m^3 (more conveniently expressed as 10^5 kg/m^3), the central temperature increases to 15,000,000 K, and the surface temperature reaches 6000 K. By stage 7, the star finally reaches the main sequence just about where our Sun now resides. Pressure and gravity are finally balanced, and the rate at which nuclear energy is generated in the core exactly matches the rate at which energy is radiated from the surface.

The evolutionary events just described occur over the course of some 40–50 million years. Although this is a long time by human standards, it is still less than 1 percent of the Sun's lifetime on the main sequence. Once an object begins fusing hydrogen and establishes a "gravity-in/pressure-out" equilibrium, it burns steadily for a very long time. The star's location on the H–R diagram will remain virtually unchanged for the next 10 billion years.

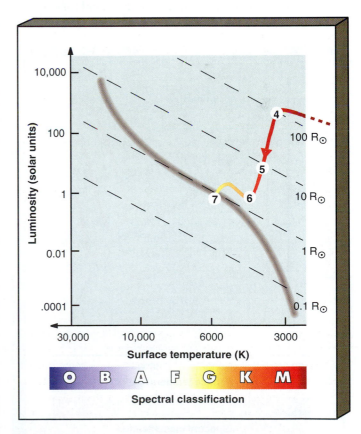

Figure 19.8 The changes in a protostar's observed properties are shown by the path of decreasing luminosity, from stage 4 to stage 6, often called the Hayashi track. At stage 7, the newborn star has arrived on the main sequence.

19.3 *Stars of Different Masses*

THE ZERO-AGE MAIN SEQUENCE

❸ The numerical values and the evolutionary track just described are valid only for the case of a 1-solar-mass star. The temperatures, densities, and radii of prestellar objects of other masses exhibit similar trends, but the numbers and the tracks differ, in some cases quite considerably. Not surprisingly, the most massive fragments within interstellar clouds tend to produce the most massive protostars and eventually the most massive stars. Similarly, low-mass fragments give rise to low-mass stars.

Figure 19.9 compares the evolutionary tracks taken by two prestellar objects—one of 0.3 and one of 3 solar masses—with the pre–main-sequence track followed by the Sun. Notice how all objects traverse the H–R diagram in the same general manner, but their luminosities and temperatures differ greatly. Cloud fragments that eventually form more massive stars approach the main sequence along a higher track on the diagram; those destined to form less massive stars take a lower track. More massive objects generally have higher luminosities and surface temperatures at any given evolutionary stage—Kelvin–Helmholtz contraction, the Hayashi track, or the main sequence—than their less massive counterparts.

The *time* required for an interstellar cloud to become a main-sequence star also depends strongly on its mass.

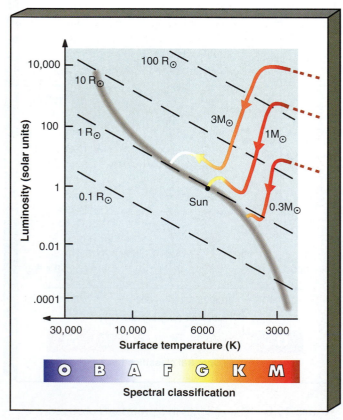

Figure 19.9 Prestellar evolutionary paths for stars more massive and less massive than our Sun.

Large cloud fragments heat up to the required 10 million K more rapidly than do less massive ones—the most massive fragments contract into stars in a mere million years, roughly 1/50 the time taken by the Sun. The opposite is the case for prestellar objects having masses less than our Sun. Cloud fragments that evolve into low-mass stars are smaller and cooler. Not only do they take a long time to become protostars, but the protostars also take their time changing into full-fledged stars. A typical M-type star, for example, requires nearly a billion years to form, some 20 times longer than it took the Sun to form.

Whatever the mass, the end point of the prestellar evolutionary track is the main sequence. The main sequence predicted by theoretical models, in which stellar properties finally settle down to stable values and an extended period of steady burning ensues, is usually called the **zero-age main sequence** (ZAMS). It agrees quite well with main sequences observed in the vicinity of the Sun and in more distant star clusters. It is important to realize that *the main sequence is itself not an evolutionary track—stars do not evolve along it*. Rather, it is just a "way station" on the H–R diagram where stars stop and spend most of their lives—low-mass stars at the bottom, high-mass stars at the top.

If all gas clouds contained precisely the same elements in exactly the same proportions, mass would be the sole determinant of a newborn star's location on the H–R diagram, and the zero-age main sequence would be a well-defined line rather than a broad band. However, the composition of a star affects its internal structure (mainly by changing the opacity of its outer layers), and this in turn affects both its temperature and its luminosity on the main sequence. Stars with more heavy elements tend to be cooler and slightly less luminous than stars that have the same mass but contain fewer heavy elements. As a result, differences in composition between stars "blur" the zero-age main sequence into a broad band instead of a narrow line.

FAILED STARS

Some cloud fragments are too small ever to become stars. The giant planet Jupiter is a good example. Jupiter contracted under the influence of gravity, and the resultant heat is still detectable, but the planet did not have enough mass for gravity to crush its matter to the point of nuclear ignition. It became stabilized by heat and rotation before the central temperature became hot enough to fuse hydrogen. Jupiter never evolved beyond the protostar stage. If Jupiter, or any of the other jovian planets, had continued to accumulate gas from the solar nebula, they might eventually have become stars. But virtually all the matter present during the formative stages of our solar system is gone now, swept away by the solar wind.

Low-mass interstellar gas fragments simply lack the mass needed to initiate nuclear burning. Rather than turning into stars, they will continue to cool, eventually becoming compact, dark "clinkers"—cold fragments of unburned matter—in interstellar space. On the basis of theoretical modeling, astronomers believe that the minimum mass of gas needed to generate core temperatures high enough to begin nuclear fusion is about 0.08 solar masses.

Vast numbers of Jupiter-like objects may well be scattered throughout the universe—fragments frozen in time somewhere along the Kelvin–Helmholtz contraction phase. Small, faint, and cool (and growing ever colder), they are known collectively as **brown dwarfs**. Our technology currently has great difficulty in detecting them, be they planets associated with stars or interstellar cloud fragments far from any star. We can telescopically detect stars and spectroscopically infer atoms and molecules, but astronomical objects of intermediate size outside our solar system are very hard to see. Interstellar space *could* contain many cold, dark Jupiter-sized objects without our knowing it. They might even account for more mass than we observe in the form of stars and interstellar gas combined.

19.4 *Observations of Cloud Fragments and Protostars*

④ The evolutionary stages we have just described are derived from numerical experiments performed on high-speed computers. Table 19-1 and the evolutionary paths described in Figures 19.6, 19.8, and 19.9 are mathematical predictions of a multifaceted problem incorporating gravity, heat, rotation, magnetism, nuclear reaction rates, elemental abundances, and other physical conditions specifying the state of contracting interstellar clouds. Computer technology has enabled theorists to construct these models, but their accuracy is only partly known, because it is difficult to test them observationally.

How then can we verify the theoretical predictions just outlined? Even the total lifetime of our entire civilization is much shorter than the time needed for a cloud to contract and form a star. We can never observe individual objects proceed through the full panorama of star birth. We can, however, observe many different objects—interstellar clouds, protostars, young stars approaching the main sequence—as they appear today at different stages of their evolutionary cycles. Each observation is like part of a jigsaw puzzle. When properly oriented relative to all the others, the pieces can be used to build up a picture of the full life cycle of a star. By observing pre–main-sequence objects at many sites in our Galaxy, astronomers have directly verified most of the prestellar stages just described.

Let us now consider in more detail some of the observational pieces that make up the modern picture of prestellar evolution.

EVIDENCE OF CLOUD CONTRACTION

Prestellar objects at stages 1 and 2 are not yet hot enough to emit much infrared radiation, and certainly no optical radiation arises from their dark, cool interiors. The best way to study the early stages of cloud contraction and fragmentation is to use radio telescopes to detect the radiation emitted or absorbed by one or more interstellar molecules. Only long-wavelength radiation can escape from these clouds to our telescopes on or near the Earth.

Figure 19.10 shows M20, the splendid emission nebula studied in Chapter 18, along with some of its surroundings. ∞ (p. 386) The brilliant region of glowing, ionized gas is not our main interest here, however. Instead, the youthful O- and B-type stars that energize the nebula alert us to the general environment where stars are forming. Emission nebulae are signposts of star birth.

The region surrounding M20 contains galactic matter that seems to be contracting. The presence of (optically)

5 pc

Figure 19.10 The beautiful emission nebula M20 (right) and its dark surroundings (left) provide examples of many phases of star formation. The nebula itself glows because of the energy of the hot young stars embedded in it. The surrounding dark regions show evidence of cloud collapse and fragmentation.

invisible gas there was illustrated in Figure 18.20, which showed a contour map of the abundance of the formaldehyde (H_2CO) molecule. These and many other kinds of molecules are widespread in the vicinity of M20, especially throughout the dusty regions below and to the right of the nebula. Their radio emission shows that they are especially abundant near the completely opaque dark region below the nebula. Further analysis of the observations suggests that this region of greatest molecular abundance is also contracting and fragmenting, well on its way toward forming a star—or, more likely, a star cluster.

The interstellar clouds in and around M20 thus provide tentative evidence for three distinct phases of star formation, as shown in Figure 19.11. The huge, dark molecular cloud surrounding the visible nebula is the stage-1 cloud. Both its density and its temperature are low, about 10^8 particles/m^3 and 20 K, respectively.

Greater densities and temperatures typify smaller regions within this huge cloud. The totally obscured regions labeled A and B, where the molecular emission of radio energy is strongest, are such denser, warmer fragments. Here, the total gas density is observed to be at least 10^9 particles/m^3, and the temperature is about 100 K. The Doppler shifts of the radio lines observed in the vicinity of region B indicate that this portion of M20, labeled "collapsing fragment" in Figure 19.11, is contracting. Less than a light year across, this region has a total mass over 1000 times the mass of the Sun—considerably more than the mass of M20 itself. It lies somewhere between stages 1 and 2 of Table 19-1.

The third star-formation phase shown in Figure 19.11 is M20 itself. The glowing region of ionized gas results directly from a massive O-type star that formed there within the past million years or so. Because the central star is already fully formed, this final phase corresponds to stage 6 or 7 of our evolutionary scenario.

EVIDENCE OF CLOUD FRAGMENTS

Other parts of our Milky Way Galaxy provide sketchy evidence for prestellar objects in stages 3 through 5. The Orion complex, shown in Figure 19.12, is one such region. Lit from within by several O-type stars, the bright Orion Nebula is partly surrounded by a vast molecular cloud that extends well beyond the roughly 5×10 parsec region bounded by the photograph in Figure 19.12(b).

The Orion molecular cloud harbors several smaller sites of intense radiation emitted by molecules deep within the core of the cloud fragment. Their extent, shown in Figures 19.12(c) and 19.12(d), measures about 10^{10} km, or 1/1000 of a light year, about the diameter of our solar system. The gas density of these smaller regions is about 10^{15} particles/m^3, much denser than the surrounding cloud. Although their temperature cannot be estimated reliably, many researchers regard these regions as objects well on their way to stage 3. We cannot determine if these regions will eventually form stars like the Sun, but it does seem certain that these intensely emitting regions are on the threshold of becoming protostars.

EVIDENCE OF PROTOSTARS

In the hunt for and study of objects at more advanced stages of star formation, radio techniques become less useful because stages 4, 5, and 6 are expected to display increasingly high temperatures. As the black-body curve of thermal emission from warm protostars and young stars shifts toward shorter wavelengths, these objects should be observable largely in the infrared.

A most interesting object within the core of the Orion molecular cloud was detected by infrared astronomers in the 1970s. Known as the Kleinmann–Low Nebula, after its discoverers, it is a strong infrared emitter with a luminosity of some $10^3 L_\odot$, lying behind the visible nebula. This compact source is outlined by contours in Figure 19.13. Most astronomers agree that this warm, dense blob is a genuine protostar, poised near the end of the Kelvin–Helmholtz contraction phase, probably around stage 4.

Until the *Infrared Astronomy Satellite* was launched in the early 1980s, astronomers were aware only of giant stars forming in clouds far away. But *IRAS* showed that stars are forming much closer to home, and some of these

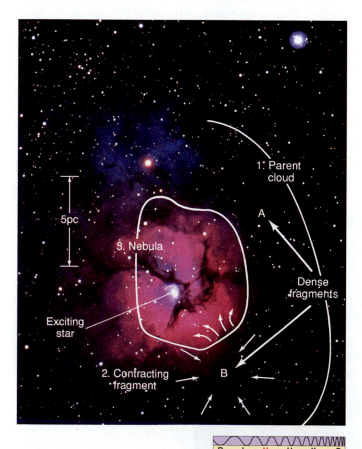

R I V U X G

Figure 19.11 The M20 region shows observational evidence for three broad phases in the birth of a star: (1) the parent cloud (stage 1 of Table 19-1), (2) a contracting fragment (between stages 1 and 2), and (3) the emission nebula (M20 itself) resulting from the formation of one or more massive stars (stages 6 and 7).

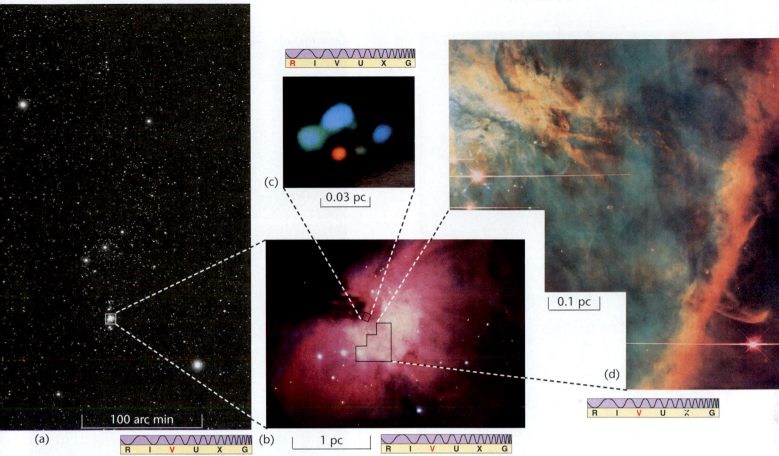

(c)

0.03 pc

0.1 pc

(d)

100 arc min

(a)

(b)

1 pc

Figure 19.12 (a) The constellation Orion, with the region around its famous emission nebula marked by a rectangle. The Orion Nebula is the middle "star" of Orion's sword. The framed region is enlarged in part (b), suggesting how the nebula is partly surrounded by a vast molecular cloud. Various parts of this cloud are probably fragmenting and contracting, with even smaller sites forming protostars. The two frames at right show some of the evidence for those protostars. (c) A false-color radio image of some intensely emitting molecular sites. (d) A real-color visible image of embedded nebular "knots" thought to harbor protostars.

protostars have masses comparable to our Sun's mass. Figure 19.14 shows a premier example of a solar-mass protostar—Barnard 5. Its infrared heat signature is that expected of an object on the Hayashi track, around stage 5.

The energy sources for some infrared objects seem to be luminous hot stars that are hidden from optical view by surrounding dark clouds. Apparently, some of the stars are already so hot that they emit large amounts of ultraviolet radiation, which is mostly absorbed by a "cocoon" of dust surrounding the central star. The absorbed energy is then reemitted by the dust as infrared radiation. These bright infrared sources are known as *cocoon nebulae*. Two considerations support the idea that the hot stars responsible for the clouds' heating have only recently ignited: (1) These dust cocoons are predicted to disperse quite rapidly once their central stars form, and (2) they are invariably found in the dense cores of molecular clouds. The central stars probably lie near stage 6.

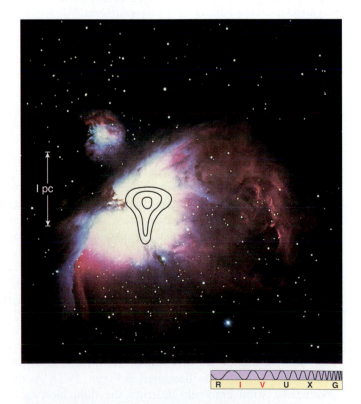

1 pc

Figure 19.13 The Kleinmann–Low Nebula, shown here as infrared contours superposed on a visible-light image, is thought to be a young protostar lying within the Orion Nebula.

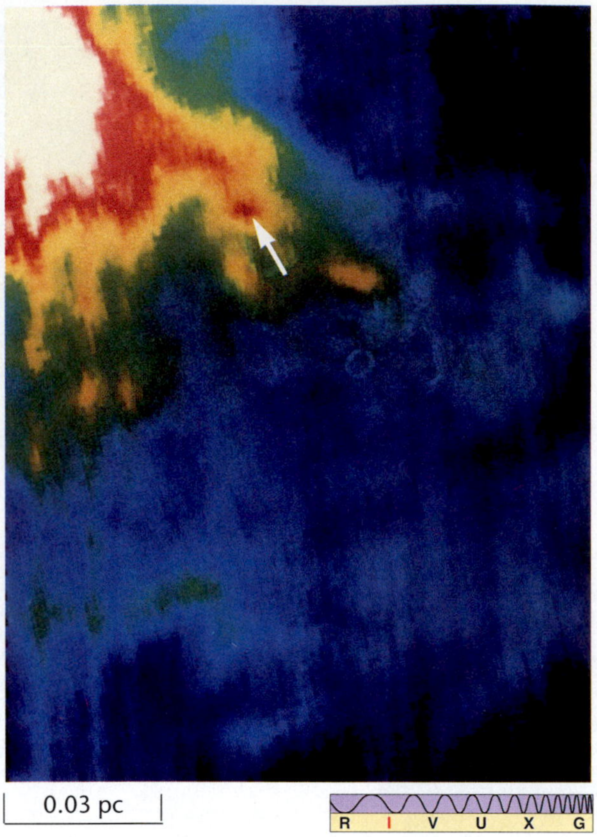

Figure 19.14 An infrared image of the nearby region containing the source Barnard 5 (indicated by the arrow). On the basis of its temperature and luminosity, Barnard 5 appears to be a protostar on the Hayashi track in the H–R diagram.

PROTOSTELLAR WINDS

Protostars often exhibit strong winds. Radio and infrared observations of hydrogen and carbon monoxide molecules, again in the Orion cloud, have revealed gas expanding outward at velocities approaching 100 km/s. High-resolution interferometric observations have disclosed expanding knots of water emission within the same star-forming region and have linked the strong winds to the protostars themselves. As mentioned earlier, these winds may be related to the violent surface activity associated with many protostars (see also *Interlude 19-1 on* p. 415).

Early in a protostar's life, it may still be embedded in an extensive disk of nebular material in which planets are forming, as discussed in Chapter 15. ∞ (p. 322) When the protostellar wind begins to blow, it encounters less resistance in the directions perpendicular to the disk than in the plane of the disk. The result is known as a *bipolar flow*—two "jets" of matter are expelled in the directions of the poles of the protostar, as illustrated in Figure 19.15. As the protostellar wind gradually destroys the disk, blowing it away into space, the jets widen until, with the disk gone, the wind flows away from the star equally in all directions. Figure 19.16 shows a real example of bipolar flow.

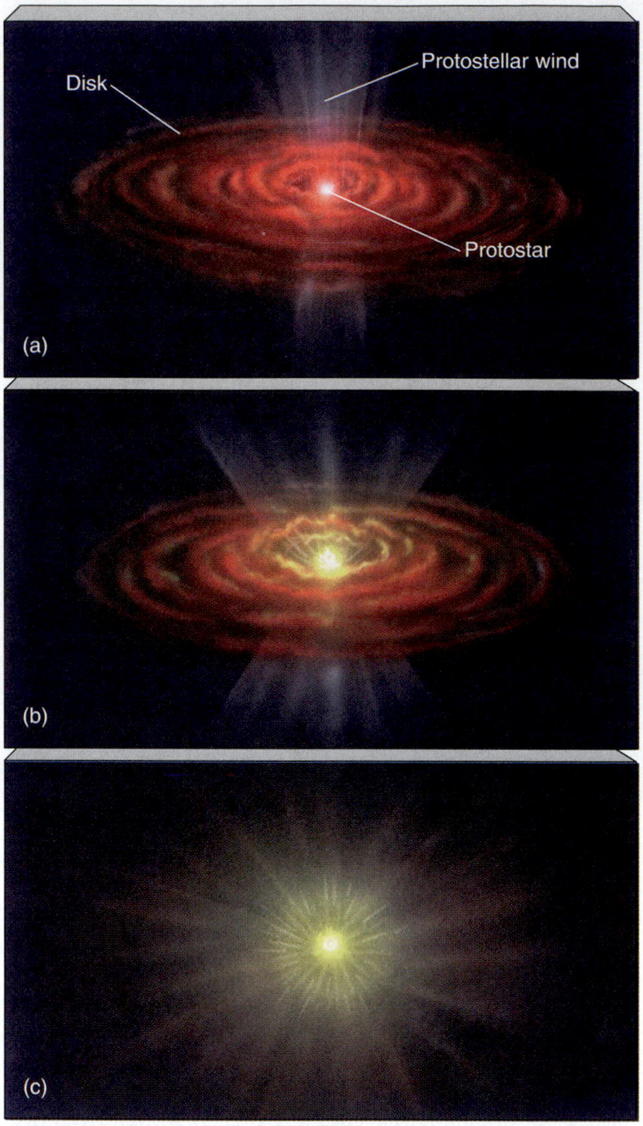

Figure 19.15 (a) When a protostellar wind encounters the disk of nebular gas surrounding the protostar, it tends to form a bipolar jet, preferentially leaving the system along the line of least resistance, which is perpendicular to the disk. (b) As the disk is blown away by the wind, the jets fan out, eventually (c) merging into a spherical wind.

19.5 *Shock Waves and Star Formation*

5 The subject of star formation is actually much more complicated than the previous discussion suggests. Interstellar space is populated with many kinds of clouds, fragments, protostars, stars, and nebulae. They all interact in a complex fashion, and each type of object undoubtedly affects the behavior of the others.

For example, the presence of an emission nebula in or near a molecular cloud probably influences the evolution of the entire region. We can easily imagine expanding waves of

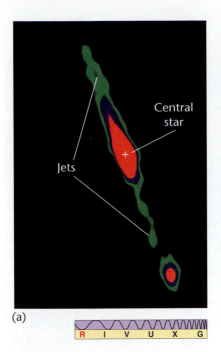

Figure 19.16 (a)This false-colored radio image shows two jets emanating from the young star system HH81-82 (whose position is marked with a cross at center). This is the largest stellar jet known, with a length of about 10,000 A.U. (The colors are coded in order of decreasing radio intensity, red, blue, green.) (b) An idealized artist's conception of a young star system, showing two jets flowing perpendicular to the disk of gas and dust rotating around the star. (See also the chapter-opening photos of more stellar jets on page 398.)

matter driven outward by the high temperatures and pressures in the nebula. As the waves crash into the surrounding molecular cloud, interstellar gas tends to pile up and become compressed. Such a shell of gas, rushing rapidly through space, is known as a **shock wave**. It can push ordinarily thin matter into dense sheets, just as a plow pushes snow.

Many astronomers regard the passage of a shock wave through interstellar matter as the triggering mechanism needed to initiate star formation in a galaxy. Calculations show that when a shock wave encounters an interstellar cloud, it races around the thinner exterior of the cloud more rapidly than it can penetrate its thicker interior.

Thus, shock waves do not blast a cloud from only one direction. They effectively squeeze it from many directions, as illustrated in Figure 19.17. Atomic bomb tests have experimentally demonstrated this squeezing—shock waves created in the blast surround buildings, causing them to be blown together (imploded) rather than apart (exploded). After shock waves cause the initial compression of an interstellar cloud, natural gravitational instabilities may divide it into the fragments that eventually form stars. Figure 19.18 suggests how this mechanism might be at work near M20.

Emission nebulae are by no means the only generators of interstellar shock waves. At least two other sources are

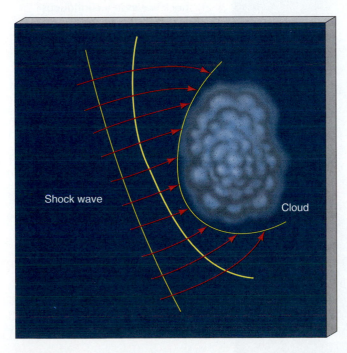

Figure 19.17 Shock waves tend to wrap around interstellar clouds, compressing them to greater densities and thus possibly triggering star formation.

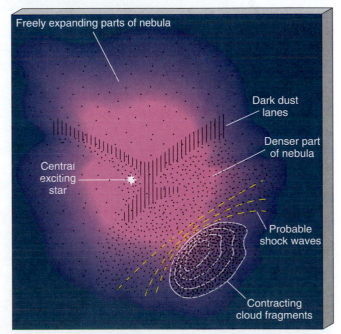

Figure 19.18 An artist's conception of a cloud fragment undergoing compression on the southerly edge of M20, as shock waves from the nebula penetrate the surrounding interstellar cloud.

available—the deaths of old stars (planetary nebulae and supernovae, to be discussed in Chapters 20 and 21) and the spiral-arm waves that plow through the Milky Way (to be discussed in Chapter 23). Supernovae are the most energetic, and probably the most efficient, way to pile up matter into dense clumps. These events are relatively few and far between, however, so the other mechanisms may be more important in triggering star formation. Although the evidence is somewhat circumstantial, the presence of young (and thus quick-forming) O- and B-type stars in the vicinity of supernova remnants does suggest that the birth of stars is often initiated by the violent, explosive deaths of others.

Wherever O- and B-type stars have recently formed, we can assume that less massive stars are still in the process of forming. It takes longer for the less massive stars to form, and thus we should not expect to see many A-, F-, G-, K-, or M-type stars directly associated with supernova remnants, provided that the star-formation mechanism really was triggered less than 1 million years ago. The neighborhoods around such remnants are probably vast stellar nurseries—the site of many invisible interstellar cloud fragments and protostars as well as the young, massive stars we see.

This scenario of shock-induced star formation is complicated by the fact that O- and B-type stars form quickly, live briefly, and die explosively. These massive stars, themselves perhaps born of a passing shock wave, may in turn create new shock waves, either through the expanding neb-

ular gas produced by their births or by their explosive deaths. These new shock waves can produce "second-generation" stars, which in turn will explode and give rise to still more shock waves, and so on, as depicted in Figure 19.19. Star formation thus resembles a chain reaction. Other, lighter, stars are also formed in the process, of course, but they are largely "along for the ride." It is the O- and B-type stars that drive the star-formation wave through the cloud. Observational evidence lends some support to this chain-reaction model. Groups of stars nearest molecular clouds do indeed appear to be the youngest, whereas those farther away seem to be older.

19.6 *Emission Nebulae and Star Clusters*

❹ We have seen how a portion of an interstellar cloud can become unstable, collapse, and fragment into stars. Let us take a moment to ask what happens next—not to the newborn stars themselves (that will be the subject of the next three chapters), but to the galactic environment in which they have just formed.

The end result of the collapse process is a group of stars, all formed from the same region of the parent cloud, along with a certain amount of unused gas and dust. How many stars form, and of what type? How much gas is left over? What does the collapsed cloud look like once the star-forma-

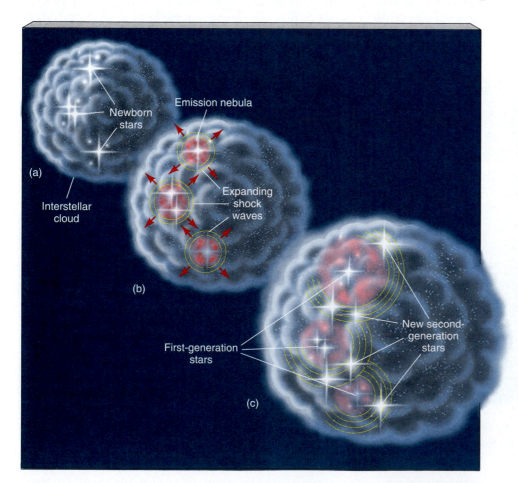

(a)

Newborn stars

Interstellar cloud

Emission nebula

Expanding shock waves

(b)

First-generation stars

New second-generation stars

(c)

Figure 19.19   (a) Star birth and (b) shock waves lead to (c) more star births and more shock waves in a continuous cycle of star formation in many areas of our Galaxy. Like a chain reaction, old stars trigger the formation of new stars ever deeper into an interstellar cloud.

tion process has run its course? At present, although the main stages in the formation of an individual star (stages 3–7) are becoming clearer, the answers to these more general questions (involving stages 1 and 2) are still sketchy and await a more thorough understanding of the cloud-collapse process.

In general, the more massive the collapsing region, the more stars are likely to form there. On the basis of observed H–R diagrams, we also know that low-mass stars are much more common than high-mass ones. However, the precise number of stars of any given mass or spectral type depends in a complex (and poorly understood) way on conditions within the parent cloud. The same is true of the *efficiency* of star formation—that is, the fraction of the total mass that actually finds its way into stars—which determines the amount of leftover gas. However, if, as is usually the case, one or more O- or B-type stars form, their intense radiation and winds will cause the surrounding gas to disperse, leaving behind a clump of young stars—a star cluster. ∞ (p. 376) Figure 19.20 shows an open cluster

in which the gas-dispersal process is almost complete.

Until recently, the existence of star clusters within emission nebulae was largely conjecture. The stars themselves could not be seen optically because they were obscured by dust. Infrared observations have now clearly demonstrated that stars really are found within star-forming regions! Figure 19.21 compares optical with infrared views of the central regions of the Orion Nebula. The optical image in Figure 19.21(a) shows only the Trapezium, the group of four bright stars responsible for ionizing the nebula. However, the false-color infrared image in Figure 19.21(b) also reveals an extensive cluster of stars (shown here as small crosses of many colors) within and behind the visible nebula. The Kleinmann–Low Nebula, discussed earlier, can be seen here as the roughly circular blue-green region in the right central portion of the infrared image. The green spot within it is thought to be a dust-shrouded B star just beginning to form its own emission nebula around it. This one image shows many separate stages of star formation.

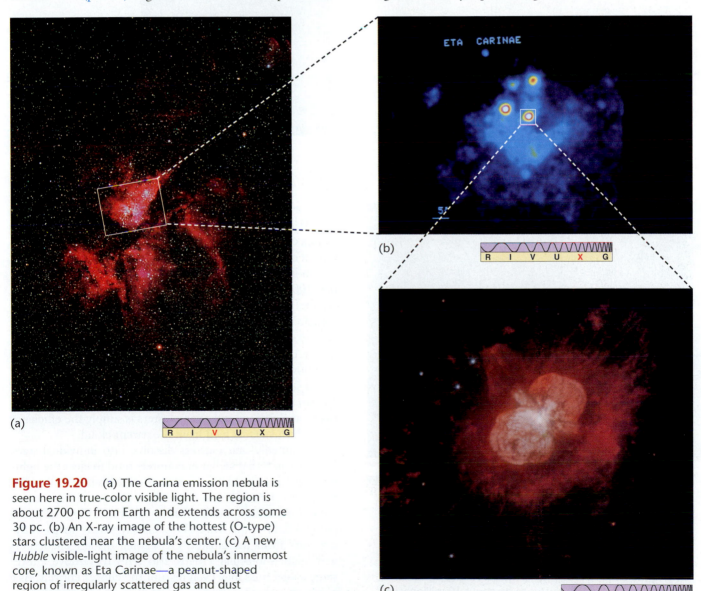

Figure 19.20 (a) The Carina emission nebula is seen here in true-color visible light. The region is about 2700 pc from Earth and extends across some 30 pc. (b) An X-ray image of the hottest (O-type) stars clustered near the nebula's center. (c) A new *Hubble* visible-light image of the nebula's innermost core, known as Eta Carinae—a peanut-shaped region of irregularly scattered gas and dust stretching across about 0.5 pc.

Figure 19.21 The central regions of the Orion Nebula seen (a) in a short-exposure visible-light image (using a filter that is transparent only to certain emission lines of oxygen) and (b) in the infrared (at roughly the same scale). The visible image shows the nebula itself and four bright O stars, known as the Trapezium. The infrared view, seen here in false color, where red is coolest and white is warmest, shows many cool stars within the nebula that are undetectable in visible light. (c) A short-exposure infrared photograph showing more clearly the four bright Trapezium stars and several faint red stars now emerging from the nebular gas.

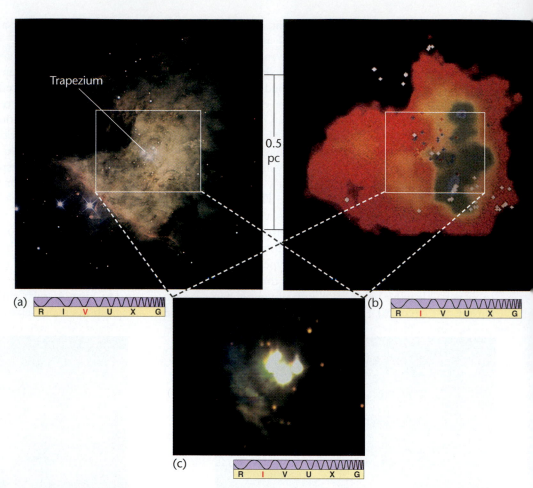

Figure 19.22 The Jewel Box cluster is a relatively young open cluster in the southern sky. Many bright stars appear in this image, but the cluster contains many more low-mass, less luminous stars. Because some red giants appear among the cluster's blue main-sequence stars, we can estimate the age of the cluster to be about 10 million years.

For every O or B giant, tens or even hundreds of G, K, and M dwarfs may form. Thus, even a modest emission nebula can give rise to a fairly extensive collection of stars. A typical open star cluster in the plane of our Galaxy, like that shown in Figure 19.22, may measure 10 pc across and contain 1000 or more stars. Less massive, but more extended, clusters are usually known as **associations**. These typically contain no more than 100 stars but may span many tens of parsecs. Associations tend to be rich in very young stars. Those containing many pre–main-sequence T Tauri stars are known as *T associations*, whereas those with prominent O and B stars, such as the Trapezium in Orion, are called *OB associations*. It is quite likely that the main difference between associations and open clusters is simply the efficiency with which stars formed from the parent cloud.

Eventually, star clusters dissolve into individual stars. One reason is that stellar encounters tend to eject the lightest stars from the cluster, just as the gravitational slingshot effect (see *Interlude 6-2* on p. 139) can propel spacecraft around the solar system. At the same time, the tidal gravitational field of the Milky Way Galaxy slowly strips outlying stars from the cluster. Occasional distant encounters with giant molecular clouds also tend to remove cluster stars; a near miss may even disrupt the cluster entirely. As a result of all these influences, most open clusters break up in

Interlude 19–1   *Evolution Observed*

Astronomical objects generally evolve over enormously long intervals of time, making it almost impossible to study their changes during a human lifetime. Even the relatively short span of 50 million years needed to form a star such as our Sun is roughly 2 million human generations—far more than have yet occurred.

Some stages of a star's evolution are nonetheless expected to occur extraordinarily rapidly by cosmic standards. One such stage is the sudden explosion of a massive star near death, to be studied in Chapter 21. Another occurs as a newly formed proto-star approaches the main sequence. Given the huge number of stars in the sky, we can detect a few objects in this evolutionary stage, despite the relatively short time they spend there.

T Tauri stars are a class of young protostars on the verge of reaching the main sequence. Their peculiar name derives from the star labeled "T" in the constellation Taurus, whose unpredictable variations in brightness have long marked it as an unusual object. (As is common in astronomy, the first known of a particular object gives its name to the entire class.) In fact, the term *star* is quite misleading. The average T Tauri star lies on the Hayashi track (somewhere around stage 5 or 6 in Figure 19.8), so T Tauri stars are really protostars—but the name has stuck. During the past half-century, astronomers have watched some T Tauri stars brighten greatly over the course of a few years, then remain at that increased level of brightness. We do not know how long this phase lasts, however, because many of these young stars have been discovered only since the 1970s, and most are still bright. The accompanying photographs record the change of one such T Tauri star.

The image at the top, taken in the 1980s, shows a region of interstellar space containing a fan-shaped gaseous nebula and a faint star located at the nebula's tip. The nebula, labeled NGC 2261, is filled with dust and gas and reflects the light of the star, which is called R Monocerotis. The same field of view, photographed three years apart in the 1960s, is shown at the bottom. The star had brightened considerably, as had the nebula (reflecting the star's increased brightness). The star, a T Tauri variable, has retained this brightness ever since.

We still lack a complete explanation for the sudden brightening of young stars like the one in the photographs. Is it caused by interstellar matter falling onto the newly formed star? Or by flares on the star's surface created by the proto-star's strong magnetic field? Or possibly by surrounding gas and dust being blown away from the new star, making its true brightness more clearly visible to us? Or are there internal changes in the star's nuclear burning rate? Whatever the cause, several T Tauri stars have undergone very definite changes in appearance on time scales shorter than a human lifetime. Astronomers around the world are monitoring them closely, hoping that further observations will reveal the full reason for their extraordinarily rapid evolution.

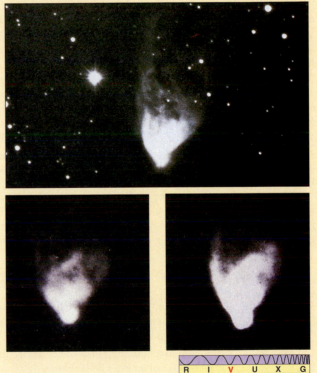

---

a few hundred million years, although the actual lifetime depends on the cluster's mass. Loosely bound associations may survive for only a few tens of millions of years, whereas some very massive open clusters such as M67, shown in Figure 19.23, are known from their H–R diagrams to be almost 5 billion years old. In a sense, only when a star's parent cluster has completely dissolved is the star-formation process really complete. The road from a gas cloud to a single, isolated star like the Sun is long and tortuous indeed!

Take another look at the nighttime sky. Ponder all that cosmic activity while gazing upward one clear, dark evening. After studying this chapter, you may find your view of the night sky greatly changed. Even the seemingly quiet nighttime darkness is dominated by continual change.

Figure 19.23 M67, one of the oldest known open clusters, has survived for almost 5 billion years—an unusually long time for a star system near the plane of the Milky Way Galaxy.

Chapter Review

SUMMARY

Stars form when an interstellar cloud collapses under its own gravity and breaks up into pieces comparable in mass to our Sun. Heat, rotation, and magnetism all compete with gravity to influence the cloud's evolution. The evolution of the contracting cloud—the changes in its temperature and luminosity—can be conveniently represented as an **evolutionary track** (p. 402) on the Hertzsprung–Russell diagram. A cold interstellar cloud containing a few thousand solar masses of gas can fragment into tens or hundreds of smaller clumps of matter, from which stars eventually form.

As a collapsing prestellar fragment heats up and becomes denser, it eventually becomes a **protostar** (p. 403)—a warm, very luminous object that emits radiation mainly in the infrared portion of the electromagnetic spectrum. At this stage of its evolution, the protostar is also known as a **T Tauri star** (p. 404), after the first object of this type discovered.

Eventually, a protostar's central temperature becomes high enough for hydrogen fusion to begin, and the protostar becomes a star. For a star like the Sun, the whole formation process takes about 50 million years. More massive stars pass through similar stages, but much more rapidly. Stars less massive than the Sun take much longer to form. The **zero-age main sequence** (p. 406) is the region on the H–R diagram where stars lie when the formation process is over.

Mass is the key property for determining a star's characteristics and life span. The most massive stars have the shortest formation times and main-sequence lifetimes. At the other extreme, some low-mass fragments never reach the point of nuclear igni-

tion. The universe may be populated with a vast number of **brown dwarfs** (p. 407)—objects that are not massive enough to fuse hydrogen to helium in their interiors.

Many of the objects predicted by the theory of star formation have been observed in real astronomical objects. The dark interstellar regions near emission nebulae often provide evidence for cloud fragmentation and protostars. Radio telescopes are used for studying the early phases of cloud contraction and fragmentation; infrared observations allow us to see later stages of the process. Many well-known emission nebulae, lit by several O-type stars, are partially engulfed by molecular clouds, parts of which are probably fragmenting and contracting, with smaller sites forming protostars.

Protostellar winds encounter less resistance in the directions perpendicular to a protostar's disk. Thus they expel two jets of matter in the directions of the protostar's poles. As the protostellar wind gradually destroys the disk, the jets widen until, with the disk gone, the wind flows away from the star equally in all directions.

Shock waves (p. 411) can compress other interstellar clouds and trigger star formation. Star birth and the production of shock waves are thought to produce a chain reaction of star formation in molecular cloud complexes.

A single collapsing and fragmenting cloud can give rise to hundreds or thousands of stars—a star cluster. Infrared observations have revealed young star clusters in several emission nebulae. Loosely bound groups of newborn stars are called stellar **associations** (p. 414). Eventually, star clusters break up into individual stars, although the process may take billions of years to complete.

SELF-TEST: True or False?

_____ **1.** Given the typical temperatures found in interstellar space, a cloud containing as few as 1000 atoms has sufficient gravity for it to begin to collapse.

_____ **2.** Both rotation and magnetic fields act to accelerate the gravitational collapse of an interstellar cloud.

_____ **3.** The time a solar-type star spends forming is relatively short compared to the time it spends as a main-sequence star.

_____ **4.** Most stars form as members of groups or clusters of stars.

_____ **5.** A stage-4 object has a luminosity about 1000 times that of the current Sun.

_____ **6.** As it evolves along the Hayashi track from stage 4 to stage 6, a prestellar object moves more or less horizontally across the H-R diagram.

_____ **7.** The rate of evolution of a stage 5 object is fast compared to the rates at previous stages.

_____ **8.** Stages 1 and 2 of star formation can be observed using optical telescopes.

_____ **9.** Shock waves produced from emission nebulae can initiate star formation in nearby molecular clouds.

_____ **10.** Shock waves for star formation can also be produced by large stars moving rapidly through the interstellar medium.

_____ **11.** In star formation, more G, K, and M type stars form than O and B type.

_____ **12.** Star clusters eventually dissipate, leaving behind individual stars like the Sun.

SELF-TEST: Fill in the Blank

1. Atoms in an interstellar cloud have random motions, with an average velocity determined by the cloud's _____.

2. A(n) _____ plots a star or protostar's changing location on the H-R diagram as the object evolves.

3. In stage 1 of prestellar evolution, a typical interstellar cloud has the following properties: temperature _____ K, size _____ parsecs, mass _____ solar masses.

4. In stage 2 of prestellar evolution, a contracting interstellar cloud _____ into smaller pieces.

5. During stage 3 of prestellar evolution, as each piece of the original interstellar cloud continues to contract, their central densities and temperatures _____.

6. At stage 4 of prestellar evolution, each piece of the interstellar cloud becomes a _____.

7. A stage 4 object is plotted in the _____ (upper/lower) _____ (right/left) part of the H-R diagram.

8. At stage 6 the central temperature of the object reaches _____ K.

9. At this temperature, a stage 6 object begins to _____.

10. At stage 7, the star has reached the _____.

11. The T Tauri phase of a star occurs during stage _____.

12. It takes a star like the Sun a total of about _____ million years to form.

13. Stars much more massive than the Sun take about _____ years to form; very low mass stars may take over _____ years.

14. Astronomers look for emissions at _____ wavelengths to identify interstellar clouds in stages 1 and 2.

15. At stages 4, 5, and 6, objects emit a great deal of radiation in the _____ part of the electromagnetic spectrum.

REVIEW AND DISCUSSION

1. Briefly describe the basic chain of events leading to the formation of a star like the Sun.

2. What is the role of heat in the process of stellar birth?

3. What is the role of rotation in the process of stellar birth?

4. What is the role of magnetism in the process of stellar birth?

5. Roughly how many atoms are needed to make a star? How much mass is this?

6. What is an evolutionary track?

7. Why do stars tend to form in groups?

8. At what point does a star-forming cloud become a protostar?

9. What event must occur in order for a protostar to become a full-fledged star?

10. What are brown dwarfs?

11. What are T Tauri stars?

12. Because stars live much longer than we do, how do astronomers test the accuracy of theories of star formation?

13. At what evolutionary stages must astronomers use radio and infrared radiation to study prestellar objects? Why can't they use visible light?

14. What is a shock wave? Name some phenomena that can produce shock waves in the interstellar medium.

15. Of what significance are shock waves in star formation?

16. Explain the usefulness of the Hertzsprung-Russell diagram in studying the evolution of stars. Why can't evolutionary stages 1–3 be plotted on the diagram?

17. Compare the times necessary for the various stages in the formation of a star like the Sun. Why are some so short and others so long?

18. In the formation of a star cluster with a wide range of stellar masses, is it possible for some stars to die out before others have finished forming? Do you think this would have any effect on the cluster's formation?

PROBLEMS

1. A certain interstellar cloud contains 10^{60} atoms. Hydrogen (mass per atom = 2×10^{-27} kg) accounts for 90 percent of the atoms, and the remainder are helium (each helium atom has four times more mass than an atom of hydrogen). What is the cloud's mass? Express your answer in solar masses ($M_\odot$ = 2×10^{30} kg).

2. Use the radius-luminosity-temperature relation ($L \propto R^2\, T^4$; see Chapter 17) to explain how a protostar's luminosity changes as it moves from stage 4 ($T = 3000$ K, $R = 2 \times 10^8$ km) to stage 6 ($T = 4500$ K, $R = 1 \times 10^6$ km).

3. What is the luminosity, in solar units, of a brown dwarf star whose radius is 0.1 solar radii and whose surface temperature is 600 K (0.1 times that of the Sun)?

4. In order for an interstellar gas cloud to contract, the average velocity of its constituent atoms must be less than half the cloud's escape velocity. From Chapter 2, the escape velocity is

$$v_{\mathrm{esc}} = \sqrt{\frac{2Gm}{r}} \ .$$

The average velocity of hydrogen atoms in a gas with temperature T kelvins is

$$v_{\mathrm{av}} = 160\sqrt{T} \ \ \mathrm{m/s}$$

(see the *More Precisely* feature on p. 136). Will a cloud of mass 1000 solar masses, radius 10 pc, and temperature of 10 K begin to collapse?

PROJECT

1. The Trifid Nebula, otherwise known as M20, is a place where new stars are forming. It has been called a "dark night revelation, even in modest apertures." An 8- to 10-inch telescope is needed to see the triple-lobed structure of the nebula. Ordinary binoculars reveal the Trifid as a hazy patch located in the constellation Sagittarius. This nebula is set against the richest part of the Milky Way, the edgewise projection of our own Galaxy around the sky. It is one of many wonders in this region of the heavens. What are the dark lanes in M20? Why are other parts of the nebula bright? There have been reports of large-scale changes occurring in this nebula in the last century and a half. The reports are based on old drawings, which show M20 looking slightly different from how it appears today. Do you think it possible that a cloud in space might undergo a change in appearance on a time scale of years, decades, or centuries?

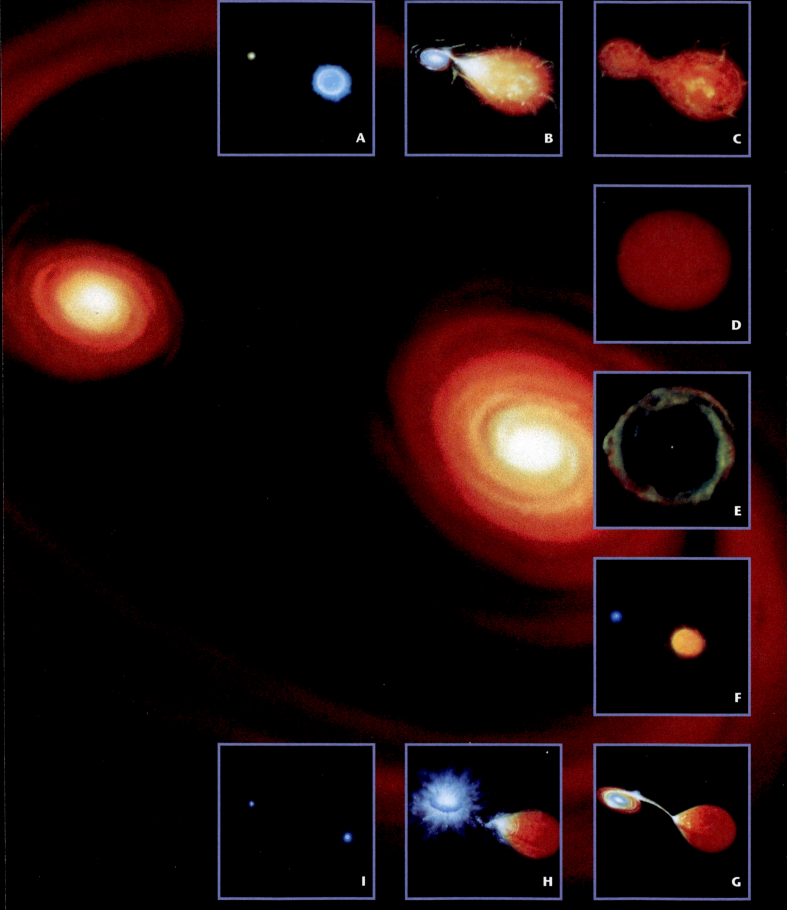

20

STELLAR EVOLUTION

From Middle Age to Death

LEARNING GOALS

Studying this chapter will enable you to:

1 Explain why stars evolve off the Main Sequence.

2 Outline the events that occur after a Sun-like star exhausts the supply of hydrogen in its core.

3 Summarize the stages in the death of a typical low-mass star, and describe the resulting remnant.

4 Contrast the evolutionary histories of high-mass and low-mass stars.

5 Discuss the observations that help verify the theory of stellar evolution.

6 Explain how the evolution of stars in binary systems may differ from that of isolated stars.

(Opposite page, background) These frames are the conceptions of noted space artist Dana Berry. The ten frames depict a sequence of the birth, evolution, and death of a binary-star system. The sequence starts with the large rendering of a 1 $M_\odot$ star and a 4 $M_\odot$ star in the process of formation and then proceeds clockwise from top left. **(Inset A)** The binary stars spend most of their lives as a yellow, 1 $M_\odot$ star much like our Sun and a blue 4 $M_\odot$ star. **(Inset B)** Nearing the end of its life, the 4 $M_\odot$ star swells, spilling gas onto its companion and forming an accretion disk around it. **(Inset C)** Continuing to flood its companion with loose gas, the accretion disk thickens, causing the two stars to resemble a dumbell. **(Inset D)** The common envelope surrounding the two stars has swollen to what appears as a single, red-giant star. **(Inset E)** The red giant has now reached the point where the gravity of the original two stars cannot contain the gas. The result is a gentle expulsion—not really an explosion—forming a planetary nebula rich in oxygen gas (hence the green color). **(Inset F)** Zooming in on the white dot at the center of the previous frame, E, we find our original binary stars. Now they are both dwarf stars—a red dwarf on the right overheated by a white dwarf on the left. **(Inset G)** An accretion bridge again joins the two stars, forming a disk of hot gas around the white-dwarf star. **(Inset H)** When the white dwarf can no longer contain the infalling plasma, a violent outburst results—a nova. This mild explosion can occur many times as members of the system continue to orbit one another. **(Inset I)** The likely end point of the system is two white dwarfs of roughly equal mass circling each other until long after their energy is exhausted.

The process of star formation produces great changes in the physical properties of a prestellar object. After reaching the main sequence, a newborn star is destined to change very little in outward appearance for more than 90 percent of its entire stellar lifetime. At the end of this period, as the star begins to run out of fuel and die, its properties once again change greatly. Aging stars travel along evolutionary tracks that take them far from the main sequence as they head toward death. In this and the next two chapters, we will study the evolution of stars during and after their main-sequence burning stages. We will find that the ultimate fate of a star depends primarily on its mass, although interactions with other stars can also play a decisive role, and that the final states of stars can be strange indeed. By continually comparing theoretical calculations with detailed observations of stars of all types, astronomers have refined the theory of stellar evolution into a precise and powerful tool for understanding the universe.

20.1 *Evolution Off the Main Sequence*

1 The main sequence of the H–R diagram is the evolutionary stage in which most stars spend most of their lives. For example, a star such as the Sun, after spending a few tens of millions of years in formation, is expected to reside on or near the main sequence for 10 *billion* years before evolving into something else. That "something else" is the main topic of this chapter. As we will see, once a star leaves the main sequence, its days are numbered. Evolving off the main sequence represents the beginning of the end for any star.

Virtually all the low-mass stars that have ever formed still exist as stars. M-type dwarfs burn their fuel so slowly that not one of them has yet left the main sequence. Some of them will burn steadily for a trillion years or more. Conversely, the most massive O- and B-type stars evolve away from the main sequence after only a few tens of millions of years of burning. Most of the massive stars that have ever existed perished long ago. Between these two extremes, many stars are observed in advanced stages of stellar evolution, their properties quite different from when they first arrived on the main sequence.

All theoretical models suggest that the final stages of stellar evolution depend critically on the mass of the star. As a rule of thumb, low-mass stars die gently, and high-mass stars die catastrophically. Depending on which astronomer you ask, the dividing line between "low mass" and "high mass" is anywhere from 5 to 10 times the mass of the Sun. In this text, we will consider stars of 8 solar masses or more to be high-mass stars.

We begin by considering the evolution of a fairly low-mass star like the Sun. The stages described in the next few sections will pertain to the Sun as it nears the end of its fusion cycle 5 billion years from now. In fact, most of the qualitative features of the discussion apply to any low-mass star, although the exact numbers vary considerably. Later, we will broaden our discussion to include all stars, large and small.

MAIN-SEQUENCE EQUILIBRIUM

Gravity is always present wherever matter exists. Only some counteracting phenomenon prevents an astronomical object from completely collapsing under its own weight. In the case of stars, that competing phenomenon is gas pressure, caused by the heat of the raging inferno at the stellar core. Figure 20.1 depicts the equilibrium in which gravity's inward pull balances pressure's outward push. Keep this simple picture in mind while you study the various stages of stellar evolution described in this chapter. Also bear in mind a simple maxim that summarizes the eventual outcome of the struggle between gravity and heat in almost every phase of a star's life: *Sooner or later, gravity wins.*

Provided that a star remains in this equilibrium state, nothing spectacular happens to it. While the star is a resident of the main sequence, its hydrogen fuel slowly fuses into helium in the core. This process is called **core hydrogen burning**. In Chapter 16, we saw how the proton–proton fusion chain powers the Sun. ∞ (p. 350) The *More Precisely* feature on p. 422 describes another sequence of nuclear reactions, of great importance in stars more

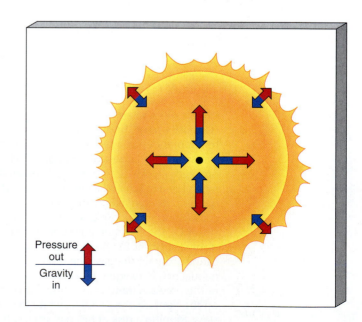

Figure 20.1 In a steadily burning star on the main sequence, the outward pressure of hot gas balances the inward pull of gravity. This is true at every point within the star, guaranteeing its stability.

massive than the Sun, that accomplishes the same basic result as the proton–proton chain, but in a very different way.

A main-sequence star's surface occasionally erupts in flares and spots, and its atmosphere ejects copious amounts of particles and photons, but for the most part the star does not experience sudden, large-scale changes in its properties. Its average temperature and luminosity remain fairly constant. (The luminosity actually increases very slowly—the Sun is now some 30 percent brighter than it was 5 billion years ago). The star might release energy indefinitely if nothing drastic occurred. But eventually something drastic *does* occur.

DEPLETION OF HYDROGEN IN THE CORE

After approximately 10 billion years of steady core hydrogen burning, a Sun-like star begins to run out of fuel. Hydrogen becomes depleted, at least in a small central region about 1/100 of the star's full size. The depletion of hydrogen is slow and steady, but the consequences are severe. It is a little like an automobile cruising effortlessly along a highway at a constant speed of 55 mph for many hours, only to have the engine cough and sputter as the gas gauge reaches empty. Unlike automobiles, though, stars are not easy to refuel.

As the nuclear burning proceeds, the composition of the star's interior changes. Figure 20.2 illustrates the increase in helium abundance and the corresponding decrease in hydrogen in the stellar core as the star ages. Three cases are shown: (a) the chemical composition of the original core, (b) the composition after 5 billion years, and (c) the composition after 10 billion years. Case (b) represents approximately the present state of our Sun.

The star's helium content increases fastest at the center, where temperatures are highest and the burning is fastest. Helium also increases near the edge of the core, but more slowly, because the burning rate is less rapid there. The inner helium-rich region becomes larger and more hydrogen-deficient as the star continues to shine. Eventually, hydrogen becomes completely depleted at the center, the nuclear fires there cease, and the location of principal burning moves to higher layers in the core. An inner core of non-burning pure helium starts to grow.

After about 10 billion years, a serious problem arises. While hydrogen burning continues in the outer core, the lack of burning at the center leads to an unstable situation. The gas pressure weakens in the helium inner core, but the inward pull of gravity does not. Gravity never lets up. Once the outward push against gravity is relaxed—even a little—structural changes in the star become inevitable.

CONTRACTION OF THE HELIUM CORE

If more heat could be generated, the star might possibly return to equilibrium. For example, were helium in the core to begin fusing into some heavier element such as carbon, all would be well once again. Energy would be created as a by-product of helium burning, and the neces-

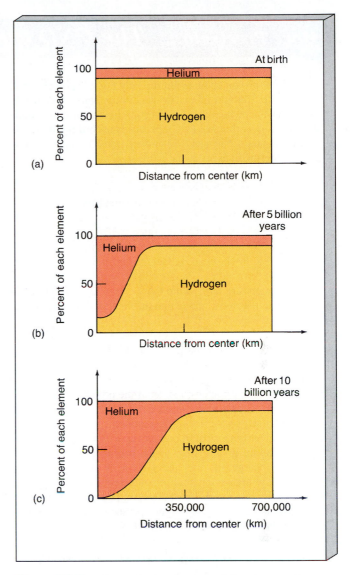

Figure 20.2 Theoretical estimates of the changes in a Sun-like star's composition. Hydrogen (yellow) and helium (orange) abundances are shown (a) at birth, on the zero-age main sequence; (b) after 5 billion years; and (c) after 10 billion years. At stage (b) only about 5 percent of the star's total mass has been converted from hydrogen into helium. This change speeds up as the nuclear burning rate increases with time.

sary gas pressure could be reestablished. But the helium there cannot burn—not yet, anyway. Despite its high temperature, the core is far too cold to fuse helium into anything heavier.

Recall that a temperature of at least 10^7 K is needed to burn hydrogen into helium. Above that temperature, colliding hydrogen nuclei—that is, protons—have enough speed to overwhelm the repulsive electromagnetic force between them. With helium, however, even 10^7 K is insufficient for fusion. Each helium nucleus, composed of two protons and two neutrons, has a net positive charge twice that of the hydrogen nucleus. As a result, the repulsive electromagnetic force between two

More Precisely...   *The CNO Cycle*

The proton–proton chain is not the only nuclear process operating in the Sun and other late-generation stars. Another fusion mechanism capable of converting hydrogen into helium, starting from carbon-12 (^{12}C), proceeds according to the following six steps. Nitrogen (N) and oxygen (O) nuclei are created as intermediate products:

1. ^{12}C + ^{1}H → ^{13}N + energy
2. ^{13}N → ^{13}C + positron + neutrino
3. ^{13}C + ^{1}H → ^{14}N + energy
4. ^{14}N + ^{1}H → ^{15}O + energy
5. ^{15}O → ^{15}N + positron + neutrino
6. ^{15}N + ^{1}H → ^{12}C + ^{4}He

These six steps are termed the *CNO cycle*. Aside from the radiation and neutrinos, notice that the sum total of these six reactions is

$$^{12}\text{C} + 4(^{1}\text{H}) \rightarrow {}^{12}\text{C} + {}^{4}\text{He}.$$

In other words, the *net* result is the fusion of four protons into a single helium-4 nucleus, just as in the proton–proton chain. The carbon-12 acts merely as a *catalyst*, an agent of change that is not itself consumed in the reaction.

The electromagnetic forces of repulsion operating in the CNO cycle are greater than in the proton–proton chain because the charges of the heavy-element nuclei are larger. Accordingly, higher temperatures are required to propel the heavy nuclei into the realm of the strong nuclear force and ignite fusion. The following figure presents a numerical estimate of the energy released in the Sun by the proton–proton

chain and the CNO cycle, each as a function of gas temperature. The proton–proton chain dominates at lower temperatures, up to about 16 million K. Above this temperature, the CNO cycle is the more important fusion process.

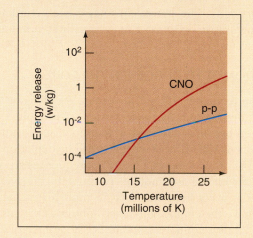

According to our theoretical models of the Sun, the temperature of the solar core is 15 million K, so these curves indicate that the proton–proton cycle is the dominant source of solar energy (notice that each step on the vertical scale corresponds to a *factor* of 100 in energy generation). The CNO cycle contributes no more than 10 percent of the observed solar radiation. However, stars more massive than our Sun often have core temperatures much higher than 20 million K, making the CNO cycle the dominant energy production mechanism.

helium nuclei is also larger, and more violent collisions are needed to fuse helium. Tremendously high temperatures are required—about 10^8 K.

A core composed of helium at 10^7 K thus cannot generate energy through fusion. As soon as the hydrogen fuel becomes substantially depleted, the helium core begins to contract because the pressure there—without nuclear burning—is too low to counteract gravity. Just as in earlier phases of stellar evolution, this shrinkage releases gravitational energy, driving up the central temperature.

The increasingly hot core heats the overlying layers. The higher temperatures—now well over 10^7 K—cause hydrogen nuclei to fuse even more rapidly than before. Figure 20.3 depicts this situation, in which hydrogen is burning at a fantastic rate in a shell surrounding the non-burning helium "ash" in the center. This phase is usually known as the **hydrogen-shell-burning** stage. The hydrogen shell generates energy faster than the original main-sequence star's hydrogen-burning core, and its energy production continues to grow. Strange as it may seem, the

star's response to the disappearance of the fire at its center is to get brighter!

Conditions in the aging star have clearly changed from the equilibrium that once characterized it as a main-sequence object. The helium core is unbalanced and shrinking, on its way to becoming hot enough for helium fusion. The rest of the core is also unbalanced, fusing hydrogen into helium at a growing rate. The gas pressure exerted by this enhanced hydrogen burning increases, forcing the intermediate layers and especially the outermost layers of the star to expand. Not even gravity can stop them. Even while the core is shrinking, the overlying layers are expanding! The star, aged and unbalanced, is on its way to becoming a red giant.

Consider the observational consequences of all this. An outside observer would see the star swell, eventually becoming nearly 100 times larger than a main-sequence star of the same spectral type. Analysis of the star's Planck curve would show that the surface was about 2000 K cooler than before. This is not to say that the period of ballooning and

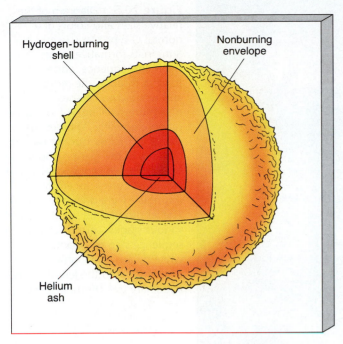

Figure 20.3 As a star's core loses more and more of its hydrogen, the hydrogen in the shell surrounding the nonburning helium ash burns ever more violently.

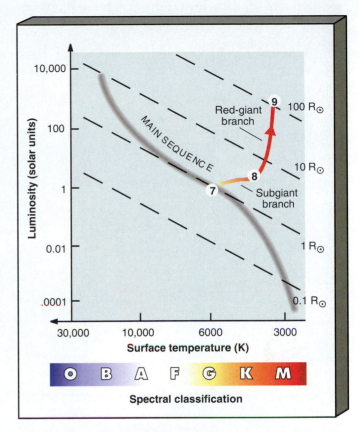

Figure 20.4 As the core of helium ash shrinks and the intermediate stellar layers expand, the star leaves the main sequence (stage 7). At stage 8, the star is on its way to becoming a red-giant star. The star continues to brighten and grow as it ascends the red-giant branch to stage 9, the top of the red-giant branch. As in Chapter 19, the diagonal lines correspond to stars of constant radius, allowing us to gauge the changes in the size of our star.

cooling of an aged star could be observed directly. The change from a normal main-sequence star to an elderly red giant takes about 100 million years to complete.

We can trace these large-scale changes on the H–R diagram. Figure 20.4 shows the path away from the main sequence, stage 7. (Recall from Chapter 19 that stage 7 corresponds to the star's arrival on the main sequence.) ∞ (p. 405) The luminosity of the giant at the point marked 8 on the figure is about $10L_\odot$ (Remember that $L_\odot$ is the Sun's luminosity.) It exceeds $100L_\odot$ at point 9. The surface temperature at stage 8 has fallen to the point at which much of the interior is opaque to the radiation from within. Beyond this point, convection carries the core's enormous energy output to the surface. One consequence is that the star's surface temperature remains nearly constant between stages 8 and 9.

Given the rising luminosity and the falling surface temperature, it follows that the star's radius must increase, and it does so to about $20R_\odot$ ($R_\odot$ is the Sun's radius) at stage 8 and eventually to about $70R_\odot$ at stage 9. The roughly constant-luminosity path from stage 7 (the main sequence) to stage 8 is often called the **subgiant branch**. The nearly vertical path followed by the star between stages 8 and 9 is known as the **red-giant branch** of the H–R diagram.

RED GIANTS

Figure 20.5 compares the relative sizes of our Sun and a stage-9 red-giant star. It also indicates the evolutionary

stages through which the Sun will evolve. The red giant is huge, having swollen to about 70 times its main-sequence size—about the size of Mercury's orbit. In contrast, its helium core is surprisingly small—only about 1/1000 the size of the entire star, making it just a few times larger than Earth.

The density in the core is now enormous. Continued shrinkage of the red giant's core has compacted its helium gas to approximately 10^8 kg/m³. Contrast this with the 10^{-3} kg/m³ in the outermost layers of the red giant, with the 5000 kg/m³ average density of the Earth, and with the 150,000 kg/m³ in the present core of the Sun. About 25 percent of the mass of the entire star is packed into its planet-sized core.

Perhaps the most famous red giant is the naked-eye star Betelgeuse in the constellation Orion (shown in Figure 17.9). Despite its great distance from Earth (about 150 pc), its enormous luminosity of $10^4 L_\odot$ makes it one of the brightest stars in the night sky.

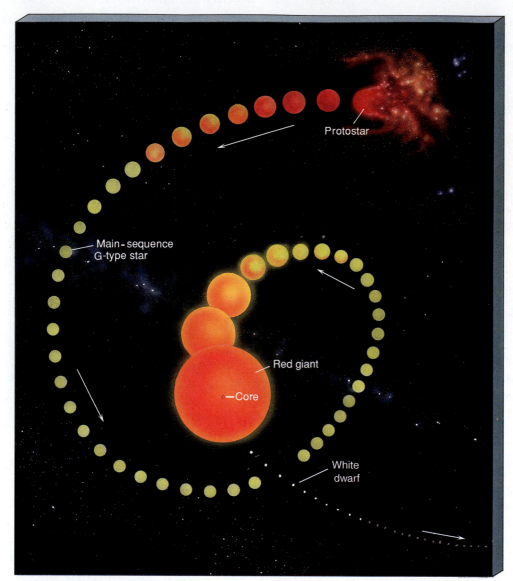

Figure 20.5 Diagram of the relative sizes and colors of a normal G-type star (such as our Sun) in its formative stages, on the main sequence, and while passing through the red-giant and white-dwarf stages. At maximum swelling, the red giant is approximately 70 times the size of its main-sequence parent; the core of the giant is about 1/15 the main-sequence size and would be barely discernible if this figure were drawn to scale. The length of time spent in the various stages—protostar, main-sequence star, red giant, and white dwarf—is roughly proportional to the length of this imaginary trek through space.

20.2 *Helium Fusion in Low-Mass Stars*

❷ Should the unbalanced state of a red-giant star continue, the core would eventually collapse, and the rest of the star would slowly drift into space. The forces and pressures at work inside a red giant would literally pull it apart. However, this simultaneous shrinking and expanding does not continue indefinitely. A few hundred million years after a solar-mass star leaves the main sequence, something else happens—helium begins to burn in the core.

By the time the central density has risen to about 10^8 kg/m^3, the temperature in the core has reached the 10^8 K needed for helium fusion. Helium nuclei then collide with one another, fusing into carbon nuclei and igniting the central fires once again. The reaction that transforms helium into carbon occurs in two steps. First, two helium nuclei come together to form a nucleus of beryllium-8 (^{8}Be). Beryllium-8 is a very unstable isotope that would normally break up into two helium nuclei in about 10^{-12} s.

However, at the densities in the core of a red giant, it is very likely that the beryllium-8 nucleus will encounter another helium nucleus before this occurs, fusing with it to form carbon-12 (^{12}C). This is the second step of the helium-burning reaction. In part, it is because of the electrostatic repulsion between beryllium-8 (containing four protons) and helium-4 (containing two) that the temperature must rise to 10^8K before this reaction can take place.

Symbolically, we can represent this next stage of stellar fusion as follows:

$$^4\text{He} + {}^4\text{He} \rightarrow {}^8\text{Be} + \text{energy},$$

$$^8\text{Be} + {}^4\text{He} \rightarrow {}^{12}\text{C} + \text{energy}.$$

Helium-4 nuclei are traditionally known as *alpha particles*. The term dates from the early days of nuclear physics, when the true nature of these particles was unknown. Because three alpha particles are required to get from helium-4 to carbon-12, the foregoing reaction is usually called the **triple-alpha process**.

THE HELIUM FLASH

For low-mass stars, there is a major complication in the helium-burning process. At the high densities found in the core, the gas has entered a new state of matter whose properties are governed by the laws of quantum mechanics rather than by those of classical physics. Up to now, we have been concerned primarily with the nuclei—protons, alpha particles, and so on—that make up virtually all of the star's mass and participate in the reactions that generate its energy. However, the star contains another important constituent—a vast sea of electrons stripped from their parent nuclei by the ferocious heat in the stellar interior. At this stage in our story, these electrons play a critical role in determining the star's evolution.

Given the conditions in the stage-9 red-giant core, a rule of quantum mechanics known as the *Pauli exclusion principle* (after Wolfgang Pauli, one of the founding fathers of quantum physics) prohibits the electrons in the core from being squeezed too close together. In effect, the exclusion principle tells us that we can think of the electrons as tiny rigid spheres that can be squeezed relatively easily up to the point of contact but become virtually incompressible thereafter. This condition is known (for historical reasons) as *electron degeneracy*, and the pressure associated with the contact of the tiny electron spheres is called **electron degeneracy pressure**. It has nothing to do with the thermal pressure (due to the star's heat) that we have been studying up to now. In a red-giant core, the pressure resisting the force of gravity is supplied almost entirely by degenerate electrons. Hardly any of the core's support results from "normal" thermal pressure.

The importance of electron degeneracy to the onset of helium burning in the core of a red giant is this: Under normal ("nondegenerate") circumstances, the core can react to and accommodate the onset of helium burning, but in its degenerate state the burning becomes unstable, with explosive consequences. In a normal star, the increase in temperature produced by the onset of helium fusion would lead to an increase in pressure. The gas would then expand and cool, reducing the burning rate and reestablishing equilibrium. In the degenerate case, however, the pressure is largely *independent* of the temperature. When burning starts and the temperature increases, there is no corresponding rise in pressure, no expansion of the gas, no drop in the temperature, and no stabilization of the core. Instead, the pressure remains more or less unchanged while the nuclear reaction rates increase and the temperature continues to rise. The temperature increases rapidly in a runaway explosion called the **helium flash**.

For a period of a few hours, the helium burns ferociously, like an uncontrolled bomb. Despite its brevity, this period of uncontrolled fusion releases a flood of new energy, enough to expand the core, lowering its density and ultimately returning it to a stable, nondegenerate state. This expansive adjustment of the core halts its gravitational collapse, returning it to equilibrium—an equilibrium reached once again between the inward pull of gravity and the outward push of gas pressure. The core, now stable, begins to burn helium into carbon at temperatures well above 10^8 K.

The helium flash terminates the giant star's ascent on the red-giant branch of the H–R diagram. Yet despite the explosive detonation of helium in the core, the flash does *not* increase the star's luminosity. On the contrary, the helium flash produces a rearrangement of the core that ultimately results in a *reduction* in the energy output. On the H–R diagram, the star jumps from stage 9 to stage 10, a stable state with steady helium burning in the core. As indicated in Figure 20.6, at this stage the surface temperature is higher than it was on the red-giant branch, whereas the luminosity is considerably less than at the helium flash. This adjustment in the star's properties occurs quite quickly—in about 100,000 years.

At stage 10 our star is now stably burning helium in its core and fusing hydrogen in a shell surrounding it. It resides in a well-defined region of the H–R diagram known as the **horizontal branch**—a "helium main sequence" of sorts, where core-helium-burning stars remain for a time before resuming their journey around the H–R diagram. The star's specific position within this region is determined mostly by its mass—not its original mass, but whatever mass remains after its ascent of the red-giant branch. The two masses differ because, during the red-giant stage, strong stellar winds

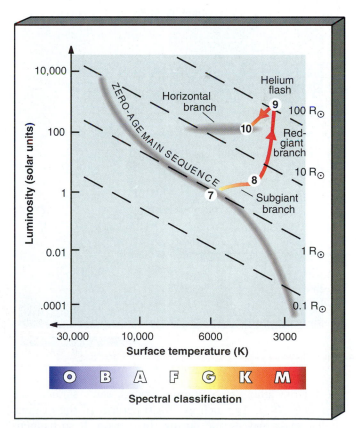

Figure 20.6 After its large increase in luminosity while ascending the red-giant branch is terminated by the helium flash, our star settles down into another equilibrium state at stage 10, on the horizontal branch.

Interlude 20–1   *Mass Loss from Giant Stars*

Astronomers now know that stars of all spectral types are active and have stellar winds. Consider the highly luminous, hot, blue O- and B-type stars, which have by far the strongest winds. Satellite and rocket observations have shown that their wind speeds may reach 3000 km/s. The result is a yearly mass loss sometimes exceeding 10^{-6} solar masses per year. Over the relatively short span of 1 million years, these stars blow a tenth of their total mass—more than an entire solar mass of material—into space. These powerful stellar winds, driven directly by the pressure of the intense ultraviolet radiation emitted by the stars themselves, hollow out vast cavities in the interstellar gas.

Observations made with radio, infrared, and optical telescopes have shown that luminous cool stars (for example, K- and M-type red giants) also lose mass at rates comparable to the luminous hot stars. Red-giant wind velocities, however, are much lower, averaging merely 30 km/s. They carry roughly as much mass into space as do O-star winds because their densities are generally much greater. Because luminous red stars are inherently cool objects (with surface temperatures of only about 3000K), they emit virtually no ultraviolet radiation, so the mechanism driving the winds must differ from that in luminous hot stars. We can only surmise that gas turbulence or magnetic fields or both in the atmospheres of the red giants are somehow responsible. The surface conditions in red giants are in some ways similar to those in T Tauri protostars, which are also known to exhibit strong winds. Possibly the same basic mechanism—violent surface activity—is responsible for both.

Unlike winds from hot stars, winds from these cool stars are rich in dust particles and molecules. Nearly all stars eventually evolve into red giants, so these winds provide a major source of new gas and dust to interstellar space. These stellar winds provide a vital link in the cycle of star formation and the evolution of the interstellar medium.

The accompanying figures show imaging and spectroscopic data acquired by the *Hubble Space Telescope* toward two stars shedding their outer atmospheres. On the left, the supergiant star AG Carinae—50 times more massive than the Sun and a million times brighter—is shown puffing out vast clouds of gas and dust. (The star, at center, is intentionally obscured to show more clearly the surrounding nebulae; the bright vertical line is also an artifact.)

On the right is the ultraviolet spectrum of a giant star called Melnick 42, located near the 30 Doradus Nebula in the Large Magellanic Cloud (LMC), the Milky Way's nearest companion galaxy, some 50 kpc away from Earth. The spectrum shows a wide carbon line in absorption, accompanied by an emission peak—a telltale sign of significant mass loss from this star. These are among many clues to a star's biography.

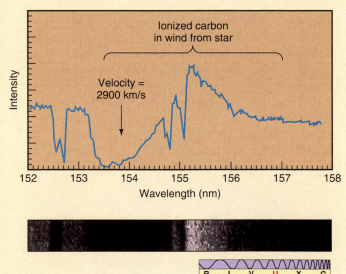

eject large amounts of matter from a star's surface (see *Interlude 20-1*). As much as 20–30 percent of the original stellar mass may escape during this period. It so happens that more massive stars have lower surface temperatures at this stage, but all stars have roughly the same luminosity after the helium flash. As a result, stage-10 stars tend to lie along a horizontal line on the H–R diagram, with more massive stars to the right, less massive ones to the left.

THE CARBON CORE

The nuclear reactions in a star's helium core burn on, but not for long. Whatever helium exists in the core is rapidly consumed. The triple-alpha helium-to-carbon fusion reaction—like the proton–proton and CNO-cycle hydrogen-to-helium reactions before it—proceeds at a rate that increases very rapidly with temperature. At

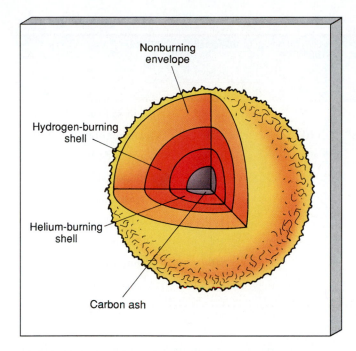

Figure 20.7   Within a few million years after the onset of helium burning, carbon ash accumulates in the inner core of a star, above which hydrogen and helium are still burning in concentric shells.

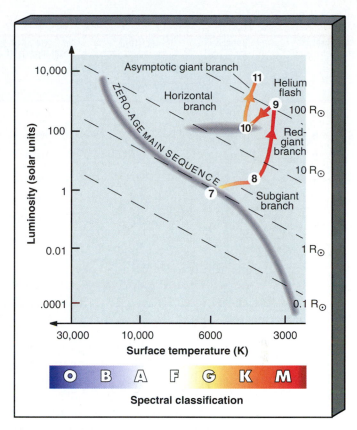

Figure 20.8   A carbon-core star reascends the giant branch of the H–R diagram—this time on a track called the asymptotic giant branch—for the same reason it evolved there the first time around: Lack of nuclear burning at the core causes contraction of the core and expansion of the overlying layers.

the extremely high temperatures found in the horizontal-branch core, the helium fuel doesn't last long—no more than a few tens of million years after the initial flash.

As the helium burns, a new inner core of carbon ash forms, and phenomena familiar to us from the earlier buildup of helium ash begin to occur. Now helium becomes depleted at the very center, and eventually fusion ceases there. In response, the carbon core shrinks and heats up a little as gravity pulls it inward, causing the hydrogen- and helium-burning rates in the overlying layers of the core to increase. The star now contains a shrinking carbon core surrounded by a helium-burning shell, which is in turn surrounded by a hydrogen-burning shell. The outer envelope of the star—the nonburning layers surrounding the core—expands, much as it did earlier in the first red-giant stage; at stage 11 the star becomes a swollen red giant for a second time. Figure 20.7 depicts the star's interior structure during this time.

The star's second ascent of the giant branch is shown in Figure 20.8. To distinguish this second track from the first red-giant stage, this phase is sometimes known as the **asymptotic giant branch**. The burning rates at the center are much fiercer this time around, and the star's radius and luminosity increase to values even greater than those reached at the helium flash on the first ascent. Our star is now a **red supergiant**. The carbon core continues to shrink, driving the hydrogen-burning and helium-burning shells to higher and higher temperatures and luminosities.

COMPARING THEORY WITH REALITY

Table 20-1 summarizes the key stages through which a solar-mass star evolves. It is a continuation of Table 19-1, except that the density units have been changed from particles per cubic meter to the more convenient kilograms per cubic meter. The numbers refer to the evolutionary stages noted in the figures and discussed in the text. Table 19-1 ended with stage 7, a main-sequence object fusing hydrogen into helium.∞ (p. 402) Table 20-1 begins at that point and then moves on to stage 8 on the subgiant branch, as the star evolves away from the main sequence and into the red-giant phase. Stage 9 is the helium flash, at the tip of the red-giant branch. Stage 10 describes an established horizontal-branch star stably fusing helium into carbon at its core, whereas stage 11 is the asymptotic giant branch, the star's final burning phase. The remaining stages (12–14) listed in the table represent the death of a low-mass star, which we will discuss in a moment.

All the H–R diagrams and evolutionary tracks presented so far are theoretical constructs based largely on computer models of the interior workings of stars. Before continuing our study of stellar evolution, let's take a moment to compare theory with reality. Figure

TABLE 20-1   *Evolution of a Sun-like Star*

STAGE	APPROX. TIME TO NEXT STAGE	CENTRAL TEMPERATURE	SURFACE TEMPERATURE	CENTRAL DENSITY	DIAMETER	OBJECT
	(yr)	(10^6 K)	(K)	(kg/m^3)	(km)	
7	10^{10}	15	6000	10^5	1.5×10^6	Main-sequence star
8	10^8	50	4000	10^7	4×10^6	Subgiant branch
9	10^5	100	4000	10^8	10^8	Helium flash
10	5×10^7	200	5000	10^7	2×10^7	Horizontal branch
11	10^4	250	4000	10^8	10^9	Asymptotic giant branch
12	10^5	300	100,000	10^9	10^5	Carbon core
	—		3000	10^{-17}	10^8	Planetary nebula*
13	—	100	50,000	10^{10}	10^4	White-dwarf star
14	—	Close to 0	Close to 0	10^{10}	10^4	Black-dwarf star

*Values refer to the envelope.

20.9 shows a real H–R diagram, drawn using the stars of the old globular cluster M3. The evolutionary stages discussed in the text and summarized in Figure 20.8 are marked. The similarity between theory and observation is very striking—stars in each of the evolutionary stages 7–11 can be seen, in numbers consistent with the theoretical models. (The points in Figure 20.9 are "shifted" a little to the left relative to Figure 20.8 because of composition differences between stars such as the Sun and stars in globular clusters—globular cluster stars tend to be slightly hotter than solar-type stars of the same mass.) Astronomers place great confidence in the theory of stellar evolution precisely because its predictions are so often found to be in excellent agreement with plots of real stars.

20.3 *The Death of a Low-Mass Star*

THE FIRES GO OUT

❸ As our red supergiant ascends the asymptotic giant branch, its envelope swells while its core, too cool for further nuclear burning, continues to contract. If the core temperature could become high enough for the fusion of carbon nuclei, or even a mixture of carbon and helium nuclei, still heavier products could be synthesized, and the newly generated energy would again support the star, restoring for a while the earlier equilibrium between gravity and heat. For the case of our solar-type star, however, this does not occur. The temperature never reaches the 600 million K needed for new nuclear reactions to occur. The red supergiant is now very close to the end of its nuclear-burning lifetime.

Before the carbon core can attain the incredibly high temperatures needed for carbon ignition, its density reaches a point beyond which it cannot be compressed further. At about

10^{10} kg/m^3, the electrons in the core once again become degenerate, the contraction of the core ceases, and its temperature stops rising. This stage represents the maximum compression that the star can achieve—there is simply not enough matter in the overlying layers to bear down any harder.

This is an extraordinarily high density. A single cubic centimeter of core matter would weigh 1000 kg on the Earth—a ton of matter compressed into a volume about the size of a grape. Yet despite the extreme compression of the core, the central temperature is "only" about 300 million K. Collisions among nuclei are neither frequent nor violent enough to fuse carbon into any of the heavier elements. Consequently, silicon, iron, gold, uranium, and the many other heavy elements are not synthesized in low-mass stars. The central fires go out when carbon has formed.

PLANETARY NEBULAE

Our aged star is now in quite a predicament. Its inner carbon core is, for all practical purposes, dead. The outer-core shells continue to burn hydrogen and helium and, as more and more of the inner core reaches its final, high-density state, the zone of nuclear burning increases in intensity. Meanwhile, the outermost layers of the star continue to expand.

Around this time, the burning becomes very unstable. The helium-burning shell is subject to a series of explosive *helium-shell flashes*. These flashes are caused by the enormous pressure there and the extreme sensitivity of the triple-alpha burning rate to small changes in temperature. The flashes produce large fluctuations in the intensity of the radiation reaching the star's outermost layers, causing them to pulsate more and more violently.

Compounding the star's problems, its surface layers are also becoming unstable. As the temperature drops to the point at which electrons can recombine with nuclei to form

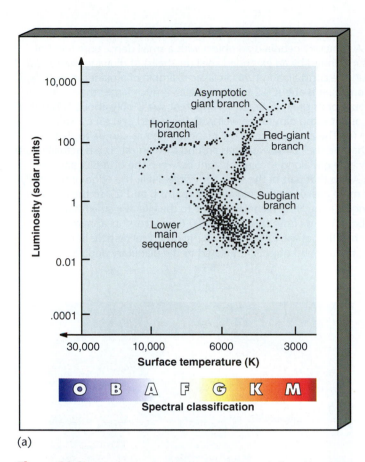

(a)

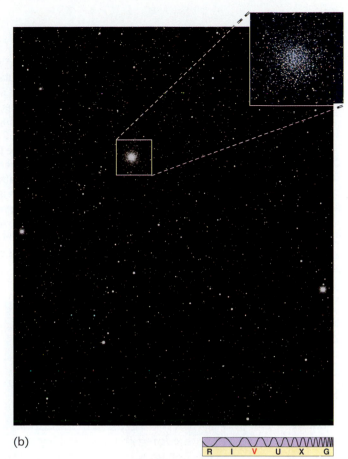

(b)

| R | I | V | U | X | G |

Figure 20.9 The various evolutionary stages predicted by theory and depicted schematically in Figure 20.8 are clearly visible in this H–R diagram (a) of an old star cluster—the globular cluster M3. The faintest main-sequence stars are not shown here because observational limitations make it difficult to determine the apparent brightness of low-luminosity stars in the cluster. (b) Wide-angle photograph showing M3 as it appears in the night sky. The inset is a more detailed view of the cluster itself; its field is a few parsecs across.

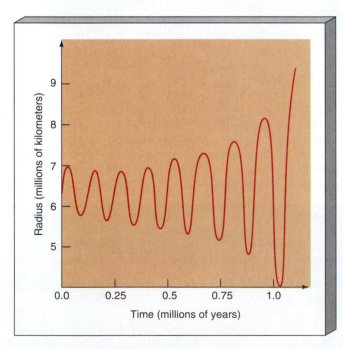

Figure 20.10 Buffeted by helium-shell flashes from within and subject to the destabilizing influence of recombination, the outer layers of a red giant become unstable and enter into a series of growing pulsations. Eventually, the envelope is ejected and forms a planetary nebula.

atoms, each recombination produces additional photons, which tend to push the outer envelope to greater and greater distances from the core. As shown in Figure 20.10, the radius of the star oscillates more and more violently. In less than a few million years, the star's outer envelope is ejected into space at a speed of a few tens of kilometers per second.

In time, a rather unusual-looking object results. We say unusual because the "star" now has two distinct parts, both of which constitute stage 12 of Table 20-1. At the center, there is a small well-defined core of mostly carbon ash. Hot and dense, only the outermost part of this core still burns helium into carbon. Well beyond the core, there is a spherical shell of cooler and thinner matter—the ejected envelope of the giant—spread over a volume roughly the size of our solar system. Such an object is called a **planetary nebula**. Some well-known examples are shown in Figures 20.11 and 20.12. In all, some 1000 planetary nebulae are known in our Galaxy.

The term *planetary* here is very misleading, for these objects have no association with planets. The name originated in the eighteenth century, when optical astronomers could barely distinguish between the myriad faint, fuzzy patches of light in the nighttime sky. With poor resolution, some of these patches did not appear as points, like stars, but instead looked more like disks—in other words, like planets. However, later observations have clearly demonstrated that the planetary nebula's fuzzy circular shape results from a shell of warm, glowing gas.

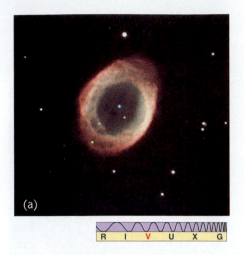

(a)

R I V U X G

Figure 20.11 A planetary nebula is an object with a small dense core (central blue-white star) surrounded by an extended shell (or shells) of glowing matter. (a) The Ring Nebula in the constellation Lyra, a classic example of a planetary nebula, is about 1500 pc from us. It is about 0.2 pc in diameter—much larger than our solar system—but because of its great distance, its apparent size is only about 1/100 that of the full Moon, and it is too dim to see well with the naked eye. (See also the images on p. 52.) (b) The appearance of the planetary nebula can be explained once we realize that the shell of glowing gas around the central core is actually quite thin. There is very little gas along the line of sight between the observer and the central star (path A), so that part of the shell is invisible. Near the edge of the shell, however, there is more gas along the line of sight (paths B and C), so the observer sees a glowing ring. (c) The Helix Nebula appears to the eye as a small star with a halo around it. About 140 pc from the Earth and 0.6 pc across, its apparent size in the sky is roughly half that of the full Moon. (All the other stars visible in the photo are foreground or background objects, unrelated to the planetary nebula.)

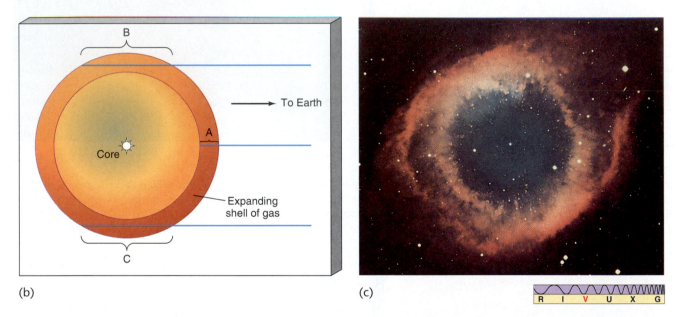

(b)

(c)

R I V U X G

Figure 20.12 (a) The Dumbell Nebula more clearly shows the shell-like structure of the expanding gases that make up a planetary nebula. (b) The Cat's Eye Nebula is an example of a much more complex planetary nebula. Intricate structures, including concentric gas shells, jets of high-speed gas, and shock-induced knots of gas are all visible. As usual, red indicates the presence of excited hydrogen. The nebula is about 1000 pc away, in the constellation Draco. It may have been produced by a pair of binary stars (unresolved at the center) that have both shed planetary nebulae.

R I V U X G

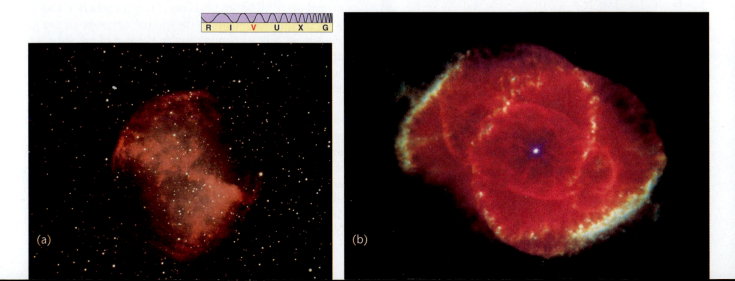

(a)

(b)

The term *nebula* is also a little confusing, because it suggests kinship with the various gaseous nebulae studied in Chapter 18. ∞ (p. 385) Although in some ways planetary nebulae resemble some emission nebulae, and both undergo similar ionization–recombination processes, these two types of objects are very different. Not only are planetary nebulae much smaller than emission nebulae, they are also associated with much older stars. Emission nebulae are the signposts of recent stellar birth. Planetary nebulae indicate impending stellar death.

The "ring" of the planetary nebula is really a three-dimensional shell completely surrounding the core. Its halo-shaped appearance is only an illusion. The shell is a complete envelope that has been expelled from around the core. However, we can see it only at the edges, where the emitting matter has accumulated along our line of sight. As shown in Figure 20.11, the shell is virtually invisible in the direction of the core. Few planetary nebulae are quite as regular as this simple picture might suggest, however. Figure 20.12 shows two systems in which the details of the gas-ejection process and interactions with the surrounding interstellar medium have apparently played important roles in determining a planetary nebula's shape and appearance.

WHITE DWARFS

The expanding envelope of a planetary nebula continues to spread out with time, becoming more diffuse and cooler,

gradually merging with interstellar space. This is one way in which interstellar space becomes enriched with additional helium atoms and possibly some carbon atoms as well. These atoms are dredged up from the depths of the core into the envelope by convection during the star's final years.

The carbon core, the stellar remnant at the center of the planetary nebula, continues to evolve. Formerly concealed by the atmosphere of the red-giant star, the core appears as the envelope recedes. The core is very small. By the time the envelope is ejected as a planetary nebula, it has shrunk to about the size of Earth. In some cases, it may be even smaller than our planet. Shining only by stored heat, not by nuclear reactions, this small star has a white-hot surface when it first becomes visible, although it appears rather dim because of its small size. The core's heat and size give rise to its new name—*white dwarf*. This is stage 13 of Table 20-1. The approximate path followed by the star on the H–R diagram as it evolves from stage-11 red supergiant to stage-13 white dwarf is shown in Figure 20.13.

Not all white-dwarf stars are found as the cores of planetary nebulae. Several hundred have been discovered "naked" in our Galaxy, their envelopes expelled to invisibility (or stripped away by a binary companion—as discussed shortly) long ago. Figure 20.14 shows an example of a white dwarf, Sirius B, that is particularly close to Earth; it is the faint binary companion of the much brighter and better-known Sirius A. ∞ (p. 362) Detailed observations show it to have the properties listed in Table 20-2. Our planet, with a radius of $0.009 R_\odot$, is actually larger than this star! With more than the mass of the Sun packed into a volume smaller

TABLE 20-2   *Sirius B—A Nearby White-Dwarf Star*

Mass	$1.1 M_\odot$
Radius	$0.008 R_\odot$
Luminosity (total)	$0.04 L_\odot$
Surface temperature	24,000 K
Average density	$3 \times 10^9 \ kg/m^3$

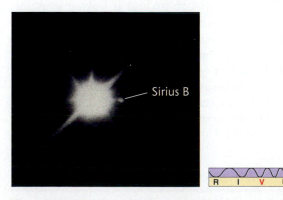

Figure 20.14   Sirius B (the speck of light at right) is a white-dwarf star, a companion to the much larger and brighter star Sirius A. (The "spikes" on the image of Sirius A are not real; they are artifacts caused by the support struts of the telescope.)

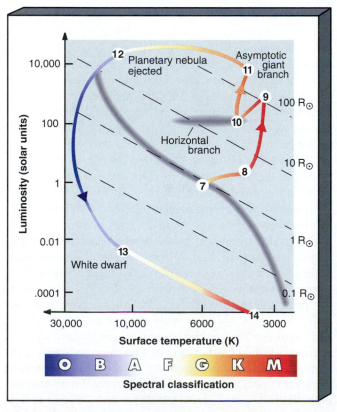

Figure 20.13   A star's passage from the horizontal branch (stage 10) to the white-dwarf stage (stage 13) by way of the asymptotic giant branch creates an evolutionary path that cuts across the entire H–R diagram.

Interlude 20–2 *Learning Astronomy from History*

Sirius A, the brighter of the two objects shown in Figure 20.14, appears twice as luminous as any other visible star, excluding the Sun. Its absolute brightness is not very great, but because its distance from us is small (less than 3 pc), its apparent brightness is very large. Sirius has been prominent in the nighttime sky since the beginning of recorded history. Cuneiform texts of the ancient Babylonians refer to the star as far back as 1000 B.C., and historians know that the star strongly influenced the agriculture and religion of the Egyptians of 3000 B.C.

Even though a star's evolution takes such a long time, we might have a chance to detect a slight change in Sirius because the recorded observations of this star go back several thousand years. The chances for success are improved in this case because Sirius A is so bright that even the naked-eye observations of the ancients should be reasonably accurate. Interestingly, recorded history does suggest that Sirius A has changed in appearance, but the observations are confusing. Every piece of information about Sirius recorded between the years 100 B.C. and A.D. 200 claims that this star was *red*. (No earlier records of its color are known.) In contrast, modern observations now show it to be white or bluish white—definitely *not* red.

Sirius has apparently changed from red to blue-white in the intervening years. The problem is this: According to the theory of stellar evolution, no star should be able to change its color in this way in such a short time. A color change such as this should take at least several tens of thousands of years, and perhaps a lot longer. It should also leave some evidence of its occurrence.

Astronomers have offered several explanations for the rather sudden change in Sirius A. These include the suggestions that (1) some ancient observers were wrong and other scribes copied them; (2) a Galactic dust cloud passed between Sirius A and the Earth some 2000 years ago, reddening the star much as the Earth's dusty atmosphere often reddens our Sun at dusk; and (3) the companion to Sirius A, Sirius B, was a red giant and the dominant star of this double-star system 2000 years ago but has since expelled its planetary nebular shell to reveal the white-dwarf star that we now observe.

Each of these explanations presents problems. How could the color of the sky's brightest star be incorrectly recorded for hundreds of years? Where is the intervening Galactic cloud now? Where is the shell of the former red giant? We are left with the uneasy feeling that the sky's brightest star doesn't fit particularly well into the currently accepted scenario of stellar evolution.

than the Earth, Sirius B's density is about a million times greater than anything familiar to us in the solar system.

BLACK DWARFS

Several tens of thousands of years are needed for the white dwarf to appear from behind the veil of expanding gas. Once a star becomes a white dwarf, its evolution, for all practical purposes, is over. It continues to cool and dim with time, following the dashed line near the bottom of the H–R diagram of Figure 20.13, finally becoming a **black dwarf**—a cold, dense, burned-out ember in space. This is stage 14 of Table 20-1, the graveyard of stars.

The cooling dwarf does not shrink much as it fades away, however. Even though its heat is leaking away into space, gravity does not compress it further. Why not? Because at the enormously high densities in the star (from the white-dwarf stage on), the resistance of electrons to being squeezed together—the same electron degeneracy that prevailed in the red-giant core around the time of the helium flash—holds the star up, even as its temperature drops almost to absolute zero. As the dwarf cools, it remains about the size of the Earth.

20.4 *High-Mass Stars*

④ High-mass stars evolve much faster than their low-mass counterparts. The more massive a star, the more ravenous is its fuel consumption and the shorter its main-se-

quence lifetime. The Sun will spend a total of some 10 billion years on the main sequence, but a 5-solar-mass B-type star will remain there for only a few hundred million years. A 10-solar-mass O-type star will depart in only 20 million years or so. This trend toward much faster evolution for more massive stars continues even after the main sequence. All evolutionary changes happen much more rapidly for high-mass stars because their larger mass and stronger gravity generate more heat, which speeds up all phases of stellar evolution.

Stars leave the main sequence for one basic reason: They run out of hydrogen in their cores. As a result, the early stages of stellar evolution beyond the main sequence are qualitatively the same in all cases: Main-sequence hydrogen burning in the core eventually gives way to the formation of a nonburning, collapsing helium core surrounded by a hydrogen-burning shell. A high-mass star leaves the main sequence on its journey toward the red-giant region with an internal structure quite similar to that of its low-mass cousin.

After the main sequence, two major divergences between low-mass and high-mass stars occur. First, when a high-mass star reaches the point at which helium begins to burn, its central density is so low that the core is *non*degenerate when helium fusion starts. As a result, the burning begins smoothly and stably, not explosively. There is no helium flash. The red giant remains a red giant as helium fuses into carbon. Second, the core *is* subsequently able to attain the 600 million K needed to burn carbon, so the evolution does not end with a carbon white dwarf. (Recall that,

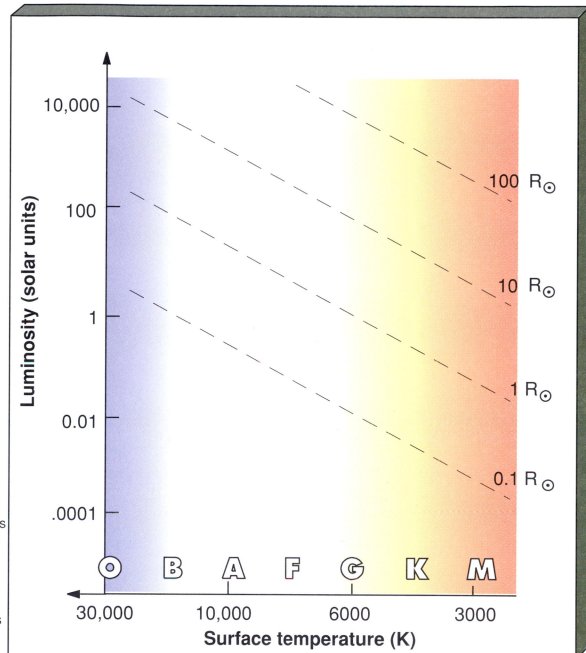

The H-R diagram plots stars by luminosity (vertical axis) and temperature, or spectral class (horizontal axis). The dashed diagonal lines are lines of constant radius.

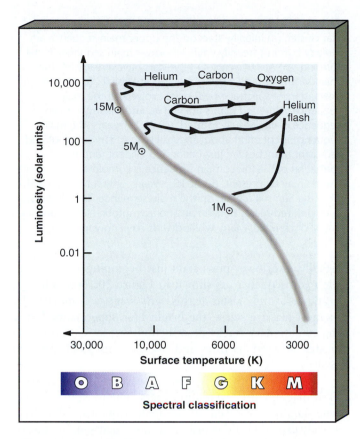

Figure 20.15 Evolutionary tracks for stars of 1, 5, and 15 solar masses (shown only up to the point of the helium flash in the low-mass cases). Low-mass stars ascend the giant branch almost vertically, whereas high-mass stars move roughly horizontally across the H–R diagram from the main sequence into the red-giant region. The most massive stars experience smooth transitions into each new burning stage. No helium flash occurs for stars more massive than about 4 solar masses. The loops in the tracks generally indicate the point at which a new burning stage begins. Some points are labeled with the element that has just started to fuse in the inner core.

the heavier the nucleus, the greater its charge and the higher the temperature needed for it to fuse.) Instead, evolution continues smoothly as the star creates heavier and heavier elements, burning ever faster as it goes.

Figure 20.15 compares evolutionary tracks for several stars of different masses, from the point at which they leave the main sequence to their arrival in the red-giant region. Whereas low-mass stars ascend the giant branch almost vertically, high-mass stars move nearly horizontally across the H–R diagram after leaving the upper main sequence. Their luminosities stay roughly constant as their radii increase and their surface temperatures drop.

The sudden "loops," or changes in direction of a star's motion, in Figure 20.15 are associated with the onset of new burning stages in the core. With no helium flash in a high-mass star, there is no sudden jump to the horizontal branch and no subsequent reascent of the asymptotic giant branch. Instead, the star simply loops back and forth at the top of the H–R diagram at roughly constant luminosity, creating heavier and heavier nuclear ash in its core. Some of the burning stages are marked on the figure. Notice that the most massive stars (those more massive than about 15 $M_\odot$) don't even reach the red-giant region before they start to fuse helium in their cores. They achieve a central temperature of 10^8 K while still quite close to the main sequence, and their evolutionary track continues smoothly across the H–R diagram, apparently unaffected by each successive phase of burning.

With heavier and heavier elements forming at an ever-increasing rate, the high-mass stars shown in Figure 20.15 are very close to the ends of their lives. We will discuss their fate in more detail in the next chapter, but suffice it to say here that they are destined to die in a violent explosion soon after carbon and oxygen begin to fuse in their cores. These stars evolve so rapidly that we are unlikely ever to "catch one in the act" of leaving the main sequence and traversing the H–R diagram. For most practical observational purposes, high-mass stars explode and die as soon as they leave the main sequence.

20.5 *Observing Stellar Evolution in Star Clusters*

5 Star clusters provide excellent test sites for the theory of stellar evolution. Every star in a given cluster formed at the same time, from the same interstellar cloud, with virtually the same composition. Only the mass varies from one star to another. This allows us to check the accuracy of our theoretical models in a very straightforward way. Having studied in some detail the evolutionary tracks of individual stars, we now consider how their collective appearance changes in time.

In Chapter 17, we saw how astronomers estimate the ages of star clusters by determining which of their stars have already left the main sequence. ∞ (p. 377) In fact, the main-sequence lifetimes that go into those age measurements represent only a tiny fraction of the data obtained from theoretical models of stellar evolution. Using the information presented in the preceding sections and starting from the zero-age main sequence, we can predict exactly how a newborn cluster should look at any later time. Ever since high-speed computers began to become available in the 1960s, this is precisely what astronomers have done. Although they cannot see into the interiors of stars to test their models, they can compare stars' outward appearances with theoretical predictions. The agreement—in detail—between theory and observation is remarkably good.

To illustrate this point, let us consider the evolution of a hypothetical star cluster somewhere in the Galaxy. We begin our study shortly after the cluster's formation, with the upper main sequence already fully formed and burning

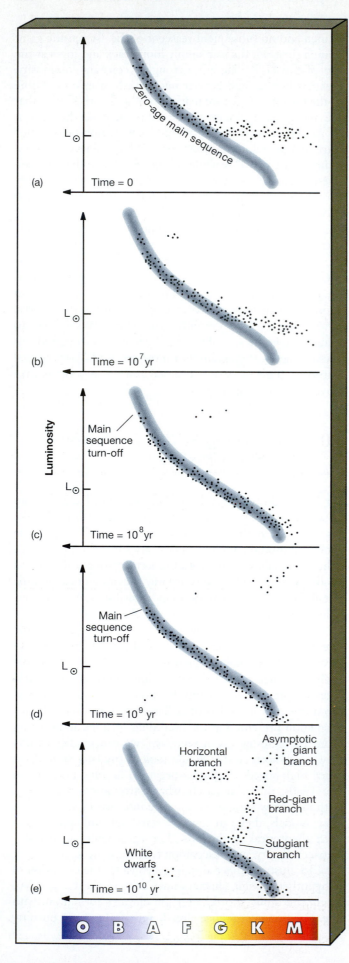

Figure 20.16 The changing H–R diagram of a hypothetical star cluster. (a) Initially, stars on the upper main sequence are already burning steadily while the lower main sequence is still forming. (b) At 10^7 years, O-type stars have already left the main sequence, and a few red giants are visible. (c) By 10^8 years, stars of spectral type B have evolved off the main sequence. More red giants are visible, and the lower main sequence is almost fully formed. (d) At 10^9 years, the main sequence is cut off at about spectral type A. The subgiant and red-giant branches are just becoming evident, and the formation of the lower main sequence is complete. A few white dwarfs may be present. (e) At 10^{10} years, only stars less massive than the Sun still remain on the main sequence. The cluster's subgiant, red-giant, horizontal, and asymptotic giant branches are all discernible. Many white dwarfs have now formed.

steadily, and lower-mass stars just beginning to arrive on the main sequence, as shown in Figure 20.16(a). The appearance of the cluster at this early stage is dominated by its most massive stars—the bright blue supergiants. Using the evolutionary tracks described above, let us follow the cluster forward in time and ask how its H–R diagram evolves.

Figure 20.16(b) shows the appearance of our cluster's H–R diagram after 10 million years. The most massive O-type stars have evolved off the main sequence. Most have already exploded and vanished, as just discussed, but one or two may still be visible as red giants. The remaining cluster stars are largely unchanged in appearance—their evolution is slow enough that little happens to them in 10^7 years. The cluster's H–R diagram shows the main sequence slightly cut off, along with a rather poorly defined red-giant region. Figure 20.17 shows the twin open clusters h and χ (the Greek letter chi) Persei, along with their combined H–R diagram. Comparing Figure 20.17(b) with such diagrams as those in Figure 20.16, astronomers estimate the age of this pair of clusters to be about 10 million years.

At any time during the evolution, the cluster's original main sequence is intact up to some well-defined stellar mass, corresponding to the stars that are just leaving the main sequence at that instant. We can imagine the main sequence being "peeled away" from the top down, with fainter and fainter stars turning off and heading for the giant branch as time goes on. Astronomers refer to the high-luminosity end of the observed main sequence as the **main-sequence turnoff**. The mass of the star that is just evolving off the main sequence at any moment is known as the *turnoff mass*.

After 100 million years (Figure 20.16c), stars brighter than type B5 or so (about 4–$5 M_\odot$) have left the main sequence, and a few more red supergiants are visible. By this time, most of the cluster's low-mass stars have finally arrived on the main sequence, although the dimmest M stars may still be in their contraction phase. The appearance of the cluster is now dominated by bright B stars and brighter red giants.

At 1 billion years, the main-sequence turnoff mass is around $2 M_\odot$, corresponding roughly to spectral type A2. The subgiant and giant branches associated with the evolu-

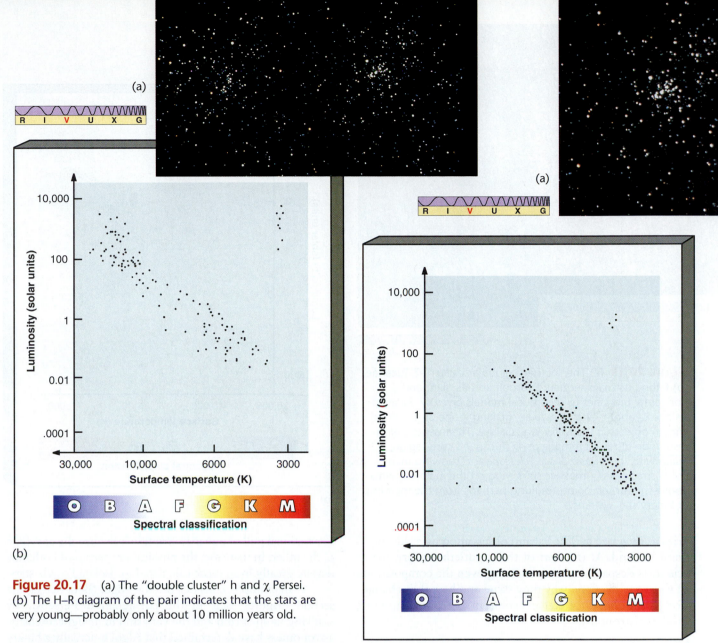

Figure 20.17 (a) The "double cluster" h and χ Persei. (b) The H–R diagram of the pair indicates that the stars are very young—probably only about 10 million years old.

Figure 20.18 (a) The Hyades cluster, a relatively young group of stars visible to the naked eye. (b) The H–R diagram for this cluster is cut off at about spectral type A, implying an age of about 500 million years.

tion of low-mass stars are just becoming visible, as indicated in Figure 20.16(d). The formation of the lower main sequence is now complete. In addition, the first white dwarfs have just appeared, although they are generally too faint to be observed at the distances of most clusters. Figure 20.18 shows the Hyades open cluster, with its H–R diagram. The H–R diagram appears to lie between Figures 20.16(c) and 20.16(d), suggesting that the cluster's age is about 5×10^8 years.

At 10 billion years, the turnoff point has reached solar-mass stars, of spectral type G2. The subgiant and giant branches are now clearly discernible (see Figure 20.16e), and the horizontal and asymptotic giant branches appear as distinct regions in the H–R diagram. Many white dwarfs are also present in the cluster. Although stars in all these evolutionary stages are also present in the 1-billion-year-old cluster shown in Figure 20.16(d), they are few in number—typically only a few percent of the total number of stars in the cluster. Also, because they evolve so rapidly, they spend very little time in these regions. Low-mass stars are much more numerous and evolve more slowly, so their evolutionary tracks are more easily detected.

Figure 20.19 shows the globular cluster 47 Tucanae. By carefully adjusting their theoretical models until the cluster's main sequence, subgiant, red-giant, and horizontal branches are all well matched, astronomers have determined its age to be roughly 14 billion years, a little older than our hypothetical cluster in Figure 20.16(e). In fact, as mentioned in Chapter 17, globular cluster ages determined this way show a remarkably small spread. ∞ (p. 377) All the globular clusters in our Galaxy appear to have formed between about 12 and 15 billion years ago.

Stellar evolution is one of the great success stories of astrophysics. Like all good scientific theories, it makes definite testable predictions about the universe, at the same time remaining flexible enough to incorporate new discov-

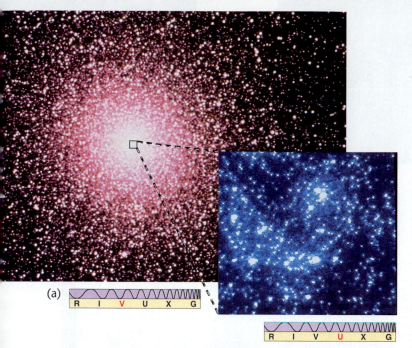

(a)

| R | I | V | U | X | G |

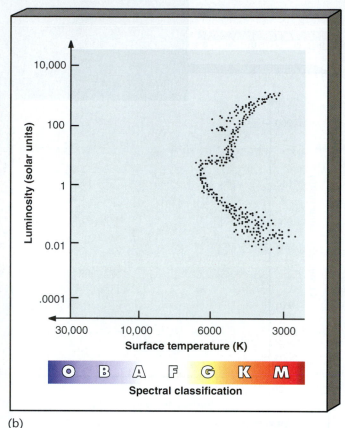

(b)

Figure 20.19 (a) The southern globular cluster 47 Tucanae. (b) Fitting its main-sequence turnoff and its giant and horizontal branches to theoretical models gives 47 Tucanae an age of about 14 billion years, making it one of the oldest known objects in the Milky Way Galaxy. The inset is a high-resolution ultraviolet image of 47 Tucanae's core region, taken with the *Hubble Space Telescope* and showing many "blue stragglers"—massive stars lying on the main sequence above the turnoff point, resulting perhaps from the merging of binary star systems.

eries as they occur. Theory and observation have advanced hand in hand. At the start of the twentieth century, many scientists despaired of ever knowing even the compositions of the stars, let alone why they shine and how they change. Today, the theory of stellar evolution is a cornerstone of modern astronomy.

20.6 *The Evolution of Binary-Star Systems*

6 We have noted that most stars in our Galaxy are not isolated objects, but are actually members of binary-star systems. However, our discussion of stellar evolution has so far focused exclusively on isolated stars. This prompts us to ask, "How does membership in a binary-star system change the evolutionary tracks we have just described?" Because nuclear burning occurs deep in the core, does the presence of a stellar companion have any significant effect? Perhaps not surprisingly, the answer depends on the distance between the two stars in question.

For a binary system whose component stars are very widely separated—that is, the distance between the stars is greater than perhaps a thousand stellar radii—the answer is that the two stars evolve more or less independently of one another, each following the track for an isolated star of its particular mass. However, if the two stars are closer, the

gravitational pull of one may strongly influence the envelope of the other. In that case, the physical properties of both may deviate greatly from those calculated for isolated single stars.

As an example, consider the star Algol (Beta Persei, the second brightest star in the constellation Perseus). By studying its spectrum and the variation in its light intensity, astronomers have determined that Algol is actually a binary (in fact, an eclipsing double-lined spectroscopic binary, as described in Chapter 17), and they have measured its properties very accurately. ∞ (p. 373) Algol consists of a $3.7M_\odot$ main-sequence star of spectral type B8 (a blue giant) with a $0.8M_\odot$ red subgiant companion moving in a circular orbit around it. The stars are 4 million km apart, with an orbital period of about 3 days.

A moment's thought reveals that there is something odd about these findings. On the basis of our earlier discussion, the more massive main-sequence star should have evolved *faster* than the less massive component. If the two stars formed at the same time (as is assumed to be the case), there should be no way that the $0.8M_\odot$ star could be approaching the giant stage first. Either our theory of stellar evolution is seriously in error or something has modified the evolution of the Algol system. Fortunately for theorists, the latter is the case.

To understand Algol, we must consider binary systems in a little more detail. As sketched in Figure 20.20, each star is surrounded by its own teardrop-shaped "zone of influence," inside of which its gravitational pull dominates the

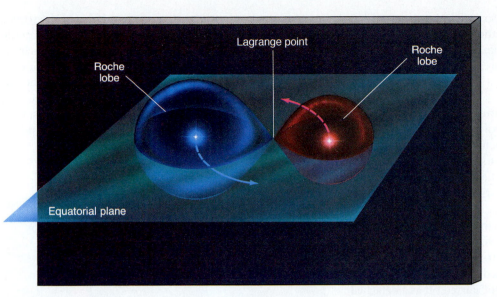

Figure 20.20 Each star in a binary system can be pictured as being surrounded by a "zone of influence," or Roche lobe, inside of which matter may be thought of as being "part" of that star. The two teardrop-shaped Roche lobes meet at the Lagrange point between the two stars. Outside the Roche lobes, matter may flow onto either star with relative ease.

effects of both the other star and the overall rotation of the binary. Any matter within that region "belongs" to the star. It cannot easily flow onto the other component or out of the system. Outside the two regions, it is possible for gas to flow toward either star relatively easily. The two teardrop-shaped regions are usually called **Roche lobes**, after Edouard Roche, the French mathematician who first studied the binary-system problem in the nineteenth century and whose work we have already encountered in the context of planetary rings. ∞ (p. 261) The Roche lobes of the two stars meet at a point on the line joining them—the inner Lagrange point (L_1), which we discussed in Chapter 14 when discussing asteroid motions in the solar system. ∞ (p. 300) This Lagrange point is a place where the gravitational pulls of the two stars exactly balance the rotation of the binary system. The greater the mass of one component, the larger is its Roche lobe and the farther from its center (and the closer to the other star) is the Lagrange point.

Normally, both stars lie well within their respective Roche lobes, and such a binary system is said to be *detached*, as in Figure 20.21(a). However, as a star evolves off the main sequence and moves toward the giant branch, it is possible for its radius to become so large that it overflows its Roche lobe. Its gas begins to flow onto the companion through the Lagrange point. The binary in this case is said to be *semidetached* (Figure 20.21b). Because matter is flowing from one star onto the other, semidetached binaries are also known as **mass-transfer binaries**. If, for some reason, the other star also overflows its Roche lobe (either because

Figure 20.21 (a) In a detached binary, each star lies within its respective Roche lobe. (b) In a semidetached binary, one of the stars fills its Roche lobe and transfers matter onto the other, which still lies within its own Roche lobe. (c) In a contact or common-envelope binary, both stars have overflowed their Roche lobes, and a single star with two distinct nuclear-burning cores results.

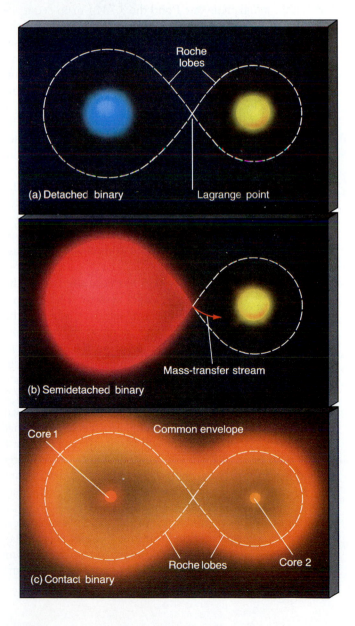

of stellar evolution or because so much extra material is dumped onto it), the surfaces of the two stars merge, and a new configuration results. The binary system then consists of two nuclear-burning stellar cores surrounded by a single continuous common envelope—a **contact binary**, shown in Figure 20.21(c).

In a binary system in which the two stars are very close together, neither star has to evolve far off the main sequence before it overflows its Roche lobe and mass transfer begins. In a wide binary, both stars may evolve all the way up the giant branch without either surface ever reaching the Lagrange point, and they evolve just as though they were isolated. Depending on the stars involved and their orbital separations, there are many different possibilities for the eventual outcome of the evolution. Let's make these ideas more definite by returning to the question of how the binary star Algol may have reached its present state.

Astronomers believe that Algol started off as a detached binary. For reference, let us label the component that is now the $0.8 M_\odot$ subgiant as star 1 and the $3.7 M_\odot$ main-sequence star as star 2. Initially, star 1 was the more massive of the two—perhaps $3 M_\odot$ or so. It thus evolved off the main sequence first. Star 2 was originally a less massive star, perhaps comparable in mass to the Sun. As star 1 ascended the giant branch, it overflowed its Roche lobe, and gas began to flow onto star 2. This had the effect of reducing the mass of star 1 and increasing that of star 2, which in turn caused the Roche lobe of star 1 to shrink as its gravity decreased. As a result, the rate at which star 1 overflowed its Roche lobe increased, and a period of unstable *rapid mass transfer* ensued, transporting most of star 1's envelope onto star 2. Eventually, the mass of star 1 became less than that of star 2. Detailed calculations show that the rate of mass transfer dropped sharply at that point, and the stars entered the relatively stable state we see today. These changes in Algol's components are illustrated in Figure 20.22.

Being part of a binary system has radically altered the evolution of both stars in the Algol system. The original high-mass star 1 is now a low-mass subgiant, whereas the roughly solar-type star 2 is now a massive blue main-sequence star. The removal of mass from the envelope of star 1 may prevent it from ever reaching the helium flash. Instead, its naked core may eventually be left behind as a *helium white dwarf*. In a few tens of millions of years, star 2 will itself begin to ascend the giant branch and fill its own Roche lobe. If star 1 is still a subgiant or a giant at that time, a contact binary system will result. If, instead, star 1 has by then become a white dwarf, a new mass-transferring period—with matter streaming from star 2 back onto star 1—will begin. In that case (as we will see), Algol may have a very active and violent future in store.

Just as molecules exhibit few of the physical or chemical properties of their constituent atoms, binaries can display types of behavior that are quite different from either of their component stars. The Algol system is a fairly simple exam-

ple of binary evolution, yet it gives us an idea of the sorts of complications that can arise when two stars evolve interdependently. A substantial fraction of all the binary stars in the Galaxy will pass through some sort of mass-transfer or common-envelope phase. In this chapter, we have seen one possible result of mass transfer involving main-sequence stars. We will return to this subject in the next two chapters, when we continue our discussion of stellar evolution and the strange states of matter that may result.

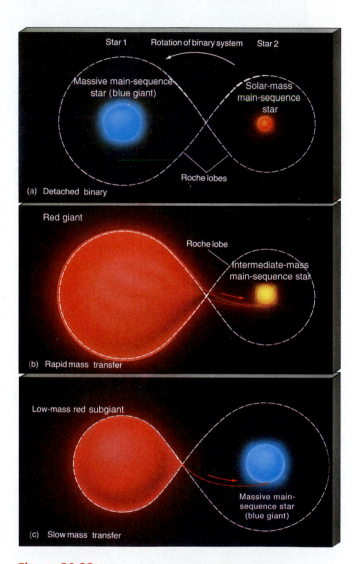

Figure 20.22   The evolution of the binary star Algol. (a) Initially, Algol was probably a detached binary made up of two main-sequence stars —a relatively massive blue giant and a less massive companion similar to the Sun. (b) As the more massive component (star 1) evolved off the main sequence, it expanded to fill and eventually overflow its Roche lobe, transferring large amounts of matter onto its smaller companion (star 2). (c) Today, star 2 is the more massive of the two, but it is on the main sequence. Star 1 is still in the subgiant phase and fills its Roche lobe, causing a steady stream of matter to pour onto its companion.

Chapter Review

SUMMARY

Stars spend most of their lives on the main sequence, in the **core-hydrogen burning** (p. 420) phase of stellar evolution, stably fusing hydrogen into helium at their centers. Stars leave the main sequence when the hydrogen in their cores is exhausted. For the Sun, which is about halfway through its main-sequence lifetime, this stage will occur about 5 billion years from now. Low-mass stars evolve much more slowly than the Sun, and high-mass stars much faster.

When the central nuclear fires cease, the helium in the star's core is still too cool to fuse into anything heavier. With no internal energy source, the helium core is unable to support itself against its own gravity and begins to shrink. The star at this stage is in the **hydrogen-shell-burning** (p. 422) phase, in which the nonburning helium at the center is surrounded by a layer of burning hydrogen. The energy released by the contracting helium core heats the hydrogen-burning shell, greatly increasing the nuclear reaction rates there. As a result, the star becomes much brighter while the envelope expands and cools. A low-mass star like the Sun moves off the main sequence on the H–R diagram first along the **subgiant branch** (p. 423), then almost vertically up the **red-giant branch** (p. 423).

As the helium core contracts, it heats up. Eventually, it reaches the point at which helium begins to fuse into carbon. The net effect of the fusion reactions is that three helium nuclei (or alpha particles) combine to form a nucleus of carbon in the **triple-alpha process** (p. 424).

In a star like the Sun, conditions at the onset of helium burning are such that the electrons in the core have become degenerate—they can be thought of as tiny, hard spheres that, once brought into contact, present stiff resistance to being compressed any farther. This **electron degeneracy pressure** (p. 425) makes the core unable to "react" to the new energy source, and helium burning begins explosively, in the **helium flash** (p. 425). The flash expands the core and reduces the star's luminosity, sending it onto the **horizontal branch** (p. 425) of the H–R diagram. The star now has a core of burning helium surrounded by a shell of burning hydrogen.

As helium burns in the core, it forms an inner core of nonburning carbon. The carbon core shrinks and heats the overlying burning layers, and the star once again becomes a red giant. It reenters the red-giant region of the H–R diagram along the **asymptotic giant branch** (p. 427), becoming an extremely luminous **red supergiant** (p. 427) star.

The core of a low-mass star never becomes hot enough to fuse carbon. Such a star continues to ascend the asymptotic giant branch until its envelope is ejected into space as a **planetary nebula** (p. 429). At that point the core becomes visible as a hot, faint, and extremely dense *white-dwarf* star. The planetary nebula diffuses into space, carrying helium and carbon into the interstellar medium. The white dwarf cools and fades, eventually becoming a cold **black dwarf** (p. 432).

Evolutionary changes happen more rapidly for high-mass stars than for low-mass stars because larger mass results in higher central temperatures. High-mass stars do not experience a helium flash and do attain central temperatures high enough to fuse carbon. They form heavier and heavier elements in their cores, at a more and more rapid pace, and eventually die explosively.

The theory of stellar evolution can be tested by observing star clusters, all of whose stars formed at the same time. As time goes by, the most massive stars evolve off the main sequence first, then the intermediate-mass stars, and so on. At any instant, no stars with masses above the cluster's **main-sequence turnoff** (p. 434) mass remain on the main sequence. Stars below this mass have not yet evolved into giants and so still lie on the main sequence. By comparing a particular cluster's main-sequence turnoff mass with theoretical predictions, astronomers can measure the age of the cluster.

Stars in binary systems can evolve quite differently from isolated stars because of interactions with their companions. Each star is surrounded by a teardrop-shaped **Roche lobe** (p. 437), which defines the region of space within which matter "belongs" to the star. As a star in a binary evolves into the giant phase, it may overflow its Roche lobe, forming a **mass-transfer binary** (p. 437) as gas flows from the giant onto its companion. If both stars overflow their Roche lobes, a **contact binary** (p. 438) results. Stellar evolution in binaries can produce states not achievable in single stars. In a sufficiently wide binary, both stars evolve as though they were isolated.

SELF-TEST: True or False?

_____ **1.** "Low mass" stars are conventionally taken to have masses less than about 8 solar masses.

_____ **2.** All the red dwarf stars that ever formed are still on the main sequence today.

_____ **3.** Once on the main sequence, gravity is no longer important in determining a star's internal structure.

_____ **4.** The Sun will get brighter as it begins to run out of fuel in its core.

_____ **5.** As a star evolves away from the main sequence, it gets larger.

_____ **6.** As a star evolves away from the main sequence, it gets hotter.

_____ **7.** As a red giant, the Sun will have a core that is smaller than it was when the Sun was on the main sequence.

_____ **8.** When helium starts to fuse inside a solar-mass red giant, it does so slowly at first; the rate of fusion increases gradually over many years.

_____ **9.** With the onset of helium fusion, a red giant gets brighter.

_____ **10.** A planetary nebula is the disk of matter around a star that will eventually form a planetary system.

_____ **11.** For a high-mass star, there is no helium flash.

_____ **12.** High-mass stars can fuse carbon and oxygen in their cores.

_____ **13.** A star cluster with an age of 100 million years will still contain many O-type stars.

_____ **14.** In a binary star system, it is never possible for the lower-mass star to be more evolved than the higher-mass companion.

_____ **15.** In a mass-transfer binary, one of the stars has filled its Roche lobe.

SELF-TEST: Fill in the Blank

1. A main-sequence star doesn't collapse because of the outward _____ produced by hot gases in the stellar interior.

2. The Sun will leave the main sequence in about _____ years.

3. While a star is on the main sequence, _____ is slowly depleted and _____ builds up in the core.

4. A temperature of at least _____ is needed to fuse helium.

5. At the end of its main-sequence lifetime, a star's core starts to _____.

6. When helium fuses, it produces _____ and releases _____.

7. Just before helium fusion begins in the Sun, the core's outward pressure will be provided mainly by electron _____.

8. As a star ascends the asymptotic giant branch, its _____ core is shrinking.

9. As a red supergiant, the Sun will eventually become about _____ times its present size.

10. The various stages of stellar evolution predicted by theory can be tested using observations of stars in _____.

11. By the time the envelope of a red supergiant is ejected, the core has shrunk down to a diameter of about _____ .

12. A typical white dwarf has the following properties: about half a solar mass, fairly _____ surface temperature, small size, and _____ luminosity.

13. As time goes by, the temperature and the luminosity of a white dwarf both _____.

14. As a star cluster ages, the luminosity of the main-sequence turnoff _____.

15. Whether being a member of a binary star system will affect the evolution of a star depends largely on the _____ of the two stars in the binary.

REVIEW AND DISCUSSION

1. Which types of stars live the longest? Why?

2. What is main-sequence equilibrium?

3. Why don't stars live forever?

4. How long can a star like the Sun keep burning hydrogen in its core?

5. Why is the depletion of hydrogen in the core of a star such an important event?

6. What makes an ordinary star become a red giant?

7. How big (in A.U.) will the Sun become when it enters the red-giant phase?

8. How long does it take for a star like the Sun to evolve from the main sequence to the top of the red-giant branch?

9. What is the helium flash?

10. What are the energy sources of a horizontal-branch star?

11. Describe an important way in which winds from red giant stars are linked to the interstellar medium.

12. How do stars of low mass die? How do stars of high mass die?

13. What is a planetary nebula? Why do many planetary nebulae appear as rings?

14. What are white dwarfs? What is their ultimate fate?

15. Do many black dwarfs exist in the Galaxy?

16. What factors determine whether the stars in a binary system will influence one another's evolution?

17. What are the Roche lobes of a binary system?

18. How did the binary system Algol come to consist of a low-mass red giant orbiting a high-mass main-sequence star?

PROBLEMS

1. (a) Use the radius-luminosity-temperature relation ($L \propto r^2 T^4$; see Chapter 17) to calculate the radius of a red supergiant with temperature 3000 K (one-half the solar value) and luminosity 10,000 solar luminosities. How many planets of our solar system would this star engulf? (b) Repeat your calculation for a 12,000 K (twice the temperature of the Sun), 0.0004 solar luminosity white dwarf.

2. The Sun will reside on the main sequence for 10^{10} years. If the luminosity of a main-sequence star is roughly proportional to the cube of the star's mass, what mass star is just now leaving the main sequence in a cluster that formed 400 million years ago?

3. A Sun-like star goes through its most rapid luminosity change between stages 8 and 9. Luminosity increases by about a factor of 100 in 10^5 years. On average, how rapidly does the luminosity change, in solar luminosities per year? Do you think this would be noticeable in a distant star?

4. Calculate the average density of a red-giant core of mass 0.25 solar mass and radius 15,000 km. Compare this to the average density of the giant's envelope, if its mass is 0.5 solar mass and its radius is 0.5 A.U. Compare each to the average core density of the Sun.

5. How many years are spent by a Sun-like star during its post-main sequence evolution? (See Table 20-1 and consider stages 8 through 12.) What is the chance of astronomers catching a star during one of its rapid periods of evolution—say between stages 8 and 9—if they examine a large number of randomly chosen post-main sequence stars? How would the odds change if all stars (including those on the main sequence) were included in the search?

PROJECTS

1. You can tour the Galaxy without ever leaving Earth, just by looking up. In the winter sky, you'll find the red supergiant Betelgeuse in the constellation Orion. It's easy to see because it's one of the brightest stars visible in our night sky. Betelgeuse is a variable star, with a period of about 6.5 years. Its brightness changes as the star expands and contracts. At maximum size, Betelgeuse fills a volume of space that would extend from the Sun beyond the orbit of Jupiter. Betelgeuse is thought to be about 10 to 15 times more massive than our Sun. It is probably between 4 and 10 million years old—and in the final stages of its evolution.

A similar star can be found shining prominently in mid-summer. This is the red supergiant Antares in the constellation Scorpius. Depending on the time of year, can you find one of these stars? Why are these stars red? What will happen to them next?

2. Find a library that has the *Astrophysical Journal*. Find an article from the late 1950s and 1960s that gives the photometry of a star cluster like the Pleiades or Hyades. Plot a color-magnitude diagram (V vs. B-V). Determine the V magnitude of the main sequence turnoff, and hence estimate the age of the cluster. Compare your age to that given in the article.

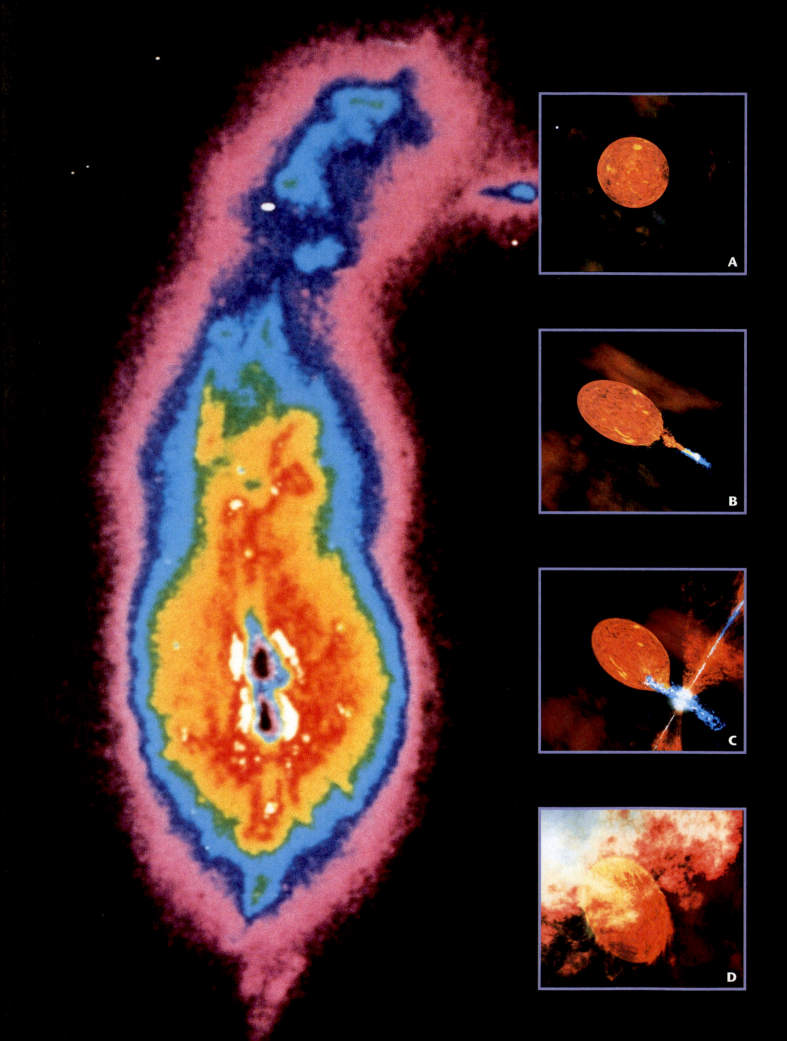

A

B

C

D

21

STELLAR EXPLOSIONS

Novae, Supernovae, and the Formation of the Heavy Elements

LEARNING GOALS

Studying this chapter will enable you to:

1. Explain how white dwarfs in binary systems can become explosively active.

2. Summarize the sequence of events leading to the violent death of a massive star.

3. Describe the two types of supernova, and explain how each is produced.

4. Describe the observational evidence that supernovae have occurred in our Galaxy.

5. Discuss the role of supernovae as distance indicators.

6. Explain the origin of elements heavier than helium, and discuss the significance of these elements for the study of stellar evolution.

(Opposite page, background) One of the best (false-color) pictures of a nova, here shown ejecting hot gas at high speeds from a binary-star system. The outer part of R Aquarii, known as a symbiotic system, resembles a luminescent geyser whose glowing plasma is twisted by the force of the explosion and channeled outward by strong magnetic fields. The geyser extends some 300 billion km, or about 30 times the size of our solar system. The double knot at the center of the image may outline the binary-star system itself—dark, because the camera overexposed the stars' light to display the fainter, erupting debris.

(Inset A) An artist's conception helps us to visualize one possible explanation of R Aquarii's periodic outbursts. Here, a white-dwarf star (upper left) orbits about a cool red-giant star; an elliptical orbit regularly brings the dwarf close to the giant.

(Inset B) The red giant spills some matter onto its companion, forming an accretion disk (lower right), drawn here as a swirling, flattened, blue vortex of hot gas.

(Inset C) As matter continues transferring, causing the process to become unstable, the disk erupts—at first gently as plasma escapes in bipolar flows (or geysers) perpendicular to the disk.

(Inset D) As the companion becomes overloaded with in-falling matter, spontaneous nuclear fusion occurs on or near its surface, resulting in a full-scale nova. The process repeats periodically, causing a "recurrent nova."

What fate awaits a star when it runs out of fuel? For a low-mass star, the white-dwarf stage is not necessarily the end of the road. The potential exists for further violent activity if a binary companion can provide additional fuel. High-mass stars—whether or not they are members of binaries—are also destined to die explosively, releasing vast amounts of energy, creating many heavy elements, and scattering the debris throughout interstellar space. These cataclysmic explosions may trigger the formation of new stars, continuing the cycle of stellar birth and death. In this chapter we study in more detail the processes responsible for these cataclysmic explosions and the mechanisms that create heavy elements during all phases of stellar evolution.

21.1 Life after Death for White Dwarfs

① Not all stars shine steadily day after day, year after year. Some change dramatically in brightness over very short periods of time—weeks, days, or even less. One type of star, called a **nova** (plural, *novae*), may increase enormously in brightness—by as much as a factor of 10,000 or more—in a very short period of time.

The word *nova* means "new" in Latin, and to early observers, these stars did indeed seem new, as they appeared suddenly in the night sky. Astronomers now recognize that a nova is not a new star at all. It is instead a white dwarf—a normally very faint star—undergoing an explosion on its surface that results in a rapid, temporary increase in luminosity. Figure 21.1 illustrates the brightening of a typical nova over a period of 3 days. Novae eventually fade back to normal, usually after a few weeks or months. On average, two or three novae are observed each year. Astronomers also know of many *recurrent novae*—stars that have been observed to "go nova" several times over the course of a few decades.

What could cause such an explosion on a faint, dead star? The energy involved is far too great to be explained by flares or other surface activity, and, as we have seen in the last chapter, there is no nuclear activity in the dwarf's interior. ∞ (p. 431) To understand what happens, we must reconsider the fate of a low-mass star after it enters the white-dwarf phase.

We noted in Chapter 20 that the white-dwarf stage represents the end point of a star's evolution. Subsequently, the star simply cools, eventually becoming a black dwarf—a burned-out ember in interstellar space. This scenario is quite correct for an *isolated* star, such as our Sun. However, should the star be part of a *binary* system, an important new possibility exists. If the distance between the two stars is small enough, the dwarf's tidal gravitational field can pull matter—primarily hydrogen and helium—away from the surface of its main-sequence or giant companion, as illustrated in Figure 21.2. The system becomes a mass-transferring binary, similar to those discussed in Section 20.6. A stream of gas leaves the companion through the Lagrange point and flows onto the dwarf. ∞ (p. 437)

As gas builds up on the white dwarf's surface, it becomes hotter and denser. Eventually its temperature exceeds 10^7 K, and the hydrogen ignites, fusing into helium at a furious rate. This surface burning stage is as brief as it is violent. The star suddenly flares up in luminosity, then fades away, as some of the fuel is exhausted and the remainder is blown off into space. If the event happens to be visible from the Earth, we see a nova. Figure 21.3(e) is a photograph of a nova apparently caught in the act of expelling mass from its surface.

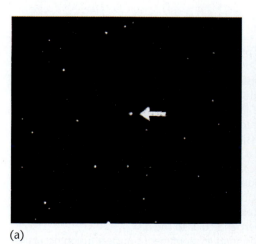

(a)

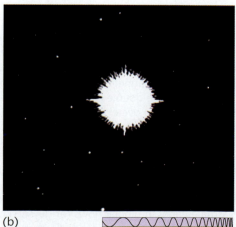

(b)

R I V U X G

Figure 21.1 A nova is a star that suddenly increases enormously in brightness, then slowly fades back to its original luminosity. Novae are the result of explosions on the surfaces of faint white-dwarf stars, caused by matter falling onto their surfaces from the atmosphere of a larger binary companion. Shown is Nova Herculis 1934 in (a) March 1935 and (b) in May 1935, after brightening by a factor of 60,000.

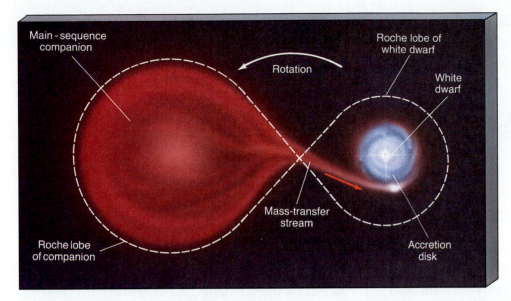

Figure 21.2 A white dwarf in a semidetached binary system may be close enough to its companion that its gravitational field can tear material from the companion's surface. Compare Figure 20.21. Notice that, unlike in the earlier figure, the matter does not fall directly onto the white dwarf's surface. Instead, as discussed a little later in the text, it forms an "accretion disk" of gas spiraling down onto the dwarf.

The initial flare-up of luminosity from a nova declines in time, and eventually the star returns to its normal, preexplosion appearance. The luminosity and temperature of a nova are not usually plotted on an H–R diagram, however. Instead, the change in luminosity is plotted in the form of a *light curve*, like that shown in Figure 21.4. Such curves show the dramatic rise in luminosity over a few days, followed by the much slower decay over the course of several months. The decline in brightness results from the expansion and cooling of the dwarf's surface layers as they are blown into space. Studies of the details of these curves provide astronomers with a wealth of information, about both the dwarf and its binary companion.

The manner in which the matter reaches the white dwarf's surface provides important observational evidence in support of this scenario. Because of the binary's rotation, material leaving the companion does not fall directly onto the dwarf. Instead, it "misses" the compact star, loops around behind it, and goes into orbit around it, forming a swirling, flattened disk of matter known as an **accretion disk** (shown in Figure 21.2). Due to the effects of viscosity (that is, friction) within the gas, the orbiting matter in the disk drifts gradually inward, its temperature increasing steadily as it spirals down onto the dwarf's surface. The inner part of the accretion disk becomes so hot that it radiates strongly in the visible, the ultraviolet, even the X-ray,

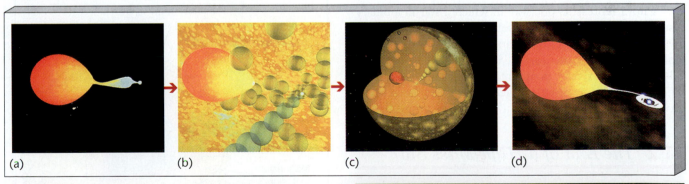

(a) (b) (c) (d)

(e)

Figure 21.3 In this artist's conception, material accumulates on a white dwarf's surface after being accreted from a companion star (a) and then ignites in hydrogen fusion as a nova outburst (b and c). Part of the surface gas is ejected into space in the form of "bubbles" of hot plasma; the rest relaxes back down onto the accretion disk (d). The real photo in (e) corresponds roughly to the events depicted by the artist in frame (c). This nova is called Nova Persei (1901). (See also the chapter opening photo.)

R I V U X G

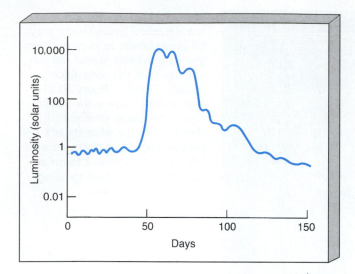

Figure 21.4 The light curve of a typical nova. The rapid rise and slow decline in the light received from the star, as well as the maximum brightness attained, are in good agreement with the explanation of the nova as a nuclear flash on a white dwarf's surface.

portions of the electromagnetic spectrum. In many systems, the disk outshines the white dwarf itself and is the main source of the light emitted between nova outbursts. X rays from the hot disk are routinely observed in many Galactic novae. The point at which the infalling stream of matter strikes the accretion disk often forms a turbulent "hot spot," causing detectable fluctuations in the light emitted by the binary system.

A nova represents one way in which a star in a binary system can extend its "active lifetime" well into the white-dwarf stage. Recurrent novae can, in principle, repeat their violent outbursts many dozens, if not hundreds, of times. But even more extreme possibilities exist at the end of stellar evolution. Vastly more energetic events may be in store, given the right circumstances.

21.2 *The End of a High-Mass Star*

❷ A low-mass star—one with mass less than about 8 times the mass of the Sun—never becomes hot enough to burn carbon in its core. It ends its life as a carbon white dwarf. A high-mass star, however, can fuse not just hydrogen and helium, but also carbon, oxygen, and even heavier elements as its inner core continues to contract and its central temperature continues to rise. The burning rate accelerates as the core evolves. Can anything stop this runaway process? Is there a stable "white-dwarf–like" state at the end of the evolution of a high-mass star? What is its ultimate fate? To answer these questions, we must look more carefully at fusion in massive stars.

FUSION OF HEAVY ELEMENTS

Figure 21.5 is a cutaway diagram of the interior of a highly evolved star of large mass. Note the numerous layers where various nuclei burn. As the temperature increases with depth, the ash of each burning stage becomes the fuel for the next stage. At the relatively cool periphery of the core, hydrogen fuses into helium. In the intermediate layers, shells of helium, carbon, and oxygen burn to form heavier nuclei. Deeper down reside neon, magnesium, silicon, and other heavy nuclei, all produced by nuclear fusion in the layers overlying the core. The core itself is composed of iron nuclei, complex pieces of matter each containing 26 protons and 30 neutrons. We will study the key reactions in this burning chain in more detail later in this chapter.

As each element is burned to depletion at the center, the core contracts, heats up, and starts to fuse the ash of the previous burning stage. A new inner core forms, contracts again, heats again, and so on. Through each period of stability and instability, the star's central temperature increases, the nuclear reactions speed up, and the newly released energy supports the star for ever-shorter periods of time. For example, in round numbers, a star 20 times more massive than the Sun burns hydrogen for 10 million years, helium for 1 million years, carbon for 1000 years, oxygen for 1 year, and silicon for a week. Its iron core grows for less than a day.

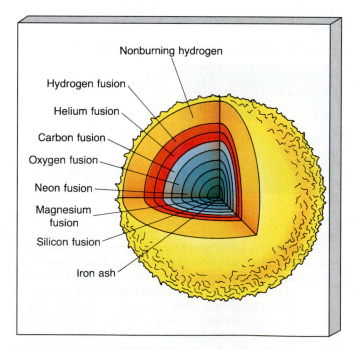

Figure 21.5 Cutaway diagram of the interior of a highly evolved star of mass greater than 8 solar masses. The interior resembles the layers of an onion, with shells of progressively heavier elements burning at smaller and smaller radii and at higher and higher temperatures.

COLLAPSE OF THE IRON CORE

Once the inner core begins to change into iron, our massive star is in trouble. Nuclear fusion involving iron does not produce energy—iron nuclei are so compact that energy cannot be extracted by combining them into heavier elements. In effect, iron plays the role of a fire extinguisher, damping the inferno in the stellar core. With the appearance of substantial quantities of iron, the central fires cease for the last time, and the star's internal support begins to dwindle. The star's foundation is destroyed, and its equilibrium is gone forever. Even though the temperature in the iron core has reached several billion kelvins by this stage, the enormous inward gravitational pull of matter ensures catastrophe in the very near future. Gravity overwhelms the pressure of the hot gas, and the star implodes, falling in on itself.

The core temperature rises to nearly 10 billion K. At these temperatures, individual photons, according to Wien's law, have tremendously high energies. ∞ (p. 64, 80) They are energetic enough to split iron into lighter nuclei and, in turn, to break those lighter nuclei apart until only protons and neutrons remain. This process is known as *photodisintegration* of the heavy elements in the core. In less than a second, the collapsing core undoes all the effects of nuclear fusion that occurred during the previous 10 million years! But to split iron and lighter nuclei into smaller pieces requires a lot of energy. After all, this splitting is just the opposite of the fusion reactions that generated the star's energy during earlier times. The process of photodisintegration *absorbs* some of the core's thermal energy—in other words, it tends to cool the core and so reduces the pressure. As nuclei are destroyed, the core of the star becomes even less able to support itself against its own gravity, and the collapse accelerates.

Now the core consists entirely of simple elementary particles—electrons, protons, neutrons, and photons—at enormously high densities, and it is still shrinking. As the core density continues to rise, the protons and electrons are crushed together, forming neutrons and neutrinos:

$$p + e \rightarrow n + \text{neutrino}.$$

This process is sometimes called the *neutronization* of the core. Recall from our discussion in Chapter 16 that the neutrino is an extremely elusive particle that interacts hardly at all with matter. ∞ (p. 350) Even though the central density by this time may have reached 10^{12} kg/m^3 or more, most of the neutrinos produced by neutronization pass through the core as if it weren't there. They escape into space, carrying away energy as they go.

The disappearance of the electrons and the escape of the neutrinos make matters even worse for the core's stability. There is now nothing to prevent it from collapsing all the way to the point at which the neutrons come into contact with each other, at the incredible density of about 10^{15} kg/m^3. At this point, the neutrons in the shrinking core play a role similar in many ways to that of the electrons in a white dwarf. When far apart, they offer little resistance to compression, but when brought into contact, they produce enormous pressures that strongly oppose further gravitational collapse. This *neutron degeneracy pressure*, akin to the electron degeneracy pressure that operates in red giants and white dwarfs (see Chapter 20), finally begins to slow the collapse. ∞ (p. 425) By the time the collapse is actually halted, however, the core has overshot its point of equilibrium, and may reach densities as high as 10^{17} or 10^{18} kg/m^3 before turning around and beginning to reexpand. Like a fast-moving ball hitting a brick wall, the core becomes compressed, stops, then rebounds—with a vengeance!

The events just described do not take long. Only about a second elapses from the start of the collapse to the "bounce" at nuclear densities. At that point, the core rebounds. An enormously energetic shock wave sweeps through the star at high speed, blasting all the overlying layers—including the heavy elements outside the iron inner core—into space. Although the details of how the shock reaches the surface and destroys the star are still uncertain, the end result is not. The star explodes, in a manner vastly more violent than the expulsion of matter in the form of a planetary nebula that marks the end of a low-mass star. The explosion is one of the most energetic events known in the universe (see Figure 21.6). For a period of a few days, the exploding star may rival in brightness the trillion-star galaxy in which it resides. This spectacular death rattle of a high-mass star is known as a **core-collapse supernova.**

Figure 21.6 A supernova called SN1987A (arrow) was exploding near this nebula (30 Doradus) at the moment the photograph on the right was taken. The photograph on the left is the normal appearance of the star field. (See *Interlude 21–2* on p. 452.)

21.3 *Supernova Explosions*

NOVAE AND SUPERNOVAE

3 Let's compare a supernova with a nova. Like a nova, a supernova is a star that suddenly increases dramatically in brightness, then slowly dims again, eventually fading from view. The exploding star is commonly called the supernova's *progenitor*. In some cases, supernovae light curves can appear quite similar to those of novae, so a distant supernova can look a lot like a nearby nova—so much so, in fact, that the difference between the two was not fully appreciated until the 1920s. But novae and supernovae are now known to be very different phenomena. Supernovae are much more energetic events, driven by very different underlying physical processes.

Well before they understood the causes of either novae or supernovae, astronomers knew of clear observational differences between them. The most important of these is that a supernova is about a million times brighter than a nova. A supernova produces a flash equaling over a billion solar luminosities, reaching that brightness within just a few hours after the start of the outburst. The total amount of electromagnetic energy radiated by a supernova during the few months it takes to brighten and fade away is roughly 10^{43} J —nearly as much energy as the Sun will radiate during its *entire* 10^{10}-year lifetime! (Enormous as this energy is, however, it pales in comparison to the energy emitted in the form of *neutrinos*, which may be 100 times greater.)

A second important difference is that the same star may become a nova many times, but a star can become a supernova only once. This fact was unexplained before astronomers knew the precise nature of novae and supernovae, but it is easily understood now that we understand how and why these explosions occur. The nova accretion–explosion cycle described earlier can take place over and over again, but a supernova destroys the star involved, with no possibility of a repeat performance.

In addition to the distinction between novae and supernovae, there are also important observational differences *among* supernovae. Some supernovae contain very little hydrogen, according to their spectra, whereas others contain a lot. Also, the light curves of the hydrogen-poor supernovae are qualitatively different from those of the hydrogen-rich ones, as illustrated in Figure 21.7. On the basis of these observations, astronomers divide supernovae into two classes, known simply as Type I and Type II. **Type-I supernovae**, the hydrogen-poor kind, have a light curve somewhat similar in shape to that of typical novae. **Type-II supernovae**, whose spectra show lots of hydrogen, usually have a characteristic "plateau" in the light curve a few months after the maximum. Observed supernovae are divided roughly evenly between these two categories.

CARBON-DETONATION SUPERNOVAE

What is responsible for these differences among supernovae? Is there more than one way in which a supernova explosion can occur? The answer is "Yes." To understand the alternative supernova mechanism, we must return to the processes that cause novae and consider the long-term consequences of their accretion–explosion cycle.

Nova explosions eject matter from a white dwarf's surface, but they do not necessarily expel or burn *all* the material that has accumulated since the last outburst. In other words, there is a tendency for the dwarf to increase slowly in mass with each new nova cycle. As the star's mass grows and the internal pressure required to support its weight rises, it can enter into a new period of instability—with disastrous consequences.

Recall that a white dwarf is held up not by thermal pressure (heat) but by the "degeneracy" pressure of electrons that have been squeezed so close together that they have effectively come into contact with one another. ∞ (p. 425) Is there a limit to the pressure that even these electrons can exert? Is there, therefore, a limit to the mass of a white dwarf beyond which electrons cannot provide the pressure needed to support the star? The answer to both questions is "Yes." Detailed calculations show that the maximum mass of a white dwarf is about 1.4 $M_\odot$, a mass often called the *Chandrasekhar mass*, after the Indian astronomer Subramanyan Chandrasekhar, whose work in theoretical astrophysics earned him a Nobel Prize in 1983.

If an accreting white dwarf exceeds the Chandrasekhar mass, the pressure of degenerate electrons in its interior becomes unable to withstand the pull of gravity, and the star immediately starts to collapse. Its internal temperature

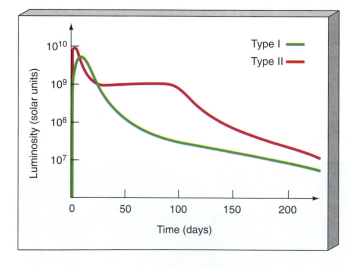

Figure 21.7   The light curves of typical Type-I and Type-II supernovae. In both cases, the maximum brightness or intensity can sometimes reach that of a billion suns, but there are characteristic differences in the fall-off of the luminosity after the initial peak. Type-I light curves somewhat resemble those of novae (Figure 21.4). Type-II curves have a characteristic bump in the declining phase.

Interlude 21-1 *Nearby Supernovae*

Only six galactic supernovae have been recorded in the past 1000 years. The accompanying figure shows their positions in the Milky Way. They are labeled by the year in which they first appeared. The supernova Cassiopeia A (CasA) apparently went unnoticed optically, although modern radio studies suggest that the first light from the explosion should have reached the Earth midway through the seventeenth century. The following combined radio–optical–X-ray image is dramatic proof that this supernova remnant, although invisible optically, makes quite an impact at other wavelengths. (Blue is radio, red is optical, and green is X ray.)

Most astronomers assume that many more stars than these six have blown up in our Galaxy. Why haven't we seen them? Possibly because they were too distant to be detected by the naked eye, or perhaps dark clouds in the galactic plane kept them from our view. Each of the Milky Way supernovae mapped in the first figure is in our "neighborhood"—that is,

in our quadrant of the Galaxy—and each is at least 100 pc above or below the galactic plane. Studies of the rate at which supernovae occur suggest that we can expect one within 100 pc of our Sun only every 500,000 years. Thus, a truly "close" supernova would be a rare event indeed. Humanity may be destined to see all supernovae from a distance.

Despite their rarity, nearby supernovae might conceivably play an important role in determining the development of life on Earth. A supernova at a distance of a few parsecs would bombard our planet with high-energy radiation for a period of several months, possibly causing substantial long-term atmospheric changes, particularly in the ozone layer. Some scientists have even gone so far as to suggest that such an event may have been responsible for episodes of mass extinction that are known to have occurred hundreds of millions of years ago, and in which, according to the fossil record, over 95 percent of all life on our planet vanished in a very short period of time. (Recall, though, that this is not necessarily the only "astronomical" explanation for mass extinctions on our planet—see *Interlude 14-1*.) ∞ (p. 300)

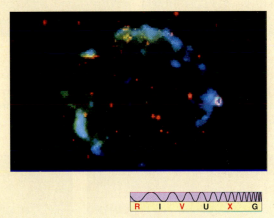

R I V U X G

---

rapidly rises to the point at which carbon can fuse into heavier elements. Carbon fusion begins everywhere throughout the white dwarf almost simultaneously, and the entire star explodes in another type of supernova—a so-called **carbon-detonation supernova**—comparable in violence to the "implosion" supernova associated with the death of a high-mass star, but very different in cause. In an alternative and (some astronomers think) possibly more common scenario, two white dwarfs in a binary system may collide and merge to form a massive, unstable star. The end result is the same—a carbon-detonation supernova.

We can now understand the differences between Type-I and Type-II supernovae. The explosion resulting from the detonation of a carbon white dwarf, the descendant of a low-mass star, is a supernova of Type I. Because this conflagration stems from a system containing virtually no hydrogen, we can readily see why the spectrum of a Type-I su-

pernova shows little evidence of that element. The appearance of the light curve (as we will soon see) results almost entirely from the radioactive decay of unstable heavy elements produced in the explosion itself.

The implosion–explosion of the core of a massive star, described earlier, produces a Type-II supernova. Detailed computer models indicate that the characteristic shape of the Type-II light curve is just what would be expected from the star's outer envelope expanding and cooling as it is blown into space by the shock wave sweeping up from below. The expanding material consists mainly of unburned gas—hydrogen and helium—so it is not surprising that those elements are strongly represented in the supernova's observed spectrum. (See *Interlude 21-2* on p. 452 for an account of a recent Type-II supernova that confirmed many basic theoretical predictions while also forcing astronomers to revise the details of their models.)

(a) Type-I Supernova

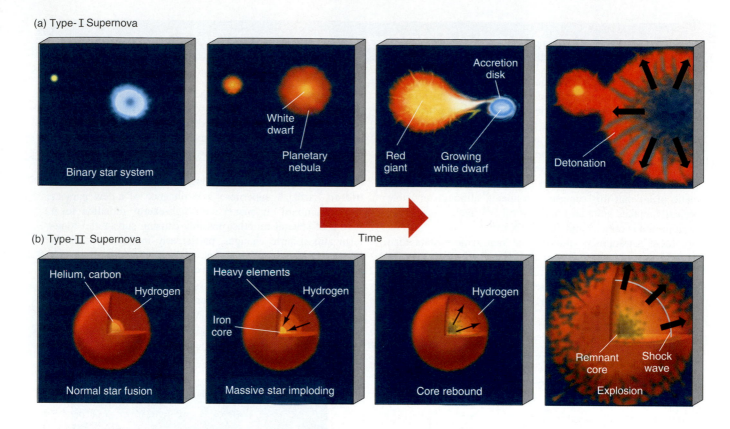

(b) Type-II Supernova

Figure 21.8 Type-I and Type-II supernovae have different causes. These sequences depict the evolutionary history of each type. (a) A Type-I supernova usually results when a carbon-rich white dwarf pulls matter onto itself from a nearby red-giant companion. (b) A Type-II supernova occurs when the core of a more massive star collapses, then rebounds in a catastrophic explosion.

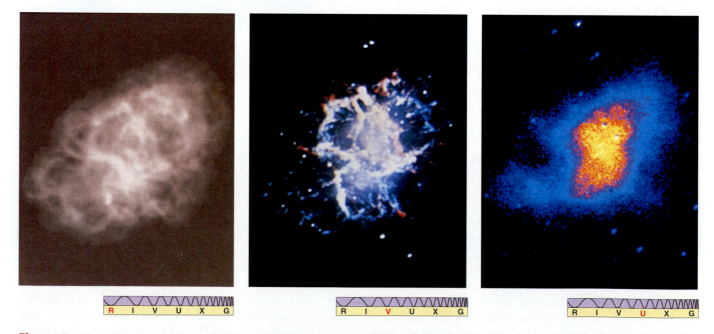

Figure 21.9 This remnant of an ancient supernova is called the Crab Nebula (or M1 in the Messier catalog). It resides about 1800 pc from the Earth and has an angular diameter about one-fifth that of the full Moon. Because its debris is scattered over a region of "only" 2 pc, the Crab is considered to be a young supernova remnant. In A.D. 1054 Chinese astronomers observed the supernova explosion itself. The center frame shows the Crab in visible light. The left and right frames, to the same scale, show the Crab Nebula in the radio and ultraviolet, respectively.

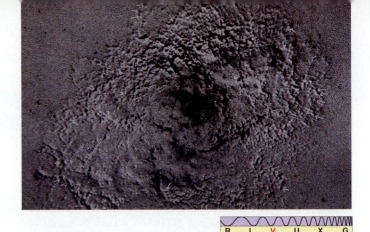

R I V U X G

Figure 21.8 summarizes the processes responsible for the two different types of supernovae.

SUPERNOVA REMNANTS

4 We have plenty of evidence that supernovae have occurred in our Galaxy. Occasionally, the explosions themselves are visible from the Earth (see *Interlude 21–1* on p. 449). In many other cases we can detect their glowing remains, or **supernova remnants**. One of the best-studied supernova remnants is known as the Crab Nebula, shown in Figure 21.9. Its brightness has greatly dimmed now, but the original explosion in the year A.D. 1054 was so brilliant that manuscripts of ancient Chinese and Middle Eastern astronomers claim that its brightness greatly exceeded that of Venus and—according to some (possibly exaggerated) accounts—even rivaled that of the Moon. For nearly a month, this exploded star reportedly could be seen in broad daylight. Native Americans also left engravings of the event in the rocks of what is now the southwestern United States.

Certainly, the Crab Nebula has the appearance of exploded debris. In fact, astronomers have proved that this matter was ejected from some central explosion. Figure 21.10 was made by superimposing a positive image of the Crab Nebula taken in 1960 and a negative image taken in 1974. If the gas were not in motion, the positive and nega-

Figure 21.10 Positive and negative photographs of the Crab Nebula taken 14 years apart do not superimpose exactly, indicating that the gaseous filaments are still moving away from the site of the explosion.

tive images would overlap perfectly, but they do not. The gas moved outward in the intervening 14 years. Knowing the total distance traveled by the gas in this time, astronomers have computed a velocity of several thousand kilometers per second for the expelled debris. Running the motion backward in time, astronomers have found that the explosion must have occurred about nine centuries ago. This date is consistent with the Chinese observations.

The nighttime sky harbors many relics of stars that blew up long ago. Figure 21.11 is another example. It shows the Vela supernova remnant whose expansion velocities imply that its central star exploded around 9000 B.C. It

R I V U X G

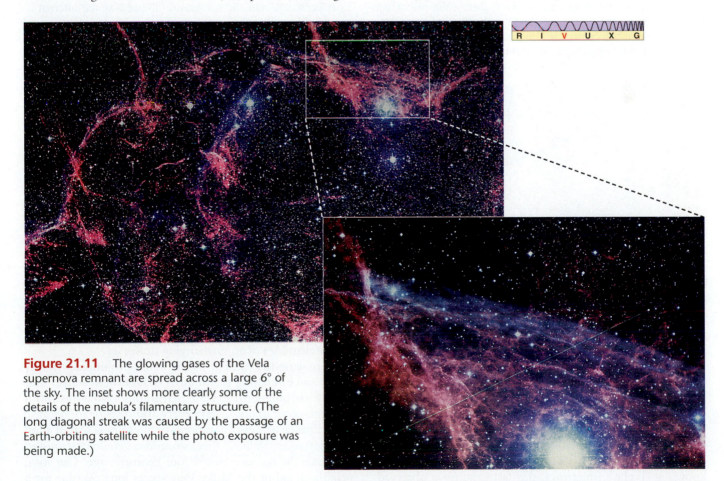

Figure 21.11 The glowing gases of the Vela supernova remnant are spread across a large 6° of the sky. The inset shows more clearly some of the details of the nebula's filamentary structure. (The long diagonal streak was caused by the passage of an Earth-orbiting satellite while the photo exposure was being made.)

Interlude 21–2   Supernova 1987A

In 1987, astronomers were treated to a spectacular supernova in the Large Magellanic Cloud. ∞ (p. 522 Observers in Chile first saw the explosion on February 24, and within a few hours nearly all Southern Hemisphere telescopes and every available orbiting spacecraft were focused on the object. It was officially named SN1987A. (The SN stands for "supernova," 1987 gives the year, and A identifies it as the first supernova seen that year.) This was one of the most dramatic changes observed in the universe in nearly 400 years. A 15-solar-mass B-type supergiant star with the catalog name of SK-69°202 detonated, outshining for a few weeks all the other stars in the LMC combined, as shown in the "before" and "after" images of Figure 21.6.

Because the LMC is relatively close to the Earth and because the explosion was detected so soon after it occurred, SN1987A provided scientists with an enormous volume of detailed information on supernovae, allowing astronomers to make key comparisons between theoretical models and observational reality. By and large, the theory of stellar evolution described in the text has held up very well. Still, SN1987A did hold some surprises.

According to its hydrogen-rich spectrum, the supernova was of Type II—the iron-core implosion–explosion type—as expected for a high-mass parent star such as SK-69°202. But a glance at Figure 20.16 (which was computed for stars in our own Galaxy) shows that, according to theory, the parent star should have been a *red* supergiant at the time of the explosion—not a blue supergiant, as was actually observed. This unexpected finding caused theorists to scramble in search of a reason, and many possibilities were considered before an expla-

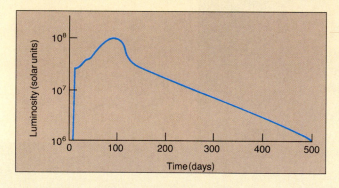

nation fitting the facts emerged. It seems that the progenitor's envelope was deficient in heavy elements compared with young stars in the Milky Way. This deficiency had little effect on the evolution of the core and on the supernova explosion itself, but it did change the star's evolutionary track on the H–R diagram. Unlike a Milky Way star of the same mass, once helium ignited in the core of SK-69°202, the star shrank in size and looped back toward the main sequence. The star had just begun to return to the right on the H–R diagram following the ignition of carbon, with a surface temperature of around 20,000 K, when the rapid chain of events leading to the supernova occurred.

The light curve of SN1987A also differed somewhat from the "standard" Type-II shape (see Figure 21.7), and the peak brightness was only about 1/10 the expected value.

For a few days after its initial detection, the supernova faded as it expanded and cooled rapidly. After about a week, the surface temperature had dropped to about 5000 K, at which point electrons and protons near the expanding surface recombined into atomic hydrogen. This recombination made the surface layers less opaque and allowed more radiation from the interior to leak out. The supernova brightened rapidly as it grew. The temperature of the expanding layers reached a peak in late May, by which point the radius of the expanding photosphere was about 2×10^{10} km—a little larger than our solar system. Subsequently, the photosphere cooled as it expanded, and the luminosity dropped as the internal supply of heat from the explosion leaked away into space.

Much of the preceding description would apply equally well to a Type-II supernova in our own Galaxy. The differences between the SN1987A light curve shown above and the Type-II light curve in Figure 21.7 are mainly the result of the (relatively) small size of the progenitor star. The peak luminosity of SN1987A was less than that of a "normal" Type-II supernova because the progenitor SK-69°202 was small and quite tightly bound by gravity. A lot of the energy emitted in the form of visible radiation (and evident in Figure 21.7) was used up in expanding SN1987A's stellar envelope, so far less was left over to be radiated into space. Thus, SN1987A's luminosity during the first few months was lower, and the early peak evident in Figure 21.7 did not occur. The peak in the SN1987A light curve at about 80 days actually corresponds to the "bump" in the Type-II light curve in Figure 21.7.

---

lies only 500 pc away from the Earth. Given its proximity, it may have been as bright as the Moon for several months. We can only speculate what impact such a bright supernova might have had on the myths, religions, and cultures of Stone Age humans when it first appeared in the sky.

Although hundreds of supernovae have been observed in other galaxies during the twentieth century, no one has ever observed with modern equipment a supernova in our own Galaxy. A viewable Milky Way star has not exploded since Galileo first turned his telescope to the heavens almost four centuries ago (see *Interlude 21-1* on p. 449). Now known as Tycho's supernova, this last supernova observed

in our Galaxy caused a worldwide sensation in Renaissance times. The sudden appearance and subsequent fading of this very bright object in the year 1604 helped shatter the Aristotelian idea of an unchanging universe.

Knowing the rates at which stellar evolutionary stages occur, and estimating the number of high-mass stars in our Galaxy, astronomers calculate that an observable supernova ought to occur in our Galaxy every 100 years or so. Because the brilliance of a nearby supernova might rival that of a full Moon, it seems unlikely that astronomers could have missed any since the last one nearly four centuries ago. Our local neighborhood of the Milky Way seems long overdue for a

That stellar evolution theory could extend to fit the facts of SN1987A is very reassuring. However, the unexpected color and size of the supernova's progenitor star underscore the importance of observational tests and checks in hammering out different theoretical models. It is just as important to know *which* model—in this case, which parent star—to use in the calculations as it is to perform the calculation correctly!

About 20 hours before the supernova was detected optically, a brief (about 13-second) burst of neutrinos was simultaneously recorded by underground detectors in Japan and the United States. As discussed in the text, the neutrinos are predicted to arise when electrons and protons in the star's collapsing core merge to form neutrons. The neutrinos preceded the light because they escaped during the collapse, whereas the first light of the explosion was emitted only after the supernova shock had plowed through the body of the star to the surface. In fact, theoretical models, consistent with these observations, suggest that vastly more energy was emitted in the form of neutrinos than in any other form. The supernova's neutrino luminosity was many tens of thousands of times greater than its optical energy output.

Despite some unresolved details in SN1987A's behavior, detection of this neutrino pulse is considered to be a brilliant confirmation of theory. This singular event—the detection of neutrinos—may well herald a new age of astronomy. For the first time, astronomers have received information from beyond the solar system by radiation outside the electromagnetic spectrum.

Theory predicts that the expanding remnant of SN1987A will be large enough to be resolvable by optical telescopes in a few years. The first accompanying photograph was taken by the *Hubble Space Telescope* in late 1990. It shows the unresolved remnant (in red) surrounded by a much larger shell of glowing gas (in yellow). Scientists reason that the progenitor star expelled this shell during its red-giant phase, some 40,000 years before the explosion. The image we see results from the initial flash of ultraviolet light from the supernova hitting the ring and causing it to glow brightly. In about 10 years, the fastest-moving debris from the remnant will strike the ring, making it a temporary but intense source of X rays.

The overexposed second photo, taken in 1994, shows the core debris indeed moving outward toward the ring. It also revealed, to everyone's surprise, two additional faint rings that might be radiation sweeping across the hourglass-shaped bub-

ble of gas. Why the gas should exhibit this odd structure, however, is unclear.

Buoyed by the success of stellar-evolution theory and armed with firm theoretical predictions of what should happen next, astronomers eagerly await future developments in the story of this remarkable object.

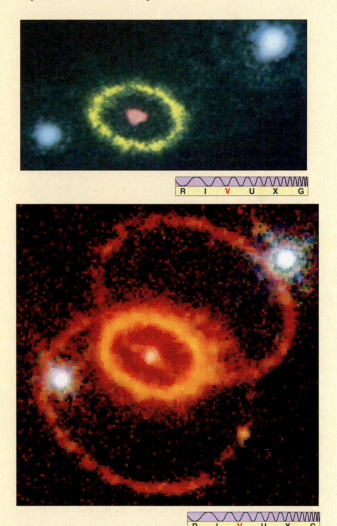

supernova. Unless massive stars explode much less frequently than predicted by the theory of stellar evolution, we should be treated to a (relatively) nearby version of nature's most spectacular cosmic event any day now.

SUPERNOVAE AS DISTANCE INDICATORS

⑤ Astronomers are often concerned with the measurement of the distances to astronomical objects. Knowing the size of the universe is essential for understanding how it works. In this regard, astronomers are especially interested in **standard candles**—cosmic objects of known absolute brightness. Once a standard candle is recognized, its absolute

brightness is immediately known, and so a measurement of its apparent brightness yields its distance. ∞ (p. 363)

Astronomers have found that supernovae are good examples of standard candles—supernovae of a given type (I or II) tend to have very similar peak luminosities (but see *Interlude 21-2* for a notable exception). When astronomers recognize a supernova and determine its type, they plot its light curve. By comparing the peak apparent brightness with the theoretical peak luminosity, they can then compute the supernova's distance from the Earth. The great advantage of these objects is that they are extremely bright, so they can be seen even when very far away. As a result, observations of supernovae are invaluable in determining cosmic distances well beyond the Milky Way Galaxy.

What is the basic reason for the similarity in peak luminosities among supernovae of a given type? The answer is that, in either case (carbon-detonation Type I or core-collapse Type II), the supernova explosion occurs only when a steadily growing stellar core reaches a well-defined critical mass. For example, regardless of how quickly or slowly a white dwarf in a binary system reaches the Chandrasekhar mass, a Type-I supernova *always* results from the explosion of a 1.4-solar-mass carbon–oxygen star. It does not matter what type of star produced the white dwarf in the first place. Similarly, the iron core of a massive star implodes and produces a Type-II supernova *only* when its mass reaches the specific value beyond which it is unable to support itself against its own gravity. Again, the conditions immediately preceding the supernova are very similar, even in progenitor stars of quite different masses.

In Type-II supernovae, the outward appearance of the explosion can be significantly modified by the amount of stellar material through which the blast wave must travel before it reaches the star's surface. Type-I supernovae, however, are particularly uniform in their properties and are now widely used in studies of galaxies far from our own.

21.4 *The Formation of the Elements*

6 Up to now, we have studied nuclear reactions mainly for their role in stellar energy generation. Now let's consider them again, but this time as the processes responsible for creating much of the world in which we live. The evolution of the elements, combining nuclear physics with astronomy, is a very complex subject, and a very important problem in modern astronomy. Let us begin by taking inventory of the composition of the universe.

TYPES OF MATTER

We currently know of 110 different elements, ranging from the simplest—hydrogen, containing 1 proton—to the most complex, discovered in 1994, with 110 protons in its nucleus. (See Appendix Table 2) All elements exist in several different *isotopic* forms, each isotope having the same number of protons but a different number of neutrons. We often think of the most common or stable isotope as being the "normal" form of an element. Some elements, and many isotopes, are radioactively unstable, meaning that they eventually decay into other, more stable, nuclei.

The 81 stable elements found on Earth make up the overwhelming bulk of matter in the universe. In addition, 10 radioactive elements—including radon and uranium—also occur naturally on our planet. Even though their half-lives (the time required for half the nuclei to decay into something else) of these elements are very long (millions or even billions of years, typically), their steady decay means that they are scarce on Earth, in meteorites, and in lunar samples. ∞ (p. 160) They are not observed in stars—there is just too little of them to produce detectable spectral lines.

Besides these 10 naturally occurring radioactive elements, 17 more radioactive elements have been artificially produced under special conditions in nuclear laboratories on Earth. The debris collected after nuclear weapons tests also contains traces of these elements. Unlike the naturally occurring radioactive elements, these artificial ones decay into other elements quite quickly (in much less than a million years). Consequently, they too are extremely rare in nature. Two other elements round out our list. Promethium is a stable element that is found on our planet only as a by-product of nuclear laboratory experiments. Technetium is an unstable element that is found in stars but does not occur naturally on Earth.

ABUNDANCE OF MATTER

How and where did all these elements form? Were they always present in the universe, or were they created after the universe formed? Since the 1950s, astronomers have come to realize that the hydrogen and most of the helium in the universe are *primordial*—that is, these elements date back to the very earliest times. All other elements in our uni-

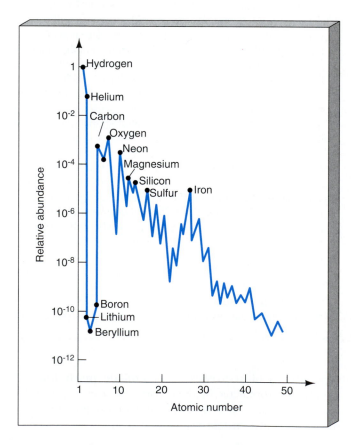

Figure 21.12 A summary of the cosmic abundances of the elements and their isotopes, expressed relative to the abundance of hydrogen. The horizontal axis shows atomic number—the number of protons in the nucleus. Notice how many common terrestrial elements are found on "peaks" of the distribution, surrounded by elements that are tens or hundreds of times less abundant. Notice especially the large peak around the element iron.

verse result from **stellar nucleosynthesis**—that is, they were formed by nuclear fusion in the hearts of stars.

A key point in understanding the creation of heavy elements is that large nuclei can be built from smaller ones by nuclear fusion. We might naturally theorize that all the heavy elements have been created in this way. In this scenario, the ultimate source of the heavy elements is the lightest and simplest of all—hydrogen. To test this idea, we must consider not just the list of different *kinds* of elements and isotopes but also their observed *abundances*, shown in Figure 21.12. This curve is derived largely from spectroscopic studies of stars, including the Sun. The essence of the figure is summarized in Table 21-1, which combines all the known elements into eight distinct groups based on the number of nuclear particles (protons and neutrons) that they contain. (All isotopes of all elements are included in both Table 21-1 and Figure 21.12, although only a few elements are marked by dots and labeled in the figure.) Any theory proposed for the creation of the elements must reproduce these observed abundances. The most obvious feature is that the heavy elements are much less abundant than most light elements.

HYDROGEN AND HELIUM BURNING

Let us now review the specific reactions leading to heavy-element production at different stages of stellar evolution. Stellar nucleosynthesis begins with the proton–proton chain studied in Chapter 16. ∞ (p. 349) Provided that the temperature is high enough—at least 10^7 K—a series of nuclear reactions occurs, ultimately forming a nucleus of ordinary helium (^{4}He) from four protons (^{1}H):

$$4\,(^1\text{H}) \;\rightarrow\; {}^4\text{He} + 2 \text{ positrons} + 2 \text{ neutrinos} + \text{energy}.$$

Recall that the positrons immediately interact with nearby free electrons, producing high-energy gamma rays through matter–antimatter annihilation. The neutrinos rapidly escape, carrying energy with them but playing no direct role

TABLE 21-1 *Cosmic Abundances of the Elements*

ELEMENTAL GROUP OF PARTICLES	PERCENT ABUNDANCE BY NUMBER*
Hydrogen (1 nuclear particle)	90
Helium (4 nuclear particles)	9
Lithium group (7–11 nuclear particles)	0.000001
Carbon group (12–20 nuclear particles)	0.2
Silicon group (23–48 nuclear particles)	0.01
Iron group (50–62 nuclear particles)	0.01
Middle-weight group (63–100 nuclear particles)	0.00000001
Heaviest-weight group (over 100 nuclear particles)	0.000000001

The total does not equal 100 percent because of uncertainties in the helium abundance. All isotopes of all elements are included.

in nucleosynthesis. The validity of these reactions has been directly confirmed in nuclear experiments conducted in laboratories around the world during recent decades. In massive stars, the CNO cycle (see *Interlude 20-1* on p. 422) may greatly accelerate the hydrogen-burning process, but the basic 4-proton-to-1-helium-nucleus reaction, illustrated in Figure 21.13, is unchanged.

As helium builds up in the core of a star, the burning ceases, and the core contracts and heats up. When the temperature exceeds about 10^8 K, helium nuclei can overcome their mutual electrical repulsion, leading to the *triple-alpha reaction*, which we saw in Chapter 20. ∞ (p. 424)

$$3\,(^4\text{He}) \;\rightarrow\; {}^{12}\text{C} + \text{energy}.$$

The net result of this reaction is that three helium-4 nuclei are combined into one carbon-12 nucleus (Figure 21.14), releasing energy in the process.

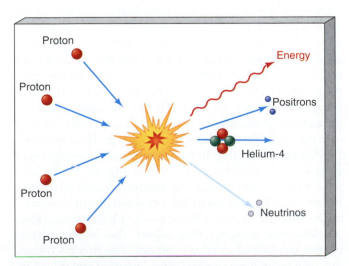

Figure 21.13 Diagram of the basic proton–proton hydrogen-burning reaction. Four protons combine to form a nucleus of helium-4, releasing energy in the process.

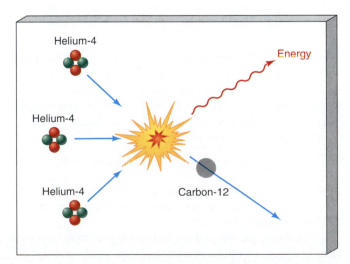

Figure 21.14 Diagram of the basic triple-alpha helium-burning reaction occurring in postmain-sequence stars. Three helium-4 nuclei combine to form carbon-12.

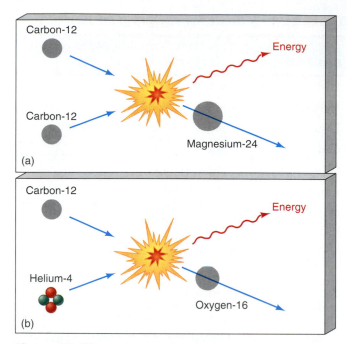

Figure 21.15 Carbon can form heavier elements (a) by fusion with other carbon nuclei or, more commonly, (b) by fusion with a helium nucleus.

CARBON BURNING AND HELIUM CAPTURE

At higher and higher temperatures, heavier and heavier nuclei can gain enough energy to overcome the electrical repulsion between them. At about 10^9 K (reached only in the cores of stars much more massive than the Sun), carbon nuclei can fuse into magnesium, as depicted in Figure 21.15(a):

$$^{12}C + {}^{12}C \rightarrow {}^{24}Mg + energy.$$

However, because of the rapidly mounting nuclear charges—that is, the increasing number of protons in the nuclei—fusion reactions between any nuclei larger than carbon require such high temperatures that they are actually quite uncommon in stars. The formation of heavier elements occurs by way of an easier path. For example, the repulsive force between two carbon nuclei is three times greater than the force between a nucleus of carbon and one of helium. Thus, carbon–helium fusion occurs at a lower temperature than carbon–carbon fusion. At temperatures above 6×10^8 K, a carbon-12 nucleus colliding with a helium-4 nucleus can produce oxygen-16:

$$^{12}C + {}^4He \rightarrow {}^{16}O + energy.$$

If any helium-4 is present, this reaction, shown in Figure 21.15(b), is much more likely to occur than the carbon–carbon reaction.

Similarly, the oxygen-16 thus produced may fuse with other oxygen-16 nuclei at a temperature of about 1.2×10^9 K to form sulfur-32,

$$^{16}O + {}^{16}O \rightarrow {}^{32}S + energy,$$

but it is much more probable that an oxygen-16 nucleus will capture a helium-4 nucleus (if one is available) to form neon-20:

$$^{16}O + {}^4He \rightarrow {}^{20}Ne + energy.$$

The second reaction is more likely because it requires lower temperatures than oxygen–oxygen fusion.

Thus, as the star evolves, heavier elements tend to form through **helium capture** rather than by fusion of like nuclei. Because these helium-capture reactions are so much more common, elements with nuclear masses of 4 units (that is, helium itself), 12 units (carbon), 16 units (oxygen), 20 units (neon), 24 units (magnesium), and 28 units (silicon) stand out as prominent peaks in Figure 21.12, our chart of cosmic abundances. Each element is built by combining the preceding element and a helium-4 nucleus as the star evolves.

SOME COMPLICATIONS

Helium capture is by no means the only type of nuclear reaction occurring in evolved stars. As nuclei of many different kinds accumulate, a great variety of reactions becomes possible. In some, protons and neutrons are freed from their parent nuclei and are absorbed by others, resulting in new nuclei with masses intermediate between those formed by helium capture. Laboratory studies confirm that common nuclei, such as fluorine-19, sodium-23, phosphorus-31, and many others, are created in this way. Their abundances, however, are not as great as those produced directly by helium capture, simply because the helium-capture reactions are much more common in stars. For this reason, many of these elements (those with masses not divisible by 4, the mass of a helium nucleus) are found in the troughs of Figure 21.12.

Around the time silicon-28 appears in the core of a star, a competitive struggle begins between the continued capture of helium to produce even heavier nuclei and the tendency of the heavier nuclei to break down into simpler ones. The cause of this breakdown is heat. By now the star's core temperature has reached the unimaginably large value of 3×10^9 K, and the gamma rays associated with that temperature have enough energy to break a nucleus apart, as illustrated in Figure 21.16(a). This is the same process of photodisintegration that will ultimately accelerate the star's iron core in its final collapse toward a Type-II supernova.

Under the intense heat, some silicon-28 nuclei break apart into 7 helium-4 nuclei. Other nearby nuclei that have not yet photodisintegrated may capture some or all of these helium-4 nuclei, leading to the formation of still heavier elements (Figure 21.1b). The process of photodisintegration provides raw material that allows the helium-capture process to proceed to greater masses. The process continues, with some heavy nuclei being destroyed and others increasing in mass. In succession the star forms sulfur-32, argon-36, calcium-40, titanium-44, chromium-48, iron-52, and nickel-56. The chain of reactions building from silicon-28 up to nickel-56 is

$$^{28}\text{Si} + 7(^4\text{He}) \rightarrow {}^{56}\text{Ni} + \text{energy}.$$

This two-step process—photodisintegration followed by the direct capture of some or all the resulting helium-4 nuclei (or alpha particles)—is often called the *alpha process.*

A further complication enters the scenario here: Nickel-56 is unstable. It rapidly decays, first into cobalt-56, then into a stable iron-56 nucleus. Any unstable nucleus will continue to decay until stability is achieved, and iron-56 is the stablest of all nuclei. Thus, the alpha process leads inevitably to the buildup of iron in the stellar core.

Iron's 26 protons and 30 neutrons are bound together more strongly than the particles in any other nucleus. Iron is said to have the greatest *nuclear binding energy* of any element. Any nucleus with more or fewer protons or neutrons has less nuclear binding energy and is not quite as stable as the iron-56 nucleus. This enhanced stability of iron explains why some of the heavier nuclei in the iron group are more abundant than many lighter nuclei (see Table 21-1 and Figure 21.12)—nuclei tend to "accumulate" near iron as stars evolve.

MAKING ELEMENTS BEYOND IRON

There are dozens of elements much heavier than iron. But if the alpha process stops at iron, how did these very heavy elements form? To form them, some nuclear process other than helium capture must have been involved. That other process is **neutron capture**—the formation of heavier nuclei by the absorption of neutrons.

Deep in the interiors of highly evolved stars, conditions are ripe for neutron capture to occur. Neutrons are produced as "by-products" of many nuclear reactions, so there are many of them present to interact with iron and other nuclei. Neutrons have no charge, so there is no repulsive barrier for them to overcome in combining with positively charged nuclei. As more and more nuclei join an iron nucleus, its mass continues to grow.

Adding neutrons to a nucleus—iron, for example—does not change the element. Rather, a more massive isotope is produced. Eventually, however, so many neutrons are added to the nucleus that it becomes unstable and then decays radioactively to form a stable nucleus of some other element. The neutron-capture process then continues. For example, an iron-56 nucleus can capture a single neutron (n) to form a relatively stable isotope, iron-57:

$$^{56}\text{Fe} + \text{n} \rightarrow {}^{57}\text{Fe}.$$

This may be followed by another neutron capture,

$$^{57}\text{Fe} + \text{n} \rightarrow {}^{58}\text{Fe},$$

producing another relatively stable isotope, iron-58. Iron-58 can capture yet another neutron to produce an even heavier isotope of iron:

$$^{58}\text{Fe} + \text{n} \rightarrow {}^{59}\text{Fe}.$$

Iron-59 is known from laboratory experiments to be radioactively unstable. It decays in about a month into cobalt-59, which is stable. The neutron-capture process then resumes: Cobalt-59 captures a neutron to form the unstable cobalt-60, which in turn decays to nickel-60, and so on.

Figure 21.16 (a) At high temperatures, heavy nuclei (such as silicon, shown here) can be broken apart into helium nuclei by high-energy photons. (b) Other nuclei can capture the helium nuclei—or alpha particles—thus produced, forming heavier elements by the so-called alpha process. This process continues all the way to the formation of iron.

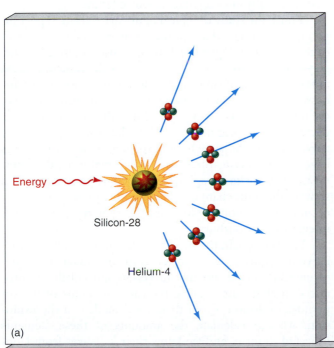

(a)

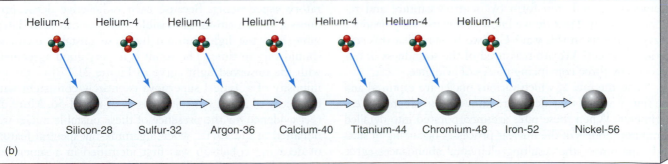

(b)

Each successive capture of a neutron by a nucleus typically takes about a year, so most unstable nuclei have plenty of time to decay before the next neutron comes along. Researchers usually refer to this "slow" neutron-capture mechanism as the *s-process*. It is the origin of the copper and silver in the coins in our pockets, the lead in our car batteries, the gold (or the zirconium) in the rings on our fingers.

MAKING THE HEAVIEST ELEMENTS

The s-process explains the synthesis of stable nuclei up to and including bismuth-209, the heaviest known nonradioactive nucleus, but it cannot account for the heaviest nuclei, such as thorium-232, uranium-238, or plutonium-242. Any attempt to form elements heavier than bismuth-209 by slow neutron capture fails because the new nuclei decay back to bismuth as fast as they form. There must be yet another nuclear mechanism that produces the very heaviest nuclei. This process is called the *r-process* (where *r* stands for "rapid," in contrast to the "slow" s-process we just described). The r-process operates very quickly, occurring literally during the supernova explosion that signals the death of a massive star.

For about the first 15 minutes of the supernova blast, the number of free neutrons increases dramatically as heavy nuclei are broken apart by the violence of the explosion. Unlike the s-process, which stops when it runs out of stable nuclei, the neutron-capture rate during the supernova is so great that even unstable nuclei can capture many neutrons before they have time to decay. Jamming neutrons into light- and middle-weight nuclei, the r-process is responsible for the creation of the heaviest known elements. The heaviest of the heavy elements, then, are actually born *after* their parent stars have died. Because the time available for synthesizing these heaviest nuclei is so brief, they never become very abundant. Elements heavier than iron (see Table 21-1) are a billion times less abundant than hydrogen and helium.

OBSERVATIONAL EVIDENCE FOR STELLAR NUCLEOSYNTHESIS

The modern picture of element formation involves many different types of nuclear reactions occurring at many different stages of stellar evolution, from main-sequence stars all the way to supernovae. Light elements—from hydrogen to iron—are built first by fusion, then by alpha capture, with proton and neutron capture filling in the gaps. Elements beyond iron form by neutron capture and radioactive decay. How do we know that stars really produce heavy elements in this way? Can we be sure that this scenario is correct? We are reassured of the soundness of our theories by three convincing pieces of evidence.

First, the rate at which various nuclei are captured and the rate at which they decay are known from laboratory experiments. When these rates are incorporated into detailed computer models of the nuclear processes occurring in stars and supernovae, the resulting elemental abundances agree extremely well, point by point, with the observational data

presented in Figure 21.12 and Table 21-1. The match is remarkably good for elements up through iron and is still fairly close for heavier nuclei. Thus, although no one has ever directly observed the formation of heavy nuclei in stars, we can be reasonably confident that the theory of stellar nucleosynthesis makes good sense in the context of nuclear physics and stellar evolution. Although the reasoning is indirect, the agreement between theory and observation is so striking that most astronomers regard it as strong evidence in support of the entire theory of stellar evolution and nucleosynthesis.

Second, the presence of one particular nucleus—technetium-99—provides direct evidence that heavy-element formation really does occur in the cores of stars. Laboratory measurements show that the technetium nucleus has a radioactive half-life of about 200,000 years. This is a very short time astronomically speaking. No one has ever found even traces of naturally occurring technetium on Earth because it all decayed long ago. The observed presence of technetium in the spectra of many red-giant stars implies that it must have been synthesized through neutron capture—the only known way that technetium can form—within the past few hundred thousand years. Otherwise, we wouldn't observe it. Many astronomers consider the spectroscopic evidence for technetium as proof that the s-process really does operate in evolved stars.

Third, the study of typical light curves from Type-I supernovae indicates that radioactive nuclei form as a result of the explosion. Figure 21.17(a) (see also Figure 21.7) displays the dramatic rise in luminosity at the moment of explosion and the characteristic slower decrease in brightness. Depending on the initial mass of the exploded star, the luminosity takes from several months to many years to decrease to its original value, but the *shape* of the decay curve is nearly the same for all exploded stars. These curves have two distinct features. After the initial peak, the luminosity first declines rapidly, then begins to decrease at a slower rate. This change in the luminosity decay invariably occurs about 2 months after the explosion, regardless of the intensity of the outburst.

We can explain the two-stage decline of the luminosity curve in Figure 21.17(a) in terms of the radioactive decay of unstable nuclei, notably nickel-56 and cobalt-56, produced in abundance during the early moments of the supernova explosion. From theoretical models of the explosion, we can calculate the amounts of these elements expected to form, and we know their half-lives from laboratory experiments. Because each radioactive decay produces a known amount of visible light, we can then determine how the light emitted by these unstable elements should vary in time. The result is in very good agreement with the observed light curve in Figure 21.17(b)—the luminosity of a Type-I supernova is entirely consistent with the decay of about 0.6 solar masses of nickel-56. More direct evidence for the presence of these unstable nuclei was obtained in the 1970s, when a gamma-ray spectral feature of decaying cobalt-56 was first identified in a supernova observed in a distant galaxy.

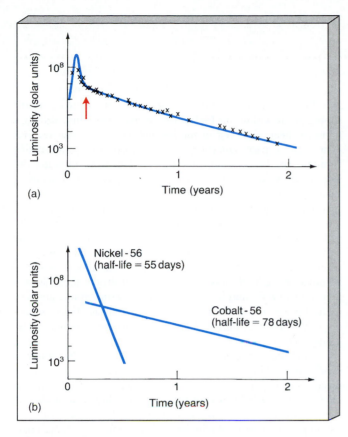

(a)

(b)

Figure 21.17 (a) The light curve of a Type-I supernova, showing not only the dramatic increase and slow decrease in luminosity, but also the characteristic change in the rate of decay about 2 months after the explosion (after the time indicated by the arrow). This particular supernova occurred in the faraway galaxy IC4182 in 1938. The crosses are the actual observations of the supernova's light. (b) Theoretical calculations of the light emitted by the radioactive decay of nickel-56 and cobalt-56 produce a light curve very similar to those actually observed in real supernova explosions, lending strong support to the theory of stellar nucleosynthesis.

21.5 *The Cycle of Stellar Evolution*

The evidence in favor of the theory of stellar nucleosynthesis is overwhelming. Theoretical calculations of stellar evolutionary paths predict that heavy elements are created deep inside stars, and spectroscopic studies of giants and stellar remnants confirm this idea. Theory likewise predicts the observed distinct differences in heavy-element abundance between the old globular cluster stars and the younger galactic cluster stars. The youngest stars contain the most heavy elements. The reason for this is that these elements are slowly produced over time, and each new generation of stars increases their concentration in the interstellar clouds from which the next generation forms. As a result, a recently formed star contains a much greater abundance of heavy elements than a star that formed long ago. In the past three chapters, we have seen all the ingredients that make up the complete cycle of star formation

and evolution in our Galaxy. Let us now briefly summarize that process, which is illustrated in Figure 21.18.

1. Stars form when part of an interstellar cloud is compressed beyond the point at which it can support itself against its own gravity. The cloud collapses and fragments, forming a cluster of stars. The hottest stars heat and ionize the surrounding gas, sending shock waves through the surrounding cloud, possibly triggering new rounds of star formation.

2. Within the cluster, stars evolve. The most massive stars evolve fastest, creating heavy elements in their cores and spewing them forth into the interstellar medium in supernova explosions. The lighter stars take longer to evolve, but they too can create heavy elements and may contribute to the "seeding" of interstellar space when they shed their envelopes as planetary nebulae.

3. The creation and dispersal of new heavy elements are accompanied by further shock waves. Their passage simultaneously enriches the interstellar medium and compresses it into further star formation.

In this way, although some material is used up in each cycle—turned into energy or locked up in low-mass stars—the Galaxy continuously recycles its matter. Each new round of formation creates stars with more heavy elements than the preceding generation had. From the old, metal-poor globular clusters to the young, metal-rich open clusters, we observe this enrichment process in action. Our Sun is the product of many such cycles. We ourselves are another. Without the heavy elements synthesized in the hearts of stars, life on Earth would not exist.

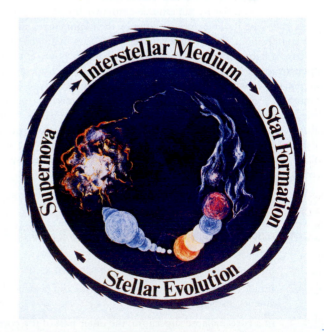

Figure 21.18 The cycle of star formation and evolution continuously replenishes the Galaxy with new heavy elements and provides the driving force for the creation of new generations of stars.

Chapter Review

SUMMARY

A **nova** (p. 444) is a star that suddenly increases greatly in brightness, then slowly fades back to its normal appearance over a period of months. It is the result of a white dwarf in a binary system drawing hydrogen-rich material from its companion. The gas builds up on the white dwarf's surface, eventually becoming hot and dense enough for the hydrogen to burn explosively, temporarily causing a large increase in the dwarf's luminosity.

The matter flowing from the companion star does not fall directly onto the surface of the dwarf. Instead it goes into orbit around it, forming an **accretion disk** (p. 445). Friction within the disk causes the gas to spiral slowly inward, heating up and glowing brightly as it nears the dwarf's surface.

Stars more massive than about 8 solar masses are able to attain high enough central temperatures to burn carbon and heavier nuclei. As they burn, their cores form a layered structure consisting of burning shells of successively heavier elements. A nonburning core of iron builds up at the center.

Iron is special in that its nuclei can neither be fused together nor split apart to produce energy. As a result, stellar nuclear burning stops at iron. As a star's iron core grows in mass, it eventually becomes unable to support itself against gravity and begins to collapse. At the enormous densities and temperatures produced during the collapse, iron nuclei are broken down into their constituent particles—protons and neutrons. The protons combine with electrons to form more neutrons. Eventually, when the core has become so dense that the neutrons are effectively brought into physical contact with one another, their resistance to further squeezing stops the collapse and the core rebounds, sending a violent shock wave out through the rest of the star. The star is blown to pieces in a **core-collapse supernova** (p. 447).

Astronomers classify supernovae into two broad categories: Type I and Type II. These classes differ by their light curves and by their composition. **Type-I supernovae** (p. 448) are hydrogen poor and have a light curve similar in shape to that of a nova. **Type-II supernovae** (p. 448) are hydrogen rich and have a characteristic bump in the light curve a few months after maximum. A Type-II supernova is a core-collapse supernova. A Type-I supernova occurs when a carbon–oxygen white dwarf in a binary system exceeds about 1.4 $M_\odot$, the *Chandrasekhar mass*—the maximum mass that can be supported against gravity by electron degenaracy pressure. The star collapses and explodes as its carbon ignites. This type of supernova is called a **carbon-detonation supernova** (p. 449).

Theory predicts that a supernova visible from Earth should occur within our Galaxy about once a century, although none has been observed in the last 400 years. We can see evidence for a past supernova in the form of a **supernova remnant** (p. 451), a shell of exploded debris surrounding the site of the explosion, and expanding into space at thousands of kilometers per second.

Supernovae of a given type tend to have quite similar maximum luminosities. Astronomers use them as **standard candles** (p. 453)—objects that are easily recognizable (in this case, by their light curves and spectra) and whose brightnesses are known. By comparing a supernova's measured apparent brightness with its known absolute brightness, we can determine its distance. Because they are so bright and hence so easily seen, supernovae are very useful for measuring distances to faraway galaxies. Type-I supernovae have the best-defined luminosities, so they are more commonly used in these measurements.

All elements heavier than helium formed by **stellar nucleosynthesis** (p. 454)—the production of new elements by nuclear reactions in the cores of evolved stars. Elements beyond carbon tend to form by **helium capture** (p. 456) rather than by the fusion of two heavy nuclei. Because of this, nuclei whose masses are a multiple of the mass of a helium nucleus tend to be more common than others.

At high enough core temperatures, photodisintegration breaks apart some heavy nuclei, providing helium-4 nuclei for the synthesis of even more massive elements, leading to a buildup of iron-56 in the core. Elements beyond iron form by **neutron capture** (p. 457) in the cores of evolved stars. With no repulsive electromagnetic barrier to overcome, neutrons can easily combine with nuclei. During a supernova explosion, rapid neutron capture occurs, producing the heaviest nuclei of all.

Comparisons between theoretical predictions of element production and observations of element abundances in stars and supernovae provide strong support for the theory of stellar nucleosynthesis.

The processes of star formation, evolution, and explosion form a cycle that constantly enriches the interstellar medium with heavy elements and sows the seeds of new generations of stars. Without the elements produced in supernovae, life on Earth would be impossible.

SELF-TEST: True or False?

_____ **1.** A nova is a sudden outburst of light coming from an old main-sequence star.

_____ **2.** Some, but not all, novae occur in binary star systems.

_____ **3.** It takes less and less time to fuse heavier and heavier elements inside a high-mass star.

_____ **4.** In a core-collapse supernova, the outer part of the core rebounds from the inner, high density core, destroying the entire outer part of the star.

_____ **5.** Most of the energy released during a supernova is emitted in the form of neutrinos.

_____ **6.** The spectrum of a type-II supernova shows the presence of lots of hydrogen.

_____ **7.** It is possible that a recurrent nova will eventually result in a core-collapse supernova.

_____ **8.** Once the process gets underway, the core of a massive star collapses in about 1 second.

_____ **9.** Different isotopic forms of an element differ only in the number of protons in the nucleus.

_____ **10.** Carbon can fuse with helium more easily than it can fuse with another carbon atom because there is more helium than carbon in stars.

____ **11.** The s-process is one in which heavy elements are formed from silicon.

____ **12.** The r-process occurs only during the first few minutes of a supernova explosion.

____ **13.** Stellar nucleosynthesis can account for the existence of all elements except hydrogen and helium.

SELF-TEST: Fill in the Blank

1. In a semi-detached binary consisting of a white dwarf and a main-sequence or giant companion, matter leaving the companion forms an _____ disk around the dwarf.

2. A nova explosion is due to _____ fusion on the _____ of a white dwarf.

3. In the collapsing core of a massive star, photodisintegration is caused by the very high _____ in the core.

4. When a proton and an electron are forced together, they combine to form a _____ and a _____ .

5. Core collapse in a massive star is eventually stopped by _____ degeneracy pressure.

6. A _____ supernova occurs when a white dwarf exceeds the Chandrasekhar mass.

7. The maximum mass for a white dwarf is roughly _____ solar masses.

8. The two types of supernova can be distinguished observationally by their spectra and by their _____.

9. Elements differing by 4 mass units, such as carbon, oxygen, neon, and silicon, are produced in stars by _____.

10. By the time silicon appears in the core of a massive star, the temperature has reached _____ , sufficient to break apart heavy nuclei.

11. Neutron capture is responsible for the formation of elements heavier than _____.

12. The first evidence of supernova 1987A was a burst of _____.

REVIEW AND DISCUSSION

1. Under what circumstances will a binary star produce a nova?

2. What is a light curve? How can it be used to identify a nova or a supernova?

3. What occurs in a massive star to cause it to explode?

4. What are the observational differences between type-I and type-II supernovae?

5. How do the mechanisms that cause type-I and type-II supernovae explain their observed differences?

6. Roughly how often would we expect a supernova to occur in our own Galaxy? How often would we expect to see a Galactic supernova?

7. What evidence is there that many supernovae have occurred in our Galaxy?

8. How are supernovae used as "standard candles"? Which type is more reliable?

9. What proof do astronomers have that heavy elements are formed in stars?

10. As a star evolves, why do heavier elements tend to form by helium capture rather than by fusion of like nuclei?

11. Why do the cores of massive stars evolve into iron, and not heavier elements?

12. How are nuclei heavier than iron formed?

13. What is the r-process? When and where does it occur?

14. Why was supernova 1987A so important?

PROBLEMS

1. The Crab Nebula is now about 1 pc in radius. If it was observed to explode in A.D. 1054, roughly how fast is it expanding? (Assume constant expansion velocity. Is that a reasonable assumption?)

2. A certain telescope could just detect the Sun at a distance of 10,000 pc. What is the maximum distance at which it could detect a nova with a peak luminosity of 10^5 solar luminosities? Repeat the calculation for a supernova with a peak luminosity 10^{10} times that of the Sun.

3. At what distance would the supernova in question 2 look as bright as the Sun? Would you expect a supernova to occur that close to us?

4. A supernova's energy is often compared to the total energy output of the Sun over its lifetime. Using the Sun's current energy output, what is its total energy output, assuming it has a 10^{10} year main sequence lifetime? How does it compare with the energy released by a supernova?

PROJECTS

1. In 1758, the French comet hunter Charles Messier discovered the sky's most legendary supernova remnant, now called M1, or the Crab Nebula. An 8-inch telescope reveals the Crab's oval shape, but it will appear faint. It is located northwest of Zeta Tauri, the star that marks the southern tip of the horns of Taurus the Bull. A 10-inch or larger telescope reveals some of its famous filamentary structure.

2. In the *Handbook of Chemistry and Physics*, available in any library reference section, look up the table of isotopes. Pick one or more isotopes and follow how they decay into a final stable isotope. For example, choose cobalt-59, formed in the s-process. Note how the isotope decays, what is emitted, and the half-life of the decays. Read the caption accompanying the table to understand the various symbols used. Try this for Uranium-235 or -238 or Plutonium-239.

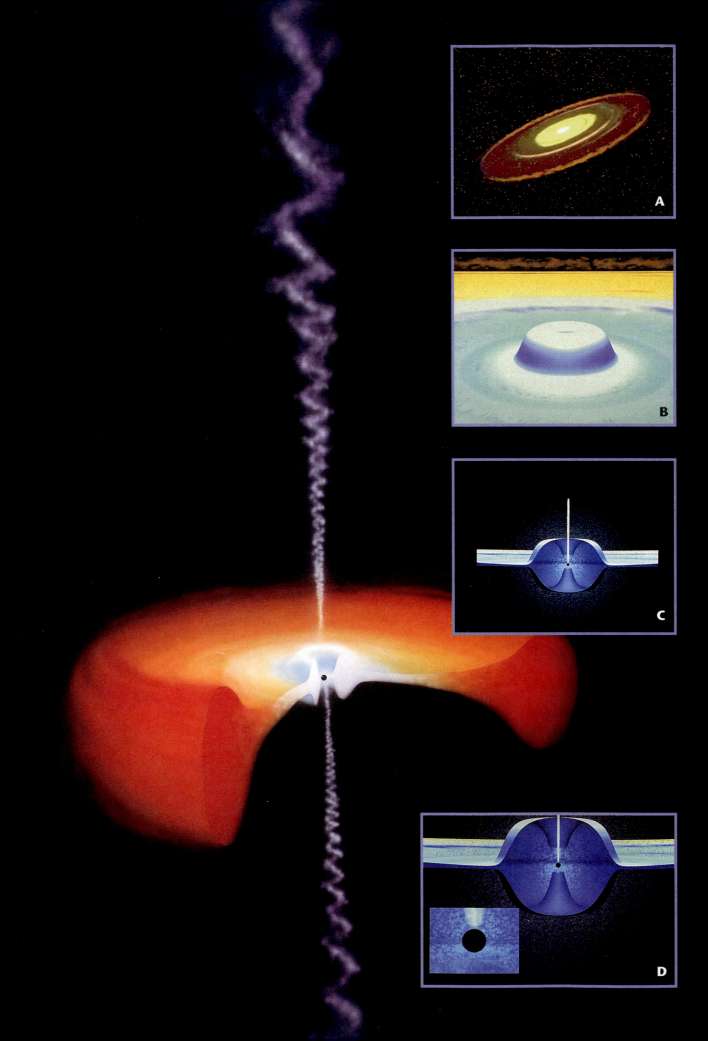

22

NEUTRON STARS AND BLACK HOLES

Strange States of Matter

LEARNING GOALS

Studying this chapter will enable you to:

1 Describe the properties of neutron stars, and explain how these strange objects are formed.

2 Explain the nature and origin of pulsars, and account for their characteristic radiation.

3 List and explain some of the observable properties of neutron-star binary systems.

4 Describe how black holes are formed, and discuss their effects on matter and radiation in their vicinity.

5 Relate the phenomena that occur near black holes to the warping of space around them.

6 Discuss the difficulties in observing black holes, and explain some of the ways in which the presence of a black hole might be detected.

(Opposite page, background) Black holes can be neither seen nor drawn easily. In this artist's conception of the vicinity of a black hole, the disk-shaped region is a rapidly whirling collection of hot gas and dust about to fall into the black hole. The hole itself is represented merely by a black dot at the center of the accretion disk.

(Inset A) An artist's view of a nearly flat, spinning torus of heated gas, called an accretion disk. For a typical stellar black hole, the disk has solar-system dimensions.

(Inset B) A closeup view of the core of the accretion disk, showing a truncated cusp of very hot gas about to make its way down into the central black hole.

(Inset C) The anatomy of a hypothetical black hole would show its accretion disk to the left and right, with the hole itself still only a black point at the very center. The white vertical spike is a geyser (or jet) of matter shot away from *near* the hole perpendicular to the innermost part of the disk.

(Inset D) A closeup view of the previous cutaway drawing of the accretion disk, the jet, and the black hole at its heart. At this scale, the stellar black hole itself (still shown as a black dot in the magnified inset) is about 30 kilometers across—and on this scale, astronomers are very uncertain of what black holes really look like.

Our study of stellar evolution has led us to some very unusual and unexpected objects. Red giants, white dwarfs, and supernova explosions surely represent extreme states of matter completely unfamiliar to us here on Earth. Yet stellar evolution can have even more bizarre consequences. The strangest states of all result from the catastrophic implosion-explosion of stars much more massive than our Sun. The almost unimaginable violence of a supernova explosion may bring into being objects so extreme in their behavior that they require us to reconsider some of our most hallowed laws of physics. They open up a science fiction writer's dream of fantastic phenomena. They may even one day force scientists to construct a whole new theory of the universe.

22.1 *Neutron Stars*

ULTRACOMPRESSION

❶ What remains after a supernova explosion? Is the entire progenitor star blown to bits and dispersed throughout interstellar space, or does some portion of it survive? For a Type-I (carbon-detonation) supernova, most astronomers regard it as quite unlikely that any central remnant is left after the explosion. The entire star is shattered by the blast. However, for a Type-II supernova, involving the implosion and subsequent rebound of a massive star's iron core, theoretical calculations indicate that part of the star might survive. The explosion destroys the parent star, but it may leave a tiny ultracompressed remnant at its center. Even by the high-density standards of a white dwarf, however, the matter within this severely compressed core is in a very strange state, unlike anything we are ever likely to find (or create) on Earth.

Recall from Chapter 21 that during the moment of implosion of a massive star—just prior to the supernova explosion itself—the electrons in the core violently smash into the protons there, forming neutrons and neutrinos. ∞ (p. 447) The neutrinos leave the scene at (or nearly at) the speed of light, accelerating the collapse of the neutron core, which continues to contract until its particles come into contact. At that point, neutron degeneracy pressure causes the central portion of the core to rebound, creating a powerful shock wave that races outward through the star, violently expelling matter into space.

The key point here is that the shock wave does not start at the very center of the collapsing core. The innermost part of the core—the region that bounces—remains intact as the shock wave it causes destroys the rest of the star. After the violence of the supernova has subsided, this ball of neutrons is all that is left. Researchers colloquially call this core remnant a **neutron star**, although it is not a star in any true sense of the word—all of its nuclear reactions have ceased forever.

NEUTRON-STAR PROPERTIES

Neutron stars are extremely small and very massive. Composed purely of neutrons packed together in a tight ball about 20 km across, a typical neutron star is not much bigger than a small asteroid or a terrestrial city (see Figure 22.1), yet its mass is greater than that of the Sun. With so much mass squeezed into such a small volume, neutron stars are incredibly dense. Their average density can reach 10^{17} or even 10^{18} kg/m^3, nearly a billion times denser than a white dwarf. A single thimbleful of neutron-star material would weigh 100 million tons—about as much as a good-sized terrestrial mountain. Even the density of a normal atomic nucleus is "only" 10^{17} kg/m^3. In a sense, we can think of a neutron star as a single enormous nucleus, with an "atomic weight" of around 10^{57}.

Neutron stars are solid objects. Provided that a sufficiently cool one could be found, you might even imagine standing on it. However, this would not be easy, as a neutron star's gravity is extremely powerful. A 70-kg (150-pound) human would weigh the Earth-equivalent of about 1 billion kg (1 million tons). The severe pull of a neutron star's gravity would flatten you much thinner than this piece of paper!

Figure 22.1 Neutron stars are not much larger than many of the Earth's major cities. In this fanciful comparison, a typical neutron star sits alongside Manhattan Island.

In addition to large mass and small size, neutron stars have two other very important properties. First, newly formed neutron stars rotate extremely rapidly, with periods measured in fractions of a second. This is a direct result of the law of conservation of angular momentum (see Chapter 15), which tells us that any rotating body must spin faster as it shrinks. ∞ (p. 321) The collapsing iron (and, later, neutron) core of an evolved star is no exception to this law. Even if the core of the progenitor star were initially rotating quite slowly (once every couple of weeks, say, as is observed in many upper main-sequence stars), it would be spinning a few times per second by the time it had reached a diameter of 20 km.

The second important property of a young neutron star is its strong magnetic field. The original field of the progenitor star is amplified by the collapse of the core because the contracting material squeezes the magnetic field lines closer together. This squeezing increases the field strength to a value on the order of a *trillion* times that of Earth's field (millions of times stronger than the fields found even in the hearts of the most violent solar flares).

In time, theory indicates, our neutron star will spin more and more slowly as it radiates its energy into space, and its magnetic field will diminish. However, for a few million years after its birth, these two properties combine to provide the primary means by which this strange object can be detected and studied.

22.2 Pulsars

❷ Can we be sure that objects as strange as neutron stars really exist? The answer is a confident "Yes." The first observation of a neutron star occurred in 1967, when Jocelyn Bell, a graduate student at Cambridge University, made a surprising discovery. She observed an astronomical object emitting radio radiation in the form of rapid *pulses*. Each pulse consisted of a 0.01-second (s) burst of radiation, after which there was nothing. Then, 1.34 s later, another pulse would arrive. The time interval between pulses was astonishingly uniform—so accurate, in fact, that the repeated emissions could be used as a very precise clock. Figure 22.2 is a recording of the radio radiation from the pulsating object Bell discovered.

Many hundreds of these pulsating objects are now known in our Milky Way Galaxy. They are called **pulsars**. Each has its own characteristic pulse period and duration. The pulse periods of some pulsars are so stable that they are by far the most accurate natural clocks known in the universe—more accurate even than the best atomic clocks on Earth. In some cases, the period is predicted to change by only a few seconds in a million years.

Most pulsars emit their pulses in the form of radio radiation. Some have been observed to pulse in the visible, X-ray, and gamma-ray parts of the spectrum as well. Whatever types of radiation are produced, these electromagnetic flashes at different frequencies are all synchronized—that is, occurring at regular, repeated time intervals—as we would expect if they arose from the same object. The period of most pulsars is usually short—ranging from about 0.03 to 0.3 s, corresponding to a flashing rate of between 3 and 30 times per second. The human eye is insensitive to such rapid flashes, making it impossible to observe the flickering of a pulsar with the naked eye or even using a large telescope. Fortunately, instruments can record pulsations of light that the human eye cannot.

A few pulsars are clearly associated with supernova remnants, although not all such remnants have a detectable pulsar within them. Figure 22.3(a) shows a pair of optical photographs of the Crab pulsar, at the center of the Crab supernova remnant. ∞ (p. 451) In the left frame, the pulsar is off; in the right frame, it is on. Figure 22.3(b) shows that the Crab also pulses in X rays. By observing the velocity and direction that the Crab's ejected matter is traveling, astronomers can work backward to pinpoint the location in space at which the explosion must have occurred and where the supernova core remnant should be located. That is precisely the region of the Crab Nebula from which the pulsating signals arise. The Crab pulsar is evidently all that remains of the once-massive star whose supernova was observed in 1054.

When Jocelyn Bell made her discovery in 1967, she did not know what she was looking at. Indeed, no one at the time knew what a pulsar was. The explanation of pulsars as spinning neutron stars won Bell's thesis advisor, Anthony Hewish, the 1974 Nobel Prize in physics. Hewish reasoned that the only physical mechanism consistent with such pre-

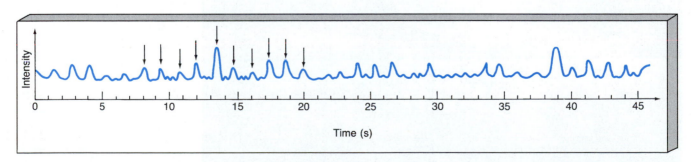

Figure 22.2   Pulsars emit periodic bursts of radiation. This recording shows the regular change in the intensity of the radio radiation emitted by the first such object known. It was discovered in 1967. Some of the pulses are marked by arrows.

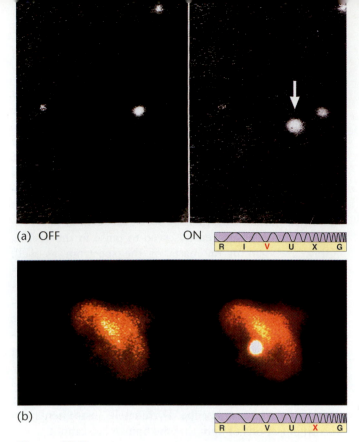

(a) OFF   ON

R I V U X G

(b)

R I V U X G

Figure 22.3 The pulsar in the core of the Crab Nebula blinks on and off about 30 times each second. (a) In this pair of closely spaced optical images, the pulsing can be seen clearly. (b) The same phenomenon is also detected in X rays.

cisely timed pulsations is a small, rotating source of radiation. Only rotation can cause the high degree of regularity of the observed pulses, and only a small object can account for the sharpness of each pulse. Radiation emitted from different regions of an object larger than a few tens of kilometers across would arrive at Earth at slightly different times, blurring the pulse profile. The best current model describes a pulsar as a compact, spinning neutron star that periodically flashes radiation toward Earth.

Figure 22.4 outlines the important features of this pulsar model. Two "hot spots" on the surface of a neutron star, or in the magnetosphere just above the surface, continuously emit radiation in a narrow "searchlight" pattern. These spots are most likely localized regions near the neutron star's magnetic poles, where charged particles, accelerated to extremely high energies by the star's rotating magnetic field, emit radiation along the star's magnetic axis. The hot spots radiate more or less steadily, and the resulting beams sweep through space, like a revolving lighthouse beacon, as the neutron star rotates. Indeed, this pulsar model is often known as the **lighthouse model**. The beams are observed as a series of rapid pulses—each time one of the beams sweeps past the Earth, a pulse is seen. The period of the pulses is the star's rotation period.

All pulsars are neutron stars, but not all neutron stars are pulsars—at least, as viewed from Earth. The reason for

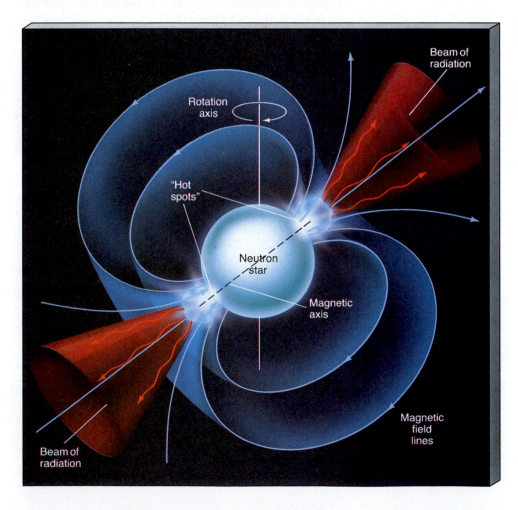

Beam of radiation

Rotation axis

"Hot spots"

Neutron star

Magnetic axis

Beam of radiation

Magnetic field lines

Figure 22.4 This diagram of the "lighthouse model" of neutron-star emission accounts for many of the observed properties of pulsars. Charged particles, accelerated by the magnetism of the neutron star, flow along the magnetic field lines, producing radio radiation that beams outward.

this is simple. The pulsar beam depicted in Figure 22.4 is relatively narrow—perhaps only a few degrees across. Unless the neutron star happens to be oriented in just the right way, the beam never sweeps across the Earth, and we never see a pulsar. However, given our current knowledge of star formation, stellar evolution, and neutron stars, pulsar observations are quite consistent with the idea that *every* high-mass star dies in a supernova, leaving a neutron star behind, and that *all* neutron stars emit beams of radiation, just like the pulsars we actually see.

22.3 Neutron-Star Binaries

3 We noted in Chapter 17 that most stars are not single, but instead are members of binary systems. ∞ (p. 373) Although many pulsars are known to be isolated (that is, not part of any binary), there is strong evidence that at least some do have binary companions, and the same is true of neutron stars in general (that is, even the ones not seen as pulsars). We will examine three types of neutron-star binaries: X-ray sources, millisecond pulsars, and pulsar planets.

X-RAY SOURCES

The late 1970s saw several important discoveries about neutron stars in binary-star systems. Numerous X-ray sources were discovered near the central regions of our Galaxy and also near the centers of a few rich star clusters. Some of these sources, known as **X-ray bursters**, emit much of their energy in violent eruptions, each thousands of times more luminous than our Sun, but lasting only a few seconds. A typical burst is shown in Figure 22.5.

This X-ray emission is thought to arise on or near neutron stars that are members of binary systems. Matter torn from the surface of the (main-sequence or giant) companion by the neutron star's strong gravitational pull accumulates on the neutron star's surface. As in the case of white-dwarf accretion (see Chapter 21), the material does not fall directly onto the surface. ∞ (p. 445) Instead, as illustrated in Figure 22.6(a), it forms an *accretion disk*. (Compare Figure 21.2, which depicts the white-dwarf equivalent.) The gas goes into a tight orbit around the neutron star, then slowly spirals inward. The inner portions of the accretion disk become extremely hot, releasing a steady stream of X rays.

As gas builds up on the neutron star's surface, its temperature rises due to the pressure of overlying material. Eventually, it becomes hot enough to fuse hydrogen. The result is a sudden period of rapid nuclear burning that releases a huge amount of energy in a brief but intense flash of X rays—an X-ray burst. After several hours of renewed accumulation, a fresh layer of matter produces the next burst. Thus, an X-ray burst is much like a nova explosion on a white dwarf, but occurring on a far more violent scale because of the neutron star's much stronger gravity.

Not all of the infalling gas makes it onto the neutron star surface, however. In at least one case—the remarkable object

known as SS433[*]—we have direct observational evidence that some material is instead shot completely out of the system at enormously high speeds. SS433 expels more than one Earth mass of material every year in the form of two oppositely directed narrow jets moving roughly perpendicular to the disk. Observations of the Doppler shifts of optical emission lines produced within the jets themselves imply speeds of almost 80,000 km/s—over 25 percent of the speed of light! As the jets interact with the interstellar medium, they emit radio radiation, as shown in Figure 22.6(b).

Jets of this sort are apparently quite common in astronomical systems in which an accretion disk surrounds a compact object (such as a neutron star or a black hole). They are believed to be produced by the intense radiation and magnetic fields near the inner edge of the disk, although the details of their formation are still uncertain. Incidentally, they are *not* the "lighthouse" beams of radiation from the neutron star itself (Figure 22.4) that can result in a pulsar. Although SS433 is the only stellar object currently known to produce

[*]*The name simply identifies it as the 433rd entry in a particular catalog of stars with strong optical emission lines.*

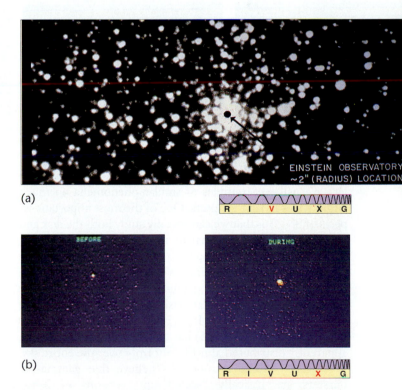

Figure 22.5 An X-ray burster produces a sudden, intense flash of X rays, followed by a period of relative inactivity lasting as long as several hours. Then another burst occurs. The bursts are thought to be caused by explosive nuclear burning on the surface of an accreting neutron star, similar to the explosions on a white dwarf that give rise to novae. (a) An optical photograph of the star cluster Terzan 2, showing a 2" dot at the center where the X-ray bursts originate. (b) X-ray images taken before and during the outburst. The most intense X rays correspond to the position of the dot shown in frame (a).

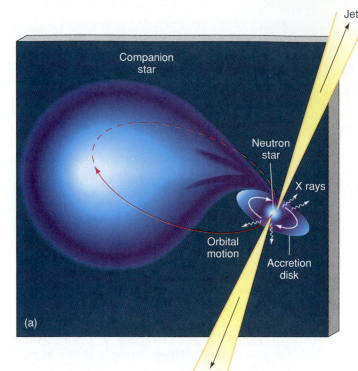

(a)

Jet

Companion
star

Neutron
star

X rays

Orbital
motion

Accretion
disk

Jet

Figure 22.6 (a) Matter flows from a normal star toward a compact neutron-star companion and falls toward the surface in an accretion disk. As the gas spirals inward under the neutron star's intense gravity, it heats up, becoming so hot that it emits X rays. In at least one instance—the peculiar object SS433—some material may be ejected in the form of two high-speed jets of gas. (b) False-color radiographs of SS433, made at monthly intervals (left to right), show the jets rotating under the gravitational influence of the companion star.

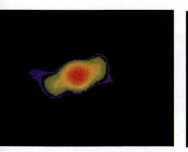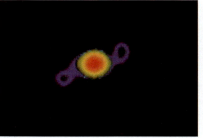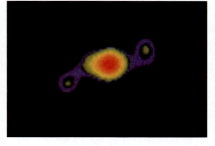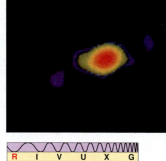

(b)

R   I   V   U   X   G

jets, we will see examples of similar phenomena on much larger scales in later chapters. One of the most important aspects of SS433 is that we can actually study both the disk and the jets, instead of simply having to assume their existence, as in more distant cosmic objects.

At around the same time as the first X-ray bursts were seen, military satellites detected the first **gamma-ray bursters**—bright, irregular flashes of gamma rays lasting only a few seconds. Most recently, the *Gamma-Ray Observatory* (see Chapter 5) has detected many such bursts, apparently distributed almost uniformly over the entire sky. ∞ (p. 120) Some astronomers believe that gamma-ray bursters are basically "scaled up" versions of X-ray bursters, in which matter from the binary companion experiences even more violent nuclear burning, accompanied by the release of gamma rays. Other explanations, based on neutron star *collisions* in very distant binary systems, have also been proposed.

MILLISECOND PULSARS

In the mid-1980s an important new category of pulsars was found—a class of very rapidly rotating objects called **millisecond pulsars**. Several dozen millisecond pulsars are currently known in the Milky Way Galaxy. These objects spin hundreds of times per second (that is, their pulse period is a few milliseconds, 0.001 s). This speed is about as fast as a typical neutron star can spin without flying apart. In some cases, the star's equator is moving at more than 20 percent of the speed of light. This speed suggests a phenomenon bordering on the incredible—a cosmic object of kilometer dimensions, more massive than our Sun, spinning almost at breakup speed, making nearly 1000 complete revolutions *every second*. Yet the observations and their interpretation leave little room for doubt.

The story of these remarkable objects is further complicated because many of them (over 40 at last count, out of a total of about 100 known in the entire Milky Way Galaxy) are found in globular clusters. This is odd because globular clusters are known to be very old—10 billion years, at least. Yet Type-II supernovae (the kind that create neutron stars) are associated with massive stars that explode within a few tens of *millions* of years after their formation, and no stars have formed in any globular cluster since the cluster itself came into being. Thus, no new neutron star has been produced in a globular cluster in a very long time. But, as we have mentioned, the pulsar produced in a super-

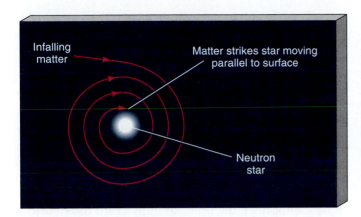

Figure 22.7   Gas from a companion star spirals down onto the surface of a neutron star. As the infalling matter strikes the star, it moves almost parallel to the surface, so it tends to make the star spin faster. Eventually, this process can result in a millisecond pulsar—a neutron star spinning at the incredible rate of hundreds of revolutions per second.

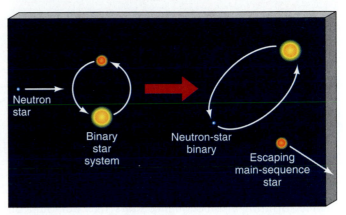

Figure 22.8   A neutron star can encounter a binary made up of two low-mass stars, ejecting one of them and taking its place. This mechanism provides a means of forming a binary system with a neutron-star component (which may later evolve into a millisecond pulsar) without having to explain how the binary survived the supernova explosion that formed the neutron star.

nova explosion is expected to slow down in only a few million years. After 10 billion years, its rotation should have all but ceased. The rapid rotation of the pulsars found in globular clusters cannot be a relic of their birth. These objects must have been spun up—that is, had their rotation rates increased—by some other, much more recent, mechanism.

The most likely explanation for the high rotation rate of these objects is that the neutron star has been spun up by drawing in matter from a companion star. As matter spirals down onto the star's surface in an accretion disk, it provides the "push" needed to make the neutron star spin faster (see Figure 22.7). Theoretical calculations indicate that this process can spin the star up to breakup speed in about a hundred million years. Subsequently, an encounter with another star may eject the neutron star from the binary, or the pulsar's radiation may destroy its companion, so an isolated millisecond pulsar results. This general picture is supported by the finding that, of the 40 or so millisecond pulsars seen in globular clusters, 10 are known currently to be members of binary systems. These numbers are quite consistent with the rate at which binaries can be broken up by encounters with other cluster members.

Thus, although a pulsar like the Crab is the direct result of a supernova explosion, millisecond pulsars are the product of a two-stage process. The neutron star was formed in an ancient supernova, billions of years ago. Only relatively recently, through interaction with a binary companion, has it achieved the rapid spin that we observe today. Once again, we see how members of a binary system can evolve in ways quite different from single stars. Notice that the scenario of accretion onto a neutron star from a binary companion is the same scenario that we just used to explain the existence of X-ray bursters. In fact, the two phenomena are very closely linked. Many X-ray bursters may be on their way to becoming millisecond pulsars.

The way in which a neutron star can become a member of a binary system is the subject of active research, because the violence of a supernova explosion would be expected to blow the binary apart in many cases. Only if the supernova progenitor lost a lot of mass before the explosion would the binary be likely to survive. Alternatively, by interacting with an existing binary and displacing one of its components, a neutron star may become part of a binary system *after* it is formed, as depicted in Figure 22.8. Astronomers are eagerly searching the skies for more millisecond pulsars to test their ideas.

PULSAR PLANETS

In January 1992, radio astronomers at the Arecibo Observatory found that the pulse period of a recently discovered millisecond pulsar lying some 500 pc from Earth varied in an unexpected but quite regular way. Careful analysis of the data has revealed that the period fluctuates on two distinct time scales—one of 67 days, the other of 98 days. The changes in the pulse period are small—less than one part in 10^7—but repeated observations have confirmed their reality. Since that time, a few other pulsars have been found with similar behavior.

The leading explanation for these fluctuations holds that they are caused by the Doppler effect as the pulsar wobbles back and forth in space. But what causes the wobble? The Arecibo group believes it is the result of the combined gravitational pulls of not one, but *two* planets, each about three times the mass of Earth! One orbits the pulsar at a distance of 0.4 A.U. and the other at a distance of 0.5 A.U. Their orbital periods are 67 and 98 days, respectively, matching the timing of the fluctuations. In April 1994, the group announced further observations that not only confirmed their earlier findings, but also revealed the presence of a *third* body, with mass comparable to the Earth's Moon, orbiting only 0.2 A.U. from the pulsar. These remarkable results constitute the first definite evidence for planets outside our solar system.

However, it is unlikely that these planets formed in the same way as our own. Any planetary system orbiting the pulsar's progenitor star was almost certainly destroyed in the supernova explosion that created the pulsar. As a result, scientists are still uncertain about how these planets came into being. One possibility involves the binary companion, which provided the matter necessary to spin the pulsar up to millisecond speeds. If the pulsar's intense radiation and strong gravity destroyed the companion, the neutron star's strong gravity might have spread its matter out into a disk (a little like the solar nebula), in whose cool outer regions the planets could have condensed. These measurements are difficult, however, and their interpretation is still controversial.

Astronomers have been searching for decades for planets orbiting main-sequence stars like our Sun, on the assumption that planets are a natural by-product of star formation. ∞ (p. 320) It is ironic indeed that the first planets to be found outside the solar system orbit a dead star and have little or nothing in common with our own world.

22.4 Disappearing Matter

④ Neutron stars are peculiar objects. Nevertheless, theory predicts that they are in equilibrium, as are most other stars. For neutron stars, however, equilibrium does not mean a balance between the inward pull of gravity and the outward pressure of hot gas. Instead, as we have seen, the outward force is provided by the pressure of tightly packed neutrons. Squeezed together, the neutrons form a hard ball of matter that not even gravity can compress further. Or do they? Is it possible that, given enough matter packed into a small

---

More Precisely...   *Einstein's Theories of Relativity*

Albert Einstein won a Nobel Prize in 1921 for his explanation of the photoelectric effect, as described in Chapter 4. ∞ (p. 80) However, he is probably best known for his two theories of relativity, the successors to Newtonian mechanics that form the foundation of twentieth-century physics.

The *special theory of relativity* (or just *special relativity*), proposed by Einstein in 1905, deals with the preferred status of the velocity of light. We have noted that the speed of light, c, is the maximum speed attainable in the universe. But there is more to it than that. In 1887, a fundamental experiment carried out by two American physicists, A. A. Michelson and E. W. Morley, demonstrated a further important and unique aspect of light—that the measured speed of a beam of light is *independent* of the motion of the observer or the source. No

matter what our velocity may be relative to the source of the light, we always measure precisely the same value for c—299,792.458 km/s.

A moment's thought leads us to the conclusion that this is a decidedly nonintuitive statement. For example, if we were traveling in a car moving at 100 km/h and we fired a bullet forward with a velocity of 1000 km/h relative to the car, an observer standing at the side of the road would see the bullet pass by at $100 + 1000 = 1100$ km/h, as illustrated in the accompanying figure. However, if we were traveling in a rocket ship at 1/10 the speed of light, $0.1c$, and we shone a searchlight beam ahead of us, the Michelson–Morley experiment tells us that an outside observer would measure the speed of the beam not as $1.1c$, as the preceding example would suggest, but as c. The

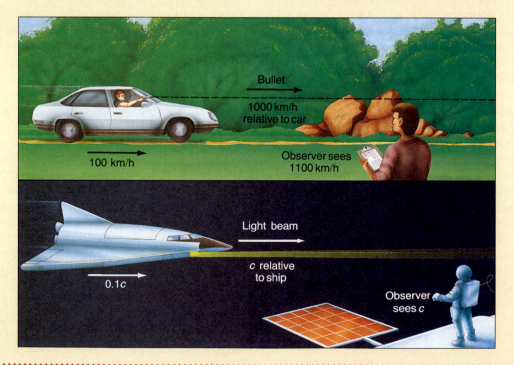

enough volume, the collective pull of gravity can eventually crush *any* opposing pressure? Can gravity continue to compress a massive star into an object the size of a planet, a city, a pinhead—even smaller? The answer, apparently, is "Yes."

THE FINAL STAGE OF STELLAR EVOLUTION

We have seen that the eventual evolution of a star depends critically on its mass. Low-mass stars leave behind a compact remnant known as a white dwarf. ∞ (p. 431) High-mass stars can also produce a compact remnant, in the form of a neutron star. The laws of physics make specific predictions about the masses of these core remnants. A white dwarf must be less than about 1.4 solar masses—the so-called Chandrasekhar mass, beyond which the electrons cannot support the core against its own gravity. ∞ (p. 448)

Similarly, a neutron star resulting from a supernova must have a mass between about 1.4 and 3 solar masses.* The lower limit of 1.4 solar masses stems from the theory of stellar evolution: The iron core of an evolved star must exceed the Chandrasekhar mass for core collapse to begin and a supernova to occur. The upper limit of 3 solar masses

*The dividing lines at 1.4 and 3 $M_\odot$ are somewhat uncertain because they ignore the effects of magnetism and rotation, both of which are surely present in the cores of evolved stars. Because these effects can compete with gravity (as discussed in Chapter 21), they influence the evolution of stars. In addition, we do not know for certain how the basic laws of physics might change in regions of very dense matter that is both rapidly spinning and strongly magnetized. However, we expect that these dividing lines will shift generally upward when magnetism and rotation are included, because even larger amounts of mass will then be needed for gravity to compress stellar cores into neutron stars or black holes.

rules that apply to particles moving at or near the speed of light are different from those we are used to in everyday life.

Special relativity is the mathematical framework that allows us to extend the familiar laws of physics from low speeds (that is, speeds much less than *c*, which are often referred to as *nonrelativistic*) to very high (or *relativistic*) speeds, comparable to *c*. Relativity is equivalent to Newtonian mechanics when objects move much more slowly than light, but it differs greatly in its predictions at relativistic velocities. For example, special relativity predicts that a rapidly moving spacecraft will appear to contract in the direction of its motion, its clocks will appear to run slow, and its mass will appear to increase. All the theory's predictions have been repeatedly verified to very high accuracy. Today special relativity is at the heart of all physical science. No scientist seriously doubts its validity.

General relativity is what results when gravity is included in the framework of special relativity. In 1915, Einstein made the connection between special relativity and gravity with the following famous "thought experiment." Imagine that you are enclosed in an elevator with no windows, so that you cannot directly observe the outside world and that the elevator is floating in space. You are weightless. Now suppose that you begin to feel the floor press up against your feet. Weight has apparently returned. There are two possible explanations for this, shown in the accompanying diagram. A large mass could have come nearby, and you are feeling its downward gravitational attraction, *or* the elevator has begun to accelerate upward and the force you feel is that exerted by the elevator as it accelerates you at the same rate. The crux of Einstein's argument is this: There is *no* experiment that you can perform within the elevator, without looking outside, that will let you distinguish between these two possibilities.

Thus, Einstein reasoned, there is no way to tell the difference between a gravitational field and an accelerated frame of reference (such as the rising elevator in the thought experiment). Gravity can therefore be incorporated into special relativity as a general acceleration of all particles. However, another major modification to the theory of special relativity

must be made. Central to relativity is the notion that space and time are not separate quantities, but instead must be treated as a single entity—*spacetime*. To incorporate the effects of gravity, the mathematics forces us to the conclusion that spacetime has to be *curved*.

In general relativity, then, gravity is a manifestation of curved spacetime. There is no such thing as a "gravitational field," in the Newtonian sense. Instead, objects move as they do because they follow the curvature of spacetime, and this curvature of spacetime is determined by the amount of matter present. We will explore some of the consequences of this view of gravity in more detail in the text.

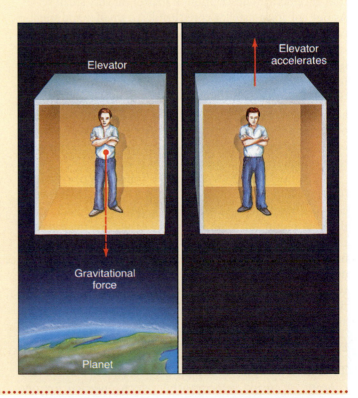

is the neutron-star equivalent of the Chandrasekhar mass—beyond 3 solar masses, not even tightly packed neutrons can withstand the star's gravitational pull.

In fact, we know of *no* force that can counteract gravity beyond the point at which neutron degeneracy pressure is overwhelmed. If enough material is left behind after a supernova, as may happen in the case of an extremely massive progenitor star, gravity finally wins once and for all, and the central core collapses forever. As the core shrinks, the gravitational pull in its vicinity eventually becomes so great that even light itself is unable to escape. The resultant object therefore emits no light, no radiation, no information whatsoever. Astronomers call this bizarre end point of stellar evolution, in which a massive core remnant collapses in on itself and vanishes forever, a **black hole**.

Can an entire star simply shrink to a point and vanish? Doesn't this violate some law of physics? Does it really make sense to talk about black holes? These questions bring us to some very fundamental issues that are presently at the forefront of modern physics. Without some agent to compete against gravity, the present laws of gravitational physics predict that a massive core remnant *will* collapse all the way to a point at which both its density and its gravitational field become infinite—a so-called **singularity**. However, we should not take this prediction of infinite density too literally. Singularities always signal the breakdown of the theory producing them. In other words, the present laws of physics are simply inadequate to describe the final moments of the star's collapse.

As it stands today, the theory of gravity is incomplete because it does not incorporate a proper (that is, a quantum-mechanical) description of matter on very small scales. As our collapsing stellar core shrinks to smaller and smaller radii, we eventually lose our ability to predict, or even describe, its behavior. However, having said that, we can at least estimate how small the core can get *before* quantum effects must become important. It turns out that by the time this stage is reached, the core is already much smaller than any elementary particle. Although the correct description of the end point of the collapse may well require a major overhaul of the laws of physics, for all practical purposes the prediction of collapse to a point is valid.

A complete analysis of Einstein's **theories of relativity** (see the *More Precisely* feature on p 470), which form the complex mathematical framework needed to understand the true nature of black holes, is far beyond the scope of this book. However, we can still usefully discuss many qualitative aspects of these strange regions of space. We can understand the essence of black holes by using the following two key facts from relativity: (1) Nothing can travel faster than the speed of light, and (2) all things, *including light*, are attracted by gravity.

ESCAPE VELOCITY

To explore how gravity attracts even light, let's consider the concept of escape velocity—the velocity needed for one object to escape from the gravitational pull of another. In Chapter 2, we noted that escape velocity is proportional to the square root of a body's mass divided by the square root of its radius. ∞ (p. 47) The escape velocity on Earth, which has a radius of about 6500 km, is nearly 11 km/s. In order for any object—a molecule, a baseball, a rocket, whatever—to be launched away from the Earth, it must move faster than 11 km/s.

Consider a hypothetical experiment in which the Earth is squeezed on all sides by a gigantic vise. As our planet shrinks under the pressure, its mass remains the same. The escape velocity increases because the radius is decreasing. Suppose that the Earth were compressed to one-fourth its present size. The proportionality just mentioned for escape velocity then predicts that the escape velocity would double (because $1/\sqrt{1/4} = 2$). Any object escaping from this hypothetically compressed Earth would need a velocity of at least 22 km/s.

Imagine compressing the Earth some more. Squeeze it, for example, by an additional factor of 1000, making its radius hardly more than a kilometer. Now a velocity of about 630 km/s would be needed to escape. Compress the Earth still further, and the escape velocity continues to rise. If our hypothetical vise were to squeeze the Earth hard enough to crush its radius to about a centimeter, the speed needed to escape its surface would reach 300,000 km/s. But this is no ordinary speed. It is the speed of light, the fastest velocity allowed by the laws of physics as we currently know them.

Thus, if by some fantastic means the entire planet Earth could be compressed to less than the size of a grape, the escape velocity would exceed the speed of light. And because nothing can exceed that speed, the compelling conclusion is that nothing—absolutely nothing—could escape from the surface of such a compressed body. Even radiation—radio waves, visible light, X rays, photons of all wavelengths—would be unable to escape the intense gravity of our reshaped Earth. With no photons leaving, our planet would be invisible and uncommunicative—no signal of any sort could be sent to the universe beyond. The origin of the term *black hole* becomes clear. For all practical purposes, such a supercompact Earth could be said to have disappeared from the universe! Only its gravitational field would remain behind, betraying the presence of its mass, now shrunk to a point.*

THE EVENT HORIZON

Astronomers have a special name for the critical radius at which the escape velocity from an object would equal the speed of light and within which the object could no longer be seen. It is the **Schwarzschild radius**, after Karl Schwarzschild, the German scientist who first studied its

---

In fact, we now know that, regardless of the composition or condition of the object that formed the hole, only three physical properties can be measured from the outside—the hole's mass, charge, and angular momentum. All other information is lost once the infalling matter crosses the event horizon. Thus, only three numbers are required to describe completely a black hole's outward appearance. In this chapter, we will consider only holes that formed from nonrotating, electrically neutral matter. Such objects are completely specified once their masses are known.

More Precisely...   *Tests of General Relativity*

Special relativity is the most thoroughly tested and most accurately verified theory in the history of science. General relativity, however, is on somewhat less firm experimental ground than is special relativity.

The problem with verifying general relativity is that its effects on the Earth and in the solar system—the places where we can most easily perform tests—are very small. Just as special relativity produces major departures from Newtonian mechanics only when velocities approach the speed of light, general relativity predicts large departures from Newtonian gravity only when extremely strong gravitational fields are involved—in effect, when orbit speeds and escape velocities become relativistic.

We will encounter other experimental and observational tests of general relativity elsewhere in this chapter. In this interlude, we will consider just two "classical" tests of the theory. These tests are solar-system experiments that helped ensure acceptance of Einstein's theory. Later, more accurate measurements have confirmed and strengthened these results. Bear in mind, however, that there are no known tests of general relativity in the "strong-field" regime—that part of the theory that predicts black holes, for example—so the full theory has never been experimentally tested.

At the heart of general relativity is the premise that everything, including light, is affected by gravity because of the curvature of spacetime. Shortly after he published his theory in 1915, Einstein noted that light from a star should be deflected by a measurable amount as it passes the Sun. The closer to the Sun the light comes, the more it is deflected. Thus, the maximum deflection should occur for a ray that just grazes the solar surface. Einstein calculated that the deflection angle should be 1.75″—a small, but detectable, amount. Of course, it is normally impossible to see stars close to the Sun. During a solar eclipse, however, when the Moon blocks the Sun's light, the observation becomes possible, as illustrated in the accompanying figure.

In 1919, a team of observers, led by the British astronomer Sir Arthur Eddington, succeeded in measuring the deflection of starlight during an eclipse. The results were in excellent agreement with the prediction of general relativity. Virtually overnight Einstein became world famous. His previous major

accomplishments notwithstanding, this single prediction assured him a permanent position as the best-known scientist on Earth!

Another prediction of general relativity is that planetary orbits should deviate slightly from the perfect ellipses of Kepler's laws. Again, the effect is greatest where gravity is strongest—that is, closest to the Sun. Thus, the largest relativistic effects are found in the orbit of Mercury. Relativity predicts that Mercury's orbit is not a closed ellipse. Instead, its orbit should rotate slowly, as shown in the (highly exaggerated) diagram. The amount of rotation is very small—only 43″ per century—but Mercury's orbit is so well charted that even this tiny effect is measurable.

In fact, the observed rotation rate is 574″ per century, much greater than that predicted by relativity. However, when other (nonrelativistic) gravitational influences, primarily the perturbations due to the other planets, are taken into account, the rotation is in complete agreement with the foregoing prediction.

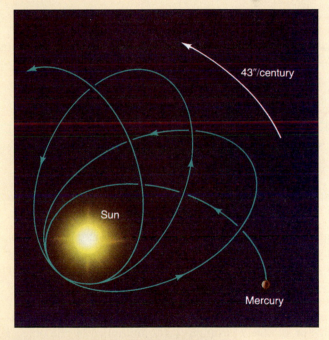

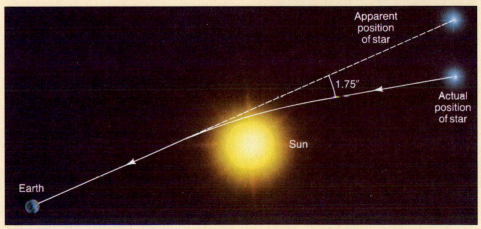

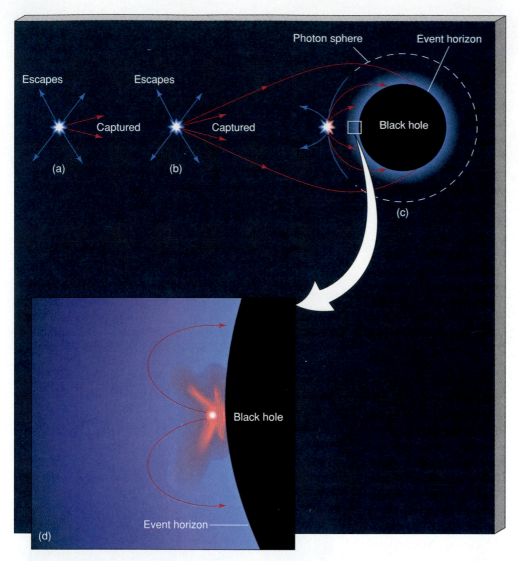

Figure 22.9 This light source, moving closer and closer to a black hole, emits radiation in all directions. (a) At large distances from the hole, most of the light (marked in blue) escapes into space. (b) As the distance decreases, more and more of the radiation is deflected by the black hole's gravity onto paths that intersect the event horizon, and this radiation is trapped (shown in red). (c) At a distance of 1.5 Schwarzschild radii, photons can orbit the hole on circular trajectories, which mark the photon sphere. (d) Eventually, when the source reaches the horizon, all the light it emits is destined to enter the hole.

properties. The Schwarzschild radius of any object is simply proportional to its mass. For Earth, it is 1 cm; for Jupiter, at about 300 Earth masses, it is about 3 m; for the Sun, at 300,000 Earth masses, it is 3 km. For a 3-solar-mass stellar core remnant, the Schwarzschild radius is about 9 km. As a convenient rule of thumb, the Schwarzschild radius of an object is simply 3 km multiplied by the object's mass, measured in solar masses. *Every* object has a Schwarzschild radius. It is the radius to which the object would have to be compressed for it to become a black hole. Put another way, a black hole is an object that happens to lie within its own Schwarzschild radius.

The surface of an imaginary sphere with radius equal to the Schwarzschild radius and centered on a collapsing star is called the **event horizon**. It defines the region within which no event can ever be seen, heard, or known by anyone outside. Even though there is no matter of any sort associated with it, we can think of the event horizon as the "surface" of a black hole.

A 1.4-solar-mass neutron star has a radius of about 10 km and a Schwarzschild radius of 4.2 km. If we were to keep increasing the star's mass, the star's Schwarzschild radius would grow, although its actual physical radius would

not. In fact the radius of a neutron star *decreases* slightly with increasing mass. By the time our neutron star's mass exceeded about 3 times the mass of the Sun, it would lie just within its own event horizon, and it would collapse of its own accord. It would not stop shrinking at the Schwarzschild radius—the event horizon is not a physical boundary of any kind, just a communications barrier. The remnant would shrink right past it to ever-diminishing size on its way to becoming a point singularity.

Thus, provided that at least 3 solar masses of material remain behind after a supernova explosion, the remnant core will collapse catastrophically, diving below the event horizon in less than a second. The core simply "winks out," disappearing and becoming a small dark region from which nothing can escape—a literal black hole in space.

PHOTON ORBITS

An alternative way of seeing the significance of the Schwarzschild radius is to consider what happens to rays of light emitted at different distances from the event horizon of a black hole. Imagine moving a light source closer and closer to the hole, as shown in Figure 22.9. Let's suppose that the source emits radiation uniformly in all directions.

At large distances from the black hole, essentially all the radiation eventually escapes into space. Only that portion of the beam that is aimed directly at the hole is captured. As the light source moves closer, however, the effect of the hole's strong gravity becomes evident. Some photons that would have missed the hole if light traveled in straight lines are instead deflected onto paths that cross the event horizon, and these photons are trapped by the black hole.

At 1.5 Schwarzschild radii, exactly half the radiation emitted by our light source escapes into space. Photons emitted perpendicular to the line joining the source to the center of the hole move in circular orbits at this radius, never escaping the hole's gravity but never crossing the event horizon. The surface on which these photons move as they travel on their circular paths is called the *photon sphere*. Closer still, the amount of deflection continues to increase, and the fraction of the radiation that escapes into space steadily decreases until, close to the event horizon, only a thin sliver of our beam is able to escape. At the event horizon itself, the sliver vanishes completely. All light rays from that surface, whatever their initial direction, are destined to enter the hole. This example clearly shows how the black hole's immense gravitational field dominates the trajectories of all particles, even photons, in its vicinity.

22.5 *Properties of Black Holes*

WARPED SPACE

5 Modern notions about black holes rest squarely on the theory of relativity. Although white dwarfs and neutron stars can be adequately described by the classical Newtonian theory of gravity, only the modern Einsteinian theory of relativity can properly describe the bizarre physical properties of black holes.

A central concept of general relativity (see the *More Precisely* feature on p. 470) is this: Matter—all matter—tends to "warp" or curve space in its vicinity. Objects such as planets and stars react to this warping by changing their paths. In the Newtonian view of gravity, particles move on curved trajectories because they feel a gravitational force. In Einsteinian relativity, those same particles move on curved trajectories because they are following the curvature of space produced by some nearby massive object. The more the mass, the greater the warping. Close to a black hole, the gravitational field becomes overwhelming and the curvature of space extreme. At the event horizon itself, the curvature is so great that space "folds over" on itself, causing objects within to become trapped and disappear.

RUBBER SHEETS AND CURVED SPACE

Some props may help us visualize the curvature of space near a black hole. Bear in mind, however, that these props are not "real," but only tools to help us grasp some exceedingly strange concepts.

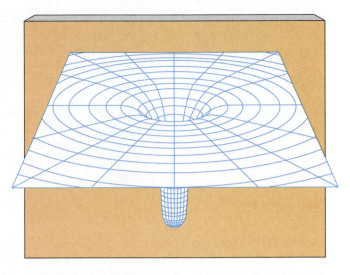

Figure 22.10 A pool table made of a thin rubber sheet will sag with a weight on it. Likewise, space is bent, or warped, in the vicinity of an astronomical object.

First, imagine a pool table with the tabletop made of a thin rubber sheet rather than the usual hard felt. As Figure 22.10 suggests, such a rubber sheet becomes distorted when a heavy weight, such as a rock, is placed on it. The otherwise flat rubber sheet sags or warps (or curves), especially near the rock. The heavier the rock, the larger the curvature. Trying to play billiards, you would quickly find that balls passing near the rock are deflected by the curvature of the tabletop.

In much the same way, both matter *and radiation* are deflected by the curvature of space near a star. For example, the Earth's orbital path is governed by the relatively gentle curvature of space created by our Sun. In the case of a very massive star, space is more severely curved. In the extreme case, a black hole curves space more than any other object.

Let's consider another analogy. Imagine a large family of people living on a huge rubber sheet—a sort of gigantic trampoline. Deciding to hold a reunion, they converge on a given place at a given time. As shown in Figure 22.11, one person remains behind, not wishing to attend. He keeps in touch with his relatives by means of "message balls" rolled out to him (and back from him) along the surface of the sheet. These message balls are the analog of radiation carrying information through space.

As the people converge, the rubber sheet sags more and more. Their accumulating mass creates an increasing amount of space curvature. The message balls can still reach the lone person far away in nearly flat space, but they arrive less frequently as the sheet becomes more and more warped and stretched—as shown in Figures 22.11(b) and (c)—and the balls have to climb out of a deeper and deeper well. Finally, when enough people have arrived at the appointed spot, the mass becomes too great for the rubber to support. As illustrated in Figure 22.11(d), the sheet pinches off into a "bubble," compressing the people into oblivion and severing their communications with the lone survivor

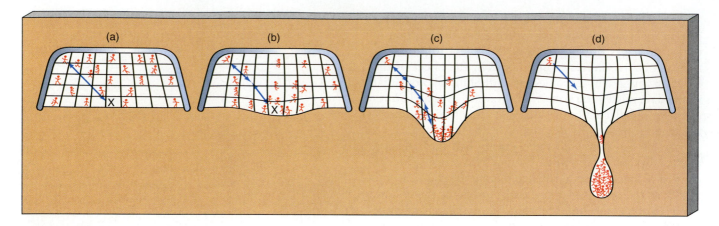

Figure 22.11 Any mass causes the rubber sheet (space) to be curved. As people assemble at the appointed spot on the sheet, the curvature grows progressively larger, as shown in frames (a), (b), and (c). The blue arrows represent some directions in which information can be transmitted from place to place. The people are finally sealed inside the bubble (d), forever trapped and cut off from the outside world.

outside. This final stage represents the formation of an event horizon around the party.

Right up to the end—the pinching off of the "bubble"—two-way communication is possible. Message balls can reach the outside from within (but at a slower and slower rate as the rubber stretches), and messages from outside can get in without difficulty. Once the event horizon (the bubble) forms, balls from the outside can still fall in, but they can no longer be sent back out to the person left behind, no matter how fast they are rolled. They cannot make it past the "lip" of the bubble in Figure 22.11(d). This analogy (very) roughly depicts how a black hole warps space completely around on itself, isolating its interior from the rest of the universe. The essential ideas—the slowing down and eventual cessation of outward-going signals and the one-way nature of the event horizon once it forms—all have clear parallels in the case of stellar black holes.

COSMIC CLEANERS? NO

Black holes are *not* cosmic vacuum cleaners. They do not cruise around interstellar space, sucking up everything in sight. The orbit of an object near a black hole is essentially the same as its orbit near a star of the same mass. Only if the object happens to pass within a few Schwarzschild radii (perhaps 50 or 100 km for a typical black hole formed in a supernova explosion) of the event horizon is there any significant difference between its actual orbit and the one predicted by Newtonian gravity and described by Kepler's laws. From a distance, the main observational difference is that an object orbiting a black hole would appear to orbit a dark, empty region of space. Neither emitted nor reflected radiation would emerge from the black hole itself.

Black holes, then, do not go out of their way to drag in matter. However, if some matter does happen to fall into one—if its orbit happens to take it too close to the event horizon—it will be unable to get out. Black holes are like turnstiles, permitting matter to flow in only one direction—inward. A black hole's mass never decreases (but see the *More Precisely* feature on p.479). Because a black hole will accrete at least a little material from its surroundings, its mass tends to increase over time. The black hole's size is proportional to its mass, so the radius of the event horizon grows with time.

COSMIC HEATERS? YES

Matter flowing into a black hole is subject to great tidal stress. An unfortunate person falling feet first into a solar-mass black hole would find herself stretched enormously in height and squeezed unmercifully laterally. She would be torn apart even before she reached the event horizon, for the pull of gravity would be much stronger at her feet (which are closer to the hole) than at her head. The tidal forces at work in and near a black hole are the same phenomenon responsible for ocean tides on Earth and the spectacular volcanos on Io. The only difference is that the tidal forces near a black hole are far stronger than any force we know in the solar system.

As shown in Figure 22.12, a similar fate awaits any kind of matter falling into a black hole. Whatever falls in—gas, people, space probes—is vertically stretched and horizontally squeezed, in the process being accelerated to high speeds. The net result of all this stretching and squeezing is numerous and violent collisions among the torn-up debris, causing a great deal of frictional heating of the infalling matter. Material is simultaneously torn apart and heated to high temperatures as it plunges into the hole.

The rapid heating of matter by tides and collisions is so efficient that, prior to submersion below the hole's event horizon, the matter emits radiation on its own accord. For a black hole of solar mass, the energy is expected to be emitted in the form of X rays. In effect, the gravitational energy of matter outside the black hole is converted into heat while that matter falls toward the hole. Once the hot

Figure 22.12 Any matter falling into the clutches of a black hole will become severely distorted and heated. This sketch shows an imaginary planet being pulled apart by the gravitational tides of the black hole.

Figure 22.13 Hypothetical robot astronauts can travel toward a black hole while performing experiments that humans, farther away, can monitor in order to learn something about the nature of space near the event horizon.

matter falls below the event horizon, its radiation is no longer detectable—it never leaves the hole. Contrary to what we might expect from an object whose defining property is that nothing can escape from it, the region surrounding a black hole is expected to be a *source* of energy. The *More Precisely* feature on p. 479 presents another, quite different, way in which a black hole can produce energy.

22.6 Space Travel Near Black Holes

How close to a black hole can we travel? Could we actually observe the event horizon at close range? One reasonably safe way to study a black hole would be to go into orbit around it, safely beyond the disruptive influence of the hole's strong tidal forces. After all, Earth and the other planets of our solar system all orbit the Sun without falling into it and without being torn apart. The gravity field around a black hole is basically no different.

However, even from a stable circular orbit, a close investigation of the hole would be unsafe for humans. Human endurance tests conducted on astronauts of the United States and the former Soviet Union indicate that the human body cannot withstand stress greater than about 10 times the pull of gravity on the Earth's surface. This breaking point would occur about 3000 km from a 10-solar-mass black hole (which, recall, would have a 30-km event horizon). Closer than that, the tidal effect of the hole would tear a human body apart.

APPROACHING THE EVENT HORIZON

Let's send an imaginary indestructible astronaut—a mechanical robot, say—in a probe toward the center of the hole, as illustrated in Figure 22.13. Watching from a safe distance in our orbiting spacecraft, we can then examine the nature of space and time near the hole. Our robot will be a useful explorer of theoretical ideas, at least down to the event horizon. After that boundary is crossed, there is no way for the robot to return any information about its findings.

Suppose, for example, our robot has an accurate clock and a light source of known frequency mounted on it. From our safe vantage point far outside the event horizon, we could use telescopes to read the clock and measure the frequency of the light we receive. What might we discover?

We would find that the light from the robot would become more and more redshifted as the robot neared the event horizon. Even if the robot used rocket engines to remain motionless, the redshift would still be detected. The redshift is *not* caused by motion. It is not the result of the Doppler effect arising as the robot falls into the hole. Rather, it is a redshift induced by the black hole's gravitational field, clearly predicted by Einstein's general theory of relativity. It is known as **gravitational redshift**.

We can explain the gravitational redshift as follows. According to general relativity, photons are attracted by gravity. As a result, in order to escape from a source of gravity, photons must expend some energy: They have to work to get out of the gravitational field. They don't slow

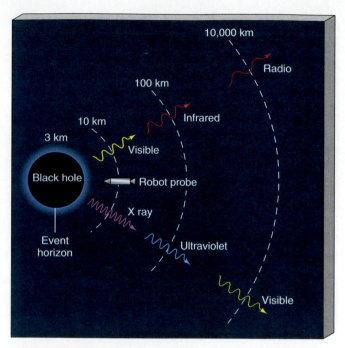

Figure 22.14   As a photon escapes from the strong gravitational field close to a black hole, it must expend energy to overcome the hole's gravity. This energy does not come from a change in the speed at which it travels (this is always 300,000 km/s, even under these extreme conditions). Rather, the photon "gives up" energy by increasing its wavelength. Thus, the photon's color changes. This figure shows the effect on two beams of radiation, one of visible light and one of X rays, emitted from a space probe as it nears a 1-solar-mass black hole.

down at all—photons always move at the speed of light—they just lose energy. Because a photon's energy is proportional to the frequency of its radiation, light that loses energy must have its frequency reduced (or, conversely, its wavelength lengthened). As illustrated in Figure 22.14, radiation coming from the vicinity of a gravitating object will be redshifted by an amount depending on the strength of the gravitational field.

Most astronomical objects' gravitational fields are far too weak to shift radiation much toward the red, although the effect can still be measured. Exceedingly delicate laboratory experiments on Earth have succeeded in detecting the tiny gravitational redshift produced by even our own planet's weak gravity. Sunlight is redshifted by only about a hundredth of an angstrom. A few white-dwarf stars do show some significant reddening of their emitted light, however. Their radii are much smaller than that of our Sun, so their surface gravity is very much stronger than the Sun's. Neutron stars should show an appreciable shift in their radiation, but it is currently impossible to disentangle the effects of gravity and magnetism on the observed signals. Only near black holes is the gravitational pull so great that the redshift should be large and easily measurable.

As photons traveled from the robot's light source to the orbiting spacecraft, they would become gravitationally

redshifted. From the standpoint of the orbiting humans, a green light, say, would become yellow and then red as the robot astronaut neared the black hole. From the robot's perspective, the light would remain green. As the robot got closer to the event horizon, the radiation from its light source would become undetectable with optical telescopes. The radiation reaching the humans in the orbiting spacecraft would by then be lengthened so much that infrared and then radio telescopes would be needed to detect it. Closer still to the event horizon, the radiation emitted as visible light from the robot probe would be shifted to wavelengths even longer than conventional radio waves by the time it reached the human observers.

Light emitted *from the event horizon itself* would be gravitationally redshifted to infinitely long wavelengths. In other words, each photon would use all its energy trying to escape from the edge of the hole. What was once light (on the robot) has no energy left upon arrival at the safely orbiting spacecraft. Theoretically, this radiation makes it to us—still moving at the velocity of light—but with zero energy. The light radiation originally emitted has become redshifted beyond our perception.

What about the robot's clock? Assuming that the distant observers in the safely orbiting spacecraft can read it, what time does it tell? Is there any observable change in the rate at which the clock ticks while moving deeper into the hole's gravitational field? We would find that, from the safely orbiting spacecraft, any clock close to the hole would appear to tick more *slowly* than an equivalent clock on board the spacecraft. The closer the clock came to the hole, the slower it would appear to run. The clock closest to the hole would operate slowest of all. Upon reaching the event horizon, the clock would seem to stop altogether. It would be as if the robot astronaut had found immortality! All action would become virtually frozen in time. Consequently, an external observer will never actually witness an infalling astronaut sink below the event horizon. Such a process would appear to take forever.

This apparent slowing down of the robot's clock is known as **time dilation**. It is another clear prediction of general relativity, and in fact it is closely related to the gravitational redshift. To see this connection, imagine that we use our light source as a clock, with the passage of a wave crest (say) constituting a "tick." The clock thus ticks at the frequency of the radiation. As the wave is redshifted, the frequency drops, and fewer wave crests pass the distant observer each second—the clock appears to slow down. This thought experiment demonstrates that the redshift of the radiation and the slowing of the clock are one and the same thing.

From the point of view of the indestructible robot, however, relativity theory predicts no strange effects at all. To the infalling robot, the light source hasn't reddened, and the clock keeps perfect time. In the robot's frame of reference, everything is normal. Nothing prohibits it from approaching within the Schwarzschild radius of the hole.

No law of physics constrains an object from passing through an event horizon. There is no barrier at the event horizon and no sudden lurch as it is crossed; it is only an imaginary boundary in space. Travelers passing through the event horizon of a sufficiently massive hole (such as might lurk in the heart of our own Galaxy, as we will see) might not even know it—at least until they tried to get out!

DEEP DOWN INSIDE

No doubt, you are wondering what lies within the event horizon of a black hole. The answer is simple: No one really knows.

Some researchers maintain that the inner workings of black holes are irrelevant. Experiments could conceivably be done by robots sent "down under" to study conditions

---

More Precisely... *Black-Hole Evaporation*

Some attempts to understand gravity on a microscopic scale suggest that black holes may not be entirely black after all. Applying what they know of subatomic physics, scientists now believe it possible that some matter and radiation can escape from a black hole. Here's how. The laws of quantum physics allow a process known as *pair creation* to occur anywhere in space: A particle and its antiparticle—an electron and a positron, say—can come into being spontaneously, literally formed out of nothing. This, of course, violates one of the most cherished laws of physics, the law of conservation of mass and energy (recall that mass and energy are equivalent, related to one another by Einstein's famous equation $E = mc^2$), but this violation is permitted so long as the "books are balanced" by the disappearance of the particles (by mutual annihilation) within a short enough period of time. In effect, the rules can be broken, so long as they are repaired before anyone notices.

Most of the time, pairs of particles appear and disappear so rapidly that energy is conserved on all macroscopic scales. However, should pair creation happen near a black hole, as illustrated in the diagram, it is possible for one of the two particles to cross the event horizon *before* it meets and annihilates its partner. The other particle would then be free to leave the scene, making the black hole appear to the outside world as a source of matter or radiation. The energy required to create the new particle ultimately comes from the black hole. Because energy and mass are equivalent, this means that the hole must decrease in mass as it radiates. Thus, black holes do not last forever—they slowly "evaporate." This possibility was first realized by Cambridge University mathematician Stephen Hawking. The radiation from a black hole is known as *Hawking radiation*.

A remarkable result, also discovered by Hawking's group, is that the spectrum of Hawking radiation is described by a Planck curve—exactly the same curve that characterizes emission from any hot body. Black holes emit black-body radiation! The temperature of the radiation turns out to be inversely related to the mass of the hole. Big black holes are very cold, whereas small black holes are hot. A hole the mass of the Sun would emit radiation at a temperature of 10^{-6}K; one the mass of a mountain—about 10^{12} kg—would have a temperature of some 10^{12}K. Knowing a black hole's temperature T and surface area A, we can calculate its luminosity L in exactly the same way as for stars: $L \propto AT^4$. ∞ (p. 361)

A black hole radiates energy (and hence mass) into space. As the hole radiates, its mass drops and its temperature increases. That is, the black hole's temperature is inversely proportional to its mass, and its area decreases as the square of the mass. The black hole, then, increases its luminosity as it evaporates. The increased luminosity, in turn, leads to a faster loss of mass. This runaway situation eventually ends violently, and the black hole explodes in a burst of gamma rays.

The lifetime of a hole depends on its mass. For a 1-solar-mass black hole, the explosion is predicted to occur after about 10^{70} years! Astronomers today hardly expect to observe such an event. Thus, the issue of evaporation is moot for the black holes described in this chapter; astronomers do not expect ever to observe either their slow decay (which begins the moment they form) or their eventual explosion.

In contrast, very small black holes, with masses of about 10^{12} kg, should have lifetimes roughly equal to the current age of the universe. Although we know of no way in which such objects could be created in the universe today, it is conceivable that conditions in the very earliest epochs of the universe might have been just right to compress pockets of matter into miniature black holes. Such black holes would have Schwarzschild radii of about 10^{-15} m, comparable to the size of a subatomic particle. If they exist, very small black holes should be exploding right now. Attempts have been made to observe the resultant gamma rays, so far without success.

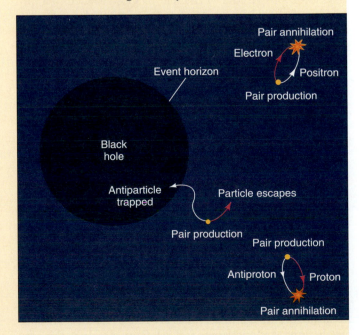

within the event horizon, but that information could never reach the rest of us outside the black hole. Theories of the insides of black holes can never be put to the experimental test. From a purely observational perspective at least, anyone's theory is as valid as anyone else's.

Other researchers point out, as we noted earlier, that relativity must be incorrect, or at least incomplete, when applied to the centers of black holes. The current laws of physics are inadequate in the vicinity of a singularity, because they lose their predictive power there. Perhaps matter trapped in black holes never actually reaches the singularity. Perhaps it just approaches this bizarre state, in a manner that we will someday understand as the subject of *quantum gravity*—the merger of general relativity with quantum mechanics—develops. However, even if this new theory succeeds in doing away with the singularity, it is unlikely that the external appearance of the hole, or the existence of its event horizon, will change. Any modifications to general relativity are expected to occur only on submicroscopic scales, not on the macroscopic (kilometer-sized) scale of the Schwarzschild radius.

Singularities are places where the rules break down, and some very strange things may occur near them. Many possibilities have been envisaged—gateways into other universes, time travel, the creation of new states of matter—but none of them has been proved, and certainly none of them has been observed. Because these regions are places where science fails, their presence causes serious problems for many of our cherished laws of physics, from causality (the idea that cause should precede effect, which runs into immediate problems if time travel is possible), to energy conservation (which is violated if material can hop from one universe to another through a black hole). Disturbed by the possibility of such chaos in science, some relativists have even proposed a "Principle of Cosmic Censorship:" Nature always hides *any* singularity, such as that found at the center of a black hole, inside an event horizon. In that case, even though physics fails, that breakdown cannot affect us outside, so we are safely insulated from any effects the singularity might have. What would happen if we one day found a so-called *naked singularity* somewhere, a singularity uncloaked by an event horizon? Would relativity theory still hold there? For now, we just don't know.

What sense are we to make of black holes? Do black holes, and all the strange phenomena that go on in and around them, really exist? The basis for understanding these weird phenomena is the relativistic concept that mass warps space—which has already been found to be a surprisingly good representation of reality, at least for the weak gravitational fields produced by stars and planets (see the *More Precisely* feature on p. 470). The larger the mass concentration, the greater the warping, and thus, apparently, the stranger the observational consequences. Although general relativity is not proven, there is presently no reason to disbelieve it, and black holes are one of its most striking predictions. As long as general relativity stands as the correct theory of gravity in the universe, black holes are real.

22.7 *Observational Evidence for Black Holes*

6 Theoretical ideas aside, is there any observational evidence for black holes? Can we prove that these invisible objects really do exist?

STELLAR TRANSITS?

One way in which we might detect a black hole, at least in principle, would be to observe it passing in front of something else. If a black hole passed in front of a much larger visible companion star, for example, we would expect to see a minute black dot glide across the otherwise bright star. Unfortunately, this would be extremely hard to see; the 12,000-km planet Venus is barely noticeable when transiting the Sun, so a kilometer-sized object moving across the image of a faraway star would be completely invisible with current equipment, or any equipment available in the foreseeable future.

Actually, this observation is not even as clear-cut as just suggested. Even if we were close enough to the star that we could resolve the 10- (or so) km disk of the black hole, the observable effect would not really be a black dot superimposed on a bright background. The background starlight would be deflected as it passed the black hole on its way to Earth, as indicated in Figure 22.15. The effect is the same as the bending of distant starlight around the edge of the Sun, a phenomenon that has been repeatedly measured during solar eclipses throughout the last several decades (see the *More Precisely* feature on p. 473). With a black hole,

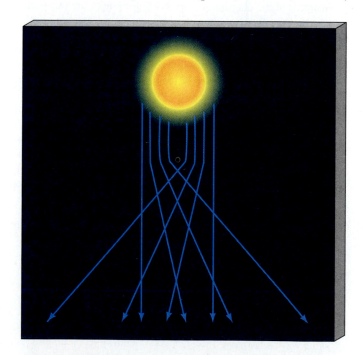

Figure 22.15 The gravitational bending of light around the edges of a small, massive black hole makes it impossible to observe the hole as a black dot superimposed against the bright background of its stellar companion.

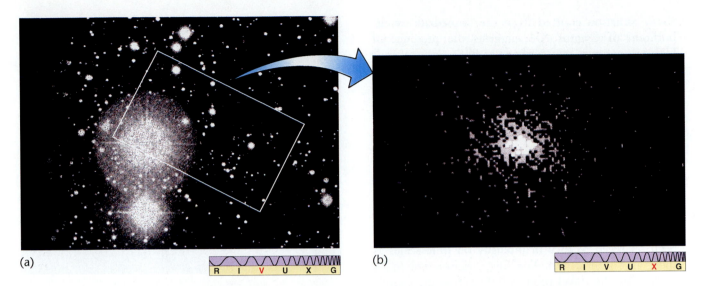

Figure 22.16 (a) The brightest star in this photograph is a member of a binary system whose unseen companion, called Cygnus X-1, is thought to be a good candidate for a black hole. (b) X rays emitted by the Cygnus X-1 source were analyzed by changing them into electronic signals that were then viewed on a video screen, from which this picture was taken. (The field of view here is outlined by the rectangle in the previous figure.)

much larger deflections would occur. As a result, our perception of a black hole in front of a bright companion star would not show a neat, well-defined black dot. Instead, the bending of light around it would make its image fuzzy. Recent studies have shown that such a blurred image would be virtually impossible to observe, even from nearby.

BLACK HOLES IN BINARY SYSTEMS

A much better way to find black holes is to look for their effects on other objects. Although black holes are invisible, they are so massive that astronomers can test for a hole's existence by studying its associated gravitational field. For example, the motion of a spacecraft, a planet, or a star could conceivably be used to probe the vicinity of a suspected hole. The hole would have the same effect on a nearby body as would a visible object of the same mass.

Our Milky Way Galaxy harbors many binary-star systems in which only one object can be seen. Recall from our study of double stars in Chapter 17 that we need to observe the motion of only one star to infer the existence of an unseen companion and measure some of its properties. ∞ (p. 373) In the majority of cases, the invisible companion is probably simply small and dim, nothing more than an M-type star hidden in the glare of an O- or B-type partner. In other cases, dust or other debris probably shroud one object, making it invisible to even the best available equipment. In either case, the invisible object is not a black hole.

A few close binary systems, however, have peculiarities that suggest that one of their members may indeed be a black hole. Some of the most interesting observations, made during the 1970s and 1980s by Earth-orbiting satellites, reveal binary systems in which the invisible member emits large amounts of X rays. The mass of the emitting

object is measured as several solar masses, so we know it is not simply a small, dim star, and radiation pressure from the binary members makes circumstellar debris an unlikely explanation for its invisibility. One particular binary system drawing much attention lies in the constellation Cygnus.

A BLACK-HOLE CANDIDATE

Figure 22.16(a) shows the area of the sky in Cygnus where astronomers have reasonably good evidence for a black hole. The rectangle outlines the celestial system of interest, some 2000 pc from the Earth. The black-hole candidate is an X-ray source called Cygnus X-1, discovered by the *Uhuru* satellite in the early 1970s. Its visible companion—a blue B-type supergiant with the catalog name HDE226868—was identified a few years later. The main observational features of this binary system are as follows:

1. Spectroscopic observations of visible radiation show that the binary system has an orbital period of 5.6 days and a diameter of 20 million km.

2. Assuming that the visible component lies on the main sequence, we know its mass must be around 30 times the mass of the Sun.

3. Knowledge of the binary's orbital period and size allows us to place a limit on the combined mass of the two components. Knowing the mass of the visible component and subtracting it from that limit, astronomers estimate the mass of Cygnus X-1 to be between 5 and 10 solar masses.

4. Other spectroscopic studies suggest that hot gas is flowing from the bright star toward an unseen companion.

5. X-ray radiation emitted from the immediate neighborhood of Cygnus X-1 suggests the presence of high-temperature gas, perhaps as hot as several million kelvins (see Figure 22.16b).

6. Rapid time variations of this X-ray radiation imply that the size of the X-ray-emitting region of Cygnus X-1 must be less than a few hundred kilometers across. The reasoning goes as follows: If the emitting region were, say, 300,000 km—1 light second—across, even an instantaneous change in intensity at the source would be smeared out over a time interval of 1 s as seen from Earth, because light from the far side of the object would take 1 s longer to reach us than light from the near side. X rays from Cygnus X-1 have been observed to vary in intensity on time scales as short as a millisecond. For this variation not to be blurred by the travel time of light across the source, Cygnus X-1 cannot be more than 1 light millisecond, or 300 km, in diameter.

These general properties suggest that the invisible X-ray-emitting companion could be a black hole. The X-ray-emitting region is likely an accretion disk formed as matter drawn from the visible star spirals down onto the unseen component. The rapid variability of the X-ray emission indicates that the unseen component must be compact—a neutron star or a black hole. The mass limit on the dark component argues for the latter, as astronomers believe that neutron stars' masses cannot exceed about three solar masses. Figure 22.17 is an artist's conception of this intriguing object. As shown, most of the gas drawn from the visible star ends up in a donut-shaped accretion disk of matter. As the gas flows toward the black hole, it becomes superheated and emits the X rays we observe, just before they are trapped forever below the event horizon.

HAVE BLACK-HOLES BEEN DETECTED?

A few other black-hole candidates are known. For example, the third X-ray source ever discovered in the Large Magellanic Cloud—called LMC X-3—is an invisible object that, like Cygnus X-1, orbits a bright companion star. LMC X-3's visible companion seems to be distorted into the shape of an egg by the unseen object's intense gravitational pull. Reasoning similar to that applied to Cygnus X-1 leads to the conclusion that the compact object LMC X-3 has a mass nearly 10 times that of the Sun, making it too massive to be anything but a black hole. The X-ray binary system A0620-00 has been found to contain an invisible compact object of mass 3.8 times the mass of the Sun.

In total, there are perhaps half a dozen known objects that may turn out to be black holes, although Cygnus X-1, LMC X-3, and A0620-00 have the strongest claims. But can we be sure that these objects are black holes? Cygnus X-1 and the other suspected black holes in binary systems

Figure 22.17 Artist's conception of a binary system containing a large, bright, visible star and an invisible, X-ray-emitting black hole. This drawing is based on data obtained from detailed study of Cygnus X-1.

all have masses relatively close to the dividing line separating neutron stars from black holes. Given the present uncertainties in both observations and theory, might they conceivably be merely dim, dense neutron stars and not black holes at all?

Most astronomers do not regard this as a likely possibility, but it highlights a problem: There is presently no observational test that can unambiguously distinguish a 10-solar-mass black hole from, say, a 10-solar-mass neutron star (if one could somehow exist). Both objects would affect a companion star's orbit in the same way; both would tear mass from its surface, and both would form an accretion disk around themselves that would emit intense X rays. The radiation from the surface of the neutron star itself, which would distinguish it from a black hole, might in some cases be so weak that it would be impossible to detect against the emission from the disk. We rule out the neutron star on purely theoretical grounds (on the basis of our understanding of neutron star masses), but we have no observational evidence to support our theory.

The argument really proceeds by elimination. Loosely stated, it goes: "Object X is compact and very massive. We don't know of anything else that can be that small and that massive. Therefore, object X is a black hole." We will see other instances later in this book of astronomers using similar reasoning to infer the existence of very massive black holes in the hearts of galaxies. Still, some astronomers are troubled that the black-hole category has in some ways become a "catch-all" for things that have no other reasonable explanation.

So have stellar black holes really been discovered? The answer is probably "Yes." Skepticism is healthy in science, but only the most stubborn astronomers (and some do exist!) would take serious issue with the reasoning that supports the case for black holes. Can we guarantee that future modifications to the theory of compact objects will not invalidate our arguments? No, but similar statements could be made in many areas of astronomy—indeed, about any theory in any area of science. We conclude that,

strange as they are, black holes have been detected in our Galaxy. Perhaps some day, future generations of space travelers will visit Cygnus X-1 or LMC X-3 and (carefully!) test these conclusions first-hand. Until then, we will have to continue to rely on improving theoretical models and observational techniques to guide our discussions of these mysterious objects.

Interlude 22-1 *Gravity Waves*

Electromagnetic waves are common everyday phenomena. Whether they are radio, infrared, visible, ultraviolet, X-ray, or gamma-ray radiation, all electromagnetic waves involve periodic changes in the strengths of electric and magnetic fields. They move through space and transport energy. Any accelerating charged particle, such as an electron within a broadcasting antenna or on the surface of a star, generates electromagnetic waves. Measurements of these waves agree perfectly with the theory of electromagnetism, which predicts them.

The modern theory of gravity—Einstein's theory of relativity—also predicts waves that move through space. A *gravity wave* is the gravitational counterpart of an electromagnetic wave. A gravity wave, or *gravitational radiation,* results from a change in the strength of a gravitational field. In principle, any time an object of any mass accelerates, a gravity wave should be emitted at the speed of light. The passage of a gravity wave should produce small distortions in the space through which it passes. Gravity is an exceedingly weak force compared with electromagnetism, so these distortions are expected to be very small—in fact, much smaller than the diameter of an atomic nucleus for the waves that might be produced by Galactic sources. Yet many researchers believe that they should be measurable. So far, no one has succeeded in detecting gravitational radiation. But detection of gravity waves would provide very strong support for the theory of relativity, so scientists are eager to search for them.

Theorists are still debating which kinds of astronomical objects should produce gravity waves detectable on Earth. Leading candidates include (1) the merger of a binary-star system, (2) the collapse of a star into a black hole, and (3) the collision of two black holes or neutron stars. Each of these possibilities involves the acceleration of huge masses, so the strength of the gravitational fields should change drastically and rapidly in each case. Other astronomical objects are also expected to emit gravity waves, but only changes involving large masses will produce waves intense enough to be observed.

Of the three candidates, the first one probably presents the best chance to detect gravity waves, at least for the present. Binary-star systems should emit gravitational radiation as the component stars orbit one another. As energy escapes in the form of gravity waves, the two stars slowly spiral toward one another, orbiting more rapidly and emitting even more gravitational radiation. This runaway situation can lead to the decay and eventual merger of close binary systems in a relatively short period of time (which, in this case, means tens or hundreds of millions of years).

Such a slow but steady decay in the orbit of a binary system has in fact been detected. In 1974, radio astronomer Joseph Taylor and his student Russell Hulse at the University of Massachusetts discovered a very unusual Galactic binary system. Both components are neutron stars, and one is observable from Earth as a pulsar. This system has become known as the *binary pulsar.* Measurements of the periodic Doppler shift of the pulsar's radiation prove that its orbit is slowly shrinking in size. Furthermore, the rate at which the orbit is shrinking is exactly what would be predicted by relativity theory if the energy were being carried off by gravity waves. Even though the gravity waves themselves have not yet been detected, the binary pulsar is regarded by most astronomers as a very strong piece of evidence in favor of general relativity. Taylor and Hulse received the 1993 Nobel Prize in Physics for their discovery.

Gravity waves should contain a great deal of information about the physical events in some of the most exotic regions of space. In 1992, funding was approved for an ambitious gravity-wave observatory called LIGO—short for Laser Interferometric Gravity-wave Observatory. This detector, which will use laser beams to measure the extremely small distortions of space produced by gravitational radiation, should be capable of detecting gravity waves from many Galactic and extragalactic sources. If successful, the discovery of gravity waves could herald a new age in astronomy, in much the same way that invisible electromagnetic waves, unknown a century ago, revolutionized classical astronomy and led to the field of modern astrophysics.

Chapter Review

SUMMARY

A core-collapse supernova may leave behind a remnant, an ultra-compressed ball of material called a **neutron star** (p. 464). The processes that form neutron stars ensure that these stars are rapidly rotating and strongly magnetized at birth.

Pulsars (p. 465) are objects that emit regular bursts of electromagnetic energy. The accepted explanation for pulsars is the **lighthouse model** (p. 466), in which a rotating neutron star sends a beam of energy into space. If the beam sweeps past Earth, we see a pulsar. The pulse period is the rotation period of the neutron star.

A neutron star in a close binary system can draw matter from its companion, forming an accretion disk. The material in the disk heats up even before it reaches the neutron star, and the disk is usually a strong source of X rays. As gas builds up on the star's surface, it eventually becomes hot enough to fuse hydrogen. As with a nova explosion on a white dwarf, when hydrogen burning starts on a neutron star, it does so explosively. An **X-ray burster** (p. 467) or even a **gamma-ray burster** (p. 468) results, depending on the temperatures reached.

The rapid rotation of the inner part of the accretion disk causes the neutron star to spin faster as new gas arrives on its surface. The eventual result is a very rapidly rotating neutron star—a **millisecond pulsar** (p. 468). Many millisecond pulsars are found in the hearts of old globular clusters. They cannot have formed recently, so they must have been spun up by interactions with other stars.

Careful analysis of the radiation received has shown that some pulsars are orbited by planet-sized objects. The origin of these "pulsar planets" is still uncertain.

The upper limit on the mass of a neutron star is about 3 solar masses. Beyond that mass, the star can no longer support itself against its own gravity, and it must collapse. No known force can prevent the material from collapsing all the way to a pointlike **singularity** (p. 472), a region of extremely high density where the known laws of physics break down. Surrounding the singularity, at a distance of a few kilometers for a solar-mass object, is a region of space from which even light cannot escape—a **black hole** (p. 472). Astronomers believe that the most massive stars form black holes, rather than neutron stars, after they explode in a supernova.

Conditions in and near black holes cannot be described by Newtonian mechanics. A proper description involves the **theories of relativity** (p. 472) developed by Albert Einstein early in the twentieth century. Even relativity theory fails right at the singularity, however.

The "surface" of a black hole is the **event horizon** (p. 474); its distance from a singularity is called the **Schwarzschild radius** (p. 472). At the event horizon, the escape velocity equals the speed of light. Within this distance, nothing can escape. Photons passing too close to a black hole are deflected onto paths that cross the event horizon and become trapped.

Relativity theory describes gravity in terms of a warping, or bending, of space by the presence of mass. The more mass, the greater the warping. All particles—including photons—respond to that warping by moving along curved paths. A black hole is a region where the warping is so great that space folds back on itself, cutting off the interior of the hole from the rest of the universe.

To a distant observer, the clock on a spaceship falling into a black hole would show **time dilation** (p. 478)—it would appear to slow down as the ship approached the event horizon. The observer would never see the ship reach the surface of the hole. At the same time, light leaving the ship would be subject to **gravitational redshift** (p. 477) as it climbed out of the hole's intense gravitational field. Light emitted just at the event horizon would be redshifted to infinite wavelength. Both phenomena are predictions of the theory of relativity. The gravitational redshifts due to both Earth and the Sun are very small, but have been detected experimentally.

Once matter falls into a black hole, it can no longer communicate with the outside. However, on its way in, it can form an accretion disk and emit X rays just as in the neutron-star case. The best candidates for black holes are binary systems in which one component is a compact X-ray source. Cygnus X-1, a well-studied X-ray source in the constellation Cygnus, is a long-standing black hole candidate. Studies of orbital motions imply that the compact objects are too massive to be neutron stars, leaving black holes as the only alternative.

SELF-TEST: True or False?

_____ **1.** The density of a neutron star is comparable to the density of an atomic nucleus.

_____ **2.** As a result of their high masses and small sizes, neutron stars have only weak gravitational pulls at their surfaces.

_____ **3.** A millisecond pulsar is actually a very old neutron star that has been recently spun up by interaction with a neighbor.

_____ **4.** Millisecond pulsars are always found in globular clusters, and nowhere else.

_____ **5.** Planet-sized bodies will never be found around a pulsar, because the supernova that formed the pulsar would have destroyed any planets in the system.

_____ **6.** Nothing can travel faster than the speed of light.

_____ **7.** All things, except light, are attracted by gravity.

_____ **8.** A black hole is an object whose escape velocity equals or exceeds the speed of light.

_____ **9.** Although visible light cannot escape from a black hole, high-energy radiation, like gamma rays, can escape.

_____ **10.** If you could touch it, the surface of a black hole, the event horizon, would be very hard.

_____ **11.** X-rays are emitted by matter accreting onto a stellar-mass black hole.

_____ **12.** Thousands of black holes have now been identified.

SELF-TEST: Fill in the Blank

1. No remnant remains after the explosion of a _____ supernova.

2. A typical neutron star is _____ km in diameter.

3. Neutron stars may be characterized as having a _____ rate of rotation and a _____ magnetic field.

4. Pulsars were discovered through observations in the _____ part of the electromagnetic spectrum.

5. Typical pulsar periods range from _____ to _____. (Give numbers and units.)

6. The pulse period of pulsar radiation tells us the _____ of the neutron star emitting the radiation.

7. All millisecond pulsars are now, or once were, members of _____ star systems.

8. X-ray bursters result from accretion of material from a binary companion onto a _____ star.

9. According to General Relativity, space is warped, or curved, by _____.

10. If the Sun were magically to turn into a black hole, Earth's orbit would _____.

11. Photons _____ energy as they escape from a gravitational field.

12. Black holes of stellar origin are believed to have been discovered in _____.

REVIEW AND DISCUSSION

1. How does the way in which a neutron star forms determine some of its most basic properties?

2. What would happen to a person standing on the surface of a neutron star?

3. Why aren't all neutron stars seen as pulsars?

4. What are X-ray bursters?

5. What is the favored explanation for the rapid spin rates of millisecond pulsars?

6. Why did astronomers not expect to find a pulsar with a planetary system?

7. What does it mean to say that the measured speed of a light beam is independent of the motion of the observer?

8. Use your knowledge of escape velocity to explain why black holes are said to be "black."

9. Why is it so difficult to test the predictions of General Relativity? Describe two tests of the theory.

10. What would happen to someone falling into a black hole?

11. What makes Cygnus X-1 a good black hole candidate?

12. Imagine that you had the ability to travel at will through the Galaxy. Explain why you would discover many more neutron stars than those known to observers on Earth. Where would you be most likely to find these objects?

13. Do you think that planet-sized objects discovered in orbit around a pulsar should be called planets? Why or why not?

PROBLEMS

1. The angular momentum of a body (see Chapter 15) is proportional to its angular velocity times the square of its radius. Using the law of conservation of angular momentum, estimate how fast a collapsed stellar core would spin if its initial spin rate was 1 revolution per day and its radius decreased from 10,000 km to 10 km.

2. Supermassive black holes are believed to exist in the centers of some galaxies. What would be the Schwarzschild radii of black holes of 1 million and 1 billion solar masses? How does the first black hole compare in size to the Sun? How does the second compare in size to the solar system?

3. A 10-km radius neutron star is spinning 1000 times per second. Calculate the speed of a point on its equator, and compare it with the speed of light. (Consider the equator as the circumference of a circle, and recall that circumference=$2\pi r$.)

4. Use the luminosity-radius-temperature relation to calculate the luminosity of a 10 km radius neutron star for temperatures of 10^5 K, 10^6 K, and 10^7 K. What do you conclude about the visibility of neutron stars? Could the brightest of these be plotted in the H-R diagram?

PROJECTS

1. Many amateur astronomers enjoy turning their telescopes on the ninth-magnitude companion to Cygnus X-1, the sky's most famous black hole candidate. Because none of us can see in X-rays, no sign of anything unusual can be seen. Still, it's fun to gaze toward this region of the heavens and contemplate Cygnus X-1's powerful energy emission and strange properties.

Even without a telescope, it is easy to locate the region of the heavens where Cygnus X-1 resides. The constellation Cygnus contains a recognizable star pattern, or asterism, in the shape of a large cross. This asterism is called the Northern Cross. The star in the center of the crossbar is called Sadr. The star at the bottom of the cross is called Albireo. Approximately midway along an imaginary line between Sadr and Albireo lies the star Eta Cygni. Cygnus X-1 is located slightly less than 0.5 degrees from this star. With or without a telescope, sketch what you see.

2. Set up a demonstration of the densities of various astronomical objects—an interstellar cloud, a star, a terrestrial planet, a white dwarf, and a neutron star. Select a common object that is easily held in your hand, something that would be familiar to anyone—an apple, for example. For the lowest densities, calculate how large a volume would contain the object's equivalent mass. For high densities, calculate how many of the objects would have to be fit into a standard volume, such as 1 cm^3. This volume is better for this than 1 m^3 because most people do not appreciate how large a volume 1 m^3 is. Present your demonstration to your class or to some other group of students. Tell them about each astronomical object and how it comes by its density.

23

THE MILKY WAY GALAXY

A Grand Design

LEARNING GOALS

Studying this chapter will enable you to:

1 Describe the overall structure of the Milky Way Galaxy.

2 Explain the importance of variable stars in determining the size and shape of our Galaxy.

3 Specify how stars in the Galactic disk differ from those in the Galactic halo.

4 Explain how and why radio astronomy is useful in mapping the structure of the Galaxy.

5 Describe the orbital paths of stars in different regions of the Galaxy, and explain how these motions are accounted for by our understanding of how the Galaxy formed.

6 Discuss some possible explanations for the existence of the spiral arms observed in our own and many other galaxies.

7 Explain what studies of galactic rotation reveal about the size and mass of our Galaxy.

8 Explain the role of dark matter in the makeup of our Galaxy, and discuss the possible nature of such matter.

9 Describe some of the phenomena observed at the center of our Galaxy.

(Opposite page, background) The varying interrelationships among the many components of matter in our Milky Way Galaxy comprise a sort of "galactic ecosystem." Its evolutionary balance might be as complex as that of life in a tidal pool or a tropical rainforest. Here, stars abound throughout the Lagoon nebula, a rich stellar nursery about 1200 pc from Earth.
(Inset A) An emission nebula, the North American Nebula, glowing amidst a field of stars, its red color produced by the emission of light from vast clouds of hydrogen atoms.
(Inset B) An open cluster, the Jewel Box, containing many young, blue stars.
(Inset C) A typical globular cluster, 47 Tucanae. This true-color image reveals its dominant member stars to be elderly red giants.
(Inset D) A true-color *Hubble* image of part of the Cygnus Loop—a supernova remnant, the remains of a colossal stellar explosion that occurred about 15,000 years ago.

*L*ooking out from Earth on a dark, clear night, we are struck by two aspects of the night sky. The first is a fuzzy band of light—the Milky Way—that stretches across the heavens. From the Northern Hemisphere, it is most easily visible in the summertime, arcing high above the horizon. Its full extent forms a great circle that encompasses the entire celestial sphere. Away from that glowing band, however, our second impression is that the nighttime sky seems more or less the same in all directions. Bunches of stars cluster here and there, but overall, apart from the Milky Way itself, the evening sky looks pretty uniform.

Yet this is only a local impression. Ours is a rather provincial view. When we consider much larger volumes of space, with dimensions far, far greater than the distances between neighboring stars, the spread of stellar and interstellar matter changes. It first becomes patchy and irregular; then, on still larger scales, new structure becomes apparent. Eventually, matter is spread so thin that stars, gas, and dust are all virtually nonexistent. We have now left the Milky Way Galaxy and entered intergalactic space—the vast, dark ocean that separates our own "island" of light and matter from its distant neighbors.

23.1 Our Parent Galaxy

1 A **galaxy** is a gargantuan collection of stellar and interstellar matter— stars, gas, dust, brown dwarfs, black holes—isolated in space and held together by its own gravity. A fairly typical galaxy—not the one we inhabit, but another, similar system—is shown in Figure 23.1. This is the Andromeda Galaxy, lying about 700 kpc (700 kiloparsecs—700,000 pc, about 2 million light years) from the Earth. We have encountered it several times already in this text. Despite its enormous distance, Andromeda is the nearest major galaxy to our own. We show it here in place of our own be-

cause our home—the Milky Way Galaxy, or just "the Galaxy," with a capital G—is too large for us to see in its entirety or to photograph. We live inside it, and we cannot step outside to take a snapshot.

Andromeda's apparent elongated shape is just a consequence of the angle at which we happen to view it. In fact, this galaxy, like our own, consists of a flattened, circular **galactic disk** of matter that fattens to a **galactic bulge** at the center. The disk and bulge are embedded in a roughly spherical ball of faint old stars known as the **galactic halo**. These three basic galactic regions are indicated in Figure 23.1 (although the halo stars are so faint that they cannot be discerned in this image). Armed with this knowledge, we

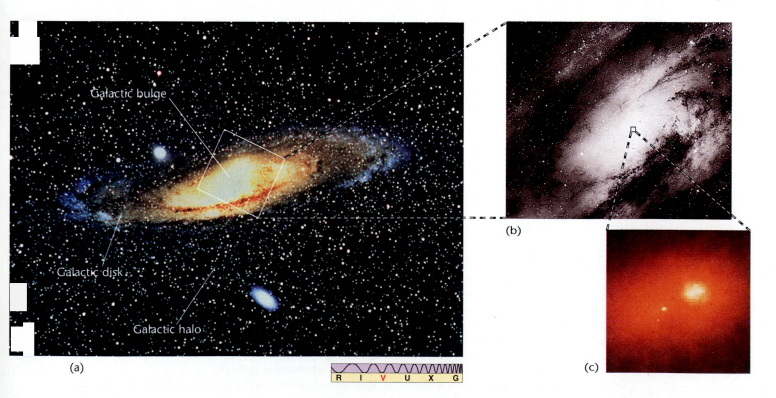

(a)

(b)

(c)

R I V U X G

Figure 23.1   (a) The Andromeda Galaxy probably resembles fairly closely the overall layout of our own Milky Way Galaxy. The disk and bulge are clearly visible in this image, which is about 30,000 pc across. (b) More detail within the inner parts of the galaxy. (c) The galaxy's peculiar—and still unexplained—double core; this inset covers a region only 15 pc across.

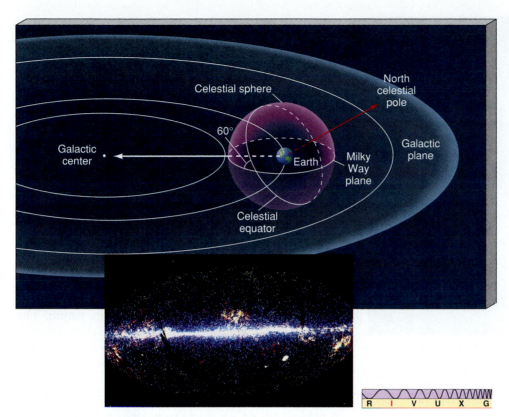

Figure 23.2   Seen from within, the flattened disk of our Galaxy appears as a band of light across the heavens, known as the Milky Way. When we gaze at the Milky Way, we are looking in the plane of our Galaxy's disk; in other directions, our line of sight is out of the plane. The plane of the celestial equator is inclined at an angle of about 60° to the Galactic plane. The inset is (mostly) the Milky Way's plane as detected by *IRAS,* the *Infrared Astronomy Satellite.*

can begin to understand and interpret the appearance of our own Galaxy from Earth, as illustrated in Figure 23.2. From our perspective within it, the Galactic disk is seen as a band of light stretching across the sky—the Milky Way.

Deciphering the structure of the Milky Way Galaxy from Earth is a difficult task—a little like trying to unravel the layout of paths, bushes, and trees in a city park without being able to leave one particular park bench. In some directions, the interpretation of what we see is ambiguous and inconclusive. In others, foreground objects completely obscure our view of what lies beyond, but we cannot move around them to get a better look. As a result, astronomers who study the Milky Way are often guided in their efforts by comparisons with more distant, but more easily observable, systems. In this chapter, we will study the various parts that make up our parent Galaxy and see some of the methods used by astronomers in piecing together this picture.

23.2 *Spiral Nebulae and Island Universes*

The brief description we have just given represents the modern picture of our Galaxy. But before the early part of the twentieth century, astronomers had a markedly different view of the stellar system we inhabit. The fact that we live in just one of many enormous "islands" of matter separated by even larger tracts of apparently empty space was completely unknown, and the clear distinction between "our Galaxy" and "the universe" simply did not exist. Like the Copernican revolution before them, the twin ideas—

that the Sun is not at the center of the Galaxy and that the Galaxy is not the center of the universe—required both time and hard observational evidence before they gained widespread acceptance.

Nineteenth-century astronomers were greatly hampered in their efforts to probe and understand the Galaxy by their inability to determine reliable distances to astronomical objects. Prior to the discovery of the main sequence in 1911 and the development of the distance-measurement technique now known as spectroscopic parallax, the locations of objects lying more than a few hundred parsecs from Earth were all but unknown. ∞ (p. 371)

As an example of the difficulties that resulted from the lack of a reliable distance scale, let's briefly reconsider the Andromeda Galaxy of Figure 23.1, or the Andromeda *spiral nebula,* as it was known in the midnineteenth century. With the observational techniques available at the time, it appeared only as a rather fuzzy, indistinct patch of light in the sky, with hints of swirling, spiral structure in it. Astronomers had no means of determining its distance, and they simply assumed that Andromeda and the many other known spiral nebulae were located in our Galaxy and were perhaps somehow similar to emission nebulae. Figure 23.3 shows a face-on view of another such system, in which the spiral structure is more plainly evident.

By the end of the nineteenth century, improved telescopes and photographic techniques had allowed astronomers to obtain much better images, showing detail comparable to Figure 23.1. However, the "nebula's" distance was still unknown. In 1888, when such images were first presented publicly, they caused great excitement among as-

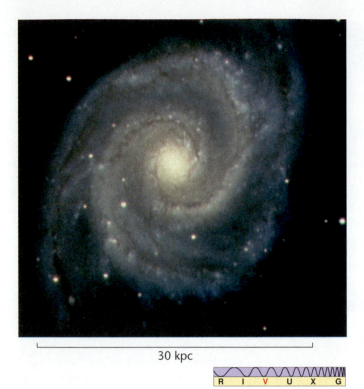

30 kpc

R I V U X G

Figure 23.3 This spiral galaxy, seen nearly face-on, is fairly similar in its overall structure to our own Milky Way Galaxy and Andromeda. It is commonly known as M51—the 51st object in the Messier catalog.

tronomers, who thought they were seeing the formation of a star from a swirling gaseous disk! Comparing Figure 23.1 with the figures in Chapter 15 (see especially Figure 15.5), we can perhaps understand how such a mistake could be made—*if* we believed we were looking at a relatively close, star-sized object. Far from demonstrating that Andromeda was distant and large, the observations seemed to confirm that it was just a small part of our own Galaxy.

Further observations soon made it clear that Andromeda is not a star-forming region. For example, Andromeda's parallax is too small to measure, indicating that it must be at least several hundred parsecs from Earth. Even at 100 pc (which we now know is vastly less than Andromeda's true distance), an object the size of the solar nebula would be impossible to resolve and simply would not look like Figure 23.1.

By the early 1900s, the questions of the size of our Galaxy and the distance to Andromeda and the other spiral nebulae were being hotly debated. One school of thought maintained that they were nebulae much smaller than, and contained within, our own Galaxy. Other astronomers held that they were "island universes" outside the Milky Way Galaxy and comparable to it in size. However, with no firm distance information, both arguments were quite inconclusive. It was not until the late 1920s, with the discovery of the next rung in our cosmic distance "ladder"—the topic of the next section—that the spirals were finally shown to lie well outside our own Galaxy. The issue was finally settled in favor of the island–universe theory.

The "spiral nebulae" are known today as **spiral galaxies**. The characteristic pinwheel-like structures that give a spiral galaxy its name are called **spiral arms**. Each arm originates close to the central bulge of the parent galaxy and extends outward throughout much of the galactic disk. Our own Milky Way Galaxy is also of the spiral type, although our location within the Galactic disk makes the spiral structure difficult to discern from the Earth.

23.3 *The Structure of the Milky Way Galaxy*

Studies of nearby stars, gas, and dust, along with comparisons with other galaxies such as Andromeda, can give us a general idea of the overall distribution of matter in our Galaxy. But the story of the spiral nebulae illustrates just how important it is for us to know the distance to an astronomical object before we can determine its nature. The growth in our knowledge of the Galaxy, and the realization that there are many other distant galaxies similar to our own, have gone hand in hand with the development of the cosmic distance scale.

VARIABLE STARS AS DISTANCE INDICATORS

2 One by-product of the laborious effort to catalog stars around the turn of the twentieth century was the systematic study of **variable stars**. These are stars whose luminosity changes with time—some quite erratically, others more regularly. Only a small fraction of stars fall into this category, but those that do are of great astronomical significance.

We have encountered several examples of variable stars in earlier chapters. Often, the variability is the result of membership in a binary system. Eclipsing binaries, novae, and Type-I supernovae are cases in point. Novae and supernovae collectively are called *cataclysmic variables* because of their sudden, large changes in brightness. In other instances, however, the variability is a basic trait of the star itself, not dependant on the star being a part of a binary system. We call such a star *intrinsically* variable.

With regard to Galactic structure, the most important intrinsic variables are the **pulsating variable stars** (which have nothing to do with pulsars, by the way). The luminosity of a pulsating variable star varies in a smooth and predictable way. Figure 23.4 shows the light curve of one such star, whose brightness rises and falls by about a factor of 2 every 3 days. This type of pulsating star is a **Cepheid variable** (or simply a "Cepheid"), after the first star of the type to be discovered—Delta Cephei, the fourth brightest star in the constellation Cepheus. Cepheid variables are recognizable by the characteristic shape of their light curves (the rapid rise followed by the slower decline in Figure 23.4b). Different Cepheids have different pulsation periods, ranging from about 1 to 100 days, but the period of any given Cepheid is essentially constant from one cycle to the next.

A related class of pulsating variables is the **RR Lyrae variable** stars (again named after the first known of their kind, in this case the variable star labeled RR in the constellation Lyra). Like the Cepheids, they have a characteristic, easily recognizable light curve (shown in Figure 23.5) that accurately repeats itself. Unlike Cepheids, however, there is little difference in period between one RR Lyrae variable and another.

Why do Cepheids and RR Lyrae variables pulsate? The basic mechanism was first suggested by the British astrophysicist Sir Arthur Eddington in 1941. The structure of any star is determined in large part by how easily radiation can travel from the core to the photosphere—that is, by the *opacity* of the interior, the degree to which the gas hinders the passage of light through it. If the opacity rises, the radiation becomes trapped, the internal pressure increases, and the star "puffs up." If the opacity falls, radiation can escape more easily, and the star shrinks. According to theory, under certain circumstances, a star can become unbalanced and enter a state in which the flow of radiation causes the opacity first to rise—making the star expand, cool, and diminish in luminosity—and then to fall, leading to the pulsations we observe.

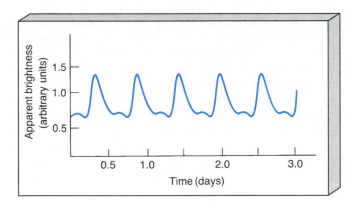

Figure 23.5 Light curve of the pulsating variable star RR Lyrae. All RR Lyrae-type variables have essentially similar light curves, with a period of less than a day.

The conditions necessary to cause pulsations are not found in main-sequence stars. However, they *do* occur in more evolved stars as they pass through a region of the Hertzsprung–Russell diagram known as the *instability strip*, shown in Figure 23.6. When a star's temperature and luminosity place it in this strip, the star becomes internally unstable. Both its temperature and its radius pulsate in a regular way, causing the variability we observe. High-mass stars

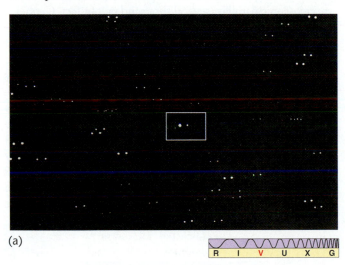

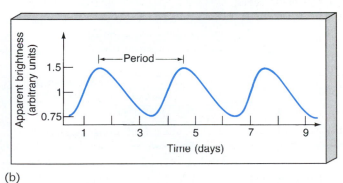

(b)

Figure 23.4 (a) A Cepheid variable star (boxed) on successive nights; two photos, one from each night, have been nearly superimposed. The star is called WW Cygni and is shown at its maximum and minimum brightness. (b) Monitoring the star over the course of a week or so yields a record of the star's changing brightness.

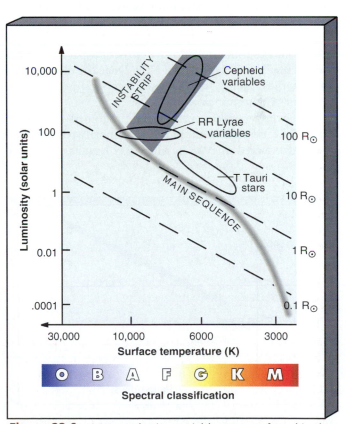

Figure 23.6 Many pulsating variable stars are found in the so-called instability strip of the H–R diagram. As a high-mass star evolves through the strip, it becomes a Cepheid variable. Low-mass horizontal-branch stars in the instability strip are RR Lyrae variables.

evolve across the upper part of the H–R diagram; when their evolutionary tracks take them into the instability strip, they become Cepheid variables. RR Lyrae variables are low-mass horizontal-branch stars that lie in the lower portion of the instability strip. Similar unstable conditions can occur within stars in other regions of the H–R diagram, particularly on the Hayashi track and the red-giant branch, and other classes of variable stars are indeed associated with such objects. However, we will confine ourselves here to the Cepheids and RR Lyrae variables just discussed. Pulsating variable stars, then, are not a special class of object. They are just normal stars experiencing a brief period of instability as a natural part of stellar evolution.

A NEW YARDSTICK

Although Cepheids and other variable stars are interesting objects in their own right, our primary goal here is to use them to obtain information about the large-scale distribution of stars in our Galaxy. The key point about Cepheid variables is this: Their *absolute* brightnesses (averaged over a complete pulsation cycle) and their pulsation periods are rather tightly connected. Cepheids that vary slowly—that is, that have long periods—have large absolute brightnesses. The converse is also true: Short-period Cepheids have small absolute brightnesses. Figure 23.7 illustrates this relationship for Cepheids found within a thousand parsecs or so of the Earth. Astronomers can plot such a diagram for relatively nearby stars because they can measure their distances using stellar or spectroscopic parallax, as described in Chapter 17. ∞ (p. 371) Once the distances are known, the absolute brightnesses (luminosities) of these stars can be determined.

This link between period and brightness, shown in Figure 23.7, is known as the **period–luminosity relationship**. It was discovered in 1908 by Henrietta Leavitt of Harvard

University (see *Interlude 23-1*, on p. 495). We know of no exceptions, and this relationship is consistent with theoretical calculations of pulsations in evolved stars. Consequently, we assume that it holds for all Cepheids, near and far. The roughly constant period of the RR Lyrae variables is also marked in Figure 23.7.

The beauty of the period–luminosity relationship is that a simple measurement of a Cepheid variable's pulsation period immediately tells us its luminosity—we just read it off the plot in Figure 23.7. The luminosity of an RR Lyrae star is even easier to determine. All such stars have basically the same absolute brightness—about 100 times that of the Sun—so once a variable star is recognized as being of the RR Lyrae type, its luminosity is known. In either case, comparing absolute and apparent brightnesses gives us an estimate of the star's distance, as described in Chapter 17. ∞ (p. 363)

Variable stars make up an important new rung in our cosmic distance ladder. The technique works well provided the star can be clearly identified and its period of variability measured. With Cepheids, this method allows astronomers to estimate distances out to several million parsecs, well beyond the range of either stellar parallax (around 100 pc) or spectroscopic parallax (several thousand pc). In fact, Cepheids can take us all the way to the nearest galaxies. Being less luminous, RR Lyrae stars are not as easily seen as Cepheids, so their useful range is not as great. However, they are much more common, so within their limited range, they are actually more useful than Cepheids.

Figure 23.8 extends our cosmic distance ladder, which we began in Chapter 1, to include Cepheid variables as a fourth method of determining distance. Because the period–luminosity relationship is calibrated using nearby stars, this latest rung inherits any and all uncertainties and errors present in the lower levels. Uncertainties also arise from the "scatter" shown in Figure 23.7. Although the

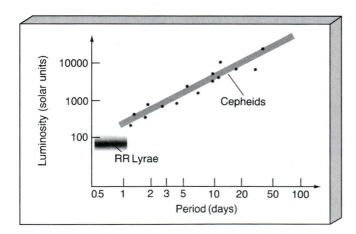

Figure 23.7   A plot of pulsation period versus average absolute brightness (that is, luminosity) for a group of Cepheid variable stars. The two properties are quite tightly linked. The pulsation periods of some RR Lyrae variables are also shown.

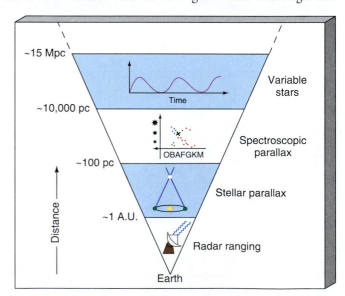

Figure 23.8   Application of the period–luminosity relationship for Cepheid variable stars allows us to determine distances out to about 15 Mpc with reasonable accuracy.

overall connection between period and luminosity is unmistakable, the individual data points do not quite lie on a straight line. Instead, there is a range of possible luminosities corresponding to any measured period.

THE SIZE AND SHAPE OF OUR GALAXY

Many variable stars—specifically, stars of the RR Lyrae type—are found in globular clusters, those tightly bound swarms of old, reddish stars that we first met in Chapter 17. ∞ (p. 377) Approximately 140 globular clusters are visible from Earth. Early in the twentieth century, the American astronomer Harlow Shapley used observations of RR Lyrae stars to make two very important discoveries. First, he proved that most globular clusters reside at great distances—many thousands of parsecs—from the Sun. Second, by measuring the directions and distances to each, he was able to determine their three-dimensional distribution in space. In this way, Shapley demonstrated that the globular clusters map out a truly enormous, and roughly *spherical*, volume of space, about 30 kpc across. But the center of the sphere lies nowhere near our Sun—in fact, it is located nearly 8 kpc away from us in the direction of the constellation Sagittarius.

In a brilliant intellectual leap, Shapley realized that this 30-kpc-wide distribution of globular clusters maps out the true extent of stars in the Milky Way Galaxy—the region that we now call the Galactic halo. The hub of this vast collection of matter, about 8 kpc from the Sun, is the **Galactic center**. As illustrated in Figure 23.9, we live in the suburbs of this truly huge ensemble of matter, in the Galactic disk (or *Galactic plane*)—the thin sheet of young stars, gas, and dust that cuts through the center of the halo. Notice, incidentally, the jump in the scale of our units. When talking about stars and "nearby" nebulae, we generally measured distances in parsecs. Now, on a galactic scale, the kiloparsec (kpc) is more appropriate. Soon, as we leave our Galaxy behind, megaparsecs (Mpc) will become the norm.

Shapley's bold interpretation of the globular clusters as defining the overall distribution of stars in our Galaxy was an enormous step forward in human understanding of our place in the universe. Five hundred years ago, Earth was considered the center of all things. Copernicus argued otherwise, demoting our planet to an undistinguished place removed from the center of the solar system. Yet even in the early twentieth century, the prevailing view was that our Sun was the center of not only the Galaxy, but also of the universe. In addition, for reasons that we will discuss in a moment, it was also believed that our Galaxy measured only a few kiloparsecs across. Shapley showed otherwise. With his observations of globular clusters, he simultaneously increased the size of our Galaxy by almost a factor of 10 over earlier estimates and banished our parent Sun to its periphery, virtually overnight!

Curiously, Shapley's dramatic revision of the size of the Milky Way Galaxy and our place in it strengthened his erroneous opinion that the spiral nebulae were part of our Galaxy and that our Galaxy was essentially the entire universe. He regarded as simply beyond belief the idea that there could be other structures as large as our Galaxy. Only in the late 1920s was the Copernican principle extended to the Galaxy itself, when American astronomer Edwin Hubble observed Cepheids in the Andromeda Galaxy and finally succeeded in measuring its distance.

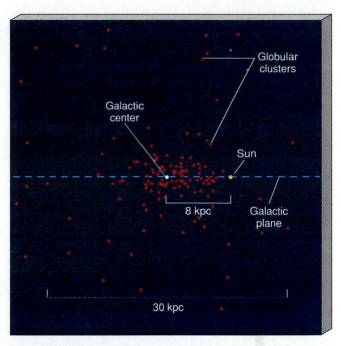

Figure 23.9 Our Sun does not coincide with the center of the very large collection of globular clusters. Instead, more globular clusters are found in one direction than in any other. The Sun resides closer to the edge of the collection, which measures some 30 kpc across. We now know that the globular clusters outline the true distribution of stars in the Galactic halo.

23.4 Disk and Halo Stars

❸ Since Shapley's time, astronomers have identified many individual stars—that is, stars not belonging to globular clusters—in the Galactic halo. The Galactic disk and halo are now recognized to be quite distinct components of our Galaxy, with very different properties. The *spatial distribution* of halo stars is unlike that of stars in the disk—material in the disk is confined to a highly flattened region in the Galactic plane that is thin compared with its diameter, whereas the halo is roughly spherical.[*] But aside from overall shape, the halo has other properties that set it apart from the disk. For one thing, the halo contains essentially *no* gas or dust—just the opposite of the disk, in which interstellar matter is common. For another, there are clear differences in both *appearance and composition* between halo and disk stars.

---

[*] *Actually, there is growing evidence that the halo is somewhat flattened in the direction perpendicular to the disk, but the degree of flattening is quite uncertain. The halo, however, is certainly much less flattened than the disk.*

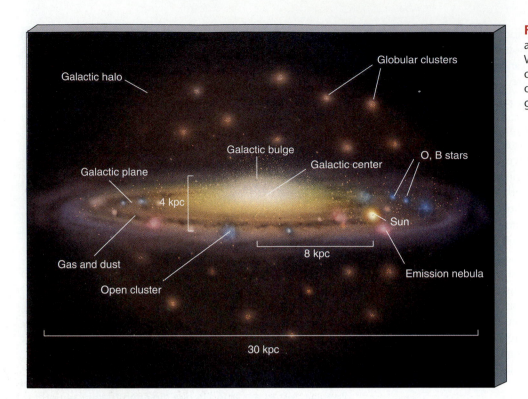

Figure 23.10 Edge-on view of a spiral galaxy, like our own Milky Way or Andromeda, showing the distributions of young blue stars, open clusters, old red stars, and globular clusters.

Stars in the Galactic bulge and halo appear distinctly *redder* than stars found in the disk. We cannot see the big picture of our own Galaxy because of our location within the disk, but careful studies of individual stars near the Sun reveal the trend. Observations of other spiral galaxies show it clearly. In distant galaxies, only the brightest stars can be distinguished as individuals—the remainder merge into a continuous blur of light—but we can say that, *on average,* their disk stars tend to be bluer in appearance than stars in their bulges (and halos too, although they are much harder to see). This difference in appearance—the bluish tint of the disk and the yellowish or reddish coloration of the bulge—is also evident in Figures 23.1 and 23.3.

Figure 23.10 sketches this twofold distribution for our Galaxy. The local blue stars, observable to distances of a few thousand parsecs, are generally confined to the Galactic plane, as are the young open star clusters and star-forming regions. In contrast, the redder stars—including those found in the old globular clusters—are more uniformly distributed throughout the disk, bulge, and halo. From a distance, our Galactic disk would appear bluish simply because main-sequence O- and B-type blue supergiants are very much brighter than G, K, and M dwarfs, even though the dwarfs dominate the total number.

The accepted explanation for this marked difference in color between the disk and the halo is this: Whereas the gas-rich Galactic disk is the site of ongoing star formation and so contains stars of all ages, all the stars in the Galactic halo are *old.* The absence of dust and gas in the halo means that no new stars are forming there, and star formation apparently ceased long ago—at least 10 billion years in the past, judging from the types of halo stars we now observe. (Recall from Chapter 20 that most globular clusters are thought to be between 12 and 15 billion years old. ∞ (p. 435)

Support for this scenario comes from studies of the spectra of halo stars, which indicate that these stars are far less abundant in heavy elements (that is, elements heavier than helium) than are stars in the disk. In Chapters 20 and 21 we saw how each successive cycle of star formation and evolution enriches the interstellar medium with the products of stellar nucleosynthesis, leading to a steady increase in heavy elements with time. ∞ (p. 459) Thus, the scarcity of these elements in halo stars compared with stars in the disk is consistent with the view that the halo formed long ago. We will discuss the reasons for this disparity between the disk and the halo in a moment, when we study the formation of our Galaxy.

Astronomers often refer to young disk stars as *Population I* stars and old disk stars as *Population II* stars. The idea of two stellar "populations" dates back to the 1930s, when the differences between disk and halo stars first became clear. It represents something of an oversimplification, as there is actually a continuous variation in stellar ages throughout the Milky Way Galaxy, not a simple division of stars into two distinct "young" and "old" categories. Nevertheless, the terminology is still widely used.

23.5 *The Galactic Disk*

OPTICAL OBSERVATIONS

In the late eighteenth century, long before the distances to any stars were known, the English astronomer William Herschel tried to estimate the size and shape of our Galaxy

Interlude 23–1   *Early Computers*

A large portion of the early research in observational astronomy focused on monitoring stellar luminosities and analyzing stellar spectra. Much of this pioneering work was done using photographic methods. What is not so well known is that most of the labor was accomplished by women. Around the turn of the century, a few dozen dedicated women—assistants at the Harvard College Observatory—created an enormous data base by observing, sorting, measuring, and cataloging photographic information that helped form the foundation of modern astronomy. Some of them went far beyond their duties in the lab to make several of the basic astronomical discoveries often taken for granted today.

This 1910 photograph shows several of those women carefully examining star images and measuring variations in luminosity or wavelengths of spectral lines. In the cramped quarters of the Harvard Observatory, these women inspected image after image to collect a vast body of data on hundreds of thousands of stars. Note the plot of stellar luminosity changes pasted on the wall at the left. The cyclical pattern is so regular that it likely belongs to a Cepheid variable. Known as "computers" (there were no electronic devices then), these women were paid 25 cents an hour.

Beginning in 1880, these workers started a survey of the skies that would be carried on for a century. Their first major accomplishment was a catalog of the brightnesses and spectra of tens of thousands of stars, published in 1890 under the direction of Williamina Fleming. On the basis of this compilation, several of these women made fundamental contributions

to astronomy. In 1897, Antonia Maury undertook the most detailed study of stellar spectra to that time, enabling Hertzsprung and Russell independently to develop what is now called the H–R diagram. In 1898, Annie Cannon proposed the spectral classification system (described in Chapter 17) that is now the international standard for categorizing stars. ∞ (p. 367) In 1908, Henrietta Leavitt discovered the period–luminosity relationship for Cepheid variable stars, which later allowed astronomers to recognize our Sun's true position in our Galaxy, as well as our Galaxy's true place in the universe.

---

simply by counting how many stars he could see in different directions in the sky. Assuming that all stars were of about equal brightness, he concluded that the Galaxy was a roughly lozenge-shaped collection of stars lying in the plane of the Milky Way, with the Sun at the center. Subsequent refinements to this approach led to essentially the same picture and some people went so far as to estimate the dimensions of the "Galaxy" as about 10 kpc in diameter by 2 kpc thick.

As we have just seen, modern astronomers hold a very different view—the Milky Way is now known to be several tens of kiloparsecs across, and the Sun lies nowhere near the center. How could the older estimate have been so flawed? The answer is that the earlier observations were made in the visible part of the electromagnetic spectrum, and astronomers failed to take into account the (then unknown) *absorption* of visible light by interstellar gas and dust. ∞ (p. 382) Only in the 1930s did astronomers begin to realize the true extent and importance of the interstellar medium.

Objects more than a few kiloparsecs away in the plane of the Milky Way are hidden from our view by the effects of interstellar absorption. The central regions of our Galaxy cannot be studied by optical techniques. The ap-

parent fall-off in the density of stars with distance in the plane of the Milky Way is thus not a real thinning of their numbers in space but simply a consequence of the rather murky environment in the Galactic disk. Because the obscuration occurs in all directions in the disk, the fall-off is roughly similar no matter which way we look, and so the Sun appears to be at the center.

In contrast, radiation coming to us from above or below the plane of the Galaxy, where there is less gas and dust along the line of sight, arrives on the Earth relatively unscathed. There is still some patchy obscuration, but the Sun happens to lie in a location where the view out of the disk is largely unimpeded by nearby interstellar clouds. Therefore, optical measurements of the disk's thickness are more accurate than are the measurements of the disk's width. Also, the observed fall-off in the density of stars above and below the Galactic plane is real.

LOCAL STRUCTURE

On the basis of optical, infrared, and radio observations of the many different types of objects—stars, gas, and dust—found within a thousand or so kiloparsecs of the Sun, astronomers have built up a fairly detailed picture of our local neighborhood. At the location of the Sun, some 8 kpc from the center, the Galactic disk is relatively thin—perhaps 300 pc thick, or about 1/100 of its diameter. Don't be fooled, though. Even if you could travel at the speed of light, it would take you 1000 years to traverse the thickness of the Galactic plane. The disk may be thin compared with the diameter of the Galaxy, but it is huge by human standards.

Actually, the thickness of the Galactic disk depends on the kinds of objects you measure. Young stars and interstellar gas are more tightly confined to the plane than are stars such as the Sun, and solar-type stars in turn are more tightly confined than are older K- and M-type dwarfs. Why? Generally, stars form in interstellar clouds close to the disk plane but then tend to drift out of the disk over time, due to their interactions with other stars and molecular clouds. Thus, as stars age, their abundance above and below the disk plane slowly increases. Note that these considerations do not apply to the Galactic halo, whose ancient stars and globular clusters extend far above and below the Galactic plane. The halo is a remnant of an early stage of our Galaxy's evolution and predates the formation of the disk.

Recently, improved observational techniques have revealed an intermediate category of Galactic stars, midway between the old halo and the younger disk, both in age and in spatial distribution. Consisting of stars with estimated ages in the range of 7–10 billion years, this so-called **thick disk** component of the Milky Way Galaxy measures some 2–3 kpc from top to bottom. Its thickness is too great to be explained by the slow drift just described. Like the halo, it appears to be a vestige of our Galaxy's distant past.

EXPLORING GALACTIC GAS

4 If we want to look beyond our immediate neighborhood and study the full extent of the Galactic disk, we cannot rely on optical observations alone. Interstellar absorption severely limits our vision. In the 1950s, astronomers developed a very important tool needed to explore the true spread of gas in our Galaxy—spectroscopic radio astronomy.

The key to observing Galactic interstellar gas is the 21-cm radio emission line produced by atomic hydrogen. ∞ (p. 393) This long-wavelength radiation is largely unaffected by interstellar dust. Thus, it travels more or less unimpeded through the Galactic disk, allowing us to "see" to great distances in this part of the electromagnetic spectrum. In addition, because hydrogen is by far the most abundant element in the interstellar medium, the 21-cm signals are strong enough that a large portion of the disk can be observed in this way. By studying the 21-cm radio lines emitted by atomic hydrogen gas throughout the Galaxy, radio astronomers have mapped out the large-scale distribution and motion of the Galaxy's interstellar clouds.

The 21-cm technique works well for regions abundant in *atomic* gas. But as noted in Chapter 18, *molecular* clouds are also widespread throughout the Galaxy. ∞ (p. 394)

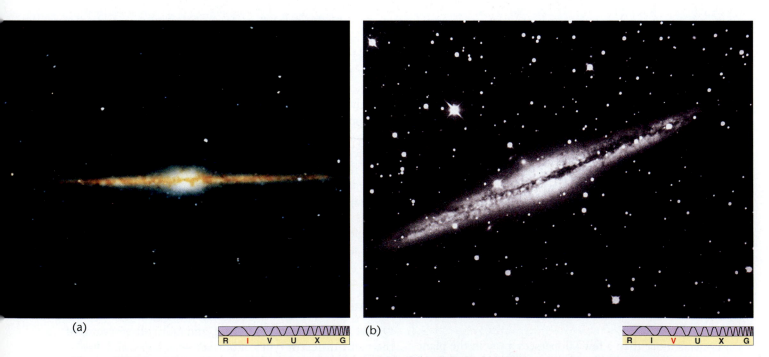

(a)   R I V U X G   (b)   R I V U X G

Figure 23.11 (a) A wide-angle infrared image of the plane and bulge of our Milky Way, as observed by the *Cosmic Background Explorer* (*COBE*) satellite. (b) An image of the galaxy NGC 891, whose shape is believed to be quite similar to the Milky Way's.

Because these clouds contain little or no atomic hydrogen, they are generally not detectable by their 21-cm emission. The problem of mapping molecular clouds is made all the more difficult because molecular hydrogen—the clouds' main constituent—is itself quite difficult to detect. Its strongest spectral lines just happen to lie in parts of the radio spectrum that are hard to observe with current radio telescopes.

Fortunately, to study the molecular gas in the Galaxy, we can use other molecules that do emit radio radiation. The *carbon monoxide* molecule is often used for this purpose, mainly because it is abundant in all molecular clouds (although it is still a million times less abundant than molecular hydrogen). Also, one of its characteristic spectral lines happens to occur in a part of the radio domain (at a wavelength of about 3 mm) that is easy to observe. Thus, although carbon monoxide accounts for only a tiny fraction of the matter in molecular clouds, it is a very useful tracer of molecular hydrogen. By measuring atomic hydrogen and carbon monoxide, we can probe most of the disk of the Milky Way.

RADIO MAPS OF THE MILKY WAY

The interstellar clouds in the Galactic disk exhibit an organized pattern on a grand scale. According to radio studies, the center of the gas distribution coincides roughly with that of the globular-cluster system, about 8 kpc from the Sun. (In fact, this figure of 8 kpc is derived most accurately from radio observations of the Galactic gas, whose center is normally taken to define the center of our Galaxy.) Radio-emitting gas has been observed out to at least 40–50 kpc from the Galactic center. Over much of the inner 20 kpc or so of the disk, the gas is confined quite closely to the Galactic plane. Beyond that, the gas distribution spreads out somewhat, to a thickness of several kiloparsecs, and shows definite signs of being "warped," possibly because of the gravitational influence of a pair of nearby galaxies (to be discussed in Chapter 24; see also Figure 23.12).

Near the center, the gas in the disk fattens markedly in the *Galactic bulge*. The high gas density in the inner part of the bulge makes it the site of vigorous ongoing star formation, and both very old and very young stars mingle there. The bulge is flattened in shape, measuring some 6 kpc across in the plane of the disk, but only about 4 kpc from top to bottom. Recent detailed studies of the motion of gas and stars in and near the bulge suggest that it may really be football-shaped, with the long axis of the football lying in the Galactic plane (see Figure 23.10).

Figure 23.11 shows two real images—(a) an infrared view of our own Galaxy and (b) an optical view of a distant galaxy like our own—that reinforce our belief in the correctness of this Galactic model.

Radio studies provide perhaps the best direct evidence that we really do live in a spiral galaxy. As shown in Figure 23.12, they show clear evidence that our Galaxy has *spiral*

30 kpc

Figure 23.12 An artist's conception of our Milky Way Galaxy seen face-on. This illustration is based on data accumulated by legions of astronomers during the past few decades, including radio maps of gas density in the Galactic disk. Painted from the perspective of an observer 100 kpc above the Galactic plane, the spiral arms are at their best-determined positions. All the features are drawn to scale (except for the oversized yellow dot near the top, which represents our Sun). The two small blotches to the left are dwarf galaxies, called the Magellanic Clouds. We will study them in Chapter 24.

arms. One of these arms, as best we can tell, wraps around a large part of the entire disk and contains our Sun. The 30-kpc diameter marked on Figure 23.10 indicates the approximate extent of both the luminous "stellar" component of our Galaxy and of the known spiral structure—about the same as the diameter of the Galactic globular-cluster distribution.

23.6 *Galactic Dynamics*

Now let's turn our attention to the *dynamics* of the Milky Way Galaxy—that is, to the motion of the stars and gas it contains. Are the internal motions of our Galaxy's members chaotic and random, or are they part of some gigantic "traffic pattern"? The answer depends on our perspective. The motion of stars and clouds we see on small scales (within a few tens of parsecs of the Sun) seems random, but on larger scales (hundreds or thousands of parsecs) the motion appears much more orderly.

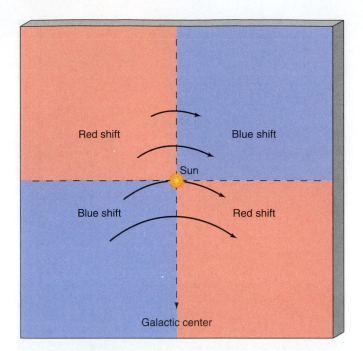

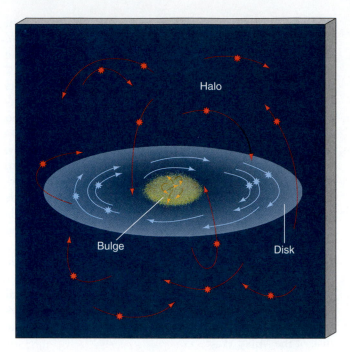

Figure 23.13   Diagram of the four Galactic quadrants in which stars and interstellar clouds show systematic Doppler motions. This information tells us that the disk of the Galaxy is spinning in a well-ordered way. These quadrants are drawn (as dashed lines) to intersect at the Sun, not at the Galactic center, because it is from the viewpoint of our own planetary system that the observations are made. The longer the arrow, the greater the angular speed of the disk material.

Figure 23.14   Stars in the Galactic disk move in orderly, circular orbits about the Galactic center. In contrast, halo stars have orbits with largely random orientations and eccentricities. The orbit of a typical halo star takes it high above the Galactic plane, through the disk, then far below the plane on the other side of the Galaxy.

DISK AND HALO ORBITS

5 We see the gas and stars in the Galactic disk only from our vantage point on Earth as we orbit the Sun. Yet, as we look around the Galactic disk in different directions, a clear pattern of motion emerges, as summarized in Figure 23.13. The spectral lines emitted from Galactic material—both nearby stars (on average) and the more widely distributed interstellar gas—in the upper right quadrant and the lower left quadrant of Figure 23.13 are *blueshifted*. At the same time, the interstellar regions sampled in the upper left quadrant and the lower right quadrant are *redshifted*. In short, some regions (the blueshifted directions) in the Galaxy are approaching the Sun, whereas others (the redshifted ones) are receding from us. The important point is that they are moving in a systematic fashion.

Careful study of this fourfold pattern of Doppler-shifted stars and gas leads to the following important conclusion: The entire Galactic disk is *rotating* about the Galactic center. In the vicinity of the Sun, the orbital speed is about 220 km/s. At the Sun's distance of 8 kpc from the Galactic center, material takes about 225 million years—an interval of time sometimes known as 1 *Galactic year*—to complete one circuit.

Furthermore, the disk does not rotate at a uniform rate. Rather, it spins *differentially*—that is, stars and gas at different distances from the Galactic center take different lengths of time to complete one orbit, as we would expect

from Kepler's laws. ∞ (p. 41)   (Contrast this with *solid-body rotation*, in which every piece of the disk would move with the same angular speed and so would have the same orbital period, like a record spinning on a turntable.) Radio observations demonstrate that the inner regions of the Galactic disk take much less time to orbit the Galactic center than do the outer parts. Similar differential rotations are observed in Andromeda (and, in fact, in all other spiral galaxies). Thus stars in the Galactic disk do not move smoothly together but ceaselessly change their positions relative to one another as they orbit the Galactic center.

This picture of orderly circular orbital motion about the Galactic center applies only to the Galactic disk. Stars in the Galactic halo (as well as stars in the thick disk and Galactic bulge) are not so well behaved. The old globular clusters and the faint, reddish individual stars that make up the halo do *not* share the well-defined rotation of the disk. Instead, their orbits are largely random.[*] Although they do orbit the Galactic center, halo objects move in all directions, their paths filling the halo's entire three-dimensional volume. At any given radius, halo stars move at speeds comparable to the disk's rotation speed, but in *every* direction, not just one—their orbits carry them repeatedly through the disk plane and out the other side. Figure 23.14 illustrates

*In fact, the stars that comprise the halo do have some net rotation about the Galactic center, but it is overwhelmed by the larger random component of their motion.

this motion and contrasts it with the much more regular orbits in the disk. Some well-known stars in the vicinity of the Sun—the bright giant Arcturus, for example—are actually halo stars that are "just passing through" the disk on orbits that take them far above and below the Galactic plane.

THE FORMATION OF OUR GALAXY

⑤ There are many differences between the Galactic disk and the Galactic halo; a few of them are listed in Table 23-1. Is there some evolutionary scenario that can naturally account for the present-day disk–halo structure we see? The answer is that there is, and it takes us all the way back to the birth of our Galaxy, some 10–15 billion years ago. Not all the details are agreed upon by all astronomers, but the overall picture is now fairly widely accepted.

When the first stars and globular clusters formed, the gas in our Galaxy had not yet accumulated into a thin disk. Instead, it was spread out over a rather irregular, and quite extended, region of space, spanning many tens of kiloparsecs in all directions. When the first stars formed, they were distributed throughout this volume. Their distribution today (in the halo) reflects that fact—it is an imprint of their birth. Many astronomers believe that the very first stars formed even earlier, in smaller systems that later

TABLE 23-1 *Overall Properties of the Galactic Disk and Halo*

GALACTIC DISK	GALACTIC HALO
highly flattened	roughly spherical
contains young and old stars	contains old stars only
contains gas and dust	contains no gas and dust
site of ongoing star formation	no star formation during the last 10 billion years
gas and stars move in circular orbits in the Galactic plane	stars have random orbits in three dimensions
spiral structure	no obvious substructure

merged to create our Galaxy. The present-day halo would look the same in either case. Figure 23.15 illustrates this view of our Galaxy's evolution.

During the past 10–15 billion years, rotation has flattened the gas in our Galaxy into a relatively thin disk, which contains virtually all the gas, dust, and young stars in the Milky Way. Physically, this process is rather similar to the flattening of the solar nebula to a disk during the

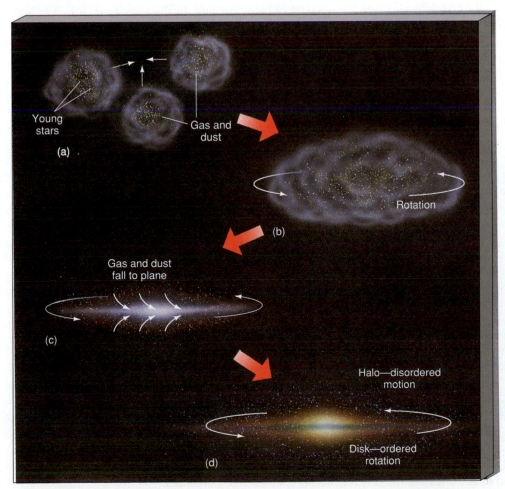

Figure 23.15 Astronomers reason that, early on, our Galaxy was rather irregularly shaped, with gas distributed throughout its volume. Possibly it formed via the merger of several smaller systems, as depicted in (a) and (b). When stars formed during these stages, there was no preferred direction in which they moved and no preferred location in which they were found. In time, rotation caused the gas and dust to fall to the Galactic plane and form a spinning disk, as in (c). The older stars were left behind, forming the halo. (d) New stars forming in the disk inherit its overall rotation and so orbit the Galactic center on ordered, circular orbits.

formation of the solar system, as described in Chapter 15, except on a vastly larger scale. ∞ (p. 320) Star formation in the halo ceased billions of years ago when the raw materials fell to the Galactic plane. Ongoing star formation in the disk gives the plane its bluish tint, but the halo's short-lived blue stars have long since burned out, leaving only the long-lived red stars that give it its characteristic pinkish glow. The Galactic halo is ancient, whereas the disk is full of youthful activity. The thick disk, with its intermediate-age stars, may represent an intermediate stage of star formation that occurred while the gas was still flattening into the plane.

The chaotic orbits of the halo stars are also explained by this theory. When the halo developed, the irregularly shaped Galaxy was rotating only very slowly, so there was no strongly preferred direction in which matter tended to move. As a result, halo stars were free to travel along nearly any path once they formed (or when their parent systems merged). As the Galactic disk formed, however, conservation of angular momentum caused it to spin more rapidly. Stars forming from the gas and dust of the disk inherit its rotational motion and so move on well-defined, circular orbits. Again, the thick disk's properties suggest that it formed while gas was still sinking to the Galaxy's midplane and had not yet reached its final (present-day) rotation rate.

In principle, the structure of our Galaxy bears witness to the conditions that created it. In practice, however, the interpretation of the observations is made difficult by the sheer complexity of the system we inhabit and by the many competing physical processes that have modified its appearance since it formed. As a result, the early stages of the Milky Way are still very poorly understood. We will return to the subject of galaxy formation in Chapter 24.

23.7 *Spiral Structure*

⑥ Studies of spiral structure in our own Galaxy and in others indicate that the spiral arms are made up of much more than just interstellar clouds. Young objects—such as O- and B-type stars, open clusters, and emission nebulae—all reside in the arms, too. In fact, these objects are generally *not* found outside the spiral arms. The obvious conclusion is that the spiral arms are the part of the Galactic disk where star formation actually takes place. The brightness of these young stellar objects is the main reason that the spiral arms of other galaxies are easily seen from afar (see Figure 23.3).

A central problem facing astronomers trying to understand spiral structure is how that structure persists over long periods of time. The basic issue is really very simple: We know that the inner parts of the Galactic disk rotate more rapidly than the outer regions. This *differential rotation* makes it impossible for any large-scale structure "tied" to the disk material to survive.

To understand why differential rotation poses a problem, consider that our Sun, at roughly 8 kpc from the Galactic center, takes about 225 million years to complete one circuit of the Milky Way. Thus our 4.5 billion-year-old solar system has cycled around the Galactic center some 20 times since being formed. In the same time, however, stars lying closer to the Galactic center have completed many more revolutions, while those farther out toward the edge of the disk have made far fewer. The result of this differential motion, illustrated in Figure 23.16, is that a spiral pattern consisting always of the *same* group of stars and gas clouds would necessarily "wind up" and disappear within a few hundred million years. Yet spiral arms clearly do exist in our own galaxy, and their prevalence in other disk galax-

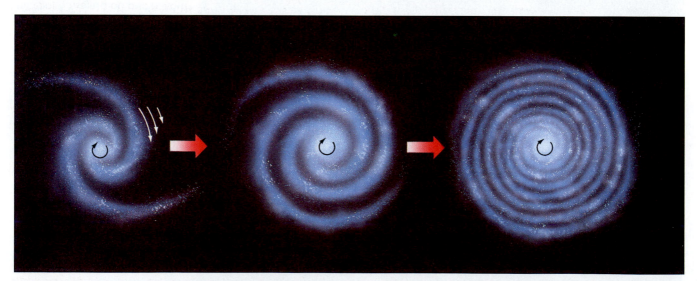

Figure 23.16 This painting illustrates the fact that the disk of our Galaxy rotates differentially— stars close to the center take less time to orbit the Galactic center than those at greater distances. If spiral arms were somehow tied to the material of the Galactic disk, this differential rotation would cause the spiral pattern to wind up and disappear in a few hundred million years. Spiral arms would be too short-lived to be consistent with the numbers of spirals actually observed.

Figure 23.17 Density-wave theory holds that the spiral arms seen in our own and many other galaxies are actually waves of gas compression and star formation moving through the material of the galactic disk. Gas (red arrows) enters the arm (white arrows) from behind, is compressed, and forms stars. The spiral pattern we see in this painting at right is delineated by dust lanes, regions of high gas density, and newly formed O and B stars. The inset photograph at left shows the spiral galaxy NGC 1566, which displays many of the features described in the text. Note, incidentally, that although the figure shows a "two-armed" spiral, astronomers are not completely certain how many arms make up the spiral structure in our own Galaxy (see Figures 23.10 and 23.12). The theory makes no strong predictions on this point.

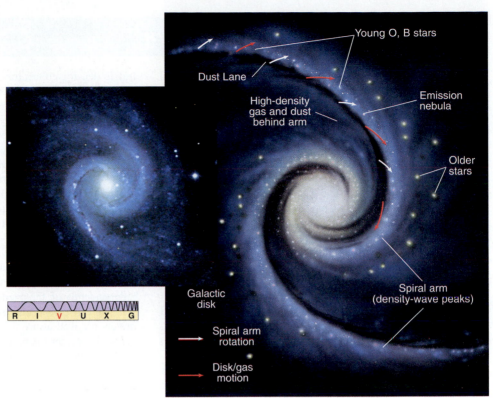

ies suggests that they last for considerably longer than this. Thus, whatever the spiral arms are, they *cannot* simply be dense star-forming regions orbiting along with the rest of the Galactic disk.

How then do the Galaxy's spiral arms retain their structure over long periods of time in spite of differential rotation? A leading explanation for the existence of spiral arms holds that they are **spiral density waves**—coiled waves of gas compression that move through the Galactic disk, squeezing clouds of interstellar gas and triggering the process of star formation as they go. ∞ (p. 411) The spiral arms we observe are outlined by the denser-than-normal clouds of gas they create, and by the new stars formed as a result of the spiral wave's passage.

This explanation of spiral structure avoids the problem of differential rotation because the wave pattern is not tied to any particular piece of the Galactic disk. The spirals we see are merely patterns *moving* through the disk, not great masses of matter being transported from place to place. The density wave moves through the collection of stars and gas comprising the disk just as a sound wave moves through air or an ocean wave passes through water, compressing different parts of the disk at different times. Even though the rotation rate of the disk material varies with its distance from the Galactic center, the wave itself remains intact, defining the Galaxy's spiral arms.

In fact, over much of the inner part of the Galactic disk (within about 15 kpc of the center), the spiral wave pattern is predicted to rotate *more slowly* than the stars and gas. Thus, Galactic material actually catches up with the wave,

is temporarily slowed down and compressed as it passes through, then continues on its way. (For a more down-to-earth example of an analogous process, see *Interlude 23-2* on page 502.) As shown in Figure 23.17, the slowly moving spiral density wave is outrun by the faster rotation of the disk. As gas enters the arm from behind, it is compressed and forms stars. Dust lanes mark the regions of highest-density gas. The most prominent stars—the bright O and B blue giants—live for only a short time, so OB associations, young star clusters, and emission nebulae are found only within the arms, near their birthsites. Their brightness emphasizes the spiral structure. Further downstream, ahead of the spiral arms, we see mostly older stars and star clusters. These have had enough time since their formation to outdistance the wave and pull away from it. Over millions of years, their random individual motions distort and eventually destroy their original spiral configuration, and they become part of the general disk population.

An alternative possibility is that the formation of stars drives the waves, instead of the other way around. Imagine a row of newly formed massive stars somewhere in the disk. As these stars form, the emission nebulae that appear around them send shock waves through the surrounding gas, possibly triggering new star formation. ∞ (p. 412) Similarly, when the stars explode in supernovae, more shocks are formed. ∞ (p. 459) As illustrated in Figure 23.18(a), the formation of one group of stars thus provides the mechanism for the creation of more stars. Computer simulations suggest that it is possible for the "wave" of star formation thus created to take on the form of a partial spi-

Interlude 23–2 *Density Waves*

In the late 1960s, American astrophysicists C. C. Lin and Frank Shu proposed a way in which spiral arms in the Galaxy could persist for many Galactic rotations. They argued that the arms themselves contain no "permanent" matter. They should not be viewed as assemblages of stars, gas, and dust moving intact through the disk—those would quickly be destroyed by differential rotation. Instead, a spiral arm should be envisaged as a *density wave*—a wave of compression and expansion sweeping through the Galaxy.

A wave in water builds up material temporarily in some places (crests) and lets it down in others (troughs). Similarly, as the spiral density wave encounters Galactic matter, the gas is compressed to form a region of slightly higher-than-normal density. Galactic material enters the wave, is temporarily slowed down and compressed as it passes through, then continues on its way. This compression triggers the formation of new stars and nebulae. In this way, the spiral arms are formed and reformed repeatedly, without wrapping up. Lin and Shu showed that the process can in fact maintain a spiral pattern for very long periods of time.

The accompanying figure illustrates the formation of a density wave in a much more familiar context—a traffic jam on a highway, triggered by the presence of a repair crew moving slowly down the road. As cars approach the crew, they slow down temporarily, then speed up again as they pass the worksite and continue on their way. The result, as might be reported by a high-flying traffic helicopter, is a region of high traffic density, concentrated around the location of the work crew and moving with it. An observer on the side of the road, however, sees that the jam never contains the same cars for very long. Cars constantly catch up to the bottleneck, move slowly through it, then speed up again, only to be replaced by more cars arriving from behind.

The traffic jam is analogous to the region of high stellar density in a Galactic spiral arm. Just as the traffic density wave is not tied to any particular group of cars, the spiral arms are not attached to any particular piece of disk material. Stars and gas enter a spiral arm, slow down for a while, then continue on their orbits around the Galactic center. The result is a moving region of high stellar and gas density, involving different parts of the disk at different times. Notice also that, just as in our Galaxy, the wave moves more slowly than, and independently of, the overall traffic flow.

We can extend our traffic analogy a little farther. Most drivers are well aware that the effects of a such a tie-up can persist long after the road crew responsible for it has stopped work and gone home for the night. Similarly, spiral density waves can continue to move through the disk even after the disturbance that originally produced them has long since subsided. According to spiral density wave theory, this is precisely what has happened in the Milky Way. Some disturbance in the past produced the wave, which has been moving through the Galactic disk ever since.

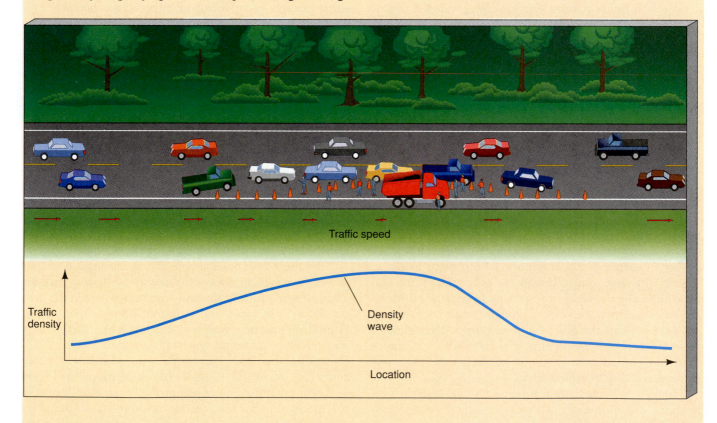

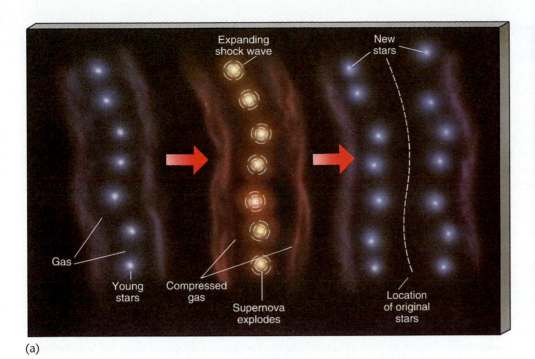

R   I   V   U   X   G

(a)   (b)

Figure 23.18   (a) Self-propagating star formation. In this view of the formation of spiral arms, the shock waves produced by the formation and later evolution of a group of stars provides the trigger for new rounds of star formation. We have used supernova explosions to illustrate the point here, but the formation of emission nebulae and planetary nebulae are also important. (b) This process may well be responsible for the "partial" spiral arms seen in some galaxies, such as NGC 300, shown here in true color. The distinct blue appearance derives from the vast numbers of young stars that pepper its ill-defined spiral arms.

ral and for this pattern to persist for some time. However, this process, sometimes known as *self-propagating star formation*, can produce only *pieces* of spirals, as are seen in some galaxies (see Figure 23.18b). It apparently cannot produce the galaxy-wide spiral arms seen in other galaxies and present in our own. It may well be that there is more than one process at work in the spectacular spirals we see.

An important question (one that unfortunately is not answered by either of the theories just described) is: Where do these spirals come from? What was responsible for generating the density wave in the first place, or for creating the line of newborn stars whose evolution drives the advancing spiral arm? Scientists speculate that (1) instabilities in the gas near the Galactic bulge, (2) the gravitational effects of our satellite galaxies (the Magellanic Clouds, to be discussed in Chapter 24), or (3) the possible asymmetry within the bulge itself may have had a big enough influence on the disk to get the process going. The first possibility is supported by growing evidence that many other spiral galaxies seem to have experienced gravitational interactions with neighboring systems in the relatively recent past (see Chapter 24). However, many astronomers still regard the other two possibilities as equally likely. For example, they point to *isolated* spirals, whose structure clearly cannot be the result of an external interaction. The truth of the matter is that we still don't know for sure how galaxies—including our own—acquire such beautiful spiral arms.

23.8 *The Mass of the Galaxy*

GALACTIC ROTATION

7   What can we learn by studying the motions of the clouds and stars that make up our Galaxy? One very important quantity that we can measure is the *mass* of the Milky Way. To see how this is accomplished, let us apply the same principles used in studies of planetary motions to observations of Galactic dynamics. Recall from Chapter 2 that Kepler's third law (as modified by Newton) connects the orbital period, orbit size, and masses of any two objects in orbit around one another. ∞ (p. 48) We expressed this law as follows:

$$\text{total mass (in solar masses)} = \frac{\text{orbit size (in A.U.)}^3}{\text{orbital period (in years)}^2}.$$

In the solar system, the total mass includes both the mass of the Sun and the mass of an orbiting planet. Because the mass of any planet is small compared with that of the Sun, we can safely ignore the planet. The mass we compute is the mass of the Sun, to good accuracy. Similarly, in applying Kepler's law to weigh the Galaxy, we can neglect the Sun's mass, which is extremely small compared to the mass of the Galaxy. Once we know the orbit of the Sun around the Galactic center, we have all the quantities we need to apply the equation. We can consider the result to be a measure of the Galaxy's mass.

We must make one important clarification at this point. In the case of a planet orbiting the Sun, there is no ambiguity about *what* mass is being measured—it is the Sun's. However, the case of the Sun orbiting the center of the Galaxy is more complicated. The Galaxy's matter is distributed over a large volume of space and is not concentrated at the Galactic center (as the Sun's mass is concentrated at the center of the solar system). Some of the Galaxy's mass lies inside the Sun's orbit (that is, within 8 kpc of the Galactic center), and some lies outside, at large distances from both the Sun and the Galactic center. The question naturally arises: Which portion of the Galaxy's mass controls the Sun's orbit? Isaac Newton answered this question three centuries ago: The Sun's orbital period is determined by the portion of the Galaxy that lies *within the orbit of the Sun*. This is the mass computed from the foregoing equation.

The distance from the Sun to the Galactic center is about 8 kpc, and the Sun's orbital period, as we have seen, is 225 million years. Substituting these numbers into the above equation, we find that the mass of the Milky Way Galaxy within the Sun's orbit is almost 10^{11} solar masses— 100 *billion* times the mass of our Sun. The Milky Way Galaxy is truly vast, both in size and in mass.

DARK MATTER

8 The mass we have just computed for the Galaxy is the mass of stars, gas, dust, and everything else residing *inside* the Sun's orbit around the Galactic center. As we have just seen (recall Figure 23.12), a lot more matter lies outside that radius, in the form of stars and interstellar gas. Yet the luminous portion of the Milky Way Galaxy—the region outlined by the globular clusters and by the spiral arms—is merely the "tip of the Galactic iceberg." There is strong evidence that our Galaxy is in reality very much larger.

To determine the mass of the Galaxy on larger scales— that is, to find how much matter is contained within spheres of progressively larger radii—we must measure the orbital motion of stars and gas farther from the Galactic center than is the Sun. Astronomers have found that the most effective way of doing this is by making radio observations of gas in the Galactic disk, because radio waves are relatively unaffected by interstellar absorption and allow us to probe to great distances, far outside the Sun's orbit. On the basis of these studies, radio astronomers have determined our Galaxy's rotation rate at various distances from the Galactic center. The resultant plot of rotation speed versus distance from the center is the called the Galactic **rotation curve**. It is shown in Figure 23.19.

Knowing the Galactic rotation curve, we can repeat our earlier calculation to compute the total mass that lies within any given distance from the Galactic center. We find, for example, that the mass within about 15 kpc from the center—the volume defined by the globular clusters and the known spiral structure—is some 2×10^{11} solar masses, about twice the mass contained within the Sun's orbit. Does most of the matter in the Galaxy "cut off" at this point, where the luminosity drops off sharply? Surprisingly, the answer is "No."

If most of the matter in the Galaxy ended at the edge of the visible structure, Newton's laws of motion would predict that the orbital speed of stars and gas beyond 15 kpc would decrease outward, just as the orbital speeds of the planets diminish with increasing distance from the Sun. The dashed line in Figure 23.19 indicates what the rotation curve would look like in that case. However, the true rotation curve is quite different. Far from falling off at larger distances, it actually *rises* slightly out to the limits of our measurement capabilities. This implies that the amount of mass contained within successively larger radii continues to grow beyond the orbit of the Sun, apparently out to a distance of at least 40 or 50 kpc. According to our earlier formula, the amount of mass within 40 kpc is approximately 6×10^{11} solar masses. In other words, roughly twice as much mass lies *outside* the luminous part of our galaxy—the part made up of stars, star clusters, and spiral arms—as lies inside!

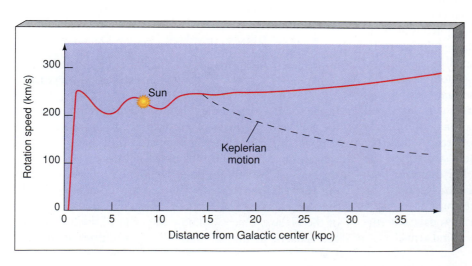

Figure 23.19 The rotation curve for the Milky Way Galaxy plots rotation speed against distance from the Galactic center. We can use this curve to compute the mass of the Galaxy that lies within any given radius. The dashed curve is the rotation curve that would be expected if the Galaxy "ended" abruptly at a radius of 15 kpc, the limit of most of the known spiral structure and the globular cluster distribution. The fact that the curve does not fall off, but in fact rises beyond that point, indicates that there must be additional matter beyond that radius.

Astronomers now believe that our Galaxy is surrounded by an extensive, invisible **dark halo**, which dwarfs the inner halo of stars and globular clusters and extends well beyond the 15-kpc radius once thought to be the limit of our Galaxy. As we will see, researchers have found evidence for dark halos in many other galaxies, leading them to suspect that most galaxies are actually much larger and more massive than their optical images suggest. But what is this mass? We do not detect enough stars or interstellar matter to account for the mass that our computations tell us must be there. We are inescapably drawn to the conclusion that *most* of the mass in our own and in other galaxies exists in the form of invisible **dark matter**, which we presently simply do not understand.

The term *dark* here does not refer just to matter undetectable in visible light. The material has (so far) escaped detection at *all* wavelengths, from radio to gamma rays. Only by its gravitational pull do we know of its existence. Dark matter is not hydrogen gas (atomic or molecular), nor is it made up of stars. Given the amount of matter that must be accounted for, we would have been able to detect it by now with present-day equipment if it were in either of those forms. Its nature and its consequences for the evolution of galaxies and the universe are among the most important questions in astronomy today.

Many candidates have been suggested for this dark matter, although none is proven. Among the strongest contenders are the *brown dwarfs* discussed in Chapter 19—low-mass prestellar objects that never reached the point of nuclear burning in their cores. ∞ (p. 407) These objects could exist in great numbers throughout the Galaxy, yet could be exceedingly hard to see. The fact that we observe so many low-mass stars strongly suggests that the even lower-mass brown dwarfs may be a very common "by-product" of star formation. *Black dwarfs*, the end result of low-mass stellar evolution (Chapter 20) are another possibility. ∞ (p. 432) Again, they would be very hard to detect, although it seems unlikely

that many stars have actually had time to reach this advanced evolutionary stage. *Black holes* also might supply the unseen mass, although their very existence is still debated, and very few candidates exist. ∞ (p. 482) However, given that they are the evolutionary product of (relatively rare) massive stars, it is hard to believe that there could be enough of them to hide large amounts of Galactic matter.

All these candidates are *stellar* in nature—they are associated with star formation or the late stages of stellar evolution. A radically different alternative is that the dark matter is made up of exotic *subatomic particles* that pervade the entire universe. Although there is (as yet) no hard experimental evidence for them, many theoretical astrophysicists believe that these particles could have been produced in abundance during the very earliest moments of our universe. If the particles have survived to the present day, there might be enough of them to account for all the dark matter we believe must be out there. This idea is hard to test, however, because any particles of this nature that might exist would be very hard to detect. Several detection experiments have been attempted—so far, without success.

THE SEARCH FOR STELLAR DARK MATTER

As noted earlier, it is unlikely that appreciable amounts of dark matter are hidden in any kind of yet unseen stars. Small, dim red-dwarf stars were once considered ideal candidates for dark matter. The reasoning was that, just as pebbles are more plentiful on a beach than rocks, then perhaps low-mass stars are very widespread, but not yet detected. However, recent observations of ancient globular clusters—where such red dwarfs might be especially abundant—have shown that faint red dwarfs are actually sparse in the Milky Way. Analysis of imagery like that shown in Figure 23.20 suggests that there is a cut-off point at about 0.2 $M_\odot$, below which nature rarely makes such dim, low-mass stars.

Recently, one very promising means of detecting stellar dark matter has begun to show results. It involves a key element of Albert Einstein's theory of general relativity (see Chapter 22, especially *Interlude 22-2*)—the prediction

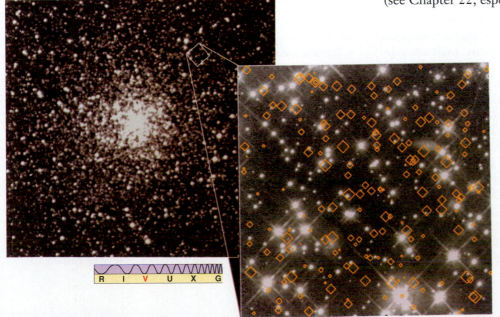

R I V U X G

Figure 23.20 Very sensitive visible observations with the *Hubble Space Telescope* have apparently ruled out faint red-dwarf stars as candidates for dark matter. The object shown here, the globular cluster NGC 6397, is one of many regions searched in the Milky Way. The inset, 0.4 pc on a side, shows a high-resolution *Hubble* view. The scores of diamonds have been overlaid at positions where red dwarfs might (statistically) have been expected if they did indeed make up the dark matter, but were not found.

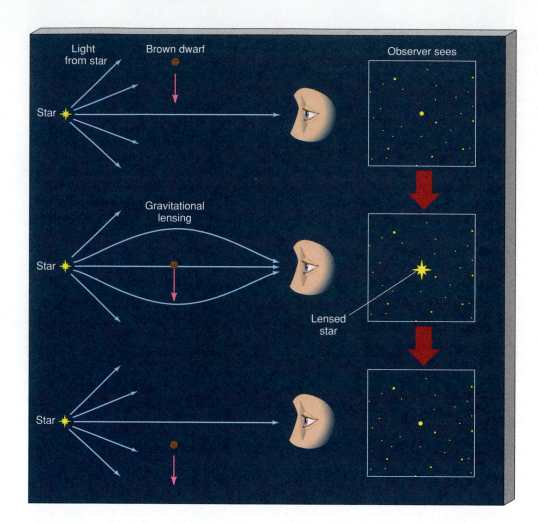

Light from star
Brown dwarf
Observer sees
Star

Gravitational lensing
Star
Lensed star

Star

Figure 23.21 Gravitational lensing by a foreground dark object such as a brown dwarf can temporarily cause a background star to brighten significantly, providing a possible means of detecting stellar dark matter.

that a beam of light can be deflected by a gravitational field, which has already been verified in the case of starlight passing close to the Sun. ∞ (p. 473) Although this effect is small, it has the potential for making otherwise invisible stellar objects observable from the Earth. Here's how.

Imagine looking at a distant star as a faint, compact object such as a brown dwarf, or a distant white dwarf, happens to cross your line of sight. As illustrated in Figure 23.21, the intervening dwarf deflects a little more starlight than usual toward the observer, resulting in a temporary, but quite substantial, *brightening* of the background star. In some ways, the effect is like the focusing of light by a lens, and the process is known as **gravitational lensing**. The foreground dwarf is referred to as a *gravitational lens*. The amount of brightening and the duration of the effect depend on the mass, distance, and velocity of the lensing object. Typically, apparent brightness increases from 2 to 5 times its usual amount for a period of several weeks. Thus, even though the dwarf cannot be seen directly, its effect on the light of the background star makes it detectable. (In Chapter 25 we will encounter other instances of gravitational lensing in the universe, but on very much larger scales.)

Of course, the probability of one star passing almost directly in front of another, as seen from Earth, is ex-

tremely small. But by observing millions of stars every few days over a period of years (using automated telescopes and high-speed computers to reduce the burden of coping with so much data), astronomers hope to see enough of these events to let them measure the mass of stellar dark matter in the Galactic halo. The technique presents an exciting, and potentially very important, new means of probing the structure of our Galaxy. In late 1993, three groups of researchers—in the United States, Europe, and Australia—each announced the detection of a lensing event. Subsequent reports from these groups seem to suggest that they are seeing too few lensing events to support the theory that brown dwarfs and faint, low-mass stars account for all of the dark matter inferred from dynamical studies. However, it is still too early to draw any definite conclusions.

Bear in mind, though, that the identity of the dark matter is not necessarily an "all-or-nothing" proposition. It is perfectly conceivable that more than one type of dark matter exists. For example, it is quite possible that most of the dark matter in the inner (visible) parts of galaxies could be in the form of brown dwarfs and very low-mass stars, while the dark matter on larger scales might be primarily in the form of exotic particles. We will return to this perplexing problem in later chapters.

23.9 *The Center of Our Galaxy*

9 Theory predicts that the Galactic bulge, and especially the region close to the Galactic center, should be densely populated with billions of stars. However, we are unable to see this region of our Galaxy—the interstellar medium in the Galactic disk shrouds what otherwise would be a stunning view. Figure 23.22 shows the (optical) view we do have of the region of the Milky Way toward the Galactic center, in the general direction of the constellation Sagittarius.

With the help of infrared and radio techniques, we can peer more deeply into the central regions of our Galaxy than we can with optical techniques. Infrared observations (see Figure 23.23a on p. 000) indicate that the Galaxy's core harbors roughly 50,000 stars per cubic parsec. That's a stellar density about a million times greater than in our solar neighborhood. Had any planets formed along with these Galactic-center stars, they would probably have been ripped from their orbits and obliterated, as these stars must experience frequent close encounters and even collisions. Infrared radiation has also been detected from what appear to be huge clouds rich in dust. In addition, radio observations indicate that a ring of molecular gas 500 pc across, containing some 30,000 solar masses of material, surrounds the central source.

High-resolution radio observations show more structure on small scales. Figure 23.23(b) shows the bright radio source Sagittarius A, which lies at the center of the circle in Figure 23.22 and within the boxed region in Figure 23.23(a)—and, we think, at the center of our Galaxy. On a scale of about 100 pc, extended filaments can be seen. Their presence suggests to many astronomers that strong magnetic fields operate in the vicinity of the center, creating structures similar in appearance (but much larger in size) to those observed on the active Sun. On even smaller scales, the observations suggest a rotating ring or disk of matter only a few parsecs across.

What could be the cause of all this activity? An important clue comes from the Doppler broadening of infrared spectral lines from the central swirling whirlpool of gas (Figure 23.23b). The extent of the broadening indicates that the gas is moving very rapidly. In order to keep this gas in orbit, whatever is at the center must be extremely massive—a million solar masses or more. Given the twin requirements of large mass and small size, a leading contender is a black hole. The hole itself is not the source of the energy, of course. Instead, the vast accretion disk of matter being drawn toward the hole by its enormous gravity emits the energy as it falls in, just as we saw (on a much smaller scale) in Chapter 22 when we discussed X-ray emission from neutron stars and stellar-mass black holes. ∞ (pp. 467, 476) The strong magnetic fields are thought to be generated within the accretion disk itself as matter spirals inward. We will see later that astronomers have reason to suspect that similar events are occurring at the centers of many other galaxies.

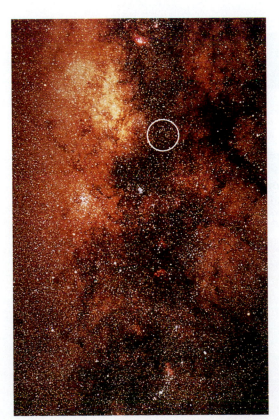

Figure 23.22 A photograph of stellar and interstellar matter in the direction of the Galactic center. Because of heavy obscuration, even the largest optical telescopes can see no farther than 1/10 the distance to the center. The M8 nebula can be seen at the extreme top center. The field is roughly 20°, top to bottom, and is a continuation of the bottom part of Figure 18.6. The circle indicates the location of the core of our Galaxy.

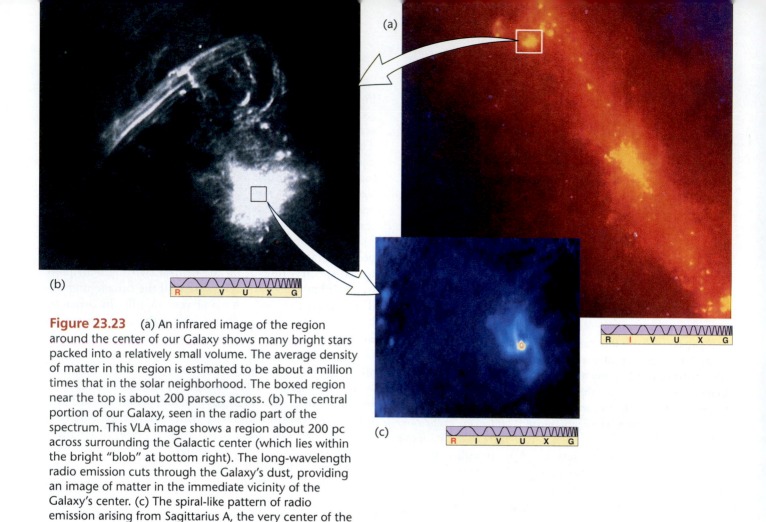

R I V U X G

R I V U X G

R I V U X G

Figure 23.23 (a) An infrared image of the region around the center of our Galaxy shows many bright stars packed into a relatively small volume. The average density of matter in this region is estimated to be about a million times that in the solar neighborhood. The boxed region near the top is about 200 parsecs across. (b) The central portion of our Galaxy, seen in the radio part of the spectrum. This VLA image shows a region about 200 pc across surrounding the Galactic center (which lies within the bright "blob" at bottom right). The long-wavelength radio emission cuts through the Galaxy's dust, providing an image of matter in the immediate vicinity of the Galaxy's center. (c) The spiral-like pattern of radio emission arising from Sagittarius A, the very center of the Galaxy. This image's scale is about 10 pc and suggests a rotating ring of matter only 5 pc across.

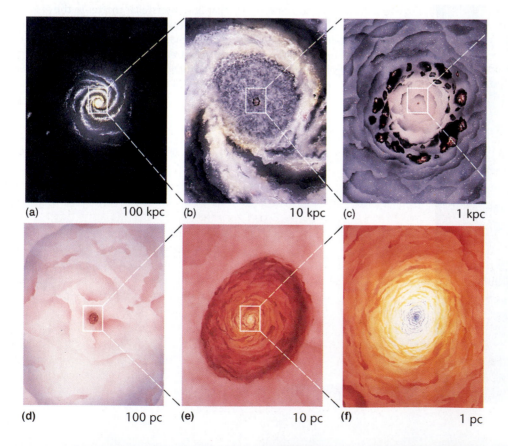

(a) 100 kpc (b) 10 kpc (c) 1 kpc

(d) 100 pc (e) 10 pc (f) 1 pc

Figure 23.24 Six artist's conceptions, each centered on the Galactic core and each increasing in resolution by a factor of 10. Frame (a) shows the same scene as Figure 23.12. Frame (f) is an artist's rendition of a vast whirlpool within the innermost parsec of our Galaxy.

Interlude 23-3   *Cosmic Rays*

In addition to stars, gas, dust, and possibly dark matter, one other type of matter populates our Galaxy. These are cosmic-ray particles, or cosmic rays for short. Cosmic rays continuously collide with the Earth, and they make up our only sample of matter from outside the solar system.

In recent decades, high-altitude balloons, rockets, and satellites have enabled astronomers to discover the chemical composition and the energies of cosmic-ray particles. Cosmic rays are actually not "rays" at all. They are subatomic particles. Nearly 90 percent are protons, the nuclei of hydrogen atoms. The nuclei of heavier atoms—helium and a long list of others—make up 9 percent. Electrons amount to 1 percent. The abundances of elements that make up the cosmic rays are close to the abundances that we find throughout the Galaxy.

Cosmic-ray particles are very energetic. By some unknown mechanism(s), cosmic rays are accelerated to extremely high velocities, much higher than any velocities we can achieve in terrestrial laboratories. Virtually all cosmic rays—even the heaviest—travel very close to the speed of light. When a galactic cosmic-ray particle hits our atmosphere, it creates a series of cascades or "showers" of many lower-energy particles. These lower-energy "secondary" particles in turn collide with objects, including humans, on the Earth's surface virtually all the time. Our bodies are being peppered with them right now.

The accompanying photograph shows an actual track of a cosmic ray particle—in this case, the nucleus of a sulfur atom (shown in red in this false-color image). Entering from the left, the particle collides with the nucleus of an atom in the photographic emulsion, producing a fluorine nucleus (green), other nuclear fragments (blue), and a multitude of subatomic particles (yellow).

On the basis of the numbers of cosmic rays hitting the Earth, astronomers estimate the number density of cosmic rays in interstellar space to be on the order of 10^{-3} particles/m^3. That makes them rare, although still more numerous than interstellar dust particles.

What is the source of our Galaxy's cosmic rays? The answer is currently uncertain. The circuitous and complex paths taken by the cosmic-ray particles while traversing the Galaxy, with its tangled magnetic field, prevent us from pinpointing their source by observing their direction of arrival at Earth. Indeed, careful searches have failed to show any preferential direction for cosmic rays. They seem to arrive from all directions with equal likelihood. Candidates for the origin of cosmic rays include (1) violent events at the Galactic center, where strong electromagnetic forces accelerate these particles to high energies; (2) supernova explosions, described in Chapter 21; and (3) the vicinity of neutron stars and black holes, as discussed in Chapter 22.

Most researchers currently favor the second of these explanations, although each of these candidate events probably contributes some particles to the observed hodgepodge of cosmic rays in our Galaxy. Some of the highest-energy cosmic rays may even originate at great distances, in the hearts of galaxies far beyond our own.

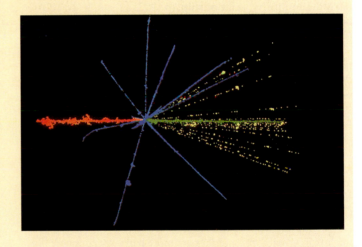

Figure 23.24 places these findings into a simplified perspective. In a series of six images, an artist has captured the important results of long-wavelength radio and infrared studies of the Milky Way's heart. Each painting is centered on the Galaxy's core, and each increases in resolution by a factor of 10.

Frame (a) renders the Galaxy's overall shape, as painted in Figure 23.12. The scale of this frame measures about 100 kpc from top to bottom. Frame (b) spans a distance of 10 kpc from top to bottom and is nearly filled by the great circular sweep of the innermost spiral arm. Moving in to a 1-kpc span, frame (c) depicts a ring of matter made mostly of giant molecular clouds and gaseous nebulae. This entire flattened, circular feature is rotating at about 100 km/s. The origin of this ring of gas is still unclear. In frame (d), at 100 pc, a pinkish region of ionized gas surrounds the reddish heart of the Galaxy. (Of course, we cannot see the colors of these regions—astronomers infer them on the basis of the gases'

temperatures and densities.) The source of energy producing this vast ionized cloud is unknown, although it is presumed to be related to activity in the Galactic center. Frame (e), spanning 10 pc, depicts the tilted, spinning whirlpool of hot (10^4 K) gas that marks the core of our Galaxy. The innermost sanctum of this gigantic whirlpool is painted in frame (f), in which a swiftly spinning, white-hot disk of gas at millions of kelvins nearly engulfs an enormously massive object too small in size to be pictured (even as a minute dot) on this scale.

If our knowledge of the Galaxy's center seems sketchy, that's because it *is*. Astronomers are still deciphering the clues hidden within its invisible radiation. We are only beginning to appreciate the full magnitude of this strange new realm deep in the heart of the Milky Way. In some respects, our research is not yet mature science. Rather, it's exploration—but absolutely fascinating exploration, enabling us to return from our telescopes with tales of new wonders at the core of our Galaxy.

Chapter Review

A **galaxy** (p. 488) is huge collection of stellar and interstellar matter isolated in space and bound together by its own gravity. Our own Galaxy, the Milky Way, and our nearest major galactic neighbor, Andromeda, are **spiral galaxies** (p. 490). Such galaxies feature flattened **galactic disks** (p. 488) that thicken into **galactic bulges** (p. 488) near the center. Spiral galaxies derive their name from the characteristic **spiral arms** (p. 490) that start near the galactic bulge and wind outward through the galactic disk. The disks are surrounded by roughly spherical **galactic halos** (p. 488) of old stars and star clusters.

Variable stars (p. 490) are stars whose luminosity changes with time. Some are quite erratic in their variations. Others, called **pulsating variable stars** (p. 490), vary in a repetitive and predictable way. Two types of pulsating variable stars that are of great importance to astronomers are the **Cepheid variables** (p. 490) and the **RR Lyrae variables** (p. 491), whose characteristic light curves make them easily recognizable.

Measuring the pulsation period of a Cepheid variable allows astronomers to determine the star's luminosity using the **period–luminosity relationship** (p. 492), a simple correlation between period and absolute brightness that holds for all Cepheids. All RR Lyrae stars have roughly the same luminosity. The brightest Cepheids can be seen at distances of millions of parsecs, extending the cosmic distance ladder well beyond our own Galaxy. RR Lyrae stars are fainter, but much more numerous, making them very useful within the Milky Way.

In the early twentieth century, Harlow Shapley used RR Lyrae stars to determine the distances to many of the Galaxy's globular clusters. He found that the clusters as a whole have a roughly spherical distribution in space, but the center of the sphere lies far from the Sun. The globular clusters are now known to map out the true extent of the luminous portion of the Milky Way Galaxy. The center of their distribution is close to the **Galactic center** (p. 493), which lies about 8 kpc from the Sun.

In the vicinity of the Sun, the Galactic disk is about 300 pc thick. Young stars, gas, and dust are more narrowly confined; older stars have a broader distribution. Intermediate between the young disk and the old halo, in both age and spatial distribution, are the stars of the **thick disk** (p. 496), which is about 2–3 kpc thick.

Astronomers use radio observations to explore the Galactic disk because radio waves are largely unaffected by interstellar dust. Regions where most of the hydrogen is in atomic form may be studied using 21-cm radiation. Regions where the gas is mostly molecular are usually studied through "tracer" molecules, such as carbon monoxide. The gas distribution fattens near the center into the Galactic bulge. Radio-emitting gas has been detected in the disk at up to 50 kpc from the Galactic center.

Disk and halo stars differ in their spatial distributions, ages, colors, and orbital motion. The halo lacks gas and dust, so no stars are forming there. All halo stars are old. The gas-rich disk is the site of current star formation and contains many young stars. Stars and gas within the Galactic disk move on roughly circular orbits around the Galactic center. Stars in the halo move on largely random three-dimensional orbits that pass repeatedly through the disk plane but have no preferred orientation.

Halo stars appeared early on, before the Galactic disk had taken shape, when there was still no preferred orientation for their orbits. As the gas and dust formed a rotating disk, stars that formed in the disk inherited its overall spin and so moved on circular orbits in the Galactic plane, as they do today.

The spiral arms in spiral galaxies are regions of the densest interstellar gas and are the places where star formation is taking place. The spirals cannot be "tied" to the disk material, as the disk's differential rotation would have wound them up long ago. Instead, they may be **spiral density waves** (p. 501) that move through the disk, triggering star formation as they pass by. Alternatively, the spirals may arise from *self-propagating star formation*, when shock waves produced by the formation and evolution of one generation of stars triggers the formation of the next.

The Galactic **rotation curve** (p. 504) plots the orbital speed of matter in the disk against distance from the Galactic center. By applying Newton's laws of motion, astronomers can determine the mass of the Galaxy. They find that the Galactic mass continues to increase beyond the radius defined by the globular clusters and the spiral structure we observe.

The rotation curves of our own and other galaxies show that many, if not all, galaxies have invisible **dark halos** (p. 505), containing far more mass than the visible galaxies themselves. The **dark matter** (p. 505) making up these dark halos is of unknown composition. Leading candidates include brown dwarfs, black holes, and exotic subatomic particles.

Recent attempts to detect "stellar" dark matter have used the fact that a dark compact object such as a brown dwarf can occasionally pass in front of a more distant star. The brown dwarf deflects the star's light and causes its apparent brightness to increase temporarily. This deflection is called **gravitational lensing** (p. 506).

Astronomers working at infrared and radio wavelengths have uncovered evidence for energetic activity within a few parsecs of the center. The leading explanation is that a massive black hole resides in the heart of our Galaxy. The activity we observe is the result of the energy released as matter falls into the hole's intense gravitational field.

SELF-TEST True or False?

_____ **1.** Cepheids can be used to determine the distances to the nearest galaxies.

_____ **2.** RR Lyrae stars are a type of cataclysmic variable.

_____ **3.** The Galactic halo contains about as much gas and dust as the Galactic disk.

_____ **4.** The Galactic disk contains only old stars.

_____ **5.** Population I objects are found only in the Galactic halo.

_____ **6.** Up until the 1930s, the main error made in determining the size of the Galaxy was due to an incorrectly calibrated method of determining stellar distances.

_____ **7.** Astronomers use 21-cm radiation to study Galactic molecular clouds.

____ **8.** Radio techniques are capable of mapping the entire Galaxy.

____ **9.** In the neighborhood of the Sun, the Galaxy's spiral density wave rotates more slowly than the overall Galactic rotation.

____ **10.** The mass of the Galaxy is determined by counting stars.

____ **11.** Dark matter is now known to be due to large numbers of black holes.

____ **12.** A million solar-mass black hole could account for the unusual properties of the Galactic center.

____ **13.** Cosmic rays are very energetic photons.

____ **14.** More than 90 percent of the mass of our Galaxy exists in the form of dark matter.

SELF-TEST Fill in the Blank

1. One difficulty in studying our own galaxy in its entirety is that we live _____.

2. The highly flattened, circular part of the Galaxy is called the Galactic _____.

3. The roughly spherical region of faint old stars in which the rest of the Galaxy is embedded is the Galactic _____.

4. Cepheids are observed to vary in _____.

5. Cepheids pulsational periods range from _____ to _____.

6. According to the period–luminosity relation, the longer the pulsational period of a Cepheid, the _____ its luminosity.

7. Harlow Shapley determined the distances to the globular clusters using _____.

8. The Sun lies roughly _____ pc from the Galactic center.

9. The orbital speed of the Sun around the Galactic center is _____.

10. The orbits of halo objects are _____ in direction.

11. The original cloud of gas from which the Galaxy formed probably had a size and shape similar to the present Galactic _____.

12. Rotational velocities in the outer part of the Galaxy are _____ than would be expected on the basis of observed stars and gas, indicating the presence of _____.

REVIEW AND DISCUSSION

1. What are spiral nebulae? How did they get that name?

2. How are Cepheid variables used in determining distances?

3. Roughly how far out into space can we use Cepheids to measure distance?

4. What important discoveries were made early in this century using RR Lyrae variables?

5. Why can't we study the central regions of the Galaxy using optical telescopes?

6. Of what use is radio astronomy in the study of Galactic structure?

7. Contrast the motions of disk and halo stars.

8. Explain why galactic spiral arms are believed to be regions of recent and ongoing star formation.

9. Describe the motion of interstellar gas as it passes through a spiral density wave.

10. What is self-propagating star formation?

11. What do the red stars in the Galactic halo tell us about the history of the Milky Way?

12. What does the rotation curve of our Galaxy tell us about its total mass?

13. What evidence is there for that dark matter in the Galaxy?

14. What are some possible explanations for dark matter?

15. Why do astronomers believe that a supermassive black hole lies at the center of the Milky Way?

PROBLEMS

1. A Cepheid variable is 100 times brighter than a typical RR Lyrae star. How much farther away could the Cepheid be used as a distance-measuring tool?

2. *HST* can see a star like the Sun at a distance of 100,000 pc. The brightest Cepheids are 10,000 times the luminosity of the Sun. How far away can *HST* see these Cepheids?

3. Calculate the total mass of the Galaxy within 20 kpc of the center if the rotation speed at that radius is 240 km/s.

4. A density wave made up of two spiral arms is moving through the Galactic disk. At the orbit of the Sun, the wave's speed is 120 km/s. Assuming that the Sun's speed is 220 km/s, calculate how many times the Sun has passed through a spiral arm since it formed 4.5 billion years ago.

PROJECTS

1. If you are far from city lights, look for a hazy band of light arching across the sky. This is our edgewise view of the Milky Way Galaxy. The Galactic center is located in the direction of the constellation Sagittarius, highest in the sky during the summer, but visible from spring through fall. Look at the band making up the Milky Way and notice dark regions; these are relatively nearby dust clouds. Sketch what you see. Look for faint fuzzy spots in the Milky Way and note their positions in your sketch. Draw in the major constellations for reference. Compare your sketch to a map of the Milky Way in a star atlas. Did you discover most of the dust clouds? Can you identify the faint fuzzy spots?

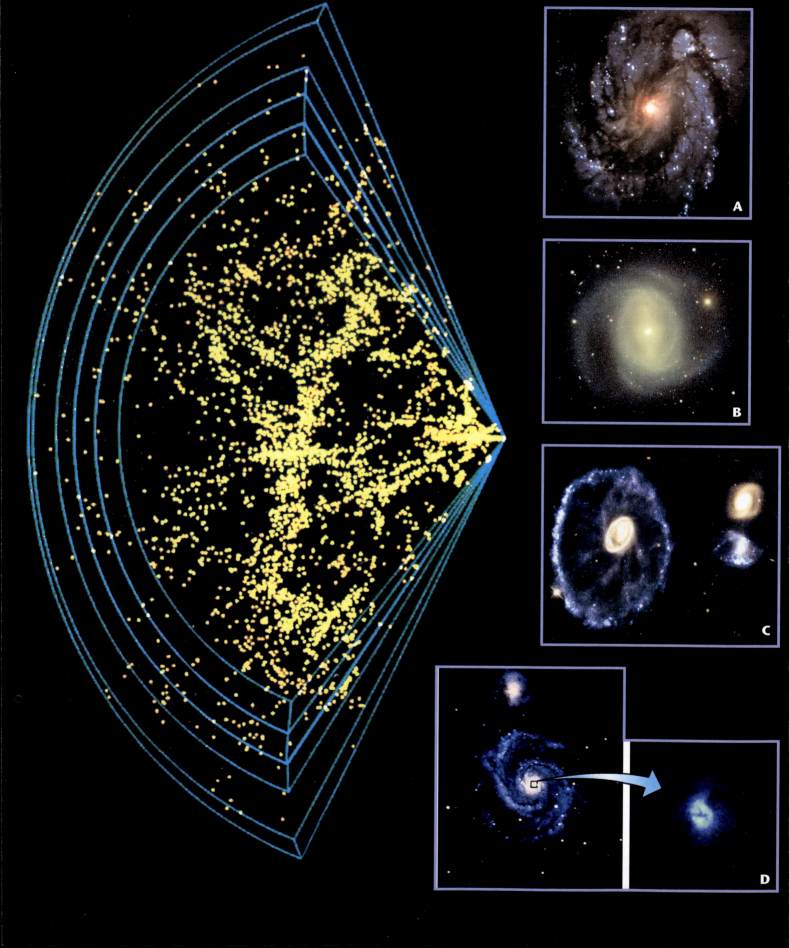

24

NORMAL GALAXIES

The Large-Scale Structure of the Universe

LEARNING GOALS

Studying this chapter will enable you to:

1 Describe the basic properties of the main types of normal galaxies.

2 Discuss the distance-measurement techniques that enable astronomers to map the universe beyond our Milky Way.

3 Summarize what is known about the large-scale distribution of galaxies throughout the universe.

4 Describe some of the methods used to determine the masses of distant galaxies.

5 Explain why astronomers think that most of the matter in the universe is invisible.

6 State Hubble's law, and explain how it is used to derive distances to the most remote objects in the observable universe.

7 Discuss some theories of how galaxies form and evolve.

(Opposite page, background) Each of the yellow dots on this map, which depicts a slice of space extending out some 150,000 kpc from Earth, represents a galaxy. Yet the nearly 4000 galaxies mapped here probably represent less than a ten-millionth of the total number in the observable universe. Even among "normal" galaxies, there is a great range of individual variation.

(Inset A) The heart of a typical galaxy, M100, is seen here in a computer-enhanced *Hubble* image. The structure of the spiral arms, with their young blue stars, is clearly visible.

(Inset B) The beautiful spiral, NGC 1433, has a bar through its center and a ring around it as well.

(Inset C) This stunning image of three galaxies probably shows the aftermath of a bull's-eye collision of one of the two galaxies at right with the one at left. The result is the Cartwheel Galaxy, about 150 Mpc from Earth, with its halo of young stars resembling a vast ripple in a pond.

(Inset D) Heart of darkness: M51, known as the Whirlpool Galaxy, is actually a pair of interacting galaxies some 10 Mpc from Earth. The breakout is a recent *Hubble* view of the the galaxy's innermost core, showing a dark X-shaped feature that is probably an accretion disk of matter girdling a black hole.

*M*uch of our knowledge of the workings of our Galaxy is based on observations of the myriad other galaxies in the universe. We know of literally millions of galaxies beyond our own—many smaller than the Milky Way, some comparable in size, a few much larger. All are vast star systems separated from us by almost incomprehensibly large distances, each a gravitationally bound assemblage of stars, gas, dust, dark matter, and radiation. Even a modest-sized galaxy harbors more stars than the number of people who have ever lived on Earth. The light we receive tonight from the most distant galaxies was emitted long before Earth even existed. By comparing and classifying the properties of galaxies, astronomers have begun to understand their complex dynamics. By mapping out their distribution in space, astronomers have traced out the immense realms of the universe. The galaxies remind us that our position in the universe is no more special than that of a boat adrift at sea.

24.1 Hubble's Galaxy Classification

① Figure 24.1 shows a vast expanse of space lying about 100 million pc from the Earth. *Every* patch or point of light in this figure is a separate galaxy—several hundred can be seen in this one photograph. Over the years, astronomers have accumulated similar images of many millions of galaxies. Let us begin our study of these enormous accumulations of matter simply by considering their appearance on the sky.

Seen through even a small telescope, images of galaxies look distinctly nonstellar. They have "fuzzy" edges, and many are quite elongated—not at all like the sharp, point-like images normally associated with stars. Although it is difficult to tell from the photograph, some of the fuzzy "blobs" of light in Figure 24.1 are actually spiral galaxies like the Milky Way and Andromeda, which we studied in

Chapter 23. Others, however, are definitely not spirals—no disks or spiral arms can be seen. Even when we take into account their different orientations in space, all galaxies do *not* look the same.

Given that galaxies display a range of shapes and sizes, can we divide them into well-defined categories, much as we classify stars? The American astronomer Edwin Hubble was the first to categorize galaxies in a comprehensive way. Working with the then recently completed 2.5-m optical telescope on Mount Wilson in California in 1924, he classified the galaxies he saw into four basic types—*spirals, barred spirals, ellipticals,* and *irregulars*—solely on the basis of appearance. Many modifications and refinements have been incorporated over the years, but the basic **Hubble classification scheme** is still widely used today. Refer to Table 24-1, p. 517 which summarizes the characteristics of the various types, as we discuss them in greater detail.

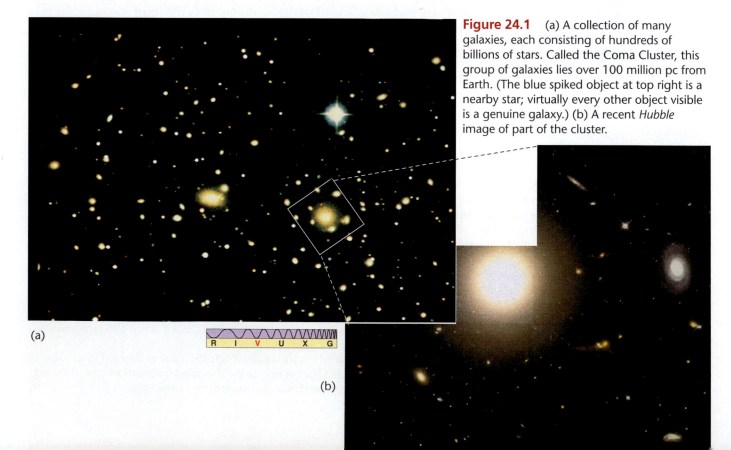

Figure 24.1 (a) A collection of many galaxies, each consisting of hundreds of billions of stars. Called the Coma Cluster, this group of galaxies lies over 100 million pc from Earth. (The blue spiked object at top right is a nearby star; virtually every other object visible is a genuine galaxy.) (b) A recent *Hubble* image of part of the cluster.

(a)

R I V U X G

(b)

Figure 24.2 Variation in shape among different spiral galaxies. As we progress from type Sa to Sb to Sc, the bulges tend to get smaller, while the spiral arms become less tightly wound.

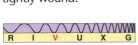

M81   Type Sa   M51   Type Sb   NGC 2997   Type Sc

SPIRALS

We saw several examples of **spiral galaxies** in Chapter 23—for example, our own Milky Way Galaxy and our neighbor Andromeda. All galaxies of this type contain a flattened galactic disk, in which spiral arms are found, and a central galactic bulge. ∞ (p. 488) The stellar density (that is, the number of stars per unit volume) is greatest in the galactic nucleus, at the center of the bulge. But within this general description, spiral galaxies can exhibit quite a variety of shapes, as illustrated in Figure 24.2.

The size of the bulge and the tightness of the spiral pattern are quite well correlated. The correlation is far from perfect, but it still provides the basis for a useful classification of galaxy properties. Tightly wrapped spirals tend to have large central bulges, and spirals with large central bulges typically have tightly wrapped arms. More open spirals exhibit smaller bulges, and spirals with smaller bulges generally have less tightly wound arms. The arms also tend to become more "knotty," or clumped, in appearance as the spiral pattern becomes more open.

In Hubble's scheme, spiral galaxies are denoted by the letter *S* and classified according to the form of the spiral arms and the size of the central bulge. *Type Sa* spiral galaxies have large bulges and tight, almost circular, spiral arms. *Type Sb* galaxies have smaller bulges and more open spiral arms. *Type Sc* spirals have small bulges and a loose, sometimes quite poorly defined spiral pattern. In the 1950s, one additional class, *Sd*, with even more poorly defined arms, was added.

Much of our description of the large-scale structure of the Milky Way Galaxy in Chapter 23 applies to spiral galaxies in general: Our Galaxy is probably a fairly typical type Sb (or perhaps Sbc—midway between Sb and Sc) spiral. The basic regions of our Galaxy—the disk, the bulge, and the halo—are found in all spirals. The halos of spiral galaxies contain large numbers of reddish old stars and globular clusters, similar to those observed in our own Galaxy and in Andromeda. Most of the light from spirals, however, comes from A- through K-type stars in the disk, giving these galaxies an overall whitish-yellowish glow. We assume that thick disks exist, too, but their faintness makes

this assumption hard to confirm—the thick disk in the Milky Way contributes only a percent or so of our Galaxy's total light. ∞ (p. 496)

Again like the Milky Way, the flat galactic disks of typical spiral galaxies are rich in gas and dust. The 21-cm radio radiation emitted by spirals betrays the presence of the interstellar gas, and detailed photographs clearly show the obscuring dust in many systems. Stars are still forming within the spiral arms, where the interstellar medium is densest, and the arms contain numerous emission nebulae and newly formed blue O- and B-type stars. Type Sc galaxies contain the most interstellar gas, and Sa galaxies the least. The Sc galaxy NGC 3184, shown in Figure 24.3, clearly shows the preponderance of interstellar dust and young blue stars tracing the spiral pattern. Spirals are not necessarily young galaxies, however. Like our own Galaxy, they are simply rich enough in interstellar gas to provide for continued stellar birth.

Many spirals are observable only at sharp angles, not face-on as in Figure 24.3. However, we do not need to see spiral structure to classify a galaxy as a "spiral." The presence of the disk, with its gas, dust, and newborn stars, is sufficient.

Figure 24.3 The type Sc galaxy NGC 3184 clearly shows young (blue) stars spread along its spiral arms.

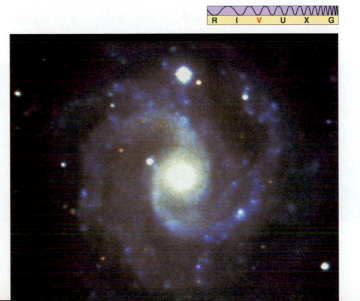

Astronomers often cannot distinguish between spirals and barred spirals, especially when a galaxy happens to be oriented with its galactic plane nearly edge-on toward the Earth, as in Figure 24.4. Because of the physical and chemical similarities of spiral and barred-spiral galaxies, some researchers do not even bother to distinguish between them. Other researchers, however, regard the differences in their structures as very important, arguing that these differences suggest basic dissimilarities in the conditions that led to the formation of the galaxies eons ago. The recent findings that the bulge of our own Galaxy may be elongated suggests—to some astronomers, at least—that the Milky Way might itself be a barred spiral, of type SBc.

Figure 24.4 The Sombrero Galaxy, a spiral system seen edge-on. Officially cataloged as M104, this galaxy has a dark band composed of interstellar gas and dust. The large size of this galaxy's central bulge marks it as type Sa, even though its spiral arms cannot be seen.

For example, the galaxy shown in Figure 24.4—the so-called Sombrero Galaxy—is thought to be a spiral because of the clear line of obscuring dust seen along its midplane.

BARRED SPIRALS

A variation of the spiral category in Hubble's classification scheme is the **barred-spiral galaxy**. The barred spirals differ from ordinary spirals mainly by the presence of an elongated "bar" of stellar and interstellar matter passing through the center and extending beyond the bulge, into the disk. The spiral arms project from near the ends of the bar rather than from the bulge (as they do in normal spirals). Barred spirals are designated by the letters *SB* and are subdivided, like the ordinary spirals, into categories SBa, SBb, and SBc, depending on the size of the bulge and the tightness of the spiral pattern. Figure 24.5 shows the variation among barred-spiral galaxies. In the case of the SBc category, it is often hard to tell where the bar ends and the spiral arms begin.

ELLIPTICALS

The next major category in the Hubble scheme contains the **elliptical galaxies**. Unlike the spirals, ellipticals have no spiral arms and, in most cases, no flattened galactic disk—in fact, they often exhibit little internal structure of any kind. As with spirals, the stellar density increases sharply in the central nucleus. Ellipticals range in shape from highly elongated to nearly circular in appearance. Denoted by the letter *E*, these systems are subdivided according to how elliptical they are. The most circular are designated E0, and the most elongated E7; intermediate elongations are classified El through E6 (Figure 24.6).

There is a large range in both the size and the number of stars contained in elliptical galaxies. The largest elliptical galaxies are much larger than our own Milky Way. These *giant ellipticals* can range up to a few megaparsecs across and contain trillions of stars. At the other extreme, *dwarf ellipticals* may be as small as a kiloparsec in size and contain fewer than a million stars. The substantial observational differences between giant and dwarf ellipticals have led many astronomers to conclude that these galaxies are members of separate classes, with quite different formation histories and stellar content. The dwarfs are by far the most common type of ellipticals, outnumbering their brighter counterparts by about 10 to 1. However, most of the *mass* that exists in the form of elliptical galaxies is contained in the larger systems.

Figure 24.5 Variation in shape among different barred-spiral galaxies. The variation from SBa to SBc is similar to that for the spirals in Figure 24.2, except that now the spiral arms begin at either end of a bar through the galactic center.

NGC 3992   Type SBa   NGC 1300   Type SBb   NGC 1365   Type SBc

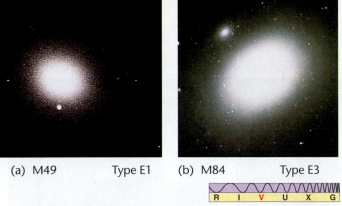

(a) M49   Type E1   (b) M84   Type E3   (a) NGC 1201   Type S0   (b) NGC 2859   Type SB0

R I V U X G   R I V U X G

Figure 24.6 Variation in shape among different elliptical galaxies. (a) The E1 galaxy M49 is nearly circular in appearance. (b) M84 is a slightly more elongated elliptical galaxy. It is classified as E3. Both these galaxies lack spiral structure, and neither shows evidence of interstellar matter.

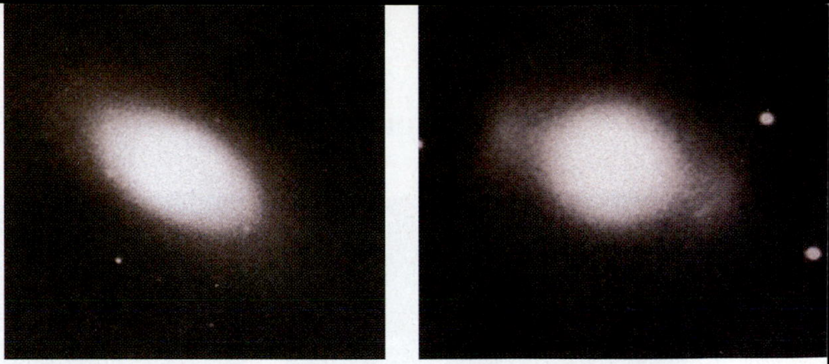

Figure 24.7 (a) S0 galaxies contain a disk and a bulge but no interstellar gas or spiral arms. They are in many respects intermediate between E7 ellipticals and Sa spirals in their properties. (b) SB0 galaxies are similar, except for a bar of stellar material extending beyond the central bulge.

Lack of spiral arms is not the only difference between spirals and ellipticals. Most ellipticals also contain little or no gas and dust. The 21-cm radio emission from neutral hydrogen gas is, with few exceptions, completely absent, and no obscuring dust lanes are seen. In most cases, there is no evidence of young stars or of any ongoing star formation. Like the halo of our own Galaxy, ellipticals are made up mostly of old, reddish, low-mass stars. Indeed, with no disk, gas, or dust, elliptical galaxies are, in a sense, "all halo." Again like the halo of our Galaxy, the orbits of stars in ellipticals are disordered, with objects moving in all directions, not in regular, circular paths as in our Galaxy's disk. Apparently all, or nearly all, of the interstellar gas within elliptical galaxies was swept up into stars (or out of the galaxy) long ago, before a disk had a chance to form, leaving no loose gas and dust for the creation of future generations of stars.

Some giant ellipticals are exceptions to most of the foregoing general statements about elliptical galaxies, as they have been found to contain disks of gas and dust in which stars are forming. Astronomers speculate that these galaxies may really be otherwise "normal" ellipticals that have collided and merged with a companion spiral system (although this explanation is not yet universally accepted). We will return later in this chapter to the important role played by galactic collisions in determining the appearance of the systems we observe today.

Intermediate between the E7 ellipticals and the Sa spirals in the Hubble classification is a class of galaxies that show evidence of a thin disk and a flattened bulge but that contain no gas and no spiral arms. Two such objects are shown in Figure 24.7. These galaxies are known as **S0 galaxies** if no bar is evident and **SB0 galaxies** if a bar is present. They look a little like spirals whose dust and gas have been stripped away, leaving behind a stellar disk. Observations in recent years have shown that many normal elliptical galaxies actually have faint disks within them, like the S0 galaxies. As with the S0s, the

TABLE 24-1 *Basic Galaxy Properties by Type*

	SPIRAL /BARRED SPIRAL (S, SB)	ELLIPTICAL (E)	IRREGULAR (Irr)
Shape and Structural Properties	Highly flattened disk of stars and gas, containing spiral arms and thickening to central bulge. Sa and SBa galaxies have largest bulges, the least obvious spiral structure, and roughly spherical stellar halos. SB galaxies have an elongated central "bar" of stars and gas.	No disk. Stars smoothly distributed through an ellipsoidal volume ranging from nearly spherical (E0) to very flattened (E7) in shape. No obvious substructure other than a dense central nucleus.	No obvious structure. Irr II galaxies often have "explosive" appearance.
Stellar Content	Disks contain both young and old stars; halos consist of old stars only.	Contain old stars only.	Contain both young and old stars.
Gas and Dust	Disks contain substantial amounts of gas and dust; halos contain little of either.	Contain little or no gas and dust.	Very abundant in gas and dust.
Star Formation	Ongoing star formation in spiral arms.	No significant star formation during the last 10 billion years.	Vigorous ongoing star formation.
Stellar Motion	Gas and stars in disk move in circular orbits around the galactic center; halo stars have random orbits in three dimensions.	Stars have random orbits in three dimensions.	Stars and gas have very irregular orbits.

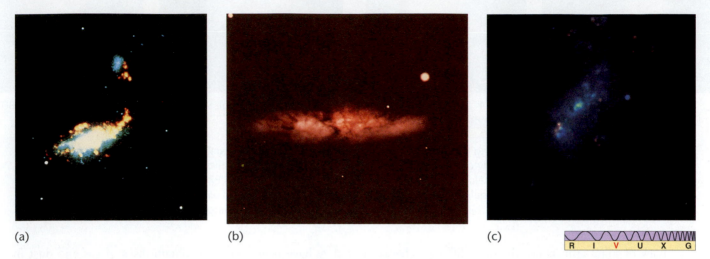

(a)   (b)   (c)

| R | I | V | U | X | G |

Figure 24.8   Photographs of some irregular (Irr II) galaxies. (a) The oddly shaped galaxies NGC 4485 and NGC 4490 may actually be physically close and be interacting with one another gravitationally. (b) The galaxy M82 likewise seems to show an explosive appearance, although interpretations remain uncertain. (c) Many irregular galaxies are small and dim, but this one, NGC 4449, is comparable in both size and luminosity to the Milky Way.

origin of these disks is uncertain, but some researchers suspect that the S0s and ellipticals actually form a continuous sequence, along which the bulge-to-disk ratio varies smoothly.

IRREGULARS

The final galaxy class identified by Hubble is a "catchall" category—**irregular galaxies**—so named largely because their visual appearance does not allow us to place them into any of the other categories just discussed. Irregulars tend to be rich in interstellar matter, but they lack any regular structure, such as well-defined spiral arms or central bulges. They are divided into two subclasses—*Irr I galaxies* and *Irr II galaxies*. The Irr I galaxies often look rather like misshapen spirals. The much rarer Irr II galaxies, in addition to their irregular shape, have other peculiarities, often exhibiting a distinctly explosive or filamentary appearance. Figure 24.8

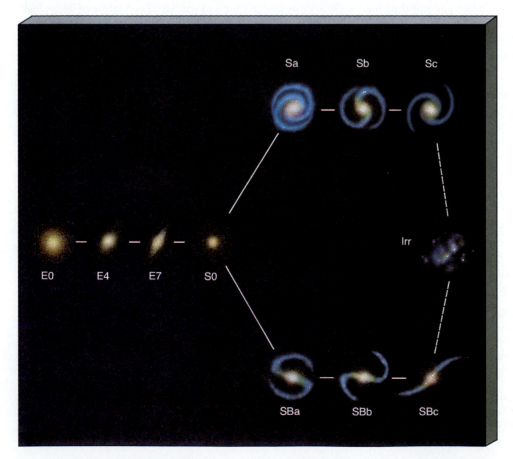

Figure 24.9   Hubble's "tuning fork" diagram, showing his basic galaxy classification scheme. The placement of the four basic galaxy types—ellipticals, spirals, barred spirals, and irregulars—in the diagram is suggestive, but no "evolutionary track" along the sequence (in either direction) is proven.

shows some examples of these strangely shaped galaxies. Their appearance once led astronomers to suspect that violent events had occurred within them. However, it now seems possible that in some (but probably not all) cases, we are actually seeing the result of a close encounter, or even a collision, between two previously "normal" systems.

Irregular galaxies tend to be smaller than spirals but somewhat larger than dwarf ellipticals. They typically contain between 10^8 and 10^{10} stars. The smallest are called *dwarf irregulars*. As with elliptical galaxies, the dwarf type is the most common irregular. Dwarf ellipticals and dwarf irregulars occur in approximately equal numbers and together make up the vast majority of galaxies in the universe. They are often found close to a larger "parent" galaxy. The **Magellanic Clouds**, a famous pair of Irr I galaxies that orbit the Milky Way, are shown to proper scale at the left of Figure 23.12 and discussed further in *Interlude 24-1* on p. 522.

AN H–R DIAGRAM FOR GALAXIES?

Having earlier found a convenient means of representing the properties of stars—namely, the Hertzsprung–Russell diagram discussed in Chapters 17 through 19—we naturally wonder if there might be something similar for galaxies. Is there anything that links the different types of galaxies in the same way that the H–R diagram relates red stars to blue stars and dwarfs to giants? When he first developed his classification scheme, Hubble arranged the basic galaxy types into the "tuning fork" diagram shown in Figure 24.9. At the time, his aim in doing this was simply to indicate similarities in appearance among galaxies, but some astronomers have since speculated that perhaps the diagram

has much deeper significance. Figure 24.9 certainly suggests an evolutionary connection of some sort among galaxy types, but does one really exist? The answer, as best we can tell, is "No." (Incidentally, Hubble himself never made any claim that such a connection might exist.)

Despite many attempts, no one has ever succeeded in explaining the observed properties of galaxies, as described by the Hubble classification scheme and presented in Figure 24.9, in evolutionary terms. Isolated normal galaxies do *not* evolve from one type to another. Spirals are not ellipticals with arms, nor are ellipticals spirals that have somehow lost their star-forming disks. In short, astronomers know of no parent–child relationship among normal galaxies. The various types of galaxies are more like cousins who trace their birth to the same ancestor—the galaxy formation process. We will return to discuss this poorly understood subject at the end of this chapter.

24.2 *The Distribution of Galaxies in Space*

3 Now that we have seen some of the basic properties of galaxies, let us ask how they are spread through the expanse of the universe beyond the Milky Way. Do they reside everywhere, scattered throughout intergalactic space all the way out to the very limits of the observable universe? We have found boundaries of sorts that mark the limits of Earth, the solar system, and our Galaxy. Is there a boundary beyond which galaxies no longer exist? Clearly, to answer this question, we must first know the *distances* to the galaxies.

The bottom four levels of Figure 24.10 show the distance-measurement techniques we have employed so far in

Figure 24.10 The inverted pyramid summarizes the various distance techniques used to study different realms of the universe. Radar-ranging, stellar parallax, spectroscopic parallax, and variable stars take us as far as the nearest galaxies in our study of the universe. To go farther, new techniques must be employed, each based on distances known by techniques at lower levels. The top level shown here will be described later in this chapter.

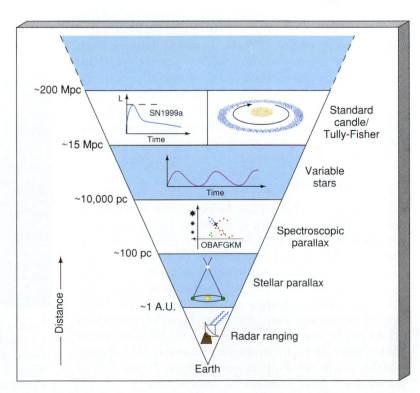

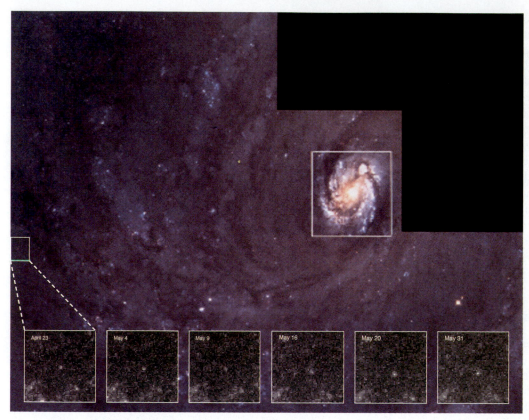

Figure 24.11 This sequence of six snapshots chronicles the rhythmic changes in a Cepheid variable star in the spiral galaxy M100. The Cepheid appears at the center of each inset, taken at the different times indicated in 1994. The star looks like a square because of the high magnification by the digital CCD camera—we are seeing individual pixels of the image. The 24th-magnitude star periodically doubles in brightness every 7 weeks. (Compare the image of M100 shown in *Interlude 5-1*. The bright, boxed core is reproduced as Inset A of the chapter opener on page 512.)

this text. First, we used radar to set the scale of the solar system, as described in Chapter 2. ∞ (p. 43) In Chapter 17 we extended our measurements confidently out to about 100 pc using stellar parallax and then proceeded on to larger distances—a few thousand parsecs (kpc)—using spectroscopic parallax. ∞ (p. 371) In Chapter 23 we expanded our range still farther—out to many millions of parsecs (Mpc)—using observations of variable stars, especially Cepheid variables, whose large luminosities make them visible to very great distances. ∞ (p. 492) In a recent series of extensive observations using telescopes on the ground and in space, astronomers have measured Cepheids in several galaxies as far away as 15 Mpc. Figure 24.11 typifies the kind of work that can be done with these variable stars in an extraordinarily rich group of galaxies—members of the Virgo cluster, discussed later in this chapter.

THE LOCAL GROUP

Figure 24.12 sketches all the known major astronomical objects within about 1 Mpc of the Milky Way. Our Galaxy appears with its three known satellite galaxies—the two Magellanic Clouds and another recently discovered companion (as yet unnamed, but designated "Sagittarius" in the figure) lying almost within our own galactic plane. The Andromeda Galaxy, lying some 700 kpc from the Milky Way, is also shown. Two galactic neighbors of Andromeda (M31) can be seen in Figure 24.13. One of them, called M33 (Figure 24.13a), is a spiral. The other, M32 (Figure 24.13b), is a dwarf elliptical, easily seen in Figure 23.1 to the bottom right of Andromeda's central bulge.

All told, some 20 galaxies populate our Galaxy's neighborhood. Three of them (the Milky Way, Andromeda, and M33) are spirals; the remainder are dwarf irregular and elliptical systems. Together, these galaxies form the **Local Group**—the next highest level of structure known in the universe. As indicated in Figure 24.12, the Local Group's diameter is roughly 1 Mpc. The Milky Way and Andromeda are by far its largest members. The combined gravity of the

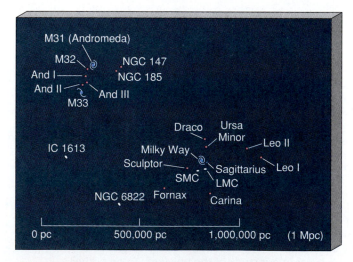

Figure 24.12 Diagram of the Local Group of some 20 galaxies within approximately 1 Mpc of our Milky Way Galaxy. Only a few are spirals; most of the rest are dwarf elliptical or irregular galaxies. Spirals and irregulars are shown in blue, ellipticals in red.

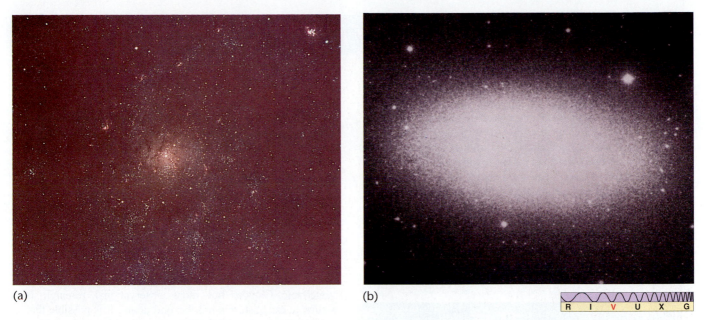

(a)

(b)

R I V U X G

Figure 24.13   The spiral M33 (a) and the dwarf elliptical M32 (b) are two well-known neighbors of the Andromeda Galaxy (M31). M32 is also visible in Figure 23.1, a larger-scale view of the Andromeda system.

galaxies in the Local Group binds them together, like stars in a star cluster, but on a millionfold larger scale. More generally, a group of galaxies held together by their mutual gravitational attraction is called a **galaxy cluster** (or *cluster of galaxies*). The Local Group is the rather small, irregularly shaped galaxy cluster that we happen to call home. Many other clusters are much larger.

Despite the great jump in scale we have just made, the principle of mediocrity continues to apply. ∞ (p. 38) Not only is Earth not the center of our solar system and the Sun not the center of our Galaxy, but our Galaxy is also not the center of the Local Group.

EXTENDING THE DISTANCE SCALE

2   In all, astronomers estimate that some *100 billion* other galaxies inhabit the observable universe. Although some reside close enough for the Cepheid technique to work (within about 20 Mpc), most known galaxies lie much farther away. Cepheid stars in very distant galaxies simply cannot be observed well enough, even through the world's largest telescopes, to measure their luminosity and period. To extend our distance-measurement ladder, then, we must find some new object to study. What individual objects are bright enough for us to observe at great distances?

Researchers have tackled this problem by using various *standard candles*—easily recognizable astronomical objects whose intrinsic brightnesses are confidently known. ∞ (p. 453) The basic idea is very simple. Once an object is identified as a standard candle—by its appearance or by the shape of its light curve, say—its luminosity can be estimated. Comparison of this luminosity with the object's *apparent* brightness then yields the object's distance. To be most useful, a standard candle must (1) have a narrowly defined luminosity, so that the uncertainty in estimating its brightness is small, and (2) be bright enough to be seen at large distances.

Astronomers have used many different standard candles over the years. Novae, emission nebulae, planetary nebulae, globular clusters, Type-I supernovae, even entire galaxies have been employed, with varying degrees of success. In fact, virtually every "bright" object discussed in this book has been considered at one time or another, so eager have astronomers been to expand their cosmic horizon. Not all of these standard candles have been equally useful, however—some have larger intrinsic spreads in their luminosities than others, making them less useful for distance-measuring purposes. Planetary nebulae and Type-I supernovae have proved particularly reliable, the latter being bright enough to use out to distances of hundreds of megaparsecs.

Just as we did with the lower rungs of our distance ladder, we calibrate the properties of each new standard candle using distances measured by more "local" techniques. In this manner, the distance-measurement process "bootstraps" its way to greater and greater distances—many hundreds of megaparsecs, by the time entire galaxies are used as standards. At the same time, the errors and uncertainties in each step continue to accumulate, so the distances to the farthest objects are the least well known.

An important alternative to standard candles was discovered in the 1970s, when astronomers found a clear correlation between the rotational speeds and the luminosities of spiral galaxies within a few tens of megaparsecs of the Milky Way. Because rotation speed is a measure of a galaxy's total mass, as we saw in Chapter 23, we should perhaps not be surprised that it is related to luminosity—the more mass a spiral galaxy has, the faster its disk rotates and the brighter it is. ∞ (p. 503) What *is* surprising is how tight the correlation is. The **Tully–Fisher relation**, as it is now known (after its discoverers), allows us to obtain a remarkably accurate estimate of a spiral galaxy's luminosity simply by observing how fast it rotates. As with Cepheid

Interlude 24-1   *The Clouds of Magellan*

Far to the south, out of viewing range of most of the Northern Hemisphere, reside two nearby galaxies called the *Magellanic Clouds*. These "Clouds of Magellan" are dwarf irregular galaxies, gravitationally bound to our own Milky Way. They orbit our Galaxy and accompany it on its trek through the cosmos. The Large Magellanic Cloud and its companion, the Small Magellanic Cloud, are visible to the naked eye from any location south of Earth's equator. Looking much like dimly luminous atmospheric clouds, they are named for the sixteenth-century Portuguese explorer Ferdinand Magellan, whose round-the-world expedition first brought word of these giant fuzzy patches of light to Europe. Figure 23.12 shows their size and position relative to the Milky Way. The accompanying figure presents them in more detail and indicates their relation to one another in the southern sky.

Studies of Cepheid variables within the Magellanic Clouds show them to be approximately 50 kpc from the center of our Galaxy. The larger cloud contains about 6 billion solar masses of material and is a few kiloparsecs across. The close-up of the Large Magellanic Cloud shows its distorted, irregular shape, although some observers claim they can discern a single spiral arm. Whatever its structure, direct observations show that this irregular galaxy contains lots of gas, dust, and blue stars (and the recent, well-known supernova discussed in *Interlude 21-2* on p. 452), indicating youthful activity and presumably current star formation. Both Clouds also contain many old stars and several old globular clusters, so we know that star formation has been going on there for a very long time.

Over the years, radio studies have hinted at a possible bridge of hydrogen gas connecting our Milky Way to the Magellanic Clouds, but more observational research is needed to establish this link beyond doubt. It is possible that the tidal force of the Milky Way tore this stream of gas from the Clouds the last time their orbits brought them close to our Galaxy. Of course, gravity works both ways, and many researchers believe that the forces exerted by the Clouds may also be responsible for distorting our Galaxy, warping and thickening the outer parts of the Galactic disk.

(a)

(b)

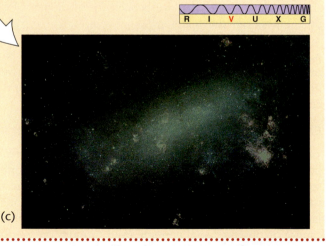

R   I   V   U   X   G

(c)

variables and standard candles, comparing the galaxy's (known) absolute brightness with its (observed) apparent brightness then yields its distance.

To see how the method is used in practice, imagine we are looking edge-on at a distant spiral galaxy (the Milky Way, say, seen from far outside). Let's suppose we are observing one particular emission line, as illustrated in Figure 24.14. Radiation from the side of the galaxy where matter is generally approaching us is blueshifted by the Doppler effect. Radiation from the other side (which is receding from us) is redshifted by a similar amount. The overall effect is that line radiation from the galaxy is "smeared out," or *broadened*, by the galaxy's rotation. The faster the rota-

tion, the greater the amount of broadening. Conversely, by measuring the amount of broadening, we can determine the galaxy's rotation speed. Once we know that, the Tully–Fisher relation tells us the galaxy's luminosity. As before, comparing the known absolute brightness (that is, the luminosity) with the measured apparent brightness allows us to determine the distance to the galaxy.

The particular line normally used in these studies actually lies in the radio part of the spectrum. It is the 21-cm line of cold, neutral hydrogen in the galactic disk. ∞ (p. 393) It is used in preference to optical lines because (1) optical radiation is strongly absorbed by dust in the disk under study and (2) the 21-cm line is normally

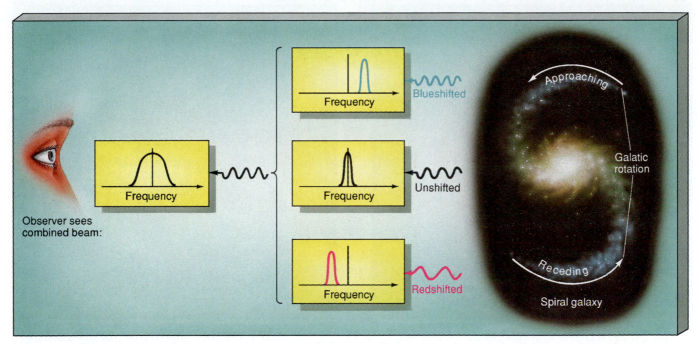

Figure 24.14   A galaxy's rotation causes some of the radiation it emits to be blueshifted and some to be redshifted (relative to what the emission would be from an unmoving source). From a distance, when the radiation from the galaxy is combined into a single beam and analyzed spectroscopically, the redshifted and blueshifted components combine to produce a broadening of its spectral lines. The amount of broadening is a direct measure of the rotation speed of the galaxy.

very narrow, making the broadening easier to observe. In addition, astronomers often use *infrared*, rather than optical, luminosities in the Tully–Fisher method to avoid absorption problems caused by dust, both in our own Galaxy and in others.

The Tully–Fisher relation can be used to measure distances to spiral galaxies out to about 200 Mpc (beyond which the line broadening becomes increasingly difficult to measure accurately). A somewhat similar connection exists for elliptical galaxies, linking the broadening of a galaxy's spectral lines (a measure of the average random velocity of the stars in the galaxy) and the galaxy's *size*. By measuring the broadening, astronomers determine the true size of the galaxy, which is then compared with the apparent size to give the distance. The important point here is that both methods bypass many of the standard candles we have described and so provide an independent means of determining distances to faraway objects. To emphasize their independence, Tully–Fisher and standard candles share the fifth rung of our distance ladder in Figure 24.10.

GALAXY CLUSTERS

Armed with these new distance-measurement techniques, we can now extend our survey of our cosmic neighborhood. Moving beyond the Local Group, the next large concentration of galaxies we come to is the Virgo Cluster. It lies about 20 Mpc from the Milky Way. Like the Local Group, the Virgo Cluster is held together by the mutual gravitational attraction of its member systems. But the Virgo Cluster does not contain a mere 20 individual galaxies; it houses approximately 2500 galaxies. Those galaxies are bound together in a tightly knit group about 3 Mpc across, each galaxy containing 100 billion or so individual stars. Figure 24.15 shows a view of the central portion of the Virgo Cluster.

R I V U X G

Figure 24.15   Part of the Virgo Cluster of galaxies, about 20 Mpc from Earth. Several large spiral and elliptical galaxies can be seen. The galaxy near the center is a giant elliptical known as M86.

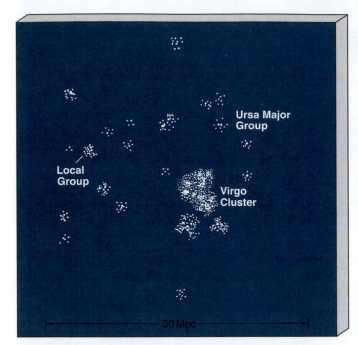

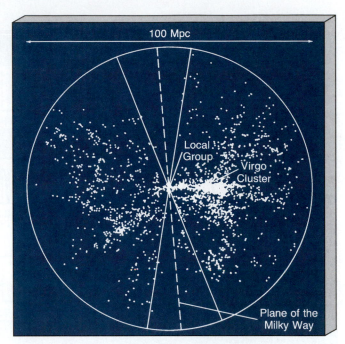

Figure 24.16   Schematic diagram of the locations of several galaxy clusters in our part of the universe. Our Milky Way is only one of these dots and our Local Group only one of the clusters of dots.

Figure 24.17   The Local Supercluster. Each of the 2200 points shown represents a galaxy, and the Sun is at the center of the diagram. The Virgo Cluster and the plane of our own Galaxy are marked. (Our Galaxy is seen edge-on. Its dust obscures our views to the top and the bottom, and two empty V-shaped regions on the map result.) The circle shown here is about 100 Mpc across.

The Virgo Cluster is not the only "nearby" group of galaxies. Figure 24.16 illustrates several well-defined clusters in our cosmic neighborhood, within about 30 Mpc of the Milky Way. Such a map clearly demonstrates that galaxies are not evenly spread throughout space—at least on scales of 50 Mpc or so.

Many thousands of galaxy clusters have now been identified and cataloged, and they come in many shapes and sizes. The largest, "rich" clusters contain many thousands of individual galaxies distributed fairly smoothly in space. The smallest, such as the Local Group, contain only a few galaxies and are quite irregular in shape. Not all galaxies are members of clusters, however. A few are apparently isolated systems, moving alone through intercluster space. Apart from these solo galaxies, is there any gaseous matter of any kind outside clusters? The answer seems to be "No." Astronomers have never found evidence of an intergalactic medium beyond any of the well-defined galaxy clusters. Evidently, when the clusters formed eons ago, they did so very efficiently, sweeping up all the matter within any given region of the universe. Matter definitely exists between the individual galaxies *within* many clusters, but the space between the galaxy clusters is apparently empty of luminous material.

CLUSTERS OF CLUSTERS

Does the universe have even greater groupings of matter, or do galaxy clusters top the cosmic hierarchy? Most astronomers now believe that the galaxy clusters themselves are clustered, forming titanic agglomerations of matter known as **superclusters**. Figure 24.17 shows the so-called *Local Supercluster*, containing the Local Group, the Virgo Cluster, and most of the other clusters shown in Figure 24.16. (The direction of view is different from that shown in the earlier figure, so some of the clusters noted there are difficult to discern here.) Each point represents a separate galaxy, and the diagram is centered on the Milky Way. The perspective is such that the disk of our Galaxy is seen edge-on and runs vertically up the page. The two nearly empty V-shaped regions at the top and bottom of the figure are not devoid of galaxies—they are simply obscured from view by the dust in our own Galaxy's plane. The total mass of the entire supercluster—which is centered in the Virgo Cluster—is about 10^{15} solar masses. By now it should come as no surprise that the Local Group is not found at the heart of the Local Supercluster (even though Figure 24.17 is centered on the Local Group). Instead, it lies far off in the periphery, some 20 Mpc from the center.

The Local Supercluster contains a huge number of individual galaxies—perhaps several tens of thousands—yet the great majority of known galaxy clusters and superclusters lie far beyond its edge. Figure 24.18 is a long-exposure photograph of one such remote cluster. Called the Corona Borealis Cluster, this rich cluster is far outside the limit of the circle shown in Figure 24.17. It is only one of many other large and distant groups of galaxies scattered throughout the observable universe. The Coma Cluster, shown in Figure 24.1, is another. On and on, the picture is

Figure 24.18 The galaxy cluster CL 0939+4713 contains huge numbers of galaxies and resides roughly a billion parsecs from the Earth. Every patch of light in this photograph is a separate galaxy. Thanks to the improved imaging capability of the *Hubble Space Telescope,* we can now discern, even at this great distance, spiral structure in some of the galaxies. In addition, we can see galaxies in collision—some tearing matter from one another, others merging into single systems.

much the same. The farther we peer into deep space, the more galaxies, clusters of galaxies, and superclusters we see. Is there structure on even larger scales? As we will see in a moment, the answer is still "yes!"

But we are moving too fast in our intellectual rush toward the limits of the universe. We are missing some interesting and important information about the nature of our extended cosmic neighborhood—within a billion or so parsecs, if we stretch our distance-measurement techniques to their limits. Let's pause for a moment to reconsider the properties of galaxies and galaxy clusters before making our final leap in distance.

24.3 *Galaxy Masses*

❹ We have quoted several galaxy masses in this chapter without giving any indication of how they were determined. Because mass is such an important physical property of any object, let's consider some of the methods used to weigh galaxies and galaxy clusters.

How can we find the masses of such huge systems? Surely, we can neither count all their stars nor estimate their interstellar content very well. Galaxies are just too

complex to take direct inventory of their material makeup. Instead, we must rely on indirect techniques. As usual, we turn to Newton's law of gravity.

MASS MEASUREMENT

Astronomers can calculate the masses of some spiral galaxies by determining their *rotation curves,* which plot rotation speed, obtained by measuring the Doppler shift of various spectral lines, versus distance from the galactic center. ∞ (p. 498) The mass within any given radius then follows directly from Newton's laws. ∞ (p. 504) Some rotation curves for nearby spirals are shown in Figure 24.19. They imply masses ranging from about 10^{11} to 5×10^{11} solar masses within about 25 kpc of the center—quite comparable to the results obtained for our own Milky Way using the same technique. Distant galaxies are generally too far away for such detailed curves to be drawn. Nevertheless, by observing the *broadening* of spectral lines—as discussed earlier in the context of the Tully–Fisher relation—we can still measure the overall rotation speed. Estimating the galaxy's size then leads to an estimate of its mass.

These methods are useful only for measuring the mass of a galaxy within about 50 kpc of the galactic center—the extent of the electromagnetic emission from stellar and interstellar material. To probe farther from the center, galactic astronomers turn to binary and multiple systems of galaxies, much as stellar astronomers study binary stars to determine stellar masses. If we knew the orbit size and period of a galaxy–galaxy binary system, the combined mass of the pair could be inferred from Kepler's third law, just as described in Chapter 17. ∞ (p. 373) But there is a problem here. Unlike the stellar case, we cannot watch the galaxies travel even a small fraction of their entire orbit.

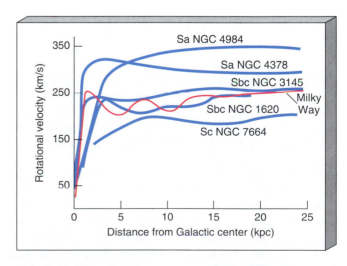

Figure 24.19 Rotation curves for some nearby spiral galaxies indicate masses of a few hundred billion times the mass of the Sun. The corresponding curve for our own Galaxy (Figure 23.19) is marked in red for comparison.

Instead, as sketched in Figure 24.20(a), the orbital period is usually simply estimated on the basis of two observations: (1) the velocities of the galaxies measured along our line of sight and (2) the present distance between the two galaxies. Masses obtained in this way, then, are fairly uncertain. However, we can combine many such measurements to obtain quite reliable *statistical* information about galaxy masses.

From investigations using these methods, we find that most normal spirals (the Milky Way included) and large ellipticals contain between 10^{11} and 10^{12} solar masses of material. Irregular galaxies often contain less mass, about 10^8 to 10^{10} times that of the Sun. Dwarf ellipticals and irregulars can contain as little as 10^6 to 10^7 solar masses of material.

We can use another statistical technique to derive the combined mass of all the galaxies within a galaxy *cluster*. As depicted in Figure 24.20(b), each galaxy within a cluster moves relative to all other cluster members. As with binary galaxies, we cannot watch galaxies move around within a cluster—typical orbit periods are billions of years. But we can estimate the cluster's mass simply by determining, using Newtonian mechanics, how massive it must be in order to bind its galaxies gravitationally. Typical cluster masses obtained in this way lie in the range of 10^{13}–10^{14} solar masses. Notice that this calculation gives us *no* information whatever about the masses of individual galaxies. It tells us only about the *total* mass of the entire cluster.

DARK MATTER IN THE UNIVERSE

3 **5** Radio observations indicate that the rotation curves of many spiral galaxies, such as those shown in Figure 24.19, remain "flat" (that is, do not decline and may even rise slightly) far beyond the visible image of the galaxy. We conclude that these spiral galaxies—and perhaps all spiral galaxies—must have invisible dark halos similar to that surrounding the Milky Way. ∞ (p. 505) Depending on how far out these halos extend (and observations appear to indicate that they reach at least as far as 50 kpc from the center of a Milky Way–sized galaxy), spiral galaxies may contain 3 to 10 times more mass than can be accounted for in the form of visible matter. Some studies of elliptical galaxies suggest similarly large dark halos surrounding them too, but these observations are still far from conclusive.

Astronomers find an even greater discrepancy when they study galaxy clusters. Calculated cluster masses range from 10 to nearly 100 times the mass suggested by the light emitted by individual cluster galaxies. Stated another way, a lot more mass is needed to bind galaxy clusters than we can see. The problem of dark matter exists, then, not just in our own Milky Way Galaxy, but also in other galaxies and, to an even greater degree, in galaxy clusters as well. It most likely applies to the entire universe. In that case, we must accept the fact that *upward of 90 percent of the universe is*

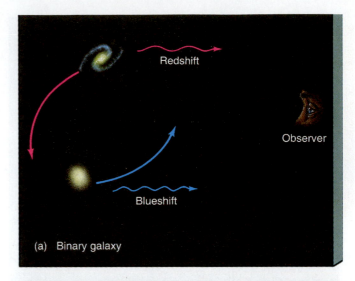

(a)   Binary galaxy

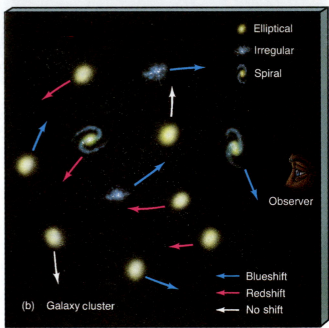

(b)   Galaxy cluster

Figure 24.20   (a) In a binary galaxy, galaxy masses may be estimated by observing the orbit of one galaxy about another. (b) In a galaxy cluster, cluster masses are measured by observing the motion of many galaxies in the cluster and then estimating how much mass is needed to prevent the cluster from flying apart.

composed of dark matter. As noted in Chapter 23, this matter is not just dark in the visible portion of the spectrum—it is invisible at *all* electromagnetic wavelengths. ∞ (p. 505)

What could the dark matter within the galaxy clusters be? Where could it hide? At present, we don't know. As discussed in Chapter 23, many solutions have been suggested, from stellar remnants of various sorts to exotic subatomic particles. ∞ (p. 505) Whatever the answer, the dark matter in clusters apparently cannot be simply the accumulation of smaller amounts of dark matter within indi-

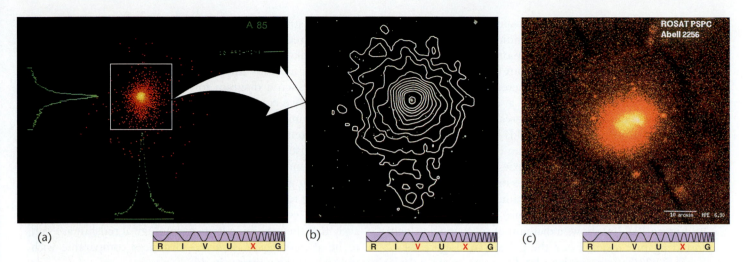

Figure 24.21   (a) X-ray image of Abell 85, an old, distant cluster of galaxies, taken by the *Einstein* X-ray satellite observatory. The cluster's X-ray emission is shown in orange. The green graphs display a smooth, peaked intensity profile centered on the cluster but not associated with individual galaxies. (b) The contour map of X rays is superimposed on an optical photo, showing its X rays peaked on Abell 85's central supergiant galaxy. Images like these demonstrated for the first time that the space between the galaxies within galaxy clusters is filled with superheated gas. (c) A ROSAT X-ray image of hot gas within another cluster of galaxies (called Abell 2256). The cluster is nearly a billion pc from Earth and measures about 3 Mpc in diameter.

vidual galaxies—even including the galaxies' dark halos, we cannot account for all the dark matter in galaxy clusters. As we look on larger and larger scales, we find that more and more of the matter in the universe is dark.

INTRACLUSTER GAS

Could the dark matter be diffuse intergalactic matter existing among the galaxies within the clusters—that is, intracluster gas? Until the late 1970s, astronomers had no observational evidence for intergalactic matter, either inside or outside galaxy clusters. Then satellites orbiting above the Earth's atmosphere detected substantial amounts of X-ray radiation in the direction of many galaxy clusters. Figure 24.21 shows false-color X-ray images of two such clusters. The X-ray-emitting region is centered on, and comparable in size to, the visible image of the cluster. These X-ray observations demonstrated for the first time the existence of large amounts of invisible hot gas—about 1 million K—within clusters. However, *no* gas has been observed outside the clusters—no "extracluster" matter has ever been found.

How much matter have the X-ray satellites found? The observations suggest that at least as much matter—and in a few cases substantially more—exists within clusters in the form of hot gas as is visible within them in the form of stars. This is a lot of material, but it still doesn't solve the dark-matter problem. To account for the total masses of galaxy clusters implied by dynamical studies, we would have to find from 10 to 100 times more mass in gas than in stars. Recently, X-ray observations of gas within one small group of galaxies actually compounded the problem. The gas was so hot that the amount of dark matter needed to

bind it to the cluster and prevent it from dispersing into intercluster space was far greater than had been previously suspected to exist!

24.4 *Hubble's Law*

❷ ❻ The motion of individual galaxies within galaxy clusters is essentially random. You might expect that, on the largest possible scales, the clusters themselves would also be found to have random, disordered motion—some clusters moving this way, some that. This is not the case at all. On the largest scales, galaxies that are not members of a cluster and galaxy clusters alike display very ordered motion.

UNIVERSAL RECESSION

In 1912, the American astronomer Vesto M. Slipher, working under the direction of Percival Lowell, discovered that virtually every spiral nebula (that is, spiral galaxy) he observed had a redshifted spectrum—it was apparently *receding* from our Galaxy. In fact, with the exception of a few nearby systems, *every* known galaxy is part of a general motion away from us in all directions. Individual galaxies that are not part of galaxy clusters are steadily receding. Galaxy clusters too have an overall recessional motion, although their individual member galaxies move randomly with respect to one another. (Consider a jar full of fireflies that has been thrown into the air. The fireflies within the jar, like the galaxies within the cluster, have random motions due to their individual whims, but the jar as a whole, like the galaxy cluster, has some directed motion as well.)

Figure 24.22 shows the optical spectra of several galaxies. These spectra are redshifted, indicating that the galaxies are steadily receding. Furthermore, the extent of the redshift increases progressively from top to bottom in the figure. Because the distance also increases from top to bottom, we conclude that there is a connection between Doppler shift and distance: The greater the distance, the greater the redshift. This trend holds for nearly all galaxies in the universe. (Two galaxies within our Local Group, including Andromeda, and a few galaxies in the Virgo Cluster display blueshifts and so have some motion toward us, but this results from their random motions within their parent clusters. Recall the fireflies in the jar.)

Figure 24.23(a) shows a diagram of recessional velocity plotted against distance for the galaxies of Figure 24.22. Figure 24.23(b) is a similar plot for some more galaxies within 1 billion pc of the Earth. Plots like these were first made by Edwin Hubble in the 1920s and now bear his name—*Hubble diagrams*. The data points generally fall close to a straight line, indicating that a simple relationship connects recessional velocity and distance: The rate at which a galaxy recedes is *directly proportional* to its distance away from us. This rule is called **Hubble's law**. We could construct such a diagram for any group of galaxies, provided that we could determine their distances and velocities. The universal recession described by the Hubble diagram is often called the *Hubble flow*.

To distinguish recessional redshift from redshifts caused by motion *within* an object—for example, galaxy orbits within a cluster or explosive events in a galactic nucleus—the redshift resulting from the Hubble flow is often called the **cosmological redshift**. Objects that lie so far away that they exhibit a large cosmological redshift are said to be at *cosmological distances*—distances comparable with the scale of the universe itself.

Hubble's law is an *empirical* discovery—that is, a discovery based strictly on observational results. Its central relationship—a statistical correlation between recessional ve-

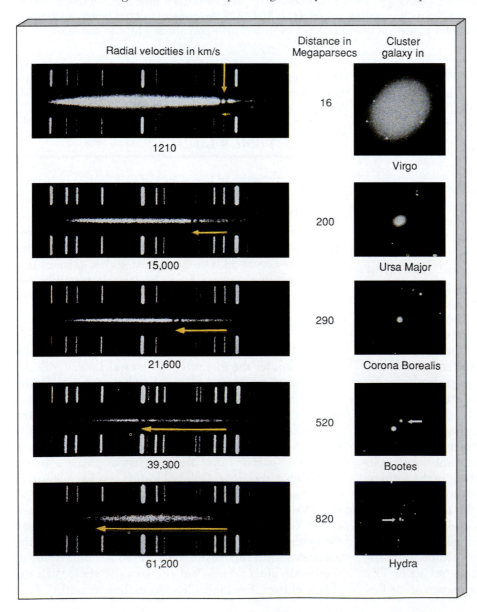

Figure 24.22 Optical spectra (on the left) of several different galaxies (on the right). The extent of the redshift (denoted by the horizontal yellow arrows) and the distance to each galaxy (in the center) increase from top to bottom.

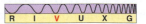

locity and distance—is well documented as far as galaxy distances can be reliably determined, but no law of nature demands that all galaxies recede, and no law of physics requires that a link exist between velocity and distance. In that sense, Hubble's "law" is not really a law at all. It is strictly a convenient way of noting the *observational* fact that any galaxy's recessional velocity is directly proportional to its distance from us.

The recessional motions of the galaxies prove that the cosmos is not steady and unchanging on the largest scales. Its contents are in constant relative motion, and the motion is not random. In fact, the universe is expanding—and expanding in a directed fashion. In short, it is evolving. But before going any further, let's be clear on *what* is expanding and what is not. Hubble's law does *not* mean that humans, the Earth, the solar system, or even individual galaxies are physically increasing in size. These groups of atoms, rocks, planets, and stars are held together by their own internal forces and are not themselves getting bigger. Only the largest framework of the universe—the ever-increasing distances separating the galaxies and the galaxy clusters—is expanding.

Hubble's law has some fairly obvious and dramatic implications. If nearly all galaxies show recessional velocity according to Hubble's law, then doesn't that mean that they all started their journey from a single point? If we could run time backward, wouldn't all the galaxies fly back to this one point, perhaps the site of some explosion in the remote past? In Chapter 26 we will explore the ramifications of the Hubble flow for the past and future evolution of our universe. For the remainder of this chapter, however, we will set aside its cosmic implications and use Hubble's law simply as a convenient distance-measuring tool.

HUBBLE'S CONSTANT

We can quantify Hubble's law to make it more useful. The constant of proportionality between recessional velocity and distance is known as **Hubble's constant**. It is denoted by the symbol H_0. The data shown in Figure 24.23 obey the following equation:

recessional velocity = Hubble's constant × distance

or

$$V = H_0 \times D.$$

The value of Hubble's constant is the slope of the straight line—recessional velocity divided by distance—in Figure 24.23(b). Reading the numbers off the graph, this comes to roughly 75,000 km/s divided by 1000 Mpc, or 75 km/s/Mpc (kilometers per second per megaparsec, the most commonly used unit for H_0). Astronomers continually strive to refine the accuracy of the Hubble diagram and the resulting estimate of H_0 because Hubble's constant is one of the most fundamental quantities of nature—it specifies the rate of expansion of the entire cosmos.

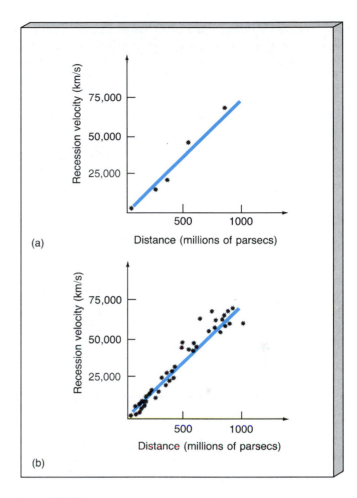

(a)

(b)

Figure 24.23   Plots of recessional velocity against distance (a) for the galaxies shown in Figure 24.22, and (b) for numerous other galaxies within about 1 billion parsecs of the Earth.

The precise value of the Hubble constant is still the subject of considerable debate among astronomers. In the 1970s, astronomers obtained a value of around 50 km/s/Mpc, using a chain of standard candles to extend their observations to large distances. However, in the early 1980s, when the (infrared) Tully–Fisher technique had become fairly well established, other researchers used it to obtain a measurement of the Hubble constant that was largely *independent* of methods relying on standard candles. From observations of galaxies within about 150 Mpc, the latter group deduced a value of $H_0 = 90$ km/s/Mpc, a result inconsistent with the earlier measurements (even allowing for the estimated uncertainties involved). For some reason, the distances obtained using the Tully–Fisher method were only about half those determined by using standard candles, and so the Hubble constant nearly doubled in value using this approach.

Subsequent determinations of the Hubble constant by other researchers, using different galaxies and a variety of distance-measurement techniques, have yielded results mostly within the range 60–90 km/s/Mpc, and most astronomers would be quite surprised if the "true" value of H_0 turned out to lie outside this range. For now, however, in the absence of any good explanation for the spread, astronomers must live with this uncertainty. Some like to "split the difference" and use $H_0 = 75$ km/s/Mpc. We will adopt this compromise value as the best current value for Hubble's constant in the remainder of the text.

THE COSMIC DISTANCE SCALE

One very important application of Hubble's law is its use as a means of determining distances. Using Hubble's law, we can derive the distance to a remote object simply by measuring the object's recessional velocity. Even within the distances accessible by other methods, Hubble's law is often the most convenient means of distance measurement.

The method works like this: An astronomer measures the redshift of the object's spectral lines. The extent of the shift is then converted to velocity, using the Doppler relationship of Chapter 3. ∞ (p. 69) Knowing the object's velocity, the astronomer then finds its distance by using the plot of Figure 24.23(b). Notice, however, that the uncertainty in the Hubble constant translates directly into a similar uncertainty in the distance determined by this method.

Using Hubble's law in this way tops our inverted pyramid of distance-measurement techniques. Sketched (for the last time) in Figure 24.24, this sixth method simply assumes that Hubble's law holds. If this assumption is correct, Hubble's law enables us to measure great distances in the universe—so long as we can obtain an object's spectrum, we can determine how far away the object is.

Many redshifted objects have recessional motions that are a substantial fraction of the speed of light. The most distant object thus far observed in the universe has the peculiar catalog name of QO051-279. Its extremely high redshift

implies a recessional velocity 93 percent that of light. At that speed, ultraviolet spectral lines are Doppler-shifted all the way into the far infrared! Hubble's law predicts that QO051-279, solely on the basis of its observed redshift, lies more than 4000 Mpc away. This object resides as close to the edge of the observable universe as astronomers have yet been able to probe.

The speed of light is finite. It takes time for light or any kind of radiation to travel from one point in space to another. The radiation that we now see from distant objects originated long ago. Incredibly, the radiation that astronomers now detect from QO051-279 left that object some 14 billion years ago, well before our planet, our Sun, or even our Galaxy came into being.

LARGE-SCALE STRUCTURE

❸ Using Hubble's law, we can complete our census of the large-scale distribution of galaxies. One of the most extensive surveys of the universe yet undertaken is being carried out by astronomers at the Center for Astrophysics (CfA) at Harvard University. Using Hubble's law as their distance indicator, these researchers are compiling a catalog of the positions and redshifts of all galaxies within about 300 Mpc of the Milky Way. This is an extremely painstaking task—even with a large telescope, it takes a long time to obtain a detailed spectrum of a distant galaxy. Rather than trying to cover the entire sky at once, the team instead has elected to map the universe in a series of wedge-shaped "slices," each 6° thick, starting in the northern sky. The first slice, shown in Figure 24.25(a), covers a region of the sky containing the Coma Cluster (Figure 24.1), which happens to lie in a direction almost perpendicular to our Galaxy's plane.

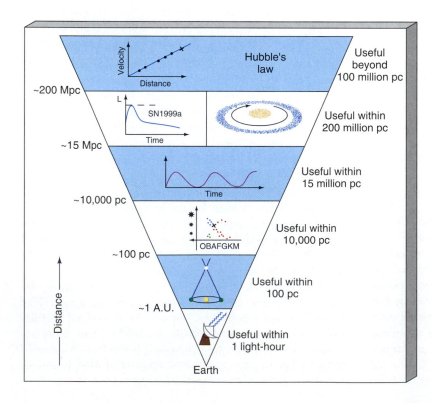

Figure 24.24 Hubble's law tops the inverted pyramid of distance techniques. This last method is used to find the distances of astronomical objects all the way out to the limits of the observable universe.

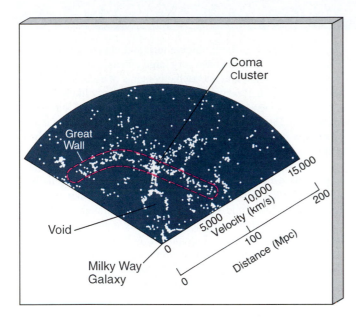

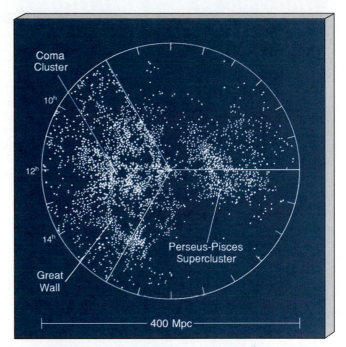

Figure 24.25 The first slice of a survey of the universe, covering 1057 galaxies out to an approximate distance of 200 Mpc, clearly shows that galaxies and clusters are not randomly distributed on large scales. Instead, they appear to have a filamentary structure, surrounding vast, nearly empty voids. The distances shown assume $H_0 = 75$ km/s/Mpc.

Figure 24.26 Combination of data from several redshift surveys of the universe reveal the extent of large-scale structure within 200–300 Mpc of the Sun. The arc on the left is the Great Wall. The empty regions are mostly areas obscured by our Galaxy. Positions for more than 4500 galaxies are plotted here. We assume a Hubble constant of 75 km/s/Mpc.

The most striking feature of maps such as this is that the distribution of galaxies on large scales is decidedly nonrandom. The galaxies appear to be arranged in a network of strings, or filaments, surrounding large, relatively empty regions of space known as **voids**. The biggest voids measure some 100 Mpc across. For a time, they were the largest objects in the universe known to astronomers. The most likely explanation for the voids and filamentary structure in Figure 24.25 is that the galaxies and galaxy clusters are spread across the surfaces of vast "bubbles" in space. The voids are the interiors of these gigantic bubbles. The galaxies only seem to be distributed like beads on strings because of the way we view them on the edges of the bubbles. Like suds on soapy water, these bubbles fill the entire universe. The densest clusters and superclusters lie in regions where several bubbles meet.

The notion that the filaments are just the intersection of the survey slice with much larger structures (the bubble surfaces) was confirmed when the next three slices of the survey, lying above and below the first, were completed (see p. 512). The region of Figure 24.25 indicated by the dashed line was found to continue through both the other slices, so we know that it covers *at least* 36° on the sky perpendicular to the outline in the figure. This extended sheet of galaxies, which has come to be known as the *Great Wall*, measures at least 70 Mpc (out of the plane of the page) by 200 Mpc (across the page). For now at least, it is one of the largest known structures in the universe. Figure 24.26 combines all the available CfA data with results of other surveys of the southern sky. (The large empty regions are obscured by our Galaxy or are not yet mapped.) The Great Wall can be seen arcing around the left side of the figure.

What might be the origin of this "frothy" distribution of galaxies? For a time, theorists tried to explain the voids as gigantic cavities hollowed out by the explosions of extremely massive stars early in the history of the universe, but this explanation does not work. The energy requirements are just too great, and the explosions would have had other observable consequences (mostly related to the radiation they would have released) that simply are not seen. Most theorists now believe that all large-scale structure in the universe (on scales larger than a few megaparsecs) is the result of small "ripples" (density fluctuations) in the early universe that became unstable and grew in time, eventually forming the large inhomogeneities we see today. We will return to this subject in Chapter 27.

Are there still larger structures in the universe? Only with more extensive surveys will we know for sure. All large-scale surveys performed to date have detected structures comparable in size to the region surveyed—in other words, they have found objects as large as the largest object they could have hoped to find. Observers and theorists alike are eager to find out how much farther this trend continues. Many theorists now believe that all structure on scales larger than a few megaparsecs traces its origin directly to the conditions found in the very earliest stages of the universe. As such, these studies of large-scale structure may be vital to our efforts to understand the origin and nature of the cosmos itself.

Contemplating the congested confines of a rich galaxy cluster (such as Virgo or Coma, with thousands of member galaxies orbiting within a few megaparsecs), we might expect that collisions among galaxies would be common. Gas particles collide in our atmosphere and hockey players collide in the rink—do galaxies in clusters collide too? The answer is "Yes."

The next two images show evidence supporting this assertion. One is an optical photograph of a small group of five galaxies in the constellation Serpens. Connecting clouds seem to link some of them, strongly suggesting that they are (or have just been) interacting with one another. The other image is a computer-enhanced view of the pair of galaxies NGC 4676 A and B (also known as "The Mice"), which show streams of gas and stars apparently generated by the encounter between the two.

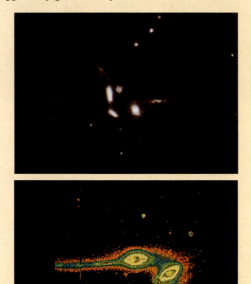

R   I   V   U   X   G

Whether these galaxies are genuinely colliding or only experiencing a close encounter cannot easily be determined. No human will ever witness an entire collision, for it would last many millions of years. However, computer simulations of these systems display formations remarkably similar to the real thing. The particular simulation shown in the third figure began with two spiral galaxies, but the details of the original structure have been largely obliterated by the collision. Notice the similarity to the real image of NGC 6240 (last figure), an object showing faint tails as well as double galactic centers only a few hundred parsecs apart. The calculations show that ultimately the two galaxies merge into one.

Direct observational evidence now indicates that galaxies in clusters apparently collide quite often. In the smaller groups, the galaxies' velocities are low enough that interacting galaxies tend to "stick together," and *mergers,* as shown in the computer simulation, are a very common outcome. In larger groups, galaxies move faster and tend to pass through one another without sticking. Since the early 1980s, it has become increasingly

clear, on the basis of both observations and numerical simulations, that collisions can have very large effects on the galaxies involved. The stellar and interstellar contents of each galaxy are rearranged, and the merged interstellar matter very likely experiences shock waves that trigger widespread bursts of star formation (see Figure 24.29). Some researchers would go so far as to suggest that *most* galaxies have been strongly influenced by collisions, in many cases in the relatively recent past.

Curiously, although a collision may wreak havoc on the large-scale structure of the galaxies involved, it has essentially *no* effect on the individual stars they contain. The stars within each galaxy just glide past one another. Although we have plenty of photographic evidence for galaxy collisions, no one has ever witnessed or photographed a collision between two stars. Stars *do* collide in other circumstances—in the dense central cores of galactic nuclei and globular clusters, or as a result of stellar evolution in binary systems—but stellar collisions are a very rare consequence of galaxy interactions.

To understand why individual stars do not collide when galaxies collide, recall that the galaxies within a typical cluster are bunched together fairly tightly. The distance between adjacent galaxies in a given cluster averages a few hundred thousand parsecs, which is only about 10 times greater than the size of a typical galaxy. Galaxies simply do not have that much room to roam around without bumping into one another. By contrast, stars within a galaxy are spread much thinner. The average distance between stars within a galaxy is a few parsecs, which is millions of times greater than the size of a typical star. When two galaxies collide, the star population merely doubles for a time, and the stars continue to have so much space that they do not run into each other. The stellar and interstellar contents of each galaxy are certainly rearranged, and the resultant burst of star formation may indeed be spectacular from afar, but from the point of view of the stars, it's still clear sailing.

R   I   V   U   X   G

24.5 *The Formation and Evolution of Galaxies*

GALAXY FORMATION

7 We saw earlier that astronomers know of no evolutionary connection among the various types of galaxies in the Hubble classification scheme. How then did those different galaxies come into being? To address this question, we must understand how galaxies formed. Unfortunately, the theory of galaxy formation is still very much in its infancy, and no definitive answers yet exist. Do we understand galaxy formation as well as, say, star formation? The answer is a resounding *no*! The theory is presently unable to answer completely even a basic question such as why spirals and ellipticals exist at all, let alone offer an explanation of the Hubble sequence (Figure 24.9) or predict the different evolutionary tracks that galaxies might take.

There are several good reasons for this lack of understanding of galaxy formation: Galaxies are much more complex than stars, they are harder to observe, and the observations are harder to interpret. We have only a partial understanding, and no observations, of the conditions in the universe immediately preceding galaxy formation, quite unlike the corresponding situation for stars, as discussed in Chapter 19. ∞ (p. 402) Finally, whereas stars almost never collide with one another, so single stars and binaries evolve in isolation, galaxies may suffer many collisions and mergers during their lives (see *Interlude 24-2* on p. 532), making it much harder to decipher their pasts. Yet despite all these difficulties, some general ideas have begun to gain widespread acceptance, and we can offer a few insights into the processes responsible for the galaxies we see.

The seeds of galaxy formation were sown in the very early universe, when small density fluctuations in the primordial matter began to grow. For now, let us begin our discussion with these "pregalactic" blobs of gas already formed. The masses of these fragments were quite small—perhaps only a few million solar masses, comparable to the masses of the smallest present-day dwarf galaxies, which may in fact be remnants of this early time. So where did the larger galaxies we find today come from? The key point in our current understanding is the realization that galaxies grow by repeated *merging* of smaller objects, as illustrated in Figure 24.27. Galactic formation is a "bottom-up" process, opposite to the "top-down" process of star formation, in which a large cloud fragments into smaller pieces that eventually become stars.

Theoretical evidence for this scenario is provided by computer simulations of the early universe, which clearly show merging taking place. Further strong support comes from recent observations that indicate that galaxies at large distances (well over 1000 Mpc away, meaning that the light we see was emitted more than 4 billion years ago) appear distinctly smaller and more irregular than those found nearby. Figure 24.28 shows one of these images. The rather vague bluish patches are separate small galaxies, each containing only a few percent of the mass of the Milky Way. Their irregular shape is thought to be the result of galaxy mergers; the bluish coloration comes from young stars that formed during the merger process.

Given that galaxies form by repeated mergers, how might we account for the differences between spirals and ellipticals? The answer is still unclear. One apparently very important factor is just when and where stars first appeared—in the original blobs, during the merger process, or later—and how much gas was used up or

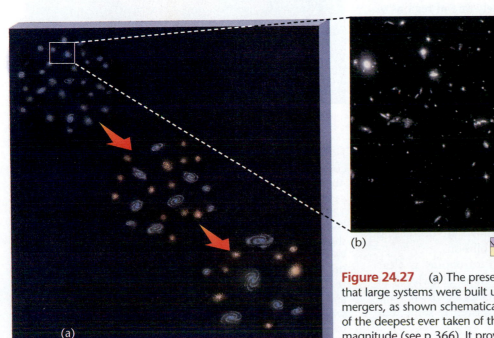

(b)

R I V U X G

Figure 24.27   (a) The present view of the formation of galaxies holds that large systems were built up from smaller ones through collisions and mergers, as shown schematically in this drawing. (b) This photograph, one of the deepest ever taken of the universe, shows objects down to 29th magnitude (see p.366). It provides "fossil evidence" for hundreds of galaxy shards and fragments, most about 3000 Mpc distant.

(a)

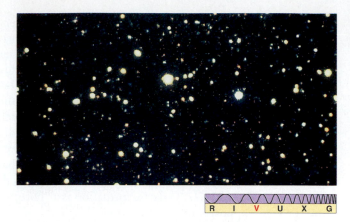

R I V U X G

Figure 24.28   Numerous small, irregularly shaped young galaxies can be seen in this very deep optical image (the 6-hour-long exposure captured objects as faint as the 26th magnitude). Redshift measurements indicate that the galaxies lie well over 1000 Mpc from the Earth. Their size, color, and irregular appearance support the theory that galaxies grew by merger and were smaller and less regular in the past. The entire field of view is only 3 by 5 arc minutes.

R I V U X G

Figure 24.29   The peculiar galaxy NGC 1275 contains a system of long filaments that seem to be exploding outward into space. Its blue blobs, as revealed by the *Hubble Space Telescope,* are probably young globular clusters formed by the collision of two galaxies.

ejected from the young galaxy in the process. If many stars formed early on and little gas was left over, an elliptical galaxy would be a likely outcome, with many old stars on random orbits and no gas to form a central disk. Alternatively, if a lot of gas remained, it would tend to sink to a central plane and form a rotating disk—in other words, a spiral galaxy would result. However, it is not known what determines the time, the place, or the rate of star formation, so whether spirals and ellipticals can form in basically the same environment, or if they tend to form in different places, is still an open question.

We do have some important clues to guide us, though. For example, spiral galaxies are relatively rare in regions of high galaxy density, such as the central regions of rich galaxy clusters. Is this because they simply tended not to form there, or is it because their disks are so fragile that they are easily destroyed by collisions and mergers, which are more common in dense galactic environments? Computer simulations suggest that collisions between spiral galaxies can indeed destroy the spirals' disks, ejecting much of the gas into intergalactic space and leaving behind objects that look very much like ellipticals. Recent observations of interacting galaxies appear to support this scenario. Figure 24.29 shows an example of this phenomenon. Additional evidence comes from the observed fact that spirals are more common at large distances (that is, in the distant past), which implies that their numbers are decreasing with time, presumably as the result of collisions.

However, nothing in this area of astronomy is clear cut. We know of numerous isolated elliptical galaxies in rather low-density regions of the universe, which are hard to explain as the result of mergers. Apparently some, but *not* all, ellipticals formed in this way.

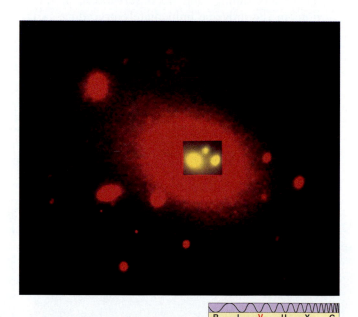

R I V U X G

Figure 24.30   This computer-enhanced, false-color composite optical photograph of the galaxy cluster known as Abell 2199 is thought to show an example of galactic cannibalism. The large central galaxy of the cluster (itself 120 kpc along its long axis) is displayed with a superimposed "window." (This results from a shorter time exposure, which shows only the brightest objects that fall within the frame.) Within the core of the large galaxy are several smaller galaxies (the three bright yellow images at center) apparently already "eaten" and now being "digested" (that is, being torn apart and becoming part of the larger system). Other small galaxies swarm on the outskirts of the swelling galaxy, almost certainly to be eaten too.

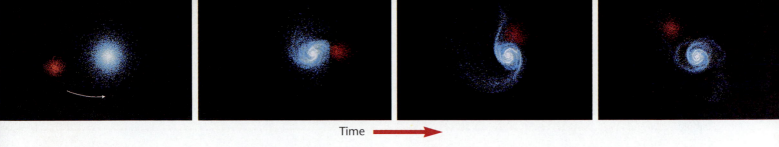

Time ➡

Figure 24.31 Galaxies can change their shapes long after their formation. In this computer-generated sequence, two galaxies closely interact over several hundred million years. The smaller galaxy, in red, has gravitationally disrupted the larger galaxy, in blue, changing it into a spiral galaxy. Compare the result of this supercomputer simulation with a photograph (Figure 24.2b) of M51 and its small companion.

MERGERS AND ACQUISITIONS

A related, and very important, question is: When did the galaxy formation process *stop*? Astronomers are divided on this issue. Some maintain that there was a fairly well-defined time in the past—given by the age of the globular clusters, for example—by which most formation was over. Others point out that many galaxies show evidence of repeated mergers and the accumulation of smaller satellite galaxies over an extended period of time—even up to the present. These astronomers suggest that the many galaxy interactions we observe today (see *Interlude 24-2* on p. 532) are just part of the same process begun when the first fragments merged. In the latter view, galaxy formation is still occurring today. In either case, astronomers have ample evidence that galaxies evolve in response to external factors, even long after they first formed.

We now know that spiral galaxies have huge, invisible dark halos surrounding them, and we strongly suspect that *all* galaxies have similar halos. Consider two galaxies orbiting one another—a binary galaxy system. As they orbit, the galaxies interact with each other's halos, one galaxy stripping halo material from the other by tidal forces. The freed matter is either redistributed within a common envelope or is entirely lost from the binary system. This interaction between the halos changes the orbits of the galaxies themselves. The galaxies tend to spiral toward one another, eventually merging. If one galaxy of the pair happens to have a much lower mass than the other, the process is colloquially termed *galactic cannibalism*. Such cannibalism might explain why supermassive galaxies are often found at the cores of rich galaxy clusters. Having dined on their companions, they now lie at the center of the cluster, waiting for more food to arrive. Figure 24.30 is a remarkable combination of images that has apparently captured this process at work.

Now consider two interacting disk galaxies, one a little smaller than the other but each having a mass comparable to our Milky Way. As shown in the computer-generated frames of Figure 24.31, the smaller galaxy can substantially distort the larger one, causing spiral arms to appear where none existed before. The entire event requires several hundred million years—a span of evolution that supercomputers can model in minutes.

The final frame of Figure 24.31 looks remarkably similar to the double galaxy shown in Figure 24.2(b). Shown there are two galaxies with sizes, shapes, and velocities corresponding very closely to those in the computer simulation. The magnificent spiral galaxy is M51, popularly known as the Whirlpool Galaxy, about 10 Mpc from Earth. Its smaller companion is an irregular galaxy that may have drifted past M51 millions of years ago. Did this smaller galaxy cause the spiral structure we see in M51? Does the model mirror reality? Perhaps. We need more evidence from other galaxies to confirm the accuracy of these and similar simulations. Still, the computer simulation does demonstrate a plausible way that two galaxies might have interacted millions of years ago and how spiral arms might be created or enhanced as a result.

It now seems that many of the most spectacular changes that occur in galaxies result from interactions with other galaxies. Astronomers have cataloged numerous **starburst galaxies**, such as the one shown in Figure 24.32, where violent events, possibly a near-collision with a neighbor, appear to have rearranged the galaxy's internal structure and triggered a sudden, intense burst of star formation in the recent past. Such close encounters are random events and do not seem to represent any genuine evolutionary sequence linking all spirals to all ellipticals and irregulars. However, it *is* clear that many galaxies and galaxy clusters have evolved greatly since they first formed long ago.

Figure 24.32 This interacting galaxy pair (IC 694, at left, and NGC 3690) shows starbursts now under way in both galaxies—hence the bluish tint. Such intense, short-lived bursts probably last for no more than a few tens of millions of years—a small fraction of a typical galaxy's lifetime.

R I V U X G

Chapter Review

SUMMARY

The **Hubble classification scheme** (p. 514) divides galaxies into several classes, depending on their appearance. **Spiral galaxies** (p. 515) have flattened disks, central bulges, and spiral arms. They are further subdivided on the basis of the size of the bulge and the tightness of the spiral structure. The halos of these galaxies consist of old stars, whereas the gas-rich disks are the sites of ongoing star formation. **Barred-spiral galaxies** (p. 516) contain an extended "bar" of material projecting beyond the central bulge.

Elliptical galaxies (p. 516) have no disk and contain no gas or dust. In most cases, they consist entirely of old stars. They range in size from dwarf ellipticals, which are much less massive than the Milky Way, to giant ellipticals, which may contain trillions of stars. **S0** and **SB0 galaxies** (p. 517) are intermediate in their properties between ellipticals and spirals. They have extended halos and stellar disks and bulges (and bars, in the SB0 case) but little or no gas and dust.

Irregular galaxies (p. 518) are galaxies that do not fit into either of the other categories. Some have a distinctly disturbed appearance and may be the result of galaxy collisions or close encounters. Many irregulars are rich in gas and dust and are the sites of vigorous star formation. The **Magellanic Clouds** (p. 519), two small systems that orbit the Milky Way, are examples of this type of galaxy.

The Milky Way, Andromeda, and several other smaller galaxies form the **Local Group** (p. 520), a small **galaxy cluster** (p. 521). Galaxy clusters consist of a collection of galaxies orbiting one another, bound together by their own gravity.

Astronomers often use standard candles as distance-measuring tools. An alternative is the **Tully–Fisher relationship** (p. 521), an empirical correlation between rotational velocity and luminosity in spiral galaxies. By measuring the rotation speed of a spiral and using this relationship, astronomers can determine the galaxy's intrinsic luminosity and hence its distance.

The nearest large galaxy cluster to the Local Group is known as the Virgo Cluster. Galaxy clusters themselves tend to clump together into **superclusters** (p. 524). The Virgo Cluster, the Local Group, and several other nearby clusters form the Local Supercluster.

The masses of nearby spiral galaxies can be determined by studying their rotation curves. For more distant spirals, masses can be inferred from observations of the broadening of their spectral lines. On larger scales, astronomers use studies of binary galaxies and galaxy clusters to obtain statistical mass estimates of the galaxies involved.

As in the Milky Way, measurements of the masses of galaxies and galaxy clusters reveal the presence of large amounts of dark matter that is presently undetectable at any electromagnetic wavelength. The fraction of dark matter apparently grows as the scale under consideration increases. Its nature is unknown.

Large amounts of hot X-ray-emitting gas have been detected among the galaxies in many clusters, but not enough to account for the dark matter inferred from dynamical studies.

Distant galaxies are observed to be receding from the Milky Way at rates that increase proportional to their distances from us. This relation between recessional velocity and distance is called **Hubble's law** (p. 528). The constant of proportionality in the law is **Hubble's constant** (p. 529). Its value is believed to lie between 60 and 90 km/s/Mpc.

Astronomers use Hubble's law to determine distances to the most remote objects in the universe. The redshift associated with the Hubble expansion is called the **cosmological redshift** (p. 528).

On very large scales, galaxies and galaxy clusters are not spread randomly throughout space. Instead, they are arranged on the surfaces of enormous "bubbles" of matter surrounding vast low-density regions, called **voids** (p. 531). The origin of this structure is thought to be closely related to conditions in the very earliest epochs of the universe.

Researchers know of no evolutionary sequence that links spiral, elliptical, and irregular galaxies together, and the process of galaxy formation is still only poorly understood. There is growing evidence that large galaxies formed by the merger of smaller ones in a process that may be continuing today.

Collisions and mergers of galaxies play very important roles in galactic evolution. Interactions between galaxies appear to be very common. A **starburst galaxy** (p. 535) may result when a galaxy experiences a close encounter with a neighbor. The strong tidal distortions caused by the encounter compress galactic gas, resulting in a widespread burst of star formation.

SELF-TEST: True or False?

_____ **1.** Barred-spiral galaxies have the same properties as normal spirals, except for the "bar" feature.

_____ **2.** Elliptical galaxies contain no flattened disk.

_____ **3.** There is no interstellar dust in elliptical galaxies, but substantial amounts of interstellar gas.

_____ **4.** Most ellipticals contain only old stars.

_____ **5.** Irregular galaxies, although small, have lots of star formation taking place in them.

_____ **6.** Spiral galaxies evolve into ellipticals.

_____ **7.** Type-I supernovae can be used to determine distances to galaxies.

_____ **8.** Every galaxy is a member of some galaxy cluster.

_____ **9.** Galaxy collisions can occur, but are extremely rare.

_____**10.** A typical galaxy cluster has a mass of about 10^{11} solar masses.

_____ **11.** Most galaxies appear to be receding from the Milky Way Galaxy.

_____**12.** Once the Hubble law is known, it can be used to determine distances to both nearby galaxies and the farthest objects in the universe.

SELF-TEST: Fill in the Blank

1. Galaxies are categorized by their _____ classification.
2. Spiral galaxies with tightly wrapped spiral arms tend to have _____ central bulges.
3. Spiral galaxies of type _____ have the least amount of gas; type _____ have the most.
4. The Milky Way, the Andromeda galaxy, and 18 other galaxies form a small cluster known as the _____.
5. In the Tully-Fisher relation, a galaxy's luminosity is found to be related to the _____ of its 21-cm line.
6. The diameter of the Local Supercluster of galaxies is about _____ .
7. When galaxies collide, the star formation rate often _____ .

8. Galaxy mass determinations from rotation curves, line broadening, and binary galaxies all make use of _____ Law.
9. Dark matter may make up as much as _____ percent of the entire universe.
10. Hubble's Law is a correlation between the redshifts and the _____ of galaxies.
11. The largest known structures in the universe, such as voids and the Great Wall, have sizes on the order of _____ .
12. By which process do galaxies form: fragmentation (large objects breaking up into small) or mergers (small objects accumulating into large)? _____

REVIEW AND DISCUSSION

1. In what sense are elliptical galaxies "all halo?"
2. Describe the four "rungs" in the distance-measurement ladder used to determine the distance to a galaxy lying 5 Mpc away.
3. Describe the contents of the Local Group. How much space does it occupy compared to the volume of the Milky Way?
4. How is the Tully-Fisher used to measure distances to galaxies?
5. What is the Virgo Cluster?
6. Describe two techniques for measuring the mass of a galaxy.
7. Why do astronomers believe that galaxy clusters contain more mass than we can see?

8. What is Hubble's law?
9. How is Hubble's law used by astronomers to measure distances to galaxies?
10. What is the most likely range of values for the Hubble constant? Why is the exact value of the Hubble constant uncertain?
11. Describe the role of collisions in the formation and evolution of galaxies.
12. Do you think that collisions between galaxies constitute "evolution" in the same sense as the evolution of stars?

PROBLEMS

1. A supernova of luminosity 1 billion times the luminosity of the Sun is being used as a standard candle to measure the distance to a faraway galaxy. From Earth, the supernova appears as bright as the Sun would appear from a distance of 10 kpc. What is the distance to the galaxy ?
2. According to Hubble's law, with $H_0 = 75$ km/s/Mpc, what is the recessional velocity of a galaxy at a distance of 200 Mpc? How far away is a galaxy whose recessional velocity is 4000 km/s? How do these answers change if $H_0 = 50$ km/s/Mpc?

3. Two galaxies are orbiting one another at a separation of 500 kpc and the orbital period is estimated to be 30 billion years. Use Kepler's Law (as stated in Chapter 23) to find the total mass of the pair.
4. In a galaxy collision, two similar-sized galaxies pass through one another with a combined velocity of 1000 km/s. If the entire event takes place over a distance of 500,000 pc, how long will the event last?

PROJECTS

1. Look for a copy of the *Atlas of Peculiar Galaxies* by Halton Arp. It is available in book form or on laser disk. Search for examples of interacting galaxies of various types: (1) tidal interactions, (2) starburst galaxies, (3) collisions between two spirals, and (4) collisions between a spiral and an elliptical. For (1) look for galactic material pulled away from a galaxy by a neighboring galaxy. Is the latter galaxy also tidally distorted? In (2) the surest sign of starburst activity are bright knots of star formation. In what type(s) of galaxies do you find starburst activity? For (3) and (4) how do collisions differ depending on the types of galaxies involved. What typically happens to a spiral galaxy after a near-miss or collision? Do ellipticals suffer the same fate?
2. Look for the Virgo Cluster of galaxies. An 8-inch telescope is a perfect size for this project, although a smaller telescope

will also work. The constellation Virgo is visible from the U.S. during much of fall, winter, and spring. To locate the center of the cluster, first find the constellation Leo. The eastern part of Leo is composed of a distinct triangle of stars, Denebola (β), Chort (θ), and Zosma (δ). Go from Chort to Denebola in a straight line east and same distance as between the two stars and you will be in the approximate middle of the Virgo Cluster. Look for the following Messier objects that make up some of the brightest galaxies in the cluster. M49, 58, 59, 60, 84, 86, 87 (giant elliptical thought to have a massive black hole at its center), 89, and 90. Examine each galaxy for unusual features; some have very bright nuclei.

25

ACTIVE GALAXIES AND QUASARS

Limits of the Observable Universe

LEARNING GOALS

Studying this chapter will enable you to:

1. Specify the basic differences between active and normal galaxies.

2. Describe the important features of Seyfert and radio galaxies.

3. Explain what drives the central engine thought to power all active galaxies.

4. Describe the observed properties of quasars, and discuss the special properties of the radiation they emit.

5. Discuss the place of active galaxies in current theories of galactic evolution.

(Opposite page, background) This artist's conception depicts one possible scenario for the "central engine" of an active galaxy. Two jets of matter are shown moving outward perpendicular to a flattened accretion disk surrounding a supermassive black hole. Such jets are seen at many places in the universe.
(Inset A) Numerical simulations—i.e., done on computers—can help us visualize key events in the universe. Here a uniform distribution of some thousand stars is spread over a dimension of several hundred parsecs.
(Inset B) As the simulation proceeds, the hypothetical stars congregate preferentially toward the core of the star cluster, forming a "luminosity cusp" of light.
(Inset C) If we could see inside the cusp, as shown in this piece of art, an accretion disk would probably be evident, swirling around a giant black hole.
(Inset D) As the accretion disk rotates, tilts, and acquires more matter—probably as whole stars fall in—jets of matter emerge to cool the environment surrounding the hole.

Our journey from the Milky Way to the Great Wall in the past two chapters has widened our cosmic field of view by a factor of 10,000, yet the galaxies that make up the structures we see show remarkable consistency in their properties. The overwhelming majority of galaxies fit neatly into the Hubble Classification Scheme, showing few, if any, "unusual" characteristics. However, sprinkled through the mix of normal galaxies, even relatively close to the Milky Way, are galaxies that are decidedly abnormal in their properties. Although their optical appearances are often quite ordinary, these abnormal galaxies emit huge amounts of energy—far more than a normal galaxy—mostly in the invisible part of the electromagnetic spectrum. Observing such objects at great distances, we may be seeing some of the formative stages of our own galactic home.

25.1   *Beyond the Local Realm*

1 Astronomers have recognized and cataloged spiral, elliptical, and irregular galaxies as far away as several hundred megaparsecs from Earth. Beyond this distance, galaxies appear so faint that it is difficult to discern their characteristic shapes, so their types are largely unknown. Nevertheless, according to their observed redshifts (and Hubble's law), we know that many galaxies lie well beyond this distance, in the farthest reaches of the observable universe. But what kinds of objects are they? Are they normal galaxies—close relatives of the galaxies that populate our local neighborhood—or are they somehow different? The answer is that they often seem to be *different* from the galaxies found in our cosmic backyard. By and large, very distant objects tend to be more active—more luminous—than those found closer to home. The most energetic objects, which can emit hundreds or thousands of times more energy per second than our entire Galaxy, are known collectively as **active galaxies**.

Not all active galaxies are distant, and only a small fraction of all distant galaxies are active. Some active galaxies are found locally, scattered among the normal galaxies that make up most of our cosmic neighborhood, while many faraway "normal luminosity" galaxies are known. As a general rule, though, active galaxies are more common at greater distances, and the most active objects lie farthest from the Earth. We might wonder whether this predominance of energetic objects with large redshifts is just an observational effect, resulting from our inability to detect relatively faint normal galaxies at great distances. After all, the apparent brightness of any astronomical object decreases as the square of its distance from us, so even with the very best telescopes, we would expect to observe only the more energetic and powerful galaxies in remote regions of space. Although this observational bias does play a role, it turns out that it can only partly explain the apparent predominance of energetic objects among those having large redshifts. We are led to conclude that bright active galaxies really are more common at great distances.

In addition to their great brightness, there is something else that is basically different about active galaxies. It seems that their *radiative character* differs fundamentally from that of normal galaxies. Most of a normal galaxy's ra-

diated energy is emitted in or near the visible portion of the electromagnetic spectrum, much like the radiation from ordinary stars. Indeed, to a large extent, the light we see from a normal galaxy *is* just the accumulated light of its many component stars. For example, our entire Milky Way has a luminosity of about 10^{37} W at optical frequencies—20 billion Suns' worth of radiation—but only 10^{31} W at radio frequencies—a million times less. By contrast, as illustrated in Figure 25.1, the radiation observed emitted by active galaxies does *not* peak at optical frequencies—far more energy is emitted at longer wavelengths than in the visible range. The radiation from active galaxies is *inconsistent* with what we would expect if it were the combined radiation of myriad stars. The radiation is thus said to be *nonstellar* in nature.

The two most important categories of active galaxies are *Seyfert galaxies* and *radio galaxies*, although other classes exist (see, for example, *Interlude 25-1* on p. 546). Even more extreme in their properties are the *quasars*. Astronomers conventionally distinguish between active galaxies and

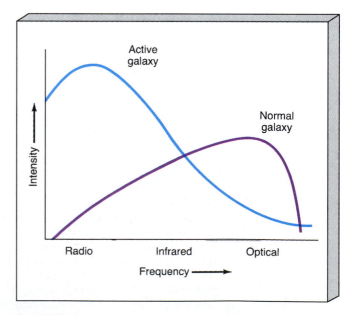

Figure 25.1   The nature of the energy emitted from a normal galaxy differs from that of an active galaxy. (This plot illustrates the general run of intensity for all galaxies of a particular type and does not represent any one individual galaxy.)

quasars based on their spectra, appearance, and distance. Active galaxies typically *look* like galaxies, whereas quasars are often so distant that little structure can be discerned. In addition, quasars are generally even brighter than active galaxies (which are themselves much brighter than normal galaxies like the Milky Way). The distinction is largely historical, however—it dates back to the days when the connection between quasars and active galaxies was not understood. As we will see later, most astronomers now believe that quasars are simply an early stage of galaxy formation, and that there is really no sharp dividing line between quasars and active galaxies. Many researchers now go so far as to include quasars in the "active galaxy" category.

Physical conditions were undoubtedly different at earlier times than they are now. Perhaps, then, we should not be surprised that remote astronomical objects, which emitted long ago the radiation we observe today, differ from nearby objects, which emitted their radiation much more recently. What *is* surprising—in fact, astounding—is the *amount* of energy radiated by some of the most luminous objects. Their tremendous power, nonstellar radiation, and abundance at great distances suggest to many astronomers that the universe was once a much more violent place than it is today.

25.2 *Seyfert Galaxies*

2 In 1943 Carl Seyfert, an American optical astronomer studying spiral galaxies from Mount Wilson Observatory, discovered the type of active galaxy that now bears his name. **Seyfert galaxies** are a class of astronomical objects whose properties lie between those of normal galaxies like the Milky Way and those of the most violent active galaxies known. This fact suggests to some astronomers that Seyferts represent an evolutionary link between these two extremes. The spectral lines of Seyfert galaxies are usually substantially redshifted—most Seyferts seem to reside at large distances (hundreds of megaparsecs) from us, although a few are as close as 20 or 30 Mpc.

Figure 25.2 shows two optical images of a typical Seyfert galaxy. A casual glance at a long-exposure photograph of a Seyfert (such as Figure 25.2a) reveals nothing strange. Superficially, Seyferts resemble normal spiral galaxies. However, closer study of Seyferts reveals some peculiarities not found in normal spirals.

First, maps of Seyfert energy emission show that nearly all of the radiation stems from a small central region known as the **galactic nucleus**. This region can be seen at the center of the overexposed core of Figure 25.2(a), and is shown in more detail in Figure 25.2(b). Astronomers suspect that a Seyfert nucleus may be quite similar to the center of a normal galaxy, such as the Milky Way or the Andromeda galaxy, but with one very important difference. The nucleus of a Seyfert is 10,000 times brighter than the center of our Galaxy. Indeed, the brightest Seyfert nuclei are 10 times more energetic than the *entire* Milky Way.

Second, Seyfert galaxies emit their radiation in two broad frequency ranges. The stars in the Seyfert's galactic disk and spiral arms produce about the same amount of *visible* radiation as those of a normal spiral galaxy. However, most of the energy from the Seyfert's nucleus is emitted in the form of *invisible* radio and infrared radiation, which cannot be explained as coming from stars—it must be nonstellar in origin.

Third, Seyfert spectral lines bear little or no resemblance to those produced by ordinary stars, although they do have many similarities to the spectral lines observed toward the center of our own Galaxy. ∞ (p. 507) Seyfert spectra contain strong emission lines of highly ionized heavy elements, especially iron. The lines are very broad, indicating either that the galaxy's gases are tremendously hot (more than 10^8 K) or that they are rotating very rapidly (at about 1000 km/s) around some central object. ∞ (pp. 88, 90) The first possibility can be ruled out, since such a high temperature would cause all the gas to be ionized, so that no spectral lines would be produced. Thus, the broadening indicates rapid internal motion in the nucleus.

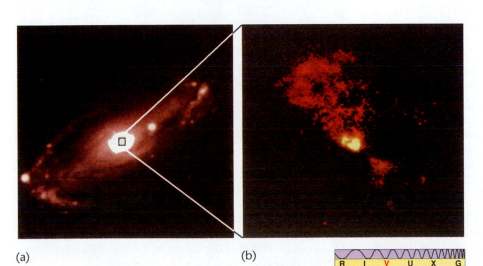

(a)   (b)

Figure 25.2 Photographs of a Seyfert galaxy (NGC 5728) (a) from the ground and (b) from Earth orbit. The enlarged view in (b) of two cone-shaped beams of light shows a group of glowing blobs near the galaxy's core, perhaps illuminated by radiation arising from the accretion disk of a black hole. This object is about 40 Mpc away—one of the closest active galaxies known.

R I V U X G

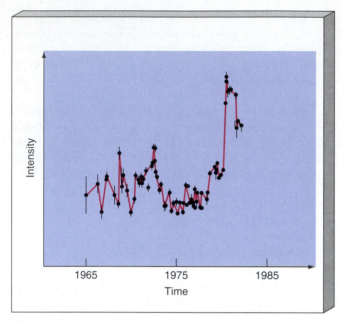

Figure 25.3 The irregular variations of a particular Seyfert galaxy's luminosity over a period of two decades. Because this Seyfert, called 3C 84, emits most strongly in the radio part of the electromagnetic spectrum, these observations were made with large radio telescopes. The optical and X-ray luminosities vary as well.

Finally, extensive monitoring of Seyfert radiation over long periods of time has shown that the energy emission often varies over time. Figure 25.3 shows an example of luminosity variations for a typical Seyfert. These radiative changes are unlike anything found in the Milky Way or any other normal galaxy. A Seyfert's luminosity can double or halve within a fraction of a year.

These rather rapid fluctuations lead us to conclude that the source of energy emissions must be quite compact. As mentioned in Chapter 22 (in the context of luminosity variations in neutron stars and black holes), for astronomers to be able to detect a coherent variation in brightness within a certain time interval, the source of radiation must be smaller in size than the distance traveled by light in that interval. ∞ (p. 482) Otherwise, the intensity variations would be blurred, not sharp, as observed. Simply put, an object cannot "flicker" in less time than radiation takes to cross it. Because the rise and fall of a Seyfert's radiation usually occurs within one year, we can confidently conclude that the emitting region must be less than one light year across—an *extraordinarily* small region, considering the amount of energy emanating from it. High-resolution interferometric radio maps of Seyfert cores generally confirm this reasoning.

Seyfert galaxies are apparently experiencing huge explosions within their cores. Their time variability and their large radio and infrared luminosities together strongly imply violent nonstellar activity. The *nature* of this activity may well resemble processes occurring in the center of our own Galaxy, but its *magnitude* is thousands of times greater than the comparatively mild events within our own Galaxy's center.

25.3 *Radio Galaxies*

Seyfert galaxies are not the only kind of active galaxy known. Radio observations have uncovered another class of extremely energetic sources that also seem to be galactic in nature. **Radio galaxies** differ from Seyferts both in the wavelength at which their energy is emitted and in the appearance and extent of the emitting region.

CORE–HALO RADIO GALAXIES

One common type of radio galaxy is often called a *core–halo* radio galaxy. As illustrated in Figure 25.4, the energy from such an object comes mostly from an extremely small central nucleus, or *core*, less than 1 pc across, with weaker emission coming from an extended *halo* surrounding it. The halo typically measures about 50 kpc across, similar in size to the surrounding visible galaxy, which is usually elliptical and often quite faint. The radio luminosity from the core can be as great as 10^{37} W—about the same as the emission from a Seyfert nucleus and comparable to the output from our entire Milky Way Galaxy at all wavelengths.

Figure 25.5 shows two optical photographs of a core–halo radio galaxy, along with false-color images of its radio and infrared emissions. This object is a giant elliptical galaxy known as M87—the eighty-seventh object in Messier's catalog. (We can be sure that this eighteenth-century Frenchman had no idea what he was really looking at. Nor perhaps do we!) M87 is roughly 20 Mpc distant, making it a prominent member of the Virgo Cluster, and one of the closest active galaxies. Its nearness and interesting activity have made it one of the most intensely studied of all astronomical objects.

Figure 25.4 Radio contour map of a typical core–halo radio galaxy. The radio emission from such a galaxy comes from a bright central nucleus, or core, surrounded by an extended, less intense halo. The radio map is superimposed on an optical image of the galaxy and some of its neighbors, shown previously in Figure 24.15.

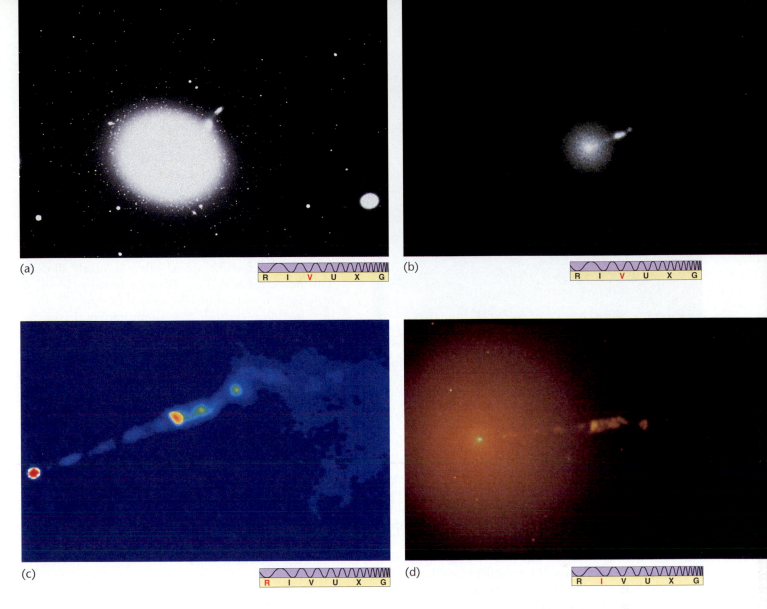

(a)

R I V U X G

(b)

R I V U X G

(c)

R I V U X G

(d)

R I V U X G

Figure 25.5 The giant elliptical galaxy M87 (also called Virgo A) is displayed here at several different wavelengths. (a) A long optical exposure of its halo and embedded central region. (b) A short optical exposure of its core and an intriguing jet of matter, on the same scale as (a). (c) A radio image of its jet, on a somewhat expanded scale compared with (b). The red dot at left marks the bright nucleus of the galaxy; the red and yellow blob near the center of the image corresponds to the bright "knot" visible in the jet in (b). (d) A recent image of the jet taken in the near-infrared, at roughly the same scale as (c).

A long time exposure (Figure 25.5a) shows a large fuzzy ball of light—a fairly normal-looking type E1 elliptical galaxy, about 100 kpc across. A shorter exposure of M87 (Figure 25.5b) captures only the brightest regions of the galaxy and reveals a compact central region only a few hundred parsecs in diameter. Beyond the core, a long thin *jet* of matter ejected from M87's center appears. The jet is about 2 kpc long and is traveling outward at a very high velocity, possible as great as half the speed of light. Computer enhancement shows that the jet is made up of a series of distinct "blobs," more or less evenly spaced along its length. This high-speed jet, which emits energy at the rate of almost 10^{35} W, has been imaged in the radio and infrared as well as in the optical regions of the spectrum (Figures 25.5c and d). Jets such as this are vital to our understanding of these energetic radio sources, as we will see in a moment.

LOBE RADIO GALAXIES

Many radio galaxies are not of the core–halo type. While they do emit most of their radiation in the long-wavelength part of the spectrum, like Seyferts and core–halo galaxies, very *little* of this emission arises from a compact central nucleus. Instead, most of the energy comes from giant extended regions called **radio lobes**—roundish clouds of gas up to a megaparsec across, lying well beyond the center of the galaxy itself. For this reason, these objects are known as *lobe radio galaxies.*

The radio lobes of all lobe radio galaxies are truly enormous. From end to end, an entire lobe radio galaxy typically is more than ten times the size of the Milky Way, comparable in size to the entire Local Group. The lobes emit no visible light, but their radio luminosity can range from 10^{36} to 10^{38} W—between one-tenth and 10 times the

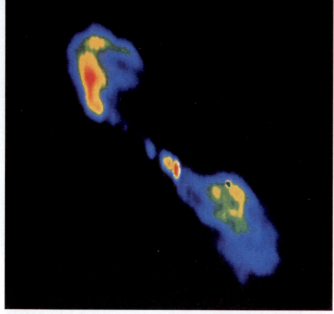

Figure 25.6 Lobe radio galaxies, such as Centaurus A shown here, have giant radio-emitting regions extending a million parsecs or more beyond the central galaxy. The lobes cannot be imaged in visible light and are observable only with radio telescopes. (The lobes are filled with false color to indicate decreasing intensity from red to yellow to green to blue.)

total energy emitted by our Galaxy. Several of these strange objects are located relatively nearby, so we can study them at close range. One such system, known as Centaurus A, is shown in Figure 25.6. It lies only 4 Mpc from the Earth.

Figure 25.7 is an optical image of Centaurus A, with another representation of Figure 25.6 superimposed to show the relation between the optical and radio emission. In visible light, Centaurus A is a rather peculiar-looking object, apparently an E2 galaxy bisected by an irregular band of dust. Numerical simulations suggest that this system is probably the result of a merger between an elliptical

galaxy and a smaller spiral galaxy about 500 million years ago. The radio lobes are roughly symmetrically placed, jutting out from the center of the visible galaxy. Note, too, that the jets are roughly perpendicular to the dust lane. The elliptical galaxy itself is very large—some 500 kpc in diameter. Relatively little radio emission is observed from the location of the optical image, however. Most of the radio radiation arises from the giant lobes well beyond it.

The lobes of radio galaxies vary in size and shape from source to source, but they maintain their alignment with the center of the optical galaxy in nearly all cases. This fact suggests that the lobes are actually "blobs" of material that were somehow ejected in opposite directions by violent events in the galactic nucleus. In the case of Centaurus A, this argument is strengthened by the presence of an additional pair of secondary lobes, smaller than the main lobes (about 50 kpc in length, marked in Figure 25.7) and closer to the visible galaxy. Both pairs of lobes share the same high degree of linear alignment. Astronomers believe that the inner lobes were expelled from the nucleus by the same basic process as the outer ones, but more recently, so they have not had time to travel as far. Still higher-resolution studies reveal the presence of a roughly 1-kpc-long jet in the center of Centaurus A, again aligned with the larger lobes.

If material is ejected from the nucleus at close to the speed of light (which seems likely), it follows that Centaurus A's outer lobes were created a few hundred million years ago, possibly around the time of the collision thought to be responsible for the galaxy's peculiar optical appearance. Apparently some violent process at the center of Centaurus A started up around that time and has been intermittently firing jets of matter out into intergalactic space ever since.

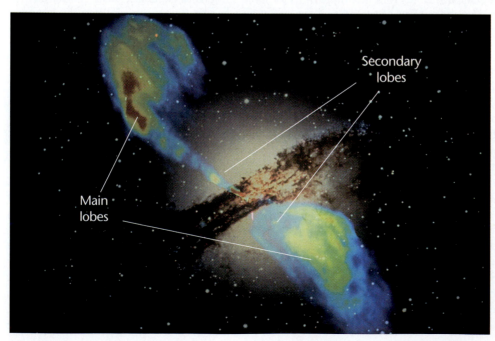

Main lobes

Secondary lobes

Figure 25.7 An optical photograph of Centaurus A, one of the most massive and peculiar galaxies known and believed to be the result of a galaxy collision some 500 million years ago. The pastel false colors mark the radio emission shown in Figure 25.6, in this case more recently acquired and with higher resolution. Note that the radio lobes emit no visible light.

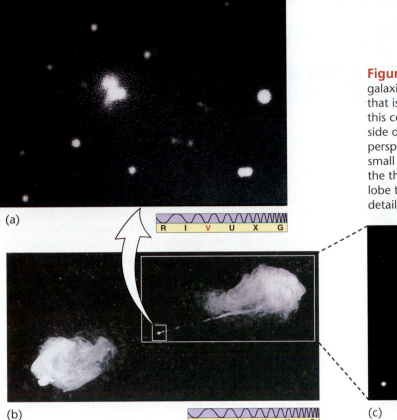

(a)

R I V U X G

Figure 25.8   (a) Cygnus A also appears to be two galaxies in collision, although it is not completely clear that that is what is really happening. (b) On a much larger scale, this cosmic object displays radio-emitting lobes on each side of the optical image. To put these images into proper perspective, the optical galaxy in (a) is about the size of the small dot at the center of the "radiograph" in (b). Notice the thin line of radio-emitting material joining the right lobe to the central galaxy. Frame (c) shows even greater detail in one of the lobes.

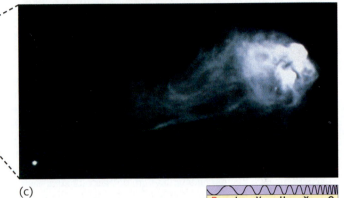

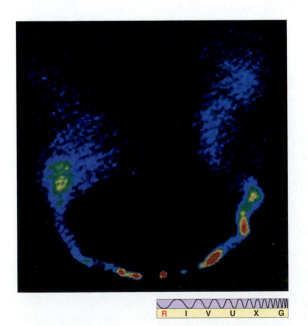

(b)

R I V U X G

(c)

R I V U X G

Further evidence in favor of the interpretation of radio lobes as material ejected from the center of a galaxy is provided by another object, Cygnus A, shown as an optical image in Figure 25.8(a) and as high-resolution radio map in Figure 25.8(b). The filamentary structure evident in the radio lobes and the thin, radio-emitting line joining the right lobe to the center of the visible galaxy (the dot at the center of the radio image) strongly suggest that we are seeing two oppositely directed jets of material running into the intergalactic medium.

In some systems, known as *head–tail* radio galaxies, the lobes seem to form a "tail" behind the main galaxy. For ex-

ample, the lobes of radio galaxy NGC 1265, shown in Figure 25.9, appear to be "swept back" by some onrushing wind. In fact, this is the most likely explanation for the galaxy's appearance. If NGC 1265 were at rest, it would be just another double-lobe source, perhaps looking quite similar to Centaurus A. However, the galaxy is traveling through the intergalactic medium of its parent galaxy cluster (known as the Perseus Cluster), and the outflowing matter forming the lobes tends to be left behind as the galaxy moves.

Radio galaxies share many characteristics with Seyfert galaxies. They emit comparably large amounts of energy, and there is good evidence that the energy source is a com-

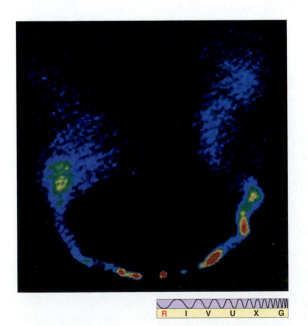

R I V U X G

R I V U X G

Figure 25.9   (a) Radiograph, in false color, of the active "head–tail" galaxy NGC 1265. (b) The same radio data, in contour form, superposed on the optical image of the galaxy. Astronomers reason that this object is moving rapidly through space, trailing a "tail" behind as it goes—a little like a comet, but on a vastly larger scale.

Interlude 25-1 BL Lac Objects

In 1929, an object thought to be a variable star was discovered in the constellation Lacerta. Astronomers gave it the two-letter code BL, so it became known as BL Lacertae, or BL Lac for short. Not until the 1970s, when it became clear that BL Lac was a strong radio source, did anyone question its classification as a star. As more objects like BL Lac were found, astronomers began to realize that they had stumbled across a whole new class of extragalactic sources. *BL Lac objects* seem starlike in a telescope and vary greatly in luminosity over time. They are also powerful compact radio sources, sometimes colloquially referred to as "blazars."

Astronomers originally placed BL Lac objects into a special class because they displayed no spectral lines. Their redshifts could not be determined and their distances were unknown. In recent years, extremely faint spectral lines strongly suggest that BL Lac objects are extremely distant, making them nearly as luminous as quasars (discussed later in the text). Further careful observations have shown that blazars reside at the centers of relatively normal elliptical galaxies. We consider these galaxies active because of the tremendous strength of their radio emission.

The weak spectral lines may be the key needed to unravel the nature of the BL Lac objects. Evidently, the thermal emission from any stars they contain, which would provide spectral lines, is swamped by their nonthermal synchrotron radiation, which does not contain spectral features. Other types of active galaxies show a mixture of thermal and nonthermal radiation, so it is more difficult to study the nonthermal radiation alone. BL Lac objects, then, may offer astronomers a chance to study the "bare machine" (which generates the nonthermal radiation) powering these active galaxies.

In some ways, the BL Lac objects seem to represent a "link" or transition phase between radio galaxies and quasars. The luminosities of the compact radio sources inside BL Lac objects span the whole range from the relatively weak cores observed in radio galaxies to the stronger ones found in quasars. Perhaps there exists an evolutionary sequence of some sort, in which the ancient quasars use up their fuel, turning into the weaker yet more erratic BL Lac objects, which subsequently become the rather inactive cores of radio galaxies. An alternative possibility is that BL Lacs—and perhaps also the weaker core–halo radio sources—*are* just lobe radio galaxies, but radio galaxies in which the jet happens to be pointing nearly straight at us. We would then see the galaxy *through* the nearer radio lobe. Only further observations and better statistics will tell us for sure which (if either) of these two possibilities is correct.

---

pact region at the center of an otherwise relatively normal-looking galaxy. In the lobe radio galaxies, that energy is fired out from the nucleus in the form of jets of matter and is ultimately *emitted* (in the form of radiation) from far beyond the galaxy itself. But the central compact nucleus is still thought to be the place where the energy is actually *produced*. Before going on to probe even more distant, and more violent, objects, let us now consider the current view of the "engine" that powers all this activity.

25.4 *The Central Engine of an Active Galaxy*

❸ The behavior of active galaxies is contrary to that expected from vast collections of stars. The lobe radio galaxies in particular, with their huge energy emission from far beyond the optical galaxy, are among the most powerful objects in the universe. Can we explain this enormous nonstellar energy output in terms of known physics? Remarkably, the answer is "Yes." The present consensus among astronomers is that, despite the great differences in appearance, these objects—Seyferts and radio galaxies—may share a common energy-generation mechanism. The energy can be reprocessed into many different forms before it is finally emitted into intergalactic space, but the engine is probably the same in either case.

As a class, active galaxies (and quasars too, as we will see) show some or all of the following properties.

1. They have *high luminosities*, generally greater than the 10^{37} W characteristic of a fairly bright normal galaxy like the Milky Way.

2. Their energy emission is *nonstellar*—it cannot be explained as the combined radiation of even trillions of stars.

3. Their energy output can be highly *variable*, implying that it is emitted from a small central nucleus much less than a parsec across.

4. They often exhibit *jets* and other signs of explosive activity.

5. Their optical spectra may show broad emission lines, indicative of *rapid internal motion* within the energy-producing region.

The principal questions then are: How can such vast quantities of energy arise from these relatively small regions of space? Why is so much of the energy radiated at low frequencies, especially in the radio and infrared? And what is the origin of the extended radio-emitting lobes and jets? Let us first consider how the energy is produced.

ENERGY PRODUCTION

To develop a feeling for the enormous emissions of active galaxies, consider for a moment an object with a luminosity of 10^{38} W. In and of itself, this energy output is not inconceivably large. The brightest giant ellipticals are comparably powerful. Thus, some 10^{12} stars—a few normal galax-

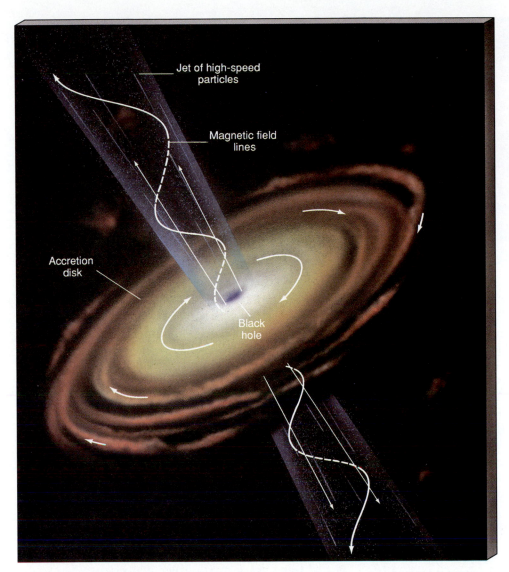

Figure 25.10 The leading theory for the energy source in active galactic nuclei (and quasars) holds that these objects are powered by material accreting onto a supermassive black hole. As matter spirals toward the hole, it heats up, producing large amounts of energy. At the same time, high-speed beams of gas may be ejected perpendicular to the accretion disk, forming the jets and lobes seen in many active objects. Magnetic fields generated in the disk are carried by the jets out to the radio lobes, where they play a crucial role in producing the observed radiation.

ies' worth of material—could *equivalently* power a typical active galaxy. The difficulty arises when we consider that in an active galaxy this energy production is packed into an object much less than a parsec in diameter!

It is difficult to imagine how several Milky Way Galaxies could be compressed into a space no larger than a parsec. Even if we could somehow squeeze that much mass into such a volume, it would immediately collapse to form a huge black hole, and none of the light it produced could escape to the outside! Thus, even neglecting its nonstellar spectrum, the total energy output of an active galaxy simply cannot be explained as the combined energy of many stars. We must think of something else.

The twin requirements of large energy generation and small physical size bring to mind our discussion of X-ray sources in Chapter 22. ∞ (p. 467) The presence of the jet in M87 and the ejection of matter to form radio lobes in Centaurus A and Cygnus A strengthen the connection. Recall that the best current explanation for those "small-scale" phenomena involves the accretion of material onto a

compact object—a neutron star or a black hole. Large amounts of energy are produced as matter spirals down onto the central object, and high-speed jets may well be a common by-product of the process. In Chapter 24, we suggested that a similar mechanism, involving a *supermassive black hole*—one with a mass of around a million suns—may also be responsible for the energetic radio and infrared emission observed at the center of our own Galaxy. ∞ (p. 507)

As illustrated in Figure 25.10, the leading model for the central engine of active galaxies is essentially a scaled-up version of the same accretion process, now involving black holes with masses between a few million and a billion times the mass of the Sun. As with its smaller-scale counterparts, infalling gas forms an accretion disk and spirals down toward the hole. It is heated to high temperatures by friction within the disk and emits large amounts of radiation as a result. In this case, however, the origin of the accreted gas is not a binary companion, as in stellar X-ray sources, but entire stars and clouds of interstellar gas that come too close to the hole and are torn apart by its strong gravity.

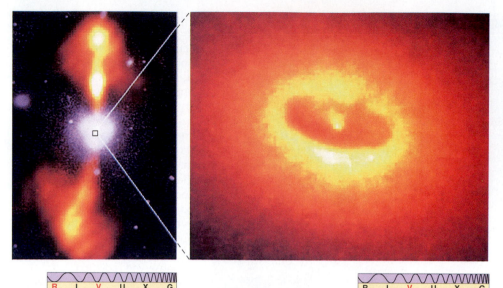

Figure 25.11 (a) A combined optical/radio image of the giant elliptical galaxy NGC 4261, in the Virgo Cluster, shows a white visible galaxy at center, from which red-orange (false color) radio lobes extend for about 60 kpc. (b) A close-up photograph of the galaxy's core reveals a 100-pc-diameter disk surrounding a bright hub that presumably harbors a black hole.

R   I   V   U   X   G

R   I   V   U   X   G

The accretion process is extremely efficient at converting infalling mass (in the form of gas) into energy (in the form of electromagnetic radiation). Detailed calculations indicate that as much as 10 or 20 percent of the total mass energy of the infalling matter can be radiated away before it crosses the hole's event horizon and is lost forever. Since the total mass energy of a star like the Sun—the mass times the speed of light squared—is about 2×10^{47} joule, if follows that the 10^{38} W luminosity of a bright active galaxy can be accounted for by the consumption of only 1 solar mass of gas per decade by a billion-solar-mass black hole. ∞ (p. 476) Less luminous active galaxies would require correspondingly less fuel—for example, a 10^{36} W Seyfert galaxy would devour only one star's worth of material every thousand years.

In this picture, the small size of the emitting region is a direct consequence of the compact nature of the central black hole. Even a billion-solar-mass hole has a radius of only 3×10^{9} km, or 10^{-4} pc—about 20 A.U.—and theory suggests that the part of the accretion disk responsible for most of the emission would be much less than a parsec across. Instabilities in the accretion disk can cause fluctuations in the energy released, leading to the variability observed in many objects. The broadening of the spectral lines observed in the nuclei of Seyferts (and in quasars) results from the rapid orbital motion of the gas in the hole's intense gravity.

Recent observations of galaxies in the Virgo Cluster by the *Hubble Space Telescope* lend strong support to this general picture. Figure 25.11 shows an image of a disk of gas and dust apparently feeding a possible black hole at the core of a giant elliptical galaxy. As expected from the theory just described, the disk is perpendicular to the huge jets emanating from the center of the active galaxy.

Hubble has also allowed astronomers to probe the fine details of the Virgo Cluster's most prominent object—the huge M87 galaxy (Figure 25.5)—and what they have found is in excellent agreement with the idea that the energy is produced by accretion onto a large black hole. At M87's distance, *Hubble*'s resolution of 0.05 arc second corresponds to a distance of about 5 pc, so we are still far from seeing the (solar-system–sized) central black hole itself, but the improved "circumstantial" evidence has convinced many doubters of the correctness of the theory. Figure 25.12 shows imaging and spectroscopic data that suggest the existence of a rapidly rotating disk of matter orbiting the galaxy's center, perpendicular to the jet. Measurements of the gas velocity on opposite sides of the disk indicate that the mass within a few parsecs of the center is approximately 3×10^{9} solar masses—we assume that this is the mass of the central black hole.

Even more compelling evidence for a supermassive black hole has recently come from studies with radio telescopes. Using the Very Long Baseline Array—a continent-wide network of ten radio telescopes—a U.S.-Japanese team was able to achieve spectacular angular resolution, in fact hundreds of times better than with the *Hubble Space Telescope*. Observations of NGC 4258, a spiral galaxy about 6 Mpc away, have uncovered a group of molecular clouds that are swirling in an organized fashion about the galaxy's core. Armed with an understanding of the Doppler effect, the astronomers have been able to detect the red and blue shifts of water-vapor spectral lines in those faraway clouds. As depicted in Figure 25.13, the pattern revealed by the clouds is that of a slightly warped and spinning disk centered precisely on the galaxy's heart. The rotation velocities imply the presence of more than 40 million solar masses all packed within a region less than 0.2 pc across—a mass density more than ten times that of any previously observed black-hole candidate.

ENERGY EMISSION

The model just described is now widely accepted as the correct picture of how active galaxies and quasars generate their enormous power. It provides a natural explanation of

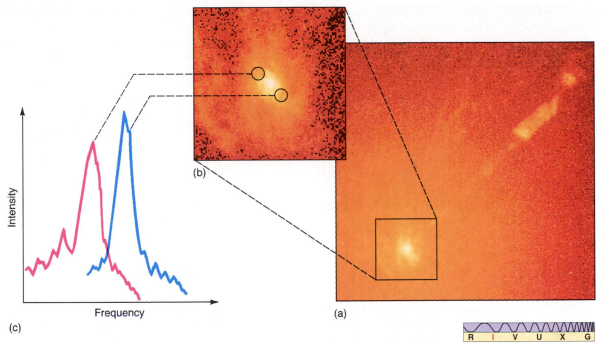

(b)

(c)

(a)

Figure 25.12 Recent imaging and spectroscopic observations of M87 support the idea of a rapidly whirling accretion disk at its core. (a) An image of the central region of M87, similar to that shown in Figure 25.5(d), shows its bright core and jet. The scale is comparable to the scale of Figure 25.5(c). (b) A magnified view of the core suggests a spiral swarm of stars, gas and dust. (c) Spectral-line features observed on opposite sides of the core show opposite Doppler shifts, implying that the material there is coming toward us on one side and moving away from us on the other. The strong implication is that an accretion disk spins perpendicular to the jet, and that at its center is a black hole having some 3 billion times the mass of the Sun.

the observed facts, and has the added advantage of being essentially similar to the processes thought to power smaller-scale phenomena, such as stellar X- and gamma-ray sources and normal galactic nuclei. Having thus accounted for the *source* of the energy, let us now turn to the way in which it is eventually *emitted* into intergalactic space.

In order to account for the details of the observed radiation spectra of some Seyfert galaxies, it is necessary to assume that the energy emitted from the accretion disk is "reprocessed"—that is, absorbed and reemitted—by gas and dust surrounding the nucleus. The jets and lobes seen in many systems consist of material (mainly protons and electrons) blasted out into space—and out of the galaxy entirely—from the inner regions of the disk. The details of how these jets form remain uncertain, but, as we have seen, there is a growing consensus among theorists that jets are a very common feature of accretion flows, large and small. The jets also contain strong *magnetic fields* (shown in Figure 25.10), possibly generated by the gas motion within the disk itself, which accompany the gas as it leaves the galaxy.

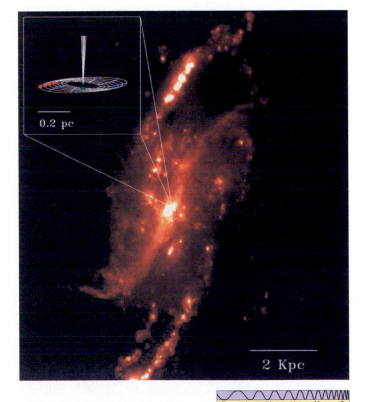

Figure 25.13 A network of radio telescopes has probed the core of the spiral galaxy NGC 4258, shown here in the light of mostly hydrogen emission. Within the innermost 0.2 pc (inset), observations of Doppler-shifted molecular clouds (designated by red, green, and blue dots) show that they obey Kepler's Third Law perfectly, and have revealed a slightly warped disk of rotating gas (shown here in artist's conception). At the center of the disk presumably lurks a huge black hole.

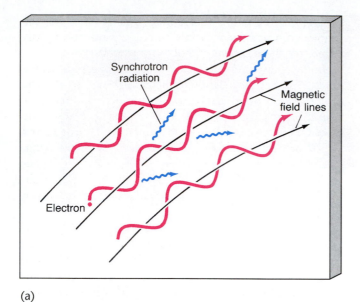

(a)

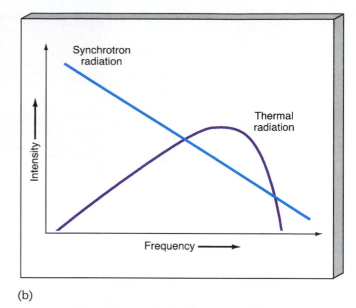

(b)

Figure 25.14 (a) Charged particles, especially fast electrons (red), emit synchrotron radiation (blue) while spiraling in a magnetic field (black lines). This process is not confined to active galaxies. It occurs, on smaller scales, when charged particles interact with magnetism in the Earth's Van Allen belts (see Chapter 7), when charged matter arches above sunspots on the Sun (see Chapter 16), in the vicinity of neutron stars (see Chapter 22), and at the center of our own Galaxy (see Chapter 23). (b) Variation in the intensity of thermal and synchrotron (nonthermal) radiation with frequency. Thermal radiation, described by a black-body curve, peaks at some frequency that depends on the temperature of the source. Nonthermal synchrotron radiation, by contrast, is most intense at low frequencies. It is independent of the temperature of the emitting object. Compare this figure with Figure 25.1.

As sketched in Figure 25.14(a), whenever a charged particle (here an electron) encounters a magnetic field, it tends to spiral around the field lines. We have encountered this idea several times previously, in a variety of different contexts (see, for example, the discussion of planetary magnetospheres in Chapters 7 and 11, or solar activity in Chapter 16. As the particles whirl around, they emit electromagnetic radiation, as discussed in Chapter 3. ∞ (p. 58) The faster the particles move, or the stronger the magnetic field, the greater the amount of energy radiated. In most cases, the fastest-moving particles are the low-mass electrons, so they are responsible for essentially all of the radiation we observe.

The radiation produced in this way—called **synchrotron radiation**—is *nonthermal* in nature. There is no link between the emission and the temperature of the radiating object, so the radiation is not described by a black-body curve. Instead, its intensity increases with decreasing frequency, as shown in Figure 25.14 (b). This is just what is needed to explain the overall spectrum of radiation recorded from active galaxies (compare Figure 25.14b with Figure 25.1). Observations of the radiation received from the jets and radio lobes in active galaxies are completely consistent with this process.

Eventually, the jet is slowed and stopped by the intergalactic medium, the flow becomes turbulent, and the magnetic field grows tangled. The result is a giant radio lobe, like those pictured in Figures 25.6–25.8, emitting virtually all of its energy in the form of synchrotron radiation. Even though the radio emission comes from an enormously extended vol-

ume of space that dwarfs the visible galaxy, the *source* of the energy is still the (relatively) tiny accretion disk—a billion billion times smaller in volume than the radio lobe—lying at the galactic center. The jets serve merely as a conduit to transport energy from the nucleus, where it is generated, into the lobes, where it is finally radiated into space.

The existence of the inner lobes of Centaurus A and the blobs in M87's jet imply that jet formation may be an intermittent process. There is also evidence to suggest that much, if not all, of the activity observed in nearby active galaxies could have been sparked by recent interaction with a neighbor. Many nearby active galaxies (Centaurus A, for example) appear to have been "caught in the act" of interacting with another galaxy, suggesting that the fuel supply can sometimes be turned on by a companion. Just as tidal forces can trigger star formation in starburst galaxies (mentioned in the previous chapter), they may also divert gas and stars into the galactic nucleus, triggering an outburst that may last for millions or even billions of years. ∞ (p. 535)

25.5 *Quasi-stellar Objects*

THE DISCOVERY OF QUASARS

❹ In the early days of radio astronomy, many radio sources were detected for which no corresponding visible object was known. By 1960, several hundred such sources were listed in the *Third Cambridge Catalog*, and astronomers were scanning the skies in search of optical counterparts. Their job was made difficult both by the low

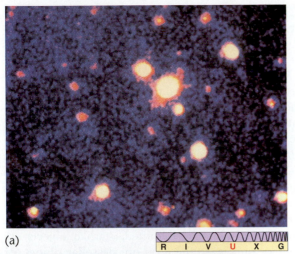

(a)

(b)

(c)

resolution of the radio observations (which meant that the observers did not know exactly where to look) and by the faintness of these objects at visible wavelengths.

In 1960, astronomers detected what appeared to be a faint blue star at the location of the radio source 3C 48 (the 48th object on the Cambridge list) and obtained its spectrum. Containing many unknown broad emission lines, the unusual spectrum defied interpretation. 3C 48 remained a unique curiosity until 1962, when another similar-looking, and similarly mysterious, faint blue object with "odd" spectral lines was discovered and identified with the radio source 3C 273. Several of these peculiar objects are shown in Figure 25.15.

The following year saw a breakthrough, when astronomers realized that the strongest unknown lines in 3C 273's spectrum were simply familiar spectral lines of hydrogen redshifted by a very unfamiliar amount—about 16 percent! This large redshift indicated a recessional velocity of about 48,000 km/s. Figure 25.16 shows the spectrum

Figure 25.15 (a) Optical photograph of 3C 275, one of the first quasars discovered. Its starlike appearance shows no obvious structure and gives little outward indication of this object's enormous luminosity. However, 3C 275 has a much larger redshift than any of the other stars or galaxies in this image; it is about 2 billion parsecs away. (b) Optical image of a field of quasars (marked), including QSO 1229+204, one of the most powerful quasars yet discovered, shown enlarged in (c). Its starlike appearance shows only a hint of structure, and gives little outward indication of the object's enormous luminosity. Like 3C 275, its distance from the Earth is about 2000 Mpc.

Figure 25.16 Optical spectrum of the distant quasar 3C 273. Notice both the redward shifts and the widths of the three hydrogen spectral lines marked as Hβ, Hγ, and Hδ. The redshift indicates the quasar's enormous distance. The width of the lines implies rapid internal motion within the quasar itself. (Note that, in this figure, red is to the right and blue is to the left.)

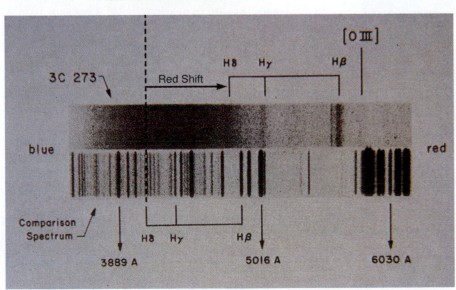

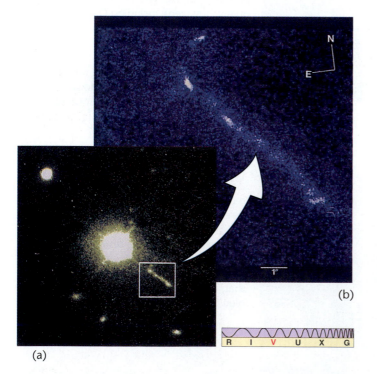

Figure 25.17 Although quasars are the most luminous objects in the universe, they are often rather unimpressive in appearance. In this optical image, a distant quasar (marked by an arrow) is seen close (on the sky) to a nearby spiral galaxy. The quasar's much greater distance makes it appear much fainter than the galaxy.

Figure 25.18 (a) The bright quasar 3C 273 displays a luminous jet of matter, but the main body of the quasar is starlike in appearance. (b) The jet extends for about 30 kpc, and can be seen better in this high-resolution image.

of 3C 273. Some prominent emission lines, and the extent of their redshift, are marked on the diagram. Once the nature of the strange spectral lines was known, astronomers quickly found a similar explanation for the spectrum of 3C 48. Its 37 percent redshift implied that it is receding from the Earth at almost one-third the speed of light.

These huge speeds mean that neither of the two objects can possibly be members of our Galaxy. Applying the Hubble law (with our adopted value of the Hubble constant $H_0 = 75$ km/s/Mpc), we obtain distances of 640 Mpc for 3C 273 and 1300 Mpc for 3C 48. Clearly not stars (with such enormous redshifts), these objects became known as *quasi-stellar radio sources* ("quasi-stellar" means "starlike"). The term has been shortened to **quasars**. We now know that not all such highly redshifted, starlike objects are strong radio sources, so the term **quasi-stellar object** (or QSO) is more common today. However, the name quasar persists, and we will continue to use it here.

OBSERVED PROPERTIES OF QUASARS

The most striking characteristic of the several hundred quasars now known is that their spectra all show large redshifts, ranging from 0.06 up to the current maximum of about 4.9. (See the *More Precisely* feature on p. 554 for an explanation of the meaning of redshifts greater than 1.) Thus, *all* quasars lie at large distances from us—the closest is 240 Mpc away, the farthest nearly 4700 Mpc (according to the table on p. 555). The majority of quasars lie more than 1000 Mpc from the Earth. We see most quasars as they existed long ago—they represent the universe as it was in the distant past.

Thus, despite their unimpressive optical appearance—see for example, Figure 25.17, which compares a quasar and a spiral galaxy that happens to lie close to it on the sky—the large distances implied by quasar redshifts mean that these faint "stars" are in fact the brightest known objects in the universe! 3C 273, for example, has a luminosity of about 10^{40} W. More generally, quasars range in luminosity from around 10^{38} W—about the same as the brightest radio galaxies—up to nearly 10^{42} W. A value of 10^{40} W, comparable to 20 trillion Suns or 1000 Milky Way Galaxies, is fairly typical. Thus quasars outshine the brightest normal and active galaxies by about a factor of 1000.

Quasars display many of the same general properties as active galaxies. Their radiation is nonthermal, and some show evidence of jets and extended emission features (although few quasars are more than a ball of luminous fuzz in visible-light images). Figure 25.18 is an optical photograph of 3C 273. Notice the jet of luminous matter, reminiscent of the jet in M87, extending nearly 3 kpc from the quasar itself. Often, as shown in Figure 25.19, quasar radio radiation arises from regions lying beyond the bright central core, much like the core–halo and lobe radio galaxies studied earlier. In other cases, the radio emission is confined to the central optical image. Quasars have been observed in the radio, infrared, optical, ultraviolet, and X-ray parts of the electromagnetic spectrum, and some have even

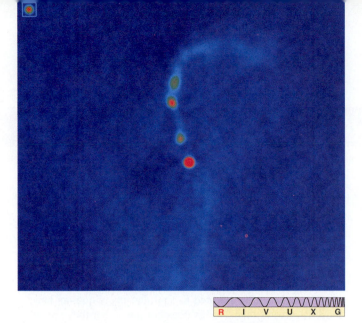

Figure 25.19 Radio image of the quasar 2300-189 showing radio jets feeding faint radio lobes. The bright (red) central object is the quasar, some 400 Mpc away.

been found to emit gamma rays. However, most quasars emit most of their energy in the infrared.

Interestingly, in addition to their own strongly redshifted spectra, many quasars also show additional absorption features that are redshifted by substantially *less* than the lines from the quasar itself. For example, the quasar known as PHL 938 has an emission-line redshift of 1.955, placing it at a distance of some 3400 Mpc, but it also shows three sets of absorption lines, with redshifts of 1.949, 1.945, and 0.613, respectively. The first two sets may well come from high-speed gas within the quasar itself (the velocity differences are only a few hundred kilometers per second), but the third is interpreted as arising from intervening gas that is much closer to us (only about 1700 Mpc away), which explains why it has a smaller redshift than the quasar itself. The most likely possibility is that this gas is part of an otherwise invisible galaxy lying along the line of sight. Quasar spectra, then, afford astronomers a means of probing previously undetected parts of the universe.

Many quasars have been observed to vary irregularly in brightness over periods of months, weeks, days, or (in some cases) even hours, in many parts of the electromagnetic spectrum. The same reasoning as we used earlier for active galaxies leads to the conclusion that the region generating the energy must be very *small*—not too much larger than our solar system in some cases. The *More Precisely* feature on p. 558 discusses another curious aspect of quasar variability that may have much to tell us about the details of the energy source.

Figure 25.20 illustrates how the light history for one quasar reveals evidence for variations in its optical radiation. In part (a), a 1937 photograph clearly shows a starlike object (marked by arrows). It has since been identified as a quasar, labeled 3C 279. In part (b), taken in 1976, the quasar has nearly disappeared. Part (c) shows the light history of 3C 279, based on photographs taken since 1930. 3C 279's great distance and measured apparent brightness

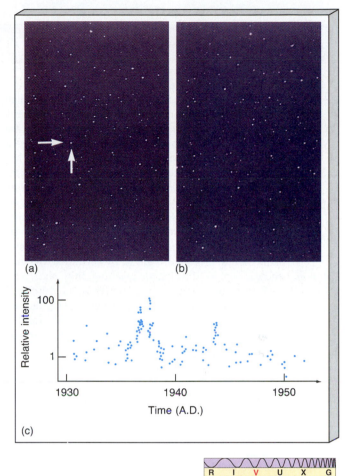

Figure 25.20 (a) Quasar 3C 279 (at the intersection of the arrows), whose luminosity in 1937 made it the most intrinsically brilliant object known in the universe. (b) The same quasar in 1976, when its luminosity was much diminished. Part (c) shows this quasar's optical variations since 1930. Its 1937 outburst later gained 3C 279 a (temporary) place in the *Guinness Book of Records* for the greatest absolute brightness of any known cosmic object.

imply that, in 1937, this faint speck of light was *intrinsically* one of the most luminous objects ever observed in the universe. At that time, its luminosity exceeded 10^{41} W. Two years later, its brightness had dropped by almost a factor of 250, making it "merely" 10 times brighter than the brightest radio galaxies.

QUASAR ENERGY GENERATION AND LIFETIMES

5 Quasars exhibit all of the properties described earlier for active galaxies—large luminosities, nonthermal emission, jets, lobes, and rapid variability (implying small size). In many respects, a quasar looks like a "souped-up" active (Seyfert or radio) galaxy, so it should come as no surprise that the best current explanation of the quasar engine is basically a scaled-up version of the mechanism powering lower-luminosity active galaxies—accretion onto a supermassive black hole residing at the galactic core.

More Precisely... *Relativistic Redshifts and Look-back Time*

When discussing very distant objects such as quasars, astronomers usually talk about their redshifts rather than their distances. Indeed, it is very common for researchers to speak of an event occurring "at" a certain redshift—meaning that the light received today from that event is redshifted by the specified amount. Of course, because of Hubble's law, redshift and distance are equivalent to one another. However, redshift is the "preferred" quantity, because it is a directly observable property of an object, whereas distance is derived from redshift using Hubble's constant, whose value is not accurately known. (In the next chapter, we will see another reason why astronomers favor the use of redshift in studies of the cosmos.)

The redshift of a beam of light is, by definition, the *fractional* increase in its wavelength resulting from the recessional motion of the source. ∞ (p. 69) Thus, a redshift of 1 corresponds to a *doubling* of the wavelength. Using the formula for the Doppler shift presented in Chapter 3, the redshift of radiation received from a source moving away from us with velocity v is given by

$$\text{redshift} = \frac{\text{observed wavelength} - \text{true wavelength}}{\text{true wavelength}}$$

$$= \frac{\text{recessional velocity } v}{\text{speed of light } c}.$$

Let's illustrate this with two examples, rounding the speed of light, c, to 300,000 km/s. A galaxy at a distance of 100 Mpc has a recessional velocity (by Hubble's law) of 75 km/s/Mpc × 100 Mpc = 7,500 km/s. Its redshift therefore is 7,500 km/s ÷ 300,000 km/s = 0.025. Conversely, an object with a redshift of 0.05 has a recessional velocity of 0.05 × 300,000 km/s = 15,000 km/s and hence a distance of 15,000 km/s divided by 75 km/s/Mpc = 200 Mpc.

Unfortunately, while it is quite correct for low velocities, the foregoing equation does not take into account the effects of relativity. As we saw in Chapter 22, the rules of everyday physics have to be modified when velocities begin to approach the speed of light, and the formula for the Doppler shift is no exception. ∞ (p. 470) In particular, while our formula is valid for velocities much less than the speed of light, when $v = c$, the redshift is not 1, as the equation suggests, but is in fact *infinite*. In other words, radiation received from an object moving away from us at nearly the speed of light would be redshifted to almost infinite wavelength.

Thus do not be alarmed to find that many quasars have redshifts greater than 1. This does not mean that they are receding faster than light! It simply means that their recessional velocities are *relativistic*—comparable to the speed of light—and the formula is not applicable. The actual formulas relating the redshift of a galaxy to its recessional velocity and its distance are quite complicated, and involve some critical—and still controversial—assumptions about the overall expansion of the universe, as we will discuss in Chapter 26. In addition (again as we will see in the next chapter), it is actually *incorrect* to think of this redshift as a Doppler shift at all. For these reasons, we will not give any formula here. Instead, we present a brief "conversion chart" between redshift, velocity and distance (shown both in megaparsecs and in millions of light years), based on reasonable assumptions and usable even for $v \approx c$. As usual, we take Hubble's constant to be 75 km/s/Mpc. Notice that, while the entries in the table agree very well with our earlier simple formulas for small redshifts (less than a few percent, corresponding to objects lying within 100 or 200 megaparsecs of Earth), they differ greatly for larger distances.

The final column in Table 25-1 lists a quantity known as the *look-back time*, which is simply how long ago an object emitted the radiation we see today. Look-back time is another measure of distance. For nearby sources, it is numerically equal to the distance in light years—the light we receive tonight from a galaxy at a distance of 100 million light years was emitted

A 10^8- or 10^9-solar-mass black hole can emit enough energy to power even the brightest (10^{38} W) radio galaxy by swallowing stars and gas at the relatively modest rate of 1 star every 10 years. To power a 10^{40} W quasar, which is 100 times brighter, the hole simply consumes 100 times more fuel—10 stars per year. The "reprocessing" mechanisms that convert the quasar's power into the radiation we actually detect—namely, the ejection of matter in jets and lobes and the reemission of radiation by surrounding gas and dust—probably operate in much the same manner as the mechanisms we described earlier for Seyferts and radio galaxies. The most likely explanation for the large luminosities of quasars is simply that there was more fuel available at very early times, perhaps left over from the formation of the galaxies in which the quasars reside. At the distances of most quasars, the galaxies themselves cannot be seen. Only their intensely bright nuclei are visible from the Earth.

In this picture, the brightest known quasars devour about 1000 solar masses of material every year. A simple calculation indicates that if they kept up this rate of energy production for the roughly 10-billion-year age of the universe, a total of 10^{13} stars would have to be destroyed. Unless the galaxies housing quasars are much larger than any other galaxy we know of, most of the quasar's parent galaxy would be completely consumed, and the universe should contain many 10^{13}-solar-mass black holes— "burned-out" quasars. We have no evidence for the existence of any such objects. One way around this problem is to suppose that a quasar spends only a fairly short period of time in this highly luminous phase—perhaps a few tens of millions of years. There is theoretical evidence to suggest that black holes tend to eat out "cavities" at the cen-

100 million years ago. But for more distant objects (for example, many normal galaxies, most active galaxies, and all quasars), the two numbers differ, and the divergence increases dramatically with increasing redshift. The reason for this apparent contradiction to the basic definition of the light year is the expansion of the universe. A galaxy that is presently located 15 billion light years from the Earth was much closer to us in the past, when it emitted the light we now see. Consequently, the light has taken considerably less than 15 billion years—in fact, only about 8 billion years—to reach us.

As a simple analogy, imagine an ant crawling across the surface of an expanding balloon at a constant speed of 1 cm/s relative to the balloon's surface. After 10 seconds, the ant might think it has traveled a distance of 10 cm, but an outside observer with a tape measure will find that it is actually more than 10 cm from its starting point (measured along the surface of the balloon), simply becasue of the balloon's expansion. Furthermore, the difference between the actual distance and 10 cm depends on the details of the balloon's expansion during the journey—the more rapid the expansion, the greater the disparity will be. In exactly the same way, the present distance to a galaxy with a given redshift depends in detail on how the universe expanded in the past. We will see in Chapter 26 that this is not well known, so the distance is in fact not well determined. For this reason, while astronomers frequently talk about redshifts, and sometimes about look-back times, they hardly ever talk of the distances to high-redshift objects. The distances given in the accompanying table will be used consistently throughout this book, but realize that they are subject to considerable uncertainty.

Finally, notice that, according to the table, the recessional velocity equals the speed of light—and the redshift becomes infinite—for objects that emitted their radiation about 8.7 billion years ago. We will examine the reasons for and implications of this fact in the next chapter.

TABLE 25–1   *Redshift and Recessional Velocity*

REDSHIFT	v/c	PRESENT DISTANCE (Mpc)	PRESENT DISTANCE (10^6ly)	LOOK-BACK TIME (millions of years)
0.000	0.000	0	0	0
0.010	0.010	40	129	129
0.025	0.025	98	320	316
0.05	0.049	193	628	613
0.10	0.095	372	1214	1158
0.20	0.180	697	2272	2080
0.25	0.220	844	2753	2473
0.50	0.385	1468	4785	3961
0.75	0.508	1952	6365	4937
1.00	0.600	2343	7638	5619
1.50	0.724	2940	9584	6493
2.00	0.800	3381	11021	7019
3.00	0.882	4000	13038	7606
4.00	0.923	4422	14415	7915
5.00	0.946	4733	15431	8101
10.0	0.984	5587	18214	8454
50.0	0.999	6879	22425	8668
100.0	1.000	7203	23482	8683
∞	1.000	7746	25253	8692

ters of their host galaxies, effectively cutting off their fuel supply through their own greed. Alternatively, as with nearby active galaxies, the high luminosities may be the result of interactions between galaxies in the early universe. The fact that quasars have been observed in some distant galaxy clusters argues in favor of this latter view. For a radically different view of quasars, however, see *Interlude 25-2* on p. 561.

QUASAR "MIRAGES"

In 1979, astronomers were surprised to discover what appeared to be a binary quasar—two quasars with exactly the same redshift and very similar spectra, separated by only a few arc seconds on the sky. Remarkable as the discovery of such a binary would have been, the truth about this pair of quasars turned out to be even more amazing. Closer study of the quasars' radio emission revealed that they were *not* in fact two distinct objects. Instead, they were two separate images of the *same* quasar! Optical views of such a so-called *twin quasar* are shown in Figure 25.21.

What could produce such a "doubling" of a quasar image? The answer is *gravitational lensing*—the deflection and focusing of light from a background object by the gravity of some foreground body. In Chapter 23 we saw how lensing by brown dwarfs or other compact objects in the halo of the Milky Way may temporarily cause the light from a distant star to be amplified, allowing astronomers to detect otherwise invisible stellar dark matter in our Galaxy. ∞ (p. 506) In the case of quasars, the idea is the same, except that the foreground lensing object is an entire galaxy or galaxy cluster, and the deflection of the light is so great (an arc second or so) that several separate images of the quasar

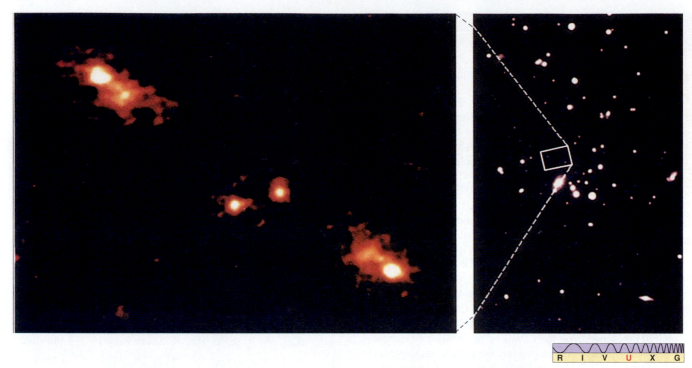

Figure 25.21 (a) This "twin" quasar (designated AC114 and located about 2 billion parsecs away) is not two separate objects at all. Instead, the two large blobs (at upper left and lower right) are images of the same object, created by a gravitational lens. These two "L" shaped blobs of light have striking symmetry. The lensing galaxy itself is probably not visible in this image—the two objects near the center of the frame are thought to be unrelated galaxies in a foreground cluster. (b) A larger perspective of AC114 in the nighttime sky.

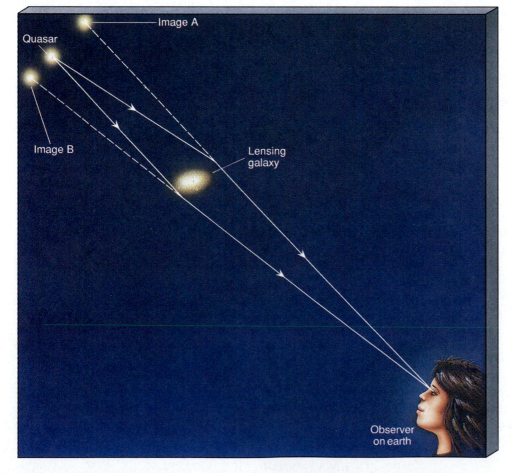

Figure 25.22 When light from a distant object passes close to a galaxy or cluster of galaxies along the line of sight, the image of the background object (here, the quasar) can sometimes be split into two or more separate images (here, A and B). The foreground object is a gravitational lens.

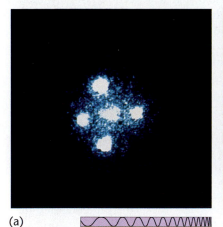
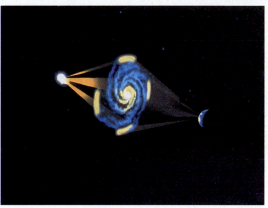

(a)   (b)

Figure 25.23   (a) The "Einstein Cross," a multiple imaged quasar. In this *Hubble* view, spanning only a couple of arc seconds, four separate images of the same quasar have been produced by the galaxy at the center. (b) An artist's conception of what might be occurring here, with the Earth at right and the distant quasar at left.

may be formed, as illustrated in Figure 25.22. About a dozen likely gravitational lenses are known. Figure 25.23 shows an image of a lensed system in which *four* images of the same quasar can be seen, neatly arranged around the central image of the lensing galaxy.

The existence of these multiple images provides astronomers with several useful observational tools. First, the lensing tends to amplify the light of the quasar, making it easier to observe. Second, because the light rays forming the images usually follow paths of different lengths, there is often a *time delay*, ranging from several days to several years, between them. This delay provides advance notice of explosive events, such as sudden flare-ups in the quasar's brightness—if one image flares up, astronomers know that in time the others will too, so they have a second chance to study the event. The time delay also permits astronomers to determine the *distance* to the lensing galaxy by carefully timing the measurements. If enough lenses can be found, this method may provide a reliable alternative means of measuring the Hubble constant that is independent of any of the techniques discussed in Chapter 24.

Third, so-called *microlensing*—lensing by individual stars in the foreground galaxy—can cause large fluctuations in a quasar's brightness. ∞ (p. 506) Microlensing allows astronomers to study the stellar content of the lensing galaxy.

Finally, by studying the lensing of background quasars and galaxies by foreground galaxy clusters, astronomers

can obtain a better understanding of the distribution of dark matter in those clusters, an issue that has great bearing on the large-scale structure of the cosmos, as we will see in the next chapter. Figure 25.24 shows the images of some faint, blue background galaxies (see Chapter 24) bent into arcs by the gravity of a nearby galaxy cluster. ∞ (p. 532) The degree of bending allows the total mass of the cluster (including the dark matter) to be measured.

Blue arcs

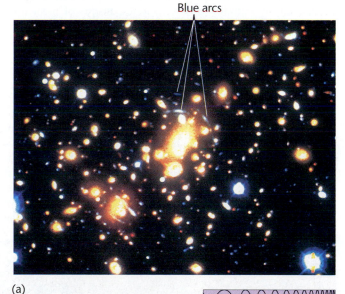

(a)

(b)

Figure 25.24   (a) The blue arcs (marked above and to the right of the central yellowish region) are images of young, blue background galaxies, warped by the gravitational field of a foreground galaxy cluster, which deflects their light and distorts their appearance. By measuring the extent of this distortion, astronomers can estimate the mass of the intervening cluster. (b) This spectacular example of gravitational lensing shows more than a hundred faint arcs from very distant galaxies. The wispy pattern spread across the intervening galaxy cluster (A2218, several billion parsecs distant) resembles a spider's web, but it is really an illusion caused by the gravitational field of the lensing cluster.

25.6 *Active Galaxy Evolution*

5 In Chapters 23 and 24 we addressed the issue of evolutionary change among normal galaxies. ∞ (p. 530) Let us now briefly consider the possibility of evolutionary links among active galaxies and between normal and active galaxies. We emphasize that this section is really mostly speculation. Although the consensus is that galaxies began to form about 8 billion years ago and that quasars were an early stage of galaxy evolution, the details of the connections among different types of active and normal galaxies are still very uncertain.

More Precisely...   *Faster-than-Light Velocities?*

The compact sources of emission within radio galaxies and quasars vary not only in intensity but also in structure. Some quasars recently mapped with very-long-baseline interferometers have displayed dramatic changes in structure, often on time scales as short as months. For example, the accompanying illustrations show three radio images of the core of the well-studied "nearby" quasar 3C 273, made several years apart. The interior of this quasar is dominated by two large blobs of gas, which move over the course of time. Knowing the distance to 3C 273 (about 640 Mpc) and measuring the angle through which the blobs moved in the course of 3 years (about 2 milliarc seconds), astronomers have calculated the blobs' velocities. Astonishingly, the result is nearly 10 times the speed of light!

The notion that the speed of light is the highest attainable velocity is central to modern physics. Scores of predictions made assuming this fact to be true have been verified to high accuracy since Einstein first published his theory of relativity early in the twentieth century (see the *More Precisely* feature on p. 473). Astronomers almost universally agree that some reinterpretation of these apparent *superluminal* (that is, faster than light) quasar motions is needed.

One alternative assumes that quasars are not at cosmological distances from us.. If quasars were relatively local, the angular motion of the blobs would not imply high physical velocities, and the problem of explaining how the individual blobs move faster than light would not exist. But if the quasars are local and not distant, the observed redshifts of their spectral lines cannot be a distance indicator, and we would be forced to find some other explanation for the large quasar redshifts. As discussed in more detail in *Interlude 25-2* on p. 560, few astronomers are prepared to make that assumption.

Several alternative solutions have been proposed to account for the apparent superluminal motion of the blobs without requiring the quasars to be local or the speeds truly faster than light. The most straightforward model suggests that the observed changes in quasar structure are not caused by actual motions of the interior blobs at all, but rather by variations of the radio intensity of *stationary* blobs. In other words, the interiors of quasars may resemble the blinking lights that sometimes appear on movie theater marquees. These marquee bulbs blink on and off in a programmed sequence that gives the illusion of motion, but the individual bulbs are of course stationary. Likewise, blinking radio sources within the quasar may suggest motion where none really exists. A variation on this model explains the motions as a different kind of illusion—a projection effect, produced by blobs moving almost precisely along our line of sight at slightly less than the speed of light. Calculations of how these blobs would look from Earth indicate that they could in fact appear to be moving faster than light.

None of the alternative models is simple. They all require peculiar geometries, and no one model is agreed upon by all researchers. We still lack a complete explanation of the puzzling phenomenon of superluminal velocities. Still, the existing models demonstrate that although the interiors of quasars are exceedingly complex and not terribly well understood, there is no compelling need to discard the laws of physics in order to explain them.

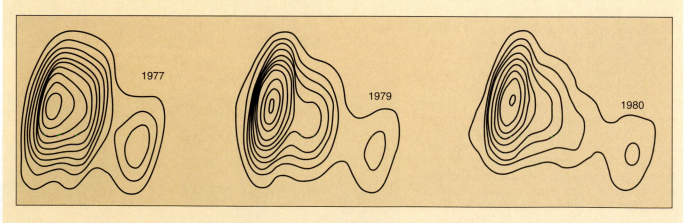

Most quasars are very distant, indicating that they were more common in the past than they are today. At the same time, "normal looking" galaxies seem to be less common at high redshift. These two pieces of evidence suggest to many astronomers that, when galaxies first formed, they probably were quasars. This opinion is strengthened by the fact that the same black-hole energy-generation mechanism can account for the luminosity of quasars, active galaxies, and the central regions of normal galaxies like our own. Large black holes do not simply vanish, at least in the 10–20 billion years that the universe has existed (see the *More Precisely* feature on p. 477). Thus, the presence of supermassive black holes in the centers of many, if not all, normal galaxies is consistent with the idea that they started off as quasars, then "wound down" to become the relatively quiescent objects we see today.

In this picture, the gradual reduction in violence from a quasar to a Seyfert galaxy to a normal spiral, for example, occurs primarily because the fuel supply is reduced as the galaxy evolves. A similar sequence might connect quasars to BL Lac objects (*Interlude 25-1* on p. 546) to radio galax-

ies to normal ellipticals. These possible evolutionary connections among the active and normal galaxies are illustrated in Figure 25.25. Adjacent objects along this sequence are nearly indistinguishable from one another. For example, weak quasars share some characteristics with some very active galaxies, and the feeblest active galaxies often resemble the most explosive normal galaxies.

If we accept this appealing (but still unproven) view, we can construct the following possible scenario for the evolution of galaxies in the universe: Galaxies formed at a redshift of around 5. The early round of massive star formation that may have expelled galactic gas and helped determine a galaxy's Hubble type—spiral or elliptical—could also have given rise to many large, stellar-mass black holes, which sank to the center of the still-forming galaxy and merged into a supermassive black hole there. Alternatively, the supermassive hole may have formed directly by gravitational collapse of the dense central regions of the protogalaxy. Whatever the cause, large black holes appeared at the centers of many galaxies at a time when there was still plenty of fuel available to power

Figure 25.25 A possible evolutionary sequence for galaxies, beginning with the highly luminous quasars, decreasing in violence through the radio and Seyfert galaxies, and ending with normal spirals and ellipticals. The central black holes that powered the early activity are still there at later times; they simply run out of fuel as time goes on.

Interlude 25-2   *Could Quasars Be Local?*

When quasars were first discovered, their large luminosities and small sizes troubled many astronomers. In the 1960s and 1970s, no known mechanism could account for the generation of 1000 Milky Ways' worth of energy within a region comparable in size to the solar system. The idea that supermassive black holes may be quite common in the cores of galaxies was unknown.

In response to these problems, some astronomers, notably the respected observer Halton Arp and the equally reputable theorist Geoffrey Burbidge, sought an alternative explanation for quasars. Instead of believing that these objects were at cosmological distances and so very luminous, these researchers argued that perhaps there was an alternative, *noncosmological* explanation for their great redshifts. Quasars could then be relatively nearby and hence much less bright. For example, if 3C 273 were 100 times closer—only 6.4 Mpc, not 640 Mpc, away—the inverse-square law says that the luminosity required to account for its observed apparent brightness would be reduced by a factor of 10,000. ∞ (p. 67) In that case, 3C 273 would radiate "only" 10^{36} W (one-tenth the energy output of our Galaxy). This is still a lot of energy, but it is perhaps more easily explainable in terms of "familiar" stellar events: the formation of high-mass stars, supernovae, and so on.

Arp has reported many examples of instances where galaxies and quasars are found close together on the sky but have very different, conflicting redshifts. Figure 25.17 is an example of such an alignment. He argues that there are simply too many of these "coincidences," where a foreground galaxy lies in nearly the same direction as a supposed background quasar, for the distant-quasar hypothesis to be correct. Instead, he claims, the quasars must be *physically* close to the galaxies, and the redshifts have some other, noncosmological, explanation. Arp and coworkers have gone further, citing instances where neighboring *galaxies* have con-

flicting redshifts, such as the system in the first image here, where one of the five galaxies (Stephen's Quintet) has a redshift very different from those of the others. Similarly, in the second image (which is shown as a photographic negative to bring out the faint structure), a small galaxy appears to be connected to a larger galaxy (NGC 7603) having a very different redshift by a faint "bridge" of gas, reinforcing the view that the two really are close together in space. Such findings call into question both the cosmological interpretation of *all* galactic redshifts and Hubble's law itself.

Most astronomers would argue that the claims of conflicting redshifts are not statistically significant. In other words, given the numbers of known galaxies and quasars, accidental superpositions on the sky should be quite commonplace, and the observed quasar–galaxy and galaxy–galaxy

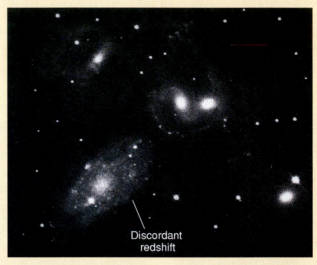

Discordant redshift

R   I   V   U   X   G

them, resulting in many highly luminous quasars. The brightest quasars—the ones we see from Earth—were those with the greatest fuel supply. The young galaxies themselves were so faint compared with their bright quasar cores that we simply cannot see them. (Unfortunately, this last point is still conjecture, and recent *Hubble* observations have failed to clarify the issue. One group of astronomers working with the telescope failed to detect the expected galactic "fuzz" around several quasars. However, another group, working with a different on-board camera, has apparently seen clear evidence for host galaxies.)

As the galaxy developed and the black hole used up its fuel, the luminosity of the central nucleus diminished. While still active, it no longer completely overwhelmed the emission from the surrounding stars. The result was an active galaxy—a radio galaxy or a Seyfert—still emitting a lot of energy, but now with a definite "stellar" component in its spectrum.

The central activity continued to decline. Eventually, only the surrounding galaxy remained visible—a normal galaxy, like the majority of those we now see around us. Today, the black holes that generated so much youthful energy lie dormant in galactic cores, producing only a relative trickle of radiation. Occasionally, two nearby normal galax-

alignments are quite consistent with pure chance. The apparent bridges may simply be photographic or image-processing defects. Usually, a lot of computer enhancement is needed to bring out the images, and there is ample opportunity for "features" to appear where none really exist. Finally, Hubble's law *is* well-established for galaxies within a few hundred megaparsecs, and some quasars have been found in galaxy clusters, *sharing* the redshift of their neighbors. Thus, at least some quasar redshifts are known to be cosmological, so the violations of Hubble's law that Arp claims exist appear only quite selectively—if they really do exist at all.

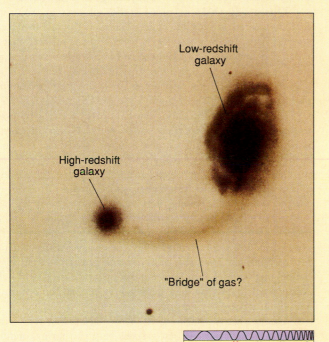

Low-redshift galaxy

High-redshift galaxy

"Bridge" of gas?

R I V U X G

If quasars are actually local, no convincing explanation has ever been advanced for their redshifts. If they are simply moving at high speeds, then we are led to the new "energy problem" of explaining how they were accelerated to such speeds. If quasars were fired out of galaxies—one alternative explanation for their great speed—we would expect some to be moving toward us, and hence blueshifted. So why are they *all* redshifted? If quasars are all "local" and come from our own Galaxy, then the Milky Way is special—a major violation of the principle of mediocrity and current scientific dogma—or the quasars must be so close (and thus quite faint) that we should have observed the motion of some of them across the sky. Arp's own supposed correlations between other galaxies and quasars would also argue against a Milky Way origin. In short, no matter where we try to put the quasars, their redshifts pose problems. An alternative attempt to explain the redshift as a gravitational redshift suffered by light as it climbs out from the vicinity of a black hole (see Chapter 22) also fails because it cannot account for the observed widths of quasar spectral lines. ∞ (p. 478)

There is no clear observational evidence for conflicting quasar redshifts and no satisfactory mechanism of producing the redshifts by noncosmological means. Furthermore, as we have seen, there really is no "quasar luminosity problem" any more. Quasar luminosities can be explained by the same mechanism that powers active galaxies, with only a modest increase in fuel consumption and certainly without posing a serious challenge to the laws of physics. Consequently, the overwhelming majority of astronomers hold that quasar redshifts are cosmological in origin and that quasars really are the most distant objects known in the universe. (Of course, there are many instances in the history of astronomy where the majority has later been proved totally wrong!)

ies may interact with one another, causing a flood of new fuel to be directed toward the central black hole of one or both. The engine starts up for a while, giving rise to the nearby active galaxies we observe.

Should this picture be correct, then many normal galaxies, including perhaps our own Milky Way Galaxy, were once brilliant quasars. Perhaps some alien astronomer, thousands of megaparsecs away, is at this very moment observing our Galaxy—seeing it as it was billions of years ago—and is commenting on its enormous luminosity, nonstellar spectrum and high-speed jets, and wondering what exotic physical process could possibly ac-

count for its violent activity!

When they were first discovered, active galaxies and quasars seemed to present astronomers with insurmountable problems. For a time, their dual properties of enormous energy output and small size appeared incompatible with the known laws of physics and threatened to overturn our modern view of the universe. Yet the problems were eventually solved, and the laws of physics remain intact. Far from jeopardizing our knowledge of the cosmos, these violent phenomena have become part of the thread of understanding that binds our own Galaxy to the earliest epochs of the universe we live in.

Chapter Review

SUMMARY

Active galaxies (p. 540) differ from normal galaxies, and are far less numerous. Active galaxies have abnormally high luminosities and spectra that are nonstellar in nature, indicating that their energy emission is not simply the accumulated light of many stars. Most of the energy from active galaxies is emitted in the radio and infrared parts of the electromagnetic spectrum. The fraction of observed galaxies that display activity increases with increasing distance from the Milky Way, indicating that galaxies were generally more active in the past than they are today.

A **Seyfert galaxy** (p. 541) looks like a normal spiral except that it has an extremely bright central **galactic nucleus** (p. 541), whose luminosity can in many cases exceed that of the rest of the galaxy. Spectral lines from Seyfert nuclei are very broad, indicating rapid internal motion. In addition, Seyfert luminosities can vary by large amounts in fractions of a year, implying that the region emitting most of the radiation is much less than a light year across.

Radio galaxies (p. 542) are active galaxies that emit most of their energy in the radio part of the spectrum. They are generally comparable to the Seyferts in total energy output. Unlike Seyferts, they are usually associated with elliptical galaxies. In a core–halo radio galaxy, the energy is emitted from a small central nucleus, again like a Seyfert. In a lobe radio galaxy, the energy comes from enormous **radio lobes** (p. 543) that dwarf the optical galaxy and lie far outside it. The lobes are usually symmetrically placed with respect to the center of the optical galaxy.

Many active galaxies have high-speed jets of matter shooting out from their central nuclei. In lobe radio galaxies, astronomers believe that the jets transport energy from the nucleus, where it is generated, to the lobes, where it is radiated into space. The jets often appear to be made up of distinct "blobs" of gas, suggesting that the process generating the energy is intermittent.

The generally accepted explanation for the observed properties of active galaxies is that the energy is generated by accretion of galactic gas onto a supermassive (billion–solar–mass) black hole lying at the center of the nucleus. As the material spirals down toward the hole, it heats up and releases enormous amounts of energy. The small size of the accretion disk explains the compact extent of the emitting region, and the high-speed orbits of gas in the

hole's intense gravity accounts for the rapid motion observed. Typical active-galaxy luminosities require the consumption of about one solar mass of material every few years.

Some of the infalling matter is blasted out into space, producing magnetized jets that create and feed the extended radio lobes. Charged particles spiraling around the magnetic field lines produce **synchrotron radiation** (p. 550) whose spectrum is consistent with the nonstellar radiation observed in radio galaxies and Seyferts.

Quasars (p. 542), or **quasi-stellar objects** (p. 542), were first discovered as starlike radio sources with unknown broad spectral lines. In the early 1960s, astronomers realized that the unfamiliar lines are actually those of familiar elements, but redshifted to wavelengths much longer than normal. Even the closest quasars lie at great distances from us. They are the most luminous objects known.

Quasars exhibit the same basic features as active galaxies, and astronomers believe that their power source is also basically the same. This source—a black hole—must be "scaled up" in a quasar to a consumption rate of many stars per year. If that is the case, the brightest quasars consume so much fuel that their energy-emitting lifetimes must be relatively short. Quasars probably represent a brief phase of violent activity early in the life of a galaxy.

Some quasars have been observed to have double or multiple images. These result from gravitational lensing, where the gravitational field of a foreground galaxy or galaxy cluster bends and focuses the light from the more distant quasar. Analysis of this bending provides a means of determining the masses of galaxy clusters—including the dark matter—far beyond the optical images of the galaxies themselves.

Quasars, active galaxies, and normal galaxies may represent an evolutionary sequence. When galaxies began to form, conditions may have been suitable for the formation of large black holes at their centers. If there was a lot of gas available at those early times, a highly luminous quasar would have been the result. As the fuel supply diminished, the quasar dimmed and the galaxy in which it was embedded became visible as an active galaxy. At even later times, the fuel supply declined to the point where the nucleus became virtually inactive, and a normal galaxy was all that remained.

SELF-TEST: True or False?

_____ **1.** Active galaxies can emit thousands of times more energy than our own Galaxy.

_____ **2.** The "extra" radiation emitted by active galaxies is due to the tremendous number of stars they contain.

_____ **3.** Active galaxies emit most radiation at optical wavelengths.

_____ **4.** Most core-halo radio galaxies are spirals.

_____ **5.** The size of a billion-solar-mass black hole is about 20 A.U.

_____ **6.** Nearby active galaxies are most likely the result of interactions between galaxies.

_____ **7.** A redshift greater than 1 means a recessional velocity greater than the speed of light.

_____ **8.** All quasars are far away.

_____ **9.** Other than a small amount of visible light, quasars emit all of their radiation at radio wavelengths.

_____ **10.** Many nearby normal galaxies may once have been quasars.

_____ **11.** Quasars emit about as much energy as normal galaxies.

_____ **12.** The quasar stage of a galaxy ends because the central black hole uses up all the matter around it.

SELF-TEST: Fill in the Blank

1. Active galaxies are more common at _____ distances.
2. Seyfert galaxies appear like normal spirals, but with a very bright galactic _____.
3. In a core-halo radio galaxy, most of the radio radiation is emitted from the _____ .
4. Lobe radio galaxies emit radio radiation from regions that are typically much _____ in size than the optical galaxy.
5. Radio lobes are always found aligned with the _____ of the optical galaxy.
6. For all types of active galaxy, the original source of the tremendous energy emitted is the galactic _____ .
7. The energy source of an active galaxy is unusual in that there is a large amount of energy emitted from a region less than _____ in diameter. (Give size and unit.)

8. The mass of the black hole responsible for energy production in the active galaxy M87 is thought to be approximately _____ solar masses.
9. The amount of mass that must be consumed by a supermassive black hole to provide the energy for an active galaxy is about _____ per _____ .
10. Quasars, in visible light, have a _____ appearance.
11. The fact that a typical quasar would consume an entire galaxy's worth of mass in 10 billion years suggests that quasar lifetimes are relatively _____ .
12. The image of a distant quasar can be split into several images by gravitational lensing, produced by a foreground _____ along the line of sight.

REVIEW AND DISCUSSION

1. Name two basic differences between normal galaxies and active galaxies.
2. Describe some of the basic properties of Seyfert galaxies.
3. What distinguishes a core-halo radio galaxy from a lobe radio galaxy?
4. What is the evidence that the radio lobes of some active galaxies consist of material ejected from the galaxy's center?
5. What conditions result in a head-tail radio galaxy?
6. Briefly describe the leading model for the central engine of an active galaxy.

7. How is the process of synchrotron emission related to observations of active galaxies?
8. What was it about the spectra of quasars that was so unexpected and surprising?
9. Why are quasars far more luminous than active galaxies?
10. How are the spectra of distant quasars used to probe the space between us and them?
11. How are BL Lac objects related to other active galaxies?
12. What evidence do we have that quasars represent an early stage of galaxy evolution?

PROBLEMS

1. Centaurus A—from one radio lobe to the other—spans about 1 Mpc. It lies at a distance of 4 Mpc from Earth. What is the angular size of Centaurus A? Compare this with the angular diameter of the Moon.
2. Assuming the same efficiency as postulated in the text, how much energy would an active galaxy generate if it consumed one Earth mass of material every day? Compare this with the luminosity of the Sun.

3. A certain quasar has a recessional velocity of 60,000 km/s and the same apparent brightness as the Sun would have if it were placed at a distance of 1 kpc. Assuming a Hubble constant of 75 km/s/Mpc, calculate the quasar's luminosity.
4. A Seyfert galaxy is observed to have broadened emission lines indicating velocities of 1000 km/s at a distance of 1 pc from its center. Assuming circular orbits, use Kepler's Laws (see Chapter 23) to estimate the mass within this 1 pc radius.

PROJECTS

Here are three observing projects that are increasingly challenging.

1. In the last chapter you were given directions for finding the Virgo Cluster of galaxies. M87, in the central part of this cluster, is the nearest core-halo radio galaxy. M87 has coordinates RA=12^h 30.8^m, dec=+12° 24′. At magnitude 8.6 it should not be difficult to find in an 8-inch telescope. Its distance is roughly 20 Mpc. Describe its nucleus; compare what you see to other nearby ellipticals in the Virgo Cluster.
2. NGC 4151 is the brightest Seyfert galaxy. Its coordinates are RA=12^h 10.5^m, dec=+39° 24′ and can be found below the Big Dipper in Canes Venatici. At magnitude 10–12 (it is variable), it should be visible in an 8-inch telescope, but it will be challenging to find. Its distance is 13.5 Mpc. As in the case of M87, describe its nucleus and compare to what you have seen for other galaxies.

3. 3C 273 is the nearest and brightest quasar. However, that does not mean it will be easy to find and see! Its coordinates are RA=12^h 29.2^m, dec=+2° 03′. It is located in the southern part of the Virgo Cluster, but is not associated with it. At magnitude 12–13 (again, it is variable), it may require a 10- or 12-inch telescope to see, but try it first with an 8-inch. It should appear as a very faint star. The significance of seeing this object is that it is 640 Mpc distant. The light you are seeing left this object over 2 billion years ago! 3C 273 is the most distant object observable with a small telescope.

If you can find the three objects listed here, you have started to become an accomplished observer!

B1

B2

B3

A1

A2

A3

26

COSMOLOGY

The Big Bang and the Fate of the Universe

LEARNING GOALS

Studying this chapter will enable you to:

1 State the cosmological principle, and explain its significance.

2 Explain how the approximate age of the universe is determined, and discuss the uncertainties involved.

3 Summarize the leading evolutionary models of the universe.

4 Discuss the factors that determine whether or not the universe will expand forever.

5 Explain the relationship between the future of the universe and the overall geometry of space.

6 Describe the cosmic microwave background radiation, and explain its importance to our understanding of cosmology.

(Opposite page, background) By virtually all accounts, the universe began in a fiery explosion some 10–20 billion years ago. Out of this maelstrom emerged all the energy that would later form the galaxies, stars, and planets (depicted here in an artist's rendering). The story of the origin and fate of all these systems—and especially of the entire universe itself—comprises the subject of cosmology.

(Inset A) These three images of spiral galaxies lying at different distances from Earth capture representative views of what the universe was like at much earlier times. At top, about 12 billion years ago, the galaxy's spiral shape is hardly recognizable. At middle, about 9 billion years ago, the spiral is still vague, though starburst activity is evident in the outer regions. At bottom, some 5 billion years ago, the galaxy's spiral features are more prominent, with much star formation in its arms.

(Inset B) Elliptical galaxies also suggest evolution with time—indeed, evolution more rapid than that of spiral galaxies. These images of different galaxies (from top to bottom, 12, 9, and 5 billion years ago, respectively) suggest that ellipticals had taken on their eventual shapes early in the universe. In contrast to spirals, the ellipticals seem to have made all their stars long, long ago.

We have reached the limits of the observable universe. Our field of view now extends for billions of parsecs into space and billions of years back in time. We have asked and answered many questions about the structure and evolution of planets, stars, and galaxies. At last we are in a position to address the central issues of the biggest puzzle of all: How big is the universe? How long has it been around? And how long will it last? What was its origin, and what will be its fate? Is the universe a one-time event, or does it recur and renew itself, in a grand cycle of birth, death, and rebirth? How and when did matter, atoms, our Galaxy form? These are basic questions, but they are hard questions. Many cultures have asked them, in one form or another, and have developed their own cosmologies—theories about the nature, origin, and destiny of the universe—to answer them. In this and the next chapter, we will see how modern scientific cosmology addresses these important issues, and what it has to tell us about the universe we inhabit. After more than 10,000 years of civilization, science may be ready to provide some insight regarding the ultimate origin of all things.

26.1 *The Universe on the Largest Scales*

THE END OF STRUCTURE

The universe shows structure on every scale we have examined so far. Subatomic particles form nuclei and atoms. Atoms form planets and stars. Stars form star clusters and galaxies. Galaxies form galaxy clusters, superclusters, and even larger structures—voids and sheets that stretch across the sky. From the quarks in a proton to the galaxies in the Great Wall, we can trace a hierarchy of "clustering" of matter from the very smallest to the very largest scales. It is natural to ask: Does the clustering ever end? Is there some scale on which the universe can be regarded as more or less smooth and featureless? Perhaps surprisingly, given the trend we have just described, most astronomers think the answer is *yes*.

We saw in Chapter 24 how surveys of the universe have revealed the existence of structures as large as 200 Mpc (megaparsecs) across. ∞ (p. 531) Yet, while they cover wide areas of the sky and enormous volumes of space, these studies are still relatively "local," in the sense that they span only a few percent of the distance to the farthest quasars (which lie nearly 5000 Mpc from Earth). Is the structure we see locally a reliable indicator of the appearance of the universe on much larger scales? Astronomers would very much like to know the answer to this question, but a major obstacle to extending these wide-angle surveys to much greater distances is the sheer observational effort of measuring the redshifts of all the galaxies within larger and larger volumes of space.

Are there any larger studies of the universe? Yes and no. There are no surveys that cover larger volumes of space, although some are underway, but deeper surveys (that is, extending to greater distances) of small regions of the sky *do* exist. An alternative to wide-angle studies is to narrow the field of view to only a few small patches of the sky, but then to study very distant galaxies within those patches. The volume surveyed then becomes a long thin "pencil beam" extending deep into space, rather than a wide swath through the local universe.

Figure 26.1 presents the results of one such survey, carried out jointly by researchers in the United States and Britain. It shows the distribution of galaxies in two directions perpendicular to the plane of our Galaxy, out to a distance of about 2000 Mpc—10 times farther than the CfA study discussed in Chapter 24. ∞ (p. 530) Although the numbers of galaxies fall off at large distances—mainly because of the simple fact that distant galaxies are very hard to see—a distinctive "on–off" pattern of galaxies, looking a little like a picket fence, can be seen. It is thought that the gaps between the "pickets" are voids like those seen closer to home and that the pickets themselves are places where the imaginary beam intersects sheets like the Great Wall. The data appear to indicate that the largest structures in the local universe are only 100–200 Mpc across—no voids or clumps of galaxies much larger than that are seen. There is presently *no* evidence for structure in the universe on scales greater than about 200 Mpc.

On the basis of a combination of sketchy data, theoretical insight, and not a little philosophical preference, cosmologists—astronomers who study the large-scale structure and dynamics of the universe—assume that the universe is roughly *homogeneous* on scales greater than a few hundred megaparsecs. In other words, if we imagine taking a huge cube—300 Mpc on a side, say—and placing it anywhere in the universe, as illustrated in Figure 26.2, the number of galaxies it enclosed would be pretty much the same—around 100,000, excluding the faint dwarf ellipticals and irregulars—no matter *where* in the universe we put it. Some of the galaxies would be clustered and clumped into fairly large structures, and some would not, but the total number would not vary much as the cube was moved from place to place. In short, the universe looks *smooth* on the largest scales.

THE COSMOLOGICAL PRINCIPLE

❶ Cosmic homogeneity is the first of two major assumptions that cosmologists make when studying the

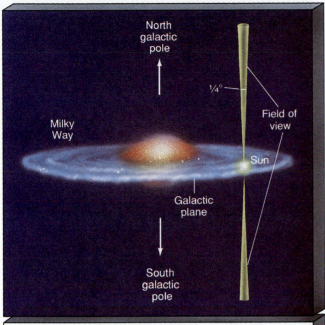

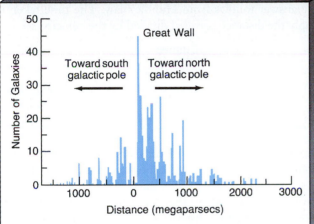

Figure 26.1 The results of a deep "pencil-beam" survey of two small portions of the sky in opposite directions from the Earth perpendicular to the galactic plane. The graph shows the number of galaxies at different distances from us, out to a distance of about 2000 Mpc. The distinctive "picket fence" appearance seems to show voids and sheets of galaxies on scales of 100 or 200 megaparsecs, but gives no indication of any larger structure.

large-scale structure of the universe. Observations suggest that it is true, but it is by no means proven. The second assumption, also supported by observational evidence and theoretical reasoning, is that the universe is *isotropic*—that is, it looks the same in any direction. Isotropy is on much firmer observational ground than homogeneity. Apart from regions of the sky that are obscured by our Galaxy, the universe *does* look much the same in all directions, at any wavelength, provided we look far enough. In other words, any deep pencil-beam survey of the sky should count about the same number of galaxies as the study mentioned previously, regardless of which patch of the sky is chosen.

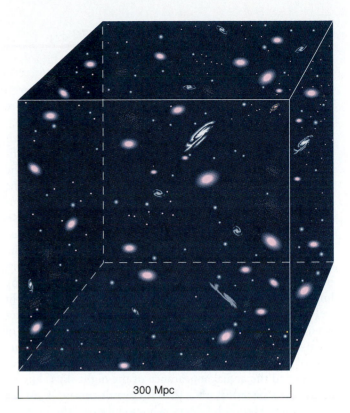

Figure 26.2 Diagram of galaxies contained within an enormous cube, 300 Mpc on a side. Cosmologists believe that regardless of where we placed this cube in the universe its contents would look similar. (Not drawn to scale; for accurate scaling see the deep-space photos in Figures 24.18, 24.27, and 24.28.)

The assumptions of homogeneity and isotropy form the foundation of modern **cosmology**—the study of the structure and evolution of the entire universe. Together, these twin pillars of cosmology are known as the **cosmological principle**. No one really knows if this principle is absolutely correct. All that we can say is that, so far, it seems consistent with observations. From this point on, we will simply assume that it holds—at least on large enough scales. Should it turn out to be incorrect—for example, were a structure a few thousand megaparsecs across to be discovered tomorrow—then some of the discussion that follows would be on very shaky ground indeed!

The cosmological principle has very far-reaching implications. For example, it implies that there can be no *edge* to the universe, because that would violate the assumption of homogeneity. Furthermore, it implies that there is no *center*, because that would mean that the universe would not be the same in all directions from any noncentral point, a violation of the assumption of isotropy. Thus, this single principle strongly limits what the overall geometry of the universe can be. The cosmological principle is the ultimate expression of the principle of mediocrity. It states not only that are we not central to the universe, but that *no one* can be central, because *the universe has no center*!

26.2 *The Expanding Universe*

OLBERS'S PARADOX

It may not have occurred to you, but every time you go outside at night and notice that the sky is dark, you are making a profound cosmological observation. Here's why.

According to the cosmological principle, the universe is homogeneous and isotropic. Let's assume that it is also infinite in spatial extent and unchanging in time—precisely the view of the universe that prevailed until the early part of the twentieth century. On average, then, the universe is uniformly populated with galaxies filled with stars. In that case, when you look up at the night sky, your line of sight must *eventually* encounter a star, as illustrated in Figure 26.3. The star may lie at an enormous distance, in some remote galaxy, but the laws of probability dictate that, sooner or later, any line drawn outward from the Earth will run into a bright stellar surface. This has a dramatic implication. No matter where you look, the sky should be as bright as the surface of a star—the entire night sky should be as brilliant as the surface of the Sun! The obvious difference between this prediction and the actual appearance of the night sky is known as **Olbers's paradox**, after the nineteenth-century German astronomer, Heinrich Olbers, who popularized the idea.

What is the resolution to this paradox? Why is it dark at night? We have accepted the cosmological principle, so we believe that the universe is homogeneous and isotropic. We must conclude, then, that one (or both) of the other

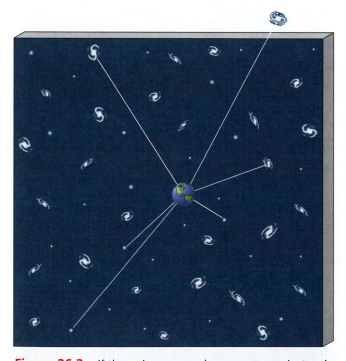

Figure 26.3 If the universe were homogeneous, isotropic, infinite in extent, and unchanging, any line of sight from the Earth should eventually run into a star, and the entire night sky should be bright. This obvious contradiction of the facts is known as Olbers's paradox.

two assumptions—that the universe is infinite in extent and unchanging in time—is false. Either the universe is finite in extent, or it evolves in time. In fact, the answer involves a little of each and is intimately tied to the behavior of the universe on the largest scales.

THE BIRTH OF THE UNIVERSE

2 We saw in Chapter 24 that all the galaxies in the universe are rushing away from us in a manner described by Hubble's law:

$$\text{recession velocity} = H_0 \times \text{distance},$$

where $H_0 \approx 75$ km/s/Mpc is the Hubble constant. We used this relation as a convenient means of determining the distances to galaxies and quasars. But it is much more than that.

Assuming that all velocities have remained constant in time, we can ask a simple question: How long has it taken for any given galaxy to reach its present distance from us? The answer follows from Hubble's law. The time taken is simply the distance traveled divided by the velocity, so

$$
\begin{aligned}
\text{time} \quad &= \quad \frac{\text{distance}}{\text{velocity}} \\[6pt]
&= \quad \frac{\text{distance}}{H_0 \times \text{distance}} \quad \text{(using Hubble's law for the velocity)} \\[6pt]
&= \quad 1/H_0 .
\end{aligned}
$$

For a Hubble constant of 75 km/s/Mpc, this turns out to be about 13 billion years. Most important, notice that the time is *independent* of the distance—galaxies twice as far away are moving twice as fast, so the time they must have taken to cross the intervening distance is the same in all cases.

Hubble's law therefore implies that, at some time in the past—13 billion years ago, according to the foregoing simple calculation—*all* the galaxies in the universe lay right on top of one another. In fact, astronomers believe that *everything* in the universe—matter and radiation alike—was confined to a single point at that instant. Then the point exploded, flying apart at high speeds. The present locations and velocities of the galaxies are a direct consequence of that primordial blast. This gargantuan explosion, involving everything in the universe, is known as the **Big Bang**. As best we can tell, it marked the beginning of the entire universe.

Thus, by measuring Hubble's constant, we can estimate the age of the universe to be $1/H_0 \approx 13$ billion years. The range of possible error in this age is considerable, both because Hubble's constant is not known precisely and because the assumption that galaxies moved at constant speed in the past is actually not a very good one—in fact, they moved faster in the past, as we will see in a moment. We will refine our estimate in a moment, but, regardless of the details, the critical fact here is that the age of the universe is *finite*.

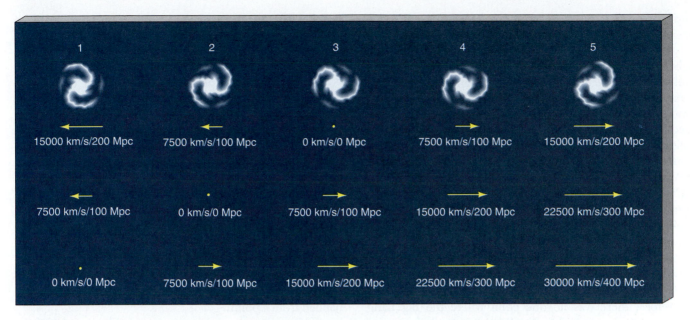

Figure 26.4   The Hubble law is the same no matter who makes the measurements. The top set of numbers are the distances and recessional velocities as seen by an observer on the middle of five galaxies, galaxy 3. The bottom two sets are from the points of view of observers on galaxies 2 and 1, respectively. In all cases, the same Hubble law holds.

This then is the explanation of why the sky is dark at night. Olbers's paradox is resolved by the evolution of the universe itself. Whether the universe is actually finite or infinite in extent is irrelevant, at least as far as the appearance of the night sky is concerned. We see only a *finite* part of it—the region lying within roughly 13 billion light years of us. What lies beyond is unknown—its light has not yet had time to reach us.

Realize that, even though it appears to place us at the center of the expansion, Hubble's law does *not* violate the cosmological principle in any way. This can be understood from Figure 26.4, which shows how observers on five separate galaxies might perceive the motion of their neighbors. For simplicity, the galaxies are taken to be equally spaced, 100 Mpc apart, and they are separating in accordance with Hubble's law with H_0 = 75 km/s/Mpc. The first pair of numbers beneath each galaxy represents its distance and recessional velocity as measured by the observer on the middle galaxy, number 3.

Now consider how the expansion looks from the point of view of the observer on galaxy 2. Galaxy 4, for example, is moving with velocity 7500 km/s to the right relative to galaxy 3, and galaxy 3 in turn is moving at 7500 km/s to the right as seen by observer 2. Therefore, galaxy 4 is moving at a velocity of 15,000 km/s to the right as seen by the observer on galaxy 2. The distances and velocities that would be measured by observer 2 are noted in the second row. Similarly, the measurements made by an observer on galaxy 1 are noted in the third row. What we find is this: *Each* observer sees an overall expansion described by Hubble's law, and the constant of proportionality—the Hubble constant—is the same in all cases. In other words, each observer measures the same expansion rate, in accordance with the cosmological principle. Far from singling out any one observer as central, Hubble's law is in fact the *only* expansion law consistent with the cosmological principle.

WHERE WAS THE BIG BANG?

The cosmological principle requires that the universe be the same everywhere, yet we have just seen that the observed recession of the galaxies described by Hubble's law, while not in violation of the principle, implies that all galaxies exploded from a point some time in the past. Wasn't that point, then, somehow different from the rest of the universe? And shouldn't there be an "edge" to the exploding debris? Doesn't that go against the assumption of homogeneity? We know *when* the Big Bang occurred. Is there any way of telling *where*?

The answer to these questions requires us to make a great leap in our view of the universe, and to shift our view of the Hubble expansion in a very important way. If we were to imagine that the Big Bang were simply an enormous explosion that spewed matter out into space, ultimately to form the galaxies we see, then the reasoning here would be quite correct. However, the Big Bang was *not* an explosion in an otherwise featureless, empty universe. The only way that we can have the Hubble expansion *and* retain the cosmological principle is to realize that the Big Bang involved the *entire* universe—not just the matter and radiation within it, but the universe *itself*. In other words, the galaxies are not flying apart into the rest of the universe. The universe itself is expanding. Like raisins in a loaf of raisin bread that move apart as the bread expands in an oven, the galaxies are just along for the ride.

Let us reconsider some of our earlier statements in light of this new perspective. We now recognize that Hubble's law actually describes the expansion of the universe. Although galaxies have some small-scale, individual random motions, on average they are *not* moving with respect to the fabric of space—any such overall motion would pick out a "special" direction in space and violate the assumption of isotropy. On the contrary, the portion of the galaxies' motion that makes up the Hubble flow is really an expansion of space itself. The expanding universe remains homogeneous at all times. There is no "empty space" beyond the galaxies into which they rush. At the time of the Big Bang, the galaxies did not reside at a point located at some well-defined place within the universe. The entire universe was a point. That point was in no way different from the rest of the universe; that point *was* the universe. Therefore, there was no one point where the Big Bang "happened"—because the Big Bang involved the entire universe, it happened *everywhere* at once.

To illustrate these ideas, imagine an ordinary balloon with coins taped to its surface, as shown in Figure 26.5. The coins represent galaxies, and the two-dimensional surface of the balloon represents the "fabric" of our three-dimensional universe. The cosmological principle applies to the balloon because every point on the balloon looks pretty much the same as every other. The frames in the figure make up a movie—a hypothetical film strip of the expanding universe. Imagine yourself as a resident of one of the three coin "galaxies" in the leftmost frame, and note your position relative to your neighbors. As the balloon inflates (as the universe expands), the other galaxies recede. Notice, incidentally, that the coins themselves do *not* expand along with the balloon, any more than people, planets, stars, or galaxies—all of which are held together by their own internal forces—expand along with the universe.

Regardless of which galaxy you choose to consider, you would note that all the other galaxies are receding. To ap-

preciate this, imagine yourself a resident on a different coin "galaxy." Now again notice the change in the positions of the galaxies while viewing the frames from left to right. The galaxies recede for any observer in the universe. Nothing is special or peculiar about the fact that all the galaxies are receding from you. Such is the cosmological principle: No observer anywhere in the universe has a privileged position. There is no center to the expansion and no position that can be identified as the location from which the universal expansion began.

Now imagine letting the balloon deflate. This would correspond to running the universe backward from the present time to the Big Bang. *All* the galaxies (coins) would arrive at the same place at the same time—at the instant the balloon reached zero size. But there is no one point on the balloon that could be said to be *the* place where that occurred. The entire balloon expanded from a point, just as the Big Bang encompassed the entire universe and expanded from a point.

This analogy has its shortcomings. The main difficulty here is that we see the balloon, which in our illustration we imagined as two dimensional, expanding into the third dimension of space. This might suggest that the three-dimensional universe is expanding "into" some fourth spatial dimension. It is not, so far as we know. At the very least, if higher spatial dimensions are involved, they are completely *irrelevant* to our theory of the universe.

BEFORE THE BIG BANG

The Big Bang was a *singularity* in space and time—an instant when the present laws of physics say the universe had zero size and infinite temperature and density. As we saw in Chapter 22, where we discussed the singularities at the center of black holes, these predictions should not be taken too literally. ∞ (p. 480) Their presence signals that, under extreme conditions, the theory—in this case, general relativity—making the predictions has broken down.

In fact, at the present moment, no theory exists to let us penetrate the singularity at the start of the universe. We

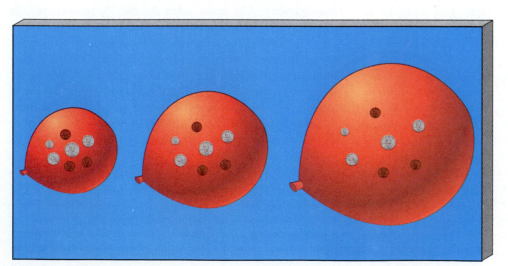

Figure 26.5   The coins taped to the surface of a spherical balloon recede from one another as the balloon inflates (left to right). Similarly, galaxies recede from one another as the universe expands. As the coins recede, the distance between any two of them increases, and the rate of increase of this distance is proportional to the distance between them. Thus, our balloon expands according to Hubble's law.

have no means of describing these earliest of times, so we have no way of answering the question, "What came *before* the Big Bang?" Indeed, given the laws of physics as we currently know them, the question itself may be meaningless. The Big Bang represented the beginning of the entire universe—mass, energy, space, *and* time came into being at that instant. With time not in existence, the notion of "before" does not exist. Consequently, some cosmologists maintain that asking what happened before the Big Bang is like asking what lies north of the North Pole! Other cosmologists disagree, arguing that one day the proper theory will explain the singularity and we can answer the question of what came before.

While we cannot extend our theory of the universe all the way back to the Big Bang itself, we can come fairly close. Theorists estimate that the "known" physics of today is adequate to describe the universe since about 10^{-43} s after the Big Bang. For the purposes of this chapter, at least, 10^{-43} is close enough to zero that we need not worry about this very early, and completely unknown, epoch.

RELATIVITY AND THE UNIVERSE

These concepts are difficult to grasp. The notion of the entire universe shrinking to a point—with *nothing*, not even space and time, outside—takes some getting used to. Nevertheless, that is the picture of the universe that lies at the heart of modern cosmology. The description of the universe itself (not just its contents) as a dynamic, evolving object is far beyond the capabilities of Newtonian mechanics, which we discussed in Chapter 2 and which we have used almost everywhere throughout this book. Instead, the more powerful techniques of Einstein's *general relativity*, with its built-in notions of warped space and dynamical spacetime, are needed.

We encountered general relativity in Chapter 22 (see especially the *More Precisely* feature on p. 470), when we discussed the strange properties of black holes. We can loosely summarize its description of the universe by saying that the presence of matter or energy causes a curvature of space (more correctly, space*time*) and that the curved trajectories of freely falling particles within warped space are what Newton thought of as orbits under the influence of gravity. The amount of curvature depends on the amount of matter present, and the orbits of particles in turn depend on the curvature. Put less formally, "Space tells matter how to move, and matter tells space how to curve." In the case of a homogeneous universe, the curvature of space must be uniform.

Must we keep this difficult notion of a warped, expanding, homogeneous, universe constantly in the forefront of our thoughts if we are to comprehend the evolution of the cosmos? Perhaps surprisingly, given that relativity is the only theory that properly describes the large-scale behavior of the universe, the answer is "No." While general relativity predicts some curious consequences for the overall *geometry* of space on large scales, as we will see later, the dy-

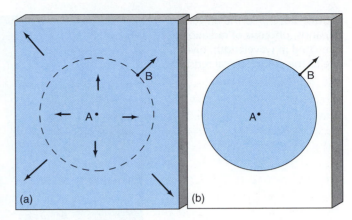

Figure 26.6 The relative motion of any two points in the expanding universe (a) can be addressed as a problem in Newtonian mechanics, even though Einstein's theory of general relativity is needed to explain why it is correct to do so. If the rest of the universe is ignored—as in part (b)—then the Newtonian calculation of B's motion relative to A would give the right answer.

namics of the universe can be understood using simple concepts that would have been thoroughly familiar to Newton.

Consider two neighboring points A and B in the expanding universe, as pictured in Figure 26.6. From the perspective of point A, every other point, including B, is rushing away from it, in accordance with Hubble's law. How does the distance between A and B change in time? Does it just keep growing, or does its rate of increase eventually slow down and stop? From a Newtonian viewpoint, we might expect that the overall gravitational pull of the universe would tend to slow the expansion, just as Earth's gravity tends to slow the upward motion of an object projected from the surface. But we have just argued that Newtonian physics is incapable of describing the motion of the universe in this way. However, a remarkable prediction of general relativity is that the Newtonian picture gives the right result! The motion of point B relative to point A is *exactly* the same as if A were at the center of a planet and B were a projectile fired vertically upward from its surface. Thus, we *can* discuss the expansion of the universe in simple Newtonian terms, although we need relativity theory to justify our doing so.

THE COSMOLOGICAL REDSHIFT

This new view of the expanding universe requires us to reinterpret the cosmological redshift. Up to now, we have explained the redshift of galaxies as a Doppler shift, a consequence of their motion relative to us. However, we have just argued that the galaxies are *not* in fact moving with respect to the universe, in which case the Doppler interpretation is incorrect. The true explanation is that, as a photon moves through space, its wavelength is influenced by the expansion of the universe. In a sense, we can think of the photon as being attached to the expanding fabric of space, so its wavelength expands along with the universe, as illus-

Figure 26.7 As the universe expands, photons of radiation are stretched in wavelength, giving rise to the cosmological redshift.

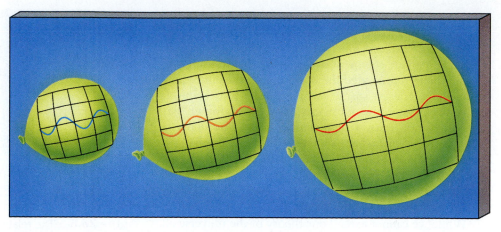

trated in Figure 26.7. While it is standard practice in astronomy to refer to the cosmological redshift in terms of recessional velocity, bear in mind that, strictly speaking, this is not the right thing to do. The cosmological redshift is a consequence of the changing size of the universe—it is *not* related to velocity at all.

Because the universe expands with time and redshift is related to that expansion, cosmologists routinely use redshift

Interlude 26–1   *The Cosmological Constant*

Even the greatest minds are fallible. The first scientist to apply general relativity to the universe was, not surprisingly, the theory's inventor, Albert Einstein. When he derived and solved the equations describing the behavior of the universe, he discovered that they predicted a universe that evolved in time. But in 1917, neither Einstein nor anyone else knew about the Hubble expansion, which would not be discovered for another 10 years. At the time, Einstein, like most scientists, believed that the universe was *static*—that is, unchanging and everlasting. The discovery that there was no static solution to his equations seemed to Einstein to be a near-fatal flaw in his new theory.

To bring his theory into line with his beliefs, Einstein tinkered with his equations, introducing a "fudge factor" now known as the *cosmological constant*. As illustrated in the accompanying figure, which shows the effect of introducing a cosmological constant into the equations describing the expansion of a critical-density universe, this factor allows many other solutions to Einstein's equations. One of these solutions describes a "coasting" universe, whose radius does in fact remain constant for an indefinite period of time, Einstein took this to be the static universe he expected.

Instead of predicting an evolving cosmos, which would have been one of relativity's greatest triumphs, Einstein yielded to a preconceived notion of the way the universe "should be," unsupported by observational evidence. Later, when the expansion of the universe was discovered and Einstein's equations—without the fudge factor—were found to describe it perfectly, he declared that the cosmological constant was the biggest mistake of his scientific career.

For many researchers—Einstein included—the main problem with the cosmological constant was (and still is) the fact that it had no clear physical interpretation. Einstein introduced it to fix what he thought was a problem with his equations, but he discarded it immediately once he realized that no problem actually existed. Scientists are very reluctant to introduce unknown quantities into their equations purely to make the results "come out right." As a result, the cosmological constant fell out of favor among astronomers for many years.

Since the early 1980s, however, the cosmological constant has made something of a comeback, and it now enjoys a measure of respectability in cosmological circles. One reason for this is that the leading theories of matter and radiation on very small scales (the "GUTs" discussed in Chapter 27) now suggest that the universe really did go through an early phase when its evolution was determined by a "cosmological constant" of sorts. As we will see in Section 27.4, this idea is now firmly entrenched in many cosmologists' models of the universe. Unfortunately, while the GUTs make the idea of a cosmological constant more acceptable, they presently offer no useful means of estimating its value.

A second reason for the cosmological constant's resurgence in popularity is the fact that it offers a possible means of resolving what may otherwise become one of the most serious contradictions in all of astronomy—the fact that the measured ages of the oldest stars in our Galaxy, 12–15 billion years (see Section 20.5), seem to be significantly greater than the age of the universe, as determined from the Hubble expansion and our best estimates of the cosmic density (see Sections 26.3 and 26.4. ∞ (p. 573) By carefully adjusting the value of the cosmological constant, theorists can construct mathematical models of the universe that did not expand so rapidly in the past, and so are older than the measured value of H_0 suggests. One such model is shown in the following figure.

With a suitably chosen cosmological constant, a Hubble constant of 75 km/s/Mpc can be reconciled with a cosmic age of 15 billion years. In fact, with a sufficiently large cosmological constant, it is even possible to avoid a Big Bang alto-

as a convenient means of expressing time (see the table on p. 555). The redshift of a photon measures the amount by which the universe has expanded since that photon was emitted. Thus, for example, when we find the wavelength of light from a quasar to have been increased by a factor of 5 (a redshift of 4—see the *More Precisely* feature on p. 554), that means that the light was emitted (and hence that we are observing the quasar) at a time when the universe was just one-fifth its present size. In general, the larger a photon's redshift, the smaller the universe was at the time the photon was emitted, and so the longer ago the event occurred.

26.3 *The Evolution of the Universe*

3 At the present moment, the universe is expanding. Will that expansion continue forever? And if not, what will happen next, and when? These are absolutely fundamental

questions concerning the fate of the entire universe. Yet, remarkably, now that we know we can confidently apply our familiar Newtonian concepts of motion under gravity to the behavior of the universe, we can address these issues by considering a simpler and much more familiar problem.

CRITICAL DENSITY

4 Consider a rocket ship launched from the surface of a planet. What is the likely outcome of that motion? There are basically two possibilities, depending on the speed of the ship. If the launch speed is high enough, it will exceed the planet's *escape velocity*, and the ship will never return to the surface. ∞ (p. 49) The speed will diminish because of the planet's gravitational pull, but it will never reach zero. The spacecraft leaves the planet on an unbound trajectory, as illustrated in Figure 26.8(a). Alternatively, if the ship's launch speed is lower than the

gether, although this possibility now seems to be quite firmly ruled out by observations of high-redshift quasars and galaxies. Notice, incidentally, that, if the cosmological constant does indeed provide the explanation of the cosmic age paradox, it also implies that the universe will expand forever, regardless of the present value of the cosmic mass density.

But before we jump to any conclusions about the cosmological constant, and overstate its potential role in solving the outstanding problems of astronomy, it is probably worth remembering the experience of its inventor and bearing in mind that—at least for now—its physical meaning remains completely unknown.

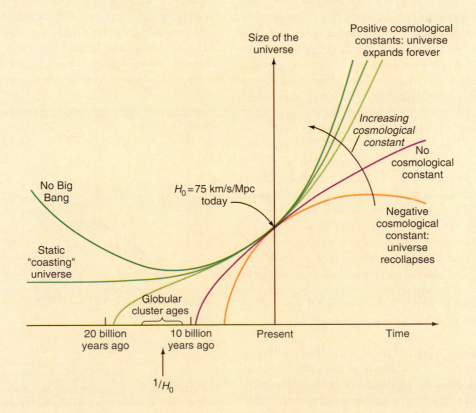

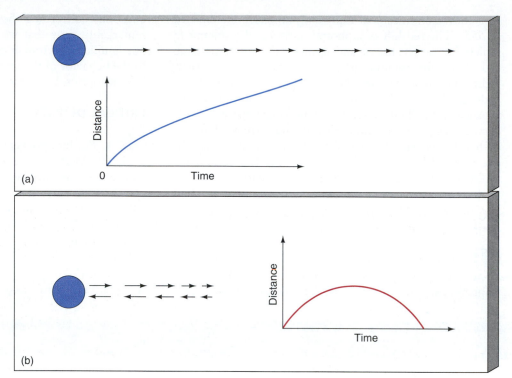

Figure 26.8 (a) A spacecraft (arrow) leaving a planet (blue ball) with a speed greater than the escape velocity leaves on an unbound trajectory (top). The graph (bottom) shows the distance between the ship and the planet as a function of time. (b) If the launch speed is less than the escape velocity, the ship eventually drops back to the planet. Its distance from the planet first rises, then falls again.

escape velocity, it will reach a maximum distance from the planet, then fall back to the surface. Its bound trajectory is shown in Figure 26.8(b).

Similar reasoning applies to the expansion of the universe. Reconsider Figure 26.6, but now imagine that *A* and *B* are galaxies at some known distance from one another, their present relative velocity given by Hubble's law. The same two basic possibilities exist for these galaxies as for our spacecraft—the distance between them can increase forever, or increase for a while and then start to decrease. What's more, the cosmological principle says that, whatever the outcome, it must be the same for *any* two galaxies—in other words, the same statement applies to the universe *as a whole*. Thus, as illustrated in Figure 26.9, the universe has only two options: It can continue to expand forever—an *unbound* universe—or the present expansion will someday stop and turn around into a contraction—a *bound* universe. These two possibilities are labeled on the figure. The third curve simply marks the dividing line between them—a *marginally bound* universe that expands forever, but at an ever-decreasing rate, analogous to our spacecraft leaving the planet with *precisely* the escape velocity. The three curves are drawn so that they all pass through the same point at the present time. All are possible descriptions of the universe, given its present size and expansion rate.

What determines which of these possibilities will actually occur? The answer is the *density* of the universe. In all cases gravity decelerates the expansion over time. The more matter there is—the denser the universe—the more "pull" there is against the expansion, just as the more mass a planet has, the less likely it is that a rocket ship can escape it. It

would take much more energy for a rocket to escape from Jupiter than from Mercury because of the large differences in the two planets' masses. In a high-density universe, there is enough mass to stop the expansion and cause a recollapse—the universe is bound. A low-density universe, conversely, is unbound and will expand forever. The dividing line between these two outcomes, the density corresponding to a marginally bound universe, is called the **critical density**. Its value depends on the Hubble constant—more matter is required to bind a more rapidly expanding uni-

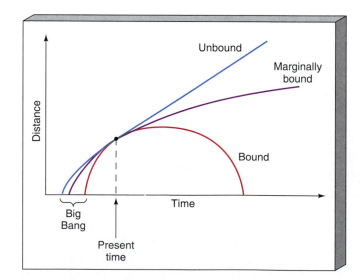

Figure 26.9 Distance between two galaxies as a function of time in each of the three possible universes discussed in the text: unbound, bound, and marginally bound. The point where the three curves touch represents the present time.

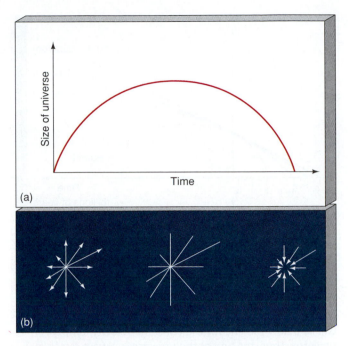

Figure 26.10 A high-density universe (a) has a beginning, an end, and a finite lifetime. The lower frames (b) illustrate its evolution, from explosion to maximum size to recollapse.

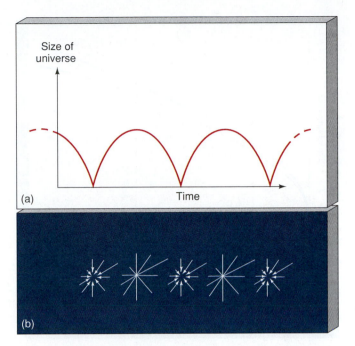

Figure 26.11 An oscillating universe has neither a beginning nor an end. Each expansion–contraction phase ends in a "bounce," which becomes the "Big Bang" of the next expansion. There is currently no information on whether or not this can actually occur.

verse. For $H_0 = 75$ km/s/Mpc, the present critical density turns out to be about 10^{-26} kg/m^3. That's an extraordinarily low density—about 6 hydrogen atoms per cubic *meter*, a volume the size of a typical household closet. In more "cosmological" terms, it corresponds to about one Milky Way Galaxy (excluding the dark matter) per cubic megaparsec.

TWO FUTURES

If the universe has a density above the critical value, then it contains enough matter to halt its own expansion, and the recession of the galaxies will eventually stop. At some time in the future, astronomers everywhere—on any planet within any galaxy—will announce that the radiation received from nearby galaxies is no longer redshifted. (The light from *distant* galaxies would still be redshifted, however, because we would see them as they were in the past, at a time when the universe was still expanding.) The bulk motion of the universe, and of the galaxies within, would be stilled—at least momentarily.

The expansion might stop, but the pull of gravity will not—gravity is relentless. The universe will begin to contract. Astronomers everywhere will announce that nearby galaxies now show blueshifts. In some ways, the contraction would be a mirror image of the expansion that preceded it. As illustrated in Figure 26.10, the universe would recollapse to a point, requiring just as much time to fall back as it took to rise.

This expansion–contraction scenario has many fascinating (and dire) implications. Following the Big Bang, the density of the universe thins to a rather small value by the

time the expansion stops. Thereafter, the density rises again, returning to its earlier huge values as matter collapses onto itself. Toward the end of the contraction phase, galaxies will collide frequently as the available space diminishes. Just as compressing the air in a bicycle pump or rubbing our hands generates heat through friction, these collisions will generate heat as well. The entire universe will grow progressively denser and hotter as the end of the contraction is neared. Near total collapse, the temperature of the entire universe will have become greater than that of a typical star. Everything everywhere will have become so bright that the stars themselves will cease to shine for want of contrasting darkness. The universe will shrink toward a superdense, superhot singularity, much like the one from which it originated—it will ultimately experience a "heat death," in which all matter and life are destined to be incinerated. Some astronomers call the final collapse of this high-density universe the "Big Crunch."

Cosmologists do not know what would happen to the universe upon reaching the point of collapse. We cannot penetrate forward in time beyond the singularity at the Big Crunch any more than we can probe backward past the Big Bang—the present laws of physics are simply inadequate to describe these extreme conditions. However, some cosmologists speculate that, with both density and temperature increasing as the contraction nears completion, the pressure might somehow be sufficient to overcome gravity, pushing the universe back out into another cycle of expansion. As depicted in Figure 26.11, the universe might not simply end—it might "bounce." A hypothetical universe having

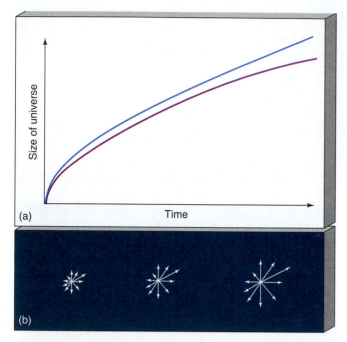

Figure 26.12   A low-density universe (a) expands forever from its explosive beginning. The lower frames (b) illustrate the continuing expansion of the universe in this case. The upper curve represents a universe with density less than the critical value. The lower curve represents a universe with density exactly equal to the critical value.

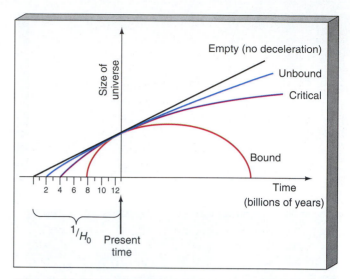

Figure 26.13   As the density of the universe increases, its deceleration increases, too. The universe contains some matter, so whatever the model, its trajectory on this graph will lie below the line for the constant-velocity, empty universe. Thus, the age of the universe is always *less* than 1 over the Hubble constant. The true age decreases for larger values of the present-day density.

many—perhaps infinitely many—cycles of expansion and contraction might be the result. Keep in mind, though, that any discussion of the universe outside of the current cycle is pure speculation.

A quite different fate awaits the universe if its density is below the critical value. In that case, its density always has been, and always will be, too small for gravity to cause it to recontract. As illustrated in Figure 26.12, this low-density universe expands forever. In this scenario, the galaxies will continue to recede forever, their radiation weakening with increasing distance. In time, an observer on Earth will see no galaxies in the sky beyond the Local Group (which is not itself expanding), even with the most powerful telescope. The rest of the observable universe will appear dark, the distant galaxies too faint to be seen. Eventually, the Milky Way and the Local Group too will peter out as their fuel supply is consumed. This universe ultimately experiences a "cold death." All radiation, matter, and life are eventually destined to freeze.

In the intermediate, critical-density case, the universe contains just enough matter eventually to halt the expansion—but only after an infinitely long time. This universe will also expand forever, its qualitative appearance resembling Figure 26.12(b).

THE AGE OF THE UNIVERSE

Earlier, when we estimated the age of the universe from the Hubble constant, we made the assumption that the expan-

sion speeds of the galaxies were *constant* in the past. However, as we have now seen, this is not the case. Gravity has slowed the universe's expansion over time, so, regardless of which evolutionary model turns out to be correct, the universe must have expanded *faster* in the past than it does today. The assumption of a constant expansion rate therefore leads to an *overestimate* of the universe's age—the universe is actually younger than the 13 billion years we calculated. That is why the ages presented in the table on p. 555 are substantially lower than 13 billion years. How much younger depends on how much deceleration has occurred.

Figure 26.13 illustrates this point. It is similar to Figure 26.9, except that we have added the line that corresponds to constant expansion at the present rate—a universe completely *empty* of matter and so experiencing no deceleration. Because all the models we have discussed lie below this line, we can see graphically that the true age of the universe is indeed less than the age obtained assuming a constant expansion speed. In the special case of critical density, the age of the universe happens to be particularly easy to calculate—it is *two-thirds* of the preceding value, or 8.7 billion years. A low-density, unbound universe is older than this (but still less than 13 billion years old); a high-density, bound universe is younger.

Impressive as it is that we can pin down the age of the universe to within a factor of two or so, these numbers indicate a potentially serious problem. Unless the universe is of *very* low density, and hence close to 13 billion years old, the age that we obtain from cosmology is less than the 12–15 billion year range implied by studies of globular clusters in our own Galaxy. ∞ (pp. 377, 433) Because the Galaxy cannot be older than the universe, and because the density of the universe appears to be at least relatively close to the critical value (as we will see in just a moment), we are forced to conclude that there may be a glaring contradiction between these two major areas of astronomy! Most astronomers would agree that H_0 probably lies between 60 and 90 km/s/Mpc. If Hubble's constant is close to 60 km/s/Mpc, and if the lower limit on the globular-cluster ages is correct, then the two age estimates might be reconciled. However, if the Hubble constant turns out to be more like 90 km/s/Mpc, then the discrepancy may become a serious embarrassment to astronomers. (See *Interlude 26-1* on p. 572 for one possible resolution of this problem.)

26.4 *Will the Universe Expand Forever?*

Is there any way for us to determine which model actually describes our universe (that is, apart from just waiting a few billion years to find out)? Will the universe end as a small, dense point much like that from which it began? Or will it expand forever? And if the expansion ceases, when will that happen? Fortunately, we live at a time when astronomers are subjecting these questions to observational tests. Unfortunately, the results of these observations are still inconclusive.

THE DENSITY OF THE UNIVERSE

One way to distinguish between the two basic cosmic models is to try to estimate the average density of matter in the universe directly, because density is the quantity that distinguishes one from the other. As noted earlier, for our "standard" Hubble constant of 75 km/s/Mpc, the critical density that separates these two possible universes is about 10^{-26} kg/m^3. Cosmologists conventionally denote the ratio of the actual density to the critical value by the symbol Ω_0 ("omega-nought"). In terms of this quantity, then, a critical universe has $\Omega_0 = 1$. A universe with Ω_0 less than 1 will expand forever; one with Ω_0 greater than 1 will recollapse.

How can we determine the average density of the universe? On the face of it, it would seem simple—just measure the average mass of the galaxies residing within some large parcel of space, calculate the volume of that space, and compute the total mass density. When astronomers do this, they usually find a little less than 10^{-28} kg/m^3 in the form of luminous matter. Largely independent of whether the chosen region contains many scattered galaxies or only a few rich galaxy clusters, the resulting density is about the same, within a factor of 2 or 3. Galaxy counts thus yield Ω_0 to be about 0.01. If this measure were correct, then the universe would expand forever.

But it is not as simple as that—there is an important additional consideration. We have noted (Chapters 23 and 24) that most of the matter in the universe is *dark*—it exists in the form of invisible material that has been detected only through its gravitational effect in galaxies and galaxy clusters. ∞ (pp. 505, 526) We currently do not know what the dark matter is, but we *do* know that it is there. Galaxies may contain as much as 10 times more dark matter than luminous material, and the figure for galaxy clusters is even higher—perhaps as much as 95 percent of the total mass in clusters is invisible. Even though we cannot see it, dark matter contributes to the average density of the universe and plays its part in opposing the expansion. Including all the dark matter that is *known* to exist in galaxies and galaxy clusters increases the value of Ω_0 to 0.2 or 0.3—still less than 1, implying an unbound universe, but getting closer to the critical value.

Unfortunately, the distribution of dark matter on larger scales is not very well known. We can infer its presence in galaxies and galaxy clusters, but we are largely ignorant of its extent in superclusters, voids, or other larger structures. However, there are indications that it accounts for an even greater fraction of the mass on large scales than it does in galaxy clusters. For example, observations of gravitational lensing by galaxy clusters suggest that dark matter may be even more extensive than is indicated by the motion of galaxies within the clusters. Furthermore, optical and infrared observations of the overall motion of galaxies in the local supercluster suggest the presence of a nearby huge accumulation of mass known as the *Great Attractor*, with a total mass of about 10^{17} solar masses and a size of 100–150 Mpc. If the current best estimates of the size and mass of this gargantuan object are correct, its average density may be quite close to the critical value.

Thus it is quite conceivable that invisible matter may account for as much as 99 percent of the total mass in the universe. In that case, the vast "voids" are not empty at all—they are huge seas of invisible matter, and the visible galaxies are merely insignificant "islands" of brightness within them. If the total amount of dark matter lurking in the darkness beyond the galaxy clusters exceeded the luminous mass by a factor of 100 or more, Ω_0 might even exceed 1, and the universe will recollapse. This is why it is so important to search for reservoirs of invisible matter beyond the galaxies. The measured value of Ω_0 has steadily increased over the past 20 years as larger and larger regions of the universe have been surveyed.

COSMIC DECELERATION

Determining the mass density of the universe is an example of a *local* measurement that provides an estimate of Ω_0. But the result we obtain depends on just how local our measurement is—the larger the scale, the larger the result we obtain. In an attempt to get around this problem, astronomers have devised alternative methods that rely instead on so-called *global* measurements, which cover a much larger portion of the observable universe. The idea is that global tests should indicate the universe's *overall* density, not just its value in our vicinity.

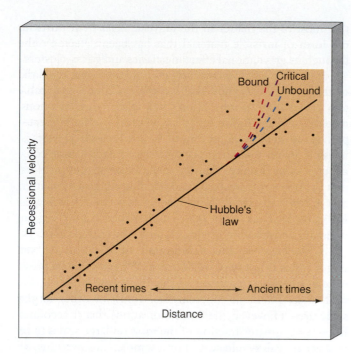

Figure 26.14 An idealized Hubble diagram, showing how we might detect evidence for a deceleration of the universe by observing a departure from the usual Hubble relationship (solid line). The dashed curves show the expected departure from the solid line for different evolving models of the universe. (The departures of the dashed curves from the solid line are exaggerated for clarity.) Uncertainties in the measured luminosities of distant galaxies (black dots) affect estimates of their distances, so this technique is too imprecise to distinguish among the various possibilities.

One such global method attempts to measure the deceleration of the universe directly by observing faraway galaxies. The universe expanded faster in the past, and will expand more slowly in the future. The greater the average density of the universe, the greater the deceleration should be, so the issue facing astronomers is to determine the *rate* at which the expansion is slowing, because this translates directly into an estimate of the density.

Astronomers cannot hope to measure the cosmic slowdown by watching the motion of any one galaxy. But, because the expansion is slowing, objects at great distances—that is, objects that emitted their radiation long ago—should appear to be receding *faster* than the Hubble law predicts. Figure 26.14 illustrates how this works. If the universal expansion were constant in time, recessional velocity and distance would be related by a straight line—the solid line in the figure. However, in a decelerating universe, the velocity is greater at large distances. Furthermore, this difference in velocities is greater for a denser universe, in which gravity has been more effective at slowing the initial expansion.

Is there any evidence for faster recessional velocities among the more distant galaxies? The answer is yes. The most distant galaxies (with distances determined by techniques *independent* of the Hubble law, of course) do indeed

have substantially greater recessional velocities than the Hubble law would predict. Unfortunately, the data are insufficient to let us pin down the fate of the cosmos. Distant galaxies are very faint, and accurate observations of their properties very difficult. As indicated in Figure 26.14, the uncertainties in their measured distances (which are much greater than the uncertainties in the velocities) mean that we simply cannot distinguish among the various possibilities. Global methods, while potentially very powerful, still await improved distance-measurement techniques before they can yield reliable results.

So, what is the ultimate fate of the universe? The answer is still not known with absolute assurance. However, while there is a large uncertainty in the value of Ω_0, most astronomers would probably agree that it lies between 0.1 and 1. As best we can tell, given the current data, the universe is destined to expand forever.

26.5 *The Geometry of Space*

We have reverted to the familiar notion of gravity, away from the more correct concept of warped spacetime, because speaking in terms of gravity makes our discussion of the evolution of the universe much easier to understand. However, general relativity makes some predictions that do *not* have a simple description in Newtonian terms. Foremost among these is the fact that space is *curved* and that the curvature is closely tied to the matter within it. Relativity asserts that mass alters the nature of spacetime. Matter effectively shapes or "warps" the geometry of space. The more matter, the greater the distortion. The degree of distortion—the curvature—must be the same everywhere (because of the cosmological principle), so there are only three distinct possibilities for the large-scale geometry of the universe, and they correspond to the possible futures we have just described. (For more information on the different types of geometry involved, see the *More Precisely* feature on p. 580.)

If the average density of the cosmos is above the critical value, space is curved so much that it bends back on itself and "closes off," making this bound universe *finite* in size. Such a universe is known as a **closed universe**. It is difficult to visualize a three-dimensional volume uniformly arching back on itself in this way, but the two-dimensional version is well known: It is just the surface of a sphere, as with the balloon we discussed earlier. Figure 26.5, then, is the two-dimensional likeness of a three-dimensional closed universe. Like the surface of a sphere, a closed universe has no boundary, yet it is finite in extent.[*] One remarkable property of a closed universe is illustrated in Figure 26.15. Just as a traveler on the surface of a sphere can keep moving forward in a straight line and eventually return to her

---

[*] *Notice that, for the sphere analogy to work, we must imagine ourselves as two-dimensional "flatlanders" who cannot visualize or experience in any way the third dimension perpendicular to the sphere's surface. Flatlanders are confined to the sphere's surface, just as we are confined to the three-dimensional volume of our universe.*

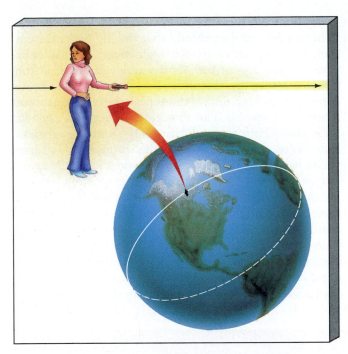

Figure 26.15 In a closed universe, a beam of light launched in one direction might return someday from the opposite direction after circling the universe, just as motion in a "straight line" upon the Earth's surface will eventually encircle the globe.

starting point, a flashlight beam shone in some direction in space might eventually traverse the entire universe and return from the opposite direction!

The surface of a sphere curves, loosely speaking, "in the same direction" no matter which way we move from a given point. A sphere is said to have *positive curvature*. However, if the average density of the universe is below the critical value, the surface curves like a saddle. It has *negative curvature*. Most people have a good idea of what a saddle looks like—it curves "up" in one direction and "down" in another, but no one has ever seen a uniformly negatively curved surface, for the simple reason that it cannot be constructed in three-dimensional Euclidean space! It is just "too big" to fit. The *More Precisely* feature on p. 580 provides a closer look at the possibility. A low-density, unbound, saddle-curved universe is infinite in extent and is usually called an **open universe**.

The third case, in which the density is precisely equal to the critical density, is the easiest to visualize. This universe, called a **critical universe**, has no curvature. It is said to be "flat," and it is infinite in extent. In this case, and *only* in this case, the geometry of space on large scales is precisely the familiar Euclidean geometry taught in high schools. Apart from its overall expansion, this is basically the universe that Newton knew.

Euclidean geometry—the geometry of flat space—is familiar to most of us because it is a good description of space in the vicinity of the Earth. It is the geometry of everyday experience. Does this mean that the universe is flat, which would in turn mean that it has exactly the criti-

cal density? The answer is "No." Just as a flat street map is a good representation of a city, even though we know the Earth is really a sphere, Euclidean geometry is a good description of space within the solar system, or even the Galaxy, because the curvature of the universe is negligible on scales smaller than about 1000 Mpc. Only on the very largest scales would the geometrical effects we have just discussed become evident.

26.6 *The Cosmic Microwave Background*

6 As we have already pointed out in Chapter 25, looking out into space is equivalent to looking back into time. ∞ (p. 554) But how far back in time can we probe? Is there any way to study the universe beyond the most distant quasar? How close can we come to perceiving directly the edge of time, the very origin of the universe?

A partial answer to these questions was discovered by accident in 1964, during an experiment designed to improve America's telephone system. As part of a project to identify and eliminate interference in satellite communications, Arno Penzias and Robert Wilson, two scientists at Bell Telephone Laboratories in New Jersey, were carrying out a study of the radio emission of the Milky Way at microwave wavelengths, using the horn-shaped antenna shown in Figure 26.16. In their data, they noticed a bothersome background "hiss" that just would not go away—a little like the background static on an AM radio station. Regardless of where and when they pointed their antenna, the hiss persisted. Never diminishing or intensifying, the weak signal was detectable at any time of the day, any day of the year, apparently filling all of space. Penzias and Wilson found that the hiss was equally intense in all directions in the sky—that is, *isotropic*—to very high accuracy.

Figure 26.16 This "sugarscoop" antenna, originally built to communicate with Earth-orbiting satellites, was used in discovering the 3K cosmic background radiation. Pictured, left to right, are Robert Wilson and Arno Penzias, who used the antenna to make the discovery.

What is the source of this radio noise? And why does it appear to come uniformly from all directions, unchanging in time? Unaware that they had detected a signal of great cosmological significance, Penzias and Wilson sought many different origins for the excess emission, including atmospheric storms, ground interference, equipment short circuits—even pigeon droppings inside the antenna! Eventually, after conversations with colleagues at Bell Labs and theorists at nearby Princeton University, the two experimentalists realized that the origin of the mysterious static was nothing less than the fiery creation of the universe itself. This discovery won Penzias and Wilson the 1978 Nobel Prize.

The radio hiss that Penzias and Wilson detected is now known as the **cosmic microwave background**. In fact, the Princeton researchers had predicted its existence and general properties a year before its discovery. Even as early as the 1940s, physicists had realized that, in addition

to being extremely dense, the early universe must also have been very *hot* and that, shortly after the Big Bang, the universe was filled with extremely high-energy thermal radiation—gamma rays of very short wavelength. (Recall from Chapter 3 that, by Wien's law, the black-body radiation curve for very high temperatures peaks at very short wavelengths. ∞ (p. 64) The Princeton researchers extended these ideas. They reasoned that, as the universe expanded and cooled, the frequency of this primordial radiation would have been redshifted from gamma ray, to X ray, to ultraviolet, eventually all the way into the radio range of the electromagnetic spectrum as the universe expanded and cooled. Figure 26.17 shows the theoretically expected change in the cosmic black-body curve. By the present time, the Princeton group argued, this redshifted "fossil remnant" of the primeval fireball should have a black-body curve corresponding to a very low temperature—no more than a few tens of kelvins—peaking in the microwave part

More Precisely...   *Curved Space*

Euclidean geometry is the geometry of flat space—the geometry taught in high schools everywhere. Set forth by one of the most famous of the ancient Greek mathematicians, Euclid, who lived around 300 B.C., it is the geometry of everyday experience. Houses are usually built with flat floors. Writing tablets and blackboards are also flat. We work easily with flat, straight objects, because the straight line is the shortest distance between any two points.

In constructing houses or any other straight-walled buildings on the surface of the Earth, the other basic axioms of Euclid's geometry would also apply: Parallel lines never meet even when extended to infinity; the angles of any triangle always sum to 180°; the circumference of a circle equals π times its diameter. If these rules were not obeyed, walls and roof would never meet to form a house!

In reality, though, the geometry of the Earth's surface is not really flat. It is curved. We live on the surface of a sphere, and on that *surface,* Euclidean geometry breaks down. Instead, the rules for the surface of a sphere are those of *Riemannian geometry,* named after the nineteenth-century German mathematician Georg Friedrich Riemann. For example, there are no parallel lines (or curves) on a sphere's surface; any lines drawn on the surface and around the full circumference will eventually intersect. The sum of a triangle's angles, when drawn on the surface of a sphere, exceeds 180°—in the 90°–90°–90° triangle shown in the figure at right, the sum is actually 270°. And the circumference of a circle is less than π times its diameter.

We see that the curved surface of a sphere, governed by the spherical geometry of Riemann, differs greatly from the flat-space geometry of Euclid. These two geometries are approximately the same only if we confine ourselves to a small patch on the surface. If the patch is small enough compared with the sphere's radius, the surface looks "flat" nearby, and

Euclidean geometry is approximately valid. This is why we can draw a usable map of our home, our city, even our state, on a flat sheet of paper, but an accurate map of the entire Earth must be drawn on a globe.

When we work with larger parts of the Earth, we must abandon Euclidean geometry. World navigators are fully aware of this. Aircraft do not fly along what might appear on most maps as a straight-line path from one point to another. Instead, they follow a "great circle"—the arc where a plane passing through the Earth's center intersects our planet's surface. On the curved surface of a sphere, such a path is always the shortest distance between two points. For example, a flight from Los Angeles to London does not proceed directly across the United States and the Atlantic Ocean, as you might expect from looking at a flat map. Instead, it goes far to the north, over Canada and Greenland, above the Arctic Circle, finally coming in over Scotland for a landing at London. This is the great circle route—the shortest path between the two cities, as you can easily see if you inspect a globe.

The "positively curved" space of Riemann is not the only possible departure from flat space. Another is the "negatively curved" space first studied by Nikolai Ivanovich Lobachevsky, a nineteenth-century Russian mathematician. In this geometry, there is an *infinite* number of lines through any given point parallel to another line, the sum of a triangle's angles is *less* than 180° (see the accompanying figure), and the circumference of a circle is *greater* than π times its diameter. Instead of the surface of a flat plane or a curved sphere, this type of space is described by the *surface* of a curved saddle. It is a hard geometry to visualize!

Most of the local realm of the *three*-dimensional universe (including the solar system, the neighboring stars, and even our Milky Way Galaxy) is correctly described by Euclidean geometry.

of the spectrum. They were in the process of constructing a microwave antenna to search for this radiation when Penzias and Wilson announced their discovery.

The Princeton researchers confirmed the existence of the microwave background and estimated its temperature at about 3K. However, proof that its frequency distribution really was described by a black-body curve was much harder to obtain. Wavelengths corresponding to the peak of the spectrum cannot be observed from the ground—the early radio measurements were made only in the low-frequency "tail" of the distribution—and the equipment necessary to make precise measurements near the peak (from

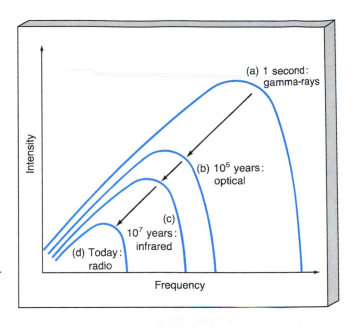

Figure 26.17 Theoretically derived black-body curves for the universe (a) 1 second after the Big Bang, (b) 100,000 years after the Big Bang, (c) 10 million years after the Big Bang, and (d) at present, approximately 13 billion years after the Big Bang.

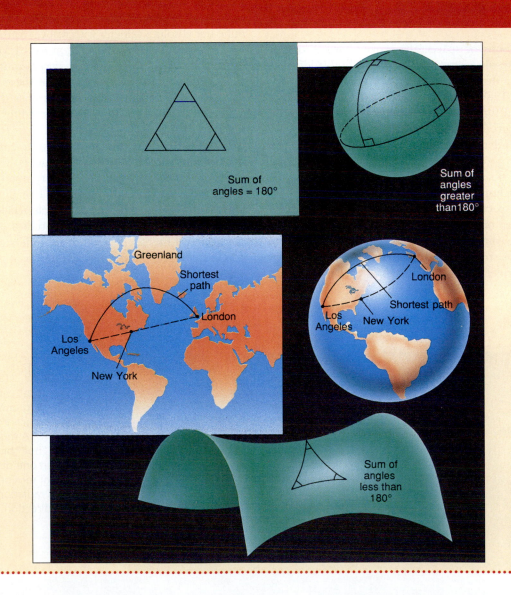

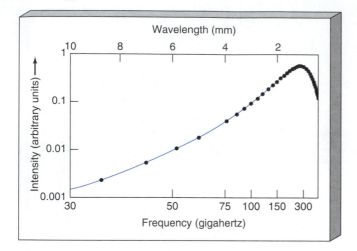

Figure 26.18   The intensity of the cosmic background radiation, as measured by the *COBE* satellite, agrees very well with that expected from theory. The curve is the best fit to the data, corresponding to a temperature of 2.735K. The experimental errors in this remarkably accurate observation are smaller than the dots representing the data points.

high-flying balloons or spacecraft) proved hard to design and build. As a result, years of painstaking observations left the shape of the distribution only poorly determined. All this changed in 1989 when the *Cosmic Background Explorer* satellite—*COBE*, for short—measured the intensity of the microwave background at wavelengths straddling the peak, from half a millimeter up to about 10 cm. The results are shown in Figure 26.18. The solid line is the black-body curve that best fits the *COBE* data. The near-perfect fit corresponds to a universal temperature of about 2.7 K.

A striking aspect of the cosmic microwave background is its high degree of *isotropy*. Its intensity is virtually constant from one direction in the sky to another. This isotropy provides strong support for the assumption of the cosmological principle. It also provides us with a novel

means of measuring the Earth's "true" velocity through space, without reference to any neighboring galaxies or galaxy clusters. If we were precisely at rest with respect to the universal expansion (like the coin taped to the surface of the expanding balloon in Figure 26.5), then we would see the microwave background as almost perfectly isotropic (actually, to about 1 part in 10^5), as illustrated in Figure 26.19(a). However, if we are moving with respect to that frame of reference, as in Figure 26.19(b), then the radiation from in front of us should be slightly blueshifted by our motion, while that from behind should be redshifted.

Thus, to a moving observer, the microwave background should appear a little hotter than average in front and slightly cooler behind. Figure 26.20 shows a *COBE* map of the microwave background temperature over the entire sky. The blue regions are hotter than average, by about 0.0034 K; the red regions cooler by the same amount. The data indicate that the Earth's velocity is about 400 km/s in the approximate direction of the constellation Leo. Even though the principle of relativity says that there is no preferred frame of reference, as the laws of physics look the same to all observers, there *is* nevertheless a way to determine our absolute velocity with respect to the universe!

When we observe the microwave background, we are looking almost all the way to the very beginning of the universe. The photons that we receive as these radio waves today have not interacted with matter since the universe was a mere 100,000 years old, when, according to our models, it was less than 1/1000 of its present size. To probe further, back to the Big Bang itself, requires us to enter the world of nuclear and particle physics. Strange as it may seem, the sciences of the very large and the very small come together in the study of the early universe. The Big Bang was the biggest and the most powerful particle accelerator of all! In the next chapter, we will see how studies of conditions in the primeval fireball may aid us in understanding the present-day structure and future evolution of the universe we live in.

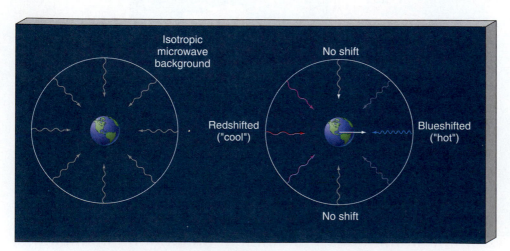

Figure 26.19   (a) To an observer at rest with respect to the expanding universe, the microwave background appears isotropic. (b) A moving observer measures "hot" blueshifted radiation in one direction (the direction of motion) and "cool" redshifted radiation in the opposite direction.

Interlude 26–2   *The Steady-State Universe*

All the cosmological models we have studied have evolution as their central theme—the universe changes with time. These models are based on Einstein's theory of relativity, and they are favored in one form or another by the overwhelming majority of cosmologists. However, other models of the universe have been proposed from time to time. Most of them do not follow directly from relativity; some do not even call for change with time. One of the most prominent and long-lived of these alternative theories was the *steady-state universe*.

The discovery in the 1920s that the universe is expanding came as a blow to the traditionalists, who preferred the idea of a static, unchanging cosmos. Even Einstein introduced the fictitious "cosmological constant" (see *Interlude 26-1*) into his equations purely to avoid the possibility of an evolving universe. The steady-state model was developed in the 1940s and 1950s by Hermann Bondi, Thomas Gold, and Fred Hoyle as an alternative to evolutionary models. For a time, it had many adherents. Motivated as much by philosophy as by science, it was an attempt to salvage as much of the old view as possible, given the new reality embodied in Hubble's law. It asserted that the universe appears the same for all observers (the cosmological principle), but it went one step further, maintaining that the universe has appeared the same *throughout all time*. This assumption is often called the *perfect cosmological principle*: To any observer at any time, the physical state of the universe is the same—the average density of the universe remains eternally constant. In this way, the steady-state model sought to avoid the thorny questions of the beginning of the universe and what happened before then. In it, the universe had no beginning and no end.

Steady-state cosmologists conceded that the universe was expanding, but because the idea of an initial explosion was unacceptable to them, they were forced to assume that an unknown repulsive force pushes the galaxies apart. Even so, the perfect cosmological principle demanded that the bulk properties of the universe—the average density of matter and the average distance between galaxy clusters—remain constant. Accordingly, to offset the dilution of the density due to the galaxies' recession, the steady-state model required the appearance of new matter in the universe. The steady-staters proposed that this *new* matter was created from *nothing*. The emergence of new matter would keep the average distance between galaxies in the future the same as in the past, preserving forever the average density of matter in the universe.

The major problem with the steady-state model was its failure to specify how the additional matter could be created. Proponents of the theory argued that *very* little new matter would be needed to offset the natural thinning of the universe as the galaxies sped apart. The creation of a single hydrogen atom in a volume equivalent to the Houston Astrodome every few years would do. The sudden appearance of such a minute quantity of matter, inside or outside a galaxy, would be quite impossible to detect, so it is not actually ruled out by either observation or experiment. Still, the sudden appearance of new matter from nothing, however little the mass involved, violates one of the most cherished concepts of modern science—the conservation of mass and energy. Matter may be created from energy, but it is very hard to understand how matter can be created spontaneously from nothing at all.

Few scientists are prepared to throw out most of the laws of physics just to avoid a philosophical difficulty. For this reason alone, the steady-state model is fatally flawed in the eyes of most astronomers. The steady-state model can be ruled out for at least two other good reasons. First, the spread of galaxies, especially the quasars, is *not* uniform throughout space. The distant quasars far outnumber those nearby. If we had lived 10 billion years ago, when quasars were the dominant cosmic objects, our view of the universe would have been much different, which is a clear violation of the perfect cosmological principle. Second, the steady-state universe offers no satisfactory explanation of the cosmic microwave background. In this theory, one simply has to imagine that the background radiation "just is," with no particular reason for its existence.

This combination of objections far outweighs any benefits the steady-state theory might bring to cosmology. After a two-decade run as a cosmological contender, the steady-state universe was abandoned because it is inconsistent with reality. This episode illustrates how personal preferences can drive good scientists to abandon the scientific method, adopting theories that have no basis in fact. In the end, however, the scientific method prevails. In science, at least, theories unsupported by data simply do not last.

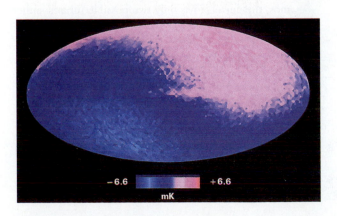

Figure 26.20   A *COBE* map of the microwave sky reveals that the microwave background appears a little hotter in the direction of the constellation Leo and a little cooler in the opposite direction. The maximum temperature deviation from the average is about 0.0034 K, corresponding to a velocity of 400 km/s.

Chapter Review

SUMMARY

On scales larger than a few hundred megaparsecs, the universe appears roughly homogeneous (the same everywhere) and isotropic (the same in all directions). In **cosmology** (p. 567)—the study of the universe as a whole—researchers usually simply assume that the universe is homogeneous and isotropic. This assumption is known as the **cosmological principle** (p. 567). It implies that the universe cannot have a center or an edge.

Olbers's Paradox (p. 568) is the fact that, if the universe were infinite and homogeneous, then any line of sight would eventually end on a star, and the night sky should be bright. The Hubble expansion implies that the entire universe is expanding. Tracing the expansion back in time indicates that, about 13 billion years ago, the universe consisted of a single point of near-infinite density and temperature which then began to expand rapidly at the time of the **Big Bang** (p. 568). That expansion continues today. However, the galaxies are not simply flying apart into the rest of an otherwise empty universe. Space itself is expanding.

General relativity provides the correct explanation of the redshifts of distant galaxies. The cosmological redshift occurs as a photon's wavelength is "stretched" by cosmic expansion. The extent of the observed redshift is a direct measure of the expansion of the universe since the photon was emitted.

Although Einstein's theory of general relativity is needed for a full description of the cosmos, the dynamics of the universe can be understood using simple Newtonian concepts. The eventual fate of the universe depends on its density. If the density is greater than the **critical density** (p. 574), then there is enough cosmic matter to stop the expansion and cause a recollapse. If the density is less than the critical value, the universe will expand forever.

Based on a Hubble constant of 75 km/s/Mpc, the age of a critical-density universe is only about 9 billion years. This age estimate is in conflict with the ages of globular clusters derived from studies of stellar evolution. Only if the density of the universe is much less than critical, or if the Hubble constant is somewhat less than 75 km/s/Mpc, can the two age estimates be reconciled.

Researchers have attempted to determine the density of the universe by measuring the luminous and dark matter detected in galaxies and on larger scales. Luminous matter contributes only about 1 percent of the matter needed to cause the universe to recollapse. When dark matter in galaxies and clusters is taken into account, the figure rises to 20 or 30 percent. Most astronomers believe that the present density is between 10 and 100 percent of the critical value, so the universe will probably expand forever.

General relativity provides a description of the geometry of the universe on the largest scales. The curvature of spacetime in a high-density universe is sufficiently large that the universe "bends back" on itself and is finite in extent, somewhat like the surface of a sphere. Such a universe is said to be a **closed universe** (p. 578). A low-density **open universe** (p. 579) is infinite in extent and has a "saddle-shaped" geometry. The intermediate-density **critical universe** (p. 579) will expand forever, but it is spatially flat.

In the early 1960s, scientists at Bell Labs discovered the **cosmic microwave background** (p. 580), a "fossil remnant" of the primeval fireball from which our universe arose. The radiation from the Big Bang has been redshifted by the expansion of the universe all the way from the gamma ray to the radio portion of the electromagnetic spectrum. The present temperature of the microwave background is about 3 K and varies little from one part of the sky to another. Its existence lends strong support to the Big Bang theory.

SELF-TEST: True or False?

_____ **1.** Cosmic homogeneity can be tested observationally, but cosmic isotropy has no observational test.

_____ **2.** Olber's paradox asks "Why is the night sky dark?"

_____ **3.** At some time in the past, everything in the universe was confined to a single point, which then exploded.

_____ **4.** Hubble's Law implies that the universe will expand forever.

_____ **5.** The Big Bang is an expansion only of matter, not of space.

_____ **6.** All points in space appear to be at the center of the expanding universe.

_____ **7.** Modern physics is capable of describing the universe as far back in time as one second after the Big Bang, but no earlier.

_____ **8.** The cosmological redshift is a direct measure of the expansion of the universe.

_____ **9.** As it travels through space, a photon's wavelength expands at the same rate as the universe expands.

_____ **10.** According to the standard Big-Bang model of the cosmos, the universe has only two possible futures: either it will continue to expand forever, or it will someday stop expanding and start to contract.

_____ **11.** A bound universe will ultimately end in a "cold death."

_____ **12.** The cosmic microwave background is the highly redshifted radiation of the early Big Bang.

SELF-TEST: Fill in the Blank

1. Pencil-beam surveys suggest that the largest structures in space are no larger than about _____ Mpc in size.

2. Homogeneous means "the same _____."

3. Isotropic means "the same in all _____."

4. Together, the assumptions of cosmic homogeneity and isotropy are known as the _____.

5. If the universe had an edge, this would violate the assumption of _____.

6. If the universe had a center, this would violate the assumption of _____.

7. A Hubble constant of 75 km/s/Mpc gives a maximum age for the universe of _____ billion years.

8. _____ is slowing down the expansion of the universe.

9. The _____ of the universe determines whether the universe will expand forever.

10. An age for the universe determined from a _____ value for the Hubble constant can not be reconciled with the ages of the oldest globular clusters.

11. Luminous matter makes up about _____ percent of the critical density.

12. By observing distant galaxies, it has been determined that the universe was expanding _____ long ago than it is today.

13. The surface of a sphere is a two-dimensional example of a _____ universe.

14. The temperature of the universe, as measured by the cosmic microwave background, is _____ K.

REVIEW AND DISCUSSION

1. What evidence do we have that there is no structure in the universe on very large scales? How large is "very large?"

2. What is the cosmological principle?

3. What is Olbers's paradox? How is it resolved?

4. Explain how an accurate measure of Hubble's constant can lead to an estimate of the age of the universe.

5. We appear to be at the center of the Hubble flow. Why doesn't this violate the cosmological principle?

6. Why isn't it correct to say that the expansion of the universe involves galaxies flying outward into empty space?

7. Where did the Big Bang occur?

8. How does the cosmological redshift relate to the expansion of the universe?

9. What measureable property of the universe determines whether or not it will expand forever?

10. Is there enough luminous matter to halt the current cosmic expansion?

11. Is there enough dark matter to halt the current cosmic expansion?

12. What is the cosmic microwave background, and why is it so significant?

13. Why does the temperature of the microwave background fall as the universe expands?

14. Which universe do you prefer philosophically: open, closed, critical, or steady-state? Why? Do you think scientists also have personal preferences?

15. Estimates of the age of the universe based on the Hubble constant are substantially lower than the estimate ages of globular clusters in our own Galaxy. How do you think astronomers should proceed in resolving this discrepancy?

PROBLEMS

1. According to the Big Bang theory described in this chapter, what is the maximum possible age of the universe if $H_0 = 50$ km/s/Mpc? 75 km/s/Mpc? 100 km/s/Mpc?

2. For a Hubble constant of 75 km/s/Mpc, the critical density is 10^{-26} kg/m^3. (a) How much mass does that correspond to within a volume of 1 cubic astronomical unit? (b) How large a cube would be required to enclose 1 Earth mass of material?

3. Assuming critical density, and using the distances presented in the table on p. 555, estimate the total amount of matter in the observable universe. Express your answer (a) in kilograms, (b) in solar masses, and (c) in "galaxies," where 1 galaxy = 10^{11} solar masses.

4. The critical density is in fact proportional to the square of the Hubble constant. If the critical density were equal to the known density of "normal" matter (not dark matter), about 10^{-28} kg/m^3, what would be the corresponding value of the Hubble constant? Is this value realistic, i.e. within the known range of values for the Hubble constant?

PROJECTS

1. Make a model of a two dimensional universe and examine the Hubble Law on it. Find a balloon that will blow up into a nice large sphere. Blow it up about half way and mark dots all over its surface; the dots will represent galaxies. Arbitrarily choose one dot as you home galaxy. Using a cloth measuring tape, measure the distances to various other galaxies, numbering the dots so you will not confuse them later. Now blow the balloon up to full size and measure the distances again, and find the new distances to each dot. Calculate the change in the distances for each galaxy; this is a measure of their velocity (= change in position/change in time; the time is the same for each and is arbitrary). Plot their velocities against their new distances as in Figure 26.14. Do you get a straight line correlation, i.e. a "Hubble" Law?

Try this again using a different dot for your home galaxy. Do you still get a Hubble Law? Does it matter which dot you choose as home? Demonstrate this to your class and show some of your results in the form of a table.

2. Go to a clear, dark country location on a night when the Moon is down. Spend an hour or more reclining comfortably, gazing up at the glittering stars. Think about the universe as revealed by modern astronomy: an expanding universe of billions of galaxies that are invisible to the eye but visible to the large telescopes and sophisticated detectors of astronomers. Can you imagine the universe of galaxies located beyond the visible stars? Can you imagine a time when those galaxies—and all that you see around you—were compressed into a single point?

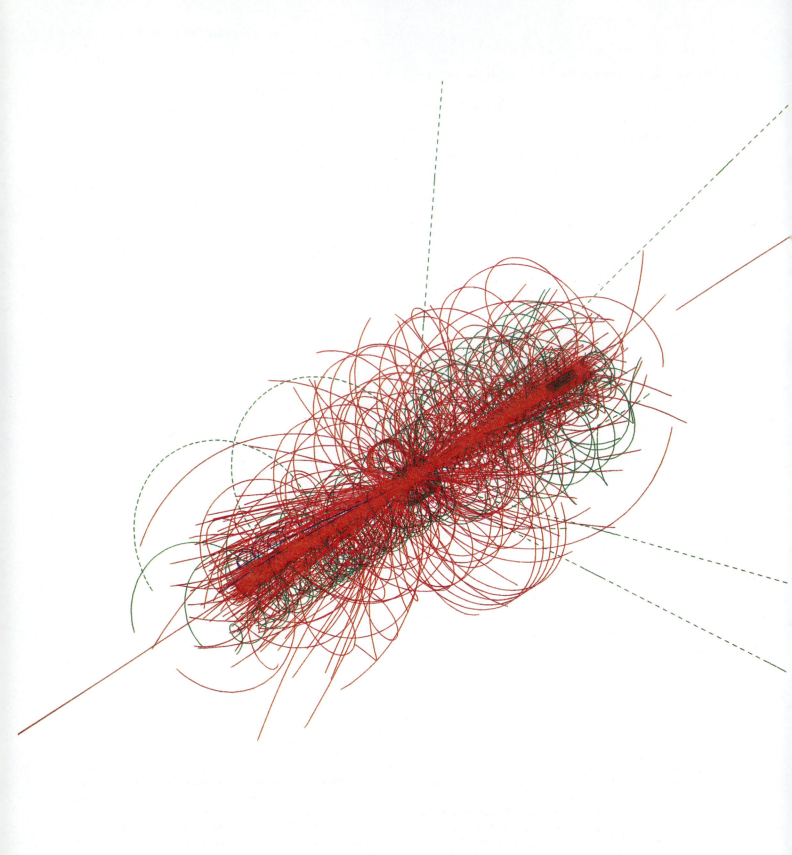

27

THE EARLY UNIVERSE

Toward the Beginning of Time

LEARNING GOALS

Studying this chapter will enable you to:

1 Describe the characteristics of the universe immediately after its birth.

2 Explain how matter emerged from the primeval fireball.

3 Enumerate the epochs in the evolutionary history of the universe, and specify the major characteristics of each.

4 Explain how and when the simplest nuclei and atoms formed.

5 Summarize the horizon and flatness problems, and discuss the theory of cosmic inflation as a possible solution to these problems.

6 Explain the formation of large-scale structure in the cosmos, and discuss the observational evidence for our theories of structure formation.

(Opposite page) Astronomers are unable to observe the universe when it was very young, because truly far-away and long-ago events were engulfed in a sea of intense radiation. In the very early universe, no galaxies, stars, or planets had yet emerged. Only subatomic particles existed—not only the protons and electrons we know today, but also, we think, various strange and exotic elementary particles predicted by current theory. Surprisingly, this melange of particles that characterized the early universe can now be studied here on Earth, in huge accelerators such as those of the Conseil Européen pour la Recherche Nucleaire (CERN) near Geneva, Switzerland. The odd image at left shows the violent collision of elementary particles, typical of events thought to have occurred about a trillionth of a second after the beginning of the universe. Two fast-moving protons (straight red lines) are shown colliding head on. Their encounter produces a multitude of charged particles, seen here curving in the magnetic field of the Large Hadron (proton) Collider at CERN. One rather massive particle thought to have played a unifying role in the early universe—a particle bizarrely named the Higgs boson—is depicted here amidst the debris, at the intersection of the straight green lines (which are yet other particles decaying in the aftermath of the violence). This image, however, is only a computer simulation; the Large Hadron Collider is still under construction, and the Higgs boson has not yet been seen by anyone. Whoever does eventually discover the Higgs (assuming it really exists) will doubtless win a free, all-expense-paid trip to Sweden—to pick up a Nobel Prize.

*W*hat was it like at the start of the universe? What conditions existed during the first few moments of the universe? And how did those conditions change to give rise to the universe we see today? These are surely basic questions, but they are hard questions. Until the twentieth century, they lay squarely in the domain of religion or philosophy. Now, after more than 10,000 years of civilization, science may be ready to provide some insight regarding the ultimate origin of all things. In studying the earliest moments of our universe, we enter a truly alien domain. As we move backward in time toward the Big Bang, our customary landmarks slip away one by one. Atoms vanish, then nuclei, then even the elementary particles themselves. In the beginning, the universe consisted of pure energy, at unimaginably high temperatures. As it expanded and cooled, the ancient energy changed into the particles that make up everything we see around us today. Modern physics has now arrived at the point where it can reach back almost to the instant of the Big Bang itself, allowing scientists to unravel some of the mysteries of our own beginning.

27.1 *Back to the Big Bang*

MATTER AND RADIATION

1 On the very largest scales, we can regard the universe as a roughly homogeneous mixture of matter and radiation. The overall density of matter is not known with certainty, but it is thought to be fairly close to the critical density of 10^{-26} kg/m³, above which the universe will eventually recollapse, and below which it will expand forever. The universe is apparently open, but barely so.

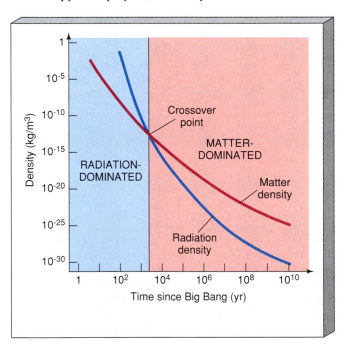

Figure 27.1 As the universe expanded, the number of both matter particles and photons per unit volume decreased. However, the photons were also reduced in energy by the cosmological redshift, reducing their equivalent mass, and hence their density, still further. As a result, the density of radiation fell faster than the density of matter as the universe grew. Tracing the curves back from the densities we observe today, we see that radiation must have dominated matter at early times—that is, at times before the crossover point.

The matter in the universe consists of the familiar building blocks of atoms—protons, neutrons, and electrons—as well as the mysterious dark matter, whose composition is still hotly debated by astronomers. Most of the radiation in the universe is in the form of the cosmic microwave background, the low-temperature (3 K) radiation field that fills all space. ∞ (p. 579) Surprisingly, although the microwave background radiation is very weak, it still contains more energy than has been emitted by all the stars and galaxies that have ever existed! The reason for this is that stars and galaxies, though very intense sources of radiation, occupy only a tiny fraction of space. When their energy is averaged out over the volume of the entire universe, it falls short of the energy of the microwave background by at least a factor of 10. For our current purposes, then, we can ignore much of the first 26 chapters of this book and regard the cosmic microwave background as the only significant form of radiation in the universe!

Is matter the dominant component of the universe, or does radiation also play an important role on large scales? In order to compare matter and radiation, we must first convert them to a "common currency"—either mass or energy. Let's choose to compare their masses. We can express the energy in the microwave background as an equivalent density by first calculating the number of photons in any cubic centimeter of space, then converting the total energy of these photons into a mass using the relation $E = mc^2$. ∞ (p. 349) When we do this, we arrive at an equivalent density for the microwave background of about 5×10^{-31} kg/m³. Thus, *at the present moment* the density of matter (about 10^{-26} kg/m³) in the universe far exceeds the density of radiation. In cosmological terminology, we say that we live in a **matter-dominated** universe.

Was the universe always matter dominated? To answer this question, we must ask how the densities of both matter and radiation change as the universe expands. Both decrease, as the expansion dilutes the numbers of atoms and photons alike. But the radiation is also diminished in energy by the cosmological redshift, so its density falls *faster* than that of matter as the universe grows (see Figure 27.1). Conversely, as we look back in time, closer and closer to the Big Bang, the density of the radiation increases faster than

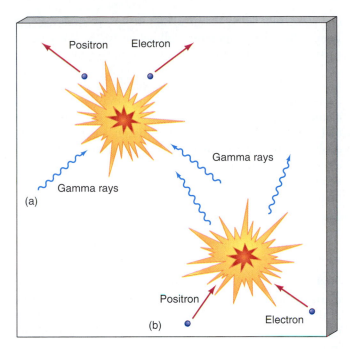

Figure 27.2   (a) Two photons can produce a particle–antiparticle pair—in this case an electron and a positron—if their total energy exceeds the mass energy of the particles produced. (b) The reverse process is particle–antiparticle annihilation, in which an electron and positron destroy each other, vanishing in a flash of gamma rays. (c) Tracks in a particle detector allow us to visualize pair creation. Here a gamma ray, whose path is invisible because it is electrically neutral, arrives from the left; it dislodges an atomic electron and sends it flying (the longest path). At the same time, it provides the energy to produce an electron-positron pair (the sprial paths, which curve in opposite directions in the detector's magnetic field because of their opposite electric charges).

that of matter. Accordingly, even though today the radiation density is much less than the matter density, there must have been a time in the past when they were equal. Before that time, radiation was the main constituent of the cosmos. The universe is said to have been **radiation dominated** then. Given our best estimates of the present densities, the crossover point—the time at which the densities of matter and radiation were equal—occurred a few thousand years after the Big Bang, when the universe was about 20,000 times smaller than it is today. The temperature of the background radiation at that time was about 60,000 K, so it peaked well into the ultraviolet portion of the spectrum.

Our discussion of the universe thus naturally breaks into two parts. For the first few thousand years after the Big Bang, the universe was small and dense and dominated by the effects of radiation. We will call this the *Radiation Era*. Some matter existed during this time, but it was a mere contaminant in the blinding radiation of the primeval Big Bang fireball. Afterwards, matter came to dominate, in the *Matter Era*. Atoms, molecules, and galaxies formed as the universe cooled and thinned toward the state we see today.

PARTICLE PRODUCTION IN THE EARLY UNIVERSE

The early phases of the Radiation Era were characterized by temperatures and densities far greater than anything we have encountered thus far, even in the hearts of supernovae. In order to fathom the early universe, we must delve a little more deeply into the behavior of matter and radiation at very high temperatures.

The key to understanding events at very early times lies in a process called **pair production**, in which two photons interact to create a *particle–antiparticle* pair, as shown in Figure 27.2(a) for the particular case of electrons and positrons. In this way, matter can be created from radiation. The reverse process can also occur—a particle and its antiparticle can *annihilate* one another to produce energy in the form of electromagnetic radiation, as depicted in Figure 27.2(b). In other words, energy in the form of radiation can be freely converted into matter in the form of particles, and matter in the form of particles can be freely converted into radiation. There are only two constraints: (1) The combined energy of the photons must be greater than the mass energy of the particle–antiparticle pair, and (2) total energy must be conserved.

The higher the temperature of a radiation field, the greater the energy of the typical photons comprising it, and the greater the masses of the particles than can be created by pair production. As an example, consider the production of the familiar electron, along with its antiparticle, the positron, as the universe expanded and cooled. At high temperatures—above about 10^{10} K—most photons had enough energy to form an electron or a positron, and pair production was commonplace. As a result, space seethed with electrons and positrons, constantly created from the radiation field and annihilating each other to form photons again. Particles and radiation are said to have been in *thermal equilibrium*—new particle–antiparticle pairs were created by pair production at the same rate as they annihilated one another.

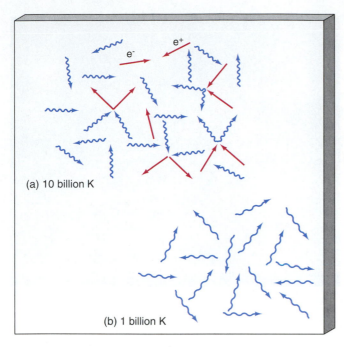

Figure 27.3 (a) At 10 billion K, most photons have enough energy to create particle–antiparticle (electron–positron) pairs, so these particles exist in great numbers, in equilibrium with the radiation. (b) Below about 1 billion K, photons have too little energy for pair production to occur.

As the temperature decreased, so did the average photon energy. By the time the temperature had fallen below a billion or so kelvins, photons no longer had enough energy for pair production to occur and only radiation remained. Figure 27.3 illustrates how this change took place. For any given particle, the critical temperature above which pair production is possible, and below which it is not, is called the particle's *threshold temperature*. The threshold temperature increases as the mass of the particle increases. For electrons, it is about 6×10^9 K. For protons, which are nearly 2000 times more massive, it is just over 10^{13} K.

Pair production in the very early universe was directly responsible for *all* of the matter that exists in the universe today. Everything we see around us was created out of radiation as the cosmos expanded and cooled. Because we are here to ponder the subject and we ourselves are made of matter, we know that some matter must have survived these early moments. For some reason, there was a slight excess of matter over antimatter at early times. A small residue of particles that outnumbered their antiparticles was left behind as the temperature dropped below the threshold for creating them. Without any antiparticles left to annihilate them, the number of particles has remained constant ever since. These survivors are said to have *frozen out* of the radiation field as the universe cooled.

The first hundred or so seconds of the universe's existence saw the creation of all of the basic "building blocks" of matter we know today—protons and neutrons froze out when the temperature dropped below 10^{13} K, when the universe was only 0.0001 s old. The lighter electrons froze

out somewhat later, about a minute or so after the Big Bang, when the temperature fell below 10^9 K. (Very heavy particles—such as those that may comprise much of the dark matter in the universe—formed even earlier than the protons.) This "matter-creation" phase of the universe's evolution ended when the electrons—the lightest known elementary particles—appeared out of the cooling primordial fireball. From that point on, matter has continued to evolve, clumping together into more and more complex structures, eventually forming the atoms, planets, stars, galaxies, and large-scale structure we see today, but no new matter has been created since that early time.

27.2 Epochs in the Evolution of the Universe

3 Let's begin our study of the early universe by summarizing in broad terms the history of the cosmos, starting at the Big Bang. Table 27-1 and Figure 27.4 present the time, density, and temperature spans of eight major stages in the development of the universe, along with a brief description of the main physical events that dominated the universe during each. The numbers in this table and figure result from pushing the known laws of physics as far back in time as we can. In the next few sections, we will expand on some of the epochs in greater detail, but let's not lose sight of the big picture and the place of each epoch in it.

THE RADIATION ERA

The universe began with an explosion from an incredibly hot and dense state. Precisely what state, we cannot say. And why it exploded, we really don't know. To understand why the universe began expanding, or even more fundamentally, why the universe exists at all, is currently beyond science—there are simply no relevant data. Although ignorant of the moment of creation itself (that is, precisely zero time), theorists nevertheless believe that the physical conditions in the universe can be understood in terms of present-day physics back to an extraordinarily short time—a mere 10^{-43} s, in fact—after the Big Bang.

Why can't theorists push our knowledge back to the Big Bang itself? The answer is that we presently have no theory capable of describing the universe at these earliest of times. Theorists believe that, under the extreme conditions of density and temperature within 10^{-43} s of the Big Bang, gravity and the other fundamental forces (electromagnetism, the strong force, and the weak force, as described in the *More Precisely* feature on p. 592) were indistinguishable from one another—a far cry from the radically different characters we see today. The four forces are said to have been *unified* at this early time—there was, in effect, only one force of nature. As mentioned in Chapter 22, the theory that combines quantum mechanics (the proper description of microscopic phenomena) with general relativity (which describes the universe on the largest scales) is known as *quantum gravity*. ∞ (p. 480) The period of time from the beginning to 10^{-43} s is often referred to as the

TABLE 27-1   *Major Epochs in the History of the Universe*

ERA	EPOCH	TIME (after Big Bang)	DENSITY (kg/m³)	TEMPERATURE (K)	MAIN EVENTS
Radiation Era		0 s	∞	∞	
	Planck				Unknown physics; quantum gravity
		10^{-43} s	10^{95}	10^{32}	
	GUT				Strong, weak, and electromagnetic forces unified
		10^{-35} s	10^{75}	10^{27}	
	Hadron				Heavy and light particles all in thermal equilibrium
		10^{-4} s	10^{16}	10^{12}	
	Lepton				Only light particles still in thermal equilibrium. Neutrinos decouple.
		10^{2} s	10^{4}	10^{9}	
	Nuclear				Deuterium and helium formed by fusion of protons and neutrons during first 1000 s.
		10^{3} yr ($\approx 3 \times 10^{10}$ s)	10^{-13}	6×10^{4}	
Matter Era		10^{3} yr ($\approx 3 \times 10^{10}$ s)	10^{-13}	6×10^{4}	
	Atomic				Matter begins to dominate. Atoms form. Electromagnetic radiation decouples.
		10^{6} yr ($\approx 3 \times 10^{13}$ s)	10^{-19}	10^{3}	
	Galactic				Galaxies and larger-scale structure form.
		10^{9} yr ($\approx 3 \times 10^{16}$ s)	3×10^{-25}	10	
	Stellar				All galaxies have formed. Stars continue to form.
		$>10^{10}$ yr ($\approx 3 \times 10^{17}$ s)	10^{-26}	3	

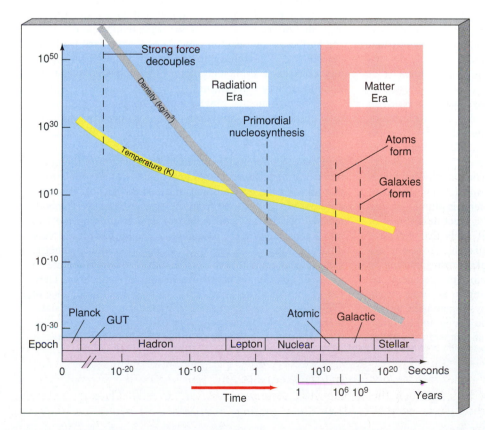

Figure 27.4 The average temperature and average density throughout the history of the universe. The epochs listed in Table 27-1 are indicated. (Numerical values for these plots are taken from Table 27-1.) Some key events in the history of the universe are also marked.

More Precisely...   *More on Fundamental Forces*

In the *More Precisely* feature on p. 352, we noted that the behavior of all matter in the universe is ruled by just three *fundamental forces*—gravity, the electroweak force (the unification of the electromagnetic and weak forces), and the strong (or nuclear) force. In terrestrial laboratories, these forces display very different properties from one another. Gravity and electromagnetism are long-range, inverse-square forces, while the strong and weak forces have very short ranges—10^{-15} and 10^{-17} m, respectively. Furthermore, the forces do not all affect the same particles. Gravity affects everything. The electromagnetic force affects only charged particles. The strong force operates between nuclear particles, such as protons and neutrons, but it does not affect electrons and neutrinos. The weak force shows up in certain nuclear reactions and radioactive decays. The strong force is 137 times stronger that the electromagnetic force, 100,000 times stronger than the weak force, and 10^{39} times stronger than gravity.

In fact, there is more structure below the level of the nucleus. Protons and neutrons are not truly "elementary" in nature, but are actually made of subparticles called quarks. (The name derives from a meaningless word coined by novelist James Joyce in his book *Finnegans Wake*.) According to current theory, there are precisely six distinct types of quark in the universe, paired with six distinct types of lepton—the electron and two related "electron-like" particles, and three types of neutrino. The most massive, and most elusive, of them—the so-called "top" quark—was discovered at the Fermi National Laboratory (Fermilab) in Illinois in April 1994. The strong nuclear force is itself just a manifestation of an even more basic force that binds quarks to one another.

On the face of it, one might not imagine that there could be any deep underlying connection between forces as dissimilar as those just described, yet there is growing evidence that they may really be just different aspects of a single basic phenomenon. In the 1960s, theoretical physicists succeeded in explaining the electromagnetic and weak forces in terms of the electroweak force. Shortly thereafter, the first attempts were made at combining the strong and electroweak forces into a single all-encompassing "superforce."

Theories that combine the strong and electroweak forces into one are generically known as *Grand Unified Theories*, or, less formally, *GUTs*. (Note that the term is plural—no one GUT has yet been proven to be "the" correct description of nature.) One general prediction of GUTs is that the three non-gravitational forces are indistinguishable from one another only at enormously high energies, corresponding to temperatures in excess of 10^{28} K. Below that temperature, the superforce splits into two, displaying its separate strong and electroweak aspects. In particle-physics parlance, we say that there is a "symmetry" between the strong and the electroweak forces that becomes broken at temperatures below

10^{28} K, allowing the separate characters of the two forces to become apparent. At "low" temperatures—less than about 10^{15} K, a range that includes almost everything we know about on Earth and in the stars—there is a second symmetry breaking, and the electroweak force splits to reveal its more familiar electromagnetic and weak nature.

The key predictions of the electroweak theory were experimentally verified in the 1970s, winning the theory's originators the 1979 Nobel prize in Physics. The GUTs have not yet been experimentally verified (or refuted), in large part because of the extremely high energies that must be reached in order to observe their predictions.

An important idea that has arisen from the realization that the strong and the electroweak forces can be unified is the notion of supersymmetry. It extends the idea of symmetry between fundamental forces to place all particles—those that feel forces (such as protons and electrons), and those that transmit those forces (such as photons and gluons)—on an equal footing. One particularly important prediction of this theory is that all particles should have so-called supersymmetric partners—extra particles that must exist in order for the theory to remain self-consistent. None of these new particles has ever been detected, yet many physicists are convinced of the theory's essential correctness. Experiments are planned that may soon provide evidence for supersymmetric partners to some of the known elementary particles.

Of particular interest to astronomers, these new particles, if they exist, would have been produced in abundance in the Big Bang and should still be around today. They are also expected to be very massive—at least a thousand times heavier than a proton. These so-called supersymmetric relics are among the current leading candidates for the dark matter in the universe (although it must be admitted that recent experimental failures to detect them has dampened some astronomers' early enthusiasm).

Efforts to include gravity within this picture have so far been unsuccessful. Gravitation has not yet been incorporated into a single "SuperGUT," in which all of the fundamental forces are united. Some theoretical efforts to merge gravity with the other forces have tried to fit gravity into the quantum world by postulating extra particles—called *gravitons*—that transmit the gravitational force. However, this is a very different view of gravity from the geometric picture embodied in Einstein's general relativity, and combining the two into a consistent theory of quantum gravity has proved very difficult. Alternative approaches start from the geometric view and attempt to explain the basic forces of nature in terms of additional curved dimensions of spacetime. They, too, encounter serious problems. At the present time, none of these theories has succeeded in making any definite statement about conditions in the very early universe. A complete theory of quantum gravity continues to elude researchers.

Planck epoch, after Max Planck, one of the creators of quantum mechanics. Unfortunately, for now at least, there is no theory of quantum gravity, so we simply cannot talk meaningfully about the universe during the Planck epoch.

By the end of the Planck epoch, the temperature was around 10^{32} K, and the universe was filled with radiation and a vast array of subatomic particles created by the mechanism of pair production. At around this time, gravity parted company with the other forces of nature—it became distinguishable from the "quantum" forces, and has remained so ever since. The strong, weak, and electromagnetic forces were still unified. The present-day theories that describe this unification are collectively known as **Grand Unified Theories**, or GUTs for short. Accordingly, we will refer to this period of time as the *GUT epoch*.

Theory indicates that, at temperatures below 10^{28} K, the strong nuclear force becomes distinguishable from the electroweak force (the unified weak and electromagnetic force). When the universe had cooled to this temperature, about 10^{-35} s after the Big Bang, the GUT epoch ended. In effect, the strong force froze out of the expanding universe at that point, just as gravity froze out at the end of the Planck epoch. Our next major subdivision of the Radiation Era covers the period when all "heavy" elementary particles—that is, all the way down in mass to protons, neutrons, and their constituent quarks—were in thermal equilibrium with the radiation. We will call this period the hadron epoch, after the general term for particles that interact via the strong force.

The universe continued to expand and cool. At a temperature of about 10^{15} K, 10^{-10} s after the Big Bang, the weak and the electromagnetic components of the electroweak force began to display their separate characters. By about 0.1 milliseconds (10^{-4} s) after the Big Bang, the temperature had dropped well below the 10^{13} K threshold for the creation of protons and neutrons (the lightest hadrons), and the hadron epoch ended. The main constituents of the universe were now lightweight particles—muons, electrons, neutrinos, and their antiparticles—all still in thermal equilibrium with the radiation. Compared with the numbers of these lighter particles, only very few protons and neutrons remained at this stage because most had been annihilated.

Electrons, muons, and neutrinos are collectively known as *leptons*, after the Greek word meaning light (that is, not heavy). Accordingly, this period in the history of the universe is known as the *lepton epoch*. Quite early on in this epoch, the rapidly thinning universe became transparent to neutrinos. These ghostly particles have been streaming freely through space ever since—most have not interacted with any other particle since the universe was a few seconds old. The lepton epoch ended when the universe was about 100 s old and the temperature fell to about 1 billion kelvins—too low for electron–positron pair production to occur. The density of the universe by this time was about 10 times that of water.

The final significant event in the Radiation Era occurred when protons and neutrons began to fuse into heavier nuclei. At the start of this period, which we will call the *nuclear epoch*, the temperature was a few hundred million kelvins, and fusion occurred very rapidly, forming deuterium and helium in quick succession before conditions became too cool for further reactions to occur. By the time the universe was about 15 minutes old, much of the helium we observe today had been formed.

THE MATTER ERA

Time passed, the universe continued to expand and cool, and radiation gave way to matter as the dominant constituent of the universe. Our next major epoch extends in time from a few thousand years (the end of the Radiation Era) to about 1 million years after the Big Bang. As the primeval fireball diminished in intensity, a crucial change occurred—perhaps the most important change in the history of the universe. At the end of the nuclear epoch, radiation still overwhelmed matter. As fast as protons and electrons combined, radiation broke them apart again, preventing the formation of even simple atoms or molecules. However, as the universe expanded and cooled, the early dominance of radiation eventually ended. Once formed, atoms remained intact. We will call this time interval the *atomic epoch*.

The last two epochs together bring us to the current age of the universe. During these late stages, change happened at a much more sedate pace. By the time the universe was about a billion years old, galaxies and large-scale structure had formed. For the first time, the visible universe departed from homogeneity on macroscopic scales. The largely uniform universe of the Radiation Era became a universe containing large agglomerations of matter. We call the period from 10^6 to 10^9 years after the Big Bang the *galactic epoch*. At its end, quasars were shining brightly in the otherwise dark sky, and the first stars were burning and exploding, helping to determine the future shape of their parent galaxies. Since then, stars, planets, and life have appeared in the universe. This final *stellar epoch* has been the subject of the first 25 chapters of this book.

27.3 The Formation of Nuclei and Atoms

HELIUM FORMATION IN THE EARLY UNIVERSE

4 We now have all the ingredients needed to complete our story of the creation of the elements, begun in Chapters 21, but never quite finished. ∞ (p. 454) The theory of stellar nucleosynthesis accounts very well for the observed abundances of heavy elements in the universe, but there are discrepancies between theory and observations when it comes to the abundances of the light elements, es-

pecially helium. Simply put, there is far more helium in the universe than can be explained by nuclear fusion in stars. No matter where they look, and no matter how low a star's abundance of heavy elements may be, astronomers find that there is a minimum amount of helium—a little less than 25 percent by mass—in all stars. The accepted explanation is that this base level of helium is *primordial*—that is, it was created during the early, hot epochs of the universe—and did not form in stars at all.

Could a large amount of helium have been created in the early universe? The answer is "Yes." The possibility of **primordial nucleosynthesis**—the production of elements heavier than hydrogen by nuclear fusion shortly after the Big Bang—was first realized in the 1940s, when physicist George Gamow pointed out that the hot, dense conditions in the early universe would have provided all of the ingredients necessary for the formation of helium. Later calculations, especially those performed in the 1970s, when the physical state of the cosmos at early times was much better understood, demonstrated that Gamow's idea was essentially correct.

Early in the nuclear epoch, the average temperature of the universe exceeded by a wide margin the 10^7 K needed to fuse hydrogen into helium through the proton–proton chain (the Sun's energy source, as discussed in Chapter 16). ∞ (p. 349) Was helium created within the primordial fireball in basically the same way that it now forms within stars? No. Helium did form, but the proton–proton chain was *not* the main route. There was an easier way, involving fusion of protons and neutrons instead.

By the end of the hadron epoch, 10^{-4} s after the Big Bang, protons and neutrons had frozen out of the cosmic fireball—the temperature of the universe had fallen below the threshold necessary to create them in quantity. By the time the temperature had dropped to about a billion kelvins (100 s after the Big Bang), there was about 1 neutron for every 5 protons in the universe. The stage was set for nuclear fusion to occur.

Protons and neutrons can combine to produce deuterium nuclei:

$$^1\text{H (proton)} + \text{neutron} \rightarrow {}^2\text{H (deuterium)} + \text{energy.}$$

Although the foregoing reaction must have occurred very frequently during the lepton epoch, the temperature was still so high then that the deuterium nuclei were broken apart by high-energy gamma rays as soon as they formed. For the same reason, the proton–proton chain could not operate because the deuterium created in its initial reaction,

$$^1\text{H} + {}^1\text{H} \rightarrow {}^2\text{H} + \text{positron} + \text{neutrino,}$$

could not survive. (Also, this reaction is much slower than the proton–neutron reaction. It plays an important role in the Sun only because there are no neutrons around to make the other reaction possible.) Although temperatures and densities before the nuclear epoch were certainly high enough for fusion to occur, the process could not get under

way because deuterium was destroyed as fast as it appeared. The universe had to wait until it became cool enough for the deuterium to survive. This waiting period is sometimes called the *deuterium bottleneck*.

Only when the temperature of the universe fell below about 900 million K, roughly 2 minutes after the Big Bang, was deuterium at last able to form and endure. Once that occurred, the deuterium was quickly converted into heavier elements by numerous reactions:

$$^2\text{H} + {}^1\text{H} \rightarrow {}^3\text{He} + \text{energy,}$$

$$^3\text{He} + \text{neutron} \rightarrow {}^4\text{He} + \text{energy,}$$

$$^2\text{H} + {}^2\text{H} \rightarrow {}^4\text{He} + \text{energy,}$$

along with many others. The result was that, once the universe passed the deuterium bottleneck, fusion occurred rapidly, and large amounts of helium were formed. In just a few minutes, most of the free neutrons were consumed, leaving a universe whose matter content was primarily hydrogen and helium. Figure 27.5 illustrates some of the reactions responsible for helium formation. Contrast it with Figure 16.28, which depicts how helium is formed today in the cores of main-sequence stars like the Sun. ∞ (p. 351)

We might imagine that fusion could have continued to create heavier and heavier elements, just as in the cores of stars. However, this did not occur. In stars, the density and the temperature both *increase* slowly with time, allowing more and more massive nuclei to form, but in the early universe the opposite was true. The temperature and density were both *decreasing* rapidly, making conditions less and less favorable for fusion as time went on. Even before the supply of neutrons was completely used up, the nuclear reactions had effectively ceased. Reactions between helium nuclei and protons may also have formed trace amounts of lithium (the next element beyond helium) by this time, but for all practical purposes, the expansion of the universe caused the fusion process to stop at helium. The brief epoch of primordial nucleosynthesis was over about 15 minutes after it began.

By the end of the nuclear epoch, some 1000 s after the Big Bang, the temperature of the universe was about 300 million K, and the cosmic elemental abundance was set. Careful calculations indicate that about 1 helium nucleus had formed for every 12 protons remaining. Because a helium nucleus is four times more massive than a proton, helium accounted for about one quarter of the total mass of matter in the universe:

$$\frac{1 \text{ helium nucleus}}{12 \text{ protons} + 1 \text{ helium nucleus}} = \frac{4 \text{ mass units}}{12 \text{ mass units} + 4 \text{ mass units}}$$

$$= \frac{4}{16} = \frac{1}{4}.$$

The remaining 75 percent of the matter in the universe was hydrogen. It would be almost a billion years before nucleosynthesis in stars would change these numbers.

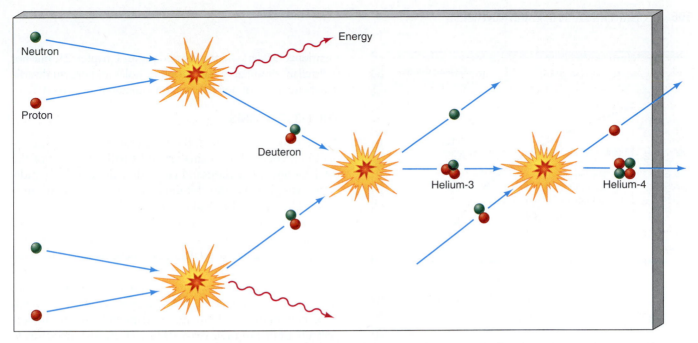

Figure 27.5 Some of the reaction sequences that led to the formation of helium in the early universe. Compare this figure with Figure 16.28, which depicts the proton–proton chain in the Sun.

It is difficult to disentangle the contributions to the present-day helium abundance from primordial nucleosynthesis and later hydrogen burning in stars, but the foregoing calculation implies that all cosmic objects should contain *at least 25* percent helium by mass. The figure for the Sun, for example, is about 28 percent. Our best hope of determining the amount of primordial helium is to study the oldest stars known, since they formed early on, before stellar nucleosynthesis had had time to change significantly the helium content of the universe. However, those stars are of low mass and quite cool, making the helium lines in their spectra very weak and hard to measure accurately. Despite this uncertainty, the observations are generally consistent with the theory just described.

Bear in mind that, while all this was going on, the matter was just an insignificant "contaminant" in the radiation-dominated universe at this early stage. Radiation outmassed matter by about a factor of 5000 at the time helium formed. The existence of helium is very important in determining the structure and appearance of stars today, but its creation was completely *irrelevant* to the evolution of the universe at the time.

DEUTERIUM AND THE DENSITY OF THE COSMOS

Although most deuterium was quickly burned into helium as soon as it formed, a small amount was left over when the nuclear reactions ceased. However, unlike helium, deuterium is not likely to be produced in stars. In fact, deuterium is *destroyed* in stars, so whatever deuterium we detect today is an *underestimate* of the amount produced in the early universe. Observations of deuterium—especially those made by orbiting satellites able to capture deuterium's strongest spectral feature, which happens to be emitted in the ultraviolet part of the spectrum—indicate a present-day abundance of about 2 deuterium nuclei for every 100,000 protons.

Primordial deuterium was produced in very small quantities, and it plays only a minor role in the evolution of stars and galaxies, so why bother to study it? The answer is that it provides cosmologists with a sensitive means of probing the density of matter in the universe that is completely *independent* of the techniques discussed in previous chapters. The more matter there was at early times, the more particles there were to react with deuterium as it formed, and the less deuterium remained when helium production stopped. But the total number of protons and neutrons present in the universe during the nuclear epoch was the *same* as it is today—protons and neutrons can combine into heavier nuclei and can even interchange identities with one another, but they cannot be created or destroyed. Consequently, the amount of deuterium formed by primordial nucleosynthesis can be directly related to the *present-day* density of matter in the universe: The denser the universe is today, the *less* deuterium must have been produced at early times.

Figure 27.6 shows the theoretical run of deuterium abundance with present-day cosmic density. Comparison of the observed deuterium abundance (marked on the figure) with the theoretical calculation implies a present-day density of *at most* 3×10^{-28} kg/m^3—only a few percent of the critical density. Furthermore, the exact amount of deuterium produced depends very sharply on the density: A little less than this maximum density and far too many deuterium nuclei would have been formed, a little more and far too few would now exist. Thus, measurements of the cosmic deuterium abundance provide us with a remarkably firm estimate of $\Omega_0 \approx 0.03$.[*]

But before we jump to the conclusion that the universe is open and will expand forever, we must make a very important qualification. Primordial nucleosynthesis as just

---

[*]*In fact, the amount of helium produced also depends somewhat on cosmic density, but the variation is not very strong, and does not provide a very accurate estimate of Ω_0.*

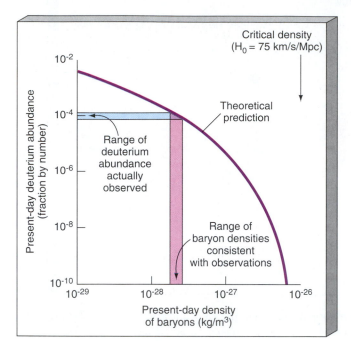

Figure 27.6 The present-day abundance of deuterium depends strongly on the amount of matter present at early times, and this, in turn, determines the present-day density of the universe. Thus, measuring the amount of deuterium in the universe gives us an estimate of the overall density of matter. The best deuterium measurements are marked; they imply that the density of matter in the universe is at most a few percent of the critical value.

described depends *only* on the presence of protons and neutrons in the early universe. Thus, measurements of the abundance of helium and deuterium tell us only about the density of so-called *baryonic* matter—matter made up of protons and neutrons—in the cosmos. Atoms, people, planets, and stars—what we have been loosely calling "normal" matter—are all predominantly baryonic in nature. Most of the mass in an atom, for example, is in its nucleus, which is composed of protons and neutrons (the orbiting electrons, which are not baryonic, make up only a tiny fraction of the atom's total mass). The point is that the foregoing arguments imply only that the present-day density of *baryonic* matter is at most 3 percent the critical value.

As we saw earlier, astronomers have concluded, on the basis of studies of the motions of galaxies in clusters and superclusters, that Ω_0 is at least 0.2 or 0.3, and may possibly be much more. ∞ (p. 577) If this reasoning turns out to be correct, and if the density of matter in the form of protons and neutrons is only a few percent of the critical value, then we are forced to admit that not only is most of the matter in the universe dark, but the dark matter *cannot* be entirely baryonic in nature. Most of the matter in the universe is apparently in the form of elusive nonbaryonic particles whose nature we do not fully understand and whose very existence has yet to be demonstrated in laboratory experiments.

For the sake of brevity, we will adopt the convention from now on that the term "dark matter" refers only to *non-baryonic* dark matter—that is, to the exotic subatomic particles

mentioned as dark-matter candidates in Chapter 23, but not to "stellar" candidates, such as black holes and brown dwarfs, which are made of normal, baryonic material. ∞ (p. 505)

THE FIRST ATOMS

As the universe expanded, the temperature fell, and the radiation accounted for a smaller and smaller fraction of the total energy. A few thousand years after the Big Bang, radiation ceased to be the dominant component of the universe. The Matter Era had begun.

By the start of the atomic epoch, matter consisted of electrons, protons, helium nuclei (formed by primordial nucleosynthesis as just described), and dark matter. The temperature was several tens of thousands of kelvins—comparable to the temperature in the atmosphere of an O-type star, and far too hot for atoms of hydrogen to exist, although some helium ions (helium atoms stripped by the high temperature of one of their two orbiting electrons) may already have formed. During the next few hundred thousand years, a major change occurred. The universe expanded by another factor of 10, the temperature dropped to a few thousand kelvins, and the electrons and nuclei combined to form neutral *atoms*. By the time the temperature had fallen to 4500 K, the universe consisted of atoms, photons, and dark matter (whose weak interactions with normal matter meant that it played no part in the atom-formation process).

The period during which nuclei and electrons combined to form atoms is often called the epoch of **decoupling**, for it was during this period that the radiation background parted company with normal matter. At early times, when matter was ionized, the universe was filled with large numbers of free electrons, which interacted frequently with electromagnetic radiation of all wavelengths. As a result, a photon could not travel far before encountering an electron and scattering off it. In effect, the universe was opaque to radiation (rather like the deep interior of a star like the Sun). Matter and radiation were strongly "tied," or *coupled*, to one another by these interactions. When the electrons combined with nuclei to form atoms of hydrogen and helium, however, only certain wavelengths of radiation—the ones corresponding to the spectral lines of those atoms—could interact with matter. ∞ (p. 82) Radiation of other wavelengths could travel virtually forever without being absorbed. The universe became nearly *transparent*. From that time on, photons passed generally unhindered through space. As the universe expanded, the radiation simply cooled, eventually to become the microwave background we see today.

As illustrated in Figure 27.7, the microwave photons now detected on Earth have been traveling through the universe ever since they decoupled. Their last interaction with matter occurred when the universe was a few hundred thousand years old and roughly 1500 times smaller (and hotter) than it is today—that is, at redshift of 1500. Thus, by observing the microwave background, we are probing conditions in the universe almost all the way back in time to the Big Bang itself, in much the same way as the study of sunlight tells us about the surface layers of the Sun.

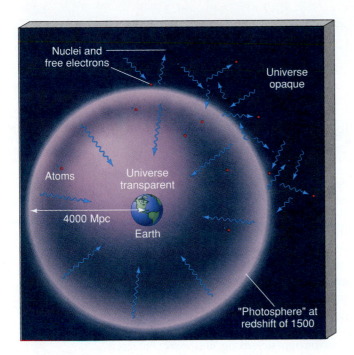

Figure 27.7   When atoms formed, the universe became virtually transparent to radiation. Thus, observations of the cosmic background radiation allow us to study conditions in the universe around a time at a redshift of 1500, when the temperature dropped below about 4500 K.

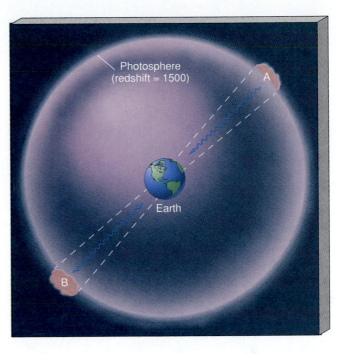

Figure 27.8   The horizon problem. The isotropy of the microwave background indicates that regions A and B in the universe were very similar to one another when the radiation we observe left them, but there has not been enough time since the Big Bang for them ever to have interacted physically with one another. Why then should they look the same?

27.4 *The Inflationary Universe*

THE HORIZON AND FLATNESS PROBLEMS

5 In the late 1970s, cosmologists trying to piece together the evolution of the universe were confronted with two nagging problems that had no easy explanation within the standard Big Bang model (basically, the sequence of events that we have just described). The resolution of these difficulties has caused cosmologists to completely rethink their view of the very early universe.

The first problem is known as the **horizon problem**, and it concerns the remarkable *isotropy* of the cosmic microwave background. Recall that the temperature of this radiation is virtually constant, at about 2.7 K, in all directions. Imagine observing the microwave background in two opposite directions on the sky, as illustrated in Figure 27.8. As we have just seen, the radiation last interacted with matter in the universe around a redshift of 1500. In observing these two distant regions of the universe, marked A and B on the figure, we are studying regions that were separated by several million parsecs at the time that they emitted this radiation. The fact that the background radiation is known to be isotropic to high accuracy means that regions A and B had very similar densities and temperatures at the time the radiation we see left them, just as required by the cosmological principle. The problem is, within the Big Bang theory as just described, there is no particular reason *why* these regions should be so similar to one another.

Just as ripples on the surface of a pond tend to spread out and merge with one another, and the cold ice cream and hot fudge of a sundae both eventually come to room temperature, density and temperature fluctuations in the early universe would gradually have been smoothed out as neighboring regions interacted. But these interactions did not occur instantaneously, any more than ripples spread instantaneously across water. Energy is carried from place to place by photons, gravitational radiation, and other wave motions, and the speed at which it moves cannot exceed the speed of light.

The point is, there simply has not been enough time since the Big Bang for light to have traveled from region A to region B. In cosmological parlance, the two regions are said to be outside each others' *horizon*. But if that is so, then why should we expect them to look the same? We cannot explain their similarity by saying that they have interacted with each other since the universe formed. Unless we are prepared simply to assume that the universe started off perfectly homogeneous—something theorists are unwilling to do, because all models of the early universe predict fluctuations at some level—then there is no good reason for regions A and B to look alike.

The second problem with the standard Big Bang model is called the **flatness problem**. Whatever the exact value of Ω_0, it appears to be quite close to 1—the density of the universe is fairly near the critical value needed for the expansion barely to continue forever. In terms of spacetime curvature, we can say that the universe is remarkably close

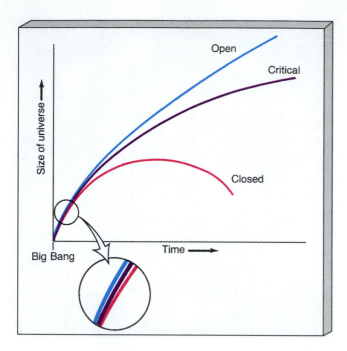

Figure 27.9 The flatness problem. If the universe deviates even slightly from the critical case, that deviation will grow rapidly in time. For the universe to be as close to critical as it is today, it must have differed from the critical density in the past by only a tiny amount.

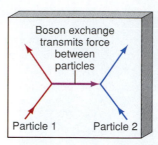

Figure 27.10 Forces between elementary particles are transmitted through the exchange of particles called bosons. As two particles interact, they exchange bosons, a little like playing catch with a submicroscopic ball.

to being flat. We say "remarkably" here because again there is no good reason *why* the universe should have formed with a density very close to critical. Why not a millionth or a million times that value? Furthermore, as shown in Figure 27.9, a universe that starts off close to, but not exactly on, the critical curve soon deviates greatly from it, so if the universe is close to critical now, it must have been *extremely* close to critical in the past. For example, if $\Omega_0 = 0.1$ today, the departure from the critical density at the time of nucleosynthesis would have been only 1 part in 10^{15}.

The standard Big Bang theory simply provides no good reason why the universe should be so nearly isotropic and flat. These observations constitute "problems" because cosmologists want to be able to explain the present condition of the universe, not just accept it "as is." They would prefer to resolve the horizon and flatness problems in terms of physical processes that could have taken a universe with no special properties and caused it to evolve into the cosmos we now see. The resolution of both problems takes us back in time almost to the instant of the Big Bang itself.

FREEZE-OUT

A central goal of modern particle physics research is to understand the four known forces of nature—*gravity* and *electromagnetism*, which we have studied in some detail already, the *strong nuclear force*, which binds protons and neutrons together to form nuclei, and the *weak nuclear force*, which plays a role in many radioactive decays. In the 1970s and 1980s, that goal was partly met when theoretical physicists suc-

ceeded in unifying three of these basic forces—electromagnetism, the strong force and the weak force—into a single all-encompassing "superforce." (Efforts to include gravity within this picture have so far been unsuccessful.) As mentioned earlier, theories that combine the three nongravitational forces into one are known as *Grand Unified Theories*, or GUTs. A general prediction of GUTs is that the three forces are indistinguishable from one another *only* at enormously high energies, corresponding to temperatures in excess of 10^{28} K. Only at lower temperatures does the superforce display its separate electromagnetic, strong, and weak characters.

An essential concept in quantum physics is the idea that forces between elementary particles are exerted, or *mediated*, by the exchange of another type of particle, generically called a *boson*. We might imagine the two particles as playing a rapid game of catch, using a boson as a ball, as illustrated in Figure 27.10. As the ball is thrown back and forth, the force is transmitted. In ordinary electromagnetism, the boson involved is the photon—a bundle of electromagnetic energy that always travels at the speed of light. The strong force is mediated by particles known as *gluons*. The electroweak theory includes a total of four bosons: the massless photon and three other massive particles, called (for historical reasons) W^+, W^-, and Z^0, all of which have been observed in laboratory experiments. Most of the particles we have encountered so far in this book—electrons, protons, neutrons, neutrinos—play "catch" with at least some of these "balls."

We saw earlier how particles "froze out" of the universe as its temperature dropped below the threshold temperature for their creation by pair production. Now that we know that the basic forces of nature are also mediated by particles, we can understand—in general terms, at least—how the fundamental forces froze out too as the universe cooled. The W and Z particles responsible for the electroweak force have masses about 100 times the mass of a proton. The threshold temperature for their production—roughly 10^{15} K—is the temperature at which the weak and electromagnetic forces parted company near the end of the hadron epoch. According to the GUTs, the particle that unifies the strong and electroweak forces is extremely massive—at least 10^{15} times the mass of the proton, and possibly much more. Because this particle is so massive, the unification of the strong and electroweak forces becomes evident only at extremely high temperatures—at least 10^{28} K.

COSMIC INFLATION

The freezing out of the electroweak force at a temperature of 10^{15} K, some 10^{-10} s after the Big Bang, had little overall effect on the cosmic expansion. By contrast, the freeze-out of the strong force, a mere 10^{-35} s after the Big Bang, produced one of the strangest events in the history of the cosmos.

As the temperature fell below 10^{28} K and the strong force appeared as a separate entity—a little like a gas liquefying or water freezing as the temperature drops—the universe briefly entered a very odd, and unstable, high-energy state which physicists call the "false vacuum." In essence, it remained in the "unified" condition a little too long, like water that has been cooled below freezing but has not yet turned to ice. This had dramatic consequences. For a short while, empty space acquired an enormous *pressure*, which temporarily overcame the pull of gravity and accelerated the expansion of the universe at an enormous rate. The pressure remained constant as the cosmos expanded, and the acceleration grew more and more rapid with time—in fact, the size of the universe *doubled* every 10^{-34} s or so! This period of unchecked cosmic expansion, illustrated in Figure 27.11, is known as the epoch of **inflation**.

Eventually, the universe returned to the lower-energy "true vacuum" state. Regions of normal space began to appear within the false vacuum, and rapidly spread to include the entire cosmos. With the return of the true vacuum, inflation stopped. The whole episode lasted a mere 10^{-32} s, but during that time the universe swelled in size by about a factor of about 10^{50}. When the inflationary phase ended, the grand unified force was gone forever. In its place were the more familiar electroweak and strong forces that operate around us in the low-temperature universe of today. With the normal vacuum restored, the universe once again resumed its (relatively) leisurely expansion, slowly decelerated by the effect of gravity. But a number of important changes had occurred that would have far-reaching ramifications for the evolution of the cosmos.

IMPLICATIONS FOR THE UNIVERSE

The inflationary epoch provides a natural solution to the horizon and flatness problems described earlier. The horizon problem is solved because inflation took regions of the universe that had already had time to communicate with one another—and so had established similar physical properties—and then dragged them far apart, well out of communications range of one another. Regions A and B in Figure 27.8 have been out of contact since 10^{-32} s after creation, but they were in contact before then. As illustrated in Figure 27.12, their properties are the same today because they were the same long ago, before inflation separated them.

Figure 27.12(a) shows a small piece of the universe just before the onset of inflation. The point that will one day

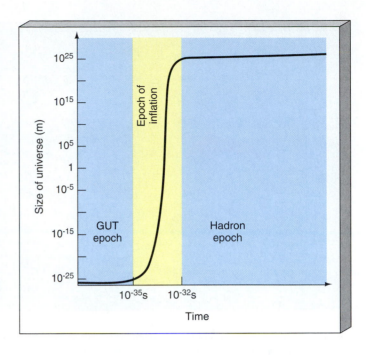

Figure 27.11 During the period of inflation at the end of the GUT epoch, the universe expanded enormously in a very short time. Afterward, it resumed its earlier "normal" expansion, except that the size of the cosmos was about 10^{50} times bigger than it was before.

become the site of the Milky Way Galaxy is at the center of the shaded region, which represents the portion of space "visible" to that point at that time—that is, there has been enough time since the Big Bang for light to have traveled from the edge of this region to its center. That entire region is more or less homogeneous because different parts of it to have been able to interact with one another, so any initial differences between them have been largely smoothed out. The points A and B of Figure 27.8 are also marked. They lie well within the homogeneous patch, so they have very similar properties. The actual size of the shaded region is about 10^{-26} m—only a trillionth the size of a proton.

Immediately after inflation, as shown in Figure 27.12(b), the homogeneous region has expanded by 50 orders of magnitude, to a diameter of about 10^{24} m, or some 30 Mpc—larger than the largest supercluster. By contrast, the visible portion of the universe, indicated by the dashed line, has grown only by a factor of a thousand, so it is still microscopic in size. In effect, the universe expanded much faster than the speed of light during the inflationary epoch, so what was once well within the horizon now lies far beyond it. In particular, points A and B are no longer visible, either to us or to each other, at this time.

Since then, the universe has expanded by a further factor of 10^{27}, so the size of the homogeneous region of space surrounding us is now about 10^{51} m (10^{28} Mpc)—10 trillion trillion times greater than the distance to the most dis-

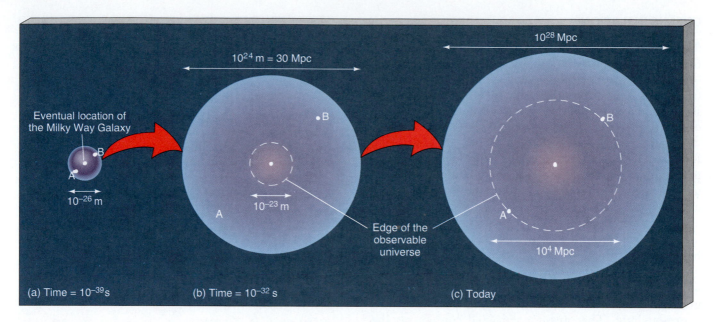

Figure 27.12   Inflation solves the horizon problem by taking a small region of the very early universe, whose parts had already had time to interact with one another and which had thus already become homogeneous, and expanding it to enormous size. In (a), points A and B are well within the (shaded) homogeneous region of the universe centered on the eventual site of the Milky Way Galaxy. In (b), after inflation, A and B are far outside the horizon (indicated by the dashed line), so they are no longer visible from our location. Subsequently, the horizon expands faster than the universe, so that today (c) A and B are just reentering our field of view. They have similar properties now because they had similar properties before the inflationary epoch.

tant quasar. As shown in Figure 27.12(c), the horizon has expanded faster than the universe, so points A and B are just now becoming visible again. As the portion of the universe now observable from the Earth grows in time, it remains homogeneous because our cosmic field of view is simply reexpanding into a region of the universe that was within our horizon long ago. We will have to wait a very long time—at least 10^{35} years—before the edge of the homogeneous patch surrounding us comes back into view.

To see how inflation solves the flatness problem, let's return once more to our earlier balloon analogy. Imagine that you are a 1-mm-long ant sitting on the surface of the balloon as it expands, as illustrated in Figure 27.13. When the balloon is just a few centimeters across, you can easily perceive the surface to be curved—its circumference is only a few times your own size. When the balloon expands to, say, a few meters in diameter, the curvature of the surface will be less pronounced, but still perceptible. However, by the time the balloon has expanded to a few kilometers across, an "ant-sized" patch of the surface will look quite flat, just as the surface of the Earth looks flat to us. Now imagine that the balloon expands 100 trillion trillion trillion trillion times, as the universe did during the period of inflation. Your local patch of the surface would now be completely indistinguishable from a perfectly flat plane, deviating by no more than 1 part in about 10^{50}.

Exactly the same argument applies to the universe. Any curvature the universe may have had before inflation

has been expanded so much that space is now virtually flat, at least on the scale of the observable universe (and, in fact, on much larger scales too). But notice that this resolution to the flatness problem—the universe looks pretty flat because it *is* flat—has a very important consequence. Because the universe is flat, the density of matter must be exactly critical: $\Omega_0 = 1$. This means that there must be a lot of invisible matter in the universe beyond the clusters and the superclusters, filling the huge voids on the largest scales. And because we just saw that primordial nucleosynthesis implies that the density of "normal" baryonic matter is at most 3 percent of the critical value, it follows that the rest of the mass—at least 97 percent of all the matter in the universe—must be in the form of nonbaryonic dark matter (whatever it may be).

When the idea of inflation was first suggested, most astronomers were very skeptical. Some still are. They point out, quite correctly, that there is *no* direct observational evidence for a cosmic density as high as the critical value. However, as we saw earlier, there is ample evidence for dark matter in the universe with as much as 20 or 30 percent of the critical density, and estimates of Ω_0 seem to increase as the scale under consideration increases. Furthermore, inflation is a clear prediction of a group of theories—the GUTs—that are becoming more and more firmly established as the standard description of matter at high energies. If the GUTs are indeed correct, then inflation *must* have occurred. Finally, inflation provides a neat solution to

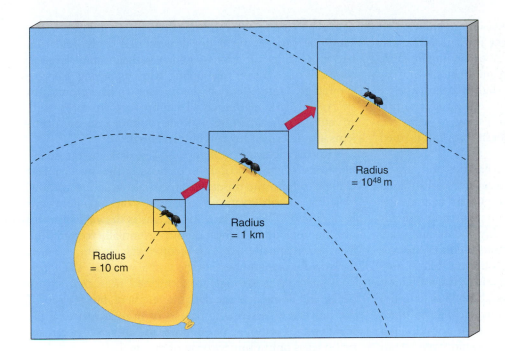

Figure 27.13 Inflation solves the flatness problem by taking a curved surface, here represented by the surface of the expanding balloon, and expanding it enormously in size. To an ant on the surface, the balloon looks virtually flat when the expansion is over.

two serious difficulties with the standard noninflationary Big Bang model. For all these reasons, inflation has become entrenched (although as an unwelcome guest in some cases) in most cosmologists' models of the universe.

27.5 The Formation of Structure in the Universe

⑥ Just as stars form from inhomogeneities in interstellar clouds, galaxies, galaxy clusters, and larger structures are believed to have grown from small density fluctuations in the matter of the expanding universe. ∞ (p. 400) Given the conditions in the universe during the atomic and galactic epochs (see Table 27-1), cosmologists calculate that regions of higher than average density which contained more than about a million times the mass of the Sun would have begun to contract. Smaller-scale fluctuations, on the other hand, would have remained stable. Since the early universe was probably subject to random density fluctuations on all scales, there was thus a natural tendency for million-solar-mass "pregalactic" objects to form. This process is illustrated in Figure 27.14. In Chapter 24, we saw a little of how these pregalactic fragments may have interacted and merged to form galaxies. ∞ (p. 535) Here, we will concern ourselves mostly with the formation of structure on much larger scales.

THE GROWTH OF PERTURBATIONS

Because the contracting matter had to "fight" the general expansion of the universe, it took time for small inhomogeneities to grow into the structures we see today. It was in

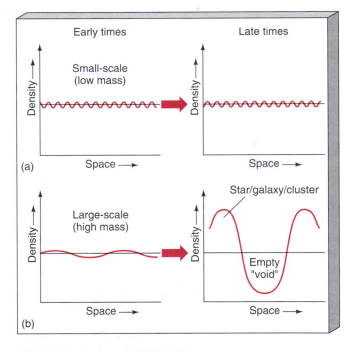

Figure 27.14 Growth of density inhomogeneities in the universe. At any instant, there is a minimum spatial scale below which density fluctuations cannot grow. (a) Regions of above-average density that are smaller than this critical length tend to oscillate, like sound waves in air, and their density does not increase with time. The amount of mass in the overdense regions is simply not great enough for gravity to overcome gas pressure. (b) For fluctuations on spatial scales larger than the critical length, collapse will occur, just as a massive enough interstellar gas cloud will start to collapse to form a star. These denser-than-average regions will tend to become even denser, leaving empty "voids" between them.

the calculation of how much time they took, and how much was available, that the theory of galaxy formation first ran into serious problems. If we imagine an expanding universe filled with normal matter—hydrogen and helium—and radiation, and if we ask how rapidly a slightly overdense region would contract, we find that the simplest idea of primordial gas clouds infalling and coalescing to form the luminous galaxies we see today just doesn't work. It is ruled out by observations. Here's why.

Any theoretical attempt to make galaxies and larger-scale structure out of normal matter is severely constrained by observations. We know that some quasars had already formed by the time the universe was one-sixth its present size—that is, at a redshift of 5, or roughly 600 million years after the Big Bang, given the assumptions used to construct the table on p. 555. Yet many theorists believe that to produce the densest galactic nuclei we see today, the formation process must have already been well established as long ago as a redshift of 10 or 20, when the universe was no more than about 100 million years old. This might not present a problem if the galaxy-growing process could have begun early enough, but there is a limit on how long ago it could have started. Calculations show that before the epoch of decoupling, at a redshift of 1500 or so, the background radiation would have prevented any large clumps of matter from contracting. The clumps had to wait until after decoupling before they could begin to grow.

Even with this shortening of the time available, structure could still form if the initial fluctuations were large enough. It turns out that, in order for inhomogeneities to grow into galaxies by a redshift of 20, the density at the epoch of decoupling would have had to vary by at least a few percent from one place to another. But since the radiation background was "tied" to the matter up until that time, any density fluctuations would have led to variations in the temperature of the cosmic background radiation—denser regions would have been a little hotter than less dense parts of the universe. The level of isotropy observed in the microwave background indicates that any density fluctuations must have been far less than 1 percent, effectively killing this whole theory of galaxy formation.

DARK MATTER

By the early 1980s, cosmologists had come to realize that galaxies could not have grown from density fluctuations in the gaseous baryonic matter of the early universe. There just wasn't enough time, given the small initial size of the fluctuations implied by the cosmic microwave background. However, the existence of dark matter, and the growing evidence that most of the universe is nonbaryonic in nature, provide an alternative. Whatever its composition, dark matter interacts only weakly with normal matter. As a result, it decoupled from the rest of the universe long ago—before the time of primordial nucleosynthesis, in fact—and its density fluctuations have been growing ever since the end of the Radiation Era, when matter first began to "con-

trol" the universe, at a redshift of 20,000. Furthermore, because the dark matter is not directly tied to the radiation background, its variations could have been quite large at the time of radiation decoupling without having a correspondingly large effect on the microwave background. In short, the dark matter could clump to form large-scale structure in the universe without running into the problems just described for baryonic material.

In this picture (see Figure 27.15), dark matter determines the overall distribution of mass in the universe and clumps to form the observed large-scale structure without violating any observational constraints on the microwave background. Then, at later times, gas is drawn by gravity into the regions of highest density, eventually forming the galaxies we actually see. This picture explains why so much dark matter is found outside the visible galaxies. The luminous material is strongly concentrated near the density peaks, and dominates the dark matter there, but the rest of the universe is essentially devoid of normal matter. Like foam on the crest of an ocean wave, the universe we can actually see is only a tiny fraction of the total.

Given that the nature of the dark matter is still unknown, theorists have considerable freedom in choosing its properties when they attempt to simulate the formation of structure in the universe. Cosmologists distinguish between two basic types of dark matter on the basis of its temperature at the time when galaxies began to form. These types are known as **hot dark matter** and **cold dark matter**, respectively, and they lead to quite different kinds of structure in the present-day universe. By performing computer simulations of model universes dominated by different combinations of dark matter and comparing the results with observations of the real universe, cosmologists try to account for the large-scale structure we see around us.

Hot dark matter consists of lightweight particles—much less massive than the electron. If neutrinos turn out to have a small mass, as many researchers suspect, they would be leading candidates for hot dark-matter particles. Simulations of a universe filled with hot dark matter indicate that large structures, such as superclusters and voids, form fairly naturally, but the computer models cannot account for the existence of structure on smaller scales. Small amounts of hot material tend to disperse, not to clump together. Attempts to produce galaxies and clusters by other means after the formation of larger objects have been only partly successful, so most cosmologists have concluded that models based purely on hot dark matter are unable to explain the observed structure of the universe.

Cold dark matter consists of very heavy particles, possibly formed during the GUT era or even before. Computer simulations modeling the universe with these particles as the dark matter easily produce small-scale structure. With the understanding that galaxies form preferentially in the densest regions and with some fine-tuning, these models can also be made to produce large-scale structure compara-

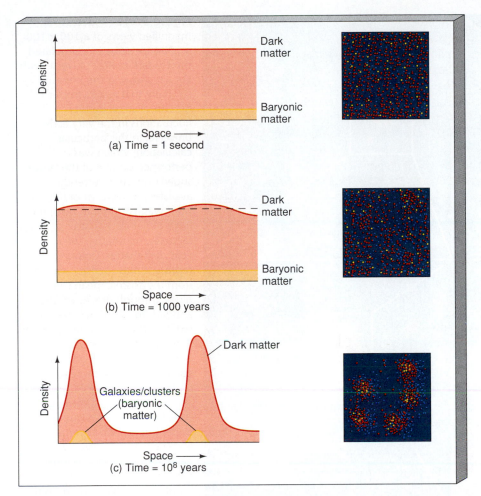

Figure 27.15 The formation of structure in the cosmos. The universe started out (a) as a mixture of (mostly) dark and "normal" baryonic matter. The dark matter began to clump quite early on (b), eventually forming large structures (c) into which the baryonic matter flowed, ultimately to form the galaxies we see today. The three frames at right represent the densities of dark matter (red) and baryonic matter (yellow) graphed at left.

ble to what is actually observed. It seems that cold dark matter is an essential ingredient in cosmologists' models if we are to understand the structure of the universe.

Perhaps the best results to date (that is, the results that agree most closely with observations) come from simulations in which a *mixture* of hot and cold dark matter is assumed. Figure 27.16 shows the results of a recent supercomputer simulation of a mixed-dark-matter universe. Compare these images with the real observations of nearby structure shown in Figures 24.25 and 24.26. Although calculations like this cannot *prove* that dark-matter models are the correct description of the universe, the similarities between the models and reality are certainly very striking.

Finally, let us note that dark matter does not *necessarily* imply inflation—not all GUTs predict the emergence of a false vacuum. A noninflated universe with $\Omega_0 = 0.5$, say, and nonbaryonic dark matter could still have formed the structure we now see. Nor does inflation necessarily imply the existence of any particular type of dark matter. Inflation and primordial nucleosynthesis together imply some sort of nonbaryonic dark matter, but say nothing about its makeup.

THE MICROWAVE BACKGROUND

One of the arguments supporting dark-matter models of the universe is the fact that they are consistent with the high degree of isotropy seen in the microwave background. Because

the dark matter does not interact directly with photons, its density fluctuations do not cause large (and easily observable) temperature variations in the radiation. However, the radiation *is* influenced slightly by the gravity of the growing dark clumps. The radiation experiences a slight gravitational redshift that varies from place to place depending on the density of the invisible matter. As a result, dark-matter models predict that there *should* be ripples (temperature fluctuations) in the microwave background, but ripples that are very small—perhaps as little as a few parts per million.

Until the late 1980s, these ripples were too small to be accurately measured, but cosmologists were confident that they were there. As observations improved, especially with the launch of the *COBE* satellite in 1989 (see Chapter 26), the anticipated temperature fluctuations remained undetected. ∞ (p. 582) Many theorists began to worry that there was something seriously wrong with this theory of structure formation too. By 1990, the observational limits on the fluctuations had fallen below 1 part in 10,000—still above the bare minimum needed for the cold dark-matter theory to survive, but getting uncomfortably close to the point where the theory would have had to be abandoned.

In 1992, cosmologists breathed a collective sigh of relief when the *COBE* team, after almost 2 years of careful observation, announced that the predicted fluctuations had been found. The temperature variations are tiny—only

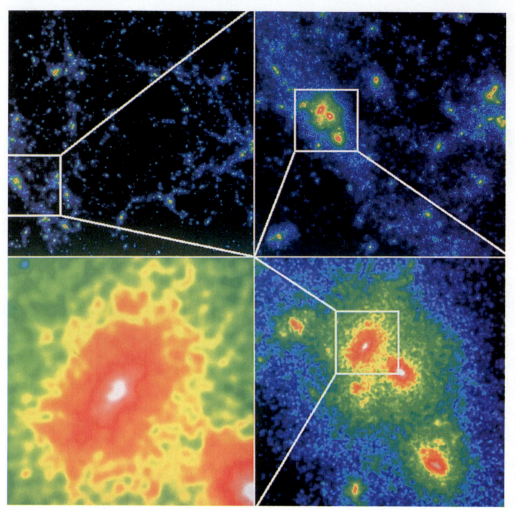

Figure 27.16 Successively magnified views of a $100 \times 100 \times 100$ Mpc cube in a simulated mixed-dark-matter universe with $\Omega_0 = 1$, showing the present-day structure that results from the growth of small density fluctuations in the very early universe. In this particular calculation, which was performed on one of the largest supercomputers currently available, 20 percent of the total mass was assumed to be in the form of hot-dark-matter particles (actually, neutrinos); the rest was cold dark matter. Colors represent mass density, ranging from the cosmic mean (dark blue), through green, yellow, and red, to 100 times the mean (white). The enlargements zoom in on one particular galaxy in one particular small group of galaxies. The last frame (at bottom left) is roughly 1.5 Mpc across. Notice both the large-scale filamentary structure evident in the top two frames and the extensive dark-matter halos surrounding individual galaxies (the galaxies are roughly the white regions in the bottom two frames).

30–40 *millionths* of a kelvin from place to place in the sky—but they are there. The *COBE* results are displayed as a temperature map of the microwave sky in Figure 27.17(a). The temperature variation due to the Earth's motion (see Figure 26.20) has been subtracted out, as has the radio emission from the Milky Way, and temperature deviations from the average are displayed.

Initially, it seemed that the fluctuations in the microwave background were not consistent with the "standard" dark-matter models that provide the best agreement with actual observations of structure in the present-day universe. The ripples seen by *COBE*, taken in conjunction with the models, appeared to imply *too little* structure on large scales—that is, the computer simulations predicted fewer superclusters, voids, Great Walls, and so on, than are actually seen. However, with some modifications to the details of the models, it now looks as though the disagreement is not as serious as it first seemed, and a growing number of cosmologists are coming to regard the *COBE* observations as confirmation of a central prediction of dark-matter theory. The simulation shown in Figure 27.16 implies temperature fluctuations in the microwave background that agree very well with the present COBE observations, but theo-

rists can (of course) go one stage further. Figure 27.17(b) presents a *prediction*, based on the same simulation, of the temperature fluctuations that should be observed by the next generation of (much higher resolution) satellite experiments. If our current understanding of the microwave background anisotropy is correct, future maps should look qualitatively like this one.

If the *COBE* results hold up—as they are checked and rechecked by collaborators and competitors alike—they may one day come to rank alongside the discovery of the microwave background itself in terms of their importance to the field of cosmology.

27.6 Toward Creation

As we mentioned earlier, our efforts to penetrate the Planck epoch are currently hampered by physicists' ignorance of how to incorporate the force of gravity into GUTs. No one has yet invented a "SuperGUT" that merges gravity with the grand unified force into a single, truly fundamental force at the energies characterizing the earliest part of the Planck epoch, although this is an area of very active research.

Figure 27.17 (a) COBE map of temperature fluctuations in the cosmic microwave background. Hotter than average regions are shown in red, cooler than average regions in blue. The total temperature range shown is ±200 millionths of a kelvin. (b) Simulated map of microwave background temperature fluctuations corresponding to the simulation shown in Figure 27.16. Dark blue and red represent temperature variations of ±200 millionths of a kelvin from the average, so this map can be compared more or less directly with the COBE map in part (a). However, the resolution here is about 0.5 degrees—twenty times sharper than the COBE map, and roughly the resolution expected in the next generation of satellite observations.

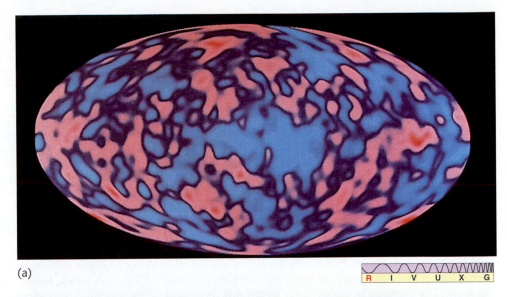

(a)

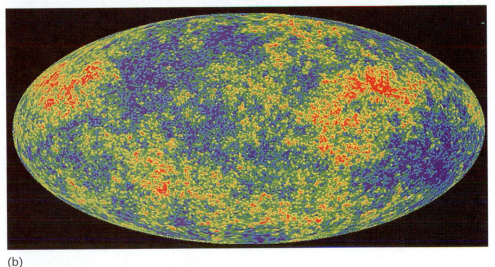

(b)

Even in the absence of a working theory, many researchers have a "gut feeling" that once we have in hand the proper description of quantum gravity, our understanding might automatically include a natural description of creation itself. Incredible though it may seem, it is conceivable that the primal energy originally emerged from literally *nothing*. Even in a perfect vacuum—a region of space containing neither matter nor energy—virtual particle–antiparticle pairs are constantly appearing and disappearing within a time span too short to observe, causing natural *quantum fluctuations* to occur in empty space. We might be living in a sort of "self-creating universe" that erupted into existence spontaneously from just such a random quantum fluctuation! This sort of "statistical" creation of the primal cosmic energy from absolutely nothing has been dubbed "the ultimate free lunch."

Whether or not this scenario proves to be the correct description of the birth of the universe, it suggests a possible explanation for the original density fluctuations that eventually grew into today's large-scale structure. They might have been microscopic quantum fluctuations that existed naturally in the universe during the GUT epoch, then grew to macroscopic size during the inflationary epoch. The fluctuations predicted by this idea agree quite well with the requirements of several cold dark-matter models.

If this view is correct—and some cosmologists believe that it is—then our Galaxy, the Sun, the Earth, even life itself, are direct consequences of a series of random events that occurred during an unimaginably short period of time some 10–15 billion years ago. Of course, these ideas are very speculative. In the strict sense, they are not really science at all, as they violate one of the central tenets of the scientific method: they are practically impossible to test experimentally. Still, whether or not you find them philosophically acceptable or intellectually pleasing, they do illustrate just how far the scope of physics and astronomy has expanded in the twentieth century. Astronomy is a subject that links the very big and the very small. Nowhere is that more evident than in our study of the most important question of all—the origin of the universe.

Chapter Review

SUMMARY

At the present time, the density of matter in the universe greatly exceeds the equivalent mass density of radiation. The universe is **matter-dominated** (p. 588). The density of matter was much greater in the past, when the universe was smaller. However, because radiation is redshifted as the universe expands, the density of radiation was greater still. The early universe was **radiation dominated** (p. 589).

During the first few minutes after the Big Bang, matter was formed out of the primordial fireball by the process of **pair production** (p. 589). In the early universe, matter and radiation were linked by this process. Particles "froze out" of the radiation background as the temperature fell below the threshold for creating them. The existence of matter today means that there must have been unequal amounts of matter and antimatter early on.

The physical state of the universe can be understood in terms of present-day physics back to about 10^{-43} s after the Big Bang. Before that, the four fundamental forces of nature—gravity, electromagnetism, the strong force, and the weak force—were all indistinguishable. There is presently no theory that can describe these extreme conditions. As the universe expanded and its temperature dropped, the forces became distinct from one another. First gravity, then the strong force, and then the weak and electromagnetic forces separated out.

Only a little of the helium observed in the universe today was formed in stars. Most of it was created by **primordial nucleosynthesis** (p. 594) in the early universe. Some deuterium was also formed at these early times, and it provides a sensitive indicator of the present density of the universe in the form of "normal" (as opposed to nonbaryonic dark) matter. Studies of deuterium indicate that normal matter can account for at most 10 percent of the critical density. The remaining mass inferred from studies of clusters must then be made of dark matter, in the form of unknown particles formed at some very early epoch.

When the universe was about 1500 times smaller than it is today the temperature became low enough for atoms to form. At that time, the (then optical) radiation background **decoupled** (p. 596) from the matter. The photons that now make up the microwave background have been traveling freely through space ever since.

According to modern **Grand Unified Theories** (p. 593), when the three nongravitational forces of nature began to display their separate characters, about 10^{-35} s after the Big Bang, a brief period of rapid cosmic expansion called the epoch of **inflation** (p. 599) occurred, during which the size of the universe increased by a factor of about 10^{50}.

Cosmologists wonder why regions of the universe that have not had time to "communicate" with one another look so similar. This is called the **horizon problem** (p. 597). Inflation solves it by taking a small homogeneous patch of the early universe and expanding it enormously in size. The patch is still homogeneous, but it is now much larger than the portion of the universe we can see today.

Cosmologists also wonder why the density of the universe seems to be so near the critical value. This is called the **flatness problem** (p. 597). Inflation implies that the cosmic density is in fact exactly critical. If this is the case, then 90 percent of the matter in the universe is dark.

The large-scale structure observed in the universe could not have formed out of density fluctuations in normal gaseous matter—there simply has not been enough time, given the twin constraints of the smoothness of the microwave background and the epoch at which the first galaxies and quasars are known to have formed. Instead, dark matter clumped and grew to form the "skeleton" of the structure now observed. Normal matter then flowed into the densest regions of space, eventually forming the galaxies we now see.

Cosmologists distinguish between **hot dark matter** (p. 602) and **cold dark matter** (p. 602), depending on its temperature at the end of the Radiation Era. In order to explain the observed large-scale structure in the universe, much of the dark matter must be cold.

In 1992, the *COBE* satellite discovered the expected ripples in the cosmic microwave background associated with the clumping of dark matter at early times.

SELF-TEST: True or False?

_____ 1. The light emitted from all the stars in the universe now far outshines the cosmic microwave background.

_____ 2. The time between the Big Bang and 10^{-43} seconds cannot be described for lack of a theory of quantum gravity.

_____ 3. All elementary particle formation was completed by the time the universe was about 1 second old.

_____ 4. About 25 percent by mass of matter in the universe is primordial helium.

_____ 5. The present-day abundance of deuterium gives information on the density of baryonic matter.

_____ 6. Decoupling refers to interactions between matter and antimatter.

_____ 7. The present-day microwave background radiation originated just before the period of decoupling.

_____ 8. After decoupling, neutral atoms could finally exist.

_____ 9. The horizon problem relates to the isotropy found in the microwave background radiation.

_____ 10. The flatness problem is the fact that the observed density of matter is unexpectedly different from the critical density.

_____ 11. Inflation was caused by the freeze-out of the strong force.

_____ 12. The universe grew in size by a factor of 100 during the inflationary period.

_____ 13. If inflation is correct, then the density of the universe is exactly the critical density.

_____ 14. Physicists have detected cold-dark-matter particles in terrestrial laboratories.

_____ 15. Galaxies had already formed by the time the universe was about 10 million years old.

SELF-TEST: Fill in the Blank

1. Comparing the mass density of radiation and matter, we find that, at the present time, _____ dominates.

2. Energetically speaking, radiation and matter were equally important approximately _____ years after the Big Bang.

3. In the process of pair production, two _____ interact to form a particle and an _____.

4. The temperature necessary to form electrons and positrons is about _____ K.

5. The temperature necessary to form protons and _____ is about _____ K.

6. When the universe was a few minutes old, nuclear fusion produced _____ and _____.

7. After about a million years, _____ had formed in the universe.

8. Elements heavier than helium were not formed primordially because, as time passed, the density and temperature of the universe _____.

9. The fact that the density of baryonic matter is so much less than the known density of matter in the universe implies that much of the dark matter must be _____.

10. The cosmic microwave background radiation last interacted with matter when the universe was _____ years old.

11. If inflation is correct, then dark matter must make up about _____ percent of all matter in the universe.

12. Which had density fluctuations first, matter or dark matter? _____.

13. Hot and cold dark matter differ in the mass of their particles. Cold dark matter consists of _____ particles.

14. Theory predicted tiny fluctuations in the _____ of the microwave background; the _____ satellite found them.

REVIEW AND DISCUSSION

1. For how long was the universe dominated by radiation? How hot was the universe when the dominance of radiation ended?

2. Why is our knowledge of the Planck epoch so limited?

3. When and how did the first atoms form?

4. Describe the universe at the end of the galactic epoch.

5. Why do all stars, regardless of their abundance of heavy elements, seem to contain at least one-quarter helium by mass?

6. Why didn't heavier and heavier elements form in the early universe, as they do in stars?

7. If large amounts of deuterium formed in the early universe, why do we see so little today?

8. How do measurements of the cosmic deuterium abundance provide a reliable estimate of Ω_0?

9. When did the universe become transparent to radiation?

10. What are GUTs?

11. How does cosmic inflation solve the horizon problem?

12. How does cosmic inflation solve the flatness problem?

13. What is the difference between hot and cold dark matter?

14. What is the connection between dark matter and the formation of large- and small-scale structures?

15. What did dark-matter models predict that was later found by the COBE satellite?

16. What additional observations have been presented in this chapter to support the cosmological principle?

PROBLEMS

1. How many times did the universe double in size during the inflationary period if its final size was 10^{50} larger than when it started?

2. Estimate the temperature needed for electron-positron pair production. The mass of an electron is 9.1×10^{-31} kg. Use $E = mc^2$ to find the energy, $E = hf$ (see Chapter 4) to find the frequency f of the photon, and finally Wien's Law to find a temperature. How does your answer compare to the threshold temperature given in the text?

3. Given that the threshold temperature for the production of electron-positron pairs is about 6×10^9 K and that a proton is 1800 times more massive than an electron, what is the threshold temperature for proton-antiproton pairs?

PROJECTS

1. Read the book *The First Three Minutes*, by Steven Weinberg. It is fairly nonmathematical in its presentation. What new results are presented in this chapter that were not known by Weinberg when he wrote the first edition of this book in 1977? How much progress has been made in understanding the very earliest epochs since that time?

2. Write a paper on the philosophical differences between living in an open, closed, or flat universe. It is quite possible that astronomers may finally determine which of these is correct within your lifetime. Does it really matter? Are there aspects of any of these that are uncomfortable to accept? Do you or others have a preference?

PARTICULATE

GALACTIC

STELLAR

PLANETARY

CHEMICAL

BIOLOGICAL

CULTURAL

BILLIONS OF YEARS AGO

12

4.6

3.3

0.6

PRESENT

28

LIFE IN THE UNIVERSE

Are We Alone?

LEARNING GOALS

Studying this chapter will enable you to:

1 Summarize the process of cosmic evolution as it is currently understood.

2 Evaluate the chances of finding life in the solar system.

3 Summarize the various probabilities used to estimate the number of advanced civilizations that might exist in the Galaxy.

4 Discuss some of the techniques we might use to search for extraterrestrials and to communicate with them.

(Opposite page) The arrow of time, from the origin of the universe to the present and beyond, spans several major epochs of history. Cosmic evolution is the study of the many varied changes in the assembly and composition of energy, matter, and life in the thinning and cooling universe.

Inserted along the arrow of time are seven "windows" outlining some of the key events in the history of the universe. As an all-encompassing and intellectually powerful world view, cosmic evolution enables us to trace a thread of understanding, linking:

- the evolution of primal energy into elementary particles,
- the evolution of atoms into galaxies and stars,
- the evolution of stars into heavy elements,
- the evolution of heavy elements into solid, rocky planets,
- the evolution of those elements into the molecular building blocks of life,
- the evolution of those molecules into life itself, and
- the evolution of advanced life forms into intelligence, culture, and a technological civilization.

*A*re we unique? Is life on our planet the only example of life in the universe? These are difficult questions, for the subject of extraterrestrial life is one on which we have no data, but they are important questions, with profound implications for the human species. Earth is the only place in the universe where we know for certain that life exists. In this chapter we take a look at how humans evolved on Earth and then consider whether those evolutionary steps might have happened elsewhere. Having done that, we will assess the likelihood of our having galactic neighbors and consider how we might learn about them if they exist.

28.1 Cosmic Evolution

1 The painting on the first page of this chapter identifies seven major phases in the history of the universe: *particulate, galactic, stellar, planetary, chemical, biological,* and *cultural* evolution. Together, these evolutionary stages make up the grand sweep of **cosmic evolution**—the continuous transformation of matter and energy that has led to the appearance of life and civilization on the Earth. The first four represent, in reverse order, the contents of this book. We now expand our field of view beyond astronomy to include the other three.

Figure 28.1 identifies some key events in the history of the universe and places them in a temporal context. From the Big Bang, to the formation of galaxies, to the birth of the solar system, to the emergence of life, to the evolution of intelligence and culture, the universe has evolved from simplicity to complexity. We are the result of an incredibly complex chain of events that spanned billions of years. Were those events random, making us unique, or are they in some sense *natural*, so that technological civilization is inevitable? Put another way, are we alone in the universe, or are we just one among countless other intelligent life forms in our Galaxy? In this chapter we will consider the development of life on Earth and try to assess the likelihood of finding intelligent life elsewhere in the cosmos.

LIFE IN THE UNIVERSE

Before embarking on our study, we need a working definition of *life*. This seemingly simple task is not an easy one— the distinction between the living and the nonliving is not as obvious as we might at first think. Though most physicists would agree on the definitions of matter and energy, biologists have not arrived at a clear-cut definition of life. Generally speaking, scientists regard the following as characteristics of living organisms: (1) they can *react* to their environment and can often heal themselves when damaged; (2) they can *grow* by taking in nourishment from their surroundings and processing it into energy; (3) they can *reproduce*, passing along some of their own characteristics to their offspring; and (4) they have the capacity for genetic change, and can therefore *evolve* from generation to generation so as to adapt to a changing environment.

These rules are not hard and fast, and there is great leeway in interpreting them. Stars, for example, react to the gravity of their neighbors, grow by accretion, generate energy, and "reproduce" by triggering the formation of new stars, but no one would suggest that they are alive. A virus

(see *Interlude 28-1* on p. 613) is crystalline and inert when isolated from living organisms, but, once inside a living system, it exhibits all the properties of life, seizing control of a living cell and using the cell's own genetic machinery to grow and reproduce. Most researchers now believe that the distinction between living and nonliving is more one of structure and complexity than a simple "checklist" of rules.

The general case in favor of extraterrestrial life is summed up in the so-called *assumptions of mediocrity*: (1) Because life on Earth depends on just a few basic molecules, and (2) because the elements that make up these molecules are (to a greater or lesser extent) common to all stars, and (3) if the laws of science we know apply to the entire universe, as we have supposed throughout this book, then—given sufficient time—life must have originated elsewhere in the cosmos. The opposing view maintains that intelligent life on Earth is the product of a series of extremely fortunate accidents—astronomical, geological, chemical, and biological events unlikely to have occurred anywhere else in the universe. The purpose of this chapter is to examine some of the arguments for and against these viewpoints.

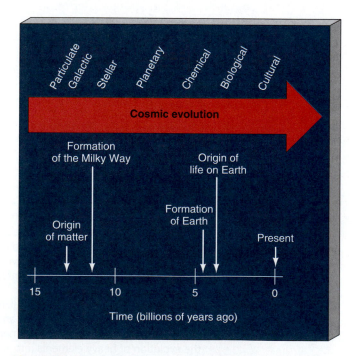

Figure 28.1 Some highlights of cosmic history are noted along this arrow of time, from the beginning of the universe to the present. Noted along the top of the arrow are the major phases of cosmic evolution.

CHEMICAL EVOLUTION

What information do we have about the earliest stages of planet Earth? Unfortunately, not very much. Geological hints about the first billion years or so were largely erased by violent surface activity as volcanoes erupted and meteorites bombarded our planet; subsequent erosion by wind and water has seen to it that little evidence has survived to the present day. Scientists believe that the early Earth was barren, with shallow, lifeless seas washing upon grassless, treeless continents. Outgassing from our planet's interior through volcanoes, fissures, and geysers produced an atmosphere rich in hydrogen, nitrogen, and carbon compounds and poor in free oxygen. As the Earth cooled, ammonia, methane, carbon dioxide, and water formed. The stage was set for the appearance of life.

The surface of the young Earth was a very violent place. Natural radioactivity, lightning, volcanism, solar ultraviolet radiation and meteoritic impacts all provided large amounts of energy that eventually shaped the ammonia, methane, carbon dioxide and water into more complex molecules known as **amino acids** and **nucleotide bases**—organic (carbon-based) molecules that are the building blocks of life as we know it. Amino acids build *proteins*, and proteins control metabolism, the daily utilization of food and energy by means of which organisms stay alive and carry out their vital activities. Sequences of nucleotide bases form *genes*—parts of the DNA molecule—which direct the synthesis of proteins, and thus determine the characteristics of the organism. These same genes also carry the organism's hereditary characteristics from one generation to the next in reproduction. In all living creatures on the Earth—from bacteria to amoebas to humans—genes mastermind life, while proteins maintain it.

The idea that complex molecules could have evolved naturally from simpler ingredients found on the primitive Earth has been around since the 1920s. The first experimental verification was provided in 1953 when scientists Harold Urey and Stanley Miller, using laboratory equipment somewhat similar to that shown in Figure 28.2, took a mixture of the materials thought to be present on Earth long ago—a "primordial soup" of water, methane, carbon dioxide, and ammonia—and energized it by passing an electrical discharge ("lightning") through the gas. After a few days, they analyzed their mixture and found that it contained many of the same amino acids found today in all living things on Earth. About a decade later, scientists succeeded in constructing nucleotide bases in a similar manner. These experiments have been repeated in many different forms, with more realistic mixtures of gases and a variety of energy sources, but always with the same basic outcomes.

While none of these experiments has ever produced a living organism, or even a single strand of DNA, they do demonstrate conclusively that "biological" molecules can be synthesized by strictly *non*biological means, using raw materials available on the early Earth. More advanced ex-

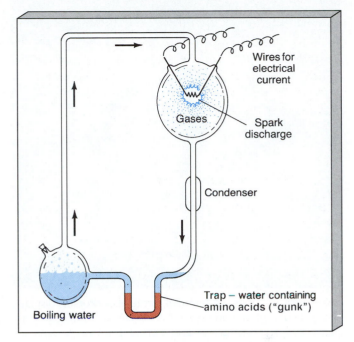

Figure 28.2 This chemical apparatus is designed to synthesize complex biochemical molecules by energizing a mixture of simple chemicals. In the diagrammed procedure, the gases are placed in the upper bulb to simulate the primordial Earth atmosphere and then energized by spark-discharge electrodes. After about a week of recycling the gases through the lightninglike electrical discharge, amino acids and other complex molecules can be found in the trap at the bottom, which simulates the primordial oceans into which heavy molecules would have fallen from the atmosphere.

periments, in which amino acids are united under the influence of heat, have fashioned proteinlike blobs that behave to some extent like true biological cells. Such near-protein material resists dissolution in water (so it would remain intact when it fell from the primitive atmosphere into the ocean) and tends to cluster into small droplets called microspheres—a little like oil globules floating on the surface of water. Figure 28.3 shows some of these proteinlike microspheres. The walls of these laboratory-made droplets permit the inward passage of small molecules, which then combine within the droplet to construct more complex molecules too large to pass back out through the walls. As the droplets "grow," they tend to "reproduce," forming smaller droplets.

Can we consider these proteinlike microspheres to be alive? Almost certainly not. Most biochemists would say that the microspheres are not life itself, but contain many of the basic ingredients needed to form life. The microspheres lack the hereditary molecule DNA. However, as illustrated in Figure 28.4, they do have similarities to ancient cells found in the fossil record. Thus, while no actual living cells have yet been created "from scratch" in any laboratory, many biochemists feel that the chain of events leading from simple nonbiological molecules almost to the point of life itself has been amply demonstrated.

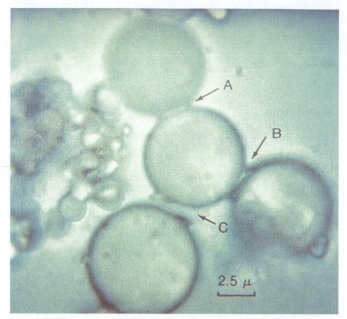

Figure 28.3 These carbon-rich, proteinlike droplets display the clustering of as many as a billion amino acid molecules in a liquid. Droplets can "grow," and parts of droplets can separate from the "parent" to become new individual droplets. The scale of 2.5 microns noted here is 1/4000 of a centimeter.

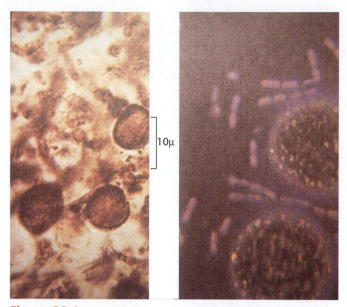

Figure 28.4 The photograph on the left, taken through a microscope, shows a fossilized organism found in sediments radioactively dated as 2 billion years old. This primitive system possesses concentric spheres or walls connected by smaller spheroids. The roundish fossils here measure about a thousandth of a centimeter. The photograph on the right, also taken through a microscope and on approximately the same scale, displays modern blue-green algae.

Recently, however, a dissenting view has emerged. Some scientists have argued that Earth's primitive atmosphere may *not* in fact have been a particularly suitable environment for the production of complex molecules. Instead, they say, there might not have been sufficient energy available to power the chemical reactions, and the early atmosphere may not have contained enough raw material for the reactions to have become important in any case. They suggest that much, if not all, of the organic material that combined to form the first living cells was produced in *interstellar space* and subsequently arrived on Earth in the form of interplanetary dust and meteors that did not burn up during their descent through the atmosphere. Interstellar molecular clouds are known to contain very complex molecules, and large amounts of organic material were detected on comet Halley by space probes when Halley last visited the inner solar system (see Section 14.2), so the idea that organic matter is constantly raining down on Earth from space in the form of interplanetary debris is quite plausible. Whether or not this was the *primary* means by which complex molecules first appeared in the Earth's oceans remains unclear. For now, the issue is unresolved.

DIVERSITY AND CULTURE

However the basic materials appeared on Earth, we know that life *did* appear. The fossil record chronicles how life on Earth became widespread and diversified over the course of time. The study of fossil remains shows the initial appearance about 3.5 billion years ago of simple one-celled organisms such as blue-green algae. These were followed about 2 billion years ago by more complex one-celled creatures, like the amoeba. Multicellular organisms such as

sponges did not appear until about 1 billion years ago, after which there flourished a wide variety of increasingly complex organisms—insects, reptiles, mammals, and humans.

The fossil record leaves no doubt that biological organisms have changed over time—all scientists accept the reality of *biological evolution*. As conditions on Earth have shifted and Earth's surface itself has evolved, those organisms that could best take advantage of their new surroundings succeeded and thrived—often at the expense of those organisms that could not make the necessary adjustments, which became extinct. What led to these changes? Chance. An organism that happened to have a certain useful genetically determined trait—for example, the ability to run faster, climb higher, or even hide more easily—would find itself with the upper hand in a particular environment. This organism was, therefore, more likely to reproduce successfully, and its advantageous characteristic would then be more likely to be passed on to the next generation. The evolution of the rich variety of life on our planet, including human beings, occurred as chance mutations—changes in genetic structure—led to changes in organisms over millions of years.

What about the development of intelligence? Many anthropologists believe that, like any other highly advantageous trait, intelligence is strongly favored by natural selection. As humans learned about fire, tools, and agriculture, the brain became more and more elaborate. The social cooperation that went with coordinated hunting efforts was another important competitive advantage that developed as brain size increased. Perhaps most important of all was the development of language. Indeed, some anthropologists have gone so

Interlude 28–1 *The Virus*

The central idea of chemical evolution is that life evolved from nonlife. But aside from insight based on biochemical knowledge and laboratory simulations of some key events on primordial Earth, do we have any direct evidence that life could have developed from nonliving molecules? The answer is "Yes." The smallest and simplest entity that sometimes appears to be alive is a virus. We say "sometimes" because viruses seem to have the attributes of both nonliving molecules and living cells. *Virus* is the Latin word for "poison," an appropriate name since viruses are often a cause of disease. Although they come in many sizes and shapes—a typical example is diagrammed at right—all viruses are smaller than the size of a typical modern cell. Some are made of only a few thousand atoms. In terms of size, then, viruses seem to bridge the gap between cells that are living and molecules that are not.

Viruses contain some proteins and genetic information (in the form of DNA or the closely related molecule RNA) but not much else—none of the material by which living organisms normally grow and reproduce. How, then, can a virus be considered alive? When alone, it cannot; a virus is absolutely lifeless when isolated from living organisms. But when inside a living system, a virus has all the properties of life. Viruses come alive by transferring their genetic material into living cells. The genes of a virus seize control of a cell and establish themselves as the new master of chemical activity. Viruses grow and reproduce copies of themselves by using the genetic machinery of the invaded cell, often robbing the cell of its usual function. Some viruses multiply rapidly and wildly, spreading the disease and—if unchecked—eventually killing the invaded organism. In a sense, viruses exist within the gray area between the living and the nonliving.

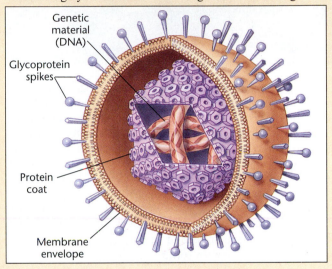

Genetic material (DNA)

Glycoprotein spikes

Protein coat

Membrane envelope

---

far as to suggest that human intelligence *is* human language. By communicating, individuals could signal each other while hunting food or seeking protection. Now our ancestors could share ideas as well as food and shelter. Experience, stored in the brain as memory, could be passed down from generation to generation. A new kind of evolution had begun, namely, *cultural evolution,* the changes in the ideas and behavior of society. Our more recent ancestors have created, within only the past 10,000 years or so, the entirety of human civilization.

To put all this into historical perspective, let's imagine the entire lifetime of Earth to be 45 years rather than 4.5 billion years. We have no reliable record of the first decade of our planet's existence. Life originated at least 35 years ago, when Earth was only about 10 years old. Our planet's middle age is largely a mystery, although we can be sure that life continued to evolve and that generations of mountain chains and oceanic trenches came and went. Not until about 6 years ago did abundant life flourish throughout Earth's oceans. Life came ashore about 4 years ago, and plants and animals mastered the land only about 2 years ago. Dinosaurs reached their peak about 1 year ago, only to die suddenly about 8 months ago (see *Interlude 14–1* on p. 300). Humanlike apes changed into apelike humans only last week, and the latest ice ages occurred only a few days ago. *Homo sapiens*—our species—did not emerge until about 4 hours ago. Agriculture was invented within the last hour, and the Renaissance—along with all of modern science—is only 3 minutes old!

28.2 *Life in the Solar System*

② Simple one-celled life forms reigned supreme on Earth for most of our planet's history. It took time—a great deal of time—for life to emerge from the oceans, to evolve into simple plants, to continue to evolve into complex animals, and to develop intelligence, culture, and technology. Have those (or similar) events occurred elsewhere in the universe? Let's try to assess what little evidence we have on the subject.

LIFE AS WE KNOW IT

"Life as we know it" is generally taken to mean carbon-based life that originated in a liquid water environment—in other words, life on Earth. Is there any reason to suppose that such life might exist elsewhere in our solar system? The answer appears to be "No." It seems that no environment in the solar system besides the Earth is particularly well suited for sustaining Earth-like life.

The Moon and Mercury lack liquid water, protective atmospheres, and magnetic fields, and so they are subjected to fierce bombardment by solar ultraviolet radiation, the solar wind, meteoroids, and cosmic rays. Simple molecules could not possibly survive in such hostile environments. The planet Venus has far too much protective atmosphere!

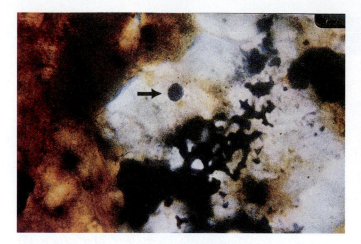

Figure 28.5 The Murchison meteorite contains relatively large amounts of organic material, indicating that chemical evolution of some sort has occurred beyond our own planet. In this magnified view of a meteorite fragment, the arrow points to a microscopic sphere of organic matter.

Figure 28.6 A trench dug by the "arm" of one of the *Viking* robots can be seen at the left. Soil samples were scooped up and taken inside the robot, where instruments tested them for chemical composition and any signs of life.

Its dense, dry, scorchingly hot atmospheric blanket effectively rules it out as a possible abode for life. The jovian planets have no solid surfaces, while Pluto and most of the moons of the outer planets are too cold. Saturn's moon Titan, with its atmosphere of methane, ammonia, and nitrogen and possibly with some liquid on its surface, is conceivably a site for life. The results of the 1980 *Voyager 1* flyby, though, suggest that its frigid surface conditions are rather inhospitable for anything familiar to us.

What about the cometary and meteoritic debris that orbits within our solar system? Comets contain many of the basic ingredients for life—for instance, ammonia, methane, and water vapor—and although comets are frozen, their icy matter warms while nearing the Sun. Indeed, some heavy molecules have been observed in comet spectra. In addition, a small fraction of the meteorites that survive the plunge to Earth's surface do contain organic compounds. The Murchison meteorite (shown in Figure 28.5), which fell near Murchison, Australia, in 1969, is a well-studied example. Located soon after crashing to the ground, this meteorite contains many of the well-known amino acids normally found in living cells. The moderately large molecules found in meteorites and in interstellar clouds are our only evidence that chemical evolution has occurred elsewhere in the universe. Most researchers regard this organic matter as prebiotic—that is, matter that could eventually lead to life but that has not yet done so.

The planet most likely to harbor life (or to have harbored it in the past) seems to be Mars. This planet does seem harsh by Earth standards—liquid water is scarce, the atmosphere is thin, and the lack of magnetism and an ozone layer allows the solar high-energy particles and ultraviolet radiation to reach the surface unabated. But the Martian atmosphere was thicker, and the surface warmer and much wetter in the past (see Section 10.4). In the hope that life might once have evolved on Mars as it did on Earth, each *Viking* lander carried a television camera to seek fossilized remnants of large plants or animals. No fossils of any kind were seen. The landers also scooped up Martian soil (see Figure 28.6) and tested for life by conducting chemical experiments designed to detect the waste gases and other products of metabolic activity (see *Interlude 10-2* on p. 228), but no unambiguous evidence for Martian life has emerged.

The consensus among biologists and chemists today is that Mars does not house any life similar to that on Earth. However, some scientists think that a different type of biology might be operating on the Martian surface. They suggest that Martian microbes capable of eating and digesting oxygen-rich compounds in the Martian soil could also explain the *Viking* results. In addition, microbial life might reside in more habitable regions on Mars, such as near the moist polar caps. After all, the two *Viking* spacecraft landed on the safest Martian terrain, not in the most interesting regions. It seems that a solid verdict regarding life on Mars will not be reached until we have thoroughly explored our intriguing neighbor.

ALTERNATIVE BIOCHEMISTRIES

It thus appears that no environment in the solar system besides Earth is particularly well suited for sustaining life. Some scientists have suggested that different types of biology may be at work out there, ones that we cannot recognize and that we do not know how to test for. What other biologies might exist?

Some scientists have pointed out that the abundant element silicon has chemical properties somewhat similar to those of carbon, and have suggested it as a possible alternative to carbon as the basis for living organisms. Ammonia (made of the common elements hydrogen and nitrogen) is sometimes put forward as a possible liquid medium in which life might develop, at least on a planet cold enough for ammonia to exist in the liquid state. Together or separately, these alternatives would surely give rise to organisms with radically different biochemistries from those we know on Earth. Conceivably, we might have difficulty even recognizing them as alive.

While the possibility of such alien life forms is a fascinating scientific problem, most biologists would argue that chemistry based on carbon and water is the one most likely to give rise to life. Carbon's flexible chemistry and water's wide liquid range are just what are needed for life to develop and thrive. Silicon and ammonia seem unlikely to fare as well as the bases for advanced life forms. Silicon's chemical bonds are weaker than those of carbon and may not be able to form complex molecules—an apparently essential aspect of carbon-based life. Also, the colder the environment, the less energy there is to drive biological processes. The low temperatures necessary for ammonia to be liquid might inhibit or even prevent completely the chemical reactions leading to the equivalent of amino acids and nucleotide bases. Still, we must admit that we know next to nothing about noncarbon, nonwater biochemistries, for the very good reason that there are no examples of them to study experimentally. We can speculate about alien life forms and try to make general statements about their properties, but we can say little of substance about them.

28.3 Intelligent Life in the Galaxy

③ Thus we look for life beyond our solar system, in the Milky Way Galaxy and in other galaxies. At such distances, though, we have little hope of detecting actual life with current equipment. Instead, we must ask: "How likely is it that life in any form—carbon-based, silicon-based, water-based, ammonia-based, or something we cannot even dream of—exists?" The word *likely* in the last sentence speaks of probabilities. Let's look at some numbers to develop statistical estimates of the probability of life elsewhere in the universe.

THE DRAKE EQUATION

An early approach to this statistical problem is usually known as the **Drake equation**, after the U.S. astronomer who pioneered this analysis:

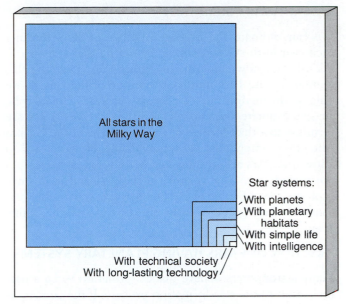

Figure 28.7   Of all the star systems in our Milky Way (represented by the largest box), progressively fewer and fewer have each of the qualities typical of a long-lasting technological society (represented by the smallest box at the lower right corner).

Realize that several of the terms in this formula are largely a matter of opinion. We do not have nearly enough information to determine—even approximately—all of the terms in the equation, so the Drake equation cannot give us a hard-and-fast answer. Its real value is that it subdivides a large and very difficult question into smaller pieces that we can attempt to answer separately. It provides the framework within which the problem can be addressed and parcels out the responsibility for the final solution among many different scientific disciplines. Figure 28.7 illustrates how, as our requirements become more and more stringent, only a small fraction of star systems in the Milky Way are likely to generate the advanced qualities specified by the combination of terms on the right-hand side of the equation.

Let's now examine the terms in the equation one by one, and make some educated guesses about their values. Bear in mind, though, that if you ask two scientists for their best estimates of any given term, you will likely get two very different answers!

| number of technological, intelligent civilizations now present in the Milky Way Galaxy | = | rate of star formation, averaged over the lifetime of the Galaxy | × | fraction of those stars having planetary systems | × | average number of planets within those planetary systems that are suitable for life | × | fraction of those habitable planets on which life actually arises | × | fraction of those life-bearing planets on which intelligence evolves | × | fraction of those intelligent-life planets that develop technological society | × | average lifetime of a techologically competent civilization. |

RATE OF STAR FORMATION

We can estimate the average number of stars forming each year in the Galaxy simply by noting that at least 100 billion stars now shine in the Milky Way. Dividing this number by the 10-billion-year lifetime of the Galaxy, we obtain a formation rate of 10 stars per year. This may be an overestimate because we think that fewer stars are forming now than formed at earlier epochs of the Galaxy, when more interstellar gas was available. However, we do know that stars are forming today, and our estimate does not include stars that formed in the past and have since exploded, so our value of 10 stars per year is probably reasonable when averaged over the lifetime of the Milky Way.

FRACTION OF STARS HAVING PLANETARY SYSTEMS

Many astronomers believe planet formation to be a natural result of the star-formation process. If the condensation theory (see Chapter 15) is correct, and if there is nothing special about our Sun, as we have argued throughout this book, we would expect many stars to have at least one planet. ∞ (p. 321) Indeed, as we saw, increasingly sophisticated observations suggest the presence of disks around other young stars. Could these disks be protosolar systems? The condensation theory suggests they are.

Is there any direct evidence for planets in orbit around other stars? Certainly, we find no indication of any planet like our own.* The light reflected by an Earth-like planet circling even the closest star would be too faint to detect even with the very best equipment. The light would be lost in the glare of the parent star. Large orbiting telescopes may soon be able to detect Jupiter-sized planets orbiting the nearest stars, but even those huge planets would be barely visible.

One possible indirect way of detecting a distant planet relies on the gravitational influence that all planets exert on their parent stars. A large planet orbiting a relatively low-mass star might produce a detectable change in that star's motion, even though the planet itself may be invisible to us. Figure 28.8 illustrates how this might look from Earth. The invisible planet pulls the star first one way and then the other during the course of its orbit. The result is a slight back-and-forth wobble in the visible star's path. Attempts to detect this wobble in Barnard's Star and other low-mass neighbors of the Sun have proved extremely frustrating, however. The amount of the back-and-forth shift is very small—it would be only a few hundredths of an arc second for a Jupiter-sized object orbiting Barnard's Star at a distance of a few A.U.—and no clear-cut detection of the expected motion has yet been made for this or any other star. Still, astronomers are confident that with

*Recall from Chapter 22 that planets have also been observed orbiting some neutron stars. However, as we saw in that chapter, those planets formed in a manner very different from Earth and the solar system. ∞ (p. 469) They are not thought to be likely candidates for the emergence of life.

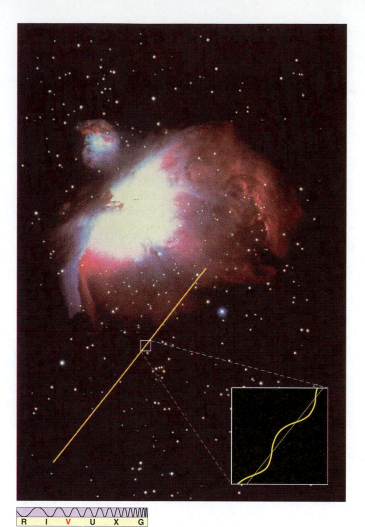

R I V U X G

Figure 28.8 Theoretically computed wobble in the path of a nearby, low-mass star having a Jupiter-sized object orbiting about it. The straight line is the path the star would take in the absence of any planets. The wavy curve (inset), shown here somewhat exaggerated for clarity, is the resulting wobble in the star's motion caused by the to-and-fro gravitational pull of an unseen planet. Even for a dwarf star some 3 pc away, the expected deviation from a straight line would be a tiny 0.005 arc second.

steadily improving resolution they will one day detect a planet by this means.

Accepting the condensation theory and its consequences, and without being either too conservative or naively optimistic, we assign a value near 1 to this term—that is, we believe that essentially all stars have planetary systems.

NUMBER OF HABITABLE PLANETS PER PLANETARY SYSTEM

Temperature, more than any other single quantity, determines the feasibility of life on a given planet. The surface temperature of a planet depends on two things: the planet's distance from its parent star and the thickness of its atmosphere. Planets with a nearby parent star (but not too close) and some atmosphere (though not too thick) should be reasonably warm, like Earth or Mars. Planets far from the star and with no atmosphere, like Pluto, will surely be cold by our standards. And planets too close to the star and with a thick atmosphere, like Venus, will be very hot indeed.

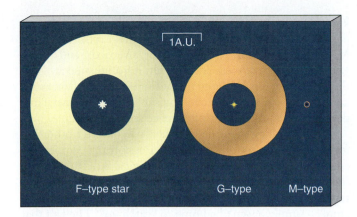

Figure 28.9 The extent of the habitable zone is much larger around a hot star than around a cool one. For a star like the Sun (a G-type star), the zone extends from about 0.85 A.U. to 2.0 A.U. For an F-type star, the range is 1.2 to 2.8 A.U. For a faint M-type star only planets orbiting between about 0.02 and 0.06 A.U. would be habitable.

Figure 28.9 illustrates that a three-dimensional zone of "comfortable" temperatures—often called a *habitable zone*—surrounds every star. It represents the range of distances within which a planet of mass and composition similar to those of the Earth would have a surface temperature between the freezing and boiling points of water. (Our Earth-based bias is plainly evident here!) The hotter the star, the larger this zone. For example, an A- or an F-type star has a rather large habitable zone. G-, K-, and M-type stars have increasingly smaller zones. O- and B-type stars are not considered here because they are not expected to last long enough for life to develop, even if they do have planets.

Three planets—Venus, Earth, and Mars—reside within the habitable zone surrounding our Sun. Venus is too hot because of its thick atmosphere and proximity to the Sun. Mars is a little too cold because its atmosphere is too thin and it is too far from the Sun. But if Venus had Mars's thin atmosphere and if Mars had Venus's thick atmosphere, both of these nearby planets might conceivably have surface conditions resembling those on the Earth.

To estimate the number of habitable planets per planetary system, we first take inventory of how many stars of each type shine in our Galaxy and calculate the sizes of their habitable zones. Then we eliminate binary-star systems because a planet's orbit within the habitable zone of a binary would likely be unstable, as illustrated in Figure 28.10. Given the known properties of binaries in our Galaxy, habitable planetary orbits would probably be unstable in most cases, so there would not be time for life to develop.

Taking all these factors into account, we assign a value of 1/10 to this term in our equation. In other words we believe that, on average, there is 1 potentially habitable planet for every 10 planetary systems that might exist in our Galaxy. Single F-, G-, and K-type stars are the best candidates.

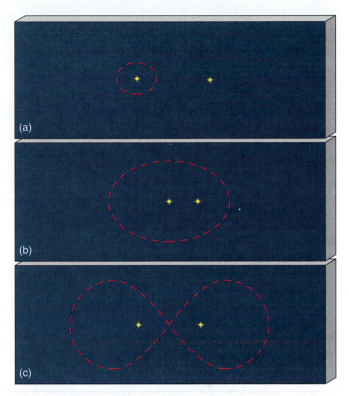

Figure 28.10 In binary-star systems, planets are restricted to only a few kinds of orbits that are gravitationally stable. The orbit in case (a) is stable only if the planet lies very close to its "parent" star, so the gravity of the other star is negligible. Case (b) shows a planet circulating at a great distance about both stars in an elliptical orbit. This orbit is stable only if it lies far from both stars. Another possible, but unstable path, (c) interweaves between the two stars in a "figure-8" pattern.

FRACTION OF HABITABLE PLANETS ON WHICH LIFE ARISES

The number of possible combinations of atoms is incredibly large. If the chemical reactions that led to the complex molecules that make up living organisms occurred completely at random, then it is extremely unlikely that those molecules could have formed at all. In that case, life is extraordinarily rare, this term is close to zero, and we are probably alone in the Galaxy, perhaps even in the entire universe.

However, laboratory experiments (like the Urey–Miller experiment described earlier) seem to suggest that certain chemical combinations are strongly favored over others—that is, the reactions are not random. Of the billions upon billions of basic organic groupings that could possibly occur on Earth from the random combination of all sorts of simple atoms and molecules, only about 1500 actually do occur. Furthermore, these 1500 organic groups of terrestrial biology are made from only about 50 simple "building blocks" (including the amino acids and nucleotide bases mentioned earlier). This suggests that molecules critical to life may not be assembled by pure chance. Apparently, additional factors are at work at the microscopic level. If a relatively small number of chemical "evolutionary tracks" is likely to exist, then the formation of complex molecules—and hence, we assume, life—becomes much more likely, given sufficient time.

To assign a very low value to this term in the equation is to believe that life arises randomly and rarely. To assign a value close to 1 is to believe that life is inevitable, given the proper ingredients, a suitable environment, and a long enough period of time. No easy experiment can distinguish between these extreme alternatives, and there is little or no middle ground. To many researchers, the discovery of life on Mars, Jupiter, Titan, or some other object in our solar system would convert the appearance of life from an unlikely miracle to a virtual certainty throughout the Galaxy. We will take the optimistic view and adopt a value of 1.

FRACTION OF LIFE-BEARING PLANETS ON WHICH INTELLIGENCE ARISES

As with the evolution of life, the appearance of a well-developed brain is a very unlikely event if only random chance is involved. However, biological evolution through natural selection is a mechanism that generates apparently highly improbable results by singling out and refining useful characteristics. Organisms that profitably use adaptations can develop more complex behavior, and complex behavior provides organisms with the *variety* of choices needed for more advanced development.

One school of thought maintains that, given enough time, intelligence is inevitable. In this view, assuming that natural selection is a universal phenomenon, at least one organism on a planet will always rise to the level of "intelligent life." If this is correct, then the fifth term in the Drake equation equals or nearly equals 1. Others argue that there is only one known case of intelligence, and that case is life on Earth. For 2.5 billion years—from the start of life about 3.5 billion years ago to the first appearance of multicellular organisms about 1 billion years ago—life did not advance beyond the one-celled stage. Life remained simple and dumb, but it survived. If this latter view is correct, then the fifth term in our equation is very small, and we are faced with the depressing prospect that humans may be the smartest form of life anywhere in the Galaxy. As with the previous term, we will be optimistic and simply adopt a value of 1 here.

FRACTION OF PLANETS ON WHICH INTELLIGENT LIFE DEVELOPS AND USES TECHNOLOGY

To evaluate the sixth term of our equation, we need to estimate the probability that intelligent life eventually develops technological competence. Should the rise of technology be inevitable, this term is close to 1, given long enough periods of time. If it is not inevitable—if intelligent life can somehow "avoid" developing technology—then this term could be much less than 1. The latter possibility envisions a universe possibly teeming with intelligent civilizations, but very few among them ever becoming technologically competent. Perhaps only one managed it—ours.

Again, it is difficult to decide conclusively between these two views. We don't know how many prehistoric Earth cultures failed to develop technology, or rejected its use. We do know that the roots of our present civilization arose independently at several different places on the Earth, including Mesopotamia, India, China, Egypt, Mexico, and Peru. Because so many of these ancient cultures originated at about the same time, it is tempting to conclude that the chances are good that some sort of technological society will inevitably develop, given some basic intelligence and enough time.

If technology is inevitable, then why haven't other life forms on Earth also found it useful? Possibly the competitive edge given by intellectual and technological skills to humans, the first species to develop them, allowed us to dominate so rapidly that other species—gorillas and chimpanzees, for example—simply haven't had time to "catch up". The fact that only one technological society exists on Earth does not imply that the sixth term in our Drake equation must be very much less than 1. On the contrary, it is precisely because *some* species will probably always fill the niche of technological intelligence that we will take this term to be close to 1.

AVERAGE LIFETIME OF A TECHNOLOGICAL CIVILIZATION

The reliability of the estimate of each term in the Drake equation declines markedly from left to right. For example, our knowledge of astronomy enables us to make a reasonably good stab at the first term, namely, the rate of star formation in our Galaxy, but it is much harder to evaluate some of the later terms, such as the fraction of life-bearing planets that eventually develop intelligence. The last term on the right-hand side of the equation, the longevity of technological civilizations, is totally unknown. There is only one known example of such a civilization—humans on planet Earth. Our own civilization has presently survived in its "technological" state for only about 100 years, and how long we will be around before a natural or human-made catastrophe ends it all is impossible to tell.

One thing is certain: If the correct value for *any one term* in the equation is very small, then few technological civilizations now exist in the Galaxy. If the pessimistic view of the development of life or of intelligence is correct, then we are unique, and that is the end of our story. However, if both life and intelligence are inevitable consequences of chemical and biological evolution, as many scientists believe, and if intelligent life always becomes technological, then we can plug the higher, more optimistic values into the Drake equation. In that case, combining our estimates for the other six terms (and noting that $10 \times 1 \times 1/10 \times 1 \times 1 \times 1 = 1$), we can say

the number of technological, intelligent civilizations now present in the Milky Way Galaxy	=	the average lifetime of a techologically competent civilization, *in years*.

Thus, if civilizations typically survive for 1000 years, there should be 1000 of them currently in existence scattered throughout the Galaxy. If they live for a million years, on average, we would expect there to be a million advanced civilizations in the Milky Way, and so on.

28.4 *The Search for Extraterrestrial Intelligence*

④ Let us continue our optimistic assessment of the prospects for life and assume that civilizations enjoy a long stay on their parent planet once their initial technological "teething problems" are past. In that case, they are likely to be plentiful in the Galaxy. How might we become aware of their existence?

MEETING OUR NEIGHBORS

For concreteness, let us assume that the average lifetime of a technological civilization is 1 million years—only 1 percent of the reign of the dinosaurs, but 100 times longer than human civilization has survived thus far. Given the size and shape of our Galaxy, we can then estimate the average *distance* between these civilizations to be some 50 pc, or about 150 light years. Thus, any two-way communication with our neighbors—using signals traveling at or below the speed of light—will take at least 300 years (150 years for the message to reach the planet and another 150 years for the reply to travel back to us).

One obvious way to search for extraterrestrial life would be to develop the capability to travel far outside our solar system. However, this may never be a practical possibility. At a speed of 50 km/s, the speed of the current fastest unmanned space probes, the trip to our nearest neighbor, Alpha Centauri, would take about 25,000 years. The journey to the nearest civilization (assuming, as earlier, that a total of 1 million such cultures exist in the Galaxy) would take almost 1 million years. Interstellar travel at these speeds is clearly not feasible. Speeding up our ships to near the speed of light would reduce the travel time, but this is far beyond our present technology.

Actually, our civilization has already launched some interstellar probes, although they have no specific stellar destination. Figure 28.11 is a reproduction of a plaque mounted on board the *Pioneer 10* spacecraft launched in the mid-1970s and now well beyond the orbit of Pluto, on its way out of the solar system. Similar information was also included aboard the *Voyager* probes launched in 1978. While these spacecraft would be incapable of reporting back to Earth the news that they had encountered an alien culture, scientists hope that the civilization on the other end would be able to unravel most of its contents using the universal language of mathematics. The caption to Figure 28.11 notes how the aliens might discover from where and when the *Pioneer* and *Voyager* probes were originally launched.

Setting aside the many practical problems of establishing direct contact with extraterrestrials, some scientists have argued that it might not even be a particularly good idea. Our recent emergence as a technological civilization implies that we must be one of the least advanced technological intelligences in the entire Galaxy. Any other civilization that we discover, or that discovers us, will almost surely be more advanced than us. Consequently, a healthy

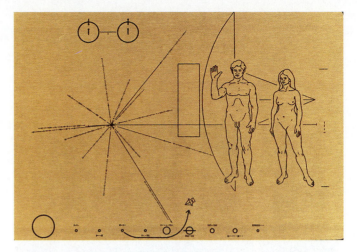

Figure 28.11 A replica of a plaque mounted on board the *Pioneer 10* spacecraft. The important features of the plaque include a scale drawing of the spacecraft, a man, and a woman; a diagram of the hydrogen atom undergoing a change in energy (top left); a starburst pattern representing various pulsars and the frequencies of their radio waves that can be used to estimate when the craft was launched (middle left); and a depiction of the solar system, showing that the spacecraft departed the third planet from the Sun and passed the fifth planet on its way into outer space (bottom). All the drawings have computer- (binary) coded markings from which actual sizes, distances, and times can be derived.

degree of caution is warranted. If extraterrestrials behave even remotely like human civilizations on Earth, then the most advanced aliens might naturally try to dominate all others. The behavior of the "advanced" European cultures toward the "primitive" races they encountered on their voyages of discovery in the seventeenth, eighteenth, and nineteenth centuries should serve as a clear warning of the possible undesirable consequences of contact. Of course, the aggressiveness of Earthlings may not apply to extraterrestrials, but given the history of the one intelligent species we know, the cautious approach may be in order.

RADIO COMMUNICATION

A cheaper, and much more practical, alternative is to try to make contact with extraterrestrials using only electromagnetic radiation, the fastest known means of transferring information from one place to another. Because light and other high-frequency radiation are heavily scattered while moving through dusty interstellar space, long-wavelength radio radiation seems to be the best choice. We would not attempt to broadcast to all nearby candidate stars, however—that would be far too expensive and inefficient. Instead, radio telescopes on the Earth listen *passively* for radio signals emitted by other civilizations. Some preliminary searches of selected nearby stars are now under way, thus far without success.

In what direction should we aim our radio telescopes? The answer to this question at least is fairly easy. On the basis of our earlier reasoning, we should target all F-, G-,

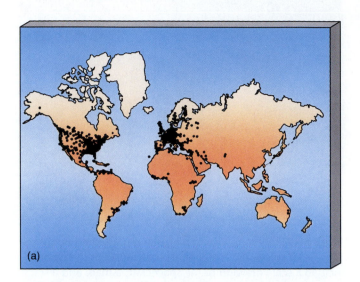

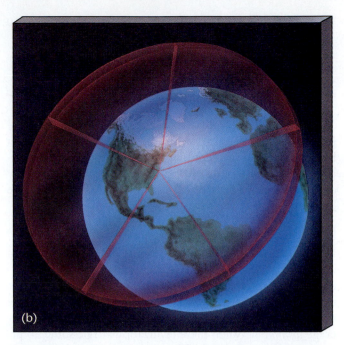

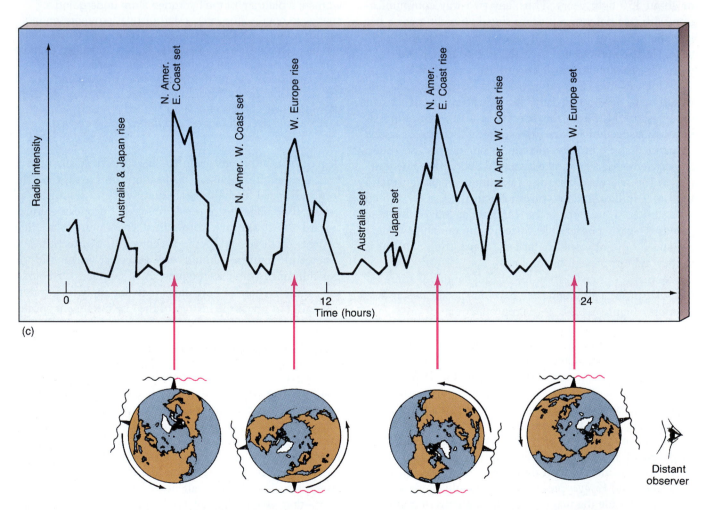

Figure 28.12   Radio radiation now leaks from Earth into space because of the daily activities of our technological civilization. (a) Most of the radiation arises from FM radio stations and television transmitters (shown here as dots), most of them found in western Europe and North America. (b) These stations broadcast their energy parallel to the Earth's surface, so they produce the strongest signal in any given direction when they happen to lie on Earth's horizon, as seen from that direction. (c) The pattern resulting from the sum of all the Earth's television stations, as viewed from a distant star. (d) Earth's radio brightness rises and falls as radio transmitters concentrated on certain parts of the surface rise and set due to our planet's rotation.

and K-type stars in our vicinity. But are extraterrestrials broadcasting radio signals? If they are not, this search technique will obviously fail. If they are, how do we distinguish their artificially generated radio signals from signals naturally emitted by interstellar gas clouds? At what frequency should we tune our receivers? This depends on whether the signals are produced deliberately or are simply "waste radiation" escaping from a planet.

Consider how Earth would look at radio wavelengths to extraterrestrials. Figure 28.12 shows the pattern of radio signals we emit into space. From the viewpoint of a distant observer, the spinning Earth emits a bright flash of radio radiation every few hours. In fact, Earth is now a more intense radio emitter than the Sun. The flashes result from the periodic rising and setting of hundreds of FM radio stations and television transmitters (marked on Figure 28.12a). Each station broadcasts mostly parallel to Earth's surface, sending a great "sheet" of electromagnetic radiation into interstellar space, as illustrated in Figure 28.12(b). (The more common AM broadcasts are trapped below our ionosphere, so those signals never leave Earth.) Because the great majority of these transmitters are clustered in the eastern United States and western Europe, a distant observer would detect blasts of radiation from Earth as our planet rotates each day (Figure 28.12c). This radiation races out into space, and has been doing so since the invention of these technologies over six decades ago. Another civilization as advanced as ours might have constructed devices capable of detecting this radiation. If any sufficiently advanced (and sufficiently interested) civilization resides within about 65 light years (20 pc) of Earth, then we have already broadcast our presence to them.

THE WATER HOLE

Now let us suppose that a civilization has decided to assist searchers by actively broadcasting its presence to the rest of the Galaxy. At what frequency should we listen for such an extraterrestrial beacon? The electromagnetic spectrum is enormous; the radio domain alone is vast. To hope to detect a signal at some unknown radio frequency is like searching for a needle in a haystack. Are some frequencies more likely than others to carry alien transmissions?

Some basic arguments suggest that civilizations might well communicate at a wavelength near 20 cm. As we saw in Chapter 18, the basic building blocks of the universe, namely, hydrogen atoms, naturally radiate at a wavelength of 21 cm. ∞ (p. 393) Also, one of the simplest molecules, hydroxyl (OH), radiates near 18 cm. Together, these two substances form water (H_2O). Arguing that water is likely to be the interaction medium for life anywhere, and that radio radiation travels through the disk of our Galaxy with the least absorption by interstellar gas and dust, some researchers have proposed that the interval between 18 and 21 cm is the best wavelength range for civilizations to transmit or listen. Called the **water hole**, this radio interval might serve as an "oasis" where all advanced Galactic civilizations would gather to conduct their electromagnetic business.

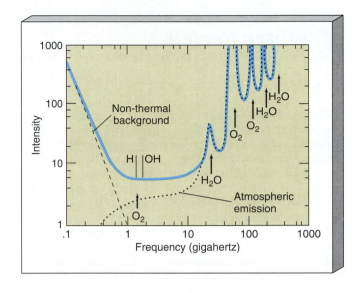

Figure 28.13 The "water hole" is bounded by the natural emission frequencies of the hydrogen (H) atom (21-cm wavelength) and the hydroxyl (OH) molecule (18-cm wavelength). The topmost solid (blue) curve sums the natural emissions of our Galaxy (dashed line on left side of diagram labeled "nonthermal background") and Earth's atmosphere (dotted line on right side of diagram denoted by various chemical symbols). This sum is minimized near the water-hole frequencies. Perhaps all intelligent civilizations conduct their interstellar communications within this quiet "electromagnetic oasis."

This water-hole frequency interval is only a guess, of course, but it is supported by other arguments as well. Figure 28.13 shows the water hole's location in the electromagnetic spectrum and plots the amount of natural emission from our Galaxy and from Earth's atmosphere. The 18- to 21-cm range lies within the quietest part of the spectrum, where the galactic "static" from stars and interstellar clouds happens to be minimized. Furthermore, the atmospheres of typical planets are also expected to interfere least at these wavelengths. Thus the water hole seems like a good choice for the frequency of an interstellar beacon, although we cannot be sure of this reasoning until contact is actually achieved. A few radio searches are now in progress at frequencies in and around the water hole. So far, however, nothing resembling an extraterrestrial signal has been detected.

The space surrounding all of us could be, right now, flooded with radio signals from extraterrestrial civilizations. If only we knew the proper direction and frequency, we might be able to make one of the most startling discoveries of all time. The result would likely provide whole new opportunities to study the cosmic evolution of energy, matter, and life throughout the universe.

Chapter Review

SUMMARY

Major phases in the history of the universe are galactic, stellar, planetary, chemical, biological, cultural, and future evolution. **Cosmic evolution** (p. 610) is the continuous process that has led to the appearance of galaxies, stars, planets, and life on Earth.

Living organisms may be characterized by their ability to react to their environment, to grow by taking in nutrition from their surroundings, and to reproduce, passing along some of their own characteristics to their offspring.

Powered by natural energy sources, reactions between simple molecules in the oceans of the primitive Earth are believed to have led to the formation of **amino acids** (p. 611) and **nucleotide bases** (p. 611), the basic molecules of life. Alternatively, some complex molecules may have been formed in interstellar space and then delivered to Earth by meteors or comets.

Organisms that can best take advantage of their new surroundings succeed at the expense of those organisms that cannot make the necessary adjustments. Intelligence is strongly favored by natural selection.

The best hope for life beyond Earth in the solar system is the planet Mars, although no evidence for living organisms has been found. Saturn's Titan may also be a possibility, but conditions there are harsh by terrestrial standards.

The **Drake equation** (p. 615) provides a means of estimating the probability of intelligent life in the Galaxy. The astronomical terms in the equation are the Galactic star-formation rate, the likelihood of planets, and the number of habitable planets. Chemical and biological terms are the probability of life appearing and the probability that it subsequently develops intelligence. Cultural and political terms are the probability that intelligence leads to technology and the lifetime of a civilization in the technological state.

Taking an optimistic view of the development of life and intelligence leads to the conclusion that the total number of technologically competent civilizations in the Galaxy is approximately equal to the lifetime of a typical civilization, expressed in years.

Even with optimistic assumptions, the distance to our nearest intelligent neighbor is likely to be many hundreds of parsecs. Space travel is not presently a feasible means of searching for intelligent life. Current programs to discover extraterrestrial intelligence involve scanning the electromagnetic spectrum for signals. So far, no intelligent broadcasts have been received.

A technological civilization would probably "announce" itself to the universe by the radio and television signals it emits into space. Observed from afar, our planet would appear as a radio source with a 24-hour period, as different regions of the planet rise and set.

The **water hole** (p. 621) is a region in the radio range of the electromagnetic spectrum, near the 21-cm line of hydrogen and the 18-cm line of hydroxyl, where natural emissions from the Galaxy happen to be minimized. Many researchers regard this as the best part of the spectrum for communications purposes.

SELF-TEST: True or False?

_____ **1.** The assumptions of mediocrity favor the existence of extraterrestrial life.

_____ **2.** The definition of life only requires that to be considered "alive" you must be able to reproduce.

_____ **3.** Miller and Urey conducted experiments in which they produced biological molecules from nonbiological molecules.

_____ **4.** Organic molecules important for life could have reached Earth's surface via comets.

_____ **5.** Organic molecules are known to exist only on the Earth.

_____ **6.** Laboratory experiments have succeeded in creating living cells from nonbiological molecules.

_____ **7.** For most of the history of the Earth, life consisted of only single-celled life forms.

_____ **8.** The *Viking* landers on Mars discovered microscopic evidence of life, but found no large fossil evidence.

_____ **9.** The rate of star formation in the Galaxy is reasonably well known.

_____ **10.** As yet there is no direct evidence that Earth-like planets orbit other stars.

_____ **11.** In estimating whether intelligence arises and develops technology, we have only life on Earth as examples.

_____ **12.** Dinosaurs existed on Earth for 10,000 times longer than human civilization has existed to date.

_____ **13.** Our civilization has launched probes into interstellar space.

_____ **14.** One disadvantage of interstellar radio communication is that we can only do it with another civilization that has a technology equal to or greater than our own.

SELF-TEST: Fill in the Blank

1. Amino acids are the building blocks of _____.

2. The naturally occurring molecules present on the young Earth included water, carbon dioxide, _____ , and _____.

3. Two sources of energy for chemical reactions available on the young Earth are _____ and _____.

4. The fossil record clearly shows evidence of life dating back _____ years.

5. Multicellular organisms did not appear on Earth until about _____ years ago.

6. The Murchison meteorite was discovered to contain relatively large amounts of _____.

7. The Drake equation estimates the number of _____ in the Milky Way Galaxy.

8. Planets in binary star systems are not considered habitable because the planetary orbits are usually _____.

9. Direct contact between extraterrestrial lifeforms may be impractical because of the large _____ between civilizations.

10. Radio communications over interstellar distances is practical because the signals travel at the speed of _____.

11. Radio waves can travel throughout the Galaxy because they are not blocked by interstellar _____.

12. Radio waves leaking away from the Earth have now traveled out a distance of _____ light-years.

13. Radio wavelengths between 18 and 21 cm are referred to as the _____.

14. A two-way communication with another civilization at a distance of 100 light-years will require _____ years.

REVIEW AND DISCUSSION

1. Why is life difficult to define?

2. What is chemical evolution?

3. What is the Urey-Miller experiment? What important organic molecules were produced in this experiment?

4. What are the basic ingredients from which biological molecules formed on Earth?

5. How do we know anything at all about the early episodes of life on Earth?

6. What is the role of language in cultural evolution?

7. Where else, besides Earth, have organic molecules been found?

8. Where—besides the planet Mars—might we find signs of life in our solar system?

9. Do we know whether Mars ever had life at any time during its past? What argues in favor of the position that it may once have harbored life?

10. What is generally meant by "life as we know it"? What other forms of life might be possible?

11. How many of the terms in the Drake Equation are known with any degree of certainty? Which factor is least well known?

12. What is the relationship between the average lifetime of Galactic civilizations and the possibility of our someday communicating with them?

13. How would Earth appear at radio wavelengths to extraterrestrial astronomers?

14. What are the advantages in using radio waves for communication over interstellar distances?

15. What is the water hole? What advantage does it have over other parts of the radio spectrum?

PROBLEMS

1. Suppose that each of the "fraction" terms in the Drake Equation turns out to have a value of 1/10, that stars form at an average rate of 20 per year, and that each star has exactly 1 habitable planet orbiting it. Estimate the present number of technological civilizations in the Milky Way Galaxy if the average lifetime of a civilization is (a) 100 years, (b) 10,000 years, (c) 1,000,000 years.

2. If the 4.5-billion-year age of the Earth were compressed to 45 years, as described in the text, what would be your age, in seconds? How long ago was the end of World War II? The Declaration of Independence? Columbus's discovery of the New World?

3. Assuming that there are 10,000 FM radio stations on Earth, each transmitting at a power level of 50 kilowatts, calculate the total radio luminosity of the Earth in the FM band. Compare this with the 10^6 W radiated by the Sun in the same frequency range.

4. Convert the water hole's wavelengths into frequencies. For practical reasons, any search of the water hole must be broken up into channels, much like you find on a television, except these channels are very narrow in radio frequency, about 100 Hz wide. How many channels must astronomers search in the water hole?

5. There are 20,000 stars within 100 light-years that are to be searched for radio communications. How long will the search take if one hour is spent looking at each star? If one day is spent per star?

PROJECTS

1. Some people suggest that if extraterrestrial life is discovered, it will have a profound effect on people. Interview as many people as you can and ask the following two questions: (1) Do you believe that extraterrestrial life exists? (2) Why? From your results, try to decide whether there will be a profound effect on people if extraterrestrial life is discovered.

2. Conduct another poll, or do it at the same time you do the first one. Ask the following question: What one question would you like to ask an extraterrestrial life form in a radio communication? How many responses do you receive that indicate the person is very "Earth-centered" in thinking? How many responses suggest a lack of understanding of how alien an extraterrestrial life form might be? Is your conclusion from the first project different or changed in any way?

3. The Drake Equation should be able to "predict" at least one civilization in our Galaxy: us. Try changing the values of various factors so that you end up with at least one. What do these various combinations of factors imply about how life arises and develops? Are there some combinations that just don't make any sense?

Appendix Tables

TABLE 1 *Some Useful Constants and Physical Measurements**

astronomical unit	1 A.U. = 1.496×10^8 km (1.5×10^8 km)
light year	1 ly = 9.46×10^{12} km (10^{13} km; 6 trillion miles)
parsec	1 pc = 3.09×10^{13} km = 3.3 ly
speed of light	c = 299,792.458 km/s (3×10^5 km/s)
Stefan-Boltzmann constant	σ [Greek sigma] = 5.67×10^{-8} W/m$^2 \cdot$ K^4
Planck's constant	h = 6.63×10^{-34} Js
gravitational constant	G = 6.67×10^{-11} Nm^2/kg^2
mass of the Earth	$M_\oplus$ = 5.97×10^{24} kg (6×10^{24} kg; about 6000 billion billion tons)
radius of the Earth	$R_\oplus$ = 6378 km (6500 km)
mass of the Sun	$M_\odot$ = 1.99×10^{30} kg (2×10^{30} kg)
radius of the Sun	$R_\odot$ = 6.96×10^5 km (7×10^5 km)
luminosity of the Sun	$L_\odot$ = 3.90×10^{26} W
effective temperature of the Sun	$T_\odot$ = 5778 K (5800 K)
Hubble constant	$H_0 \approx$ 75 km/s/Mpc
mass of the electron	m_e = 9.11×10^{-31} kg
mass of the proton	m_p = 1.67×10^{-27} kg

The rounded-off values used in the text are shown in parentheses.

Conversions between English and Metric Units

1 inch	=	2.54 centimeters (cm)	1 mile	=	1.609 kilometers (km)
1 foot (ft)	=	0.3048 meters (m)	1 pound (lb)	=	453.6 grams (g) or .4536 kilograms (kg) (on Earth)

TABLE 2 *Periodic Table of the Elements*

1	I A	II A											III A	IV A	V A	VI A	VII A	VIII A
1 **H** 1.0080 Hydrogen																		2 **He** 4.003 Helium
2	3 **Li** 6.939 Lithium	4 **Be** 9.012 Beryllium											5 **B** 10.81 Boron	6 **C** 12.011 Carbon	7 **N** 14.007 Nitrogen	8 **O** 15.9994 Oxygen	9 **F** 18.998 Fluorine	10 **Ne** 20.183 Neon
3	11 **Na** 22.990 Sodium	12 **Mg** 24.31 Magnesium	III B	IV B	V B	VI B	VII B		VIII B		I B	II B	13 **Al** 26.98 Aluminum	14 **Si** 28.09 Silicon	15 **P** 30.974 Phosphorus	16 **S** 32.064 Sulfur	17 **Cl** 35.453 Chlorine	18 **Ar** 39.948 Argon
4	19 **K** 39.102 Potassium	20 **Ca** 40.08 Calcium	21 **Sc** 44.96 Scandium	22 **Ti** 47.90 Titanium	23 **V** 50.94 Vanadium	24 **Cr** 52.00 Chromium	25 **Mn** 53.94 Manganese	26 **Fe** 55.85 Iron	27 **Co** 58.93 Cobalt	28 **Ni** 58.71 Nickel	29 **Cu** 63.54 Copper	30 **Zn** 65.37 Zinc	31 **Ga** 69.72 Gallium	32 **Ge** 72.59 Germanium	33 **As** 74.92 Arsenic	34 **Se** 78.96 Selenium	35 **Br** 79.909 Bromine	36 **Kr** 83.80 Krypton
5	37 **Rb** 85.47 Rubidium	38 **Sr** 87.62 Strontium	39 **Y** 88.91 Yttrium	40 **Zr** 91.22 Zirconium	41 **Nb** 92.91 Niobium	42 **Mo** 95.94 Molybdenum	43 **Tc** (99) Technetium	44 **Ru** 101.1 Ruthenium	45 **Rh** 102.90 Rhodium	46 **Pd** 106.4 Palladium	47 **Ag** 107.87 Silver	48 **Cd** 112.40 Cadmium	49 **In** 114.82 Indium	50 **Sn** 118.69 Tin	51 **Sb** 121.75 Antimony	52 **Te** 127.60 Tellurium	53 **I** 126.90 Iodine	54 **Xe** 131.30 Xenon
6	55 **Cs** 132.91 Cesium	56 **Ba** 137.34 Barium	57 TO 71	72 **Hf** 178.49 Hafnium	73 **Ta** 180.95 Tantalum	74 **W** 183.85 Tungsten	75 **Re** 186.2 Rhenium	76 **Os** 190.2 Osmium	77 **Ir** 192.2 Iridium	78 **Pt** 195.09 Platinum	79 **Au** 197.0 Gold	80 **Hg** 200.59 Mercury	81 **Tl** 204.37 Thallium	82 **Pb** 207.19 Lead	83 **Bi** 208.98 Bismuth	84 **Po** (210) Polonium	85 **At** (210) Astantine	86 **Rn** (222) Radon
7	87 **Fr** (223) Francium	88 **Ra** 226.05 Radium	89 TO 103															

Atomic number — 2
Symbol of element — **He**
Atomic weight — 4.003
Name of element — Helium

57 **La** 138.91 Lanthanum	58 **Ce** 140.12 Cerium	59 **Pr** 140.91 Praseodymium	60 **Nd** 144.24 Neodymium	61 **Pm** (147) Promethium	62 **Sm** 150.35 Samarium	63 **Eu** 151.96 Europium	64 **Gd** 157.25 Gadolinium	65 **Tb** 158.92 Terbium	66 **Dy** 162.50 Dysprosium	67 **Ho** 164.93 Holmium	68 **Er** 167.26 Erbium	69 **Tm** 168.93 Thullium	70 **Yb** 173.04 Ytterbium	71 **Lu** 174.97 Lutetium
89 **Ac** (227) Actinium	90 **Th** 232.04 Thorium	91 **Pa** (231) Protactinium	92 **U** 238.03 Uranium	93 **Np** (237) Neptunium	94 **Pu** (242) Plutonium	95 **Am** (243) Americium	96 **Cm** (247) Curium	97 **Bk** (249) Berkelium	98 **Cf** (251) Californium	99 **Es** (254) Einsteinium	100 **Fm** (253) Fermium	101 **Md** (256) Mendelevium	102 **No** (254) Nobelium	103 **Lw** (257) Lawrencium

TABLE 3A *Planetary Orbital Data*

PLANET	SEMI-MAJOR AXIS (A.U.)	SEMI-MAJOR AXIS (10^6 km)	PERIHELION (A.U.)	PERIHELION (10^6 km)	APHELION (A.U.)	APHELION (10^6 km)	ECCENTRICITY e
Mercury	0.39	57.9	0.31	46.0	0.47	69.8	0.206
Venus	0.72	108.2	0.72	107.5	0.73	108.9	0.007
Earth	1.00	149.6	0.98	147.1	1.02	152.1	0.017
Mars	1.52	227.9	1.38	206.6	1.66	249.1	0.093
Jupiter	5.20	778.3	4.95	740.9	5.45	815.7	0.048
Saturn	9.54	1427.0	9.01	1347	10.51	1507	0.056
Uranus	19.19	2869.6	18.31	2738	20.07	3002	0.046
Neptune	30.06	4496.6	29.76	4452	30.36	4542	0.010
Pluto	39.53	5914	29.73	4447	49.33	7381	0.248

PLANET	MEAN ORBITAL VELOCITY (km/s)	SIDEREAL PERIOD (tropical years)	SYNODIC PERIOD (days)	INCLINATION TO THE ECLIPTIC (degrees)	GREATEST ANGULAR DIAMETER AS SEEN FROM EARTH (arc seconds)
Mercury	47.89	0.24	115.88	7.00	13
Venus	35.03	0.62	583.92	3.39	64
Earth	29.79	1.00	—	0.01	—
Mars	24.13	1.88	779.94	1.85	25
Jupiter	13.06	11.86	398.88	1.31	50
Saturn	9.64	29.46	378.09	2.49	21
Uranus	6.81	84.01	369.66	0.77	4.1
Neptune	5.43	164.8	367.49	1.77	2.4
Pluto	4.74	248.6	366.73	17.15	0.11

TABLE 3B *Planetary Physical Data*

PLANET	EQUATORIAL RADIUS (km)	EQUATORIAL RADIUS (Earth = 1)	MASS (kg)	MASS (Earth = 1)	MEAN DENSITY (kg/m³)	SURFACE GRAVITY (Earth = 1)	ESCAPE VELOCITY (km/s)
Mercury	2,439	0.38	3.30×10^{23}	0.055	5430	0.38	4.3
Venus	6,052	0.95	4.87×10^{24}	0.82	5250	0.90	10.4
Earth	6,378	1.00	5.97×10^{24}	1.00	5520	1.00	11.2
Mars	3,397	0.53	6.42×10^{23}	0.11	3930	0.38	5.0
Jupiter	71,492	11.21	1.90×10^{27}	317.9	1330	2.53	60
Saturn	60,268	9.45	5.69×10^{26}	95.18	710	1.07	36
Uranus	25,559	4.01	8.68×10^{25}	14.54	1240	0.90	21
Neptune	24,764	3.88	1.02×10^{26}	17.13	1670	1.14	24
Pluto	1,123	0.18	1.46×10^{22}	0.0025	2290	0.07	0.07

PLANET	SIDEREAL ROTATION PERIOD (solar days)*	AXIAL TILT (degrees)	SURFACE MAGNETIC FIELD (Earth = 1)	MAGNETIC AXIS TILT (degrees relative to rotation axis)	ALBEDO**	SURFACE TEMPERATURE*** (K)	NUMBER OF MOONS
Mercury	58.6	7.0	0.011	<10	0.11	100 to 700	0
Venus	−243.0	177.4	<0.001		0.65	730	0
Earth	0.9973	23.45	1.0	11.5	0.37	290	1
Mars	1.026	23.98	<0.002		0.15	180 to 270	2
Jupiter	0.41	3.08	13.89	9.6	0.52	124	16
Saturn	0.43	26.73	0.67	0.8	0.47	97	20
Uranus	−0.69	97.92	0.74	58.6	0.50	58	15
Neptune	0.72	29.6	0.43	46.0	0.5	59	8
Pluto	−6.387	118	?		0.6	40 to 60	1

* A negative sign indicates retrograde rotation.
** Fraction of light reflected from surface.
*** Temperature is effective temperature for Jovian planets.

TABLE 4   *The Twenty Brightest Stars*

NAME	STAR	SPECTRAL TYPE* A	B	PARALLAX (arc seconds)	DISTANCE (pc)	APPARENT VISUAL MAGNITUDE* A	B
Sirius	αCMa	A1V	wd**	0.37	2.7	−1.46	+8.7
Canopus	αCar	F0Ib-II		0.033	30	−0.72	
Rigel Kentaurus	αCen	G2V	K0V	0.77	1.3	−0.01	+1.3
Arcturus	αBoo	K2III		0.091	11	−0.06	
Vega	αLyr	A0V		0.13	8.0	+0.04	
Capella	αAur	GIII	M1V	0.071	14	+0.05	+10.2
Rigel	βOri	B8Ia	B9	—	250	+0.14	+6.6
Procyon	αCMi	F5IV-V	wd**	0.29	3.5	+0.37	+10.7
Betelgeuse	αOri	M2Iab		—	150	+0.41	
Achernar	αEri	B5V		0.050	20	+0.51	
Hadar	βCen	B1III	?	0.011	90	+0.63	+4
Altair	αAql	A7IV-V		0.20	5.1	+0.77	
Acrux	αCru	B1IV	B3	0.008	120	+1.39	+1.9
Aldebaran	αTau	K5III	M2V	0.063	16	+0.86	+13
Spica	αVir	B1V		0.013	80	+0.91	
Antares	αSco	MIIb	B4V	0.008	120	+0.92	+5.1
Pollux	βGem	K0III		0.083	12	+1.16	
Formalhaut	αPsA	A3V		0.14	7.0	+1.19	+6.5
Deneb	αCyg	A2Ia		—	430	+1.26	
Mimosa	βCru	B1IV		—	150	+1.28	

NAME	VISUAL LUMINOSITY* (SUN = 1) A	B	ABSOLUTE VISUAL MAGNITUDE* A	B	PROPER MOTION (arc seconds/yr)	TRANSVERSE VELOCITY (km/s)	RADIAL VELOCITY (km/s)
Sirius	23.5	0.003	+1.4	+11.6	1.33	17.0	−7.6***
Canopus	1510		−3.1		0.02	2.8	+20.5
Rigel Kentaurus	1.56	0.46	+4.4	+5.7	3.68	22.7	−24.6
Arcturus	115		−0.3		2.28	119	−5.2
Vega	55.0		+0.5		0.34	12.9	−13.9
Capella	166	0.01	−0.7	+9.5	0.44	29	+30.2***
Rigel	4.6×10^4	126	−6.8	−0.4	0.00	1.2	+20.7***
Procyon	7.7	0.0006	+2.6	+13.0	1.25	20.7	−3.2***
Betelgeuse	1.4×10^4		−5.5		0.03	21	+21.0***
Achernar	219		−1.0		0.10	9.5	+19
Hadar	3800	182	−4.1	−0.8	0.04	17	−12***
Altair	11.5		+2.2		0.66	16	−26.3
Acrux	3470	2190	−4.0	−3.5	0.04	24	−11.2
Aldebaran	105	0.0014	−0.2	+12	0.20	15	+54.1
Spica	2400		−3.6		0.05	19	+1.0***
Antares	5500	115	−4.5	−0.3	0.03	17	−3.2
Pollux	41.7		+0.8		0.62	35	+3.3
Formalhaut	13.8	0.10	+2.0	+7.3	0.37	12	+6.5
Deneb	5.0×10^4		−6.9		0.003	6	−4.6***
Mimosa	6030		−4.6		0.05	36	

*A and B columns identify individual components of binary systems.
** "wd" stands for "white dwarf."
*** Average value of variable velocity.

TABLE 5 *The Thirty Nearest Stars*

NAME	SPECTRAL TYPE*		PARALLAX	DISTANCE	APPARENT VISUAL MAGNITUDE*	
	A	B	(arc seconds)	(pc)	A	B
Sun	G2V				−26.72	
Proxima Centauri	M5		0.772	1.30	+11.05	
Alpha Centauri	G2V	K0V	0.750	1.33	−0.01	+1.33
Barnard's Star	M5V		0.545	1.83	+9.54	
Wolf 359	M8V		0.421	2.38	+13.53	
Lalande 21185	M2V		0.397	2.52	+7.50	
UV Ceti	M6V	M6V	0.387	2.58	+12.52	+13.02
Sirius	A1V	wd**	0.377	2.65	−1.46	+8.3
Ross 154	M5V		0.345	2.90	+10.45	
Ross 248	M6V		0.314	3.18	+12.29	
ε Eridani	K2V		0.303	3.30	+3.73	
Ross 128	M5V		0.298	3.36	+11.10	
61 Cygni	K5V	K7V	0.294	3.40	+5.22	+6.03
ε Indi	K5V		0.291	3.44	+4.68	
Grm 34	M1V	M6V	0.290	3.45	+8.08	+11.06
Luyten 789-6	M6V		0.290	3.45	+12.18	
Procyon	F5IV-V	wd**	0.285	3.51	+0.37	+10.7
Σ 2398	M4V	M5V	0.285	3.55	+8.90	+9.69
Lacaille 9352	M2V		0.279	3.58	+7.35	
G51-15	MV		0.278	3.60	+14.81	
Tau Ceti	G8V		0.277	3.62	+3.50	
BD + 5°1668	M5		0.266	3.77	+9.82	
Lacaille 8760	M0V		0.260	3.86	+6.66	

NAME	VISUAL LUMINOSITY* (Sun = 1)		ABSOLUTE VISUAL MAGNITUDE*		PROPER MOTION	TRANSVERSE VELOCITY	RADIAL VELOCITY
	A	B	A	B	(arc seconds/yr)	(km/s)	(km/s)
Sun	1.0		+4.85				
Proxima Centauri	0.00006		+15.5		3.86	23.8	−16
Alpha Centauri	1.6	0.45	+4.4	+5.7	3.68	23.2	−22
Barnard's Star	0.00045		+13.2		10.34	89.7	−108
Wolf 359	0.00002		+16.7		4.70	53.0	+13
Lalande 21185	0.0055		+10.5		4.78	57.1	−84
UV Ceti	0.00006	0.00004	+15.5	+16.0	3.36	41.1	+30
Sirius	23.5	0.003	+1.4	+11.2	1.33	16.7	−8
Ross 154	0.00048		+13.3		0.72	9.9	−4
Ross 248	0.00011		+14.8		1.58	23.8	−81
ε Eridani	0.30		+6.1		0.98	15.3	+16
Ross 128	0.00036		+13.5		1.37	21.8	−13
61 Cygni	0.082	0.039	+7.6	+8.4	5.22	84.1	−64
ε Indi	0.14		+7.0		4.69	76.5	−40
Grm 34	0.0061	0.00039	+10.4	+13.4	2.89	47.3	+17
Luyten 789-6	0.00014		+14.6		3.26	53.3	−60
Procyon	7.65	0.00055	+2.6	+13.0	1.25	2.8	−3
Σ 2398	0.0030	0.0015	+11.2	+11.9	2.28	38.4	+5
Lacaille 9352	.013		+9.6		6.90	117	+10
G51-15	0.00001		+17.0		1.26	21.5	—
Tau Ceti	0.45		+5.7		1.3	23.7	−16
BD + 5° 1668	0.0015		+11.9		3.8	59.8	+26
Lacaille 8760	0.028		+8.7		2.4	42.4	+21

* *A and B columns identify individual components of binary systems.*
** *"wd" stands for "white dwarf."*

Glossary

A

A ring One of three Saturnian rings visible from Earth. The A ring is farthest from the planet and is separated from the B ring by the Cassini division. (p. 259)

absolute brightness The apparent brightness a star would have if it were placed at a standard distance of 10 parsecs from Earth. (p. 362)

absorption line Dark line in an otherwise continuous bright spectrum, where light within one narrow frequency range has been removed. (p. 77)

acceleration The rate of change of velocity of a moving object. (p.45)

accretion Gradual growth of bodies, such as planets, by the accumulation of other, smaller, bodies. (p. 322)

accretion disk Flat disk of matter spiraling down onto the surface of a star or black hole. Often, the matter originated on the surface of a companion star in a binary system. (p. 445)

active galaxy The most energetic galaxies, which can emit hundreds or thousands of times more energy per second than the Milky Way. (p. 540)

active optics Collection of techniques now being used to increase the resolution of ground-based telescopes. Minute modifications are made to the overall configuration of an instrument as its temperature and orientation change, to maintain the best possible focus at all times. (p. 107)

active region Region of the photosphere of the Sun surrounding a sunspot group, which can erupt violently and unpredictably. During sunspot maximum, the number of active regions is also a maximum. (p. 346)

active Sun The unpredictable aspects of the Sun's behavior, such as sudden explosive outbursts of radiation in the form of prominences and flares. (p. 341)

adaptive optics Technique used to increase the resolution of a telescope by deforming the shape of the mirror's surface under computer control while a measurement is being taken, to undo the effects of atmospheric turbulence. (p. 108)

aesthenosphere Layer of the Earth's interior, just below the lithosphere, over which the surface plates slide. (p. 162)

amino acids Organic molecules which form the basis for building the proteins that direct metabolism in living creatures. (p. 611)

amplitude The maximum deviation of a wave above or below zero point. (p. 56)

angular momentum problem The fact that the Sun, which contains nearly all of the mass of the solar system, accounts for just 0.3 percent of the total angular momentum of the solar system. This is an aspect of the solar system that any acceptable formation theory must address. (p. 327)

angular resolution The ability of a telescope to distinguish between adjacent objects in the sky. (p. 102)

annular eclipse Solar eclipse occurring at a time when the Moon is far enough away from the Earth that it fails to cover the disk of the Sun completely, leaving a ring of sunlight visible around its edge. (p. 18)

aphelion The point on the elliptical path of an object in orbit about the Sun that is most distant from the Sun (p. 41)

Apollo asteroid *See* Earth-crossing asteroid.

apparent brightness The brightness that a star appears to have, as measured by an observer on Earth. (p. 67)

arc degree Unit of angular measure. There are 360 arc degrees in one complete circle. (p. 14)

association Small grouping of (typically 100 or less) stars, spanning up to a few tens of parses across, usually rich in very young stars. (p. 414)

asteroid One of thousands of very small members of the solar system orbiting the Sun between the orbits of Mars and Jupiter. Often referred to as "minor planets." (p. 129, 298)

asteroid belt A region of the solar system, between the orbits of Mars and Jupiter, in which most asteroids are found. (p. 129, 298)

astronomical unit (A.U.) The average distance of the Earth from the Sun. Precise radar measurements yield a value for the A.U. of 149,603,500 km. (p. 13)

astronomy Branch of science dedicated to the study of everything in the universe that lies above Earth's atmosphere. (p. 4)

asymptotic giant branch Path on the H-R diagram corresponding to the changes that a star undergoes after helium burning ceases in the core. At this stage, the carbon core shrinks and drives the expansion of the envelope, and the star becomes a swollen red giant for a second time. (p. 427)

atmosphere A layer of gas confined close to a planet's surface by the force of gravity. (p. 146)

atom Building block of matter, composed of positively charged protons and neutral neutrons in the nucleus, surrounded by negatively charged neutrons. (p. 81)

aurora Event which occurs when atmospheric molecules are excited by incoming charged particles from the solar wind, then emit energy as they fall back to their ground states. Aurorae generally occur at high latitudes, near the north and south magnetic poles. (p. 154)

autumnal equinox Date on which the Sun crosses the celestial equator moving southward, occurring on or near September 22. (p. 10)

B

B ring One of three Saturnian rings visible from Earth. The B ring is the brightest of the three, and lies just past the Cassini division, closer to the planet than the A ring. (p. 259)

barred-spiral galaxy Spiral galaxy in which a bar of material passes through the center of the galaxy, with the spiral arms beginning near the ends of the bar. (p. 516)

baryonic matter Matter that is composed primarily of baryons—protons and neutrons. "Normal" matter. (p. 596)

baseline The distance between two observing locations used for the purposes of triangulation measurements. The larger the baseline, the better the resolution attainable. (p. 23)

belt Dark, low-pressure region, where gas flows downward in the atmosphere of a jovian planet. (p. 237)

Big Bang Event that cosmologists consider the beginning of the universe, in which all matter and radiation in the entire universe came into being. (p. 568)

binary-star system A system which consists of two stars in orbit about their common center of mass, held together by their mutual gravitational attraction. Most stars are found in binary-star systems. (p. 373)

black-body curve The characteristic way in which the intensity of radiation emitted by a hot object depends of frequency. The frequency at which the emitted intensity is highest is an indication of the temperature of the radiating object. Also referred to as the Planck curve. (pp. 58, 62)

black dwarf The end-point of the evolution of an isolated, low-mass star. After the white dwarf stage, the star cools to the point where it is a dark "clinker" in interstellar space. (p. 432)

black hole A region of space where the pull of gravity is so great that nothing-not even light-can escape. A possible outcome of the evolution of a very massive star. (p. 472)

blue giant Large, hot, bright star at the upper left end of the main sequence on the H-R diagram. Its name comes from its color and size. (p. 369)

blue shift Motion-induced changed in the observed wavelength from a source that is moving toward us. Relative approaching motion between the object and the observer causes the wavelength to appear shorter (and hence bluer) than if there were no motion at all. (p. 68)

blue supergiant The very largest of the large, hot, bright stars at the uppermost left end of the main sequence on the H-R diagram. (p. 369)

Bohr model First theory of the hydrogen atom to explain the observed spectral lines. This model rests on three ideas: that there is a state of lowest energy for the electron, that there is a maximum energy, beyond which the electron is no longer bound to the nucleus, and that within these two energies the electron can only exist in certain energy levels. (p. 81)

brown dwarf Remnants of fragments of collapsing gas and dust that did not contain enough mass to initiate core nuclear fusion. Such objects are then frozen somewhere along their pre-main-sequence contraction phase, continually cooling into compact dark objects. Because of their small size and low temperature they are extremely difficult to detect observationally. (p. 407)

brown oval Feature of Jupiter's atmosphere that appears only at latitudes near 20 degrees N, this structure is a long-lived hole in the clouds that allows us to look down into Jupiter's lower atmosphere. (p. 239)

C

C ring One of three Saturnian rings visible from Earth. The C ring lies closest to the planet and is relatively thin compared to the A and B rings. (p. 259)

carbon-detonation supernova See type-I supernova. (p. 449)

Cassegrain telescope A type of reflecting telescope in which incoming light hits the primary mirror and is then reflected upward toward the prime focus, where a secondary mirror reflects the light back down through a small hole in the main mirror, into a detector or eyepiece. (p. 100)

Cassini Division A relatively empty gap in Saturn's ring system, discovered in 1675 by Giovanni Cassini. It is now known to contain a number of thin ringlets. (p. 259)

catastrophic theory A theory that invokes statistically unlikely accidental events to account for observations. (p. 319)

celestial coordinates Pair of quantities-right ascension and declination-similar to longitude and latitude on Earth, used to pinpoint locations of objects on the celestial sphere. (p. 13)

celestial equator The projection of the Earth's equator onto the celestial sphere. (p. 9)

celestial sphere Imaginary sphere surrounding the Earth, to which all objects in the sky were once considered to be attached. (p. 9)

center of mass The "average" position in space of a collection of massive bodies, weighted by their masses. In an isolated system this point moves with constant velocity, according to Newtonian mechanics. (p. 48)

Cepheid variable Star whose luminosity varies in a characteristic way, with a rapid rise in brightness followed by a slower decline. The period of a Cepheid variable star is related to its luminosity, so a determination of this period can be used to obtain an estimate of the star's distance. (p. 490)

chaotic rotation Unpredictable tumbling motion that non-spherical bodies in eccentric orbits, such as Saturn's satellite Hyperion, can exhibit. No amount of observation of an object rotating chaotically will ever show a well-defined period. (p. 271)

charge-coupled device (CCD) An electronic devise used for data acquisition, composed of many tiny pixels, each of which records a buildup of charge to measure the amount of light striking it. (p.106)

chromatic aberration The tendency for a lens to focus red and blue light differently, causing images to become blurred. (p. 97)

chromosphere The Sun's lower atmosphere, lying just above the visible atmosphere. (p. 332)

closed universe Geometry that the universe as a whole would have if the density of matter is above the critical value. A closed universe is finite in extent, and has no edge, like the surface of a sphere. It has enough mass to stop the present expansion, and will eventually collapse. (p. 578)

cold dark matter Class of dark-matter candidates made up of very heavy particles, such as supersymmetric relics. (p. 602)

collecting area The total area of a telescope that is capable of capturing incoming radiation. The large the telescope, the greater its collecting area, and the fainter the objects it can detect. (p. 101)

color index A convenient method of quantifying a star's color by comparing its apparent brightness as measured through different filters. If the star's radiation is well described by a blackbody spectrum, the ratio of its blue intensity (B) to its visual intensity (V) is a measure of the object's surface temperature. (p. 364)

color-magnitude diagram A way of plotting stellar properties, in which absolute magnitude is plotted against color index. (p. 368)

coma An effect occurring during the formation of an off-axis image in a telescope. Stars whose light enters the telescope at a large angle acquire comet-like tails on their images. (p. 305)

coma The brightest part of a comet, often referred to as the "head." (p. 305)

comet A small body, composed mainly of ice and dust, in an elliptical orbit about the Sun. As it comes close to the Sun, some of its material is vaporized to form a gaseous head and extended tail. (pp. 129, 304)

condensation nuclei Dust grains in the interstellar medium which act as seeds around which other material can cluster. The presence of dust was very important in causing matter to clump during the formation of the solar system. (p. 321)

condensation theory Currently favored model of solar system formation which combines features of the old nebular theory with new information about interstellar dust grains, which acted as condensation nuclei. (p. 321)

conservation of mass and energy A fundamental law of modern physics which states that the sum of mass and energy must always remain constant in any physical process. In fusion reactions, the lost mass is converted into energy, primarily in the form of electromagnetic radiation. (p. 349)

constellation A human grouping of stars in the night sky into a recognizable pattern. (p. 6)

contact binary A binary star system in which both stars have expanded to fill their Roche lobes and the surfaces of the two stars merge. The binary system now consists of two nuclear burning stellar cores surrounded by a continuous common envelope. (p. 438)

continuous spectrum Spectrum in which the radiation is distributed over all frequencies, not just a few specific frequency ranges. A prime example is the black-body radiation emitted by a hot, dense body. (p. 75)

convection Churning motion resulting from the constant upwelling of warm fluid and the concurrent downward flow of cooler material to take its place. (p. 150)

convection zone Region of the Sun's interior, lying just below the surface, where the material of the Sun is in constant convection motion. This region extends into the solar interior to a depth of about 20,000 km. (p. 332)

Copernican revolution The realization toward the end of the sixteenth century that Earth is not at the center of the universe. (p. 35)

core The central region of the Earth, surrounded by the mantle. (p. 146) The central region of any planet or star. (p. 332)

core-collapse supernova *See* type-II supernova.

core hydrogen burning The energy burning stage for main sequence stars, in which the helium is produced by hydrogen fusion in the central region of the star. A typical star spends up to 90% of its lifetime in hydrostatic equilibrium brought about by the balance between gravity and the energy generated by core hydrogen burning. (p. 420)

corona The tenuous outer atmosphere of the Sun, which lies just above the chromosphere, and at great distances turns into the solar wind. (pp. 208, 332)

coronal hole Vast regions of the Sun's atmosphere where the density of matter is about 10 times lower than average. The gas there streams freely into space at high speeds, escaping the Sun completely. (p. 341)

cosmic distance scale Collection of indirect distance-measurement techniques that astronomers use to measure the scale of the universe. (p. 23)

cosmic evolution The collection of the seven major phases of the history of the universe, namely galactic, stellar, planetary, chemical, biological, cultural, and future evolution. (p. 610)

cosmic microwave background The almost perfectly isotropic radio signal that is the remnant of the Big Bang explosion. (p. 580)

cosmological principle Two assumptions which make up the basis of cosmology, namely that the universe is homogeneous and isotropic on large scales. (p. 567)

cosmological red shift The component of the redshift of an object which is due only to the Hubble flow of the universe. (p. 528)

cosmology The study of the structure and evolution of the entire universe. (pp. 32, 567)

crater Bowl-shaped depression on the surface of a planet or moon, resulting from a collision with interplanetary debris. (p. 176)

critical density The cosmic density corresponding to the dividing line between a universe that recollapses and one that expands forever. (p. 574)

critical universe Geometry that the universe would have if the density of matter is exactly the critical density. The universe is infinite in extent, and has zero curvature. The expansion will continue forever, but approach an expansion speed of zero. (p. 579)

crust Layer of the Earth which contains the solid continents and the seafloor. (p. 146)

D

D ring Collection of very faint, thin rings, extending from the inner edge of the C ring down nearly to the cloud tops of Saturn. This region contains so few particles that it is completely invisible from Earth. (p. 262)

dark dust cloud A large cloud, often many parsecs across, which contains gas and dust in a ration of about 1012 gas atoms for every dust particle. Typical densities are a few tens or hundreds of millions of particles per cubic meter. (p. 389)

dark halo Region of a galaxy beyond the visible halo where dark matter is believed to reside. (p. 505)

dark matter Term used to describe the mass in galaxies and clusters whose existence we infer from rotation curves and other techniques, but which has not been confirmed by observations at any electromagnetic wavelength. (p. 505)

declination Celestial coordinate used to measure latitude above or below the celestial equator on the celestial sphere. (p. 14)

decoupling Event in the early universe when atoms first formed, and after which photons could propagate freely through space. (p. 596)

deferent A construct of the geocentric model of the solar system which was needed to explain observed planetary motions. A deferent is a large circle encircling the Earth, on which an epicycle moves. (p. 34)

degree A unit of angular measure. There are 360 degrees in a complete circle. (p. 14)

density A measure of the compactness of the matter within an object, computed by dividing the mass by the volume of the object. Units are kilograms per cubic meter (kg/m^3), or grams per cubic centimeter (g/cm^3). (p. 131)

deuteron An isotope of hydrogen in which there is a neutron bound to the proton in the nucleus. Often called "heavy hydrogen" because of the extra mass of the neutron. (p. 350)

differential rotation The tendency for a gaseous sphere, such as jovian planet or the Sun, to rotate at a different rate at the equator than at the poles. For a galaxy or other object, a condition where the angular speed varies with location within the object. (p. 234)

differentiation Variation in the density and composition of a body, such as Earth, with low density material on the surface and higher density material in the core. (p. 159)

diffraction The ability that waves have to bend around corners. The diffraction of light establishes its nature as a wave. (p. 56)

Doppler effect Any motion-induced change in the observed wavelength (or frequency) of a wave. (p. 68)

Drake Equation Expression that gives an estimate of the probability that intelligence exists elsewhere in the galaxy, based on a number of supposedly necessary conditions for intelligent life to develop. (p. 615)

dust grain An interstellar dust particle, roughly 10-8 m in size,, comparable to the wavelength of visible light. (p. 382)

dust lane A lane of dark, obscuring interstellar dust in an emission nebula or galaxy. (p. 387)

dust tail The component of a comet's tail that is composed of dust particles. (p. 306)

dwarf Any star with radius comparable to, or smaller than, that of the Sun (including the Sun itself) (p. 362)

dynamo theory Theory that explains planetary and stellar magnetic fields in terms of rotating, conducting material flowing in an object's interior. (p. 154)

E

E ring A faint ring, well outside the main ring system of Saturn, which was discovered by Voyager, and is believed to be associated with volcanism on the moon Enceladus. (p. 262)

Earth-crossing asteroid An asteroid whose orbit crosses that of the Earth. Earth-crossing asteroids are also called Apollo asteroids, after the first of the type was discovered. (p. 299)

earthquake A sudden dislocation of rocky material near the Earth's surface. (p. 156)

eccentricity A measure of the flatness of an ellipse, equal to the distance between the two foci divided by the length of the major axis. (p. 41)

eclipse Event during which one body passes in front of another, so that the light from the occulted body is blocked. (p. 171)

eclipse season Times of the year when the Moon lies in the same plane as the Earth and Sun, so that eclipses are possible. (p. 21)

eclipsing binary Rare binary-star system that is aligned in such a way that from Earth we observe one star pass in front of the other, eclipsing the other star. (p. 373)

ecliptic The apparent path of the Sun, relative to the stars on the celestial sphere, over the course of a year. (p. 10)

electric field A field extending outward in all directions from a charged particle, such as a proton or an electron. The electric field determines the electric force exerted by the particle on all other charged particles in the universe; the strength of the electric field decreases with increasing distance from the charge according to an inverse-square law. (p. 58)

electromagnetic radiation Another term for light, electromagnetic radiation transfers energy and information from one place to another. (p. 54)

electromagnetic spectrum The complete range of electromagnetic radiation, from radio waves to gamma rays, including the visible spectrum. All types of electromagnetic radiation are basically the same phenomenon, differing only by wavelength, and all move at the speed of light. (p. 60)

electromagnetism The union of electricity and magnetism, which do not exist as independent quantities, but are in reality two aspects of a single physical phenomenon. (p. 58)

electron An elementary particle with a negative electric charge, one of the components of the atom. (p. 57)

electron degeneracy pressure The pressure produced by the resistance of electrons to compression once they are squeezed to the point of contact. (p. 425)

element Matter made up of one particular atom. The number of protons in the nucleus of the atom determines which element it represents. (p. 84)

ellipse Geometric figure resembling an elongated circle. An ellipse is characterized by its degree of flatness, or eccentricity, and the length of its long axis. In general, bound orbits of objects moving under gravity are elliptical. (p. 41)

elliptical galaxy Category of galaxy in which the stars are distributed in an elliptical shape on the sky, ranging from highly elongated to nearly circular in appearance. (p. 516)

emission line Bright line in a specific location of the spectrum of radiating material, corresponding to emission of light at a certain frequency. A heated gas in a glass container produces emission lines in its spectrum. (p. 75)

emission nebula A glowing cloud of hot interstellar gas. The gas glows as a result of a nearby young star which is ionizing the gas. Since this gas is mostly hydrogen, the emitted radiation falls predominantly in the red region of the spectrum, because of a dominant hydrogen emission. (p. 386)

emission spectrum The pattern of spectral emission lines, produced by an element. Each element has its own unique emission spectrum. (p. 76)

Encke Division A small gap in Saturn's A ring. (p. 259)

epicycle A construct of the geocentric model of the solar system which was necessary to explain observed planetary motions. Each planet rides on a small epicycle whose center in turn rides on a larger circle (the deferent). (p. 34)

escape velocity The speed necessary for ran object to escape the gravitational pull of an object. Anything that moves away from the object with more than the escape velocity will never return. (p. 49)

event horizon Imaginary spherical surface surrounding a collapsing star, with radius equal to the Schwarzschild radius, within which no event can be seen heard, or known about by an outside observer. (p. 474)

evolutionary theory A theory which explains observations in a series of gradual steps, explainable in terms of well-established physical principles. (p. 319)

evolutionary track A graphical representation of a star's life, as a path on the H-R diagram. (p. 402)

excited state State of an atom when one of its electrons is in a higher energy orbital than the ground state. Atoms can become excited by absorbing a photon of a specific energy, or by colliding with a nearby atom. (p. 81)

extinction The dimming of starlight as it passes through the interstellar medium. (p. 382)

F

F ring Faint narrow outer ring of Saturn, discovered by Pioneer in 1979. The F ring lies just inside the Roche limit of Saturn, and was shown by Voyager to be made up of several ring strands apparently braided together. (p. 263)

flare Explosive event occurring in or near an active region on the Sun. (p. 346)

flatness problem One of two conceptual problems with the Standard Big Bang model, which is that there is no natural way to explain why the density of the universe is so close to the critical density. (p. 597)

fluidized ejecta The ejecta blankets around some Martian craters, which apparently indicate that the ejected material was liquid at the time the crater formed. (p. 221)

focus One of two special points within an ellipse, whose separation from each other indicate the eccentricity. In a bound orbit, objects move in ellipses about one focus. (p. 41)

forbidden line A spectral line seen in emission nebulae, but not seen in laboratory experiments, because collisions kick the electron in question into some other state before emission can occur. (p. 388)

force Action on an object that causes its momentum to change. The rate at which the momentum changes is numerically equal to the force. (p. 44)

fragmentation The breaking up of a large object into many smaller pieces (for example, as the result of high-speed collisions between planetesimals and protoplanets in the early solar system). (p. 323)

Fraunhofer lines The collection of over 600 absorption lines in the spectrum of the Sun, first categorized by Joseph Fraunhofer in 1812. (p. 77)

frequency The number of wave crests passing any given point in a given period of time. (p. 56)

full moon Phase of the Moon in which it appears as a complete circular disk in the sky. (p. 15)

fusion Mechanism of energy generation in the core of the Sun, in which light nuclei are combined, or fused, into heavier ones, releasing energy in the process. (p. 348)

G

galactic bulge Thick distribution of warm gas and stars around the galactic center. (p. 488)

galactic cannibalism A galaxy merger in which a larger galaxy consumes a smaller one. (p. 535)

galactic center The center of the Milky Way, or any other, galaxy. The point about which the disk of a spiral galaxy rotates. (p. 493)

galactic disk Flattened region of gas and dust that bisects the galactic halo in a spiral galaxy. This is the region of active star formation. (p. 488)

galactic halo Region of a galaxy extending far above and below the galactic disk, where globular clusters and other old stars reside. (p. 488)

galactic nucleus Small central high-density region of a galaxy. Nearly all of the radiation from an active galaxy is emitted from the nucleus. (p. 541)

galaxy Gravitationally bound collection of a large number of stars. The Sun is a star in the Milky Way Galaxy. (p. 488)

galaxy cluster A collection of galaxies held together by their mutual gravitational attraction. (p. 521)

Galilean satellites The four brightest and largest moons of Jupiter (Io, Europa, Ganymede, Callisto), named after Galileo Galilei, the 17th century astronomer who first observed them. (p. 243)

gamma ray Region of the electromagnetic spectrum, far beyond the visible spectrum, corresponding to radiation of very high frequency and very short wavelength. (p. 59)

gamma-ray burster Object that radiates tremendous amounts of energy in the form of gamma ray, possibly due to the accretion of matter onto a neutron star from another star in a binary orbit. The accreting matter eventually reaches temperatures high enough to begin violent nuclear burning on the neutron star's surface. (p. 468)

geocentric model A model of the solar system which holds that the Earth is at the center of the universe and all other bodies are in orbit around it. The earliest theories of the solar system were geocentric. (p. 32)

giant A star with a radius between 10 and 100 times that of the Sun. (p. 361)

globular cluster Tightly bound, roughly spherical collection of hundreds of thousands, and sometimes millions, of stars spanning about 50 parsecs. Globular clusters are distributed in the halos around the Milky Way and other galaxies. (p. 377)

Grand Unified Theories Theories which describe the behavior of the single force that results from unification of the strong, weak, and electromagnetic forces in the early universe. (p. 593)

granulation Mottled appearance of the solar surface, caused by rising (hot) and falling (cool) material in convective cells just below the photosphere. (p. 336)

gravitational field Field created by any object with mass, extending out in all directions, which determines the influence of that object on all others. The strength of the gravitational field decreases as the square of the distance. (p. 47)

gravitational lensing The effect induced on the image of a distant object by a massive foreground object. Light from the distant object is bent into two or more separate images. (p. 506)

gravitational red shift A prediction of Einstein's general theory of relativity. Photons lose energy as they escape the gravitational field of a massive object. Because a photon's energy is proportional to its frequency, a photon that loses energy suffers a decrease in frequency, which corresponds to an increase, or redshift, in wavelength. (p. 477)

gravity The attractive effect that any massive object has on all other massive objects. The greater the mass of the object, the stronger is its gravitational pull. (p. 46)

Great Dark Spot Prominent storm system in the atmosphere of Neptune, located near the equator of the planet. The system is comparable in size to the Earth. (p. 280)

Great Red Spot A large, high-pressure, long-lived storm system visible in the atmosphere of Jupiter. The Red Spot is roughly twice the size of the Earth. (p. 236)

greenhouse effect The partial trapping of solar radiation by a planetary atmosphere, similar to the trapping of heat in a greenhouse. (p. 15)

ground state The lowest energy state that an electron can have within an atom. (p. 81)

H

heliocentric model A mode of the solar system which is centered on the Sun, with the Earth in motion about the Sun. (p. 35)

helioseismology The study of conditions far below the Sun's surface through the analysis of internal "sound" waves that repeatedly cross the solar interior. (p. 334)

helium capture The formation of heavy elements by the capture of a helium nucleus. For example, carbon can form heavier elements by fusion with other carbon nuclei, but it is much more likely to occur by helium capture, which requires less energy. (p. 456)

helium flash An explosive event in the post-main-sequence evolution of a low-mass star. When helium fusion begins in a dense stellar core, the burning is explosive in nature. It continues until the energy released is enough to expand the core, at which point the star achieves stable equilibrium again. (p. 425)

helium precipitation Mechanism responsible for the low abundance of helium of Saturn's atmosphere. Helium condenses in the upper layers to form a mist, which rains down toward Saturn's interior, just as water vapor forms into rain in the atmosphere of Earth. (p. 259)

high-energy telescope Telescope designed to detect radiation in X rays and gamma rays. (p. 118)

highlands Relatively light-colored regions on the surface of the Moon which are elevated several kilometers above the maria. Also called terrae. (p. 175)

homogeneity Assumed property of the universe such that the number of galaxies in an imaginary large cube of the universe is the same no matter where in the universe the cube is placed. (p. 566)

horizon problem One of two conceptual problems with the standard Big Bang model, which is that some regions of the universe which have very similar properties are too far apart to have exchanged information in the age of the universe. (p. 597)

horizontal branch Region of the H-R diagram where post-main sequence stars again reach hydrostatic equilibrium. At this point, the star is burning helium in its core, and hydrogen in a shell surrounding the core. (p. 425)

hot dark matter A class of candidates for the dark matter in the universe, composed of lightweight particles, such as neutrinos, much less massive than the electron. (p. 602)

H-R diagram A plot of luminosity against temperature (or spectral class) for a group of stars. (p. 368)

Hubble Classification scheme Method of classifying galaxies according to their appearance, developed by Edwin Hubble. (p. 514)

Hubble's constant The constant of proportionality which gives the relation between recessional velocity and distance in Hubble's law. (p. 529)

Hubble's law Law that relates the observed velocity of recession of a galaxy to its distance from us. The velocity of recession of a galaxy is proportional to its distance. (p. 528)

hydrogen envelope An invisible region engulfing the coma of a comet, usually distorted by the solar wind, and extending across millions of kilometers of space. (p. 305)

hydrogen shell burning Fusion of hydrogen in a shell that is driven by contraction and heating of the helium core. Once hydrogen is depleted in the core of a star, hydrogen burning stops and the core contracts due to gravity, causing the temperature to rise, heating the surrounding layers of hydrogen in the star, and increasing the burning rate there. (p. 422)

hydrosphere Layer of the Earth which contains the liquid oceans and accounts for roughly 70 percent of Earth's total surface area. (p. 146)

I

image The optical representation of an object produced when light from the object is reflected or refracted by a mirror or lens. (p. 97)

inertia The tendency of an object to continue in motion at the same speed and in the same direction, unless acted upon by a force. (p. 44)

inflation Short period of unchecked cosmic expansion early in the history of the universe. During inflation, the universe swelled in size by a factor of about 10^{50}. (p. 599)

infrared Region of the electromagnetic spectrum just outside the visible range, corresponding to light of a slightly longer wavelength than red light. (p. 55)

infrared telescope Telescopes designed to detect infrared radiation. These telescopes are designed to be lightweight so that they can be carried above most of Earth's atmosphere by balloons, airplanes, or satellites. (p. 115)

intensity A basic property of electromagnetic radiation that specifies the amount or strength or the radiation. (p. 62)

intercrater plains Regions on the surface of Mercury that do not show extensive cratering, but are relatively smooth. (p. 185)

interference The ability of two or more waves to interact in such a way that they either reinforce or cancel each other. (p. 57)

interferometer Collection of two or more radio telescopes working together as a team, observing the same object at the same time and at the same wavelength. The effective diameter of an interferometer is equal to the distance between its outermost dishes. (p. 114)

interferometry Technique in widespread use to dramatically improve the resolution of radio maps. Several radio telescopes observe the object simultaneously, and a computer analyzes how the signals interfere with each other. (p. 113)

interplanetary space The space between the objects in the solar system. (p. 128)

interstellar dust Microscopic dust grains that populate the space between stars, having their origins in the ejected matter of long-dead stars. (p. 321)

interstellar medium The matter between stars, composed of two components, gas and dust, intermixed throughout all of space. (p. 382)

inverse-square law The law that a field follows if its strength decreases with the square of the distance. Fields that follow the inverse square law rapidly decrease in strength as the distance increases, but never quite reach zero. (p. 46)

Io plasma torus Doughnut-shaped region of energetic ionized particles, emitted by the volcanoes on Jupiter's moon Io, and swept up by Jupiter's magnetic field. (p. 246)

ion tail Thin stream of ionized gas that is pushed away from the head of a comet by the solar wind. It extends directly away from the Sun. Often referred to as a plasma tail. (p. 306)

ionized State of an atom that has had at least one of its electrons removed. (p. 81)

ionosphere Layer in Earth's atmosphere above about 100 km where the atmosphere is significantly ionized, and conducts electricity. (p. 151)

irregular galaxy A galaxy which does not fit into any of the other major categories in the Hubble classification scheme. (p. 518)

isotopes Nuclei containing the same number of protons but different numbers of neutrons. Most elements can exist in several isotopic forms. A common example of an isotope is deuterium, which differs from normal hydrogen by the presence of an extra neutron in the nucleus. (p. 350)

isotropy Assumed property of the universe such that the universe looks the same in every direction. (p. 567)

J

jovian planet One of the four giant outer planets of the solar system, which resembles Jupiter in physical and chemical composition. (p. 132)

K

Kepler's Laws of Planetary Motion Three laws, based on precise observations of the motions of the planets by Tycho Brahe, which summarize the motions of the planets about the Sun. (p. 40)

Kirchhoff's Laws Three rules governing the formation of different types of spectra. (p. 78)

Kirkwood gaps Gaps in the spacings of semi-major axes of orbits of asteroids in the asteroid belt, produced by dynamical resonances with nearby planets, especially Jupiter. (p. 301)

Kuiper Belt A region in the plane of the solar system outside the orbit of Neptune where most short-period comets are thought to originate. (p. 304)

L

Lagrange point One of five special points in the plane of two massive bodies orbiting one another, where a third body of negligible mass can remain in equilibrium. (p. 270)

light *See* electromagnetic radiation.

light curve The variation in brightness of a star with time. (p. 373)

lighthouse model The leading explanation for pulsars. A small region of the neutron star, near one of the magnetic poles, emits a steady stream of radiation which sweeps past Earth each time the star rotates. Thus the period of the pulses is just the star's rotation period. (p. 466)

light-year The distance that light, moving at a constant speed of 300,000 km/s, travels in one year. One light year is about 10 trillion kilometers. (p. 4)

line of nodes The line of intersection of the Moon's orbit with the ecliptic plane. (p. 21)

lithosphere Earth's crust and a small portion of the upper mantle that make up Earth's plates. This layer of the Earth undergoes tectonic activity. (p. 162)

Local Group The small galaxy cluster that includes the Milky Way Galaxy. (p. 520)

luminosity One of the basic properties used to characterize stars, luminosity is defined as the total energy radiated by star each second, at all wavelengths. (p. 334)

luminosity class A classification scheme which groups stars according to the width of their spectral lines. For a group of stars with the same temperature, luminosity class differentiates between supergiants, giants, main-sequence stars and subdwarfs. (p. 372)

lunar eclipse Celestial event during which the moon passes through the shadow of the Earth, temporarily darkening its surface. (p. 17)

lunar phase The appearance of the moon at different points along its orbit. (p. 15)

M

Magellanic clouds Two small irregular galaxies that are gravitationally bound to the Milky Way Galaxy. (p. 519)

magnetic field Field which accompanies any changing electric field, and governs the influence of magnetized objects on one another. (p. 58)

magnetosphere A zone of charged particles trapped by a planet's magnetic field, lying above the atmosphere. (p. 146)

main sequence A well-defined band on an H-R diagram, on which most stars tend to be found, running from the top left of the diagram to the bottom right. (p. 369)

main-sequence turnoff Special point on an H-R diagram for a cluster. If all the stars in a particular cluster are plotted, the lower mass stars will trace out the main sequence up to the point where stars begin to evolve off the main sequence toward the red giant branch. The point where stars are just beginning to evolve off is the main-sequence turnoff. (p. 434)

mantle Layer of the Earth just interior to the crust. (p. 146)

mare Relatively dark-colored and smooth region on the surface of the Moon. (p. 175)

mass A measure of the total amount of matter contained within an object. (p. 44)

mass-luminosity relation The dependence of the luminosity of a main-sequence star on its mass. The luminosity increases roughly as the mass raised to the third power. (p. 374)

mass-radius relation The dependence of the radius of a main sequence star on its mass. The radius rises roughly in proportion to the mass. (p. 374)

mass-transfer binary *See* semi-detached binary.

matter-dominated universe A universe in which the density of matter exceeds the density of radiation. The present-day universe is matter-dominated. (p. 588)

meteor Bright streak in the sky, often referred to as a "shooting star," resulting from a small piece of interplanetary debris entering Earth's atmosphere and heating air molecules, which emit light as they return to their ground states. (p. 310)

meteor shower Event during which many meteors can be seen each hour, caused by the yearly passage of the Earth through the debris spread along the orbit of a comet. (p. 310)

meteorite Any part of a meteoroid that survives passage through the atmosphere and lands on the surface of the Earth. (p. 310)

meteoroid Chunk of interplanetary debris prior to encountering Earth's atmosphere. (p. 310)

meteoroid swarm Pebble-sized cometary fragments dislodged from the main body, moving in nearly the same orbit as the parent comet. (p. 310)

micrometeoroid Relatively small chunks of interplanetary debris ranging from dust particle size to pebble-sized fragments. (p. 310)

millisecond pulsar A pulsar whose period indicates that the neutron star is rotating nearly 1000 times each second. The most likely explanation for these rapid rotators is that the neutron star has been spun up by drawing in matter from a companion star. (p. 468)

molecular cloud A cold, dense interstellar cloud which contains a high fraction of molecules. It is widely believed that the relatively high density of dust particles in these clouds plays an important role in the formation and protection of the molecules. (p. 394)

molecular cloud complex Collection of molecular clouds that spans as much as 50 parsecs and may contain enough material to make millions of Sun-sized stars. (p. 395)

molecule A tightly bound collection of atoms held together by the electromagnetic fields of the atoms. Molecules, like atoms, emit and absorb photons at specific wavelengths. (p. 85)

moon A small body in orbit about a planet. (p. 128)

N

nebula General term used for any "fuzzy" patch on the sky, either light or dark. (p. 386)

nebular theory One of the earliest models of solar system formation, dating back to Descartes, in which a large cloud of gas began to collapse under its own gravity to form the Sun and planets. (p. 319)

neutrino Virtually massless and chargeless particle that is one of the products of fusion reactions in the Sun. Neutrinos move at close to the speed of light, and interact with matter hardly at all. (p. 350)

neutrino oscillations Possible solution to the solar neutrino problem, in which the neutrino has a very tiny mass. In this case, the correct number of neutrinos can be produced in the solar core, but on their way to Earth, some can "oscillate," or become transformed into other particles, and thus go undetected. (p. 353)

neutron An elementary particle with roughly the same mass as a proton, but which is electrically neutral. Along with protons, neutrons from the nuclei of atoms. (p. 84)

neutron capture The primary mechanism by which very massive nuclei are formed in the violent aftermath of a supernova. Instead of fusion of like nuclei, heavy elements are created by the addition of more and more neutrons to existing nuclei. (p. 457)

neutron star A dense ball of neutrons that remains at the core of a star after a supernova explosion has destroyed the rest of the star. Typical neutron stars are about 20 km across, and contain more mass than the Sun. (p. 464)

new Moon Phase of the moon during which none of the lunar disk is visible. (p. 15)

Newtonian mechanics The basic laws of motion, postulated by Newton, which are sufficient to explain and quantify virtually all of the complex dynamical behavior found on Earth and elsewhere in the universe. (p.44)

Newtonian telescope A reflecting telescope in which incoming light is intercepted before it reaches the prime focus and is deflected into an eyepiece at the side of the instrument. (p. 100)

north celestial pole Point on the celestial sphere directly above the Earth's north pole. (p. 9)

nova A star that suddenly increases in brightness, often by a factor of as much as 10,000, then slowly fades back to its original luminosity. A nova is the result of an explosion on the surface of a white dwarf star, caused by matter falling onto its surface from the atmosphere of a binary companion. (p. 444)

nucleotide base An organic molecule, the building block of genes that pass on hereditary characteristics from one generation of living creatures to the next. (p. 611)

nucleus Dense, central region of an atom, containing both protons and neutrons, and orbited by one or more electrons. (p. 81)

nucleus The solid region of ice and dust that composes the central region of the head of a comet. (p. 305)

O

Olber's paradox A thought experiment suggesting that if the universe were homogeneous, infinite, and unchanging, the entire night sky would be as bright as the surface of the Sun. (p. 568)

Oort cloud Spherical halo of material surrounding the solar system, out to a distance of about 50,000 A.U. where most comets originate. (p. 305)

opacity A quantity that measures a material's ability to block electromagnetic radiation. Opacity is the opposite to transparency. (p. 60)

open cluster Loosely bound collection of tens to hundreds of stars, a few parsecs across, generally found in the plane of the Milky Way. (p. 376)

open universe Geometry that the universe would have if the density of matter were less than the critical value. In an open universe there is not enough matter to halt the expansion of the universe. An open universe is infinite in extent. (p. 579)

outflow channel Surface feature on Mars, evidence that liquid water once existed there in great quantity, believed to be the relics of catastrophic flooding about 3 billion years ago. Found only in the equatorial regions of the planet. (p. 221)

ozone layer Layer of the Earth's atmosphere at an altitude of 20 to 50 km where incoming ultraviolet solar radiation is absorbed by oxygen, ozone, and nitrogen in the atmosphere. (p. 150)

P

pair production Process in which two photons of electromagnetic radiation give rise to a particle-antiparticle pair. (p. 589)

parallax The apparent motion of a relatively close object with respect to a more distant background as the location of the observer changes. (p. 24)

parsec The distance at which a star must lie in order that its measured parallax is exactly 1 arc second, equal to 206,000 A.U. (p. 358)

partial eclipse Celestial event during which only a part of the occulted body is blocked from view. (pp. 17, 18)

penumbra (I) Portion of the shadow cast by an eclipsing object in which the eclipse is seen as partial. (p. 18) (ii) The outer region of a sunspot, surrounding the umbra, which is not as dark, and not as cool as the central region. (p. 342)

perihelion The closest approach to the Sun of any object in orbit about it. (p. 41)

period The time needed for an orbiting body to complete one revolution about another body. (p. 42)

period-luminosity relation A relation between the pulsation period of a Cepheid variable and its absolute brightness. Measurement of the pulsation period allows the distance of the star to be determined. (p. 492)

permafrost Layer of permanently frozen water ice believed to lie just under the surface of Mars. (p. 221)

photoelectric effect Experiment concerning the detection of electrons from a metal surface, whose speed off the surface was dependent on the frequency of light striking the surface. The theoretical explanation rests on viewing light as made up of photons, or individual "bullets" of energy. (p. 80)

photometer A device that measures the total amount of light received in all or part of the image. (p. 101)

photometry Branch of observational astronomy in which intensity measurements are made through each of a set of standard filters. (p. 364)

photon Individual packet of electromagnetic energy that makes up electromagnetic radiation. (p. 79)

photosphere The visible surface of the Sun, lying just above the uppermost layer of the Sun's interior, and just below the chromosphere. (p. 332)

pixel One of many tiny picture elements, organized into an array, making up a digital image. (p. 106)

Planck curve see black-body curve (p. 62)

planet One of nine major bodies that orbit the Sun, visible to us by reflected sunlight. (p. 128)

planetary nebula The ejected envelope of a red giant star, spread over a volume roughly the size of our solar system. (p. 429)

planetary ring system Material organized into thin, flat rings encircling a giant planet, such as Saturn. (p. 259)

planetesimal Term given to objects in the early solar system that had reached the size of small moons, at which point their gravitational fields were strong enough to begin to influence their neighbors. (p. 322)

plate tectonics The motions of regions of the Earth's lithosphere, which drift with respect to one another. Also known as continental drift. (p. 161)

polarization The alignment of the electric fields of emitted photons, which are generally emitted with random orientations. (p. 384)

positron Atomic particle with properties identical to those a negatively charged electron, except for its positive charge. This positron is the antiparticle of the electron. Positrons and electrons annihilate one another when they meet, producing pure energy in the form of gamma rays. (p. 350)

precession The slow change in the direction of the axis of a spinning object, caused by some external influence. (p. 21)

primary atmosphere The chemical components that would have surrounded the Earth just after it formed. (p. 153)

prime focus The point in a reflecting telescope where the mirror focuses incoming light to a point. (p. 96)

primordial nucleosynthesis The production of elements heavier than hydrogen by nuclear fusion in the high temperatures and densities which existed in the early universe. (p. 594)

Principle of Cosmic Censorship A proposition to separate the unexplained physics near a singularity from the rest of the well-behaved universe. The principle states that nature always hides any singularity, such as a black hole, inside an event horizon, which insulates the rest of the universe from seeing it. (p. 480)

prominence Loop or sheet of glowing gas ejected from an active region on the solar surface, which then moves through the inner parts of the corona under the influence of the Sun's magnetic field. (p. 346)

proper motion The angular movement of a star across the sky, as seen from Earth, measured in seconds of arc per year. This movement is a result of the star's actual motion through space. (p. 359)

proton An elementary particle, carrying a positive electric charge, a component of all atomic nuclei. The number of protons in the nucleus of an atom dictates what type of atom it is. (p. 57)

proton-proton chain The chain of fusion reactions, leading from hydrogen to helium, that powers main-sequence stars. (p. 350)

protoplanet Clump of material, formed in the early stages of solar system formation, that was the forerunner of the planets we see today. (p. 320)

protostar Stage in star formation when the interior of a collapsing fragment of gas is sufficiently hot and dense that it becomes opaque to its own radiation. The protostar is the dense region at the center of the fragment. (p. 403)

protosun The central accumulation of material in the early stages of solar system formations, the forerunner of the present-day Sun. (p. 320)

Ptolemaic model Solar system model, developed by the sec-ond century astronomer Claudius Ptolemy, perhaps the best geocentric model to be proposed. It predicted with great accuracy the positions of the know planets, using more than 80 circles to model the (then know) planets. (p. 34)

pulsar Object that emits radiation in the form of rapid pulses with a characteristic pulse period and duration. Charged particles, accelerated by the magnetic field of a rapidly rotating neutron star, flow along the magnetic field lines, producing radiation that beams outward as the star spins on its axis. (p. 465)

pulsating variable star A star whose luminosity varies in a predictable, periodic way. (p. 490)

Q

quantization The fact that light and matter on small scales behave in a discontinuous manner, and manifest themselves in the form of tiny "packets" of energy, called quanta. (p. 81)

quarter moon Lunar phase in which the moon appears as a half disk. (p. 15)

quasar Star-like radio source with an observed redshift that indicates extremely large distance from Earth. (p. 542)

quasi-stellar object (QSO) *See* quasar.

quiet Sun The underlying predictable elements of the Sun's behavior, such as its average photospheric temperature, which do not change in time. (p. 341)

R

radar Acronym for Radio Detection And Ranging. Radio waves are bounced off an object, and the time at which the echo is received indicates its distance. (p. 43)

radial motion Motion along a particular line of sight, which induces apparent changes in the wavelength (or frequency) of radiation received. (p. 68)

radiation A way in which energy is transferred from place to place in the form of a wave. Light is a form of electromagnetic radiation. (p. 54)

radiation darkening The effect of chemical reactions that result when high-energy particles strike the icy surfaces of objects in the outer solar system. The reactions lead to a build-up of a dark layer of material. (p. 288)

radiation-dominated universe Early epoch in the universe, when the density of radiation in the cosmos exceeded the density of matter. (p. 589)

radiation zone Region of the Sun's interior where extremely high temperatures guarantee that the gas is completely ionized. Photons are only occasionally diverted by electrons, and travel through this region with relative ease. (p. 332)

radio Region of the electromagnetic spectrum corresponding to radiation of the longest wavelengths. (p. 59)

radio galaxy Type of active galaxy that emits most of its energy in the form of long-wavelength radiation. (p. 542)

radio lobe Roundish region of radio-emitting gas, lying well beyond the center of a radio galaxy. (p. 543)

radio telescope Large instrument designed to detect radiation from space in radio wavelengths. (p. 108)

radioactivity The release of energy by rare, heavy elements when their nuclei decay into lighter nuclei. (p. 159)

radius-luminosity-temperature relation A mathematical proportionality, arising from Stefan's Law, which allows astronomers to indirectly determine the radius of a star once its luminosity and temperature are known. (p. 361)

red dwarfs Small, cool faint stars at the lower-right end of the main sequence on the H-R diagram, whose color and size give them their name. (p. 369)

red giant A giant star whose surface temperature is relatively low, so that it glows with a red color. (p. 361)

red-giant branch The section of the evolutionary track of a star that corresponds to continued heating from rapid hydrogen shell burning, which drives a steady expansion and cooling of the outer envelope of the star. As the star gets larger in radius and its surface temperature cools, it becomes a red giant. (p. 423)

red giant region The upper right hand corner of the H-R diagram, where red-giant stars are found. (p. 371)

red shift Motion-induced change in the wavelength of light emitted from a source moving away from us. The relative recessional motion causes the wave to have an observed wavelength longer (and hence redder) than it would if it were not moving. (p. 68)

red supergiant An extremely luminous red star. Often found on the asymptotic giant branch of the H-R diagram. (p. 427)

reddening Dimming of starlight by interstellar matter, which tends to scatter higher-frequency (blue) components of the radiation more efficiently than the lower-frequency (red) components. (p. 382)

reflecting telescope A telescope which uses a carefully designed mirror to gather and focus light from a distant object. (p. 96)

refracting telescope A telescope which uses a lens to gather and focus light from a distant object. (p. 96)

refraction The tendency of a wave to bend as it passes from one transparent medium to another. (p. 60)

residual cap Portion of Martian polar ice caps that remains permanently frozen, undergoing no seasonal variations. (p. 224)

retrograde motion Backward, westward loop traced out by a planet with respect to the fixed stars. (p. 32)

revolution Orbital motion of one body about another, such as the Earth about the Sun. (p. 10)

right ascension Celestial coordinate used to measure longitude on the celestial sphere. The zero point is the position of the Sun on the vernal equinox. (p. 14)

rille A ditch on the surface of the Moon where molten lava flowed in the past. (p. 185)

ringlet Narrow region in Saturn's planetary ring system where the density of ring particles is high. Voyager discovered that the rings visible from Earth are actually composed of tens of thousands of ringlets. (p. 262)

Roche limit Often called the tidal stability limit, the Roche limit gives the distance from a planet at which the tidal force, due to the planet, between adjacent objects exceeds their mutual attraction. Objects within this limit are unlikely to accumulate into larger objects. The rings of Saturn occupy the region within Saturn's Roche limit. (p. 261)

Roche lobe An imaginary surface around a star. Each star in a binary system can be pictured as being surrounded by a tear-shaped zone of gravitational influence, the Roche lobe. Any material within the Roche lobe of a star can be considered to be part of that star. During evolution, one member of the binary star can expand so that it overflows its own Roche lobe, and begins to transfer matter onto the other star. (p. 437)

rotation Spinning motion of a body about an axis. (p. 9)

rotation curve Plot of the orbital speed of disk material in a galaxy against its distance from the galactic center. Analysis of rotation curves of spiral galaxies indicates the existence of dark matter. (p. 504)

RR Lyrae star Variable star whose luminosity changes in a characteristic way. All RR Lyrae stars have more or less the same period. (p. 491)

runaway greenhouse effect A process in which the heating of a planet leads to an increase in its atmosphere's ability to retain heat and thus to further heating, quickly causing extreme changes in the temperature of the surface and the composition of the atmosphere. (p. 202)

runoff channel River-like surface feature on Mars, evidence that liquid water once existed there in great quantities. They are found in the southern highlands, and are thought to have been formed by water that flowed nearly 4 billion years ago. (p. 221)

S

S0 galaxy Galaxy which shows evidence of a thin disk and a bulge, but which has no spiral arms and contains little or no gas. (p. 517)

SB0 galaxy An S0-type galaxy whose disk shows evidence of a bar. (p. 517)

scarp Surface feature on Mercury believed to be the result of cooling and shrinking of the crust forming a wrinkle on the face of the planet. (p. 185)

Schwarzschild radius The distance from the center of an object such that, if all the mass compressed within that region, the escape velocity would equal the speed of light. Once a stellar remnant collapses within this radius, light cannot escape and the object is no longer visible. (p. 472)

scientific method The set of rules used to guide science, based on the idea that scientific "laws" be continually tested, and replaced if found inadequate. (p. 38)

seasonal cap Portion of Martian polar ice caps that is subject to seasonal variations, growing and shrinking once each Martian year. (p. 224)

seasons Changes in average temperature and length of day that result from the tilt of Earth's (or any planet's) axis with respect to the plane of its orbit. (p. 10)

secondary atmosphere The chemicals that composed the Earth's atmosphere after the planet's formation, once volcanic activity outgassed chemicals from the interior. (p. 153)

seeing A term used to describe the ease with which good telescopic observations can be made from Earth's surface, given the blurring effects of atmospheric turbulence. (p. 105)

seeing disk Roughly circular region on a detector over which a star's pointlike images is spread, due to atmospheric turbulence. (p. 105)

seismic wave A wave that travels outward from the site of an earthquake through the Earth. (p. 156)

semi-major axis One half of the major axis of an ellipse. The semi-major axis is the way in which the size of an ellipse is usually quantified. (p. 41)

Seyfert galaxy Type of active galaxy whose emission comes from a very small region within the nucleus of an otherwise normal-looking spiral system. (p. 541)

shepherd satellites Satellites whose gravitational effects on a ring preserve its shape, such as the two satellites of Saturn, Prometheus and Pandora, whose orbits lie on either side of the F ring. (p. 264)

shield volcano A volcano produced by repeated nonexplosive eruptions of lava, creating a gradually sloping, shield-shaped low dome. Often contains a caldera at its summit. (p. 207)

shock wave Wave of matter, which may be generated by a star, which pushes material outward into the surrounding molecular cloud. The material tends to pile up, forming a rapidly-expanding shell of dense gas. (p. 411)

sidereal day The time needed for a star on the celestial sphere to make one complete rotation in the sky. (p. 10)

sidereal month Time required for the Moon to complete one trip around the celestial sphere. (p. 15)

sidereal year The time required for the constellations to complete once cycle around the sky and return to their starting points, as seen from a given point on Earth. (p. 11)

singularity A point in the universe where the density of matter and the gravitational field are infinite, such as at the center of a black hole. (p. 472)

solar constant The amount of solar energy reaching the Earth per unit area per unit time, approximately 1400 W/m2. (p. 333)

solar core The region at the center of the Sun, with a radius of nearly 200,000 km, where powerful nuclear reactions generate the Sun's energy output. (p. 332)

solar cycle The 22-year period that is needed for both the average number of spots and the Sun's magnetic polarity to repeat themselves. The Sun's polarity reverses on each new 11-year sunspot cycle. (p. 344)

solar day The period of time between the instant when the Sun is directly overhead (i.e. at noon) to the next time it is directly overhead. (p. 10)

solar eclipse Celestial event during which the new Moon passes directly between the Earth and Sun, temporarily blocking the Sun's light. (p. 18)

solar interior The region of the Sun between the solar core and the photosphere. (p. 332)

solar maximum The starting point of the sunspot cycle, during which only a few spots are seen. They are generally confined to narrow regions, one in each hemisphere, at about 25-30 degrees latitude. (p. 344)

solar minimum The starting point of the sunspot cycle, during which only a few spots are seen. They are generally confined to narrow regions, one in each hemisphere, at about 25—30 degrees latitude. (p. 344)

solar nebula The swirling gas surrounding the early Sun during the epoch of solar system formation, also referred to as the primitive solar system. (p. 319)

solar neutrino problem The discrepancy between the theoretically predicted flux of neutrinos streaming from the Sun as a result of fusion reactions in the core and the flux which is actually observed. The observed number of neutrinos is only about half the predicted number. (p. 352)

solar system The Sun, and all the planets that orbit the Sun-Mercury, Venus, Earth, Mars, Jupiter, Saturn, Uranus, Neptune, and Pluto. (p. 128)

solar wind An outward flow of fast-moving charged particles from the Sun. (p. 341)

south celestial pole Point on the celestial sphere directly above the Earth's south pole. (p. 9)

spectral class Classification scheme, based on the strength of stellar spectral lines, which is an indication of the temperature of a star. (p. 367)

spectrograph Instrument used to produce detailed spectra of stars. Usually, a spectrograph records a spectrum on a photographic plate, or more recently, in electronic form on a computer. (p. 74)

spectroscope Instrument used to view a light source so that it is split into its component colors. (p. 74)

spectroscopic binary A binary-star system which from Earth appears as a single star, but whose spectral lines show back-and-forth Doppler shifts as two stars orbit one another. (p. 373)

spectroscopic parallax Method of determining the distance to a star by measuring its temperature and then determining its absolute brightness by comparing with a standard H-R diagram. The absolute and apparent brightness of the star give the star's distance from Earth. (p. 371)

spectroscopy The study of the way in which atoms absorb and emit electromagnetic radiation. Spectroscopy allows astronomers to determine the chemical composition of stars. (p. 77)

speed of light The fastest possible speed, according to the currently known laws of physics. Electromagnetic radiation exists in the form of waves or photons moving at the speed of light. (p. 58)

spin-orbit resonance State that a body is said to be in if its rotation period and its orbital period are related in a simple way. (p. 179)

spiral arm Distribution of material in a galaxy in a pinwheel-shaped design apparently emanating from near the galactic center. (p. 490)

spiral density wave (i) A wave of matter formed in the plane of planetary rings, similar to ripples on the surface of a pond, which wrap around the rings forming spiral patterns similar to grooves in a record disk. Spiral density waves can lead to the appearance of ringlets. (p. 262) (ii) A proposed explanation for the existence of galactic spiral arms, in which coiled waves of gas compression move through the galactic disk, triggering star formation. (p. 501)

spiral galaxy Galaxy composed of a flattened, star-forming disk component which may have spiral arms and a large central galactic bulge. (pp. 490, 515)

standard candle Any object with an easily recognizable appearance and known luminosity, which can be used in estimating distances. Supernovae, which all have the same peak luminosity (depending on type) are good examples of standard candles and are used to determine distances to other galaxies. (p. 453)

Standard Solar Model A self-consistent picture of the Sun, developed by incorporating the important physical processes that are believed to be important in determining the Sun's internal structure, into a computer program. The results of the program are then compared with observations of the Sun, and modifications are made to the model. The Standard Solar Model, which enjoys widespread acceptance is the result of this process. (p. 334)

star A glowing ball of gas held together by its own gravity and powered by nuclear fusion in its core. (p. 332)

star cluster A grouping of anywhere from a dozen to a million stars which formed at the same time from the same cloud of interstellar gas. Stars in clusters are useful to aid our understanding of stellar evolution because they are all roughly the same age and chemical composition, and lie at roughly the same distance from Earth. (p. 376)

starburst galaxy Galaxy in which a violent event, such as near-collision, has caused a sudden, intense burst of star formation in the recent past. (p. 535)

Stefan's Law Relation that gives the total energy emitted per square centimeter of its surface per second by an object of a given temperature. Stefan's Law shows that the energy emitted increases rapidly with an increase in temperature, proportional to the temperature raised to the fourth power. (p. 65)

stellar nucleosynthesis The formation of heavy elements by the fusion of lighter nuclei in the hearts of stars. Except for hydrogen and helium, all other elements in our universe result from stellar nucleosynthesis. (p. 454)

stellar occultation The dimming of starlight produced when a solar system object such as a planet, moon or ring, passes directly in front of a star. (p. 285)

stratosphere The portion of Earth's atmosphere lying above the troposphere, extending up to an altitude of 40 to 50 km. (p. 150)

subgiant branch The section of the evolutionary track of a star that corresponds to changes that occur just after hydrogen is depleted in the core, and core hydrogen burning ceases. Shell hydrogen burning heats the outer layers of the star, which causes a general expansion of the stellar envelope. (p. 423)

summer solstice Point on the ecliptic where the Sun is at its northernmost point above the celestial equator, occurring on or near June 21. (p. 10)

sunspot An Earth-sized dark blemish found on the surface of the Sun. The dark color of the sunspot indicates that it is a region of lower temperature than its surroundings. (p. 342)

sunspot cycle The fairly regular pattern that the number and distribution of sunspots follows, in which the average number of spots reaches a maximum every 11 or so years then fall off to almost zero. (p. 344)

supercluster Grouping of several clusters of galaxies into a larger, but not necessarily gravitationally bound, unit. (p. 524)

supergiant A star with a radius between 100 and 1000 times that of the sun. (p. 361)

supergranulation Large-scale flow pattern on the surface of the Sun, consisting of cells measuring up to 30,000 km across, believed to be the imprint of large convective cells deep in the solar interior. (p. 336)

supernova Explosive death of a star, caused by the sudden onset of nuclear burning (type I), or an enormously energetic shock wave (type II). One of the most energetic events of the universe, a supernova may temporarily outshine the rest of the galaxy in which it resides. (p. 447)

supernova remnant The scattered glowing remains from a supernova that occurred in the past. The Crab Nebula is one of the best-studied supernova remnants. (p. 451)

synchrotron radiation Type of nonthermal radiation caused by high-speed charged particles, such as electrons, emitting radiation as they are accelerated in a strong magnetic field. (p. 550)

synchronous orbit State of an object when its period of rotations is exactly equal to its average orbital period. The Moon is in a synchronous orbit, and so presents the same face toward Earth at all times. (p. 178)

synodic month Time required for the Moon to complete a full cycle of phases. (p. 16)

T

T Tauri star Protostar in the late stages of formation, often exhibiting violent surface activity. T Tauri stars have been observed to brighten noticeably in a short period of time, consistent with the idea of rapid evolution during this final phase of stellar formation. (p. 404)

tail Component of a comet that consists of material streaming away from the main body, sometimes spanning hundreds of millions of kilometers. May be composed of dust or ionized gases. (p. 305)

telescope Instrument used to capture as many photons as possible from a given region of the sky and concentrate them into a focused beam for analysis. (p. 96)

temperature A measure of the amount of heat in an object, and an indication of the speed of the particles that comprise it. (p. 62)

terrae *See* highlands.

terrestrial planet One of the four innermost planets of the solar system, resembling the Earth in general physical and chemical properties. (p. 132)

theories of relativity Einstein's theories, on which much of modern physics rests. Two essential facts of the theory are that nothing can travel faster than the speed of light, and that everything, including light, is affected by gravity. (p. 470)

thick disk Region of a spiral galaxy where an intermediate population of stars resides, younger than the halo stars, but older than stars in the disk. (p. 496)

tidal bulge Elongation of the Earth caused by the difference between the gravitational force on the side nearest the Moon and the force on the side farthest from the Moon. The long axis of the tidal bulge points toward the Moon. (p. 147) More generally, the deformation of any body produced by the tidal effect of a nearby gravitating object. (p. 147)

tidal force The variation in one body's gravitational force from place to place across another body-for example, the variation of the Moon's gravity across the Earth. (p. 148)

tides Rising and falling motion that bodies of water follow, exhibiting daily, monthly and yearly cycles. Ocean tides on Earth are caused by the competing gravitational pull of the Moon and Sun on different regions of the Earth. (p. 146)

time dilation A prediction of the theory of relativity, closely related to the gravitational reshift. To an outside observer, a clock lowered into a strong gravitational field will appear to run slow. (p. 478)

total eclipse Celestial event during which one body is completely blocked from view by another. (p. 17)

transition zone The region of rapid temperature increases that separates the Sun's chromosphere from the corona. (p. 341)

transverse motion Motion perpendicular to a particular line of sight, which does not result in Doppler shift in radiation received. (p. 68)

triangulation Method of determining distance based on the principles of geometry. A distant object is sighted from two well-separated locations. The distance between the two locations and the angle between the line joining them and the line to the distant object are all that are necessary to ascertain the object's distance. (p. 22)

triple-alpha process The generation of Carbon-12 from the fusion of three helium-4 nuclei (alpha particles). Helium-burning stars occupy a region of the H-R diagram known as the horizontal branch. (p. 424)

Trojan asteroid One of two groups of asteroids which orbit at the same distance from the Sun as Jupiter, 60 degrees ahead and behind the planet. (p. 300)

troposphere The portion of Earth's atmosphere from the surface to about 15 km. (p. 150)

Tully-Fisher relation A relation used to determine the absolute luminosity of a spiral galaxy. The rotational velocity, measured from the broadening of spectral lines, is related to the total mass, and hence the total luminosity. (p. 521)

21-centimeter radiation Radio radiation emitted when an electron in the ground state of a hydrogen atom flips its spin to become parallel to the spin of the proton in the nucleus. (p. 393)

type-I supernova One possible explosive death of a star. A white dwarf in a binary system can accrete enough mass that it cannot support its own weight. The star collapses and temperatures become high enough for carbon fusion to occur. Fusion begins throughout the white dwarf almost simultaneously and an explosion results. (p. 448)

type-II supernova One possible explosive death of a star, in which the highly evolved stellar core rapidly implodes and then explodes, destroying the surrounding star. (p. 448)

U

ultraviolet Region of the electromagnetic spectrum, just outside the visible range, corresponding to wavelengths slightly shorter than blue light. (p. 55)

ultraviolet telescope A telescope that is designed to collect radiation in the ultraviolet part of the spectrum. The Earth's atmosphere is partially opaque to these wavelengths, so ultraviolet telescopes are put on rockets, balloons or satellites to get high above most or all of the atmosphere. (p. 118)

umbra (i) Central region of the shadow cast by an eclipsing body. (p. 18) (ii) The central region of a sunspot, which is its darker and cooler part. (p. 342)

unbound An orbit which does not stay in a specific region of space, but where an object escapes the gravitational field of another. Typical unbound orbits are hyperbolic in shape. (p. 49)

universe The totality of all space, time, matter and energy. (p. 4)

V

Van Allen belts At least two doughnut-shaped regions of magnetically trapped charged particles high above the Earth's atmosphere. (p. 154)

variable star A star whose luminosity changes with time. (p. 490)

visible spectrum The small range of the electromagnetic spectrum that human eyes perceive as light. The visible spectrum ranges from about 4000 to 7000 angstroms, corresponding to blue through red light. (p. 60)

vernal equinox Date on which the Sun crosses the celestial equator moving northward, occurring on or near March 21. (p. 10)

visible light The small range of the electromagnetic spectrum that human eyes perceive as light. The visible spectrum ranges from about 400 to 700 nm, corresponding to blue through red light. (p. 54)

visual binary A binary star system in which both members are resolvable from Earth. (p. 373)

void Large, relatively empty region of the universe around which superclusters of galaxies are organized. (p. 521)

volcano Upwelling of hot lava from below Earth's crust to the planet's surface. (p. 159)

W

water hole The radio interval between 18 cm and 21 cm, the wavelengths at which hydroxyl (OH) and hydrogen (H) radiate, respectively, in which intelligent civilizations might conceivably send their communication signals. (p. 621)

wave A pattern that repeats itself cyclically in both time and space. Waves are characterized by the velocity with which they move, their frequency, and their wavelength. (p. 55)

wave period The amount of time required for a wave to repeat itself at a specific point in space. (p. 55)

wavelength The length from one point on a wave to the point where it is repeated exactly in space, at a given time. (p. 55)

weird terrain A region on the surface of Mercury of oddly rippled features. This feature is thought to be the result of a strong impact which occurred on the other side of the planet, and sent seismic waves traveling around the planet, converging in the weird region. (p. 186)

white dwarf A dwarf star with a surface temperature that is hot, so that the object glows white. (p. 362)

white dwarf region The bottom left-hand corner of the H-R diagram, where white dwarf stars are found. (p. 370)

white light Visible light that contains approximately equal proportions of all colors. (p. 370)

white oval Light-colored region near the Great Red Spot in Jupiter's atmosphere. Like the red spot, such regions are apparently rotating storm systems. (p. 239)

Wien's Law Relation which gives the connection between the wavelength at which a black-body curve peaks and the temperature of the emitter. The temperature is inversely proportional to the peak wavelength, so the hotter the object, the bluer its radiation. (p. 64)

winter solstice Point on the ecliptic where the Sun is at its southernmost point below the celestial equator, occurring on or near December 21. (p. 10)

X

X ray Region of the electromagnetic spectrum corresponding to radiation of high frequency and short wavelengths, far outside the visible spectrum. (p. 59)

X-ray burster X-ray source that radiates thousands of times more energy than our Sun, in short bursts that last only a few seconds. A neutron star in a binary system accretes matter onto its surface until temperatures reach the level needed for hydrogen fusion to occur. The result is a sudden period of rapid nuclear burning and release of energy. (p. 467)

Z

zero-age main sequence The region on the H-R diagram, as predicted by theoretical models, where stars are located at the onset of nuclear burning in their cores. (p. 406)

zodiac The twelve constellations through which the Sun moves as it follows its path on the ecliptic. (p. 11)

zonal flow Alternating regions of westward and eastward flow, roughly symmetrical about the equator of Jupiter, associated with the belts and zones in the planet's atmosphere. (p. 238)

zones Bright, high pressure regions, where gas flows upward, in the atmosphere of a jovian planet. (p. 237)

Photo Credits

Index

A0620–00, 482
Aberration of starlight, 39
Absolute brightness, 362, 371, 376, 378, 492
Absolute magnitude, 367
Absolute zero, 63
Absorption lines, 76–78, 82–84, 92
   effect of dark dust clouds, 390–91
   solar, 337–38
   of stars, 365, 382
Acceleration, 45, 50
Accretion disk, 445–46, 460, 467, 548–49
Accretion process:
   in active galaxies, 547–48
   in quasars, 553–55
   in solar system formation, 322–24, 328
Active galaxies, 540–41, 562
   central engine of, 546–50
   energy emission from, 548–50
   energy production in, 546–48
   evolution of, 558–61
   internal motion of, 546
   luminosity of, 546
   radiative character of, 540
   radio galaxies, 540, 542–46
   Seyfert galaxies, 540–42, 549, 559–62
Active optics, 107, 124
Active prominence, 346
Active sites, on Earth's surface, 159–61, 166–67
Active Sun, 341–47, 354
Adams, John, 277
Adaptive optics, 108, 124, 361
Advanced X-ray Astrophysics Facility, 120
Aesthenosphere, 162–63, 165, 168
Aldebaran, 364
Algol, 436–38
Alpha capture, 458
Alpha Centauri, 358–61, 364, 376
Alpha particle, 424
Alpha process, 457
Alpha ring, of Uranus, 288–89
Amalthea, 242–43
Amino acids, 611, 614, 622
Amor asteroid, 298
Ancient astronomy, 30–32
Andromeda Galaxy, 54, 489–90, 520
Angstrom, 13, 60
Angular measurement, 14–15
Angular momentum, 321
Angular momentum problem, 327–28
Angular resolution, 102–4, 109, 113, 124
Angular size, 131
Annual displacement, 359
Annular eclipse, 18–19, 21, 26
Antiparticle, 350
Aphelion, 41, 179–80
Apollo asteroid, 298–99
Apollo program, 188
Apparent brightness, 67, 70, 362–63, 366, 371, 376, 382, 521
Apparent magnitude, 367
Arc degree, 14

Archival data, 368
Arc minute, 14–15
Arc second, 14–15, 358
Arcturus, 364
Arecibo telescope, 110–12
Ariel, 283–84
A ring, of Saturn, 259–60, 264, 272
Aristarchus of Samos, 33
Aristotle, 32, 35, 39, 44–45
Armstrong, Neil, 188
Arp, Halton, 560–61
Associations (star clusters), 414–16
Assumption of mediocrity, 610
Asteroid, 129, 133, 142, 298–304, 314, 318
   collision with Earth, 298–301
   earth-crossing asteroids, 298–99, 314
   formation of, 323
   orbital resonance, 300–302
   orbit of, 298
   physical properties of, 303–4
   stray asteroids, 310–11
Trojan asteroids, 300–302, 314
Asteroid belt, 129–30, 142, 298–99, 314, 326
Astrology, 8
Astronomical unit, 13, 42–44, 50
Astronomy, 6, 8–9, 26
Atmosphere of Earth, 136–37, 146, 149–53, 168, 201
   blockage of radiation by, 61–62
   circulation of, 349
   color of sky, 152
   composition of, 61, 137
   opacity of, 61–62, 70
   origin of, 153
   stability of, 136–37
   structure of, 150–51
   surface heating, 151–52
   windows in, 62
Atmosphere of Jupiter, 236–40
Atmosphere of Mars, 225–26
Atmosphere of Mercury, 174–75
Atmosphere of Moon, 174–75
Atmosphere of Neptune, 279–81
Atmosphere of Saturn, 254–58
Atmosphere of Sun, 337–41
Atmosphere of Uranus, 279–81
Atmosphere of Venus, 198–203
Atom, 92
   Bohr model of, 81
   excited state of, 81–82
   formation of, 593–96
   structure of, 81–85
Atomic epoch, 591, 593, 596
Aurora, 154, 349
Aurora australis, 154
Aurora borealis, 154
Autumnal equinox, 10–11, 26

Balmer series, 88–89
Barnard 5, 409
Barnard's Star, 358–59, 364

Barred-spiral galaxy, 514, 516–17, 536
Baryonic matter, 596
Basalt, 159, 183
Baseline, of triangle, 23–24, 26
Belinda, 283
Bell, Jocelyn, 465
Belts, of Jupiter, 237–38, 250
Beryllium-8, 424
Bessel, Friedrich, 358
Beta ring, of Uranus, 288–89
Betelgeuse, 13, 15, 364
Bianca, 283
Big Bang, 568–71, 584
Big Crunch, 575
Big Horn Medicine Wheel, 30–31
Binary asteroid, 304
Binary galaxy, 525–26, 535
Binary pulsar, 483
Binary quasar, 555–58
Binary star system, 373–74, 378, 444
   black holes in, 481–82
   contact binaries, 438–39
   detached binaries, 437–38
   eclipsing binaries, 373, 378
   evolution of, 436–38
   gravitational radiation from, 483
   mass-transfer binaries, 437, 439, 444–46
   neutron-star binaries, 467–70
   semidetached binaries, 437
   spectroscopic binaries, 373, 378
   visual binaries, 373, 378
Biological evolution, 612–13
Bipolar flow, 410
Black-body curve, 62, 64–66, 70, 363, 479
   for universe, 580–81
Black-body spectrum, 62
Black dwarf, 428–32, 439, 444, 505
Black hole, 474
   in binary system, 481–82
   candidates for, 481–83
   center of, 479–80
   at center of Milky Way, 507
   as dark matter, 505
   evaporation of, 479
   final stage of stellar evolution, 472
   matter falling into, 476–77
   object orbiting in, 476
   observational evidence for, 480–83
   properties of, 475–77
   space travel near, 477–80
   stellar transit of, 480–81
   supermassive black hole, 547–48, 553, 561
Blazar, 546
BL Lac objects, 546, 559
Blue giant, 369–70, 378
Blueshift, 68–70, 87, 90, 498, 522, 575
Blue straggler, 436
Blue supergiant, 369, 378
Bode, Johann, 135
Bohr, Niels, 81, 88
Bohr atom, 81, 92

Bondi, Hermann, 583
Boson, 598
Bound universe, 574, 578
Brackett series, 89
Brightness, 362, 366–67
B ring, of Saturn, 259–60, 264, 272
Brown dwarf, 407, 416, 505–6
Brown oval, 239–40, 250
Bruno, Giordano, 38
Burbridge, Geoffrey, 560

Caldera, 208
Calendar, 8, 30
Callisto, 242–44, 248–49
Calypso, 265, 270
Cannon, Annie, 495
Canopus, 364
Capricornus, 11
Capture theory, of origin of Moon, 189
Caracol temple, 30–31
Carbon:
    carbon-based life, 613–14
    carbon-helium fusion, 456
    formation of, 455–56
Carbonaceous meteorite, 313
Carbon burning, 456
Carbon core, of star, 426–29
Carbon-detonation supernova, 448–51, 460
Carbon monoxide, in molecular clouds, 497
Cassegrain, Guillaume, 100
Cassegrain telescope, 100, 124
Cassini, Giovanni Domenico, 259
Cassini Division, 259, 262–64, 272
Cassiopeia A, 449
Cataclysmic variables, 490
Catalyst, 422
Catastrophic theory, 319, 326–28
Cat's Eye Nebula, 429
Celestial coordinates, 13–15, 26
Celestial equator, 9, 14, 26
Celestial mechanics, 139
Celestial pole, 9, 26
Celestial sphere, 9, 26
Celsius temperature scale, 63
Center of mass, 48, 50
Centimeter-gram-second system, 13
Cepheid variable, 490–92, 510, 520
Ceres, 303
Chandrasekhar, Subramanyan, 448
Chandrasekhar mass, 448, 460, 471–72
Chaotic rotation, 271–72
Charge-coupled device, 106, 124
Charged particle interactions, 57–58
Charon, 292–93, 327
Chemical evolution, 611–12, 614
Chromatic aberration, 97–98, 124
Chromosphere, of Sun, 332–33, 339–40, 354
Closed universe, 578–79, 584
Cloud cover:
    of Neptune, 280
    of Saturn, 256, 258
    of Venus, 199, 201
CNO cycle, 422, 455
COBE satellite, 582–83, 603–5
Cocoon nebula, 409
Cold dark matter, 602–3, 606
Collecting area, 101–2, 124

Color index, 364, 378
Color-magnitude diagram, 368, 378
Coma, 101, 305, 314
Coma Cluster, 5, 514, 524
Comet, 129, 133, 142, 304–10, 314, 318
    cometary fragments, 310–12
    cometary impact, 308–10
    formation of, 323
    orbit of, 304–5
    physical properties of, 308
    structure of, 305–6
Comet Kohoutek, 306
Comet Shoemaker-Levy 9, 308–10
Comparative planetology, 128, 141–42
Compton Observatory, 121
Condensation nuclei, 321, 328
Condensation theory, 319–24, 328
Conservation of mass and energy, 349, 354, 583
Constellations, 6–9, 26
    as navigational guides, 7–8
    primitive calendars, 8
    summer, 11–12
    winter, 11–12
Constructive interference, 57
Contact binary, 438–39
Continental drift, 161–65, 168
Continuous spectrum, 75, 78, 92
Convection, 150, 168, 225
    convection zone, of Sun, 332–36, 354
    plate drift and, 164–65
Convection cell, 150
Co-orbital satellite, 270
Copernican principle, 40
Copernican revolution, 35–39, 50
Copernicus, Nicholas, 4, 35–36
Cordelia, 283, 289
Core-collapse supernova, 447, 460
Core-halo radio galaxy, 542–43
Core-hydrogen burning, 420–21, 439
Corona, solar, 18, 332–33, 340–41, 347, 354
Coronae (volcanic structure), 208
Coronal holes, 341, 354
"Coronium," 340
Cosmic distance scale, 22–26, 371–72, 489–90, 530
    (*See also* Distance indicators)
Cosmic evolution, 610, 622
Cosmic microwave background, 579–84, 588, 596–97, 603–4
Cosmic rays, 509
Cosmological constant, 572–73, 583
Cosmological distance, 528
Cosmological principle, 566–67, 574, 584
Cosmological redshift, 528, 536, 571–73
Cosmology, 32, 50, 565–84
Coudé room, 100–101
Crab Nebula, 450–51, 465
Crater-chain pattern, 184
Craters:
    of Mars, 220–21
    of Mercury, 184–86
    of Moon, 176, 180–82, 191
    of Venus, 206–9
Crescent Moon, 15–16
Cressida, 283
C ring, of Saturn, 259–60, 264, 272

Critical density, 573–77, 584
Critical universe, 579, 584
C-type asteroid, 303
Current sheet, 242
Curved space, 475–76, 578–81
    negative curvature, 579–81
    positive curvature, 579
Cygnus X-1, 481

Dark dust clouds, 389–92, 396
Dark halo, 505, 510, 535
Dark matter, 557–58, 592, 596, 602–3
    cold dark matter, 602–3, 606
    distribution on larger scales, 577
    in galaxies, 526–27
    hot dark matter, 602–3, 606
    in Milky Way Galaxy, 504–5, 510
    mixed-dark-matter universe, 603
    search for, 505–6
Daughter theory, of origin of Moon, 189
Declination, 14–15, 26
Decoupling, epoch of, 596, 602, 606
Deferent, 33–35, 50
Degree, 14, 26
Deimos, 216–17, 227–29
Delta ring, of Uranus, 288–89
Deneb, 11
Dense interstellar clouds, 391
Density, uncompressed, 132
Density wave, 502
Descartes, René, 319
Desdemona, 283
Designate, 286
Destructive interference, 57
Detached binary, 437–38
Detector, telescope, 101
Deuterium:
    density of cosmos and, 595–96
    formation in early universe, 594
Deuterium bottleneck, 594
Deuteron, 350, 354
Differential rotation, 234, 250, 332, 344, 498, 500
Differentiation:
    of Earth, 159, 168
    of solar system, 324–26
Diffraction, 56–57, 70, 102–3, 109
Diffraction grating, 74–75
Diffraction-limited resolution, 103
Dinosaurs, extinction of, 300–301
Dione, 265, 267–70
Disk stars, 493–94, 498–500
Distance indicators, 22–24, 519–23, 530
    supernova, 453–54
    variable stars, 490–92
Diurnal motion, 10
DNA, 611, 613
Doppler, Christian, 68
Doppler effect, 67–70, 87–90, 92, 179
Double-line spectroscopic binary, 373
Drake equation, 615–18, 622
D ring, of Saturn, 262, 272
Dumbell Nebula, 429
Dust, lunar, 183–84
Dust grain, 382, 384, 396
    (*See also* Interstellar dust)
Dust lanes, 387, 396, 501

Dust storm, on Mars, 226
Dust tail, 306, 314
Dwarf elliptical galaxy, 516, 526
Dwarf irregular galaxy, 519, 526
Dwarf stars, 361–62, 378
Dynamo theory, 154

Earth:
asteroid collisions with, 298–301
atmosphere of, 61–62, 136–37, 146, 149–53, 168, 201, 349
axis of, 21–22
biological evolution on, 612–13
chemical evolution on, 611–12
core of, 146, 158
crust of, 146, 158, 168
density of, 131, 146
differentiation of, 159, 168
escape velocity, 49, 131
gravity on, 46
interior of, 156–59
magnetosphere of, 146, 153–56, 168
mantle of, 146, 158–59, 168
mass of, 47, 131, 146
meteoroid collisions with, 311–12
moon of (*See* Moon)
orbit of, 42, 129–31
precession of, 21–22, 26
radius of, 131, 146
revolution of, 10, 26
rotation of, 9–10, 14, 22, 26, 148–49, 178
size of, 6, 25
solar radiation, 150–51
solar-terrestrial relations, 349
surface of, 151–52, 159–67, 611
temperature of, 151–52
tides on, 146–49
volcanic activity on, 153, 159–62
weather on, 349
Earth-crossing asteroid, 298–99, 314
Earthquake, 156–62, 164, 168
shadow zone, 158
Earthshine, 172
Eccentricity of orbit, 41–42, 50
Eclipse, 17–21
favorable configuration for, 20–21
lunar (*See* Lunar eclipse)
solar (*See* Solar eclipse)
unfavorable configuration for, 20–21
Eclipse seasons, 21
Eclipsing binary, 373, 378
Ecliptic, 10, 129
Eddington, Sir Arthur, 473, 491
Effective temperature, 333
Einstein, Albert, 78, 349, 470, 572
Einstein Observatory, 119
Ejecta blanket, 181–82, 208
Electric field, 58–59, 70
Electromagnetic spectrum, 59–62, 70
Electromagnetism, 54, 58–59, 70, 353, 592, 598
Electron, 57, 70, 81, 350
orbitals of, 81–82
quantized, 81–82
Electron degeneracy, 425
Electron degeneracy pressure, 425
Electronic transition, 82, 86

Electron spin, 392
Electron volt, 88
Electroweak force, 592, 598–99
Elements, 84
cosmic abundance of, 455
formation of, 454–59
Ellipse, 40–41
Elliptical galaxy, 514–18, 523, 526, 536, 544, 546, 548
Elliptical orbit, 40–42
Emission lines, 75–78, 82–86, 92, 340, 388
Emission nebula, 385–89, 396, 407–15
Enceladus, 265, 267–70
Encke Division, 259–60, 272
Endeavour space shuttle, 111
Energy:
conservation of, 349, 354, 583
production in Sun, 349
Energy flux, 65, 67, 361–62
Epicycle, 33–35, 50
Epimetheus, 265, 270
Epsilon ring, of Uranus, 288–89
Equinox, 10–11
Erathosthenes, 25
E ring, of Saturn, 262, 272
Escape velocity, 49–50, 131, 136–37, 472, 573–74
Eta ring, of Uranus, 288–89
Euclidean geometry, 580–81
Europa, 242–44, 247
Event horizon, 472–79
Evolutionary theory, 319, 328
Evolutionary track, 402, 416
Excited state, of atom, 81–82
Exponent, 8
Extinction, of starlight, 382
Extraterrestrial life, 610
alternative biochemistries, 614–15
carbon-based, 613–14
fossil evidence for, 228–29
intelligent life, 618
on Mars, 228–29
in Milky Way, 615–18
search for, 228–29, 614, 619–21
in solar system, 613–15
technological civilization, 618–19
Eyepiece, 97, 100

Fahrenheit temperature scale, 63
Failed stars, 406–7
False vacuum, 599, 603
Fault, 164
Fireball, 310
Flare, solar, 346–47, 349, 354
Flatness problem, 597–600, 606
Fleming, Williamina, 495
Fluidized ejecta, 221, 230
Focal length, 96
Focus of ellipse, 41, 50
Forbidden lines, 388–89, 396
Force, 44
continuous, 46
instantaneous, 46
Formaldehyde, interstellar, 395
Fossil record, 166, 612
Fragmentation, in solar system formation, 322–24, 328

Fraunhofer lines, 76, 78, 84, 92, 337–38
Freeze-out, 598–99
F ring, of Saturn, 263–64, 272
Full Moon, 15–16, 26
Fundamental forces, 592

Galactic bulge, 488, 497, 503, 510, 515
Galactic cannibalism, 533–34
Galactic center, 493, 510
Galactic cluster, 376
Galactic disk, 488, 493–500, 510, 515, 522
Galactic epoch, 591, 593
Galactic gas, 496–97
Galactic halo, 488, 493, 496–500, 510
Galactic nucleus, 541, 550, 562
Galactic plane, 493
Galactic year, 498
Galatea, 286
Galaxy, 4–5, 488, 510
active galaxies (*See* Active galaxies)
barred-spiral galaxies, 514–17, 536
binary galaxies, 525–26, 535
collisions and mergers of, 532–35
dark matter in, 526–27
distance scale, 521–23
distribution of, 519–25
elliptical galaxies, 514–18, 523, 526, 536, 544, 546, 548
evolution of, 532–35
formation of, 532–35, 602
H-R diagram for, 519
Hubble classification of, 514–19, 536
Hubble's law (*See* Hubble's law)
irregular galaxies, 514, 517–19, 526, 536
large-scale distribution of, 530–31
Local Group, 520–21
mass of, 525–27
radio galaxies, 559–60, 562
recession of, 527–31, 568–70, 578
size of, 523
spiral galaxies, 490, 510, 514–17, 523, 526, 534–36, 559
starburst galaxies, 535–36
Galaxy cluster, 521, 523–24, 536, 577
dark matter in, 526
intercluster gas, 527
mass of, 526
Galaxy supercluster, 524, 536
Galilean moons, 234, 242–44, 250
Galileo, 4, 36–39, 129, 175, 259
Galileo probe, 141
Galle, Johann, 277
Gamma-ray burster, 468
Gamma-Ray Observatory, 120, 468
Gamma-ray radiation, 55, 61, 70
Gamma-ray telescope, 118–20, 122–24
Gamma ring, of Uranus, 288–89
Gamow, George, 594
Ganymede, 242–44, 248–49
Gaseous planet, 132
Gas exchange experiment, 228
Gaspra, 303–4
Gemini, 11
Gene, 611, 613
General relativity, 471, 473, 478, 480, 571–72, 578
Geocentric universe, 32–34, 50

Giant elliptical galaxy, 516–17
Giant stars, 361–62, 378, 426
Gibbous Moon, 15–16
Global measurement, 577–78
Globular cluster, 377–78, 468–69, 493
Gluon, 598
Gold, Thomas, 583
GONG project, 334
Grand Unified Theories (GUTs), 572, 592–93, 598, 600, 603
Granite, 159
Granulation, solar surface, 336, 354
Gravitational constant, 45, 47
Gravitational deformation, 146–48
Gravitational field, 47, 50
Gravitational lens, 506
Gravitational lensing, 506, 510, 555–58, 577
Gravitational radiation, 483
Gravitational redshift, 477–78
Gravitational slingshot, 139–41
Graviton, 592
Gravity, 46, 50, 353, 592, 598
    atmosphere and, 136–37
    Newton's laws of, 45–47
    star formation and, 400–401
Gravity waves, 483
Great Attractor, 577
Great circle route, 580–81
Great Dark Spot, 280–81, 294
Great Red Spot, 236, 238–39, 250
Great Wall, 531
Greek astronomers, 32–33, 128, 172–73, 366
Greenhouse effect, 151–52, 168, 201–3
    runaway, 202–3, 211
Ground state, 81–82, 88, 92
Gum Nebula, 451–52
GUT epoch, 591, 593

Habitable planets, 616–18
Habitable zone, 617
Hadron epoch, 591, 593–94
Hale telescope, 100–101
Half-life, 160
Hall, Asaph, 227
Halley, Edmund, 306
Halley's Comet, 306–8, 612
Halo stars, 493–94, 498–500
h and χ Persei, 434–35
Harvard College Observatory, 495
Hawking, Stephen, 479
Hawking radiation, 479
Hayashi track, 404–5
Head-tail radio galaxy, 545
Heat, 63
    differentiation of solar system, 324–25
    star formation and, 400–401
    Heavy elements:
    fusion in high-mass stars, 446
    production of, 455
Helene, 265, 270
Heliocentric universe, 35–36, 50
Helioseismology, 334, 354
Helium:
    cosmic abundance of, 455
    formation in early universe, 593–95
    fusion in low-mass stars, 424–28
    helium-capture reaction, 456, 460

Helium-burning reaction, 424–28, 455
Helium flash, 425–26, 428–29
Helium precipitation, 259, 272
Helium-shell flash, 428–29
Helium white dwarf, 328
Helmholtz, Hermann von, 404
Herschel, William, 276
Hertz, Heinrich, 56
Hertzsprung, Ejnar, 368
Hertzsprung-Russell diagram, 367–71, 378
    asymptomatic giant branch of, 427–29, 439
    black-dwarf region of, 431–32
    development of, 495
    for galaxies, 519
    high-mass stars on, 433
    horizontal branch of, 425, 428–29
    instability strip, 491–92
    main sequence (See Main sequence)
    protostars on, 404
    red-giant region of, 370–71, 378, 423, 439
    star clusters, 375–77, 434–36
    of star formation, 400–405
    stellar evolution, 427–28
    subgiant branch of, 423, 439
    white-dwarf region of, 370–71, 378, 431
Hewish, Anthony, 465–66
High-Energy-Astronomy-Observatories, 119
High-energy telescope, 118–20, 122–24
Highlands, of Moon, 175, 182–83, 190–91
High-mass stars:
    collapse of iron core, 447
    evolution of, 432–33
    fusion of heavy elements, 446
High-resolution astronomy, 104–9
Hipparchus, 366
History of astronomy:
    ancient astronomy, 30–32
    Copernican revolution, 38–39
    dimensions of solar system, 42–44
    Galileo's observations, 36–39
    geocentric universe, 32–34
    Harvard College Observatory, 495
    heliocentric model, 35–36
    Kepler's laws, 40–42
    Newton's laws, 44–47
    telescope development, 128–29
Homogeneity, cosmic, 566
Hooke, Robert, 238
Horizon, 597
Horizon problem, 597, 599, 606
Hot dark matter, 602–3, 606
Hot longitudes, 180
Hour (angular unit), 14
Hoyle, Fred, 583
Hubble, Edwin, 105, 493, 528
Hubble classification, of galaxies, 514–19, 536
Hubble diagram, 528
Hubble flow, 528
Hubble's constant, 529–30, 536, 568, 572, 577
Hubble's law, 527–31, 536, 554, 560–61, 568–70, 578
Hubble Space Telescope, 105–7, 110–11, 118
Hulse, Russell, 483
Huygens, Christian, 259
Hydrogen:
    atomic structure of, 88–89
    conversion to helium, 421–22

core-hydrogen burning, 420–21, 439
    cosmic abundance of, 455
Hydrogen-burning process, 455
Hydrogen envelope, of comet, 305, 314
Hydrogen-shell-burning stage, 422, 439
Hydrosphere, 146, 168
Hyperbola, 49
Hyperion, 265, 270–71

Iapetus, 265, 267–70
Ida, 303–4
Image, telescope, 97, 101–2
Impact theory, of origin of Moon, 189
Inertia, 44–45, 50
Inferior conjunction, 196
Inflation, cosmic, 597–601
Inflationary epoch, 599, 606
Infrared astronomy, 115–17
Infrared Astronomy Satellite, 408–9
Infrared radiation, 55, 61, 70
Infrared Space Observatory, 117
Infrared telescope, 115–17, 122–24
Instability strip, of H-R diagram, 491–92
Intelligent life:
    development of intelligence, 612–13
    extraterrestrial, 618
Intercloud medium, 393
Intercluster gas, 527
Intercrater plains, Mercury, 184–85, 191
Interference, 56–57, 70
Interferometry, 113–14, 124
    speckle interferometry, 361
Intergalactic matter, 527
International Ultraviolet Explorer, 118
Interplanetary debris, 129, 133, 182, 298–314
Interplanetary space, 142
Interstellar clouds, 394
    contraction, in star formation, 400–403, 407–8
    fragments, in star formation, 400, 403, 407–10
Interstellar dust, 320–22, 328, 382–84, 394, 426, 515, 517
Interstellar gas, 382, 426, 515, 517
    21-centimeter radiation, 392–94
    composition of, 384
    density of, 383–84
    nebular spectra, 388
    temperature of, 383–84
Interstellar matter, 382–85, 533
Interstellar medium, 382, 396
Interstellar molecules, 394–95, 612
Interstellar space, IUE mission, 393
Inverse-square law, 46–47, 50, 67
Io, 242–47
Ion, 81
Ionization, 87, 92, 387
Ionosphere, 61, 151, 168
Ion tail, 306, 314
Iridium, 300–301
Iron, in stellar core, 457
Iron core, collapse of, 447
Irregular galaxy, 514, 517–19, 526, 536
    Irr I galaxy, 518–19
    Irr II galaxy, 518–19
Islamic astronomy, 32
Island universes, 489–90, 577

Isotope, 160, 350, 354, 454
Isotropy, 566, 570, 579, 582, 597, 603

Jansky, Karl, 108
Janus, 265, 270
Jets:
    of active galaxy, 543, 546, 548–50
    of quasars, 552
Jewel Box cluster, 414
Joule, 65, 80
Jovian planets, 132–34, 142, 323, 325
Juliet, 283
Jupiter:
    atmosphere of, 236–40
    brightness of, 234
    color of, 236–37
    cometary impact on, 308–10
    density of, 131
    energy emission from, 240–41
    escape velocity, 131
    formation of, 406
    interior of, 240–41
    magnetosphere of, 234–35, 241–43, 246
    mass of, 131, 234
    moons of, 37, 131, 141, 234, 242–49
    orbit of, 42, 129–31, 234, 300–302
    radius of, 131, 234
    ring of, 249
    rotation of, 131, 234–36
    spacecraft exploration of, 138–41
    temperature of, 240–41
    weather on, 236, 238–40

Keck telescope, 114
Kelvin, Lord, 404
Kelvin-Helmholtz contraction phase, 404
Kelvin temperature scale, 62–63
Kepler, Johannes, 4, 40–42
Kepler's laws, 40–42, 48, 50
Kirchhoff, Gustav, 78
Kirchhoff's laws, 77–79, 82–84, 92
Kirkwood, Daniel, 301
Kirkwood gaps, 301, 314
Kleinmann-Low Nebula, 408–9, 413
Kruger 60, 374
Kuiper, Gerard, 283, 285
Kuiper Belt, 304–5, 325

Labeled release experiment, 228
Lagrange, Joseph Louis, 300
Lagrange points, 270, 272, 300–302, 438
LaPlace, Pierre Simon de, 319
Larissa, 286
Laser, 180
Laser-ranging, 162, 180
Lassell, William, 283, 285
Leavitt, Henrietta, 492, 495
Lens, of telescope, 96, 99
Leo, 11
Lepton, 592
Lepton epoch, 591, 593–94
Leverrier, Urbain, 277
Life:
    definition of, 610
    origin of, 153
    (See also Extraterrestrial life)
Light, 70

radiation and, 54–55
    speed of, 58–59, 67, 70, 470, 472, 558
    wavelength of, 60, 67–69
Light curve, 373, 378, 445–46, 452, 458–59,
        490–91
Lighthouse model, 466
Light year, 4, 13, 26
Limb, of Sun, 337
Lin, C.C., 502
Linear momentum, 321
Line of nodes, 21
Lithosphere, 162–63, 168
LMC X-3, 482
Lobe radio galaxy, 543–46
Local Group, 520–21, 524, 528, 536
Local measurement, 577
Local Supercluster, 524
Logarithmic scale, 61
Long-period comet, 305
Look-back time, 554–55
Lowell, Percival, 223, 290
Low-mass stars:
    death of, 428–32
    helium fusion in, 424–28
Luminosity:
    of active galaxy, 546
    period-luminosity relationship, 493, 495,
        510
    of quasar, 561
    of stars, 361–63, 371, 374
    of Sun, 333–34, 354
Luminosity class, 372, 378
Lunar eclipse, 17, 21
    partial, 17, 26
    total, 17, 26
Lunar lander, 188
Lunar Orbiter, 188
Lyman series, 88

Magellanic Clouds, 519–20, 522, 536
Magellan spacecraft, 138
Magnetic field, 58, 70
    of active galaxies, 549–50
    interstellar, 385
    of neutron stars, 465
    spectral-line broadening and, 90
    of Sun, 343–47
Magnetic pole, 153
Magnetopause, 156, 242
Magnetosphere:
    of Earth, 146, 153–56, 168
    of Jupiter, 234–35, 241–43, 246
    of Neptune, 281–83
    of Saturn, 259
    of Uranus, 281–83
    of Venus, 210–11
Magnitude scale, 366–67
Main sequence, 369–75, 378, 402, 428–29
    star formation, 405
    stellar evolution off, 420–32
    zero-age main sequence, 406, 416
Main-sequence equilibrium, 420–21
Main-sequence turnoff, 434–35, 439
Maria, 175–77, 182–83, 190–91
Mariner missions, 134–35, 137–38, 177
Mars:
    atmosphere of, 225–26

brightness of, 216
canals on, 218, 223
craters on, 220–21
density of, 131, 216–17
earth-based observations of, 217–18
escape velocity, 131
mass of, 131, 216–17
moons of, 131, 216–17, 227–29
oppositions of, 216–17
orbit of, 42, 129–31, 216
permafrost on, 221, 223, 230
polar caps on, 224
radius of, 131, 216–17
rotation of, 131, 217
running water on, 221–24
search for life on, 228–29, 614
spacecraft exploration of, 134–38, 224–25,
        228–29
surface of, 217–25
temperature of, 224–26
Valles Marineris, 220–21
volcanic activity on, 219–20
weather on, 225–26
Mars Observer, 138
Mars Pathfinder, 229
Mass, 44, 46
    center of, 48, 50
    conservation of, 349, 354, 583
Mass extinction, 449
Mass-luminosity relationship, 374, 378
Mass-radius relationship, 374, 378
Mass-transfer binary, 437, 439, 444–46
Matter:
    abundance of, 454–55
    created from radiation, 589–90
    disappearing, 470–75
    types of, 454
Matter-dominated universe, 588, 606
Matter Era, 589, 591, 593, 596
Maunder minimum, 346, 349
Maury, Antonia, 495
Maxwell, James Clerk, 260
Measurement:
    angular, 14–15
    astronomical, 13
    (See also Distance indicators)
Mercury:
    atmosphere of, 174–75
    craters on, 184–86
    density of, 131, 174
    escape velocity, 131, 174
    evolution of, 190–91
    gravity on, 174
    interior of, 186–87, 327
    magnetic field of, 186
    mass of, 131, 174
    orbit of, 42, 129–31, 172–74, 179, 473
    radius of, 131, 174
    rotation of, 131, 178–80
    surface of, 175–77, 184–86
    transit of Sun, 43
    volcanic activity on, 185, 191
Messier, Charles, 385
Metallic elements, 158
Meteor, 129, 181, 311, 314
Meteorite, 133, 310–14, 318
Meteoritic impact:

on Earth, 182
on Moon, 180–82, 190
on Venus, 208
Meteoroid, 133, 149, 180–81, 310–14
collisions with Earth, 311–12
orbit of, 310
Meteoroid swarm, 310–11, 314
Meteor shower, 310–11, 314
Michelson, A.A., 470
Microlensing, 557
Micrometeoroid, 182, 310, 312, 314
Microwave background, cosmic, 579–84, 588, 596–97, 603–4
Midocean ridge, 164–67
Milky Way Galaxy, 5, 123, 385–86
  center of, 507–9
  dark matter in, 504–5, 510
  disk stars, 493–94, 498–500
  dynamics of, 497–500
  extraterrestrial life in, 615–18
  formation of, 499–500
  galactic disk, 494–97
  halo stars, 493–94, 498–500
  mass of, 503–6
  radio map of, 497
  rotation of, 503–4
  size and shape of, 6, 493
  spiral structure, 500–503, 515
  structure of, 490–93
Miller, Stanley, 611
Millisecond pulsar, 468–69
Mimas, 263, 265, 267–70
Minute (angular unit), 14
Mira, 361, 370
Miranda, 283–85, 327
Mirrors, of telescope, 96–104, 110
Mixed-dark-matter universe, 603
Modeling, 23
Molecular clouds, 394–96, 496–97, 612
Molecular velocity, 136–37
Molecules, 85–86, 92
  interstellar, 394–95, 612
Month, 15
Moon, 128, 131
  atmosphere of, 174–75
  craters on, 180–82, 191
  density of, 174, 186
  distance from Earth, 180
  escape velocity, 174
  evolution of, 190
  exploration of, 188
  gravity on, 46, 174
  interior of, 186–87
  lunar eclipse, 17, 21, 26
  mass of, 174
  orbit of, 172
  origin of, 189–90
  phases of, 15–17, 26
  properties of, 131
  radius of, 174
  revolution about Earth, 15, 177–78
  rotation of, 131, 177–78
  surface of, 175–77, 180–84, 188, 190
  tides on Earth and, 146–49
  volcanic activity on, 183–85, 190
Moonlets, of Saturn, 262
Moons, 142

of asteroid Ida, 304
formation of, 324
of Jupiter, 37, 131, 141, 234, 242–49
of Mars, 131, 216–17, 227–29
of Neptune, 131, 277, 285–87
of Pluto, 131, 292
revolution of, 318
of Saturn, 131, 139, 254, 263–71
of Uranus, 131, 283–85
Morley, E.W., 470
Motion, Newton's laws of, 44–46
Mountain building, 163–64
Murchison meteorite, 614

Naiad, 286
Naked singularity, 480
Nanometer, 13, 61
National Radio Astronomy Observatory, 108
Navigation, interplanetary, 139
Neap tide, 148
Nebula, 396
  (See also specific types)
Nebular contraction, 319–20
Nebular emission spectra, 388
  forbidden lines, 388–89
Nebular theory, 319–20, 328
"Nebulium," 388
Negative charge, 58
Neptune:
  atmosphere of, 279–81
  cloud cover on, 280
  density of, 131
  discovery of, 277
  escape velocity, 131
  interior of, 281–83
  magnetosphere of, 281–83
  mass of, 131, 278
  moons of, 131, 277, 285–87
  orbit of, 42, 129–31, 277
  radius of, 131, 278
  rings of, 290
  rotation of, 131, 278–79
  weather on, 280–81
Nereid, 285–87, 327
Neutrino, 350, 354, 447–48, 453, 455, 464, 592, 594
  solar (See Solar neutrino)
Neutrino oscillations, 353–54
Neutrino pulse, 453
Neutrino telescope, 352–53
Neutron, 84, 464
Neutron capture, 457–58, 460
Neutron degeneracy pressure, 447, 464, 472
Neutronization, 447
Neutron-star binary, 467–70
Neutron stars, 464–65, 478, 482
  final stage of stellar evolution, 471–72
  gamma-ray bursters, 468
  pulsars, 465–67
  X-ray bursters, 467
Newborn star, 405
New Moon, 15–16, 26
Newton, Isaac, 44–47, 60, 76, 80, 100
Newton (unit of force), 45
Newtonian mechanics, 44, 47, 50
Newtonian telescope, 100, 124
Newton's laws, 44–47

laws of gravity, 45–47
laws of motion, 44–46
laws of planetary motion, 47
North celestial pole, 9, 26
North Pole, 153
Nova, 444–46
  supernova and, 448
Nuclear binding energy, 457
Nuclear epoch, 591, 593–95
Nuclear fusion, 348–49, 354, 455, 458
Nucleotide bases, 611, 622
Nucleus of atom, 81, 84
  formation of, 593–96
Nucleus of comet, 308, 314
Nutation, 22

OB association, 414
Oberon, 283–85
Observation, 38–39
Olbers's Paradox, 568–69, 584
Omega Centauri, 377
Oort, Jan, 305
Oort Cloud, 305, 314, 318, 325–27
Open cluster, 376–78
Open universe, 579, 584
Ophelia, 283, 289
Optical doubles, 373
Orbit:
  escape from, 49
  semi-major axis of, 48
  (See also specific planets and moons)
Orbital period, 42, 48, 50
Orbital resonance, 262–64
Organic molecules, interstellar, 612
Orion, 6–7, 11, 13, 364
Orion Nebula, 112–13, 116–17, 389, 408–9, 413–14
Oscillating universe, 575–76
Oscillations, solar, 334
Outflow channel, 221–22, 230
Outgassing, 153, 226
Ozone layer, 61, 150–51, 153, 168, 449

Pair production, 479, 589–90, 606
Paleomagnetism, 167
Pallas, 303–4
Pandora, 264
Pangaea, 165–67
Parabola, 49
Parallactic angle, 24, 26
Parallax, 24, 26, 33, 358
  spectroscopic, 371–72, 378, 519–20
  stellar, 39, 358–59, 519–20
Parsec, 13, 358, 378
Particle-antiparticle pair, 589
Paschen series, 89
Pauli, Wolfgang, 425
Pauli exclusion principle, 425
Pencil-beam survey, 566–67
Penumbra, 18–19, 26, 342
Penzias, Arno, 579–80
Perfect cosmological principle, 583
Perihelion, 41, 179–80
Period-luminosity relationship, 493, 495, 510
Permafrost, on Mars, 221, 223, 230
Pfund series, 89
Phobos, 216–17, 227–29

Phoebe, 265, 270
Photodisintegration, 447, 456–57
Photoelectric effect, 78–80, 92
Photography, stellar, 368
Photometry, 101, 124, 364, 378
Photon, 79, 82, 92
    orbits of, 474–75
    redshift of, 571–73
Photon sphere, 475
Photosphere, 332–33, 337, 354, 403
Piazzi, Giuseppe, 298
Pickering, E.C., 368
Pioneer spacecraft, 136, 138, 619
Pixel, 106
Planck, Max, 62, 80
Planck curve (*See* Black-body curve)
Planck epoch, 590–93, 604
Planck's constant, 80
Planetary motion:
    geocentric universe, 32–34
    Kepler's laws of, 40–42, 48, 50
    Newton's laws of, 47
Planetary nebula, 428–31, 439, 521
Planetary ring system, 259–65, 272
Planetary system, 128–29
    comparative planetology, 128, 141–42
    fraction of stars having, 616
    habitable planets in, 616–17
    planetary properties, 130–32
    Planetismal, 322–23, 325–26, 328
    Planets, 4, 128, 142
    brightness of, 33
    discovery of, 129
    formation of, 322–24
    jovian planets, 132–34, 142
    orbits of, 40–42, 135, 318
    pulsar planets, 469–70
    revolution of, 318
    rotation of, 318
    spacecraft exploration of, 134–41
    terrestrial planets, 132–34, 142
    (*See also* individual planets)
Plasma torus, 243, 246
Plate tectonics, 161–65, 168
Pleiades cluster, 376
Pluto, 129–30
    density of, 131, 293
    discovery of, 290–91
    escape velocity, 131
    mass of, 131, 292–93
    moon of, 131, 292
    orbit of, 42, 129, 291, 293
    origin of, 293
    radius of, 131, 292–93
    rotation of, 131, 292
Polar caps, on Mars, 224
Polaris, 8–9
Polarities, magnetic, 343
Polarized light, 384–85, 396
Population I stars, 494
Population II stars, 494
Portia, 283
Positive charge, 58
Positron, 350, 354, 455, 594
Precession, 21–22, 26
Pressure broadening, 90
Pressure waves, 156

Prime focus, 96, 124
Primordial element, 454
Primordial nucleosynthesis, 594–96, 606
Primordial soup, 611
Progenitor, of supernova, 448
Prograde motion, 32–33, 172, 198
Prometheus, 264
Prominence, solar, 346, 354
Proper motion, 359, 378
Proteins, 611, 613
Proteus, 286
Protomoon, 324
Proton, 57, 70, 81, 84, 509
Proton capture, 458
Proton-proton chain, 349–54, 422, 455, 594
Protoplanet, 319, 323–24, 326, 328
Protostar, 403–4, 407–10, 415–16
    evolution of, 404–5
    on H-R diagram, 404
Protostellar winds, 410
Protosun, 319, 324, 328
Proxima Centauri, 358, 376
Ptolemaic model, 34–36, 50
Puck, 283
Pulsar, 465–67
    binary pulsars, 483
    millisecond pulsars, 468–69
Pulsar planet, 469–70
Pulsating variable stars, 490–92, 510
P-waves, 156–58
Pyrolitic release experiment, 228

QO051–279, 530
Quantum fluctuations, 605
Quantum gravity, 480, 590, 604
Quantum mechanics, 79, 480
Quark, 592
Quarter Moon, 15–16, 26
Quasar, 162, 540–41, 562, 583, 602
    binary quasars, 555–58
    discovery of, 550–52
    energy generation in, 553–55
    evolution of galaxies, 558–61
    interior of, 558
    lifetime of, 553–55
    local quasars, 560–61
    luminosity of, 561
    "mirages," 555–58
    properties of, 552–53
Quasi-stellar objects, 550–58, 562
Quiescent prominence, 346
Quiet Sun, 341, 354

Radar, 43–44, 50
Radar ranging, 519
Radial motion, 68, 70
Radial velocity, 87
Radiation:
    astronomical applications, 65–66
    atmospheric blockage of, 61–62
    black-body spectrum, 62
    charged particle interactions, 57–58
    created from matter, 589
    diffraction and interference, 56–57
    distribution of, 62–66
    Doppler effect, 67–69
    electromagnetic spectrum, 59–62

    electromagnetic waves, 58–59
    intensity of, 62, 70
    light and, 54–55
    particle nature of, 78–79
    Stefan's law, 64–65
    wave properties of, 55–58
    Wien's law, 62–65
Radiation darkening, 285, 294
Radiation-dominated universe, 589, 606
Radiation Era, 589–93
Radiation frequency, 79–80
Radiation zone, of Sun, 335, 354
Radioactive dating, 160–61
Radioactive heating, 159
Radioactivity, 159–60, 168, 454, 458
Radio astronomy, 108–13, 121
    search for extraterrestrials, 619–21
    spectroscopic radio astronomy, 496
    Radio galaxy, 540, 542–46, 559–60, 562
    core-halo radio galaxies, 542–43
    head-tail radio galaxies, 545
    lobe radio galaxies, 543–46
Radio interferometry, 113–14
Radio lobe, 543–46, 562
Radio map, of Milky Way, 497
Radio telescope, 108–13, 121–24
Radio waves, 55, 61, 70
Radius-luminosity-temperature relationship,
    361, 369, 378
Ranger missions, 188
Rayleigh, Lord, 152
Rayleigh scattering, 152
Recessional motion, of galaxies, 527–31, 578
Recessional velocity, redshift and, 554–55
Recurrent nova, 444–46
Reddening, 382, 396
Red dwarf, 369–70, 378, 505
Red giant, 361, 370–72, 378, 423–26, 434, 439
Redshift, 68–70, 87–88, 498, 522, 528, 530,
    551–53, 575, 602
    cosmological redshift, 571–73
    gravitational redshift, 477–78
    noncosmological explanation for, 560–61
    recessional velocity and, 554–55
    relativistic redshift, 554–55
Red supergiant, 427–28, 434, 439
Reflecting telescope, 96–98, 103, 124
Reflection nebula, 386
Refracting telescope, 96–98, 103, 124
Refraction, 60, 70, 96
Regolith, 183–84
Regression of the line of nodes, 21
Relativistic redshift, 554–55
Relativity, 470, 472, 571
Residual polar cap, 224, 230
Resolving power, of telescope, 102–3
Retrograde motion, 32–33, 50, 197–98, 287,
    292
Revolution:
    of Earth, 10, 26
    of Moon, 15, 177–78
Rhea, 265, 267–70
Rho Ophiuchi, 390
Riemann, Georg Friedrich, 580–81
Riemannian geometry, 580–81
Rigel, 13, 15, 364, 376
Right ascension, 14–15, 26

Rille, volcanic, 184–85, 191
Ringlets, of Saturn, 262
Ring Nebula, 429
Rings:
of Jupiter, 249
of Neptune, 290
of Saturn, 254–55, 259–65, 272
of Uranus, 288–89
Roche, Edouard, 261, 437
Roche limit, 260–62, 264, 272
Roche lobe, 437–39
Rocks, lunar, 183–84
Rosalind, 283
ROSAT, 120–21
Rotation:
of earth, 9–10, 14, 22, 26, 148–49, 178
of galactic disk, 498, 500
of Jupiter, 131, 234–36
of Mars, 131, 217
of Mercury, 131, 178–80
of Milky Way Galaxy, 503–4
molecular, 86
of Moon, 131, 177–78
of Neptune, 131, 278–79
of neutron stars, 465
of Pluto, 131, 292
of Saturn, 131, 254
spectral-line broadening due to, 90
of Sun, 332–33, 344
of Uranus, 131, 278–79, 327
of Venus, 131, 197–98, 327
Rotation curve, Galactic, 504, 510, 525
Rounding off numbers, 8
r-process, 458
RR Lyrae stars, 491–93, 510
Runaway greenhouse effect, 202–3, 211
Runoff channel, 221–22, 230
Russell, Henry Norris, 368

S0 galaxy, 517–18, 536
Sagittarius, 11
Saturn:
atmosphere of, 254–58
cloud cover of, 256, 258
color of, 254–56
density of, 131, 254
escape velocity, 131
interior of, 258–59
internal heating, 258–59
magnetosphere of, 259
mass of, 131, 254
moons of, 131, 139, 254, 263–71
orbit of, 42, 129–31, 254
radius of, 131, 254
rings of, 254–55, 259–65
rotation of, 131, 254
temperature of, 256, 258
weather on, 256–58
SB0 galaxy, 517–18, 536
Scarp, 185, 191
Schiaparelli, Giovanni, 178, 223
Schmidt, Bernhard, 101
Schmidt camera, 101
Schmidt telescope, 101
Schmitt, Harrison, 188
Schwarzschild, Karl, 472
Schwarzschild radius, 472–74

Scientific method, 38–39
Scientific notation, 8
Seasonal polar cap, 224, 230
Seasons, 26, 278–79
Second (angular unit), 14
Seeing, 105, 124
Seeing disk, 105, 124
Seismic waves, 156–58, 168
Self-propagating star formation, 503, 510
Semidetached binary, 437–38
Semi-eccentricity of ellipse, 41
Semi-major axis of ellipse, 41, 50
Seyfert, Carl, 541
Seyfert galaxy, 540–42, 549, 559–60, 562
Shapley, Harlow, 493
Shear waves, 157
Shepherd satellites, 262–64, 272, 289
Shield volcano, 207–8, 211, 220
Shock waves, 411, 416, 501
star formation and, 410–12
Shooting star (*See* Meteoroid)
Short-period comet, 304
Shu, Frank, 502
Sidereal day, 10, 26, 108
Sidereal month, 15–17
Sidereal year, 11, 22, 26
Silicon-28, in stellar core, 456
Silicon-based life, 614–15
Single-line spectroscopic binary, 373
Singularity, 472, 480, 570–71
Sirius, 11, 364, 376
Sirius A, 421–32
Sirius B, 362, 370, 431–32
Sister theory, of origin of Moon, 189
Sky, color of, 152
Slipher, Vesto M., 527
Solar constant, 333, 354
Solar cycle, 344–46, 354
Solar day, 10, 26
Solar disk, 332, 354
Solar eclipse, 18–21, 26, 339–40, 473
annular, 18–19, 21
partial, 18
total, 18–19, 21
Solar maximum, 344, 354
Solar minimum, 344, 354
Solar nebula, 319–21, 328
Solar neutrino, 350–53
Solar neutrino problem, 352, 354
Solar system, 142
contents of, 128
differentiation of, 324–26
dimensions of, 42–44
as double-star system, 241
evolution of, 319
extraterrestrial life in, 613–15
Galilean moons as model of, 242–44
layout of, 129–30
size of, 6, 129
spacecraft exploration of, 134–41
Solar system formation, 317–27
angular momentum problem, 327
catastrophes in, 326–27
condensation theory of, 319–24
irregularities in, 319, 327
modeling of, 318–19
Solar transit, 43

Solar winds, 133, 154–56, 174, 210, 241–42, 259, 306, 326–27, 332–33, 341, 354
South celestial pole, 9, 26
South Pole, 153
Space, geometry of, 578–79
Spacecraft exploration:
of Jupiter, 138–41
of Mars, 134–38, 224–25, 228–29
of Moon, 188
near black holes, 477–80
of Venus, 135–37, 209–10
Spacetime, 471, 473, 571
Special relativity, 471, 473
Speckle interferometry, 361
Spectral-line analysis, 86–91
line broadening, 87–91, 522, 525
line intensity, 87
Spectral lines, 74–78, 81–82
absorption lines, 76–77
astronomical applications, 78
emission lines, 75–76
Kirchhoff's laws, 77–79
of molecules, 85–86, 394
of stars, 365–68, 378
width of, 372
Spectrometer, 74, 101, 124
Spectroscope, 74, 92, 100
Spectroscopic binary, 373, 378
Spectroscopic parallax, 371–72, 378, 519–20
Spectroscopic radio astronomy, 496
Spectroscopy, 74–92
Spectrum, 60, 74
Speed of light, 58–59, 67, 70, 470, 472, 558
Spherical aberration, 110
Spicule, solar, 339
Spin-orbit resonance, 179–80, 191
Spiral arms, 490, 497, 500–503, 510, 515
Spiral density wave, 262, 501–3, 510
Spiral galaxy, 490, 510, 514–17, 523, 526, 534–36, 559
Spiral nebula, 489–90
Spring tide, 148
s-process, 458
SS433, 467–68
Stable element, 454
Standard candle, 453, 460, 519, 521, 523
Standard Solar Model, 334–35, 354
Starburst galaxy, 535–36
Star cluster, 375–78, 412–15
associations, 414–16
evolution in, 433–36
globular cluster, 377–78, 468–69, 493
H-R diagram for, 434–36
open cluster, 376–78
Star formation:
cloud contraction in, 400–403, 407–8
cloud fragments in, 400, 403, 407–10
efficiency of, 413
emission nebulae and, 412–15
failed stars, 406–7
gravitational competition, 400–401
H-R diagram of, 400–405
Kelvin-Helmholtz contraction, 404
main sequence, 405
newborn star, 405
protostar, 403–4 (*See also* Protostar)
self-propagating, 503, 510

Star cluster (con't)
    shock waves and, 410–12
    star clusters and, 412–15
    stars like the Sun, 400–405
    zero-age main sequence, 406
Stars, 4, 332, 354
    absorption lines of, 365
    age of, 572
    brightness of, 67, 362–63, 366–68
    classification of, 364–67
    collisions of, 533
    color of, 363–64
    day-to-day changes in, 10
    death of, 412
    determining composition of, 86
    distance to, 358–59
    evolution of (See Stellar evolution)
    explosions of, 444–59
    formation of, 616
    H-R diagram (See Hertzsprung-Russell diagram)
    lifetimes of, 374–75
    luminosity of, 361–63, 371–74
    mass of, 372–75
    motion of, 359–61
    radius of, 374
    seasonal changes in, 10–11
    size of, 361–62
    spectral analysis of, 365–68
    temperature of, 65–66, 363–67
    twinkling of, 105
    (See also specific types of stars)
Steady-state universe, 583
Stefan, Josef, 65
Stefan-Boltzmann constant, 65
Stefan's law, 64–65, 70, 151, 338, 361
Stellar epoch, 591, 593
Stellar evolution, 415, 420–38
    binary systems, 436–38
    contraction of helium core, 421–23
    core hydrogen burning, 420–21
    cycle of, 459
    death of low-mass star, 428–32
    depletion of hydrogen in core, 421
    final stage of, 471–72
    helium fusion in low-mass stars, 424–28
    high-mass stars, 432–33
    historical records of, 432
    H-R diagram, 427–28
    hydrogen-shell-burning stage, 422, 439
    main-sequence equilibrium, 420–21
    off main sequence, 420–32
    in star clusters, 433–46
Stellar nucleosynthesis, 454–60, 593–95
Stellar occultation, 288, 294, 480–81
Stellar parallax, 39, 358–59, 519–20
Stellar winds, 426
Stonehenge, 30–31
Stratosphere, 150, 168
Stray asteroid, 310
Strong nuclear force, 353, 592, 598–99
S-type asteroid, 303
Subatomic particles, 505, 509
Subduction zone, 164
Subgiant branch, of H-R diagram, 423, 439
Summer constellations, 11–12
Summer solstice, 10–11, 26

Sun:
    active regions of, 346–47, 354
    active Sun, 341–47, 354
    angular momentum of, 327
    atmosphere of, 337–41
    composition of, 339
    corona of, 18, 332–33, 340–41, 347, 354
    day-to-day changes in, 10
    density of, 334–35
    edge of, 337
    energy production in, 333–34, 348
    Fraunhofer lines, 76
    gravity of, 47
    habitable zone surrounding, 617
    interior of, 332–37, 348–51, 354
    luminosity of, 333–34, 354
    magnetic field of, 343–47
    mass of, 13, 332
    modeling structure of, 334–35
    nuclear fusion in, 348–49
    planetary motion and, 47
    proton-proton chain on, 349–51
    quiet Sun, 341, 354
    radius of, 13
    rotation of, 332–33, 344
    seasonal changes in, 10–11
    solar cycle, 344–46
    solar neutrinos, 350–53
    solar spectrum, 337–39
    solar-terrestrial relations, 150–51, 349
    structure of, 332–33
    surface of, 65
    temperature of, 65, 333–35, 340
    tidal influence on Mercury, 179–80
    tides on Earth and, 148
    T-Tauri phase, 326
    weight of, 47
Sunspot cycles, 344–47, 349, 354
Sunspots, 37, 342–44, 354
Superforce, 598
Supergiant stars, 361, 378, 426–29
Supergranulation, solar, 336, 354
SuperGUT, 604
Superior conjunction, 196
Superluminal velocity, 558
Supermassive black hole, 547–48, 553, 561
Supernova, 412, 471
    ancient Chinese data on, 31
    carbon-detonation supernova, 448–51, 460
    core-collapse supernova, 447, 460
    as distance indicators, 453–54
    explosions of, 448–54, 509
    formation of, 458
    Milky Way, 449
    1987A, 452–53
    nova and, 448
    Type I, 448–50, 454, 458–60, 464, 521
    Type II, 448–50, 452–54, 460, 464
Supernova remnant, 449–53, 460, 465
Supersymmetric relics, 592
Supersymmetry, 592
Surveyor missions, 188
S-waves, 156–58
Synchronous orbit, 178–79, 191
Synchrotron radiation, 550, 562
Synodic month, 17

Tail, of comet, 305–6, 314
T association, 414
Taylor, Joseph, 483
Technetium-99, 454, 458
Technology, extraterrestrial, 618–19
Tectonic fracture, 221
Telescope, 124
    active optics, 107, 124
    adaptive optics, 108, 124, 361
    atmospheric blurring, 104–6
    Cassegrain telescope, 100, 124
    design of, 99–101
    full-spectrum coverage, 121–23
    high-energy telescope, 118–20, 122–24
    high-resolution telescope, 104–9
    history of, 128–29
    image processing, 106–7
    images and detectors, 101
    infrared telescope, 115–17, 122–24
    interferometry, 113–14
    light-gathering power of, 101–2
    neutrino telescope, 352–53
    Newtonian telescope, 100, 124
    radio telescope, 108–13, 121–24
    reflecting telescope, 96–98, 103, 124
    refracting telescope, 96–98, 103, 124
    resolving power of, 102–3
    Schmidt telescope, 101
    size of, 101–4
    ultraviolet telescope, 118, 122–24
Telesto, 265, 270
Temperature, 62–66, 70
    atmosphere, 136
    spectral-line intensity and, 87
Terminator, of Moon, 175–76
Terrestrial planets, 132–34, 142, 323–26
Tethys, 265, 267–70
Thalassic, 286
Tharsis bulge, 219, 221
Theory, 38–39
Thermal energy, 63
Thermal equilibrium, 589
Thermal motion, of atoms, 88–89
Thick disk, 496, 500, 510, 515
Threshold temperature, 590
Tidal bulge, 147–49, 168, 178, 260, 287
Tidal deformation, 178
Tidal field, 264
Tidal force, 148, 260, 476
Tidal stability limit, 261
Tides:
    on Earth, 146–49
    effect on Earth's rotation, 148–49
Time delay, 557
Time dilation, 478
Titan, 139, 265–67
Titania, 283–85
Titius, Johann, 135
Titius-Bode law, 130, 135
Tombaugh, Clyde, 291
Transition zone, of Sun, 333, 340, 354
Transverse motion, 68, 70
Triangulation, 22–24, 26, 42
Trigonometry, 23, 25, 32
Triple-alpha process, 424, 426, 439, 455
Triton, 277, 285–87, 327
Trojan asteroid, 300–302, 314

Trophosphere, 150, 168, 225, 236
Tropical year, 10, 22
T Tauri star, 326, 404, 415–16
47 Tucanae, 435
Tully-Fisher relation, 521–23, 536
Turbulence:
    atmospheric blurring, 105
    spectral-line broadening and, 89–90
Turnoff mass, 434, 436
21-centimeter radiation, 392–94, 396, 496–97, 522–23
Twin quasar, 555–58
Tycho Brahe, 40
Tycho's supernova, 452

Ultracompression, 464
Ultraviolet astronomy, 118
Ultraviolet Explorer, 393
Ultraviolet radiation, 55, 61, 70
Ultraviolet telescope, 118, 122–24
Umbra, 18–19, 26, 342
Umbriel, 283–85
Unbound orbit, 49–50
Unbound universe, 574
Uncompressed density, 132
Universe:
    age of, 568, 572, 576–77
    Big Crunch, 575
    birth of, 568–69
    black-body curve for, 580–81
    bound universe, 574, 578
    closed universe, 578–79, 584
    cosmological principle, 566–67, 584
    critical density, 573–75
    critical universe, 579, 584
    deceleration of, 577–78
    definition of, 4
    density of, 574, 577, 595
    end of structure, 566–67
    evolution of, 573–77, 583, 610–13
    expansion of, 568–78
    formation of structure in, 601–4
    future of, 575–76
    geometry of space, 578–79
    inflationary, 597–601
    island universes, 489–90
    large-scale structure in, 566–67
    life in, 609–21
    matter-dominated universe, 588, 606
    Matter Era, 591, 593, 596
    open universe, 579, 584
    oscillating universe, 575–76
    radiation-dominated universe, 589, 606
    Radiation Era, 589–96
    relativity and, 571
    steady-state universe, 583
    unbound universe, 574
Unpolarized light, 384–85
Uranus:
    atmosphere of, 279–81
    density of, 131
    discovery of, 276–77
    escape velocity, 131
    interior of, 281–83
    magnetosphere of, 281–83
    mass of, 131, 278
    moons of, 131, 283–85

orbit of, 42, 129–31, 276
radius of, 131, 278
rings of, 288–89
rotation of, 131, 278–79, 327
seasons on, 278–79
weather on, 280–81
Urey, Harold, 611

Van Allen belts, 154, 168, 349
Variable stars, 368, 510
    as distance indicators, 490–92, 519–20
    pulsating variables, 490–92, 510
Vega, 11, 364
Venera missions, 135–36, 209–10
Venus:
    atmosphere of, 198–203
    brightness of, 196
    craters on, 206–9
    density of, 131, 196
    escape velocity, 131
    greenhouse effect on, 201–3
    magnetic field of, 210–11
    mass of, 131, 196
    meteoritic impact on, 208
    orbit of, 42, 129–31, 196
    phases of, 37–38
    radius of, 131, 196
    rotation of, 131, 197–98, 327
    spacecraft exploration of, 135–37, 209–10
    surface of, 199, 203–10
    temperature of, 199, 201, 210
    transit of Sun, 43
    volcanic activity on, 206–9
Vernal equinox, 10–11, 22, 26
Vesta, 303
Vibration, molecular, 86
Viking missions, 138, 228, 614
Virgo Cluster, 523–24
Virus, 613
Visible light, 54–55, 59–60
Visible spectrum, 60, 70
Visual binary, 373, 378
Voids, 531, 536, 577
Volcanic activity:
    on Earth, 153, 159–62
    on Enceladus, 268–70
    on Io, 246
    on Mars, 219–20
    on Mercury, 185, 191
    on Moon, 183–85, 190
    shield volcanos, 207–8, 211, 220
    on Venus, 206–9
Volcano, 159
Voyager missions, 139, 262, 619

Warm longitudes, 180
Warped space, 475, 571, 578
Water, on Mars, 221–24
Water hole, 621–22
Watt, 65
Wave, 55–56
    amplitude of, 56, 70
    crest of, 55
    frequency of, 56, 70
    trough of, 55
    velocity of, 56
    wavelength, 55–56, 60

wave period, 55, 70
Weak force, 353, 592, 598
Weather:
    on Earth, 349
    on Jupiter, 236, 238–40
    on Mars, 225–26
    on Neptune, 280–81
    on Saturn, 256–58
    on Uranus, 280–81
Wegener, Alfred, 165
Weird terrain, 186
White dwarf, 362, 370–72, 378, 428–34, 439, 444–48, 454, 471, 478
White oval, 239, 250
Wien's law, 62–65, 67, 70, 117
Wilson, Robert, 579–80
Winds, 150
Winter constellations, 11–12
Winter solstice, 10–11, 26
Wispy terrain, 268
Wollaston, William, 76

X-ray burster, 467–68
X-ray radiation, 55, 61, 70, 341
X-ray telescope, 118–20, 122–24

Yerkes Observatory, 99

Zero-age main sequence, 406, 416
Zodiac, 11, 26
Zonal flow, 238, 250, 256–57
Zones, of Jupiter, 237–38, 250

These star maps show the brighter stars and the prominent constellations as they appear on the dates and at the times indicated. To use these maps, face the south and hold the book overhead with top of the map toward the north and the right-hand edge toward the west. The brightest stars are indicated by the star symbol (☆) and the names are indicated. (Star maps courtesy of Robert Dixon, *Dynamic Astronomy,* 6th ed., Prentice Hall, 1992.)

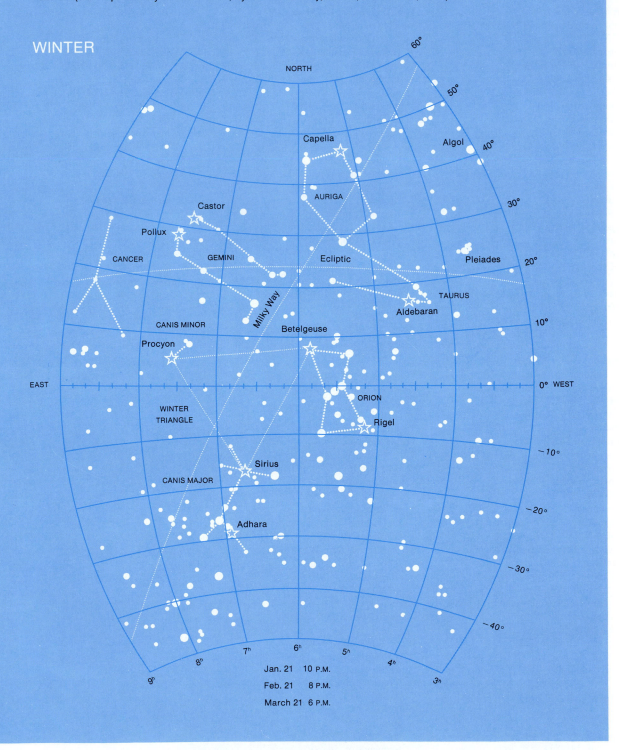

WINTER

Jan. 21   10 P.M.

Feb. 21   8 P.M.

March 21   6 P.M.

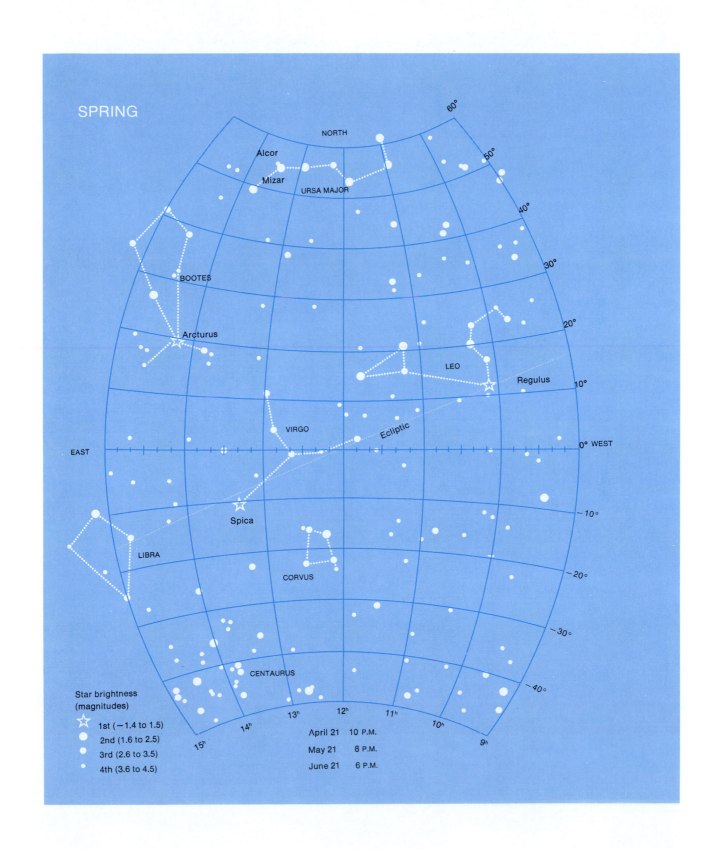

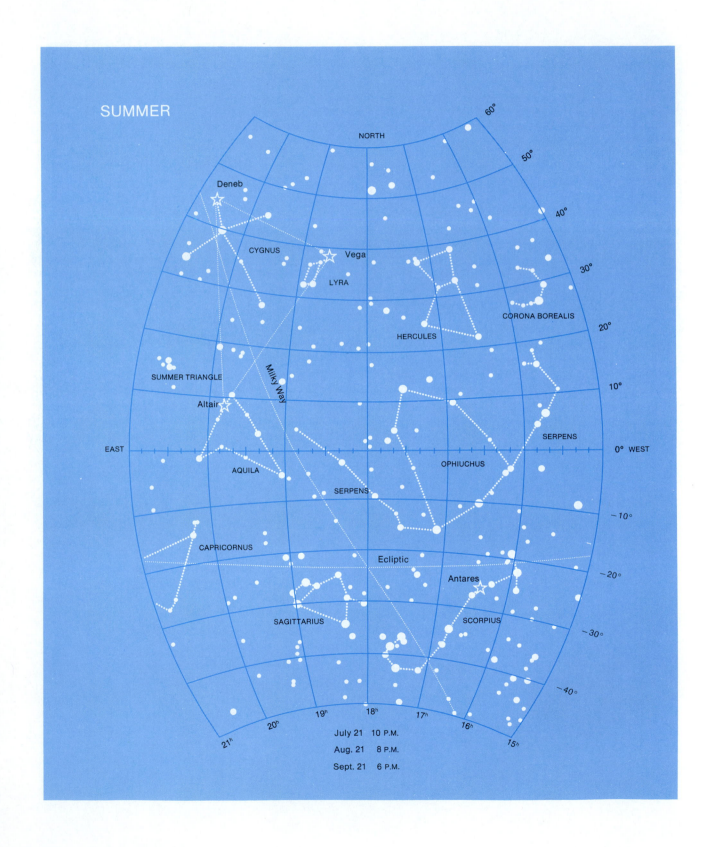

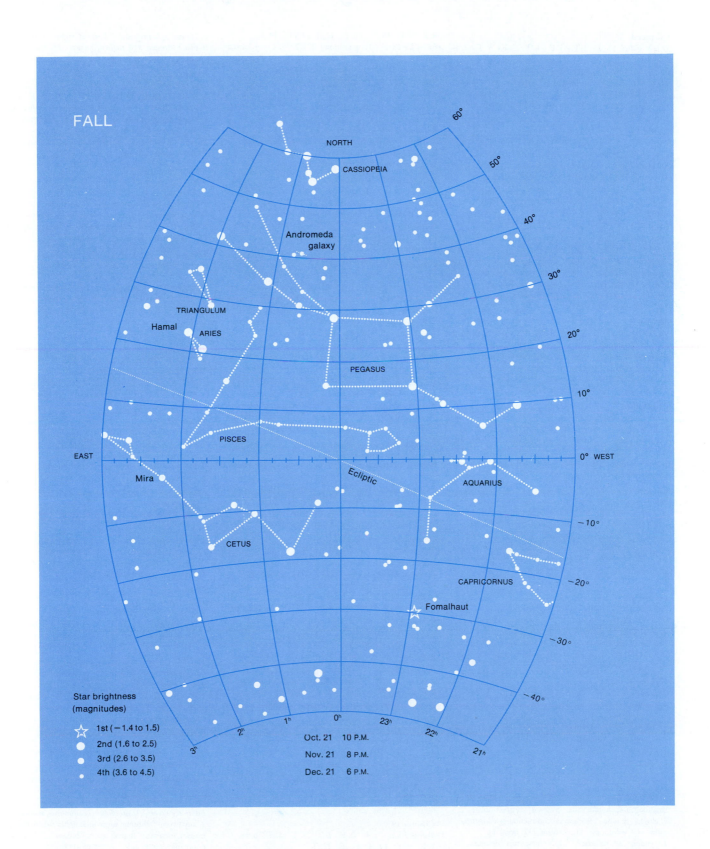

Answers to Self-Test Questions

Chapter 1
True or False? 1. T 2. F 3. F 4. T 5. F 6. T 7. F 8. F 9. F 10. F 11. T 12. F 13. F 14. T
Fill in the Blank 1. galaxy 2. distance 3. axis 4. celestial sphere 5. stars 6. ecliptic 7. winter solstice, lowest 8. celestial equator 9. 1/60 10. quarter 11. lunar 12. angular size 13. distance from the sun or radius of orbit 14. Earth

Chapter 2
True or False? 1. F 2. T 3. T 4. F 5. F 6. F 7. T 8. T 9. T 10. F 11. F 12. T 13. T 14. T
Fill in the Blank 1. calendar 2. Chinese 3. Arab 4. retrograde 5. scientific method 6. Copernicus 7. Earth's 8. moons, phases, sunspots 9. ellipse, circle 10. square, cube 11. radar 12. force 13. product, square 14. masses

Chapter 3
True or False? 1. T 2. F 3. T 4. F 5. F 6. T 7. F 8. F 9. F 10. T 11. F 12. F 13. T 14. T
Fill in the Blank 1. 300,000 2. wavelength 3. frequency 4. electric, magnetic 5. 4000, 7000 Å or 400, 700 nm 6. red 7. radio, infrared, visible 8. temperature 9. 0 10. 273 11. 6000 12. the 1200 K object 13. hot 14. 3 15. shorter

Chapter 4
True or False? 1. T 2. F 3. T 4. T 5. F 6. F 7. F 8. T 9. T 10. F 11. F 12. T 13. F 14. F
Fill in the Blank 1. prism 2. continuous 3. Sun 4. dense 5. cool 6. particle 7. photoelectric effect 8. positive, negative 9. electrons 10. absorbs 11. emits 12. difference 13. infrared 14. radio

Chapter 5
True or False? 1. F 2. F 3. T 4. F 5. F 6. T 7. T 8. F 9. T 10. T 11. F 12. T 13. F 14. F
Fill in the Blank 1. refracting 2. reflecting 3. reflecting 4. area or diameter 5. wavelength 6. atmosphere 7. one 8. digital 9. resolution 10. reflecting 11. interferometer 12. infrared

Chapter 6
True or False? 1. F 2. T 3. T 4. F 5. F 6. F 7. F 8. F 9. F 10. T
Fill in the Blank 1. planets 2. asteroids 3. Mercury, Pluto 4. average random motion 5. higher 6. rocky, icy 7. terrestrial 8. icy, outer 9. U.S., Mercury 10. radar 11. Mars 12. Pioneer 10 13. Voyager 2 14. gravitational

Chapter 7
True or False? 1. F 2. F 3. T 4. T 5. F 6. T 7. T 8. F 9. T 10. T 11. F
Fill in the Blank 1. 6500 (6378) 2. crust 3. liquid water 4. difference 5. nitrogen, oxygen 6. convection 7. infrared 8. raise or increase 9. volcanoes 10. life 11. auroras 12. solid, liquid 13. granite, basalt 14. molten 15. plate tectonics

Chapter 8
True or False? 1. T 2. F 3. T 4. F 5. F 6. T 7. T 8. F 9. F 10. F 11. T 12. T
Fill in the Blank 1. laser ranging 2. dawn, sunset 3. 1/4, 1/3 4. heavy elements 5. larger 6. closer to the Sun 7. maria 8. impacts 9. radar 10. poles 11. 10 12. mantle 13. magnetic field

Chapter 9
True or False? 1. T 2. T 3. F 4. T 5. F 6. F 7. T 8. T 9. F 10. F 11. T 12. T 13. F 14. F
Fill in the Blank 1. spacecraft 2. density 3. slow or retrograde 4. carbon dioxide 5. trace 6. greenhouse 7. closer 8. radar 9. elevated continental-sized formations 10. volcanoes 11. coronae 12. break up 13. young 14. resurfaced 15. temperature, pressure

Chapter 10
True or False? 1. T 2. F 3. T 4. F 5. F 6. T 7. F

Chapter 11 (continued top of column 2)
8. F 9. F 10. F 11. T 12. T 13. F 14. F
Fill in the Blank 1. opposition 2. 1/2 3. 25 4. carbon dioxide 5. cratered 6. bulge 7. volcanoes 8. gravity 9. permafrost or frozen water 10. rain water 11. flooding 12. billion 13. water 14. crust

Chapter 11
True or False? 1. T 2. F 3. F 4. F 5. T 6. F 7. F 8. T 9. F 10. T 11. T 12. T 13. F 14. F 15. F
Fill in the Blank 1. density 2. hydrogen, helium 3. rotation 4. hurricanes 5. twice 6. zones, bands 7. twice 8. liquid 9. hydrogen 10. 4, 16 11. Ganymede 12. water 13. volcanoes

Chapter 12
True or False? 1. T 2. F 3. F 4. F 5. T 6. F 7. F 8. F 9. T 10. F 11. F 12. T 13. T 14. F
Fill in the Blank 1. ring 2. ammonia 3. gravity 4. helium precipitation 5. 3 6. A, B 7. B 8. Roche limit 9. shepherd 10. nitrogen 11. organic 12. temperature 13. E 14. Lagrange

Chapter 13
True or False? 1. F 2. F 3. T 4. T 5. T 6. T 7. F 8. T 9. T 10. T 11. T 12. T 13. F 14. F
Fill in the Blank 1. twice 2. perpendicular 3. methane 4. Uranus 5. Neptune 6. magnetic field 7. Miranda 8. 9. retrograde 10. eccentricity 11. atmosphere 12. Triton 13. Neptune 14. Triton

Chapter 14
True or False? 1. F 2. T 3. F 4. T 5. T 6. T 7. F 8. F 9. F 10. F 11. T 12. T
Fill in the Blank 1. rocky 2. Mars, Jupiter 3. hundreds, 100 4. Jupiter 5. resonance 6. icy 7. gas 8. eccentric 9. a few, 1 A.U. or 100s of million of kilometers 10. meteoroid swarm 11. comets 12. asteroids, comets 13. meteor 14. meteorites 15. 4.6 billion years

Chapter 15
True or False? 1. T 2. F 3. T 4. T 5. F 6. F 7. T 8. F 9. F 10. F 11. T 12. F 13. T 14. T 15. T
Fill in the Blank 1. nebular 2. interstellar dust 3. collisions 4. gravity 5. gas 6. 1000 to 2000 K 7. icy 8. fragmentation 9. comets 10. comets 11. Jupiter 12. T Tauri 13. size 14. lower 15. higher

Chapter 16
True or False? 1. T 2. F 3. F 4. T 5. F 6. T 7. F 8. F 9. F 10. T 11. T 12. F 13. F
Fill in the Blank 1. photosphere 2. chromosphere, corona 3. convective, radiative, core 4. granules 5. photosphere 6. hydrogen 7. helium 8. 98 or 99 9. ionized 10. cooler 11. 11, 22 12. core 13. 4, helium, neutrinos, gamma-rays 14. few

Chapter 17
True or False? 1. T 2. T 3. F 4. F 5. F 6. T 7. F 8. F 9. F 10. T 11. F 12. T 13. F
Fill in the Blank 1. 2 A.U. 2. spectrum, Doppler 3. proper motion, distance 4. luminosity, temperature 5. white dwarfs 6. temperature 7. ionized 8. G2 9. spectral type, luminosity 10. main sequence 11. binary 12. decrease

Chapter 18
True or False? 1. F 2. T 3. T 4. F 5. F 6. F 7. F 8. T 9. F 10. F 11. F 12. T 13. T 14. T
Fill in the Blank 1. gas, dust 2. similar or larger 3. dimming, dust 4. low 5. hydrogen, helium 6. 8,000 7. emission nebula 8. 10 or 20 9. atomic 10. spin, hydrogen 11. 20 12. radio 13. hydrogen 14. millions of

Chapter 19
True or False? 1. F 2. F 3. T 4. T 5. T 6. F 7. F 8. F 9. T 10. F 11. T 12. T
Fill in the Blank 1. temperature 2. evolutionary track 3. 10, 10 or more, 1,000 or more 4. fragments

Chapter 19 (continued top of column 3)
5. increase 6. protostar 7. upper, right 8. 10 million 9. fuse hydrogen 10. main sequence 11. 5 or 6 12. 50 13. one million, one billion 14. radio 15. infrared

Chapter 20
True or False? 1. T 2. T 3. F 4. T 5. T 6. F 7. T 8. F 9. F 10. F 11. T 12. T 13. F 14. F 15. T
Fill in the Blank 1. pressure 2. 5 billion 3. hydrogen, helium 4. 100 million K 5. contract 6. carbon, energy 7. degeneracy 8. carbon 9. 100 10. clusters 11. the Earth 12. high, low 13. decrease 14. decreases 15. separation

Chapter 21
True or False? 1. F 2. F 3. T 4. T 5. T 6. T 7. T 8. T 9. F 10. F 11. F 12. T 13. T
Fill in the Blank 1. accretion 2. hydrogen, surface 3. temperature 4. neutron, neutrino 5. neutron 6. Type I 7. 1.41 8. light curves 9. helium capture 10. 3 billion K 11. iron 12. neutrinos

Chapter 22
True or False? 1. T 2. F 3. F 4. F 5. T 6. T 7. F 8. T 9. F 10. F 11. T 12. F
Fill in the Blank 1. Type I 2. 20 3. high, strong 4. radio 5. 0.03 to 0.3 seconds 6. rotation 7. binary 8. neutron 9. matter 10. be unchanged 11. lose 12. binary stars

Chapter 23
True or False? 1. T 2. F 3. F 4. F 5. F 6. F 7. F 8. T 9. T 10. F 11. F 12. T 13. F 14. T
Fill in the Blank 1. in side it 2. disk 3. halo 4. luminosity 5. 1, 100 6. brighter 7. RR Lyrae stars 8. 8,000 9. 220 km/s 10. random 11. halo 12. higher, dark matter

Chapter 24
True or False? 1. T 2. T 3. F 4. T 5. T 6. F 7. T 8. F 9. F 10. F 11. T 12. F
Fill in the Blank 1. Hubble 2. large 3. Sa, Sc 4. Local Group 5. broadening 6. 100 Mpc 7. increases 8. Kepler's Third 9. 90 10. distance 11. 100 to 200 Mpc 12. mergers

Chapter 25
True or False? 1. F 2. F 3. F 4. F 5. T 6. T 7. F 8. T 9. F 10. T 11. F 12. T
Fill in the Blank 1. large 2. nucleus 3. core 4. larger 5. nucleus 6. nucleus 7. one parsec 8. 1 to 3 billion 9. one solar mass, decade 10. starlike 11. short 12. galaxy or cluster of galaxies

Chapter 26
True or False? 1. F 2. T 3. T 4. T 5. F 6. T 7. F 8. T 9. T 10. T 11. F 12. T
Fill in the Blank 1. 200 2. everywhere 3. directions 4. Cosmological Principle 5. homogeneity 6. isotropy 7. 13 8. gravity 9. density 10. high 11. one 12. faster 13. closed 14. 2.7

Chapter 27
True or False? 1. F 2. T 3. F 4. T 5. T 6. T 7. F 8. T 9. F 10. F 11. F 12. T 13. T 14. F 15. F
Fill in the Blank 1. matter 2. a few thousand 3. photons, antiparticle 4. 109 5. neutrons, 1013 6. deuterium, helium 7. neutral atoms 8. decreased 9. nonbaryonic 10. a few hundred thousand 11. 90 12. dark matter 13. heavy 14. temperature, COBE

Chapter 28
True or False? 1. T 2. F 3. T 4. T 5. F 6. F 7. T 8. F 9. F 10. T 11. T 12. T 13. T 14. T
Fill in the Blank 1. proteins 2. ammonia, methane 3. any two of lightning, ultraviolet light, radioactivity, volcanism, meteoritic impact 4. 3.5 billion 5. 1 billion 6. organic compounds 7. technical civilizations 8. unstable 9. distances 10. light 11. dust 12. 65 13. water hole 14. 200

The Distance Pyramid

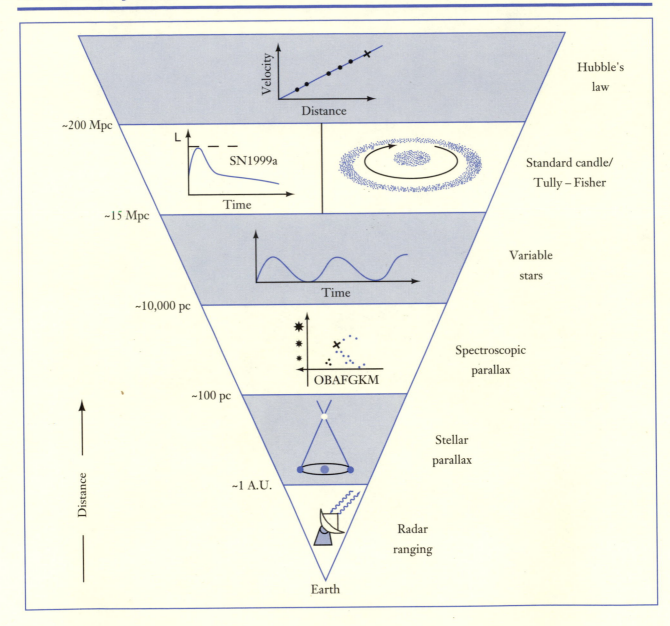